ISBN 978-0-266-66390-4
PIBN 10877129

PHYSIOLOGICAL CHEMISTRY

A TEXT-BOOK AND MANUAL FOR STUDENTS

BY

ALBERT P. MATHEWS, Ph.D.

PROFESSOR OF PHYSIOLOGICAL CHEMISTRY, THE UNIVERSITY OF CHICAGO

SECOND EDITION

ILLUSTRATED

NEW YORK

WILLIAM WOOD AND COMPANY

MDCCCCXIX

1919

First Edition, September, 1915
Reprinted, February, 1916
Reprinted, March, 1916

Second Edition, September, 1916

PRINTED IN THE U. S. A.

THE QUINN & BODEN CO. PRESS
RAHWAY, N. J

TO
MY WIFE

PREFACE

I hope that this book may raise in the minds of those who read it more questions than it answers. Enormous as the science of physiological chemistry, or bio-chemistry, has grown to be, covering as it does the whole of the chemical and physico-chemical phenomena of living nature, only a beginning has as yet been made in it. To few of its fundamental questions can we now give an answer. The great discoveries remain for the future. To arouse interest in the subject, to stimulate curiosity and inquiry, are the main objects of every teacher. I hope that in the pages which follow I have not hit too wide of this mark.

Of so large a subject one can be personally familiar with but a small part. It is difficult to estimate the value of work done in fields other than those in which one has worked. It may be that the emphasis has not always been put in the right place. Some parts of the subject have been treated far more fully than others, and, possibly, more fully than their importance deserves. The chapters on the chemistry of the carbohydrates, fats and proteins and the physical chemistry of the cell are longer than is usual. But a thorough knowledge of this part of the subject is essential to a comprehension of physiology and pathology. On the other hand, this has necessitated a briefer treatment than they deserved of some other matters. I have not been able to consult the whole of the vast literature of biochemistry and I know that many valuable and suggestive papers have probably escaped my attention. At the end of each chapter there will be found a short list of papers hearing on the subject dealt with in that chapter. Many of these should be read by students, and material may be taken from them for conferences. Most of these papers are recent. They have been chosen not because they are necessarily better than older papers, for the reverse may be the case, but because in them the older literature is cited and they reflect the more modern point of view. While I have expressed opinions here and there, I have, as far as space permits, given definite experiments rather than conclusions only, so that the reader may judge the evidence for himself.

In the preparation of the practical work I have been assisted by my colleague, Professor F. C. Koch, whose aid is gratefully acknowledged. For the derivation of the scientific words and their meanings I have relied on the excellent Medical Dictionary of Stedman. I have drawn freely for tables and cuts on other works.

UNIVERSITY OF CHICAGO,
May, 1915.

PREFACE TO THE SECOND EDITION

The publishing of a second edition enables me to correct mistakes, typographical and other, which had got into the first; to add a few references to papers appearing in the past year; to improve in a few places the practical directions, and to recast slightly the discussion of colloids so as to bring more sharply into the foreground their surface properties.

I am much indebted to Professor C. C. Stewart of Dartmouth College for his courtesy in placing at my disposal his library and laboratory in the preparation of the book during two summers' stay in Hanover; and to Dr. Ben Nicolet of the Mellon Institute, Dr. L. B. Mendel, Dr. Max Morse, and several other friends who have been so kind as to call my attention to mistakes and misstatements.

CHICAGO, May, 1916.

CHAPTER PAGE

PART III.

PART I

THE CHEMISTRY OF PROTOPLASM AND THE CELL

CHAPTER I.

THE GENERAL PROPERTIES OF LIVING MATTER.

The various objects on the surface of the earth may be divided into two great classes, the living and the lifeless; the former being characterized by the possession of certain properties which the latter lack. The first of the distinctive properties of living matter is the power of movement; and of movement having an internal rather than an external origin. These movements are either from place to place as in animals; or movements of growth and foliage as in plants. It is by the property of movement that we instinctively distinguish living and lifeless. A second property is that of growth; growth not by the apposition of particles to the outside of the living thing, but growth from within, by the intercalation of substances within the organism. Another, the most characteristic, and the only property it is certainly known that some of the simpler organisms possess, organisms too small to be seen, is that of reproduction. Such organisms are called living because they are capable of indefinite multiplication. Finally we have two properties which often require special apparatus for their detection, but which are, none the less, fully as fundamental as the others, the properties, namely, of respiration and irritability. All living things respire, that is they consume oxygen, liberate energy by combustion or oxidative changes, and they give off a gas, carbon dioxide; and they are irritable; that is they respond in some way, either by a change in the rate of reproduction, in movement, in growth, or in some other of their functions when their surroundings change. We cannot directly observe that many of the smaller forms of life are irritable, but we believe from analogy that they must be so.

These five properties, movement, growth, reproduction, respiration and irritability, are, hence, those properties possessed by living things, and not possessed, or at least not all of them, by any non-living thing. Their possession defines a living thing. When we speak of life we mean this peculiar group of phenomena; and when we speak of explaining life, we mean the explanation of these phenomena in the terms of better known processes in the non-living.

How it happens that living things have these properties which are lacking in the non-living has only within comparatively recent times become a subject of scientific investigation. For many centuries the problem was regarded as solved. Since living things are apparently lifeless things plus something else, it was assumed that there was in living things a spirit, an energy, an entelechy, or a demon, which did not exist in lifeless matter, and to this spirit, or entelechy, all of these peculiar vital properties were ascribed. It was not until the end of the eighteenth and the beginning of the nineteenth century that this explanation was doubted, and only since then has the attempt been made to discover the origin of the vital properties.

To the solution of this problem many men have contributed and it is perhaps invidious to pick out anyone for special mention, but physiological chemistry certainly took a long stride forward, if indeed it may not be said to have originated, about 1775-1793 in the work of that great man of science, Lavoisier. In that beautiful series of papers published in the *Memoirs of the French Academy*, papers which should be read by every student of the science as true examples of real scientific work, embodying the happiest combination of imagination and experimental verification, Lavoisier showed that the heat of the body, that peculiar property of the living body, was due to the burning, or combustion, of its substances,—a burning analogous in all respects to the combustion of a candle, or of a piece of coal. Animal heat and animal respiration were thus correlated, and the living energy was seen to have its origin in the combustion of hydrogen and carbon.

It remained, however, for the histologists to show what was the real physical substratum of the living phenomena, and this grew immediately out of the discovery of the compound microscope. Living things, in their outward form, are extremely diverse, but when they are examined microscopically it is found that all are composed of microscopic units called cells. Within these cells there is a substance of a peculiar and unique nature found nowhere else; a substance called by Dujardin, who first described it in animals, *sarcode;* and by von Mohl, who saw it in plants, *protoplasm* (*protos*, first; *plasma*, form). This sarcode, or protoplasm, Dujardin described as a sticky, viscid, clear, or slightly granular, substance, which would adhere to a glass rod and could be pulled out in long thin strands, much as candy can be pulled out. In it was a more refractive, spherical body called the nucleus, discovered by Robert Brown in 1831. It was not, however, until about 1861 that sarcode and protoplasm were recognized as essentially identical in all plants and all animals, and the conclusion drawn that it was the real living basis, the physical basis of life. Max Schulze especially contributed to the establishment of this conception.

The recognition of the fact that all living things had in them a substance essentially identical in its main features in all cells provided at once a basis for those peculiar and common properties of living things.

Irritability, respiration, growth, metabolism, movement are the properties of living matter, or protoplasm. It is the chemistry of this substance and its products with which the science of physiological chemistry, or biochemistry, has to deal.

The physical appearance and consistence of this living matter varies in different cells, sometimes being jelly-like in its rigidity; at other times, or in other locations, decidedly fluid. It may be seen in many vegetable cells, such for example as the fine stamen hairs of the spider-lily, Tradescantia, or in Nitella, to be in active movement, the protoplasm keeping up a circulation within the cells; its flowings may carry unicellular organisms from place to place; even in the cells of higher animals, as in the eggs of one of the tunicates, the external layer of the protoplasm appears fluid and may flow about the egg; and in the nerve cell of the vertebrate brain its movements are supposed to make and break those fine, inter-cellular connections at the basis of memory, association and thought. On the other hand, protoplasm may be quite jelly-like and semi-rigid and highly elastic, as in the epithelial and muscle cells of vertebrates; and it may be now rigid and now fluid as its state changes with its condition of activity. These facts have been established, in part, by Kite's microscopic dissection of cells by very fine glass needles.

The optical appearance of living matter is that of a clear, transparent ground substance in which are imbedded a great number of granules of different sizes and often of different densities and different tints. It is generally believed, because of its uniformity and universality, that the clear ground substance with the nucleus is the living substance itself, and that the granules represent raw materials, or secretory, or waste substances. The granules are generally colorless, but they may be colored as in pigment cells, or in the blood cells of the sea-urchin, Arbacia, where they are a beautiful deep red. They may be either spherical, or rod shaped, ellipsoidal, or crystalline. When stains enter living matter they may combine with and color the granules, but the ground substance does not appear ever to color while it is living. Finally living matter is always probably very slightly alkaline in reaction, but it becomes acid on dying.

Living matter, therefore, is a substance found in all living things, essentially the same in all, but differing somewhat in its physical appearance and chemical composition in each particular kind of cell. The physical and psychological complex of phenomena to which is given the collective name of "life" is associated always, so far as we know, with this substance, although each individual property may be independent of it; and it is the problem of the science of physiology to discover, to analyze these phenomena and, if possible, to find how they arise from the physical-chemical-psychic constitution of protoplasm.

How the differentiation into living and lifeless arose on the earth is still unknown, but most physiological chemists are of the opinion that since living matter is to-day being constantly made out of lifeless, and we have no reason to believe that the course of events was different in this respect in the past, that living originated from lifeless; and, probably, not at one step, but as the result of a series of transformations taking in the first instance a very long time. It must be remembered, too, in considering the gap between living and lifeless, that while this appears to be wide and profound, if we consider the higher organisms such as man himself, it is not so profound if we consider the very simplest forms of life. Living forms exist so minute as to be almost, or quite, beyond the realm of microscopic vision; such forms can have only the simplest structure, since their volume is so small that they can contain only a small number of molecules of the size of those in living matter. The difference between these forms and lifeless matter would seem to be reduced almost to a simple chemical difference. In fact, the differences between living and lifeless appear on closer examination to be quantitative rather than qualitative.

Living matter is nearly always in movement, movements of growth, of active streaming or of changes of shape; and since to move objects, such as nuclei, requires that work be done, and since energy is that which does work, living matter must be the seat of energy transformations. It might be supposed that this energy, or capacity for work, was due to some peculiar, non-physical, vital force or spirit, but experiment has now clearly demonstrated that this is not the case, but that this energy comes ultimately from light and immediately from the union of the living matter, or its constituents, with oxygen. The law of conservation of energy in living things is the most fundamental law of biology. Living matter is, indeed, a machine for the transformation of chemical and other forms of potential energy into various forms of kinetic energy, or into the chemical energy of new compounds.

The kinetic energy of living things may appear as heat, as mass movements, as light or as electrical energy. Thus all forms of living matter are exothermic; they constantly produce heat, so that their temperature is more or less above that of their environment. The chemical transformations of living things are necessarily, for the most part, exothermic. In some cases, however, the energy appears as light rather than heat. This is the case, for example, in the luminous organs of the fire-fly; and probably in the phosphorescent organs of the Ctenophores and in Noctiluca; in these forms combustion produces light, and the liberation of heat is reduced to a minimum, so that the light of the fire-fly may be said to be the most efficient lamp in existence, in the sense of there being least waste of energy as heat.

Another form of energy set free by living things is electrical. Electrical disturbances occur in all cells when combustion takes place in them, but in some instances nearly the whole of the energy appears to take this form instead of heat. This is well illustrated in the electrical organ of the Torpedo, in which stimulation causes a strong electrical current, so that this organ, made of modified muscle, is a very efficient battery and a study of its physiology may ultimately show how fats, sugars or other carbon compounds, or carbon itself, may be burnt with the liberation of electrical energy in place of heat. But the most striking example of this kind is found, probably, in the nerve impulse, which though it is accompanied by, or is due to, the production of a large amount of carbon dioxide, and is hence a direct or indirect oxidation, nevertheless appears to generate no heat, but only a well-marked electrical current of momentary duration. On the other hand, the muscle cell has developed a mechanism by which much of the energy appears to be used in producing molar movements; although here the larger proportion still appears as heat.

Finally in all these cases some of the energy is re-transformed, with some consumption of heat, into the potential energy of new chemical compounds, forming thus new combustible substances.

Thus far a very important manifestation of living things has been omitted, namely, the psychical phenomena which accompany the energy transformations in our brains, and which we must believe arise in some way from simple phenomena of the same kind perhaps occurring in every chemical transformation. These psychical phenomena are omitted because it has not yet been possible to show that consciousness, or intellectual activities, represent any portion of the transformed energy; and they are, at present, not supposed to be in the chain of physical cause and effect. They are generally regarded, in other words, as outside, or concomitant, or epiphenomena, which occur parallel with the physical changes, and which appear to be dependent upon them, but which do not themselves produce or influence such changes. It cannot be denied, however, that this is a most unsatisfactory solution of the most interesting of all problems, since if consciousness has this position it becomes difficult to attack the problem as all other physical problems have been attacked. It is perhaps wiser to wait until more light has been thrown upon this subject. Negative evidence, the failure to detect loss of energy accompanying consciousness changes, is not a satisfactory basis for any firm conclusion. It may prove to be the case, although the evidence is certainly not favorable at present, that consciousness, or rather the psychical basis of it, should be put together with heat, light and electricity as one of the accompanying manifestations of energy transformations in living and, presumably, in lifeless things also.

It is very important to remember in the course of the transformation of potential into kinetic energy in living matter that the kinetic energy may appear in various forms, and that if it appears in some other form than heat, the heat which one might expect to appear does not do so, but it is replaced by light, electrical currents, movements, possibly psychic energy, if there is such a thing, or some other form of energy of movement.

Since living matter is constantly giving off energy in these different forms, it must be receiving it from some source, or creating it. Careful experiments, which will be cited later in the book, prove that living matter does not create energy, but that in it energy is simply transformed from one kind to another, as it is elsewhere in the universe. Living matter must then get its energy from some source. This source is the food and the oxygen of the air. The chemical system consisting of oxygen and foods contains potential energy. This system is formed, with its potential energy, by the action of chlorophyll, the green coloring matter of plants and the protoplasm of plants. Sunlight acting on these green plant parts in the presence of carbon dioxide and water brings about a separation of the carbon and oxygen of the carbon dioxide. The energy of the sunlight is transformed in this process. The carbon, with a small part of the oxygen, becomes converted into various food substances (carbohydrates, etc.) and the oxygen accumulates in the air. This separation of carbon and oxygen requires that work should be done and consequently the expenditure of energy, and this energy is obtained from the light absorbed by the green leaves. All the energy of living things comes, therefore, in the long run from the sun. The food and oxygen thus separated contain between them potential energy, since, under favorable conditions, not well understood but such as exist in living matter, they will combine again to form carbon dioxide and water and set free, in so doing, the energy required for their previous separation. The energy of living things, whether it appears as heat, light, electrical disturbances or movements of masses, is due then directly, or indirectly, to the combustion of the carbon and hydrogen of the body by the oxygen of the air. Living matter is a combustion engine, with cylinders and connecting rods of molecular dimensions and provided, possibly, with an electrical sparking device not so dissimilar in principle from that of an internal combustion or explosion engine. The discovery of the origin of the energy of living protoplasm in the combustion of carbon and hydrogen was one of the greatest, if not the greatest and most fundamental, discovery in chemical biology; and it is considered more at length in Chapter VI.

But it must not be thought from what has preceded that combustive changes are the only kinds of chemical changes occurring in living matter. The fact is quite otherwise. There are also, in the first place, reducing reactions. In order that any substance may oxidize it must also be reducing. A reducing substance is one which has the power of combining with oxygen. Now all the food and organic substances of protoplasm have the power of combining with oxygen under appropriate conditions, hence living matter is seen to be made of reducing substances. If it happens that there is not sufficient free oxygen for these reducing substances to unite with when they enter into an actively reducing condition, and how they come to enter such a condition will be considered presently, those which are the stronger reducing steal the oxygen away from other weaker reducing bodies which have got a little; or the reducing particles, finding no oxygen to unite with and being in a condition to unite with something, join, or condense, together, two or more parts of molecules uniting to form new substances; and in this way, probably, the fats are formed from the sugars. Since no cell ever has a sufficient supply of oxygen to oxidize all the reducing substances set free or active, and since, indeed, it cannot continue to exist if ever the oxygen becomes thus plentiful, all living matter has a steady reducing action and there are a great many reducing reactions, as well as oxidations, going on in cells. It is, indeed, as we shall see, this play of oxidation and reduction which accounts for many of the synthetic transformations in protoplasm. Furthermore, since the absorption of oxygen must be proportional to the surface of the cell, whereas the requirement goes proportional to the mass, the size of cells must be regulated or fixed in some way to secure the proper balance between oxidation and reduction.

A very large class of chemical transformations in protoplasm consists of hydrations, as would be anticipated in a medium containing, as protoplasm does, 80 per cent. of water. By a hydration is meant the union of water with a substance. When this union takes place some substances become unstable, for some reason not understood by the writer, and fall into fragments. This process of decomposition with the taking on of water is called hydrolytic decomposition, or cleavage (Gr. *hydōr*, water; *lysis*, separation). And among the disintegrative, or catabolic (*Kata*, down) chemical changes, this is one of the most important. All digestive changes are of this kind.

Besides oxidations and reductions, condensations and hydrolyses, there is finally another great class of chemical reactions known as dehydration syntheses. It is a singular fact that protoplasm, although it is four-fifths water, nevertheless synthesizes complex substances such as proteins, carbohydrates and fats by a process which involves the liberation of water and which is ordinarily duplicated outside the cell by means of high temperature, or by strong water-attracting substances, such as phosphorus pentoxide or sulphuric acid. These dehydration syntheses taking place in such a wonderfully aqueous medium have been a great puzzle. It has been suggested by Drechsel that many of them are dehydrations produced not by a simple taking out of water, but by a reduction followed by an oxidation. There is reason to believe this explanation in some instances to be well founded, although syntheses of the more complex of these bodies by this method have not yet been produced outside the cell. The subject requires further investigation. One fact strongly in its favor is that such syntheses are retarded if the respiration of the cell is reduced by deprivation of oxygen, by anesthetics or in other ways; or if the reducing power of the cell is destroyed by the supply of too much oxygen.

There is still another feature of cell chemistry which must strike even the most superficial observer, and that is the *speed* with which growth and the chemical reactions occur in it. Everyone knows that sugar dissolved in water does not rapidly oxidize to carbon dioxide, but remains intact for a long period; but in the cell it oxidizes with surprising speed, liberating heat, light, or doing work by the energy set free. It has been found that if glucose is dissolved in water and exposed to air, particularly in the light, it undergoes a very slow oxidation and decomposition. The difference between its behavior in and out of the cell is a difference of speed of decomposition, rather than a difference in kind. A similar fact is seen in the behavior of starch. Starch boiled with water does not easily take on water and split into sweet glucose, but in the plant cell it changes into sugar under appropriate conditions very rapidly. How does it happen then that the chemical changes of the foods go on so rapidly in living matter and so slowly outside? This is owing to the fact, as we now know, that living matter always contains a large number of substances, or compounds, called enzymes (Gr. *en*, in; *zymē*, yeast; in yeast) because they occur in a striking way in yeast. These enzymes, which are probably organic bodies, but of which the exact composition is as yet unknown, have the property of greatly hastening, or as is generally said, catalyzing, various chemical reactions. The word catalytic (*Kata*, down; *lysis*, separation) means literally a down separation or decomposition, but it is used to designate any reaction which is hastened by a third substance, this third substance not appearing much, if at all, changed in amount at the end of the reaction. Living matter is hence peculiar in the *speed* with which these hydrolytic, oxidative, reduction or condensation reactions occur in it; and it owes this property to various substances, catalytic agents, or enzymes, found in it everywhere. Were it not for these substances reactions would go on so slowly that the phenomena of life would be quite different from what they are. Since these catalytic substances are themselves produced by a chemical change preceding that which they catalyze, we might, perhaps, call them the memories of those former chemical reactions, and it is by means of these memories, or enzymes, that cells become teachable in a chemical sense and capable of transacting their chemical affairs with greater efficiency. Whether all

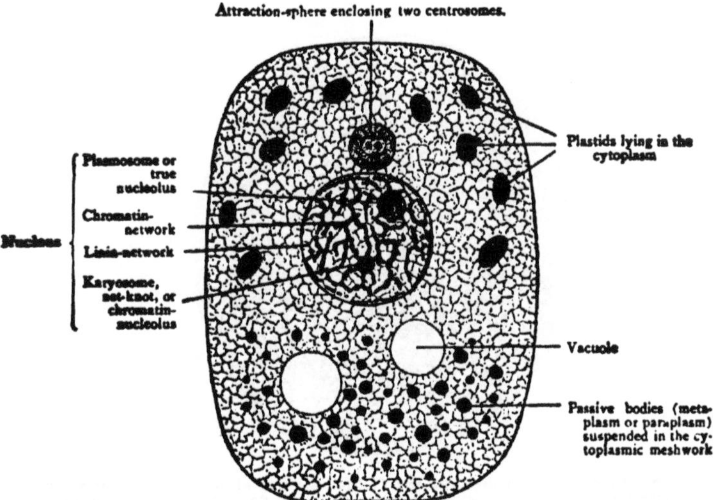

FIG. 1.—Diagram of a cell according to Wilson, illustrating the organization and specialization of the cell.

our memories have some such basis as this we cannot at present say, since we do not yet know anything of the physical basis of memory.

Living reactions have one other important peculiarity besides speed, and that is their "*orderliness.*" The cell is not a homogeneous mixture in which reactions take place haphazard, but it is a well-ordered chemical factory with specialized reactions occurring in various parts. If protoplasm be ground up, thus causing a thorough intermixing of its parts, it can no longer live, but there results a mutual destruction of its various structures and substances. The orderliness of the chemical reactions is due to the cell structure; and for the phenomena of life to persist in their entirety that structure must be preserved. It is true that in such a ground-up mass many of the chemical reactions are presumably the same as those which went on while structure persisted, but they no longer occur in a well-regulated manner; some have been checked, others greatly increased by the intermixing. This orderliness of reactions in living protoplasm is produced by the specialization of the cell in different parts shown in Figures 1 and 2. Thus the nuclear wall, or membrane, marks off one very important cell region and keeps the nuclear sap from interacting with the protoplasm. Profound, and often fatal, changes sometimes occur in cells when an admixture of nuclear and cytoplasmic elements is artificially produced by rupture of this membrane. Other localizations and organizations are due to the colloidal

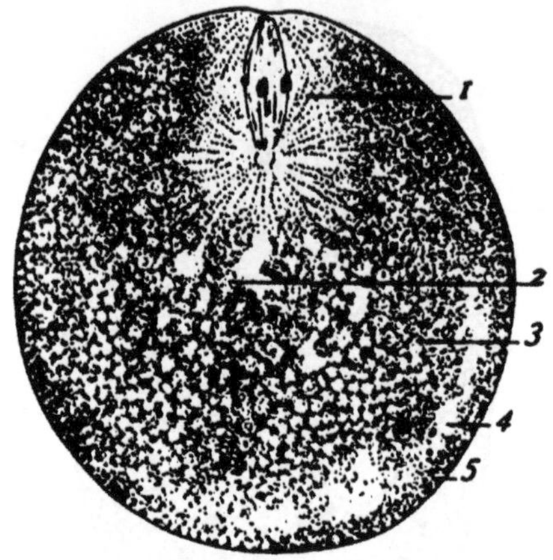

FIG. 2.—Section of a dividing egg cell (Lillie) showing alveolar or granular structure of protoplasm at 3, the spindle with chromatin at 1, and finely granular protoplasm at 2. The peripheral layer at 5 is different from the parts lying inward.

nature of the cell protoplasm and possibly to its lipoid character. By a colloid is meant, literally, a glue-like body; a substance which will not diffuse through membranes and which forms with water a kind of tissue, or gel. It is by means of the colloids of a protein, lipoid or carbohydrate nature which make up the substratum of the cell that this localization of chemical reactions is produced; the colloids furnish the basis for the organization or machinery of the cell; and in their absence there could be nothing more than a homogeneous conglomeration of reactions. The properties of colloids become, therefore, of the greatest importance in interpreting cell life, and it is for this reason that they have been studied so keenly in the past ten years. The colloids localize the cell reactions and furnish the physical basis of its physiology; they form the cell machinery.

The general chemical composition of living matter. Water.—It is little short of astounding that living matter with all its wonderful properties of growth, movement, memory, intelligence, devotion, suffering and happiness should be composed to the extent of from 70 to 90 per cent. of nothing more complex or mysterious than water. Such a fact as this is most perplexing, especially when all experiment shows that this water is playing a profoundly important part in the generation of the vital phenomena. Any interference with the amount normally present makes a change at once in the activities of the cell. In fact, we might say that all living matter lives in water, as Claude Bernard put it. For not only is this obviously true in the lower and simpler forms of animals and plants, which are little more than naked masses of protoplasm living in water, but it is no less true of the higher forms, since in all of them an internal medium, or environment, of a liquid nature, the lymph, the blood, or sap, is found which is the immediate environment of the cells. Water is the largest and one of the most important constituents of living matter; and if organisms are carefully examined the most various devices are found to assure the regulation of the water content of the cells of the body. The younger, the more vigorous, the more alive, the more actively growing, the more impressionable cells are, the more watery are they. Perhaps more than anyone else the French physiologist, Dubois, has emphasized the important rôle of water in life. Table I gives the proportion of water found in various kinds of tissues.

TABLE I. AMOUNT OF WATER IN VARIOUS TISSUES.

Organ	Percentage of water	Organ	Percentage of water
Brain. White matter	68	Liver (human)	76
Brain. Gray matter	84	Cartilage (hyaline)	67
Brain. Embryonic	91	Thymus (calf)	77
Muscle (mammalian)	73	Kidney (child)	78
Muscle (fish)	80	Suprarenal gland	80
Electrical organ	92	Dentine	10

Salts, and inorganic elements.—One would very naturally expect that living matter might contain some very rare, peculiar and costly metal, or substance, like radium, to which its properties might be attributed. But quite the contrary seems to be the case. Besides water, the inorganic constituents of protoplasm are salts, and they are among the commonest salts on the surface of the earth. Sodium, potassium, magnesium, calcium, iron, sulphates, chlorides, phosphates and carbonates are essential to life and are found in practically all living matter. The amount of these various inorganic elements differs somewhat in different cells and tissues, but they occur in all. Other common elements are sometimes present, such as iodine, manganese, copper, zinc, barium or silicon, but these are generally confined to special plants and animals. About 1 per cent. of the weight of the protoplasm is composed of the salts or inorganic metals and acids mentioned (Figure 3). Furthermore, these salts are not mere inert substances, they are not simply absorbed with the water and tolerated, but they are in combination, in part at least, with the organic matter of the protoplasm. They are not simply clinkers clogging the grates of the protoplasmic fires, but they are active in the production of the vital phenomena. Indeed, some have gone so

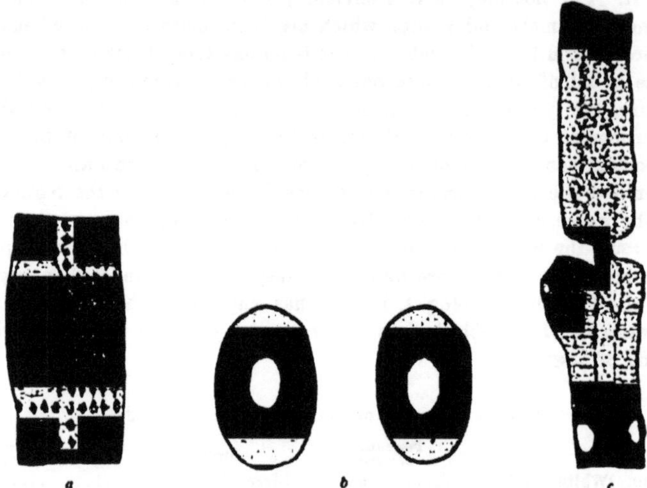

FIG. 3.—The distribution of potassium in cells after Macallum. (a) striated muscle; (b) nucleated blood corpuscles; (c) nerve fiber. The black precipitate represents the potassium.

far as to believe, as we shall see, that by means of the electrical charges they bear when in solution they vitalize the colloidal, organic substratum of the cell and make it alive. Any change in their relative proportions at once affects the activity of the cell; thus by increasing or diminishing the proportion of sodium, calcium or potassium skeletal muscle may be made to twitch rhythmically or to remain at rest; nerve impulses may be set up in motor nerves, or the irritability of the nerve raised or lowered; chromophores of fish scales may be contracted or expanded; and the activities of all cells increased or diminished. Magnesium sulphate acts much as an anesthetic on mammals, but paralyzes, also, the endings of the motor nerves in the muscles. Furthermore, by increasing the total amount of salt in protoplasm many cells may be stimulated and egg cells of some animals caused to develop parthenogenetically without the aid of sperm.

Thus in some instances 94 per cent. of living matter consists of nothing more unusual or remarkable than water and the commonest salts.

It is certainly not without significance that living matter is so watery and contains the salts of the sea. It would appear probable from this that living matter originated either in the sea itself or, perhaps, in some pool of water which contained, possibly in dilute form, the common salts. It has been suggested that it was in some slowly-drying volcanic pool where concentration could take place, and where cyanides and other similar reactive organic compounds might have been formed by the vigorous electrical discharges accompanying the eruptions, that living matter first appeared. We would thus have sprung from the thunderbolts of Jove, if this theory is true; but we are, at any rate, the children of the sun and the sea, of Apollo and Aphrodite.

The organic matter.—The remainder of living matter, 10 to 25 per cent. by weight, is organic. This organic matter is found to consist of, or may be divided for purposes of convenience into, four great groups of substances: 1, substances of the fat group soluble in alcohol and ether, called lipins; 2, substances of the sugar group, carbohydrates; 3, substances containing nitrogen, carbon, hydrogen and oxygen, called proteins; 4, various simple substances such as urea, creatinine, inosite, phenols, etc., called extractives, because they are soluble in water by which they may be extracted from the cell when the latter is first coagulated. In muscle the relative proportion of these substances is as follows: protein, 19 per cent.; carbohydrate, 0.3 per cent.; lipin, 3 per cent.; salts, 3 per cent.; water, 75 per cent. In the following chapters the chemistry of each of these great groups of organic substances, beginning with the carbohydrates, will be discussed.

CHAPTER II.

THE CARBOHYDRATES.

Occurrence.—All living organisms, except the most simple, which are nothing more than naked masses of protoplasm, consist of both living and lifeless matter, the lifeless having been formed or secreted by the living. This lifeless matter forms the greater part of the supporting framework, or serves as reserve food. In plants these supporting tissues, or reserve foods, consisting of the cellulose or woody parts, the starches, mucilages or gums, such as that which exudes from the bark of the cherry-tree, are composed of the elements carbon, hydrogen and oxygen and belong to a great group of substances known as sugars or carbohydrates. The supporting tissues of animals, unlike those of plants, contain a large proportion of nitrogen and belong generally to the group of proteins, although chitin, which forms the hard shell of crabs and other invertebrates, contains a large amount of carbohydrate (glucosamine). But it is not only as the supporting tissues of plants and animals that carbohydrates occur. They are found, also, in the living matter itself, making part of the chromatin of the nucleus, or distributed as glycogen or sugar, free or combined, through the cytoplasm; and it is, indeed, largely by the combustion of carbohydrate that we derive our energy. Since substances of this class are the simplest of the colloidal materials of cells, and are among the most abundant organic constituents of living things; since they are formed from the inorganic compounds of carbon dioxide and water, and in the long run all the energy of living matter comes from them, and since both the fats and proteins originate from them, a study of the organic constituents of protoplasm may best begin by a study of their composition and chemical nature.

Definition.—The carbohydrates are compounds of carbon, hydrogen and oxygen occurring in animals and plants. They get their name from the fact that in the majority, though not in all, the hydrogen and oxygen are in the proportion of two atoms to one, that is, they are in the same proportion as in water; and, indeed, by the action of heat, or of strong dehydrating agents, they are split into carbon and water, as in the process of making charcoal or in the charring of sugar. The formula of glucose, a typical carbohydrate, is $C_6H_{12}O_6$. But while in the majority of the naturally occurring members of this group the hydrogen and oxygen are in this proportion, in some cases, as in rhamnose, $C_6H_{12}O_5$, a methyl pentose, they have not this proportion. Many substances, also, have hydrogen and oxygen in this proportion which are not carbohydrates, such as lactic acid, $C_3H_6O_3$, or acetic acid, $C_2H_4O_2$, which differ from the carbohydrates in their chemical properties. Many of the carbohydrates have a sweet taste, although some substituted members of the group among the glucosides are intensely bitter, and polysaccharides may be tasteless. When pure they are white; some, like cane sugar, crystallize; others, like starch, are colloidal and do not crystallize.

The chemical properties of carbohydrates characterize them as well as, or better than, their composition. All of the simpler ones readily oxidize. They are, hence, reducing substances and a large part of the reducing powers of protoplasm are due, in the long run, to these substances. They reduce ammoniacal silver nitrate, or alkaline solutions of mercury, copper, gold or bismuth salts. On the other hand, they have oxidizing properties too. They will absorb nascent hydrogen, uniting with it and oxidizing the substance from which the hydrogen is taken. The simultaneous possession of these and other properties shows that they contain aldehyde, or ketone, groups, $-\overset{H}{\overset{|}{C}}=O$ or $=C=O$, in the molecule. Either of these groups can take up hydrogen yielding an alcohol; or by oxidation go over into a carboxyl group, $R-\underset{\underset{O}{\|}}{C}-OH$. The simplest carbohydrates, therefore, are aldehydes or ketones, and they form accordingly two groups: aldoses and ketoses. Their reaction in aqueous solution is neutral to the usual indicators, but they possess, nevertheless, very weak acid and basic characters, being very weak amphoteric compounds. Thus they contain some hydrogen which may be replaced by a metal, such as lead, or sodium, and they are thus able to neutralize, to a slight extent, the causticity of sodium hydrate. They are to this extent acids, though they lack the acid taste. This acid property is due to the fact that they contain alcohol groups, all alcohols behaving like very weak acids, since the alcohol hydrogen may be, in part, replaced by a metal. They are, however, very weak acids. The number of hydrogen ions in their solutions is very small, smaller than in solutions of carbon dioxide of equal concentration. The dissociation constant of every sugar is very small. By the dissociation constant is meant the value K, where $K=\frac{C_1 \times C_2}{C_3}$; C_1 is the concentration of hydrogen ions, C_2 the concentration of the sugar anion and C_3 the concentration of the undissociated molecule. The dissociation constant of glucose at 18° is 5.9×10^{-13} (Osaka) or 3.6×10^{-13} (Madsden); that of saccharose is 1.14×10^{-13} (Madsden), or 2.4×10^{-13} (Michaelis and Rona); maltose is 18×10^{-13} (Michaelis and Rona); and levulose is 8.8×10^{-13} (Michaelis and Rona). With bases such a sugar as glucose will react according to the following equation:

$$C_6H_{12}O_6 + NaOH = C_6H_{11}O_6Na + H_2O$$

The sugars are, then, alcohols as well as aldehydes or ketones. They are polyhydric alcohols having one alcohol group attached to each carbon atom, but that of the aldehyde or ketone group.

Their basic properties are due to the oxygen of the aldehyde. By the aldehyde oxygen they have the property of uniting with acids to form so-called oxonium salts, but this union is easily dissociated, the basicity being very weak.

$$C_6H_{12}O_6 + HCl = C_6H_{12}O_6.HCl$$

The carbohydrates may, then, be defined thus: They are compounds of carbon, hydrogen and oxygen, the oxygen and hydrogen being often but not always in the proportion to form water; and, further, they are aldehyde or ketone derivatives of polyhydric alcohols.

The aldehyde structural formula for dextrose is

```
    OH  OH  OH  OH  OH  H
    |   |   |   |   |   |
H — C — C — C — C — C — C = O ;
    |   |   |   |   |
    H   H   H   H   H
```

and the formula for the ketose, levulose, is

```
    OH  OH  OH  OH  O   H
    |   |   |   |   ||  |
H — C — C — C — C — C — C — OH.
    |   |   |   |       |
    H   H   H   H       H
```

Classification.—It is convenient to divide the carbohydrates into three great classes according as their molecules contain one, two or several saccharide (simple carbohydrate) groups. These classes are the monosaccharides, the disaccharides and the polysaccharides. The members of the first two groups are generally crystalline bodies; but many, though not all, of the last group are colloidal in aqueous solution. The more important monosaccharides found in nature are d-glucose, or grape sugar, or dextrose as it is also called; d-levulose, or fruit sugar; galactose; xylose; arabinose; mannose; and d-ribose. The disaccharides are saccharose, or sucrose, as cane sugar is also called; lactose, or milk sugar; and maltose, or malt sugar. The common polysaccharides are cellulose, gums, dextrins, starches and glycogen.

The monosaccharides are in their turn classified by the number of carbon atoms, or more properly by the aldehyde, ketone and alcohol groups they contain into bioses, trioses, tetroses, pentoses, hexoses, heptoses, octoses, nonoses, etc. Of these the first six are found in nature, but the hexoses are the more abundant. Each of these groups from the trioses on is subdivided into two groups, the aldoses and ketoses, according as they are aldehydes or ketones. Thus mannose, dextrose and galactose are hexose aldoses having the general formula, $C_6H_{12}O_6$; levulose is a ketose hexose; ribose and xylose are pentose aldoses, $C_5H_{10}O_5$; of the trioses, glycerose, $C_3H_6O_3$, is an aldose, while dioxyacetone is a ketose.

- CARBOHYDRATES
  - I. MONOSACCHARIDES
    - 1. Bioses. Aldose. *Glycolaldehyde.*
    - 2. Trioses.
      - Aldoses. *Glycerose.*
      - Ketoses. *Dioxyacetone.*
    - 3. Tetroses.
      - Aldoses. *Erythrose.*
      - Ketoses. *d-Erythrulose.*
    - 4. Pentoses.
      - Aldoses. *Arabinose, xylose, ribose.*
      - Ketoses. *l-Arabinulose.*
    - 5. Hexoses.
      - Aldoses. *Dextrose, galactose, mannose.*
      - Ketoses. *Levulose, sorbose.*
    - 6. Heptoses. Aldoses. *d-Mannoheptose.*
  - II. DISACCHARIDES
    - 1. Lactose. (Glucose + galactose.)
    - 2. Maltose. (Glucose + glucose.)
    - 3. Saccharose. (Glucose + levulose.)
    - 4. Trehalose. (Glucose + glucose.)
    - 5. Melibiose. (Galactose + glucose.)
  - III. POLYSACCHARIDES
    - 1. Trisaccharides.
      - *Melitose* (Raffinose) in molasses.
      - *Melizitose.* (Pinus larix.) (Levulose + glucose + galactose.)
    - 2. Tetrasaccharides. *Lupeose* in peas; *stachyose.* (Lupeose consists of two molecules of galactose, one of glucose, and one of levulose.)
    - 3. Colloidal polysaccharides.
      - *Dextrins.*
      - *Glycogen.*
      - *Cellulose.*
      - *Starch.*
      - *Mucilages.*
      - *Gums.*
      - *Inulin.*

Monosaccharides. Structural formulas. Isomerism. Optical properties.

a. *Hexoses.*

Analysis of glucose, galactose and mannose shows that they all contain the same proportion of carbon, hydrogen and oxygen; a proportion corresponding to the formula: $C_6H_{12}O_6$. They have also the same chemical properties showing that all of them are aldehydes and polyhydric alcohols. When chemical compounds have the same chemical atoms in their molecules in the same proportions they are called *isomers;* or are said to be isomeric with each other. Thus lactic acid, $C_3H_6O_3$, and dioxyacetone, $C_3H_6O_3$, are isomers. When, in addition to having the same number of atoms of the same kind in the molecule, these atoms are arranged in the same general way so that the chemical nature of the substances is the same, then those substances are said to be *stereo-isomeric*, a word which means "having a like form" (Greek, *stereos*, solid). Since mannose, galactose and dextrose are all of them aldoses and polyhydric alcohols their molecules must be, on the whole, very similar; they are, therefore, stereo-isomers. Their molecules differ only in their forms and we may now examine how these molecules may differ in their shape. This brings us to one of the most important subjects in the whole of physiological chemistry, namely, the subject of the shapes of molecules; in the pages which follow we shall find many examples illustrating the importance of molecular form in vital processes of all kinds.

The proof that the atoms in a molecule occupy definite positions, so that the molecule has a definite shape, was one of the most beautiful and fundamental discoveries of Pasteur, made while he was still a very young man, in 1848; and since this discovery is at the bottom of all the beautiful science of molecular form which has been built upon it, and as the importance of this molecular property is showing itself in every field of biological work, it is fitting that we consider Pasteur's work at some length. Pasteur had been greatly interested in crystalline form. Why do substances crystallize in definite shapes? Among the substances of an organic nature which gave very fine, large crystals, tartaric acid and its salts were noteworthy. Now there were two kinds of tartaric acid known to Pasteur, the ordinary tartaric acid, the acid of wine, which Biot had shown to be dextro-rotatory, i.e., its solutions had the property of rotating the plane of polarization of polarized light to the right; and another kind of tartaric acid found by Kastner and called racemic acid (*L. racemus*, a bunch of grapes) of the same composition as the other but which had no action at all on polarized light. It and its salts were inactive. Pasteur undertook to study the crystalline forms of these two acids. He expected to find that racemic acid would have a different crystalline form from the ordinary dextro-rotatory tartaric acid. He found, however, that when the sodium-ammonium salt of the inactive (racemic) acid was crystallized below 28° crystals of the same shape as those of the corresponding salt of the dextro acid appeared. On looking at the crystals more closely, however, he found that there were in reality among the crystals of sodium-ammonium racemate crystals of two different kinds which are illustrated in Figure 4. These crystals were exactly alike with the exception of a small facet, o^1, and the corresponding facet diagonally opposite to it. These two facets were so placed in these two kinds of crystals that the crystals would not correspond if superimposed one on the other. In the one kind of crystal the facet was on the right side as it was in the dextro-tartaric acid; while in the other form of crystal it was on the left side. The crystals were not symmetric, they were asymmetric and, as it were, mirror images of each other.

He separated these two forms of crystals and thinking that they might show different optical properties he dissolved them and examined the solutions in the polariscope. To his great joy, he found that the solution of the one form now rotated the plane of polarization to the right; while the solution of the other form rotated it to the left. This great discovery showed at once that crystalline form must depend on molecular form, because in the solution the molecules were separated and the crystalline form had disappeared, but the asymmetrical action on light persisted. The action of the solution on light showed that the indi-

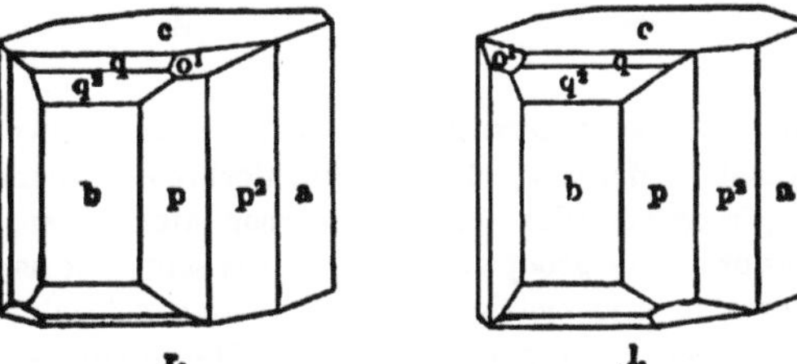

FIG. 4.—Two forms of crystals of levo and dextro tartaric acid (Landolt).

vidual molecules must be of two different forms, a dextro-rotatory and a levo-rotatory form. The molecules of tartaric acid must be asymmetrical, just as the crystals were asymmetrical. The discovery, of course, cleared up at once the difference between the two kinds of tartaric acid. It showed that there were at least three different forms of tartaric acid, the dextro-rotatory, the levo-rotatory and the third, or racemic, form which was composed of equal amounts of the other two kinds and which was inactive on light. Pasteur afterwards discovered a fourth, the meso-tartaric acid. By this discovery of Pasteur we know that the shapes of molecules may be asymmetric, and that the atoms of these molecules do not easily rearrange themselves, for if they did the molecule would readily pass from the one form to the other. It is one of the most fundamental discoveries in physics or chemistry.

The difference in shape of the molecules of the two forms of tartaric acid was made more precise many years later, practically coincidently, in 1874 by LeBel and van 't Hoff. They actually pictured the possible arrangement of the atoms in the molecule by which the asymmetry was produced. If the carbon atom is represented as lying at the center of a tetrahedron, of which the apices represent the position of the four atoms attached to the carbon atom, it becomes possible to picture the different arrangements of the atoms causing the asymmetry. This is illustrated in Figure 5. If the four atoms or atomic groups attached to the carbon atom are all different, as they are in the case of iodo, chlor, brom, methane, CHIClBr, then it is possible to arrange these atoms in two different ways, as is shown in the figure, the two tetrahedrons not being superimposable, but being mirror images of each other. If, however, two of the atom groups attached to the carbon are the same,

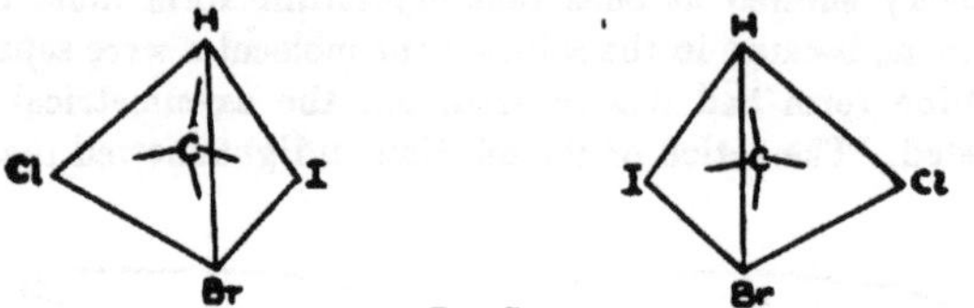

Fig. 5.

then it is impossible so to arrange them that the tetrahedrons will not be superimposable. Methane, chlor- or dichlorbrom-methane can have but one form, a symmetrical one. A carbon atom, then, with four different atoms or atomic groups attached to it is said to be asymmetrical, since it produces an asymmetrical crystalline and molecular form, and an asymmetrical action on polarized light. The atomic groups about such a carbon atom may have two different arrangements. Asymmetric carbon atoms in the sugar molecules illustrated on page 28 are printed in black-face type. Not all compounds with asymmetric carbon atoms rotate the plane of polarized light, since in some, of which mesotartaric acid is an example, compensation may occur, some atoms rotating the plane of polarized light in one direction; while others rotate it in an opposite direction: the total effect of the molecule on light being nil. Most compounds with asymmetric carbon atoms, however, exist in two forms, one dextro- the other levo-rotatory.

The various forms of tartaric acid (stereo-isomers) may be represented as follows, the asymmetric carbon atoms being printed in black-face type:

```
      COOH              COOH              COOH
       |                 |                 |
   H — C — OH       HO — C — H         H — C — OH
       |                 |                 |
  HO — C — H         H — C — OH        H — C — OH
       |                 |                 |
      COOH              COOH              COOH
 d—Tartaric acid   l—Tartaric acid   Meso-tartaric acid (Inactive).
 \_________________________________/
        Racemic acid (Inactive).
```

All compounds having an asymmetric carbon atom in them may exist, therefore, in two different forms, these forms being stereo-isomeric forms and also optical antipodes. One of these optical isomers rotates the plane of polarized light in the one direction, just as much as its antipode rotates it in the other direction. The physical and chemical properties, such as the melting points and solubility in symmetrical solvents, of these two antipodes are almost or quite the same. Stereoisomers which are not optical antipodes generally have different melting and boiling points and solubilities. The separation of the optical antipodes can be accomplished by picking out the crystals in the way Pasteur did in a few instances; or by the different solubilities of their compounds with other optically active substances; or by the action of moulds, yeasts or other living organisms which often destroy one, but not the other antipode. The mould, penicillium glaucum, destroys the dextro- but not the levo-tartaric acid.

In the figure which has been given of the possible shape of the molecule (Figure 5), one might suppose that the atoms in the molecule were far apart, in which case it would be difficult to see why the molecule should keep its form. The figure is, however, probably incorrect in this particular. The attraction between the atoms of a molecule is so great that they probably lie closely packed together and with very little freedom of movement beyond that of minute vibration about a center. The amount of this vibration and the space at the disposal of the atoms becomes somewhat greater as the temperature rises, since there is good reason for believing that molecules expand with a rise in temperature, although the expansion is not very great. The pressures due to molecular and atomic attractions on the surfaces of molecules are enormous. Thus the pressure called the internal pressure of a liquid or a gas, which is due to molecular cohesion, or the attraction between the molecules, is, at zero centigrade in ether, about 2,000 kilograms per square centimeter, and it increases considerably at temperatures below this. Now the attraction between the atoms within the molecule is certainly many times greater than the attraction between the molecules, although it is not yet known just how great it is. By this attraction, therefore, the atoms within the molecules will be under a compression certainly of many thousands of kilograms per square centimeter in addition to the cohesive pressure. It is not impossible that the pressure driving together the atoms of a molecule may be more than a hundred thousand kilograms per square centimeter. It is not probable that this pressure is distributed evenly over the molecule, since some atoms are held more firmly than others. So great a pressure as this must certainly drive the atoms of the molecule very close together so that, at relatively low temperatures at least, the molecules must have the properties of rigid solids with the atoms having very little power of movement. Theoretically, however, they will always have some movement at temperatures above absolute zero, and so the molecules above this temperature are not absolutely incompressible. Of course at higher temperatures as the molecules separate this pressure is reduced and in some cases the attraction between particular atoms of a compound is less than that stated. Greater mobility of the atoms exists in such molecules so that the atoms may shift their positions, undergoing what is known as a tautomeric change. It is not surprising, however, that subjected to such high pressures the atoms of a molecule generally arrange themselves in the position of greatest stability and if, temporarily, they take unstable positions, they may undergo rearrangement. Such molecular rearrangements are by no means uncommon. The racemization of optically active compounds is such a process of atomic rearrangement.

Molecular form is of fundamental importance throughout living nature. Most naturally occurring organic compounds are asymmetric and usually only one of two possible isomers occurs in any organism. Of the optical isomers of any amino acid or carbohydrate only one generally will serve to nourish an organism, or, if both are foods, one is usually better used than the other. The enzymes, or catalytic agents, will only act on compounds of a very particular molecular form. Yeast will ferment d-glucose, d-mannose or d-fructose, all of which have the same configuration of the last three carbon atoms, but it will not ferment l-fructose, or l-glucose, or l-mannose, or l-galactose. The phenomena of immunity, such as specific antitoxins, precipitins and anaphylaxis, also involve molecular form. Protein which has been racemized by the action of sodium hydrate will no longer cause anaphylaxis. In the very accurate and specific adjustment of the spermatozoon to the ovum, an adjustment so accurate that a spermatozoon will usually only fertilize the eggs of its own species, it is probable that the form of the molecules of sperm and eggs are in some manner related or adjusted to each other. In fact, the whole living world is an asymmetric world; the development of different species and varieties probably depends on asymmetric molecules, since animal forms, like the forms of crystals, must, in the last analysis, be but the expression of the forms of the molecules of which the protoplasm is composed.

Molecular asymmetry may be most easily detected by means of the action of the molecules on polarized light. When polarized light, that is light which has passed through a Nicol's prism, passes through a solution of a substance of which the molecules are asymmetrical it is acted upon, so that the plane of polarization of the light on emergence from the solution does not coincide with the plane of polarization of the entering light. The plane of polarization has been rotated to one side or the other, the degree to which it is rotated depending on the kind of molecules and the number of molecules the light has passed. It is dependent, in other words, upon the concentration and the length of the tube. It is also dependent upon the wave length of the light. The plane of polarization of blue light is rotated, for some substances, about twice as much as that of yellow light by the same molecules. Hence one uses always monochromatic light and the degree of rotation is generally expressed for sodium light for a concentration of one gram of substance in a cubic centimeter of solution and for a tube one decimeter in length. This angle is called the *specific rotatory power* of the substance. Temperature also affects the degree of rotation. In general the higher the temperature, the lower the rotation. It is usual to give the specific rotation at or near 20° C. The specific rotatory power as just described (α) is written as follows:

$$(\alpha)_D^{20^\circ}$$

Since for many substances the specific rotatory power varies, also, with the concentration of the solute and the character of the solvent, it is desirable to give these data also. (α) in the above formula is the angle of rotation which the plane of polarization of the D line of the spectrum (sodium) would undergo in passing through 1 dm. of a solution containing one gram of substance to one cubic centimeter at 20° C.

The specific rotatory power is calculated from the angle of rotation produced by a solution of known strength in a tube of known length. The formula is as follows:

$$(\alpha)_D^{20^\circ} = \frac{a.100}{l.c.}; \text{ or } (\alpha)_D^{20^\circ} = \frac{a.100}{l.p.d.}$$

a being the observed angle of rotation at 20° C.; l, the length of the tube in decimeters; c, the number of grams of active substance in 100 c.c. of solution; p, the number of grams of active substance in 100 grams of solution; and d, the density. pd=c. (α) is the specific rotatory power.

Just how molecules with asymmetric carbon atoms rotate the plane of polarization of light is not yet understood. It would seem necessary for the light to pass through all the molecules in one direction, in order that the actions of the different molecules should coincide and not neutralize each other. If this is so, polarized light must orient the molecules and perhaps the molecular asymmetry enables the light waves to do this. If the molecules of an asymmetric substance in solution, or in a liquid, are thus oriented by light so that all the molecular axes coincide, then the conditions in such a solution might approximate to those in a crystal; the magnetic properties of the molecules, if they have any, should coincide and might be detectable. This very interesting and fundamental problem remains for future investigation. The reason why a rise in temperature diminishes the rotatory power would also be clear; since by heat the molecular vibration increases and presumably it would be more difficult to hold the molecular axes in line. Thus increasing the temperature should diminish the rotatory power for the same reason that increasing the temperature of iron diminishes its magnetism, i.e., by destroying molecular orientation.

The rotation of the plane of polarization may be due to the fact that the vibrations of some of the valence electrons occur more easily in some planes than in others.

The Polariscope.—The Polariscope is used to measure the rotatory power. Figure 6. In this instrument the light of a sodium flame, produced by heating sodium chloride or bromide, is first passed through a light filter of potassium

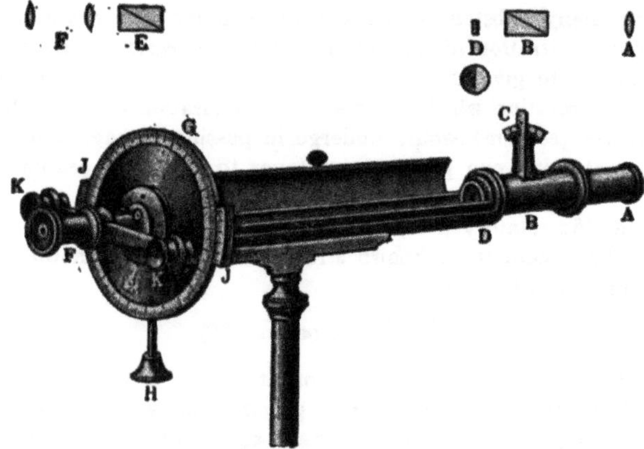

Fig. 6.—Polariscope (Landolt). *A*, lens; *B*, polarizing Nicol prism; *C*, arm to rotate the polarizer; *D*, quartz plate; *E*, analyzing prism mounted so that it rotates with the circle *G* which is marked in degrees; *F*, the observing telescope; *J*, the vernier for reading the rotation, and *K*, telescopes for increasing the accuracy of reading. The tube containing the solution goes between the analyzer and polarizer, the cover of this space being shown open.

bichromate to remove extraneous rays, and then is plane polarized by passing through a Nicol prism or a Glan-Thompson prism of Iceland spar, called the polarizer (B, Figure 6). The light then passes through the solution and then through another Nicol prism, E, called the analyzer, which is so mounted that it can be rotated about an axis. The light on emerging from the polarizer is plane polarized in a plane at right angles to the optical section of the Nicol prism. When this light passes through a solution of an active substance such as glucose, the plane of polarization is rotated, or bent, at an angle to the right or to the left. If the analyzing prism is so placed that its optical section corresponds to that of the polarizer, the light passes through it to the eye without change; if, however, its optical section is at an angle with that of the polarizer the light from the latter is split into two rays, one of which is reflected, so that only part of the light passes to the eye and the field is less light than when the optical sections coincided; and when the angle of the optical section of the analyzer is at right angles to that of the polarizer, no light at all comes through it, all being reflected, at the plane of section of the prism, to the side where it is absorbed by black surfaces. The field is then dark. If the

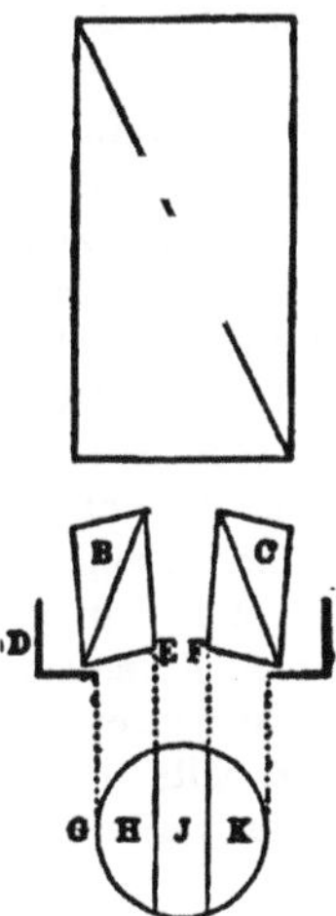

FIG. 6A.—Arrangement of the analyzer with two accessory Nicols to give a three divided field (Landolt).

analyzer is placed at the point of total absorption of the light this may be taken as the zero point. If now an active solution is placed between the analyzer and the polarizer the plane of polarization of the light emerging from the polarizer is twisted to one side, hence the vibrations of the light entering the analyzer are no longer in the plane of the optical axis, in which case they would be totally reflected, but they are at an angle with that so that more or less of the light comes through. It is necessary to rotate the analyzer to one side or the other to again produce the complete absorption of the light. If it is necessary to rotate to the right, the substance is said to be dextro-rotatory. In order to make the polariscope more sensitive it is common, in the better instruments, to introduce close to the analyzer and between it and the solution two small prisms, Nicols, so placed that they project with a sharp edge partly across the circular field. These prisms are fixed, and their edges are focussed by the observing telescope. This has the effect of dividing the field of view into three parts as shown in Figure 6A. At the zero point these three fields should have the same illumination. The advantage of this is that the zero end point is more sharply determined, since the shade of the three fields may be matched very exactly. Some instruments have three prisms in addition to the Nicol polarizer, giving a four divided field. These instruments are called two, three or four shadow instruments respectively. In using the polariscope it is essential that the light should be uniform in the field, of a maximum brightness, it should be carefully centered through the apparatus and into the eye, and the polarizer, C, should be turned to a minimum angle which it is possible to read clearly. If colored solutions are to be examined, it is necessary to select a colored light which is not absorbed by the solution. A mercury lamp is useful as a source of light when combined with the proper light filters.

We may now return to the problem of the way in which we shall represent on a plane surface the fact that several aldose sugars of the general formula $C_6H_{12}O_6$ are known. How shall the different structures of these molecules be pictured? A careful study of the possible arrangements of the atoms in the molecule shows that there are eight different aldose, hexose, stereo-isomeric carbohydrates possible, depending on the arrangement of the hydrogens and hydroxyls in the chain, and that there are two optical antipodes of each of these stereo-isomers, making sixteen possible aldose hexoses in all. Not all of these have been found in nature. It will be seen that there are four asymmetric carbon atoms in each hexose molecule. The number of possible stereo-isomers of any substance may be found from the formula: Number$=2^n$, where n is the number of asymmetric carbon atoms in the molecule. Some of the structural formulas of the sixteen aldose hexoses and ketose hexoses are given below. Their different structures are represented on a plane surface by writing the formulas with the aldose group at the top and the alcohol and hydrogen atoms variously placed at the sides of the carbon atoms.

d-glucoce.	l-glucose.	d-galactose.	l galactose.
COH	COH	COH	COH
HCOH	HOCH	HCOH	HOCH
HOCH	HCOH	HOCH	HCOH
HCOH	HOCH	HOCH	HCOH
HCOH	HOCH	HCOH	HOCH
CH_2OH	CH_2OH	CH_2OH	CH_2OH

l-talose.	d-talose.	d-mannose.	l-mannose.
COH	COH	COH	COH
HOOH	HOCH	HOCH	HCOH
HCOH	HOCH	HOCH	HOOH
BCOH	HOCH	HCOH	HOCH
HOCH	HCOH	HCOH	HOCH
CH_2OH	CH_2OH	CH_2OH	CH_2OH

l-gulose.	d-gulose.	l-tagatose.	d-tagatose.
COH	COH	CH$_2$OH	CH$_2$OH
HCOH	HOCH	CO	CO
HCOH	HOCH	HCOH	HOCH
HOCH	HCOH	HCOH	HOCH
HCOH	HOCH	HOCH	HCOH
CH$_2$OH	CH$_2$OH	CH$_2$OH	CH$_2$OH

l-idose.	d-idose.
CHO	CHO
HOCH	HCOH
HCOH	HOCH
HOCH	HCOH
HCOH	HOCH
CH$_2$OH	CH$_2$OH

l-sorbose.	d-sorbose.	d-levulose.	l-levulose.
CH$_2$OH	CH$_2$OH	CH$_2$OH	CH$_2$OH
CO	CO	CO	CO
HOCH	HCOH	HOCH	HCOH
HCOH	HOCH	HCOH	HOCH
HOCH	HCOH	HCOH	HOCH
CH$_2$OH	CH$_2$OH	CH$_2$OH	CH$_2$OH

b. *Pentoses. Isomerism.*

There are three asymmetric carbon atoms in each pentose, so that there are possible 2^3, or eight possible isomers of the aldoses. The structural formulas are as follows:

d-arabinose.	l-arabinose.	d-xylose.	l-xylose.
COH	COH	COH	COH
HO—C—H	H—C—OH	HO—C—H	H—C—OH
H—C—OH	HO—C—H	H—C—OH	HO—C—H
H—C—OH	HO—C—H	HO—C—H	H—C—OH
CH$_2$OH	CH$_2$OH	CH$_2$OH	CH$_2$OH

COH	COH	COH	COH
H—C—OH	HO—C—H	HO—C—H	H—C—OH
H—C—OH	HO—C—H	HO—C—H	H—C—OH
H—C—OH	HO—C—H	H—C—OH	HO—C—H
CH_2OH	CH_2OH	CH_2OH	CH_2OH
d-ribose.	l-ribose.	d-lyxose.	l-lyxose.

Of these pentoses, d-ribose, xylose and arabinose are of most interest to biologists, d-ribose being found in some nucleic acids (guanylic and yeast); and arabinose occurring in the gum associated with, or making part of, the enzyme, amylase. Xylose (Gr. *Xylon*, wood) is a pentose obtained by the hydrolysis of straw or wood. The pentoses generally occur in nature in gums and tetra, or polysaccharides. Xylose has been found as a constituent of the cephalopod muscle and other tissues (Henze). There are also ketose pentoses.

c. *Heptoses.*

A heptose, d-mannoheptose, called volemose, was found by Bourquelot in Persea gratissima and Lactuarius volemus, and by Allard in the dry residue of Primulaceæ. An unknown heptose, osazone melting at 195° and formed only on long heating, was isolated from human urine (Rosenberger).

Dissociation of the monosaccharides. Reactions of the monosaccharides of biological interest.—The monosaccharides, while comparatively stable in the test tube, are very unstable in living matter. They break up there and are converted into fats, proteins and other substances. How they are rendered so unstable by the protoplasm is unknown and is a very interesting problem on which many men are at present working. It will help us to understand the possible causes of instability in living matter, if we study how this instability or decomposition may be produced outside the body; and into what kinds of substances the carbohydrates break up when they are thus decomposed. Among the agents which we may use to produce decomposition of the mono- or disaccharides, alkalies and acids are the simplest.

Action of alkalies on monosaccharides.—All the monosaccharides and some of the disaccharides are unstable in alkaline solution and decompose into a variety of substances. If a solution of glucose, levulose, galactose, maltose or lactose is made alkaline, it turns a yellowish-brown color and acquires a smell of caramel. If heated, this change goes on more rapidly and the solution quickly turns brown in some cases, or yellow in others. This behavior is the basis of Moore's test for sugars. The stronger the alkali the more rapid is the change. If, however, air has free access to the alkaline solution being shaken with it or drawn rapidly through it, and if the alkali is not too strong, the brown color does not develop, but a rapid oxidation occurs, causing at times a faint phosphorescence and always liberating heat.

Chemical examination of the brown liquid shows that the monosaccharides and many of the disaccharides have undergone profound decomposition even if the amount of alkali is small. In strong alkali a great number of acids are produced having six, five, four, three, two or one carbon atoms in them. Moreover, volatile substances appear in the absence of oxygen, which give the iodoform test like ethyl alcohol but which are more probably glycolaldehyde, or oxyacetone, or glyoxal. Condensation products are also formed, in the absence of oxygen, leading to the development of the brown color due to humus and caramel substances.

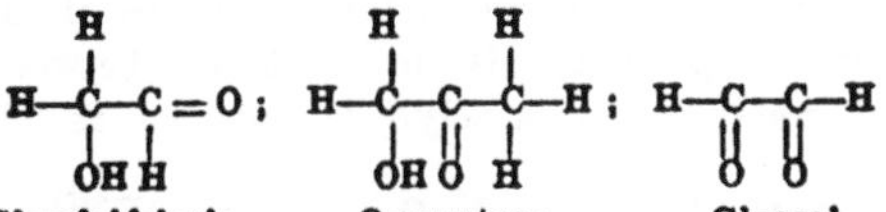

```
     H                H     H
     |                |     |
  H—C—C=O ;       H—C—C—C—H ;    H—C—C—H
     |  |             |  ‖  |         ‖  ‖
    OH  H            OH  O  H         O  O
 Glycolaldehyde.     Oxyacetone.      Glyoxal.
```

If the alkali is very weak a molecular (tautomeric) rearrangement of the sugar molecule occurs, accompanied by very little or no decomposition of the carbon chains (Nef). Thus d-glucose, d-mannose or d-levulose have the same configuration of the molecule except in the first two carbon atoms of the chain, as may be seen in the structural formulæ on page 28. If any one of these sugars is dissolved in weak alkali and allowed to stand, all the other sugars of this group appear in time in the solution. Thus there is the formation of a ketose, d-levulose, from an aldose, d-glucose, sugar. On the other hand, the sugars of the galactose series, such as tagatose, sorbose or talose, do not appear. Only those sugars appear which involve a change in structure of the first two or three carbon atoms of the chain, thus showing that the molecule is most unstable and reactive at this end. There is, as it were, a gradient of reactivity in the molecule from the aldehyde end extending downward, resembling, superficially at any rate, the gradient in reactivity in an earthworm, which is most reactive at the head end. The transformation of an aldehyde to a ketose sugar, and from the one isomer to the other, probably takes place with the intermediate formation of an enol modification as follows:

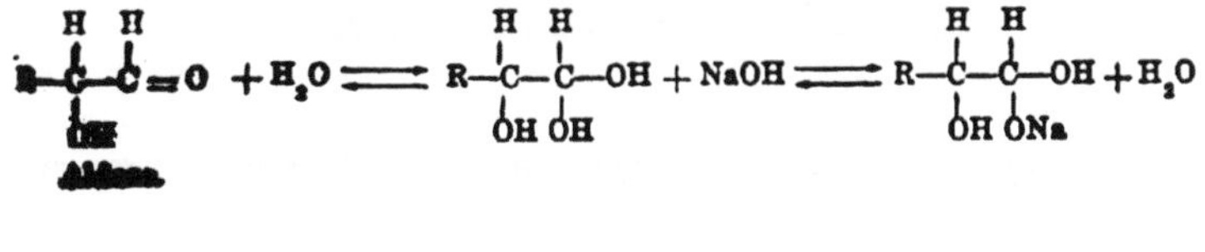

```
  H  H                 H  H                       H  H
  |  |                 |  |                       |  |
R—C—C=O + H₂O ⇌  R—C—C—OH + NaOH ⇌  R—C—C—OH + H₂O
  |                    |  |                       |  |
  OH                   OH OH                      OH ONa
 Aldose.
```

$$\mathrm{R{-}\overset{H}{\underset{OH}{C}}{-}\overset{H}{\underset{ONa}{C}}{-}OH} \rightleftharpoons \mathrm{R{-}\overset{H}{C}{-}\overset{H}{C}{-}OH}\ (\text{with } \mathrm{O} \text{ bridging the two C}) \rightleftharpoons \underset{\text{Enol.}}{\mathrm{R{-}\overset{OH}{C}{=}\overset{H}{C}{-}OH}} \rightleftharpoons \underset{\text{Ketose.}}{\mathrm{R{-}\overset{O}{C}{-}\overset{H}{\underset{H}{C}}{-}OH}}$$

Their great instability in alkaline solutions makes it necessary in evaporating sugar solutions to be sure that the solution is exactly neutral.

The explanation of this behavior of the sugars is very interesting. The decomposition of glucose may be taken as a type of all. The first thing which happens when mixed with an alkali like sodium hydrate is that a union occurs and a salt is formed. One of the hydrogens of the glucose behaves as an acid hydrogen and is, hence, believed to be slightly ionized. It is this hydrogen which is replaced by sodium. The hydrogen thus replaced may be the hydroxyl hydrogen just behind the aldol group, the α hydroxyl, or else one of the hydrogens of the aldol group. An aldehyde easily opens up its double bonds between carbon and oxygen and adds water to form a polyhydric alcohol, as follows:

$$\mathrm{R{-}\underset{H}{C}{=}O} \longrightarrow \mathrm{R{-}\underset{H}{\overset{/}{C}}{-}O{-}} \longrightarrow \mathrm{R{-}\underset{H}{C}{<}^{OH}_{OH}}$$

With sodium hydrate there is formed the salt either:

$$\mathrm{R{-}\overset{H}{\underset{O-Na}{C}}{-}\overset{H}{C}{=}O}\ ; \qquad \text{or, } \mathrm{R{-}\overset{H}{\underset{OH}{C}}{-}\overset{H}{\underset{OH}{C}}{-}O{-}Na.}$$

This salt is unstable and the molecule now first forms enols and then breaks apart into a number of pieces, double bonds appearing first between the carbons in the manner described on page 36 and then disruption occurring at the double bonds, thus:

$$\mathrm{HO{-}\overset{H}{\underset{H}{C}}{-}\overset{H}{C}{=}\overset{OH}{C}{-}\underset{H}{C}{=}\overset{OH}{C}{-}\underset{H\ \ OH}{C}{-}O{-}Na};$$

and this is perhaps followed by subsequent decomposition into such pieces as:

$$\underset{1.}{\mathrm{HO{-}\overset{H}{\underset{H}{C}}{-}\overset{H}{C}{=}}}\ ; \quad \underset{2.}{\mathrm{{=}\overset{OH}{C}{-}\ \overset{H}{C}{=}}}\ ; \quad \underset{3.}{\mathrm{{=}\overset{OH}{C}{-}\ \overset{H}{\underset{OH}{C}}{-}O{-}Na}}\ ;$$

$$\underset{4.}{\mathrm{HO{-}\overset{H}{\underset{H}{C}}{-}\overset{H}{C}{=}\overset{OH}{C}\ {-}\overset{H}{C}{=}}}\ ; \quad \underset{5.}{\mathrm{{=}\overset{OH}{C}{-}\underset{H}{C}{=}\overset{OH}{C}{-}\overset{OH}{\underset{H}{C}}{-}O{-}Na.}}$$

By this dissociation pieces of varying numbers of carbon atoms are probably formed. These dissociated pieces are very reactive in their nascent state when the free bonds on the carbon are open. They undergo intermolecular changes into acids, aldehydes or alcohols; they have strong reducing properties and if oxygen is present they unite with it to form aldehydes and acids; but if sufficient oxygen is not present to oxidize each piece as rapidly as it is set free, the particles interact, condensation occurs, caramel and resinous substances are produced which cause the brown color.

The part which is hypothetical in the foregoing explanation is the composition of the fragments which are first formed under the action of the alkali. There is no doubt that a salt is first formed and that this salt is unstable and decomposes with an unsaturated enol state intervening. We infer the nature of the fragments from the composition of the final products.

This decomposition, or fragmentation, is probably closely similar to the decomposition of the sugars in living matter; and there, as here, if sufficient oxygen is present, as it probably is on the periphery of cells, sugar will be burned or oxidized to lactic, carbonic, formic, glyceric, tartaric or tartronic acids; while if oxygen is not present in sufficient amounts to burn these reactive pieces as rapidly as they are set free, and this will probably be the case in the interior of the cells, the pieces will reduce substances near them or each other, or they will condense with ammonia or with each other transforming into amino acids, aromatic substances, fatty acids and other products of the metabolism of the sugars. The important thing, however, to note in this and to remember, for we shall return to it in discussing the metabolism of the sugars and indeed of other substances in the body, is *that the decomposition or rearrangement of the molecule into reactive pieces is a preliminary to metabolic transformations.*

SOME ACIDS FORMED FROM THE CARBOHYDRATES BY OXIDATION.

HCOOH	Formic
HOCOOH	Carbonic
COOH—COOH	Oxalic
CH_3—CHOH—COOH	Lactic
CH_3—CO—COOH	Pyruvic
COOH—CHOH—COOH	Tartronic
COOH—CH_2—CH_2—COOH	Succinic
COOH—CH_2—CHOH—COOH	Malic
COOH—CHOH—CHOH—COOH	Tartaric
COOH—CHOH—CHOH—CHOH—CH_2OH	Ribonic
COOH—CHOH—CHOH—CHOH—CHOH—COOH	Saccharic
COOH—CHOH—CHOH—CHOH—CHOH—CH_2OH	Gluconic

The ionic theory explains the reason why the molecule is so unstable in the salt form, whereas it is so stable in the form of the free monosaccharide. This explanation is as follows: The sugar molecule itself ionizes in aqueous solution very little. This is shown by the fact that solutions of the sugars in water are non-conductors; the avidity of the sugar as an acid is very low. There are, hence, at any instant of time in the solution of a monosaccharide very few $C_6H_{11}O_6^-$ ions. The sugar is a weaker acid than carbonic; it is about as weak as boric, or hydrocyanic acid. The salt formed by the addition of sodium hydrate, however, ionizes easily, hence the salt is widely dissociated, just as sodium acetate is much more widely dissociated than acetic acid. Just why the sodium salt ionizes more than the hydrogen salt is not yet known, but it may be that it is connected with the power of the sodium to unite with water molecules through its reserve or extra valences, this power being absent in the hydrogen which apparently has very few such reserve valences. At any rate, whatever the reason may be, sodium does ionize more. As a result the oxygen atom of the carbohydrate from which the sodium is separated is left with a free negative charge and this may be supposed to exert an influence over the whole molecule, since the bonds between the atoms are electrical in nature, but the effect is strongest in the two or three terminal carbons. As a result the molecule loses water and double bonds appear at several places in the molecule. The double bond between carbon atoms is not stronger than a single bond, but it is weaker. Why it is weaker is not certainly known, but it may be because when two bonds are present the atoms can separate without electro-static stresses being set up between them, since each atom takes a positive and negative charge with it thus: $C^{-+}_{+-}C \rightarrow C^{+}_{-}$, and C^{-}_{+}. We shall find the same facts of the instability of unsaturated carbon compounds illustrated in the fats and indeed in other cases, double bonded compounds being generally more reactive than single bonded. A very striking example of a molecular rearrangement due to ionization is shown by the indicator phenol-phthalein in which the negative ion undergoes rearrangement to a red quinonoid substance.

If the alkali is very weak so that only the terminal and α carbon atoms are involved (Nef), condensation of the monosaccharides to form di- and polysaccharides may occur. Thus cane sugar may be synthesized from d-glucose. There is at first under the influence of the alkali a transformation of some of the glucose to d-fructose and then the condensation of some of these molecules with glucose to make cane sugar.

Action of alkalies on di- and polysaccharides.—The action of alkalies on disaccharides varies with the nature of the sugar. Cane sugar, for example, is a very weak acid, it has no free aldehyde group and it is stable in an alkaline solution. Consequently it does not break into fragments and accordingly cane sugar reduces an alkaline copper salt solution only at a very slow rate. The case is, however, quite different for those disaccharides like lactose and maltose which contain free aldehyde groups. They are almost as unstable as glucose in alkaline solution; their solutions turn brown or yellow very quickly. This is probably the reason, for example, that soda biscuits become yellow if too much soda is used in their preparation. The decomposition of the lactose of the milk by the alkali causes a yellow color. Solutions of lactose and glucose and maltose of the same molecular concentration and the same alkalinity absorb oxygen at almost the same rate. The character of the fragments into which the disaccharides decompose is still unknown. Many of the polysaccharides are quite stable in alkaline solution. This is the case, for example, with glycogen or animal starch. It is prepared by destroying all the protein matter of the cells by cooking the tissue with 30 to 40 per cent. potassium hydrate, a procedure which leaves the glycogen quite unaffected. Starch also is quite resistant to alkalies.

Action of acids on monosaccharides.—Not only do the carbohydrates decompose spontaneously in alkaline solution, but they do so in acid as well. The molecule is, indeed, most stable in the neutral form and least stable as a salt. The decomposition of the monosaccharides in acid solution is slower than in alkali and the decomposition is not so complete, so that the various steps can be followed and the intermediate products, or some of them, can be isolated. But it is as yet uncertain whether the decomposition in the acid is in all particulars identical with the early stages of the alkaline decomposition or not. Advantage is taken of the difference in the decomposition of hexose and pentose sugars in acid solution to distinguish between them.

If acid is added to the solution of a hexose and if the reaction is quickened by boiling, it will be found that the hexose decomposes. If the acid is strong, brown or black humus substances are produced; if the distillate is collected it is found to contain formic acid, carbon monoxide, a little hydroxymethylfurfural, and in the solution remaining in the flask considerable quantities of levulinic acid, $CH_3—\overset{\overset{O}{\|}}{C}—CH_2—CH_2—COOH$, are to be found. If, however, a pentose is subjected to the same treatment, it distills over almost quantitatively as furfuraldehyde or furfural, and this we detect by the colored condensation products it yields with aniline acetate, orcine, phloroglucine, resorcine and other substances. Pentoses may, therefore, be readily distinguished from hexoses by the large quantity of furfural yielded on distilling the former with acid, and by the levulinic acid formed under the same circumstances from a hexose. Berthelot gives the following figures illustrating the decomposition of hexoses on heating with phosphoric acid:

Glucose heated with phosphoric acid for several hours:

CO_2	2.07%
CO	1.19
Formic acid	11.90
Levulinic acid	39.88
Humic acid	23.60
	78.64
Loss (in part H_2O)	21.36

This different behavior of hexoses and pentoses is the basis of the orcine, aniline acetate and most tests for pentoses. Bran and straw contain large amounts of polysaccharides containing a pentose (xylose), which are hydrolyzed by the acid. The free pentose is then converted into furfural. Bran heated with acid yields accordingly a great deal of furfural.

As has already been stated the decomposition of the monosaccharides by acids is closely analogous to that by alkalies, but is not so rapid or profound, and some of the intermediate products may be isolated. The acid first unites with the carbohydrate molecule, forming a salt. This is shown by the fact that the addition of a carbohydrate solution to a weak acid solution diminishes its acidity. The acids probably unite with the double bonded oxygen of the aldehyde, if this is present, or with the oxygen of the alcohol groups. It is well known that the aldehydes, ketones and other organic compounds containing oxygen have very weak basic properties, due to the oxygen they contain. They form with acids what are known as *oxonium salts*, analogous to the ammonium salts of nitrogen. Oxygen is at times tetravalent; it opens up two residual valences and is able thereby to unite with acids just as ammonia does. Thus we probably have in the sugars when treated with hydrochloric acid solution the compounds:

```
                    H               H
R—C=O ;   R—C=O—Cl ;   R—C=O          + Cl̄
  |         |            |   \
  H         H            H    +
```

The chloride thus formed ionizes, the chlorine being negative, the rest of the molecule positive and, possibly as a result of this free positive charge, molecular rearrangement takes place, water is eliminated and, in the case of pentose, furfural is formed thus:

```
  OH H  OH OH         +              H  H
   |  |  |  |        /               |  |
H—C—C—C—C—C=O—H + Cl̄ ——→ HC=C—C=C—C=O + 3H₂O + HCl
   |  |  |  |                 |_____O_____|   |
   H  OH H  H                                 H
```

In alkaline solution furfural is unstable and rapidly decomposes as already described, but in acid solution it does not readily decompose, but may be distilled. In the case of a hexose the reaction may go in various ways, hydroxymethylfurfural being formed in small quantities, perhaps as an unstable intermediate product, and many decomposition and condensation products, i.e., levulinic acid, humic acid, etc. Hydroxymethylfurfural has the following formula:

$$\begin{array}{c} \quad H \qquad\quad H \quad H \qquad\quad H \\ HO\overset{|}{C}-C=\overset{|}{C}-\overset{|}{C}=C-\overset{|}{C}=O \\ \underset{H}{|} \quad \lfloor\!\_\_\_\; O \;\_\_\_\!\rfloor \end{array}$$

Hydroxymethylfurfural.

It is probably the substance which is responsible for the color in the Molisch test and in the Seliwanoff reaction.

The behavior of the carbohydrates in acids and alkalies shows that there are two points of attack in the molecule, two points at which the molecule may unite with protoplasm or enzymes, namely, the residual or extra valences on the oxygen, where acids unite; or the double bonds in the aldehyde, where the alkali unites. It is possible, perhaps probable, that the decomposition of a monosaccharide by an enzyme is analogous to that produced by acids and alkalies, the enzyme upsetting the equilibrium of the molecule in the same manner as it is upset by acids or alkalies, the easiest change produced being in the aldehyde or α carbon groups.

It must not be concluded from this discussion that substances are always more unstable in the ionic or salt form than they are in the undissociated form. In some cases, at any rate, the reverse is the case. Thus cysteine, one of the amino acids, unites with oxygen with great speed in the neutral or undissociated form, whereas it is very stable in the acid solution where it exists as a salt, and oxidizes at a slow rate in an alkaline solution. And many other examples of this sort might be given. The free organic acids, undissociated, are indeed often less stable than their salts.

Action of acids on polysaccharides.—The di- and polysaccharides decompose readily into monosaccharides by hydrolysis when treated with acids, and in this respect they are less stable in acid than in alkaline solution. The various disaccharides break up with varying speed in acid solution and those which are the most resistant to the action of alkali are the most sensitive to the action of acid. Thus saccharose, which decomposes very slowly in the alkali, is the easiest of the disaccharides to invert with acid; and lactose and maltose, which are so unstable in alkali, hydrolyze much more slowly than cane sugar in acids. Starch and glycogen are quickly hydrolyzed by acid, but are very resistant to alkalies. It is by the action of acids on starch that commercial glucose is prepared. The probable reason why the polysaccharides are so sensitive to acids and so inert to alkalies is that, as their acidity is reduced by polymerization, they react very little with alkalies to form salts; on the other hand, their basicity is increased, causing them to unite more readily with acids. The reactions are written as follows:

$$C_{12}H_{22}O_{11} + H_2O + HCl \rightarrow 2C_6H_{12}O_6 + HCl.$$
$$(C_6H_{10}O_5)_n + nH_2O + HCl \rightarrow nC_6H_{12}O_6 + HCl.$$

Oxidation.—The monosaccharides are readily oxidized; and in acid, but particularly in alkaline, solution they are fairly strong reducing agents, oxidizing themselves to acids. Their reducing powers furnish one means of detecting them and measuring the quantity present in a solution. An alkaline solution of any monosaccharide will reduce methylene blue, cupric, silver, ferric, mercuric, gold or bismuth salts; and either in acid or alkaline solution they will reduce bromine, chlorine, permanganates, peroxides or atmospheric oxygen. They burn spontaneously with the liberation of heat, or even some light, in alkaline solutions, and with the liberation of electrical energy, heat and possibly light in protoplasm. By their combustion various acids are formed, carbonic, lactic, tartaric, malic, malonic, tartronic, oxalic, gluconic and saccharic. By very mild oxidation the osone is first formed, p. 43.

Fehling's Solution. They are quantitatively estimated, or qualitatively detected, by their reducing action on alkaline cupric tartrate solutions. Such a solution is that of Fehling. This consists of two solutions, A and B, which are kept separate until they are used. A consists of an aqueous solution of cupric sulphate containing 34.64 grams, $CuSO_4.5H_2O$, in 500 c.c.; B is made by dissolving 125 grams of potassium hydrate and 173 grams of sodium potassium tartrate (Rochelle Salts) in water and making up to 500 c.c. Just before using, equal quantities of the two solutions are mixed and heated to boiling and an equal, or smaller, quantity of the solution supposed to contain the sugar is added and the solution boiled in a qualitative test for two or three minutes. The solutions are kept separate until the time of using because they interact slowly, the tartrate slowly reducing the copper. This fact suffices to show the incorrectness of the statement sometimes made that the reducing action of the monosaccharides is due entirely to the aldehyde groups they contain. There are no such groups in the tartrates.

The various constituents of Fehling's solution act as follows: The sodium hydrate unites with the carbohydrate, decomposing it, in the manner already discussed, into a number (3—4) of reactive fragments, these fragments being the actual reducing substances. The tartrate is added to unite with the cupric hydrate formed by the interaction of sodium hydrate and cupric sulphate, which in the absence of the tartrate would be precipitated as a blue, gelatinous precipitate, turning into the black oxide on heating. This precipitate would obscure any cuprous oxide which might be formed by reduction. The tartrate by uniting with the cupric hydrate holds it in solution; the following reaction taking place:

```
Na—O—C=O                          NaO—C=O
      |                                 |
    HC—OH                             HC—O—CuOH
      |                                 |
    HC—OH + Cu (OH)₂ ——→              HC—O—CuOH + H₂O
      |                                 |
K—O—C=O                            KO—C=O
  Rochelle salt.
```

The cupric hydrate-tartrate compound is a deep-blue color and is soluble. It is in equilibrium with a very small quantity of $Cu(OH)_2$ which remains in solution; and this is in equilibrium with a very minute amount of free Cu ions and OH ions. The cupric ion having two free positive charges, which it gives up very readily, is the oxidizing agent and it is reduced to the cuprous state by the active reducing fragments formed from the sugar molecule by the alkali. Cuprous hydrate, which is thus formed, is a yellow substance which is very insoluble. It either precipitates as such or it loses water and forms the bright red, insoluble cuprous oxide, Cu_2O. This cuprous oxide may be filtered from the solution by an asbestos filter, washed and weighed direct, or it may be oxidized to the black cupric oxide and weighed as such; or dissolved in nitric acid and the copper determined by electrolytic deposition; or dissolved in ferric alumn and the ferrous salt formed titrated by permanganate. In this way the amount of reduction occurring in the solution may be determined. The best method is the last (often called Bertrand's). It will be seen that the tartrate-copper-hydrate acts as a reservoir of cupric hydrate, which is distributed all through the solution, and as soon as the cupric ions are reduced and the equilibrium thus upset, new cupric hydrate is at once dissociated from the tartrate compound. It thus happens that although at any instant of time there is only a minute amount of the intermediate substance, the cupric ion, present, yet since the ion is formed instantaneously the reaction can proceed at a rapid rate. If it took an appreciable time for the dissociation of the copper tartrate compound this solution could not be used in the way it is. This is a very good example, in all likelihood, of the manner in which a large transformation can take place through an intermediate stage, yet the intermediate stage itself be present at any instant of time only in the most minute amounts. One molecule of glucose will under certain conditions reduce four or five molecules of cupric hydrate.

Inasmuch as the reaction is not complete, but the oxidation continues at a slow rate for a long time, the malonic, tartaric and gluconic acids being slowly oxidized, the quantitative determinations have to be made under strictly comparable conditions of concentration, time of heating, etc., and the method is at best rather unsatisfactory. For the quantitative determination more in detail, see Part III, experiment 241.

Glucose, levulose or galactose will reduce copper sulphate without being made alkaline, but the time required is much longer, the reason being that the number of active sugar particles in such solutions and the number of hydroxyl ions is enormously less than in the alkaline solutions. On the other hand, the number of cupric ions is larger in the acid solution, but this is not sufficient to counterbalance the diminution in the active reducing particles formed from the sugar. If cupric sulphate is heated with levulose for some time the cupric oxide is reduced in large part to metallic copper, beautiful reddish crystals of copper being formed. The reaction in this case goes so slowly that the cuprous oxide is not formed rapidly enough to precipitate, but remains in solution and is further reduced to the metallic form. Copper acetate solutions with more or less acetic acid added to them may also be used in place of Fehling's solution. One such mixture is that of Barfoed, which is incorrectly used sometimes to distinguish between monosaccharide and disaccharide sugars. All sugars, however, reduce copper acetate with or without the addition of acetic acid if given time enough, and the separation of the different sugars by this test is purely quantitative, depending on the velocity with which various sugars oxidize. It is not qualitative. Of the various sugars levulose reduces in acid solution the most rapidly, since it is the strongest acid among the sugars; then galactose, glucose, maltose, lactose and cane sugar follow in the order named. Since there is quite a decrease in acidity or stability between glucose and maltose a glucose solution reduces Barfoed's quite a good deal faster than maltose under similar conditions of concentration and heating, so that with a strict control of these factors, Barfoed's solution may be used to distinguish qualitatively between these two groups of sugars, the monosaccharides on the one hand and the di- and polysaccharide sugars on the other; but a strong solution of maltose will reduce more rapidly than a weak solution of dextrose, so that as the test is ordinarily performed without control of the concentration of the sugar used, and time of heating, it is indecisive. It is sometimes stated that the disaccharides must be inverted first before reducing, but the reducing action of the disaccharides in alkaline solution, in which they are not inverted, shows that inversion is not a necessary prerequisite for reduction. The weak reducing powers of saccharose are due to the stability of the sugar in alkaline solution, but it will reduce Fehling's solution at a very slow rate. Glycerine also will reduce Fehling's solution, but the rate is so slow as not to introduce an appreciable error as the estimation is ordinarily carried out.

By oxidation of glucose by bromine in acid solution the sugar is converted quantitatively into gluconic acid, the monocarboxylic hexonic acid, and by further oxidation into saccharic acid. By nitric acid both of these acids are obtained from glucose, and galactonic and mucic acids from galactose.

Other substances reduced by the carbohydrates with a color change are described in the practical part. Among these may be mentioned white bismuth subnitrate which is reduced to black bismuth, or the suboxide; picric acid which is reduced to red picramic acid:

$$\underset{\text{Picric acid.}}{C_6H_2.OH.(NO_2)_3} \longrightarrow \underset{\text{Picramic acid.}}{C_6H_2.OH.NH_2.(NO_2)_2}$$

silver nitrate or mercury salts to black metals or mirrors. All of these changes are produced by reduction and may be caused by other substances than sugars.

The reduction of carbohydrates.—Carbohydrates not only have the power of oxidizing themselves at the expense of other bodies, thus acting as reducing substances, but they also may oxidize other substances, being themselves reduced thereby. When oxidized the reactions liberate heat; they are exothermic and it is by means of such reactions that the body gets its energy. But if the carbohydrates are reduced, the reactions are endothermic; heat is absorbed to be given out again when the substance is finally burned. Both kinds of reactions go on in living matter and in our own bodies in the decomposition of the carbohydrates of the food. The greater part of the carbohydrate is oxidized and the heat and energy are set free by which we live, but some of the fragments of the carbohydrate molecule do not oxidize at once; they are reduced, not oxidized; and by this means various important substances, pre-eminently the fats, are formed in the body. The carbohydrates have the power of absorbing nascent hydrogen and they are changed thereby in part to alcohols. Such alcohols are quite widespread in nature. Thus levulose and mannose are transformed into d-mannitol, a hexatomic alcohol; d-glucose into d-sorbite. Similarly the fragments into which the glucose molecule falls may take up hydrogen and be converted to the long carbon chains of fatty acids such as palmitic, $C_{16}H_{32}O_2$, and stearic, $C_{18}H_{36}O_2$, acids. But the exact steps of this process are still unknown. By means of this oxidizing property the carbohydrates play a great part in what is known as anærobic respiration. Many bacteria and animal tissues have the property of continuing to live and give off carbon dioxide in the absence of air. They can carry out a great number of reactions which are in part oxidative without atmospheric oxygen. It has been found that the injection of glucose enables fish and other animals to get on for a much longer time without atmospheric oxygen than when they have no glucose; and glucose is particularly favorable for the anærobic life of many bacteria. Glucose thus plays, in virtue of its oxidizing properties, a very important part in cell life. Tissues which contain glycogen or glucose will generally live longer in the absence of air than tissues which do not contain these substances. It has been suggested that in these cases the tissue oxidizes itself from the oxygen of the water, the nascent hydrogen set free being taken up by the glucose. Whether this explanation of the favorable action of glucose on anærobic respiration is correct or not is doubtful. It may be that one molecule is reduced to an alcohol, while the other sugar molecule is oxidized to an acid. This would be possible if the amount of heat set free by the oxidation of aldehyde to acid was greater than the amount of heat rendered latent by the reduction of aldehyde to alcohol.

Relation to hydrocyanic acid.—An interesting property of the carbohydrates is their power of uniting with hydrocyanic acid. If glucose be dissolved in water and potassium cyanide added to it, not all the hydrocyanic acid can be distilled off from the glucose on acidification. Some has combined with the glucose. Glucose exists, for example, in the white of hen's eggs to about 0.5 per cent. It was found that the alcoholic extract of egg-white had the power of binding hydrocyanic acid, due to the presence of this glucose. The importance of this power of hydrocyanic acid from a chemical point of view is that it enables one to build up the sugars carbon atom by carbon atom. A heptose may in this way be formed from glucose as follows:

$$CH_2OH-CHOH-CHOH-CHOH-CHOH-CHO + HCN \longrightarrow CH_2OH-CHOH-CHOH-CHOH-CHOH-CHOH-CN$$

$$CH_2OH.(CHOH)_5CN + 2H_2O \longrightarrow CH_2OH.(CHOH)_5.COOH + NH_3$$

$$CH_2OH.(CHOH)_5COOH + 2H \longrightarrow \underset{\text{Heptose.}}{CH_2OH(CHOH)_5COH} + H_2O$$

By the action of acids the nitrile is saponified to the acid which by reduction yields the aldehyde, glucoheptose. Physiologically the reaction is of interest because hydrocyanic acid unites with some substance in the cell which is of great importance in respiration; and it indicates that glucose should be a good antagonist to the cyanides. Whether this reaction plays any part in the building up of sugar in the plant cells is very doubtful.

Oximes.—Just as by the action of hydrocyanic acid the sugars may be built up carbon atom by carbon atom, so by the action of hydroxylamine the sugar of the next smaller number of carbon atoms may be obtained. Hydroxylamine, HNHOH, which corresponds to hydrogen peroxide, HOOH, and like it has reducing and oxidizing properties, acts on glucose to form the glucose-oxime thus:

$$CHOH)_4.COH + HNHOH \longrightarrow CH_2OH(CHOH)_4CH{:}NOH + H_2O$$
Glucose oxime.

vhen heated with acetic anhydride loses water and is con·he acetylated glucosenitrile; which when warmed with am·rer nitrate loses hydrocyanic acid and is converted into an entose from which the pentose may be set free.

ı.—A reaction very important for distinguishing monosac·ne formation of the osazones. If phenyl hydrazine and glu·be warmed the following reaction takes place:

$$C_6H_5NH.NH_2 \longrightarrow CH_2OH.(CHOH)_4.CH = N.NHC_6H_5 + H_2O$$
henyl hydrazine. Phenyl-gluco-hydrazone.

$$HOH)_4.CH = N.NHC_6H_5 + 2C_6H_5NH.NH_2 \longrightarrow$$
$$_2 + CH_2OH.(CHOH)_3.C(= N.NHC_6H_5).CH = N.NHC_6H_5 + NH_3$$
ne. Phenyl-glucosazone.

osazones are generally needle-shaped small crystals, yellow ne melting points of the different osazones differ somewhat glucosazone, 205° C.; galactosazone, 192-195°; maltosazone, zone, 200°.

azones which are formed in the reaction are generally not but in the case of mannose the hydrazone is insoluble and ss, plate-like crystals (m. p. 195°) which make the identifi· sugar very easy. If heated long enough, however, mannose yellow osazone also. The crystals of the various osazones istinguished easily by their shapes. Some of the substi· hydrazines give more insoluble osazones, the methyl or hydrazine being often employed. Since the sugars may ed from their osazones, this is one method which may be ir separation from a solution. It will be found that the hich the various sugars react with phenyl hydrazine differs, order of their ease of oxidation, levulose reacting with the and lactose most slowly. Levulose, mannose and glucose l osazones, showing that the last four carbon groups in ie same in configuration. Glucosamine readily yields In reaction 2 above the sugar is oxidized to an osone which th the third molecule of phenyl hydrazine.

zone is treated with hydrochloric acid and water, glucosone d, which on reduction yields d-fructose. Glucosone is an one of the formula: $CH_2OH.(CHOH)_3.CO.CHO$. By this n convert an aldose into a ketose. Glucosone is one of the of oxidation of glucose by mild oxidizing agents.

tion of the osazones is made in practice by using phenyl hydrazine n place of the free phenyl hydrazine. The hydrochloride is used for oth because the salt is more stable than the free base and because it is more soluble. Owing to hydrolytic dissociation of the hydrochloride, free hydrochloric acid is produced. This checks the speed of the reaction by reducing the dissociation of the sugar and, moreover, the reaction of the phenyl hydrazine with the sugar involves the free base of the hydrazine and not the ion of the salt. Accordingly to reduce the acidity of the hydrochloride of phenyl hydrazine and to increase the amount of free base it is customary to add sodium acetate to the phenyl hydrazine hydrochloride. This makes sodium chloride and phenyl hydrazine acetate, and since acetic acid is a very weak acid the phenyl hydrazine acetate undergoes hydrolytic dissociation into the acid and free base and the reaction goes much better than if the sodium acetate is not used.

By the use of methyl-phenyl hydrazine (Neuberg) the ketoses can be distinguished from the aldoses. The ketoses form osazones with this reagent, while the aldoses form only hydrazones. According to Betti the aldoses may also be separated from the ketoses by the use of β-naphthol-benzylamine,

$$\begin{matrix} C_6H_5 & \\ & \searrow CH-NH_2 \\ C_{10}H_6OH & \nearrow \end{matrix}$$

This combines easily with the aldoses to form a crystalline product, but not with ketoses. Glucose can thus be readily separated from levulose.

Reactions of carbohydrates with ammonia to form nitrogen containing compounds.—Glucose and other carbohydrates react readily with ammonia in weakly alkaline solutions to form nitrogen-containing substances such as the amino-acids. See page 186. By this reaction the proteins probably originate from the carbohydrates.

Synthesis in Plants.—All the energy of the animal body has come in the long run from the sun in the form of light energy. We are the children of the sun. The great synthetic agency on the earth's surface has not been heat, but light. Living matter could not have come into existence when the temperature of the earth was very different from what it is at present, for the activities of all living things are limited practically to the very narrow temperature range from about zero centigrade to about 70°. Light is to-day actively bringing about the syntheses underlying all vital processes and, in the absence of evidence to the contrary, we may believe that this has been the case since before the origin of living matter. Heat produces its syntheses in a very rough manner by the mechanical shocks of the rapidly vibrating molecules, or by the vibrations of the atoms in the molecules; but light, by a finer mechanism, its shorter waves, picks out the very bonds of the atoms themselves, the valence electrons, and tears them away from or injects them into the atoms. Light is thus an enormously more powerful chemical agent than heat. It seems from the work of Drude and his successors that it is the valence electrons which have the property of vibrating with light, and these are responsible for the refractive and color properties of substances. Of the light waves it is particularly the short ultra-violet rays which are most powerful.

The synthesis of the monosaccharides from carbon dioxide and water is at present carried on only in the chlorophyll or other pigment-bearing plants such as the red or blue-green algæ; but, as we shall see, chlorophyll is not necessary for this synthesis; it only makes possible the utilization of parts of the spectrum which otherwise cannot be used. The synthesis occurs in the chlorophyll bodies composed of chlorophyll and protoplasm; in these bodies carbon dioxide and water unite in the sunlight to form carbohydrates and liberate at the same time oxygen. If green leaves are exposed to carbon dioxide and sunlight starch appears in them. The same leaves in the dark make no starch and that which existed already disappears. Chlorophyll is not necessary for the transformation of a monosaccharide into a polysaccharide since glucose turns into starch in the roots, or tubers.

The method of the synthesis of the monosaccharides while in some particulars still obscure has been greatly illuminated by the discovery that light itself, and particularly ultra-violet light, even in the absence of all chlorophyll or protoplasm, will make formaldehyde and oxygen from a mixture of carbon dioxide and water. This has been the work of Berthelot and Gaudichon. For this synthesis ultra-violet light, that is tight of very short wave length, beyond the spectrum visible to our eyes, is most effective. If a mixture of carbon dioxide and water is illuminated by the light of a mercury arc in a tube of pure silica, this light being particularly rich in ultra-violet rays and silica not absorbing them as does glass, it has been found that the mixture very quickly contains hydrogen, oxygen, carbon monoxide, carbon dioxide, formaldehyde and hydrogen peroxide. The reaction goes to the point of equilibrium; it is never complete. If one start with a carbohydrate solution, or with formaldehyde and oxygen, the reaction is reversed to the same mixture, as if one started with carbon dioxide and water. The reaction which ensues may be written thus:

$$\text{Light} + 2CO_2 + 2H_2O \longrightarrow O_2 + CO + CH_2O + H_2O_2$$

The energy of the ultra-violet light has been absorbed and appears as the potential energy of the system "oxygen-formaldehyde." It is generally stated that the energy goes into the formaldehyde as potential chemical energy, but this is not strictly true: the energy is really represented by the system, "O_2—H_2," and by the system, "Oxygen-formaldehyde," for by the reunion of these substances the energy is set free. In a still narrower sense the energy may be said to be locked up in the oxygen molecule; the oxygen atom in this condition being actually less stable (in the presence of water) owing to its having one negative electron less than in the ionic form, or when it is united with hydrogen or carbon.

Oxygen seems particularly sensitive to the action of these light waves, a great many organic compounds being easily oxidized in light and in the presence of oxygen, but being quite stable in the dark in the presence of oxygen; or in light, in the absence of oxygen. Cholesterol, for example, behaves in this way. The light syntheses are also independent, or largely independent, of temperature, so that light syntheses and decompositions will occur at low as well as at high temperatures. In short, the chemist has in light an enormously more efficient and finer agent for chemical syntheses than he has in heat.

How very powerful these ultra-violet radiations are is made apparent by their action on the skin. An exposure of the skin to the light of a quartz mercury tube for thirty seconds is often sufficient to cause as severe a sunburn as one will ordinarily have after a day's exposure to the sun; and the destructive action of the rays on the retina or cornea is said to be so great that glass spectacles, preferably darkened, must always be worn in conducting these experiments. Blindness is easily produced by the rays. Their powerful irritant action on the skin is made use of therapeutically in treating skin diseases by sunlight or by the Finsen method. The irritant action of these rays greatly aggravates the skin eruption in smallpox. It is an interesting fact that although these rays irritate the cornea in so destructive a manner they do not give rise to a sensation of light. We cannot see by them. There are, however, creatures which can. Ants perceive the longer of these ultra-violet radiations; but on the other hand they cannot perceive light at the other end of the spectrum which is readily perceived by us, consequently ants under a brownish-red glass which absorbs the violet end of the spectrum are in darkness to them, whereas we can watch them without difficulty. The ultra-violet rays from the sun are for the most part absorbed in passing through the atmosphere, being absorbed by the oxygen and moisture of the air. But since carbon dioxide is always present in the air it is interesting to reflect that the fundamental synthesis of formaldehyde is probably going on all the time in the air at the expense of these rays; and H_2O_2 and hydrogen are constantly forming. Before the carboniferous period, when the coal now locked in the earth was partly in the air as carbon dioxide, this synthesis may have been far more extensive than it is to-day. Ultra-violet light is then absorbed by carbon dioxide and water and the energy thus absorbed becomes potential energy represented chiefly by the system oxygen-formaldehyde.

The amount of ultra-violet light reaching the earth's surface after passing the atmosphere is small, and the advantage of chlorophyll appears to be that it makes possible the utilization of the longer light vibrations, which reach the earth in greater abundance, or with more energy in them. Chlorophyll absorbs in the red or orange end of the spectrum and has hence a green color. The energy of the absorbed red light is then used in the synthesis of formaldehyde, just as the ultra-violet light is used in the absence of chlorophyll. It has been suggested that cholorophyll acts the part of a transformer, absorbing the long waves and either giving off shorter ultra-violet vibrations or in an indirect way bringing the same synthesis to pass. Chlorophyll seems to act much the part of the light sensitizers added to photographic plates to make them sensitive to the red end of the spectrum. The exact manner of action of the chlorophyll is, however, not yet clear. It has been shown by Priestly and Usher that chlorophyll extracted from leaves has the property of forming formaldehyde from sunlight, carbon dioxide and water, so that living matter is not necessary for this fundamental synthesis. It is, no doubt, not without significance that chlorophyll solutions are fluorescent and that uranium salts seem to have a similar photodynamic action on carbon dioxide.

Formaldehyde thus formed is itself the simplest of the carbohydrates. If made very faintly alkaline it has been found that it transforms itself spontaneously by condensation into a mixture of sugars called formose or acrose. This condensation goes probably as follows:

$$2H_2C=O \longrightarrow C_2H_4O_2.$$

Further condensation leads to the hexoses. This transformation is all down hill as far as the energy goes. That is, the reaction is exothermic, one molecule of a hexose liberates less energy on oxidation than do six molecules of formaldehyde.

The transformation of formaldehyde into a true monosaccharide is not, however, produced in plants by an alkaline reaction for plant protoplasm is almost neutral in reaction. It has a hydrogen ion concentration of about 2×10^{-8} normal. The condensation of formol would be slow under these conditions. Moreover plants make different sugars so that the transformation must be directed or regulated in some way. It is necessary, in order that the light synthesis should go steadily forward, that the oxygen and the formaldehyde should be removed from the reaction as quickly as possible so that equilibrium cannot be established. An additional reason for the removal of the formol is that the latter is very toxic and it must be changed to some non-toxic substance. Plants possess, to accomplish these ends, a catalytic agent of some kind as yet unknown which converts the formaldehyde almost as quickly as it is formed into hexose or pentose, so that at any instant of time there is not more than a trace of the aldehyde in the leaves. As has been said.

hydrogen peroxide is formed in the light reaction, but as all cells possess a catalase to convert this into oxygen and water, oxygen is at once freed and escapes. Hydrogen peroxide, if it accumulates, would destroy the chlorophyll. It is probable that some animal cells, as well as plants, are able to convert formaldehyde into a sugar, since some animals have chlorophyll and can form starch. Volvox and Euglena viridis have this power. It has recently been stated that turtle's liver can convert formaldehyde into glycogen, but the proof is not yet convincing.

It is probable that not all the formol produced by light synthesis, or which may be produced in the reverse process of the decomposition of the carbohydrates, is used in making carbohydrate. Formol is extremely reactive, combining with amino-groups (NH_2) wherever they are unsubstituted to make methylene derivatives, $R—N=CH_2$, which, by reduction, become methyl derivatives. Many methylated bases are found in plants and some also in animals. Such bases are choline, stachydrine, betaine, creatine. It is very interesting, also, that similar methylations happen in the animal body when pyridine and some other compounds are ingested, a fact which would indicate that formaldehyde might be an intermediate product of carbohydrate metabolism in animals, as well as in plants. So far as the author knows it has not yet been isolated from animal tissues as it has from plants, but formic acid is found in the brain and elsewhere.

The chemical composition of chlorophyll, which has this wonderful function in the world, is not yet known. It is easily extracted from leaves by ether, but is then mixed with various lipin impurities. When decomposed it yields, like hemoglobin, the coloring matter of the blood, pyrrol derivatives. It is evidently related more or less closely to the hematin of the hemoglobin, hemopyrrol being identical with phytopyrrol. Unlike hemoglobin it contains no iron, but the plant must have iron for its synthesis. Its close relationship to hemoglobin is further established by the discovery of the plant chromoproteins, phycoerythrin and phycocyan, which are crystalline conjugated proteins like hemoglobin, but they are found in plants and are closely related to their chlorophyll. Chlorophyll contains magnesium in place of iron in its molecule and it has been suggested that its powers depend on the presence of this element. But there is no good reason for this supposition except that magnesium is used by the chemist for certain condensations, although under widely different conditions from those prevailing in plant tissues. Hemoglobin also absorbs light, but it is the light chiefly at the violet end of the spectrum, although there are some bands in the green. By this absorption the blood pigment is supposed to protect the delicate tissues from the irritant action of the more refractive rays. It has recently been suggested that the iron which is always present in the chloroplasts of plant cells plays a very important part in the synthesis of the formaldehyde as well as of the chlorophyll. There can be no doubt that many colorless plant tissues, if exposed to light and if they contain iron, are able to synthesize chlorophyll.

Special properties of various carbohydrates. A. monosaccharides. —*d-Levulose or d-Fructose*. This is one of the most unstable of the monosaccharides. It is a levo-rotatory, ketose hexose. It shows the property of mutarotation. The d- does not mean that it is dextro-rotatory, but that it is related to d-glucose, which is dextro-rotatory. When first dissolved the solution exhibits a much greater rotatory power than after standing. The specific rotation of the fresh solution is $[\alpha]_D^{20} = -104.0^\circ$ but it falls gradually to $[\alpha]_D^{20} = -92.3^\circ$. The final rotation is obtained quickly by the addition of a small amount of sodium carbonate, or other weak alkali. It is one of the sweetest of the sugars, being sweeter than glucose. It is widespread in nature, being a constituent of many di- and polysaccharides. It is one of the constituents of cane sugar, melitose and lupeose, and inulin, a polysaccharide of the dahlia bulb, is composed of levulose. It reduces alkaline, or neutral, solutions of copper salts at a rate faster than does glucose, but the total amount of cuprous oxide formed per molecule of sugar is less than in glucose. It is prepared readily by the hydration of inulin by water at 100°, or by the hydrolysis of cane sugar by sulphuric acid. To prepare it from cane sugar the mixture of glucose and levulose known as invert sugar is freed from the acid, concentrated, cooled and calcium hydrate added. The levulose is precipitated as the calcium salt, which is filtered, suspended in water and decomposed by carbonic acid. Pure levulose is white and crystallizes in small needles (m. p. 95°), or in a dense mass, which, on standing, slowly decomposes, particularly in the light, and becomes a faint yellow color. Levulose is easily fermentable by yeast, since the last four carbon atoms have the same configuration as d-glucose, and for this reason it was called by Fischer d-levulose, although levo-rotatory. d- and l-levulose have been synthesized by the action of weak alkali from formaldehyde. The mixture was called α-acrose. When heated with hydrochloric acid, solutions of levulose turn a deep orange-red and by this reaction they may be easily distinguished from glucose solutions. Levulose may also be distinguished by the reaction of Seliwanoff described on page 865. It forms glucosazone, m. p. 205° C. Levulose together with other sugars is formed spontaneously from glucose by the action of very weak alkali.

The mutarotation of levulose is probably due, like that of glucose, to the fact that it exists in solution in two forms, a ketone and a lactone form, the ketone form being very much stronger rotatory than the other. The two forms suggested are the following:

```
    CH₂OH           CH₂OH
     |               |
     C=O          HO—C———┐
     |               |   |
  HO—C—H          HO—C—H |
     |               |   |
   H—C—OH          H—C—OH|
     |               |   |
   H—C—OH          H—C—O—┘
     |               |
    CH₂OH           CH₂OH
```

Ketone form. Lactone form.

d-Glucose or Dextrose. Also called grape sugar. $C_6H_{12}O_6$. This is the chief constituent of commercial glucose and is formed by the action of dilute acids on starch; it is one of the constituents of cane sugar and is found free in the sap of most plants and in the blood of many animals. Together with levulose it occurs in fruits and it may undoubtedly be called the most important monosaccharide. The glucose found in nature is d-glucose; it is dextro-rotatory, the specific rotatory power being either $+105°$, or $+22°$, as it exists in two varieties. Ordinary glucose in water solution has the specific rotatory power of $+52.5°$, being a mixture of the other two varieties, α glucose and γ glucose. This mixture is sometimes called β-glucose. In a fresh solution only the first, or α, variety exists, but it slowly transforms itself into the γ variety until a point of equilibrium is reached. This happens when there is present .368 parts of the first and .632 parts of the second. This is the explanation of the mutarotation which such solutions show. The point of equilibrium is reached only after a day at ordinary temperature and in neutral reaction; but the addition of even a small amount of alkali brings it to pass in a few minutes. The two forms of the glucose are supposed to have the structural formulas shown as follows:

```
     COH            H—C—OH
     |                |  \
   H—C—OH           H—C—OH┐
     |                |   |
  HO—C—H           HO—C—H |
     |                |   |
   H—C—OH           H—C—O—┘
     |                |
   H—C—OH           H—C—OH
     |                |
    CH₂OH            CH₂OH
```

α-d-glucose. γ-d-glucose.

The osazone melts at 205° C. Dextrose crystallizes with one molecule of water at ordinary temperatures having a melting point of 86°; from concentrated solutions at higher temperatures it crystallizes anhydrous, m. p. 146°.

Commercial glucose, or corn syrup, as it is often called in America, is made by the action of dilute acid, generally hydrochloric, on potato starch abroad, but out of corn starch in America. Sulphuric acid was at one time used in the hydrolysis and still is in some localities. The syrup known as commercial glucose contains often a large proportion of dextrins as well as glucose. There have been in the past cases of arsenical poisoning from the use of glucose prepared from sulphuric acid which contained arsenic. One such case occurred in England and resulted in the death of several persons. If commercial glucose contained only dextrose and dextrins, no objection could be taken to its use as a food, except on the ground that it is less sweet than cane sugar, and it may be used as an adulterant of cane syrup. Unfortunately, however, the purity of the commercial article, as is shown by the accident just referred to where it was used in brewing, is not always above suspicion. A further difficulty arises from the fact that purified starch may not be used as the raw product, and the possibility exists that other substances of a non-carbohydrate nature may find entrance to the final product.

The fermentation of glucose by yeast, with the production of alcohol, carbon dioxide and some other substances in small amounts, is one of the most important reactions with which the physiological chemist has to deal and it has of recent years been the subject of many investigations, but its mechanism is still obscure. The discovery of the mechanism of this process would possibly reveal the manner in which the dextrose molecule is broken down in the course of metabolism and the determination of this fact is one of the most fundamental problems of metabolism and nutrition. Many practical advantages would probably come from its discovery. The fermentation is brought about by an enzyme, *zymase*, which does not dissolve out of the yeast cell as long as it is alive and is accordingly called an endoenzyme. It was obtained by Buchner by grinding the yeast with sand, mixing it with diatomaceous earth and pressing the liquid out with an hydraulic press. It may also be obtained by treating yeast with acetone, or the vapors of methyl alcohol, and in other ways which cause the discharge of the fluid contents of the yeast cell. The nature of this fermentation change in glucose will be referred to again in the chapter on the metabolism of the carbohydrates.

d-Galactose. This monosaccharide, a hexose aldose, is found both in animals and plants. In the animal body it is made in the mammary glands and forms one of the constituents of lactose, or milk sugar, the other constituent being glucose; it possibly occurs also in the glycoprotein, mucin, of the saliva; and it is an important constituent of phrenosin and kerasin, substances making part of the myelin sheaths of nerves. It appears in the nervous system at the time myelination begins. In plants it occurs in the hemi-cellulose of the endosperm of

Molinia cœrulea which yields on hydrolysis galactose, mannose and arabinose; and also in the seed coats of peas and garden beans (Phaseolus vulgaris) which yield fructose, galactose and arabinose. Schulze states that carbohydrates yielding mucic acid when oxidized, and, hence, containing galactose, are almost as widespread in plant seeds as cane sugar. Lupeose, which is probably a tetrasaccharide from the seeds of Lupinus luteus and angustifolius, by hydrolysis yields one-half of the sugar as galactose. Lupeose is closely similar to stachyose and to a carbohydrate from the seeds of Cicer aretrium. Galactose (m. p. 162-164°) is a weaker acid than levulose, but a little stronger than glucose. In a copper acetate solution the oxidation goes at a slightly greater speed than that of glucose, but far slower than levulose. On oxidation in an acid medium by bromine, or chlorine, or nitric acid, it yields galactonic acid, $C_6H_{12}O_7$, and mucic acid, $C_6H_{10}O_8$. It is by the latter reaction that it can be distinguished. It is fermented very slowly by ordinary yeast. It is interesting that mucic acid corrodes teeth in a manner closely similar to the corrosion occurring in the mouths of some people. It may be the active principle of this corrosion. Galactosazone melts at 192-195°. The specific rotation of galactose is

$$(a)_D^{20°} = +81.5 \text{ (10\% solution); when first dissolved } (a)_D^{20} = +134.5°.$$

Glucosides. These are bodies widespread in nature, found both in animals and plants, which are characterized by the fact that on hydrolysis by acids they yield glucose, or some other monosaccharide. They are in reality ethers, that is compounds of some alcohol with the aldehyde group of the glucose or carbohydrate. Since they contain no free aldehyde they yield no osazones, nor do they reduce Fehling's solution without previous inversion. The simplest glucosides are the methyl and ethyl glucosides produced by the union of methyl or ethyl alcohol and glucose, in the presence of hydrochloric acid. Other sugars as well as glucose yield such unions. Thus there are pentosides in some of the simpler nucleic acids and glucosides in the more complex. The pentose in the former case is d-ribose and the other component guanine or adenine. The formula of methyl glucoside is the following (Fischer):

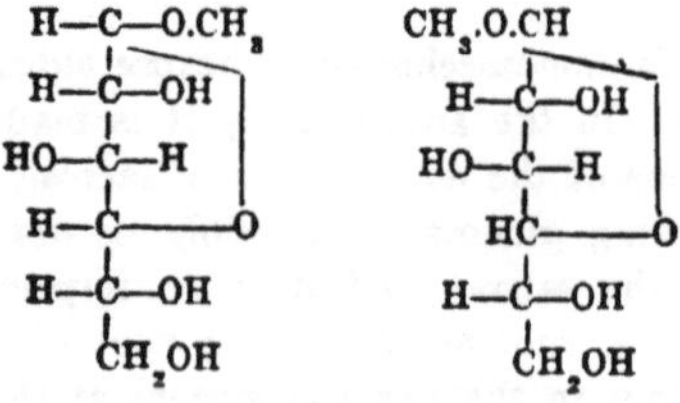

α-Methyl-glucoside. β-Methyl-glucoside.

There are, as will be seen from the formula, two classes of glucosides: the α and the β. The former are split by maltase; the latter by the enzyme, emulsin. Lactose is a β-glucoside. The following table[1] will give some idea of the variety and importance of this group of substances. Many of them form the active principles of some of the best-known drugs.

1. ***Ethylene derivatives.***

Sinigrin. The potassium salt of myronic acid, $C_{10}H_{16}NS_2KO_9H_2O$. Found in black mustard and horse radish. Has a burning taste. Decomposes on hydrolysis into glucose, allyl mustard oil, and potassium bisulphate. Fermented by the enzyme myrosin, found in mustard. The formula is as follows:

$$C_6H_{11}O_5.S.C(\text{—O—}SO_2OK)\ N\text{—}C_3H_5 + H_2O \longrightarrow C_6H_{12}O_6 + C_3H_5NCS + KHSO_4$$

Dextrose. Allyl mustard oil.

Sinalbin, $C_{30}H_{42}N_2S_2O_{15}$. Found in white pepper. It hydrolyses into mustard oil, $HO.C_6H_4.CH_2.NCS$, glucose and sinapin sulphate, a compound of choline and sinapinic acid and sulphuric acid. Its formula is as follows:

$$C_6H_{11}O_5.S.C.NCH_2.C_6H_4OH$$
$$\quad \searrow O.SO_2.OC_{16}H_{24}O_5N$$

Sinapin is $(CH_3O)_2C_6H_2(OH).CH{:}CH{:}CO.O.C_2H_4.N(CH_3)_3OH$

Choline.

2. ***Benzene derivatives.*** Arbutin, $C_{12}H_{16}O_7$, composed of glucose and hydroquinone, is found in the bearberry. It has a diuretic action, and some antiseptic powers.

Salicin, also called saligenin, is found in the willow. $C_{13}H_{18}O_7$. Ptyalin and emulsin hydrolyse it to glucose and saligenin, ortho-oxybenzylalcohol, $C_6H_4(OH).CH_2OH$. Populin, $C_{20}H_{22}O_8$, a benzoyl salicin, is found in the bark of the poplar, Populus tremula.

3. ***Styrolene derivatives.*** Derivatives of styrolene, $C_6H_5.CH{:}CH_2$.

Coniferin, $C_{16}H_{22}O_8$. Cambium of conifers. By emulsin hydrolyzed to glucose and coniferyl alcohol.

Phlorhizin, $C_{21}H_{24}O_{11}$. Root bark of cherry and other fruit trees. Yields glucose and phloretin, $C_{15}H_{14}O_5$, the phloroglucin ester of paraoxyhydratropic acid. Produces glucosuria in mammals. $C_{15}H_{14}O_5 + H_2O = OH.C_6H_4.CH(COOH).CH_3 + C_6H_6O_3$.

p-oxyhydratropic acid. Phloroglucin.

4. ***Anthracene derivatives.***

In this group occur many of the purgatives. The sugar may be rhamnose in place of glucose. Chrysophanic acid and emoidine of rhubarb occur as glucosides, or rhamnosides. Similar substances are found in Frangula and Jalap. Digitoxin, saponin and strophanthin are also glucosides.

[1] Abbreviated from a similar table in the Encyclopædia Britannica, article "Glucosides."

Amygdalin, $C_{20}H_{27}NO_{11}$, a glucoside in the bitter almond, is decomposed by maltase into glucose and mandelic nitrile glucose, and the latter substance by emulsin, a ferment found with the amygdalin, into glucose, benzaldehyde and hydrocyanic acid.

Saponin, a glucoside from Sapindus utilis, yields on hydrolysis d-fructose, arabinose, rhamnose and sapogenin, $C_{18}H_{28}O_3$. The saponin of horse-chestnuts yields sapogenin and arabinose, d-glucose and d-fructose (Winterstein and Blau).

We may mention also the fact that glucose enters in similar glucoside union to make many important animal substances, as, for example, some of the phosphatides which we shall consider more at length.

TABLE II. ILLUSTRATING THE HYDROLYZING ACTION OF VARIOUS ENZYMES ON DIFFERENT GLUCOSIDES.

Glucosides acted upon and hydrolyzed by

I Invertin	II Maltase	III Emulsin
Saccharose	Maltose	Amygdaline
Raffinose	Methyl-d-glucoside-α	Coniferine
Gentianose	Ethyl-d-glucoside-α	Piceine
	Benzyl glucoside	Salicine
	Glycerine glucoside-α	Helicine
	Amygdaline	Esculine
	Trehalose	Arbutine
	Methyl-d-fructoside	Lactose
		Methyl-d-galactoside-β
		Benzyl-glucoside (one isomer)
		Glyceryl-glucoside " "
		Methyl-d-glucoside-β

B. Disaccharides.—Hexose disaccharides have the formula $C_{12}H_{22}O_{11}$ and they are characterized by yielding two molecules of monosaccharides when hydrolyzed. The hydrolysis may be produced either by proper ferments or by the action of acids. The following are some of the hexose disaccharides which have been obtained, together with their place of occurrence and the monosaccharides which they yield:

Disaccharide	Occurrence	Yields on hydrolysis:
Cane sugar. (Sucrose) (Saccharose)	Sugar cane; beets (Saccharum officinarum)	Levulose Dextrose
Maltose	Germinating barley Digestion of starch	Dextrose Dextrose
Lactose	Milk	Dextrose Galactose
Trehalose	Various fungi. Boletus edulis. Ergot. Trehala	Dextrose Dextrose
Melibiose	From melitose in molasses Australian manna	Galactose Dextrose

Disaccharides have been artificially synthesized. Thus two molecules of glucose unite in the presence of acids to form a small amount of isomaltose, a disaccharide which is like maltose but differs from it in being amorphous and not fermenting with yeast. Cane sugar has been synthesized by Marchlewski from potassium fructosate and aceto-chloro-glucose, and also by Nef from weakly alkaline glucose solutions. In fact, it is probable that in every solution of glucose a small amount of maltose or isomaltose is formed spontaneously, but in the absence of acid, or the ferment maltase, the reaction goes very slowly.

Cane sugar. Sucrose or Saccharose. $C_{12}H_{22}O_{11}$. This is commercially the most important of the disaccharides. It is a crystalline sugar, very sweet to taste, which reduces Fehling's solution so slowly that it is generally said not to reduce it at all. It is readily oxidized in acid solution. It is dextro-rotatory, the specific rotatory power being $[\alpha]_D^{20^\circ} = +66.67^\circ$ (varies with concentration). It melts at 160°, and at 200° is changed to a brown mass of caramel by the loss of water. It yields saccharates with lime, strontia or barium hydrate. Cane sugar is inverted, that is changed into a mixture of glucose and levulose, by the action of acids, or the enzyme invertin, which is found in beets, in the intestinal secretions of mammals, in many plants and in yeast, Saccharomyces cerevisiæ. The fact that it does not reduce Fehling's solution and forms neither a hydrazone nor an osazone, leads to the conclusion that the aldehyde and ketone groups cannot be free but must be substituted, and the following structural formula has been proposed for it:

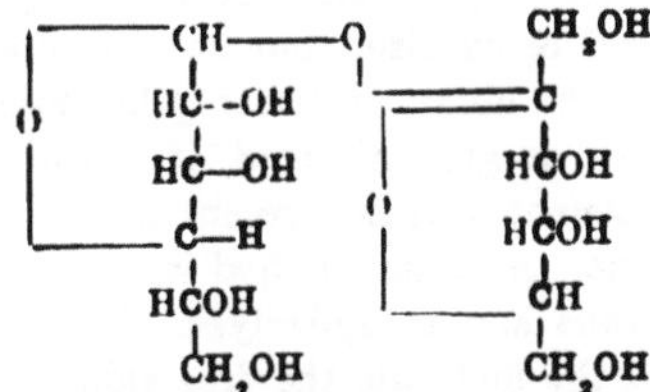

Fischer formula for cane sugar.

When cane sugar is hydrolyzed the superior rotating power of the levulose, as compared with the glucose, causes the total rotation of the mixture, which is called invert sugar, to be levo-rotatory, in place of dextro-rotatory. For this reason the sugar is said to be inverted. In studying the rate of rotation in the polariscope it is important to remember that dextrose has a far greater mutarotation than levulose. The glucose first set free from the levulose has a very high dextro-rotation, α-glucose, so that the inversion progresses faster than one would suppose from the polarimetric determination. This error can be avoided by adding to the invert mixture a small amount of sodium carbonate, which causes the glucose to take its stable rotatory power almost at once, and the reading will, therefore, enable one to determine the amount of inversion which has taken place before the alkali is added. The specific rotatory power of invert sugar is $[\alpha]_D^{20°}=-19.84°$ (c=5). Cane sugar does not combine with phenyl hydrazine.

Lactose. $C_{12}H_{22}O_{11}+H_2O$. This is the sugar of milk, but it is also found at times in the urine of pregnant animals and in the amniotic liquid of the cow. It is formed in the mammary glands. It is not nearly so sweet as cane sugar and unlike it, it reduces Fehling's solution as readily as glucose. It forms, also, an osazone melting at 200° C. When hydrolyzed by acids it breaks into galactose and glucose, and it hydrolyzes at a much slower rate than does cane sugar. It is also hydrolyzed by an enzyme, lactase, found in the intestinal mucosa. It shows mutarotation. When first dissolved $[\alpha]_D^{20°}=+87°$ (82.9?). Its final specific rotatory power is $[\alpha]_D^{20°}=+52.53°$. Lactose does not ferment with yeast nor can it be used as a food by animal tissues. It must be hydrolyzed first. It is hydrolyzed and fermented by lactase in the alimentary canal, and by several of the bacteria, for example by the bacillus coli communis, but not by the typhoid bacillus, and this constitutes, therefore, one way of distinguishing these two species. It is an interesting question whether lactose is a glucose-galactoside or a galactose-glucoside. In other words, is the aldehyde group of the galactose or of the glucose concerned in the union of the two molecules. This point can be determined in two ways. We may oxidize the lactose gently. By this the free aldehyde group will be oxidized, but not the substituted one. Then by hydrolysis there will be obtained either galactonic acid, if the galactose aldehyde was free; or gluconic acid, if the glucose aldehyde was free. The other component will be present as a hexose and it can be separated as the osazone. Another method is to form the osazone of the sugar and then the osone and to hydrolyze. The osone obtained indicates in which part of the molecule the free aldose group was (Fischer and Armstrong). By these methods it has been found that lactose is a glucose-galactoside. That is, the aldehyde of the galactose is substituted by the glucose molecule, the free aldose being the glucose. On the other hand, melibiose is a galactose-glucoside. The structural formula of lactose is hence probably that shown on the next page.

Lactose is easily distinguished from glucose and maltose by its nonfermentability by yeast and by its yielding mucic acid when heated with nitric acid. It is split by emulsin, but not by ptyalin. Since the free aldehyde group of lactose is in the glucose molecule, bromine water,

oxidizes this group, forming lactobionic acid, $C_{12}H_{22}O_{12}$, which on hydrolysis is converted into d-gluconic acid and galactose.

```
          H                       H
          |                       |
  ┌───────C───────────O───────────C—H
  │       |                       |
  │    H—C—OH                 HO—C—H
  │       |                       |
  │   HO—C—H                  HO—C—H
  │       |                       |
  └──O—C—H                     H—C—OH
          |                       |
       H—C—OH                 HO—C—H
          |                       |
         CH₂OH                   COH
     d-galactose.             d-glucose.
```

Formula of lactose.

Maltose. $C_{12}H_{22}O_{11}+H_2O$. This is one of the commonest and most important of the disaccharide sugars. It is formed by the action of amylase on starch, and since both the enzyme and starch are almost universal in the plant world, and the enzyme and animal starch, or glycogen, widespread in the animal world, maltose is very common in animal and plant tissues. It crystallizes readily in white needles which contain one molecule of water. It reduces Fehling's solution, but a given weight of glucose reduces more copper hydrate than the same weight of maltose. It forms an osazone, more soluble than glucosazone, which melts close to that of glucose at 206° C. Maltose is dextro-rotatory, its specific rotation being $[\alpha]^{20°}_{D}=+137$, computed for $C_{12}H_{22}O_{11}$ and after standing. The beginning rotation is +121. On hydrolysis by acids, or by the enzyme maltase, it splits into two molecules of dextrose. It, like cane sugar, is an α-glucoside and is, accordingly, not hydrolyzed by emulsin. (Contrary to lactose.) Maltose may be distinguished from glucose by the fact that when heated with dilute acids the reducing power of the maltose solution increases, whereas that of glucose undergoes no change. Maltose is not so sweet as cane sugar and it readily ferments with yeast, showing that yeast must have a maltase in it. It is possible that maltose may be directly utilizable by animal tissues, but this is still in doubt The formula of maltose is

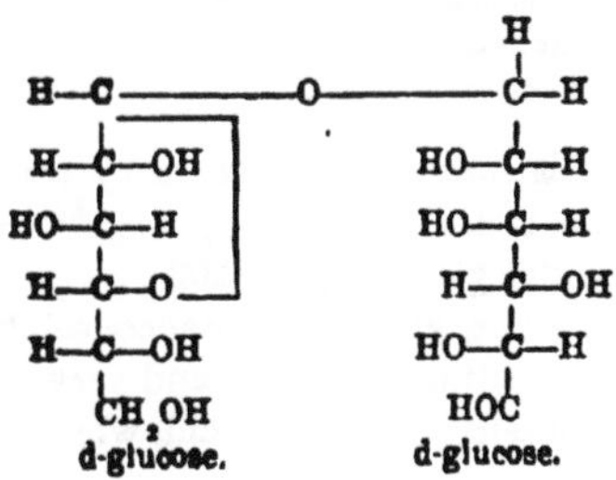

C. Polysaccharides which are colloidal in aqueous solution, or which are insoluble.—The more important of the polysaccharides are starch, or amylum, cellulose, glycogen, dextrins, various gums, mucilages and inulin. They are all soluble in water with the exception of cellulose, and they form emulsoid colloids, which may be precipitated by salts. Only a very brief account of some of their properties need be given here.

The formation of polysaccharides from monosaccharides is probably an attribute of all living matter, just as the formation of proteins from amino acids is an attribute of all living matter without exception. This transformation apparently takes place very easily. Thus a small amount of alkali will cause a condensation of some molecules of monosaccharides, and small amounts of acid will make some iso-maltose from glucose. We do not know, however, how this transformation is produced in living matter. Whether the condensation is due to a simple dehydration, or whether the phenomena of oxidation and reduction play a part in it. It seems in animal tissue, at any rate, that the synthesis of glycogen from glucose cannot take place if the cell is anesthetized, a fact which is taken by some to indicate that the respiration of the cell is in some way involved in this condensation. The whole subject is a fertile field for investigation.

Starch. Starch, or amylum, is a polysaccharide of glucose which it yields on hydrolysis with acids. It may be composed of maltose groups, since it yields maltose on hydrolysis by ptyalin, an enzyme. The formula is generally given as $(C_6H_{10}O_5)_n$, but it is very hard to prepare starch entirely free from phosphoric acid. It is possible that phosphoric acid is in union with the starch molecule. The size and composition of the starch molecule is still very uncertain. A recent determination of the number of hexose molecules in the molecule of some of the polysaccharides using a colorimetric method gave the following results:

Starch	7	hexose groups.
Glycogen	8-9	" "
Erythro-dextrin	4	" "
Achroodextrin	4	" "
Raffinose	3	
Maltose	2	
Lactose	2	

Soluble starch, which is formed from starch by the action of hot water, dilute acids or amylase, is dextro-rotatory, $(\alpha)_D^{17.6°}=+190.24$ (c=3.995). Brown, Morris and Millar give $(\alpha)_D=+202°$.

Gums. These are white, gummy substances soluble in water but generally precipitated by Fehling's solution and very widespread in nature. They nearly always contain phosphoric acid, which it is impossible to get entirely separated from the organic matter without hydrolysis. The phosphoric acid appears to be in union with the gum molecule and it is not impossible that it plays a very important part in the synthesis of the gum. These may be complex phosphatides. The gums, like the starches, form electro-negative colloidal solutions. Gums are found chiefly in the plant world, although there is some evidence that they occur also in the animal world. There is one always associated with the enzyme, invertin. It is a white gum, easily obtained from brewer's yeast. It yields on hydrolysis both mannose and glucose. Many of the gums contain either rhamnose, or arabinose, or other pentoses. Thus the gum associated with the enzyme amylase yields arabinose on hydrolysis. The mucilages resemble the gums except that they are more hygroscopic and their solutions will not filter. Gum arabic is one of the best-known gums. Many of the gums contain galactose.

Cellulose. This is the main constituent of wood. This polysaccharide forms the main part of the wall of plant cells. It is also found in some animals such as the tunicates. There are probably many different celluloses, but the composition of the more complex members of the group is still unknown. Cellulose is insoluble in all ordinary solvents, but dissolves in ammoniacal copper sulphate solution. Cellulose does'not reduce Fehling's solution, but on hydrolysis with acids it yields glucose and some other sugars which do reduce. Nitric acid converts cellulose into the nitro-derivative known as guncotton, a very explosive substance. Hemi-celluloses are found in many seeds and young plant tissues and they serve either as reserve foods or as supporting tissues. They yield on hydrolysis galactose, arabinose or rhamnose, mannose and, some of them, fructose. They are probably simpler in composition than the celluloses. Concentrated sulphuric acid dissolves cellulose, which may be precipitated from it by the addition of water. The cellulose is changed by this process into a compound which gives a blue color with iodine (amyloid).

REFERENCES.

Photosynthesis. *B. J. Moore:* The presence of inorganic iron compounds in the chloroplasts of the green cells of plants considered in relation to natural photosynthesis and the origin of life. Proceedings of the Royal Society, London, Series B, 87, p. 557, 1914. Ibid., 87, p. 163, 1913. References will be found to the work of Berthelot and Gaudichon, Priestly and Usher and others.

Products of decomposition of various sugars. *J. U. Nef:* Dissoziationsvorgänge in der Zuckergruppe. Annalen der Chemie, 403, p. 204, 1913. References to Nef's earlier work will be found here, and also to that of Lobry de Bruyn and Alberda van Ekenstein.

Rate of oxidation of various sugars in alkalies and acids. *Mathews:* The spontaneous oxidation of sugars. Jour. Biol. Chem., Vol. VI, p. 3, 109.

Mathews and Bunzel: The mechanism of the oxidation of glucose by bromine in neutral and acid solutions. Journal of the American Chemical Society, Vol. XXXI, p. 464, 1909.

Articles in the Encyclopædia Britannica on *Sugar and Glucosides.*

Molecular form. Life of Louis Pasteur by Vallery-Radot. Translation. *Pasteur:* Recherches sur la dissymétrie moléculaire des produits organiques naturels. Leçons de chimie professée en 1860. Paris, 1861. (Translated into German. Ostwald's Classiker der exakten Wissenschaften. No. 28, Leipzig, 1891.)

Le Bel: Sur les relations qui existent entre les formules atomiques des corps organiques et le pouvoir rotatoire de leur dissolutions. Bull. Soc. Chem. (2), 22, p. 337, 1874. Various other papers later in the same publication. Also in the Comptes rendus, pp. 114, 304 and 417, 1892.

Simple sugars and glucosides. *E. Frankland Armstrong.* In the Series of Biochemical monographs. Longmans, Green & Co., London.

Betti: Sulla distinzioni degli aldosi dei chetosi. Gazz. chim. Ital., 1912, p. 288. Abstract. J. de Phar. et de Chim. (7), 6, p. 222, 1912.

Neuberg: Berichte, 34, p. 2626. *Hill:* A method of isolating maltose when mixed with glucose. Pro. Chem. Soc., 17, 1901, p. 45.

CHAPTER III.

THE LIPINS. FATS, OILS, WAXES, PHOSPHATIDES, STEROLS.

It will be recalled that all living matter contains a larger or smaller amount of organic substances which are soluble in alcohol, ether and other fat solvents. These substances help to give to protoplasm its properties of containing large amounts of water but not dissolving; and also the power of taking up readily and in large amounts chloroform, ether and other substances soluble in fats but not readily soluble in water. They are among the fundamental and ever-present constituents of living matter. They may be given the group name of lipins (Greek, *lipos*, fat). In this chapter the amount, chemical nature, origin and some of the general properties of the more important of these substances will be considered, while certain of them found chiefly in some special tissue like the brain will be dealt with more completely in the chapters dealing with the chemistry of the organs.

Properties.—While the group of lipins contains such widely different chemical substances as the aromatic essential oils, like clove oil, the true neutral fats, like mutton tallow, the sterols, which are aromatic alcohols, and the phosphatides, or phospholipins, which contain large amounts of phosphoric acid, the members of the group all possess two or three properties in virtue of which they are called lipins. These properties are their greasy, or fat-like feel, their solubility in chloroform and fat solvents, and their insolubility in water. They constitute, then, a very heterogeneous group, chemically and physiologically. The following classification based on that proposed by Gies will give a general view of these bodies:

LIPINS. CLASSIFICATION.

LIPINS. Constituents of protoplasm having a greasy feel soluble in alcohol-ether and insoluble in water.

1. (a) **Fats.** Neutral esters of glycerol and fatty acids which are solid at 20° C.
   (b) **Fatty acids.**
2. **Fatty oils.** Neutral esters of glycerol and fatty acids liquid at 20° C.
   - A. **Drying oils.** Harden on exposure to light and air. Linseed oil.
   - B. **Semi-drying oils.** Thicken slowly on exposure to light and air. Cottonseed oil.
   - C. **Non-drying oils.** Remain liquid on exposure to light and air. Olive oil.

3. **Essential oils.** Volatile, generally odoriferous substances of oily and of varied chemical nature, being aldehydes, acids, terpenes, alcohols, etc. Oil of cloves, turpentine, wintergreen, etc.
4. **Waxes.** Esters of sterols and fatty acids. Beeswax, carnauba wax. Sperm oil. Spermaceti.
5. **Sterols.** Alcohols, generally of terpene group, solid at ordinary temperatures. Cholesterol, phytosterol, cetyl alcohol, myricyl alcohol, etc. (Oxidation products terpenic acids.)
6. **Phospho-lipins.** Phosphatides. Fatty substances yielding on hydrolysis phosphoric acid and fatty acids.
   A. Mono-amino-monophospholipins. Lecithin, cephalin.
   B. Di-amino-monophospholipins.
   C. Mono-amino-diphospholipins.
7. **Glyco-lipins.** Fatty substances yielding on hydrolysis fatty acids and a carbohydrate, generally glucose or galactose. Contain no phosphorus. Cerebron. Kerasin. Phrenosin.
8. **Sulpho-lipins.** Fatty substances yielding on hydrolysis fatty acids and sulphuric acid. Sulphatide of brain. Protagon. Nature still undetermined.
9. **Amino-lipins.** Fatty substances containing amino nitrogen and fatty acids. Contain no phosphorus. Not well characterized. Bregenin.

Historical.—The fats were the first of the three great classes of food-stuffs to have their composition determined. This was the work of the French chemist, Chevreul, in 1814. He found that fats were saponifiable by alkalies into glycerol and fatty acid. He discovered and named cholesterin, or solid bile, which was a non-saponifiable, fat-like substance. Chevreul is notable, also, for living longer than any scientist since Aristotle, dying at the great age of 102, in the year 1889.

Amount.—The amount of lipins, that is of ether-alcohol soluble, fat-like substances, found in different cells and tissues is widely variable, and the character of the lipin is peculiar to each tissue, but in general a tissue composed chiefly of living matter, but not serving as a depot of fats, contains about 1-10 per cent. by weight of lipins. Of the total organic matter of a rapidly-growing tissue, such for example as an embryo pig, about 1.6 per cent is lipin. In some tissues, however, the amount is much greater. In the sperm, eggs, brain and supra-renal capsules of mammals, for example, the lipin makes about 7.58 to 19.51 per cent. of the wet weights; and in those tissues which are the store-house of fat, such as the subcutaneous fatty tissue, the mesenteric fat, or bone marrow, even as much as 90 per cent. by weight may be lipin.

The fats and fatty oils.—*Composition.* The fats and fatty oils are the neutral esters of the tri-hydric alcohol, glycerol, and certain higher fatty acids such as palmitic, $C_{16}H_{32}O_2$, stearic, $C_{18}H_{36}O_2$, oleic, $C_{18}H_{34}O_2$, etc. They are, therefore, compounds of carbon, hydrogen and oxygen, but they contain far less oxygen in proportion to the carbon than do the carbohydrates. The fats differ from the oils physically in that they are solid at ordinary temperatures, whereas the oils are liquid at ordinary (18-25° C.) temperatures; and chemically they differ in that the fatty acids of the fats are, for the most part or wholly, saturated; while in the oils some of them are unsaturated. A typical fat is tri-palmitin, or tri-stearin. These two constitute the chief parts of the fats of mammals. These fats have the following composition:

$$\begin{array}{l}
\mathrm{H{-}\overset{H}{\overset{|}{C}}{-}O{-}\overset{O}{\overset{\|}{C}}{-}(CH_2)_{14}{-}CH_3} \\
\mathrm{H{-}\overset{|}{C}{-}O{-}\underset{O}{\underset{\|}{C}}{-}(CH_2)_{14}{-}CH_3} \\
\quad | \\
\mathrm{H{-}\underset{H}{\underset{|}{C}}{-}O{-}\underset{O}{\underset{\|}{C}}{-}(CH_2)_{14}{-}CH_3}
\end{array}
\qquad\qquad
\begin{array}{l}
\mathrm{H{-}\overset{H}{\overset{|}{C}}{-}O{-}\overset{O}{\overset{\|}{C}}{-}(CH_2)_{16}{-}CH_3} \\
\mathrm{H{-}\overset{|}{C}{-}O{-}\underset{O}{\underset{\|}{C}}{-}(CH_2)_{16}{-}CH_3} \\
\quad | \\
\mathrm{H{-}\underset{H}{\underset{|}{C}}{-}O{-}\underset{O}{\underset{\|}{C}}{-}(CH_2)_{16}{-}CH_3}
\end{array}$$

Tri-palmitin, or palmitin. Tri-stearin, or stearin.

A typical oil is tri-olein, which is the chief constituent of olive oil and which is composed of glycerol and three molecules of the unsaturated acid, oleic acid.

$$\begin{array}{l}
\mathrm{H{-}\overset{H}{\overset{|}{C}}{-}O{-}\underset{O}{\underset{\|}{C}}{-}(CH_2)_7{-}CH{=}CH{-}(CH_2)_7{-}CH_3} \\
\quad | \\
\mathrm{H{-}C{-}O{-}\underset{O}{\underset{\|}{C}}{-}(CH_2)_7{-}CH{=}CH{-}(CH_2)_7{-}CH_3} \\
\quad | \\
\mathrm{H{-}\underset{H}{\underset{|}{C}}{-}O{-}\underset{O}{\underset{\|}{C}}{-}(CH_2)_7{-}CH{=}CH{-}(CH_2)_7{-}CH_3}
\end{array}$$

It is obvious that by substituting other fatty acids many different fats can be made. A great number of these exist in nature; and in fact the different tissues of the same animal, or the corresponding fats of different animals, differ either in the nature or proportion of the fatty acids in the fats. Thus cold-blooded animals, such as fishes and amphibia, have fats which are fluid at ordinary temperatures. They are, in reality, oils and contain much unsaturated fatty acid, either oleic or an analogous acid. The fat of the earthworm contains, for example, butyrin, 4.47 per cent.; olein, 87.42 per cent.; stearin and palmitin, 8.11 per cent. In naturally occurring fats, such as lard, lard oil, tallow or milk fat, there are always present the glycerides of various fatty acids. No natural fat as it occurs in the tissue contains only a single glyceride, and the glycerides may contain more than one kind of fatty acid. In lard and tallow, the glycerides are chiefly those of stearic and palmitic acid, but there is also present some olein. Lard oil, which is obtained from lard by cooling until the tristearin and palmitin have partially crystallized out and then pressing the warm mass, thus expressing the oil, consists largely of triolein. From beef fat, or tallow, a similar oil is obtained in the same way and this is called oleo oil, or oleomargarine, and is used in the manufacture of artificial butter.

It is important to remember that most naturally occurring oils, or oils pressed from fats, contain some cholesterol, or phytosterol, and often phospho-lipins. They are not pure oils.

Butter yields about 7 per cent. of volatile fatty acids which are mainly butyric, $C_4H_8O_2$, and caproic acid, $C_6H_{12}O_2$; but there are also small amounts of caprylic, $C_8H_{16}O_2$, and capric, $C_{10}H_{20}O_2$, acids. Of the non-volatile fat of butter about 30 to 40 per cent. is olein and 60 to 70 per cent. palmitin, with a little stearin. Butter also contains a small amount of a phospho-lipin and some cholesterol. Small amounts of other fatty acids, possibly derived from the phosphatide, such as arachidic and lauric, $C_{20}H_{40}O_2$ and $C_{12}H_{24}O_2$, have also been found in butter. The natural yellow coloring matter of butter is mainly carotin, with some xanthophyll, and is derived from the green fodder.

Oleomargarine, or butterine, is a buttery substance made mainly from oleomargarine oil, which is generally churned with milk. When made from clean materials it is a wholesome food, but the experiments of Mendel and Osborne suggest that its nutritive value is not equal to that of butter, since it did not promote growth of young rats as did butter.

Table IV contains the formulas, boiling and melting points of various fatty acids occurring in the natural fats. There are many others which are not included in this table.[1]

TABLE IV.

1. *Acids of acetic series.* $C_nH_{2n}O_2$

		Pressure	Boiling	Melting ° Cent.	Found in
Acetic	$C_2H_4O_2$	760	119°	17	Spindle tree oil.
Butyric	$C_4H_8O_2$	760	162.3°	-6.5	Butter fat.
Isovaleric	$C_5H_{10}O_2$	760	173.7°	-51	Porpoise; dolphin.
Caproic	$C_6H_{12}O_2$	760	202°	-8	Butter fat. Cocoanut oil. Palm nut oil.
Caprylic	$C_8H_{16}O_2$	761	236°	16.5	Butter fat. Cocoanut oil. Palm nut oil.
Capric	$C_{10}H_{20}O_2$	760	268°-270°	31.3	Butter fat. Cocoanut oil. Palm nut oil.
Lauric	$C_{12}H_{24}O_2$	100	225°	43.6	Laurel oil. Cocoanut.
Myristic	$C_{14}H_{28}O_2$	100	250.5°	53.8	Mace butter. Nutmeg.
Palmitic	$C_{16}H_{32}O_2$	100	271.5°	62.62	Palm oil. Lard.
Stearic	$C_{18}H_{36}O_2$	100	291°	69.32	Tallow.
Arachidic	$C_{20}H_{40}O_2$			77	Arachis oil.
Behenic	$C_{22}H_{44}O_2$			85-84	Ben oil.
Lignoceric	$C_{24}H_{48}O_2$			80.5	Arachis oil.

[1] For other data see table pp. 10-11. Leathes, The Fats. Longmans, Green, 1910.

	2. *Acryllic or oleic acid group.* $C_nH_{2n-2}O_2$				
Tiglic	$C_5H_8O_2$	760	198.5°	64.5	Croton oil.
Physetoleic	$C_{16}H_{30}O_2$			30	Caspian seal oil.
Oleic	$C_{18}H_{34}O_2$	100	285.5°	14	Most oils.
		10	223°		
Rapic	$C_{18}H_{34}O_2$				Rape.
Erucic	$C_{22}H_{42}O_2$	30	281°	35-34	Rape; fish oils.
	3. *Linolic series.* $C_nH_{2n-4}O_2$				
Linolic	$C_{18}H_{32}O_2$			50.5	Maize; cottonseed.
Tariric	" "				Oil of Pic.
Elaeomargaric	" "			48	Tung oil.
	4. *Linolenic acid series.* $C_nH_{2n-6}O_2$				
Linolenic	$C_{18}H_{30}O_2$				Linseed oil.
Iso "	" "				" "
	5. *Series* $C_nH_{2n-8}O_2$				
Clupanodonic	$C_{18}H_{28}O_2$				Fish, liver; blubber.
	6. *Hydroxylated acids.*				
Ricinoleic	$C_{18}H_{34}O_3$	15	25C°	4-5	Castor oil.
Quince oil acid	$C_{18}H_{34}O_3$				
	7. *Dihydroxylated acids.*				
Dihydroxy stearic acid	$C_{18}H_{36}O_4$			141-143	Castor oil.

Physical Properties. While the neutral fats and oils are not crystalline in the cells, the temperature being high enough to keep the fat liquid, they may often be crystallized on cooling their hot alcohol solutions. The crystals are long needles. They are quite insoluble in water for some reason as yet unknown. The affinity of fat and water is small. The molecular weight is large, for tristearin being 896. Fats and the fatty acids and soaps reduce the surface tension of water. The lipin collects in the surface film so that the concentration in the film is greater than in the water as a whole. For this reason an old surface of water contaminated by oil or soap contains more soap or oil than a new surface and has a lower surface tension. The oils have the property of spreading over water in an extremely thin layer of molecular dimensions. This layer, according to Lord Rayleigh, has a thickness of only 2.7×10^{-7} cms., a diameter but little greater than that of a fat molecule. It is possible that this property of collecting in surface films and of lowering the surface tension may be of value in some vital mechanics of protoplasm; but the whole matter of surface tension and, above all, the surface tension of the surface of contact of water and such a mixture as that of protoplasm is still too obscure to permit of any definite statement concerning this possible function of the fats.

The specific gravities of all oils and fats are lower than that of water.

Rape oil having a specific gravity of 0.915 is one of the lightest. The non-drying oils, like olive or sweet oil, range from 0.916-0.920; the drying oils are about 0.930; the ordinary fats 0.930; castor oil and cocoa butter have the highest specific gravity, i.e., 0.970. The lower the cohesion the more liquid will be the fat, the larger will be the space at the disposal of its molecules, the greater their freedom of movement and hence the lower the specific gravity.

Fats have no sharp melting points, since none of them are pure glycerides, all being mixtures. But even the pure glyceride tri-stearin shows a curious behavior. It melts at 55°, solidifies and then again melts at 71.5-75°. This is due possibly to a tautomeric rearrangement of the molecule. The drying oil, linseed oil, remains fluid to the lowest temperature —28° C. On heating, oils slowly decompose and at about 300° C. acrolein is given off. It is by the appearance of acrolein on heating that neutral fats may be distinguished from fatty acids.

Glycerol. There are only two or three reactions of glycerol which need mention at this time. One of the most important of its reactions is the formation of acrolein from it by heating. If glycerol is heated to 300°, and particularly if it is heated with some substance such as acid potassium sulphate, or P_2O_5, which has an affinity for water, it loses two molecules of water and is converted into the unsaturated aldehyde, acrolein. Acrolein has a very irritating action on the mucous membrane and may be detected by its sharp odor. The reaction of its formation may be written as follows:

$$\begin{array}{ccc} CH_2OH & & CH_2 \\ | & & \| \\ CHOH & -2H_2O \longrightarrow & CH \\ | & & | \\ CH_2OH & & CHO \\ \text{Glycerol.} & & \text{Acrolein.} \end{array}$$

By oxidation glycerol is converted into glycerose, a mixture of dioxyacetone and glycerine aldehyde, and finally into glyceric acid, CH_2OH-$CHOH$-$COOH$. Glycerol has a sweet taste. It reduces Fehling's solution only at a very slow rate. It readily esterifies with phosphoric and other acids.

$$\begin{array}{ccccccccccc} CH_2OH & & CH_2OH & & CH_2OH & & CH_2OH & & COOH & & COOH \\ | & & | & & | & & | & & | & & | \\ CHOH & \longrightarrow & CHOH & ; & CO & \longrightarrow & CHOH & \longrightarrow & CHOH & \longrightarrow & C(OH)_2 \\ | & & | & & | & & | & & | & & | \\ CH_2OH & & COH & & CH_2OH & & COOH & & COOH & & COOH \\ \text{Glycerol.} & & \text{Glycerylaldehyde.} & & \text{Dioxyacetone.} & & \text{Glyceric acid.} & & \text{Tartronic acid.} & & \text{Mesoxalic acid.} \end{array}$$

Fatty oils. The true fatty oils are divided into three general classes: into those which harden on exposure to air, light and moisture; those which dry slowly, and those which do not dry. The first are known as drying oils; the second as semi-drying; the last as permanent or non-drying oils. The first are used in painting, and linseed oil is a typical, and probably the most valuable, drying oil; while cottonseed oil dries very much more slowly, and olive oil is a non-drying oil. The chemical difference between these kinds of oils lies in the different stabilities of the fatty acids they contain. The drying oils have very unstable fatty acids; while the non-drying have relatively stable acids. The very unstable fatty acids of the drying oils are very reactive and as might be expected they are found, among other places, in the lipins of the brain, heart, liver (cod-liver oil); and in other organs of which the metabolism is very high; whereas the stable acids of the fats and non-drying oils are found in locations of lowered metabolism and in the depots of fats in fat tissues. Owing to their reactivity, the drying oils absorb oxygen and become rancid much more rapidly than the non-drying. The difference in stability of the fatty acids in the drying oils, such as linseed oil, and the non-drying, such as oleic, is due to the presence of more double-bonded carbon atoms, called unsaturated carbon atoms, in the linseed-oil group. Oleic acid contains only a single double-bonded couple of carbon atoms, whereas the drying oils contain acids having two or three such couples. The formula of oleic acid is $C_{18}H_{34}O_2$, or $CH_3.CH_2.CH_2.CH_2.CH_2.CH_2.CH_2.CH_2.CH{=}CH.CH_2.CH_2.CH_2.CH_2.CH_2.CH_2CH_2.COOH$, the double bond being in the middle of the chain. This is shown by the fact that on oxidation it yields nonylic acid, $C_9H_{18}O_2$, and azelaic acid, $C_9H_{16}O_4$. In linseed oil there occur such unsaturated acids as linolenic, $C_{18}H_{30}O_2$, with three pairs of double-bonded carbons, and linolic, $C_{18}H_{32}O_2$, with two pairs of double-bonded carbons.

Resemblance of the chemistry of painting to some biological processes. It is interesting to consider the many curious resemblances of the chemical processes involved in painting with protoplasmic respiration, memory and growth. The use of linseed oil in painting depends on the fact that it spontaneously oxidizes in the air, especially in the light, and decomposes, the decomposition products forming a resinous hard mass of a composition still largely undetermined. Linseed oil has the power, then, of spontaneous oxidation; it respires. It takes up oxygen and it gives off carbon dioxide and other volatile substances. Light, and particularly ultra-violet light, accelerates this respiration just as it does that of protoplasm. Moreover growth, or rather synthesis, occurs, for in the condensation following the decomposition substances are formed more complex than the linolenic acid. Heat moreover is set free. It is, in other words, a veritable metabolism which the linseed oil undergoes. But the relationship to protoplasmic metabolism does not end here. In living matter there are substances which hasten the oxidation, catalytic substances, or oxidases as they are called. In the presence of these substances oxidation goes on much more rapidly than in their absence. Cells use various substances as oxidative accelerators. Manganese salts are used by some; copper or iron salts by others; or organic oxides and peroxides, like the quinones, by others. The painter uses similar substances to accelerate the decomposition and drying of the oil. The oil is sometimes boiled in iron or copper vessels and dryers of various kinds are added to help the oxidation, such as manganese dioxide, litharge, manganous borate or iron salts; or he uses an organic oxidizer, turpentine, which in the light picks up oxygen with great ease and carries it

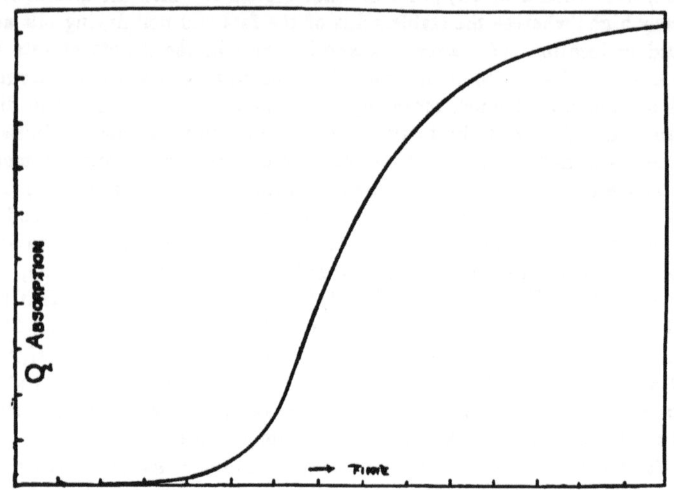

Fig. 7.—Curve showing the rate of absorption of oxygen by linseed oil. Ordinate: mms. of negative pressure due to absorption of oxygen; abscissa: time in days.

over to the oil. In this respect, then, the metabolism of paint resembles that of protoplasm.

But most remarkable of all is the fact that the oil may be taught to oxidize itself and it remembers its lesson for some time. If linseed oil is exposed to light, or ultra-violet rays, in the presence of air in a closed flask provided with a mercury manometer, for the first 24 to 36 hours nothing seems to happen; but then slowly the oil begins to oxidize and it oxidizes at a constantly accelerating pace so that the oxygen is used up in the flask and the negative pressure may be measured by the mercury manometer. The curve of the rate of absorption of oxygen is at first convex downward, not, like most chemical reactions, concave. Figure 7. It is the curve of an autocatalysis. It is as if the oil had to be taught by the light to oxidize itself and learned to oxidize better and better.

Now it may be shown that the oil remembers. If after 60 hours' illumination when the oil is oxidizing let us say at a fairly rapid rate, the illumination is discontinued and the oil put in the dark, the oxidation goes on at a slower pace. If after a period of a few hours in the dark the oil is again illuminated by the light, it will be found that the oxidation no longer waits 24 hours before beginning, but now the stimulation by the lamp is effective within an hour or more; the oil acts as if it remembered the teaching by a previous illumination and now oxidizes at a more rapid rate. However, oil can also forget. If left in the dark 24 hours or more after being taught to oxidize, it has forgotten and now the teaching must be done all over again, a long illumination being necessary before the oxidation begins. We do not usually speak of the long latent period of the oxidation as a period of teaching, but it is called in chemistry a period of "inductance"; and we do not say that the oil is learning to oxidize itself, and doing it better and better, but we say that it shows phenomena of autocatalysis; nor do we say that it forgets again in the dark, but that the intermediary, autocatalytic agent has disappeared; but when organisms show the same kind of phenomena we speak of teaching, of latent periods, of stupidity, of good or bad memories. And it is not impossible by any means that the phenomena of memory, shown in greatest perfection by the mammalian cerebrum, may have at the bottom some such basis as this, and the persistence within certain cells of substances of an autocatalytic nature which have remained from a previous stimulation. Perhaps the brain cells remember longest because they most carefully maintain intact, or preserve, these labile autocatalytic substances. It may be mentioned that the whole of growth is an autocatalytic process. There are always left over in the cell, at the close of a period of feeding, substances, enzymes, derived from the metabolism of the foods, which hasten the metabolism of the next succeeding feeding and hasten growth. It is because of the presence of these autocatalytic substances that foods change into protoplasm so much more rapidly in cells than outside of them.

Methods of identification of oils and fats.—The identification of the oils and fats is a matter of commercial as well as of scientific interest. A partial identification can be made by means of the melting point and by the determination of the iodine, hydrogen, oxygen, saponification, Reichert-Meissl and acetyl numbers.

The melting points, as already stated, are not sharp and definite, as fats are not pure substances but always mixtures, but nevertheless something may be learned from the melting points. A low melting point means either that the fats contain saturated acids of short carbon chains, or that there are long but unsaturated chains, as in oils. The melting point of a fat as it occurs in the tissue is generally, or always, lower than the usual temperature of the tissue in which it is found. Thus in cold-blooded animals, such as fishes, the fats are found to be really oils, liquid at 20°. In mammals the fat of the hoofs which are exposed to cold is generally oil (Neat's foot oil), while that about the kidneys melts at a high temperature and contains more tri-stearin.

The melting point of tri-stearin is 55°, permanent 71.5°
tri-palmitin 50.5°, " 66.5°
tri-olein − 6°

Iodine number. The unsaturated fatty acids have the property of adding atoms at the double bond. Thus iodine, bromine, oxygen and hydrogen may be added here. Hence these fats have reducing powers, in that they readily oxidize themselves at the double bond going over into the hydroxy-acid, or even into a ketone acid; they also have oxidizing powers, in that they will take up hydrogen and be converted into the more stable saturated fats. By reduction or by oxidation with bromine or iodine, the number of unsaturated bonds may be discovered, since each such bond adds two atoms of iodine or bromine, two of hydrogen and presumably two oxygen atoms.

$$C_8H_{17}-\underset{H}{C}=\underset{H}{C}-C_8H_{15}O_2+I_2 = C_8H_{17}-\overset{I}{\underset{H}{C}}-\overset{I}{\underset{H}{C}}-C_8H_{15}O_2.$$

Thus if it is desired to know whether linseed oil has been adulterated with cottonseed oil, it is only necessary to determine how much iodine in centigrams a gram of the oil will take up from a solution of iodine; or how much hydrogen or oxygen it will unite with. The first figure is especially important and is easily determined (see page 872); it is called the iodine number; but recently the two last, the hydrogen and oxygen numbers, have also become important aids in the identification of the fats.

TABLE V. IODINE VALUES OF VARIOUS OILS, FATS AND WAXES.

I. Vegetable.	
1. *Drying oils.*	
Linseed	175-205
Hempseed	148
Walnut	145
Sunflower	119-135
2. *Semi-drying.*	
Soja bean	122
Maize, corn	113-125
Beech nut	111-120
Cottonseed	108-110
Sesame	103-108
Rape	94-102
Brazil nut	90-106
3. *Non-drying.*	
Apricot kernel	96-108
Peach "	93-109
Almond	93-100
Olive	79-88
Grape seed	96
Castor	83-86
II. Animal oils.	
Fish	140-173
Menhaden	161

Sardine	161-193	2. *Semi-drying.*	
Herring	124-142	Horse	75-85
Cod liver	167		
Shark "	115	3. *Non-drying.*	
Whale oil	121-136	Goose fat	70
Dolphin	99-126	Lard	50-70
Horses' foot	74-90	Bone	45-56
Neat's foot	67-73	Beef tallow	38-46
Egg oil	68-82	Mutton tallow	25-46
		Butter	26-38
III. Vegetable fats.			
Laurel oil	68-80	**V. Waxes.**	
Palm oil	53	Sperm oil	81-90
Mace butter	40-52	Carnauba wax	13
Cacao butter	32-41	Wool	102
Borneo tallow	15-31	Bees	8-11
Palm kernel oil	13-14	Spermaceti (Cetin)	0-4
Cocoanut oil	8-9	Insect wax (Coccus ceriferus)	0-1.4
Japan wax	4-10		
		Iodine number of various fatty acids.	
IV. Animal fats.		Oleic acid	90.07
1. *Drying.*		Erucic	75.15
Ice bear	147	Linolic	181.42
Rattlesnake	106	Linolenic	274.1
		Clupanodonic	367.7

The hydrogen number. The discovery of a practicable method of adding hydrogen to unsaturated oils is a matter of great commercial importance, since by this means ill-smelling, or ill-tasting, or cheap vegetable oils, such as cottonseed or even linseed oils, may be converted into substances closely resembling true stable animal fats, which at present are far more expensive. The ill-taste and smell of the oil is due to the decomposition products which arise from the fatty acids in consequence of their unstable nature. By hydrogenation the true fats are formed. A method recently employed on a commercial scale for this hydrogenation has been by the use of hydrogen and finely divided nickel oxide which has the important advantage over finely divided nickel, in that it is said not to be so easily poisoned by the impurities in the oils. The only disadvantage in the use as foods of these cheap vegetable lards is that they usually contain small amounts of nickel which it has been impossible thus far to eliminate. Since nickel, when absorbed, is a toxic substance, the presence of even very small amounts of nickel in any food must be regarded as undesirable.

Ozonides. The unsaturated fats absorb oxygen. The heat generated by this process may cause the spontaneous inflammation of cotton. In an acetic acid solution oleic acid absorbs an equimolecular quantity of ozone. An ozonide is thereby formed, $C_{18}H_{34}O_5$, a viscid, almost colorless, transparent liquid, heavier than water, which does not combine with iodine and which is stable at 80-90°. At 120° this decomposes; 5 per cent. being converted into a gaseous mixture of the following composition: CO_2, 2.7-7 per cent.; CO, 71-88 per cent.; CH_4, 16.5 per cent.; H_2, 5.4 per cent. There is also formed a mobile oil (Molinari and Sonciri). Triolein forms the ozonide, $C_{57}H_{104}O_6.3O_3$; each molecule of the fatty acid adding a molecule of ozone at the double bond. This is a viscid, colorless oil decomposing at 136°, and when saponified by potassium alcoholate it yields glycerol, azelaic, $C_9H_{16}O_4$, and nonylic, $C_9H_{18}O_2$, acids, and also oxystearic acids, $C_{18}H_{36}O_3$ and $C_{18}H_{32}O_6$.

Fatty acids are also rendered unstable by partial oxidation, and when in this condition they have a more intense physiological action. Thus castor oil may owe its cathartic action to the ricinoleic acid, $C_{18}H_{34}O_3$, an oxidized oleic acid which is especially active when in a free state, as it is in the intestine. An oxidized stearic acid, ketostearic acid, is found in mushrooms of the genus Lactarius, $CH_3(CH_2)_{11}—CO—(CH_2)_4.COOH$ (Bouzant and Charaux), and dihydroxy stearic acid, possibly formed by oxidation of oleic acid, in castor oil.

Saponification. The property of fats of being hydrolyzed by water and alkalies is of practical importance. That fats when treated with alkalies make soaps is generally known, but the exact mechanism of the process is not yet clear. If a neutral fat is heated with acid, preferably in an alcoholic solution, or by alkali, or even by superheated steam, it is split into the fatty acids of which it was composed and glycerol. This is the process of saponification.

$$
\begin{array}{l}
H-\overset{H}{\underset{|}{C}}-O-\overset{O}{\overset{\|}{C}}-(CH_2)_{16}-CH_3 \\
H-\overset{|}{\underset{|}{C}}-O-\overset{O}{\overset{\|}{C}}-(CH_2)_{16}-CH_3 \\
H-\overset{|}{\underset{H}{C}}-O-\overset{O}{\overset{\|}{C}}-(CH_2)_{16}-CH_3
\end{array}
+ 3KOH \longrightarrow
\begin{array}{l}
H-\overset{H}{\underset{|}{C}}-OH \\
H-\overset{|}{\underset{|}{C}}-OH \\
H-\overset{|}{\underset{H}{C}}-OH
\end{array}
+ 3KO\overset{O}{\overset{\|}{C}}-(CH_2)_{16}-CH_3
$$

If alkalies are used, the fatty acids set free at once unite with the alkali to form the alkali salts of the fatty acids which are called soaps. The characteristic of a soap is its slippery feel and its property of making a foam when dissolved in water and shaken. Not all substances with these physical properties are soaps, however. Many plants, such as the soap-bark tree, contain saponaceous or soap-like substances, as far as these physical properties are concerned, but which are not soaps in a chemical sense. A soap, therefore, is a salt of a higher fatty acid which, when dissolved in water, foams on shaking and the solution has a low surface tension. Sodium acetate is the salt of a fatty acid, but it is not a soap because it lacks the physical properties of soap.

The speed of saponification of a fat has been found to be proportional to the number of hydroxyl, or hydrogen ions in the solution. For this reason saponification by ammonium hydrate is much slower than by sodium hydrate; and by water, $C^{20°}_{OH} = .8 \times 10^{-7}$, is still slower. The process of saponification forms then one method of measuring the concentration of such ions.

But while the fact of hydrolysis or saponification is quite evident, when we ask the further question of the mechanism of the process we get at once into very badly-known territory. What is the point of attack of the enzyme, or the alkali, or the acid on the fat molecule? The most probable answer is that it is either the doubly bonded oxygen atom, or that linking glycerol and fatty acid together, which is the point at which union with the saponifying agent takes place. It seems that one molecule can influence another only when it is united with it. If this is true, the saponifying agent must unite with the fat molecule. Since one of the oxygen atoms of esters is almost certainly quadrivalent, there being two extra valences in such molecules, union probably occurs here, at least in the case of lipase. Union with acids probably occurs as follows:

$$\underset{\underset{O=}{\|}}{R{-}O{-}C}{-}R' + HCl \rightleftharpoons \underset{\underset{HOCl}{\|}}{R{-}O{-}C}{-}R'$$

Probably the last compound is unstable and breaks into RCl and R'-COOH. RCl reacts with the water to form ROH and HCl.

The hydrogen or hydroxyl ions appear to act catalytically because they are left at the end of the reaction with their concentration unchanged, but in reality they have united with the ester, decomposed it and are again set free.

Saponification numbers of various fats. The saponification number is the number of milligrams of KOH required to neutralize the fatty acids contained in one gram of oil or fat. It serves to tell whether the fatty acids in the fat have a high or low molecular weight, for it is clear that the smaller the molecular weight the more molecules of the acids there will be in the gram of fat. Most fats and oils having chiefly palmitic, oleic or stearic acid in them have saponification numbers of about 195, but some which have more complex acids are lower than this. Thus for the rape-oil group, rape oil containing erucic acid, $C_{22}H_{42}O_2$, the saponification number is about 175. Similarly, lipins which contain carnaubic acid, $C_{24}H_{48}O_2$, will have still lower saponification numbers. On the other hand, butter which contains caproic, caprylic and butyric acids, all of these of low molecular weight, has a high saponification number, the saponification number of butter fat being 227. The saponification number of oleomargarine is about 195, so that it constitutes an easy method of distinguishing between butter and oleomargarine. Cocoanut and palm oil and some of the blubber oils, such as porpoise-jaw oils, have also a considerable quantity of volatile acid of low molecular weight and the saponification number is between 240 and 260.

Reichert-Meissl number. Another useful method of discovering the true nature of a fat or oil is to determine the amount of volatile fatty acid in it. For this determination a weighed amount of the fat is saponified with alcoholic potash or by potassium hydrate, the mixture is acidified to set free the fatty acids, phosphoric or sulphuric acid being used, as they are non-volatile, and then the mixture after adding pumice stone is distilled. The amount of fatty acid distilling over is titrated with N/10 KOH. The amount of N/10 KOH required to neutralize the volatile fatty acids from five grams of oil or fat is called the Reichert-Meissl number. Ordinary fats have a Reichert number of 0 when the saponification number is 195. For butter fat the number is 25-30; cocoanut oil 6-7; and for palm kernel oil 5-6. If the saponification number is high, the Reichert number is, also, generally high.

Acetyl number. Some oils, such as castor oil, contain oxidized fatty acids and, since the number of these hydroxyl groups is known if the oil is pure, the determination of this number is of use in detecting adulteration. These hydroxyl groups are detected by acetylation, and hence we have the acetyl number, which is the number of milligrams of KOH necessary to neutralize the acetic acid saponified from 1 gram of acetylated fat.

Separation of the fatty acids. Another not difficult method for learning something of the chemical nature of a fat consists in saponifying the fat and then determining the proportion of oleic acid present. After saponification the fatty acids are dissolved in alcohol and lead acetate in alcoholic solution is added. The lead salts of the fatty acids are precipitated, filtered, dried and extracted with ether. Lead oleate is soluble in ether, while the palmitate and the stearate are insoluble. The determination of the relative amount of oleic acid present is thus simple. By treating the ether solution of the lead oleate with hydrochloric acid, lead chloride is precipitated. On evaporating the ether oleic acid remains behind.

Physiological value of fats. Fats are of use to the body in several ways. Being poor heat conductors, a layer of fat beneath the skin helps to conserve bodily heat; by their physical properties they contribute to the physical constitution of protoplasm; and finally they are the heat-producing foods par excellence. They are burned or oxidized in the cells, setting free much heat, nine large calories per gram of fat. An

dation is, therefore, very important in under-

rn in the body with ease, but outside the body ledly inactive at body temperature. It is not ıer the fatty acids are burned in metabolism e from the glycerol, or whether they can be r union. The general presence of lipases, ats, in all cells and the fact that the form of mulates in cells is generally the stable form, is stored, leads to the inference that they are g. The fatty acids are, then, probably united stens their oxidation.

u how the higher fatty acids are oxidized, but lation occurs first in the β carbon atom, i.e., xyl, where it has been shown to occur in the ric. It has been found by Dakin that hydro- ch the same kind of oxidations as the body. nd living matter oxidize the fatty acids by he beta carbon atom is first oxidized, giving reactions may be represented as follows:

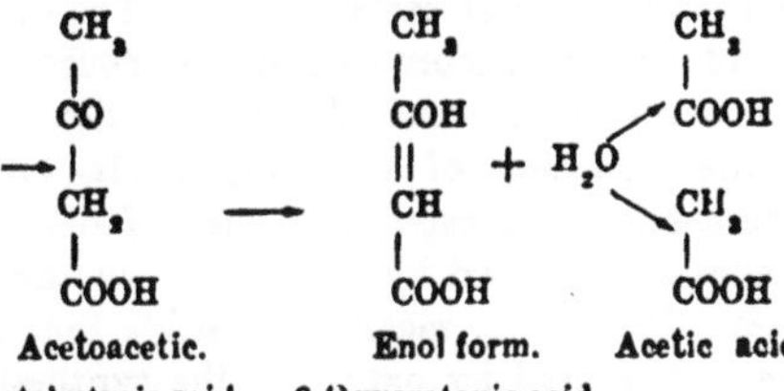

Acetoacetic. Enol form. Acetic acid.
etobutyric acid. β-Oxycrotonic acid.
ybutyric acid.
utanon acid.

n the β carbon are very unstable and in the se; they are far less stable than those oxidized ons. By reduction the enol form will go over acid as follows:

CH₃ | COH || CH | COOH + H₂ ⟶ CH₃ | CHOH | CH₂ | OOOH

l form. β-hydroxybutyric acid.

The oxidation of the long carbon-chain acids is believed to follow this same process, the β carbon being oxidized each time to the ketone and then this splitting-off acetic acid leaving the fatty acid with two less carbons. The long chain is thus split up, two carbons going each time. Every fatty acid with an even number of carbons in the chain will thus ultimately produce the β-ketone butyric acid. It is not certain, however, that oxidations may not occur, also, in other carbons of the chain. The process may be pictured as follows:

$$
\begin{array}{ccccccc}
CH_3 & & CH_3 & & CH_3 & & \\
| & & | & & | & & \\
(CH_2)_{12} & & (CH_2)_{12} & & (CH_2)_{10} & & \\
| & & | & & | & & \\
CH_2 & +O_2 \longrightarrow & CO & +H_2O \longrightarrow & CH_2 & + & CH_3 \\
| & & | & & | & & | \\
CH_2 & & CH_2 & & CH_2 & & COOH \\
| & & | & & | & & \\
COOH & & COOH & & COOH & & \\
\text{Palmitic acid.} & & \beta\text{-Keto-palmitic.} & & \text{Myristic acid.} & & \text{Acetic acid.}
\end{array}
$$

It is not impossible that the synthesis of the fatty acids may be brought about by a reverse process of reduction, two carbon atoms being added each time. This would explain why it is that all the naturally occurring fatty acids have an even number of carbon atoms. The oxidation of the unsaturated fatty acids follows a more complicated course and a variety of splitting products are formed, some of them volatile.

The course of the oxidation of the fatty acids will well repay further study. The aceto acetic ester, the ester of diacetic acid, is a very reactive substance widely used by chemists in synthesizing a great variety of substances. It is by no means impossible that some ester of diacetic acid is also widely used by nature in the synthesis of the substances in the cell.

Origin of fats. There appear to be two distinct problems in the formation of the fats: the first is the origin of the long carbon chains making the fatty acids; the second, the nature of the synthesis of these chains with glycerol. While neither of these problems has as yet been solved, we may consider what is known about them.

From the fact that the long carbon chains in the natural fats always contain an even number of carbon atoms, there can hardly be any other conclusion than that the synthesis is not made carbon atom by carbon atom, but by the addition of two, four or six carbon atoms at a time. From the decomposition by oxidation where two carbons are split off at a time the inference may be drawn that the synthesis also involves pieces of two carbons. Since the fats are formed by reduction from the sugars, the synthesis obviously involves reductions. What the pieces are which are thus added it is impossible to say positively at the present time, but they may possibly be either ethylidene or glyoxal, glycolic aldehyde or acet aldehyde. But this is not the only possibility.

Whatever may be the nature of the carbon fragments giving rise to the fatty acids, there is no doubt of two facts: that they arise by the process of the fermentative decomposition of the sugars common to all tissues; and, in the second place, that these pieces act as oxidizing agents, being themselves reduced. The fats accordingly represent, in essentials, reduced, condensed carbohydrates. That processes of reduction play a part in fat formation is shown by the chemical composition of the fats. The fatty acids contain little oxygen; they are carbon-hydrogen compounds except for the carboxyl group. The system oxygen-fatty acid contains, therefore, far more potential energy than the system oxygen-carbohydrate. One gram of fat burned produces nine calories of heat as compared with four for a gram of carbohydrate. That the fats originate from the carbohydrates is shown not only by practical human experience that a carbohydrate diet is fattening, but by direct experiment. The transformation of starch into oil takes place commonly in plants and many bacteria form acetic, propionic, lactic or butyric acid by carbohydrate fermentation.

Since the fats when oxidized yield more energy than an equal weight of the sugars from which they are derived, it is probable, since most reactions are in their entirety exothermic, that the reduction of some of the sugar to fat coincides with the oxidation of some of the sugar to carbon dioxide. In this way the gain in energy in the system fat-oxygen represents some of the energy set free by the total combustion of some of the sugar. Thus it may be inferred that to make one molecule of palmitic acid, $C_{16}H_{32}O_2$, perhaps four, or more, molecules of $C_6H_{12}O_6$ are destroyed, some being oxidized as shown by the schematic reaction:

$$6C_6H_{12}O_6 + 13O_2 \longrightarrow 20CO_2 + C_{16}H_{32}O_2 + 20H_2O + n \text{ Cals.}$$

The condensation of the fatty acid thus formed with glycerol to make neutral fat is one of the condensations so common in protoplasm and of which the synthesis of the polysaccharides from the monosaccharides is a type, but the nature of which is still so obscure. If glycerol and fatty acid are mixed, some spontaneous condensation occurs with the elimination of water. The rate at which this spontaneous condensation occurs is hastened by lipase, the fat-splitting ferment. It has accordingly been suggested that the synthesis of the fats from glycerol and fatty acids in the cells may be brought to pass by the action of the lipase. However reasonable this explanation appears at first glance, there are certain grave objections to it. The point of equilibrium of fatty acid, glycerol and neutral fat is one in which very little neutral fat coexists with a great deal of glycerol and fatty acids. Now in protoplasm much neutral fat is found and only traces of the glycerol and fatty acid, unless these are formed by autolysis. This fact alone makes it impossible to ascribe the synthesis to lipase unless some other condition be also assumed by which the neutral fat is removed as quickly as it is formed from the influence of the lipase and away from the system in which it has been part. A second reason is the fact that the synthesis goes on most rapidly in those cells which contain little or no lipase. There seems to be no lipase, for example, in the mammary glands in which fats are rapidly formed. Practically conclusive evidence against the lipase theory is the fact that the synthesis of the fats, like other syntheses, is dependent upon oxygen, and the lipase action is independent of oxygen. The syntheses depend on the vitality of the cell. They do not take place in an etherized cell. It is clear that this synthesis must be left for future work to decipher.

The origin of the glycerol is not difficult. It is closely related to the carbohydrates, and glycerose is one of the products of their decomposition. By reduction glycerose yields glycerol. Glycerose is a mixture of dioxyacetone and glycerine aldehyde, $CH_2OH\text{-}CHOH\text{-}COH$.

The question may finally be asked of the location in the cell of the neutral fat or oil. In many cells the fat or oil may be seen in the form of fine round droplets distributed through the cytoplasm. This is the case in egg cells and in some plant cells. In fat tissue the cell is almost composed of one large droplet of fat, inclosed in the cell membrane. But in living protoplasm itself fat also occurs, but in such a fine state of subdivision, or union with the other constituents, presumably phospholipins, that it cannot be detected microscopically. That the fat is there is shown by chemical examination and by its appearance in cells, particularly after they have been exposed to chloroform vapors. After anesthesia by chloroform of only one hour's duration, the liver, kidney and heart cells may show by staining with Sudan 3 an abundant presence of fat in small droplets, whereas the total fat is found by analysis to be no greater than before. The anesthetic evidently causes a change in its distribution and perhaps in its saturation, so that it aggregates into visible droplets. Fat may be stained in cells by Sudan 3, or by Nile blue. Only lipins with unsaturated acids stain with osmic acid.

Essential oils.—These oils are a heterogeneous chemical group with widely different compositions but having three properties in common: they are volatile, they make a temporary grease spot on paper, and they are combustible. Most of them also have an odor. Among these oils are clove oil, cedar oil, oil of cardamom, oil of peppermint, oil of wintergreen, to mention only a few. They are widely distributed among plants. Some of them, perhaps the majority, are aromatic aldehydes or ketones belonging to the terpenes of which the oil of turpentine is an example; others are aliphatic compounds such as the essential oil of Rita graveolus, which is methyl-n-nonyl ketone, $CH_3.CO.(CH_2)_7.CH_3$; others are acids. Those of the aromatic group which are terpenes are related to the sterols and the resins. The essential oils, while of considerable importance in pharmacy and therapeutics and in commerce as flavoring extracts, are of less importance in animal physiology, none having been isolated from the human body.

The terpenes constitute a group of very great commercial and practical importance, camphor and menthol, $C_{10}H_{20}O$, belonging in this group. The true terpenes have one or several hexa-carbon rings with two side chains:

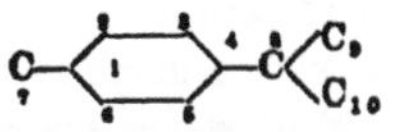

[Structural formula: hexa-carbon ring numbered 1–6, with side chain C_7 at position 1 and, at position 4, carbon 8 bearing C_9 and C_{10}.]

Oil of turpentine is chiefly pinene, $C_{10}H_{16}$. The terpenes are generally benzene derivatives, but the benzene nucleus is usually only partially unsaturated and contains only one or two double bonds. Some of them have two or more rings, some only one. India rubber is a complex terpene. This can be formed by the condensation of isoprene, a diolefine:

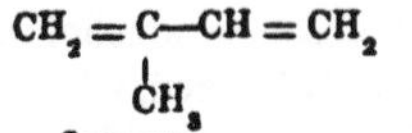

$$CH_2 = C—CH = CH_2$$
$$|$$
$$CH_3$$

Isoprene.

[Structural formulas:]

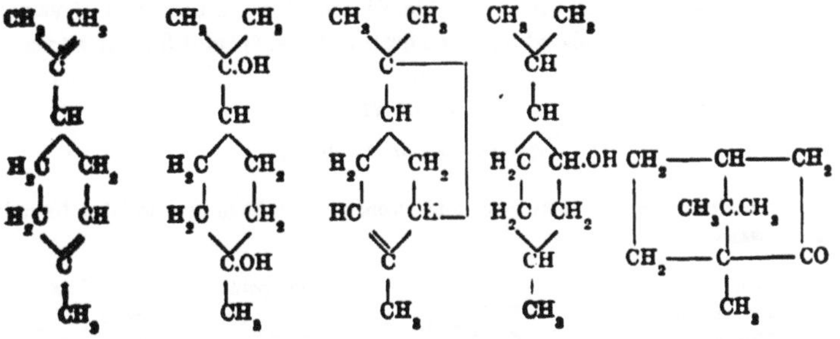

Limonene, $C_{10}H_{16}$.

Terpin, $C_{10}H_{20}O_2$.

Pinene, $C_{10}H_{16}$.

Menthol, $C_{10}H_{20}O$.

Camphor, $C_{10}H_{16}O$.

Terpenes and some oxygen derivatives.

The aliphatic terpenes are of great theoretical importance because they stand intermediate between the fatty acids on the one hand and the resins and aromatic terpenes and cholesterol on the other. Such an olefine terpene is myrcene, $C_{10}H_{16}$, found in the oil of bay. It is probably $CH_2{=}CH{-}CH{=}CH.CH_2.CH_2.CH{:}C(CH_3)_2$. While these substances occur in plants, it is not impossible that similar substances may appear as intermediate products of metabolism in animals. Thus d-citronellol, $C_{10}H_{19}OH$, is $CH_3.C(:CH_2).CH_2.(CH_2)_3.CH(CH_3).CH_2OH$. On oxidation it yields acetone and β-methyl adipic acid. Geraniol, $C_{10}H_{18}O$, yields on oxidation such products as acetone and levulinic acids. Moreover these citrals, such as geraniol, readily condense to six carbon rings on heating with acids, thus showing one possible origin of the aromatic rings. Santalol is a sesquiterpene alcohol of unknown composition, $C_{15}H_{26}O$. Cholesterol, other sterols and bile acids are probably terpene derivatives, but they are solids at ordinary temperatures and differ widely in chemical and physical appearance from the essential oils.

The waxes.—Waxes are of both animal and vegetable origin. They are the esters of fatty acids with a mono-hydric solid alcohol, or sterol. They contain no glycerol. Most waxes are solids at ordinary temperatures, but at least one, sperm oil, is liquid. Most naturally occurring waxes are mixtures of the esters of various fatty acids with various sterols. Lanolin, or wool fat, is such a wax; another is bees-wax. The waxes generally have a different texture from the fats, but their solubilities are similar. They are saponifiable like the fats, but often with more difficulty. The saponification is easily carried out in petroleum ether containing a little absolute alcohol by the addition of metallic sodium. Sodium alcoholate is formed, which saponifies the wax. Among the waxes the cholesterol esters of the blood deserve special mention.

Composition of the waxes. Table VI contains the formulas of various fatty acids and alcohols which have been isolated from different waxes.

TABLE VI.

Composition of the Waxes.

Formulas of various fatty acids and alcohols which have been isolated from different waxes.

I. *Acids. Saturated.*	Formula	Melting point	Wax
Ficocerylic	$C_{13}H_{26}O_2$	57° C.	Gundang.
Myristic	$C_{14}H_{28}O_2$	53.8°	Wool.
Palmitic	$C_{16}H_{32}O_2$	62.62°	Bees. Spermaceti.
Carnaubic	$C_{24}H_{48}O_2$	72.5°	Carnauba. Wool.
Cerotic	$C_{26}H_{52}O_2$	77.8°	Bees. Wool. Insect.
Melissic	$C_{30}H_{60}O_2$	91°	Bees.
Psyllostearylic	$C_{30}H_{60}O_2$	94-95°	Psylla.
II. *Acryllic series.*			
Physetoleic	$C_{16}H_{30}O_2$	30°	Sperm oil.
Doeglic (?)	$C_{19}H_{36}O_2$		" "
Lanopalmic	$C_{16}H_{32}O_3$	87-88°	Wool.
Cocceric	$C_{31}H_{62}O_3$	92-93°	
Lanoceric	$C_{30}H_{60}O_4$	104-105°	Wool.

III. *Alcohols. Sterols.*			
Pisang ceryl	$C_{16}H_{34}O$	78°	Pisang.
Cetyl (Ethal)	$C_{16}H_{34}O$	50°	Spermaceti.
Octadecyl	$C_{18}H_{38}O$	59°	"
Carnaubyl	$C_{24}H_{50}O$	68-69°	Wool.
Ceryl	$C_{26}H_{54}O$	79°	" Chinese.
Myricyl (Melissyl)	$C_{30}H_{62}O$	85-88°	Bees. Carnauba.
Psyllostearyl	$C_{33}H_{66}O$	68-70°	Psylla.
Lanolin alcohol	$C_{12}H_{24}O$	102-104°	Wool.
Ficoceryl	$C_{17}H_{28}O$	198°	Gundang.
Cocceryl	$C_{30}H_{62}O_2$	101-104°	Cochineal wax.
Cholesterol	$C_{27}H_{44}O$	148.4-150.8°	Wool.
Iso-cholesterol	$C_{26}H_{44}O$	137-138°	Wool.

The sterols.—The sterols (*stereos*, solid; *ol*, the ending signifying an alcohol), literally solid alcohols, are a very important group of the lipins. As their name signifies, they are alcohols solid at ordinary temperatures. They are readily soluble in ether and $CHCl_3$, easily crystallized, the crystals having a mother-of-pearl glance and a greasy feel. Cholesterin, or cholesterol, as it is now more generally called, was the first member of the group to be discovered and is the most important member. Phytosterol, stercorin or coprosterin, cetyl alcohol are others.

Cholesterol.—This, the most important substance in the group, was isolated from gall stones in 1785 by Fourcroy, when it was confounded with adipocere. Its true nature as a non-saponifiable, fat-like body was discovered by Chevreul in 1814, who named it cholesterin from the Greek *chole*, bile; and *stereos*, solid. It is most readily obtained from gall stones by pulverizing them, and extracting with boiling alcohol containing a small amount of potassium alcoholate. On cooling, or if necessary concentrating first by evaporation of the alcohol, cholesterol crystallizes out as white, shining, plate-like rhombic crystals generally having one corner broken. Figure 8. The angles of the sides are generally 76° 30′ or 87° 31′. The form varies with the solvent. The pure crystals melt at 148.5° C. A solution of the crystals in $CHCl_3$, or ether, is levorotatory $[\alpha]^{20°}_{D} = -31.12°$ for a 2 per cent. ether solution. Cholesterol is volatile at 300° in vacuo, subliming unchanged.

Cholesterol is insoluble in water, acids or alkalies; slightly soluble in soap solutions and much more soluble in solutions of bile salts. It dissolves readily in hot or cold ether, carbon bisulphide, chloroform, benzene, various hydrocarbons, acetone and hot alcohol. It is slightly soluble in cold alcohol. It is readily soluble in oleic acid and fluid fats. Its solutions react neutral and have neither taste, color, nor smell. It cannot be saponified, but is very slowly decomposed by concentrated alkalics.

If gall stones are not available, cholesterol can be most easily prepared by extracting mammalian brains, such as sheep or hog brains, with cold acetone. The brains, freed as far as possible from adherent blood vessels and membranes, are ground in a meat chopper and three times their weight of acetone is added. After six to twelve hours of extraction the acetone, which is diluted with the water of the tissue, is filtered off and a new portion of acetone added and extracted for twenty-four hours or more while heating on the water bath. On filtering off the acetone and concentrating by distillation on the steam bath impure

Fig. 8.—Crystals of cholesterol.

Fig. 9.—Crystals of phytosterol from soil (Schreiner and Shorey).

cholesterol separates out. It may be purified by recrystallizing from hot alcohol, or other solvents. The cholesterol may be obtained, also, by making one extraction with cold acetone to remove most of the water and then extracting the brain mass in a large Soxhlet with acetone.

Color reactions. Cholesterol is remarkable for its power of forming pigments. Advantage is taken of this property for the purpose of its qualitative detection by the Salkowski, Liebermann, Lifschütz and Neuberg methods. These reactions are not characteristic of cholesterol.

Salkowski reaction. Cholesterol, or the substance to be examined, is dissolved in 5 c.c. of chloroform; an equal volume of concentrated sulphuric acid is poured underneath the chloroform. In the presence of cholesterol the chloroform solution very soon becomes colored a cherry red, while the sulphuric acid is dark red with a brilliant green fluorescence. Poured into a porcelain evaporating dish, the chloroform becomes violet, green and finally yellow. Cholesterol esters give this reaction, but only after a delay, probably consumed in setting free the cholesterol. This reaction is not so delicate or specific as the acetic anhydride reaction.

Acetic anhydride reaction. *(Liebermann-Burchard.)* A little cholesterol is dissolved in 2 c.c. of chloroform, then ten drops of acetic anhydride and one or two drops of concentrated sulphuric acid are added. In the presence of cholesterol a violet, changing quickly to a blue-green, is obtained. If more chloroform is added in proportion to the acetic anhydride, the development of the green is delayed. This reaction is the best and most delicate of the qualitative reactions.

Dry cholesterol reaction. (*Schiff's.*) If a few crystals of cholesterol in a porcelain dish are moistened with concentrated HCl, 2-3 volumes, containing one volume of dilute $FeCl_3$; or with a drop of concentrated sulphuric acid containing a trace of $FeCl_3$; and then evaporated carefully to dryness over a flame, a reddish violet residue changing to a bluish violet is obtained. With concentrated nitric acid alone cholesterol crystals give, on evaporation of the acid, a yellow spot changing to orange when moistened with an alkali (xantho reaction). This is not peculiar to cholesterol, but is given by many aromatic substances.

Oxycholesterol reaction. (*Lifschütz reaction.*) A few mgs. of cholesterol are dissolved in 2-3 c.c. of glacial acetic acid and oxidized by the addition of a few particles of benzoyl peroxide. The solution is brought once or twice to the boiling point. To the cooled solution of oxycholesterol four drops of concentrated sulphuric acid are added and shaken. The color becomes a bright red, then blue and finally green. Spectroscopically it shows the typical absorption band in the red characteristic of oxycholesterol. One part of cholesterol in 10,000 may be detected by this method. Oxycholesterol gives this reaction without the addition of the benzoyl peroxide.

Methyl furfural reaction. (*Neuberg and Rauschwerger's reaction.*) —A trace of rhamnose or methyl furfural is added to the alcoholic solution of cholesterol and then concentrated sulphuric acid is cautiously added. At the contact surface a red ring forms at once and by shaking, while cooling the solution, the whole fluid colors itself a beautiful red and shows, after dilution with alcohol, an absorption band which begins before E and ends at the B line. The color is permanent for several days. Phytosterol also gives this reaction and so do the bile acids. (Pettenkofer's reaction. Camphor derivatives, etc.)

Quantitative determination. For the quantitative determination of the amount of cholesterol in a tissue, the latter is repeatedly extracted with boiling alcohol and then by ether; the extracts are united and evaporated to dryness on the steam bath; the residue is then repeatedly extracted in dry ether and the ether portions united. The ether is distilled and the residue is dissolved in benzene. To the benzene is now added an amount of metallic sodium equal in weight to the residue and absolute alcohol is added cautiously in small amounts until the sodium just dissolves. When the sodium is dissolved saponification of the fats is continued on the water bath with a reflux condenser for two hours. The solution is then cooled, and filtered from the soaps, and the soaps carefully extracted with water-free benzene, the extracts being added to the filtrate. The benzene is now washed repeatedly with water to remove the sodium alcoholate until the wash water reacts neutral. The benzene is now distilled and the residue is dissolved in hot 94 per cent. alcohol and made up to known volume. Ten c.c. of the alcoholic solution of cholesterol are now taken and mixed hot with a 1 per cent. solution in 90 per cent. alcohol of crystallized digitonin as long as a precipitate forms. The precipitate of digitonin-cholesterol, after standing a few hours, is filtered through a weighed asbestos Gooch crucible, washed with cold alcohol and ether, dried at 100-110° and weighed. To the weight is added .0016 gm. for each 10 c.c. of solution as a correction for the amount of the digitonin cholesterol left in solution, the compound not being absolutely insoluble. The united weights multiplied by the factor 0.2431 gives the weight of cholesterol in the amount of solution taken.

Quantitative determination. Colorimetric. (Grigant's.)—The gravimetric method described is probably superior to the colorimetric, but digitonin cannot always be easily obtained. For the colorimetric determination the tissue is heated in 1 per cent. sodium carbonate in 50 per cent. alcohol for twenty-five minutes on the steam bath. After cooling the dried residue is extracted with ether to remove the cholesterol, the ether is evaporated and the cholesterol is determined colorimetrically by the acetic anhydride method by comparing the color with standard tubes made with known amounts of pure cholesterol at the same time.

Amount of cholesterol in different tissues.—Cholesterol is relatively more abundant in the brain than in any other organ of the body. It forms an important constituent of the medullary sheath of the nerve fibers, but it probably also occurs in the axis cylinder, since it is found in practically all cells. More of it is found in the white matter of the pons or corpus callosum than in the gray matter. In the human brain Kirschbaum and Linnert found by the digitonin method in the cortex 1.15 per cent. of cholesterol; in the white matter, 2.47 per cent.; in the cerebellum, 1.31 per cent.; in the pons and medulla oblongata, 4.03 per cent., and in the whole brain, 2.69 per cent. These figures are computed on the weight of the fresh undried tissue. Dimitz found 3.35-4.26 per cent. in the spinal cord. Thudichum, by measurement of the non-saponifiable residue of the brain, found in the gray matter 1.95 per cent.; in the white substance, 3.26 per cent. In the dry ox brain Lifschütz got 11-12 per cent. as nearly pure cholesterol. Ninety-eight per cent. of this was free cholesterol, the other 2 per cent. existed in the ester form. Egg-oil, prepared from the yolks of eggs, contains about 5.5 per cent. of raw cholesterol, one-fifth being oxycholesterol ester. Trade lecithin (Merck) contains esters of cholesterol and oxycholesterol about 2-2.5 per cent. Cholesterol esters are also present in the kidneys of the ox, and other glandular organs; there is no oxycholesterol in the liver. Cholesterol is present in the vernix caseosa, 16.8 per cent. of the total cholesterol being oxycholesterol. In human blood serum the total cholesterol content varied from 1.17-2.95 mgs. per c.c. (Weston and Kent); or 1.97 gram in the liter (Tscovesco). Esters of cholesterol form doubly refracting masses in atheromatous arteries.

Chemistry of cholesterol.—The empirical formula for cholesterol is $C_{27}H_{43}OH$ or $C_{27}H_{43}OH$ (Mauthner and Suida). When dissolved in glacial acetic or propionic acid it easily esterifies. It is, therefore, an alcohol and there is but one alcohol group. It adds bromine, two atoms to the molecule, and also two atoms of iodine, or hydrogen. It has, therefore, at least one double bond. According to Mauthner, however, it adds two molecules of ozone, or six atoms of oxygen, which would indicate two double bonds. When oxidized it yields, first, a ketone, cholesteron, a neutral body also called oxycholesterol, $C_{27}H_{45}O$; and on further oxidation a series of cholesterinic acids, dicarboxylic and tricarboxylic. It is only very slowly acted upon by potassium hydrate in solution, but when fused with KOH it is attacked below the point of fusion. It is readily oxidized by benzoyl peroxide, by potassium permanganate, or bichromate in glacial acetic solution. By reduction, it absorbs two atoms of hydrogen and goes into hydrocholesterin. It is unstable in light in the presence of oxygen and changes to a darker color, a lower melting point, and to oxycholesterol, giving then Lifschütz' reaction. (Schulze; Winterstein.) In lanolin, the fat of sheep's wool, various decomposition products of cholesterol as well as oxy- and unchanged and iso-cholesterol are found. Esters are readily formed, most of them being crystalline. The palmitic, stearic, oleic, benzoic and salicylic esters, among others, have been prepared. By fusing cholesterol with propionic acid, the propionic acid ester is formed. Many of these esters when cooling after melting show a beautiful play of colors from violet, blue-green, orange and carmine red. This may be used for detecting cholesterol. If gently fused with propionic anhydride, two to three drops in a dry test tube, and a little of the fused mixture taken out on a glass rod, the play of colors may be seen.

From a study of the decomposition products and the properties of the substance, the conclusion has been tentatively reached that cholesterol is a mono-hydroxy-secondary alcohol, with a terminal vinyl group; with a double bond; and probably containing four saturated hexa-carbon rings. The structure of the nucleus has not yet been made out. The formula as far as known is, then, according to Mauthner:

$$\begin{array}{l} CH_3 \\ \quad > CH.CH_2.CH_2{-}C_{17}H_{26}CH{:}CH_2 \\ CH_3 \\ \qquad\qquad\qquad\qquad H_2C \quad CH_2 \\ \qquad\qquad\qquad\qquad\quad CH(OH) \end{array}$$

It belongs in the general group of the terpenes and it is probably closely related to the bile acids.

Physiological importance.—Cholesterol is one of the most important physiological substances. In pathology, also, it plays a considerable rôle. It forms one of the chief constituents of gall stones, and deposits in the walls of arteries. It may have some value as a remedy. As we have seen, it forms a weak molecular union with saponaceous substances like digitonin. It acts in a similar way with other hemolytic substances, such as saponin, and other glucosides. It neutralizes their toxic action and thus protects the blood corpuscles of the body. The red cells are constantly being attacked by hemolyzing substances and dissolved. If the rate of destruction surpasses the rate of new formation anemia is produced. Cholesterol has the property of protecting the corpuscles from many of these dissolving substances, such as the bile salts among others. It is, hence, being tried as a remedy in hemoglobinuria and anemia. It probably plays a very important rôle in the blood in this way. Another property of cholesterol, possibly of no less importance, is that it neutralizes, or checks, the action of lipolytic enzymes. It may in this way, perhaps, protect the lipins of the cell from digestion, or help regulate the rate of their digestion. It has recently been found by Faust and Abel that derivatives of cholesterol are among the most powerful of heart poisons. These investigators have isolated from the skins of toads crystalline substances, evidently oxidation products of cholesterol, which have most powerful digitalis-like actions on the heart. Their artificial formation from cholesterol may put in the hands of the physician a valuable remedy of the digitalis class. Cholesterol, or its degradation products, aids the other lipins in giving to cells their power of holding large quantities of water without losing their peculiar semifluid characters, and without dissolving. As a constituent of wool fat, or lanolin, it helps form a valuable salve menstruum for the physician, lanolin taking up 80 per cent. of water in which water soluble drugs may be dissolved and applied as a salve. As a constituent of waxes and the sebum of the skin it protects the epidermal structures; it forms one of the most abundant lipins in the brain and occurs in nearly all living tissues; it is believed to be the mother substance from which the bile acids are derived and so plays indirectly an important part in the absorption of fats from the intestine; it probably gives rise, also, to some of the lipochrome pigments and possibly to some of the odoriferous substances of plants and animals.

The emulsions formed from water and cholesterol esters, or cholesterol in union with, or mixed with, other lipins, particularly the amino-lipins, are extremely interesting and should be carefully studied. Some of the acids formed by the oxidation of cholesterol are poisons having a toxic and hemolytic action comparable with some of the snake venoms. By repeated oxidation of cholesterol Windaus obtained an acid, $C_{27}H_{40}O_8$,

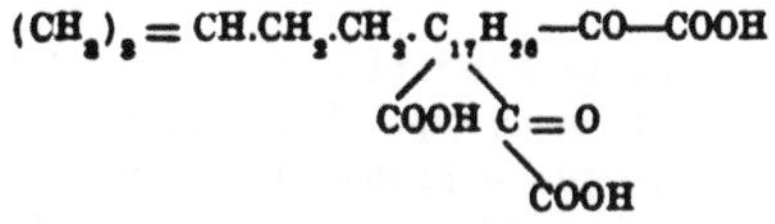

$$(CH_3)_2 = CH.CH_2.CH_2.C_{17}H_{26}—CO—COOH$$

$$COOH \quad C = O$$

$$COOH$$

more hemolytic than the bile acids. The acid $C_{27}H_{40}O_8$ was as powerful as many saponins in dissolving red blood cells. It caused local necrosis like that caused by many snake poisons. It probably plays an important part in the Wasserman reaction. Cerebro-spinal liquid of syphilitics generally contains more cholesterol than normal (Pighini).

Other sterols.—Other sterols isolated from animal and plant tissues are the following:

Substance	Formula	M.P.	Rotation $(\alpha)^{20}_D$	Where found
Phytosterols	$C_{27}H_{46}O$	135°-144°		Various plants
Sitosterol	$C_{26}H_{44}O$ or $C_{27}H_{45}OH\ H_2O$	136.5° (Ritter) 137.5° (Burian)	—26.4 (Ether) —33.9 ($CHCl_3$)	Wheat embryos Calabar bean.
Psylla alcohol	$C_{33}H_{67}OH$	69°-70°		Psylla wax. Bumble-bee wax
Isocholesterol	$C_{26}H_{46}O$	140.1° (137°-138°) (Dried at 80°)	$(CHCl_3)^{17°}$ +59.1 +60.0	Wool fat
Taraxasterol	$C_{27}H_{47}OH$			Taraxicum officinale (Power and Browning)
Homo " "	$C_{25}H_{39}OH$			
Clutyianol	$C_{29}H_{46}O(OH)_4$			Clutyia similis (Tutin and Clewer)
Spongosterol	$C_{27}H_{48}O$	124°	$[\alpha]^{15}_D$—19.59°	Sponge
Bombicesterol		148°	—34° (Ether)	Silkworm chrysalis
Stigmasterol	$C_{30}H_{48}O$	170°	—45.0° ($CHCl_3$)	Calabar bean Rape oil; cacao butter
Ergosterol		160°		Boletus edulis
Cucurbitol	$C_{24}H_{40}O_4$	260°		Watermelon seed
Sterorol (Koprosterol)	$C_{27}H_{48}O$	95-96°	+24°	Human feces
Hippocoprosterol	$C_{27}H_{54}O$	78.5-79.5°		Horse feces

Phospholipins, or phosphatides.—Although the fats have been discussed before the phosphatides because of the simple structure of the former, were we to place them in the order of their importance in the production of vital phenomena, there can be little doubt that the phosphatides should have been considered first, as among the most important substances in living matter. For they are found in all cells, and it is undoubtedly their function to produce, with cholesterol, the peculiar semifluid, semisolid state of protoplasm. The latter holds much water in it, but does not dissolve. Indeed it might be said that the phosphatides with cholesterol make the essential physical substratum of living matter.

Definition. This physical substratum of phospholipin differs in different cells and probably in the same type of cells in different animals, but everywhere, from the lowest plants to the highly differentiated brain cells of mammals and of man himself, it possesses certain fundamental chemical and physical properties. In all cases the phospholipin substratum is soluble in alcohol containing some water. The phospholipins may be separated from the protein, salt and other insoluble substances by extraction with 85 per cent. ethyl alcohol. Many, but not all, are soluble when pure in absolute ether, or in absolute alcohol, but all, or nearly all, may be extracted with impurities by 85 per cent. alcohol. It is extremely difficult to remove the phospholipins completely from the protein, with which a part appears to be in combination, but by repeated, fifteen to twenty, extractions with boiling 85 per cent. or 90 per cent. alcohol and ether they may be practically completely removed.

Most of the so-called fat of tissues is in reality phospholipin and cholesterol. It is usually obtained by ether or alcohol extraction of the dried tissue. The total alcohol-ether extract (fat) of the pig's liver was found by MacLean and Williams to contain 84 per cent. of phospholipin and 16 per cent. only was neutral fat and cholesterol, or substances soluble in acetone.

There is probably always more than one phospholipin in a cell and when they are separated from each other and from fat and sterol some of them, such as cephalin, are almost insoluble in absolute alcohol even when boiling; and others are insoluble in ether; but in a mixture, those which are the more soluble hold those which are less soluble in solution, so that their separation by differential solubility is a matter of difficulty.

The phospholipins are so called because they always contain phosphoric acid; in addition they always contain some higher fatty acids; and the typical ones an organic base, which is sometimes choline, but in many its nature is unknown. Most, but apparently not all, contain also glycerol. They resemble the fats and oils, then, in containing fatty acids and glycerol, but they differ from them in having, in addition, phosphoric acid and some nitrogen base. It is just this composition which explains their peculiar relation to water, for by their fatty parts they are prevented from dissolving in the water, while in virtue of the choline or other base and phosphoric acid in their molecules they have a great affinity for it.

The exact chemical composition of none of the phospholipins has been definitely established, but the constitution of at least two of them is probably approximately known. These two are lecithin and cuorin. The reason why more of them are not known is owing to the lack of accurate chemical quantitative methods for their preparation and analysis, and the great difficulty of securing stable, pure substances for analysis. But lecithin and cuorin may be taken as types of phospholipins which are widespread.

Classification of the phospholipins (phosphatides).—The following classification was suggested by Thudichum for the phosphatides and is coming more and more into use. The phospholipins are divided into the following groups:

1. Mono-amino-monophospholipins. Lecithin, cephalin.
2. Mono-amino-diphospholipins. Cuorin.
3. Di-amino-monophospholipins. Sphingomyelin.
4. Tri-amino-monophospholipins. Carnaubon.

This classification is of course of a temporary nature until the real composition of the members of the group is known, when a more definite classification can be made.

General method for the separation of phospholipins from cells.

Dry the tissue at low temperature. If possible freeze it immediately after removal from the animal to stop at once the action of possible phosphatideases. Shave the frozen tissue in a Kossel knife machine and allow it to dry in a frozen state in a vacuum desiccator over Al_2O_3 or $CaCl_2$. Extract the dry tissue twice with cold acetone at room temperature. This removes most of the fat and sterol. Suck off the acetone; free the powder from acetone by evaporation at room temperature; and extract with absolute ether at room temperature. The ether will remove any cephalin, or lecithin. Some cerebrin and other similar cerebrosides or glycolipins, which when pure are insoluble in ether, may also go into solution in the ether when the latter has lecithin in it. Most of the glycolipins, however, will remain behind. The ether extraction may be followed by a boiling alcohol extraction, which will take out most of the glycolipins, if any are present, and the remnant of lecithin and other phospholipins. The ether extract had best be treated separately from the alcoholic. Evaporate the ether under diminished pressure at low temperature; take up the extract in cold absolute ether and allow to stand twenty-four hours or longer, in a tall cylinder until a white precipitate (glycolipin, sulphatide, etc.) has settled out. The clear solution is then sucked off from the precipitate and to it two volumes of good acetone are added. This precipitates the phospholipins and leaves most of the fat and cholesterol in solution. Redissolve in ether and reprecipitate, after settling out the white substance; and repeat this until the ether dissolves entirely clear. By the addition now of three volumes of absolute alcohol to each volume of ether the lecithin substances may be separated from the cephalin, or cuorins which are precipitated.

The substances thus separated as lecithin and cephalin are still impure mixtures of several phospholipins. The method for the preparation of pure lecithin is given in the chapter on the brain. It consists in the precipitation of the lecithin as the cadmium salt by alcoholic cadmium chloride, the fractioning of the precipitate by means of benzene and the final separation of the lecithin as the chloride by H_2S and of the free lecithin by the use of Millon's base.

If it is not possible to freeze the tissue it should be placed at once in a large amount of 95% alcohol, cut into small pieces and extracted twice with cold alcohol, to remove the water, and then boiled out repeatedly with 85% alcohol. The phospholipins are recovered by distilling off the alcohol and allowing to cool. They may be separated in the method described under the brain.

Lecithin.—This is one of the phospholipins of the yolk of the hen's egg. The name is from the Greek, *lekythos*, meaning yolk of egg. It was first isolated and the name given by Gobley[1] in 1846. Diakonow, in Hoppe-Seyler's laboratory, worked out what is believed to be its composition in 1867, although it is very improbable that he ever had pure lecithin. On hydrolyzing it with barium hydrate, glycerol, phosphoric acid, stearic acid, oleic acid and the base choline (neurine, according to Thudichum) were obtained. He wrote its formula $C_{44}H_{90}NPO_9+H_2O$ and regarded it as distearyl lecithin. Its structural formula was believed to be as follows:

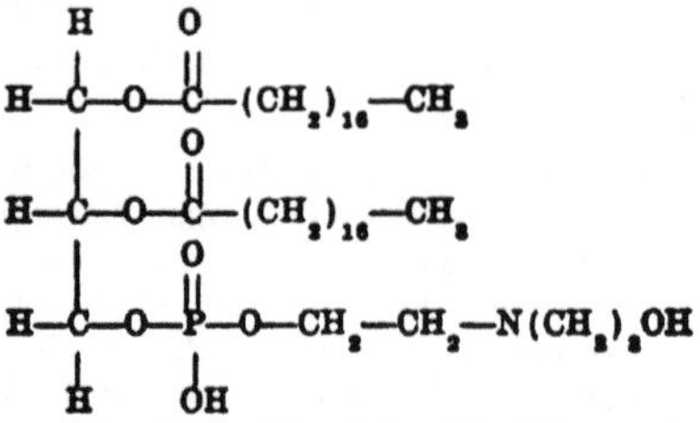

```
   H   O
   |   ||
H—C—O—C—(CH2)16—CH3
   |   O
   |   ||
H—C—O—C—(CH2)16—CH3
   |   O
   |   ||
H—C—O—P—O—CH2—CH2—N(CH3)3OH
   |   |
   H   OH
```

Hypothetical formula of lecithin.

Its molecular weight has not been directly determined. The molecular weight with the above formula would be 807. It was very quickly found by Hoppe-Seyler and his pupils that lecithin-like bodies were present in all cells and that they might contain palmitic, oleic and other acids; but until recently it was believed that the base was always choline and the alcohol glycerol. It is now clear that there are a group of substances differing widely in chemical composition, of which lecithin is but one. This group we have called the phospholipins. Lecithin in ether solution will not diffuse through a rubber membrane. Its molecular weight may possibly be double that cited.

[1] Gobley: *Journal de pharmacie et de chimie*, Vol. XI, p. 409; XII, p. 5; XVII, p. 401; XVIII, p. 107.

The hydrolytic decomposition of lecithin is supposed to follow the reaction:

$$\underset{\text{Lecithin.}}{C_{42}H_{84}NPO_9} + \underset{\text{Water.}}{4H_2O} \longrightarrow \underset{\text{Oleic acid.}}{C_{18}H_{34}O_2} + \underset{\text{Palmitic acid.}}{C_{16}H_{32}O_2} + \underset{\text{Glycerol.}}{C_3H_8O_3} + \underset{\text{Phosphoric acid.}}{H_3PO_4} + \underset{\text{Choline.}}{C_5H_{15}NO_2}$$

There is one atom of nitrogen to one of phosphorus, so that lecithin is a mono-amino-monophospholipin. N:P::1:1.

The actual analyses of egg lecithin have given figures for the fatty acid content which correspond closely to the theoretical; and approximately the theoretical amount of phosphorus and nitrogen has been found, but the quantitative determination of the choline is still impossible and only some 80 per cent. of the theoretical amount has actually been recovered. The chief difficulties in the way of a quantitative determination of choline have been two: the first is that choline is not entirely stable in barium hydrate which is generally used for the hydrolysis, and some nitrogen of an unknown nature always remains in the fatty acid residue when this method of hydrolysis is used; the second is that the precipitation of choline by platinum chloride is not quantitative, particularly in the presence of other substances such as glyceryl-phosphoric acid. Moreover, unless special methods are used, the platinum compound nearly always contains potassium. The first difficulty can be avoided by the use of the method of alcoholysis, that is hydrolysis by hydrochloric acid in absolute alcohol. But the second is still unsurmounted.

Pure lecithin is a white substance, crystallizing in very thin plates which mass together to form a waxy mass. It is readily soluble in alcohol and ether and can be separated from other phospholipins which may accompany it only by precipitating it with cadmium chloride and the use of a complicated method of purification. See page 569. Few people have prepared pure lecithin. According to Thudichum, lecithin always contains oleic acid. The point of union of the choline is still disputed and the exact structure of the molecule is uncertain. The formula given is, perhaps, the most probable. Thudichum believed that the fatty acids were united with the phosphoric acid, but this does not seem very probable.

On hydrolysis lecithin yields oleic, palmitic or stearic acids, choline or neurine, glyceryl-phosphoric acid, and free glycerol and phosphoric acid. Many substances have in the past been called lecithins which were either very impure lecithin or phospholipins of a similar character. Lecithin is fairly stable and may be kept as the cadmium salt or even as free lecithin for a long period unchanged. It is much more stable than the cephalins (kephalins).

It is probable that there are a number of different lecithins, since the character of the fatty acid may vary in different members of the group. A phospholipin yielding choline, fatty acids, glycerol and phosphoric acid may be called lecithin. It is doubtful, however, whether the phospholipin of the egg, which is properly called lecithin, is found in any other tissue. We might call those phospholipins which yield choline the choleo-phospholipins.

Choline. This is the base found in egg lecithin. It is trimethyl, oxyethyl ammonium hydrate and has the following formula:

$$\begin{array}{l} CH_3 \diagdown \quad \diagup CH_2.CH_2OH \\ CH_3-N \\ CH_3 \diagup \quad \diagdown OH \end{array}$$

Choline, $C_5H_{15}NO_2$.

It is related to neurine and muscarine, the latter being an oxidized choline found in poisonous mushrooms such as Agaricus muscarius.

$$\begin{array}{l} CH_3 \diagdown \quad \diagup CH_2.COH \\ CH_3-N \\ CH_3 \diagup \quad \diagdown OH \end{array} \qquad \begin{array}{l} CH_3 \diagdown \quad \diagup CH=CH_2 \\ CH_3-N \\ CH_3 \diagup \quad \diagdown OH \end{array}$$

Muscarine, $C_5H_{13}NO_2$. Neurine, $C_5H_{13}NO$.

Choline was discovered by Strecker in ox-bile, hence the name choline, from the Greek *chole,* bile. It is found in many different cells, as phospholipins which are said to yield choline are widespread in nature. In some instances, however, the substances identified micro-chemically as choline may not have been choline, but trimethyl amine (Dorée and Golla). Neurine is found in the nervous system of the ox, but not free. It is probably derived from choline in the process of preparation. It is difficult to see how neurine could form a part of a phospholipin except as a salt through the hydroxyl. Choline is also related to the betaïnes. Betaïne is the anhydride of trimethyl amino-acetic acid. The formula is as follows:

$$\begin{array}{l} CH_3 \diagdown \\ CH_3-N.\ CH_2-C=O \\ CH_3 \diagup \quad \diagdown O \diagup \end{array}$$

Betaïne, $C_5H_{11}NO_2$.

Betaïne gets its name from the fact that it occurs free in the sap of the sugar beet, Beta vulgaris. It is also found in many other plants and the homologues of betaïne are very common in plants, and one, the α-oxy-γ trimethyl butyro betaïne, is found in muscle (carnutine). Large amounts of betaïne occur in the fresh muscles and salivary glands of cephalopods.

Choline is a fairly strong base, alkaline in reaction and absorbing CO_2 from the air. It crystallizes readily as the double salt of platinum chloride $(C_5H_{14}NOCl)_2\ PtCl_4$ in reddish yellow plates. Thudichum states that the corresponding neurine salt crystallizes much more beautifully if there is some potassium chloride present as an impurity. Choline is precipitated by phospho-tungstic acid. It is unstable as the free base, but stable as the salt. It will be noticed that in lecithin, if the ordinary formula is correct, the hydroxyl of the choline is free. This may be substituted by other anions such as chlorine, making lecithin chloride. This is very important from the point of view of the action of inorganic salts on protoplasm, for at this point the anions of the salts may act on the living matter by changing the state of aggregation of the lecithin by substituting this hydroxyl of the lecithin. Moreover, by the interaction of the hydrogen of the phosphoric acid with the hydroxyl of the choline anhydrides may be formed.

Choline has a pronounced physiological action and its esters are even more powerful. It is said by most observers to cause a lowering of the blood pressure. Modrakowski and Popielski, however, maintain that pure choline raises the blood pressure in non-anesthetized animals. The latter recrystallized the platinum double salt and picked out the larger, better-formed crystals in order, as they thought, to get a pure product. The reason for the discrepancy between these different observers in the action of choline is not clear. Choline decomposes readily. They ascribe the lowering of the blood pressure obtained by others to impurities produced by decomposition. On the other hand, carefully prepared, synthetic choline still lowers the blood pressure. Most investigators tested the action on anesthetized animals. Perhaps the difference lies here; possibly the larger crystals picked out by Popielski may have been impurities or choline esters, since Thudichum has shown for neurine that a slight impurity of potassium greatly improves the power of crystallization of the platinum salt. Choline, as the free base, is said to decompose readily, yielding trimethyl amine; and trimethyl amine has been found to be a normal constituent of human blood, urine and other fluids and tissues. If a lecithin, or other phospholipin, is hydrolyzed with barium hydrate some nitrogen evaporates and part of it goes as trimethyl amine. The curious fact is that the amount of trimethyl amine obtained does not seem to depend on the time of heating, which would indicate the possibility either that trimethyl amine is not derived from the choline but from some other base, or else that it is preformed in small amounts in the lipin. The subject needs further investigation. As an example of the contradictory evidence on some of these points by the best investigators, the observation of Thudichum on cephalin may be cited. According to Thudichum cephalin, which he purified with the greatest care, more carefully than most other observers, contains quantities of neurine which was identified as the platinum double salt. He found also amino-ethyl alcohol and another base. Now according to Koch there is probably no neurine or choline in cephalin and this has been confirmed by Foster. What, then, was the base which was so much like neurine as to mislead so sharp-sighted an investigator as Thudichum? What weight shall be attributed to the reports of others as to the presence or absence of choline or neurine in animal tissues or phospholipins?

The decomposition of choline by heating into trimethyl amine and glycol is represented by the following equation:

$$(CH_3)_3N(OH)CH_2.CH_2OH \longrightarrow (CH_3)_3N + CH_2OH.CH_2OH$$

Choline. Trimethyl amine. Glycol.

It is the opinion of several observers that choline and similar substances of the class of betaïnes, which are quite common in the plant and probably in the animal world, originate by the methylation of some of the amino acids.

Quantitative determination of choline.—The following method was used by Kinoshita for the quantitative determination of the amount of choline in tissues. One-half to one kilo of the organ freed from connective tissue and fat is hashed and mixed with a double weight of water acidified with acetic or hydrochloric acid and boiled. The extract is filtered through a cloth and the extraction three times repeated. The fluid of the extract is allowed to cool, the fatty layer removed and after acidification with hydrochloric acid to prevent the decomposition of the choline it is concentrated first over the free flame and then on the water bath; the concentrate is poured into a 3-liter, round-bottom, Jena glass flask and by vacuum distillation on the water bath brought to dryness. The residue is then extracted three times in the same flask with boiling ether and the ether poured off. The choline is then removed by three extractions with boiling 95% alcohol and the united alcohol extracts brought by distillation in vacuo on the water bath to dryness. The residue is extracted with absolute alcohol, the alcohol evaporated and again extracted with absolute alcohol and the last absolute alcohol extract is filtered. The choline in the absolute alcohol solution is now precipitated by the addition of an absolute alcohol solution of $PtCl_4$ or $HgCl_2$. The precipitate after twenty-four hours' standing in the cold is suspended in water and decomposed with H_2S and then without filtering it is evaporated to dryness in a porcelain dish on the water bath. The residue is taken up in a little water, filtered and the residue carefully washed. The solution is evaporated to a small bulk and the gold double salt formed by the addition of $AuCl_3$. The gold chloride-choline crystals are weighed.

Choline may be precipitated from its solution by KI_3; by $PtCl_4$; or by an alcoholic

$HgCl_2$ solution. KI_3 will precipitate when added to an aqueous choline chloride solution. A 10% $HgCl_2$ solution in alcohol, or a 10% solution of $PtCl_4$ in alcohol when added to an alcoholic solution of choline chloride, are about equally efficacious as precipitant. A .002-.001% solution of choline chloride will give with either reagent a distinct precipitate in 10 to 15 minutes; and a .0006-.0003% solution will give a noticeable precipitate after 30 minutes. Even a .00005% solution will give a slight precipitate after 2 hours. Sublimate solution in absolute alcohol added to the alcoholic solution

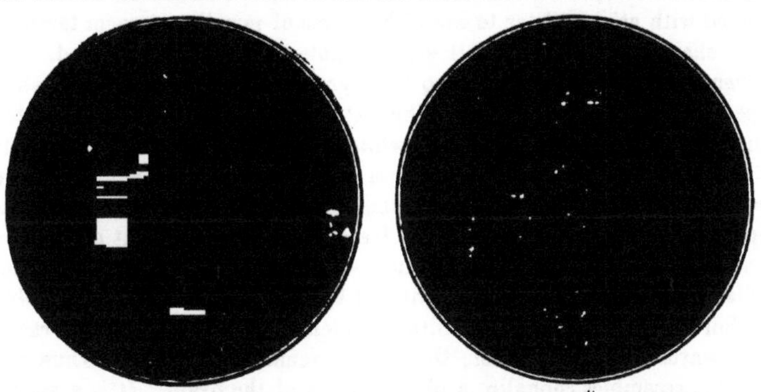

FIG. 10.—*A*. Crystals of choline periodide. *B*. Crystals undergoing disintegration and formation of oily drops (Rosenheim).

of choline chloride precipitates far more completely than the aqueous solution (Kinoshita).

None of the micro-chemical tests which have been proposed for the detection of choline are reliable. Most of these reactions, such as the Florence reaction, are given by neurine, trimethyl amine and other bodies (Kinoshita).

Amount of choline in various tissues.—The quantitative determination of the choline in various tissues has yielded the following results. This choline is not free but is split off by heating the tissue with hydrochloric acid.

Amount of choline in per cent. of the wet weight of the tissue. Direct determination by the gold salt (Kinoshita).

TISSUES OF OX

	Per cent. choline.		Per cent. choline.
Pancreas I	.015	Spleen	.012-.030
" IIa	.031	Muscle	.016-.032
" IIb	.022	Liver	.010-.017
" IIc	.021	Kidney	.020-.029
Small intestine	.022-.030	Lung	.016-.011

Cuorin.—$C_{71}H_{128}NP_2O_{21}$. This as the name implies is a phospholipin isolated from the beef heart. In its solubilities and other properties it resembles cephalin.

Method of preparation. The method used by Erlandsen for its preparation is as follows: The ox heart freed from fat, blood vessels, etc., is ground in a meat grinder and spread on glass plates before a fan to dry. It loses about 70 per cent. of its weight as water evaporates. It is then ground in a mill and dried in vacuo over sulphuric acid at 40°. (The author's experience is that it is better not to use sulphuric acid in vacuo as a drying agent because of the vapor pressure of the acid. $CaCl_2$ or Al_2O_3 are not open to this objection.) The powdered mass is then extracted with ether, 1 liter to each 500 grams of powder, at room temperature, changing daily until all ether-soluble material is extracted. The extracts are evaporated in vacuo to a syrup. The ether extract is about 7 per cent. of the dry powder. This extract is re-extracted with absolute ether, and allowed to stand. A white substance (probably glycolipin, A.P.M.) separates out. The ether is decanted from the white residue; precipitated with water-free acetone. It is redissolved in ether and reprecipitated several times. This frees the cuorin from fat and cholesterol and some other substances such as the white substance. It is now redissolved in ether and reprecipitated by four volumes of cold absolute alcohol at 0°-2° C. The precipitate, which is crude cuorin, is extracted with warm absolute alcohol; the cuorin remains insoluble. Thus prepared it resembles cephalin, a phospholipin of the brain. It is a yellowish brown, transparent, almost odorless, glassy substance: it is very hygroscopic and becomes sticky and finally fluid on standing in the air. It melts, undergoing decomposition. It contains no sulphur. It is easily soluble in ether, chloroform, carbon bisulphide and petroleum ether. At high temperatures it is soluble in glacial acetic acid, acetic ester and amyl alcohol, and precipitates out on cooling. It is insoluble in hot and cold methyl and ethyl alcohol and acetone. In water it sinks at once to the bottom, but swells and finally dissolves to a turbid emulsion, which reacts neutral. The emulsion becomes quite clear on adding alkali. Analysis of cuorin gave C, 61.63 per cent.; H, 9.03 per cent.; N, 1.015 per cent.; P, 4.46 per cent.; O, 23.86 per cent. The ratio of P:N::2:1. It is hence a mono-amino-diphosphatide. The iodine number is 101. It contains three fatty acid groups per molecule of the average composition of $C_{19}H_{34}O_2$. Melting point 47-48°. Iodine number of the acids, 130.1. The fatty acids contain acids of the linolic and linolinic groups, and are, therefore, doubly or trebly unsaturated. This is a point of much interest in view of the keen metabolism of the heart tissue. Glyceryl-phosphoric acid was isolated. The base was not choline, but an unknown base, the nature of which has not been determined. Cuorin resembles cephalin in many ways. It is very improbable that the cuorin thus prepared is a single substance and further work, using the methods of Thudichum for the separation of the cuorin into its constituents, must be done. There are other phospholipins in the heart than cuorin.

We shall have occasion to examine the composition of other members of the phospholipins in the chapter on the brain.

Physical properties of phospholipins.—The physical characters of the phospholipins, by whatever method they are prepared, are extremely interesting. They are generally light yellow to brown, waxy or salve-like substances, not crystalline, although some have been crystallized. They dry in vacuo over a drying agent to a hard mass, which may easily be pulverized. The powder so obtained is generally very hygroscopic and when exposed to the air it takes up quantities of water and turns to a dark-brown, thick, semifluid mass. It is probably partially oxidized at the same time, and it is chemically changed, so that after thus becoming partially fluid and standing some time, there is a change in solubility; generally the solubility in ether is diminished and that in water increased. The phospholipins, unlike the fats, form permanent emulsions, or colloidal solutions, if shaken with water and many of these emulsions are unstable, being easily precipitated by various common salts. One of their most striking peculiarities is that if rubbed with water or shaken with water, but not too long, and then examined under the microscope, they will be seen in isolated shreds or strings. Water enters the masses of phospholipin, which do not usually take a spherical shape but often the most regular and striking forms, reminding one irresistibly of the structure of cells. Often they take the form of long fibers with a hollow core and highly refractive walls, looking much like a medullated nerve fiber with the axis cylinder in the center; or they may take forms of flasks with long perfectly regular, whip-like processes looking like enlarged, exploded nettle cells. These forms are called myelin forms because of the resemblance they bear to the myelin sheaths of nerves. They form liquid crystals, the molecules having a definite orientation. In this condition they may be doubly refracting. Liquid crystals are easily deformed by pressure and fuse when brought into contact. A mixture of cephalin, amidolipin and cholesterol is particularly apt to assume these forms. Phrenosin and kerasin, two glycolipins of the brain, behave similarly. The power of making these myelin forms differs in different phospholipins, and it is possible that the presence of certain impurities, such as cholesterol, or its esters, may influence the process. It is possible that cells owe their hygroscopic character, their clear refractive appearance, their organization and power of taking definite shapes in part to this phospholipin groundwork. Indeed, if one rubs together in a mortar some phospholipin containing cholesterol, some white of egg and some olive oil, there is obtained a mixture of substances which mimics exactly the physical appearance and many of the vital staining reactions of protoplasm.

Auto-oxidation.—Another very important chemical property of the phospholipins is that they undergo auto-oxidation, that is they oxidize spontaneously when exposed to the air. The fact that this fundamental substratum of living matter has this power of taking up oxygen, of burning itself and, possibly, inducing oxidation in substances dissolved in it, although this possibility has not yet been investigated, is of the greatest importance for the theory of the mechanism of respiration, since all protoplasm has also this power of reduction, or auto-oxidation. This property of the phospholipins can be very easily observed, as it is accompanied by a darkening of the phosphatide preparations on exposure to air, and a change in their solubilities. The phospholipin from the heart, cuorin, and the brain cephalin are particularly active in this way, but nearly all phospholipins show something of this same property of taking up oxygen. Erlandsen observed the following increase in weight in some cuorin left in a desiccator over sulphuric acid.

	Weight of cuorin
December 29.	0.3705 gram.
February 6.	0.4039
November 20.	0.3949

The iodine number of the fresh cuorin was about 101, but after standing in the desiccator it fell to 22.33. The original preparation showed, on analysis, a composition of $C_{71}H_{123}NP_2O_{21}$; but after standing these numbers changed to $C_{71}H_{125}NP_2O_{30}$. At the same time the substance became insoluble in ether. It is this ease of oxidation which is one of the complicating circumstances in the isolation of unchanged cuorin and cephalin.

This power of auto-oxidation is due, in large measure at least, to the unsaturated fatty acids in these phospholipins. Thus cuorin and cephalin have acids of the linolenic acid series which contain three pairs of unsaturated carbon atoms, $C_{18}H_{30}O_2$; and nearly all the phospholipins contain some oleic or other unsaturated acids. They may possible contain also unsaturated groups in the basic constituents, but this, while in some instances probable, is not yet certain.

It is very suggestive that the oxidized phospholipin has a much greater affinity for water than the unoxidized. Possibly this process may play a part in the cell mechanics, since most movements in cells are apparently due to this changing affinity for water. By oxidation the phospholipin might be made to take up water, by reduction to lose it.

Other functions of phospholipins.—Besides their fundamental function of making, with the sterols, a water-containing, semifluid, highly-reducing, auto-oxidizable, semilipoid crystalline substratum, which contributes to or makes possible the vital phenomena, the phospholipins are also concerned in some of the phenomena of immunity. Flexner and Noguchi observed that the hemolytic action of cobra venom toward certain red blood corpuscles was dependent upon the presence of some substance in the serum. Kyes found that this serum substance was extractable with cold alcohol and it proved to be a phospholipin. He then showed that cobra venom united with lecithin (?) to form a substance which he called a lecithide, and these lecithides were extremely hemolytic,—they dissolved blood corpuscles with great power. Lecithin emulsions have also some hemolytic power themselves. The phosphatides may, therefore, by uniting with toxins, or possibly with the foods, influence the power of the latter to unite with and affect the activity of protoplasm. Either the phospholipins or glycolipins have the power of binding, or uniting with tetanus toxin. This function of the phospholipins is being investigated at the present time. The phospholipins appear to play an important part in the Wasserman reaction for syphilis.

Method of hydrolysis of the phospholipins.—The phospholipins (phosphatides) are generally decomposed by heating an emulsion with barium hydrate under a condenser for several hours. By this method, however, the fatty acids or soaps are generally a deep brown and contain some nitrogen, a partial decomposition of the base having occurred. Other methods suggested and tried have been by using a methyl alcohol solution of barium hydrate; or by suspending the phospholipin in water and boiling with 5% HCl. The best and cleanest method, however, for all alcohol soluble lipins, is to dissolve or suspend them in absolute alcohol, to saturate the solution with dry hydrochloric acid gas, stopper the flask and allow it to stand twenty-four hours at room temperature. Then boil on the water bath for 3-5 hours, using a reflux condenser. By this method the fatty acids are esterified with alcohol as rapidly as they are set free, and they remain in solution. The mixture is then evaporated with the addition of water on the water bath; the esters are insoluble in water and may be easily separated by filtration. In the acid filtrate the base, often choline, which is more stable as the salt than the free base, may be separated by precipitation with platinum chloride leaving the glycerol and phosphoric acid in solution.

A useful method of determining the character of the phospholipin is that of Herzig and Meyer as developed by Foster in the writer's laboratory. It consists in heating the phospholipin with hydriodic acid and ammonium iodide and collecting the isopropyl and methyl iodides in silver nitrate and weighing the silver iodide. The isopropyl iodide comes over at 112-120° and represents the amount of glycerol present. Two methyl groups come from the choline, and possibly three, at 240-310°, as methyl iodide. From the silver iodide collected at these two temperatures an idea may be had of the amount of glycerol and choline present, and the phospholipin partially identified. Cephalin contains glycerol about 10%, but no base yielding methyl iodide, i.e., neither neurine nor choline.

Other phospholipins.—Among the other phospholipins a few may be mentioned at this place, namely, heart lecithin, carnaubon, jecorin, glucose lecithin, cephalin and sphingomyelin. The last two are found in brain, the first in the ox heart, carnaubon in the kidney and the other two in the liver. They will be taken up more in detail under the chemical composition of the organs in which they occur, and at this place we will give only a summary of their composition to give some idea of the composition of the group. Especial attention is called to the fact that many of the members of the group, like jecorin, contain a carbohydrate radicle in the molecule. It is found that those which contain such a group are more soluble in water than the rest, and they are less soluble in ether. Some of them are quite insoluble in absolute ether, but will dissolve in ether containing water. They are also more hygroscopic than egg lecithin. Their chemical nature is still quite obscure. Such carbohydrate-containing phospholipins have been found not only in the livers of many animals, but also in the egg cells of Echinoderms, and they are probably widespread in the animal and plant kingdom. Several have been isolated from plants. Their group name is glyco-phospholipins.

The character of the base in the phospholipins is probably very variable. Amino-ethyl alcohol, or oxyethyl amine was isolated by Thudichum from cephalin, and it has since been found in many phospholipins. Serine and oxyamino-butyric acid were isolated by MacArthur from cephalin. Vidin, $C_9H_{26}N_2O_2$, was obtained from a phospholipin in Lupinus albus. Many other nitrogenous bases probably occur in the different phospholipins and the investigation of these bodies is one of the most promising fields of physiological chemistry.

Glycolipins.—The glycolipins, or cerebrosides, are phosphorus-free lipins which contain a carbohydrate and fatty acids. Phrenosin, kerasin and cerebron are typical members of the group. Phrenosin is a white, crystalline lipin found in the medulla of the fibers of the vertebrate nervous system. It occurs in considerable quantities in the brain after medullation has begun. The properties of this substance or group of substances are described in the chapter on the brain. The glycolipins on hydrolysis yield a sugar, which is often galactose, various fatty acids and a nitrogen-containing alcohol. The molecular weight is undetermined. While these bodies have been found in the brain, it is probable that they are not confined to that tissue, since phospholipins containing a sugar are widespread. It is not impossible that these may be compounds or mixtures of a glycolipin with a phospholipin.

Localization of the phospholipins in the cell.—The proportion of phospholipin in the sperm is high and it is found chiefly, or altogether, in the tails. The nuclear heads appear to be almost, or quite, free of alcohol-ether soluble materials. The heads and tails of the herring sperm were examined, the heads being separated by suspension of the sperm in water and centrifugalization. The heads being heavier are thrown off from the tails. This observation would indicate that the nucleus of the ripe sperm had little or no lipin material in it. The conclusion that the lipins are, for the most part, in the cytoplasm is confirmed, also, by examination of the red blood corpuscles. Mammalian erythrocytes have no nuclei, and yet they contain phospholipins and sterols. The localization of the phospholipin in the cell may also be studied by the use of staining reagents. The phospholipins are acids or salts of acids. There is a free acid hydroxyl in the phosphoric acid, if the ordinary formula is correct. The phospholipins are generally present in cells as salts, the hydroxyl being occupied by sodium, potassium, calcium, ammonium or organic bases. Being salts they have the power of forming salts with the basic dyes. They do this and they stain intravitam by means of the basic dyes. If an emulsion is made of lecithin, or other phosphatide, egg albumin and olive oil, it will be found that the granules of lecithin or phospholipin take up neutral red, or other basic dye, most easily and the granules stain a deep red. It does not do to conclude that all the granules of cells which stain red in neutral red are phospholipin, because other colloids of an electro-negative, or anionic, character will stain in the same way, but certainly some of the fine granules in cells look very much like lecithin; they stain in the same way and they are soluble in ether. Furthermore, the density of lecithin is greater that that of water, so that if cells are centrifugalized we should expect these granules to go to the base of the cell and the oil to come to the part of the cell nearest the axis of the centrifuge where the centrifugal force is least. It is found that these small granules which have the staining reaction and appearance and solubility of lecithin also behave the same way in the centrifuge. It is probable, therefore, that the phospholipin is distributed all through the cytoplasm; and that in egg cells, in which there is a storage of lecithin or other phospholipin as reserve food, the lipin occurs also as granules in the cytoplasm. Many believe that the surface of the cell has a character more particularly lipoid or lipin in character than the interior. They consider the blood corpuscles to be essentially but bags of lipin and sterol filled with hemoglobin, and they attribute the ease of penetration of ether, alcohol and other substances into cells to their solubility in this membrane. In the author's opinion there is no satisfactory evidence as yet of the existence of any such particularly lipin-like outer membrane and experiments indicate that the lipin character of the cytoplasm extends throughout it. The entrance of basic dyes can as easily be explained by their power of union with protein as by their power of union with phospholipin. Kite's work shows that the permeability of the cell is the same all the way through. From this it follows that if the entrance of substances into cells depends upon the presence of lipoids in the membrane, these same lipoids must occur all through the cell. There are, also, far too large quantities of lipins in cells to be confined to a surface layer of molecular thickness. Phospholipins are found, also, in many secretions, such as that of the stomach glands, milk, pancreatic secretion,

etc. It is not impossible that many of the granules called zymogen granules in cells which stain readily in neutral red may owe this property to the fact that they contain phospholipins. The mito-chondria are probably in part phospholipin.

REFERENCES. LIPINS.

1. Classification. *Gies and Rosenbloom:* Biochemical Bulletin, I, p. 51, 1912.
2. Oils and fats. Encyclopædia Britannica, XI edition.
3. Waxes. " " "
4. The fats. *Leathes:* Longmans, Green & Co. London, 1910.
5. Cuorin. *Erlandsen:* Untersuchungen über die lecithinartigen Substanzen. Zeits. f. physiol. Chem., 51, p. 71, 1907.
6. Egg yolk. Phospholipins in egg yolk. *Stern and Thierfelder:* Zeits. f. physiol. Chem., 53, 1907, p. 371.
7. Nature of base in lecithin. *MacLean:* Biochemical Journal, iv, p. 455, 1900.
8. Monamino-diphosphatide in egg yolk. *MacLean:* Biochemical Journal, iv, p. 168, 1909.
9. Glyco-phospholipins. *Winterstein:* Zeitschrift f. physiol. Chem., lviii, p. 500, 1909.
10. Acid hydrolysis of phospholipins. *Moruzzi:* Zeit. f. physiol. Chem., lv, p. 352, 1908.
11. Choline in animal tissues and fluids. *Webster:* Biochem. Journal, iv, p. 331, 1909.
12. Trimethyl amine in blood, urine and cerebrospinal fluid. *Dorée and Golla:* Biochem. Journal, v, p. 306, 1911.
13. Quantitative determination of choline. *MacLean:* Zeits. f. physiol. Chem., lv, p. 360, 1908.
14. Quantitative determination of choline and betaine. *Stanek:* Zeits. f. physiol. Chem., xlviii, p. 334, 1906.
15. Nature of union of choline in lecithin. *Malengreau and Prigent:* Zeits. f. physiol. Chem., 77, p. 107, 1912.
16. Amino-ethyl alcohol in phospholipins. *Trier:* Zeits. f. physiol. Chem., 73, p. 383, 1911.
17. Oxyethyl amine in egg lecithin. *Eppler:* Zeits. f. physiol. Chem., 87, p. 233, 1913.
18. Origin of betaines. *Schulze and Trier:* Zeits. f. physiol. Chem., 81, p. 53, 1912.
19. Vidine. *Njegorau:* Zeits. f. physiol. Chem., 76, p. 1, 1911.
20. Alcoholic saponification of lecithin. *Kollet:* Zeits. f. physiol. Chem., 61, p 210, 1910.
21. Purification of phosphatides. *MacLean:* Biochem. Jour., vi, p. 355, 1912.
22. Carnaubon. *MacLean:* Biochem. Jour., vi, p. 354, 1912.
23. Acetyl choline in ergot. *Ewins:* Biochem. Jour., viii, p. 44, 1914.
24. Staining reactions of lipins. Weigert reaction. *Lorrain Smith:* Journal of Pathology and Bacteriology, XI, p. 415; 12, p. 1, 1907. Biochem. Jour., 8, 1914.
25. Role of phospholipins and cholesterol in immunity and hemolysis. *Browning:* Jour. of Pathology and Bacteriology, xiv, p. 484, 1909; xvi, p. 225.
26. Cobra poison and lecithin. Toxolecithides. *Morgenroth and Kaya:* Biochem. Zeits., 25, p. 88, 1910.

and lipins. *Browning, Cruickshank and McKenzie:*
5, 1910, p. 76.
Vernon: Biochem. Zeits., 47, p. 374, 1912.
: Biochemical Journal, vi, p. 100, 1911.
ve determination. *Windaus:* Zeits. f. physiol. Chem.,

ve determination. *Ritter:* Zeits. f. physiol. Chem.,

ve determination. *Lifschütz:* Biochem. Zeits., 25

Cerebral spinal fluid. *Pighini:* Zeits. f. physiol.
).
Lapworth: Journal Pathology and Bacteriology, 15,

of hemoglobinuria. *Pringsheim:* Münchener med.
1757, 1912.
Darmstaedter and Lifschütz: Berichte d. d. Chem.

ol and physiology. Various articles by *Dorée and*
lings of the Royal Society, London, ser. B, 1913-1914.
Alicard: Jour. de Physiol. et Pathol. gen. 16, p. 624,

cells under various conditions of diet. *Terroine:*
th gen. 16, pp. 583, 607, 1914.
activity. *Mayer and Schaeffer:* Jour. de Physiol. et
, 1914. See also p. 23.
f lecitho-proteins. *Mayer, Rathery and Schaeffer:*
thol. gen., 16, pp. 583, 607, 1914.
in hen's egg. *Mueller:* Jour. Biol. Chem., 21, p. 24,

CHAPTER IV.

THE PROTEINS.

Occurrence.—If living matter of any kind, whether plant or animal, be extracted with alcohol and ether to remove the fats and lipins, the greater part of the organic matter remains behind as a white amorphous mass. If this mass be rapidly extracted with water, or dilute alcohol, various salts (chlorides, phosphates, carbonates) and a small amount of organic matter, called extractives, are removed, together with a little of an acid, nucleic acid. There remains the greater part of the organic matter insoluble. If this organic matter be analyzed it is found to contain nitrogen. It consists in most instances of about 52 per cent. of carbon; 7 per cent. hydrogen; 15 per cent. nitrogen; 1-2 per cent. of sulphur; 0.2-3 per cent. of phosphorus; and 23 per cent. of oxygen. This organic, nitrogenous, amorphous matter since it makes by far the greater portion of the organic substance of living matter proper, exclusive of cell walls, is called *protein*,[1] a word meaning "of the first importance." Protein substances are found, like the fats and carbohydrates, nowhere else in nature than in living matter, or as the products of the action of living matter. In plants more or less carbohydrate is associated with the protein; but in animals by far the greater part of the tissue solids are generally protein.

The proteins, therefore, very early attracted the attention of investigators, whether biologists or chemists, and the opinion has from the first been held that they must play a predominant rôle in the production of the vital phenomena. Probably the fact that the composition of the fats and carbohydrates was relatively simple, was easily understood and early made out, in its general features at any rate, whereas the proteins were apparently more complex and have until recently kept the mystery of their composition inviolate, strengthened this view. Some have even gone so far as to suppose that the protein in living matter is alive and endowed with properties other than ordinary chemical and physical properties. Whether the protein, which we are able to separate from cells for analysis, has the same composition it had while in living matter is a question which is unanswerable, as long as we know nothing of its

[1] The name protein, from the Greek πρωτεῖος, pre-eminence, was given by Mulder, *Jour. f. prak. Chem.*, 16, p. 129, 1839. The word proteid has now been dropped and protein substituted for it.

composition while in the living mixture. But while the statement is often categorically made that the materials which we examine chemically are dead materials, and therefore incapable of throwing light on the composition of living matter, it must be remembered that it is the part of those who assert that they differ to show that this is the case. And this at present cannot be done. While, therefore, it is conceivable, and in itself not unlikely, that the protein may contain, while it is in the living mixture, unstable groups, i.e., aldehyde or nitrile groups, which are made stable in the process of extraction, no one has as yet proved this to be the case; nor, indeed, has any strong evidence been produced showing it to be the case. While some speak of living, as distinct from lifeless protein, no one can as yet point to any single chemical property found in living and not found in dead protein; although the differences between living matter and dead protein are profound. Whatever the truth may prove to be we must, at any rate, attack the dead protein, that is protein separated from non-protein substances, for following this path has already led to many brilliant and fortunate discoveries. These discoveries have revolutionized the science of nutrition and have reacted on all branches of biological science, from botany and anthropology to the diagnosis and treatment of disease. Let us see, therefore, what substances have been isolated from this amorphous, colorless mixture of protein bodies constituting so essential a part of the framework of living matter; and what properties the substances so isolated possess.

Methods of extraction.—The separation and exact analysis of these protein bodies is not yet complete. It was necessary to devise methods for the separation of the various proteins, for in any one cell there is a mixture of such bodies; and the elucidation of the structure of the molecules was most difficult, owing to their complexity and instability. It was found that the solubilities and some other properties of the proteins were easily modified by acids and alkalies, by heat, by alcohol, by some metal salts and by many enzymes and oxidizing agents. Subjected to the action of any of these agencies many proteins change their nature, as is evident from their change in solubility; they are said to be *denaturized.* Proteins so modified are called *derived proteins.* For the separation from the cell of natural, or unmodified, protein it was found necessary, in most instances, to dissolve them out with water, with neutral solutions of sodium chloride, or the sulphates of the alkali metals, and to avoid alkaline or acid reactions, high temperatures or alcohols.

It is very difficult indeed to get from living matter itself individual, pure proteins. The living matter is such a mixture of different kinds of proteins that it is only occasionally that a form of protoplasm is obtained which consists largely, or exclusively, of a single protein. One of the very few cases in which probably pure individual proteins have been separated from a living animal cell is the protamine which may be separated from the head of the fish sperm. The spermatozoon is a very highly differentiated cell, the head being composed of pure nuclear matter of the simplest kind known. Moreover the protein in the head is of a very peculiar kind, being very strongly basic and soluble in strong or dilute acids, so that its separation from other proteins is not difficult, and there is apparently in some sperm but one protein present. Elsewhere in the animal kingdom one can at times, owing to the peculiar nature of the protein, secure proteins which are probably pure substances. Hemoglobin, the red protein of the blood, crystallizes very easily in characteristic crystals, so that this protein is probably pure. In the egg cells of selachians crystalline substances exist which are probably proteins. In animals it is only in special cells such as the cells of the thyroid gland (colloid matter) or the egg cells, and in very few of these, that a storage of a particular kind of reserve protein exists, so that this one kind of protein is in such excess that it may be separated from the others. In general, animals do not store proteins, as they do carbohydrates and fats, except in the form of living matter. The condition is different in plants and it is from plants that crystalline proteins which are probably pure individuals may be most easily had for analysis. In seeds and nuts there is a large amount of reserve or stored protein. Many of these proteins are of a crystalline form in the seeds or nuts. Moreover the living protoplasm in these structures is reduced to a minimum. It is found chiefly in the embryo, which in some cases, as in wheat, may be separated by milling processes from the rest of the grain; and in any case, it is generally present in very small amount compared to the large amount of reserve protein present. There is another feature of these plant reserve proteins, besides their presence in large amounts and their ease of crystallization, which makes them of particular value for the study of the nature of the proteins. They are very stable proteins and do not easily change their condition; or they do not change their condition and solubility so easily as the protein in living matter, in the protoplast proper. In order that substances may accumulate in cells they must be rather resistant; the reactive substances are short-lived. This is a general rule for all the substances of the cells. Thus the fat which is laid down as reserve fat is generally stable saturated fat; in the carbohydrates it is the stable polysaccharides or the most stable disaccharide, cane sugar, which accumulate as a reserve. The same thing is true for the phospholipins. The phospholipin stored as a reserve in the egg cell is the most stable form, lecithin. The phospholipins isolated from active living matter are very unstable with highly unsaturated acids in them. This same rule is true for the proteins and in seeds and nuts the large amount of reserve protein present is unusually stable and so best adapted to withstand the crude manipulation of the chemist. The stable proteins used by animals for supporting tissues, such as horn or elastin, are extremely insoluble. A very easily crystallizable substance obtained from nuts is the protein, excelsin, of the brazil nut. This occurs crystalline in the nut itself. Another crystalline reserve protein is edestin, which occurs in hemp seed. And there are many others. These proteins may be obtained crystalline by the simple expedient of dissolving them out of the seeds or nuts with 10 per cent. sodium chloride and dialyzing the salt out of the solution. The protein will crystallize in the dialyzing tube. The investigation of these plant proteins is largely the work of the American chemist, Osborne, who has thus carried forward the work well begun many years ago by the German investigator, Ritthausen.

The separation of proteins from plant seeds and nuts is generally best made by extracting the ground nuts, or seeds, with distilled water, or by 10 per cent. sodium chloride; or 75 per cent. alcohol. A few proteins exist in some plant seeds which are insoluble either in water or absolute alcohol, but which are soluble in dilute alcohol. Gliadin of wheat and zein of corn are examples of such proteins. The separation of the mixture of the proteins obtained by extraction with salt solution is best made either by dialysis, which will precipitate some of them and dialyze away impurities; or by precipitating them by saturating their solutions with ammonium sulphate which precipitates all the proteins. The redissolved precipitate may be separated by fractional precipitation with ammonium sulphate, the globulins, being less soluble than the albumins in ammonium sulphate, are precipitated by half saturation. Since acids change the plant proteins very easily, both by their own action and by their power of setting free digestive enzymes which are present in all forms of living matter, it is often wise to add just enough dilute barium hydrate to keep the reaction neutral to phenolphthalein during the extraction.

While many of the proteins are thus extracted with neutral salt solutions or water, there generally remains a residue which can only be dissolved by the use of alkali, which may and probably does alter the composition of the protein. In the cereals, for example, water or neutral salt solution dissolves very little of the protein.

The extraction of the proteins from animal tissues is a very much more difficult matter than from vegetable. In the first place the animal cells have more reactive protein in them and generally there are present powerful digestive enzymes, the so-called autolytic enzymes, which are apt in a prolonged extraction to change, more or less, the protein of the cell. These enzymes are easily checked by the maintenance of a neutral or faintly alkaline reaction. Extraction always removes, also, some of the phospholipins with the proteins and the separation of these from the proteins without coagulation of the latter is very difficult, and it is doubtful if it is often possible. The methods of extraction employed differ so widely in different tissues that it is hard to make any general statements. It is sometimes advisable to extract the ground or finely-cut tissue directly with 10 per cent. sodium chloride solution so as to bring everything possible into solution; and sometimes it is better to dry the tissue and after removing the lipin by some method which does not coagulate the protein, to extract then with salt solution and fraction the extract. In a few cases the proteins can only be separated by the use of alkali, as in the nucleo-proteins; and in other cases, when strongly basic proteins are present, acids must be used (protamines and histones).

Probably the best general method for obtaining proteins in a least modified form from animal tissues is the following: The organ or tissue, of which it is desired to examine the proteins, is frozen while still living. It can be frozen either by liquid carbon dioxide or liquid air or by any other method which chills it at once to a very low temperature so that all chemical change is at once inhibited and the development of an acid reaction prevented. It is then quickly shaved into a snow in a Kossel knife machine, which is a large microtome, the snow being kept below the freezing point. The cooled snowy mass of the organ is placed in vacuum desiccators over some good drying agent. Aluminum oxide is one of the best, but any other non-volatile, water-absorbing medium may be used; and the desiccator is evacuated to a very high vacuum, .01-.001 mm. of Hg., by a good air pump. In the vacuum all conduction of heat is prevented and the organ dries while still frozen. When thoroughly dry the ground mass is fairly stable. It may now be extracted by a good, dry petroleum ether. It is better to use petroleum ether rather than ether for this purpose, since this consists chiefly of the light, saturated hydrocarbons like pentane and these are less reactive than ether and the chlorine substitution products of the hydrocarbons such as chloroform, which are apt to contain impurities and reactive decomposition products. By this extraction nearly all the phospholipins, cholesterol and fat are taken out. The last portions of phosphatides can be removed only by repeated boilings with absolute alcohol, a process which coagulates the protein and hence cannot be used if it is desired to keep the proteins soluble. After the extraction with petroleum ether the protein residue is dried at room temperature and then finally ground in a good mill. It is best now to sieve it through a fine wire sieve to remove fibers of connective tissue, blood vessels, etc., if an animal organ has been used. By this sifting process the sample is made as homogeneous as possible.

The further treatment of the sample will depend on what substances are desired. One method, however, is to extract those proteins which are soluble in dilute, 10 per cent. sodium chloride. A considerable part of the proteins of many organs goes into solution by this process, but generally much remains undissolved. This mixture may be separated into several fractions, for example, by half saturating it with ammonium sulphate, or by saturating it with magnesium sulphate, a portion of the proteins known as globulins are thrown out and may be removed by filtration, redissolved in weak salt solution and reprecipitated. In some such way as this, the exact method which it is best to follow varying with the different organs, the various proteins making up the protein residue of cells may be examined. It may be said that in all cells thus far examined, a number of different bodies may thus be isolated from each kind of living matter and the character of the proteins thus isolated is peculiar to the kind of cells from which they are obtained. Thus one finds different proteins in the liver and spleen, or in the livers of different species of animals. The protein mass of any one cell is, therefore, made up of a group of bodies, which may be separated one from another by their solubilities in water, salt or dilute alcohol.

Composition.—If a little of this protein mass after extraction with alcohol and ether be mixed with soda lime in a dry test-tube and heated, it will be found to yield ammonia, which we may detect by its odor. By this reaction it is found that the proteins contain nitrogen. If some be heated by itself it chars to a black mass, showing that it contains carbon and is organic. If the combustion is carried out in a dry tube, moisture will condense on the walls of the test-tube, showing that the protein contains hydrogen; if some of the mass is fused with sodium carbonate or hydrate, or even suspended in dilute sodium hydrate and thoroughly boiled, and to the solution lead acetate is added, it will be found that a black precipitate of lead sulphide will be formed, proving that the proteins contain unoxidized sulphur; and, finally, if another portion be fused with sodium carbonate and potassium nitrate, the fusion mass dissolved in nitric acid, and ammonium molybdate added and warmed, a yellow precipitate of phosphomolybdic acid proves the presence of phosphorus.

A quantitative estimation of the amounts of the elements in different proteins has shown some variation, but the typical protein substances contain, of carbon about 50 per cent.; nitrogen, 16 per cent.; hydrogen, 7 per cent.; oxygen, 22 per cent.; sulphur, 0·3 per cent.; phosphorus, 0·4 per cent.

COMPOSITION OF VARIOUS PLANT AND ANIMAL PROTEINS.

	C	H	N	S	O	
Amandin, almond	51.30	6.90	18.90	0.43	22.47	
Corylin, hazel-nut	50.72	6.86	19.03	0.83	22.56	
Excelsin, Brazil-nut	52.23	6.95	18.26	1.09	21.47	
Edestin, hemp-seed	51.36	7.01	18.65	0.88	22.10	
Globulin, cotton-seed	51.71	6.80	18.30	0.62	22.51	
Vignin, cow-pea	52.64	6.95	17.25	0.42	22.74	
Glycinin, soy-bean	52.01	6.89	17.47	0.71	22.92	
Legumin, lentil, vetch	51.72	6.95	18.04	0.39	22.90	
Phaseolin, kidney-bean	52.57	6.97	15.84	0.33	24.29	
Conglutin, blue lupine	51.13	6.86	18.11	0.32	23.10	
Vicilin, lentil	52.29	7.03	17.11	0.17	23.40	
Legumelin, lentil	53.31	6.71	16.08	0.97	22.93	
Gliadin, wheat, rye	52.72	6.86	17.66	1.03	21.73	
Glutenin, wheat	52.34	6.83	17.49	1.08	22.26	
Globulin, wheat	51.03	6.85	18.30	0.69	23.13	
Hordein, barley	54.29	6.80	17.20	0.85	20.86	
Animal proteins:						
Collagen	50.75	6.47	17.86			Hoffmeister
Gelatin (commercial)	49.38	6.80	17.97	0.7	25.13	Chittenden
Gelatin, ligaments	50.49	6.71	17.90	0.57	24.33	Richards and Gies
Elastin, yellow elastic	53.95	7.03	16.67	0.38	21.97	Schwarz
Mucin, submaxillary	48.84	6.80	12.32	0.84	31.20	Hammarsten
Mucoid, tendon	48.76	6.53	11.75	2.33	30.63	Hammarsten
Serum globulin	52.71	7.01	15.85	1.11	23.32	Chittenden and Gies
Serum albumin	53.08	7.10	15.93	1.90	21.96	Michel
Fibrin	52.68	6.83	16.91	1.10	22.48	Hammarsten
Hemoglobin, ox	54.66	7.25	17.70	0.45	19.54	Fe = 0.4 Hufner
Casein, cow	53.00	7.00	15.70	0.8	22.65	P = 0.85
Nucleo-histone, thymus	43.69	5.60	16.87	0.47		P = 5.23
Nucleo-protamine, herring	41.20	5.75	21.06	0.00	25.90	P = 6.07
Clupein, herring	47.93	7.59	31.68	0.00	12.78	

The composition of the plant proteins of this table have been taken from Osborne's work on the plant proteins.

Properties.—A further study of these substances shows that they form *colloidal solutions.* If they are dissolved in water or dilute salt solution and the solution is placed in a bag of parchment paper and the bag is placed in the solvent, the proteins do not pass outside the bag. They cannot pass through parchment, or collodion, or animal membranes by diffusion. Dissolved substances which do not pass through membranes of this kind were called by the English physicist, Graham, colloids, or glue-like bodies, because glue, which is also a protein, cannot diffuse. Most proteins, therefore, are colloids and we may infer from the fact that they are colloids and so non-diffusible that their molecules in aqueous solvents are large. Another property they possess, also indicating a large molecular size when in solution at any rate, is the opalescence of many of their solutions. Moreover they are generally amorphous, although a few, like hemoglobin, crystallize easily and others can be made to crystallize with some difficulty.

Another property of nearly all proteins is that when they are treated with a mixture of the digestive juices of the intestine and pancreas of mammals, or when they are heated for some time with acids, they are decomposed into a mixture of substances called *amino-acids* which crystallize. In other words proteins are digestible by certain enzymes and on digestion yield amino-acids.

They have besides certain powers of developing *colored products* when treated with various reagents, which are discussed farther on; and they may be precipitated by the salts of heavy metals and they form insoluble compounds with many acids such as tannic or picric or phosphotungstic. These reactions are used for their easy detection, or even their quantitative determination. And, finally, they will neutralize in a marked degree the causticity of acids, and to some extent that of alkalies, showing, in this way, that they are amphoteric bodies, being both bases and acids. These various properties served to define the proteins long before we had any definite information of their chemical composition, and we may recapitulate them here as a definition of the proteins:

Definition.—The proteins are organic substances found in nature in living matter, or associated with it, and always produced by it. They consist of carbon, hydrogen, nitrogen, oxygen, generally, but not always, some sulphur and sometimes they contain phosphorus. The proportions of these constituents are approximately: C, 50 per cent.; H, 7 per cent.; N, 16 per cent.; O, 25 per cent.; P, 0·3 per cent.; S, 0·3 per cent. Their molecules are large, at least in aqueous solution, so that they are colloidal and do not pass through parchment paper; they give certain color and precipitation reactions, being precipitated by alkaloidal precipitating agents; chemically they are both bases and acids uniting with acids to diminish their acidity, and with bases to diminish their causticity; they are digestible by various enzymes and are broken up by acids when in solution and yield, when thus digested or decomposed, a mixture of crystalline, simple substances known as amino-acids and most, if not all, of these are α-amino acids.

All of these properties of the proteins are explained, or accounted for, by the hypothesis that the protein molecule is composed of a series of amino-acids united through their carboxyl and amino groups. This hypothesis was first well grounded by Kossel by his analytical studies of the simplest proteins, the protamines; and confirmed by Curtius and Emil Fischer, who synthesized bodies having these same properties by the condensation of amino-acids in the manner supposed by the hypothesis of Kossel.

Classification.—By the means which have just been discussed a great variety of protein substances, all having certain properties in common in virtue of which they have been called proteins, have been isolated from living cells, tissues and secretions, and separated one from another by their solubilities and chemical properties. While the chemical composition of none of these substances is as yet accurately known, it has been found necessary and convenient to divide them into classes, the basis of classification being chemical so far as possible, and where a chemical classification is not possible, solubility has been made the basis. Two such classifications have been proposed, one by the English Society of Physiologists, the other by the American Society of Biochemists. The two classifications are similar in their principal features.

American classification.—Proteins are defined as nitrogenous, organic substances consisting wholly, or in part, of amino-acids united by their carboxyl and amino groups. They are divided into three main classes:

1. Simple proteins.
2. Compound or conjugate proteins.
3. Derived proteins.

The first two classes are natural proteins; the last includes the artificial and proteins modified by reagents.

I. THE SIMPLE PROTEINS.—These are naturally occurring proteins which on being treated with enzymes or acids break up only into α-amino acids or their derivatives. They differ from the conjugate proteins in that the latter break up not only into amino-acids, but also into other non-protein substances. The simple proteins are separated into the following groups by their solubilities and other properties:

A. **Albumins.** Simple proteins, coagulable by heat, soluble in water and dilute salt solutions. *Ovalbumin, serum albumin.*

B. **Globulins.** Simple proteins, heat coagulable, insoluble in water, but soluble in dilute solution of salts of strong bases and acids. *Serum globulin, edestin.*

C. **Glutelins.** Simple proteins, heat coagulable, insoluble in water or dilute salt, but soluble in very dilute acids or alkalis. *Glutenin.*

D. **Prolamines.** Simple proteins, insoluble in water, soluble in 80 per cent. alcohol. *Gliadin, hordein, zein.* Found in grains.

E. **Albuminoids.** Simple proteins, insoluble in dilute acid, alkali, water or salt solutions. *Elastin, keratin, collagen.*

F. **Histones.** Simple proteins, not coagulable by heat, soluble in water and in dilute acid; strongly basic, and insoluble in ammonia. *Histone* from birds' corpuscles and thymus.

G. **Protamines.** Simple proteins, strongly basic, non-coagulable by heat, soluble in ammonia, and yielding large amounts of diamino-acids on decomposition. *Sturin, salmin, clupein,* etc. Found in ripe sperm of fishes.

II. CONJUGATED PROTEINS.—These are compounds of simple proteins with some other non-protein group. The other group is generally acid in nature. They are subdivided into the following classes:

A. **Chromoproteins.**[1] The prosthetic group (Gr. *prosthesos,* additional) is colored. It may be hematin as in hemoglobin, or the colored radicles of phycoerythrin or phycocyan. *Hemoglobin, hemocyanin, phycoerythrin, phycocyan.*

B. **Glyco- or glucoproteins.** The prosthetic group contains a carbohydrate radicle. In mucin and cartilage it may be chondroitic acid. *Mucin, ichthulin, mucoids.*

C. **Phosphoproteins.** Proteins of the cytoplasm. The prosthetic group is not known, but it contains phosphoric acid, but not nucleic acid or a phospholipin. *Casein, vitellin.*

D. **Nucleoproteins.** Proteins of the nucleus. The chromatin. The prosthetic group is nucleic acid. *Nuclein, nucleohistone.*

E. **Lecithoproteins.** Found in the cytoplasm and limiting membrane. Prosthetic group is lecithin or a phospholipin. No lecithoprotein has as yet been isolated. They probably exist.

F. **Lipoproteins.**[2] Existence is still doubtful. The prosthetic group is a higher fatty acid.

III. DERIVED PROTEINS.—This group is an artificial one, but it includes all the various decomposition products of the naturally occurring proteins which are produced by the action of reagents, or enzymes, or physical agents such as heat; and also artificially synthesized proteins. It is divided into various groups by solubility and also somewhat according to the degree of decomposition.

A. PRIMARY PROTEIN DERIVATIVES.

a. **Proteins.** Derived proteins. The first products of the action of acids, enzymes or water. Insoluble in water. *Edestan, myosan.*

b. **Metaproteins.** The further action of acids and alkalies produces metaproteins. These are soluble in weak acids or alkalies but insoluble in neutral solutions. *Acid albumin (acid metaprotein); alkali albumin.*

c. **Coagulated proteins.** Insoluble protein products produced by the action of heat or alcohol.

B. SECONDARY PROTEIN DERIVATIVES.

a. **Proteoses.** Hydrolytic decomposition products of proteins. Sol-

[1] The subdivision here called chromoproteins is called "Hemoglobins" in the usual American classification. The word Chromoprotein has so many points of superiority over that recommended by the Society that it has been adopted here. It was adopted by the English classification.

[2] Artificial lipoproteins are easily made. This subdivision has been added by the author to the classification adopted by the Society.

uble in water, not coagulable by heat, precipitated by saturatin solutions with ammonium sulphate.

b. **Peptones.** Hydrolytic decomposition products of protein ble in water, not coagulable by heat, not precipitated by saturati ammonium sulphate; generally diffusible and giving the biuret r

c. Peptides. These are compounds of amino-acids of which t position is known. Many are synthetic. The amino-acids are through the amino and carboxyl groups. They may or may r the biuret reaction. They are not heat coagulable. They are di-, tri-, tetra-, penta-peptides, etc., according as they contain twc eral amino-acids in the molecule.

ENGLISH CLASSIFICATION.

I. SIMPLE PROTEINS.

1. Protamines.
2. Histones.
3. Globulins.
4. Albumins.
5. Glutelins.
6. Gliadins. (Prolamins.) (Soluble in 80 per cent. insoluble in water.)
7. Sclero-proteins. (Forming the skeletal struct animals.)
8. Phosphoproteins. Caseinogen.

II. CONJUGATED PROTEINS.

1. Chromoproteins.
2. Nucleoproteins.
3. Glucoproteins.

III. HYDROLYZED PROTEINS.

1. Metaproteins.
2. Albumoses or Proteoses.
3. Peptones.
4. Polypeptides.

The definitions adopted of these groups are essentially those given in the American classification. It will be noticed that the classification places the phosphoproteins among the simple prot

Decomposition products of the proteins.—Before we take up detail the peculiar characters of the different proteins included protein mass from cells, we may consider, first, those propertie all such masses show, whatever their origin, since these are the mental or peculiar properties common to the proteins as a class, a were observed before the composition of the proteins was kn the reactions explainable. The first of these properties, nam chemical composition, has already been referred to. The elemen posing them are among the most abundant on the earth. Besides their elementary composition, all such protein masses show certain color reactions which are convenient for their identification; and they all yield certain crystalline amino-acids when cooked with acid. We may begin with a study of these decomposition products, since the way followed to get an insight into the composition of such a large molecule as that of protein is to break it into smaller pieces in various ways and from the character of these pieces, which are sufficiently small to permit the determination of their composition, to imagine how they are united in the larger molecule; we then put this imagination to the test, as to its validity, by artificially uniting the pieces in the way they are supposed to occur and see if the artificial product thus obtained corresponds, in its properties, with the natural product whose composition was sought. If it does correspond the structure of the molecule is considered solved.

The main difficulty, or one difficulty, in such studies on chemical structure, comes from the possibility that the substances isolated in the decomposition may not have been preformed in the molecule, but may have arisen secondarily during the reaction; but this difficulty can be avoided, in part at least, by breaking up the molecule in a variety of ways; if the products are the same whatever method is used, it may be confidently believed the fragments actually pre-existed in the molecule. In the case of the proteins the problem was in this particular unusually simple. The proteins may be broken up in a variety of ways, but they always yield the same end products, from which it is certain that these fragments pre-exist, or at least the greater part of them, in the protein molecule. The proteins may be broken up by heating them for several hours at a boiling temperature with 10 per cent. sulphuric acid, or 30 per cent. hydrochloric acid; or by the action of superheated steam acting for many days; or by the action of barium hydrate either cold or heated; or they may be decomposed at ordinary temperatures in a solution of nearly neutral reaction by digestive enzymes isolated from animal or plant tissues. The result is always, in its main features, the same. If treated in any of these ways the colloidal character of the protein disappears and it is converted into a mass of simple substances, which are crystallizable and which can be separated from each other. These substances are amino-acids. From these experiments the conclusion was drawn that the amino-acids must pre-exist in the molecule, since it is very unlikely that in these various conditions of acidity and temperature always the same products would be formed did they not pre-exist. Furthermore, by oxidation of the protein by permanganate, or other oxidizing agents, oxidized products of the amino-acids may be had. The conclusion that the amino-acids are preformed in the molecule is also strengthened by the character of the decomposition which has taken place. It is found that the total weight of the amino-acids from a given weight of protein is greater than that of the protein from which they are derived; the proportion of carbon and nitrogen in the decomposed mass is less; and that of hydrogen and oxygen is more. This shows that in the process of decomposition the elements of oxygen and hydrogen have been added; and further study shows that water has been taken up and combined in the process of decomposition, and that water alone will bring the decomposition to pass. From this it is clear that all these decompositions produced by acids, alkalies or enzymes are in reality produced by water, they are hence called hydrolyses (Gr. *Hydōr*, water; *lysis*, to loosen) or hydrolytic decompositions. Since this is a mild process, no great energy transformations accompanying the decomposition, it is an additional evidence that the amino-acids pre-exist in the protein molecule and that they are separated from each other by the entrance of water.

The amino-acids which have been identified are nearly all α-amino acids, but it is not impossible that others than α acids may still be found. An α-amino acid is an organic acid in which one hydrogen of the α carbon atom, that is the carbon atom just behind the carboxyl, is substituted by an amino group. An amino group is NH_2.

The following amino-acids have been isolated from the cleavage of simple proteins: glycocoll, alanine, valine, caprine, or glycoleucine, leucine, iso-leucine, aspartic acid, glutamic acid, tyrosine, phenyl alanine, tryptophane, histidine, arginine, lysine, proline, cystine, oxyproline, diamino-acetic acid, dioxy-diaminosuberic acid, $C_8H_{16}N_2O_6$, and diamino-trioxydodecanoic acid, $C_{12}H_{26}N_2O_5$. Both of these latter were found in caseine. The composition of these acids is shown in the following structural formulas. A β-alanine, that is β-amino-propionic acid has been isolated from carnosine, a di-peptide found in muscle.

Structure of amino-acids isolated from simple proteins.

I. *Aliphatic, mono-amino, mono-carboxylic acids.*

H	H	H	H	CH_3 CH_3
H—C—NH_2	H—C—H	H—C—H	H—C—H	C—H
O=C—OH	H—C—NH_2	H—C—H	H—C—H	H—C—NH_2
	O=C—OH	H—C—NH_2	H—C—H	O=C—OH
		O=C—OH	H—C—NH_2	
			O=C—OH	
Glycocoll (α-aminoacetic acid)	**Alanine** (α-amino propionic acid)	—— (α-amino butyric acid)	—— (α-amino valerianic acid)	**Valine** (Iso-propyl α-amino acetic acid)
$C_2H_5NO_2$.	$C_3H_7NO_2$.	$C_4H_9NO_2$.	$C_5H_{11}NO_2$.	$C_5H_{11}NO_2$.

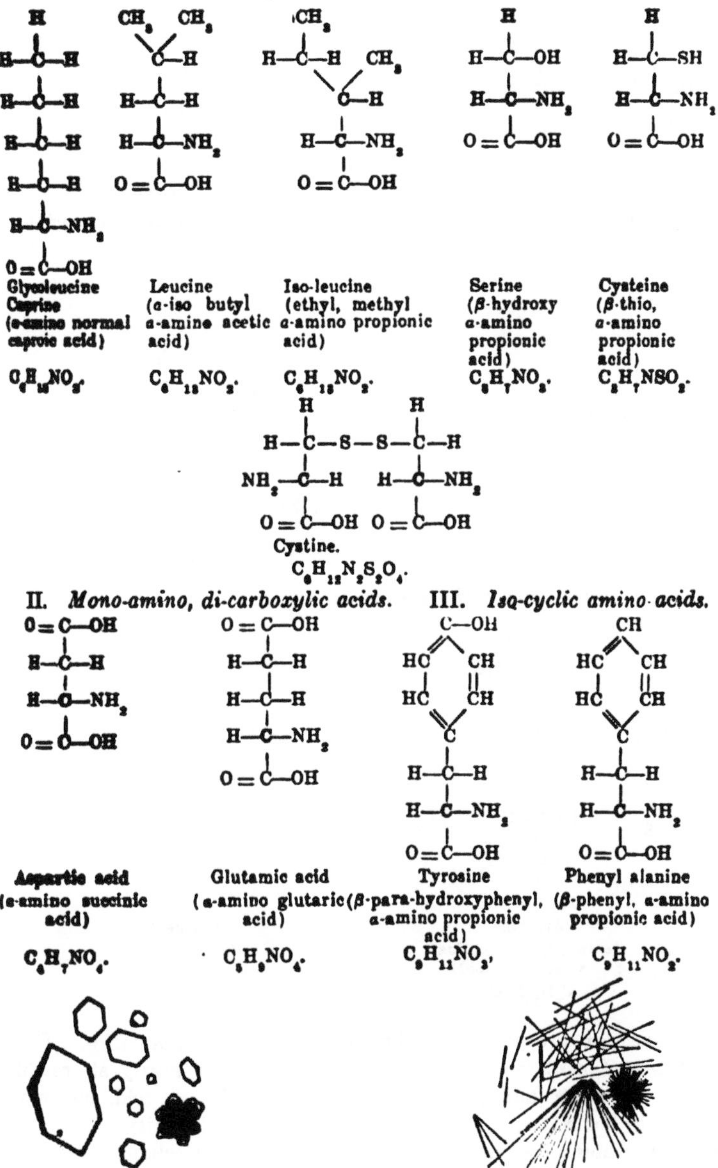

FIG. 11.—Crystals of cystine.

FIG. 12.—Crystals of tyrosine.

IV. *Hetero-cyclic amino-acids.*

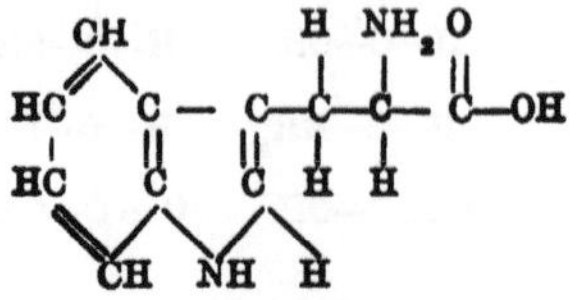

```
  CH          H  NH2 O
 //  \        |   |  ||
HC    C  —  C—C—C— C—OH
|     ||    ||  |   |
HC    C     C   H   H
  \\ / \  / \
   CH   NH   H
```

Tryptophane
(α-amino, β-indole propionic acid)
$C_{11}H_{12}N_2O_2$.

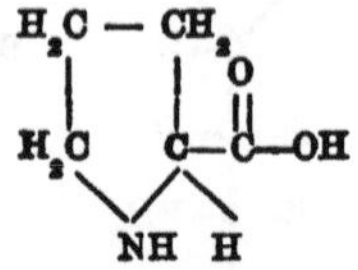

```
H2C — CH2
 |      |    O
 |      |    ||
H2C     C—C—OH
   \   / \
    NH    H
```

Proline
(α-pyrrolidine carboxylic acid)
$C_5H_9NO_2$.

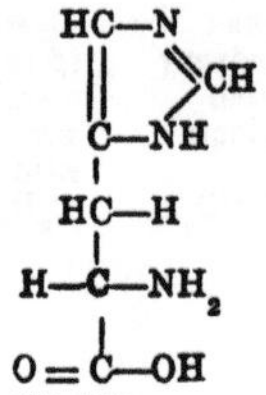

```
  HC—N
  ||   \CH
  C—NH
  |
  HC—H
  |
H—C—NH2
  |
O=C—OH
```

Histidine
(α-amino, β-imidazole propionic acid)
$C_6H_9N_3O_2$.

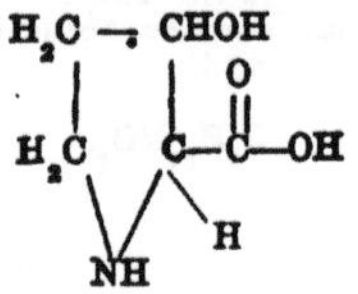

```
H2C — CHOH
 |      |    O
 |      |    ||
H2C     C—C—OH
   \   / \
    NH    H
```

Oxy-proline
(The position of the hydroxyl is uncertain)
$C_5H_9NO_3$.

V. *Diamino-mono carboxylic acids.*

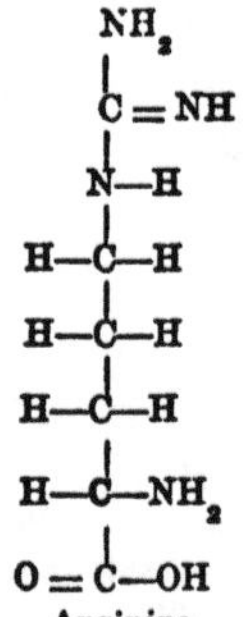

```
  NH2
  |
  C=NH
  |
  N—H
  |
H—C—H
  |
H—C—H
  |
H—C—H
  |
H—C—NH2
  |
O=C—OH
```

Arginine
(α-amino, δ-guanidine valerianic acid)
$C_6H_{14}N_4O_2$.

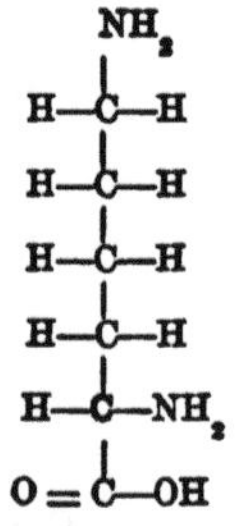

```
  NH2
  |
H—C—H
  |
H—C—H
  |
H—C—H
  |
H—C—H
  |
H—C—NH2
  |
O=C—OH
```

Lysine
(αξ di-amino, caproic acid)
$C_6H_{14}N_2O_2$.

Important properties of the amino-acids.—1. *Solubility. Compounds with acids and bases.* The amino-acids are nearly all readily soluble in water, but tyrosine is sparingly soluble in cold, but more soluble in hot water, while cystine is soluble with difficulty in both hot and cold water. Cysteine, on the other hand, is very soluble. They are all readily soluble in dilute acids and alkalies except cystine, which does not dissolve easily in dilute ammonia. Leucine, while it is quite soluble in cold water, redissolves very slowly. The amino-acids are insoluble in ether. The reaction of all the mono-carboxylic-mono-amino acids is amphoteric to litmus; the diamino acids and histidine and arginine react

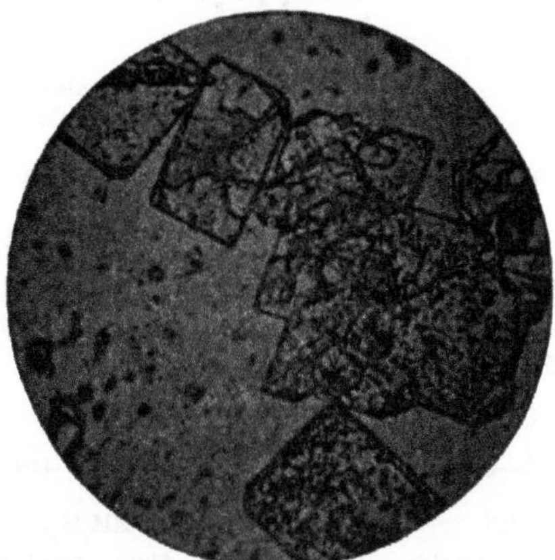

FIG. 13.—Crystals of histidine dichloride (Schreiner and Shorey).

alkaline in solution and absorb CO_2 from the air; whereas the mono-amino, dicarboxylic acids (aspartic, glutamic) are acid in reaction. Possessing both carboxyl and amino groups, they unite both with acids and bases and behave thus both as bases and acids. They form two series of salts. As bases they react like substituted ammonias to form hydrochlorides with hydrochloric acid. The salts thus formed ionize into the amino-acid as the cation and chlorine as the anion. As acids they unite with bases, such as sodium hydrate, to form the sodium salt of the amino-acid, which on ionizing yields sodium as the cation and the amino-acid as the anion. These reactions may be written as follows:

```
   H                      H                          H
   |                      |                          |
H—C—H                  H—C—H  /H                  H—C—H  /H
   |                      |   //H                    |   //H
H—C—NH₂  +  HCl  —→    H—C—N—H        ⇌          H—C—N—H   +   C̄l
   |                      |   \Cl                    |   \+
O═C—OH                 O═C—OH                     O═C—OH

Alanine.           Alanine hydrochloride.         Alanine    Chlorine
                                                   cation.     anion.
```

The positive charge, it will be noticed, is on the nitrogen atom of the amino group.

With bases the reaction is as follows:

$$\begin{array}{c}H\\|\\H—C—H\\|\\H—C—NH_2\\|\\O=C—O—H\end{array} + NaOH \longrightarrow \begin{array}{c}H\\|\\H—C—H\\|\\H—C—NH_2\\|\\O=C—O—Na\end{array} + H_2O \rightleftharpoons \begin{array}{c}H\\|\\H—C—H\\|\\H—C—NH_2\\|\\O=C—O^-\end{array} + \overset{+}{Na}$$

Alanine. Sodium alanate. Alanine anion. Sodium cation.

The negative charge is on the oxygen atom when the amino-acid is present as an anion.

As they are both weak acids and weak bases, both of their salts undergo hydrolytic dissociation, so that the sodium salts are alkaline in reaction, due to sodium hydrate; and the hydrochlorides are acid, there being always present some free hydrochloric acid.

$$CH_3—\underset{\underset{NH_2HCl}{|}}{\overset{\overset{H}{|}}{C}}—\overset{\overset{O}{\|}}{C}—OH + H_2O \rightleftharpoons CH_3—\underset{\underset{NH_2H_2O}{|}}{\overset{\overset{H}{|}}{C}}—\overset{\overset{O}{\|}}{C}—OH + HCl$$

Alanine hydrochloride. Alanine free base.

$$CH_3—\underset{\underset{NH_2}{|}}{\overset{\overset{H}{|}}{C}}—\overset{\overset{O}{\|}}{C}—O—Na + H_2O \rightleftharpoons CH_3—\underset{\underset{NH_2}{|}}{\overset{\overset{H}{|}}{C}}—\overset{\overset{O}{\|}}{C}—O—H + NaOH$$

Sodium alanate. Alanine.

This property of the amino-acids of uniting both with bases and acids is of very great importance in determining the behavior of the proteins, which act similarly.

2. *Union with salts of metals.*

By their amino groups the acids will unite, also, with such salts as mercuric chloride or nitrate, silver nitrate, platinum chloride and cupric chloride to form double salts, and some of these compounds are crystalline. All of these metals come below hydrogen in the order of their solution tensions, that is in the electromotive series of the metals. The reaction with cupric chloride may be written as follows:

$$CH_3—CH_2—CH.NH_2—COOH + CuCl_2 \longrightarrow CH_3—CH_2—CH.NH_2CuCl_2—COOH.$$

Advantage is taken of this power of the amino-acids to combine with copper and mercuric salts to precipitate them from their solutions. Mercuric acetate in the presence of carbonates is one of the best reagents to use for this purpose, although the precipitation is probably not quantitative and many other compounds than the amino-acids will precipitate with this reagent. One makes the solution weakly alkaline by the addition of sodium carbonate and then adds alternately a little of a 25 per cent. solution of mercuric acetate, and a few drops of sodium carbonate to keep the reaction alkaline, as long as a white precipitate is obtained and then sufficient of the reagents so that on stirring the precipitate has a faint yellowish-red color. Then 5-8 volumes of 98 per cent. alcohol are added.

The amino-acids also have the property of forming compounds with ordinary salts such as sodium chloride.

3. *Condensation of aldehydes with the amino group.*

The amino groups will also condense with aldehydes, particularly with formaldehyde, with the elimination of water according to the reaction:

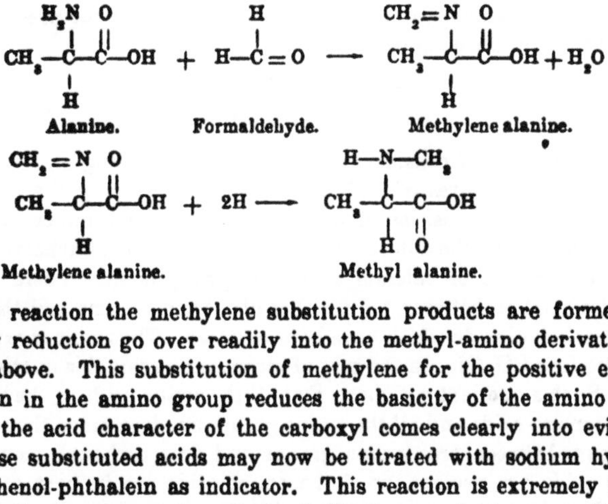

$$\underset{\text{Alanine.}}{CH_3\text{—}\overset{H_2N}{\underset{H}{C}}\text{—}\overset{O}{\overset{\|}{C}}\text{—}OH} + \underset{\text{Formaldehyde.}}{H\text{—}\overset{H}{C}{=}O} \longrightarrow \underset{\text{Methylene alanine.}}{CH_3\text{—}\overset{CH_2{=}N}{\underset{H}{C}}\text{—}\overset{O}{\overset{\|}{C}}\text{—}OH} + H_2O$$

$$\underset{\text{Methylene alanine.}}{CH_3\text{—}\overset{CH_2{=}N}{\underset{H}{C}}\text{—}\overset{O}{\overset{\|}{C}}\text{—}OH} + 2H \longrightarrow \underset{\text{Methyl alanine.}}{CH_3\text{—}\overset{H\text{—}N\text{—}CH_3}{\underset{H}{C}}\text{—}\underset{O}{\underset{\|}{C}}\text{—}OH}$$

By this reaction the methylene substitution products are formed and these by reduction go over readily into the methyl-amino derivatives as shown above. This substitution of methylene for the positive element hydrogen in the amino group reduces the basicity of the amino group so that the acid character of the carboxyl comes clearly into evidence, and these substituted acids may now be titrated with sodium hydrate, using phenol-phthalein as indicator. This reaction is extremely important, partly because such methyl substitution products are quite common in animals and plants and they are probably produced in this way, and partly because this reaction is the basis of the Sorensen method, which is one of the best methods for determining in a simple way the amount of amino-acids in a mixture. See p. 914. Other aldehydes do not react so readily as formaldehyde with the amino group.

4. *The carbamino reaction of amino-acids.*

Another very interesting and important reaction of the amino-acids is their union with calcium and carbonic acid to form carbamino compounds, a reaction discovered and studied by Siegfried and a reaction of considerable biological importance, since it occurs in the living organism. When ammonia and carbonic acid unite, two compounds may be formed, namely, ammonium carbonate, $(NH_4)_2CO_3$, and ammonium carbamate, the ammonium salt of carbamic acid, $NH_4OCO.NH_2$.

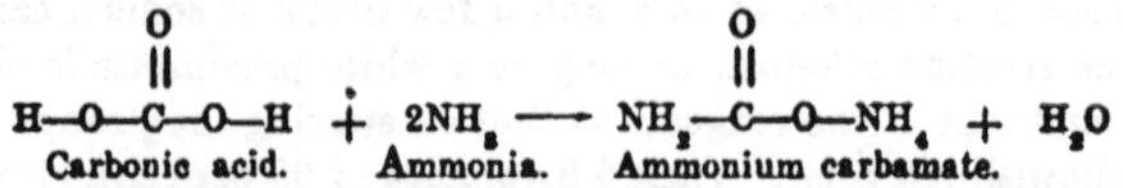

$$\underset{\text{Carbonic acid.}}{H{-}O{-}\overset{\overset{\large O}{\|}}{C}{-}O{-}H} + \underset{\text{Ammonia.}}{2NH_3} \longrightarrow \underset{\text{Ammonium carbamate.}}{NH_2{-}\overset{\overset{\large O}{\|}}{C}{-}O{-}NH_4} + H_2O$$

The esters of carbamic acid are quite important. They are called urethanes. The ethyl ester, ordinary urethane, is a useful hypnotic.

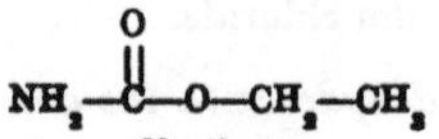

$$\underset{\text{Urethane.}}{NH_2{-}\overset{\overset{\large O}{\|}}{C}{-}O{-}CH_2{-}CH_3}$$

The carbamino reaction of the amino-acids is similar to this union of ammonia and CO_2 to make carbamic acid.

Alanine, for example, unites with carbon dioxide in the presence of a lime salt to form the compound

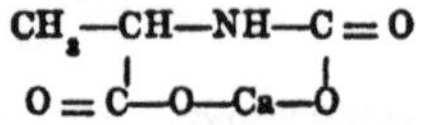

$$\begin{array}{l} CH_3{-}CH{-}NH{-}C{=}O \\ \quad\;\; | \qquad\quad\;\; | \\ O{=}C{-}O{-}Ca{-}O \end{array}$$

This reaction is of considerable importance in getting an idea of the composition of mixtures of amino-acids and in studying the course of the hydrolytic decomposition of the proteins. One determines the nitrogen and, by boiling the filtered solution containing the carbamino compound, the amount of calcium carbonate precipitated. There is thus obtained a relation between the number of nitrogen atoms and the molecules of carbon dioxide which have been in union. The result is expressed as the quotient CO_2/N. In the case of monoamino acids this quotient will be one. Such a quotient will mean that all the amino-acids in the mixture are monoamino acids and all the amino groups are free, since the reaction only occurs between free amino groups and free carboxyl groups. If diamino acids are present, or if the amino-acids are in part combined in peptide unions, the quotient will be less than one, since the amount of nitrogen is greater than the number of free amino groups. It is possible by studying this quotient to follow the course of the digestion of the proteins and determine when the reaction is complete. There are, however, easier methods which will be shown later.

5. *Deamidization by oxidation.*

The amino group is detached only with great difficulty in acid hydrolysis, or by the action of alkalies. One would think it might easily be replaced by hydroxyl to form the hydroxy acid, but this is not the case. If it were, it would be impossible to isolate the amino-acids from the proteins by acid hydrolysis. The amino-acids are most stable in the form of their acid salts. Most of them are quite stable, too, in alkaline solution, but cystine and cysteine will lose their sulphur to a large extent in alkaline solution and arginine decomposes into ornithine and urea. The fact that the amino nitrogen is so firmly attached and the amino-acids so stable in acid solution enables the conclusion to be drawn that the ammonia, which always appears in small amounts when a protein is hydrolyzed in acid solution, cannot have come from the amino acids, but must have had some other linking in the protein molecule. As a matter of fact it represents *acid amide nitrogen* and not amino-acid nitrogen. It was in a free carboxyl group. It is called *amide* nitrogen. Its amount is variable, as we shall see. But while the amino group is not readily detached by a simple hydrolysis, it can be readily removed by an oxidation. By various oxidizing agents such as hydrogen peroxide, or permanganate, the amino group is displaced and the corresponding ketonic acid is formed. This reaction is reversible. The reaction may be written as follows:

$$\underset{\text{Alanine.}}{\begin{array}{c} CH_3 \\ | \\ H{-}C{-}NH_2 \\ | \\ O{=}C{-}OH \end{array}} + O \rightleftharpoons \underset{\text{Pyruvic acid.}}{\begin{array}{c} CH_3 \\ | \\ C{=}O \\ | \\ O{=}C{-}OH \end{array}} + NH_3$$

The reaction in the left direction goes on only in the presence of reducing agents. Both these reactions are of very great importance, since they occur in the metabolism of amino-acids in living matter and show the way in which amino-acids may form from the decomposition products of the sugars and ammonia; and how the proteins may be converted, with the loss of ammonia, into carbohydrates. The ammonia thus formed in the body by this oxidation of the amino-acids has a great many functions which are discussed on page 248. By further oxidation the ammonia thus set free may be oxidized to nitrogen gas. This may be done by bromine, or chlorine, or nitrous acid in the following reaction:

$$NH_3 + \underset{\text{Nitrous acid.}}{HO.NO} \longrightarrow 2H_2O + N_2$$

This last reaction is used in determining the quantity of amino-acids or the amino nitrogen of a protein in the van Slyke method. The nitrogen gas evolved by treatment of the protein or its hydrolyzed products by nitrous acid is collected and measured and the number of amino groups in a solution of the acids is thus determined with very great ease. The same method, using bromine in place of nitrous acid, is employed clinically for the determination of urea and some other forms of nitrogen in the urine. Although the amino group is replaced by oxygen with ease, it is not replaceable with hydrogen. That is, if a protein is hydrolyzed with hydrochloric acid and tin, so that nascent hydrogen is set free, the hydrogen does not displace the amino group to form the fatty acid. In fact, tin is not infrequently added to reduce the decomposition caused by oxidation during an hydrolysis.

6. *Formation of lactams and piperazine nuclei.*

Another important peculiarity of these α acids is the fact that they form anhydrides, two molecules combining, with the greatest readiness. Thus it is only necessary to evaporate solutions of leucine or other amino-acids to produce some condensation to imides, such as leucinimide; diketopiperazine nuclei are thus formed:

$$\begin{array}{l}(CH_3)_2=CH—CH_2—CH.NH—\overset{\displaystyle O}{\overset{\|}{C}} \\ \qquad\qquad\qquad O=\underset{|}{C}—HN—\underset{|}{C}H—CH_2—CH=(CH_3)_2\end{array} ; \quad \begin{array}{c}CH_2—CH_2 \\ HN \qquad\quad NH \\ CH_2—CH_2\end{array}$$

Leucinimide (Diisobutyl-diketopiperazine). Piperazine.

It is considered probable that heterocyclic rings of this kind occur in the protein molecule, but their presence has not yet been proved. Such a condensation may, however. give rise to some of the cyclic amino-acids from straight chain amino-acids.

When the amino group is in the γ or δ position as compared with the carboxyl, as it is in glutamic acid, for example, $COOH—\overset{\alpha}{C}H_2—\overset{\beta}{C}H_2—\overset{\gamma}{C}HNH_2—COOH$, an internal condensation within the molecule occurs with the greatest ease. These anhydride substances correspond to the lactones and are called lactams. It is an interesting fact that although the amino-acids themselves are without special physiological or toxic action, these lactams are powerful poisons, producing strychnine-like convulsions. By this lactam formation some of the cyclic amino-acids may be formed from the straight chain amino-acids.

This may, for example, be the origin of proline. Glutamic acid undergoes a condensation of this kind to form the lactam, pyrrolidon carboxylic acid, which by reduction will yield pyrrolidine carboxylic acid, or proline.

$$\begin{array}{l}O=C—CH_2—CH_2—CH.NH_2—COOH \longrightarrow \\ \quad | \\ \quad OH\end{array}$$

$$\begin{array}{c}CH_2—CH_2 \\ O=\underset{|}{C} \quad \underset{|}{C}H—COOH \\ NH\end{array} + 2H_2 \longrightarrow \begin{array}{c}CH_2—CH_2 \\ \underset{|}{C}H_2 \quad \underset{|}{C}H—COOH \\ NH\end{array} + H_2O$$

Pyrrolidon carboxylic acid. Proline.

7. *The taste of amino-acids.*

The amino-acids being very weak acids and having amino groups which resemble, in their properties, the hydroxyl groups of the carbohydrates, have also, in many cases, a sweet taste. Thus glycocoll, as its name implies (Gr. *glykys*, sweet; *kolla*, glue; the sweet substance from glue), is very sweet. Alanine is also sweet and so is caprine, or glycoleucine. Leucine, however, is tasteless and iso-leucine is bitter.

It is not yet known why some substances are sweet and others bitter. The problem is identical with the general problem of the nature of stimulation and the character of the irritable response of living matter. In this case it is essentially the question of how the taste buds of the sweet-perceiving nerves are stimulated.

8. *Optical properties of amino-acids.*

The amino-acids obtained from proteins by hydrolysis by acids or enzymes are all except glycine optically active. Most are levo-rotatory. The α carbon is always asymmetric. Since the total rotatory action of any collection of amino-acids obtained by acid hydrolysis or by digestive enzymes is different from, and generally greater than, the rotation of the protein from which they come, the change in the rotatory power of a digesting protein enables us to follow the rate of digestion.

The amino-acids obtained from the proteins by hydrolysis by alkalies, or which have been obtained by acid hydrolysis from protein which has been treated for a time with dilute alkali, are for the most part inactive. The amino-acids obtained from such alkali-treated proteins are as a rule, but not exclusively, composed of equal amounts of the dextro- and levo-rotatory forms of the acids. Since in the proteins probably only one of the optically active forms of an amino-acid occurs, as is shown by the activity of those acids when freed by enzymes or acid hydrolysis, it follows that by the action of alkali the other optically active form has been produced. This process by which from one optically active isomer the opposite optically active isomer is produced is called "racemization." The name is derived from racemic acid (page 22) which is composed of equal quantities of the two optical antipodes of tartaric acid. Now it is found that the amino-acids when free do not easily or rapidly racemize when treated by dilute alkali. The racemization produced by the action of dilute alkali on protein must, therefore, occur while the amino-acids are combined in the molecule of the protein. The way this is produced is very interesting and very valuable information about which amino-acids have free carboxyl groups on them in the protein molecule can be obtained by taking advantage of this ease of racemization by alkali of the amino-acids which have the carboxyl combined. The explanation as given by Dakin is as follows:

The amino-acids of the proteins are linked through their amino and carboxyl groups in the following way as illustrated by alanyl-alanine.

Dipeptides like this do not easily undergo racemization; but plex polypeptides do.

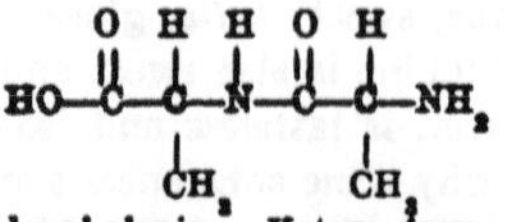

```
       O  H  H  O  H
       ‖  |  |  ‖  |
   HO—C—C—N—C—C—NH2
          |        |
         CH3      CH3
```

Alanyl-alanine. Ketone form.

This is the ketone form. Now by intramolecular rearrange hydrogen of the α carbon atom passes over to the oxygen of t group to make the enol form, thus:

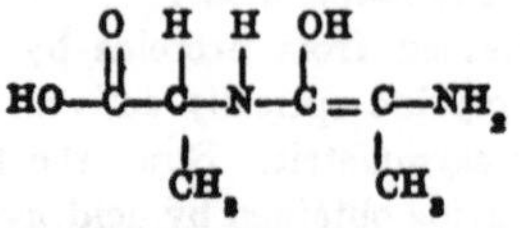

```
       O  H  H  OH
       ‖  |  |  |
   HO—C—C—N—C=C—NH2
          |     |
         CH3   CH3
```

Enol peptide form.

The ketone and enol forms are probably in equilibrium other, some enol going back to ketone and some ketone to enol.

By this enol formation the carbon atom of the second alaı which had been asymmetrical now becomes symmetrical. Wl the ketone form with its asymmetric carbon is regenerated, tl acid with the double bond will form equal amounts of the tw isomers, since there will be nothing to determine that one ra the other isomer shall be formed. It is clear that the alkali mus producing the enol form in the peptide groups and in this way it its racemization. Furthermore, since only those amino-acids w their carboxyl groups combined can undergo the enol formatioı find out, by studying the optical activity of various amino-a duced from protein by acid hydrolysis after the protein had bee with dilute alkali, which of the acids have been racemized and Those which were racemized must have had their carboxyl grouj those which were not probably had their carboxyls free.

Amounts of various amino-acids in different proteins.— ous proteins differ in the amounts of the different amino-aci they yield on hydrolysis. While it is not yet possible to detern of the amino-acids quantitatively, yet this is possible for som and approximations may be made for the others. The quantitat mination of arginine, lysine and histidine in a mixture of a can be made. The protein is hydrolyzed by prolonged cookir per cent. sulphuric acid or concentrated HCl and the bases, ε ration of the humus materials by filtration and decolorization coal, are precipitated by phosphotungstic acid. The further s three acids is accomplished by the different solubiliti

In the [illegible] it will be [illegible] that the [illegible] are [illegible] composition. Thus the [illegible] of [illegible] composed of glycocoll and alanine, whereas in [illegible] 87 per cent. of the molecule is [illegible]. The alcohol soluble proteins [illegible] contain [illegible] large amounts of proline and glutamic acid. Tyrosine, tryptophane and [illegible] amino-acids are present in some and absent in others. The [illegible] which is obtained by distilling the [illegible] alkaline solutions of the [illegible] products is [illegible] to the amount of glutamic acid in the molecule, these proteins [illegible] a large amount of glutamic [illegible]

While all proteins yield amino-acids the relative amounts of the different ones vary in each protein. Thus salmin, a protamine found in the head of the salmon sperm, yields arginine to over 85 per cent. of its weight; others yield large amounts of leucine, or glycocoll or glutamic acid; some yield no tyrosine, or cystine, or tryptophane. The proteins differ from each other chemically, therefore, in the amount of the different amino-acid nuclei they contain.

PROTAMINES COMPOSITION.

PER CENT. OF TOTAL NITROGEN PRESENT AS NITROGEN OF DIFFERENT AMINO ACIDS.

	Salmin[2]—Salmon—Salmon salar	Clupein[4]—Herring—Clupea harengus	Sturin[4]—Sturgeon—Accipenser sturio	Scombrin[4]—Mackerel—Scomber scomber	Percin[1]—Yellow perch—Perca flavescens	Thynnin[1]—Tunny fish—Thynnus thynnus	Pike[1] Perch—Stizostedion vitreum	Xiphin[1]—Sword fish—Xiphias gladius	Esocin[1]—Esox lucius	Salmin[1]—Chinook Salmon—Oncorhynchus tschawytscha	Coregonin[1]—White fish—Coregonus albus	Salvelin[1]—Lake trout—Salvelinus namaycush α	Salvelin β	Cyclopterin[4]—Cyclopterus lumpus	Cyprinin[4]—Carp—Cyprinus carpio α	Cyprinin β	Crenilabrin[5]—Cunner—Crenilabrus pavo
Histidine	0	0	11.8	0	5.6	0	6.7	0	0	0	0	0	0	0	0		0
Arginine	87.8	88.0	68.5	88.6	78.1	79.5	76.8	81.5	86.8	86.2[3]	87.3	88.9	88.5	67.7	8.7	28.0	42.8
Lysine	0	0	8.4	0	0	0	0	0	0	0	0	0			30.8	6.6	11.0
Monoamino nitrogen	9.2				9.8	11.0	10.7	14.0	11.3	9.4	9.4	7.1	6.7				25.1
Tyrosine	0	0	0	0	0.0	0.6	0	+	0	0	0	0	0	2.2		1.5	+
Alanine	0	+	+	+													
Valine	1.65	+			+	+											
Serine	3.25	+															
Proline	4.3	+		3.8	+	+											
Ammonia	0	0	0	0	0	0	0	0	0	0	0	0					0
Leucine			+														
Tryptophane	0	0	0	+				0	0	0	0	0		+			0

[1] Kossel: Zeit. physiol. Chem., **88**, 163 (1913), Kossel and Edlbacher ibid **88**, 186 (1913).
[2] Kossel and Dakin: Zeit physiol Chem., **40**, p 311, 565 (1904), **41**, 407 (1904).
[3] Compare with A. E Taylor: Jour biol Chem., **3**, 389 (1908).
[4] Kossel: Zeit physiol. Chem., **44**, 347 (1905).
[5] Kossel: Zeit. physiol. Chem., **69**, 138 (1910).

In the accompanying tables it will be noticed that the different proteins have a widely different composition. Thus the proteins of silk have half their molecule composed of glycocoll and alanine; whereas in salmin 87 per cent. of the molecule is arginine. The alcohol soluble proteins like zein contain very large amounts of proline and glutamic acid. Tyrosine, tryptophane and other amino-acids are present in some and absent in others. The amount of ammonia which is obtained by distilling the weakly alkaline solutions of the hydrolytic products is seen to be roughly parallel to the amount of glutamic acid in the molecule, those proteins like gliadin having a large amount of glutamic acid, have also a large amount of ammonia, whereas the protamines which lack glutamic acid in their molecules have no ammonia. This fact means that glutamic and possibly aspartic acids have an amide group in the free carboxyl. Atten-

PROTEINS, COMPOSITION.

PER CENT. OF AMINO-ACIDS ISOLATED FROM VARIOUS PROTEINS.

	Gliadin [1]—Rye	Gliadin [2]—Wheat	Hordein [3]—Barley	Zein [1]—Maize	Legumin [4]—Vetch	Legumin [4]—Cow pea	Salmin [10]—Salmon	Edestin [5]—Hemp	Excelsin [5]—Brazil nut	Keratin [6]—Sheep's horn	Squash seed [5]—Globulin	Amandin [4]—Almond	Ovalbumin [11]—Cryst.
Glycocoll	0.13	0.02	0.00	0.00	0.39	0.38	0.00	3.80	0.60	0.45	0.57	0.51	0.00
Alanine	1.33	2.00	0.43	9.79	1.13	2.08	0.00	3.60	2.33	1.6	1.92	1.40	2.22
Valine	?	0.21	0.13	1.88	1.36	?	4.3	6.20 [19]	1.51	4.5	0.26	0.16	2.50
Leucine	6.30	5.61	5.67	19.55	8.80	8.00	0.00	14.50	8.70	15.3	7.32	4.45	10.71
Proline	9.82	7.06	13.73	9.04	4.04	3.22	11.0	4.10	3.65	3.7	2.82	2.44	3.56
Phenylalanine	2.70	2.35	5.03	6.55	2.87	3.75	0.0	3.09	3.55	1.9	3.32	2.53	5.07
Aspartic acid	0.25	0.58	Not found	1.71	3.21	5.30	0.0	4.50	3.85	2.5	3.30	5.42	2.20
Glutamic acid	37.8	43.66	43.19	26.17	18.30	16.97	0.0	18.84	12.94	17.2	12.35	23.14	9.10
Serine	0.06	0.13	?	1.02	?	0.53	7.8	0.33	0.0	1.1	?	?	?
Cystine	?	0.45	?	?	Undet.	Undet.	0.0	1.00	?	7.5	0.23	?	?
Tyrosine	1.19	1.20	1.67	3.55	2.42	1.55	0.0	2.13	3.03	3.6	3.07	1.12	1.77
Arginine	2.22	3.16	2.16	1.55	11.06	11.71	87.4	14.17	14.14	2.7	14.44	11.85	4.91
Histidine	0.39	0.61	1.28	0.43	2.94	1.69	0.0	2.19	1.47	?	2.63	1.58	1.71
Lysine	0.00	0.00	0.00	0.00	3.99	4.98	0.0	1.65	1.64	0.2	1.99	0.70	3.76
Ammonia	5.11	5.11	4.87	3.64	2.12	2.05		2.28	1.80	?	1.55	3.70	1.34
Tryptophane	Pres.	Pres.	Pres.	0.00	Pres.	Pres.	0.0	Pres.	Pres.		Pres.	Pres.	
Total	68.31	71.46	78.16	85.27	62.65	62.22	110.5	82.38	60.21		55.77	59.00	30.08

	Vitellin [12]—Hen's egg	Muscle [13]—Scallop	Cocoon [7]—Ocortiens talonsis	Indian [14] Tussah silk	Italian [14] Silk cocoons	Japanese [15] Silk cocoons	Globin [8]—Horse haemoglobin	Gelatin [17]	Casein [16]—Cow	Elastin [18]	Histone [19]—Thymus
Glycocoll	1.00	0.00	27.19	1.5	38.50	35.00	0.00	16.5	0.00	25.75	0.5
Alanine	0.75	......	18.8	9.8	20.00	22.6	4.2	0.8	1.5	6.58	3.46
Valine	1.87	......	—	—	—	—	......	1.0	7.2	1.4	......
Leucine	9.87	8.78	0.75	4.8	0.75	0.7	29.0	2.1	9.4	21.38	11.80
Proline	4.18	2.28	3.2	8.0	0.8	0.7	2.3	5.2	6.7	1.74	1.46
Phenylalanine	2.54	4.90	1.8	0.8	1.2	1.3	4.2	0.4	3.2	3.89	2.20
Aspartic acid	2.13	3.47	0.25	2.8	1.0	1.3	4.4	0.56	1.4	......	......
Glutamic acid	12.95	14.88	2.35	1.8	0.25	0.07	1.7	1.88	15.55	0.76	3.68
Serine	?	......		5.4	1.9		0.6	0.4	0.5	......	......
Cystine	?	......					0.3	0.0	0.07	?	
Tyrosine	3.37	1.95	0.00	1.0	9.0	9.7	1.3	0.00	4.5		5.2
Arginine	7.46	7.38					5.4	7.62	4.84 [20]	0.3	15.5
Histidine	1.90	2.02					11.0	0.4	2.50		1.5
Lysine	4.81	5.77					4.3	2.75	5.95		6.9
Ammonia	1.25	1.08							1.61		
Tryptophane		Pres.					Pres.	0.0	1.5		
Total	54.02	32.51									

[1] Osborne and Clapp · Amer. Jour. Physiol., **9**, 498 (1907); 477 (1907).
[2] " " " " " " **19**, 475 (1907).
[3] " " " " " " **19**, 53 (1907).
[4] " Hey l: " " " **2**, 362 (1908).
[5] " Clapp: " " " **20**, 470 (1907).
[6] Abderhalden and Voitinovici: Zeit. physiol. Chem., **52**, 348 (1907).
[7] " " Landau · " " " **71**, 443, (1911).
[8] " " Zeit. physiol. Chem., **37**, 484 (1903); **51**, 397 (1907).
[9] " " " " " **37** (1903); **40**, (1904); Kossel & Patten: Zeit. physiol. Chem., **38**, 39 (1903); Osborn and Harris, Jour. Amer. Chem. Soc., **25**, 323 (1903); Osborne and Gilbert. Amer Jour. Physiol., **15**, 333 (1906).
[10] Levene and Van Slyke: Proc. Soc. Biol. Exp. Med., **6**, 54 (1909).
[11] Kossel and Dakin · Zeit. physiol Chem., **40**, 311, 565 (1904); **41**, 407 (1904).
[12] Osborne, Jones and Leavenworth · Amer Jour Physiol., **24** (1909); Osborne and Jones, ibid., **24**, 153 (1909).
[13] Osborne and Jones Amer. Jour. Physiol., **24**, 161 (1909).
[14] Roose: Zeit. physiol. Chem., **68**, 272 (1910).
[15] Suwa " " " **68**, 275 (1910).
[16] Strauch " " " **71**, 315 (1911).
[17] Fischer, Levene and Aders: Zeit. physiol. Chem., **35**, 70 (1902), Levene and Beatty, ibid., **49**, 252 (1906).
[18] Abderhalden Zeit. physiol. Chem., **44**, 28 (1905); Hart · Zeit. physiol. Chem., **33** (1901); Morner: Zeit. physiol. Chem., **34** (1901); Osborn and Guest: Jour. Biol. Chem., **9**, 333 (1911).
[19] Abderhalden and Rona · Zeit. physiol. Chem., **41** (1904).
[20] By the "group" method Van Slyke, Jour Biol. Chem., **16**, 531 (1913), obtained 7.6% total N as arginine N; 10.0% lysine N; and 6.21% histidine N.

tion may be directed also to the fact that proteins corresponding in kin in related seeds have very similar but not identical compositions. Th legumin of the vetch and the pea illustrate this fact. The protamines ar the proteins which have the fewest different amino-acids in their mole cules; and those acids which are present are chiefly basic acids.

The total amount of the amino-acids found rarely equals 100 pe cent. of the protein molecule. It is in fact seldom more than two thirds of the protein. Salmin alone yields 110 per cent. of amino-acids The weight of the amino-acids recovered is greater than the weight of th salmin hydrolyzed, for the reason that a molecule of water has bee added between each two amino-acids in the process of hydrolysis. Wer the methods of determination accurate all the proteins should show mor than a hundred per cent. of the weight of the protein as amino-acids The 30-40 per cent. of the protein molecule not accounted for in mos proteins might be due to the losses in analysis, or to the presence of othe unknown amino-acids in the decomposition products. It is the opinio of Osborne, who has particularly studied this question, that the defi ciency is chiefly due to the losses in analysis, since from a known amoun of amino-acids he could recover only about 60 per cent. It is probabl also that there are some unknown amino or other acids in the residue There are reasons for thinking that some of the sulphur may be in anothe form than cystine.

The structure of the protein molecule.—Since all methods o hydrolysis, whether by water, by the mild action of enzymes at body o room temperature, by acids and alkalies, yield amino-acids, it is saf to conclude that these nuclei are not secondary products of decomposi tion, but that they pre-exist in the protein molecule. That the protein are indeed made up of amino-acids linked through the carboxyl group o one acid and the α-amino group of another is now certain. This resul is largely due to the work of A. Kossel on the composition of the basi proteins found in the cell nuclei of the sperm of the salmon and sturgeon Kossel discovered that the protamine, salmin, a strongly basic protei which can be separated from the head of the salmon sperm, yielded o hydrolysis nearly 90 per cent. of its weight as the single amino-aci arginine; in the case of the sturgeon protamine, sturin, two other amino acids were present, namely lysine and histidine. From this and othe considerations he drew the conclusion that the proteins were made up o amino-acids linked through their amino and carboxyl groups, many o them at any rate having a protamine-like nucleus to which the differen amino-acids were attached, the number and kind of these amino-acids be ing variable in different proteins. This conception allied the proteins t the scheme of the carbohydrates. In this view proteins corresponded t the polysaccharides; the amino-acids to the various monosaccharides; an

Kossel named those amino-acids with six carbon atoms, namely histidine, lysine and arginine, "hexone bases" to bring out this similarity. By the work of Emil Fischer and Curtius this conception of the constitution of proteins was proved to be correct by the synthesis on the basis of Kossel's theory of various bodies of a protein nature.

The amino-acids are linked together in the protein molecule in the following way through their amino and carboxyl groups. The union of a molecule of alanine with one of leucine may be pictured as follows:

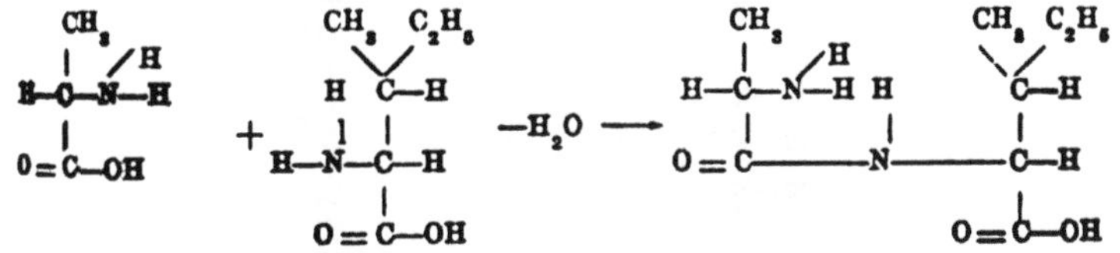

```
 CH3              CH3  C2H5                      CH3                 CH3  C2H5
 |  /H              \  /                         |  /H                 \  /
H—C—N—H           H  C—H                      H—C—N—H  H                C—H
 |             +  |  |       —H2O ——→            |        |              |
O=C—OH           H—N—C—H                      O=C———————N——————————C—H
                     |                                                    |
                   O=C—OH                                              O=C—OH
Alanine.        Iso-Leucine.                        Alanyl-leucine.
```

This leaves a free amino group at one end of the chain and a free carboxyl at the other, at which other amino groups can be attached. A series of amino-acids put together in this way to form a polypeptide as it is called, in this case a decapeptide, is shown on page 132.

The resemblance of this union to that of the polysaccharides is very close. By looking at the formula of the disaccharide, maltose, on page 57, it will be seen that the two monosaccharide molecules are attached to each other through an oxygen atom. The carbons of the different monosaccharide groups do not unite directly with each other. In the formula of a polypeptide just given, the different amino-acids are united through a nitrogen atom. The carbons of the monopeptides do not unite directly to make a polypeptide, any more than do the carbons of the monosaccharides to make a polysaccharide. A further resemblance lies in the fact that in each case the synthesis involves the loss of a molecule of water between each two monopeptide groups, or monosaccharide groups. The main difference apparently lies in the fact that there are a far larger number of amino-acids used in the synthesis of the proteins, or polypeptides, than of monosaccharides to make polysaccharides. No protein has as yet been discovered which yields only a single amino-acid, although salmin yielding 88 per cent. of arginine does not come far from it. Inulin, however, is supposed to yield only levulose when it is hydrolyzed; and glycogen is supposed to yield only glucose. Many of the other carbohydrates, however, are composed of several different monosaccharides. No doubt as means of separation of the monosaccharides improve, it will be found that the polysaccharides contain more kinds of monosaccharides than is at present believed.

The evidence that the amino-acids in the proteins are linked through

```
      1          2           3           4           5           6           7           8           9          10
      H O     H  H O      H  H O      H  H O      H  H O      H  H O      H  H O      H  H O      H  H O      H  H O
      | ‖     |  | ‖      |  | ‖      |  | ‖      |  | ‖      |  | ‖      |  | ‖      |  | ‖      |  | ‖      |  | ‖
H2N—C—C——N—C—C——N—C—C——N—C—C——N—C—C——N—C—C——N—C—C——N—C—C——N—C—C——N—C—C—OH
      |          |           |           |           |           |           |           |           |           |
     CH3        HCH         HCH         HCH         HCH         HCH         HCH         HCH         HCH          H
                / \          |           |           |           |           |           |          / \
            CH3   C2H5       OH         OCOH         C          HCH         SH          HCH      CH3   CH3
                                                    / \\         |                       |
                                                  HC   CH       HCH                     HCH
                                                  ‖     |        |                       |
                                                  HC   CH        NH                    HCNH2
                                                    \ //         |                       |
                                                     C           C                       H
                                                     |          // \
                                                     OH       NH   NH2
```

Alanyl—leucinyl—seryl—aspartyl—tyrosinyl—arginyl—cysteinyl—lysinyl—valyl—glycin
1 2 3 4 5 6 7 8 9 10

the amino and carboxyl groups is the fact that they have been synthesized into protein-like bodies by such union, and the further fact that the number of the free amino and carboxyl groups in a protein molecule is very small, showing that both amino and carboxyl groups are combined.

Synthesis of the proteins.—The synthesis of protein-like substances from the amino-acids has been accomplished in several ways. 1. By dehydration. By heating leucine and glycocoll in the presence of phosphorus pentoxide Grimaux and later Pickering obtained colloidal bodies with many of the properties of the proteins. 2. By the condensation of glycocoll Curtius obtained a base, the biuret base, which is now known to be triglycyl-glycine ethyl ester. 3. The first systematic attempts at synthesis which were successful were those of Emil Fischer on the basis of Kossel's theory of the nature of the protein molecule, and these attempts have led to the successful synthesis of a great number of artificial polypeptides, some having the general nature of the albumoses, being digestible by trypsin and erepsin and giving the color reactions of the proteins. The methods used in the synthesis are as follows:

The carboxyl and amino groups are not of themselves sufficiently reactive to combine rapidly, more rapidly than they dissociate. It is necessary to make one of them at least more reactive, so that the velocity of the reaction which is leading to their synthesis is greater than the velocity of their decomposition by hydrolysis. This greater reactivity is secured for the carboxyl group by substituting the hydroxyl with chlorine, to make the acid chloride. This can be done by treatment of the amino-acid by phosphorus pentachloride. There is thus formed from alanine, or glycine, the hydrochloride of the acid chloride:

```
           H
           |
     CH3—C—NH2HCl
           |
       O=C—Cl
```

This will now unite with a molecule of an amino-acid, or a polypeptide, liberating hydrochloric acid thus:

```
     H                                           H
     |                                           |
CH3—C—NH2HCl        CH3                  CH3—C—NH2HCl       CH3
     |                |                        |                |
 O=C—Cl     +   H—N—C—H      ——→       O=C————————————N—C—H + HCl
                H   |                                    |   |
                O=C—OH                                   H   |
                                                          O=C—OH
Alanyl chloride     Alanine.              Alanyl-alanine hydrochloride.
hydrochloride.
```

By treating the alanyl-alanine with phosphorus pentachloride it may be converted into the acid chloride in its turn and it will then unite with the amino group of some other amino-acid, for example, leucine:

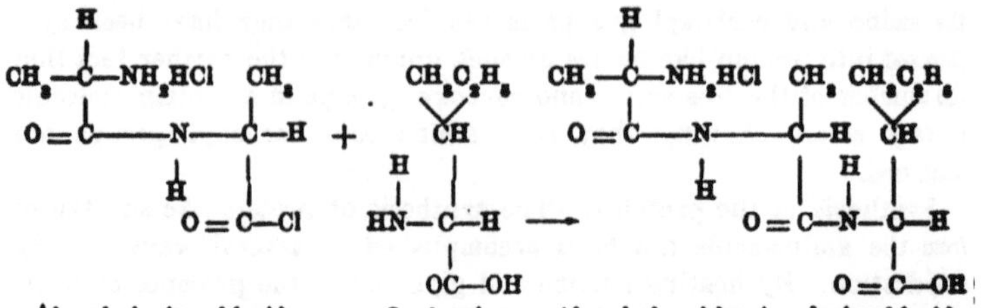

```
       H                                          H
       |                                          |
CH₃—C—NH₂HCl  CH₃        CH₃C₂H₅     CH₃—C—NH₂HCl  CH₃  CH₃C₂H₅
       |          |             \/                  |           |      \/
O=C———N———C—H  +       CH           O=C———N———C—H   CH
          |       |          H  |                 |         |  H  |
          H       |          |  |                 H         |  |  |
            O=C—Cl      HN—C—H   ——→           O=C—N—C—H
                               |                                 |
                            OC—OH                          O=C—OH
```

Alanyl-alanine chloride. Iso-leucine. Alanyl-alanyl-leucine hydrochloride.

Two tri-peptides or even more complex peptides may in this way be condensed into a hexa- or other poly-peptide.

Another method used by Fischer consisted in adding an amino-acid to the amino group of the terminal acid of a peptide, using a bromine substitution product of a fatty acid chloride; and then after union with the amino group replacing the bromine by an amino group by treating with ammonia. The process is then repeated. Suppose it is desired to make an alanyl-leucine. The leucine is treated with brompropionyl chloride and then the reaction product with ammonia as follows:

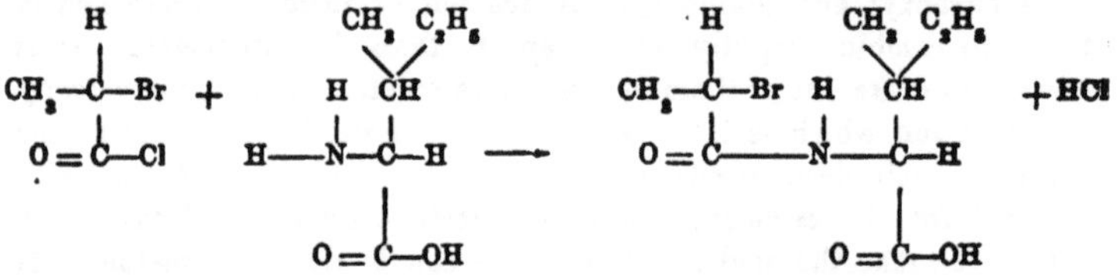

```
     H              CH₃  C₂H₅              H          CH₃  C₂H₅
     |                 \/                  |              \/
CH₃—C—Br  +       H   CH          CH₃—C—Br  H   CH         +HCl
     |            |   |                |    |   |
O=C—Cl     H———N——C—H   ——→   O=C———N———C—H
                      |                          |
                   O=C—OH                     O=C—OH
```

Brompropionyl chloride. Iso-leucine. Brompropionyl-leucine.

2. CH_3—CHBr—CO—NH—CH(C_4H_9)—COOH + NH_3 ——→
CH_3—$CHNH_2$—CO—NH—CH(C_4H_9)—COOH + HBr

Brompropionyl-leucine. Alanyl-leucine.

The process may now be repeated with the alanyl-leucine and either brompropionylchloride or some other similar compound may be united to the di-peptide and converted into the amino compound by the action of ammonia. So a tri-peptide may be made. By the use of these methods a great number of artificial polypeptides have been made by Fischer, Abderhalden, Curtius and their co-workers. One of the most complex of these polypeptides contained 18 amino-acid groups, namely three leucine and 15 glycocoll groups. It was l-leucyl-triglycyl-l-leucyl-triglycyl-l-leucyloctoglycylglycine. $NH_2CH(C_4H_9)CO.(NHCH_2CO)_3.NHCH(C_4H_9)CO.(NHCH_2CO)_3.NHCH(C_4H_9)CO.(NHCH_2CO)_8NHCH_2COOH$.

These complex artificial polypeptides have the properties of the derived proteins. They are like albumoses. They give the biuret and other reactions of the proteins, which are given by the various amino-acids of which they may have been composed, such as the tyrosine or tryptophane reactions. They are precipitated by mercuric chloride and phosphotungstic acid. And some of them are digestible by trypsin and erepsin. They are optically active, also, like the natural bodies. One of them produced an anaphylaxis reaction. It has not yet been possible to form a protein which is coagulated by heat; nor has any artificial protein been made which is identical with the naturally occurring proteins. On the other hand, many of the di- and tri-peptides which appear in the artificial hydrolysis of the naturally occurring proteins have been synthesized artificially. The final synthesis of the natural proteins is probably only a question of industry and time.

Other linkings in the molecule.—It must not be supposed that the NH—CO— grouping is the only method of linking amino-acids in the protein molecule, although it is undoubtedly the principal one. Another is certainly by means of the cysteine sulphur. This union is brought about by oxidation and released again by reduction. This linking may be of great importance in determining the reactivity of living protoplasm, since oxidations and reductions are constantly taking place in it. Thus if two molecules of cysteine are oxidized, and in neutral or in the faintest alkaline reaction the oxidation goes spontaneously very rapidly in the air, they are converted into one molecule of cystine.

The reaction is as follows:

$$\begin{array}{ccccccccc}
\mathrm{H} & & & & \mathrm{H} & & \mathrm{H} & & \mathrm{H} \\
| & & & & | & & | & & | \\
\mathrm{HC{-}SH} & & & & \mathrm{HS{-}C{-}H} & & \mathrm{HC{-}S} & \mathrm{{-}{-}S} & \mathrm{{-}C{-}H} \\
| & & & & | & & | & & | \\
\mathrm{HCNH_2} & + & \mathrm{O} & + & \mathrm{HCNH_2} & \longrightarrow & \mathrm{HCNH_2} & & \mathrm{HCNH_2} \quad + \; \mathrm{H_2O} \\
| & & & & | & & | & & | \\
\mathrm{COOH} & & & & \mathrm{COOH} & & \mathrm{COOH} & & \mathrm{COOH} \\
\text{Cysteine.} & & & & \text{Cysteine.} & & & \text{Cystine.} &
\end{array}$$

It is possible, although it has not yet been shown to be the case, that if two proteins each containing cysteine are oxidized, a more complex cystine protein would be the result. By reduction this could be broken up again. There seems to be evidence from certain color reactions with sodium nitro-prusside, with which cysteine gives a beautiful red color, that some natural proteins contain cysteine, while others contain cystine. It would seem not impossible that this union might join molecules of protein into more complex groups; and possibly the fibers of the aster in cell division might be formed in this way. The author found that these fibers would only form in the sea-urchin egg in the presence of oxygen and they at once broke up and disappeared when oxygen was withdrawn from the egg. At any rate we have in the cysteine sulphur one of the most reactive points of the protein molecule. Heffter and the writer have particularly tried to bring it int relationship with cell processes. The union is as follows:

$$R—CH_2—S—S—CH_2—R'$$

It would seem that the protein with cysteine in the molecule migh change its state of solution when it became cystine, and this migh alter the state of viscidity of the protoplasm, or possibly even it affinity for water.

Another linking which is possible is an ester union through th hydroxyl of the serine, or oxyproline, or tyrosine with carboxyl Whether such unions exist is still unknown. Another linking is tha typefied by guanidine and ornithine in forming arginine. The linkin is of the following kind: $NH{=}C(NH_2)—NH—CH_2—R$. So far as i known this union occurs only in arginine.

This part of the subject should not be left without reference t another very suggestive fact. None of the artificial polypeptides ar digestible by pepsin, though many of them digest with trypsin o erepsin. This matter is discussed on page 404. This fact may mean tha there are other kinds of unions between the polypeptide groups whicl go to make up the protein molecule than unions between the amino an carboxyl groups as just stated. Pepsin might act on these unions. O the other hand, it might be that the failure of pepsin to digest th protamines or the artificial polypeptides was owing to the fact that th pepsin acts only on certain specific amino-acid junctions and that w have not yet happened to test these particular junctions with th enzyme. The fact that during peptic digestion the free amino group increase in numbers (page 362) bears out the latter supposition.

A very curious relationship has recently been found by Kossel i the protamines of the fish sperm and may be mentioned in this connec tion. He finds that in these proteins there are always approximately or exactly, two molecules of a basic amino-acid like arginine, histidin or lysine, to each molecule of a mono-amino acid. This fact suggest that possibly the protamine may be made of a series of tri-peptide Similar tri-peptides have been isolated by Siegfried in the course of th slow hydrolysis of various proteins and called by him, kyrins. It ha been suggested by Taylor that the protamine, salmin, may be made u of these tri-peptides, or protones, united as follows:

Arginine		Arginine		Arginine		Arginine		Arginine		Arginine
|		|		|		|		|		|
Serine	—	Serine	—	Proline	—	Proline	—	Proline	—	Valine
|		|		|		|		|		|
Arginine		Arginine		Arginine		Arginine		Arginine		Arginine

The first cleavage of the molecule by hydrolysis would consist in th setting free of the tri-peptides which would then be separately broke up. This view, while it is in consonance with many facts, cannot yet be said to be well grounded.

Number of free amino and carboxyl groups in proteins.—That there are only a few free amino groups in the protein molecule is shown by a variety of reactions. Acids, for example, combine with the amino, NH_2, groups, but not with the imino, NH—, groups; or if they unite with the latter the union is a very weak one and dissociation occurs. The basicity of the group, NH—, is no doubt reduced by the neighboring C=O group. At any rate, the acid-combining power of the protein molecule is generally only two to four molecules of hydrochloric acid to what we believe to be a single molecule of protein. Thus edestin, a crystalline protein from hemp seed, forms two series of salts, a mono- and a di-chloride (Osborne). As digestion takes place and the amino groups become free, the power of taking up acid greatly increases. Kossel has shown that the amount of acid taken up by protamine is in direct relation to the amount of the free amino groups it has. In general, proteins with more lysine and arginine combine with more acid. This indicates that one of the amino groups in each of these acids is uncombined in the molecule; in other words, that only the α-amino group is bound in both arginyl and lysyl.

Another proof that there are few free amino groups is the power of union with formalin. Formaldehyde unites with the free amino groups to form water and methylene addition products (see page 121). It does not react with the imino, NH, groups. Now it is found that the amount of formalin bound or taken up is small in the intact proteins, but undergoes a steady increase as hydrolysis progresses. Indeed by means of formol titration the progress of a hydrolysis can be most easily followed. It is found that the rate of increase of the acid-combining power and of formalin binding in such a hydrolysis go parallel. Still another method for the detection of the amount of free amino groups, and from a quantitative standpoint perhaps the best, is the method of Van Slyke, which depends on the fact that nitrous acid reacts with free amino groups liberating nitrogen gas which can be collected and measured (for reaction see page 123). It is found that the amount of nitrogen deplaceable from a protein by nitrous acid is a very small fraction (5 to 8%) of the total nitrogen, but that as hydrolysis proceeds and the amino groups become free, the amount steadily increases. All of these methods, then, prove beyond question that the amino-acids have most of their amino groups combined and that they are, therefore, probably linked through the amino groups.

That the carboxyl groups also are in combination in the protein and few of them free is shown, in the first instance, by the fact that the power of the protein to combine with alkali increases as the hydrolysis proceeds. It may be shown also by the method of Dakin. By the action of dilute alkali on protein a decrease in the rotatory power results and the subsequent acid hydrolysis of the protein thus acted upon yields the racemic form of nearly all the amino-acids (page 126). There are only a few amino-acids in such hydrolyses which have not been racemized by the alkali. Now since the acids having free carboxyl groups do not racemize, the fact that most of them are racemized by alkali treatment of the proteins shows that the great majority of the carboxyls must have been united with something in the molecule.

Since the great majority of both the carboxyl and amino groups of the protein molecule are combined, it is probable that they have combined with each other.

Molecular weight of the proteins.—We may now ask the question how large is the molecule of protein? How many of these amino-acids does a molecule have in it? This is a very difficult question to answer for the majority of the proteins, but for a few of them it may be answered with a considerable degree of probability. There is no doubt that the molecular size of the great majority of the proteins, of all the natural proteins, is very large. This is shown by the fact that they will not diffuse through parchment paper. They are colloidal in aqueous solution. This means that the diameter of their molecules is certainly more than 1 $\mu\mu$. Even protamine, which is in many ways the simplest of the proteins, is colloidal. It might be, however, that the proteins were colloidal in water but not in other solvents. Soap is colloidal in water, but not in alcohol. The molecular size of the proteins in other solvents than water has hardly been investigated. It is possible that in water several simple protein molecules might aggregate by processes known as association to form large complexes, just as many simple substances, such as alcohol or acetic acid, associate to form double or triple molecules. The molecular size of the proteins dissolved, for example, in formamide, if they will dissolve in it, should be investigated. While it is possible for the reason just stated that the large molecular size of the proteins when dissolved in water does not necessarily mean that the individual molecules of the protein are large, there are other reasons which make such a conclusion practically inevitable. There are several different ways in which the molecular weight may be determined both by indirect and direct methods. The results obtained by these two methods are in very good agreement. We will consider the indirect methods first.

Calculation from the sulphur content. The crystalline form of several of the proteins is so distinct and constant that we may assume that these represent chemical individuals. On repeated precipitations they do not change their form or composition. Many of these proteins have the sulphur largely in the form of cysteine. Perhaps it is altogether in that form in some. It is probably present in other forms than cysteine, perhaps as cystine in others. If there is one molecule of cystine in a molecule of protein there must be two atoms of sulphur to each protein molecule. Two atoms of sulphur have a molecular weight of 64. If there is 1 per cent. of sulphur in the molecule, the molecular weight of such a protein would be at least 6,400. If there was 0.5 per cent. S, the molecular weight would be 12,800. The following computations of the molecular weights and formulas of various plant and animal proteins were made by Osborne from the sulphur on the basis that there were two or more atoms of S to the molecule.

MOLECULAR WEIGHT, COMPOSITION AND POSSIBLE EMPIRICAL FORMULAS OF PROTEINS.

Composition								Formula							Molecular weight
C	H	N	S	Fe	P	O		C	H	N	S	Fe	P	O	
[illegible]	[illegible]	18.90	0.4[illegible]	......	......	23.471	Amandin..............	638	1080	202	2	..		209	14922
51.72	[illegible]	18.04	[illegible]	......	......	22.906	Legumin...............	718	1158	214	2	..	..	228	16642
[illegible]	7.[illegible]	16.13	0.600	......	......	20.78	Zein.....................	736	1161	184	3	..		208	15963
[illegible]	6.[illegible]	17.21	0.847	......	......	20.53	Hordein..................	675	1014	181	4	..	..	194	14660
[illegible]	[illegible]	18.[illegible]	0.[illegible]	......	......	22.10	Edestin.	622	1080	198	4			201	14580
[illegible]	6.[illegible]	17.66	1.027	......	......	21.733	Gliadin	685	1068	196	5	...	..	211	15568
[illegible]	6.[illegible]	18.[illegible]	1.0[illegible]	......	......	21.514	Excelsin	642	1018	192	5	...	..	198	14738
							Animal proteins								
[illegible]	7.[illegible]	16.[illegible]	0.42	......	......	20.51	Globin....................	700	1098	184	2	....	..	196	15274
[illegible]	6.[illegible]	16.[illegible]	1.10	......	......	22.48	Fibrin....................	645	1004	178	5	. .	.	207	14706
[illegible]	7.[illegible]	15.[illegible]	1.11	......	......	23.[illegible]	Serglobulin, horse....	628	1002	180	5	....	..	209	14310
[illegible]	6.[illegible]	16.[illegible]	1.[illegible]	......	......	22.[illegible]	Fibrinogen...............	679	1062	183	6	....	..	207	15276
[illegible]	7.[illegible]	15.51	1.616	......	......	23.024	Ovalbumin...............	696	1125	175	8	...	.	220	15708
[illegible]	7.1[illegible]	15.77	1.7[illegible]	......	......	23.13	Lactalbumin............	644	1064	166	8	....	..	214	14792
[illegible]	7.[illegible]	15.[illegible]	1.[illegible]	......	......	22.14	Seralbumin, horse......	662	1051	171	9	....	..	207	14989
[illegible]	6.[illegible]	15.[illegible]	2.[illegible]	......	......	22.95	Seralbumin, human....	684	1045	178	11	..		225	15697
[illegible]	7.[illegible]	17.[illegible]	0.[illegible]	0.3[illegible]	......	20.165	Oxyhemoglobin, horse.	758	1181	207	2	1	..	210	16655
[illegible]	7.11	16.[illegible]	0.5[illegible]	0.3[illegible]	......	21.036	Oxyhemoglobin, dog ..	758	1185	195	3	1	..	219	16667
[illegible]	7.[illegible]	15.78	0.[illegible]	......	0.86	22.37	Casein.....................	708	1130	180	4	....	4	224	15982
[illegible]	7.12	16.[illegible]	1.0[illegible]	......	0.82	23.942	Ovovitellin...............	671	1112	182	5	....	4	227	15698

From the foregoing figures it is clear that if the proteins in question are individuals their molecular weight is certainly high. In the case of hemoglobin it will be seen that the computation of the molecular weight on the assumption that there is one molecule of cystine gives the same result for horse hemoglobin as the assumption of one atom of iron in the molecule; for dog hemoglobin, however, it is necessary to assume that there are three sulphur atoms in the molecule, which would mean one molecule of cystine and one molecule of some other sulphur compound, possibly cysteine. The molecular weight might, of course, be some multiple of these figures.

Computation of the molecular weight of hemoglobin from the oxygen compound. That the molecular weight of oxyhemoglobin is approximately that indicated in the foregoing table is shown, also, by a calculation of the molecular weight from the number of grams of oxygen or carbon monoxide taken up by a gram of hemoglobin, assuming that each molecule of hemoglobin combines with one molecule of the gas. The molecular weight of the carbon-monoxide hemoglobin is given by the ratio 28 : x :: a : 1, where a is the weight of carbon monoxide combined in one gram of carbon-monoxide hemoglobin, and x the molecular weight of the hemoglobin. Hüfner found that one gram of the carbon-monoxide hemoglobin contains 1.338 c.c. of CO computed at 0° and 760 mm. pressure, or .0016745 gram. From this the molecular weight of the carbon-monoxide hemoglobin is computed as 16,669 (.0016745 : 1 :: 28 : M). This figure agrees almost exactly with that computed from the sulphur and iron. It is also in agreement with the direct determination of the molecular weight made from the osmotic pressure. Computing from the heat of formation of one gram of oxyhemoglobin from hemoglobin and oxygen, Barcroft and Hill found the molecular weight to be 15,200.

Direct determination of molecular weight by the osmotic pressure method. The determination of the molecular weight of proteins cannot be made by the boiling-point method because most of the proteins coagulate or change on boiling. The freezing-point method also is not sufficiently accurate for such large molecules. There are two methods which may be used: the osmotic-pressure method, and the measurement of the vapor pressure at lower temperatures than boiling by the method recently introduced by Menzies. The determination of the molecular weight of hemoglobin has been made by measuring the osmotic pressure of solutions of known strength of hemoglobin. The only real difficulty in this method consists in getting perfectly tight membranes which are truly semipermeable, that is membranes which readily pass the solvent but not the solute through them. Hüfner and Gansser used the apparatus in Figure 14. The solution of hemoglobin is brought into the diffusion shell of Schleicher and Schull which is closed and connected with a mercury manometer. The diffusion shell is then placed in water and by osmosis the water enters the solution, forcing the mercury up until the pressure becomes so high that it presses just as much water out as that which enters. The principle of the method is that a solution which contains in a liter an amount of the substance equal in grams to the molecular weight will have a pressure at 0° of 22.41 atmospheres. A half-molecular solution which has only an amount of substance equal to half a molecular weight has half this pressure, and so on. It is only necessary then to measure the osmotic pressure of the hemoglobin at 0° or some other temperature to find what fraction this is of 22.41 atmospheres or the corresponding osmotic pressure at the temperature employed for the hemoglobin, and divide the weight of hemoglobin dissolved in one liter of solution by this fraction to get the molecular weight. The formula is as follows:

$$M = \frac{22.41\,(1 + 0.00366t)\,760.c}{p'}$$

In this formula (1+0.00366t) is the temperature correction, since the osmotic pressure increases with the temperature. t is the temperature at which the determination is made; c is the concentration of the solute in grams in one liter of solution; and p′ is the osmotic pressure of the

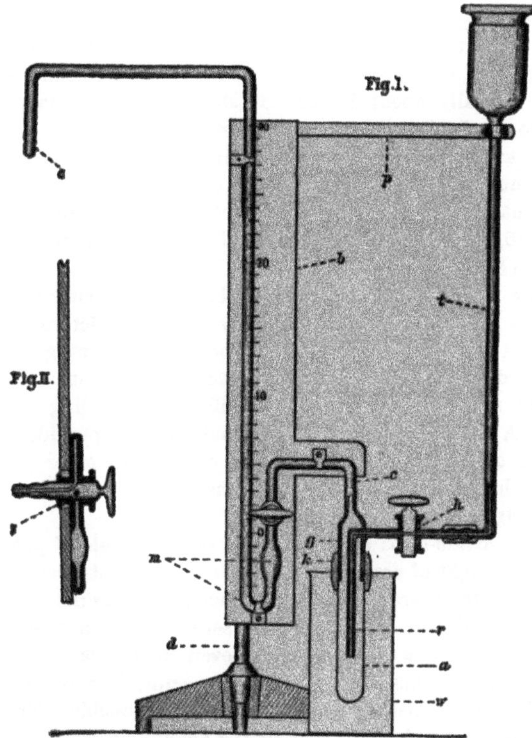

Fig. 14.—Osmometer for determining the osmotic pressure of oxyhemoglobin solutions (Hüfner and Gansser). *a*, diffusion cell containing oxyhemoglobin solution run in through *t*, *h*, and *r*; *b*, manometer for measuring osmotic pressure; *w*, beaker containing water. Fig. II. Detail of cock *e*.

solution if it had not been diluted from the volume v to v′ by the entrance of water. The correction is of course a small one. It took several hours for the pressure to reach its maximum and it remained at this maximum for several hours. Some of the results obtained are given in the following table:

c	t	p	v	v'	p'=pv'/v	M	Kind of hemoglobin.
		mm. Hg.			mm. Hg.		
52.72	10°	62.7	23.5	23.6	62.97	14,780	
	10	58.5			58.75	15,840	Horse Hb.
108.0	1	109.0		23.7	109.9	16,790	Ox Hb.
109.2	1	114.9			115.9	16,110	
216.0	1	198.0		23.9	201.0	18,370	
216.0	1	224.0			227.8	16,210	

The mean value of all the determinations gave for horse hemoglobin the molecular weight of 15,115 and for ox Hb 16,321.

This direct determination confirms fully the determinations by the other method and leaves very little doubt that the molecular weight of hemoglobin is really about 16,693. It should be mentioned, however, that Weymouth Reid by the osmotic-pressure method got 48,000 and Roaf 32,000 as the molecular weight of oxyhemoglobin. It is probable from these numbers that in both these cases some association of the hemoglobin had occurred giving Roaf double molecules and Reid triple molecules. 16,693 is the minimum molecular weight if there is one atom of iron to each molecule.

But while these results are so concordant and striking there is one fact which is not apparently in harmony with this determination; or at any rate it is as yet unexplained. The molecular weight of casein when dissolved in formamide was found to be only about 400. In the same solvent starch had a weight of 645, corresponding to a tetra-saccharide. A molecular weight of 400 would be a tri-peptide. Further investigation of the molecular weight of casein and other proteins in this solvent should be made. In water there is no doubt but that the molecular weight is far higher than this.

How many amino-acids would there be in a molecule of protein? If the molecular weight of casein is 16,000 it must have at least 120 amino-acids in it since the average weight of a molecule of amino-acid is about 130. Some 15 different acids have been separated from casein, so that on the average there would be about seven molecules of each kind.[1] If the molecule has this size and so many acids, it will be seen that there may be an astonishing number of caseins possible. They might differ from each other in the order or the amount in which the amino-acids occur in the molecule; or the acids might be isomers. One might have leucine and another iso-leucine. In fact, the number of amino-acids is so great that by modifying the proportion of those present in different proteins, or by modifying the arrangement of them in the molecule, or by the introduction of optical isomers practically an infinite

[1] Recent indirect determinations of the molecular weight of casein by Van Slyke indicate that the molecular weight of casein is about half this amount, or about 8,000.

number of combinations is possible. It is this great diversity, combined of course with the diversity in the lipins and carbohydrates, which has made possible the very large number of different kinds of organisms on the earth.

Crystallized proteins.—A very interesting crystallized protein has been obtained (Katake and Knoop) from the milk of Antiaris toxicaria, the poisonous Upas tree of Java, by extraction with 85 per cent. alcohol, drying the extract and then cooking out the extract with 0.8 per cent. acetic acid. On evaporating the extract the protein crystallizes out in needles and prisms. Recrystallized from hot water, the crystals are eventually obtained of uniform appearance containing 15.73 per cent. of water. The ash-free crystals are small, solid polyhedra. They react acid in solution. They give all protein reactions including sulphur, except Molisch. The solution is not precipitated by picric or nitric acids, nor by ferrocyanide and acetic acid, but is precipitated by phosphotungstic acid. Dissolved in glacial acetic acid, the substance shows the Tyndall phenomenon of scattering a beam of light. It is, therefore, colloidal in this solution. The rotation is $(\alpha)_D = -19.25°$. The composition was C, 48.02; H, 5.71; N, 15.65; S, 7.20; O, 23.47. It contains more sulphur than any other protein. If there is only one molecule of cystine present in the molecule, the minimum molecular weight would be 900. It certainly yields on hydrolysis cystine, lysine, glycocoll, alanine, proline and valine.

	C	H	N	S	O
Computed for $(C_{36}H_{50}N_{10}S_2O_{13})_n$	48.27;	5.63;	15.69;	7.16;	23.25
Found	48.02;	5.71;	15.65;	7.20;	23.42

Water of crystallization found 15.73 per cent. Computed for the above formula with $9H_2O$, 15.35 per cent.

Distribution of nitrogen in the protein molecule.—The analysis of the proteins by hydrolysis and the quantitative isolation of the various amino-acids is exceedingly laborious and requires a very large amount of material. Shorter methods have been devised to give a general idea of the nitrogen distribution between various amino-acids and which are applicable to as little material as 2 grams. The best of these methods is the group method perfected by Van Slyke. The total nitrogen of the protein molecule may be divided into four main groups, namely, ammonia nitrogen, amino N, imino N and basic N. These groups are determined in the following way: During the acid hydrolysis of the proteins the acid amide nitrogen is split off as ammonia. It is determined by getting rid of the acid of the hydrolysate, making the solution faintly alkaline with lime, and distilling off the NH_3 under diminished pressure. The material freed from ammonia and filtered to remove excess lime and some melanine is precipitated, after acidification, with phosphotungstic acid. This precipitates the basic amino-acids and cystine. The nitrogen determined in this precipitate is called the basic nitrogen. A portion of the filtrate from this precipitate, after removal of excess phosphotungstic acid and neutralization, is treated with nitrous acid by the Van Slyke method. This liberates the nitrogen preser free amino groups. The nitrogen is collected and measured. This i amino N. It comes from the monoamino acids. Another portion o filtrate has the total nitrogen determined by Kjeldahl and the differ between this and the amino nitrogen gives the imino nitrogen, na that in proline, oxyproline and one-half of tryptophane nitrogen. method has been still further refined by Van Slyke to permit a dete nation of the different basic amino-acids and some of the others. Son the results he has obtained in the examination of different proteins embodied in the accompanying table. The nitrogen which is evo when non-hydrolyzed proteins are treated by nitrous acid comes f the ε-amino group of lysine, which is thus shown to be free in the m cule. The α-amino group of lysine is combined.

PERCENTAGE OF THE TOTAL NITROGEN OF VARIOUS PROTEINS PRESENT IN VA AMINO-ACIDS (Van Slyke).

	Gliadin	Edestin	Hair (Dog)	Gelatin	Fibrin	Hemo-cyanin	O
Ammonia N	25.52	9.99	10.05	2.25	8.32	5.95	
Melanin N	0.86	1.98	7.42	0.07	3.17	1.65	
Cystine N	1.25	1.49	6.60	0.0?	0.99	0.80	
Arginine N	5.71	27.05	15.33	14.70	13.86	15.73	
Histidine N	5.20	5.75	3.48	4.48	4.83	13.23	
Lysine N	0.75	3.86	5.37	6.32	11.51	8.49	
Amino N of the filtrate	51.98	47.55	47.5	56.3	54.3	51.3	
Non-amino N of the filtrate (proline, oxyproline, ½ tryptophane)	8.50	1.7	3.1	14.9	2.7	3.8	
Sum	99.77	99.37	98.85	99.02	99.58	100.95	1

Color reactions of the proteins.—The proteins yield colored prod when acted upon by various reagents, and these colors are utilize detecting the presence of protein matter in solutions and body fl and in determining easily the presence or absence of some amino- from the molecule. The most characteristic of these reactions, th the reaction given by all native proteins and by the larger numb the derived products, is the biuret reaction. It is not, howeve delicate as some of the others.

The biuret reaction. If a solution of a protein is made alka preferably by sodium or potassium hydrate, and a drop or two of d cupric sulphate solution is added, well mixed and allowed to stan room temperature, or if it is gently heated, the clear fluid above precipitate which may be formed has, if a protein is present, a v tinge. The reaction is most delicate when made at room tempera but it may be hastened by heating, only in some cases the c is destroyed by heat. The shade of the color varies from a reddish v in the case of some peptones, or simple peptides, to a blue violet in n other proteins. Sometimes in the presence of certain gums which are precipitated by the copper, the color may be on the precipitate, but this is the exception.

The reaction is called the biuret reaction for the reason that it is given, also, by biuret, a substance $NH_2—CO—NH—CO—NH_2$ formed by the condensation of two molecules of urea (hence biurea, or biuret) with the elimination of ammonia.

$$O=C\langle^{NH_2}_{NH_2} + O=C\langle^{NH_2}_{NH_2} \longrightarrow \underset{\substack{\| \quad | \quad \| \\ O \quad H \quad O}}{NH_2—C—N—C—NH_2} + NH_3$$

Biuret is easily made by heating a few crystals of urea in a dry test-tube to a little above their melting point and cooling when the odor of ammonia is perceived. The biuret may be detected by the biuret test. The fact that biuret gives this reaction shows that the reaction is not peculiar to the proteins. Many other substances give this reaction. Schiff has shown that any diacid amide in which the two amide groups are not attached to the same carbon will give the reaction. Thus oxamide, $NH_2—CO—CO—NH_2$, or malonamide react. One of the amide groups must be unsubstituted, but the other may be substituted as it always is in the protein molecule. Thus $NH_2—CO—CO—NHR$ will give the reaction. Asparagine, the amide of aspartic acid, gives a blue-violet biuret reaction. In this case we have $COOH—CH_2—CHNH_2—CONH_2$, which is not a diacid amide. The reaction may, however, be due to the formation of an amino compound by a kind of lactone (lactam) formation thus

$$O=\overset{\lfloor\_\_\_\_\_\_\_\_\_\_\_\_ NH}{C—CH_2—\overset{|}{C}H—CONH_2}.$$

We would thus have two acid amide groups, one of them free. Similarly leucine amide gives with sodium hydrate and cupric sulphate a red salt-like compound in red crystals (Bergell and Busch). Succinimide also forms, in similar circumstances in the presence of potassium hydrate, reddish needles fairly stable in the solid form and of the composition

$$K_2Cu\left(N\langle^{CO—CH_2}_{CO—CH_2}\right)_4 6H_2O;$$

but which are readily decomposed in aqueous solution by acids. The rubidium and cæsium salts are red violet; the sodium salt pale blue; the lithium salt ultramarine. All of these are supposed to be derived from the hypothetical acid

$$Cu\left(N\langle^{CO—CH_2}_{CO—CH_2}\right)_4 H_2$$

Schiff isolated the biuret potassium compound in red needles to which he ascribed the formula:

```
          OH                  OH
          |                   |
         NH2 ——— Cu ———————— NH2
O=C <                               > C=O
       > NH  OH          NH <
O=C <  /             HO           > C=O
         NH2            > NH2
          \           K
           K
```

Among other substances giving the reaction are urobilin, a coloring matter derived from the bile and found in the urine (Stokvis, Salkowski). It may be mentioned that strongly basic proteins which are already alkaline in their aqueous solution, such as the protamines and protones (Gota), will give the biuret reaction without any addition of alkali.

From the foregoing we may conclude that the proteins give this reaction because they contain at least one acid amide group and other substituted amide groups attached to neighboring carbon atoms. If the proteins are deamidized, that is if the free amide groups are split off by the action of strong acid, the product which remains does not give the biuret reaction, although it is still a protein, digestible by trypsin and other enzymes and giving other protein reactions. All native proteins, therefore, since they give the reaction, contain some acid amide nitrogen. The biuret reaction, unlike all the other color reactions, is not a reaction for any specific amino-acid, but rather is dependent on the constitution of the proteins.

The color of the biuret test is due probably to the copper atom. Many copper compounds are blue and others, like the metal itself or cuprous oxide, are red. It is probable that in the blue compounds the copper atom is in a different state from what it is in the red form, possibly being partially reduced, consequently the valence electrons have a different period of vibration so that the light absorption is changed. As this state of the atom may be induced by a great number of compounds, it is clear that the biuret test cannot be a specific test for proteins, or for any particular class of bodies.

Since the color change depends on an alteration of the state of the copper atom, it may be anticipated that other metals having several stages of oxidation and different colors and which combine with amino groups may also give a similar reaction. This is the case. Pickering found that cobalt salts also might be used for the biuret test, and the reaction is even more delicate than with copper. Zinc, iron and manganese gave no color change.

Millon's reaction. This reaction consists in the development of a red color, when a protein is heated or allowed to stand some time in contact with a mixture of mercuric nitrite and nitrate. If a few drops of Millon's reagent is added to a solution, or suspension, of many proteins and this is heated, the protein is precipitated and the precipitate after a time turns red. To make Millon's reagent dissolve 1 part by weight of mercury in 2 parts concentrated nitric acid and dilute with twice its bulk of water, allow the precipitate to settle and use the supernatant liquid. The protein does not need to be in solution for this reaction and it may hence be used for the detection of proteins in sections of tissues. The color is not deep enough for a good microscopic stain. In place of Millon's solution, which contains a good deal of free acid, Nasse recommends that an aqueous solution of mercuric acetate be used, to which at the time of using there is added a few drops of a 1 per cent. solution of sodium or potassium nitrite. It is usually not necessary to acidify, but the addition of a little acetic acid to the above solution is sometimes advantageous.

The Millon reaction is given by all organic compounds containing a monohydroxy benzene nucleus. It is hence given by phenol, salicylic acid and many other substances. It is not given by a di- or tri-hydroxy phenol unless one of the hydroxyls is substituted, as in esters or ethers. Since the only group thus far recognized of the protein molecule which contains a monohydroxy benzene ring is the tyrosine group, this reaction when applied to proteins detects the presence of this group. As not all proteins contain tyrosine, for example pure gelatin and various protamines, not all proteins give the reaction. It is a good deal more delicate than the biuret reaction and the presence of proteins when the dilution is great may be detected by this and the xantho-proteic reaction, when the biuret test quite fails to show their presence.

The character of the colored compound formed has been studied by Vaubel. The color probably involves the state of oxidation of the mercury atom, since many mercury compounds are red (cinnabar).

Millon's reaction is interfered with by hydrogen peroxide, chlorides and by alcohol. If these are present it is necessary to use an excess of reagent.

Xantho-proteic reaction. This, as the name says, is the yellow reaction of proteins (Greek, *xanthos*, yellow). In contact with nitric acid most proteins develop a lemon-yellow color which changes to an orange when the solution is made alkaline. The protein either in solution or suspension is heated with a few drops of concentrated nitric acid to 5 c.c. of water in the test-tube for from one to three minutes, cooled and ammonia or sodium hydrate added to an alkaline reaction.

This reaction is due to the benzene nuclei in the molecule. The reaction is given by tyrosine, phenyl alanine and by tryptophane, the three amino-acids contained in proteins having benzene nuclei. Tryptophane gives the reaction most intensely and easily; then tyrosine; whereas phenyl alanine requires a longer heating, or more nitric acid. Proteins which lack these three groups, for example salmin, sturin and clupein among the protamines, do not give the xantho-proteic reaction.

The mechanism of the reaction consists in the formation of a mononitro benzene, or possible a dinitrobenzene. The nitrated benzenes such as picric acid, $C_6H_2(NO_2)_3OH$, are light yellow in acid solution, but a deep orange in the salt form. Such nitro derivatives are formed in the course of the reaction. These nitrobenzenes are all toxic and are sometimes used as dyes (Martius Yellow) for coloring macaroni and other foodstuffs, although their use is forbidden in most countries. The yellow color is probably due to the NO_2 groups (vibration periods of the electrons of the valences of the nitrogen or oxygen), since some of the nitrogen oxides are brown or reddish yellow.

Tryptophane reactions. Tryptophane, containing as it does the indole group, is the chromogenic radicle of the protein molecule *par excellence.* Tryptophane and tyrosine are the protein nuclei which give rise in their metabolism to most of the body pigments, such as the blood pigment (pyrrol nucleus), bile pigments (pyrrol from tryptophane), melanins and reds from tyrosine, etc. Tryptophane, as its name implies, i.e., the bright (Gr. *phanos,* bright) substance formed in the course of tryptic digestion, readily yields, like indole, a series of bright colors, reds, violets, blues, when oxidized. There are a number of color reactions which depend on the presence of tryptophane and among these is the Adamkiewicz reaction.

Adamkiewicz reaction. If to a few c.c. (2-3) of a protein solution one adds an equal quantity of glacial acetic acid and then 4-5 c.c. of concentrated sulphuric acid, at the zone of contact a violet ring forms in the presence of a protein containing tryptophane. If the tube is shaken, the violet color generally develops all through the solution if not too much sulphuric acid has been used. This reaction depends on the presence of aldehydes in the glacial acetic acid. It has been found (Hopkins and Cole) that most samples of glacial acetic acid which have stood some time contain some glyoxylic acid, HCO.COOH. It is said that some samples of glacial acetic acid will not give Adamkiewicz reaction, although the writer has never seen any such. The test may be performed, therefore, by using glyoxylic acid in place of glacial acetic. The glyoxylic acid is easily made by reducing oxalic acid with powdered magnesium. An equal volume of this acid (Hopkins-Cole reagent) is added to the solution in the place of the glacial acetic and the test performed otherwise in the same manner. The rôle of the glyoxylic acid is not explained, but it possibly consists in hastening the oxidation of the tryptophane or condensing with it in the presence of acid to give the color.

Other aldehydes may be used in this test besides glyoxylic acid. *Formaldehyde* has been suggested by Rosenheim and Acree. In fact, this reaction is used for the detection of formaldehyde in milk and is of very great delicacy. Casein, the protein in milk, contains relatively a large amount of tryptophane in its molecule. If a little formaldehyde is added to milk and the milk does not stand long enough for the formaldehyde to have united with the free amino groups of the milk proteins, the addition of strong hydrochloric acid containing a trace of iron, or of sulphuric acid with iron, leads to the development of a violet color. It has been recently suggested by Cole that perhaps the Adamkiewicz reaction is due to the presence of formaldehyde in the glacial acetic acid rather than to the glyoxylic acid. Perhaps other aldehydes will act similarly.

Liebermann's reaction. Another color reaction involving tryptophane is that of Liebermann when carried out in the manner originally prescribed by him. Liebermann found that protein treated first with alcohol and ether and then with hydrochloric acid developed often a violet or bright blue color. This reaction is probably due to the presence of aldehydes in the alcohol and ether (Cole) which combine with the protein and on subsequent heating with strong hydrochloric acid develop the tryptophane reaction. If the protein contains both tryptophane and sugar, it is not necessary to treat it with alcohol or ether first, since by the action of the strong acid on the carbohydrate aldehydes are formed which give a colored reaction product with some of the protein groups and presumably with the tryptophane. See page 36.

Other tryptophane reactions. Bromine. Tryptophane when free, but not when united in the protein molecule, gives in a faintly alkaline solution with bromine or chlorine water a beautiful violet color. This reaction was discovered by Claude Bernard as distinguishing tryptic from peptic digestion. Adamkiewicz' reaction is given both by the free and linked tryptophane. The color in the bromine test is possibly due to the formation of indigo, since indole gives a similar reaction. By this bromine reaction one can follow the course of the splitting off of tryptophane from the protein molecule during the process of digestion.

Tryptophane will also give colored products in the presence of aromatic aldehydes (Rohde). If a little p-dimethyl-amino-benzaldehyde is dissolved in concentrated sulphuric acid and run beneath a solution of protein in a test-tube, a red-violet ring at the zone of junction develops. A similar reaction occurs with vanillin, or benzaldehyde sulphuric acid and protein. These reactions are given also by free indole groups as well as by tryptophane; p-nitro-benzaldehyde gives an intense, stable, green color; vanillin,

$$C_6H_3(COH)(OCH_3)(OH),$$

a beautiful red, becoming violet by dilution; para-dimethyl-amino benzaldehyde,

$$C_6H_4(CHO)N(CH_3)_2$$

a red becoming violet. In the spectrum a wide absorption band in the orange between λ 615-570 and a second in the green between λ 555-540 are to be seen. The method of making this test is as follows: To 6 c.c. of the protein solution or suspension in a test-tube add 5-10 drops of a 5 per cent. solution of p-dimethyl-amino-benzaldehyde in 10 per cent. sulphuric acid and then add concentrated sulphuric acid drop by drop, with frequent shaking until color appears. If the albumin is very dilute this method is not sensitive enough. In that case put concentrated sulphuric acid containing 1 per cent. dissolved aldehyde beneath the solution and see if a colored ring of contact develops. In the ring method it is possible to detect tryptophane in 0.003 per cent. concentration. Casein reacts in about 0.15 per cent. concentration, so that tryptophane must make about 2 per cent. of the casein molecule.

Triketo-hydrindene hydrate reaction. Ninhydrin reaction. A very sensitive reagent for most amino-acids, proteins, peptones and some other substances is triketo-hydrindene hydrate. A blue color develops on boiling. The test is given by amino-acids which have at least one free carboxyl and a free amino group. Ninhydrin is

$$C_6H_4\langle{}^{CO}_{CO}\rangle C(OH)_2.$$

A description of the test is given on page 883. The reaction is positive with proteins, proteoses and with all the amino-acids with the exception of proline, oxyproline, pyrrolidon carbonic acid. It is positive also with asparagine and glutamine, amino-oxy-valerianic, diamino propionic, sarkosine and alanyl alanine. It is negative with proline, oxyproline, glucosamine, guanine, allantoine, leucinimide, urea. The albumins give a very blue color, as do also all polypeptides, all α-amino acids and β-alanine. Ammonium carbonate gives a red coloration, and histidine after a while becomes a Burgundy red. Glycocoll will give the reaction in 1:10,000 solution. By means of this valuable reagent it is possible to show the presence of amino-acids in fresh urine and in the protein-free blood serum.

Carbohydrate reaction. Many proteins contain a carbohydrate nucleus in their molecule. This may be detected by *Molisch's* reaction. The principle of the reaction consists in converting the carbohydrate into aldehyde decomposition products (furfural, formol, pyruvic aldehyde, etc.) by the action of strong acid and then the detection of these by some aromatic substance. The method usually employed is that of Molisch. To the solution (5-6 c.c.) to be examined 1-2 drops of a 10 per cent. alcohol solution of α-naphthol, (OH), are added and then a few c.c. of concentrated sulphuric acid is poured carefully down the side of the tube. A violet ring develops at the zone of contact in the presence of carbohydrates. The α-naphthol in the presence of sulphuric acid condenses with the aldehydes formed from the carbohydrate by the action of the acid to form colored compounds. If the protein contains a good deal of carbohydrate and also tryptophane, it may not be necessary to add the α-naphthol, the tryptophane taking its place. Thus egg-white contains a good deal (0.5 per cent.) of glucose. If a little egg white is boiled in water with strong hydrochloric acid a violet color develops without any addition of α-naphthol. In this case the aldehyde is generated from the glucose by the acid, and the proteins furnish the tryptophane. Liebermann's reaction is sometimes tried in this form. Molisch's reaction for carbohydrates appears later as Pettenkofer's test for bile acids. In this case the carbohydrate is added and the chromogen is supplied by the bile acids.

Sulphur reaction. Reference may also be made here to two or three sulphur reactions. Sulphur occurs in the protein molecule in the reduced form either as cysteine or cystine. If a protein containing either cystine or cysteine is boiled with sodium hydrate, the sulphur is in part split off as the sulphide. If a little lead acetate is added either before or after heating, a brown or black color develops and ultimately a black precipitate of lead sulphide settles out.

Some proteins, and particularly those from actively metabolic cells, probably contain cysteine in place of cystine in the molecule and, as we have already noticed elsewhere, this difference may be of great importance in cell life (see Heffter and Arnold). If a protein which contains cysteine is dissolved in water and 2-4 drops of a fresh 4-5 per cent. solution of sodium nitroprusside and then a few drops of ammonia are added, an intense purple-red color appears at once. The color disappears on the addition of acetic acid. This reaction, however, is not specific or characteristic. The color is given by other substances than cysteine, for example by other sulphides, by acetone, creatinine etc., but cysteine is the only substance likely to be present in protei which will give the reaction. Proteins of the supporting tissues of th body generally contain cystine; those of active organs cysteine.

Precipitation reactions of the proteins.—Both for the purpose o detecting the presence of proteins in solution and of removing then from solution their precipitation reactions are important. Probably al natural proteins contain a small number of free amino groups and fre carboxyl groups. They are hence both basic and acid. By means o these groups they can unite and form salts, many of which are insolubl with both acids and bases. Among the acids giving more or less insolu ble compounds with proteins are tannic, metaphosphoric, picric, picro lonic, phosphomolybdic, phosphotungstic, tri-iodo-hydriodic, chromic an bichromic acids, and many acid dyes; and among the bases are the metal copper, iron, manganese, aluminum, lead, mercury, nickel, platinum gold; organic bases such as quinine, strychnine and many other alka loids, some basic proteins, such as protamines and histones; and basi dyes such as thionin, fuchsin and methylene blue or neutral red. Th acids which precipitate are generally those which precipitate alkaloid also. A great deal of confusion exists in the literature on this subjec of precipitation of proteins because of a failure to realize that thes precipitates are true chemical compounds. They are sometimes calle without any good reason "adsorption" compounds, indicating that the belong to that hypothetical class of physical unions of which so littl of a definite nature is known, but which is supposed to depend on surfac tension. The whole behavior of the proteins shows these precipitates t be true compounds.

The reactions are as a matter of fact almost certainly simple sal formations. Whenever the precipitation is to be made by a reagent o which the precipitating part is in the anion or negative group of th molecule, the solution must, for all except the basic proteins such a histone and protamine, be acid in reaction. The basic proteins may b precipitated either in neutral or even slightly alkaline reactions fo the reason given below. If, however, the precipitating substance is metal, or base, the precipitation either does not take place at all or no so completely unless the solution be slightly alkaline. The reason fo this is as follows: The precipitating agents of the first class mentione are the free acids, or the salts of acids, and the part of their molecul which precipitates is the negative part, or the anion. In this group ar all the acids mentioned above and many others not there included, suc as bichromic, chromic, ferrocyanic, etc. The precipitates which ar formed have been found always to be the protein salts of the precipi tating acids. They are protein bichromate, tannate, picrate, picrolonate ferrocyanide, etc.; and in the case where basic precipitating substances are used the precipitates always carry down the base and they are generally the salts of the protein, such as quinine or lead proteinate. If a colored base is used to precipitate, the fact that the precipitate is colored shows that the base has gone down with the protein. In a few cases, such as precipitation with mercury, platinum or copper salts, the union of the salt is with the amino group, as will presently be shown.

The reason why the protein must be in an acid solution to precipitate with the alkali salts of the acids mentioned is that the protein must be electro-positive to unite with the electro-negative radicle of the salt; and it must be in an alkaline medium to precipitate with the bases, because the protein must be electro-negative to unite with the electro-positive bases.

In acid solutions proteins become electro-positive; and in alkaline solution they become electro-negative. This was shown by Hardy. If egg white be dialyzed against distilled water until free from salts and then boiled, it becomes opalescent, but the protein is not precipitated; it remains in colloidal solution. If, now, to this solution a little acid is added and an electric current is sent through the solution, the protein collects in a tough, white mass at the cathode; while, if the solution is made very faintly alkaline, the protein collects at the anode. The fact that the protein moves in the electric stream proves that it carries an electric charge; that it moves to the negative electrode, or cathode, in an acid solution shows it to be electro-positive; and to the anode in an alkaline solution proves it is there electro-negative. The electric sign of the protein molecule is different in an acid from what it is in an alkaline solution.

Some rather extraordinary explanations have been given of this change of sign, which is a matter of fundamental importance in understanding cell metabolism, vital and ordinary staining, etc. Thus it was suggested that as the hydroxyl ion moves faster than the sodium or potassium ion it hurries on ahead of the sodium and hitting the protein molecule first buries itself in that molecule, thus making the molecule electro-negative; and in acids, the hydrogen ion goes first, is entombed in its turn and makes the molecule of protein electro-positive. There is no need, however, for this fanciful explanation which has nothing to recommend it except its picturesque nature. The real explanation is probably quite different. By means of the free amino groups the proteins are basic and they combine with the acid by these groups, forming thereby salts like substituted ammonias thus:

$$R—CHNH_2 + HCl \longrightarrow R—CHNH_2.HCl.$$

R is the rest of the protein molecule. The salt $R—CHNH_3Cl$ now ionizes

into $RCHNH_3^+$ and $\bar{Cl}$. The chlorine is electro-negative and the rest of the molecule is electro-positive. Hence in acid solutions the proteins, with the exception of some very acid ones like casein, are always electro-positive.

In alkaline solution the free carboxyls unite with the alkali to form salts:

$$RCOOH + NaOH \longrightarrow RCOONa + H_2O.$$

RCOONa now ionizes into $RCOO^-$ and Na^+. Thus the protein becomes electro-negative.

The reactions with the precipitating reagents now become clear. They are as follows:

1. Protein + $CH_3.COOH$ ⟶ Protein acetate.
2. Protein acetate + Na bichromate ⟶ Protein bichromate + $NaOCOCH_3$.
   Precipitate.
3. Protein + NaOH ⟶ Na proteinate.
4. Na proteinate + Pb acetate ⟶ Lead proteinate + $NaOCO.CH_3$.
   Precipitate.

But, while this is the rule for most of the proteins, there are certain ones which appear at first glance to be exceptions. For example, the protamines and histones may be precipitated by colored acid dyes, or by sodium picrate, or bichromate in neutral, or even faintly alkaline solution. The reason for this is that these proteins are so strongly basic, having so many basic amino-acids in their molecules, that they are electro-positive even in a neutral solution in which they exist as the free bases. They may even be positive in faintly alkaline media. They do not change to the electro-negative state until some excess of alkali has been added. Similarly some of the acid proteins, such for example as some of the vegetable proteins which contain a large amount of glutamic acid in the molecule and are hence fairly strong acids, may be precipitated in neutral or even faintly acid solution by the basic precipitating reagents. For these proteins do not at once become electro-positive as soon as the reaction becomes faintly acid. Casein is a protein of this kind. Another complication is introduced by the affinity of all metals below hydrogen in the scale of solution tension, such as mercury, gold, copper and platinum, for amino groups. These metals will not only form simple salts with the proteins by displacing the hydrogen from the carboxyl group, but they will also form addition compounds or double salts by union with the amino groups. It will be found, therefore, that mercuric chloride will precipitate even in a faintly acid medium, and so will the others of this group. This, however, is not an exception to the rule stated, but an additional kind of chemical union between the precipitating agent and the protein. In most of these cases, also, the precipitation is found to be more complete in a faintly alkaline than in a faintly acid medium.

One of the best ways of completely separating the proteins from a solution is by using basic lead acetate. Mercuric chloride in a faintly alkaline solution may, however, also be used.

With this brief account of the properties of the proteins we may pass to the consideration of some few which are of particular interest in the cell. We shall not now consider all the different kinds of proteins, leaving the individual members of the group to be treated more at length in connection with the organs or fluids of the body in which they occur. There is one group, however, which is colored and of very general interest, as members of this group are found both in plants and animals. These are the chromoproteins. They occur in the cytoplasm of cells.

Chromoproteins.—There are two groups of chromo, or colored, proteins which may occur in the cytoplasm: the hemo-chromoproteins obtained from blood, of which the hemoglobins are the best examples; and, second, the phyco-chromoproteins which are obtained from seaweed. These latter are very interesting proteins because in a way they are intermediate between hemoglobin and chlorophyll. The chromatic group of hemoglobin is an iron containing pyrrol complex called hematin; and the iron free part of hematin resembles chlorophyll, which also yields pyrrols on decomposition. It is very interesting, therefore, as showing the close relation between hemoglobin and chlorophyll that a chromoprotein closely resembling hemoglobin in several ways and particularly in its ease of crystallization has been isolated from the red and blue-green algæ. The red coloring matter of the Floridiæ, *phykoerythrin*, and the blue coloring matter of the blue-green algæ, *phycocyan* (*phykos*, seaweed; *cyan*, blue; *erythros*, red), crystallize most readily. The substances are obtained from seaweed just as hemoglobin is obtained from the corpuscles of the blood by laking in distilled water. Ammonium sulphate (30 grams to 100 c.c. solution) is then added and the phykoerythrin and the phykocyan precipitate. They are globulins. If the precipitate is redissolved by the addition of water and the salt dialyzed out, the protein crystallizes out in the dialyzing tube in microscopic crystals. Phykoerythrin is coagulated by boiling; it is soluble in weak alkalies and neutral salts, but insoluble in distilled water. It is precipitated by acetic acid, but redissolves in an excess. It is precipitated by $(NH_4)_2SO_4$, $MgSO_4$ and alcohol. It quickly loses its color in the light, particularly in an alkaline solution. The analyses gave C, 50.82; H, 7.01; N, 15.37; S, 1.60; O, 25.20. It is free from ash and resembles chlorophyll in containing no iron.

Distribution of protein substances between the cytoplasm and the nucleus.—The proteins of the cell nucleus are sharply differentiated from those of the cell cytoplasm. In the nucleus many of the proteins, in some cases all of them, are nucleoproteins, characterized by the presence in the molecule of nucleic acid. The simple proteins in the nucleus are often more basic than the general run of proteins and sometimes they are very basic proteins, such as the protamines and histones. The occurrence of these proteins is, however, the exception rather than the rule. The composition of the nuclear proteins will be considered presently. The proteins of the cytoplasm are less well characterized and of very diverse character. They include both the proteins of the living proto-

FIG. 15.—Phycoerythrin crystals (Kylin).

plast and lifeless secretory or reserve proteins of a varied nature. They are often globulins, that is simple proteins insoluble in water, but soluble in dilute salt solution. Thus in the cytoplasm of muscle there are the simple proteins, myosin and myogen and myosin fibrin; in the thyroid gland, the thyreoglobulin of the colloid material which is found in the cytoplasm is a globulin. On the other hand, albumins are found there also. In the white blood corpuscles a simple protein corresponding to serum albumin has been found. In many cells of the body there occurs in the cytoplasm, also, a globulin coagulating at the low temperature of 56°, which is the temperature of coagulation of fibrinogen. It is generally believed, too, that phosphoproteins are found in the cytoplasm, and this is certainly the case in some cells. Thus casein is found in the cytoplasm of the milk glands and vitellin in the cytoplasm of the hen's egg and some other eggs. Both of these bodies are phosphoproteins. There is some reason for believing that in the living protoplasm the protein may be in union with phospholipins, carbohydrate and possibly fats. It is not possible, however, to make a definite statement on this point. The decomposition products of protein metabolism probably also occur there.

We may then say that in the nucleus are found the nucleoproteins; whereas in the cytoplasm of the cell these are probably lacking (see page 173). The protoplast of the cytoplasm consists in all likelihood of a mixture of simple albumins and globulins, coagulable by heat, and phospholipins, and some of these simple proteins may be and probably are in loose physical or chemical union with phospholipin, fat and carbohydrate. In other cells one finds mucin, which is a glycoprotein. These proteins do not occur free for the most part, but in union with inorganic salts, salts of sodium, potassium, calcium and magnesium preponderating. These cytoplasmic proteins in the living cell are predominantly electronegative, but occasionally electro-positive protein may be present, as in the red blood corpuscles in which the hemoglobin is electro-positive. Since the whole of the protein world is at some time in the cytoplasm of cells, it will be seen that this part of the cell is wonderfully diverse in its chemical nature. The general features of the living protoplast, as distinct from secretory granules, reserve proteins or structural elements, are, however, so similar in all cells that it is probable that in its fundamental chemical constitution it is everywhere closely alike, although differing in some particulars. What this constitution is, is the great unsolved problem of physiological chemistry.

CHEMISTRY OF THE CELL NUCLEUS.

Morphology. If living cells are examined under the microscope, all except the simplest animal cells (Monera) and the bacteria may be seen to contain within the granular protoplasm a clear, almost or quite homogeneous, more refractive area. This area, called the nucleus and first described by Robert Brown in 1831, is generally spherical or ellipsoidal in shape, though at times it is quite irregular in outline. Figure 1, p. 11. Sometimes it is separated from the surrounding protoplasm by a distinct visible membrane; at other times no membrane may be seen in the living cell, though it is probably always present. In size, the nucleus may fill almost the entire cell, as in cells of the thymus gland or the sperm head, or it may be a very small part of the total bulk of the cell, as in many eggs and muscle cells.

Generally no structure can be seen within the *living* nucleus, but in some cases, as in the germinal vesicle of many eggs, there may be seen, in addition to a distinct membrane, spherical or irregularly shaped more dense portions which are known morphologically as nucleoli. When the cell divides by caryokinesis there may also be seen, in the most favorable cases, as in the testes of grasshoppers, and in some transparent eggs, the spindle fibers and the chromatic masses called by morphologists chromosomes. In general, however, as long as the cell is alive no other structure may be seen within the nucleus than the nucleolus.

The physical structure. The physical consistence of the nucleus has been found by Kite to vary greatly in different cells. By his very ingenious method of microscopic cell dissection by means of extremely fine glass needles (diameter 1 μ or less), Kite has found that most nuclei are separated from the protoplasm by a very tough distinct nuclear membrane. Within this membrane one generally finds either a liquid (sol) or a fairly viscid gel in which no structure, except sometimes the nucleolus, is to be discovered by his methods. The nucleus of an amœba, for example, or the nucleus of an immature starfish egg, contains a liquid, and when the nuclear wall is ruptured the contents escape into the surrounding cytoplasm, mixing with the latter and setting up most interesting chemical changes within it, discussed further on page 180. But the nuclei of most differentiated cells which he examined, such as epithelial, liver or pancreas cells of the amphibian, Necturus, or the frog, or rabbit, are quite jelly-like. They may be cut into several pieces, each piece retaining its form and in this case not mixing with the cytoplasm. It is indeed altogether probable that the physical state of the nuclear contents is not constant in any cell, but varies from fluïd to gel under various conditions. This is indicated, for example, by the experiments of Calkins and Miss Peebles in their cutting to pieces of infusoria. At times the cutting could be made as if through a jelly, the pieces not losing their contents when cut; and at other times the protoplasm was so liquid that it readily escaped through the cut. Kite has made similar observations on Amœba proteus and they have been made also by Gruber. One of the constituents of the nucleus is nucleic acid and this has quite remarkable powers of forming gels; and it may be that this jelly-like consistence of many nuclei is due to the presence of this substance.

One of the most important observations of Kite is that it is impossible by his method of dissection to find in living nuclei any more dense masses, or networks, which might correspond with the chromatin network, or chromosomes to be seen in fixed and stained nuclei. Whether these pre-exist in the cell nucleus when it is alive, or whether they first appear as the result of the action of fixing agents, may seem doubtful from this observation; but the extreme and detailed regularity of these morphological pictures in fixed cells (Figure 2, p. 12), and their steady development during karyokinesis, make it unlikely that they are produced by the fixing agent. It seems more probable that they exist in the living nucleus, though perhaps not quite in the form revealed in the nuclear corpse, even though they can neither be seen nor found by dissection. It may be remarked, indeed, that the dissection method, by the enormous stimulation of the cell which it entails and the mechanical mixing of the parts of the cell, must render an interpretation of the results obtained by it somewhat uncertain.[1]

Function. There is no question but that the nucleus, forming as it does so universal a constituent of cells, is of fundamental importance to cell life. The sperm head, which alone enters many eggs, the tail being left outside, and which is able to produce the development of an organism resembling in many most minute particulars the parent organism from which it came, is often composed exclusively of a nucleus. So the nucleus must play a great part in inheritance. Inheritance is equally from father and mother, and it can hardly be a coincidence that the embryo contains an equal share of nuclear material from father and mother, whereas the cytoplasmic material is obtained almost exclusively from the mother.

The importance of the nucleus is shown very clearly in many experiments which have been performed on unicellular organisms. If an amœba, or other protozoon organism, be cut into two parts, one of which contains the nucleus, while the other lacks it, it is found that while both pieces may continue in motion and may capture food, it is only the part with the nucleus which is able to grow and reconstitute the cell; the protoplasm without the nucleus cannot regenerate the nucleus and in a short time it dies and disintegrates. This experiment shows that both nucleus and cytoplasm are necessary for growth and development.

Similar facts showing the great importance of the nucleus in the growth and synthesis of new protoplasm are beautifully illustrated in gland and vegetable cells. If vegetable cells are plasmolyzed, that is shrunk from the cell wall by the action of hypertonic salt solutions, it sometimes happens that the protoplasm becomes divided within the cell into a nuclear containing and a nuclear free portion; it is only the former which makes a new cell wall and grows to a new cell. In many gland cells the protoplasm during glandular rest in whole, or in large part, becomes differentiated into secretory material, generally taking the form of granules. The nucleus remains with only a very small quantity of cytoplasm around it. Now, when the cell secretes, these granules are discharged or dissolved, and the new undifferentiated protoplasm which takes their place appears always first close to the nucleus, as if it were being formed here.

[1] Chambers has recently found that the chromosomes may appear quite suddenly in nuclei.

There can be no doubt from all these facts that the nucleus plays a very important part in the synthesis of the cell protoplasm. It appears as if under favorable conditions the nucleus might be able to make the cytoplasm about it, but no one has as yet succeeded in proving this. It might be tested by growing the spermatozoa in such conditions that they would make themselves into cells provided with cytoplasm. Perhaps it might be proved, also, by isolating nuclei by means of Kite's method. At present all that can be said definitely is that both protoplasm and nucleus appear to be necessary for growth and development. Many chemical transformations, probably most of them, occur in the extra nuclear part of the protoplasm. But the nucleus is nevertheless of fundamental importance.

Chemical composition. Method of obtaining nuclei for chemical analysis.—The chemical composition of an organ of such vital importance in inheritance and cell life is a matter of very great interest. What knowledge we have of it is owing more particularly to Miescher and above all to Kossel. There are several ways in which the chemical nature of the nucleus may be studied. We may study cells consisting chiefly of nuclei, such as leucocytes, and contrast their composition with that of cells consisting chiefly of cytoplasm, such as muscle, or egg cells, or red blood corpuscles of mammals. Substances which are found in predominating amounts in the first group of cells we would be justified in inferring came from the nuclei. Another method, although one to be used with great caution in interpreting observations, is the use of microchemical stains. The best method is to separate the nucleus from the cytoplasm and to study the chemical composition of each separately.

The first method was that used by Miescher, with whom our knowledge of the composition of the nucleus begins in 1876. It had been known that living tissues all contained large amounts of phosphoric acid in different combinations. This acid early attracted the attention of chemists, some of whom even went so far as to say "ohne Phosphor keine Gedanke" ("without phosphorus no ideas"). And we are coming to realize more and more clearly the fundamental rôle phosphoric acid plays in all vital phenomena. It was soon found that the phosphoric acid was present in at least two forms. One part could be extracted by alcohol and was in organic union. It was present in the lecithin discovered by Gobley. Another part could be extracted by cold water from the tissues already extracted with alcohol. This part consisted of inorganic phosphates. After removing these two forms of phosphoric acid there remained a considerable proportion of the phosphoric acid in the protein residue of the cell. Hoppe-Seyler put his pupil Miescher at the task of finding out what compound of phosphoric acid remained in this residue. Miescher worked chiefly with pus, which in those days of septic surgery could be readily obtained. He found that most of this remnant of phosphoric acid-containing material could be extracted with dilute alkalies and reprecipitated by acetic acid. It was in organic union with proteins. Since pus cells consisted chiefly of nuclei with very small amounts of cytoplasm and this material constituted the greater part of the residue, there could be little doubt that it came from the nucleus and for this reason it was called by Miescher "nuclein." It was quickly found that nuclein was a constituent of all cells examined. Thus Hoppe-Seyler found it in yeast; it was isolated from sperm, spleen and a great variety of tissues. This nuclein contained varying amounts of phosphorus varying from 0.9-4 per cent. One of the easiest ways of preparing such a nucleoprotein is to extract a tissue with dilute alkali; or even to boil it with water, some of the nuclein goes into solution in the boiling water. Shortly after this Kossel found that if this nuclein was boiled with acids, it yielded a number of xanthine bases, of which the formulas will be given presently, such as xanthine, hypoxanthine, guanine and adenine, a new base which he discovered and named adenine (Gr. *adēn*, gland) because he isolated it first from the pancreatic gland. The discovery that the xanthine bases could be obtained from nuclein was a discovery of fundamental importance, for it indicated that these bases, which are found in human urine, and uric acid, which belongs in the same group of substances, must come from the nuclein of the body and not from the ordinary albumin, as had been supposed up to that time. In 1887 Altmann, an histologist, took a long step forward when he succeeded in isolating from Miescher's nuclein by digesting it with pepsin-hydrochloric acid an organic acid, containing 8-9 per cent. of phosphorus, which was free from albumin, all the albumin tests being negative. He called this acid *nucleic acid.*

Before examining the constitution of this important acid discovered by Altmann, a word may be said about another method of determining the constitution of the nucleus. The best method is to examine the heads of spermatozoa. These, in the fishes and most animals, consist wholly, or almost wholly, of nuclear material; and while they undoubtedly represent very highly specialized nuclei, nevertheless they are still nuclei. This method of studying nuclear composition was found by Miescher.

If the ripe testes of a fish such as the salmon, which Miescher studied, or the herring, are taken and ground to a pulp and then strained through cheesecloth the sperm go through; the connective tissue remains behind. It is an additional advantage that in fishes the sperm all ripen at the same time so that a homogeneous product is had. The unripe sperm have a different, more complex, composition from the ripe. The sperm mass is then suspended in normal salt solution, or in a dilute magnesium sulphate solution, and centrifugalized. By this means they are freed from the liquid in which they are suspended in the testes. After one or two washings of this kind, the sperm are suspended in distilled water and centrifugalized very rapidly. In the distilled water the tails swell, and the heads are so much heavier and denser that they are separated from the tails by the centrifugal force and accumulate at the bottom of the tube as a pure white mass. Above this mass of heads, there may be seen in the centrifugal tube a slimy tenacious layer of swollen tails more gray in color than the heads. This layer of tails is coherent and may be easily lifted out. Above this again is a layer of water, opalescent, and containing the greater part of the lecithin, cholesterol and much protein in solution. After several washings and centrifugalizing in distilled water the heads are clean from tails. Under the microscope they look perfectly normal. They are not changed in shape nor apparently in size. They appear to have lost none of their constituents. They constitute pure nuclear matter. It is of course possible that they have lost some material in the washing in spite of the fact that they do not appear to have done so. Thus far only two kinds of sperm have been examined in this way, the salmon by Miescher, and the herring by the author and Steudel. If these pure white sperm heads are now extracted by alcohol and ether only traces (.1-.01 per cent.) of alcohol-ether soluble substances are found in them. From this it appears either that the lecithin and lipoids have been extracted by the distilled water, or else that they are confined chiefly to the middle pieces and tails, and that they are not found in the nucleus. The small amount found was so variable as to suggest that it may have come from remnants of tails, which had not been completely separated from the heads.

The other kind of nucleus which has been obtained free and pure for analysis is that of the red blood corpuscles of hens. The corpuscles, treated in the same manner as the sperm, swell, they are laked, and the nuclei become free and may be accumulated by centrifugal action. These nuclei have been recently examined by Ackermann.

THE COMPOSITION OF CHROMATIN.—The sperm head consists wholly, or almost entirely, of chromatin. This chromatin consists of a nuclein. In the heads of salmon sperm the chromatin is salmin nucleate; in the herring it is clupein nucleate. See page 178. In all cells it has been found that the chromatin consists of two parts: an acid part, nucleic acid, discovered by Altmann, and a basic part which is always some member of the simple proteins, but a different protein in every kind of cell which has been examined thus far. We will consider first the composition of the acid part of the nucleus, or nucleic acid, and then the basic or protein part of the molecule.

Nucleic acid.—*Method of isolation.* Nucleic acid may be obtained from tissues without necessarily isolating the nuclei first. It is most easily obtained by the Kossel-Neumann method. Perfectly fresh tissue must be taken and as quickly as possible after its removal from the body it is ground in a meat chopper and thrown into boiling water slightly acidified with acetic acid to destroy the enzymes. The reason for the necessity of haste is that there are present in most cells enzymes, called nucleases, which very rapidly attack and partially decompose the nucleic acid. The residue is ground as fine as possible and then brought into twice its weight of a boiling solution of sodium hydrate and sodium acetate (1.6 per cent. NaOH and 10 per cent. Na acetate) and extracted for from ½-2 hours at boiling temperature. By this treatment the nucleic acid is dissolved and extracted from the cells. The mass is then neutralized with acetic acid, centrifugalized and, if necessary, filtered hot. The filtrate is now concentrated and the filtered solution is poured into alcohol, about three volumes of 95 per cent. to one of the solution. The nucleic acid is precipitated as the sodium salt. It may be purified by resolution and reprecipitation. By this method (Neumann's) from 1 kg. of dry thymus gland 180-200 grams of nucleic acid are obtained.

Nucleic acid.—*Physical and chemical properties.* The sodium salt of nucleic acid thus prepared is soluble in water. If dissolved in hot water to a concentration of 5 per cent. it gelatinizes firmly, on cooling, to a clear, slightly opalescent gel. This property has already been mentioned in connection with the jelly-like consistence of some nuclei, and the solidity of the chromosomes. When dry the salt is pure white, amorphous, having neither taste nor smell. It gives no protein tests; the biuret, Millon, xanthoproteic and tryptophane reactions are negative. If added to a solution of protein containing a little free acetic acid, it precipitates the protein, forming thereby an artificial nuclein. It does not reduce Fehling's solution; it is not crystalline in any of its salts. It is optically active, dextro-rotatory, the rotatory power being $(\alpha)_D =$ +154.2. The substance is fairly stable with alkalies, but on long boiling (2 hours) in alkaline solution it goes over into a β-nucleic acid, which no longer gelatinizes, and which has a different per cent. of composition from the first. It is very unstable in the form of the free acid and is readily hydrolyzed into its constituents. The free acid is white like the salt, unstable in the light, turning a reddish or brownish red color when exposed in the powder form. It is fairly soluble in hot water, but much less soluble in cold. It is insoluble in alcohol, ether and similar solvents.

Per cent. of composition of nucleic acid. The very great ease with which the purines are split off from the molecule and the necessity of using acid at some stage of the separation makes it very difficult to obtain nucleic acids which are entirely normal. All of the older analyses in which the nucleic acid was precipitated by free acid are almost certainly incorrect. The following analyses are some which have been obtained:

Origin	C	H	N	P	Observer
Sperm of Alosa	36.27	5.00	15.96	8.11	Levene and Mandel.
Human placenta	37.44	4.32	15.32	9.67	Kikkoje.
Spermatozoa (Muraenoesox).	37.50	4.36	16.04	9.73	Inouye.

Various acids have given percentages of composition which differ somewhat among themselves. The relation of P:N is as 4 atoms to 14 or 15.

The most probable formula according to Steudel is $C_{43}H_{57}N_{15}P_4O_{30}$, which requires that the molecule should be composed of four hexose molecules, two purines, two pyrimidines and four molecules of phosphoric acid. By the action of endocellular enzymes nucleic acid is very quickly partially digested, which accounts for many of the discordant results of analyses.

Decomposition. We will first consider the composition of the true nucleic acids, or polynucleotides, as they are called, such as are found in the nuclei of all cells thus far examined, leaving the simpler, or mononucleotides, such as guanylic, or inosinic acid, for later consideration.

The true nucleic acids thus prepared by Neumann's method are extremely unstable if heated in the presence of acids; or even if left in an acid solution for a short time at room temperature. They decompose on prolonged heating with 3 per cent. sulphuric acid, or by heating under pressure with acetic or other acids, into orthophosphoric acid, various basic substances, i.e., guanine, adenine, cytosine, thymine, uracil, and either into a pentose or levulinic and formic acids. A method which gives the guanine and adenine in almost quantitative amounts and which is very simple is that of Steudel, who treats the copper salt with half-concentrated nitric acid.

It was Kossel who showed that the nucleic acids split under acid hydrolysis into the purine bases, orthophosphoric acid, levulinic acid, or a pentose, and the pyrimidine bases which he discovered and named. He found that the purine bases, some phosphoric acid and levulinic acid appeared very easily; the remnant of the molecule consisting of phosphoric acid, carbohydrate and pyrimidine bases was isolated by Kossel and Neumann and called thymic acid. The pyrimidine bases are far more difficult to detach from the molecule than the purines.

The work of Steudel and Levene has shown that in the nucleic acid itself there are two purine bases, adenine and guanine. These bases are heterocyclic compounds, and may be regarded as derivatives of the substance, purine. Caffeine, the active principle of coffee and tea, is a purine.

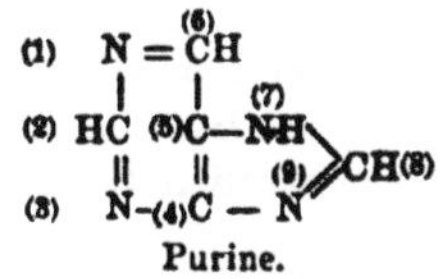

```
          (6)
(1)  N = CH
     |    |   (7)
(2) HC (5)C—NH
     ||   ||  (9)  >CH(8)
(3)  N—(4)C — N
```

Purine.

Products of hydrolysis.—*Chemistry of the purines. Guanine.* This rine base, $C_5H_5N_5O$, or 2-imino-6-oxypurine, or 2-amino-6-oxypurine, es its name to the fact that it was first isolated from guano. Its aphic formula is either

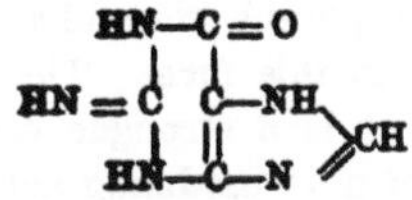

```
      HN—C=O                      HN—C=O
      |   |                        |   |
HN = C   C—NH            ;or  H2N—C   C—NH
      |   ||   >CH                 ||  ||   >CH
      HN—C—N                       N—C—N
```

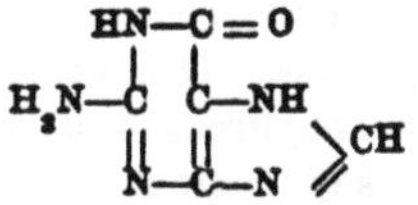

s may be seen in the graphic formula it contains the radicles urea, anidine and tartronic acid. It is a fairly strong base, precipitated by amonia from its aqueous solutions, a peculiarity which makes it easy separate it from adenine. It is soluble in acids and in strong alkalies. is precipitated by silver nitrate either in neutral or an ammoniacal lution, and forms double salts with the nitrate. Its nitrate crystallizes adily. The nitrate is insoluble in strong (half-concentrated) nitric id. Guanine crystallizes readily from a dilute solution as the picrate. forms a crystalline compound with bichromates.

Free guanine is found in various deposits in tissues. Thus it is found the free state in the concretions about the joints of hogs suffering from -called guanine gout. It occurs free in the scales and skins of the bony shes; and in the swim bladder, to which it gives the peculiar pearly-hite appearance. It is easily isolated from these sources by extracting ith dilute acid and precipitating with ammonia. On oxidation it yields anthine, uric acid, allantoine, urea, oxalic acid and other substances.

Guanase is an enzyme found in various organs of the body, in the iver, spleen, lungs, etc., which hydrolyzes guanine with the formation f ammonia **and xanthine**:

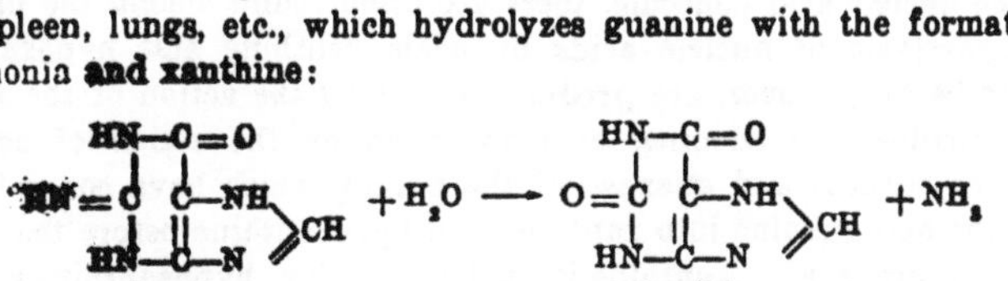

```
      HN—C=O                            HN—C=O
      |   |                              |   |
HN = C   C—NH         + H2O  ——>   O = C   C—NH         + NH3
      |   ||   >CH                       |   ||   >CH
      HN—C—N                             HN—C—N
      Guanine.                           Xanthine.
```

Adenine. This base, $C_5H_5N_5$, or 6-amino purine, was discovered by Kossel in the cleavage products of the nuclein of the ox pancreas and alled adenine (Gr. *ad n*, gland) because of its origin from a gland. Its mpirical formula is that of a polymer of hydrocyanic acid, and indeed hydrocyanic acid and cyanogen spontaneously change into substances which are allied to the purines. Adenine has been found to be formed in the nucleic acid molecule and it occurs in all true nu acids, polynucleotides, where it has been looked for. It is not pre tated by ammonia, hence its separation from guanine. Its struct formula is:

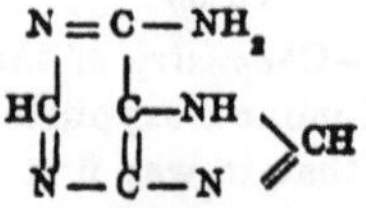

```
N = C — NH2
|     |
HC    C—NH \
||    ||     >CH
N — C — N  //
```

Adenine.

It is precipitated like guanine by picric or metaphosphoric acid and f crystalline picrates. It is usually separated in this form. The mel point of the anhydrous base is 360-365°. It is a stronger base guanine. It is quite stable in the presence of mild oxidizing agents, is easily decomposed by acids in the presence of a reducing agent far better yield is obtained by Steudel's method of hydrolysis of nucleic acid by nitric acid, than by hydrolysis with hydriodic, or c non-oxidizing or reducing acid reagents. The fact that the base unstable in the presence of reducing agents may have some bearin cell physiology, since the nucleus is probably always situated at a p in the cell where reductions are strongest. Hydrochloric acid at 180- C. decomposes it into carbon dioxide, glycocoll, ammonia and fo acid. Many cells, perhaps all, contain a ferment known as "adena discovered by Jones, which by hydrolysis converts adenine into hypo thine as follows:

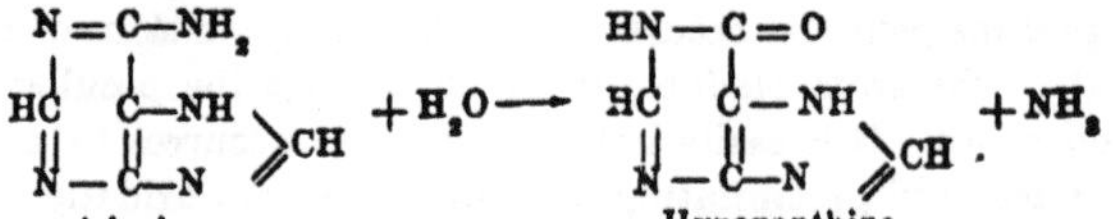

```
N = C—NH2                      HN — C = O
|    |                          |     |
HC   C—NH \                     HC    C — NH \
||   ||     >CH  + H2O —→       ||    ||       >CH  + NH3
N — C—N   //                    N — C—N     //
Adenine.                        Hypoxanthine.
```

Xanthine and *Hypoxanthine.* Besides these purines, which pre in the nucleic acid molecule, there are often found among the prod of hydrolysis of nucleic acids by acids xanthine and hypoxant These bases, however, are produced either by the action of the aci the guanine and adenine, or more often by the action of enzy such as adenase and guanase of the tissues, which have converted adenine and guanine into xanthine and hypoxanthine before the nu acid was prepared. Xanthine is 2,6-dioxy purine; hypoxanthine is 6- purine.

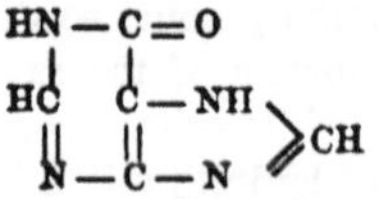

```
HN — C = O                 HN — C = O
|     |                     |     |
HC    C — NH \          O = C     C — NH \
||    ||       >CH          |     ||       >CH
N — C — N   //              HN — C — N  //
```

Hypoxanthine. $C_5H_4N_4O$; 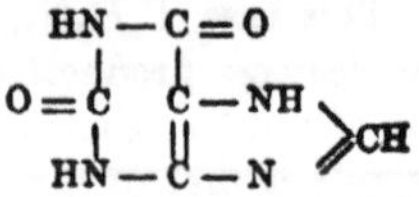 Xanthine. $C_5H_4N_4O_2$.

Xanthine owes its name (Gr. *xanthos*, yellow) to the yellow reaction it gives when heated to dryness in a porcelain dish with nitric acid. The yellow spot moistened with sodium hydrate turns first red and then purple red on heating, in distinction from uric acid. It was discovered in urinary calculi in 1817 by Marcet. On dry heating it decomposes into hydrocyanic acid, carbon dioxide and ammonia. It is both an acid and a base. It owes its acid properties to the fact that by a tautomeric rearrangement of the molecule the enol form appears:

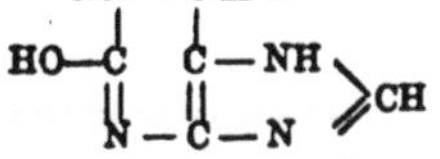

```
      HN — C = O
      |    |
  HO—C    C — NH
      ||   ||       >CH
      N  — C — N
```

Enol form of xanthine.

The hydrogen of the hydroxyl is replaceable by metals.

Hypoxanthine, literally little, or less, xanthine, is a reduced xanthine. It was formerly called sarkine. It is 6-oxypurine, having the following formula:

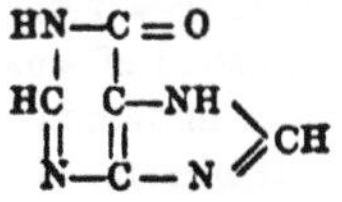

```
  HN—C = O
  |    |
  HC   C—NH
  ||   ||     >CH
  N—C — N
```

Hypoxanthine.

Hypoxanthine forms small colorless needles. It does not give the xanthine reaction with nitric acid, nor does it give the Weidel reaction. With hydrochloric acid and zinc a solution of hypoxanthine becomes first a ruby red and then turns brownish red on addition of alkali. Hypoxanthine is soluble in dilute alkalies and is not precipitated by ammonia. When treated with ammonia and an excess of silver nitrate, a crystalline compound having, when dried at 120°, a constant composition of $(C_5H_2Ag_2N_4O)H_2O$ separates out. Use is made of this in the quantitative separation. Hypoxanthine picrate is little soluble. Hypoxanthine as well as other purines which have the nitrogen at number 7 or the carbon in 8 unsubstituted give red azo compounds with diazo-benzolsulphonic acid in alkaline or neutral solution. The compound is probably of the following nature:

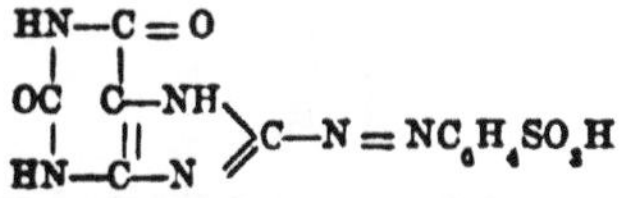

```
  HN—C = O
  |    |
  OC   C—NH
  |    ||      >C—N = NC6H4SO3H
  HN—C—N
```

Some pyrimidines give this reaction also. All purines are precipitated by cupric sulphate and a reducing substance such as sodium bisulphite. They form insoluble cuprous compounds. This is the basis of their quantitative determination by the Krüger-Schmidt method. Hypoxanthine is present in nearly all cells. It is a constituent of inosinic acid of muscle.

Pyrimidine bases.—Nucleic acid yields two or three pyrimidine bases when it is hydrolyzed long enough, but probably only two of them are preformed in the molècule, uracil being formed from the cytosine during the hydrolysis. These bases were discovered by Kossel. They are thymine, cystosine and uracil.

Thymine. This is 2,6-dioxy, 5-methyl pyrimidine. The structural formula is as follows:

```
    HN—C=O
     |  |
  O=C   C—CH3      The empirical formula is C5H6N2O2.
     |  ||
    HN—CH
```

It was first isolated from the hydrolytic products of thymic acid obtained from the thymus gland, hence its name. The pyrimidines are found very generally in cells not only in nucleic acid but as glucosides. Vicin and convicin discovered by Ritthausen and Preuss are hexose glucosides of pyrimidines. Thymine crystallizes from cold water, in which it is little soluble, in the form of clusters of small leaves or needles. (m.p. about 321°.) Thymine sublimes undecomposed. It is not readily precipitated by ammonia and silver nitrate. It is precipitated by phosphotungstic acid.

Cytosine. This is 2-oxy, 6-amino pyrimidine or

```
     N=C—NH2
     |  |
  O=C   C—H      or C4H5N3O
     |  ||
    HN—C—H
```

The free base is little soluble in water and crystallizes in thin plates with a mother-of-pearl glance. It is precipitated by silver nitrate in the presence of an excess of barium hydroxide, and by phosphotungstic acid. It gives the murexide reaction with chlorine water and ammonia. Like uracil it also gives a violet color when treated with bromine until cloudy and then baryta water added (Wheeler and Johnson).

Uracil is 2,6-dioxy pyrimidine.

```
    HN—C=O
     |  |
    OC  CH
     |  ||
    HN—CH
```

The reactions of this base are much like those of cytosine, but it is not precipitated by phosphotungstic acid. It is only imperfectly precipitated by silver nitrate and baryta water. It crystallizes from water in clusters of needles. It is nearly insoluble in alcohol and ether. Unlike thymine it does not sublime undecomposed, except on very careful heating. Generally decomposition takes place with the formation of red vapors.

Carbohydrate group.—All true nucleic acids, or polynucleotides, of animal origin thus far examined have been found to contain a hexose group, or several of them; whereas the nucleic acid from yeast and that from wheat, called tritico-nucleic acid, contain a pentose. Kossel discovered that on hydrolysis the thymus nucleic acid yielded levulinic acid and formic acid. It does not yield a reducing sugar. The production of levulinic and formic acid indicated clearly the presence of a hexose, since, as we have seen in the chapter on carbohydrates, the hexoses yield these bodies when heated with acid. On the other hand, he found in yeast nucleic acid on hydrolysis no levulinic acid, but a reducing sugar which gave large quantities of furfural when distilled. This showed the carbohydrate in this nucleic acid to be a pentose. Nucleic acids from fish sperm, thymus, spleen, liver, testes, pancreas, supra-renals, brain, lining of the alimentary canal and kidneys have all been found to yield levulinic acid and hence contain hexoses in the molecule. The nature of this hexose is still uncertain. It gives a saccharic acid (episaccharic acid) of as yet undetermined nature when the nucleic acid is hydrolyzed with nitric acid (Steudel). It has recently been suggested (Feulgen) that it is of the nature of glucal, an aldehyde derivative of glucose, $C_6H_{10}O_4$. Glueal is an unstable non-toxic substance. (Unpublished observations.) When a pentose is present it is d-ribose.

That the substances thus obtained constitute all that there are in the nucleic acid molecule is made probable by the recent work of Steudel and Levene. Steudel by means of his nitric acid method of hydrolysis obtains nearly a quantitative yield of the purine bases. The phosphoric acid is easy to determine, but the determinations of the carbohydrate and the pyrimidines are still far from being quantitative. Steudel gives the following result of an attempt at a quantitative analysis. It is assumed that the molecule contains four phosphoric-acid groups; two purines; two pyrimidines; and four carbohydrate nuclei. He found 28.95 per cent. of the total nitrogen as guanine nitrogen; 38.42 per cent. as adenine nitrogen; 11.47 per cent. as cytosine; and 13.11 per cent. as thymine nitrogen, making a total of 92 per cent. of the whole nitrogen. As the methods are not exactly quantitative, it is clear that these four bases are probably the only ones present. The amounts of the bases isolated and computed were as follows:

	Computed	Found
Guanine	10.72	9.01
Adenine	9.58	10.68
Cystosine	7.86	4.26
Thymine	8.93	8.33

As some of the cytosine is unavoidably converted into uracil by the hydrolysis, the agreement must be considered as very satisfactory.

To determine the carbohydrate he weighed the levulinic acid formed and computed from the figures of Conrad and Gultzeit how much carbohydrate this amount of levulinic acid represented. His complete analyses of thymus and sperm nucleic acids were as follows:

	Computed for $C_{43}H_{57}N_{15}O_{30}P_4$	Found
Guanine	10.88	8.7
Adenine	9.73	10.5
Thymine	9.08	8.2
Cytosine	9.15	4.2
Phosphoric acid	20.46	20.31
Hexose	51.	57.
	111.20	108.9

The decomposition may be represented as follows:

$$\underset{\text{Nucleic acid.}}{C_{43}H_{57}N_{15}P_4O_{30}} + 8H_2O = \underset{\text{Guanine.}}{C_5H_5N_5O} + \underset{\text{Adenine.}}{C_5H_5N_5} + \underset{\text{Thymine.}}{C_5H_6N_2O_2} + \underset{\text{Cytosine.}}{C_4H_5N_3O} + \underset{\text{Hexose.}}{4C_6H_{12}O_6} + \underset{\text{Metaphosphoric acid.}}{4HPO_3}$$

The agreement is as good as could be expected. Nucleic acid consists, then, of these few building stones and 50 per cent. of the molecule is carbohydrate. The nature of this carbohydrate of the animal nucleic acids has not yet been determined beyond the fact that it is a hexose. It is possibly not always the same hexose.

Structure of the molecule.—We may now take up the problem of the way in which these smaller molecules are united to build up the big. Kossel very early suggested that nucleic acid was composed of a polymerized metaphosphoric acid to which the bases and carbohydrates were attached and structural formulas based on his findings were proposed by Bang and Osborne and Harris. In these suggestions the backbone of the molecule consisted of four molecules of phosphoric acid to which the bases and carbohydrates were attached. The real structure of the molecule has been elucidated largely by the work of Neuberg on the simple nucleic acid, inosinic acid, and of Bang and Levene and Jacobs on guanylic acid. Since these last two acids have contributed to our understanding of the structure of the nucleic acid molecule, we may stop and consider them here, although they are possibly not constituents of the nuclei.

Guanylic acid.—This is an acid belonging to the general group of nucleic acids, but being less complex than those found in the cell nuclei. It was isolated by Bang from the ox pancreas and was found by him to contain no other base than guanine, whence its name, phosphoric acid and a pentose. Bang thought it contained glycerol, but this was incorrect. This acid is found in the ox pancreas in addition to the real nucleic acid which we have been considering. It has been obtained also from the liver and spleen and from yeast. It is best obtained from Hammarsten's nucleoproteid in the following way:

If the fresh pancreas of the ox is hashed and boiled with water, Hammarsten found that a nucleoproteid went into solution in the water, from which it could be obtained by slightly acidifying with acetic acid, the nucleoproteid being precipitated. The gland residue from which this nucleoproteid has been extracted will yield the true nucleic acid of the type of those already considered, if treated by Neumann's method. The guanylic acid is separated from the nucleoproteid precipitate by redissolving in sodium hydrate, reacidifying, precipitating and filtering. The filtrate is poured into alcohol. The guanylic acid precipitates as a powder. This guanylic acid Steudel showed contained no glycerol, no levulinic acid, but only guanine, phosphoric acid and a pentose. Its constitution was worked out by Levene, who succeeded in isolating from it both a compound of guanine and pentose, a pentoside, or a nucleoside as he called it, guanosine; and on the other hand a phosphoric acid pentose compound. These facts showed that the pentose was united both to the guanine and to the phosphoric acid and that its composition was as follows:

```
                                                  O = C—NH
       O      H   H   H   H   H                       |   |
       ||     |   |   |   |   |          ,NH — C   C = NH
HO — P — O — C — C — C — C — C — C<          ||   |
       |      |   |   |   |   |          \\N  — C—NH
       |      |   |   OH  OH  |
       OH     H   O___________|
```

Guanylic acid.

While its molecular weight has not been directly determined, the compounds it forms leave little doubt that it is but a single molecule, a mononucleotide as Levene and Jacobs call it. The character of the pentose was long in doubt, but the authors just mentioned have shown that it is d-ribose, a levo-rotatory, aldose pentose of the arabinose type not previously known to occur in animals. The point of union of the sugar with the guanine is not yet certain, but it is either in purine 7 or 8 as is figured, and probably the latter, although Burian thought the union was in number 7. *Guanosine* is, therefore, a pentoside. It may be mentioned that the position of the attachment of the phosphoric acid in the sugar is also uncertain. Guanosine, $C_{10}H_{13}N_5O_5$, does not reduce Fehling's solution until it is decomposed. $[\alpha]_D^{20} = -60.52$.

Guanylic acid has also been separated from ox liver and Jones succeeded in getting it from yeast nucleic acid by a quick digestion by an enzyme, tetra-nucleotidase, found in the pig pancreas. Guanylic acid is dextro-rotatory.

Inosinic acid.—This is an acid similar to guanylic acid, but it is composed of a molecule of hypoxanthine, a pentose and phosphoric acid. It was isolated from Liebig's beef extract and is supposed to occur in muscle. Whether it does pre-exist in the muscle is probable, but not certain. It was the study of this acid by Neuberg which really gave the key to the structure of the nucleic acids. Neuberg thought it had the formula

```
                 O
                 ‖
Hypoxanthine—O—P—O—pentose
                 |
                 OH
```

But Levene and Jacobs isolated from it a compound called *inosine*, a union of pentose and hypoxanthine, showing that inosinic acid must have a formula similar to guanylic acid. It is not, however, identical in its structure. From yeast another pentoside was isolated, an adenine pentoside called *adenosine*. *Guanosine* had already been isolated by Schulze from plants and called by him *vernin*.

Nucleic acid.—Levene and Jacobs have also isolated other fragments of the molecule of yeast and thymo-nucleic acid. They conclude from their work and that of Steudel that the structure of thymus nucleic acid is probably

```
            ┌─────────────────┐
                                        OC—NH
   H    H   O   H    H        |          |    |
   |    |   |   |    |        |      NH—C    C—NH₂
H—C —  C—C—C—C—C—C<          ‖    ‖
   |    |   |   |    |        |      N —C—N
   |   OH  H  OH  OH  H
   O
   |
HO—P=O
   |          O  OH
   |          ‖ /
   O     O—P—OH
   |     |    |  ┌──────────┐
   |     |    |   H    H   O
H—C  —  C—C—C—C—C  — Thymine.
   |     |    |   |    |   |
   H     H    H  OH   O   H
                       |
   H     H    H  OH   |   H
   |     |    |   |    |   |
H—C  —  C—C—C—C—C  — Cytosine.
   |     |    |   |    |   |
   |     |    |   H    H   O
   O    O—POH ──────────────┘
   |      / \
   |     O   OH
HO—P=O
   |                              H₂N—C=N
   O   OH   H   OH  OH  H            |    |
   |    |    |   |    |   |     NH——C    C—H
H—C—C—C—C—C—C—C<           ‖    ‖
   |    |    |   |    |   |     N——C—N
   H    H    O   H    H   |
                 └─────────┘
```

Thymo-nucleic acid (Levene and Jacobs).

This would correspond with Steudel's formula, $C_{43}H_{57}N_{15}O_{30}P_4$. Such a nucleic acid would be a tetra-nucleotide.

While the facts seem to bear out this formula, in its main features at any rate, it cannot be said that it is as yet conclusively established. The exact point of attachment of the phosphoric acid to the sugar is still obscure. The great difficulty of hydrolyzing the di-nucleotide, thymic acid, seemed to indicate that the union between the pyrimidine nucleo. tides was not through phosphoric acid, but was an ether-like union. It will be noticed that the molecule as written in the Levene-Jacob's formula is hexabasic.

Another possible formula would be the following:

```
  HO                 O----------------          NH2C = N
   |         H   H   |   H   H       |             |   |
   |         |   |   |   |   |       |          NH-C   CH
O = P--O--  C - C - C - C - C - C - C <          |    |
   |         |   |   |   |   |   |             N--C--N
   |         H  OH   H  OH  OH   H
   O                 ----------------
   |         H   H   O   H   H       |
   |         |   |   |   |   |       |
O = P--O--  C - C - C - C - C - C -- Thymine
   |         |   |   |   |   |   |
   |         H  OH   H  OH  OH   H
   O                 ----------------
   |         H   H   O   H   H       |
   |         |   |   |   |   |       |
O = P--O--  C - C - C - C - C - C -- Cytosine
   |         |   |   |   |   |   |
   |         H  OH   H  OH  OH   H
   O                 ----------------         O = C--NH
   |         H   H   O   H   H       |             |    |
   |         |   |   |   |   |       |          NH-C   CNH2
O = P--O--  C - C - C - C - C - C - C <          ||   ||
   |         |   |   |   |   |   |             N--C--N
  OH         H  OH   H  OH  OH   H
```

Nucleic acid. A possible formula.

Does nucleic acid exist outside the nucleus?—There are several very interesting questions as yet unsolved concerning the location in the cell of the nucleic acid. It seems probable, though there is nothing really known about it, that guanylic and inosinic acid may be in the cytoplasm of the cells in which they occur, though they may be in the nucleus. It is possible that they do not exist free in the cell, but are united with the true nucleic acid and are set free by endocellular enzymes. Nothing is really known about their function or location. Their staining reaction will probably resemble that of the real nucleic acids. Guanylic acid gelatinizes much as the nucleic acids, and it was this property that caused Bang to maintain that it must be more complex than a single nucleotide. Inosinic acid is probably the source of the hypoxanthine of muscle and it is very interesting that this substance is increased during muscular contraction.

There can be little doubt that the true nucleic acids, that is the poly-nucleotides, like thymus nucleic acid, are found only in the nucleus. This was first indicated by the work of Kossel, who determined the amount of purine bases obtainable from different tissues. The amount ran proportional to the amount of nuclear material present; it was high in embryonic tissue; in the thymus; and low in muscle. It is shown also by the fact that no nucleic acid is found in some unfertilized eggs where the nuclei are very small in proportion to the cytoplasm, and none in the mammalian red blood cells which lack nuclei. On the other hand, nucleic acid is found wherever nuclei occur, as in the red corpuscles of bird's blood which are nucleated. It has never been shown positively to be a constituent of the cytoplasm, but it is certain that it is found in the nucleus. It is probable, therefore, that it is confined to the nucleus, but there are some facts which may be urged against this conclusion. For example, some believe that nucleic acid is found in the cytoplasm, because not all the cytoplasmic phosphoric acid in organic union is split off from its union by sodium hydrate. If the substance in the cytoplasm was a vitellin, or casein-like compound, it would presumably have been split off. Nucleic acid, unlike the phosphoproteins, does not split off its phosphoric acid when treated by alkali hydrates. And recently nucleic acid has been found in the sea-urchin's egg, where the nuclei are very small. The author got a substance with some of the properties of nucleic acid in some quantity from unfertilized eggs of the sea-urchin. It could not be positively identified, however, as the quantity was too small. In all these cases, then, it is still uncertain whether the substances described were really nucleins, and the probability is that they did not contain true nucleic acid. Further work, however, is necessary on this subject before a definite statement can be made that nucleic acid is never found in the cytoplasm. It is certain, however, that most of the phosphoric acid compounds in the cytoplasm are not nucleic acids.

Are all nucleic acids the same?—The question whether all animal nuclei contain the same, or different, nucleic acids cannot be answered definitely, since only two of the animal nucleic acids have been accurately examined, namely that of the sperm of herring and from the thymus gland of calves. These two acids appear to be identical. They contain the same bases in the same proportions and they have the same physical properties. Until the nature of the carbohydrate is discovered it is impossible to say whether they contain the same carbohydrate, but all indications are that these two nucleic acids are identical. Since they come from such widely different sources, it would indicate that probably the same nucleic acid is found in totally different kinds of cells, a conclusion of the utmost importance in interpreting the probable rôle of nucleic acid in the cell. All other nucleic acids of animal origin, except guanylic and inosinic acids, have been found to yield the same splitting products when hydrolyzed, so that they must be closely similar to thymus nucleic acid, if they are not identical with it.

On the other hand, only two plant nucleic acids have been carefully examined. These are triticonucleic acid from wheat, and yeast nucleic acid. These are apparently identical, and they differ from the animal nucleic acids in having d-ribose, a pentose sugar, in the place of a hexose. They may also differ in other particulars. The composition of neither of these acids is exactly known, and particularly the molecular weight has not been determined. Steudel's analyses indicate that yeast nucleic acid may be a tri-nucleotide and not a tetra-nucleotide, as Levene thinks. No one has as yet isolated yeast nucleic acid which on analysis would yield figures for carbon, phosphorus and nitrogen comparable with a tetra-nucleotide. But this may be due to the fact that yeast contains a nucleotidase, and possibly if some of the yeast cells are dead when analyzed a partial digestion of the nucleic acid may have taken place. Only fresh, living, active yeast should be used for the preparation of this acid.

Another possibility which complicates the question of the identity of the nucleic acids is that in the nucleus we may have a polymer of a tetranucleotide, as Steudel has suggested for the sperm head. He found, namely, that the viscosity of the solution of the herring sperm heads in alkali was greater than an equivalent solution of protamine nucleate; and he inferred from this a different state of aggregation of the nucleic acid outside and inside the cell. It is of course possible that some other factor than that suggested was responsible for the observed result.

The tentative conclusion may with all reserve be drawn from the foregoing facts, that the nucleic acids of different nuclei of animal tissues are certainly closely similar if they are not identical; but that they differ in their carbohydrate radicles from such plant nucleic acids as have been examined. It is possible that the hexose component will not be found to be the same everywhere. Their similarity would clearly indicate that nucleic acids have the same function in all cells. If they intervene actively in cell metabolism, it must be in connection with some fundamental cell property such as growth, irritability or respiration which is common to all cells. It may be, however, that it has only the function of a supporting structure, or aids in keeping the physical viscosity of the nucleus what it has to be. In favor of this view it may be mentioned that it is a fairly stable substance, otherwise it could not accumulate, and its most probable function would appear to the writer to be that it serves as a colloidal, gelatinous substratum in the nature of an organic skeleton to which the specifically active, more labile, albuminous constituents, possibly of a catalytic nature, may be attached. Forming a firm union with the acid, these active substances may be thus confined to, or located in, the nucleus from which they may at times get free. But nothing positive as to its function can be stated without further investigation.

It is of interest to recall, in view of the foregoing statement, that all so-called nuclear stains of a basic nature, with the exception of the mordanted stains such as iron hematoxylin, combine with the nucleic acid. In thus following the chromatin and chromosomes by means of these stains, cytologists, if the view stated above of the significance of nucleic acid is correct, may be following the inert skeletal material of the nucleus, while the active albuminous material is entirely neglected for the reason that it does not gel and does not stain with basic dyes. All theories of inheritance based on the behavior of the nucleic acid of the nucleus, that is the behavior and number of the chromosomes, must be accepted only with the greatest reserve, until the function of this substance may be shown to be something more than a skeletal substance. We have as yet no chemical evidence that the different chromosomes have different nucleic acids in them, but such evidence as we have is contrary to this view. If the chromosomes do differ chemically, as perhaps their individual and peculiar shapes and sizes may indicate, it is more probable, as we shall shortly see, that they differ in their protein or basic rather than in their acid moieties.

THE BASIC CONSTITUENTS OF THE NUCLEUS.—Nucleic acid is either a hexa- or tetra-basic acid, probably the former; and it forms a series of salts. We have now to ask the question with what basic substances is nucleic acid united in the chromatin? Are the bases organic or inorganic?

It is probable that some inorganic bases, i.e., calcium, are present; it is certain that organic bases of a protein nature are always present. The only nuclei carefully examined in a clean form, free from cytoplasm, are the sperm heads, and possibly the nuclei of birds' corpuscles. These always yield some calcium phosphate when dissolved or ashed. It seems certain that calcium is generally present. MacCallum, from cytological, microchemical studies, has concluded that nuclei contain no potassium, since around the outside of the nucleus he generally obtains a deposit of potassium-cobalto nitrite by his method, but none in the nucleus. But to his conclusion it may be objected that the place where the precipitate forms is not necessarily indicative of the location of the soluble salt. There is, indeed, very little evidence of what inorganic salts or bases we have in the nucleus itself. This question must be left for further work. It appears, from some recent work, that iron, contrary to an earlier view, is not present in all nuclei.

The organic bases which occur in some chromatins are among the most interesting substances in the cell, whether considered from the physiological or chemical point of view. Our knowledge of these bases, the study of which gave Kossel the clew to the constitution of the proteins, we owe chiefly to Kossel and Miescher and pre-eminently to the former. These bases are protein in nature and consist either of basic proteins called protamines or histones, or of other more complex proteins.

The protamines.—If the sperm heads of the salmon, sturgeon, herring and other fishes are extracted with 10 per cent. sulphuric acid, or hydrochloric acid, there goes into solution about 19 per cent. of the dry, alcohol- and ether-extracted heads. The nucleic acid remains behind more or less altered and insoluble. Three extractions of the heads with 10 per cent. sulphuric acid for about half an hour at a time will take out practically all of the removable base. The substance which goes into solution as a sulphate is of a protein nature; when precipitated by alcohol as the sulphate it is a white, somewhat hygroscopic, amorphous powder, giving, in the case of the herring, salmon and sturgeon sperm, no Millon, or xanthoproteic, or tryptophane reaction, but a good biuret reaction. This substance was named protamine by its discoverer, Miescher, who obtained it from salmon sperm (Gr. *protos*, first, *amine*). The protamine from salmon is called salmin.

General properties. The protamines, although individually different, have the following properties in common: In the free state all are strong bases, alkaline to litmus, and not precipitated by ammonia. They give a splendid biuret test, but Millon, xanthoproteic or Adamkiewicz reactions are in many cases negative, but in some protamines positive. They are digestible by trypsin, but not by pepsin-hydrochloric acid; they are readily soluble in water, but not in alcohol, and their sulphates separate as an oil when the saturated aqueous solution is shaken with ether. They are not coagulated or changed by heating. They precipitate proteins by uniting with them in ammoniacal solution, and this is a very delicate test for them. In this respect they act like metallic bases. Unlike most proteins, they are precipitated from a neutral solution by neutral solutions of sodium picrate, ferrocyanide or tungstate, and they may even be precipitated in faintly alkaline solutions. The reason for this peculiarity has already been explained. They are such strong bases that their molecules are electro-positive even in faintly alkaline solutions. On analysis they consist of carbon, hydrogen, oxygen and nitrogen, but they contain no sulphur. The elementary analyses of some are as follows:

	C	H	N	O	Pt	Cl	
Clupein	47.93	7.59	31.68	12.78	—	—	Free base.
Salmin	22.96	4.32	14.83	6.7	24.73	26.56	Plat. chloride salt.
Sturin	24.32	4.49	14.20	8.47	23.10	25.42	

The formula for salmin is probably $C_{30}H_{57}N_{17}O_6$; that for sturin, $C_{36}H_{69}N_{19}O_7$. The molecular weight is not yet determined.

The protamines differ from all other proteins in the small number of different amino-acids they yield on hydrolysis and in the character of these acids. Kossel found that salmin, one of the simplest, yielded 87 per cent. of its molecule as arginine, and it was this discovery which suggested to him the constitution of the proteins. The composition of the hydrolytic cleavage products of numerous protamines is given on page 128.

Does the sperm chromatin consist exclusively of protamine nucleinate?—The chromatin of the sperm head is supposed to be the bearer of the hereditary qualities and zoölogists have pictured it as composed of individual units, biophores or determinants, each of which represents some specific unit-character of the adult. If this hypothesis were true, we should expect the sperm chromatin to be extremely complex; more complex indeed than any chromatin in the body, since it is supposed to represent them all. As a matter of fact, chemical examination shows this chromatin in the fish sperm to be the simplest found anywhere. The heads of the herring sperm do not contain any tyrosine; they give no Millon, xanthoproteic or tryptophane test. They contain no coagulable protein. They have the following composition after extraction with alcohol and ether:

		Average
C	40.99—41.48	41.20
H	5.62— 5.83	5.75
N	20.78—21.44	21.06
P	5.87— 6.33	6.07

Steudel has recently confirmed these figures. Accepting his formula for the composition of nucleic acid, $C_{43}H_{57}N_{15}P_4O_{30}$, and Kossel's formula for clupein, or salmin, $C_{30}H_{57}N_{17}O_6$, there would be required for protamine nucleate:

	Computed for $C_{73}H_{112}N_{32}O_{36}P_4$	Found
C	40.97	41.24
H	5.33	5.27
N	20.95	21.09
P	5.80	6.02
O	26.95	26.37

This formula requires 64.8 per cent. nucleic acid and 35.2 per cent. protamine. He actually isolated 93 per cent. of the calculated amounts of each of these substances and the deficit was undoubtedly due to the fact that the methods are not entirely exact. There can be no doubt, therefore, that the chromatin of herring sperm when fully ripe consists of a neutral salt of protamine nucleate. Miescher found very similar relationships in the salmon sperm, the head consisting largely or wholly of salmin nucleate.

Nature of the union of protamine and nucleic acid. The ease with which the protamine may be separated from the nucleic acid by acids or alkalies indicates clearly that the two are in a salt-like union. Probably the union is between the free amino groups at the end of the chain of the arginine and the acid radicle of the nucleic acid (Steudel). By extracting first with alkali these free amino groups of the arginine of the salmin are decomposed, ammonia being set free and ornithine remaining. If, now, the compound is acidified a reunion of the nucleic acid and protamin does not take place. This is the probable basis of the Neumann method of preparing nucleic acid. But there can be equally little doubt that we often have other than salt unions between the protein and nucleic acid. It is impossible to extract all the protein from the nuclei of all cells by acid. The union is too firm.

Other basic constituents. Histone. In the sperm of the sea urchin, Arbacia, the author isolated by acid extraction a basic protein resembling histone in some particulars and protamine in others. About 11 per cent. by weight of the alcohol and ether extracted, dried whole sperm was extracted by acid. The arbacin sulphate contained 15.91 per cent. of nitrogen, whereas protamine sulphate contains about 25.13 per cent. In this experiment the sperm heads were not separated from the tails. The substance was not a true histone, for it did not precipitate with ammonia, except very incompletely. Nucleic acid was also isolated. Arbacin was strongly basic and gave the Millon test. Only a small proportion of the protein could be extracted by acid from the sperm, indicating that not all of it was in a salt union, or else that the tails made a very considerable proportion of the whole.

The chromatin of both thymus gland and bird's blood corpuscles contains a basic, simple protein, *histone,* in a salt union with nucleic acid. This fact was also discovered by Kossel. These nuclei have been recently obtained and studied by Ackermann.

The method of isolating the nuclei has already been given (page 162). The dried nuclei after alcohol and ether extraction contained 3.93 per cent. P; 17.20 per cent. N. If Steudel's formula for nucleic acid is used in place of the formula employed by Ackermann, it is computed from the phosphorus that the nuclei contain 43.96 per cent. nucleic acid and 56.04 per cent. of histone, if they contain only histone nucleate. From Steudel's formula nucleic acid contains 15.18 per cent. of N. Hence in 100 grams of the nuclei containing 17.2 grams of nitrogen, 6.67 grams are in the nucleic acid and 10.53 grams in the histone. Since histone contains 18.3 per cent. N, the nuclei must contain 57.5 per cent. of histone. Both nitrogen and phosphorus indicate, therefore, that the nuclei contain 43-44 per cent. of nucleic acid and 56-57 per cent. of histone. Ackermann actually extracted by hydrochloric acid (1 per cent.) 63.9 per cent.

(53.9?) of histone, leaving 46.1 per cent. insoluble nucleic acid instead of about 44 per cent. Some purine bases undoubtedly went into solution and the residue contained only 7.79-7.99 per cent. of P and 15 per cent. of N, so that some histone may have been left unextracted. Although these figures do not check exactly, the method not being quantitative, it is clear, nevertheless, that these nuclei consist chiefly, if not entirely, of histone nucleate, and contain no other protein substance in any quantity. If the molecular weight of nucleic acid is 1,387 and that of histone about 1,600, which is the simplest formula which can be ascribed to it, a molecule of chromatin might be simply histone nucleate containing one molecule of each substance.

It is greatly to be desired that studies similar to these should be made on other tissues so that we may have a more accurate knowledge of the composition of the chromatin of as many cells as possible. Only when this is done will physiological chemistry be able to contribute to the vexed and vexing question of chromosomal inheritance.

Concerning the nature of the simple protein united with nucleic acid in other nuclei than these few kinds, nothing is known. Basic proteins corresponding to histone and protamine have not been isolated from other cells than those mentioned.

Enzymes in the nucleus.—Many nuclei, and particularly the large germinal vesicles of starfish eggs when unripe (Asterias vulgaris, etc.) contain very little of the morphological substance called chromatin. The greater part of these nuclei consists of a liquid sap which contains protein matter, if we may conclude from the fine precipitate produced in it by fixing agents such as mercuric chloride. No one has yet obtained this nuclear sap for chemical analysis, but there is no question that its admixture with the extra-nuclear cytoplasm produces marked chemical changes in the latter and greatly stimulates cell respiration. Delage, Loeb and the author have particularly studied the changes so produced. If unripe or immature eggs in which the germinal vesicle is intact are placed in sea-water, some of the eggs rupture the nuclear membrane and the nuclear sap mixes with the cytoplasm. Some eggs do not rupture the nucleus spontaneously, but they may be made to do so artificially by shaking. Eggs in which the nuclear sap has penetrated the cell cytoplasm behave very differently from eggs in which the nuclear sap remains in the nucleus. If rupture of the membrane takes place, the eggs become very sensitive to oxygen and they will only live about 10-18 hours in oxygenated sea-water. At the end of that time the protoplasm becomes opaque and seems filled with a multitude of spherules, the protoplasm being disintegrated into these spherules. If, however, the nuclear sap does not penetrate the cell cytoplasm and the nuclear membrane remains intact, or if the eggs after the nucleus and cytoplasm are mixed are placed in an atmosphere of hydrogen, or if they are slightly poisoned by potassium cyanide which prevents oxidation, the eggs remain alive for several days. It is very clear from this experiment that when the nuclear wall is ruptured either naturally or by mechanical means, the eggs become very sensitive to oxygen and, if not protected by fertilization, they will rapidly die in the presence of oxygen. The most probable explanation of these facts is that substances are present in the nuclear sap which when mixed with the protoplasm cause the mixture to undergo auto-oxidation leading, if not checked, to death. A simple, though perhaps not a correct, way of stating these facts is that the nuclear sap contains oxidases, or substances which stimulate respiration.

The change in the cytoplasm produced by this admixture of nuclear sap is also made visible in other ways than by oxidative changes. Sometimes spermatozoa penetrate eggs which do not maturate and in which the nuclear wall remains intact. In that case no typical aster develops about the advancing sperm, but only the faintest radiations about the sperm nucleus. This may be the case even though the sperm is lying close to the germinal vesicle. If, however, it enters an egg which has lost the nuclear wall so that the nuclear sap can escape, the typical big asters develop at once about the sperm, provided the eggs have oxygen.

Similar facts have been recorded by Delage. If a piece of protoplasm cut from an egg in which the nucleus is intact be entered by a spermatozoon, no division figure is developed. If, however, a sperm enters a piece cut from an egg in which the nuclear membrane has been ruptured, then the large normal sperm aster develops. It is clear that the change in the cytoplasm produced by the nuclear admixture enables the sperm to produce its typical effects. Inasmuch as these astral figures are dependent for their existence upon a supply of oxygen and disappear if the eggs are placed in hydrogen, reappearing again when they are returned to oxygen, their behavior again indicates the important part the nuclear sap plays in respiration. Yatsu found that nucleus-free pieces of Cerebratulus eggs, if cut off from the eggs before maturation occurred, would not develop asters when treated by strong magnesium chloride solutions, whereas similar pieces cut after maturation would develop them.

A very similar phenomenon illustrating the importance of the nuclear sap is shown in the first segmentation of the egg of the sea urchins, Arbacia and Toxopneustes. Wilson and the author observed that a marked pause in the segmentation process occurs just before the segmentation. The nuclear wall of the big segmentation nucleus is at that time intact. The large segmentation asters fade out, except near the nucleus. Suddenly the nuclear wall breaks at the two poles close to the asters. It appears to be dissolved or digested away. By this means the nuclear sap and the asters may come into contact; and coincident with this, the great radiations of the asters burst out in full magnificence, their streamers, like a miniature aurora borealis, flung wide throughout the cell, and cell division is rapidly consummated. Just at this time, too, there is a sudden outburst of carbon dioxide and the cell becomes extremely sensitive to ether, cyanides, acids and other poisons, a fact clearly indicative that the protoplasm is in a very reactive and unstable condition.

All these facts indicate in no uncertain manner that substances are present in the nuclear sap which on entering the cytoplasm produce chemical changes there. Not only are respiratory changes stimulated many fold, but also digestion seems to be inaugurated. Autolytic enzymes also evidently become active, either because they are set free from the nucleus, or because the nuclear materials activate, directly or indirectly, the inactive enzymes of the cytoplasm. Many yolk granules are dissolved and the nucleolus also dissolves and disappears; the nuclear membrane suffers a like fate and the chromatin itself, which has been more voluminous and less avid for basic dyes, diminishes in bulk and increases its staining power as if a considerable amount of protein had been digested or separated from it. It is also well known that the unfertilized eggs of hens keep much better and do not undergo autolytic digestion as do the fertilized eggs. These phenomena speak for the presence in the nucleus of oxidases and digestive enzymes. Since during cell division these enzymes are set free and at the same time the chromatic elements are in many cases plainly losing substance, it is possible that these two facts should be correlated and the conclusion drawn that in the resting condition of the nucleus enzymes of various kinds stick to, or combine with, the nucleic acid and are thus accumulated, made resistant, more stable and rendered inert, and that during caryokinesis, and possibly at other times also, they are split off from the acid, become free in the sap, enter the cytoplasm and rejuvenate the cell by digesting its accumulated colloidal material. Possibly the guanylic acid may, as an extra nuclear material, combine with the trypsin of the pancreas to make the inactive trypsinogen. Possibly there are also within the nucleus some of the nucleases which digest nucleic acid itself.

These few remarks will serve to illustrate the attractiveness, the importance and the obscurity of the field of the composition and function of the cell nucleus. Possibly they may stimulate some to the investigation of a subject of which the importance is only commensurate with our ignorance. We may in this connection recall the fact that it has been suggested (Gautier) that immediately about the nucleus there takes place something of the nature of an anærobic fermentation of the food materials, by which CO_2 is produced and many active fragments are formed which later in the periphery of the cell are oxidized by the entering oxygen, or condense to other compounds.

The formation and destruction of nuclear material.—We may close this chapter on the composition of the nucleus with a brief review of what is known concerning the formation and destruction of nuclear material. From what substances does a cell make nucleic acid, or protamine or histone? And what are the substances formed from its disintegration?

Origin of the proteins and nucleic acid. We will consider first the origin of nucleic acid, since this is the simpler problem. The question is then this: From what substances and in what manner is nucleic acid formed in cells? There are certain aspects of this question which can be definitely answered. There is evidence that the source of the phosphoric acid is inorganic phosphates. It is known that phosphates are necessary ingredients of the foods of all animals and all plants. Indeed this acid has quite a peculiar position in the cell. It not only enters into the composition of many of the proteins and of all nucleic acids of which it appears to form the backbone as it were, but in the phospholipins it no doubt contributes powerfully to the production of vital phenomena. It plays an important part in the maintenance of the neutrality of the protoplasm and in the activity of many enzymes. The acid owes its fundamental importance in metabolism probably to its power of polymerizing in the form of metaphosphoric acid, HPO_3, and, in the second place, to its power of forming ester unions with carbohydrate and other substances. It has in this regard a power only second to that of boric acid. By this power it forms the basis of nucleic acid, for at the bottom of this acid is the ester of phosphoric acid with either a pentose or an unknown hexose. This same property of forming esters with carbohydrates is shown at its best in the case of inosite, which is found probably in all cells combined with several molecules of phosphoric acid as in phytin, which is the hexa-phosphoric acid ester. (See page 610.) This part of the molecule of nucleic acid offers no difficulty for an understanding of the method of its formation, although we are not yet certain of the exact steps in the process. The formation of the purine and pyrimidine bases, however, is a somewhat more difficult problem. It has recently been discussed by Johnson. Since the pyrimidines are the simpler bases, we will assume that the purines are formed from them.

There is no doubt that all cells, animal as well as plant, can make their purines without being fed purines. Whether they can all make pyrimidine also is not entirely certain, but there is no doubt that plants have this power and it is probable that animals have it also. In milk or in the yolk and white of egg there are neither purines nor pyrimidines in more than extremely small amounts, and yet the developing organism nourished by these foods makes both of these substances at a very rapid rate. Birds and reptiles, too, can certainly form purine, uric acid, from amino-acids of various kinds, so that there is no question but that they have the power of synthesizing these bases from the simplest compounds, and probably from carbohydrates and ammonia.

Hydrocyanic acid, HCN, is one of the most reactive of substances. It is found combined in a great many plants. Its great importance in the synthesis of living matter was clearly recognized by Gautier. Hydrocyanic acid dissolved in water and allowed to stand gives rise to many substances found in living matter. Urea, alanine, carbamic acid, cyanates and, according to Gautier, substances related to xanthines or really xanthines, although this is denied by Fischer, appear in it. It has been repeatedly suggested that this substance may have been a very important contributor to the formation of living matter in the first instance. Three molecules of hydrocyanic acid will condense to form the amino-malonic nitrile,

$$3HCN \longrightarrow \begin{array}{c} CN \\ | \\ H_2N—C—H \\ | \\ CN \end{array}$$

Amino-malonic nitrile.

This nitrile might condense directly with urea to form a pyrimidine; or it might be hydrolyzed first to form the amino-malonic acid,

$$H_2N.CH(CN)_2 + 4H_2O \longrightarrow NH_2.CH(COOH)_2 + 2NH_3$$

Amino-malonic acid.

The acid might then condense with urea or guanidine to give a pyrimidine. If the condensation is with the nitrile, a diamino pyrimidine is the result; if with the acid, uramil, an amino-oxy-pyrimidine results:

$$\begin{array}{l} NH_2 \\ | \\ CO \\ | \\ NH_2 \end{array} + \begin{array}{l} NC \\ | \\ HCNH_2 \\ | \\ NC \end{array} \longrightarrow \begin{array}{l} NH—CO \\ |\quad\quad | \\ CO \quad CNH_2 \\ |\quad\quad \| \\ NH—CNH_2 \end{array} + NH_3$$

Dioxy-diamino pyrimidine.

$$\begin{array}{l} NH_2 \\ | \\ CO \\ | \\ NH_2 \end{array} + \begin{array}{l} HOOC \\ \quad\quad | \\ \quad\quad CHNH_2 \\ \quad\quad | \\ HOOC \end{array} \longrightarrow \begin{array}{l} NH—CO \\ |\quad\quad | \\ CO \quad CHNH_2 \\ |\quad\quad | \\ NH—CO \end{array} + 2H_2O$$

Uramil.

The condensation of either of these bodies with another molecule of urea to form a purine is analogous to numerous syntheses in living matter, although we do not know just how they are produced. The reaction may be represented as follows:

$$\begin{array}{l} NH-CO \\ |\quad\;\; | \\ CO\;\; CHNH_2 \\ |\quad\;\; | \\ NH-CO \end{array} + \begin{array}{l} NH_2 \\ \\ NH_2 \end{array}\!\!>CO \longrightarrow \begin{array}{l} NH-CO \\ |\quad\;\; | \\ CO\;\; C-NH \\ |\quad\;\; \| \\ NH-C-NH \end{array}\!\!>CO + NH_4OH$$

Uric acid.

The synthesis could as readily go through alloxan which might be formed from glyoxalcarbonic acid formed from the carbohydrate decomposition. It has been shown that glucose when it decomposes in weakly alkaline solution forms some glyoxalcarbonic acid. With ammonia this will condense with formic aldehyde to make an imidazole as follows:

$$\begin{array}{l} COH \\ | \\ CO \\ | \\ COOH \end{array} + \begin{array}{l} NH_3 \\ \\ NH_3 \end{array} + H_2CO \longrightarrow \begin{array}{l} HC-NH \\ \;\,\| \\ \;\,C-N \\ \;\,| \\ \;\,COOH \end{array}\!\!>CH$$

Glyoxalcarbonic acid. Imidazolylcarbonic acid.

A similar condensation might occur with urea:

$$\begin{array}{l} COH \\ | \\ CO \\ | \\ COOH \end{array} + \begin{array}{l} NH_2 \\ | \\ CO \\ | \\ NH_2 \end{array} \longrightarrow \begin{array}{l} N=CH \\ |\quad\;\; | \\ OC\;\;\; CO \\ |\quad\;\; | \\ HN-CO \end{array} + H_2O$$

Trioxypyrimidine.

The trioxypyrimidine by oxidation could give alloxan which by condensation with urea might yield a purine.

Another possible source of pyrimidine would be by oxidation of arginine to guanidine propionic acid and the condensation of this body to an amino pyrimidine:

$$\begin{array}{l} NH_2\quad\;\; COOH \\ | \qquad\quad\;\; | \\ NHC \qquad CH_2 \\ | \qquad\quad\;\; | \\ NH- \;\;\; CH_2 \end{array} \longrightarrow \begin{array}{l} NH-CO \\ |\qquad\; | \\ NHC\;\;\; CH_2 \\ |\qquad\; | \\ NH-CH_2 \end{array} + H_2O$$

Guanidine propionic acid.

This formation would be analogous to the formation of creatinine from creatine, page 704.

While the exact course of the formation in the cell is thus obscure, there are no great difficulties in imagining how the condensation might occur in the presence of ammonia, or urea or hydrocyanic acid or formamide and the reactive decomposition products of the carbohydrates. Whatever may be the exact steps in the process, it may be regarded as probable that they, like the amino-acids, are formed by the condensation of ammonia with the reactive decomposition products of the carbohydrates. Essentially, therefore, speaking broadly, the proteins and the nucleins arise by the condensation of the decomposition products of the carbohydrates with ammonia. It may be added further that, in order that the proper decomposition products shall be formed from the carbohydrates, it is necessary that the reaction shall be guided or directed, and that this is probably accomplished by the presence in cells of accelerating agents, or enzymes, which hasten one reaction or another, the particular reaction differing in different cells, so that the proper decomposition products shall occur in the proper amounts.

Origin of the amino-acids. The amino-acids of the animal body are obtained chiefly as the products of the digestion of plant proteins, but the animal organism has certainly the power of making some of them from ketonic acids, like pyruvic acid and ammonia, a subject discussed on p. 818. To what extent animals have this power of making amino-acids from ketonic acids and ammonia, or in any other way, is still being investigated and no certain answer to the problem can be given at the present time. While it appears that animal protoplasm has in general the same chemical properties as plant, there is no doubt that this power of manufacture of amino-acids which is so noteworthy a property of plant life is reduced certainly to a very subordinate power in the animal, for it appears necessary to supply most animals with ready-made amino-acids. The plant amino-acids are almost certainly derived in the long run and in large measure from ammonia and carbohydrates. By the fermentation of glucose, or when glucose is decomposed by alkalies and presumably by the processes of plant metabolism, various ketonic aldehydes, such as pyruvic aldehyde, are produced. Pyruvic acid, CH_3—CO—COOH, is thus formed or glyoxylic acid, HCO—COOH. Ammonia, derived from the nitrates which are reduced in the plant protoplasm, condenses with these compounds to form imino compounds which by reduction yield amino-acids, thus

$$CH_3\text{—CO—COOH} + NH_3 \longrightarrow CH_3\text{—CNH—COOH} + H_2O$$
$$CH_3\text{—CNH—COOH} + H_2 \longrightarrow CH_3\text{—CHNH}_2\text{—COOH}$$

Alanine.

$$HCO\text{—COOH} + NH_3 \longrightarrow HCNH\text{—COOH} + H_2O$$
$$CHNH\text{—COOH} + H_2 \longrightarrow CH_2NH_2\text{—COOH}$$

Glycocoll.

By a similar reaction guanidine, one of the constituents of arginine and guanine, may arise from urea and ammonia:

$$O = C(NH_2)_2 + NH_3 \longrightarrow HNC(NH_2)_2 + H_2O$$

Urea. Guanidine.

The origin of proline from glutamic acid has already been indicated (p. 124). The exact method in which the other amino-acids arise in the plant is still uncertain, but it is probable that they are formed for the most part from the degradation products of the sugars uniting with ammonia. Light, or at least chlorophyll, is not necessary for this synthesis, since many of the bacteria and moulds which are free from chlorophyll can make many different amino-acids from a single source of ammonia such as asparagine and some carbohydrate. Imidazole groups may be formed by long contact of ammonia, glucose and oxygen, or an oxidizing agent, from glyoxal carbonic acid:

$$\begin{array}{l} COH \\ | \\ C=O \\ | \\ COOH \end{array} + \begin{array}{l} NH_3 \\ \\ NH_3 \end{array} + H_2CO \longrightarrow \begin{array}{l} HC-NH \\ \| \\ C-N \\ | \\ COOH \end{array} \!\!>\! CH + 3H_2O$$

REFERENCES. Proteins and Chemistry of Nucleus.

1. **General Works:** The Vegetable Proteins. *Osborne:* Monographs on Biochemistry. Longmans, Green and Co. 1909. London.
Chemical Constitution of Proteins. *Plimmer:* Part 1, 1912; Part 2, 1913. Monographs on Biochemistry. Edited by Plimmer and Hopkins. London.
General Characters of Proteins. *Schryver:* Monographs on Biochemistry, 1910.
Protamines and Histones. A. *Kossel:* Monographs on Biochemistry. Longmans, Green and Co. 1914.
Chemistry of Proteins. *Mann.*

2. **Nucleic acid.** *Composition. General.* Nucleic Acids. Their Chemical Properties and Physiological Behavior. W. Jones: Monographs on Biochemistry. Longmans, Green and Co. 1914.

3. " " *Composition. First work. Altmann:* Arch. f. Anat. u. Physiol. p. 526, 1889. Kossel: ibid., 1893, p. 157.

4. " " *Composition. Levene:* Journal of the Amer. Chem. Soc., 32, p. 231, 1910.

5. " " *Decomposition by nitric acid.* Steudel: Zeit. f. physiol. Chem., 49, p. 406, 1906; 48, p. 425, 1906; 53, p. 14, 1907.

6. " " *Formula.* Thymus nucleic acid. Levene and Jacobs: Jour. Biol. Chem., 12, p. 411, 1912.

7. " " *Yeast nucleic acid.* Levene and Jacobs: Ber d. d. chem. Gesell. 42, p. 2474, 1909; 43, p. 3150, 1910; 44, p. 1027, 1911. Kowalewsky: Zeit. f. physiol. Chem., 69, p. 240.

8. " " *Formation of guanylic acid from yeast nucleic acid.* Jones: Jour. Biol. Chem. 12, p. 31, 1912.

9. " " *Inosinic acid.* Levene and Jacobs: Ber. d. d. chem. Gesell. 44, p. 746, 1911.

10. " " *Guanylic acid.* Steudel: Zeit. f. physiol. Chem. 80, p. 40, 1910. Levene and Jacobs: Jour. Biol. Chem. 12, 1912.

11. " " *Pyrimidine nucleosides.* Levene and LaFarge: Ber. d. d. chem. Gesell. 45, p. 608. Johnson and Chernoff: Jour. Biol. Chem. 14, p. 307, 1913.

12. " " *Purine hexose compound.* Mandel and Dunham: Jour. Biol. Chem. 11, p. 85, 1912.

13. Nucleic acid. *Guanosine hexoside from thymus nucleic acid.* Levene and Jacobs: Jour. Biol. Chem. 12, p. 377, 1912.
14. " " *Carbohydrate group in molecule.*. Steudel: Zeit. f. physiol. Chem. 55, p. 407, 1908; 56, p. 212, 1908. Feulgen: Zeit. f. physiol. Chem. 92, p. 154, 1914.
15. " " *Pentose group.* Levene and Jacobs: Ber. d. d. chem. Gesell. 43, p. 3147, 1910.
16. " " *Wheat nucleic acid.* Levene and Jacobs: Ber. d. d. chem. Gesell. 43, p. 3164, 1910.
17. " " *Relation to glucosidal enzymes.* Tschermorutzky: Zeit. f. phys. Chem. 80, p. 298, 1912.
18. " " *Decomposition by enzymes.* See chapter on uric acid.
19. Nucleoproteid, of pig's liver. Scaffadi: Zeit. f. physiol. Chem., 58, 1908-9, p. 272.
20. Nucleus. Function. Gruber: Biologische u. expt. Untersuchungen an Amoeba proteeus. Archiv. f. Protistenkunde 25, pp. 316-374, 1912.
21. " Chemistry. Kossel: Ueber die chemische Beschaffenheit des Zellkerns. Münchener med. Wochenschrift, 2, 1911. Nobel prize address, 1910.
22. " Chemistry. Mathews: *Sperm nucleus.* Zeits. f. physiol. Chem. 23, 1897. *Birds' blood corpuscles.* Ackermann: Zeits. f. physiol. Chem. 43, 1904-5, p. 299. *Nuclei of thymus.* Abderhalden and Kashiwado: Ibid., 81, 1912, p. 285. *Sperm.* Steudel: Zeits. f. physiol. Chem. 83, 1913, p. 72.
23. " Calcium content. Horkammer: Biochem. Zeitschrift, 39, 1912, p. 271.
24. " Iron content. Masing: Zeits. f. physiol. Chem. 66, 1910, p. 262.
25. " Nucleic acid in eggs. Tscherroutzky: Zeits. f. physiol. Chem., 80, 1912, p. 194. Plimmer and Scott: Journal of Physiology, 38, p. 247. Levene and Mandel: Zeits. f. physiol. Chem. 49, p. 262. Masing: Ibid., 75, 1911, p. 135; 67, 1910, p. 161.
26. Nucleic acid. Purines. Origin of purines in plants. Johnson: Jour. Amer. Chem. Soc., 36, 1914, p. 337.
27. Pyrimidine compounds. Physiological action. Kleiner: Jour. Biol. Chem., 11, 1912, p. 443.
28. Relation of nucleic acid to stains. Feulgen: Zeits. f. physiol. Chem., 80, 1912, p. 73.
29. Nature of free amino groups in proteins. Van Slyke and Birchard: Jour. Biol. Chem., 16, p. 539, 1913. Kossel and Cameron: Zeits. f. physiol. Chem., 76, 1912, p. 457. Kossel and Gawrilow: Ibid., 81, 6. 274, 1912. Skraup: Annalen der Chemie, cccli, p. 379, 1906. Abderhalden and Van Slyke: Zeits. f. physiol. Chem., 74, p. 505, 1911.
30. Racemization of proteins. Dakin and Dudley: Casein. Jour. Biol. Chem., 15, p. 263, 1913; 13, p. 357, 1912. Kossel: Zeits. f. physiol. Chem., 72, p. 486, 1911; 78, p. 402, 1912; 84, p. 1, 1913.
31. Racemized protein not digestible by enzymes. Dakin and Dudley: Jour. Biol. Chem., 15, p. 271, 1913.
32. Racemized protein does not cause anaphylaxis. Ten Broeck: Jour. Biol. Chem., 17, p. 369, 1914.
33. Anaphylaxis by synthetic polypeptide. Abderhalden: Zeits. f. physiol. Chem., 81, 1912.

34. **Conversion of amino-acids into ketonic aldehydes.** Dakin: Jour. Chem. Soc., 1913.

35. **Protamin.** Kossel: Zeits. f. physiol. Chem., 69, 1910, p. 138.

36. **Amines from amino-acids.** Bickel and Pawlow: p-oxyphenyl ethyl amine. Biochem. Zeitschrift 47, 1912.

37. **Aporrhegmas.** Ackermann and Kutscher: Zeits. f. physiol. Chem., 69, p. 265, 1910.

38. **Agmatine.** Kossel: Zeits. f. physiol. Chem., 68, p. 170, 1910.

39. **p-Oxyphenyl amino butyric acid.** Goldschmidt: Monatshefte f. Chemie., 33, p. 1379, 1912.

40. **Dioxytyrosin.** Guggenheim: Zeits. f. physiol. Chem., 88, 1913, p. 276.

41. **Quantitative determination of hexone bases.** Weiss: Zeits. f. physiol. Chem., 52, p. 109, 1908. Van Slyke: Jour. Biol. Chem., 10, 1911, p. 15.

42. **Molecular weight of hemoglobin.** Hüfner and Gansser: Archiv. f. Physiol., 1907, p. 209.

43. **Heat coagulation of oxyhemoglobin.** Chick and Martin: Journal of Physiology, 40, p. 404, 1910.

44. **Casein. Isoelectric point.** Michaelis: Biochem. Zeitschrift 47, p. 260, 1912.

45. **Gelatin. Isoelectric point.** Michaelis and Grineff: Biochem. Zeits. 41, 1912, p. 373.

46. **Separation of gelatin from other substances.** Berrar: Biochem. Zeits. 47, 1912, p. 189

47. **Precipitation by half saturation with $(NH_4)_2SO_4$.** Wiener: Zeits. f. physiol. Chem., 74, 1911, p. 29

48. **Carbamino reaction.** Siegfried: Zeit. f. physiol. Chem., 44, p. 85, 1905; 46, pp. 401-414, 1905; 54, pp. 423, 437, 1908; 58, p. 84, 1908. Ergebnisse der Physiologie, 9, pp. 334-350, 1910.

49. **Amides and imides of amino-acids.** Bergel and Boll: Zeit. f. physiol. Chem., 76, p. 464, 1912. (Earlier papers cited here.)

50. **Biuret reaction.** Schiff: Berichte d. d. chem. Gesell., 29, p. 298, 1896.

51. **Anhydrides of amino-acids.** Grimaux: Sur des colloides azotés. Bull. soc. chim. 1882, (2), 38, p. 64. Schiff: Ueber Polyaspartsauren. Annalen der Chem., 310, p. 301, 1899. Schützenberger: Essai sur la synthèse des materiès proteiques. C. Rend. 112, p. 198, 1891.

52. **Polypeptides by partial hydrolysis of proteins.** Abderhalden: Zeits. physiol. Chem., 65, p. 417, 1910. See also various articles by Siegfried on "Kyrins" in the Zeits. f. physiol. Chem.

53. **Physical Chemistry of Proteins.** Ueber die Verbindungen der Proteine mit anorganischen Substanzen und ihre Bedeutung für die Lebensvorgänge. Robertson: Ergebnisse der Physiol., Asher and Spiro, 10, 1910, pp. 216-361.

54. **Diazo Reaction.** Pauly: Zur Kenntnis der Diazoreaktion des Eiweisses. Zeit. f. physiol. Chem., 94, p. 284, 1915.

CHAPTER V.

THE PHYSICAL CHEMISTRY OF PROTOPLASM.

Thus far we have considered the general composition of living matter and the chemical nature and origin of the carbohydrates, fats and proteins which make up the larger part of the organic basis of the cell, furnish energy for its vital activities and form its machinery. Knowledge of the chemical composition of these bodies does not enable us to understand how they can produce vital phenomena. For this it is necessary to understand not only their chemical composition, but also their physics or dynamics. In this chapter the physical chemistry of the cell will be considered, since physical chemistry is the science which deals with the explanations of chemical reactions.

Water.—The most abundant element of the cell is water. 70-93 per cent. of the protoplasm is water. To understand vital mechanics, knowledge must be had of the properties and possibilities of water. What is it doing in the cell? What does water contribute to the complex of life? What is water? It is a singular fact that the exact composition of this abundant substance, a *sine qua non* of life, is not yet known. That water decomposes into hydrogen and oxygen and that there are very nearly, if not exactly, two volumes of hydrogen liberated to one of oxygen is common knowledge. Also, it is certain that water is formed by the union of hydrogen and oxygen. The simplest formula which can be written for water is H_2O, H—O—H, and this is generally given as its formula, but there are many facts which show that water as it exists in the liquid and solid form and probably in the form of its vapor even at 365°, which is its critical temperature, has a more complex formula. Its high critical temperature, cohesion, refractive index and boiling point all show that the formula is more complex than H_2O. The molecule of water would be very light were the above formula true; it should boil at a low temperature, and have a low surface tension. Instead it has a very high surface tension, much higher than any of the hydrocarbons. Hence it is certain that the formula is more complex, at least at temperatures lower than 400° C. That the formula is some multiple of H_2O is shown also by the following circumstance: Eötvös found that if the surface tension is multiplied by the $^2/_3$ power of the volume of a gram mol. of a liquid the result, which is the surface energy of a gram mol., was equal, for all normal non-associating substances, to 2.27 (T_c-T) ergs, T_c being the critical temperature and T the absolute temperature at which the surface tension was measured. For all substances which associated, that is substances in which polymerization occurred, the product was less than 2.27 (T_c-T). Now water was found to have a surface tension energy which was less than half 2.27 (T_c-T) and the coefficient instead of being 2.27 fell lower and lower as the temperature was lower. Since all liquids in which the molecules do not remain the same but coalesce to form larger molecules as the temperature falls behave in this way, it is clear that water is also more complex than H_2O at temperatures below the critical and that the degree of complexity increases as the temperature falls. Ramsay and Shields computed from the surface tension that the formula at the boiling point must be about $(H_2O)_3$, and in ice about $(H_2O)_4$. Eötvös had also come to this result earlier. Determination of the freezing point of solutions of water in other solvents leads to the formula $(H_2O)_2$. Water is indeed one of the most associated liquids known. The molecular weight and the valence of the molecule at the critical temperature can also be determined from the cohesion, and this determination shows that the molecule at the critical temperature is at least $(H_2O)_2$. From some of these and other facts, Armstrong has concluded that the molecule of water in the liquid form is probably $(H_2O)_4$; and that by the condensation of the simple molecule H_2O, which he has named hydrone, into a ring or chain compound like the polymethylenes water is formed. It is probable that not all the molecules are thus associated, but that some dissociation takes place so that some free hydrone probably exists even in liquid water. The following kinds of molecules, then, probably exist in liquid water at 20-40° C.:

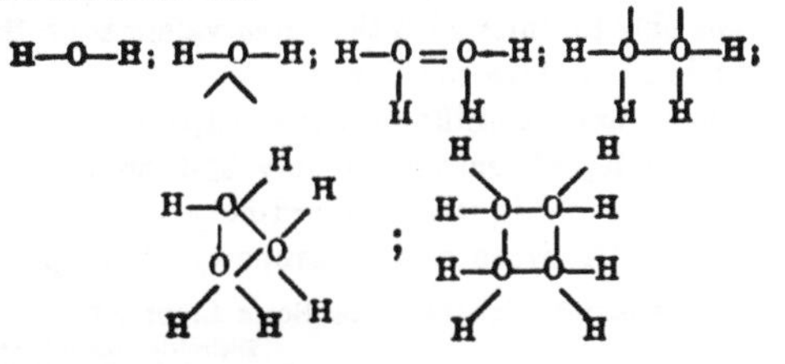

H—O—H; H—O—H; H—O=O—H; H—O—O—H;

```
                    H       H        H
        H—O<  H           \       /
          |  \/       ;    H—O—O—H
          O /O             H—O—O—H
         /   \  \            /    \
        H    H   H          H      H
```

The cause of this great association of water is probably the extra valences of the oxygen. Oxygen may be tetravalent here. Now, hydrogen differs from all other elements thus far studied in the fact that its valence is almost or entirely fixed and unchangeable; it has in it almost none of those reserve, or extra, valences, which appear in all the other elements. Chlorine, for example, may be univalent, trivalent, pentavalent or heptavalent. The result is that when hydrogen is united with a single other atom the extra valences which may occur on the other atom cannot be satisfied by union with those of the hydrogen; there is, hence, nothing else for them to unite with than the other similar valences on another molecule, thus producing molecular unions and association. Oxygen is certainly at times quadrivalent and hence the oxygen atom of hydrone may have, in addition to the valences uniting it with the hydrogen, two extra valences.

A physical property of water of very great biological importance, which is probably correlated with this association, is the high specific heat of water. It takes more heat to raise the temperature of a gram of water one degree than is required to raise the temperature of a gram of any other substance, either solid or liquid, one degree. This high specific heat of water is due in part to the fact that there are in a gram of water a large number of molecules, but chiefly to the fact that the dissociation of the water into hydrone consumes heat and the association accordingly liberates heat. At any rate, whatever may be its cause, this high specific heat is of value to the cell, since when heat is liberated in the course of the vital reactions the temperature of the cell does not rise very greatly; the water acts as if it were a buffer, taking up the heat liberated and giving it off gradually. Thus this property of water is of importance in preventing violent temperature changes which might lead to uncontrollable decompositions in the cell.

Another very remarkable property of water is its power of solution. No other solvent surpasses water. All kinds of substances dissolve in it: salts, carbohydrates, proteins and even fat solvents to some extent. Its power of solution, also, contributes much to the possibilities of life. This power of solution has not yet been explained, but it is probable that it, also, is correlated with, or due to, the extra valences on the oxygen atoms, which are perhaps able to unite with the extra valences on the dissolving molecules, and thus to produce solution.

Water has also a higher specific inductive capacity, or dielectric constant, than any other liquid, except possibly hydrogen dioxide. It is a good insulator. It does not in itself, at ordinary temperatures, conduct the current readily. In virtue of this property it happens that when

DIELECTRIC CONSTANTS OF SOME LIQUIDS.*

	Dielectric constant or specific inductive capacity
Water	77.0
Formic acid	63.0
Methyl alcohol	33.7
Ethyl alcohol	25.9
Propyl alcohol	22.0
Ammonia, liquid	22.0
Amyl alcohol	16.0
Chloroform	5.0
Ether	4.4
Carbon bisulphide	2.6
Benzene	2.3

* Jones: *Elements of Physical Chemistry*. Macmillan. 1902, p. 146.

electrical disturbances occur in a cell they are not instantly compensated, so that oppositely charged particles may coexist in water. It is probably because of this property that water forms such a good ionizing medium. At any rate, this property may account for the undoubted fact, whatever explanation we may choose to give of that fact, that substances dissolved in water interact with greater ease and speed than when dissolved in any other medium. *It has the property then,* so important for the cell, *of accelerating all kinds of chemical reactions.* Thus hydrogen and oxygen will not unite, except at very high temperatures, unless some water is present; hydrochloric acid and sodium hydrate react vigorously in the presence of water, but not when they are quite dry; chlorine and hydrogen do not form hydrochloric acid, except at very high temperatures, unless water be present; and everyone knows that the rusting of iron does not occur unless water is there too. Water has, then, this fundamental property of making reactions go on between bodies dissolved in it or wet by it. This property is believed by many to be correlated with its ionizing powers, and with the fact that its solutions conduct electrical current more than those of any other solvents. And this property brings us to the consideration of the salt solution in protoplasm.

Salts.—All protoplasm contains a solution of salts and these salts are of the nature of those of the sea. What, then, is a salt solution? How can that in protoplasm be assisting in the production of vital phenomena? Just as it is not yet known with certainty what the composition of liquid water is, so it is not known what is the exact state of affairs in a salt solution. No fact shows more clearly the limitations of chemical and physical knowledge at the present time than that one cannot say positively just what is a solution of common table salt in water. It is known, however, that salt solutions have certain properties and these may be dealt with even in the absence of their explanation. One of these properties of most fundamental importance is that aqueous solutions of the common salts conduct the electrical current. This fact was studied by that inspiring British physicist, Michael Faraday. He found that if a current flows through a solution of sodium chloride, for example, the sodium moved down with the current to the cathode, or negative electrode, and the chlorine moved up against the current to the positive electrode, the anode. Since the metal part of the salt moved down with the current, he called such wandering metals *cations,* from the Greek *kata,* down, and *ion,* going; and the negative, or metalloid part of the molecule, was called an *anion* (Gr. *ana,* up). Now it is clear that if the sodium moves down with the current it must be positively charged, and the chlorine moving up must be negatively charged, since only particles with charges on them move in an electrical field. Faraday did not know where the sodium got its charge. He thought that these ions did not pre-exist in the solution, but that the action of the current separated the neutral sodium-chloride molecule into a positive and negative particle. On the other hand, it was later suggested by Clausius that the ions did pre-exist, since no energy seemed to be consumed in the separation. This view of Clausius was put on a much firmer foundation and introduced as a powerful and fruitful theory into chemistry by the Swedish physicist, Arrhenius, in the year 1881. The basis of this theory of Arrhenius of the pre-existence of the ions, the so-called ionic theory, was that the osmotic pressures of solutions of electrolytes was higher than the osmotic pressure of equally concentrated solutions of non-electrolytes. The osmotic pressure and the vapor pressure are functions of the number of dissolved molecules in a given volume. It was found that a molar solution of sodium chloride depressed the freezing point, or raised the boiling point, of water more than a molar solution of sugar. Arrhenius brought this fact into relation with the anomalous pressures of some gases. It is found, for example, in heating nitrogen tetroxide, N_2O_4, that the product of the pressure by the volume increases more than it should, according to Boyle's law, $PV=RT$, and the explanation of this is that some of the N_2O_4 dissociates into two molecules of NO_2. Arrhenius suggested that the greater osmotic pressure and lower vapor pressure of electrolyte solutions, as compared with equally concentrated solutions of non-electrolytes, was due to the fact that the salt dissociated, also, like vapors of chlorine, bromine or iodine, and that the pieces into which it dissociated were the electrically charged ions of Faraday and Clausius. This theory, it will be noticed, explained at once the anomalous conductivity, the low freezing and high boiling points and the higher osmotic pressure of salt solutions. The ionic theory thus introduced has proved to be one of the most fruitful theories of chemistry. It has explained more facts, which without it were quite unexplainable, than probably any other chemical hypothesis except the atomic theory; and while some are disposed to criticise it and there are some facts which are, at first glance, difficult to explain by the theory, there can be no question of the enormous usefulness of the theory whether in its present form it is exactly true or not.

We may perhaps pause for a moment to consider a few of the more important evidences of the truth of this fundamental theory so illuminating for physiology. It enables us to understand the avidity of acids and bases. There was no explanation of the variation in the strength of acids and bases before this theory. It was known that hydrochloric acid was much more powerful and active than acetic or lactic acid. The ionic theory explained this at once. Acids, on the ionic theory, are bodies which dissociate in solution so as to form hydrogen ions. This dissociation may be represented as follows:

$$HCl \rightleftharpoons \overset{+}{H} + \overset{-}{Cl} \qquad CH_3.COOH \rightleftharpoons \overset{+}{H} + \overset{-}{O}.CO.CH_3$$

$$HNO_3 \rightleftharpoons \overset{+}{H} + \overset{-}{NO_3} \qquad C_6H_5OH \rightleftharpoons \overset{+}{H} + \overset{-}{OC_6H_5}$$

$$H_2SO_4 \rightleftharpoons \overset{+}{H} + \overset{-}{HSO_4} \qquad HCN \rightleftharpoons \overset{+}{H} + \overset{-}{CN}$$

All acids, then, have hydrogen ions in their solutions; their acidity is due to this; and their activity is proportional to the number of such ions there are in unit volume. This conclusion may be tested by comparing the conductivities of acids with their chemical or physiological activity. The amount of current which can be ferried by the ions between the electrodes in a solution in unit time will evidently be a function of the number of ions and their speed. It is found that a solution of hydrochloric acid will carry in a given time much more electricity across than a solution of acetic acid of the same concentration. There is no reason to believe that the speed of the hydrogen ion differs in the two cases; and while the acetic ion moves at a slower pace than the chlorine ion, its velocity has been determined and it is found that the difference is not sufficient to account for the difference in conductivity. There seems to be but the single possibility that the number of hydrogen ions is greater in the solution of hydrochloric acid than in that of acetic; hence, if the strength of the acid is proportional to the number of hydrogen ions, hydrochloric acid should be much stronger than acetic and in the same proportion as is determined by their conductivities. This was found to be the case. All acids split cane sugar into glucose and levulose; invert it, in other words. The speed with which they do this is different in different acids. It is a function of the number of hydrogen ions which are in the solution, so that if the speed of hydrolysis is measured the relative number of hydrogen ions in different acids of the same concentration can be determined and they should be approximately in the same proportion as the figures for the conductivities and other powers of the acids. This is found to be the case, as is shown in the accompanying figures:

Acid	Inversion coefficient	Equivalent conductivity at 18° (0.1n except when otherwise noted)
Hydrobromic	1.114	360
Hydrochloric	1.00	351
Nitric	1.00	350
Trichloracetic	0.754	323 (n/32)
Sulphuric	0.536	225
Oxalic	0.186	117
Phosphoric	0.0621	46.8
Monochloracetic	0.0484	72.4 (n/32)
Formic	0.0153	29.3 (n/32)
Acetic	0.0040	4.6

It is a general law that solutions freeze at a lower temperature than the pure solvent. It has been found by a further study of this phenomenon that the depression of the freezing point of dilute solutions is proportional to the concentration of the dissolved substance, that is to the number of molecules in a given volume. A solution as concentrated as

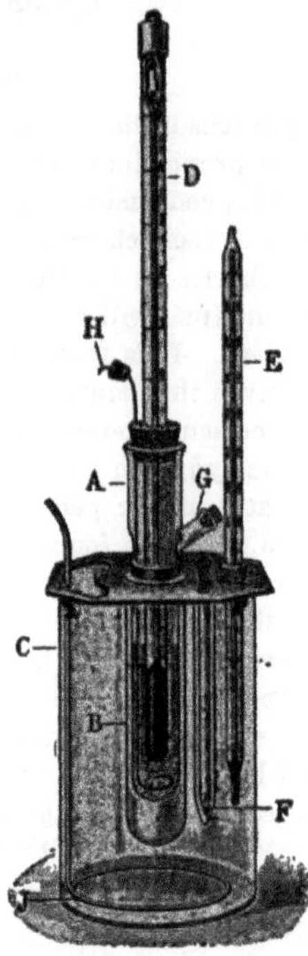

FIG. 16.—Beckmann freezing-point apparatus. *A*, tube containing liquid to be frozen; *D*, thermometer; *H*, stirrer; *G*, side tube for introducing ice crystals, etc.; *B*, large outer test tube; *C*, jar containing freezing mixture; *J*, stirrer for same.

a one-tenth gram mol. solution, that is a solution which contains 6.06×10^{22} solute molecules in a liter volume, depresses the freezing point of water 0.186°, so that a solution of glucose which contains 18.0 grams of glucose in one liter will freeze at —0.186° C. A solution half as concentrated will freeze at —.093° C. In this way by taking the freezing point of a solution by means of an accurate thermometer measuring to hundredths or thousandths of a degree, it is possible to tell how many molecules there are in a liter of any solution. It is found that a 0.205 M solution of calcium chloride does not depress the freezing point approximately .370°, as one would expect were there only $CaCl_2$ molecules present, but it depresses it 1.012°. The most probable interpretation of this fact is that the solution contains more particles than had been supposed. But to get a larger number of particles it is necessary to split the calcium-chloride molecules into Ca and Cl particles. About 91 per cent. of the molecules must have dissociated into Ca and Cl ions. If the number of such particles is computed from the freezing point, it is found to be about the same as that which is computed on the ionic theory from the conductivity. As in this case no electricity is used and it is unlikely that depressing the temperature could cause such a dissociation, this fact lends support to the view that some substances dissociate into particles and these particles are the ions, or electrically charged particles, already mentioned.

There is one circumstance which strongly corroborated the truth of the ionic theory, namely, that a great number of facts which were formerly wholly unexplained were at once explicable; and new facts could be predicted and found to be true by experiment. It resulted in an entirely new development of electro-chemistry and quantitative analysis was put by it on a firm theoretical foundation. For all these reasons we may repeat what was already said, that no more clarifying, fruitful theory has appeared in chemistry than the electrolytic dissociation theory. Inasmuch, however, as there are some who do not yet accept the theory as positively established, for reasons into which we cannot go at this place, it must be accepted provisionally only, as the most probable explanation of the facts which has yet been proposed. The conception of the chemical union of solvent and solute may eventually considerably modify the ionic theory.

When a salt dissolves in water then, as it does in living matter, there are these reasons for believing that it breaks, in part, into electrically charged particles which, like so many tiny electrodes, each bearing one or more electrical charges, float about in the protoplasm and become thereby capable of doing many things. Living matter contains before it is stimulated, then, a large number of electrically charged particles, and it is clear that if in any way an accumulation of positive particles in one place and of negative in another could be produced, and if the negative and positive particles had different actions on the vital processes, momentous changes might thus be brought about in living matter. This is what happens when an electric current is sent through protoplasm. Moreover, it is clear that if the nature of these little electrodes is changed so that instead of carrying one charge each carries two or three, or if they carry them at a different potential, the electrical equilibrium of the protoplasm might be upset as surely as if a separation of opposite electricities had occurred. The ionic theory, then, is at present fundamental to an understanding of the nature of electrical and chemical stimulation and depression of protoplasm; of the action of salts and drugs on living matter; and it also enables us to see how if by any reaction taking place in living matter a change in the distribution of positive and negative ions could be produced something in the nature of a condenser might be formed which, under suitable conditions, would discharge. Later on, under the heading of colloids, the relation of these charges on the ions to the physical state of the protoplasm will be considered. It may be stated, also, that oxidation in protoplasm is accom-

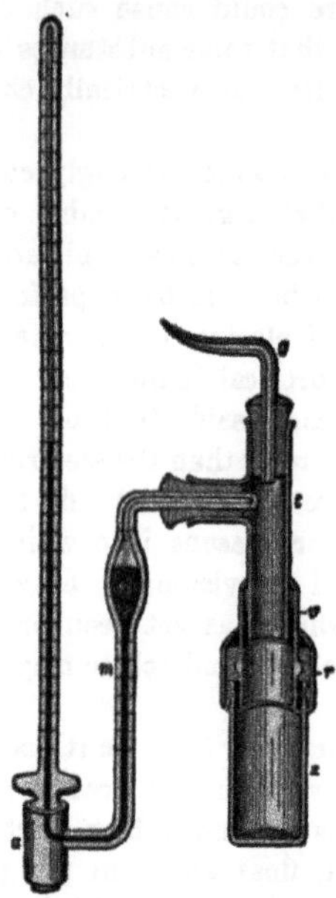

FIG. 17.—Porous cup and manometer for measuring osmotic pressure as used by Pfeffer. *m*, manometer; *z*, porous clay cup with ferrocyanide in its pores. In making the determination this is put into a beaker of water.

panied by such an electrical disturbance which in its turn probably acts as a stimulus to the surrounding parts of the protoplasm, the stimulus being propagated in this way.

Another property of salt solutions of great interest is their high internal pressure. The internal pressure of salt solutions, or even of water alone, is very high. By the internal pressure is meant the cohesive pressure due to the attraction of the molecules for each other. This pressure in such a liquid as ether, which is very labile and volatile and of a low internal pressure, is about 2,000 kilograms per square cm. at zero degrees; and in water it is certainly far greater than this, being probably between 5,000 and 10,000 atmospheres. The addition of salt to water increases this pressure still higher, and the more salt there is added the greater the internal pressure becomes. The internal pressure being so high, the spaces between the water molecules are very small. It is this internal pressure which is probably at the basis of osmotic pressure.

Osmotic pressure.—This is another property of solutions of great importance in vital phenomena, since it is one of the factors controlling the amount of water in protoplasm and its turgor. It was found by the English physicist, Graham, that if solutions of two different substances, or two differently concentrated solutions of the same substance, were separated by a membrane, either animal or vegetable, the substances in solution would in some instances pass through the membranes and sometimes they would not. Using parchment paper, or bladder, as the membrane he divided all substances into two classes: those which passed through he called crystalloids, and those which did not were called colloids. The process of passage of solvent, or solute, through a membrane is called osmosis. It has been found possible to prepare membranes which are freely permeable to water, but which oppose a resistance to the passage of the crystalloid solute; such a membrane is said to be semipermeable, since only the solvent goes through. The botanist, Pfeffer, prepared such a membrane by precipitating the gelatinous copper ferrocyanide in the pores of a porous clay cup. If potassium ferrocyanide is put within the cup of which the pores are filled with water and the cup is immersed in a 3 per cent. copper sulphate solution for 24-48 hours, a gelatinous precipitate of cupric ferrocyanide occurs at the junction of the solutions within the porous wall. This precipitate is permeable to water and some ordinary salts, but it does not permit cane sugar to pass through it. If a cup thus prepared, or prepared by the electrolysis method of Morse and Horn, holding a solution of cane sugar is immersed in water, sugar cannot go out, but water can and does enter. If the cup is closed by a mercury manometer, water will continue to pass into the cup, expanding the solution and forcing the mercury of the manometer upward until a certain pressure is reached, when the manometer becomes stationary and the solution takes up no more water. This pressure is known as the osmotic pressure of the sugar solution. It is the pressure which is just sufficient to prevent the solution from increasing in volume when separated from the solvent by a semi-permeable membrane. Before considering the cause of this passage of water inward, the relation of the amount of the pressure to the concentration of the solution may be discussed.

Pfeffer made an osmometer of the nature of that just described (Figure 17) and measured the amount of the osmotic pressure of sugar solutions of various concentrations and at different temperatures. Some of the results he obtained are given in the following tables. It will be observed that the osmotic pressure increases with the temperature and

Cane sugar C=Concentration per cent.	P (t = 13.°5-14.°7) Osmotic pressure cms. Hg.	$\frac{P}{C}$
1	53.5 cms	53.5
2	101.6 "	50.8
4	208.2 "	52.0
6	307.5 "	51.2

Temperature	P 1% cane sugar
6.8°	50.5 cms
13.2	52.1 "
14.2	53.1 "
22.0	54.8 "
36.0	56.7 "

with the concentration; and also that the amount is proportional to the concentration and is high. Thus a 0.1 molecular solution, 34.2 grams saccharose in a liter or about 3.1 per cent., has an osmotic pressure of 2.24 atmospheres at 0°; a .05 molecular of 1.12 atmospheres and so on. This rule only holds for dilute solutions. Concentrated solutions have a higher pressure than that calculated.

Since it is not always possible to find semi-permeable membranes with which to measure osmotic pressure directly, recourse must often be had to indirect methods. The pressure may be determined by taking the freezing point of the solution. A 0.1 molecular solution depresses the freezing point of water 0.186°. This has an osmotic pressure of 2.24 atmospheres at zero degrees. If the freezing point is depressed only half of the foregoing amount, the solution must be .05 molecular and the osmotic pressure is only 1.12 atmospheres. Ordinarily, therefore, instead of measuring the osmotic pressure, the freezing point may be taken, a correction made for the concentration change produced by the ice which has separated and the osmotic pressure calculated. Of course the calculation is made on the assumption, which is not always correct, that the degree of dissociation and association does not markedly change with the temperature; this is virtually true for most common salts. A very useful table for calculating the osmotic pressure from the freezing point is that of Harris and Gortner, on page 201.

The van't Hoff law of the correspondence of osmotic and gas pressure only holds for dilute solutions. It does not hold strictly even for a solution of sugar 0.1 mol. in strength and higher solutions have osmotic pressures greater than that calculated. (Morse; Berkeley and Hartley; Garrey.) Thus the freezing point of a molecular cane-sugar solution is not —1.86° C. as calculated from the freezing point of a 0.05 molecular solution, but it is —2.775°. The osmotic pressure in place of being the theoretical amount of 22.4 atmospheres at 0° is actually about 33.3 atmospheres. The deviation becomes greater at higher concentrations. It does not disappear entirely if we calculate the concentration on the basis of the pressure being that which would be exerted by the gas when calculated for the volume occupied by the solvent only. The osmotic pressure of the sea-water at Woods Hole is that of a solution freezing at —1.81° C. or about that of a 3/4 molecular cane-sugar solution (256.6 grams per liter. Garrey).

TABLE OF OSMOTIC PRESSURES IN ATMOSPHERES FOR DEPRESSION OF THE FREEZING POINT TO 2.99° C. (Harris and Gortner).

Δ	Hundredths of degrees, Centigrade									
	0	1	2	3	4	5	6	7	8	9
0.0	0.000	0.121	0.241	0.362	0.482	0.603	0.724	0.844	0.965	1.085
0.1	1.206	1.327	1.447	1.568	1.688	1.809	1.930	2.050	2.171	2.291
0.2	2.412	2.532	2.652	2.772	2.893	3.014	3.134	3.255	3.375	3.496
0.3	3.616	3.737	3.857	3.978	4.098	4.219	4.339	4.459	4.580	4.700
0.4	4.821	4.941	5.062	5.182	5.302	5.423	5.543	5.664	5.784	5.904
0.5	6.025	6.145	6.266	6.386	6.506	6.627	6.747	6.867	6.988	7.108
0.6	7.229	7.349	7.469	7.590	7.710	7.830	7.951	8.071	8.191	8.312
0.7	8.432	8.552	8.672	8.793	8.913	9.033	9.154	9.274	9.394	9.514
0.8	9.635	9.755	9.875	9.995	10.12	10.24	10.36	10.48	10.60	10.72
0.9	10.84	10.96	11.08	11.20	11.32	11.44	11.56	11.68	11.80	11.92
1.0	12.04	12.16	12.28	12.40	12.52	12.64	12.76	12.88	13.00	13.12
1.1	13.24	13.36	13.48	13.60	13.72	13.84	13.96	14.08	14.20	14.32
1.2	14.44	14.56	14.68	14.80	14.92	15.04	15.16	15.28	15.40	15.52
1.3	15.64	15.76	15.88	16.00	16.12	16.24	16.36	16.48	16.60	16.72
1.4	16.84	16.96	17.08	17.20	17.32	17.44	17.56	17.68	17.80	17.92
1.5	18.04	18.16	18.28	18.40	18.52	18.64	18.76	18.88	19.00	19.12
1.6	19.24	19.36	19.48	19.60	19.72	19.84	10.96	20.08	20.20	20.32
1.7	20.44	20.56	20.68	20.80	20.92	21.04	21.16	21.28	21.40	21.52
1.8	21.64	21.76	21.88	22.00	22.12	22.24	22.36	22.48	22.60	22.72
1.9	22.84	22.96	23.08	23.20	23.32	23.44	23.56	23.68	23.80	23.92
2.0	24.04	24.16	24.28	24.40	24.52	24.63	24.75	24.87	24.99	25.11
2.1	25.23	25.35	25.47	25.59	25.71	25.83	25.95	26.07	26.19	26.31
2.2	26.43	26.55	26.67	26.79	26.91	27.03	27.15	27.27	27.39	27.51
2.3	27.63	27.75	27.87	27.99	28.11	28.23	28.34	28.46	28.58	28.70
2.4	28.82	28.94	29.06	29.18	29.30	29.42	29.54	20.66	20.78	29.90
2.5	30.02	30.14	30.26	30.38	30.50	30.62	30.74	30.86	30.08	31.09
2.6	31.21	31.33	31.45	31.57	31.69	31.81	31.93	32.05	32.17	32.29
2.7	32.41	32.53	32.65	32.77	32.89	33.00	33.13	33.25	33.36	33.48
2.8	33.60	33.72	33.84	33.96	34.08	34.20	34.31	34.43	34.56	34.68
2.9	34.79	34.91	35.04	35.16	35.27	35.39	35.51	35.03	35.75	35.87

A convenient form of apparatus for determining the freezing point of blood, vegetable saps, milk or other animal juices is that shown in Figure 18, described by Bartley:

"The apparatus consists of a Dewar tube, A, 22 cm. high and 6 cm. inside diameter, set in a wooden base. This is fitted with a rubber stopper having three holes. Into the large hole is fitted a heavy glass test tube 20 cm. long and 3 cm. wide passing down to near the bottom of the vessel A. Two other holes are for small brass or copper tubes, one (C) terminating just below the rubber stopper and the other (B) passing to the bottom of A and coiled around two or three times. These coils are perforated with a series of small holes. Inside of the test tube passing through the rubber stopper is a second test tube of about the same length and 2.5 cm. in diameter, held in place by a section of rubber tubing drawn over it and separating the two tubes by a narrow space. In operation, this space is filled with alcohol. A delicate thermometer (F) with a platinum wire coiled loosely around its lower end completes the apparatus. In the apparatus as here figured, and as used by the author, the stirrer (E) is operated by a toy motor (D) run by an ordinary dry cell. This can be dispensed with, if desired, and the stirrer operated by hand, although this mechanical contrivance makes the apparatus almost automatic.[1]

To use the apparatus, fill the tube A about one-third full of ether or carbon disulphide. Insert the rubber stopper tightly, connect the shorter metal tube with a Richards aspirator pump, attached to the water service. The liquid to be frozen is placed in the inner test tube. There should be enough liquid to cover the mercury bulb of the thermometer, when the latter is lowered to the bottom of the tube.

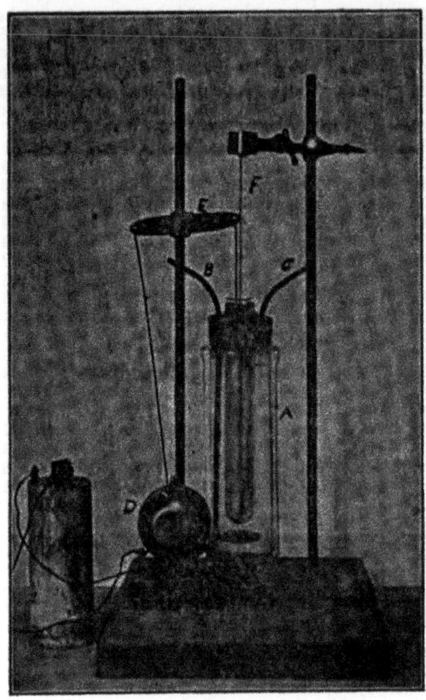

FIG. 18.—Bartley freezing-point apparatus.

The water is then started through the Richards respirator pump, which draws air through the ether in a series of bubbles, causing it to evaporate.

Owing to the well-known principle of the Dewar tube, applied in the popular thermos bottle, almost all the heat used to vaporize the ether is derived from the thin layer of alcohol between the two test tubes and from the liquid under examination. There is no frosting of the outer vessel, the whole system remains clear and transparent and the thermometer can easily be read at all times.

When the temperature reaches zero, the stirrer is started. It will be observed that the temperature steadily sinks to —2° C. o —3° C. before freezing begins, i.e.,

[1] Bartley: *Archives of Diagnosis*, 1913.

two or more degrees below the true freezing point of the liquid. Then, suddenly, freezing occurs and the temperature reading rises to a fixed point and remains there for some minutes. When this point is reached the water is shut off and an accurate reading taken. This is the freezing point of the liquid. There is no necessity of adding ice to start the freezing, as is usually done in other forms of apparatus. The whole process is automatic and all the observer need do is to regulate the flow of water running through the pump and read the thermometer. It is advisable, when the temperature reaches zero, to draw the air through the ether more slowly until freezing takes place, by partly shutting off the flow of water. For accurate work the Beckmann adjustable thermometer should be used. The thermometer is the most important and most expensive part of the apparatus."

The osmotic pressure may then be defined as that pressure which is just sufficient to prevent any increase of volume of a solution when it is separated from its solvent by a truly semi-permeable membrane.

Using the measurements of Pfeffer, van't Hoff discovered that for dilute solutions the osmotic pressures were equal to the pressure which a true gas would exert if the same number of molecules were contained in a space as large as that at the disposal of the solute molecules. Thus a one-tenth gram mol. of sugar in a liter space at 0° exerts an osmotic pressure of 2.24 atmospheres per square cm. One-fifth of a gram of hydrogen gas in the same space and at the same temperature would have the same pressure. Moreover, the temperature coefficient is the same both for the osmotic pressure and the gas pressure. In the case of a gas it is known to be 1/273, or .00366 per degree. Pfeffer found for the osmotic pressure of sugar approximately the same value.

A 1 per cent. solution of cane sugar contains one gram in 100.6 c.c. The same number of molecules of hydrogen in the same space, or .0581 grams per liter at 0° exerts a pressure of .646 atmospheres. Van't Hoff gives the following table comparing gas and osmotic pressure:

Temperature	Osmotic pressure of cane sugar	Gas pressure of hydrogen gas
6.8°	0.664	0.665 atmosphere
13.7	0.691	0.681
15.5	0.684	0.686
36.0	0.746	0.735

These facts were all determined empirically, but the explanation has not yet been given to the satisfaction of all. At first the conceptions of the molecular kinetic theory of gas pressure were carried over bodily to explain osmotic pressure. The pressure in the case of a gas is due to the bombardment of the walls by the rapidly moving molecules of the gas; the osmotic pressure was ascribed to the bombardment of the semi-permeable membrane by the dissolved molecules. A more probable explanation of the pressure is the following: The vapor pressure over a salt solution is less than over pure water. This is shown either by direct measure of the vapor pressure or by a boiling-point determination. The boiling point of a solution is that temperature at which the vapor pressure becomes equal to the external, generally the atmospheric, pressure. It is found that it is necessary at atmospheric pressure to heat salt solutions to temperatures higher than 100° C. before they begin to boil; from which we conclude that their vapor pressures at 100° and below are less than an atmosphere and lower than that of pure water. It is also found that the increase in the boiling point is proportional to the molecular concentration of the dissolved substance for all substances which vaporize at a temperature higher than does water. Why is the vapor pressure of a salt solution lower than that of water? Various reasons may be assigned. One is that the attraction between salt molecules and between water and salt is greater than that of water for water. Hence the internal pressure of the solution is higher than that of water alone. Now at the same temperatures all molecules possess the same average kinetic energy; that is, the product of the mass by the square of the average velocity is a constant for all molecules at any given temperature, the heavier molecules moving more slowly, the lighter faster. The mean kinetic energy of the water molecules in water and salt solution is the same, but the cohesive attraction is greater in the salt than in the water. Only those molecules which have a kinetic energy above the mean value are able to escape from this cohesive attraction of the liquid into the vapor. Since the cohesive energy is greater in the salt solution, there will be, on the average, fewer molecules able to escape this attraction in unit time. Hence, when equilibrium is reached and just as many molecules in the vapor are coming into the liquid as escape from the latter, this equilibrium will be attained when fewer molecules are in the vapor space in the case of the salt solution and hence the vapor pressure over the salt solution will be lower than over the water.

If two receptacles are closed except for a glass tube connecting them and the one is partly full of water, the other partly full of salt solution, the vapor pressure over the salt solution will be lower than that over the water. The water will gradually distill over into the salt solution. The conditions are not different if the two solutions are brought into contact; for now the attraction, or cohesion, of the salt solution molecules for water is greater than that of the water molecules for water, and the water molecules will gradually penetrate the salt solution until equilibrium is attained, when the solution becomes homogeneous. If we put a semi-permeable membrane between the solution and the solvent and then exert a pressure on the salt solution, molecules of solvent may be forced outward, by filtration, through this membrane. By increasing the pressure, the number of solvent molecules thus leaving the salt solution may be increased until a point is reached at which the numbers thus forced out by pressure, added to those which are leaving as vapor, equals the number leaving pure water when in equilibrium with its vapor. This pressure will thus just suffice to prevent more water entering the solution than is leaving and such a pressure is called the osmotic pressure.

The cause of the osmotic pressure is evidently ultimately the attraction of a physical or chemical nature between the solvent and the solute molecules. It is the cohesive or internal pressure of the solution.

Since salt solutions and all things in solution exert osmotic pressure, protoplasm has a decidedly higher osmotic pressure than water. The amount of this pressure varies in different cells, but for the mammalian tissues it is supposed to be about that of a 0.9 per cent. NaCl solution, since in such a solution the tissue neither gains nor loses weight. This is about 7.1 atmospheres. For the cells of apples, the juice obtained by pressing the apples has an osmotic pressure of about 17 atmospheres. It is partly by means of osmotic pressure that plant and animal cells preserve their turgor and keep the cell wall stretched; and it is by changes in turgor that movements are produced in many plants, i.e., the sensitive plant, and possibly in our own brain cells.

The determination of the osmotic pressure of animal and plant cells may be directly made by immersing them in solutions of salts or substances which do not penetrate them and determining whether they shrink or swell or remain unaltered. That solution in which they neither swell nor shrink is supposed to have an osmotic pressure equal to that of the cell contents. This method was used by the botanist, de Vries, to determine the osmotic pressure of plant cells and also the concentration of various salts all of which left the size of the cells unaffected. He used cells of many plants, among others of Tradescantia, the spider lily. Algæ serve as well. Normally the cell contents are under high pressure, due to turgor which keeps the protoplasm applied to the cellulose wall, but if the cell is put into a solution of which the solute does not penetrate the cell, and if the osmotic pressure is high, the protoplasm shrinks away from the cellulose wall. It is said to be *plasmolyzed,* and the method is called the *plasmolysis method*. By this method the osmotic pressure of various plant cells was determined. Some vegetable saps have an osmotic pressure of 14 atmospheres.

This method has several serious sources of error. The plant cell is not a bag of liquid with a semi-permeable wall, but probably a jelly-like substance. Furthermore, this gel is one of the most unstable substances known. It is living matter, and the activities of living matter are wonderfully dependent on different kinds of salts and other substances. It is not surprising, therefore, that the method has given only approximate results, although these results have been of great value, since it was from de Vries' osmotic measurements made by this method that van 't Hoff and Arrhenius drew part of their material for the laws of osmotic pressure and dissociation.

Animal cells presumably have an osmotic pressure approximately equal to that of the circulating liquids like the blood, which is somewhat more than seven atmospheres. The freezing point of blood serum is about —0.6°, which would be an osmotic pressure of 7.2 atmospheres as shown in the table. The red blood corpuscles of mammals are often used for osmotic pressure determinations. The concentration of a solution is determined in which the corpuscles have the same volume (in the hematokrit, see page 921) that they usually have in the serum. The osmotic pressure of the serum is hence equal to that of the solution. For mammalian corpuscles it is about that of a 0.9 per cent. NaCl solution. Solutions of this osmotic pressure are said to be *isosmotic* or *isotonic*. Stronger solutions which shrink the corpuscles are *hypertonic;* weaker, which swell them, are *hypotonic*. Although these corpuscles have little chemical activity, they are gels like the plant cells and their use for determining osmotic pressure is hence very limited.

Surface tension.—Besides the properties of osmotic pressure and ionization and the physical properties which have been mentioned, salt solutions, such as occur in protoplasm, or in fact all liquids, possess certain properties at the surfaces which separate them from other substances of a gaseous, liquid, or solid nature. Such surfaces are supposed to and probably do exist in protoplasm between the more solid and the more liquid parts of the protoplasm; and the physical properties of such surfaces of separation become at least worthy of attention in any examination of the physical properties of protoplasm. It is clear that where a liquid comes in contact with another substance of a different kind, the molecules of the periphery of the liquid are no longer under similar attractions in all directions. It will seldom or never happen that the attraction between the molecules of the two substances in contact will be precisely the same as that between the molecules of each substance. The result of this will be that the molecules in the surface film of the liquid will be attracted with a different force outward than they are inward. Their freedom of movement, therefore, will no longer be precisely the same in all directions, as it is in the interior of the liquid, but will be restricted in certain directions. The surface of a liquid thus comes to possess different properties from the interior; and, since the molecular freedom of movement is restricted in a certain direction, the surface perpendicular to this direction acquires the property of a solid, since a solid is a liquid in which the freedom of movement of the molecules is reduced. The surface has a certain resistance to rupture owing to the inability of the molecules to move freely out of the plane of the surface; and this resistance to rupture of the surface film is called the surface tension. Wherever there are surfaces of separation of liquids, or of liquids from solids in protoplasm, such surface films will be found; and their surface tension becomes then a very important matter in the physiology of the cell.

Method of determining the surface tension.—The surface tension of a liquid can be determined in several different ways, of which only a brief outline can be given here. The most accurate is perhaps the so-called ripple method of Lord Rayleigh, which consists in measuring the speed of propagation of a series of ripples set up in a pan of the liquid. There is a relation between the velocity and the surface tension. This is applicable to pure liquids. Another equally accurate method is that devised by Michelson of measuring directly, by means of a balance, the tension of a double surface film of a given length. This method is not applicable to volatile liquids. There are two methods which are more convenient, but which are not so accurate. One is the measurement of the height to which a liquid will rise in a capillary tube of known bore. From this height the surface tension in dynes per cm. may be calculated by the formula: Surface tension $= \gamma = \frac{1}{2}\, grh\, (D_l - D_v)$. r is the radius of the tube in cms.; g, the acceleration due to gravity; h is the height to which the liquid rises; and D_l and D_v the densities of liquid and vapor. The drawback to this formula and this method of the measurement of the surface tension is that it involves the assumption that the angle of contact of the liquid with the wall of the tube is zero, so that the cosine of the angle is unity. While this is very nearly approximated to in water at low temperatures, it is probably not true at higher temperatures and particularly for liquids which have a lower tension than water; hence all determinations of the surface tension by the capillary method are open to the suspicion of being too low, the error increasing with the temperature. Another method of determining the tension is the drop method. The drop weight which any surface film can support is dependent on the surface tension. The number of drops which are formed from a given volume of liquid is determined by means of a stalagmometer, Figure 19, and if the density of the liquid is known, the weight of each drop may be calculated from the weight of the liquid divided by the number of drops. The surface tension of water being taken as unity, the surface tension of any other liquid measured in the same stalagmometer may be found from the formula:

$$\gamma_1 = \frac{zs_1}{z_1}$$

γ_1 is the surface tension of the liquid sought; z and z_1 the number of drops of equal volumes of water and solution; s_1 the specific gravity of the unknown liquid of which γ_1 is the surface tension.

This method, while not so accurate as some others, is nevertheless most applicable for the determination of the surface tension of animal and plant liquids. It has been refined in the hands of I. Traube and Morgan. Another accurate method, also avoiding the error of the angle of contact of liquid with the solid, is that of Eötvös, which is particularly applicable for the accurate determination of the tension at the junction of liquid and saturated vapor. It involves only the measure of certain

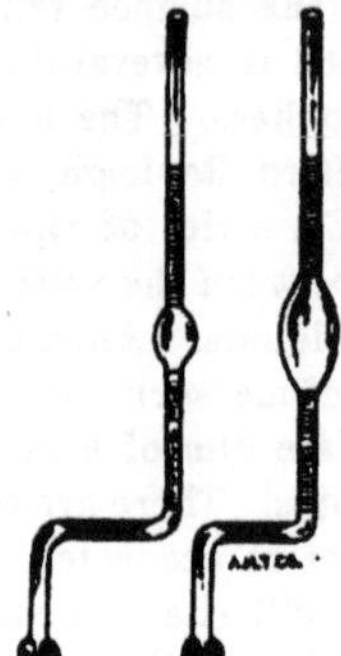

FIG. 19.—Traube stalagmometers for determining surface tension.

angles determined by reflected light and is carried out in sealed tubes. It may be used for the determination of the surface tension of condensed gases. There are also other methods, but these are the more important.

It is found by the use of the ripple method that the surface tension of pure water is 73.24 (74 by Rayleigh) dynes per cm. at 18°. The addition of any of the common salts increases the tension, as is shown in the table:

SURFACE TENSION OF SODIUM CHLORIDE, POTASSIUM CHLORIDE AND $ZnSO_4$ 18°.

Concentration	NaCl	KCl	½$ZnSO_4$
.1 M	73.42	73.48	73.40
.2	73.51	73.60	73.60
.3	73.55	73.75	73.75
.5	74.10	74.20	74.20
.7	74.40	74.50	74.50
1.0	74.80	75.00	75.10

These results are represented by the formula, $T_s = T_w + kC$. The value of k was for NaCl, 1.53; KCl, 1.71; ½ Na_2Co_3, 2.00; ½ K_2Co_3, 1.77; ½ $ZnSO_4$, 1.86. T_w is the tension of water.

It will be noticed that the addition of each salt has a specific effect. That is, the tension is not increased to the same extent by the same concentration of each salt. The surface tension is also a linear function of the concentration, at least within certain limits. The compressibilities of the solutions decrease, in homologous salts, as the surface tension increases, showing that the internal pressure of the solution due to cohesion is also increased by the action of the salt.

Fats and soaps and bile salts decrease the surface tension of water. The least trace of grease has a marked effect on the surface tension of water, if the tension is measured by the ordinary methods, where the surface is not fresh. But if the jet method is used for the determination of the tension, it is found that the perfectly fresh surface of the water has its tension changed very little by the addition of soap. It is only if there has been a chance for the surface to stand for a few moments that the surface tension is found to fall rapidly. The reason for this is that the concentration of the soap in the surface film increases up to a certain point with the time and thereby makes the surface tension steadily lower. It is a very important fact to remember, in considering the surface tension of substances which may exist in two or more states, that the state with the lower surface tension will accumulate in the surface.

The surface tension of water may be used to test the presence of oil in the skin. If camphor is placed on the surface of perfectly pure water, it darts hither and thither on the surface until by its solution in the water it has lowered the surface tension of the water a certain amount. If there is an extremely small amount of grease on the surface, and there is generally enough grease in the air of an ordinary laboratory to spoil the surface of water very quickly, the camphor stands still. Now it is found if a glass rod be touched to the skin beside the nose and then touched to the water, it makes the camphor still, provided the skin has a normal amount of oil. A quantitative determination of the oiliness of the skin in different localities can be made by this method. It has been found that the ingestion of boric acid in sufficient quantity so completely prevents the secretion of oil, causing all the hair of the body to come out, that the camphor no longer becomes still if the rod is rubbed by the side of the nose and then touched to the water. Lord Rayleigh has calculated how thin the film of oil must be to prevent the movement of camphor on water and he has found that it is about of the order of magnitude of a single molecular diameter of the oil. In other words, the oil is a layer only a molecular diameter thick. The reason why the camphor darts about on the surface is that by the solution of a little of the camphor under the piece there is a local lowering of the surface tension so that the surface yields at this point and is stretched by the superior tension of the surface elsewhere. This jerks the camphor away with it. It is a good demonstration of how rapid movements may be produced through the influence of surface tension. Many believe that the movements of protoplasm and even muscle contraction are due to surface-tension changes. But this may be discussed later. When the concentration of the soap or oil in the water is sufficient to lower the surface tension to the point where the addition of camphor can lower it no further, then the camphor stays still.

A more difficult question is raised if it be asked how it comes that sodium chloride added to water increases the surface tension and that soap or fat lowers it. Perhaps it follows from the fact that the cohesion of salt is greater, and that of fat is less, than that of water. What the cohesive pressure of soap or oil may be is unknown, but it certainly is a good deal less than that of water. Water probably has an internal pressure of about 10,000 kilograms per square cm. at 15°. No other liquid has at this temperature as high an internal pressure as this. The surface tension of a pure liquid is a function of the internal pressure, and the surface tension of water is accordingly higher than that of oil. Acetic acid also lowers the surface tension of water, and here again the surface tension of the acid is less than that of water. Salt, on the other hand, has an internal pressure so great that the substance is a solid at the ordinary temperatures; it is much higher than water. We may say, then, that those substances with a lower surface tension than water will move into the surface film and those of a higher surface tension will move away from the film.

The accumulation of substances in the surface film.—The eminent American mathematical chemist, Willard Gibbs, drew the conclusion from that general principle of energetics and thermodynamics which says that systems always endeavor to take that state, or form, in which their potential energy is at a minimum, that if any substance lowered surface tension it would accumulate in the surface film, and that if it raised surface tension it would be less concentrated in the surface film than elsewhere. This prediction was experimentally confirmed. It is easy to see why this should be. If a substance by its presence in the surface film is going to increase the surface energy it is evident, from the law of conservation of energy, that this increase of energy can only be obtained by the doing of work. The substance in order to move into the film must then do work. This is as if there was an obstacle to its moving into the film and hence there will be fewer molecules moving into the film against this pressure than are moving in other directions. Hence the concentration will be less in the film. Just the contrary will be the case for substances which by their presence in the surface diminish the surface tension. For these substances the surface film acts as a trap. Once in it they find an obstacle to their leaving it, since by their departure the energy of the surface will be increased, and hence to leave it they must do work. It is found, as a matter of fact, that this diminution or increase of concentration in the surface film actually occurs, although the amount is not usually very great. In sodium oleate solution, Milner found an excess of 0.4 mg. oleate per square meter in the surface film.

The difference of concentration between the surface film and the rest of the solvent may be of considerable importance in protoplasm. Thus it is suggested that in the surface of contact of protoplasm with water, lipin substances will accumulate and thus make a kind of intermediate layer of a lower surface tension and of a fatty nature. But, inasmuch as the whole substratum of the cell is of a fatty or lipin nature, it is difficult to see how the surface tension of the junction of fat and water could be changed by the passage of more lipin into the film; and, as a matter of fact, there is no good evidence that there is such a layer about the protoplasm. It is probable that often the protoplasm is not a liquid at its surface at all but a gel-like solid.

Quite apart from the accumulation of soluble substances in the surface film due to the general principle of maximum stability just mentioned, we often find that solid substances will accumulate in the surface. If finely divided substance be placed in an emulsion and the emulsion afterwards separates from the liquid, as an oil or ether emulsion may separate from water, the material in suspension is carried along with the emulsion and thus separated from the liquid. This method may sometimes be used to purify solutions from finely divided precipitates which filter badly. The accumulation of these solids in the surface is not due to the principle of Gibbs, just stated. They get into the surface by movements accidentally carrying them there by the shaking when the emulsion is made. Once there they are kept there by the surface film which is like a solid membrane. They are supported at the surface because they come to lie actually outside the water and between that and the ether. They are supported there just as flowers of sulphur are supported at the surface of water and they are mechanically carried up by the rising oil or ether. In protoplasm substances may get caught in this same way at the surface boundary of protoplasm and water or possibly even between boundaries in the cell and thus, perhaps, materials for the making of shells or membranes may be accumulated (Macallum). But this process is sometimes confused with that indicated by Gibbs, whereas it is only remotely related to it.

There can be no doubt that there is a certain tension of the surface of the water which touches the protoplasm. The water at least is liquid. But the same cannot be said of the protoplasm. It was long believed that the movement of the amœba was due to these surface tension forces in the protoplasm. The internal protoplasm of the amœba is certainly at times liquid, for example the protoplasm which rolls out to form a pseudopod; but the rest of the protoplasm in the external layer, according to Kite, is solid and gel-like and it can be cut off and cut into pieces.

It is difficult to see how surface-tension changes of the water should cause the movements in the interior protoplasm. Moreover, the movements begin, according to Harrington and Leaming, not in the periphery but in the interior. Jennings, who has very carefully studied the movements of the amœba, concludes that whatever the cause of these movements may be they are certainly not due to surface tension. When the pseudopod moved forward the surface went forward too, not backward as it should have done if the pseudopod was formed as the result of the lowering of surface tension at the point of rupture. An examination of the movement of the amœba from the side instead of from the top shows that the amœba walks on pseudopods as if they were legs and that the motion is not surface tension. According to the observations of Kite, the movements seem more probably due to the liquefaction or taking up of water by the cell protoplasm, this differing in differing regions and causing the movements in the protoplasm. It is very doubtful, therefore, whether the movements of an amœba are due to surface tension any more than those of a fish are due to surface tension. It is very difficult to apply to such a complex organized half-gel and half-sol substance such as protoplasm is, the conclusions derived from the study of pure liquids in the simplest conditions. The application is extremely hazardous.

When one comes to consider the protoplasm as a whole, it is impossible to say to what extent it is made up of small chambers of capillary dimensions; to what extent it has a structure of such a kind that capillarity should play a large part in it. The granules and droplets of protoplasm are many of them solid, not liquid, and they are imbedded not in a liquid but in a more or less solid gel. It is impossible to say to what extent surface forces are active in such a semisolid medium. It is, therefore, at least too early to speak of surface tension as determining the distribution of substances in the cell, as has been done by some observers.

Surface tension plays an important physiological rôle in its relation to the absorption of water by the cell colloids. If a colloidal or gel-like substance such as gelatin, perfectly dry, be put in contact with water it absorbs a considerable quantity of the latter. This absorption is due to the chemical affinity of the water for the gel substance. By the penetration of water into the gel there is produced an enormous surface of contact of the water and the gel particles. Now, if the gel be acted upon by any substance which increases its affinity for water, which increases its power of union with the water molecules, the attraction for the water is increased and consequently the surface tension of the water at the surface boundary is lowered and the surface will be increased. In other words, more water will be absorbed. On the other hand, if we add to the liquid any substance which increases the surface tension of the water, the surface tension will be increased and the surface of contact will be reduced, the gel will lose water. It is because of this fact that the movement of water into and out of the structures of protoplasm becomes possible. Acids, for example, enormously increase the attraction of proteins for water, consequently acids will lead to the taking up of water by the protoplasm, as they are found to do. Salts, on the other hand, may have an opposite action. The movements of all kinds of living cells are probably due to this swelling, or dehydration, of the protoplasmic gel; and we may, therefore, consider it briefly. Since the protoplasmic gel is made of colloids, we may begin by a study of these substances.

Colloids.—The microscopic examination of living matter shows that the cell is not alike in all its parts; it is not homogeneous, but it has a definite structure. This structure is due to the colloids of the cell. There are numerous coarse granules of various sizes and kinds; a nucleus; nucleolus; and a clear, more homogeneous matrix in which very fine granules are revealed under the highest powers of the microscope, and particularly when photographed by ultra-violet light. The details of this structure appear somewhat different in different cells, but it has been suggested by Bütschli, after long investigation both of fixed and living protoplasm, that, including that part which appears to be homogeneous, protoplasm has in reality a foam-like structure, the compartments of the foam being very small and the walls extremely thin. Within the cavities of the foam a solution is supposed to exist. This conception is probably not strictly accurate, but there is no doubt of the organization and heterogeneity of the cell protoplasm whatever the exact nature of its finest structure. The cell is an organized structure; it is not formless. If the structure of the cell is destroyed, if the nuclear membrane or cell membranes are ruptured by mechanical means, as by cutting or grinding the cell, or by the penetration of ice crystals in freezing and thawing, or by stirring up the protoplasm so as to bring about a thorough mixture of its various parts, there is a great outburst of chemical activity evidenced by the formation of acid and the liberation of carbon dioxide, and cell life stops. Organization is, therefore, essential for metabolism. The different substances of the cell must be kept apart, localized in different regions. If they are mixed, they react with and destroy each other.

The cell is in fact not a single room in which all the chemical processes occur in a higglety-pigglety manner, as they occur in a beaker, but it is rather a well-organized chemical factory with different chemical processes occurring in different regions and in which substances are being elaborated as fast as they are required. How their production is regulated will be discussed farther on.

This division of labor within the cell, this separation into different compartments is due to the fact that protoplasm is not primarily a solution, or is so only in part, but it is a jelly-like substance or technically a gel. It is a semisolid substance consisting of solid and water in intimate admixture or union. This gel structure of protoplasm is due to the fact that the organic substances of which it is in part composed have very large molecules, or are large particles, so that they have little velocity of translation, but cohere together. Such substances are known as colloids, and it is in virtue of the colloidal nature of the products elaborated from the foods by the cell's chemical processes that life is possible.

The colloidal substances in protoplasm contributing to its structure are the proteins, carbohydrates and lipins. These together form the vital, organized substratum of the cell, containing in its interstices the water and substances of simple molecular kind, the extractives, salts and various other organic bodies in true solution. As the whole organization of the cell depends on the colloids and vital activity is so dependent upon their affinity for the water or solution present, an affinity which is easily modified by salts, metabolic products, acids, anesthetics and other drugs and by digestive enzymes, an examination of the general properties of colloids and the colloidal state and particularly of colloidal proteins is necessary for the understanding of vital processes.

Properties of colloids. All substances in solution were divided into two great groups about the middle of the nineteenth century by the English physicist, Graham; into substances which would diffuse through parchment paper or other membranes wet by water, substances generally crystalline in nature, which he named crystalloids; and into substances which would not diffuse through parchment or other similar membranes, substances which he called colloid (Gr. *kolla*, glue; *eidos*, appearance) or glue-like bodies, because they behaved like glue in this respect. Among the colloidal, or glue-like bodies, were albumins, gum arabic, glue itself, starch and many other animal and plant substances. Besides the property of not diffusing through paper, these colloids had several properties in common. Most of them, but not all, were amorphous, non-crystalline bodies; they formed viscous solutions which when sufficiently concentrated would set, or gel. When in aqueous solution, Graham called them *hydrosols;* when gelled, *hydrogels.*

It is now known that many colloidal bodies may be crystalline. For example, the chromoproteins hemoglobin, phycoerythrin and phycocyan are all readily crystallized; and many other colloidal proteins such as edestin, excelsin, serum albumin and ovalbumin may be obtained crystalline; but nevertheless it is true that in most cases special conditions are necessary for the crystallization of colloids, and when crystalline the crystals are small and of microscopic dimensions; and many colloids have never been crystallized. A great many crystalloids, also, even substances like common salt, may be obtained in a colloidal form.

It is in virtue of these three properties—non-diffusibility, of forming viscous solutions and real gels—that the colloids are able to act as true organizers of the cell's activity.

The peculiar and distinctive properties of colloidal solutions are due to the large size of the particles which are dispersed. Owing to this large size surface tension phenomena between solute and solvent come into play at the boundaries of the particles, and these phenomena, which are lacking in ordinary solutions, give to colloidal solutions properties which ordinary solutions lack.

Colloids may be arbitrarily defined as substances of which the particles in solution have a diameter ranging from 1-100 $\mu\mu$. One μ is the one thousandth part of a millimeter. They grade into the diffusible crystalloids on the one hand, and suspensions on the other. An idea of the size of a colloidal particle may be obtained from the fact that a molecule of ether has a diameter of about 3×10^{-8} cms. or about .000,000,3 mm. One $\mu\mu$ is .000,001 mm. The shortest visible waves of violet light have a wave length of about 400 $\mu\mu$. That the size of colloidal particles is large is shown not only by their non-diffusibility, but also by the fact that they may at times be seen in the ultra microscope; that they scatter light and the light so reflected from their surfaces is polarized (Tyndall phenomenon); and by the fact that they may be centrifugalized out of solution.

The size of colloidal particles cannot be directly determined by microscopic measurement because of the diffraction halos which surround them and indeed which make them visible. The particles themselves cannot be seen. The diameter of a particle of sodium oleate can be calculated approximately by measuring the thinnest spots of the films of solutions of sodium oleate. These are found to be about 6×10^{-7} cms. in thickness. Since the thinnest films are at least three times the diameter of a molecule, a molecule or particle of sodium oleate cannot have a diameter greater than 2×10^{-7} cms. or 2×10^{-6} mms. The smallest particles which are visible in the ultra microscope are said to be about 5 $\mu\mu$ or 5×10^{-6} mms. The ultra microscope can show particles, therefore, which have a diameter little larger than three particles of sodium oleate. Linear dimensions found for some colloidal particles are: Gold 6-130 $\mu\mu$; silver 50-77 $\mu\mu$; platinum 44 $\mu\mu$.

In the ultra microscope light enters the solution from the side instead of from beneath as in the ordinary microscope. The light strikes the colloidal particle and is reflected upward to the eye. One sees the colloidal particles, when they are sufficiently large, as bright specks on a dark field. These bright points are usually in active Brownian movement. The smallest particles cannot be seen in this way. For example, the particles of casein in solution are colloidal, but they do not appear in the ultra microscope. When a casein solution clots, however, the particles become visible and may be seen to grow.

Fig. 20.—Lantern and microscope arranged for ultra-microscopic observation. Cardioid condenser on the microscope.

When it is remembered that some forms of living matter exist, sub-microscopic germs of disease, which are filterable through a porcelain filter, but scarcely visible in the ultra microscope, it is probable that their dimensions can hardly be larger than a very few molecules of a protein

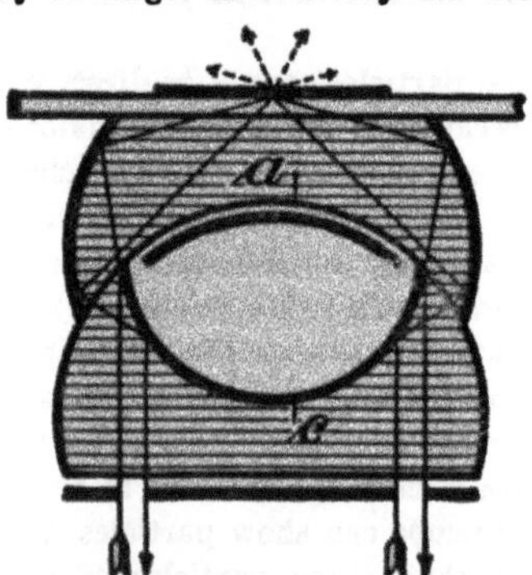

Fig. 21.—Cardioid condenser for ultra-microscopic vision. The rays of light illuminate the liquid on the glass slide from the side.

colloid. Their organization must, hence, be extremely simple and can hardly be other than that of a chemical substance.

Colloidal substances readily separate from crystalloids if brought into parchment paper immersed in the solvent. The crystalloids pass through; the colloids remain behind. This process of separation is called *dialysis* (*dia*, through; *lysis*, to loosen). Colloidal solutions may be purified in this manner.

The Tyndall phenomenon. Most colloidal particles are sufficiently large to show the Tyndall phenomenon. By this is meant that colloidal solutions have the property of scattering a beam of light passing through the solution, so that the path of the light rays in the solution becomes visible, just as in passing through a dusty atmosphere. This is known as the Tyndall phenomenon, since Tyndall used this method to determine when the dust particles had subsided out of the air in his famous experiments on artificial biogenesis. The light which is thus scattered from the particles, or reflected from their surfaces, is found to be elliptically polarized like other reflected light. Since the blue rays are the more easily reflected, colloidal solutions often show a blue opalescence.

Suspensoids and emulsoids. For convenience, but not because there is any sharp line of demarcation between them, for on the contrary they grade one into the other, colloids are divided into two classes: into suspensoids and emulsoids. The colloidal solutions of metals are typical suspensoids. They are easily precipitated from their solutions by the action of salts; they do not gel; and they form generally rather dilute unstable sols; the emulsoids, on the other hand, of which protein colloids, starch, gum arabic and gelatin are types, have the property of forming semisolid or solid gels; that is, solid systems containing a great deal of water. Most of the emulsoids, however, will flock and not gel if the solutions be sufficiently dilute, so that the distinction is not a fundamental one. The colloids in protoplasm are emulsoid colloids.

Suspensoids	Emulsoids
Collodial metals	Gum arabic
Kaolin	Proteins
Antimony sulphide	Starch
Cadmium sulphide	Gelatin
Arsenious sulphide	Silicic acid
	Soap
	Agar-agar
	Nucleic acid

Colloidal particles are electrically charged. A fundamental fact about aqueous colloidal solutions is that the particles bear electrical charges, the charge of opposite sign being in the water contiguous to the colloid. That the colloids are electrically charged may be shown by placing electrodes connected with a battery in a colloidal solution. The colloidal particles move with or against the current. Since only electrically charged particles are thus transported, the colloidal particles must be charged. The various colloids may be divided into those which move to the anode, and are, hence, electro-negative; and those which move to the cathode, and are, accordingly, electro-positive.

Electro-negative	Electro-positive
Arsenious sulphide.	Ferric hydrate.
Antimony sulphide.	Basic proteins, histones and protamines.
Gold.	Proteins in acid solution.
Platinum.	Oxyhemoglobin.
Copper and other metals.	
Most natural proteins in neutral or slightly alkaline solution.	
Lecithin and phosphatides.	
Gum arabic.	
Glycogen and starch.	
Nucleic acid.	
Soaps.	

How many charges there are on a single colloidal particle has not been determined, so far as I know. Some writers speak as if there were a complete electric double layer all about the particle. There is probably but

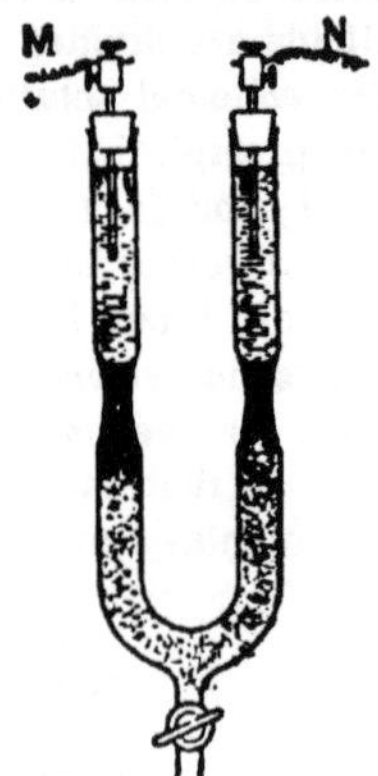

FIG. 22.—Apparatus for the study of cataphoresis of colloids. Non-polarisable electrodes are in the top compartments. The colloidal solution is brought into the U tube below the gelatin plugs. In the figure it may be seen that the colloid is accumulating below the plug on the anode side and is leaving the cathode chamber. The colloid is electro-negative.

a single charge on some soap colloids, but the number undoubtedly is much greater in others.

Origin of the electrical charges. The origin of these electrical charges, of the existence of which there can be no doubt, was at first obscure. It was originally suggested that the particles owed their charges to the faster speed of migration of the hydrogen or hydroxyl ions of water, the ion which was going faster would presumably strike the colloid first (see p. 153) and in this way give it a positive charge in acid solutions, where hydrogen ions predominate, and a negative in alkaline, where the hydroxyl ions are predominant. It is, however, generally recognized that this explanation is incorrect; and there can be little or no doubt that they acquire their charges like any other ions by the process of ionic dissociation. The colloidal particle sends into the water one ion, metal or metalloid, and it retains the opposite charge. This process may be illustrated by glass. Glass in contact with water becomes electrically negative and the water positive, the reason being that glass, which is a silicate, sends potassium or sodium ions into the water, thus making the glass electro-negative, and the water, containing the ion, positive. It is quite possible to substitute the sodium ionized from the glass by another metal. If, for example, a glass bottle contains a solution of copper sulphate it will be found, if the sulphate is poured out, that some sodium from the glass has gone into the copper sulphate solution and some of the copper remains attached to the glass so firmly that it is very difficult to remove it with water. It is necessary to treat the glass bottle with acid in order to free it from copper. A copper silicate has been formed in place of the sodium silicate on the surface of the glass. It is for this reason that in trying physiological experiments glass utensils, which have had mercury or copper salt solutions in them, must be washed with the greatest care.

If, instead of using the glass as a bottle, finely pulverized glass, or glass wool, is placed in a copper sulphate solution, the surface of contact is so much larger than in the case of the bottle that the power of combining with the metal is greatly increased, so that the effect of the glass

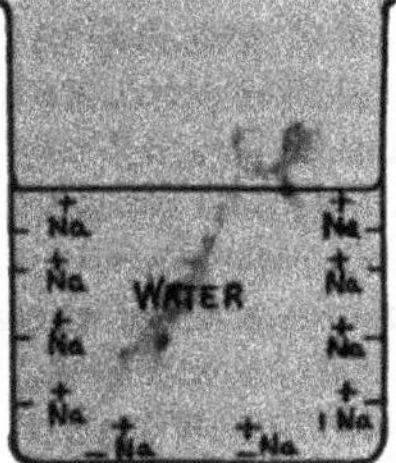

FIG. 23.—Beaker of water showing negative charges on the glass and the water made electro-positive by the sodium ions in it.

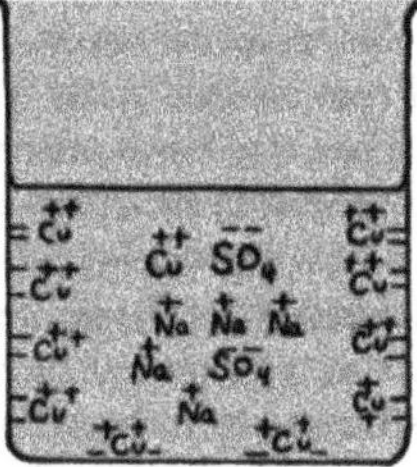

FIG. 24.—Showing how copper in a copper sulphate solution replaces the sodium of the glass and is absorbed by the beaker wall.

in modifying the concentration of the copper in the solution becomes plainly noticeable. Thus quite large quantities of copper may be separated on the glass and removed from the solution. A very interesting experiment was tried by True and Ogilvie illustrating this combining power of glass. They placed sprouted pea seedlings in copper sulphate solution just concentrated enough to be toxic to the seedlings, as shown by the wilting of the rootlet. If now some powdered glass or glass wool, or even filter paper, was placed in the bottom of the tube containing the copper and the seedlings, so much copper was taken out of the solution by the filter paper or glass that it was no longer toxic and the seedlings grew.

It is easy in imagination to carry the process of subdivision of the glass further, until, instead of the pulverized glass, a colloidal silicate is obtained, the particles becoming so small that their surface compared

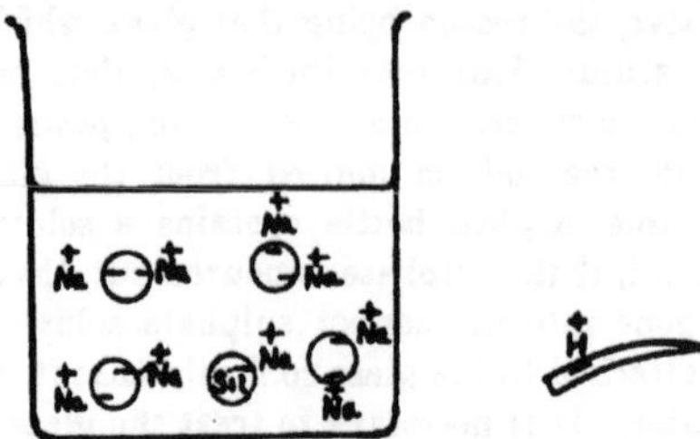

FIG. 25.—Colloidal sodium silicate showing the silicate particles electro-negative with the positive sodium ions in the water close by.

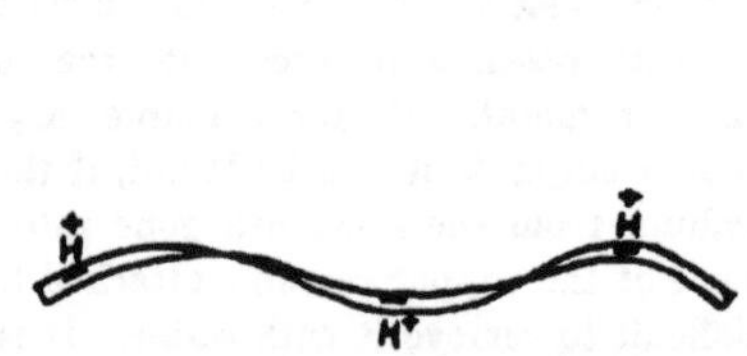

FIG. 26.—To illustrate the possible way in which cotton fibers become electro-negative in water by sending some hydrogen ions into the water.

to their bulk is so large that the particles remain suspended in the water. As the number of particles of silicate in the surface increases by subdivision, the number of bonds between the water and the silicate increases also. The action of the glass in detoxicating the copper solution thus appears to increase proportional to the surface of separation of water and glass and processes of this kind are often referred to as surface phenomena, and treated as if they differed in kind from other chemical reactions. There is, however, no difference in principle between the condition of the soluble glass as a colloidal silicate in solution in the water and the solid glass bottle containing the water in it. In each case the glass is charged by the process of ionization; and the union of metal and glass is a true chemical union. But it will be seen that if the glass should be filtered off and analyzed the proportion of copper and glass would not be found fixed, as in ordinary chemical compounds, but varying in every degree with the size of the surface of the glass.

Let us suppose, now, that instead of the glass being a solid, so that it retains its shape and the water is forced by its affinity for the glass to spread itself over the surface, the glass molecules were freer to move, suppose the glass were a liquid. It is clear that if the attraction between the water and glass were sufficiently great, the surface energy would be most reduced by the most complete and extensive contact possible between the water and the glass. Hence the potential energy of the system would be least when the surface of contact was most extensive. The system would proceed as far as possible in the direction of reducing its surface energy and the glass would, accordingly, dissolve in the water and the water in the glass. Ordinarily in a colloidal solution the division of the colloid proceeds to that point at which the surface energy is a minimum.

Other colloidal solutions obtain their charges in the same way as the glass. Thus colloidal silicic acid sends a hydrogen ion into the water. It is probable that all carbohydrate materials of an insoluble nature or of a nature to form colloidal solutions do the same. Figure 26. The carbohydrates are very weak acids, as discussed in Chapter II. In some

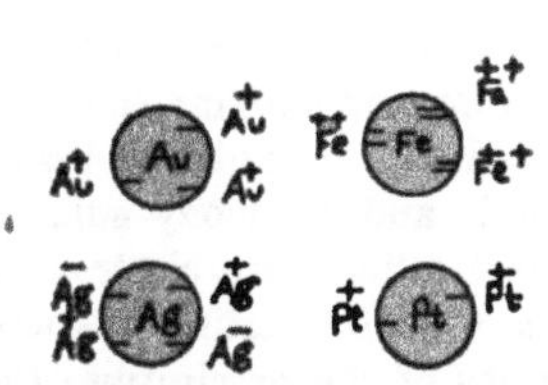

FIG. 27.—Showing how various colloidal metals are electro-negative since they send a few positive metal ions into solution. Each particle has an electric double layer at the surface.

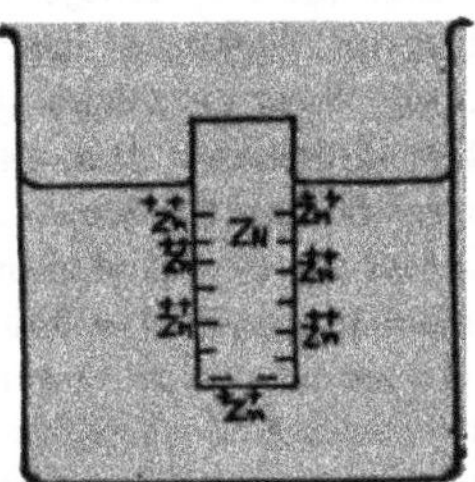

FIG. 28.—Showing how a zinc plate in water becomes electro-negative while positive ions are in the water contiguous to the plate.

cases, however, there may be metal ions present in place of the hydrogen, as in gum arabic. Filter paper or cellulose or glycogen, like glucose and cane sugar, probably send hydrogen ions into the water, the insoluble residue taking the negative sign. Filter paper will thus combine with and hold positive colloids filtered through it. Gum arabic is electro-negative. The positive ion in this case may be calcium, since this is always present in gum arabic. The colloidal metals are all electro-negative. This is due to the fact that all the metals, in the presence of pure water, send some positive ions of the metal, the number varying with the solution tension of the metal, into the solution, and accordingly the colloidal metal particle is electro-negative. Since the water is electro-positive, due to the presence in it of the positive metal ion, there is, along the surface of a plate of metal in water, an electric double layer. The colloidal solution of ferric hydrate is made by shaking ferric hydrate in ferric chloride solution and then dialyzing out the ferric chloride. The hydrate is electro-positive because by the affinity of the iron of the hydrate for the iron of the ferric chloride, the ferric hydrate is held in solution. The ferric hydrate has thus the charge of the positive ferric ion and the negative ion in this case is chlorine, some of which always remains in the colloidal solution. A similar

$$1.\quad Fe(OH)_3 + FeCl_3 = \overset{+++}{Fe}(Fe(OH_3))_n + 3\overset{-}{Cl}$$

Colloidal ferric hydrate.

$$2.\quad (As_2S_3)_n + H_2S = H_2S(As_2S_3)_n = \overset{+}{H} + \overset{-}{HS}(As_2S_3)_n$$

Colloidal arsenious sulphide.

phenomenon occurs in the case of arsenious sulphide sol. In this case the solution is prepared by passing H_2S into arsenious acid. The arsenious sulphide does not precipitate, but remains as a yellowish red solution. If this solution is dialyzed a long time, to get rid of the sulphureted hydrogen, and is then evaporated and analyzed, it is found that there is always more sulphur than is sufficient to form the sulphide of arsenic. It is clear, then, that the arsenious sulphide, as it forms, unites with the hydrogen sulphide, probably by the affinity of the sulphur of the two sulphides for each other. The hydrogen sulphide ionizes, separating hydrogen as a positive ion and sulphur as a negative, so that the arsenious sulphide thus is united with the anion and is accordingly electro-negative. Cadmium sulphide and antimony sulphide behave in the same way, so that all of these colloids are electro-negative. In addition cadmium sulphide shows a very interesting phenomenon, which recalls some of the specific reactions of the precipitins; the cadmium sulphide sol is extremely sensitive to cadmium salts and is precipitated by very small amounts of these salts,—by a dilution of 1:250,000 $CdSO_4$, as compared with 1:20,000 of $MnSO_4$.

Perhaps one of the most interesting and instructive cases of a colloidal solution is that of soap in water. In alcohol a soap is quite normal, the molecules being monomolecular and not colloidal; but in water there is a true colloidal solution. The explanation of this throws light on many puzzling colloidal phenomena. In water soap is hydrolyzed, that is, some of the soap molecules react with the water to form sodium hydrate and the free fatty acid, the fatty acids being very weak acids.

Stearic acid is not by itself soluble in water, but just as atoms unite readily in a physical or chemical union with other atoms of the same kind, so stearic acid unites with the stearic ion of that portion of the soap which has not been hydrolyzed. Every sodium atom has, therefore, attached to it one stearate ion, but united with this stearate ion, in either a physical or a loose chemical union, are two or three molecules of stearic acid. If now the soap is salted out of the solution, this mixture of sodium stearate and stearic acid separates in the form of soap. The soap thus forms an electro-negative colloidal particle consisting of several molecules of palmitic, or stearic, acid, and this is held in solution because of the great attraction of the sodium ion for water. The alcoholic soap

$$1.\quad Na{-}O{-}\overset{\overset{\displaystyle O}{\|}}{C}{-}(CH_2)_{16}{-}CH_3 \rightleftharpoons \overset{+}{Na} + \overset{-}{O}{-}\overset{\overset{\displaystyle O}{\|}}{C}{-}(CH_2)_{16}{-}CH_3$$

Sodium stearate Stearate ion

$$2.\quad Na{-}O{-}\underset{\underset{\displaystyle O}{\|}}{C}{-}(CH_2)_{16}{-}CH_3 + H_2O \rightleftharpoons NaOH + H{-}O{-}\underset{\underset{\displaystyle O}{\|}}{C}{-}(CH_2)_{16}{-}CH_3$$

Stearic acid

$$3.\quad \overset{+}{Na} + \overset{-}{O}{-}\underset{\underset{\displaystyle O}{\|}}{C}{-}(CH_2)_{16}{-}CH_3 + 2HO{-}\underset{\underset{\displaystyle O}{\|}}{C}{-}(CH_2)_{16}{-}CH_3 \rightleftharpoons$$

$$\overset{+}{Na} + \left\{ \begin{array}{l} \overset{-}{O}{-}\overset{\overset{\displaystyle O}{\|}}{C}{-}(CH_2)_{16}{-}CH_3 \\ 2HO{-}\underset{\underset{\displaystyle O}{\|}}{C}{-}(CH_2)_{16}{-}CH_3 \end{array} \right\}$$

Colloidal soap.

solution is normal, and not colloidal, for the reason that hydrolysis does not occur in the alcohol and the stearic acid, if formed, has so much greater an affinity for alcohol than for water that it does not form molecular complexes. The cleansing power of soap depends upon this same principle of affinity between the palmitic or stearic acid colloidal particle and the fatty acids of the neutral fats. When soap is put on the skin, the fats of the skin, like the palmitic acid of the soap, adhere to the latter, and the whole is suspended in water because of the attraction of the sodium for the water and the electro-static affinity between the sodium and the palmitate or stearate ion. Very large, loose physico-chemical aggregates may be built up in this way. Thus vaseline, a hydrocarbon, does not readily combine with soap, but it does have an affinity for oil and oil for soap. Thus by rubbing vaseline with oil it is easily removed by soap, the oil acting as an intermediate body. Probably such unions as these contribute to the formation of protoplasm; the union between fat, phospholipin and cholesterol may be of this nature.

These examples will suffice to show how colloidal particles get the electrical charges they have in solution and that they are produced by processes of ionization.

Precipitation of colloids by salts. Many colloids, particularly the suspensoids, are very easily precipitated from their solutions by salts of any kind; but all colloids aggregate into larger or smaller particles, or change their surface of contact with water when they are in the gel state, under the action of salts. This is one of the fundamental changes which salts can produce in living matter, and since there is good reason for thinking that the mechanics of living matter involves this process, er, a careful study of it has been made. The by salts is of great practical importance not yeing, in the treatment of sewage, in mining, , but also in therapeutics and physiology.

neutral salts to the solutions of colloids gentation. But some colloids are dissolved by e amount of the salt necessary to precipitate he colloid, but toward all colloids the salts it the same order of precipitating efficiency. st of which are taken from the work of Linder the first to investigate these phenomena, put venient form, show the minimum amount of ut a precipitation of the colloid in a given

rious salts on arsenious sulphide sol (Linder and ity. The figures represent the relative concentration pitate. The limiting concentration for precipitation ir.

Bivalent cations		Monovalent cations	
rCl_2	20.0	HCl	954
$r(NO_3)_2$	20.9	HBr	909
$CaCl_2$	21.3	HI	933
$CaBr_2$	21.3	HNO_3	933
$CaSO_4$	26.0	H_2SO_4	1,980
$ZnSO_4$	27.3	H_2SO_3	3,640
$ZnCl_2$	21.8	H_3AsO_4	5,100
$FeCl_2$	23.1	H_3PO_4	4,430
$FeSO_4$	31.8	NH_4Cl	1,010
$CoCl_2$	20.9	NH_4Br	1,200
$CoSO_4$	31.9	KCl	1,590
$NiCl_2$	24.6	KBr	1,640
$MnSO_4$	32.8	NaCl	1,680
$CuSO_4$	14.8	NaBr	1,770
$BaCl_2$	19.1	$NaNO_3$	1,900
$Ba(NO_3)_2$	18.6	$LiNo_3$	1,770
		Tl_2SO_4	13

positive colloid. Albumin from Picea excelsa in 0.1 e figures indicate the molecular concentration of the

alt	Concentration	Salt	Concentration
$_4Br$	0.230	$NaNO_3$	0.116
Br	0.200	KNO_3	0.136
r	0.206	$\frac{1}{2}H_2SO_4$	0.0714
I	0.069	$\frac{1}{2}(NH_4)_2SO_4$	0.0376
	0.098	$\frac{1}{2}Na_2SO_4$	0.0274
O_3	0.137	$\frac{1}{2}K_2SO_4$	0.0402
I_4NO_3	0.135		

It will be seen from the examples cited that salts containing monovalent metals (cations) require stronger solutions to precipitate electro-negative colloids than salts containing bivalent metals; and these in turn require stronger solutions than salts of trivalent metals. The valence of the ion of the opposite charge to the colloid appears to determine, or to be a powerful factor in, the precipitation of a colloid. It will be seen that apparently, at any rate, valence is of more importance than the chemical nature of the ion. Furthermore, the valence of the salt ion which is of the same sign as the colloid, the anion in the case of electro-negative colloids, appears to exert no influence on the precipitation. For example, while calcium chloride is much more powerful as a precipitating salt than the chloride of sodium, sodium chloride and sodium sulphate have about the same precipitating power. These facts have been found to be very general. Toward electro-positive colloids the valence of the anion is important.

It is necessary to examine the character of the precipitate formed if an insight is desired into the mechanism of precipitation by salts. Does the salt go down with the colloid or not? It has been found in all cases which are thus examined that the ion of the opposite charge to the colloid, that is the precipitating ion, always is found in the precipitate. Thus, when antimony sulphide is precipitated by sodium chloride, there is always sodium in the precipitate; if by potassium chloride, there is potassium in the precipitate, and so on. If a protein is salted out of solution by a sulphate, sulphuric acid is always found attached to the protein after dialysis of the salt. These general phenomena have been formulated in the following rules:

1. The precipitating agent is always the ion of the opposite sign to that of the colloid. That is, if the colloid is negative, the precipitating ion is always the cation; if positive, the anion.

2. The precipitating power of the precipitating ion is a function of its valence. Bivalent ions are much more powerful than monovalent; polyvalent more powerful than bivalent.

3. Some of the precipitating ion is always precipitated with the colloid.

4. The valence of the ion of the same sign as the colloid is of no importance in the precipitation.

5. The ion of the same sign appears to exert an influence antagonistic to the precipitating action of the ion of opposite sign.

How does the valence of the ion act? The fact that the precipitation is a function of the number of valences is of great significance, because the valences are probably electrical in nature. The electrical state of the ion thus appears to be of more importance than its chemical nature. Attempts have been made to explain how the valence might act. There have been two explanations given of the way in which an increase of valence might increase the precipitating action of a salt. The first is that of Whetham and Hardy. Since it is the valence, or the number of electrical charges on the ion, which is of importance in precipitating, Hardy suggested, and Whetham computed, that there was a far greater chance of two charges arriving simultaneously in the neighborhood of a colloid particle when both charges were on the same ion, than when they were on separate univalent ions. For a trivalent ion the chances were very much better that the three charges should arrive simultaneously if all were on one ion, than if each were on a separate ion. Their idea was far removed from that of a chemical union between the ion and the colloid. Hardy supposed that the electrical double layer about the colloidal particle was destroyed by the approach of the precipitating ion, and the solution was in this way made unstable. This interpretation was rendered very unlikely when it was found that the precipitating ion went down with the colloid in union with it. Furthermore, the author has shown that the ion of the same sign as the colloid exerts a dissolving action, making the colloidal solution more stable, but valence plays no part in its action. Compare, for example, the sodium and potassium salts in Posternak's work. It always takes a higher concentration of a potassium salt to precipitate than of a sodium salt of the same acid. If it were simply a question of the opposite action of electrical charges, the same reasoning should hold for the ions of the same sign; and the efficiency of a polyvalent ion of the same sign in holding a colloid in solution should also be greater commensurately than that of a monovalent ion. Since valence is of importance in one case, that in which the ion unites with the colloid, but is without importance for the ion which does not unite with the colloid, the writer has suggested that bivalent and trivalent ions are more effective in precipitating because they unite two or three or more colloidal aggregates into very large aggregates of the following kind: Ca-colloid-Ca-colloid-Ca-colloid-Ca-colloid. The aggregates are nearly always obviously larger when the precipitation is by a polyvalent ion. Since ions of the same sign do not unite with the colloid, the number of charges they bear is of no effect.

How does the ion of opposite sign precipitate? This is a very fundamental question and one to which no definite answer can as yet be given. It is essentially the question of solubility. It is not known why sodium sulphate is soluble and barium sulphate insoluble. There is a surface of contact between the colloidal particle and the water or salt solution. The action of the salt on the surface tension of the water may, therefore, be considered first. All of these salts raise the surface tension of the water, as may be seen in the figures cited on page 208.

Any agent which raises the surface tension of the water will, if it have no other action, cause the system water-colloid to reduce the surface of contact in order to reduce the potential energy of the surface to a minimum. This factor then will result in the flocking, or the coalescence, of the colloidal particles into larger aggregates of smaller surface. An examination of the effects of salts on surface tension shows that they arrange themselves somewhat in the same order as they do in their precipitating powers. This does not explain the fact; it simply expresses another fact. It does not say exactly how this coalescence is brought about. Another factor may be this: The colloid is rendered stable by the electric double layer; the effect of this double layer is to reduce the tension of the surface, because it sets up electro-static stresses across the surface between colloid and solution. Now, it is generally true that salts of bivalent metals ionize somewhat less readily than monovalent. Hence, if a monovalent ion is replaced by a bivalent, the ionization will be reduced, the electric double layer reduced, the surface tension increased and hence a reduction of surface will occur, if it can occur. The water in contact with the uncharged particle has, of course, the highest surface tension when there is no union or attraction between the water and the colloid. The consequence is that the undissociated particle is the least soluble particle. The ion, or charged colloidal particle, is more soluble, because the double layer reduces the surface tension.

In some of the colloids, as in the globulins for example, which are soluble in dilute salt solutions but not in water, the addition of a little salt is able to cause the colloid to dissolve. This can only be explained by supposing that the salt acts on the colloid so as to increase its affinity for water, so that by this the surface tension is reduced more than it is raised by the direct action of the salt on the water. This might be accomplished in the following way:

$$\text{Colloid} \rightleftharpoons \overset{-}{\text{Colloid}} + \overset{+}{\text{H}}$$

$$\overset{-}{\text{Colloid}} + \overset{+}{\text{H}} + \overset{+}{\text{Na}} + \overset{-}{\text{Cl}} \rightleftharpoons \overset{-}{\text{Colloid}} + \overset{+}{\text{Na}} + \overset{+}{\text{H}}\overset{-}{\text{Cl}}$$

$$\overset{-}{\text{Colloid}}\overset{+}{\text{Na}} + \text{HCl} \longrightarrow \overset{-}{\text{Colloid}}\text{HCl} + \overset{+}{\text{Na}}$$

In this case the HCl formed is probably united with the colloid and may ionize itself, making the colloid positive at one place and negative at the other. At any rate, the tension of the surface between NacolloidHCl and water is reduced below that of the globulin alone, and solubility is increased. The addition of more salt precipitates. The same thing may happen in arsenious sulphide, which is also rendered more soluble by very small amounts of salt, but precipitated by larger. Here normally the positive ion is hydrogen. By the replacement with sodium in sodium chloride the ionization will be increased. However, H_2S is so weak an acid that this action will soon cease, then the addition of more NaCl will reduce the ionization and the colloid will be precipitated as the sodium salt. Addition of the salt pushes back the ionization and the sodium salt of the colloid is accordingly precipitated.

$$\overset{+}{Na} + \overset{-}{colloid} \longrightarrow Nacolloid$$

Soluble → Very little soluble.

According to this explanation the weaker the acid of the sodium salt used, the larger should be the amount of salt necessary to precipitate. It should take more sodium iodide than of chloride to precipitate, and this is found to be the case.

Influence of solution tension. There is another factor in the precipitating power of an ion or salt besides the valence. It is found that hydrogen chloride is always more effective in precipitating an electro-negative colloid than sodium chloride; and differences exist between the lithium, rubidium, cæsium, potassium and sodium salts, all of which are monovalent. Similar differences exist between calcium, magnesium, barium and strontium, or between aluminum, ferric and chromic salts. These differences have been studied by the author and they are illustrated by the preceding tables. Thus Posternak found that the limiting precipitating concentrations of NaCl, NaBr and NaI were .325, .200 and .069 M. respectively. The anions have the same valence presumably, but the precipitating action of the iodide is greater. In Linder and Picton's work, KCl had a precipitating power represented by 1/1590; while HCl was 1/954.

It is a matter of general experience also that the heavy, and particularly the noble, metals precipitate albumin colloids more effectively than do the alkaline or alkaline earth metals of the same valence. Cobalt, cupric and mercuric chlorides are far more powerful precipitants of the colloids than are the alkaline earths. To explain this difference the author has pointed out that the metals arrange themselves in the order of their solution tensions. In other words, that besides the number of charges carried by the ion, the efficiency of the ion is measured by the *voltage* of the ion; that is, by the tendency of the ion to give up its charge, or by the amount of available energy in the ion. It thus happens that although silver is monovalent it is a better precipitant of the colloids than is calcium. There are always two factors in energy, the volumetric or capacity factor, i.e., the quantity factor; and the intensity factor. Just as in an electric current the amount of work it can do is measured by the amount of the current and the voltage, so in an ion the work it can do is measured by the number of charges and by the voltage or intensity factor of the charge. The silver ion holds its positive charge far less firmly than does sodium. Silver, as an ion, attempts to get rid of its charge and go over into the metallic state; so that ionic silver is a good oxidizing agent. When it is reduced a large amount of energy is set free. In the silver ion there is a large amount of available potential energy as compared with metallic silver.

It is for this reason also that silver salts are so toxic and poisonous, whereas the metal is so inert. In the case of sodium the reverse is the case. Metallic sodium has more energy in it than the ion. It is this difference in energy content that makes the properties of ionic sodium so different from those of the metal, one being a necessary food for the body, the other a terrible caustic which destroys all living matter with which it comes in contact. Copper, ferric iron, lead, gold, platinum, arsenic, bismuth and mercury all resemble silver in the respect that they have more energy in the ionic than in the metallic form. Of the anions, chlorine, bromine, fluorine and iodine have very much more available energy in the atomic than in the ionic form and consequently these substances are, as elements, strong oxidizing agents. Fluorine with the most energy is the most caustic and toxic; chlorine, bromine and iodine following in the order named, iodine being least toxic.

It would take us too far afield, however, to discuss farther in a book of this character this relationship of the solution tension, or energy content of the ion, to its precipitating power, and we may now pass on from a consideration of the effect of salts on colloidal solutions to the properties and nature of gels.

Structure of gels.—Many colloidal solutions, particularly solutions of emulsoids, and some crystalloid solutions have the property of solidifying as a whole without the separation of the solute and solvent into visibly distinct zones or phases. A sufficiently concentrated solution of gelatin will set when cool into a jelly. A gel has the properties of a solid, in that it holds its shape and resists shearing stresses; it is more or less elastic. The molecules of which it is composed are not like those of a solution free to move about. Their motion is in some way restricted as in a solid. A gel is never homogeneous. It consists always of two distinct phases or substances, one of these is a liquid and it may generally be separated from the other by pressure, leaving behind the more solid phase of the solute. Since protoplasm has the property of changing very readily from a liquid to a gel state, the study of the structure and physics of gels becomes very interesting for the physiologist. A gel may be defined as a disperse system of a solid consistence and consisting of a liquid and a more solid phase, or of two liquid phases. It is a solid disperse system in which the degree of disperson is not to molecular fineness. Some colloids form gels with great ease. Gelatin is a typical example of this; agar-agar is another example. A solution of sodium nucleate gels very easily. These colloids are called hydrophil colloids, meaning that they have an attraction for water. On the other hand, some colloids, such for example as colloidal metals, do not form gels. They flock out of solution. Closely allied to the gels are the emulsions. These are systems of several substances having at times a solid consistence. They differ from the true colloidal solutions and gels in that the degree of dispersion is not so fine. A soap foam is an emulsion of this character; cream is another.

Many gels are converted into a sol state by warming or other treatment, and gel again on cooling. These are called reversible gels. Some gels are not reconvertible into sols. They are irreversible. Gelatin or agar-agar form reversible gels; blood when it clots forms an irreversible gel; a strong solution of coagulable protein forms on heating an irreversible gel also. It will not redissolve on further heating. Protoplasm appears to form reversible gels.

What, then, is the structure of a gel? What has happened when a colloidal solution gels? How is the water held in the gel? Why do some colloids flock out of solution and some gelatinize? Since gels are solids, not liquids, it is clear that in some way or other the freedom of movement of the solvent molecules is restricted in the gel as compared with the state in the liquid. How is this loss of freedom produced? To examine the structure of gels an ultra microscope, that is a dark field microscope, is best. In this case the light enters from the side instead of from underneath and the finest particles are shown as bright spots on a dark field. The process of gelatinization has been studied with such a microscope. It has been found that the structures of various gels may be quite different in details. Some gels are formed of very minute, or rather very thin, acicular crystals which penetrate the gel in all directions, and which hold the saturated solution from which the crystals have been deposited entangled between them. Such a gel made of a crystalloid, not a colloid, is very easily obtained by dissolving a good deal of caffeine in water and allowing it to cool. The caffeine crystallizes out in a mass of very long, extremely thin, acicular crystals, and the whole makes a gel, so that the test-tube may be inverted without any liquid escaping. This experiment shows that crystalloids may form gels as well as colloids. Tyrosine when quite pure often forms similar gels. To form such a crystalline gel it is apparently necessary that the crystals should come out in a very minute dimension, at least one or two dimensions must be minute. The crystals may be very long. Among other examples of colloidal substances which gel by the formation of very long, extremely thin acicular crystals, the clotting of the blood may be mentioned. The crystals of fibrin are shown in Figure 29. The corpuscles and liquid are entangled between these crystals. Most gels, however, are not crystalline in structure. A typical gel of a non-crystalline form is that of casein. If rennin is added to a solution of casein under suitable conditions the casein is converted into an insoluble form, paracasein, which forms a clot or jelly. This process has been watched under the ultra microscope and the appearances are reproduced in Figures 30 and 31. In the solution nothing can be seen. The field is homogeneous and dark. As the clotting begins there is at first a diffuse, very faint light in the

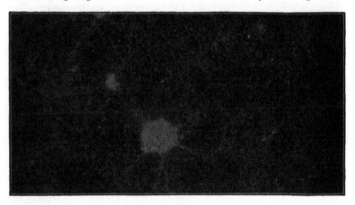

FIG. 29.—The clotting of fibrin showing the long, acicular crystals. A crystalline gel as seen in the ultra-microscope (Stubel).

field, but no visible, distinct points. As the clotting goes on distinct points, very minute and in active motion, appear first, these grow larger and gradually cease to move. Finally the gel is seen to consist of a very large number of small clumps or specks distributed throughout the gel and having no Brownian movement. The gelatinization appears in this case to be due to the formation of an insoluble precipitate which does not flock out of the solution but remains in situ and which holds the liquid between the particles. The liquid between the particles will hold other substances in solution and will also necessarily consist of a saturated solution of the substance, in this case paracasein, which has been in part precipitated. Most gels are of this general character. Silicic acid, for example, forms a gel of this nature and so does sodium nucleate. Still another form of gel is sometimes obtained of the nature of a very fine emulsion. A mixture of gelatin, water and alcohol will form such a gel as shown by Hardy. If the concentration of the gelatin was 36 per cent., the gelatin formed solid walls or alveoli containing a dilute solution of the gelatin in the cavity; if the gelatin was 13.5 per cent., the more concentrated phase separated out as spherical drops surrounded by the more dilute phase. A gel may be formed of three liquids.

The most probable explanation of the formation of a gel is that the liquid solvent, or more liquid phase, is made solid by surface forces. It has already been explained that at the surface of boundary of two liquids or of liquid and solid the attraction of the molecules in the two directions outward and inward is different, so that the surface layer of liquid mole-

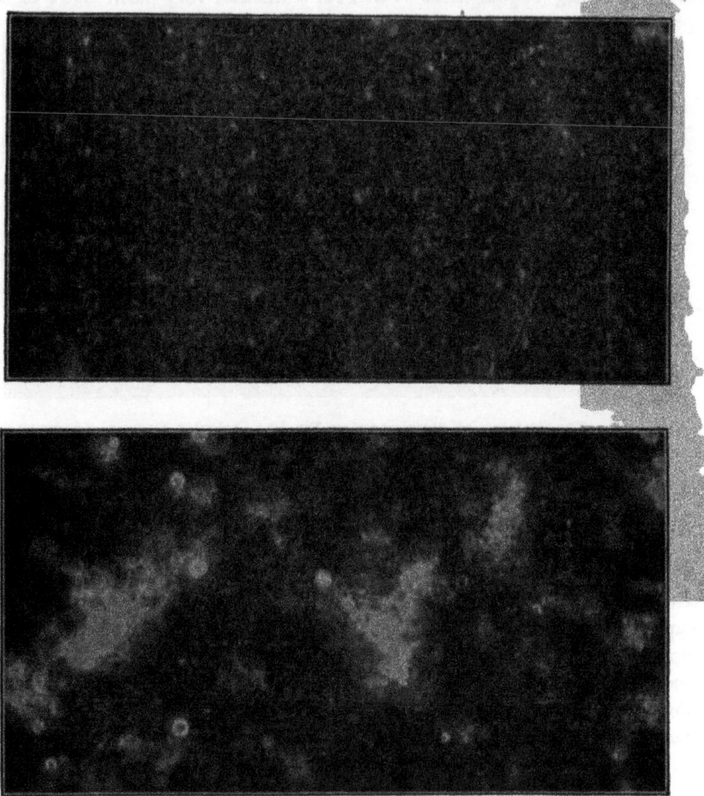

FIGS. 30 AND 31.—Two stages in the clotting of casein as seen in the ultra microscope. In Fig. 31 the casein particles have aggregated into coarser clumps (Stubel).

cules have their freedom of movement reduced; they are really converted into a solid of two dimensions. In order to make a solid gel it is only necessary that the amount of liquid in the surfaces shall be large compared to the amount of liquid not under the action of unequal attractions. In other words, the proportion of liquid in the form of surface film must be sufficiently high. To accomplish this it is only necessary to have a very fine state of subdivision of the precipitate, together with a certain concentration of the precipitate and an attraction between the precipitate and the solvent. It is a matter of indifference whether the finely divided precipitate is crystalline or colloidal; all that is necessary is that the surface of the particles be very large compared to their bulk and that there be enough of such fine crystals, or other finely divided matter, so that the amount of free liquid shall not be too great. If the amount of liquid is too great, the particles, surrounded by their surface layers of water, will separate out. Emulsions do not differ in principle from gels, nor do foams. Although a soap foam is composed of a gas and a liquid soap solution, yet if the liquid be distributed in the form of surface films it is changed to solid supporting lamellæ and the foam is a solid. The essential thing about a gel is, therefore, that the liquid which is present, whatever its character, be present for the greater part as surface films.

All surface films, as we have seen, have a contractile action. There is always a tension in such films. This is shown very clearly in the contraction of a small soap bubble when one stops blowing it and removes the pipe from the mouth without stopping the vent. It is this contractile action which constitutes or measures or is the expression of the surface tension. It is not surprising, hence, that gels which really owe their solidity to surface films of liquid generally contract. In this contraction they press out some of the liquid and this liquid is always, naturally, a saturated solution of the material of one phase of the gel, generally of the solid matter, and it often contains other substances in solution such as salts. This process of contraction of gels with the separation of some liquid of this character is called "*Syneresis.*" It is a true process of secretion, probably identical in character with the processes of secretion by cells. The contraction of an agar-agar tube with the formation of the so-called water of condensation is known to every bacteriologist; and the contraction of clotted blood with the formation of serum which is squeezed out of the jelly is known to many. Most housewives know how annoying this tendency of contraction of gels is, since the liquid thus expressed is often a good culture medium for moulds and bacteria.

Protoplasm may be regarded either as a very viscid sol, or as a gel. Its structure is that of a microscopic emulsion. In other words, it has the structure of a gel, or when it is more liquid, of a sol. Like other gels it is able to contract and thus to press out a solution. It can take up water. It is probable that while a good deal of the solidity of protoplasm may be due to the formation in the cell of tough fibers or crystals of protein or other matter, a part of its solidity, and often a large part, is due to the fact that the liquid in the cell is distributed very largely in the form of surface films between the granules, microsomes or droplets of various kinds found in living matter. The liquid between the various granules is bound, probably, in the form of surface films of a contractile kind. This structure of protoplasm by which it is allied to an emulsion was particularly studied by Bütschli. Figure 32 shows how closely the structure resembles that of an emulsion. The real living part of the protoplasm in its more solid moments is probably a microscopic foam or emulsion. At other times it may be distinctly fluid. This will happen when the films are broken, or when the surface tension is diminished, or

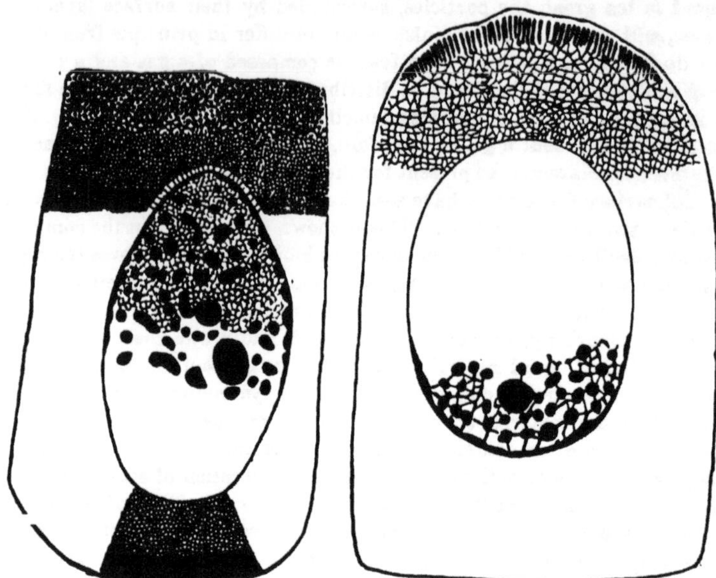

FIG. 32.—Illustrating the foam-like structure of fixed protoplasm (Hardy).

when there is too much water in the protoplasm in proportion to that present in the form of surface films. The recent, very important work of Clowes on emulsions should be read in this connection.

Absorption of water by gels.—The absorption of water by gels is extremely important in physiology, because the protoplasm of the cells of the higher animals is a gel and many of the physiological activities appear to be due to the absorption and loss of water by various parts of the protoplasm. Thus the contraction of muscle cells is apparently due to the passage of water into and out of the fibrillar elements, sarcostyles, as is discussed on page 624; in secretion there is evidently an alternate absorption and loss of water by the cell, the process being so regulated that the loss of water takes place from another part of the cell than that in which it is taken in; the abstraction of water from the gel of the nerve fibers causes the development of a nerve impulse; and in plants the movements of leaves, petals and other parts of the plant are due, generally, to turgor changes in the different cells. In fact, it appears that the mechanics, that is the physical part of the protoplasmic activity, is very largely a matter of the orderly absorption and loss of water by the cell colloids. In this connection attention may be called to the concentration of the chromatin of the nucleus into very dense chromatic masses during cell division as a probable example of the concentration of a gel by the loss of water, accompanying a physiological process. The

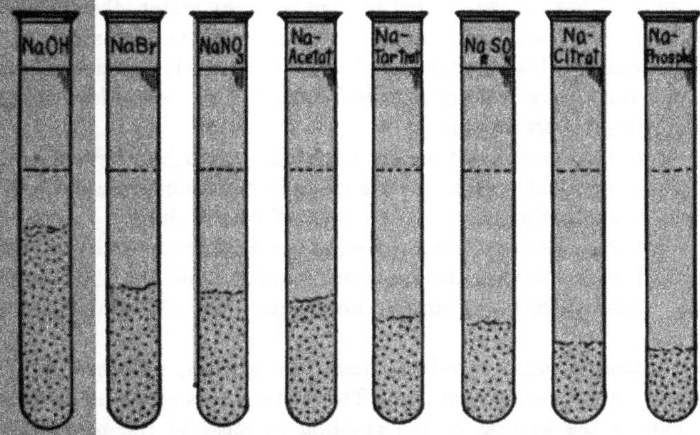

FIG. 32A.—The swelling of fibrin in different salt solutions. Equal amounts of fibrin in each tube. Fibrin is represented by the punctate masses at the bottom of the tubes (from Bechold).

great importance of a thorough understanding of the physics of the process of the absorption and loss of water by gels, of swelling or imbibition processes in other words, for the understanding of vital actions will be apparent from this statement; but in addition it may be mentioned that from the point of view of pathology and therapeutics the subject is no less important, since the process undoubtedly plays a very great rôle in œdema, in the swelling of the brain after a blow on the head and in the occasional swelling of organs like the kidneys to such an extent that the circulation is impeded.

Hofmeister was one of the first to recognize the great importance of this process of swelling in animal physiology, and he has contributed much to our knowledge of the conditions of the process. Any protein gel can be used to illustrate the action of salts, acids, etc., on the water content, but perhaps gelatin and fibrin are the most easily obtained and have been most studied. Small muscles may also be used, for in them the gel is protoplasm itself, but muscles have the drawback that they are the seat of various chemical processes, which complicate and make obscure the interpretation of the results. The method employed by Hofmeister, and generally followed by others, is to make a fairly strong gel of gelatin by dissolving the latter in hot water. The gelatin is poured into a flat-bottom vessel in a thin layer, and after hardening it is cut into cubes of about the same size. These cubes are weighed and then placed in solutions of salts of various strengths and kinds, or of acids, or other substances, and then, after varying times, they are removed, the adhering water removed by blotting paper and the cubes weighed. If the salt or acid has caused the gel to take up more water, the cube will now be heavier; if loss of water has occurred, it will be lighter.

"By swelling," Hofmeister says, "one understands the taking up of liquid by a solid body without chemical change." Three different processes may play a part in this:

1. A porous mass may take up liquid in previously formed capillaries and spaces filled with air. This is capillary imbibition and is illustrated by the absorption of water by a mass of porous clay.

2. A porous mass may take up liquid by osmosis into previously built spaces filled with soluble substances or liquids. This is imbibition by osmosis. This occurs in all animal and plant cells which have a permeable skin.

3. A homogeneous, pore-free mass, like gelatin, absorbs a liquid undergoing an increase in volume. This is the molecular imbibition of Fick. This is the process of the swelling of most of the proteins, of lecithin and other phospholipins and most of the cell structures like the chromosomes, cellulose fibers, dried peas, etc.

In animal or plant cells all three forms of swelling may coexist. Thus a muscle brought into a solution takes by capillarity some water into the spaces between the fibers. A second part is taken by osmosis, not all substances in the muscle being diffusible through the surface layer; and a third by molecular imbibition. Of the three forms, the capillary depends on surface tension; imbibition is a peculiar kind of process which possibly belongs in the group of adsorption processes and may also be of a surface tension nature at the bottom.

Three laws of swelling which have been discovered are these: a. Many substances capable of swelling take up a definite amount of water, not an indefinite amount. There is a swelling maximum. b. The volume of the swollen body is smaller than the combined volumes of the original substance and the liquid taken up. Swelling is accompanied by a contraction in volume (Quincke). c. Swelling is regularly accompanied by the evolution of heat (Duvernoy; Wiedemann and Lüdeking).

Hofmeister used small squares of gelatin or agar not over 0.2 mm. thick and weighing when dried at 100° C. from .01 to .05 gram. These were placed for a time in water warmed to a certain temperature and weighed from time to time to discover the rate at which water was taken up and the swelling maximum. It was found that the weight of water absorbed could be represented by the following formula:

If W is the weight of water taken up by one part of the dry substance in t minutes it was found that

$$W = P(1 - \frac{1}{1 + \frac{c}{d}t})$$

P is the largest amount of water which can be taken up by one unit of substance, or the swelling maximum; c is a constant computed from the experiment; d, the thickness of the plate in the maximum swollen state measured in mm.; and t, the time in minutes. P was found by leaving the plates in water for from 1,000 to 2,000 minutes. P depends on the affinity of the particles for water and the cohesion of the particles for themselves. The velocity is proportional to P—W.

The following results were obtained with agar plates of a thickness of 0.60 mm. P=5.62935; c/d=.15348.

t	W found	W computed
5	2.44	2.44
10	3.35	3.41
15	3.86	3.92
20	4.31	4.25
30	4.69	4.63
40	4.84	4.84
50	5.07	5.03
60	4.98	5.08
1440	5.52	5.60

It will be seen from these figures that the water is absorbed with great quickness. The smaller and the thinner the piece, that is the larger the surface in proportion to the bulk, the more rapid will be the swelling. A completely dry agar plate 0.206 mm. in thickness in the first minute absorbed 75 per cent. of the total amount it could take up. If the size of the plate was that of a red blood cell, i.e., .002 mm., in the first minute it would have reached the maximum, or at least 98 per cent. of the maximum absorption. This is a point of very great importance in the interpretation of muscle contraction where the taking up and loss of water must take place with great speed, if the contraction is to be ascribed, as it seems that it must be ascribed, to this process. It will be seen, also, that when a cell is placed in water or other solution, the first effect must be on the delicate membrane surrounding the cell which will almost at once, because of its great thinness, alter its state of swelling to that of the solution in which it is placed. It is seen, then, that the first effect of a solution is thus on the state of swelling of the peripheral membrane and it is not impossible that this is of importance in determining the ease or difficulty of penetration of substances into cells, and is a matter of importance in the making of the fertilization membrane of eggs.

The second problem investigated by Hofmeister had to do with the question, How does the swelling behave when the gelatin is in a salt solution and not in water? He made cubes of concentrated gelatin and determined the amount of ash and water in several of them. The gelatin cubes were then brought into the solution to be tested and on the next or later days were taken out and after drying were weighed. He at times determined also the amount of salt taken up. The figures in the table show the number of times the weight of solution in the plates is that of the dry substance. At the start the water was 4.42 times the dry substance. The cubes were placed in solutions of the salts which contained one gram mol. of the salt to a thousand grams of water.

	After				
Salt	48 h.	72 h.	96 h.	11 days	25 days
Na Acetate	2.69	2.76	2.99	3.07	3.25
Water	6.39	6.74	8.13	9.37	Putrefaction
KCl	6.89	7.49	9.25	10.05	11.86
NaCl	6.82	7.49	9.82	10.81	13.07
NH_4Cl	9.39	10.33	12.83	14.01	16.47

In solutions of LiCl, $MgCl_2$, $CaCl_2$, NaBr, $NaClO_3$, the gelatin dissolved in 24 hours.

In double normal solutions in 48 hours, cubes, which before contained 5.39 parts of water to one part gelatin, contained the following amounts:

Alcohol	8.70
$C_6H_{12}O_6$	9.29
Na Acetate	9.83
Sucrose	10.77
Water	12.12
NH_4Cl	15.95
KCl	18.84
NaCl	22.19

These experiments show that the amount of water taken up by a gel from a salt solution depends on the nature of the salt present, being reduced by some salts and increased by others. Most salts, however, increase the swelling of the gelatin. This is exactly parallel to the solution of a globulin, which will only dissolve if some salt is present. Hofmeister interpreted his results to mean that all salts had, as a primary effect, an increase in the water-absorbing power, but that some had a greater affinity for water themselves and that there was a strife between the gelatin and salt particles for the water. It is, however, possible that the greater power of the ammonium chloride may be due to the presence of a small amount of acid in its solution and the inhibitory action of the acetate to the presence of a little hydroxyl in the solution. Acids in small amounts greatly increase the swelling power of protein colloids. The amount of salt taken up was in general proportional to the quantity in the solution, but the absorption of the salt and the water did not run perfectly parallel. Thus, at first, the salt content increased more rapidly than the water content. While it is clear that some kind of a chemical affinity must exist between the swelling body and the liquid, since rubber swells in ether but not in water and gelatin shows the reverse effect, Hofmeister concludes nevertheless that the attraction is not that of an ordinary chemical kind, since in the latter a definite proportion of the reacting substances is found. The selective absorption of the gelatin for various solutes was made very obvious by placing the gelatin cubes in solutions of methyl violet. In dilute solutions most of the color passed out of the solution into the gelatin. He concludes that the gelatin particles must have a far greater attraction for the color molecules than for the water. It is probable, since salts unite with amino-acids, that they also combine chemically with gelatin.

Similar experiments have been carried out by Pauli and others, among whom special mention may be made of Fischer for his application of the results to the study of the cause and treatment of œdema. Pauli studied the action of various salts on the fusion points of gels. He found that the iodides lowered the point of fusion most, then came the bromides, chlorides, acetates, tartrates and sulphates in the order named.

While the facts as they stand enable us to forecast the influence of salts on the colloidal substratum of living matter and thus furnish a partial explanation of how salts are affecting vital processes, yet it is instructive to inquire somewhat more deeply into the physics of this process. Why do salts in dilute solution cause swelling, and in stronger solution, shrinking? It is probable that the manner of action is the same as the precipitation of a colloid from solution. If the colloid is sufficiently dilute, under the influence of the salt it falls out of the water: if it is too concentrated, as it is in the gel form so that it can no longer precipitate, the salt causes the water to fall out of it. The same result may be had if a piece of fibrin is placed in water; it takes up some water, but it takes up far more when there is a little acid present. Some salts and acids increase the affinity of protein colloids for water and these cause swelling and dissolve protein colloids; others diminish the affinity, and these cause a loss of water. Anything which increases this affinity lowers the tension of the surface of contact of the water and the colloid and hence this surface is increased, the gelatin swells, water penetrates it; on the other hand, any salt which increases the surface tension of contact between colloid and water will cause a diminution of the surface of contact, and the colloid will lose water. The way in which acids thus act is that they unite with the protein to form a salt which ionizes readily. This salt thus causes the gelatin particles to be surrounded by an electrical double layer. By the electro-static affinity across this layer the affinity of the water and the gelatin is increased, and accordingly the surface tension of the water in contact with the gelatin is reduced, and if the gelatin is not too rigid an increase of surface will take place. All acids, therefore, in more than quite minute amounts, cause proteins to absorb water. If, however, the amount of acid is very small and the protein is in a very dilute alkaline solution, the first action will be in the direction of a loss of water. The production of acid in a tissue Fischer believes to be the prime cause of œdema. The action of salts is more difficult to understand, but it is probable, as Hofmeister believed, that the first effect of the salt is to make a weak union with the gelatin, leading it to take up more water or rendering it more soluble. We may infer from the fact that iodides cause solution of gelatin that the gelatin colloid is, in water, chiefly electro-negative, that which direct observation confirms.

The swelling of solid bodies following or accompanying imbibition of water is often very great. Thus peas will swell to double their size, and Irish moss (Chondrus crispus) swells to treble its size when dry. In all such cases the cohesion of the swollen body is much less than that of the same body before imbibition. Thus Reinke found that dried Laminaria absorbed 300 per cent. of water and its ductility increased sixty times, while the breaking strain fell to one-tenth its value. Solid substances which take up water in visible pores without swelling undergo no change in cohesion. Thus a dried brick will absorb a good deal of water in its pores, but the cohesion of the brick is not altered thereby. There is probably no difference in principle between the imbibition of water in visible capillary pores and true swelling where there are no visible pores and the imbibition is molecular or micellar. The sole difference is probably that in the case of imbibition without swelling the attraction between the water and the solid substance is not sufficiently great to overcome the attraction of the molecules of solid for each other. The molecules of the solid cannot be separated by the penetrating water and the body does not swell. The water is able to displace the air from the capillary spaces and spreads itself accordingly over all free surfaces. In the case of swelling gelatin, or woody fibers, or liquid crystals, the attraction between the molecules of the substance and the water molecules is greater than between the molecules of the substance, so that the latter are separated. Swelling of solid bodies in this way passes imperceptibly into swelling so considerable that the solid becomes a liquid solution. Thus gum arabic and peptone at first swell as solids and then dissolve in the water taken up. Katz has shown that the processes of molecular imbibition in solids are at bottom identical with the taking up of water by glycerol, sulphuric acid or other liquids which absorb water with the liberation of heat.

The heat liberated by these swelling processes may be considerable. Katz found that dried cellulose liberated the following amounts of heat on swelling. W is the quantity of heat in gram calories liberated when 1 gram of the dried substance takes up i grams of water. Formula: $W=Ai/(B+i)$.

Cellulose. $A = 11.6$; $B = 0.030$.

i	W deter.	W calc.	Δ
0	0	0	
0.014	3.5	3.7	—0.2
0.041	6.9	6.7	—0.2
0.054	7.6	7.5	—0.1
0.074	9.0	8.3	—0.7
0.261	10.5	10.4	—0.1

Adsorption by colloids.—Finely divided solids often have the power of uniting with substances in solution so that if the solid is filtered off the solute remains very largely in the precipitate of the finely divided solid. Thus colors filtered through charcoal or diatomaceous earth often remain in the charcoal or earth and the solution is freed from them. If permanganate solution is filtered through clean sand, it is said that the first water which comes through is almost colorless. There are many other examples known to chemists, and one of them has already been mentioned: namely, the taking up of color from a solution of methyl violet by gelatin. This separation of the solute from a solvent by substances in suspension is generally said to occur by means of adsorption. By many observers it is referred to surface forces at the boundary of liquid and solid. It is probable that many of these so-called adsorptions are in reality due to true molecular union between the solid and the substance adsorbed. This is the case, for example, between the filter paper and copper adsorbed from a solution of copper sulphate. Probably the majority of adsorption compounds are in reality chemical unions. They differ, however, from ordinary chemical unions in the fact that the two combining substances are not united in definite proportions, but the amount of substance which is adsorbed is more or less proportional to the amount of surface.

That surface forces may, also, be responsible for some of these processes of adsorption is, however, possible and on the whole probable. It was shown by Willard Gibbs that substances which lower surface tension will accumulate in the surface film. That is, their concentration will be larger there than in the liquid behind. If there is a very large amount of surface, it is possible that quite a good deal more substance will be taken up than by an equal amount of pure solvent not in the surface form. Suppose at the boundary of liquid and more solid portions of protoplasm the surface tension might be lowered by the passage into the film of some substance present in the protoplasm. In that case the distribution of the substance might be altered by the presence of the film. It might be more concentrated immediately about the granules, for example, than in the protoplasm between the granules. It might be said

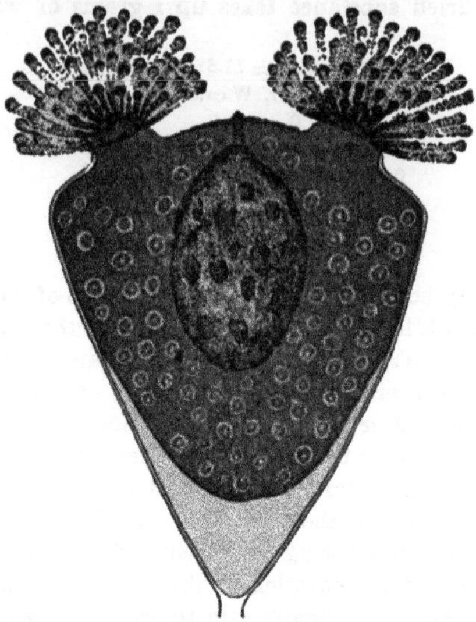

FIG. 33.—Distribution of potassium in Acineta. The black granules represent potassium (Macallum). Macallum believes this distribution is determined by surface forces.

to be concentrated by adsorption. Some physiologists believe that this principle of distribution of substances in protoplasm by means of surface forces of adsorption plays a large part in determining the distribution of substances in cells. Macallum, for example, has studied the distribution of potassium in cells from this point of view. He has concluded that probably surface tension forces do determine the distribution of potassium in Acineta. The curious distribution of potassium in these cells is shown in Figure 33. It is extremely difficult to be sure in these cases that the substance is really distributed by surface forces rather than by the operation of the usual chemical means of forming insoluble precipitates. To what extent adsorption, as distinct from ordinary chemical unions, determines the taking up of substances by living matter must be left open for the present. There can be little doubt, however, that many so-called adsorption phenomena are in reality nothing more than chemical unions. As the tension of water in contact with a solid cannot be measured, it is impossible to say whether any substance raises or lowers this tension and, hence, whether the accumulation is due to its lowering tension, or because it unites chemically with the particle. For potassium to accumulate in the surface in the way mentioned above, it is necessary to suppose that it unites with the substance which the liquid bathes and so reduces the tension by increasing the attraction across the surface; or else that it is present in union with some substance which reduces the tension and so accumulates in the surface.

Brownian movement.—The fine particles of a suspensoid or emulsoid colloid when they are large enough may be seen to be in a state of violent agitation. The particles dance about a fixed point, first in one direction and then in another. This movement was studied by Robert Brown in 1828 and has since been known as the Brownian movement. The vigor of the movement increases with the temperature and with the smallness of the particles. It has been variously explained, but it is now ascribed to the bombardment of the particles by the molecules of liquid in which they are immersed. It represents the true dance of the molecules. By this bombardment they are knocked first in one direction and then in another. In protoplasm the particles are not generally in Brownian movement, but when the protoplasm liquefies the movement begins at once. It would appear from this that the protoplasm of many cells at any rate is rather in a gel than a sol state.

Osmotic pressure of colloidal solutions.—As is to be anticipated, colloidal solutions possess osmotic pressure. They must do so if they have any affinity for water. The colloidal particles are, however, so large that their number is very small even in fairly concentrated solutions when compared with the number of particles of a non-colloidal substance. Indeed, the freezing-point method, or the boiling-point method of determining osmotic pressure, which depends on the vapor tension of the solution, is not sufficiently precise to measure the osmotic pressure accurately, and indeed for some time it was doubted whether colloids really exerted an osmotic pressure. By directly measuring this pressure in an osmometer, however, there is no doubt of its existence. The osmotic pressure of hemoglobin, a colloid, is given on page 142. Since osmotic pressure is but the expression of the affinity of particles and water, it is immaterial whether the particles are in a solid state or in solution. The swelling pressure of gelatin in water is but the expression of its osmotic pressure. The protoplasmic colloids, whether gel or sol, have then an osmotic pressure, but it is customary to call it osmotic pressure only when the colloid particles are in solution.

It is rather interesting to consider the number of particles which are necessary to produce a given osmotic pressure. Suppose a gram of a pure crystalline protein is dissolved in 100 c.c. of solution. The osmotic pressure is measured and found to be 76 mm. of mercury. What is the molecular weight of the protein and what is the number of particles in this solution? The number of molecules in a gram mol. of any substance has been measured in various ways and found to be 6.05×10^{23}. A gram mol. of glucose in a liter of solution has a theoretical osmotic pressure of 22.4 atmospheres at 0° C. In each cubic centimeter there must be 6.05×10^{20} molecules. The solution just mentioned had an osmotic pressure of 76 mm. Hg, that is 1/10th of an atmosphere. It must contain in a c.c. $6.05 \times 10^{19}/22.4$ particles or 2.7×10^{18}. Since the solution contained 0.01 gr. in a c.c., the weight of each molecule is easily calculated. A gram mol. per liter gives a calculated but not an actual osmotic pressure of 22.4 atmospheres. The actual osmotic pressure is more than this (p. 200). This solution containing 10 grams per liter gives an osmotic pressure 1/224th of that of a gram mol. Hence 10 is 1/224th of the molecular weight, and the molecular weight would be 2240. It may be seen from this that an almost inconceivably great number of molecules may be present in a solution and yet exert a very small osmotic pressure.

The electrical phenomena of protoplasm.—Every activity of living matter, whether it is an act of secretion, a muscle contraction such as the rhythmic contraction of the heart, or a nerve impulse, is accompanied by an electrical disturbance in the protoplasm. This electrical disturbance may be detected by connecting two points in the tissue which are not equally active with the electrodes coming from an electrometer or galvanometer. In some cases, as for example the electrical eels, Gymnotus or Malopterurus, or the electrical fishes, such as the Torpedo, the electrical disturbance may be so heavy as to produce a strong shock to animals touching or being in the neighborhood of the discharging animal, and the discharge is of use in stunning enemies or prey. Every living cell is, therefore, a battery and the study of protoplasm as an electrical machine may help elucidate its vital properties.

For the detection of these electrical currents non-polarizable electrodes should be used. These are made usually by immersing a zinc wire or zinc electrode in a strong solution of zinc sulphate, the zinc sulphate is kept from the tissue by interposing between the sulphate solution and the tissue a solution of inert saline, that is N/6 NaCl solution; a string from this solution connects with the tissue. Current passing in either direction through these electrodes does not produce any opposing electromotive force. For the detection of the current various devices are used, such as the oscillograph, the thread galvanometer or the capillary electrometer, for the operation of which special memoirs may be consulted.

By the means just mentioned it is found that the part of protoplasm or tissue which is active is always electro-negative to the part which is more at rest. Every excitation of the protoplasm causes a momentary negativity in the part excited so that a momentary current flows through the galvanometer from the less active to the more active part; and within the tissue the current is completed by the current flowing from the more active to the less active part of the protoplasm. This momentary electric discharge is called the current of action, or the negative variation, or by Waller the "blaze" current. Waller called it a blaze current to indicate that it is as if, following the stimulation, the protoplasm suddenly blazed up in a conflagration. It is usually of very short duration, but if the difference of activity is constant the electrical difference is also constant.

The cause of the electrical phenomena.—The mechanism of the production of this current of action, or of the blaze current, is still unknown. Three or four more or less probable suggestions have been made to explain it. The first, and in many ways the most important of these suggestions, is that of DuBois-Reymond, that protoplasm is composed of polarized electro-motive molecules being electro-negative at the ends and positive in the center. The current is produced by the rotation of these molecules. If cohesion is of a magnetic nature, as appears probable, it is not unlikely that such magnetically polarized molecules really exist in living matter.

The second view is that of the polarizable membrane. It is supposed that protoplasm has in it membranes which are permeable for cations but not for anions and so generates an electro-motive force at the boundaries. Excitation lets the anions out, too, and so produces an electrical disturbance.

The third is that the current is caused by a change in the surface of contact between colloids and water, as it is in the capillary electrometer between mercury and acid.

The fourth view is that it is caused by the production of acids setting free hydrogen ions, making a concentration chain effect.

The fifth view is that it is produced by oxidation and reduction, that it is a burning with the energy set free as electrical energy rather than heat, as it is in a battery.

It is impossible at present to decide which of these theories is correct, or whether any one is correct.

The reaction of protoplasm and the preservation of neutrality.—So far there has been considered the physical chemistry of the structure of living matter, its properties which are due to its colloids and the surface tension (and electrical properties); those properties, in other words, have been considered which have to do with the machinery of protoplasm. We shall now take up briefly the physical chemistry of some of the fundamental chemical properties of the cell, dynamical physical chemistry in other words. One of the simplest and most important of these chemical properties is the reaction, whether alkaline or acid, of the protoplasm and the methods used to preserve neutrality.

Whether the reaction is acid, alkaline or neutral may be determined with greater or less certainty in a variety of ways, all of which show that protoplasm is generally very faintly alkaline in reaction, but that it becomes acid after death and when it works in a medium with too little oxygen. The reaction of living matter may be determined by the use of indicators and in various indirect ways. Many stains are good indicators: that is, they have a strong color change when they pass from an alkaline to an acid reaction. Such a stain is neutral red, which is a deep orange red in an alkaline reaction and a pink in an acid. Neutral red penetrates nearly all cells with ease and stains the granules in the cells and sometimes other structures. It always takes the color it has in an alkaline reaction; moreover, it is found that when protoplasm is made acid it no longer stains with neutral red. Another stain is cyanamine, which has been introduced by Bensley and Harvey. This stain is red in an alkaline solution and blue in an acid solution. It is found that it stains cells or parts of cells red. Other stains may also be used. Thus acid fuchsin will not stain protoplasm as long as it is alkaline, but as soon as it is acid it takes up the stain. By this means it is found that if acid fuchsin is injected into a frog and then the blood vessel to one leg is ligated while that to the other leg is intact, and then if the muscle of the ligated side is made to contract, that muscle will be found, on taking off the skin, to be a bright red, while the other has scarcely a tint of red. This proves that acid is formed in muscle during contraction and accumulates there, if there is not a supply of oxygen. Ordinarily the protoplasm is neutral or alkaline, since it does not stain. The same experiment has been tried with the electrical organ of the Torpedo and with other tissues and the same facts found for the difference between the reaction of active and passive protoplasm. Miss Greenwood fed various stains to Stentor, Paramœcia and other protozoa and found the reaction of the protoplasm to be alkaline to litmus and the contractile and digestive vacuoles to be acid.

There are several difficulties in this use of stains for the determination of reaction. The first is that the stain may itself be poisonous and by killing the cell bring about the production of acid, since dead protoplasm is generally acid in reaction. Another difficulty is that only those stains will accumulate in protoplasm which combine with some element in the cell. Not all stains penetrate easily. Moreover, the cell is strongly reducing so that many stains, like methylene blue, are reduced to a colorless substance as long as the cell is alive; the color can only be seen in those parts of the cell where the reducing or oxidizing powers are least;

where, in other words, the vitality of the cell is least. But the most serious difficulty, and one which is usually overlooked by cytologists, is that the stain is not, properly speaking, a reagent for acids and alkalies at all, when used in a heterogeneous system like the cell. In all cases the difference of color shown by the stain is between the color of the salt and the free base or acid. Now, let us suppose that there are in protoplasm substances, electro-negative colloids, which will form insoluble salts with neutral red. If these colloids do not change their sign with a slight acidity, but remain electro-negative, they will form a true salt with the stain and give the reaction of the salt; as if the reaction were acid, even though the reaction is neutral or faintly alkaline. The same is true of any acid stain. If there are electro-positive colloids anywhere in the cell, these will react with the acid dyes, forming a true salt and giving the salt color, which in a true solution would be held to be the sign of an acid. For example, silk exposed to acid, but thereafter thoroughly washed so that it is not acid in reaction, will stain a brilliant red in acid fuchsin. Stains can be used, then, for the detection of the acidity of protoplasm only with the greatest caution, but nevertheless the evidence, as far as it goes, is to the effect that the protoplasm is usually very faintly alkaline.

Another method of determining the alkalinity of protoplasm is to study the alkalinity of the liquid in which the cells live. It is not likely that a cell will have a reaction very different from the medium in which it lives, although this is at times possible. The blood may be regarded as a living tissue. The reaction of the blood plasma and the lymph of mammals and vertebrates generally is, like that of the sea-water, very weakly alkaline. The blood is alkaline to litmus, but acid to phenolphthalein. The exact reaction of the blood may be determined by the methods given in Chapter XII. By these methods it is found that the plasma has a concentration of hydroxyl ions of about 7×10^{-7} N.

Changes produced by a change in reaction in the cell. The physical and chemical changes which ensue in protoplasm when its reaction is rendered less alkaline, or more acid, are extremely important and profound. The first effect of the production of acid in the cell is seen in a reduction in the rate of oxidation. The respiratory oxidation of the cell is wonderfully affected by a rise or fall of acidity. By an increase in acidity it is checked; by a decrease in acidity it is greatly stimulated. It is always found that the appearance of acid, or any reduction of alkalinity, is accompanied by a failure to oxidize certain cell products which under normal circumstances are oxidized. Thus acidity in the mammal is always accompanied by the appearance of acetone bodies in the urine which may disappear at once, if alkalies, such as sodium bicarbonate, are given. A second important change is the appearance in the cell of autolytic enzymes of various kinds, both proteolytic and carbohydrate splitting enzymes. Invertin, amylase and other digestive and deamidizing enzymes are set free. The appearance of these enzymes is in the nature of an adaptive change, the cell by means of them seeking to recover its metabolic balance. By means of the carbohydrate enzymes it obtains glucose and levulose from glycogen or cane sugar, or other carbohydrate reserves, and by these it is able to carry on its oxidations and respiration under conditions of lack of oxygen and acidosis which it cannot otherwise withstand. Thus in the liver any acidosis is at once accompanied by the transformation of some glycogen into glucose for the protection of the respiration of the liver. The proteolytic enzymes attack the proteins of the cell, setting free ammonia from them by means of which the cell acids are neutralized. Thus every acidosis in the body is at once accompanied by the increase of proteolysis in the cells and the appearance of ammonia, so that by many physiologists the appearance of more ammonia in the urine is regarded as characteristic of acidosis. The autolytic digestion of proteins is distinguished from the digestion by the digestive ferments of the alimentary tract by the production of vastly more ammonia in the former process.

A third marked change produced by the action of acids in cells is a slowing down of all the activities of the cell. This is caused by the period of recovery from activity becoming much prolonged: that is, fatigue comes on much sooner when there is a reduced alkalinity. It is for this reason that athletes are able to withstand fatigue longer if they load the blood with oxygen first. By the action of the muscles in exercise lactic acid is produced, this is at once oxidized, or otherwise removed, if plenty of oxygen is present, but it accumulates and produces fatigue if too little oxygen is present; and by its accumulation it checks oxidation, having thus an autocatalytic action.

Another physical change produced by the accumulation of acid in cells in small amounts is an accumulation of water in the cells and œdema. The colloids of protoplasm take up more water in the presence of acid; this may interfere with the circulation in the case of those organs which, like the kidneys, are inclosed in a tough inelastic capsule. A vicious cycle is thus inaugurated, acid causing swelling, the swelling checking the blood and oxygen supply, thus increasing the production of acid and increasing swelling. In many cases the ingestion of alkalies in such conditions produces a sudden and marvelous improvement.

Regulation of the reaction of protoplasm. With such grave consequences following from any permanent derangement of its reaction, it is not surprising that cells have various methods of preserving that reaction most suitable to their activities. The means employed are in part organic and in part inorganic. The organic consist in the setting free of the autolytic enzymes already mentioned by means of which ammonia, for the neutralization of the acid is set free, and glucose, to enable the cell to survive the period of depressed oxidation caused by the acid. There is in addition the power of the protein constituents of the cell to combine with acid, thus neutralizing it. We have, also, in many cells another mechanism which may also be brought into relation with this process, but which is usually held in abeyance as long as carbohydrate food is present; this is the mechanism of the carboxylase. On the digestion of the proteins amino-acids are set free. By means of these the acids are in part neutralized by the formation of salts, just as by the ammonia, but by the action of the carboxylases the carboxyl group of the amino-acid is split off according to the following equation,

$$CH_3.CHNH_2.COOH \longrightarrow \underset{\text{Ethyl amine.}}{CH_3.CH_2NH_2} + CO_2$$

and a base is left with much greater acid-combining powers than the amino-acid itself. These amines play, also, in the mammalian organism another important rôle, since some of them are strong stimulants of the heart or vaso-dilators, thus increasing the circulation through the parts and hastening the removal of the acid.

In the simpler organisms, but also playing an important part in the higher, are the inorganic safeguards against acidosis. These are primarily the carbonates and phosphates. There are considerable amounts of both these salts in all protoplasm and they have this very important function among others. They act as buffers, in that they can take up a good deal of acid with a very slight change in acidity. Sodium acetate has a similar power. They owe these powers to the fact that both carbonic acid and phosphoric acid are very weak acids, so that when acids such as lactic, or oxybutyric, or almost any other acid, are produced in the cell, the acid so formed at once takes the base away from the carbonate, making bicarbonate, which is scarcely at all acid. If still more acid is formed, carbon dioxide is set free, which is a very weak acid and which escapes from the cell with the greatest ease. The phosphates, also, play a part, although they are held rather in reserve, as phosphoric acid is stronger than carbonic. There is in the cell a mixture of disodium hydrogen phosphate and dihydrogen sodium phosphate. The acid produced is neutralized by the base taken from the former. The former is alkaline, the latter slightly acid in reaction. H_2NaPO_4 ionizes very little into H and $HNaPO_4$, so that the number of hydrogen ions is only very slightly increased. Moreover, there are some $NaHPO_4$ ions from the Na_2HPO_4 which still further check the ionization. The equilibrium of this reaction has been studied by Henderson.

To sum up, then, the production of acid in cells: The immediate effect of this acid is to cause the protein gel to absorb water at the point of formation of the acid and to become more fluid. Possibly the liquefaction of the protoplasm of the amœba at certain regions has this cause. The activities of protoplasm are almost certainly due to this action, in part at any rate. This acid, if it is not oxidized, accumulates. It checks oxidation. It is in part neutralized by the following reactions, which are written for lactic acid, $C_3H_6O_3$:

I. $C_3H_6O_3 + Na_2HPO_4 \longrightarrow NaC_3H_5O_3 + NaH_2PO_4$

II. $C_3H_6O_3 + Na_2CO_3 \longrightarrow NaC_3H_5O_3 + NaHCO_3$

III. $C_3H_6O_3$ + ProteinNa $\longrightarrow$ HProtein + $NaC_3H_5O_3$

IV. 1. Protein + Protease $\longrightarrow$ Amino-acids + Protease
2. $C_3H_6O_3$ + Amino-acid $\longrightarrow$ Amino-acid $C_3H_6O_3$

V. 1. Amino-acid + Deamidase $\longrightarrow$ NH_3 + Oxyacid + Deamidase
2. $NH_3 + C_3H_6O_3 \longrightarrow NH_4C_3H_5O_3$

VI. 1. Amino-acid + Carboxylase $\longrightarrow$ Amine + CO_2 + Carboxylase
2. $C_3H_6O_3$ + Amine $\longrightarrow$ Amine $C_3H_6O_3$

Catalysis.—The mechanism of protoplasm certainly involves the passage of water into and out of the cell or of the cell elements. This movement of water back and forth is in its turn probably to be ascribed to a varying affinity of the protoplasmic colloids, whether protein, carbohydrate or lipin, for water and in part to the varying osmotic pressure of the cell contents. It has been suggested, and the evidence is on the whole favorable to the view, that this varying affinity of the colloids for water is due in large measure to the varying reaction of the protoplasm, often a variation of a strictly local nature, due to the production of acid in the course of the cell metabolism. This consideration led to the discussion of those chemical processes by which acid is produced and got rid of and by which the colloids of the cell are made and broken down. These chemical changes are the source of the energy which moves the water and animates the machine. While these chemical processes are very diverse in their nature and may be considered in each tissue in turn, they are all alike in that they proceed at a rate much superior to that at which they go on outside of the cell when under similar conditions of temperature. This superior rate of reactions in protoplasm is due to the fact that these reactions are hastened or catalyzed, and it is this process which we have now to examine.

The word catalysis is from the Greek *kata*, meaning down, and *lysis*, to loosen. Literally a down loosening, it has come to mean the hastening of a chemical reaction by a third substance, the catalyst, which emerges at the end of the reaction unchanged in amount, or nearly unchanged, since all substances are more or less unstable, and which accordingly has appeared to act only by means of its presence. But while it appears to have acted by its presence only, there is no doubt that in many, if not in all cases, it has actually entered into the reaction at some stage or other, but has become free again. There are a great number of reactions of this kind known outside of cells. For example, hydrogen and oxygen gas do not combine at a measurable speed at ordinary temperatures, although they will combine at higher temperatures. It is, however, to be inferred that they do unite at ordinary temperature, but at so slow a rate that it is not measurable in the times so far studied. But if this mixture of gases is passed over finely divided platinum, union takes place and at a rate so great that it heats the platinum. In this case the platinum is the catalyst. Another example of a catalytic reaction is the union of sulphur dioxide and oxygen to sulphuric acid in the lead chamber process of sulphuric acid manufacture. The presence of nitric trioxide, N_2O_3, hastens this reaction, the nitric trioxide appearing at the end in unchanged amount. Water is one of the most important catalytic agents known. Thus perfectly dry ammonia, NH_3, and HCl will not unite with measurable speed, nor will ammonium chloride dissociate into NH_3 and HCl in the absence of water. The presence of a very minute amount of water catalyzes, or hastens both reactions. Water is necessary for the rusting of iron, or for the union of chlorine and hydrogen. In fact, an enormous number of reactions are catalyzed by water.

To understand how catalytic agents may hasten reactions, we must first consider the factors which determine the velocity of ordinary reactions.

The velocity of a chemical reaction is directly proportional to the chemical affinity, and inversely proportional to the chemical resistance. There is as yet no good means of measuring chemical affinity. It invòlves two factors, mass and attraction. Chemical reactions take place in the direction of doing the maximum of external work. This is simply another way of saying that the reaction is always in such a direction that the total potential energy of the system reacting is reduced to a minimum: in other words, the reaction as a whole, but not necessarily in all its parts, always goes in the direction of greater stability under the conditions of the reaction. By the velocity of a chemical reaction is meant the amount of substance transformed divided by the time. If a gram of cane sugar is inverted in an hour, the velocity of the reaction is 1/60th of a gram a minute. The time required for any chemical transformation is evidently the sum of two periods, namely, the time required for the two or more reacting molecules to come into contact, since chemical actions only take place between molecules in contact or more probably they only take place when they are united, that is they only occur within molecules. It is necessary, then, in order that the reaction shall take place, for the reacting species of molecules to come into contact and unite into a single molecule. The second period of the total time taken is for the molecular rearrangement to take place which constitutes the reaction.

These two periods of time may be illustrated by the reaction by which sulphuric acid is made in the lead chamber process. The first part of the reaction consists in the time necessary for the formation of the intermediate compound, nitrosyl-sulphuric acid, and the second period the time required for the decomposition of this compound:

1. $2SO_2 + H_2O + 2O + N_2O_3 \longrightarrow 2(SO_2.OH.NO_2)$
2. $2(SO_2.OH.NO_2) + H_2O \longrightarrow 2H_2SO_4 + N_2O_3$

Now the first part of the time, i.e., that required for the molecules to meet, will be the shorter the more molecules there are in the space in which they are confined; and evidently the speed of the reaction will be proportional to the number of molecules of each reacting species in the space, or in other words to the concentration of each of the reacting species. This general law of chemical reactions by which the velocity is proportional to the concentration is known as the law of mass action and is sometimes called the law of Guldberg and Waage. It may be put in the form of an equation as follows:

$$\text{Velocity} = V = \frac{\text{Amt. Transformed}}{\text{Time}} = K\,C_a \times C_b$$

K being the constant of proportion and C_a and C_b being the concentrations respectively of the two kinds of reacting molecules *a* and *b*. Since in the absence of any special means of keeping the concentration constant C_a and C_b must of course diminish as the molecular species *a* and *b* are used up in the reaction, it is obvious that the velocity of such a reaction is not constant but must diminish with the time. If, however, a very minute interval of time was taken, the velocity would remain approximately constant during that time interval. If dx is a very minute amount of substance transformed in the very minute time, dt, then the velocity at any instant, t, will be dx/dt and this will be proportional to the amount of substance actually present, and this will be equal to the amount A at the start of the reaction minus the amount x transformed during the period, t, or $dx/dt = K(A-x)$; this is the differential equation of the velocity of a reaction in which only a single molecular species A is undergoing a change in concentration. It applies, for example, to the hydrolysis of cane sugar, the water, which is the other molecular species entering the reaction, not materially changing its concentration, being present in great excess. Since dx and dt are too minute for direct measurement, it is necessary to add a great number of these together to get a time interval and an amount of x which can be measured. This addition is the process of integration; and if the foregoing equation be integrated, or added, there is obtained the velocity equation

$$\text{Log}A - \text{Log}(A - x) = Kt; \text{ or } \text{Log}(A/(A-x)) = Kt$$

t being the time from the beginning of the reaction, A the concentration of the substance at the start and A—x the concentration at the end of the interval t.

Temperature coefficient. The velocity of this part of the reaction, namely the time required for collision of the reacting molecules, is not only a function of the concentration of the molecules, but also, very naturally, it is a function of the speed of their movement; the velocity of the reaction, or rather of this part of it, must hence increase with the temperature. Most chemical reactions increase in speed with the temperature, however, at a very much greater rate than can be accounted for by the increase of velocity of the molecules. For most chemical reactions at temperatures of from 10-40° C. the rate of the reaction doubles or trebles with a rise in temperature of 10°, but the rate of increase is not constant, being greater than twice or thrice at lower temperatures and less at higher. If only the velocity of the molecular movement was concerned in this increase in the reaction velocity, the rate should increase uniformly with the temperature. For example, the average kinetic energy of all molecules at 20° C. (293° Abs.) is $2.015 \times 10^{-16} \times 293$ ergs, and at 30° it is $2.015 \times 10^{-16} \times 303$ ergs. Since the kinetic energy, $\frac{1}{2} MV^2$, is proportional to the square of the velocity of movement, the velocity of the reaction, which is proportional to the speed of molecular movement, at 30° should be to that at 20° as the square root of 303/293 or 1.017. It is evident from this calculation, as the actual increase is 2-3 times this amount, that chemical reactions are increased by a rise of temperature in some other way than exclusively by the increase of velocity of molecular movement. This brings us to the second period in a chemical reaction: namely, the time required for the molecular rearrangement.

Chemical resistance. If it be admitted that rearrangement only takes place within the molecule, if in other words it is admitted that molecules really interact only when combined, a longer or shorter period of time will be necessary for the molecular rearrangement to take place by which the reaction is consummated and the new molecular species are formed. Now this intramolecular rearrangement can only spontaneously occur in the direction of greater molecular stability; that is, of less potential energy. This brings us to the question of the resistance to chemical reactions. While very little is known about this, it is not impossible that it consists, in a measure at any rate, in the stability of the molecular form of the molecule. The atoms are probably packed very closely together in a molecule. The resistance to movement, or the internal molecular friction opposing the molecular rearrangements, may be either high or low, but it is often high. The length of time union between molecular species persists before rearrangement takes place, and the reaction is ended, is extremely variable. In some cases the time is long and the intermediate substances are hence so stable that they may be isolated in quantities; in other cases the resistance to rearrangement is so slight that the reaction takes place almost instantaneously and these intermediate compounds are so unstable that they cannot be isolated and often their existence can only be inferred from the character of the transformation. It is, for example, very difficult often to prove that molecular oxygen unites with the substance oxidized before the reaction is consummated, but oxyhemoglobin is such an intermediate substance which in the absence of either alkaline or acid reaction is fairly stable and may be isolated. Since a rise in temperature increases the motion of the atoms in the molecule and thus increases their lability, it shortens the period of the reaction by shortening the time taken up in the intermediate stage and so hastens the reaction. It accelerates by diminishing the resistance, but does not so greatly affect the chemical affinities. Heat having this double action accelerates chemical reactions more than physical. Nearly all vital reactions or activities are doubled or trebled by a rise of temperature of 10° between the limits 10-40° C.

The action of a catalyst may be pictured in very much the same way as the action of heat, in that the chemical resistance is reduced, so that the time required for the intermediate stage of the reaction is greatly shortened and hence the reaction is hastened. For example, it is probable that the reaction in the union of hydrogen and oxygen to form water involves the intermediate formation of hydrogen peroxide, thus

1. $H_2 + O_2 \longrightarrow H_2O_2$
2. $2H_2O_2 \longrightarrow 2H_2O + O_2$

Finely divided platinum has the property of rendering hydrogen peroxide so unstable that it decomposes with great speed, so that the total time required in the reaction is enormously shortened. There can be very little doubt, also, of the manner in which this hastening is produced. It is found that any substance which will unite with the platinum, and thus presumably occupy the bonds of the platinum where the hydrogen peroxide ordinarily takes hold, will poison or prevent the action of the platinum. Thus hydrogen sulphide or carbon bisulphide are true poisons of this catalysis. It is probable, therefore, that in the presence of platinum there are these reactions:

1. $H_2 + O_2 \longrightarrow H_2O_2$
2. $2H_2O_2 + Pt \longrightarrow 2H_2O_2Pt$
3. $2H_2O_2Pt \longrightarrow 2H_2O + O_2 + Pt$

Another reaction of this same type where the action is hastened by the formation of an intermediate union between the catalyst and an intermediate product of the reaction is that of the formation of ether from alcohol by the action of sulphuric acid.

her fundamental property of catalytic agents is that, in many any rate, they do not change the point of equilibrium of revers- tions. A great many reactions, possibly all of them, never go ely to an end. They apparently come to rest, but it is found that more or less of the unchanged, reacting substance still present is happens. Such reactions are said to be reversible. It is char- c of the common reversible reactions that very little energy e takes place in them. Such a reversible reaction is that gen- ited between acetic acid and ethyl alcohol. If acetic acid and are mixed, union takes place between them with the formation of etate. The reaction apparently comes to an end when there are 3 molecules each of alcohol and acetic acid and 66 of the ethyl If ethyl acetate be dissolved in water, it will break up into acetic l alcohol until the three reacting molecules are present in the same on as before. This point is called the point of equilibrium of tion.

$$C_2H_4O_2 + C_2H_5OH \rightleftharpoons C_2H_5.O.CO.CH_3 + H_2O$$

ction goes in the left-handed direction if ethyl acetate is dis- n water, and in the right-handed if a start is made with alcohol l. At the time of equilibrium the reaction has not stopped, but on in such a way that the number of molecules of acetate break- s just equal to the number being formed in any time interval.

reaction may be catalyzed by lipase, an enzyme or catalytic agent n cells, but it is found that while the point of equilibrium is in a shorter time, it has not changed the relative concentration eacting molecules. The fact that the point of equilibrium of the is not changed by the catalyst means that the catalyst must te the reactions in each direction equally, otherwise, in any of time, there would be more molecules of acid and alcohol unit- n of ethyl acetate breaking up, or vice versa, and the point of ium would be shifted. On the theory of the catalysis being due formation of an intermediate unstable stage, this behavior is understood, since the reaction has to pass through this stage in er direction it is going. This fact, that the catalysts catalyze both reactions in a reversible reaction, is known as *the reversible of enzymes*. It was first observed in the case of the enzyme . This catalyzes the union of glucose to form isomaltose and nd of maltose and water to form glucose. It has since been shown other reversible reactions and other catalysts. Kastle and Loe- showed the reversible reaction of lipase, and Taylor reported the le synthesis of protamine by a proteolytic enzyme.

intermediate body composed of catalyst and reacting molecules is generally unstable, but it is conceivable that when it is formed it might in some way be rendered more stable. If this were the case, we should have a complex formed of enzyme and various species of reacting molecules which might be very complex and stable within narrow limits. It is possible that the synthesis of amino-acids to make proteins, and other syntheses, are brought about in this way in protoplasm, the protoplasm being essentially composed of the enzymes united with the substances upon which they act. The substance making the intermediate stage stable might be called an anti-ferment.

The catalytic agents of cells are known as enzymes, a word meaning literally *in yeast,* from the Greek, *en,* in; *zyme,* leaven. An enzyme is an organic catalytic agent found in, or isolated from, living matter. These catalytic agents are very numerous and it is to them that the activity of living protoplasm in a chemical sense is due. Some of these enzymes are easily isolated from cells; they are exocellular; such are the various digestive enzymes, pepsin, trypsin, invertin, ptyalin, maltase, and the alcoholic enzyme, zymase. Others, however, have not yet been isolated, and the more fundamental reactions of living matter, such as the oxidations or the preliminary fragmentations of the molecules, are apparently due to some enzymes of a very unstable kind which are firmly tied to the structure of the cell. It may be that for these reactions the simultaneous presence of two or more contiguous or loosely-bound enzymes may be necessary, so that by separating them their action is lost. At any rate, it has so far been impossible to bring about the synthesis produced by the cell if the structure of the cell is first destroyed.

There are a great variety of enzymes found in cells; among them are those which produce hydrolyses, such as the digestive enzymes, and by whose action the synthetic formation of various colloidal constituents may be explained; oxidases; peroxidases; catalases; and various enzymes producing fermentations such as zymase. Among the hydrolytic enzymes may be mentioned invertin, maltase, laccase, amylase, dextrinase, cytase, emulsin, myrosin, pepsin, trypsin, erepsin, probably rennin, and other proteases; the esterases, such as various lipases; deamidizing enzymes, such as adenase, guanase; arginase; nucleases; and glyoxalase. It is possible that these enzymes form part of the organized protoplast, and that their hydrolytic activity is checked by the presence of anti-enzymes, particular conditions of alkalinity and so on.

The physical chemistry of oxidation.—Since the whole of the energy used in the production of living phenomena comes immediately or secondarily from the oxidation of the foods, an understanding of the process of oxidation is necessary before vital processes can be understood; a short account of the theories of the nature of oxidation may be given at this place. There are two kinds of oxidations going on in living matter:

namely, those taking place at the expense of the oxygen of the air, and those in which the oxidation is produced by easily reducible food substances or their metabolic fragments. The first process is called aërobic respiration; the second anaërobic. In aërobic respiration the oxidizing agent is the oxygen of the air.

The term oxidation, which literally means a process of souring, from the Greek, *oxys*, acid, includes in chemistry not only processes which involve the transfer of oxygen, but it is used to signify any process which results in the increase of the number of positive valences, or the diminution of negative valences of a compound or element, whether this is produced by oxygen or some other agent. Thus the reaction between ferric chloride and potassium iodide by which iodine is set free is called an oxidation, the iodide being oxidized by the ferric chloride. The reaction is as follows:

$$\overset{+++}{Fe} + 3\overset{---}{Cl} + \overset{+}{K} + \overset{-}{I} \longrightarrow \overset{++}{Fe} + 3\overset{---}{Cl} + \overset{+}{K} + I$$

It will be seen from the ionic reaction that the oxidation has really involved the passage of a positive charge of electricity from the ferric atom, which is the oxidizing reagent, to the iodine atom; or the passage of a negative electrical charge from the iodine ion to the ferric ion, thus reducing it. It will be noticed that there cannot be an oxidation without a corresponding reduction. A similar reaction is that between nitric acid and silver, leading to the formation of silver nitrate. This may be written as follows from the ionic point of view:

$$\overset{+}{NO_2} + \overset{-}{OH} + Ag \longrightarrow \overset{+}{Ag} + \overset{-}{OH} + NO_2$$

$$\overset{+}{Ag} + \overset{-}{OH} + \overset{+}{H} + \overset{-}{NO_3} \longrightarrow \overset{+}{Ag} + \overset{-}{NO_3} + H_2O$$

In this reaction it will be seen that the oxidizing reagent is the ion, NO_2, which has a positive charge held at a very high potential, and that this is the cause of the oxidation of the metallic silver to the positively charged Ag ion. It may at first seem unlikely that nitric acid should dissociate in this way into NO_2 and OH, but it is not by any means impossible that such a dissociation in small amounts takes place. The weaker the acid is, the more likely it is to dissociate somewhat as a base as well as an acid. Water, for example, functions both as a base and an acid. Boric acid is nearly as basic as it is acid. Hypochlorous acid, HClO, is also a very weak acid and probably dissociates in this way in part into $\overset{+}{Cl} + \overset{-}{OH}$; the positive chlorine being the active agent. It will be observed that in all the oxidations of this type a hydroxyl group combines with the oxidized substance. Another reaction, an oxidation which does not involve oxygen, is the oxidation of zinc by an acid. In this case there is the reaction

$$Zn_2 + 2\overset{++}{H} + 2\overset{--}{Cl} \longrightarrow \overset{++}{Zn} + 2\overset{--}{Cl} + H_2$$

In this reaction the hydrogen is the oxidizing agent giving up a positive charge to the zinc which is thus oxidized. Similarly all processes of oxidation, could we trace them out, would be found to involve the transfer of a negative electron from one element to another, the one which receives it being reduced and the element losing the negative charge being thus rendered more positive and being said to be oxidized.

Whether the foregoing picture of the process of oxidation be in all particulars right or not, it is beyond question that the oxidation does involve, in all cases in which the process can be watched, the transfer of positive and negative electrical particles or electrons, and that this is the essence of the process. Moreover, the more easily a substance gives up a negative charge the more active will it be as a reducing agent; and similarly the more easily it gives up a positive charge, or acquires a negative, the more active will it be as an oxidizing agent. Oxygen acts as an oxidizing agent because it has a great tendency to take away a negative charge from other substances and go over into the form of an oxygen ion, or of electro-negative oxygen.

The great importance of this theory from the point of view of physiological chemistry is that it shows at once that every oxidation in protoplasm is at the bottom an electrical process involving the transfer of electrical charges. In other words, an electrical disturbance of some kind, albeit possibly within molecular dimensions, must occur in every combustion in the protoplasm. It thus furnishes a point of attack of the origin of the electrical disturbances which are so characteristic of living matter of all kinds and enables an understanding of the disappearance of these currents when the respiration of the protoplasm is prevented.

While ordinarily the transfer of the electrical charge from one atom to another releases, directly or indirectly, in the form of heat, energy which had been potential, under suitable conditions this energy does not take the form of heat, that is of indiscriminate molecular vibrations, but of a steady migration of the ions, the positive in one direction, the negative in another, so that we have an electrical current which can do work. This happens in the particular arrangement which is called a battery. If a piece of zinc is placed in a solution of sulphuric acid, it dissolves with the liberation of hydrogen gas and of much heat. In this case the oxidizing substance is the hydrogen ions of the acid, and the oxidized substance is the zinc which escapes into solution as a positive ion. The heat may be due to the violent separation of the zinc and hydrogen after the transfer of the charge from one to the other. If, however, the zinc be placed in a solution of zinc sulphate, and this is in contact through a porous cup or directly with a solution of copper sulphate in which is a plate of copper, and if the copper and the zinc are connected with a wire, the zinc dissolves as it did before, and copper is deposited, but there is no appearance of heat or very little, but instead an electrical current is produced from the zinc to the copper in the solution and in the opposite direction outside. This reaction is a true oxidation-reduction reaction. In this case the zinc dissolves extremely slowly when the battery is not short-circuited, that is when the zinc and copper are not connected, for the reason that there are in a zinc sulphate solution so very few hydrogen ions to oxidize the zinc; but as soon as the connection is made with the copper, the copper ions in the solution which have a greater oxidizing potential than the hydrogen are able to give up their charges to the copper plate and these charges are conducted by the wire to the zinc plate, thus oxidizing the zinc so that it can go into solution as a positive ion. This is, as it were, an oxidation at a distance, the oxidizing and reducing agents not being in direct union, but indirectly through the copper plate and wire. There is, of course, some loss of energy as heat produced by the movement of the ions through the solution and in part by the movement of the electrons through the wire, but the loss is not great. The important point is that the processes in a battery which give rise to the electrical phenomena of the battery are oxidation-reduction processes.

It is not inconceivable, although it has not yet been possible to prove it, that the electrical phenomena of protoplasm might arise directly in this way from the oxidation-reduction phenomena of protoplasm. They may, however, have an indirect relation to the oxidation, as has been pointed out.

In order that the oxygen of the air shall oxidize it must first go into solution. It is only in the presence of water that oxygen has the power of oxidizing rapidly. The first question, then, is the mechanism of the oxidation by oxygen. What happens to oxygen when it goes into solution in water? This raises a most fundamental question, to which no definite answer can at present be given; there have been several answers.

The first view is that of van't Hoff. According to him, the oxygen probably ionizes when it goes into solution in water, splitting into a small number of $\overset{+}{O}$ and $\overset{-}{O}$ ions. It is the $\overset{+}{O}$ ions which have the oxidizing power. It has not yet been possible to prove that such an ionization takes place, although something similar appears to happen in many gases at high temperatures, dissociation into atoms taking place. It is also suggested by others that all processes of ionization involve a union between solvent and ionizing substance and this view, while not entirely incompatible with the foregoing, will, if substantiated, cause some modification of the explanation.

Another view is that of Traube, according to which the oxygen always unites with the water to form hydrogen peroxide, which is the real oxidizing agent. The reaction might be written as follows:

$$2H_2O + O_2 \longrightarrow 2H_2O_2$$

This view leaves unexplained the cause of the oxidizing power of the hydrogen peroxide.

There is a growing amount of evidence that all true solution is a process of chemical union between the solvent and solute, possibly through the extra valences on the molecules of the two kinds. It may be that the oxidation is similar to that of the oxidation by chlorine or bromine. When bromine dissolves in water it is known to form hypobromous acid, bromates and bromide. The reactions might be written as follows on the basis that the interaction of the water and the bromine can only take place when a chemical union exists between them:

1. $2(Br_2) + 2H_2O \longrightarrow (2H_2O)(Br_2)_2$
2. $(2H_2O)(Br_2)_2 \longrightarrow 2HBr + 2HOBr$
3. $HBrO \longrightarrow \overset{+}{H} + \overset{-}{OBr}$; or $HBrO \longrightarrow \overset{+}{Br} + \overset{-}{OH}$

Or the reaction might be with the water molecules $(H_2O)_3$

1. $2Br_2 + (H_2O)_3 \longrightarrow Br_4(H_2O)_3$
2. $H_6O_3Br_4 \longrightarrow 3HBr + HOBr + H_2O_2$

Since the ion, BrO, has little or no oxidizing power, and the power of the hydrogen ion is very much less than the solution possesses, the oxidizing agent would be the positive bromine obtained from the bromine hydrate.

The oxidizing power of oxygen may be represented in the same way:

1. $2O_2 + 3H_2O \longrightarrow H_6O_7 \longrightarrow 2OOH + H_2O_2 + H_2O$
2. $OOH \longrightarrow \overset{+}{O} + \overset{-}{OH}$

The oxidizing agent would be the oxygen hydrate. The above reaction would be in case the oxygen is monovalent in the gaseous form. If it were bivalent, the reaction would be

$$2O_2 + 3H_2O \longrightarrow O(OH)_2 + 2H_2O_2$$

These reactions would account for the general appearance of hydrogen peroxide during the reaction. They are, of course, in the case of oxygen, purely hypothetical, since neither $O(OH)_2$ nor OOH have been isolated. There are two facts, however, which are undoubted: one is that the action of the oxygen is enormously more rapid in the presence of water, and that hydrogen peroxide is formed accompanying many oxidations. It would seem, therefore, that a union of some kind between the water and the oxygen certainly precedes the oxidizing process. The solubility of oxygen clearly indicates this also, since the solubility is greater than that of a completely indifferent gas such as hydrogen, or helium. There is no doubt, either, that in the case of the halogens the acids corresponding to hypobromous acids are always formed when they dissolve in water; and it is equally certain that the oxidizing power of the metals, such as

$Cu(OH)_2$, is due to the presence of hydrates in the solutions. The facts are, then, that the exact behavior of oxygen in water is uncertain, so that it is impossible to say just what the oxidizing principle really is. It is worth noting that, if there is in protoplasm a substance which will combine with oxygen in the way supposed for the water, it will form just such a union with the oxygen. Hemoglobin is such a substance.

Summary.—We may summarize as follows the results of the study of the physical chemistry of protoplasm made in this chapter. Protoplasm, that is the real living protoplast, consists of a gel, or sol, which is composed of the colloids of an unknown nature which include protein, lipin and carbohydrate. Whether these colloidal particles consist of one large colloidal compound in which enzymes, protein, phospholipin and carbohydrate are united to make a molecule which may be called a biogen, cannot be definitely stated, but it seems probable that something of the sort is the case. This colloid exists in the form of a gel. That is, it always contains a large amount of water and this water has in it salts. The gel of the protoplasm is not often uniform, but it is differentiated physically and chemically in different parts of the cell. The cell is not isotropic, as the morphologists say. The movements of the protoplasm and so the activity of the cell, the vital activity, appear to be due to the varying affinity of this gel, or of particular parts of it, for water, by which water is caused to enter, or leave it. This affinity for water may be modified in various ways. It may be modified by salts, which exist usually in loose or more firm chemical union with the colloids. Some salts if introduced into the protoplasm will cause the protoplasm to take up more water, others to lose water. It may be modified by a change in the reaction of the cell, by the production of acid. And, above all, it is modified by the chemical changes occurring in the colloid itself, for this colloid is very unstable. The last is the cause undoubtedly of most of the rhythmic and other activities of protoplasm. The protoplast, which is probably the continuous phase, undergoes oxidations, and in virtue of the changes thus produced the affinity for water by the colloid is changed. It may be simply a local affinity alteration, such as we see in the streaming amoeba, in which the protoplasm suddenly appears to become more liquid in one region or another of the cell. A stimulus, on this view, is anything which alters the affinity for water on the part of the protoplast, and as this affinity is a very delicate adjustment it may be altered by a great variety of means. Hence stimuli may be either chemical, physical or mechanical, since by all of these means we may produce chemical changes which will alter the affinity of the cell for water.

Every activity of this protoplast is accompanied by an electrical disturbance, the blaze current, or current of action. The way in which this electrical disturbance is produced is still entirely dark; and its significance, or rather its possible function, is equally dark. But men have imagined three possible ways in which it might be produced: It might be produced by the change in the surface of contact of the colloid and water, as happens in a capillary electrometer when the surface of contact of mercury and acid is altered; it might be produced by the appearance of acid as a result of chemical decomposition, the hydrogen ions in some way setting up a concentration chain effect by their greater velocity of movement, the colloids assisting by forming semi-permeable membranes, thus interposing resistances to the passage of the negative ion; or the electrical disturbance might be the direct result of the oxidation, every oxidation involving a minute current when the positive charge is passed from the oxidizing to the oxidized body. How such an effect could be propagated to a distance beyond the molecule has not been explained. Evidently the explanation of the mechanism of the production of this electrical disturbance must be left to the future.

All the chemical processes in the cell, so necessary for the quick response to a stimulus and to recovery from the effects of a stimulus, are accelerated by the presence in the cells of accelerators of these reactions, and these accelerators are called enzymes. The nature of none of these is definitely known, and the protoplast itself, or the biogens, appear to be the most important of these accelerators. The enzymes, there are reasons for thinking, are not distributed evenly through the cell, but exist in definite locations, so that the changes in one part differ from those in another, thus producing a physiological division of labor and a physiological diversity no less marked than the morphological diversity.

Finally the structures of cells are so characteristic and definite as to show that the cell is organized in some way or other. The suggestion has been made that this organization must in the long run be caused by the molecules of which the protoplast is composed, just as the form of a crystal is produced by the molecules of which it is composed. It must hence be the expression of the molecular form of the biogens.

REFERENCES. Osmotic Pressure.

1. *Findlay:* Osmotic Pressure. Monographs on Inorganic and Physical Chemistry. Longmans, Green and Co. London, 1913.
2. *Van 't Hoff:* Zeit. physikal. Chem. 1, p. 481, 1887.
3. *Traube:* Arch. f. Anat. Physiol. u. wissensch. Med. p. 87, 1867.
4. *Pfeffer:* Osmotische Untersuchungen, 1877.
5. *Graham:* Phil. Trans. 144, p. 117, 1854.
6. *Kahlenberg:* Jour. Physical Chem., 10, p. 141, 1906.
7. *Berkeley and Hartley:* Phil. Trans. A, 206, 486, 1906. A. 209, 177 and 319.
8. *Morse:* American Chem. Journal, 48, p. 29, 1912. Earlier papers in the same journal. Osmotic pressure of concentrated solutions.
9. *Garrey:* Some cryoscopic and osmotic data. Biol. Bull., 28, 77, 1915.

10. *Callendar:* Pro. Roy. Soc. A, 80, 466, 1908.
11. *Stern:* Zeit. physikal. Chem. 81, 441, 1912.
12. *I. Traube:* Phil. Mag., 8, 704, 1904. Surface tension and osmotic pressure.
13. *Armstrong:* Proceedings Roy. Soc. A, 78, 264, 1906.
14. *Taylor:* Chemistry of Colloids. Longmans, Green and Co., London, 1915.

PROTOPLASM AS A PHYSICAL SYSTEM.

1. *Rhumbler:* Das Protoplasma als physikalisches System. Ergeb. d. Physiol. 14, 1914, pp. 475-617.

PHYSICAL STRUCTURE OF PROTOPLASM.

1. *d'Arsonval:* Rélations entre la tension superficielle et certains phenomènes d'origine animale. Arch. de Physiol. (6), 1, p. 460, 1880.
2. *Bechhold:* Strukturbildung in Gallerten. Zeits. physikal. Chem., 52, p. 185, 1905.
3. *Bechhold:* Die Kolloide in der Biologie und Medizin. Dresden, 1912.
4. *Bernstein:* Chemotropische Bewegungen eines Quecksilbertropfens. Ar. ges. Physiol. 80, p. 628, 1900.
5. *Bernstein:* Die Kräfte der Bewegung in der lebenden Substanz. Naturwiss. Rundschau, 16, 1901, pp. 413, 429, 441.
6. *Bernstein:* Die Energie des Muskels als Oberflächenenergie. Ar. ges. Physiol. 85, p. 271, 1901.
7. *Bernstein:* Bemerkung zur Wirkung der Oberflächenspannung im Organismus. Eine Entgegung. Anatom. Hefte. 27, p. 823.
8. *Berthold:* Studien über Protoplasmamechanik. Leipzig, 1886.
9. *Bethe:* Zellgestalt, Plateausche Flüssigkeitsfigur und Neurofibrille. Anat. Anz. 40, p. 209, 1911.
10. *Biedermann:* Vergleichende Physiologie der irritablen Substanzen. Ergeb. Physiol. 8, p. 26, 1909.
11. *Biedermann:* Die Aufnahme, Verarbeitung und Assimilation der Nahrung. Winterstein's Handbuch der vergl. Physiol. 2, 1911.
12. *Buglia:* Hängt die Resorption von der Oberflächenspannung der resorbierten Flüssigkeit ab? Bioch. Zeits., I, p. 1, 1909.
13. *Bütschli:* Untersuchungen über mikroskopische Schäume und das Protoplasma. Leipzig, 1892.
14. *Bütschli:* Ueber Strukturen künstlicher und natürlicher quellbarer Substanzen. Verh. d. naturhist.-mediz. Ver. Heidelberg. N.F. 1895, pp. 360-368.
15. *Bütschli:* Untersuchungen über die Mikrostruktur künstlicher und natürliche Kieselsaure Gallerten. Ibid., 6, p. 287, 1900.
16. *Danilewsky:* Versuche über die elektrische Pseudoirritabilität toter Substanzen. Arch. Anat. Physiol. 1906, p. 413.
17. *Dellinger:* The cilium as a key to the structure of contractile protoplasm. Jour. Morph. 20, p. 171, 1909.
18. *Fischer:* Fixierung, Färbung u. Bau des Protoplasmas. Jena, 1899.
19. *Fitz-Gerald:* On the change of superficial tension of solid-liquid surfaces with temperature. Scientific writings, p. 307, 1878.
20. *Freundlich:* Kapillarchemie. Eine Darstellung der Chemie der Kolloide und verwandter Gebiete. Leipzig, 1909.
21. *Gaidukow:* Dunkelfeldbeleuchtung und Ultramikroskopie in der Biologie und in der Medizin. Jena, 1910.
22. *Gebhardt:* Knochenbildung und Kolloidchemie. Arch. Entwickelungsmechanik, 32, 1911, p. 727.

PART II.

THE MAMMALIAN BODY CONSIDERED AS A MACHINE. ITS GROWTH, MAINTENANCE, ENERGY TRANSFORMATIONS AND WASTE SUBSTANCES.

We have now to inquire more in detail concerning the nature of the processes by which the animal body grows, and maintains itself from the foods; what the source is of the heat it so constantly produces; whence comes its power of moving and doing work; what is the nature of the waste products it forms, and the causes of the variations which they show in diverse conditions of diet and health. All these processes are included in the province of the science of nutrition. It is, then, the nutrition of animals and in particular of mammals which we shall consider in the following chapters. We shall take up, first, the production of heat in the body and then pass on to the study of the processes involved in the maintenance of the animal organization and in its development and energy transformations.

In order that organisms shall maintain their form unchanged, that they shall transform into their own substance this mass of material different from themselves on which they subsist,—that they shall make themselves out of their foods,—it is necessary that there shall be some kind of an organizing force at work. Organization is at the bottom of everything living, not only of the material side of our existence, but of the mental as well. The organism may be but an enlarged, complex and semifluid crystal. That we shall remember experiences, that we shall continue to exist as individuals, something of a material kind must persist, or be reconstituted from moment to moment. What is that something? What is the nature of the organizing principle of living things? These are fundamental questions, but they are questions to be solved by the experiments of the future. We cannot answer them at present.

The body resembles a magnet. The body in many particulars resembles a magnet. A magnet has existence as a magnet only as long as the materials of which it is composed are organized; only as long as the molecules are oriented or organized into a definite relation to each other. The magnet is itself, then, an organism. It has the power, like the body, of organizing other materials like itself, so that if new iron is brought to it, it makes the new iron into a part of the magnet by organizing the ultimate particles of the new iron to correspond to the organization or orientation of its own molecules. It has the power of growth. It will pick the right materials, fine iron particles, out of a mixture of iron and other substances, rejecting the useless and assimilating the substances like itself. It may be cut in two, like many organisms, and each part becomes a magnet like that from which it came. If heated to a point at which its organization is lost; if heated, in other words, to a point at which the orientation of the molecules in consequence of their rapid movement is lost, the magnetism disappears. Its property of magnetism is lost. The magnet dies as an individual. How closely similar to all this, at least in its superficial aspect, is the life history of the mammalian organism. Beginning life as a very minute organism, it assimilates to itself, out of the mixture of foodstuffs brought to it, substances like itself. It grows like the magnet. Its power of organization reminds one irresistibly of the magnet's powers of organization. Like the magnet, the organism cannot assimilate all kinds of things, but only those things of a certain special kind depending on the shape and nature of the molecules. Like the magnet, too, it can only continue to exist below a certain temperature. If heated to its critical temperature the organism, like the magnet, ceases to exist as an organism, although composed of the same elements as before the heating. In the heated magnet, although it ceases to exist as a magnet, the property of magnetism still exists concealed from our eyes in the molecules of which it is composed. Each molecule preserves the property of magnetism. Is it not probable that so in an organism the psychic and other properties of vitalism are really inherent in and continue to exist in the particles of which the molecules and atoms are composed, but that they are concealed from our dull vision? The organism at death may be resolved into a multitude of psychic elements, of elements possessed of vitalism, but this vitalism, this psychism, is now in a multitude of small molecular and atomic units and the organism as an organism ceases to exist. The law of the conservation of energy, namely, that energy can neither be created nor destroyed, but may be organized to appear in different forms, is one of the fundamental laws of physics, chemistry and physiology. The biologist may some day add its counterpart in the law of the conservation of psychism. The property of psychism may be a fundamental property of every atom; nay, of every electron. But it is only when organized that this property appears to us in the form of an individual, a larger psychic unit.

The magnet does not appear to be undergoing vigorous chemical change. In this respect it differs from the animal body, which is the seat of such changes. But this difference is probably but a difference in degree. The atoms of which the iron magnet is composed, there is reason now for thinking, are not everlasting and changeless, but they may be having a metabolism of their own, new electrons being absorbed from the ether, and old electrons being given off to the ether. All is in flux in nature. Stability is but an appearance. Our brief lives are like the fraction *dt* in a differential equation, infinitely brief in the tim of the universe. Things appear constant when observed for so short a time.

We cannot pursue speculations of this nature concerning those great problems awaiting the future biologists. It is for us to clear the ground. And we may comfort ourselves with two reflections as we consider how hopeless the solution of those fundamental problems appears at the present time: the first is that there is really nothing insignificant in nature, however small our problems may look to us. The problem most trivial in appearance, if followed but a little way, brings us to the most profound mysteries of nature; and the second is that, however small a fact may appear to be, any discovery made is multiplied by an infinite factor in the course of time, since it is multiplied by infinite time, so that every discovery is in reality infinitely valuable.

With these preliminary observations, which are not in many ways very a propos, we may pass now to the consideration of the immediate problems of how organisms maintain themselves, what the nature is of the processes by which the material stream which flows through them is organized into the body itself; whence come from this organization those properties of life and consciousness which we may include under the term vitalism?

CHAPTER VI.

ANIMAL HEAT.

Animal heat.—The mammalian body is generally warmer than its surroundings. It has a temperature of about 37.5° C., which is kept nearly constant irrespective of the outer temperature. It produces heat like a stove; it moves and expends energy in moving. Whence comes the heat of the body; and what is the source of the energy used in movements, in doing work? What are the natures of the processes by which heat is liberated? These questions, or their prototypes, were among the most puzzling which men asked when they began to inquire concerning the nature of things; and very crude were the answers which were given in the twilight of learning. Men imagined that heat was an essence, or substance, which found its way into and out of the body; or that it was one of the four cardinal elements of which things were made, earth, air and water being the other three. Men knew nothing concerning the nature of heat. They did not know what was happening when a piece of wood burned, so that their ideas about animal heat were necessarily crude and generally wide of the mark. Even near the end of the eighteenth century such brilliant and far-seeing men as Lavoisier and Laplace still considered heat to be a substance which could be added to or taken away from bodies; although they also suggested that it might be a mode of motion of the finer particles of which matter was composed.

History of the discovery of the origin of animal heat.—In the last part of the seventeenth and first part of the eighteenth century men began more and more to try experiments to discover something about nature, for " it is only by experiment that we see in the full light of day." Some experiments were tried by Mayow, Boyle and Priestley. They put small animals into confined spaces and discovered that they soon died. They found also that if when one animal was dead they introduced a second into the same jar without renewing the air, the second died in a shorter time than the first. The air in the jar was no longer able to support life. Moreover, if a candle was put under a similar jar, it was extinguished after a time. If, then, a second lighted candle was introduced without renewing the air, the light at once went out. These observations showed that animals and candles behaved in the same way in a closed space. Life and light were extinguished. The experiment was then tried of seeing whether a mouse would live in the air exhausted by a candle and whether a candle would burn in air in which a mouse had suffocated. In other words, did a candle and a mouse change the air in the same way? It was found that mice would not live in air in which candles could not burn. There was evidently something similar in the burning of a candle and the breathing of an animal. Black, who studied what had happened to the air, says in 1757: " I have convinced myself that the change produced in healthful air by the act of respiration consists principally, if not entirely, in converting a part of it into fixed air [now called carbon dioxide], because I found that in breathing by means of a tube into water of lime or into a solution of caustic alkali, I precipitated the lime and caused the alkali to lose its causticity." Shortly after this Priestley made a kind of air by heating mercuric oxide, which he called dephlogisticated air, but which is now called by Lavoisier's name of oxygen. He collected this air and put a lighted candle into it. It burned better than in ordinary air. He then put a mouse into some more of the dephlogisticated air and it lived longer than in the same quantity of ordinary air. This " dephlogisticated air " was beneficial alike to burning candles and mice. Priestley also found that it changed dark venous blood into red arterial blood, a change which was known to occur in the lungs.

Lavoisier.—The question now turned on finding out what the candle did to the air when it burned in it. This problem was solved by the great French chemist Lavoisier about 1776. He proved by a series of simple and convincing experiments that Priestley's " dephlogisticated air " was a gas present in ordinary air and forming about one-fifth of its volume. When metals were heated in air they became heavier, not because they lost the light spirit of phlogiston which had been driven away by the heat, but because they combined with this oxygen. And he proved also that the so-called " fixed air " was carbon dioxide, a compound of carbon and oxygen. The burning of a candle consisted, then, in the combination of the carbon of the candle with the oxygen of the air to form carbon dioxide, and in this process heat was liberated. Since carbon, or phosphorus, or sulphur burning in this air formed acids, he called the new or dephlogisticated air, the acid-maker, " oxygen " (Gr. *oxys*, sour or sharp; *gennao*, I produce). He then proceeded to find if animals affected air in the same way as burning candles. He found that they did. In 1777 he published the observation that animals placed under a jar until dead changed a portion of the oxygen of the air to carbon dioxide. To re-establish the air so that it would support life, it was necessary to absorb the carbon dioxide and restore an equal part of oxygen. He says: " If one augments, or if one diminishes, in a given quantity of air, the quantity of eminently respirable air (oxygen) that it contains, one augments or one diminishes in the same proportion the quantity of metal that one can calcine in it and up to a certain point the time that an animal can live in it." Having thus clearly demonstrated that both burning candles and living mice took oxygen out of the air and added carbon dioxide to it, he goes on to say, in regard to the place in the body where the conversion of oxygen to carbon dioxide takes place, " I have been forced to two conclusions equally probable and between which my experiments do not permit me to decide. Either the portion of air eminently respirable (oxygen) contained in atmospheric air is converted into (carbon dioxide) fixed air in passing through the lungs; or it makes an exchange in this organ, on the one hand the eminently respirable air is absorbed, and on the other hand the lung returns for it a portion of carbonic acid almost equal to it in volume."

Having thus shown the identity of the chemical changes produced by combustion and respiration, he at once inferred that, since a candle liberates heat when it burns and animals also produce heat, this animal heat probably came from the combustion of the body. In his Memoir to the French Royal Academy in 1777 he says: " I have shown that the pure air after having entered the lungs comes out in part in the state of fixed air (carbonic acid). The pure air in passing the lungs undergoes a decomposition analogous to that which takes place in the combustion of coal. But in the combustion of coal there is a disengagement of heat (materie de feu), of which there should be a similar disengagement in the lungs in the interval of inspiration and expiration, and it is this heat, without doubt, which distributed by the blood throughout the animal economy gives rise to the constant temperature of about 32.5° Réaumur. This idea appears perhaps hazardous at first glance, but before rejecting or condemning it, I ask that it be considered that it is founded on two constant and incontestable facts: namely, on the decomposition of air in the lungs, and on the liberation of the matter of fire (heat) which accompanies all decomposition of pure air, that is to say, all passage of pure air to the state of fixed air." As further proof of this hypothesis he points to the fact that those animals which are warmest, such as the birds, produce the most carbon dioxide in a given time in proportion to their weight.

In order to establish this revolutionary and beautiful hypothesis that animal heat was due to the combustion of the carbon of the body, he attempted to measure the amount of heat produced by an animal and the amount of carbon burned and thus see whether the combustion of this amount of carbon would account for the heat produced. It was necessary for him to perfect first a means of measuring heat, so with the aid of his great compatriot Laplace, one of the greatest of French physicists, he perfected the ice calorimeter which had been made by Black. This calorimeter consisted of a double-walled can, as shown in Figure 34.

In the space between the walls cracked ice was placed. This prevented any heat from reaching the interior of the can from the outside. In the interior there was placed a cage which would hold a guinea pig, a rabbit or other small animal, and this cage was packed in ice. The heat given off by the body of the animal melted some of the ice in the interior compartment and the water from the melted ice was collected and weighed. Since the melting of each gram of ice requires an amount of heat sufficient to raise the temperature of a gram of water from 0° to 80°, and the amount of heat necessary to raise the temperature of

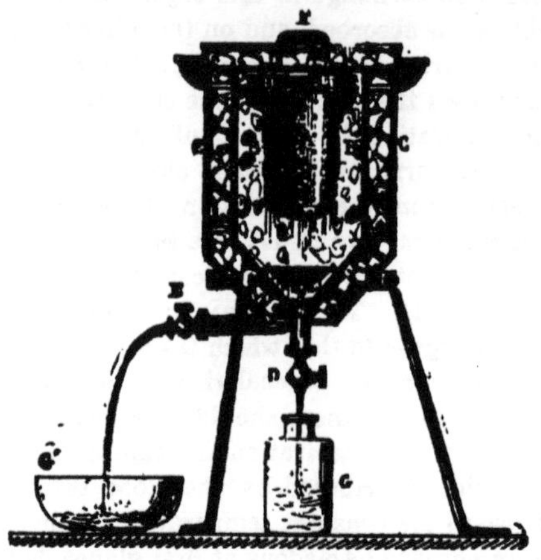

Fig. 84.—Ice calorimeter of Lavoisier and Laplace. The animal was placed in the cage, *A*, and the heat of its body melted ice in *B*, the water being collected in *G* and weighed. The outer layer of ice in *C* prevented heat from entering.

water one degree was taken as the unit amount of heat and called one calorie, to melt a gram of ice requires 80 calories of heat. Of course, arrangements had to be made for the ingo and outgo of air for respiration. Before or after being in the calorimeter the guinea pig was placed in another apparatus in which the carbon dioxide exhaled could be measured. The result of this experiment was as follows when translated into modern units: In 10 hours the guinea pig burned 3.33 grams of carbon, computed from the carbon dioxide exhaled. The heat liberated by burning this amount of carbon Lavoisier and Laplace determined to be sufficient to melt 326 grams of ice to water at 0°. The guinea pig when placed in the calorimeter for a given time liberated sufficient heat to melt in 10 hours 402 grams of ice. This should be a little larger than the heat generated by the combustion of the carbon, for the pig was at a lower temperature when emerging from the calorimeter than on entering it. Some of the preformed body heat had been lost. Estimating that the heat thus lost from the body would melt 61 grams of ice, there remain 341 grams of ice melted, as compared with the 326 grams which should have been melted by the combustion of the 3.33 grams of carbon of the body. Thus he arrived at the conclusion that at least 96 per cent. of the heat of the body was to be accounted for by the combustion of the carbon. The experiment may be tabulated as follows:

Amount of carbon burned by a guinea pig in 10 hours, 3.33 grams.
Heat liberated by burning this amount of carbon melts 326.5 grams of ice.
Pig in calorimeter for 10 hours melted 402.27 grams ice.
Reduction of temperature of body accounts for 61.19 grams ice.
Heat produced by guinea pig in 10 hours hence melted 341.08 grams ice.
Heat produced by burning carbon in per cent. of total = 326/341 = 96 per cent.

Having thus proved that animal heat was due, in large part, to the union, or combustion, of carbon and oxygen, he proceeded to find out if the oxygen combined with anything else than the carbon. Measuring the amount of oxygen consumed and the oxygen in the carbon dioxide, he found that only 81 per cent. of the oxygen combined with the carbon; the other 19 per cent. must combine with something else, namely hydrogen, to form water. If the figures just given were corrected by the addition of the heat formed by the combustion of the hydrogen, there was a slight excess of heat computed over that found. There was also a source of error in the experiment. He had measured the heat produced when the pig was exposed to a temperature of 0°, but the carbon dioxide produced while the pig was exposed to 18° C. Perhaps the body burned more carbon and produced more heat when it was cold than it did at room temperature. This thought led him to other fundamental discoveries. He undertook to determine whether more oxygen was consumed at low temperatures than at high; and whether more oxygen was consumed and more heat liberated when work was done. His experiments in this direction were never published in full, owing to the interruption in the publication of the Academy Memoirs caused by the French revolution, and to the death of Lavoisier on the guillotine in 1794, but an abstract had been published. These conclusions are fundamental for the science of nutrition. They are as follows:

1. A man in repose and fasting at an external temperature of 32.5° R. consumed per hour 24.002 liters O_2.

2. A man in repose and fasting at an external temperature of 15° R. consumed per hour 26.66 liters oxygen.

3. A man during digestion consumed per hour 37.689 liters oxygen.

4. A man fasting, doing work necessary to raise in 15 minutes a weight of 7.343 kilos to a height of 199.776 meters, consumed per hour 63.477 liters oxygen.

5. A man during digestion doing a similar amount of work consumed per hour 91.248 liters oxygen.

Here we have the fundamental observations that the consumption of oxygen and excretion of carbon dioxide are increased by cold, diminished by warmth; and increased by work and by digestion. Furthermore, the quantitative data are given for man himself.

Lavoisier sums up his work in the following words: "Respiration is only a slow combustion of carbon and hydrogen, which is similar in all respects to that which takes place in a lamp or lighted candle; and from this point of view the animals which respire are truly combustible bodies which burn and consume themselves." "In respiration, as in combustion, it is the air which furnishes oxygen and heat, but since in respiration it is the substance of the animal itself, it is the blood, which furnishes the combustible, if the animals do not habitually repair by their aliment that lost by respiration, the oil would be lacking just as in a lamp; the animal would perish like a lamp extinguished when it lacks nourishment. The proof of this identity of respiration and combustion is deduced immediately from experiment." "In concluding these reflections upon the results which have preceded, one sees that the animal machine is principally governed by three regulatory principles: respiration, which consumes hydrogen and carbon and furnishes heat; transpiration, which augments or diminishes, following the necessity of getting rid of more or less heat; and digestion, which returns to the blood that which has been lost by transpiration and respiration."

I have dwelt so long on the work of this great scientist because the discoveries he made and expressed in such clear and beautiful language are the fundamental discoveries of our science. Of that wonderful group of brilliant Frenchmen who lived at the end of the eighteenth and the beginning of the nineteenth century he was one of the greatest. He terminates his last Memoir by a consolatory reflection wrung from him, no doubt, by the distracted state of his country and his own threatening future:

"We will terminate this memoir by a consolatory reflection. It is not indispensable, in order that one should merit well from humanity and pay a tribute to his country, that he should be called to public functions and aid in the organization and regeneration of empires. The physician, also, in the silence of his study and laboratory, can exercise his patriotic functions; he can hope by his work to diminish the mass of evils which afflict the human species, to augment its happiness and well-being; and can he contribute by new ways which he discovers to the prolongation of some years, of some days even, of the average human life, he, also, may aspire to the glorious title of benefactor of humanity."

Where in the body is the heat produced?—The heat and the energy of the body were thus proved by Lavoisier to come from the combustion of the materials of the body. We may now consider where this burning takes place by which carbon dioxide and water are formed and heat set free. Lavoisier left this question undecided. He did not know whether the combustion took place in the lungs, or whether only an exchange of gases, of oxygen for carbon dioxide, took place in the lungs, but the combustion somewhere else. This question was settled by the work of various men. Lagrange objected if the combustion took place in the lungs the amount of heat the body gives off is so great that if all were liberated there it would injure the lung tissue. Spallanzani showed that all kinds of animals, whether they had lungs or not, gave off carbon dioxide and consumed oxygen and liberated heat. He took the lungs out of a frog and found that the animal still breathed through the skin. Edwards drew hydrogen gas through blood and found that the blood gave off carbon dioxide and oxygen to the hydrogen, thus showing that it already contained both of these gases before it reached the lungs. Finally Magnus in 1838 received some blood under the receiver of an air pump and demonstrated that there was a large amount of gas in blood and that venous blood contained more CO_2 and less oxygen than arterial blood. These facts showed that only an exchange between the gases of the blood and the outside air took place in the lungs; the combustion took place somewhere else. This conclusion was confirmed by measuring the temperature of the blood going to the lungs and that of the blood coming away. The blood coming away was cooler, not hotter. It should have been warmer if the combustion took place in the lungs. On the other hand, blood coming from the muscle was hotter, not cooler, than that going to it. Bertholet showed that muscles became hotter during contraction. Consequently by 1850 it was believed that the combustion took place either in the blood of the capillaries in the organs of the body or in the organs themselves. Finally by the work of Pflüger and Hoppe-Seyler, among others, it was shown in 1870 that the blood had little power of combustion and that combustion took place in the tissues, and indeed in the living matter of the tissues. The union of oxygen and living matter, or combustion, takes place and carbon dioxide and heat are produced, then, in the living cells themselves. The real respiration occurs here. It is the living matter which is burning, not the food substances circulating in the blood.

Origin of the heat set free on combustion.—Where is the heat before it appears as heat? Is it in the foods, or in the oxygen, and in what form is it? This was long a puzzling question and in some ways it still is puzzling. It may be answered tentatively and in part as follows: Both carbon and hydrogen have a great attraction for oxygen. For some reason, not at present understood, an atom of oxygen and an atom of carbon in certain conditions attract each other so strongly that to separate them much work is required. A similar attraction exists, also, between hydrogen and oxygen. Why the union between hydrogen and oxygen should be so much firmer than between other elements is not yet clear; but there is no doubt about the fact. When, therefore, they are separated, energy is consumed; and this energy is represented by the separate positions or distance apart of the atoms. It is energy of position or potential energy. It is supposed to be a condition of strain in the ether which fills all space. Light, working in the chlorophyll parts of plants, is able to bring this separation to pass. The light energy disappears and is represented by the potential energy of the system carbohydrate-oxygen. Now for some reason carbohydrate, which contains both carbon and hydrogen, does not combine readily with oxygen in spite of the great attraction between the oxygen and carbon atoms. It is as if there was some resistance in the way of their union; but under the conditions prevailing in protoplasm this resistance, whatever its nature, disappears, and now the carbon and oxygen atoms rush together with great violence, drawn by their mutual attraction. In the violence of their impact they rebound and vibrate vigorously back and forth. Sometimes this vibration is so fast and so vigorous as to give rise to light. This happens in the phosphorescent substances; in other cases the vibration is communicated to the surrounding molecules and is gradually dissipated in longer molecular vibrations, and this we call heat.

Comparison of the amount of heat liberated with the amount calculated from the oxidation of the carbon and hydrogen. The conservation of energy.—Having shown that the heat and energy of the body come from the oxidation of the substances of the body and chiefly the carbon and hydrogen, for as we shall see nitrogen leaves the body in an unoxidized form, chiefly as urea, $CO(NH_2)_2$, we may now consider the no less important question: Is all the energy of the body due to oxidation? Has the body no other source of energy? Is there an exact balance of the energy set free by the combustion in the body and the amount given off from it? Energy may leave the body as heat, or be rendered latent, that is changed to potential energy, in the evaporation of water or in the doing of work. Does the law of conservation of energy hold in living things? Has man no other source of energy than his foods and oxygen? For there might be some kind of radiant energy penetrating the universe which is neither light nor heat, nor X-rays, but of a period of vibration of such a character that it might be absorbed by man's body and transformed into intellectual or physical work. Has man any such source of energy? This, as will be seen, is a very important question and one which has, perhaps, not yet been absolutely settled.

If there were a kind of ray of radiant energy which could freely penetrate the bones of the skull and surrounding tissues, so that these were transparent to it, but which was absorbed by some of the constituents of the brain cells so as to produce metabolic changes in them, it might be possible to affect the brains of men, or portions of the brains, or other organs, directly without the interposition of matter. It might in this way be possible, if the wave length was such as to influence only certain specific parts of the brain, to set going processes which might result in definite ideas in trained minds. There is hardly any question that each substance has its own absorption spectrum, although that spectrum may lie in the ultra-violet or involve wave lengths far too short for the substances of the retina to perceive. There would seem to be no theoretical objection to the possibility of such an absorption of radiant energy on the part of the brain or other tissues of the body; but there is as yet no evidence for such a source of energy.

Work of Dulong and Depretz.—Lavoisier, in the rough experiments which had sufficed to lead him to such correct and brilliant conclusions, had not been able to strike an exact balance between the income and outgo of energy. He measured as heat about 104 per cent. of that calculated from the combustion of the carbon; if he added to the calculated amount the amount of heat computed from the combustion of the hydrogen, then he found too little. The result was not satisfactory when left in this form. Accordingly the French Academy offered a prize for an investigation of this matter. Two investigations were undertaken, one by Depretz and the other by Dulong. Both of these observers improved on Lavoisier's methods in that they measured the heat produced and the carbon and hydrogen burned at the same time, and the animal was kept at or near room temperature. They weighed the carbon dioxide and water given off while the animal was in the calorimeter. They made use of a number of small animals, dogs, rabbits, birds, etc. They made what has since been called a respiration calorimeter, the heat being absorbed by water instead of ice. Figure 35 shows Dulong's calorimeter. The results obtained by these observers are summed in the table after correcting their results for more modern values for the heat of combustion of carbon and hydrogen.

Dulong: Animals used: dog, rabbit, pigeon, cat, fowl. For every 100 calories registered by the calorimeter that calculated from the respiration accounted for:

Minimum	79.2 cals.
Maximum	99.4 "
Mean	90.6 "
Depretz:	
Minimum	84.2 "
Maximum	101.8 "
Mean	92.3 "

As will be seen, they were unable to get a complete agreement between the heat calculated from the products of respiration and that measured, but they came sufficiently close to show that certainly 92 per cent. of the heat of the body was due to the combustion. These observers made one observation which turned out to be of great importance. They found that in dogs the volume of carbon dioxide given off was a smaller pro-

FIG. 35.—Respiration calorimeter of Dulong. The air entered at *D*, and after passing through the coil, *S*, passed out at *D'*. The can containing the rabbit is immersed in the water of the calorimeter, as shown.

portion of the volume of the oxygen consumed than in rabbits and fowls. They suggested that this might be due to a difference in the character of the food. The ratio of the volume of carbon dioxide exhaled to that of the oxygen consumed (c.c. of CO_2 exhaled ÷ c.c. O_2 consumed) is called the ***respiratory quotient***. This suggestion of Dulong's has proved to be correct and has contributed much to the determination of the character of the substance burning in the body.

Respiratory quotient. Regnault and Reiset.—Dulong's suggestion, that the respiratory quotient was dependent on the character of the food, was substantiated by Regnault and Reiset in what is one of the fundamental investigations in nutrition. They made use of the apparatus in Figure 36, which is essentially a closed respiration system in which oxygen consumption is measured directly. They measured more accurately the ratio of carbon dioxide produced to the oxygen consumed, and the relation of the respiratory activity to the size of the animal.

They made two fundamental discoveries. First, that as required by the theory of Lavoisier, those animals of the smallest size, and consequently of the largest surface in proportion to their weight so that the radiation of heat was a maximum, consumed more oxygen per gram per hour and gave off relatively more carbon dioxide than larger animals of the same

FIG. 36.—Respiration apparatus of Regnault and Reiset. The oxygen was supplied as consumed from the containers *N*, *N'* and *N''* and the CO_2 was absorbed by the absorption bulbs, *C* and *C'*, which were alternately moved up and down. The system was closed.

kind. This observation of the relation of respiration to surface was extended by Rubner many years later to the heat production. Smaller animals produce more heat per gram than larger. Second, they showed that there was a relationship between the amount of the oxygen consumed and the amount of carbon dioxide exhaled and the character of the food. They thus discovered what is now known as the dependence of the respiratory quotient on diet, a possibility suggested by Dulong.

Regnault and Reiset. CO_2/O_2 ratio and its dependence on diet:

Animal	Diet	$c.c.CO_2/c.c.O_2$
Rabbit	Carrots and vegetables	0.92
Dog	Bread	0.93
	Lean meat	0.74
	Fat meat	0.69
Fowl	Grains	0.93
	Bread	0.97
	Meat	0.68

Those foods consisting chiefly of carbohydrate gave a high respiratory quotient; those of protein were intermediate; and foods containing much fat produced the lowest quotient.

Cause of the variation of the respiratory quotient with diet.—The reason why the respiratory quotient varies with the nature of the materials being oxidized in the body is the following: Carbohydrates, having the general formula $C_nH_{2n}O_n$, already have in their molecules sufficient oxygen to oxidize all of the hydrogen to water. This leaves only the carbon to be oxidized by the oxygen consumed in respiration. Each molecule of oxygen consumed combines then with an atom of carbon to make one molecule of carbon dioxide. Since both carbon dioxide and oxygen are gases, they follow Avogadro's law, according to which the same number of molecules of every gas occupies the same space at the same temperature and pressure. Accordingly the volume of the carbon dioxide exhaled equals the volume of oxygen inhaled and the respiratory quotient for a carbohydrate is unity.

$C_6H_{12}O_6—6H_2O \longrightarrow C_6$; $6C + 6O_2 \longrightarrow 6CO_2$; vol. $6CO_2$/vol. of $6O_2 \longrightarrow 1$.

There is not sufficient oxygen in the molecule of fats to oxidize the hydrogen of the molecule, consequently a portion of the oxygen consumed will not reappear in the form of carbon dioxide, but in that of water. The respiratory quotient will hence be less than unity when fat is burning in the body.

1. $C_{57}H_{110}O_6 - 6H_2O \longrightarrow C_{57}H_{98}$
   Tristearine.
2. $C_{57}H_{98} + 81.5\ O_2 \longrightarrow 57CO_2 + 49H_2O$
3. Vol. of $57CO_2$/vol. $81.5\ O_2 \longrightarrow 0.7$

In burning tristearine, then, only 57 volumes of carbon dioxide are exhaled for each 81.5 volumes of oxygen consumed, the respiratory quotient when fat is burning is then 57/81.5 or about 0.7. For triolein it would be a little higher than this, since there are six less hydrogen atoms in the molecule.

For protein the respiratory quotient lies between the two, since the protein molecule contains a larger proportion of oxygen than the fats and less than the carbohydrates. For a protein the ratio CO_2/O_2 is about 0.8.

In the preceding computations it has been assumed that the oxidation is complete to carbon dioxide and water. In no case, however, does the oxidation in the body go completely to carbon dioxide and water. Small fragments of the molecules partly oxidized are to be found in the urine. This error is practically negligible in the case of the carbohydrates. In round numbers the respiratory quotient of a carbohydrate is taken as unity; of a fat as .71 and of a protein as .8. If the respiratory quotient is high, carbohydrate is supposed to be oxidizing in the body. If it is low, fat is burning. It is clear that in anaërobic respiration, and all tissues may have some anaërobic respiration for a short time, carbon dioxide is liberated, although no atmospheric oxygen is consumed. For short periods, therefore, the amount of carbon dioxide exhaled may be greater than that of oxygen inhaled, giving a respiratory quotient greater than unity. For other short periods oxygen may be stored, that is combined without the production of an equivalent part of carbon dioxide, the oxidation not being complete. In experiments extending over 24 hours or several days, however, these irregularities more or less completely neutralize each other, since in the long run the oxygen of the CO_2 comes

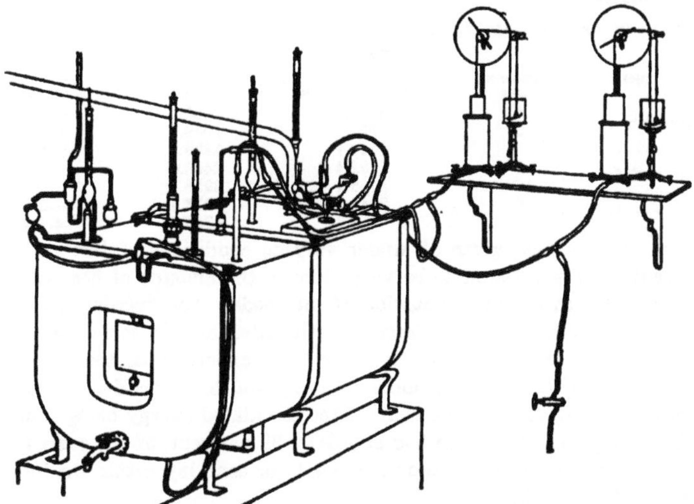

Fig. 37.—Rubner's respiration calorimeter for animals.

directly or indirectly from the oxygen of the air, at least in the higher animals.

Proof of the conservation of energy in the body. Rubner.—It was still uncertain from these determinations whether the income and outgo of energy of the body could be balanced. It was necessary, if a complete energy balance was to be had, to measure not only the energy of the food taken, but that still left in the feces and urine. The final proof that such a balance exists and that the law of conservation of energy holds for the highest animals was not given until 1892. It was the work of the German physiologist Rubner.

In 1892 Rubner had constructed for him a very accurate respiration calorimeter. Figure 37. The calorimeter was kept as nearly as possible at a constant temperature and at the temperature of the room; and the heat produced by an animal in the calorimeter raised the temperature of the air passing through the apparatus and could thus be measured.

The calorimeter was large enough to receive a dog and permitted the simultaneous examination of the water and carbon dioxide produced. The urine and feces were collected and examined for the amount of potential or unconsumed energy they contained, and the experiments were continued each time for several days. The amount of potential energy contained in the food was carefully determined. With this apparatus Rubner was able to get the results a few of which are shown in the table.[1]

COMPARISON OF THE HEAT ACTUALLY PRODUCED WITH THAT COMPUTED FROM THE METABOLISM.

Diet	Number of days	Heat calculated from metabolism	Heat directly determined	Difference—per cent.
Starvation	5	1296.3	1305.2	—1.42
	2	1091.2	1056.6	
Fat	5	1510.1	1498.3	—0.97
Meat and fat ...	8	2492.4	2488.0	
	12	3985.4	3958.4	
Meat	6	2249.8	2276.9	—0.42
	7	4780.8	4769.3	+0.43

The amount of heat given off under varying conditions of fasting and different diets was found to be very close to the amount of heat calculated as set free by the combustion of the foods. The maximum variation was some 1.5 per cent. from the calculated and in some cases the two values agreed almost exactly. By these experiments, then, Lavoisier's theory of the source of animal energy and heat was demonstrated to be true. Instead of 92 per cent. of the calculated energy being found, with the improvement in method and data, 99 per cent. of the energy of the body was accounted for and the other 1 per cent. lay within the limits of error of the method.

From these experiments Rubner established the fundamental law of animal thermodynamics: *Energy is neither created nor destroyed in the animal body. The energy appearing as free energy is exactly equal to the energy of the materials burned. Conservation of energy is as true within the animal body as elsewhere in nature.* The animal behaves in all respects like a machine. But the error of 1 per cent. was still too large. It was desired to test the law on man himself and, moreover, to carry out on him experiments in nutrition. The ultimate aim of all physiological inquiry is to enlarge man's knowledge of himself and to enable him to control the living world. It might still have been true that man had some unknown source of energy different from other animals, although this was unlikely. For the purpose of studying the nutrition of man, calorimeters had to be constructed which should be large enough to accommodate a man. The best of these calorimeters was first constructed in this country in Middletown (Conn.) in the laboratories of Wesleyan University.

[1] Lusk: *Science of Nutrition*. 2nd edition, p. 42.

The Atwater-Rosa-Benedict calorimeter.—The following description of the Atwater-Rosa-Benedict calorimeter is taken from the report of Benedict and Milner.[1] Figure 38. "The respiration apparatus is in principle essentially the same as that of Regnault and Reiset, that is a so-called closed circuit apparatus. The same current of air is kept in circulation through the chamber, the carbon dioxide and water vapor imparted to it by the subject being removed as it is withdrawn from the

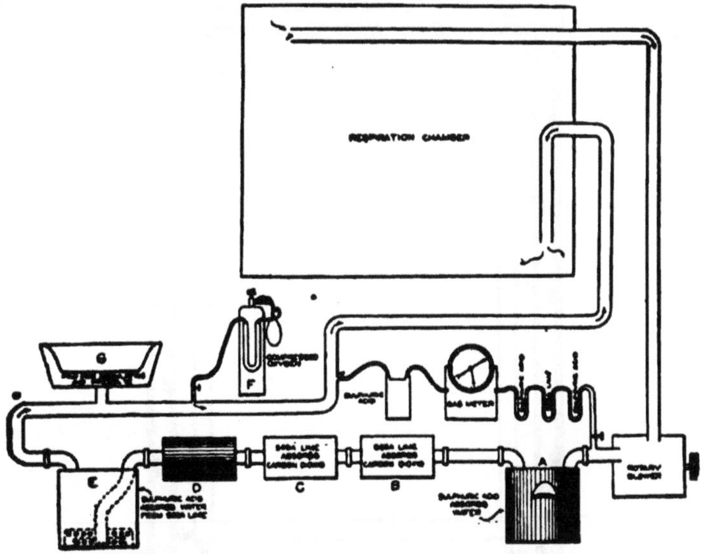

FIG. 38.—Diagram of Atwater-Rosa-Benedict respiration calorimeter for human beings (Benedict and Milner).

chamber and oxygen restored to it as it is returned to the chamber." The quantities of carbon dioxide and water removed from the air and of oxygen supplied to it are ascertained as follows: The carbon dioxide is absorbed by the soda lime tubes C and B in the figure, which are weighed on an accurate balance: the water is absorbed by the sulphuric acid absorber just before the soda lime tubes; the amount of oxygen is determined by weighing the bomb containing compressed oxygen before and after the period of observation. Of course, corrections have to be made for the water condensed in the chamber about the water pipes which carry away the heat; and for the composition of the air left in the chamber at the end of the experiment. The air is caused to circu-

[1] Benedict and Milner: Experiments on the Metabolism of Matter and Energy in the Human Body, 1903-1904. U. S. Dept. Agriculture, Bulletin 175.

late by the rotary blower which goes at such a speed that 75 liters of air a minute are drawn from the chamber. To provide for the expansion or contraction of the air due to changes of barometric pressure, or of slight changes of temperature in the apparatus, the pan G, which is provided

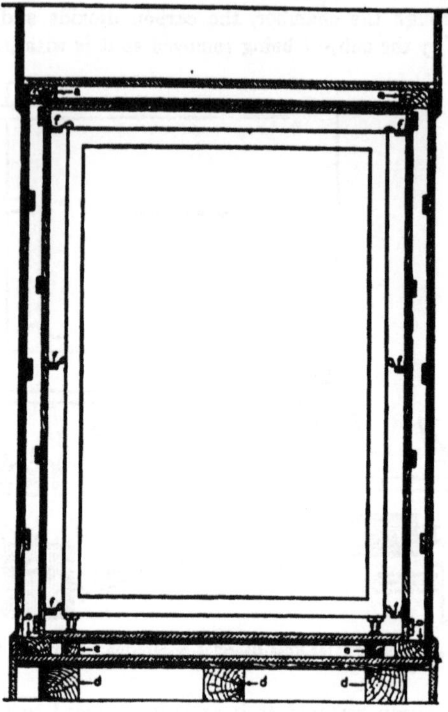

FIG. 39.—Vertical cross section of the calorimeter showing wooden outer walls and zinc and copper inner walls and air spaces (Atwater and Benedict).

with a rubber disk, which rises or falls as the pressure is greater or less than that in the outer air, is put into the circuit.

The calorimeter.—"The device employed in these investigations combines a respiratory apparatus and a calorimeter in the same construction, and the determination of the heat output of the subject is made concurrently with the measurements of the respiratory products and oxygen. Figure 39. The apparatus is so devised and manipulated that the passage of heat through the walls of the chamber is prevented, and the heat evolved by the subject cannot escape in any other way than that provided for carrying it out and measuring it. A small quantity leaves the chamber as latent heat of water vapor, but the major portion is sensible heat absorbed by a current of cold water passing through a coil of pipe within the chamber." "The principle of construction of the calorimeter as a whole is much like that of an ordinary refrigerator, namely, a chamber surrounded by a series of confined air spaces. The walls, ceiling and floor of the calorimeter chamber, which is of course the respiration chamber, are copper. On all sides of the chamber the copper is attached to a wooden framework, to the outside of which is fastened a zinc shell concentric with that of the copper, a dead air space, about three inches across, separating the two metal shells. About three inches outside of the zinc is a concentric shell of wood, and an equal distance from this is the outer wooden structure. The space between the zinc and the inner wooden shell is the 'inner air space,' and that between the two wooden shells is the 'outer air space,' both of which are thus designated hereafter. The wooden casing surrounding the chamber on all sides, and especially the air spaces with the devices for heating and cooling them, afford means for controlling the temperature of the zinc wall and protecting it against fluctuations in temperature of the air surrounding the outer wooden casing."

Gain or loss of heat through the metal walls of the chamber is prevented by keeping the zinc wall at the same temperature as the copper, in which case there will be no exchange of heat between them. For this purpose provision is made for heating or cooling the inner air space, and thus heating or cooling the zinc. The heating is accomplished by passing a current of electricity through a German silver resistance wire installed in the space, the amount of heat being controlled by a rheostat on the observer's table. For cooling, a current of water is passed through a small brass pipe in the same space.

Similar provision is made for heating or cooling the outer air space to aid in protection against changes in temperature of the air of the laboratory. "The attempt is made, of course, to keep the temperature of the laboratory as near as possible to that of the interior of the calorimeter. In order to know whether to heat or cool the zinc wall, it is necessary to know whether it is hotter or colder than the copper. For this purpose thermo-electric elements of iron and German silver wire are installed between the two metal walls in such a way that one end of each element is in thermal contact with the copper and the other with the zinc wall, and are connected with a delicate galvanometer on the observers' table. The difference between the temperature of the copper wall at one end of the elements and that of the zinc wall at the other end is indicated by the deflections on the galvanometer, one direction showing that the zinc is cooler, the other that it is warmer than the copper, and accordingly whether to heat or cool."

The arrangements are made in such a way that it is possible to heat the bottom, top or sides separately.

"In order that the chamber will neither gain nor lose heat by changes

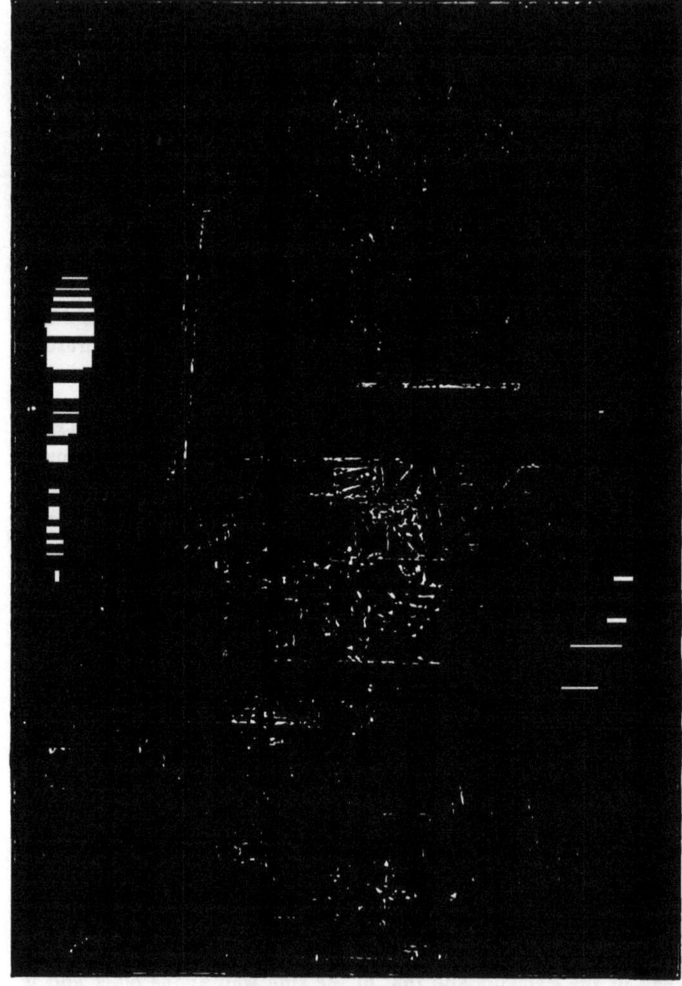

FIG. 40.—General view of the respiration calorimeter in the laboratory. The calorimeter with its window open stands on the right of the picture. The observer controlling the temperature sits at the table in front of the calorimeter with the galvanometer before him (Benedict and Milner).

in temperature of the air current, the temperature of the ingoing air is kept the same as that of the air leaving the chamber. A system of thermal junctions is installed with one end in the ingoing and the other in the outgoing air, and provision is made for heating or cooling the ingoing air as indicated by the galvanometer deflections."

Removal of heat from the chamber. A portion of the heat generated within the chamber is carried out as latent heat of water vapor in the air current. The major portion, however, is removed by a current of

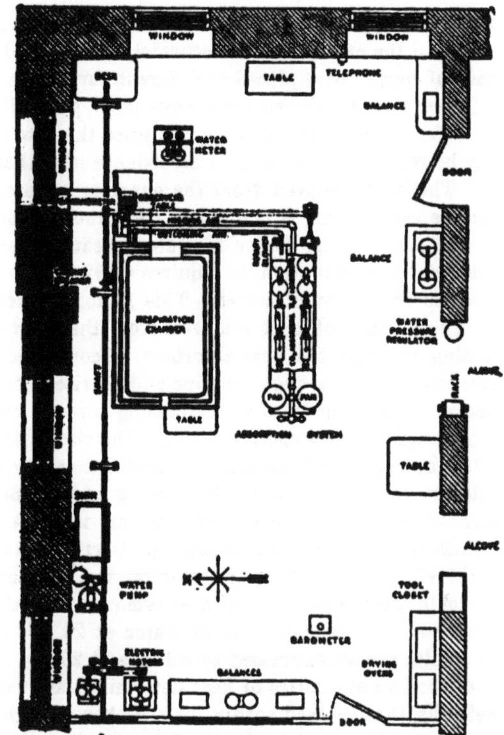

FIG. 40A.—Diagram of the respiration calorimeter laboratory (Benedict).

cold water which passes through a pipe extending around the chamber near the ceiling and absorbs the heat. To increase the heat-absorbing area a large number of copper disks are soldered along the pipe.

It has been stated that an effort is made to extract the heat just as fast as it is produced, and thus maintain a comparatively constant temperature in the chamber. To this end it is necessary that the rate of absorption may be accurately controlled. This is accomplished in three ways: First, by increasing or diminishing the rate of flow of water through the pipe; second, by raising or lowering the temperature of the water entering the heat-absorber; and third, for the finer regulation, exposure of the absorbing area to the heat may be increased or diminished by raising or lowering the metal shields which partially surround the pipe and disks. As the shields are raised they become filled with cold air and act as insulators; when they are lowered a greater absorption area is exposed.

Determination of the quantity of the heat evolved. " The temperature of the air within the calorimeter is accurately determined by means of resistance coils of copper wire placed in different parts of the chamber. The variations in temperature shown are always small, generally amounting to not over a few hundredths of a degree, since the rate of abstraction of heat may be regulated so closely in accordance with that at which it is produced. The heat removed from the chamber as latent heat of water vapor in the air current is computed from the weight of water absorbed from the air and the factor for latent heat of vaporization. According to the best available data, it requires 0.592 calorie (kilogram calorie) to vaporize one gram of water. This factor is used in these computations. The amount of heat removed from the chamber by the cold water passing through the heat-absorbers is computed from the amount of water that passes through the pipe and its rise in temperature during its passage. The quantity of water passing through the absorber is determined by weighing on a special device. The rise in temperature is determined by observation of carefully calibrated mercury thermometers whose bulbs are immersed, one in the ingoing and the other in the outgoing water. These are read every two to four minutes, according to the rate at which the temperature is changing. Corrections in readings are made for the effect of pressure of water on the bulb of the thermometer. In calculating the quantity of heat removed from the chamber by the water current, the specific heat of water at 20° C. is taken as unity, and the results are all expressed as calories at 20°.

" The sum of the two quantities of heat determined as just described comprises practically that evolved within the chamber, though allowance is made for certain small quantities involved in the change of temperature of the calorimeter, and in the passage of objects into or out of the chamber through the food aperture at a temperature different from that of the chamber."

Appointments of the respiration chamber. The respiration chamber is 7 feet 6 inches long, 4 feet wide and 6 feet 6 inches high, the dimensions being such as to allow a man to stand or lie down at full length, or even to move about to a limited extent. In one end is an opening through which a subject enters or leaves the chamber at the beginning or end of an experiment. During an experiment this is closed by glass, tightly sealed in place and serves as a window, admitting ample light for reading and writing. In the opposite end of the chamber is a smaller opening known as the food aperture, through which receptacles for food, drink, excreta and other articles may be passed in the course of an experiment. Within the chamber are a chair, a table and a bed, all of metal and all of which may be folded and put aside when not in use. For experiments in which muscular work is performed, a device by means of which the amount of work done may be measured is provided. There is a telephone for communication with persons on the outside. The arrangement of the furniture, bed, shelving, etc., in the space available is such as to make the subject comfortable during an experiment which may continue for two weeks.

Testing the accuracy of the apparatus. To test the accuracy of the calorimeter and the respiration apparatus alcohol is burned in the calorimeter. The results of such a test are given in the following table:

RESULT OF A TYPICAL ALCOHOL CHECK EXPERIMENT.

Date	Duration		Alcohol Burned	Carbon Dioxide			Water		
				Found	Required	Per cent. Theory	Found	Required	Per cent. Theory
	Hrs.	min.	Grams	Grams	Grams		Grams	Grams	
Dec. 16-17	2	30	50.22	86.02	87.30	98.54	58.47	58.16	100.53
	5	15	101.76	174.27	175.84	99.11	115.73	117.15	98.79
	11	25	212.59	369.63	369.53	100.03	248.24	246.20	100.83
	2	05	40.14	70.96	69.77	101.70	46.52	46.49	100.07
Total	21	15	404.71	700.88	702.44	99.78	468.96	468.00	100.21

Date	Oxygen consumed			Heat		
	Found	Required	Per cent. Theory	Found	Required	Per cent. Theory
	Grams	Grams		Calories	Calories	
Dec. 16-17	96.48	95.23	101.31	295.10	288.98	102.11
	183.02	191.83	95.41	582.18	582.10	100.00
	405.37	403.14	100.55	1223.20	1223.29	100.00
	78.27	76.12	102.83	233.13	230.98	100.02
Total	763.14	766.32	99.59	2333.61	2325.35	100.36

What, then, has been the result obtained by this most accurate calorimeter in answer to the question with which this chapter started; namely, Does the amount of latent and active heat and external work done equal the amount of heat which is calculated from the oxidation of the protein, carbohydrate and fat of the body and that of the food? The following table taken from Benedict and Milner summarizes all the experiments which have been tried up to the date of publication of the bulletin by this calorimeter. It was believed that by taking a large number of observations the accidental errors would very largely eliminate each other. By the net income is meant the heat calculated from the combustion of the foods and the material of the body, when there has been a change of weight, less the heat of combustion of the feces and urine. By the net outgo is meant the total outgo, including the heat measured by the calorimeter, the heat made latent by the evaporation of water and the heat equivalent of the external work done.

TOTAL INCOME AND OUTGO OF ENERGY.

Subjects and kinds of experiments. Diet	Duration	Net income	Net outgo	Difference in terms of net income	
REST EXPERIMENTS	Days	Calories	Calories	Calories	Per cent.
H. F. Carbohydrate	3	5,678	5,712	34	0.6
B. F. D. "	3	6,441	6,683	242	3.7
A. L. L. "	4	10,458	10,301	157	1.5
Total, 5 rest experiments	10	22,577	22,696	119	0.5
Total, 22 previous rest experiments	67	151,638	152,051	413	0.3
Total all experiments	77	174,215	174,747	532	0.3
WORK EXPERIMENTS					
J. C. W. Fat diet	3	15,423	15,852	429	2.7
J. C. W. Carbohydrate	3	16,378	16,304	74	0.5
B. F. D. "	1	4,463	4,565	102	2.2
A. L. L. "	3	14,257	14,464	207	1.5
A. L. L. Fat diet	3	14,792	14,671	121	0.8
A. L. L. " " severe work	1	7,185	7,137	48	0.7
Total 6 work expts.	14	72,498	72,993	495	0.7
Total 23 previous expts.	76	346,167	345,641	526	0.2
Total all work experiments	90	418,665	418,634	31	0.0

The difference between the total calories computed and those actually measured in the whole series of experiments, both those reported here and those previously reported, amounted to only 501 calories in a total of 592,880 computed, or a difference of only 0.1 per cent. The individual experiments show, to be sure, a larger deviation, particularly when they are for short periods of a day or two, but this deviation is certainly within the limits of error of the method and probably depends on the use of factors not exactly correct for the computation of the potential energy of the foods used in the body; or they come from errors in the estimation of the amount of energy in the feces. It is difficult to determine exactly the feces which correspond to the periods under examination, when these periods are short.

It may be said, therefore, as the result of these and other experiments, that the conclusion of Lavoisier is correct and that the combustion of the materials of the body is the sole source of energy of the human body, as far at least as our methods permit us at present to determine. The law of the conservation of energy applies to the human machine as well as to the inanimate world.

How much energy does the combustion of a gram of various foods yield the body; do all foods serve equally well for the production of the heat and energy of the body? Having shown that the source of the energy and heat is the oxidation of the materials of the tissues and that there is an equivalence between the heat and work actually produced and that computed from the carbon dioxide and water formed by the combustion, we may ask the farther question whether proteins, carbohydrates and fats serve equally well as sources of energy. If we wish to do muscular work, is the energy for that work obtained with equal ease from the fat, carbohydrate, or protein? Have the foods an isodynamic value as heat and energy producers? This question may in its solution also throw light on the other question of whether the energy of the muscles comes mainly from carbohydrates, or fats, or proteins.

This question is one which may be attacked and partially solved by the use of the calorimeter, and in order that the method of attack may be understood I have presented in some detail the results of a most carefully conducted experiment of this kind reported by Benedict and Milner. The object of this experiment was to discover whether fats or carbohydrates serve best as sources of muscular energy. To solve this question a man entered the calorimeter and, while on a diet in which most of the energy was in the form of fat, he did a heavy day's work by riding a stationary bicycle which was geared to a dynamo so that the amount of the work done could be measured. A second period of observation followed of three days' duration during which he did as nearly as possible the same amount of work, but in this his main source of energy was in the carbohydrate of the diet.

Many other interesting facts will appear also from this experiment, namely, the amount of energy in the feces and urine, the proportion of the total energy consumed which appears in the form of work done, in other words the efficiency of the human machine, the volume of oxygen consumed per day by a man, the weight of carbon dioxide and water given off. The methods of computation of the results are no less interesting.

The subject after a preliminary digestion experiment of four days' duration entered the calorimeter at 10 P.M. The experiment in the calorimeter began at 7 the next morning and continued for three days. The object of the first experiment was to test the value of fat as a source of energy as compared with carbohydrate. The diet for this period was, therefore, a fat diet. He rode the bicycle for six hours a day. The feces were marked off by taking charcoal before entering the chamber and at the end of the period. The feces between the passage of the two charcoal markers corresponded to the period under observation. The daily régime was as follows: The subject was called at 7 A.M. He took his pulse and temperature, voided his urine, and weighed himself. He then collected the water from the drip on the heat-absorbers in dry, weighed bottles and weighed the absorbers to ascertain the amount of water adherent to them. He had breakfast and began work at 8.15. The food was carefully analyzed and weighed. All water that he drank was measured. The feces were collected in a closed can. The perspiration in his clothing was measured by weighing his underclothing when dry and after exercise. The nitrogen secreted in the perspiration was also determined by examining the wash water from the underclothing. It amounted to about 0.5 gram per day. The diet was very simple and consisted of bread, ginger snaps, shredded wheat, butter, milk and cream and an infusion of cereal coffee.

The diet contained approximately 110.9 grams of protein, 450 grams of carbohydrate and 350 grams of fat. The total heat of combustion of the foods eaten was on the average 5,550 calories per day. Of this amount the total energy from the fat, since this was butter fat, was 9.3 calories per gram or a total of 3,100 to 3,350 calories or about 60% of the total energy intake. These results more in detail for the first day are given in the accompanying table.

Total Daily Weight, Composition, and Heat of Combustion of Food, Feces and Urine.

Kind of Material	Total weight	Water	Water free substance								Heat
			Protein	Fat	Carbo hydrates	Ash	Nitrogen	Carbon	Hydrogen	Oxygen	
Food	Gms.	Gms.	Gms.	Gms.	Gms.	Gms.	Gms.	Gms.	Gms.	Gms.	Cals.
Bread	440	170.7	36.5	6.6	222.2	4.0	6.42	122.67	17.82	118.39	1210
Butter	45	4.0	.6	38.8	......	1.6	.09	29.11	4.73	5.47	361
Ginger snaps	60	4.5	3.1	5.4	45.6	1.4	.55	25.45	3.68	24.42	255
Shredded wheat	40	3.3	3.9	.7	31.5	6	.68	16.53	2.26	16.63	161
Sugar	45				45			18.95	2.92	23.13	178
Cereal coffee	1,200	1189.2	1.2	......	9.6	......	.12	5.64	.84	4.20	46
Total basal ration	1,830	1371.7	45.3	51.5	353.9	7.6	7.86	218.35	32.25	192.24	2211
Supplemental: 1st day											
Milk	1,000	858.0	37.0	44.0	53.0	8.0	6.00	75.80	11.80	40.40	853
Cream	1,020	705.8	28.6	238.7	40.8	6.1	4.59	213.79	32.23	57.49	2484
Total: 1st day	3,850	2935.5	110.9	334.2	447.7	21.7	18.45	507.94	76.28	290.13	5548
" 2d day	3,900	2970.1	112.3	345.9	449.7	22.0	18.68	518.42	77.86	292.94	5549
" 3d day	3,960	3011.7	113.9	359.9	452.1	22.4	18.95	531.00	79.76	296.19	5599
Excreta											
Total feces: 3 days	719.1	573.8	33.8	36.7	47.5	27.3	5.39	80.40	12.22	19.99	927
Av. per day	239.7	191.3	11.3	12.2	15.8	9.1	1.80	26.80	4.07	6.66	309
Urine: 1st day	1,495.6	1425.8	......	......	......	14.1	14.31	11.57	2.92	26.90	147
" 2d day	973.0	927.5	......	......	......	9.1	15.49	12.52	3.16	5.23	130
" 3d day	950.1	914.3	......	......	......	9.0	17.17	13.96	3.52	1.15	138
Total urine	3,427.7	3267.6	......	......	......	32.2	46.97	38.05	9.60	33.28	415

It will be noticed by inspecting the last column of this table that of the 5,550 calories contained in the food, 309 calories per day appeared in the feces and on the average about 138 calories in the urine, making all told about 8% of the total food calories which were not set free in the body but which were re-excreted. About 92% of the heat of the food was utilized or was available to the body.

It will be remembered that Lavoisier found that the oxygen consumption of a man doing work was about 90 liters per hour. The average in the table on page 293 for the whole day was about 41 liters per hour. Lavoisier found for a man at rest but digesting about 37 liters per hour. The respiratory quotient of 0.815 is rather high for a diet containing so much fat. It is due to the fact that there was a good deal of carbohydrate also present and the respiratory quotient when carbohydrates are burned is very nearly 1. When fat alone is burned the whole respiratory quotient should be 0.71.

WATER AND CARBON DIOXIDE GIVEN OFF AND OXYGEN CONSUMED. THE OUTGO OF WATER AND CARBON DIOXIDE AND THE CONSUMPTION OF OXYGEN.

	Water of respiration and perspiration	Carbon dioxide		Carbon	Oxygen consumed		Respiratory quotient e/h
		d Grams exhaled	e Liters d x .509	f In CO_2 d x 3/11	g Grams	h Liters g x 0.7	
1st day	3,702.27	1,775.61	903.96	484.25	1,545.24	1,081.67	0.836
2d "	4,179.78	1,813.56	923.28	494.60	1,652.56	1,156.79	0.798
3d "	3,907.14	1,697.41	864.15	462.91	1,517.80	1,062.46	0.813
Total	11,789.19	5,286.58	2691.39	1,441.76	4,715.60	3,300.92	0.815

Amount of heat eliminated.—The manner in which the heat is computed for the various factors involved in the calorimeter is shown in Table C.

TABLE C

SUMMARY OF CALORIMETRIC MEASUREMENTS.

	a Heat measured in terms of 20° cals	b Change of temperature of calorimeter	c Capacity correction of calorimeter b x 60	d Correction due to temperature of food and dishes	e Heat used by vaporization of water	f Total heat determined
	Calories	Degrees	Calories	Calories	Calories	Calories
1st day	4,594.50	—0.01	—0.60	—37.97	739.17	5,295.10
2d "	4,657.59	+.28	+16.80	—42.33	801.35	5,433.41
3d "	4,443.16	—.20	—12.00	—44.33	737.05	5,123.88
Total	13,695.25	+.07	4.20	—124.63	2,277.57	15,852.39

It will be seen from this table that by far the greater part of the heat given off from the body is measured directly by the outgoing water of the heat-absorbers, column a. The other principal measurement is that of the water vapor in the air streaming from the calorimeter. From the amount of water in the air the amount of heat used in vaporizing it may be computed. These figures are given in column e. The total output of heat of the subject doing hard work was about 5,300 calories per day.

Comparison of the income and outgo of material and energy.—"From the data given in the preceding tables the amounts of material and energy received and given off by the body in different ways may be compared, and gain or loss by the body estimated. Such a comparison is made in the tables which follow." "The computation of the amounts of materials gained or lost by the body are given for purposes of illustration in the data for the first day of the experiment."

The difference between the total weight of the income and that of the outgo was 122.06 grams. It will be observed on comparing the total income of elements with the total outgo that there was a gain of 1.69 grams of nitrogen, 20.80 grams of hydrogen, and 125.75 grams of oxygen, and a loss of 14.68 grams of carbon and 1.50 grams of ash. "In the lower section of the table are shown the amounts of protein, fat, carbohydrates, and water gained or lost by the body on the first day of the experiment as computed from the gains and losses of the elements. The gains or losses

TABLE D.

GAINS AND LOSSES OF BODY MATERIAL ON FIRST DAY OF METABOLISM EXPERIMENT 56.

	Total weight *a*	Nitrogen *b*	Carbon *c*	Hydrogen *d*	Oxygen *e*	Ash *f*
Income	Grams	Grams	Grams	Grams	Grams	Grams
Oxygen from air	1,545.24	....	....	....	1,545.24	....
Water in drink	1,950.00	....	....	218.20	1,731.80	....
Water in food	2,935.50	....	....	328.48	2,607.02	....
Solids in both	914.50	18.45	507.94	76.28	290.13	21.70
Total	7,345.24	18.45	507.94	622.96	6,174.19	21.70
Outgo						
Water in feces	191.30	....	....	21.41	168.89	....
Solids in feces	48.40	1.80	26.80	4.07	6.63	9.10
Water in urine	1,425.80	....	....	159.55	1,266.25	....
Solids in urine	69.80	14.31	11.57	2.92	26.90	14.10
Water of respiration and perspiration ..	3,702.27	.65*	....	414.21	3,287.41	....
CO_2 of respiration ..	1,775.61	....	484.25	....	1,291.36	....
Total	7,213.18	16.76	522.62	602.16	6,048.44	23.20
Gain (+) or loss (—)	+ 132.06	+ 1.69	— 14.68	+ 20.80	+ 125.75	— 1.50
Ash of protein gained (+)	.16					
Gains and losses of body material						
Protein	+ 10.14	+ 1.69	+ 5.35	+ .71	+ 2.23	+ .16
Fat	+ 43.82	....	+ 33.35	+ 5.17	+ 5.30	
Glycogen	— 120.17	....	— 53.36	— 7.45	— 59.36	
Water	+ 199.93	....	....	+ 22.37	+ 177.56	
Ash	— 1.50	....	....	....	....	— 1.50
	+ 132.22	+ 1.69	— 14.00	+ 20.80	+ 125.73	— 1.34

* Nitrogen of perspiration.

of these different compounds are computed in the following way by means of formulæ derived from the following data regarding the elementary composition of these compounds:

TABLE E.

ASSUMED PERCENTAGE OF COMPOSITION OF BODY MATERIALS.

	Nitrogen Per cent.	Carbon Per cent.	Hydrogen Per cent.	Oxygen Per cent.	Ash Per cent.
Protein	16.67	52.80	7.00	22.00	1.53
Fats		76.10	11.80	12.10	
Carbohydrates		44.40	6.20	49.40	
Water			11.19	88.81	

In the above table the carbohydrate is assumed to have the composition of glycogen. The ash of the protein may be disregarded and the following equations may be derived from the above data, letting p = protein; t = fat; r = carbohydrates; and w = water, and indicating the names of the elements by the initial letter, as is customary:

$$0.1667\,p = N$$
$$0.4440\,r + 0.7610\,t + 0.5280\,p = C$$
$$0.1119\,w + 0.0620\,r + 0.1180\,t + 0.0700\,p = H$$
$$0.8881\,w + 0.4940\,r + 0.1210\,t + 0.2200\,p = O$$

"The solutions of these equations in terms of nitrogen, carbon, hydrogen and oxygen give the following formulæ:

Protein = 6.0 N.

Fat = 0.005 C + 9.693 H — 1.221 O — 2.476 N

Carbohydrates = 2.243 C — 16.613 H + 2.093 O — 2.892 N

Water = 1.248 C + 7.920 H + 0.128 O + 0.460 N

"If the quantity of each element gained or lost by the body as expressed in grams in Table D is substituted in these equations for the corresponding sign of the element and the indicated calculations are performed the results obtained will be the quantities of the different compounds gained or lost. For example, according to Table D the body gained 1.69 grams of nitrogen on the first day of the experiment. This value would therefore be substituted for N in the first of the above equations as follows:

Protein = 6 N = 6 (+ 1.69) = + 10.14."

The gains and losses of protein, fat, glycogen, and water as thus computed from the gains and losses of the elements for the first day of experiment 56 are shown in the lower section of the first column of Table D.

The energy income and outgo in this experiment totaled as follows for the three days:

Available energy from the foods	15,354	calories
Energy by the combustion of body material lost	69	"
Total income of energy	15,423	"
Total energy outgo, work and heat observed	15,852	"
Excess of heat measured	429	"

This was the first experiment with the respiration calorimeter in its revised form and the difference of about 2.7% between the computed and measured heat is much larger than in other experiments. In making these computations of the energy equivalent of the materials of the body which had lost in weight the following factors were used:

Heat of combustion of protein ..... 5.65 calories per gram.
" " " " fat .... 9.54 " " "
" " " " carbohydrates 4.19 " " "

These are of course the heats as determined in the bomb calorimeter.[1] The value for the protein is that of fat-free muscular tissue from which the nitrogenous non-proteid compounds have been removed.

Having thus illustrated the method by the detailed experiment just quoted, the question may now be asked whether it is possible to with-

[1] It must be noted that if we consider the fats and carbohydrates in the food alone and not those in the body material catabolized since the absorbability is not perfect we do not use the figures given here, but rather the relative efficiency of the foods are shown in the following table which expresses how many calories of energy are actually available to the body for each gram of foodstuff eaten:

Protein	4.0 cals. per gram.
Carbohydrate	4.0 " " "
Fats	8.9 " " "
Alcohol	6.9 " " "

draw the fat in the diet mentioned and replace it with an equivalent amount of carbohydrate, equivalent that is in the amount of energy it contains. For this purpose the subject after a few days' interval re-entered the calorimeter and for a period of three days did nearly the same amount of work, but the cream was now withdrawn from his diet and cane sugar containing the same amount of energy substituted for it. On the fat diet the fat furnished 60 per cent. of the total energy; on the carbohydrate diet the carbohydrates furnished about 4,000 calories or nearly 80 per cent. of the total. The total energy outgo was, as before, 5,500 calories per day. Of this amount the equivalent of 602 calories was in the form of external work.

The respiratory quotient while on the carbohydrate diet was 0.92 on the average, on one day being as high as 0.946. Now if the carbohydrate cannot be used as efficiently to produce energy as the fats, then it would be necessary for the body to call upon its own tissues to supply the energy, and proteins or fats should be catabolized. As a matter of fact, the experiment showed that the body lost on the average per day the following amounts of material: Protein, 1.12 grams; fat, 21.48 grams; carbohydrates, 26.76 grams, and water was gained to the extent of 199.63 grams.

In the fat experiment there was a loss of body fat, but a gain of protein and carbohydrate. The change was, however, small. From this it might appear that the fats were slightly better sources of energy for the body than the carbohydrates, but as a matter of fact these two substances are really nearly equivalent. The slight difference was due to the energy output being somewhat greater in the second experiment. From this experiment, then, the conclusion may be drawn that the body can get its heat and the energy for its external work from either fats or carbohydrates and that they replace each other in the energy balance of the body approximately according to their calculated heat equivalents. *They are, in other words, isodynamic.*

We have now to consider the proteins. Are they also equivalent to the others as energy-producers? Does it make no difference so far as the body is concerned whether the energy is supplied from protein or carbohydrate? It is at the outset clear that proteins and fats suffice for the greater part of the food of many people, namely the Esquimaux. Further, carnivorous animals generally have very little carbohydrate in their diet, and yet they are capable of great exertion. Moreover, in the calculations which have been made in the experiment cited, it is assumed that the protein is oxidized and produces heat and energy in proportion to its content. It is not, of course, possible to substitute the proteins completely by fats or carbohydrates, since the latter contain no nitrogen, but it is possible to supply the whole needs of the body for a time at least with protein and a small amount of fat or carbohydrate. The proteins can replace the latter as sources of energy as shown by Rubner. 100 grams of fat are equivalent to 227 grams of protein or carbohydrate in the food, according to Rubner; or 223 grams, according to Atwater.

But, while the foods thus have an isodynamic action in this sense, they are of course by no means equivalent physiologically. Each has a specific action in nutrition; the proteins affect the production of heat directly apparently by stimulating it; they are necessary to the formation of many of the internal secretions, and they, or the foods which contain them, may have a direct action on the nervous system. The digestibility is different. There are also other substances in foods besides the proteins, fats or carbohydrates which are of great importance, of vital importance indeed, although present in small amounts. These will be considered presently.

Growth.—Thus far the foods have been considered only as the source of energy. That energy was shown to be necessary, first, to supply the heat necessary to keep the temperature of the body high enough to permit the chemical processes to occur with sufficient speed and to keep the protoplasm in the necessary condition of viscosity; second, to furnish the energy which is transformed in doing external and internal work, for energy is consumed in moving the blood, in secreting the urine, as well as in all the bodily movements; and third, to furnish the energy used in the syntheses of specific materials in the body, for not all the chemical transformations in the body are exothermic decompositions; there are also synthetic, reducing, endothermic reactions by which some heat is rendered latent and in virtue of which the specific materials of the body are produced. But foods are needed not only because with oxygen they supply energy; they must also construct the machine. They are the raw materials from which the body is made. It is probable, although it is not certain, that much of the materials which are oxidized in the body are built into the real living matter before they are oxidized. In other words, it is the living matter which is burning and not the foods as such. Foods may or may not contain energy, but they must have the raw materials out of which the body is formed. They must, in other words, nourish the body. And it is from this point of view that we shall now consider them.

Summary.—In this chapter we have considered the income and outgo of the energy of the body considered as a whole. The body is warmer than its surroundings. It is constantly giving off heat. The amount of heat given off in the course of the day by a man at rest is about 2,500 calories, and during working days it is greater than this. Energy leaves the body not only as heat, but some is converted into potential energy of work done and some into the potential energy involved in the separation of the water molecules to the form of a vapor. The source of the heat of the body was long a puzzle. It was found by the French chemist, Lavoisier, that it came from the combustion of the materials of the body, and in particular from the combustion of the carbon and hydrogen of the body by the oxygen of the air. Oxygen enters the body by the lungs and combines with the carbon and hydrogen in the tissues giving carbon dioxide and water. This combustion does not take place in large measure in the lungs or in the blood, but chiefly in the tissues and in the muscles in greater part, since these are the most bulky tissues of the body.

The measurement of the heat given off is made by means of a respiration calorimeter. The most perfect of these instruments has been constructed in this country. By this calorimeter it was shown that the amount of heat given off in perceptible heat, or in other forms such as latent heat of vaporization, is equivalent to the amount of heat set free by the combustion of the carbohydrates, fat and protein burning in the body. The source of animal heat is, hence, the combustion of these materials and the amount of heat is that calculated from the combustion. So far as the measurements go, they indicate that there is a balance between the income and outgo of energy and that the law of conservation of energy holds for human beings as well as for other animals. The human body does not have, so far as can be determined, any other source of energy than the combustion of its materials. Respiration and heat production are thus related.

The amount of heat produced and of carbon dioxide given off per kilo body weight varies at different ages and is greater in children than in adults. It is generally larger in small animals than in large of the same kind. There is, hence, a more rapid respiration and heat production in the young than in adults, and this is true even for the heat production per unit of surface area of the body. In other words, the metabolism of young is higher than that of older people.

Some light may be thrown on the nature of the materials burning in the body by the study of the respiratory quotient. This is the ratio between the volume of carbon dioxide exhaled and the volume of oxygen consumed. The respiratory quotient is unity for the combustion of carbohydrate, about 0.8 for proteins and 0.71 for fats. It may, for short spaces of time, be larger than one, or smaller than .7.

By the use of the respiration calorimeter it has been found that the carbohydrates and fats can very largely replace each other in the metabolism of the body and that in the absence of much of either of these the energy required may come from the proteins. The latter, however, have also a direct action on the body in that they increase the heat production and have in this way a specific action different from the fats and sugars. The explanation of this action is as yet obscure.

Finally, while we speak of the foods as oxidizing in the body and while for practical purposes they may be treated as if they were undergoing oxidation in this way, it is more probable that it is not the foods which are oxidizing, but the living matter; and that the foods only oxidize as they are combined in the living protoplasm. When life ceases, the oxidation of the food stops. It is impossible thus far to isolate from dead protoplasm substances, or enzymes, which have the power of oxidizing the foods apart from living matter. The oxidation of the purine bases is a possible exception to this statement. The question of the nature of the combustion is considered more fully farther on.

REFERENCES. ANIMAL HEAT.

1. *Black:* Quoted from Gavarret: De la chaleur produite par les êtres vivantes, p. 163. Paris, 1855.
2. *Lavoisier:* Experiences sur la respiration des animaux. Histoire et Mémoires de l'Acad. de Sci. Paris. 1777, p. 187; 1780.
3. *Lavoisier:* Sur la combustion en general. Ibid., 1777, p. 599.
4. *Lavoisier et Laplace:* Sur la chaleur. Ibid., 1780, p. 355.
5. *Lavoisier:* Ibid., 1780, p. 401.
6. *Lavoisier et Seguin:* Première mémoire sur la respiration des animaux. Ibid., 1789.
7. *Dulong:* Mémoire de l'Institut, 1844, 18, p. 341; Annales de chimie et de physique (3), I, 1841.
8. *Depretz:* Annales de chim. et de phys. (2), vol. 25, p. 337.
9. *Regnault et Reiset:* Annales de chim. et de phys. (3), 1849, vol. 26, p. 26.
10. *Spallanzani:* Mémoire sur la Respiration.
11. *Rubner:* Zeitschrift f. Biologie, 30, p. 73 (1894).
12. *Laulanie:* Archives de Physologie, 1898, p. 748.
13. *Atwater and Benedict:* Metabolism of Matter and Energy in the Human Body. U. S. Dept. Agriculture, O. E. S. Bulletin No. 136, 1903.
14. *Benedict and Milner:* Ibid., Bulletin 175, 1907.
15. *Benedict:* Carnegie Institution Publications, Washington, No. 42.
16. *Atwater:* The nutritive value of alcohol. Mem. National Acad. of Science, vol. 8, 1902.

CHAPTER VII.

THE RAW MATERIALS OR FOODS.

Foods.—The body consists of water, various salts, of proteins, carbohydrates and fats, and various other organic substances which are formed from the latter in the course of their decomposition in the body. These organic constituents of the body are burning or oxidizing as long as life lasts, and it is their combustion which gives rise to the heat of the body; to the energy appearing as work done, or transformed into such other forms of energy as chemical, or electrical. By this combustion of its substances simple compounds are produced, such as carbon dioxide, water, urea and many other substances which are constantly leaving the body in the excretions, either through the skin, by the lungs, kidneys or alimentary canal. The water leaving the body by the skin or kidneys carries out always some of the salts with it in solution, so that the body is constantly losing carbon, nitrogen, oxygen, hydrogen, salt and water. The body may be likened to a lamp which as long as combustion continues continues to give light and heat. So the body as long as it continues to burn, or, as it is called in living things, to respire, continues to give off heat and those various forms of energy found in living things. And just as the lamp, if the supply of oil runs low, burns its wick, one of its constituent parts, and is ultimately extinguished; so in the body, if the supply of food or raw material for combustion is exhausted, it burns in part its own substance and then life is extinguished. But, while this analogy with a lamp or candle is in many ways most apt, it must not be pushed too far, for the body not only burns the fuel supplied it, but manufactures or repairs the wasting of the machinery of the body. It is as if the lamp were able to make new wick material from the oil so as to make good the loss of wick material which occurs daily.

Food. Definition.—The term food has been variously defined. It has been said to be a substance containing available potential energy, that is energy which can be utilized by the body. But by this definition some poisons are foods, since, like alcohol, morphine or caffeine, they are partly oxidized in the body and thus setting free their energy contribute to its work or its store of heat; and other substances such as water, salts or oxygen which are absolutely necessary to the body are not foods, since they carry no energy into the body. Obviously such a definition is of very little value. A food has one characteristic, it has the power of repairing waste, or contributing substance necessary for the growth of the body. The body is not a stove which burns fuel. Foods do not burn as they do in a stove, being contained in the material but not made part of it. It is true that some oxidation of this kind may occur, but not the combustion of the foods. It is the living matter which is burning, not the foods which are in it. Foods may or may not bring available energy into the body; but they all repair waste and provide the raw materials for growth. All substances which have this power are foods; those which lack it, although they may be burned in the body and set free energy, are not foods. Thus water, salts, oxygen are true foods. All things necessary for the construction of the living machine are foods. Poisons, such as alcohol, caffeine, morphine or other substances which may be oxidized in the body, but which are incapable of repairing waste, —these are not foods. All substances which do not nourish living matter and which are not chemically inert are harmful to it. So a poison may be defined as a substance, not a food, which acts chemically on the body. Too much of one kind of food or too little of another may be harmful, but such substances do not thereby become poisons. They are still foods. On the other hand, poisons may change living matter in a manner to benefit it temporarily, or restore it when it has been disturbed, but such substances are none the less poisons, although they may be temporarily used for valuable ends. No material has as yet been found which is able to improve protoplasm so that it will live longer than it does under its normal circumstances. All substances not foods shorten life when they find entrance to protoplasm, provided they are taken in sufficient amount to affect it at all.

The food materials must contain, then, the materials out of which the body is made, and we may at this time examine the constitution of various foods as raw materials for the manufacture of the body.

Water.—Human beings normally require 3-5 liters of water per day, although life is possible on less. Some of this is obtained in the foods; fruit and many vegetables consist of 80-90 per cent. of water, but most is taken in beverages of various kinds. Water that we drink is never pure. It always contains salts of various kinds in solution and these salts may have a physiological action. Thus hard and soft waters may differ in their action on the bowels; in localities using hard waters goiter is often endemic, though it is very doubtful whether the hardness of the water due to the lime salts in it has anything to do with the goiter. It is more probable that the cause of the goiter is an organic substance or germ. Thus in the region of the Great Lakes goiter is very common. Nearly all the dogs in Chicago are more or less goiterous. In general, also, hard waters are constipating, soft relaxing. But this effect again is not due to the water, but to the salts. Some waters, too, are radio-active and at the present time there is a good deal of evidence indicating that this circumstance is of importance in the beneficial action of such waters in many diseased states. How the radio-activity produces any physiological action is still quite obscure. No one has as yet shown that living matter is at all radio-active, so that we have no reason for thinking that radium in minute traces may be a necessary part of our foods. Still it is not impossible that this may be the case. Minute traces of substances are frequently of extraordinary importance. Thus it was recently found by Berthelot that manganese in extremely small amounts was necessary for the development of the gonidia in Aspergillus. The amount of manganese which was necessary was so small that it was very difficult to provide water and containers and foods which did not have these minute traces, so that growth was impossible. This fact makes one very cautious in concluding that minute traces of substances which occur in living matter may not have an influence on it. While the favorable action of various naturally radio-active waters is to be ascribed in part to the salubrity of the climate, to the more restricted diet generally enjoined on those who visit these watering-places for health, to enforced exercise, to diminished intestinal putrefaction due to the laxative action of the water or to drinking more water, since many people usually drink too little, yet some of the action it seems most probable is almost certainly due to the radio-activity of the water.

Diversity of foods. Importance.—The various food materials of civilization have enormously multiplied in kind with the improvement of transportation and the introduction into all countries of the products of other countries. The temperate zone now obtains the vegetables and fruits of the tropics; and America not only has its native vegetables, such as the potato and maize, but it cultivates also those of nearly all other lands. The diet of various nations, and in particular of Americans, has thus become more varied than that of any nation or group of people previously existing. This is due in part, so far as America is concerned, to the favorable location of the country, reaching as it does from the tropics to the north and having no customs duties; and in part it is due to the efforts of the Department of Agriculture to introduce into the country all those fruits and cereals which are cultivated elsewhere and which custom has shown to be valuable foods. As a varied diet is more certain to contain all those food materials which the body needs than is a restricted or monotonous diet, the general result of this expansion in the kinds of foods consumed is certain to be good; for what one foodstuff may lack in one particular is certain to be contained in another.

With the wonderful development of chemistry, also, food substances are now manufactured from materials which were formerly wasted. Thus while the human body cannot make glucose from cellulose, the chemist can accomplish this. A great many prepared foods supposedly more nutritious or partially digested or suitable for special ends, or having a great many other real or imagined advantages, are being consumed in constantly increasing amounts. Mention may be made of various infant foods designed to eke out or replace mother's milk; breakfast cereals of all kinds; bran and wheat mixtures for laxative foods; and so on. This development of the chemical manufacture or chemical treatment of foods is a matter of the very greatest importance to public health, since while the end sought is in many ways desirable, the foods by chemical treatment are often altered in their chemical composition, or substances not naturally in the foods may be introduced from the chemicals used in manufacture. The physiological effects of even small amounts of certain compounds are so important that it is very necessary to guard against the admixture of any toxic or harmful substance which may be present in the chemicals used even in very small amounts. Thus some years ago sulphuric acid containing arsenic was used in the manufacture of glucose from starch. The arsenic contaminated the glucose; the glucose was used in the manufacture of beer and many people in England were poisoned, some fatally, by the arsenic. It is important to remember, also, that in the chemical treatment of foods there is the possibility of forming in them toxic substances not in the food itself. Often the effects of such substances in small quantities are not immediately apparent and it is extremely difficult to trace such injurious effects as there may be to their true source in the altered food. It is for this reason that any manufactured or chemically changed food which has not had years and even generations of tests behind it, as all natural foods have had, is to be regarded with a certain amount of suspicion, unless the methods used in its manufacture are such as practically to preclude injury. Physicians and scientific men have particularly to be on their guard against hasty indorsements of artificial foods. For example, physicians and scientific men usually mean by the word glucose the pure hexose sugar, dextrose, and when the problem of the use of glucose is presented to them they think only of the pure sugar, which is of course an admirable food. But the manufacturer uses the term glucose, or did use it for a long period, for the very impure mixture obtained by the partial hydrolysis of starch. This substance not only contained pure glucose, but many other substances such as dextrins, maltose, etc.; it occasionally contained some nitrogenous substances, and if made with impure materials it might contain some deleterious substances such as arsenic. Before indorsing a food of any kind, a scientific man or physician would do well to consult the government authorities whose business it is to investigate such foods, their constitution, methods of manufacture, etc., and to inform people of their desirability.

Amount of food required by a human adult.—The human adult requires roughly 60-150 grams of protein a day and sufficient fat and carbohydrate, or protein, in addition to cover his energy requirements. All three of these substances are found in all the foods, but in widely varying proportions. Vegetables contain as a rule a great excess of carbohydrate, and meat an excess of protein. Many vegetables contain much water. The following tables give the proportion of food matters in the principal foodstuffs. Attention is particularly given to the various meats and such vegetables as potatoes, turnips, squash, peas and beans, corn, wheat, oatmeal, and the fruits: apples, pears and bananas.

COMPOSITION OF SOME FOOD MATERIALS. BULLETIN 28. E. O. S. DEPT. OF AGRICULTURE.

Foods	Refuse	Water	Protein N x 6 25	Fat	Total carbohydrates	Ash	Fuel value per pound
Beef.							
Brisket, medium fat as purchased. Average	23.3	41.6	12.0	22.3		0.6	1165
Chuck, including shoulder. Medium fat. As purchased. Average	15.2	57.9	16.6	10.1		0.8	735
Flank. All analyses. Average. As purchased	5.5	56.1	18.6	19.9		0.8	1185
Loin. Medium fat. As purchased. Average	13.1	58.2	17.1	11.1		0.9	785
Round. Fat. As purchased	12.0	54.0	17.5	16.1		0.8	1005
Heart. As purchased	5.9	53.2	14.8	24.7		0.9	1320
Corned beef. Brisket	21.4	40.0	14.4	19.4		4.5	1085
Veal: Fresh.							
Leg, lean. As purchased	9.1	66.8	19.4	3.7		1.1	520
Leg. Cutlets. As purchased	3.4	68.3	20.1	7.5		1.0	690
Mutton: Fresh.							
Leg. Medium fat. As purchased	18.4	51.2	15.1	14.7		0.8	900
Loin, without kidney or tallow. As purchased	14.8	40.4	13.1	31.5		0.6	1575
Pork: Fresh.							
Ham. Medium fat. As purchased	10.7	48.0	13.5	25.9		0.8	1345
Loin chops, medium fat	19.5	41.8	13.4	24.2		0.8	1270
Tenderloin as purchased	00	66.5	18.9	13.0		1.0	900
Ham. Smoked	12.2	35.8	14.5	33.2		4.2	1670
Salt pork, clear fat	00	7.9	1.9	86.2		3.9	3670
Bacon, smoked	8.7	10.4	9.5	59.4		4.5	2685
Sausage. Bologna	3.3	55.2	18.2	19.7	0.3 (?)	3.8	1170
Pork sausage		39.8	13.0	44.2	1.1	2.2	2125
Chickens. Broilers as purchased	41.6	43.7	12.8	1.4		0.7	295
Goose. Edible portion		46.7	16.3	36.2		0.8	1830
" As purchased	17.6	38.5	13.4	29.8		0.7	1505
Fish. Mackerel. Whole. As purchased	44.7	40.4	10.2	4.2		0.7	365
Mackerel. Edible portion		73.4	18.7	7.1		1.2	645
Eggs. Uncooked. Edible portion		73.7	14.8	10.5	0.3?	1.0	720
" " As purchased	11.2	65.5	13.9	9.3		0.9	635
Eggs. Boiled whites		86.2	13.0	0.2	0.5	0.6	250
" " yolks		49.5	16.1	33.3		1.1	1705

Refuse	Water	Protein N x 6.25	Fat	Total carbohydrates	Ash	Fuel value per pound
	11.0	1.0	85.0		3.0	3605
	91.0	3.0	0.5	4.8	[illegible].7	165
	31.6	28.8	35.9	0.3	.4	2055
	72.0	20.9	1.0	4.3	.8	510
	35.2	7.1	17.7		.0	1435
	39.3	2.6	29.5	1.8	.8	1700
	87.0	3.3	4.0	5.0	.7	325
	68.2	9.6	9.3	11.2	.7	780
	26.9	8.8	8.3	54.1	.9	1520
	9.5	1.2	83.0		.3	3525
	12.5	9.2	1.9	75.4	1.0	1655
	12.6	7.1	[illegible].3	[illegible]	0.[illegible]	1645
	.3	16.1	.2	7.	1.	1860
	7.7	16.7	.3	6.	2.	1850
	84.5	2.8	.5	1	0.	285
	12.3	8.0	.3	9.	0.	1630
	11.4	13.8	.9	1.	1.	1675
	11.9	10.9	1.1	75.6	0.5	1655
	13.1	.2 3	1.1	73.0	0.5	1635
	10.9	11.0	1.4	76.3	0.4	1685
	8.[illegible]	10.5	1.4	77.9	2.1	1700
	10	3.4	0.9	4.1	1.3	1665
	78.	3.0	1.5	5.8	1.3	415
	35.0	9.1	1.6	53.3	1.0	1225
	38.4	8.7	0.9	49.7	1.3	1140
	4.8	11.3	10.5	70.5	2.9	1965
	5.9	9.8	9.1	73.1	2.1	1925
	42.5	3.1	9.8	42.8	1.8	1270
	18.2	0.4		81.2	0.2	1520
	94.0	1.8	0.2	3.3	0.7	105
	58.9	9.4	0.6	29.1	2.0	740
50.0	29.4	4.7	0.3	14.6	1.0	370
	68.5	7.1	0.7	22.0	1.7	570
	95.3	0.8	1.1	1.9	0.9	95
	88.6	2.3	0.1	7.4	1.6	[illegible]
	91.5	1.6	0.3	5.6	1.0	[illegible]
	3.5	7.7	3.6	80.3	4.9	1790
	75.4	3.1	1.1	19.7	0.7	470
	95.4	0.8	0.2	3.1	0.5	80
	92.9	1.2	0.3	5.1	0.5	130
	8.4	25.7	1.0	59.2	5.7	1620
	9.5	24.6	1.0	62.0	2.9	1655
	74.6	7.0	0.5	16.9	1.0	465
	75.5	2.5	0.1	20.9	1.0	440
	75.1	2.6	3.0	17.8	1.5	505
	60.0	1.8	0.7	27.4	1.1	570
	88.3	1.4	0.5	8.0	0.8	215
	94.	0.9	.4	.9	0.5	105

Foods	Refuse	Water	Protein N x 6.25	Fat	Total carbohydrates	Ash	Fuel value per pound
Fruits.							
Apples. Edible portion		84.6	0.4	0.5	14.2	0.3	290
Bananas. " "		75.3	1.3	0.6	22.0	0.8	460
Blackberries		86.3	1.3	0.0	10.9	0.5	270
Oranges. Edible portion		86.9	0.8	0.2	11.6	0.5	240
Pears. Edible portion		84.4	0.6	0.5	14.1	0.4	295
Nuts.							
Almonds. Edible portion		4.8	21.0	54.9	17.3	2.0	3030
Brazil nuts		5.3	17.0	66.8	7.0	3.9	3265
Hickory nuts		3.7	15.4	67.4	11.4	2.1	3345
Peanuts		9.2	25.8	38.6	24.4	2.0	2560

Milk.—Milk is a food for the young and growing. It is a most interesting food, since it represents the answer Nature has given to the question as to the best food for developing mammals. After a very long period of experimentation in the monotremes, marsupials and lower placental mammals, the milk of the higher placentals was evolved. It is probable that there are good reasons for the presence in milk of most, if not all, of its constituents. Milk is the secretion of the mammary glands and these are modified skin glands. The constituents of the milk are for the greater part manufactured in the gland itself; some of them, however, come directly from the blood. The organic substances in the milk are peculiar to it. They are not found in other tissues of the body. Human milk is alkaline to litmus; cows' amphoteric. The specific gravity is 1.026-1.035.

The composition of the milks of the different mammals varies, but they are alike in their fundamental qualities. In all, the organic matter consists for the most part of protein, carbohydrate and lipin, and each of these is peculiar to milk. The carbohydrate is lactose; the lipin is peculiar because of the proportion of fatty acids of low molecular weight it contains and also for its special properties in nutrition; the protein is largely casein. The composition of human and cows' milk is as follows (Meigs):

	Fat Per cent.	Lactose Per cent.	Protein Per cent.	Ash Per cent.
Human milk	2—4	6—7.5	0.7—1.5	0.15—0.3
Cow's milk	2—4	3.5—5	2.5—4	0.66—0.77

Besides the substances mentioned in the foregoing table, both cows' and human milk contain organic substances which have little or no nitrogen, which are soluble in ether and alcohol, as well as water, but of which the chemical nature is still unknown. The amount of these substances in human milk at the beginning of lactation is about 1 per cent., and in the middle period of lactation about 0.5 per cent. Cows' milk at the middle of lactation contains about 0.3 per cent. These substances are possibly of great importance in nutrition.

The composition of the milk varies at different periods of lactation.

The milk secreted during the first three or four days after delivery is called *colostrum* milk. According to Keenig colostrum of cows has the following composition:

Water	74.67
Solids	25.33
Casein	4.04
Albumin and globulin	13.60
Fat	3.59
Lactose	2.67
Salts	1.56

The colostrum is richer in proteins and total solids than the later milk. It contains many cells, milk corpuscles, which contain nuclear material, whereas the later milk is practically free from such material. The fat, also, is different in that it has a higher melting point, a higher iodine number and it contains more cholesterol and lecithin. The protein material is different. It contains, probably in the colostrum corpuscles, more coagulable proteins, so that unlike ordinary milk it coagulates on heating. These same differences between colostrum and later milk are shown also in human milk, but they are not so marked. Meigs gives the following figures of the human colostrum and later milk, comparing the total nitrogen, the casein, globulin and albumin, and unknown substances (figures in per cent. of weight of whole milk):

	Total N	N in casein and globulin precipitated by magnesium sulphate	N in albumin precipitated by acetic acid and heat	Nitrogen not precipitated by either $MgSO_4$ or heat
Colostrum Human	0.2962	0.1572	0.0514	0.0876
Early milk "	0.2186	0.1377	0.0267	0.0542
Cows' milk (Guernsey)	0.5623	0.4925	0.0482	0.0216

The main differences between human milk and cows' milk are these: Human milk contains considerably more lactose than cows' milk and more substances of an unknown nature which contain little or no nitrogen; it contains very much less protein than cows' milk. The greater proportion of lactose in human milk may be correlated with the very much greater brain development of human beings. There is a very rapid myelinization of the fibers of the brain occurring shortly after birth in the first six weeks of life. In myelin there is a large amount of galactolipins of the nature of phrenosin, or cerebrosides of various kinds. Galactose is one of the constituents of this material. It may be that the larger amount of lactose in human milk is to supply this need. No other place of formation of galactose in the body is known than the mammary glands and these are of course very rudimentary in the infant. The replacement of lactose by cane sugar in milk, or in milk substitutes, would seem to be open to serious criticism on this account. It must be remembered that the mammary glands have been evolved primarily to make a food best suited for rapidly developing human beings; even undiluted cows' milk cannot be tolerated by very young children, although its composition is so similar to human milk. Inasmuch as we are quite ignorant of the composition of some constituents of milk which may be very important, the attempt to find a suitable substitute for human milk is seen to be very largely a groping in the dark.

Proteins of milk.—The proteins of milk are casein (English call it caseinogen) and lactalbumin. These two proteins do not contain all the nitrogen particularly in the early stages of lactation, when quite a large percentage of the nitrogen is in the form of substances, extractives, etc., which are not precipitated by $MgSO_4$ and not by heat after slight acidification. They are not precipitated either by $Al(OH)_3$ and are presumably non-colloidal.

Casein. This is the most abundant and characteristic protein of milk. It is probable that each kind of milk has its own kind of casein in it. In cows' milk there is about 3-4 per cent. of casein; in human milk between 0.5-1.5 per cent. This protein is a phosphoprotein. It is a conjugate, or compound, protein containing phosphoric acid. It is a strong acid probably forming three or four series of salts and having 4 or 6 carboxyls free. Its molecular weight, according to Van Slyke, is about 8,000. Casein has several peculiar properties. It is not coagulated by heating; it is precipitated by faint acidification with acids, but soluble in excess; it clots or coagulates when it is treated by a proteolytic enzyme, and in particular by the enzyme found in calves' stomachs, or the stomachs of other young mammals. Its percentage of composition is: C, 53.00; H, 7.00; N, 15.70; S, 0.8; O, 22.65; P, 0.85. These figures are for cow's casein. When hydrolyzed it yields the amino-acids shown in the table on page 129. It is very unstable and hydrolyzes with great ease. It will be noticed that it is remarkable for the very small amount of cystine it contains, for the absence of glycocoll and for the large amount, relatively, of tryptophane. Furthermore, the quantity of unknown splitting products is larger in casein than in many other proteins. The known products recovered do not total more than 65 per cent. of the molecule. Among the acids found in casein, but not found in other proteins, may be mentioned diamino-trioxydodecanoic acid, $C_{12}H_{26}N_2O$, found by Fischer and Abderhalden; dioxydiaminosuberic acid, $C_8H_{16}N_2O_6$, found by Skraup; caseanic acid, $C_9H_{16}N_2O_7$, and caseinic acid, $C_{12}H_{24}N_2O_5$. The nature of these is uncertain (Skraup). Caseinic acid may be identical with dioxyaminododecanoic acid, although it appears to differ from it in its optical rotation. Caseins from various animals differ in the per cent. of nitrogen and phosphorus. Mares' and asses' caseins contain more nitrogen (16.44-16.28 per cent.) than cows'. Asses' casein has more phosphorus (Tangl and Csókás). Casein contains calcium. The amount varies. $(\alpha)_D = -97.8$ to

—111.8° in a N/10—N/5 NaOH (Long). In neutral solution it is lower, i.e., —80°.

The preparation of casein is described on page 891. The clotting is discussed on page 376.

Lactalbumin. This is a crystallizable protein present in small amounts in milk and resembles seralbumin, but differs from it in its rotation, $(\alpha)_D$ =—37°. It coagulates at about the same temperature as seralbumin, i.e., 72-80°. Its composition is: C, 52.19; H, 7.18; N, 15.77; S, 1.73; O, 23.13. The amount of this protein in human milk is about one-fifth of the casein; and in cows' milk it is about one-tenth of the amount of the casein.

Lipins of milk.—The composition of butter fat has already been given on page 64. It is remarkable and different from the fat of most other organs in the large proportion of butyric, caproic and caprylic acids in it; in the proportion, in other words, of the volatile fatty acids. Milk fat contains, in addition to the glycerine esters of the usual fatty acids and those just mentioned, small amounts of both lecithin and cholesterol. Koch and Woods found in milk 0.036—0.049 per cent. of phosphatide computed from the phosphoric acid of the alcohol soluble phospholipins as lecithin; and 0.027-0.045 per cent. of a phospholipin computed as cephalin. This latter is insoluble in alcohol. The total of phospholipins was 0.072-0.086 per cent. For human milk " lecithin " was 0.041 per cent.; cephalin, 0.037 per cent., and the total, 0.078 per cent. The amount of cholesterol has not been determined by the digitonin method so far as can be found, but Marsh found by the Ritter method 0.021 per cent. The character of the phospholipin should be further studied.

The importance of these lipins in diet has been emphasized by the work of Stepp, who found that mice could not be nourished on food materials which had been carefully extracted by alcohol and ether, whereas the addition of this extract to the food restored them to health. The active substance was not fat, cholesterol, lecithin or salts. Similar observations have been made by Hopkins, who found that the addition to an incomplete ration of a very small amount of milk powder enabled mice to thrive when on a diet containing too little of some amino-acid. Similar observations have been made also by McCollum and Osborne and Mendel, who found that the active principle in milk was contained in the alcohol-ether extract and that similar active substances were present in the oils of fish livers, such as cod-liver oil. The nature of the constituent, or constituents, is still quite uncertain. They may belong to the general group of vitamines which are discussed on page 833. The small amount of lecithin in milk, which is designed as a food for rapid growth, is in striking contrast with the large amount of lecithin of the yolk of hens' eggs. This fact indicates that the infant is able to make the necessary phospholipins from the inorganic phosphates. There are other evidences that the phospholipins are readily formed by most kinds of protoplasm from inorganic material, fats and amino-acids.

Lactose.—The presence of lactose in milk to the extent of from 4 per cent. in cows to 7 per cent. in human milk is significant for the reasons already stated. The physical properties of this sugar are treated on page 56; the method of isolating it on page 867. The mammary glands have the power of making galactose from glucose.

Other organic constituents of milk.—Milk contains in small quantities a considerable number of organic substances. Among these particular mention may be made of *citric acid,* which is present in cows' and human milk to the extent of .05-.1 per cent. Among other constituents may be mentioned one found recently in human milk by Meigs and Marsh, which crystallizes readily, which makes about 1 per cent. of the milk, and which gives a strong test for unoxidized sulphur. The composition of this substance is unknown. Biscaro and Balloni have isolated a substance called *orotic acid* (Gr. *oros,* milk), which was precipitated by basic lead acetate. It had the composition $C_8H_{11}N_2O_4 \cdot 2H_2O$. The nature of the acid is unknown. Siegfried has isolated carniferrin, related to nucleon or phosphocarnic acid, from milk by his iron method. This yielded on decomposition carnic acid, fermentation lactic acid, succinic acid and oxylic acid ($C_{18}H_{28}N_4O_8$). The latter yielded leucine on hydrolysis with hydrochloric acid. There are also present in milk small amounts of urea, some purine nitrogen (in 1 liter cows' milk 0.004-0.006 gram purine N) and various other substances in small amounts. Some observers have described peptone or proteose bodies in milk, but it is doubtful whether they occur in perfectly fresh milk. Cows' milk as ordinarily obtained contains very large quantities of bacteria which rapidly change its constitution. To get sterile milk is difficult and it can be obtained only by passing a sterile catheter past the sphincter into the ducts. It must be remembered, however, that bacteria, although they are present in great quantities in ordinary city milk, often 1,000,000 or more per c.c., are not normal constituents, but have fallen in during the act of milking and that they multiply in the milk with great rapidity. As Sedgwick has put it: "A cow secretes milk, not bacteria."

Inorganic substances in milk.—Milk must contain most of the substances required for the early development of a child. It must provide substance for the growth of bone, hence it must and does contain large amounts of calcium and phosphoric acid. Moreover, all growing organisms need potassium, since potassium in nearly all forms of living matter is present in larger quantities than sodium. Rapidly growing cells need potassium. Iron is present in milk in very small amounts. This is in contrast to the egg where there is an abundant supply of iron in the yolk. But Bunge has shown that the iron is stored in the liver of the fœtus, this organ at birth having a rich supply which is called upon during development to make good the lack in the milk. The ash of milk contrasted with the ash of the new-born pup, showing how closely they resemble each other, is given by Bunge as follows:

	In 1000 parts of ash:	
	Of dog milk	Of new-born dogs
K_2O	149.8	114.2
Na_2O	88.0	106.4
CaO	272.4	295.2
MgO	15.4	18.2
Fe_2O_3	1.2	7.2
P_2O_5	342.2	394.2
Cl	169.0	83.5

In human beings there does not seem to be so close a correspondence between the ash of milk and of the new-born (Camerer and Söldner).

Coloring matter.—The milk of cows fed on green fodder is more or less yellow, while the milk of cows fed on dry hay and other forms of dry fodder is not yellow. Guernsey cows give particularly yellow milk and their fat is very yellow. This natural coloring matter is derived entirely from the green fodder and consists of carrotin and xanthophyll.

Hulls of the milk globules.—It has been generally supposed that the fat globules were surrounded by a very fine sheath of casein kept or condensed there by the action of surface tension. Recent investigations indicate that this is incorrect. By allowing the milk globules to rise through a long column of water they can be separated from the casein and other constituents of the milk. They are collected, the fat extracted with ether and the residue, which constitutes the hulls, analyzed. The result has been to show that the hulls may give no biuret reaction at all. When hydrolyzed they yield glycocoll, which is not present either in casein or lactalbumin. The amount of nitrogen is very low and variable, 4.04-5.70 per cent. The ash varies from 4.57-45.28 per cent., and P between 0.18 and 0.57 per cent. It is clear that it is not a homogeneous substance nor is it composed of casein.

Milk glands.—The composition of the mammary glands has been very little studied. A nucleic acid similar in its general properties to that of other organs has been isolated from it by Mandel and Levene. This acid contained pentose, so that it is probable that these glands contain not only the usual tetranucleotide, but also an acid analogous to, or identical with, guanylic acid. By treating the glands in the same way as the pancreas is treated for the extraction of a nucleoproteid by Hammarsten (see page 171) a nucleoproteid was isolated by Odenius of the following composition: C, 46.89-47.15 per cent.; H, 6.04-6.15 per cent.; N, 17.26-17.29 per cent.; S, 0.875-0.904 per cent.; P, 0.275-0.278 per cent. and 0.962-0.922 per cent. Mandel and Levene isolated a substance of a reducing character which contained sulphur, and they called it glucothionic acid. S, 2.65 per cent.; N, 4.38 per cent. There are probably present substances of the sulpholipin or glycolipin type and these may yield the reducing substances obtained by several observers on hydrolysis of the organic matter. The extractives are about the same as the other organs so far as they have been studied. The composition of the substances thus far isolated throws singularly little light on the method of formation of the milk constituents or on the metabolism, and the gland should be further investigated.

Metabolism of the glands.—There is very little that can be said about this. The raw material from which the galactose is made is undoubtedly glucose. The metabolism undergoes a remarkable rhythmic alteration dependent upon the sexual organs. The mammæ are in the nature of secondary sexual organs. That is, although they are present in a rudimentary form in the male, they do not develop, but remain rudimentary. In the female they begin their development with the ripening of the ovaries and the formation of ova. It is clear that their development must be stimulated in some way by the ovary. Further research of this very interesting problem has shown that the corpora lutea have the power of stimulating the growth and activity of the mammary glands. An extract of the corpora has a remarkable action on these organs. It is found, too, that an extract of the uterine wall at the time when it is undergoing regressive metamorphosis after parturition has the power, when injected into a female animal of the same kind, of exciting the activity of the mammæ so that the glands become full of milk. On the other hand, an extract of fœtus has an inhibiting action. It will be noticed, too, when the hypophysis is considered, that the extract of this organ may cause a discharge of milk from a pregnant female. The development of the gland appears to depend, then, upon the internal secretion of the ovaries. The development ceases and the gland atrophies after the discharge of ova from the ovary ceases and there are no more corpora lutea. On the other hand, the cause of the differentiation of the skin tissue into milk-gland tissue does not depend upon this internal secretion, for it occurs in the male also. It would appear that in this, as in so many other ways, males have in them often many of the potentialities of females, and vice versa, and that whether male or female characters predominate depends upon the development of ovaries or testes. This is a matter, however, somewhat foreign to the present discussion, but which will no doubt occupy a much larger space in the biochemistries of the future.

Enzymes.—The presence of enzymes in the milk has been much disputed, the contradictory results being due possibly to the rich bacterial flora of most milks examined. Human milk is said to contain small quantities of amylolytic enzyme, and as this is found in the blood it probably comes from the blood. The presence or absence of lipase is disputed. There is no lipase in the glands (Bradley) and this indicates that lipase is not the cause of the synthesis of the fats in the gland. Catalase and peroxydases are present in milk, but catalase is found almost universally in the organized world. A proteolytic enzyme of an ereptic nature has been described (Russell and Babcock) and given the very inappropriate name of *galactase*. This does not appear to be always present. An enzyme which ferments lactose to lactic acid, alcohol and CO_2 has been described by Stoklasa in both human and cows' milk.

Souring of milk.—This is produced by the action of bacteria of various kinds which attack the lactose, forming from it lactic acid. The acid thus set free precipitates the casein and thus clots the milk. Some bacteria are able to clot milk in a neutral reaction. These clot because they secrete a proteolytic enzyme like rennin which forms paracasein from the casein.

Sterilized milk as a food.—Many very young children do not thrive when fed exclusively on boiled, diluted cows' milk, whereas they will assimilate fresh, diluted cows' milk. The character of the change produced in the milk by heating, which is responsible for this malnutrition, is still unknown. On heating, small amounts of sulphureted hydrogen are split off from the milk. This is probably one reason for the change in taste and odor produced by cooking. Whether it has anything to do with the change in nutritive qualities is uncertain. It may be that cooking destroys some vitamine in the milk. If infants can have a portion of human milk, they generally assimilate without trouble cooked, diluted cows' milk to which lactose has been added.

Foreign substances in milk.—Various substances pass directly from the blood to the milk. Thus salts of various kinds taken by the mother may affect the child. Cases are on record of infants developing bromism from the effects of bromides taken by the mother during nursing. Various odoriferous substances pass readily into milk, as everyone knows. The diet of the mother is hence of importance to the child. It is possible that various internal secretions from the mother's body may pass by means of the milk to the child and may affect it.

Various kinds of milk.—Of the various kinds of milk which are available as foods for children, asses' milk approaches most closely in composition to that of human beings. The composition of goats', asses' and mares' milk is as follows:

albumin. It resembles ovalbumin in elementary composition, but will not crystallize and coagulates at a lower temperature, i.e., 50-60°. It contains about 16 per cent. N and 1.7 per cent. of sulphur.

Ovomucin. This is separated in the globulin fraction by precipitation by half saturation with ammonium sulphate (Eichholz). It is a glycoprotein. It rapidly becomes insoluble in water on dialysis. It yields about 11 per cent. of glucosamine.

Ovomucoid. This makes about 10 per cent. of the total solids of hen's egg white. It is a typical mucoid substance. It contains 12.65 per cent. nitrogen and 2.20 per cent. of sulphur. According to Seeman, it contains 34.9 per cent. of glucosamine. It is not precipitated by half saturation with ammonium sulphate or by complete saturation with sodium chloride or magnesium sulphate. It does not coagulate on boiling and it may be prepared by this means. All coagulable proteins are removed by slightly acidifying with acetic acid, coagulating by heat, and filtering. The concentrated filtrate is precipitated with alcohol. This behavior and the low content of nitrogen may mean that the substance is allied to chondroitic acid, or contains a large amount of this substance. Further work on the composition of this body is needed.

Ash. The ash of the white contains about 30 per cent. of Na_2O, the same amount of K_2O, about 29 per cent. of Cl and small amounts of Ca, Mg, phosphoric acid and other substances.

The yolk.—The yolk is very much more complex than the white. It contains a very large amount of oil or fat, egg oil, and the protein is chiefly vitellin. The solids are about 50 per cent., and these consist of protein, 16 per cent.; fat, 23 per cent.; lecithin, 11 per cent.; cholesterol, 1.5 per cent., and salts 3 per cent. There is some variation in these figures in different eggs. The lecithin and egg oil make the greater part of the solids. The phospholipins are not all lecithin; bodies similar to cephalin and some other myelin-like phospholipins have been isolated (Thierfelder). The yolk generally makes about one-third of the whole egg. It weighs, on the average, about 20 grams. There should be about two grams of phospholipins in one egg yolk.

Ovovitellin. This is a typical phosphoprotein, or nucleoalbumin. According to Osborne and Campbell, ovovitellin as extracted from egg yolk by salt solution is a mixture of various compounds of lecithin and vitellin proper. The P content varies from 2.5-4.2 per cent. The amino-acids it yields are given on page 129, and its composition on page 139. The ovovitellin is insoluble in water, but soluble in dilute salt solutions and very dilute alkalies. It behaves much like a globulin, but it differs from a globulin in that it contains phosphoric acid in some union or other, and when digested with pepsin HCl it yields an acid rich in phosphorus, about 8 per cent. P, which was at first supposed to be nucleic acid, but is not. It yields no purine bases. The vitellin is extracted from the yolk by shaking the fat out with ether and then treating the residue with 10 per cent. NaCl and precipitating the filtrate by diluting with water. The lecithin is separated by boiling alcohol. Pure vitellin has not been obtained. The nature of the phosphoric acid residue is unknown and needs further investigation.

Hematogen. Egg yolk contains a substance which on digestion with pepsin HCl yields a nuclein-like substance which contains iron. This iron compound is supposed to be the mother substance of the hemoglobin. It was named hematogen to show this relationship. The composition of this substance is as follows: C, 43.5; H, 6.9; N, 12.6; P, 8.7; S, trace; Fe, 0.455; Ca, 0.352; Mg, 0.126. The substances analyzed are probably mixtures. There is no doubt that egg yolk contains all the necessary ingredients for the manufacture of hemoglobin, since this substance is formed during development.

Lipins. Egg oil. This yields a large amount of oleic and palmitic acid. Liebermann obtained: palmitic acid, 38.0 per cent.; oleic, 40 per cent., and stearic acid, 15.21 per cent. The composition of this fat is dependent upon diet. Thus, if chickens are fed fish, the eggs have a very strong taste of fish; if they are fed on rape seed and the young rape plants, the boiled yolks are a deep black and have a disagreeable taste. Fat soluble dyes such as Sudan 3 will be laid down in the yolk if it is ingested by the birds. The yolk is laid down in rings. Each ring means a time interval of 24 hours. The deposition of the yolk is slow at night, or some reabsorption may occur so that the appearance of the fat is changed and the yolk is in layers (Riddle). It takes about four days when a hen is laying rapidly to produce the greater part of the yolk. The period can be greatly lengthened by cold or poor nutrition. The phospholipins include lecithin and lipins resembling cephalin and aminomyelin.

Mineral substances. The following analysis shows the composition of the mineral constituents in parts per thousand:

Na_2O	51.2 - 65.7	FeO	11.9 - 14.5
K_2O	80.5 - 89.3	P_2O_5	638.1 - 667.0
CaO	122.1 - 132.8	SiO_2	5.5 - 14.0
Mgo	20.70 - 21.1		

REFERENCES. FOODS.

1. Milk. *Meigs and Marsh:* The comparative composition of human milk and of cows' milk. Jour. Biol. Chem., 16, p. 147, 1913. (Literature cited in this paper.)

2. *Difference between cooked and raw milk.* Schardinger reaction. *Romer:* Biochem. Zeits., 40, p. 5, 1912. *McCollum and Davis:* J. Biol. Chem., 23, p. 247, 1915.

mechanism becomes capable of directly or indirectly affecting digestion. So cooking, because of the development of pleasant and appetizing tastes and smells in foods, aids digestion more than by its direct digestive action on the foods.

In still a third way cooking may profoundly aid digestion by killing parasites or bacteria which otherwise would gain a foothold in the alimentary canal and thus change the nature of the digestive processes. The meat of chickens and pigs is particularly apt to become infected with bacteria which produce poisonous products, or which become lodged in the intestine and growing there produce toxines. Many of the headaches, attacks of constipation, diarrheas, feelings of malaise, fatigue, etc., from which we suffer, are due to such infections, which are, as often as not, so ill defined in their onset, so general in their symptoms and so gradual in their disappearance that the origin of the ill feeling is often incorrectly referred to the nervous system, or to any other rather than to the right source.

Auto-digestive processes occur, too, in many foods. Thus meat and eggs contain digestive enzymes which come into activity particularly in meat when its reaction becomes acid. These ferments digest the tissue and even in the absence of bacteria gradually destroy the structure and render the constituents soluble. The ripening of fruits; the changes in tenderness and taste of meat of various kinds, or of glandular organs used as food; the change in taste of vegetables with age; all these are due to such auto-digestive processes, which are thus more or less important adjuncts to the digestive juices of the body.

Saliva. Origin.—The first of the digestive juices formed by the body with which the food comes in contact is the saliva. This fluid is secreted by the salivary glands, the parotid, submaxillary and sublingual, and by the mucous membrane of the mouth.

The saliva is formed by the protoplasm of the cells of these glands with the admixture of salts, water and some other substances derived from the blood. The protoplasm of the inner end of the cells of the gland and sometimes the greater part of the cell body is transformed into saliva. Saliva is thus nothing else than transformed protoplasm. During the time of active secretion raw material is brought to the gland by the rapid blood stream, since vasodilation takes place at that time and the flow of blood to the gland is greatly increased. Out of this raw material crude protoplasm is formed which afterwards, during the period of glandular rest, either forms mucin and the other salivary constituents, or it is itself transformed into these substances by a process of differentiation. The mucin thus formed accumulates in the cells in the form of granules and is discharged from the cell during the process of secretion.

The character of the saliva secreted by the different glands varies. The sublingual secretion is very thick and viscid, due to the presence in it of a great deal of mucin; the submaxillary, also, generally contains a great deal of mucin and its saliva is very slimy; the parotid, on the other hand, has a saliva with less mucin, or none at all. Its secretion is generally more watery, although it has a good deal of protein in it, and it is the secretion of this gland, more than the others, which has a digestive action on starches. Differences exist, also, in the salivas secreted by the same gland in different species of animals. The parotid saliva of the horse, rabbit and sheep is generally more fluid than that of the dog, which indeed may be very viscid.

Nervous control of the secretion.—Each of the salivary glands is controlled by a double set of nerves, by a nerve which dilates the arterioles of the gland, a nerve which comes directly from the brain; and by a nerve which comes from the sympathetic system, and which when stimulated produces constriction of the arterioles of the gland. It is by means of the actions of these two nerves that the nutrition of the gland is controlled and its activities regulated. The quantity of saliva secreted on stimulation of the cerebral nerves (chorda tympani of the submaxillary and sublingual, and the auriculo-temporalis branch of the fifth nerve, which receives fibers from the glossopharyngeal nerve for the parotid) is much greater and the saliva is more watery than when the sympathetic nerves are stimulated. By some authors, notably by Heidenhain, this difference was ascribed to differences in the manner of action of the nerves on the cells of the gland. Heidenhain suggested that the cerebral nerve acted upon the gland cells in such a manner as to cause the secretion proper, that is the discharge of water and salts, whereas the sympathetic acted upon the cells so as to make the cell contents soluble, so that the secretion was a great deal more concentrated in the latter case. By other physiologists the difference in the character of the secretion is ascribed to the difference in the state of the blood vessels accompanying secretion, to the vasodilation when the watery juice is secreted and to the constriction when the more concentrated juice is formed; by still others (Eckhard, Bernard and some others) the difference in composition of the two kinds of salivas is ascribed to the fact that in part the nerve acts on the cells of the gland and the blood vessels, and in part it acts on the basement membrane, or the basket cells, which are supposed to be contractile and to press out of the gland some of the stored secretion. The sympathetic causes its secretion mainly by the latter method, the chorda by stimulating secreting cells, blood vessels and contractile sheath. It is impossible to go here into a discussion of this question which is still unsettled.

Composition of mixed saliva.—The composition of mixed human saliva was found by Frerichs to be as follows:

Water	99.41%
Solids	.59
Mucin and epithelium	.213
Soluble organic matter	.142
Inorganic salts	.219
KCNS	.01 - .00

In 1,000 parts of the mineral ash there were found the following constituents:

K	457.2
Na	95.9
CaO and traces of Fe_2O_3	50.11
MgO	1.55
SO_3	63.8
P_2O_5	188.48
Cl	183.52

The specific gravity is 1.002-1.008. The freezing point is $\Delta = -0.28$ to $-0.43°$. The freezing point is higher than that of the blood. The viscosity of submaxillary-sublingual saliva may be 18-35 times that of water. It is extremely variable. The great richness of the ash in potassium will be noticed. In the blood, sodium greatly surpasses potassium in amount. This potassium is not free potassium chloride, but the greater part is present in organic combination.

The reaction of the saliva of the healthy is generally, if not always, in its fresh state, alkaline to litmus, but acid to phenol-phthalein. This would give it a reaction about that of the blood, or approximately $2 \times 10^{(-8)}$ normal concentration of hydrogen ions. To neutralize it to litmus there must be added for 10 c.c. saliva 4.5-14.9 c.c. of N/100 acid; and to make it sufficiently alkaline to redden phenol-phthalein, about the same amount of N/100 alkali is necessary. It will not infrequently be found, however, in testing a large number of individuals, that some have the mixed saliva acid to litmus. This is most often due to an acidity produced by the action of bacteria on the fragments of food left between the teeth, particularly carbohydrate food, but it sometimes has another explanation, for the saliva is found to be acid to litmus when collected directly from the duct by a cannula. This acidity may be correlated with age, state of health or diet and should be further studied. The reaction of the saliva is of considerable importance in the hygiene of the mouth, for it is a general impression of dentists that the acids thus produced by the action of the bacteria on the saliva and food remnants are an important element in the decay and corrosion of teeth, or in the deposition of the tartar on the teeth. The subject needs a thorough investigation.

Amount of saliva secreted per day.—It was estimated by Bidder and Schmidt that about 1,500 c.c. saliva were normally secreted per day, but the amount is no doubt subject to wide variation, depending on the amount of water consumed, the length of time the food is chewed, the character of the food, its dryness and so on, and the estimate is at best an approximation only. Since the salivary glands weigh together only about 66 grams in the adult, it is obvious that the amount of saliva secreted is many times the weight of the glands. The secretion may be greatly increased by many poisons, such as pilocarpine, by mercury salts or by some organic toxins which produce a salivation very similar to mercury salivation. It is increased by smoking. It was not infrequent in the days of mercury treatment of disease to give mercury freely until three or four liters of saliva, or even more, were secreted daily. In mercury poisoning one of the symptoms is the copious flow of saliva and the other manifestations of salivation, such as the sore gums and lips.

Functions of the saliva.—In man the saliva has two main functions, i.e., to aid swallowing and to assist in the digestion of starches. These functions are subserved by two different constituents of the saliva. The former by the water and mucin; the latter function by the ptyalin. These are, therefore, the most important constituents and their chemistry may now be briefly considered.

Chemistry of mucin.—The saliva is a modified skin secretion, since the salivary are modified skin, or epidermal glands. It is interesting, therefore, that they contain mucin, a glycoprotein, which is such a constant constituent of the skins of the amphibia and the fishes and which occurs even in the skin of mammals in a disease of the thyroid known as myxœdema. Mucin, or mucoid substances, are found not only in the salivary glands and in the skin, but similar glycoproteins may be isolated from the tendons, connective tissues, from cartilage, from amyloid, and they occur not infrequently in the invertebrates. The general composition of various mucins is shown in the following figures:

	C	H	N	S	O	Author
Snail's mucin.......	50.32	6.84	13.05	1.75	27.44	Hammarsten
Tendon mucoid......	48.76	6.53	11.75	2.33	30.63	Chittenden and Gies
Submaxillary mucin..	48.84	6.80	12.32	0.84	31.30	
Salivary mucin......	48.26	6.91	10.70	1.38-1.45		Müller

They contain on the average less carbon and nitrogen and more oxygen and sulphur than the simple proteins.

The mucins are conjugated proteins, they are glycoproteins. That is, when they are hydrolyzed by acids they yield a carbohydrate radicle. The exact composition of the mucin of the submaxillary or of any other true slimy mucin has not yet been determined, but extensive studies have been made by Gies, Levene and others on the mucoid of the tendons. The relation between the chrondroproteins (mucoids) and mucin is still uncertain, but the former contains more sulphur than the latter. The mucins have the general properties of the nucleins, that is they are readily soluble in dilute alkali, are readily precipitated by dilute acid, but are soluble in stronger acid. They are prepared by extracting tissues with water, or with dilute alkali and precipitating with acetic acid. Such preparations generally contain nucleoprotein. They are acid bodies and exist in the tissues as salts. The mucin of the submaxillary gland probably exists as the potassium salt. Being acid bodies, like the nucleins, they form compounds with basic dyes, or at least some of the mucins thus combine, and one of the principal mucin stains is thionin, which is such a basic dye. According to Levene, submaxillary mucin has some of its sulphur as ethereal sulphate.

On decomposition of tendon mucoid by hydrolysis by acids, Levene found that it split into sulphuric acid, galactose, galactosamine; and by stronger acid into leucine, tyrosine, levulinic acid and acetic acid. It contains, therefore, a hexose (shown by the levulinic acid) and sulphuric acid. It is thus clear that the mucoids must be closely related to the cartilages. The organic matrix of cartilage consists of two proteins, one of which is a conjugated protein formed by a simple protein united with chondroitic acid, that is with cartilage acid, since the word chondroitic means cartilage. This cartilage acid, or chrondroitic acid, is an acid found in small amounts in the urine. When hydrolyzed by acid it breaks up according to the scheme shown on page 325.

The final products of decomposition of the chrondroitic acid are therefore glucosamine, or levulinic acid from it, sulphuric acid, acetic acid, glycuronic acid. These are just the splitting products which Levene found among the decomposition products of tendon mucoid. It seems, therefore, that the prosthetic group of the mucoid molecule is chrondroitic acid, or a closely similar acid. He was not able to isolate the chrondroitic acid itself. Mucin is related, therefore, to cartilage through the mucoids. The difference between them is probably to be found in the protein part of the molecule and the proportion of chondroitic acid. Mucin also occurs chiefly as the potassium salt, whereas cartilage mucoid is probably present chiefly as the calcium salt.

The relation of mucin to cartilage is extremely interesting from the evolutionary standpoint. It shows that mesodermal and the ectodermal tissues are possibly not so unlike in their chemical nature. A very interesting fact, also, is the resemblance in composition of mucin, which occurs in such quantities in the skin of the lower vertebrates, to chitin, the hard covering of the arthropods. When chitin is hydrolyzed it yields glucosamine and acetic acid. It is supposed to be a polymerized monoacetyl glucosamine. There is always present, also, some sulphuric acid, but this is generally regarded as an admixture of inorganic sulphate. In other words, chitin, which is the main constituent of the external and internal hard parts of the arthropoda, is chemically related to the cartilage or the organic matrix of the internal skeleton of the vertebrates and to mucin, which is such an important constituent of the skins of vertebrates. These facts lend support to the view that the chitin of the invertebrate, perhaps by combination with glycuronic and sulphuric acids, formed the matrix of the mucin, mucoid and cartilage of the vertebrates. These facts are of interest in the light of the theory of Gaskell and Patten that the arthropods were the ancestors of the vertebrates.

$$\underset{\text{Chondroitic acid.}}{C_{18}H_{27}NSO_{17}} + H_2O = H_2SO_4 + \underset{\text{Chrondroitin.}}{C_{18}H_{27}NO_{14}}$$

Chrondroitin in its properties resembles gum arabic. It yields acetic acid and chrondrosin.

$$\underset{\text{Chondroitin.}}{C_{18}H_{27}O_{14}N} + 3H_2O = \underset{\text{Acetic acid.}}{3C_2H_4O_2} + \underset{\text{Chondrosin.}}{C_{12}H_{21}NO_{11}}$$

On further hydrolysis chondrosin, which is a reducing substance, splits as follows:

$$\underset{\text{Chondrosin.}}{C_{12}H_{21}NO_{11}} + H_2O = \underset{\text{Glycuronic acid.}}{C_6H_{10}O_7} + \underset{\text{Glucosamine.}}{C_6H_{11}(NH_2)O_5}$$

```
                              HC——————O——————CH
                               |‾‾‾‾‾|  |‾‾‾‾‾|
OH                           HCOH   |  |  HOCH                              OH
|           CH3               |     O  O   |           CH3                   |
SO2         |               HOCH    |  |  HCOH         |                    SO2
|           CO                |     |  |   |           CO                    |
O           |                 HC————|  |———CH          |                     O
|   OHH  H  NH·  H            |            |       H   HN  H           OH    |
CH2—C—C—C—C   —C—O—CH                     HC—O—C———C—C———C— —C—CH2
    H  | OHH   |             |            |       |   H  OH |H    H
       |——O————|            COOH        HOOC     |————————O——|
```

Proposed formula for chondroitic acid. $(C_{28}H_{46}N_2S_2O_{29})$
(Levene and LaFarge.)

While the composition of the salivary mucin is still uncertain, there are several facts which indicate that it is closely related to the mucoids. It has been shown (Müller), for example, that of the total sulphur in the compound 35 per cent. is in the form of oxidized sulphur, and (Levene) that part of it is present as ethereal sulphate. The carbohydrate radicle makes 36.9 per cent. of the salivary mucin, and about 24 per cent. of the submaxillary mucin (Müller). This radicle is probably at least in part glucosamine. Moreover, it is probably a monoacetyl hexosamine. This relates the mucin at once to chitin, which is a polymer of an acetylated glucosamine. On cooking with alkali it splits off a complex carbohydrate, called animal gum by Landwehr, which does not itself reduce Fehling's solution but only after hydrolysis. This again would relate it to chitin. It appears, then, that the prosthetic group of mucin, if not chondroitic acid, is at any rate related to it and resembles chitin in many ways. The protein part of the molecule is still not well investigated. Mucin which has been warmed with alkali gives a very red color with dimethylaminoazobenzaldehyde (Ehrlich). This reaction is given also by acetylated glucosamine when it is treated with alkali. Chitosan from chitin does not give this reaction.

Dry mucin is a white amorphous substance, which swells in water and dissolves on the addition of a little alkali. As precipitated by acetic acid its reaction is acid. It gives the protein reactions, but it does not coagulate on heating a neutral solution. It is completely precipitated by saturation with ammonium sulphate, but not by magnesium sulphate.

Preparation of mucin.—Mucin is prepared from the submaxillary gland of the dog by extracting the finely hashed organ with water and filtering. HCl of 25 per cent. is then added to the filtrate until this contains 0.15 per cent. of HCl. On the addition of the acid there is at first a precipitate, but this soon redissolves. The clear solution may be then diluted with 2-3 volumes of water when the mucin precipitates out. Tendon mucoid may be prepared from the ox Achilles tendon by extracting the chopped tendon first with water and then 10 per cent. NaCl to free it from protein. It is then extracted with $Ca(OH_2)$. The extract is filtered; the mucoid is precipitated with acetic acid and purified by repeated solution in dilute alkali and reprecipitation with acid.

B. Digestive action of saliva.—That human saliva has the power of converting starch into sugar was discovered by Leuchs in 1831. If a little saliva is mixed with starch paste, it may be seen that the paste very quickly becomes clear like water; if a little of it is then poured out, it will be found that, whereas it was before very thick, it now pours easily; in other words, its viscosity has been greatly reduced; if a little of the clear solution is added to some Fehling's solution in a test-tube and boiled, it will be found that the Fehling's solution is reduced, whereas it was not reduced either by the starch alone, or by the saliva alone; by the action of the saliva a reducing substance has been formed; finally, if after a short time one adds to the mixed solution of starch and saliva a solution of iodine in potassium iodide, it will be found that the iodine no longer forms the blue color which it formed in the starch solution before the addition of the saliva, but it either gives no color at all or it develops a red color. These facts show that saliva changes both the chemical and physical properties of starch solutions and that the starch disappears and a reducing substance appears in its place. If the starch solution thus acted upon by saliva is tasted, it will be found to have become sweet. From these observations we infer that saliva changes starch into some kind of reducing sugar.

Nature of the changes involved in the digestion of starch.—The further study of the changes which have occurred in the starch solution have not yet succeeded in entirely clearing up the matter. The question may first be asked, What is the nature of the sugar produced? This is certainly in greater part the disaccharide, maltose, $C_{12}H_{22}O_{11}$. There appears, however, particularly toward the end of the action, to be some glucose present also. But besides the maltose there are also formed in the course of the digestion substances of a very much lower reducing power, or of no reducing power at all, and substances which give with iodine a red coloration, or no coloration at all. These substances are called "dextrins," because they are dextro-rotatory. The name was given by Biot, the father of polariscopic investigations. The dextrin which gives with iodine a red coloration is called erythrodextrin, or the red dextrin (Gr. *erythros*, red). The dextrin which gives no color with iodine is called achroodextrin, or the colorless dextrin (Gr. *achroos*, colorless). These substances are colloidal in nature. While all observers are agreed that these various substances are formed in the course of the digestion, there is no agreement as to the exact order in which they appear and their relation to each other. The difficulty in the way of a solution of the problem arises in part from the difficulty of separating quantitatively the various substances formed, but also because the substrate, the raw material which is digested, is not a pure substance. There is certainly in every starch grain some substance of the nature of cellulose which is far more resistant to the action of the enzyme than starch proper. Furthermore, the ferment solutions may contain more than one enzyme. Probably a great deal of experimental work will be required before this matter is entirely cleared.

The products of the digestion of starch by malt extract, which has a digestive property similar to, but not in all points identical with, ptyalin, were found by Musculus and Gruber to be the following:

(a) Soluble starch. This was insoluble in cold water, but soluble at 50°. Iodine colored its solution wine red, but the dry substance blue. The reducing power was 6% of that of glucose. $(\alpha)_D = +218°$.

(b) Erythrodextrin. Soluble in cold water. Both when dry and in solution gives a red color with iodine. Not a pure substance.

(c) Achroodextrin α. No color with iodine $(\alpha)_D = +210°$. Reducing power 12% that of glucose. Less easily changed to sugar than erythrodextrin.

(d) Achroodextrin, β. $(\alpha)_D = +190$. Not digested by malt diastase. Reducing power 12% that of glucose.

(e) Achroodextrin, γ. $(\alpha)_D = +150°$. Not digested by diastase. Reducing power 28% that of glucose.

(f) Maltose, $C_{12}H_{22}O_{11}$. $(\alpha)_D = +150°$ Reducing power 66% that of glucose. It is capable of fermentation by yeast. It is not attacked by malt diastase.

(g) Glucose, $C_6H_{12}O_6$. $(\alpha)_D = +56°$. Fermentable. Reducing power 100%.

According to Brown and Herron, the simplest formula for soluble starch is $(C_{12}H_{20}O_{10})_{10}$. This was believed to split off maltose under the action of malt diastase and to give erythrodextrin $(C_{12}H_{20}O_{10})_8$. This losing another molecule of maltose went into a second erythrodextrin, and this after the loss of another molecule of maltose went into achroodextrin. The final mixture contained 81 per cent. of maltose and 19 per cent. of achroodextrin. In their opinion the dextrins had no power of reduction when they were entirely free from maltose.

Since the maltose comes off almost instantaneously when starch solution is mixed with saliva, it is generally believed that the course of the reaction is a gradual etching away of the starch molecule by splitting off successive molecules of maltose by hydrolysis, as pictured in the following scheme:

Starch + H_2O = Soluble starch + Maltose
Soluble starch + H_2O = Erythrodextrin + Maltose
Erythrodextrin + H_2O = Achroodextrin α + Maltose
Achroodextrin α + H_2O = Achroodextrin β + Maltose
Achroodextrin β + H_2O = Maltose + Maltose.

We may accept this scheme as the best one possible with our present knowledge, although there are many grave difficulties in the way of its acceptance as a real picture of the process. The main fact is, however, that dextrins and maltose are produced by the action of saliva on starch. There is also in human saliva some maltase which converts some of the maltose to glucose.

Ptyalin.—The question may now be asked, What is it in the saliva which enables it to act thus on starch? A few simple experiments show that the active substance is destroyed or loses its activity if the saliva is heated, and that it is precipitated from the saliva by alcohol. It is then not heat stable, and it is probably an organic substance. Since a very little of it can convert a very large amount of starch into sugar by hydrolysis, it evidently belongs in the group of substances known as catalysts, or enzymes. It is called "ptyalin," a name given by Berzelius to the peculiar organic matter of the saliva, but now confined to the ferment. The word comes from the Greek, *ptyalon*, spittle. Ptyalin is, therefore, the enzyme in saliva which digests starch. It is hence an "amylase," that is an enzyme which hydrolyzes starch, or amylum. It is not certain whether there is but a single enzyme in ptyalin. There may be several,—an amylase converting starch to dextrin, and a dextrinase which converts dextrin to maltose. There is hardly a doubt that in the pancreas the amylase is such a mixture, and it is probable that this is the case in the saliva also. Such amylolytic enzymes are widespread in nature, being found not only in all vegetable cells which contain starch, but also widely distributed in animals, being found in the blood, saliva, pancreatic secretion, in the liver and probably in the muscle also. They are also very often present in moulds, yeasts and bacteria, although common brewer's yeast does not contain amylase. These amylolytic (literally starch-loosening) enzymes are sometimes called *diastases* in English, American and German literature, but the French use the term diastase as a synonym for enzyme in general. The French usage is the more correct. The name diastase was given to the starch-splitting enzyme in malt by Payen and Persoz to indicate that it was different, or stood in a class by itself. The word means " to stand apart," Gr. *dia*, apart, or through. With this confusion in its significance it is perhaps better to drop the word entirely and replace it by the word enzyme as the group name, and amylase as a specific name of the starch-splitting ferments. Malt amylase was first discovered by Dubrumfaut in 1823, and isolated and carefully studied by Payen and Persoz in 1833. It was found by Kraussman and Krauch in leaves; by Baranetsky in the tubercles of potatoes in repose; by Duclaux in moulds such as Aspergillus niger. It was found in the saliva by Leuchs in 1831, and extracted from saliva by Miahle in 1845; the latter year the amylase of the pancreas was discovered by Bouchardat and Sandras.

While all these different enzymes have in common a similar action in producing a soluble sugar, maltose, from starch, they differ among themselves so that they constitute not one enzyme, but a group of enzymes. The amylase of the saliva is much more sensitive to heat than that of malt. The amylase of the pancreas, amylopsin, converts the starch into dextrins more rapidly than the dextrins into maltose. And there are other differences in their actions.

Composition and manner of action of ptyalin.—The composition of ptyalin has not been much studied, owing in large part to the difficulty of procuring sufficient material and of separating the enzyme from mucin. But the composition of other amylases, such as that from malt, has been often investigated, as well as that from the pancreas. The best method for the extraction from malt or the pancreas is to extract the tissues or material with two to four volumes of dilute alcohol, 20-30 per cent., and then to precipitate the ptyalin by the addition of three volumes of absolute alcohol. This is the method followed by Lintner in the study of malt diastase. It may also be extracted with water and precipitated by saturating with magnesium sulphate. Still another method was followed by Fraenkel and Hamburg. They fermented malt extract by yeast to get rid of any maltose or other fermentable sugar; filtered from the yeast through a porcelain filter; precipitated the protein very carefully by basic lead acetate; and fermented the filtrate with a nitrogen-hungry yeast to free from the amino-acids and peptones; again filtered through a porcelain filter; and evaporated to dryness in a vacuum at a low temperature. The per cent. of composition of various amylases obtained by various observers is as follows:

	From	C	H	N	S	Ash	O and S	P
Lintner	Malt	44.33	6.98	8.92	1.07	4.79	34.98	
Tegorow		40.24	6.78	4.70	0.70	4.60	42.23	1.45
Osborne	Hemp seed	52.50	6.72	16.10	1.90	0.66	22.78	

Opinions vary widely as to the nature of the active principle. Armstrong has suggested that the enzymes are of the nature of the substances they act on, that glucoside-splitting enzymes, for example, are glucosides. Lintner thinks that the active principle is a carbohydrate and the nitrogen present in most preparations is an impurity. On the other hand, there was no carbohydrate in Osborne's preparation, so that the active principle can hardly be a carbohydrate. Fraenkel and Hamburg found that their very active preparations gave no protein reactions, except the faintest Millon and xanthoproteic. The carbohydrate reaction was strong. They concluded with Lintner that it was a carbohydrate. Wroblewski says that he has prepared a very active substance from malt extract which gave no carbohydrate reactions, but was of the nature of an albumose. The following facts, however, are admitted by all investigators: Nearly all preparations, if not all, contain phosphoric acid which it is very difficult or impossible to eliminate and to keep the activity of the enzyme. There is a growing consensus of opinion that in most enzyme decompositions, at least those of the carbohydrates, phosphoric acid is playing some unexplained rôle. There is practical agreement, also, that generally the enzyme is found to be accompanied by, or to be part of, a gum, and this gum contains phosphoric acid. This gum is an araban, a pentose gum. In all organs from which amylase has been isolated pentoses have been found on hydrolysis. All observers agree that the active principle is a colloid. It will not diffuse. No preparation of active amylase has been obtained free from nitrogen and the protein reactions, whereas some have been obtained free from carbohydrate reactions. In view of these facts we may tentatively conclude that, while the nature of the enzyme is still quite unknown, the indications are that the active part of the molecule is a protein, probably colloidal, and that this active principle is usually combined with a colloidal, carbohydrate gum. It is possible that it is only active when united with the gum, but the probability is, the writer thinks, that the gum acts only as a bearer of the active principle, making it more stable, and that the principle is more active when free from the gum. In other words, that the compound, gum-active principle, is the proenzyme, or zymogen as it is called, from which the active principle may be set free.

There is no doubt that the active principle, whatever its nature, combines with starch and other carbohydrates. This is shown by the fact that it is much more stable in the presence of these bodies. The enzyme is more resistant to heat, ultra-violet light, alcohol and other poisons when carbohydrates are present in the solution.

Determination of the activity of amylolytic enzymes.—The activity of any preparation of amylase may be measured in various ways. As soon as the starch solution becomes clear, the further course of the digestion may be followed by the diminution in the rotatory power. In quantitative determinations by this method it is necessary to correct for the mutarotation of the maltose by the addition of a little alkali before polarization. The alkali stops the digestion also. Another method is to fill glass tubes of an internal diameter of about 2 mm. with a thick starch paste, cut the tubes in small lengths, place them horizontally in a small flask with the solution of amylase to be tested and measure the length of the starch column dissolved at different time intervals. Another method is to determine the length of time which is required for the blue iodine reaction of a starch paste of known concentration at a known temperature and with a known amount of ferment to disappear. This method is better than measuring the rate of increase of the reducing power of the solution by Fehling's solution, for the latter may involve several enzymes, whereas the iodine reaction probably involves only the amylase proper. Another method which again involves the dextrinases as well as the amylases is to measure the rate of change of the viscosity of the solution by a viscosimeter. In all these measurements, if the activity of different preparations of the enzyme are to be compared, it is necessary to be sure that the conditions of activity, such as the acidity, proportion of inorganic salts, etc., are in all cases at the optimum for the enzyme being tested.

Conditions of activity of the ptyalin.—The optimum temperature for the action of ptyalin or saliva on starch is 40-45°. The action is permanently lost if saliva is heated rapidly to 75°; and more gradually lost if heated for longer periods at lower temperatures. While the reaction of the saliva is very faintly alkaline to litmus, the optimum reaction for the digestion of starch is a very faint acidity. Thus the addition of carbonic acid hastens the action, as illustrated by the table on page 332 (Chittenden and Painter). The degree of acidity must, however, be very slight. The most favorable concentration of hydrogen ions was found to be: $N \times 10^{-6.7}$.

If more acid than this is present, if sufficient is present to turn congo red violet, a hydrogen ion concentration of $N \times 10^{-4}$, the activity is stopped. The digestion will still proceed in faintly alkaline solution, but it is already partially inhibited if the reaction is alkaline to phenolphthalein and stronger amounts of alkali quickly bring the reaction to an end. How acids accelerate the ptyalin action, and they act in exactly this same manner toward another enzyme, invertin, is not yet known. One suggestion is that the acids form the active principle from the inactive zymogen. Perhaps they set free the active principle from the gum; perhaps they may act in other ways by combining with the enzyme, or the starch. This matter has to be left for future investigation.

	Starch changed	Relative activity
When no gas passed through	24.15%	100
Air	25.02	103.6
O_2	27.72	114.7
CO_2	28.82	116.8
H_2S	25.23	104.4
H_2	22.86	94.6

The presence of some salt is necessary for the action. Carefully dialyzed saliva loses much of its activity, and this is restored by the action of dilute salts. Phosphates particularly favor some of the amylases. How the salts act is not clear, but possibly they increase the fineness of subdivision of the enzyme, just as they help to dissolve globulins, and thus in effect increase its concentration. Salts of the heavy metals may be very toxic. Thus uranium salts even in a dilution of .0001 per cent. retard and 0.008 per cent. of uranyl nitrate completely inhibits. Silver nitrate and mercuric chloride are also highly toxic. The toxic action of these salts indicates that the active principle is probably a protein. If the acidity is already beyond the optimum, the addition of amino-acids greatly improves the rate of digestion, since these substances combine with the acids and thus help establish the optimum acidity. It will be observed that the optimum conditions for the action of saliva are those present in the medium for which it was designed, if we may use a teleological expression. When the food is swallowed we have in the stomach a very faint acidity, some salts and proteins with their decomposition products.

The products of digestion inhibit the action of ptyalin.—As is the case with most enzymes, the speed of conversion of the substrate is reduced by the presence of the hydrolytic products of the digestion. It will be found that if two starch solutions of the same strength and the same temperature are taken and to one some maltose is added and not to the other, and if to each the same amount of saliva is added, the one without the maltose loses the iodine-starch reaction before the one with the maltose. The action of the ptyalin on the starch is reduced by the presence of the maltose. This fact of the retardation of the velocity by the products of the digestion is generally true for all hydrolytic enzymes. Now in the stomach and intestines this retarding action is avoided by the reabsorption of the maltose by the walls of the intestine as rapidly as it is formed so that it does not remain to vex the ptyalin by its presence.

The effect of the removal of the maltose in increasing the speed of digestion of starch is shown very clearly in the following experiment of Lea, in which the course of the digestion of the starch was followed by the iodine reaction. In one case the digestion was carried on in a beaker, and in the other in a dialyzing tube immersed in running water so that the maltose was dialyzed out.

Time	Dialysed starch solution	Not dialysed
11.00	Pure blue	Pure blue
11.20	Trace of violet	" "
11.40	Red with violet tinge	Violet
12.15	Colorless	Red with violet
12.00	"	Very faint red
1.15	"	Colorless

This retarding action of maltose and other sugars on the rate of digestion of starch by ptyalin was at first supposed to be due to the mass action of the maltose. The reaction by which the digestion of the starch is produced was supposed to be a reversible one.

$$\underset{\text{Starch.}}{(C_{12}H_{20}O_{10})_n} + H_2O \rightleftharpoons \underset{\text{Maltose.}}{n\,C_{12}H_{22}O_{11}}$$

The addition of maltose might be supposed to reverse the reaction and thus to check the rapidity of the disappearance of the starch. This explanation is incorrect, or at least insufficient. The fact is that the maltose combines with the enzyme, thus removing it from acting on the starch. This can be shown by the addition of sugars other than maltose to the reacting mixture. All will be found to retard the reaction, but maltose and glucose retard it most.

Law of action of the ptyalin.—If a dilute solution of starch is used and the enzyme added in more than a minimum amount, the rate of digestion will be found to double with a doubling of the amount of enzyme. In other words, the speed of the reaction goes proportional to the concentration of the enzyme. Chittenden and Smith, using 100 c.c. 1 per cent. starch solution with varying amounts of saliva which had been diluted 50 to 100 times, obtained in 30 minutes at 40° C. the following per cent. of sugar:

Saliva	0.5 c.c.	1.0 c.c.	2.0 c.c.
Sugar per cent.	3.60	7.23	16.92

The increase in sugar was nearly proportional to the increase in the saliva. If, however, tubes of starch paste are used, so that the enzyme must always act on the starch in the presence of a strong solution of the products of its activity which diffuse away very slowly, then the rate goes approximately proportional to the square root of the concentration of the enzyme. That is, if of two tubes one is placed in an enzyme solution four times as concentrated as the other, the rate at which the starch is digested in the two tubes will not be as one is to four, but approximately as one is to two. The rate will be doubled, not increased four times. One will digest two millimeters while the other is digesting one. This is called the law of Schütz and Borissow. Amount digested $\div$ by the time $= K\sqrt{C_{ferment}}$. $C_{ferment}$ is the concentration of the ferment.

Time of appearance of the ptyalin in development.—Only the parotid of the new-born contains amylase; the submaxillary lacks this action. It appears in the submaxillary and the pancreas only at the end of the second month of life. At birth the activity of the parotid is little more than that of many other tissues.

Variation in different types of animals.—Not all salivary glands have in them ptyalin or an amylase. There is very little or none present, for example, in the saliva of various carnivores, such as the dog and cat. The parotid glands of rodents are said to be most active. In the horse mixed saliva is said to have a very powerful diastatic action, whereas saliva collected from the parotid ducts is said to be inactive (Ellenberger and Hofmeister). There is a possibility that a co-ferment or kinase is necessary for the action of the enzyme and this may be absent in some glands. The late Dr. G. W. Cook told the writer that if the human mouth is carefully washed out with a sterile solution of water, or dilute antiseptic, the saliva collected from the ducts may be inactive, whereas the saliva which has been in contact with the mucous membrane of the mouth is very active. It is possible that bacteria are necessary for the setting free of the amylase, as Ellenberger and Hofmeister suggested for the horse, or it may be that the mucous membrane contributes a kinase to assist in the action. This matter should be carefully investigated. It might clear up some of the contradictory statements in the literature. Various observers have found that a carbohydrate diet increases the amylase and maltase content of the saliva of human beings and dogs. Others have failed to find such an increase.

Other enzymes in the saliva.—Besides the amylolytic enzyme, human saliva contains a substance, an erepsin, which will split the tripeptide, l-leucyl-glycyl-d-alanine and probably other peptides; and maltase, an enzyme which will convert maltose into glucose, the amount of which is, however, small and its origin unknown. It may be derived from the maltase of the blood. Saliva also contains catalase, which catalyzes hydrogen peroxide setting free oxygen; and an oxidase. Saliva blues guaiac tincture, and develops color with p-phenylendiamin, a-naphthol and m-toluylendiamin. It is probable that other enzymes in small quantities will be found in it.

Potassium sulphocyanate and other excretory substances.—Potassium sulphocyanate is found in most human saliva in small quantities. It is generally present in larger amounts in the saliva of smokers shortly after smoking. It is in the nature of an excretory substance and is not supposed to have any function. Whenever cyanides are ingested, as they are in some fruits or in tobacco smoke, they are excreted as sulphocyanates, which are excreted in practically all of the secretions. It is possible, also, although it has not been proved, that hydrocyanic acid may be formed in small amounts in the oxidation of protein in the body, since nitriles appear in the oxidation of proteins by permanganate. Many other excretory substances may be found in the saliva in larger or smaller amounts, depending on their concentration in the blood. Thus urea, uric acid, small amounts of glucose and other excretory substances have been isolated from saliva. Iodides pass out in the saliva very quickly and, since nitrites may also be excreted here or be formed in the mouth by the action of bacteria from nitrates excreted in the saliva, the iodine of ingested iodides may be set free and produce irritation of the throat and mouth.

Importance of salivary digestion.—The importance of the digestive action of the saliva on starch was long underestimated, for it was thought that the action was checked at once on entrance of the food into the stomach. We now know that this is not the case, but that the food collects in the fundus of the stomach in a large mass, in the interior of which the reaction is very faintly acid, the optimum conditions of ptyalin action prevail and that for a half hour or longer, depending on the size of the meal, the ptyalin may continue to digest the starches. Its action is permanently checked as soon as the hydrochloric acid accumulates sufficiently to give a reaction with congo for free acid. In another way, also, the saliva may be an important aid to digestion. It is stated by several competent observers that the food matters when thoroughly mixed with saliva digest very much faster with gastric juice than food matters not so mixed, and this is not supposed to be due entirely to the finer state of division of the food. The presence of saliva, or its digestive products, may also act as a stimulus to the secretion of gastric juice. For all these and other reasons a thorough mastication of the food undoubtedly conduces to good digestion.

Composition and metabolism of the salivary glands.—The composition of the salivary glands has been very little investigated, and what is known about them throws very little light on their metabolism. Besides the mucin which the sublingual and submaxillary glands contain in small but undetermined quantities, they contain also a nucleoprotein and nucleic acid. Xanthine and guanine have been isolated from a nucleoprotein which is obtained by extracting the mucin first by water, and then the nucleoprotein by dilute alkali, such as lime water, and precipitating by acetic acid. The nucleoprotein thus obtained (Holmgren) contained 15.11 per cent. of N and some phosphorus. On peptic digestion it yielded a nuclein containing 2.9 per cent. P. It is said (Horsley) that after thyroid extirpation in apes mucin is increased in the glands, and even appears in the parotid gland where it is normally absent. Ptyalin or a ptyalinogen exists in unknown amounts in the parotid gland.

The consumption of oxygen by the gland increases when it is in a state of activity (Barcroft). Like the other glands, it has an unusually high rate of oxygen consumption per gram of tissue: namely, .028 c.c. O_2 per gram per hour. Muscles use .004 c.c. O_2 per gram per hour.

REFERENCES. Salivary Digestion.

1. Composition of mucin. *Levene:* Zeit. f. physiol. Chem., 31, 1900, p. 395.
   Richard and Gies: American Journal of Physiology, 7, p. 93, 1902.
2. Amylase. Discovery in malt. *Payen and Persoz:* Mémoir sur la diastase. Annales de chimie, 53, 1883.
   Discovery in saliva. *Leuchs:* Kastner's Archiv f. d. ges. Naturlehre, 1831.
3. Amylase. Composition. *Wroblewski:* Ber. d. d. chem. Gesell. 31, p. 1128, 1898.
   Osborne: Jour. Amer. Chem. Soc., 17, p. 587, 1895.
   Fraenkel and Hamburg: Hofmeister's Beiträge z. chem. Physiol. 8, p. 389, 1906.
4. Digestive decomposition of starch. *Musculus and Gruber:* Zeits. f. physiol. Chem., 2, p. 177, 1878.
   Brown and Millar: Jour. of the Chem. Soc., 75, 1899.
   Moritz and Glendenning: Jour. Chem. Soc., 61, 1892, p. 689.
   Nasse: Pflüger's Archiv f. d. ges. Physiol., 14, p. 473, 187.
   v. Mehring: Zeit. f. physiol. Chem., 5, 1881, p. 185.
   Brown and Herron: Liebig's Annalen, 199, 1880, p. 65.
5. Effects of temperature. *Chittenden and Martin:* Transactions Connecticut Acad. Sci. 7, 1885.
6. Effects of acids and other substances on speed of digestion. *Chittenden and Ely:* Jour. of Physiol., 3, p. 327, 1882.
   Wood: Amer. Chem. Jour., 15, 1893, p. 663; 16, 1894, p. 313.
   Cole: Jour. of Physiol., 30, p. 202, 1904.
   McGuigan: Amer. Jour. Physiol., 10, p. 444, 1904.
   Chittenden and Smith: Trans. Conn. Acad. Sci., p, p. 343, 1884.
7. Inhibiting action of digestive products. *Lea:* Jour. of Physiol., 11, p. 226, 1890.
8. Proteolytic enzyme in saliva. *Koelker:* Jour. Biol. Chem., 8, p. 145, 1910.
   Warfield: J. Hopkins Bulletin 22, 1911, p. 100.
   Koelker: Zeit. f. physiol. Chem., 76, 1911, p. 27.

9. **Amylase in saliva of carnivora.** *Carlson and Ryan:* Amer. J. of Physiol., 22, 1908, p. 1.

10. **Excretory substances in saliva.** Uric acid. *Boucherson:* C. R. de l'Acad. de Sci., Paris, 100. p. 1308

11. **Activity of mixed saliva of horses.** *Ellenberger and Hofmeister:* Arch. f. wiss. u. prakt. Thierheilkunde 12, p. 265, 1881.

CHAPTER IX.

DIGESTION IN THE STOMACH.

After a more or less thorough mastication and mixing with the saliva, the food is swallowed. It passes down the œsophagus into the stomach, where it remains for a period of from one to five hours, undergoing a process of solution by the action of the juices secreted by this organ. At first it lies in the stomach in a mass in the left side or fundus of the stomach. Small portions are gradually separated from this mass by the contractions of the stomach, partially digested in the pyloric region, and as soon as they are reduced to a state of fine division or solution passed on through the pylorus into the intestine for further digestion and absorption there. We may now proceed to examine in detail the nature of the processes involved in gastric digestion.

Morphology.—The stomach is an enlargement of the alimentary canal found in all vertebrates, which functions both as a reservoir and as a digestive organ. It has an acid secretion and contains hydrochloric acid. The shape of the stomach differs in different animals, varying from a slight tubular enlargement of the alimentary canal, such as is found in the frog, to a many-chambered organ. In most of the vertebrates, including the mammalia, and particularly in the Carnivora and Primates, there is only a single cavity, but this may be partially separated into two by a muscular constriction, more or less pronounced, between the fundus and the pyloric portions. The mucous membrane of these two portions is often plainly different in its macroscopic appearance, and histological examination shows it to be different, also, in its finer structure. The glands in the pyloric region have in them none of the so-called parietal cells. There is also a difference in the chemical nature of the juice they furnish.

The most complex form of stomach is found in ruminants. In cows and other animals of this class the stomach consists of several chambers. These divisions are called respectively the paunch, or rumen, which receives the food as it is first swallowed; the reticulum; the omasum, or psalterium; and the abomasum. The first two cavities have an alkaline secretion and here the digestion is carried on by means of bacteria and saliva. When the bolus of food, after regurgitation from the reticulum, is chewed and reswallowed, it passes into the second stomach, or psalterium, and this corresponds to the fundus portion of the human stomach.

the paunch and reticulum corresponding to an enlargement of the œsophagus. The psalterium and the abomasum, or rennet sack, have an acid secretion like the human stomach. The fourth stomach, or abomasum, corresponds to the pyloric end of the human stomach. In some of the ruminants there is, in addition, a division of the paunch for carrying water. This is brought to a high state of perfection in the camel.

When empty the human stomach is normally contracted on itself so that its walls are in contact and the reaction of its mucous membrane is either very faintly acid or even neutral. When water is swallowed into an empty stomach it runs rapidly through the stomach into the intestine, where it is absorbed. But when food is swallowed the stomach begins to relax and it relaxes sufficiently to adapt itself to the amount of food taken, the powers of relaxation being really remarkable. But while the stomach is normally contracted and empty between meals, it sometimes happens that it re-expands, after emptying itself, and becomes filled with a juice acid in reaction. Such a dilated stomach is a source of much discomfort, and is pathological.

General Physiology of the Human Stomach.—What actually happens in the human stomach when food is swallowed and during digestion was first accurately observed by the American army surgeon, Dr. Beaumont, in the stomach of the Canadian coureur du bois, Alexis St. Martin. These observations, though old, are still in many respects the best observations on the human organ, and they have since been many times confirmed. The nature of the wound in St. Martin and his robust health enabled a very careful scrutiny of a perfectly healthy organ. Alexis St. Martin was wounded in the left side by the accidental discharge of a musket close to him. The charge carried away some of the lower rib, and a part of the wall of the stomach and abdomen. The injured lung protruded from the wound. On healing, the wall of the stomach and the abdominal wall grew together so as to leave a hole or fistula in the fundus part of the stomach, through which food could be introduced, or gastric juice extracted, and the inside of the stomach directly observed. After a while a flap of mucous membrane grew down over the wound and acted as a valve, preventing the escape of the contents of the stomach, unless the valve was pushed to one side. The accident occurred on the island of Mackinac, at the head of Lake Michigan, in 1822. The post surgeon was Dr. Beaumont and he at once seized this most fortunate opportunity for a study of gastric digestion. He hired St. Martin to submit to observation and he made out many of the fundamental facts of our knowledge of stomachal digestion. He observed that the mucous membrane of the empty stomach was pale pink and covered by a thin layer of mucus of a neutral or slightly alkaline reaction, but that as soon as the stomach was excited, either by the introduction of food into it or by mechanical irritation, the membrane became a bright red and appeared to be gorged with blood. Its temperature rose a degree or so at the same time, and if it was watched it could be seen that at the openings of the stomach glands clear drops of colorless juice welled up and, running together, trickled down the sides of the organ. His description of the organ was as follows:

"The inner coat of the stomach, in its natural and healthy state, is of a light or pale pink color, varying in its hues according to its full or empty state. It is of a soft or velvet-like appearance, and is constantly covered with a thin, transparent, viscid mucus, lining the whole interior of the organ. Immediately beneath the mucous coat, and apparently incorporated with the villous membrane, appear small spheroidal- or oval-shaped granular bodies, from which the mucous fluid appears to be secreted. On the application of aliment the size of the vessels is increased, the color heightened and vermicular movements excited. The gastric glands begin to discharge a clear, transparent fluid, which continues rapidly to accumulate as aliment is received for digestion. This fluid is invariably distinctly acid. The mucus of the stomach is less fluid and more viscid, and sometimes a little saltish, but does not possess the slightest character of acidity. On applying the tongue to the mucous coat of the stomach in its empty, unirritated state, no acid taste can be perceived. When food or other irritant has been applied to the membrane, the acid taste is immediately perceptible."

By irritating the stomach with a feather, a thermometer tube or a rubber catheter he got several ounces of juice secreted and collected for examination. The human gastric juice thus obtained was a perfectly clear, generally colorless solution, like so much water in appearance, except that it was now and then mixed with some bile. It had in it a little mucus. It was salty and sour to the taste, turned purple cabbage red and effervesced with carbonates. Some of the juice was analyzed by Professor Silliman, of Yale University, who found that it contained a good deal of hydrochloric acid. Beaumont found that the juice thus collected had the power of dissolving meat put in it when kept warm outside the body, and he thus confirmed Spallanzani's observations of the chemical or solvent nature of the juice. The collection of several ounces of juice was accompanied by no other sensation than that of faintness and a sinking at the pit of the stomach. Among the other very interesting properties of this juice was its power of preventing putrefaction. Meat in it did not putrefy; it had a preservative action, and Beaumont found that applied to wounds it kept them clean and fresh, helping them to heal by first intention and stopping suppuration.

He also examined the time which it took to digest meals: i.e., the time necessary for the stomach to empty itself after a meal. He found that pork took longer to digest than beef, and in general fatty food took longer than lean. Thus after a meal of mutton chops and potatoes the stomach was empty in two hours, whereas after a hearty meal of potatoes, roast pork, vegetables and pudding, it took four or five hours to empty itself. Among the most valuable of Beaumont's observations were those on the optical appearance of the mucous membrane under varying conditions of health, after drinking ardent spirits and coffee, and during constipation. He was able to see what effect drugs had on the mucosa and, above all else, he was able to find out to what extent the condition of the mucosa could be inferred from external symptoms. Thus he found that after hard drinking of ardent spirits the mucosa was inflamed, small ulcers appeared in it and it secreted a large amount of mucus, but a very small amount of gastric juice, and this almost without digestive action. The same conditions prevailed in fever, so that gastric digestion almost ceased. In constipation, too, when the tongue was coated, the mucosa was bright red, dry and secreted little juice. All of these symptoms could be changed in the latter case by the use of a good cathartic, such as calomel, the mucosa recovering its normal condition. Particularly instructive was the fact that the condition of serious disturbance of the mucosa could not be inferred from any outward symptoms.

Beaumont also tried many other experiments. He introduced various foods, such as vegetables or meat done up in cloth bags, into the stomach, leaving them there for some time and afterwards withdrawing and examining them to see what substances would be affected by the gastric digestion. Beaumont did not observe any secretion of gastric juice in St. Martin purely as the result of smelling food, or tasting it, or from the fact that it was meal time. It secreted only when he ate or when the stomach was mechanically irritated. He observed no appetite juice, in other words.

It seems hardly probable that, had there been such an appetite secretion, it would have escaped so acute an observer, but apparently it did not occur to him to try to see whether the mere sight of food would stimulate the secretion. Pawlow has questioned whether the juice Beaumont got was due to mechanical irritation and has suggested that it was in reality an appetite juice due to hunger, but there is no reason to suspect that Beaumont's observations were faulty in this particular. Pawlow's criticism was based chiefly on the behavior of the dog's stomach. Richet in another case of gastric fistula in a human being later showed that it was not necessary for the food to enter the stomach, but that the taking of sugar or other foods into the mouth would start the secretion of juice.

Beaumont also made a suggestion as to the nature of the stimulus leading to the sensation of hunger. He suggested that the turgidity of the gastric glands caused this sensation, which, if it be acute, is often ascribed to the pit of the stomach. This suggestion remained the most probable until the recent observations of Boldyreff, and those of Cannon, and later of Carlson, that a sensation of hunger is accompanied by, and these observers believe is caused by, the rhythmic contraction of the empty stomach.

Manner of obtaining gastric juice.—a. *From fistulas in human beings.* Human gastric juice unmixed with food has been obtained since Alexis St. Martin from several individuals in whom, owing to a blocking of the œsophagus, a fistula or artificial opening had been made into the stomach for the purpose of feeding. Among the commonest causes of such strictures of the œsophagus which prevent swallowing have been the drinking of caustic liquids, such as strong acids or alkalies, either accidentally or purposely. It may happen that the health of such persons is not very good, so the composition of the juice obtained from them may not be altogether normal. An examination of such cases shows, as with St. Martin, that the stomach when empty is contracted and contains no juice, but only an alkaline mucous secretion. They can when hungry often cause the stomach to secrete by chewing food. This appetite secretion, as it is called, has provided samples of pure gastric juice of human origin for examination.

b. *By stomach tube.* Impure juice mixed with saliva and food remnants can be obtained from normal individuals either by causing vomiting or drawing off the stomach contents with a stomach tube after eating a meal. This is the clinical method for determining whether the secretion is normal in amount and quality. The test meal, as it is called, is taken in the morning when the stomach is empty, only a light supper or none at all having been eaten the evening before. There are several test meals. That of Ewald consists of a buttered roll and a cup of weak tea; another such breakfast consists of a roll or a couple of pieces of toast, a glass of water, or a cup of tea, and some bacon. This is thoroughly chewed. After three-quarters of an hour a stomach tube, a flexible rubber tube, is swallowed and a portion of the contents are drawn out.

c. *From animals by artificial fistulas.* To obtain juice from animals, the method usually employed is to make a gastric fistula: i.e., an opening is made into the stomach, generally at the fundus end, and the walls sewed into the abdominal walls, leaving an opening into the interior. If the walls of the stomach are brought out between the muscle fibers, the pressure of these may act as a valve and prevent the escape of gastric contents. The gastric juice flows from the cannula introduced into the fistula when the dog is teased with the sight and smell of food or when he is fed pieces of meat. Such juice may have saliva mixed with it. A great step forward in the study of gastric secretion was taken by Pawlow in his introduction of the œsophageal fistula and his formation of a small stomach pouch separate from the main stomach. In making an œsophageal fistula the œsophagus is divided and the two ends are then sutured to the skin. The result is that when a dog eats, the swallowed food falls out of the upper end of the fistula and neither it nor the saliva enters the stomach. Meanwhile, from the stomach fistula juice may be drawn off if it is secreted.

1. *Appetite secretion.* Richet discovered in human beings that it was not necessary to swallow food to arouse the stomach to activity. The taste or smell of food was sufficient to provoke a flow. Pawlow found the same facts for dogs. Advantage may be taken of this fact to procure pure juice. It is only necessary for hungry dogs to see or smell solid food, or even to hear some sound, such as a special note, which had always been sounded as they were about to be fed and so had become associated with the idea of feeding, to start the stomach secreting. This is the appetite secretion and it begins after a very definite latent period of about five minutes if the dog is hungry. There is no appetite secretion if the dog is not hungry.

2. *Juice obtained by sham feeding.* Another method of getting the juice is by means of sham feeding. This provides large amounts of juice. A dog with an œsophageal fistula as well as a gastric fistula is given meat to eat. He swallows it, but it falls out of the opening in the œsophagus. A dish put under the œsophageal fistula receives the swallowed food and the dog will eat it over and over again. By this means a constant stimulus is provided to the taste buds in the mouth and to the gums and teeth and the nerves of smell. Under this stimulus the stomach secretes a considerable amount of juice, as may be seen in the experiment, page 344. By this means pure canine gastric juice was obtained. It is a clear, limpid liquid, looking like water, but having a doggy smell which is distasteful to human beings. This smell is, however, eliminated by aërating the juice, and such juice, acid and very active, is now obtainable in Russia for therapeutic purposes. The dogs kept for this purpose are large and healthy. They are fed two hours a day in this sham feeding and work cheerfully at producing gastric juice. The flow continues only during the eating, and grows less as the sham feeding proceeds.

d. *The stomach pouch.* By the foregoing method it was possible to obtain pure juice and to study its secretion in the absence of food in the stomach. It was very desirable to study also the effect of having food in the stomach on the composition and amount of the juice, and particularly to see whether the character of the juice remained the same no matter what kind of food was eaten. To do this and yet get juice free from food admixture, Pawlow devised the small separated stomach

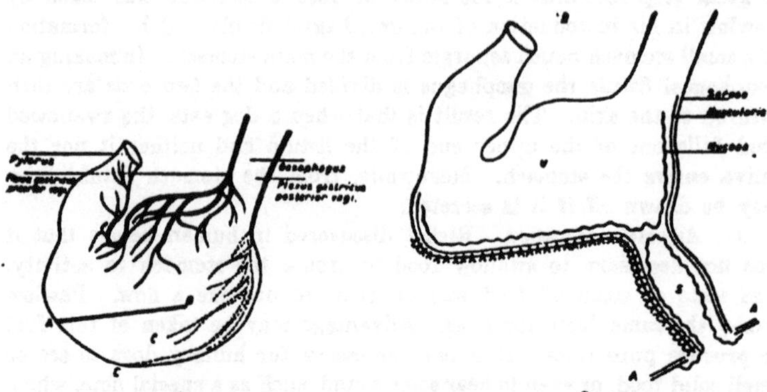

Fig. 41.—Manner of making a Pawlow pouch. The incision is made along the line *A—B* (1). The complete pouch is shown in (2); its cavity, *S*, is separated from the general cavity of the stomach, *V*, by a double layer of mucous membrane (Pawlow).

or stomach pouch, which in its processes completely mirrors the large stomach which has the food in it.

Sham Feeding Experiment.

Time	Amount of juice collected from stomach	
12 h. 5'	3.2 c.c.	Before feeding.
15	3.4	
20	3.5	
25	4.0	
30	3.0	
35	2.4	
45	1.8	Meat feeding begun.
50		
55	10.8	
1 h. 00	15.4	
05	17.8	Stopped feeding.
10	16.0	
15	12.0	
20	10.8	
25	8.6	
30	7.6	
35	6.2	
40	6.2	
45	2.8	

The idea of cutting out a part of the stomach and sewing it into a pouch discharging its secretion to the exterior was not original with Pawlow. It had been tried by Heidenhain and by Klemenciewicz. They made the mistake, however, of cutting clear through the walls of the stomach and thus cutting off all the nerves to the small pouch. The secretion they obtained was on this account far from normal. Pawlow made the pouch in another way so as to leave the nerve supply intact. The method is illustrated in Figure 41.

The nerves of the stomach are the two vagi which come down the œsophagus and spread out, one on the posterior and one on the anterior surface of the organ. They run on the surface and then plunge into the muscular coat to get to the mucosa. There are, also, sympathetic nerves which come in from the splanchnic nerves from the pylorus end of the stomach. To make the small pouch, a V-shaped cut is made completely through all the coats of the stomach in the direction and location shown in the figure. The mucous membrane and the muscular wall are now sewed together in such a way that a complete floor is made for the large stomach, so that food passes normally through it and into the pylorus. The V-shaped piece is now sewed into a pouch and the end of it brought to the surface, passing through the muscles of the abdominal wall, the muscle fibers being separated and not divided. They thus press the mouth together and act as a sphincter. This little pouch has its vagus connections intact and it is separated from the large stomach by a double layer of mucous membrane. If desired, a fistula can also be made into the large stomach. Various foods can now be allowed to enter the large stomach, and the amount and character of the juice secreted by the small stomach under these varying conditions can be studied. It was found that the secretion in the large and small stomachs began and stopped at the same time, and that the introduction of a large amount of food into the large stomach caused the secretion of a large amount of juice in the small, and that when the amount of food was halved, the amount of secretion in the small stomach fell to one-half. This established the very important fact that *the amount of juice secreted is normally proportioned to the amount of food to be digested* and that the small stomach was a faithful image of the secretion in the large stomach.

Character of pure gastric juice of the dog.—The pure gastric juice obtained from dogs or human beings, when unmixed with bile, is a perfectly clear, limpid, not viscid, colorless juice. Its specific gravity lies between 1.002 and 1.0059. It is levo-rotatory. In a decimeter tube the juice collected by Madame Schumova-Simonowski rotated —0.7-—0.73°; that collected by Rosemann, —.109°-—.168°. The freezing point was —.490-0.638°. The freezing point of the first juice secreted was —0.643°, and after two hours —.628°. As the freezing point of the blood is generally —0.571°, the juice is generally hypertonic, not hypotonic, to the blood. The secretion of the gastric juice raises the concentration of the blood when the juice is discharged to the exterior through a fistula.

Thus in an experiment of Rosemann the freezing point of the blood before secretion was —0.589° and after —0.600°. The dry residue is 0.26-0.653 per cent. The ash was 0.133 per cent. (0.105-0.204 per cent.). The organic matter varied from 0.176-0.434. The ash contained iron, calcium and phosphoric acid, and the usual salts of blood serum. The organic substance gives both the Millon and biuret test, although Madame Schumova-Simonowski stated that it did not give the biuret test when it was perfectly fresh, but only after standing. Ammonium chloride is present only in small amounts. There was no lactic acid. The acidity computed as hydrochloric acid was from 0.46-0.58 per cent. The juice contained, as a mean, 0.6137 per cent. of chlorine. Of this 0.5322 per cent. was present in hydrochloric acid; 0.0653 per cent. was in the ash as chlorides; and the remainder was 0.0162 organically combined. Since the per cent. of chlorine in dog's blood is 0.295-0.275 per cent., the gastric juice is twice as concentrated in chlorine as the blood. During secretion, therefore, on one side of the mucous membrane the concentration of the chlorine is 0.54-0.64 per cent; on the other, 0.41 per cent. in the plasma; and in the membrane itself 0.09 per cent.

The hydrogen ion concentration of pure human gastric juice from a case of stomach fistula was found by Dr. Menten in the author's laboratory, using the gas-chain method, to be about equivalent to that of N/10 HCl, or even a little more acid than this. In the stomach, however, the juice is generally mixed with food substances so that its acidity is lower than this. See page 368.

COMPOSITION OF GASTRIC JUICE.

Dog (Rosemann)		Human (Bidder and Schmidt)	
Water	99.74 - 99.36	Water	99.44
Dry residue	0.26 - 0.64	Organic	.32
Organic	0.17 - 0.43	HCl	0.20
Inorganic	0.10 - 0.21	$CaCl_2$	0.0061
Hydrochloric acid	0.46 - 0.58	NaCl	0.146
Chlorine	0.61	KCl	0.055
Chlorine in HCl	0.53	NH_4Cl	——
Chlorine in chlorides	0.077	$Ca_3(PO_4)_2$, $Mg_3(PO_4)_2$, $FePO_4$	0.0125
Chlorine in organic	0.016		
Ash	0.127		
Water sol. ash	0.120	**Human Appetite Juice (Carlson)**	
Na	0.025	Sp. gravity	1007
K	0.030	Total acidity	0.45 - 0.50
Cl	0.067	Freezing point	0.580 - 0.530
SO_3	0.0012	Total solids	0.48 - 0.61%
Insol. ash	0.0023	Organic	0.34 - 0.47
Ca	0.00022	Inorganic	0.11 - 0.14
Mg	0.00022	Free acidity	0.35 - 0.45
P_2O_5	0.0006	Total chlorine	0.49 - 0.56%
Fe	Trace	NH_3	0.051 - 0.074%

Amount of gastric juice secreted.—The amount of juice secreted is subject to great variations, depending on the amount of food eaten, the state of health and so on. By means of the small stomach pouch just described Pawlow found that the amount secreted by the small stomach was directly proportional to the amount of food eaten. If he doubled the size of a meal, the amount of juice was doubled. This proportionality is of course true only within limits. It was estimated by Bidder and Schmidt that in human beings 2-3 liters of gastric juice were secreted per day, and this is probably not far from the truth. In an experiment by Rosemann the stomach of a large dog weighing 24,100 grams secreted during a sham feeding in the first two hours at the rate of 200-300 c.c. per hour; it then fell to 100 c.c. per hour. It was found by Pawlow that when food entered the stomach the secretion was larger than when there was only a sham feeding. We may, perhaps, estimate that in this dog at least 500 c.c. of juice would be secreted for the digestion of each meal. If a man's stomach secretes at the same rate, this would give about 1,500 c.c. for the digestion of a meal.

In these fistula dogs the amount of chlorine and water secreted in the juice may form a very considerable proportion of that contained in the whole body. Thus in the dog experiment just quoted the total amount of juice secreted in the sham feeding in 2½ hours amounted to one-half the total volume of the blood in the body. The secretion of this amount of liquid caused great thirst. In various researches Rosemann found the total chlorine secreted in 3¼ hours to be from 4.4-5.5 grams. In a dog weighing 26 kilos there are in the blood about 5.4 grams of chlorine, so that as much chlorine was secreted in the gastric juice as the total amount present in the blood. Such a loss of chlorine would almost certainly affect the activities of all the cells of the body. The total chlorine in the body of a dog of this weight was estimated as 20.8 grams, so that approximately one-fourth of it was eliminated by the gastric juice in three hours.

The gastric juice of human beings contains on the average about 0.3 per cent. of hydrochloric acid; that of dogs about 0.53 per cent. If a liter or 1,500 c.c. of such juice is secreted in each person for the digestion of a meal, this would mean the secretion of from three to five grams of hydrochloric acid each meal time. Since this acid is poured out before the secretion of the alkaline bile and pancreatic juice, it causes a reduction in the acidity of the other juices of the body and an increase in their alkalinity. It produces, in other words, an alkaline tide in the body Thus the urine is always reduced in its acidity during the digestion of a meal and may even become alkaline. It is not impossible that this alkaline tide may be part of the cause of the sensation of well-being which accompanies the eating of a meal. A slight change of alkalinity of the cells very materially affects their activity.

Variation of the character of the juice with the diet.—Among the interesting facts found by Pawlow and his pupils by the use of the small stomach was that the character of the juice varied with the character of the diet. The juice was most powerfully proteolytic when the dog had bread to eat and weakest when fed milk. The figures on page 380 illustrate the variation of the juice. It was also found that a sudden increase in the rate of secretion not only increased the rate but also the concentration of the juice, confirming a fact observed by Heidenhain. The explanation of the difference in the activity of the juice on these different diets has not been given. The juice which is secreted at the beginning of an experiment is usually more active than that after the secretion has gone on for a time, presumably owing to the fact that during secretion some of the stored pepsinogen has been exhausted. Some of the variations noted may be due to the fact that not all the glands of the stomach secrete at the same time. It is probable that the number entering into activity may vary with the size of the meal; some may be at rest while others are secreting. If now by giving meat one arouses those which have been resting, it is possible that their secretion would be more concentrated and more active than the secretion which has been derived from the glands which had hitherto been in activity and which had been partially exhausted.

How is the secretion of the juice produced and controlled? Gastric hormones.—There is no doubt that the secretion of the gastric juice is under the control of nerves, the vagi and the splanchnics. The stimulation of these nerves under proper conditions causes secretion of the juice. Secretion is greatly reduced in the dog after section of the nerves. If we ask the further question of how the nerves cause secretion, we ask a question not yet answered. There seems to be evidence that either owing to the stimulation of the mucosa by the nerves or other agencies, i.e., the food, there is produced in the stomach cells a substance which when injected into the blood of another animal or intramuscularly causes the secretion in that animal of gastric juice. Such a substance as this is called a hormone, meaning " I rouse to activity " (Gr. *hormân*, I rouse, or set in motion). As we shall see in other cases, there is some reason for thinking that such substances are normally produced perhaps in all cells under the influence of nerves and it is these substances which directly stimulate the cell to activity. The work on the gastric hormones is yet incomplete, but the observations of Edkins that such gastric hormones exist have been confirmed in the author's laboratory (Keeton and Koch). These hormone substances in the gastric gland are extracted from the mucosa by hot acid and are hence probably substances of a basic nature. They are heat stable. The gastric hormone does not cause secretion in the salivary glands, but only in the stomach. The question of the relation of gastrin to secretin, the active secretory hormone of the intestine, is still uncertain. They appear to have many properties in common. Gastrin injected subcutaneously into dogs having Pawlow pouches causes a copious secretion of typical gastric juice in the pouch (Keeton and Koch). (See also Emsmann, Tomazewski and Ehrmann, who have studied gastrin.)

Digestive actions of gastric juice.—1. *Action on proteins.* It was found by Spallanzani that gastric juice exerted a solvent action on meats. Pieces of meat immersed in the juice gradually dissolve until only a few fragments of fibers remain undigested. The cause of this solvent action was early studied. It was at first supposed that it was due to the hydrochloric acid which had been discovered in the juice by Prout in 1823, but experiment showed that pieces of meat immersed in solutions of hydrochloric acid as strong as that of the gastric juice did not dissolve at all, or at most they dissolved at a rate so slow as not to be comparable to the action of the juice. There must be something else in the juice to bring about this digestion. It was discovered by Schwann, one of the founders of the cell theory, that if the juice was boiled first it lost its digestive action, although its acidity remained. The active principle, therefore, is destroyed by heat. Since the juice contains a good deal of organic matter and this is coagulated by heat, it was supposed that the active principle was organic in nature and Schwann proposed that it be called *pepsin* (Greek, *pepsis*, digestion). Pepsin is, hence, the active principle in the juice which digests meat, or, more generally, proteins, in an acid medium.

2. **The clotting of milk.**—Another property of the gastric juice which was very early discovered is that it causes milk to coagulate, or clot, so that it becomes jelly-like. Since acid by itself produces a similar physical change in milk, as is shown in the souring of milk, it might be supposed that this power of the gastric juice is due to the acid it contains: but this is not the case. Carefully neutralized gastric juice, particularly that of young animals, will still produce clotting if added to neutral milk. The milk does not change its reaction in the process. Even an aqueous extract, neutral in reaction, of the calf's stomach will clot milk, although the action is faster in a faintly acid medium. If the gastric juice or these extracts are boiled, they lose their power of coagulation. Like pepsin, the active principle is found stored in the mucous membrane. This power of the juice is accordingly due to an active principle or enzyme, and this enzyme is called rennet, or *rennin*, or *chymosin*. The mechanism of this clotting and the question of the identity or difference of pepsin and rennin will be discussed presently. Since this

power of clotting milk is most pronounced in the stomachs of very young animals, it is evidently in the nature of an adaptation to milk as an article of diet, and the advantages possibly secured by it are discussed farther on.

3. Action on carbohydrates.—Toward carbohydrates gastric juice has a very unimportant action. Starch and dextrins it does not act upon, but saccharose or cane sugar, which is the easiest of the disaccharides to invert, is very slowly split into levulose and glucose by the hydrogen ions, or the acid of the stomach. There is, however, under normal circumstances when no regurgitation has taken place from the intestine, no enzyme in the gastric juice capable of digesting carbohydrates. Lusk found that the inversion of cane sugar was no more rapid than could be ascribed to the acid of the juice. Since lactose and maltose are inverted or digested by acid much more slowly than cane sugar, the action on these sugars is even less important than that on cane sugar. All three are digested by enzymes found in the intestine. To what extent cane sugar is inverted in the stomach will depend on the length of time it remains there after free acid appears. This time is normally so short that probably little inversion occurs. The concentration of free hydrogen ions at the best is not more than .08 N, and in a mixed meal is less than this. Such weak acid inverts slowly.

4. Action on fats.—Gastric juice has quite marked powers of digesting fats which are already emulsified, such as the fats of milk and of yolk of egg, but it has almost no power on non-emulsified fats such as those in meat or butter. The power of the juice to digest emulsified fats was described long ago by Ogata and confirmed by many observers, but it was for some reason long omitted from the text-books. It has recently been confirmed by the work of Volhard and his pupils. Thus the fats of milk and yolk of eggs are split or digested into fatty acid and glycerine to about 50-80 per cent. in the stomach, whereas non-emulsified fat is split only to the extent of .5-2 per cent. in the stomach. The greater action on the fats of milk and yolk of eggs is probably due to the fact that the area of contact between fat and water is much greater owing to their emulsion. This power of the gastric juice to split fats is still under investigation. There is no doubt that the action is due to a fat-splitting enzyme in the gastric juice, or lipase as it is called. For some time it was doubtful whether this lipase was secreted by the stomach, or whether it had regurgitated into the stomach from the intestine, the juices in the intestine containing much lipase, but experiment has shown that lipase pre-exists in the stomach mucosa, and also that the stomach lipase differs in the optimum acidity of its digestive action from that of the intestine (Davidson).

5. Antiseptic action of the juice.—In practically all of the verte-

brates the secretion in the stomach is strongly acid with hydrochloric acid. The end secured in having an acid rather than an alkaline reaction in this reservoir is presumably to check fermentation. Hydrochloric acid kills all forms of living things except spores. Many animals swallow their food alive. Frogs eat each other, or snakes and other living animals. The tissues of these animals live well enough in an alkaline or neutral digestive medium, but in the strong acid of the stomach they are attacked, killed and digested. But probably of even greater importance is the additional protection secured against parasites of all kinds. Animals are constantly eating with their food bacteria, moulds, protozoa or other parasites which are killed by the gastric juice. Of the vast numbers of bacteria we swallow the great majority are killed in the stomach, and people having a copious and active gastric juice are less liable to infection by typhoid and cholera than those with less acid juice. The digested strongly acid material called chyme as it escapes from the stomach is almost sterile and contains few living bacteria. Any reduction of the acidity is apt to be followed by a bacterial or yeast fermentation in the stomach which may produce irritating organic acids and gas. As long as hydrochloric acid remains in normal amount in the stomach, one never finds more than traces of lactic, butyric or other acids which are produced by fermentation, but in the absence of hydrochloric acid these generally appear.

Action of the juice on proteins.—1. *Pepsin. Its origin. Pepsinogen.* We may now proceed to consider more in detail the nature and source of the active principles of peptic digestion, namely the pepsin and the hydrochloric acid, and the character and conditions of their action on proteins.

Pepsin is stored in the mucous membrane of the stomach. It was found by Schwann that if the mucous membrane of the pig's or dog's stomach, which is alkaline or neutral in reaction, is extracted with dilute (0.4 per cent.) HCl or extracted by water and then the water extract mixed with acid to make about 0.3 per cent., the extract has digestive powers similar to that of gastric juice.

This experiment showed that pepsin, or a substance which gave rise to it, was stored in the mucosa of the stomach, but since the aqueous extract of the membrane was neutral, it was clear that the hydrochloric acid was not stored in the mucosa. The pepsin was probably made in the mucosa between meals, but the hydrochloric acid must be made only at the moment of secretion. Further facts about this pepsin were discovered by Langley. He found that it was extremely sensitive to alkalies. If, for example, the juice be neutralized by sodium hydrate or carbonate and then made acid again, it will be found to have lost much of its activity. If it be made plainly alkaline, it is entirely inactive when reacidified, although the reacidification is made at once. The pepsin has been destroyed or rendered inactive by the alkali. It is necessary, if one wishes to neutralize the juice without destroying the activity of most of the pepsin, to add $CaCO_3$ to it and then a very weak alkali, such as sodium acetate or milk of lime, very cautiously. Langley found that if an aqueous extract be made of the mucosa of the stomach, this extract might be made slightly alkaline for a short time, but would still be active if made acid again; but if an acid extract be made of the stomach, this could not be made weakly alkaline without permanently destroying its activity. It appeared, then, that the extract made with water was more resistant to the action of alkali than the extract made with acid. Langley interpreted this to mean that the pepsin did not exist as such in the mucous membrane, but that there was an antecedent and more stable substance from which the active pepsin was formed by the action of acids. This more resistant antecedent substance he called "*pepsinogen*," since it formed pepsin when acted on by acids. Another evidence that the substance in the mucous membrane differs from pepsin is the fact that if carbonic anhydride gas be passed through a neutral, aqueous extract of the frog's œsophagus, which contains pepsinogen, most of the pepsinogen is destroyed, but if the extract is first acidified and then neutralized and carbonic anhydride passed through it, it is not destroyed. Pepsinogen is, then, more sensitive to carbon dioxide and less sensitive to alkalies than pepsin. It has since been shown that the loss of activity of the pepsin is not permanent, as Langley thought, when the juice is carefully neutralized and then made alkaline. The activity of the pepsin of such juice may be reobtained if acid is very carefully added nearly to the neutral point and then after standing twenty hours the reaction made acid. But there is no doubt of the difference in resistance of pepsin to CO_2 and alkalies after and before it has been treated with acid. If his explanation is correct, we may then assume that pepsin exists in an inactive state in the mucosa as pepsinogen and is set free by the acid. Since it is quite a common thing for the enzymes to be in union in the cells with some colloidal matter, the union being far more stable than the free enzyme, it is probable that the explanation of Langley is the true one. The bald statement is often made in the literature that pepsinogen and not pepsin exists in the mucosa, but it is well to remember that this is as yet an inference and, while quite probable, should not be stated as a fact.

The fact that pepsinogen is so very sensitive to carbonic acid gas, whereas pepsin is not at all sensitive to it, is very interesting and may throw light on the nature of the change which occurs when pepsinogen is converted into pepsin. The reaction may be the carbamino reaction of Siegfried (see page 121). If it is the carbamino reaction, then the change produced by acid in pepsinogen by which it is converted into pepsin might be similar to the conversion of creatine to creatinine by the action of acids. The pepsinogen might be represented thus:

$$\begin{array}{l} R—CHNH_2 \\ \quad | \\ \quad CH_2—COOH \end{array} \qquad \begin{array}{l} R—CHNH \\ \quad | \quad \diagdown \\ \quad CH_2—C{=}O \end{array}$$

Pepsinogen. Pepsin.

No importance is to be attached to the relative positions of the carboxyl and amino groups in the above formula, but only that under the action of the acid an anhydride may be formed like creatinine. The carbonic acid in the presence of a calcium salt would combine as follows with the pepsinogen, but not with the pepsin:[1]

$$\begin{array}{l} R—CHNH—COO—Ca \\ \quad | \qquad\qquad\qquad | \\ \quad CH_2 - - CO—O \end{array}$$

Carbamino pepsinogen.

2. ***What cells of the mucous membrane form the pepsin?*** The question which followed immediately on the discovery of the fact that pepsin or its antecedent, pepsinogen, existed in the mucous membrane in noteworthy quantities was that of the location of this substance. A histological examination of the mucosa of the stomach showed it to contain a large number of simple tubular glands. There are in these glands in the fundus end of the stomach in mammals three kinds of cells: namely, mucous cells near the neck of the gland, chief or adelomorphic cells, and parietal or delomorphic cells. The latter are also called oxyntic cells. Oxyntic means acid, and adelomorphic means "of uncertain shape" (Gr. *adelos*, uncertain; *morphe*, shape). In the pyloric end of the stomach the glands have no parietal cells and they differ in some other particulars from those of the fundus end. The mucous cells secrete mucin; both the parietal and chief cells contain granules presumably of a protein nature, but the granules are larger in the chief cells than in the parietal, and these two kinds of cells differ in their affinities for certain dyes. The granules are more abundant at certain times than at others and it is believed, though it has not been demonstratively proved, from analogy with other glands which can be observed in the living state, that these granules, elaborated by the cell, are secreted or dissolved and pass out from the cell when the latter secretes. They are probably protein in nature, judging from their solubilities and relation to reagents,

[1] The reaction would probably occur in the absence of calcium, giving simply the free carbamic acid compound. If ammonia is present a uramido compound might be formed.

$$\begin{array}{l} R—CH—NH—COOH \\ \quad | \\ \quad CH_2—COOH \end{array} \text{, or } \begin{array}{l} R—CH—NH—CO—NH_2 \\ \quad | \\ \quad CH_2—COOH. \end{array}$$

and it is natural to suppose that they constitute a portion at least of the organic matter found in the gastric juice. Many have gone farther than this and called them plainly zymogen granules and assumed that they represent so much pepsinogen. They are in fact ordinarily called ferment granules, but it must be remembered that this is an inference, and whether the pepsin is represented by the granules or by some substances in solution in the protoplasm it is quite impossible to say at the present time, although it is possible that some of the granules may contain, or be, pepsin or pepsinogen itself. There is no question from the work of Harvey and Bensley that the secretion of the parietal cells is alkaline and protein in character. These observers discovered an intravital stain, cyanamine, which is blue when acid and red when alkaline and which is taken up with avidity by the secretion in the fine canaliculi of the parietal cells. The duct contents stain a clear red by this stain and the blue is found only close to the neck of the gland in the lumen and in the cells of the neck and the cells of the foveolæ. The secretion of the chief cells is also of an alkaline nature.

It has been found that pepsin is secreted in frogs by some of the cells in the lower part of the œsophagus, and the opinion is widely held, although it cannot be said that the evidence is satisfactory, that pepsin is secreted mainly by the chief cells. Until there is some microchemical test which will enable a detection of pepsin, it is impossible to say which kind of cell secretes pepsin. It is of course little likely, but it is by no means excluded by experiment that some might be secreted in the mucous cells.

3. *The nature of pepsin.* Curiosity as to the nature of the digestive principle, pepsin, is of course very keen and many attempts have been made to discover something of its chemical composition. It was early found that a small amount of this substance was able to digest a very large amount of meat, an amount enormously greater than its own weight. This at once classified it with the enzymes or catalytic substances, for this is their main characteristic. But while the amount is large it is not unlimited. If a large piece of meat is put into gastric juice, the digestion will stop long before the digestion is complete, and it proceeds if more pepsin is added. Moreover, pepsin is itself unstable, and gastric juice even when kept in the dark and in the ice box slowly loses its activity.

Pepsin is colloidal in nature, or if not colloidal itself it is attached to colloids. If gastric juice is placed in parchment paper tubes and dialyzed, the pepsin stays in the tube while acids and salts go out; moreover, it will be found that when dialyzed in this way against water, or very weak acid, a precipitate appears in the tube and this precipitate contains most of the pepsin. Under the ultra microscope, also, pepsin solutions are seen to contain actively-moving, colloidal particles.

Unfortunately very few of the investigators who have attempted to separate pepsin from the juice have made any quantitative determinations of the activity of the substances they have obtained, but have contented themselves with the statement that the powder was very active, or digested a piece of fibrin in a certain length of time. In the absence of decisive, carefully carried-out measurements of the activity of the powders called pepsin, it is impossible to know whether the powders described as pepsin contained much or little of the active principle.

Brücke made in the phosphoric acid extract of the mucosa an insoluble precipitate of $Ca_3(PO_4)_2$. This carried down some pepsin with it. This was dissolved in hydrochloric acid and then an ether-alcoholic solution of cholesterol was poured into the solution. The cholesterol was precipitated and again carried down some of the pepsin. The cholesterol was removed from the precipitate with ether, leaving the pepsin behind. The small amount of powder thus obtained was of unknown activity. It gave no xanthoproteic reaction, nor was it precipitated by tannic acid or potassium ferrocyanide and acetic acid.

Sundberg, in 1885, extracted the mucosa with saturated sodium-chloride solution which prevented the solution of much protein. The solution was very powerful, but contained only a trace of protein. This solution he precipitated with $Ca_3(PO_4)_2$, redissolved in HCl and dialyzed. He states that the solution was more active than before, but gave no reactions for protein. It contained ash and nitrogen. Pekelharing in 1896 and 1897 made an HCl extract of the mucosa and dialyzed it. A precipitate formed in the tube. This was many times redissolved in acid and precipitated by dialysis. He finally obtained a precipitate which had a powerful digestive action. It was a yellow powder giving the albumin reactions and containing about 1 per cent. P_2O_5, some of which was soluble in alcohol. It contained a nucleoprotein. 1/100 mg. of this powder dissolved in .2 per cent. HCl was able to digest a piece of fibrin in one hour and a piece of coagulated egg albumin 2 mm. thick and 1 cm. long in a few hours. The substance gave no xanthoproteic reaction, but was certainly a protein and he thought probably a nucleoprotein.

Nencki and Sieber examined the natural pepsin precipitated from natural gastric juice of dogs on standing in the cold. This pepsin contained Cl, .475 per cent.; P, .104 per cent.; Fe, .16 per cent. in the dry pepsin. After washing with alcohol the ash was .399 per cent.; Fe, 0.115 per cent.; P, 0.059 per cent.; Cl, .188 per cent. of dry pepsin. The elementary analysis was C, 51.26 per cent.; H, 6.74 per cent.; N, 14.33 per cent.; S, 1.5 per cent. These are the figures of a protein. It lost power on washing in alcohol. Lecithin made about 10 per cent. of the pepsin. They concluded that the molecule was very complex and labile and that it contained protein, lecithin, chlorine, iron, phosphorus and a pentose. These conclusions are, however, quite incorrect for the active principle. Pekelharing subsequently obtained a pepsin which was active, but which was free from phosphorus and contained, hence, neither nucleic acid nor lecithin, but still contained protein.

It will be clear from this résumé that practically nothing is known of the nature of the active principle and that the subject is badly in need of investigation. It seems, on the whole, probable from all its reactions that the pepsin is either a protein itself or united with a protein. The fact that it is so easily extracted with acid and that it is so unstable in alkaline media indicates that it is of a basic nature and hence probably contains nitrogen. Pepsin is destroyed in a moist state when heated to 57-58°, which is the temperature at which the proteins present coagulate (Langley).

4. *Different pepsins exist.* Just as each animal has its own specific proteins which resemble those of others but which are nevertheless different from them, so there appear to be different pepsins. By pepsin, as a group name, is meant a proteolytic enzyme which acts in an acid, but not in a neutral or alkaline medium. That there is more than one kind of pepsin is indicated by the fact that the optimum concentration of acid is not always the same. For example, that of the dog has an optimum with a hydrogen ion concentration of .05 N, while human pepsin has its optimum at a lower concentration, namely .03 N. They differ also in their resistance to heat. The pepsin of cold-blooded animals works as well at 10° as at 40° C., whereas that of warm-blooded animals works far better at the latter temperature. They show differences, also, in their speed of digestion of different proteins. Thus the pepsin of the calf's stomach digests casein at a rapid rate, but coagulated egg albumin very much more slowly. The pepsin of the pig's stomach digests both proteins rapidly.

5. *Pepsin in embryonic development.* Pepsin is present in the gastric mucosa of new-born cats and dogs, but this pepsin only begins to attack coagulated egg albumin toward the end of the third week of life. It appears at an earlier time in the rabbit. In children pepsin is already present at birth and casein is changed by it to proteose. Whether this pepsin is identical with that of adults is not yet certain. Casein is digested with greater ease than many other proteins. Extract of the cow's stomach has digestive power in an acid solution from the third fetal month. The stomachs of children have less acid in the juice they secrete than is found in adults and the pepsin in their stomachs works in more weakly acid media than in adults. Pepsin may, in fact, be

regarded as an enzyme which appears rather late in the course of evolution and also late in the course of embryonic development, and it is preceded by other enzymes of the nature of erepsin which are active in a weaker acid. No invertebrate has a true pepsin active in strongly acid media and inactive in neutral or alkaline. The proteolytic enzymes of invertebrates are usually of the tryptic, ereptic or autolytic class. Some, however, are said to be active in weakly acid media.

6. *Conditions of activity of pepsin.* The most important of these conditions is the acidity of the mixture. The most favorable concentration of acid is that which is generally found in the stomach of the animal of which the pepsin is being examined. For human beings it is an acidity of about 0.3 per cent. hydrochloric acid, or an hydrogen ion concentration of about 0.03 N. In dogs the concentration of both free acid and of hydrogen ions is greater and the optimum is higher: namely, at about 0.56 per cent. hydrochloric acid, or about 0.05 N hydrogen ions. Other acids are able to replace the hydrochloric acid, but none are better, and probably none quite so good as the hydrochloric acid which is normally present. The action of the acid involves not only the hydrogen ion, but also the anion, since when various acids are taken with the same hydrogen ion concentration it is not found that all the acids aid the pepsin equally well. The sulphate ion appears to be particularly deterrent. Perhaps it acts on the protein to prevent its ready solution. It appears that pepsin unites with protein which is being digested and advantage may be taken of this fact to separate pepsin from a solution. If fibrin or elastin is immersed in a slightly acid solution containing pepsin, the pepsin will stick to the protein and when the fibrin or elastin is taken out of the solution pepsin goes with it. If, then, the elastin is thoroughly washed with water to get rid of all adherent substances and then brought into hydrochloric acid of the right concentration, the pepsin attacks and digests the protein. The course of the digestion may be followed very easily by the polariscope, since the products of proteolytic digestion are optically active and it is only necessary to filter off some of the solution from time to time and examine the progressive change in the rotatory power of the solution. The rate of solution may be followed, also, by means of the ninhydrin, or other delicate reaction for the presence of amino-acids or other polypeptides in the solution. Free acid is not necessary for the action of the pepsin. If a protein is first soaked in acid so that it combines with the acid and is then brought into a neutral solution of pepsin, it will be found that pepsin will digest the protein, although the reaction of the solution when tested with congo red shows no free acid present. In fact, it is probably not the concentration of the hydrogen ions which is important in the digestion, but the combined acid. It is the acid

which is combined with the protein and pepsin which enables the pepsin to act. Neither pepsin nor hydrochloric acid will digest rapidly except in the presence of the other.

The optimum H ion concentration is difficult to determine. It lies between 1.7×10^{-2} N and 3×10^{-2}. When the concentration of H ions falls to 10^{-3} N, Michaelis states that the digestion is already bad, and at $N \times 10^{-4}$ it almost ceases. It is probable that the optimum will be found to differ with different pepsins, since in a baby's stomach the digestion goes on with a much weaker acidity than this.

7. *Law of peptic activity.* The question may now be asked what relation, if any, exists between the amount of pepsin, the amount of digestive products and the time of the digestion. What is the relation between the rate of digestion and the quantity of pepsin? If the quantity of pepsin in unit volume be doubled, will the rate of digestion be doubled, or will it increase more or less than this? One reason for finding out any such proportion between rate and amount of pepsin is that it may enable us to compute the amount of pepsin present, or the relative amount compared with some standard, by measuring the rate of digestion. By the rate of digestion is meant the amount of protein digested divided by the time in minutes or hours during which the digestion has proceeded. This gives the average rate of digestion per minute or hour during the period of observation.

It happens that we can only measure differences in the amount of digestion after an hour or more interval. The rate may not remain constant during this time. It is, therefore, much more convenient to take a very short interval of time, an interval so short that the rate does remain practically constant throughout it. This time which is relatively very minute we represent ordinarily by the symbol dt. During this interval of time the amount digested, let us say of coagulated albumin, is a very small amount and it is represented by the symbol dx. The rate of digestion for this small time interval is, then, the quantity digested divided by the time, or dx/dt. We can make various guesses concerning the relation of this velocity of digestion, dx/dt, to the concentration of the enzyme. The simplest guess is that the rate is proportional to the amount of the ferment present, or that $dx/dt = KC_fC_s$. Where K is a constant of proportion to be determined by experiment, and C_f and C_s are respectively the concentration of the ferment in unit volume and the concentration of the protein or substrate. If C_s remains constant, the velocity should be constant also, provided of course that C_f does not change. If, however, C_s is changing, that is if there is less and less protein left undigested, then the velocity should of course steadily diminish as C_s diminishes. Let us suppose, for example, that this is the case. If a is the concentration of the protein at the start and x is the amount digested during the interval t, then a—x is the concentration at the time t. The equation becomes then $dx/dt = KC_f\,(a-x)$. By integrating this expression we sum up all the amounts digested in the successive intervals which have elapsed since the beginning of the digestion and that enables us to compare the actual digestion as measured with that computed in this way. Integrating, or summing the expression, we have, $\log(a/(a-x)) = KC_f t$. If on the other hand C remains constant, we should have x, the amount digested, $=KC_f\,t$. That is, the amount digested should be proportional to the time t.

One of the most useful methods of measuring the rate of digestion is that of Mett's tubes, which has already been described (p. 331). The length of the column of coagulated egg albumin digested away from the end of the tube is carefully measured. This method is certainly far from ideal. The cross-section of the protein exposed to be acted upon by the pepsin remains approximately constant in this method, so that this substrate is constant; but the concentration of the ferment does not remain constant. The fragments of protein which go into solution are not fully digested. They combine with and neutralize in this manner some of the pepsin, so that the concentration of the pepsin immediately at the end of the cylinder of protein is reduced. They also combine with the acid and reduce the concentration of the acid at this point, and in both these ways check the rate to a maximum. Nevertheless, by this method a certain relation has been found to exist between rate and amount of ferment: namely, the amount of coagulated protein digested is proportional to the square root of the amount of enzyme, $x = tK\sqrt{C_f}$. This is known as the law of Schütz and Borrissow. The time is taken always as 24 hours. It expresses only a relation within very narrow and special conditions, but it is a useful method none the less. This relation between rate of digestion and enzyme concentration is not followed if the conditions are changed. For example, Gross has measured the amount of pepsin by the rate of conversion of casein into caseoses which are not precipitated by acids. In this method the casein and the enzyme are both in solution, but the casein is changing its quantity rapidly. He measured the time it took various concentrations of pepsin to convert all the casein into caseoses and he found that if he doubled the amount of pepsin he halved the time. This corresponds to our first equation $\log(a/(a-x)) = KC_f t$. For in this case the concentration of the casein at the start, a, and the concentration at the time t, or (a—x), are always taken the same and are constant, so that we have simply $\log(a/(a-x)) = \text{Constant} = KC\,t$. That is, the concentration of the ferment multiplied by the time is a constant. If, therefore, the ferment is doubled, the time will be halved. A very similar law has been found to hold also for the liquefaction of gelatin by various proteolytic enzymes of bacterial and other origins by Jordan. The time required for liquefaction is inversely proportional to the concentration of the enzyme present.

Among other simple but not very accurate methods devised for the study of the rate of digestion are various colorimetric methods. These generally involve the setting free of some dye from fibrin by digestion. Carmine and acid fuchsin are sometimes used. The amount of digestion is estimated by the color of the solution. Fibrin may be brought into a slightly acid (acetic acid) solution of acid fuchsin in which it stains an intense red. The color is fixed more firmly by heating to 80° C. A small piece of colored fibrin is placed in 5-10 c.c. of the juice to be examined and the rate of digestion is determined by the deepening of the color of the solution. For two reasons this method is not accurate. In the first place, it is impossible to get two pieces of fibrin of exactly the same surface of contact between it and the enzyme solution. And, in the second place, the combining power of the fibrin for congo red or any other color increases as digestion proceeds and more molecules are set free. One has thus a varying relation between the amount of color and the amount of fibrin in the solution.

8. *Products of the peptic digestion of proteins.* The action of the gastric juice is to dissolve the proteins partially or wholly. To bring this to pass a chemical change is produced in the protein. If fibrin is dissolved in gastric juice, the substance in solution is no longer fibrin. It has been changed to a mixture of derived proteins called meta-protein and proteoses by the action of the juice. Whether amino-acids and di- and tri-peptides are produced at the same time or not has been the subject of much controversy, owing in part to the fact that the digestion has at times been carried out with pure juice and at times by extracts of the mucous membrane. In these extracts there are present sometimes other enzymes than pepsin. The digestion of a protein by pepsin hydrochloric acid is in its main features as follows: If a mass of fibrin is dissolved in gastric juice, there remains a small undissolved residue which is supposed to represent nuclein of white blood cells entangled in the fibrin, or a lecitho-protein (Wooldridge). From this it appears that gastric juice will not dissolve nuclein or some lecithoproteins. If the clear filtrate from any such undissolved digestion residue is neutralized carefully with NaOH or $NaHCO_3$, there is generally precipitated at or near the neutral point, if the digestion has not lasted long, a more or less copious flocculent precipitate which will redissolve if more alkali is added. This substance (1) is a *meta-protein* known as *syntonin* or *acid albumin*. The greater part of the digested fibrin remains in solution even at the neutral point. The solution gives a violet-red biuret test.

If the solution freed from the syntonin is half saturated with ammonium sulphate, a white, flocculent or sticky precipitate separates out, which often clings to a glass rod or sticks to the side of the beaker. This substance (2) is known as *primary*, or *proto-proteose* or *protalbumose*. If more ammonium sulphate is added, another precipitate begins to form and at complete saturation of the warm solution there separates a second substance, or rather a mixture of substances (3) which are known as *deutero*, or *secondary proteoses* or *albumoses*. These two fractions, 2 and 3, make by far the greater part of the weight of the original fibrin, and the *process of digestion evidently consists in converting the protein into acid albumin, proto- and deutero-albumose*. It will be found by the examination of the digested mixture from which the albumoses have been separated by ammonium sulphate that it still gives a good biuret test. The separation of these biuret-giving, proteolytic products which are soluble in saturated ammonium sulphate is a matter of some difficulty. They are all called *peptones* (4). Whether some amino-acids or di- and tri-peptides are produced, or not, is still a matter of dispute, but probably they are not set free by the pepsin, even though they may appear in the digestive mixtures. It is, perhaps, more probable that they are the result of the action of the *ereptic protease*, sometimes and perhaps always present. There is a general agreement, therefore, that by peptic digestion acid albumin, proteoses and peptones are formed.

We should not leave this part of the subject, however, without a reference to the very careful work of Zuntz on the extent of peptic decomposition of proteins, since his results raise many questions of interpretation of which the solution may be of great value. Zuntz found that protein fragments of small size, precipitated by phosphotungstic acid and not giving the biuret test, were formed very early in the course of digestion.

a. *The proteoses or albumoses.* The proteoses or albumoses are a group of derived proteins. There are three divisions of the group: namely, the primary proteoses, including the proto-proteoses and hetero-proteoses, and the secondary, or deutero-proteoses. The primary proteoses are precipitated by half saturation of their solutions by ammonium sulphate. Proto-proteose is soluble both in hot and cold water and dilute salt solution. The hetero-proteoses, while not coagulated by heat, undergo a precipitation when heated, but the precipitate redissolves readily in a little acid or alkali. The hetero-proteoses are insoluble in water, but soluble in dilute salt solutions. The secondary proteoses are only precipitated by full saturation. It is probable that each of these groups is a mixture of many different fragments of the protein molecule. They can be further fractioned by ammonium sulphate, but we need not go into that matter at this place.

It is found that the different proteoses differ in the kind and amount of the amino-acids that they contain. They differ also from each other depending on the protein from which they come, the deutero-gelatose from gelatin being different from the deutero-caseose from casein. The protalbumoses differ from the deutero-albumoses, also, in their reaction to other precipitating reagents than ammonium sulphate. Thus potassium ferrocyanide and acetic acid precipitate the proto, but not the deutero-albumose; copper sulphate will precipitate the former, but not the latter, and other differences have been noted by Kutscher. It is probable, from his results, that the latter have more of the basic amino-acids in them than the former and that the protalbumose is more acid than the deutero-albumose.

9. *Character of the linkings in the protein molecule attacked by pepsin.* Although it is uncertain whether some small amounts of amino-acids are or are not set free by the prolonged action of pepsin hydrochloric acid, there is no doubt that the action is incomparably more rapid in the first stages of digestion by which the protein is split into albumoses and peptones than in the further decomposition of these substances. Another very significant fact is that no artificial polypeptides have yet been made which can be digested by pepsin, although many are digested by erepsin and trypsin. Moreover, pepsin has no action at all on certain proteins which are easily digested by trypsin and erepsin. It has no action on the protamines, salmin and clupein. In this case we have linkings between the basic amino-acids and proline, serine and valine which pepsin cannot break. Since it will not act either on any of the artificial polypeptides, it seems either that pepsin must act only on some special amino-acid linkings, or else that there are in the protein molecule linkings uniting the albumoses of a different nature than amino-acid linkings. It may be remarked in this connection, however, that in the course of peptic digestion the number of free amino groups undoubtedly increases, as is shown presently, and this would indicate that if the digestion is really due to the pepsin and not to some admixed enzyme of an ereptic nature, like that, for example, of the saliva, that the pepsin must have acted on some amino-linkings. The further investigation of this problem will possibly throw light on the synthesis of albumoses to make complex proteins. In any case the action of pepsin shows that the possibility that the protein molecule is not a single long chain of amino-acids, but may be composed of larger groups of two, three or a larger number of amino-acids, just as starch is composed of a number of maltose groups, cannot be disregarded. The protein digestion is thus apparently similar to that of the polysaccharides. By digestion of the polysaccharide, starch, a number of dextrin groups are first formed, and these are subsequently transformed into monosaccharides. This

is accomplished by two or three ferments: amylase, which attacks the starch and carries it over into dextrin; dextrinase, which attacks the dextrin and carries it to maltose; and maltase, converting maltose to glucose. Amylase, dextrinase and maltase correspond in a general way with the three proteolytic enzymes pepsin, trypsin and erepsin. That the pepsin hydrochloric acid decomposes by hydrolysis is shown by the investigations of Chittenden and Kühne. If a protein is analyzed before and after digestion, it is found that the amount of hydrogen and oxygen in the split products is much greater than in the original protein. The total weight, also, of the solid protein left on evaporation is increased, showing that the elements of water are taken on.

During the course of peptic digestion there is a breaking of amino-carboxyl linkings so that the number of free, that is unsubstituted, amino groups greatly increases. This inference is drawn from two facts: first, the capacity of the digestive products of combining with formaldehyde steadily increases, as is shown by the Sorensen titration; and, second, coincident with this is a steady increase in the power of combining with acid, as is shown by the steady fall in the amount of uncombined or free hydrochloric acid as determined by the Günzberg reagent. The following experiments illustrate these facts:

100 c.c 1% gluten in N/10HCl plus 50 c.c 1% Parke-Davis pepsin in N/10HCl in the thermostat at 38°.

Titration	Free HCl (Günzberg)	Formol titration
At once	88*	6
After 4 days	84	12
After 12 days	77	17

* The figures represent c.c. of N/10 NaOH necessary to neutralize 100 c.c. of the mixture. 100 c.c., for example, contains 88 c.c. free N/10 HCl. On the addition of formol the acidity increases equivalent to 6 c.c. N/10 acid, phenol-phthalein being the indicator.

50 c.c. saturated gluten solution in N/10 HCl plus 50 c.c. 1% Parke-Davis pepsin in N/10 HCl in thermostat.

Titration		
At once	91	13
After 4 days	84	21

The experiments show that the increase in the formol titration is almost equal to the decrease in the free acid. This fact is also brought out in the following experiments which give also the change in the titration by congo red:

Titration	Günzberg acidity Free HCl	Congo acidity	Congo-Günzberg acidity	Formol titration
At once	72	88	16	18
After 2 hours	64	86	22	25
" 1 day	56	86	30	32
" 60 days	29	89	60	58
" 135 days	20	88	68	67

The congo titration remains practically constant, since this gives both free and combined acid; and the difference between Günzberg and congo acidity represents the amount of hydrochloric acid which has been bound by the amino groups set free. It will be seen that this number is almost the same as that of the formol titration which titrates the number of such amino groups which have been set free in another way.

These experiments show very clearly that whatever the special linkings may be which are broken by the pepsin, or other enzymes in the commercial pepsin, that they involve amino groups. It will be seen, too, that the free acidity decreases markedly, and since the activity of the pepsin is dependent on the acidity this factor of diminishing acidity will probably alter the rate of digestion. In any experiment in which the activity of the pepsin itself is to be studied the acidity should be kept constant.

Such a substance as Witte's peptone, which is in reality largely a mixture of albumoses, is still capable of partial digestion by pepsin, as is shown in the following experiment:

50 c.c. 4% Witte's peptone plus 50 c.c. 1% Parke-Davis pepsin in N/10 HCl plus 25 c.c. water at 38° C.

Titration	Free acidity (Günzberg)	Free and combined (Congo red)	Formol
At once	15	38	24
After 2 days	9	37	29
After 9 days	—10	29	42

There is a noticeable increase in the free amino groups, as shown by the rise in the formalin and decrease in the Günzberg acidity.

If the formation of other products than peptones and albumoses recorded by Zuntz is really due to pepsin, the following scheme may perhaps indicate the course of peptic digestion of a simple protein.

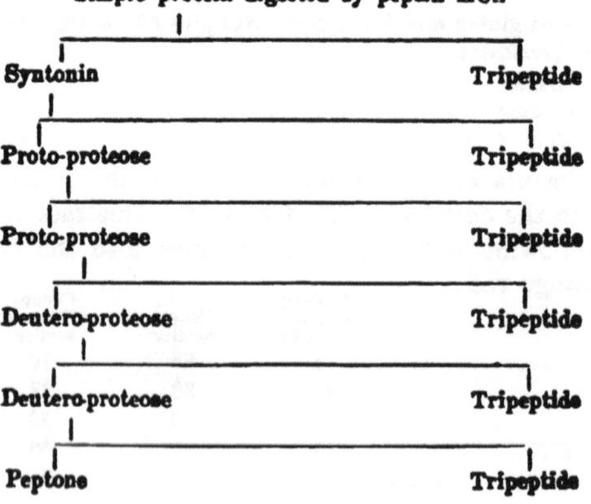

10. *Energy changes in digestion.* Peptic digestion does not involve a noticeable loss of energy, a matter of great importance for the organism, since it indicates that neither the synthesis nor the decomposition of the proteins from the amino-acids involves much energy dissipation.

If the heat produced by burning a certain weight of the protein before digestion is compared with that produced by burning the same weight of the digested products, allowing for the increase of weight due to the water added by hydrolysis, the difference is negligible. Energy is not lost, then, by the hydrolytic decomposition of the proteins, or the loss is extremely little, nor is energy required for their synthesis by dehydration.

11. *Fate of the pepsin.* What becomes of the pepsin which is secreted by the stomach and finds its way into the intestine? Is it destroyed there by the action of the bacteria and enzymes of the small and large intestine? or is it reabsorbed? and, if it is reabsorbed, what becomes of it? Is it picked out from the blood by the stomach mucosa and used over again or is it taken up and destroyed by the blood or body cells?

These are questions to which only most imperfect answers can be given. In fact, it is not known what becomes of most of the pepsin passing with the food into the intestine. It was long thought that by the action of the bile, duodenal and pancreatic secretions, all of which are alkaline, it would be destroyed and that its action would cease on entering the intestine and the pepsin itself be perhaps digested. This is perhaps the fate of most of it. But it has been found that the reaction of the chyme in the upper part of the intestine remains weakly acid sometimes clear to the large intestine, at other times only a short distance below the pylorus, but in any case the neutralization gives very little or no free alkali. Recent experiments by Abderhalden and Meyer have shown that active pepsin is to be found in the intestinal contents of the dog's duodenum, jejunum and ileum. By mixing these contents with elastin to which the enzyme sticks and then placing the elastin in hydrochloric acid solution, the elastin was found to be digested, proving the presence of active pepsin throughout the canal. There can be little doubt that the peptic digestion does not cease as soon as the chyme passes into the intestine, but continues for some time, and that the pepsin is not killed at once, at any rate, by trypsin and erepsin.

The further fate of this pepsin is not known. Very little active pepsin is found in the feces. There is little doubt that some of it is absorbed, for the blood always contains some anti-pepsin, a fact which proves that pepsin has found entrance to the blood either from the peptic glands of the stomach or from the alimentary canal. More probably it enters from the latter. Pepsin is found. also, in small amounts in the urine, which proves that some of it is circulating in the blood. These facts make it possible that there may be a circulation of pepsin analogous to the circulation of the bile, the pepsin being in part reabsorbed, carried to the stomach and there picked out and resecreted. The conditions of this resecretion are of course much less favorable than for the analogous circulation of the bile, since before reaching the stomach the blood from the intestine must pass through many other organs, all of which may act upon the pepsin. How extensive and whether or not there is such a circulation of the pepsin cannot at present be said and the question should be investigated. While it is not impossible that a larger proportion of the pepsin which is secreted in the stomach may be reabsorbed in the intestine than is at present believed, the probability is that most of the pepsin is destroyed in the intestine either by the other digestive enzymes or by the bacteria. The whole question is badly in need of further investigation.

Hydrochloric acid. Methods of its quantitative determination.— The activity of the pepsin depends on the acidity of the juice. It becomes, therefore, important from a clinical point of view to determine what the amount of acid is in the juice and whether it is hydrochloric or some other acid. Generally the secretions of the hydrochloric acid and the pepsin run closely parallel to each other, so that if the juice is weak in pepsin it is also weak in hydrochloric acid. In hyperacidity there is usually a hyperpeptic activity also. There may, however, be achlorhydria, that is an absence of hydrochloric acid, with some peptic action. The close parallelism of the secretions of these two substances may mean that they come from the same cells or are even formed together in one action. The acid is generally lacking in carcinoma of the stomach, and this is a fact of considerable diagnostic importance. It is desirable to have, therefore, methods for the easy and accurate determination of the stomach acid. The examination is usually made in the following way:

In the morning when no meal, or but a light one, has been eaten the evening before, and the stomach has, therefore, been resting for 12 hours and should be empty, a test breakfast is eaten of either a roll and a cup of weak tea, or five Albert or other crackers (biscuits) well chewed. After 45 minutes the patient either vomits, or a stomach tube is swallowed and a sample of the gastric juice is drawn for examination. It is wise not to be content with a single sample, but to examine the contents on more than one occasion, since the acidity of the contents is not always the same and it may happen that the material withdrawn has not been thoroughly mixed with the juice.

A few c.c. of the unfiltered juice is titrated with N/10 NaOH, using phenol-phthalein as an indicator. This indicator is so delicate that it will detect all acids, even very weak organic acids, and that also which is temporarily combined with the protein. Hence this is the maximum acidity which can be present. The number of c.c. of N/10 NaOH necessary to neutralize 100 c.c. of the unfiltered contents is called the *total acidity*.

If this number turns out to be normal, and human normal gastric contents should be sufficiently acid so that 100 c.c. will require about 74 c.c. of the N/10 alkali, the next process is to discover whether the acid is hydrochloric acid or not, or at least to see whether the hydrochloric acid is present in a free state. This is best determined by Günzberg's reagent or by Töpfer's. The latter introduced dimethylamino-azo-benzene as an indicator for the free hydrochloric acid, since it only changes color when the concentration of hydrogen ions is at least $N \times 10^{-6}$. Any organic acid occurring in the stomach would need to be present in very high concentration, 0.5 per cent. or more, to produce this concentration. This indicator is pink when acid and yellow when alkaline. If now 10 c.c. of filtered juice are taken and a drop or two of the indicator solution added and then titrated with N/10 NaOH, the result, if the contents are normal, should give a figure corresponding to about .03 NHCl. The Günzberg reagent should also be used as described on page 368. An accurate determination of the number of hydrogen ions in the juice, and this is the only true test of its actual activity or avidity, can best be made at the present time by means of the gas-chain method explained on page 538. This method is too difficult for clinical work. The actual acidity of the juice as determined by this method varies greatly in different conditions. In the fasting stomach Tangl obtained, by drawing some of the contents with a sound 10-12 hours after the last meal at 8 A.M., from 2-25 c.c. All 13 samples, with one exception, were acid to litmus. Only one was alkaline. The concentration of hydrogen ions in gram equivalents per liter were: .035; .022; .0001; .00042; 12×10^{-8}; .085; .013; .029. The acidity of normal gastric contents should be about .035. The acidity ran between 10^{-2} and 7×10^{-2} N hydrogen ions. The optimum concentration of hydrogen ions for the digestion of proteins by pepsin was found by Sorensen's gas-chain method to be 0.020-0.060 N. Christiansen found the optimum for human juice to be .020-.033; for the pig, 0.052; and for the dog, 0.053 N.

While the direct determination of the hydrogen ion concentration is thus too difficult for clinical use, except in a well-equipped laboratory, a very close approximation may be made by the use of the Günzberg reagent. This reagent consists of a mixture of phloroglucin, 2 grams; vanillin, 1 gram; alcohol, 100 c.c. The reagent should be kept well protected from the light. One drop of the reagent is placed on a white porcelain plate, or evaporating dish, and dried cautiously over the flame, or better on the water bath. Then add one drop of the juice to be tested to the clear yellow spot left by the evaporation and warm again carefully. If hydrochloric acid is present in the free state, a red color develops on heating. If it is heated too much, it will be brown. By diluting the gastric juice a limit will be found which just gives the reaction. This limit contains 0.0004 NHCl. The red color is due to the production of a red crystalline substance, $C_{20}H_{18}O_8$, which is formed by the union of two molecules of phloroglucin, $C_6H_6O_3$, and one molecule of vanillin, $C_8H_8O_3$, with the elimination of water. If acid is tested in the cold with this reagent, it will be found to react only with N HCl; weaker acid gives no color.

The following table illustrates the findings in a large number of cases after a test breakfast (Ewald's). The figures in all except the first column represent the number of c.c. of N/10 NaOH necessary to neutralize 100 c.c. of the gastric contents when the different indicators are used. The Günzberg figures are not determined by direct titration, but in the manner indicated, i.e., by dilution. They express the amount of free hydrochloric acid there is in 100 c.c. of juice. The figure 20 under Günzberg means that 100 c.c. of this juice contained 20 c.c. of free N/10 HCl.

No. of exper.	C_H by gas chain	C_{HCl} computed from C_H and expressed as titration number.	C_{HCl} corrected for dissociation	Günzberg	Töpfer	Congo red	Litmus	Phenol-phthalein
1.	.0003	0	0	....	....	4	......	20
2.	.033	33	34	....	....	41	60	66
3.	.004	4	4	....	....	23	......	47
4.	.036	36	38	....	....	62	......	85
5.	.004	4	4	Trace	....	17	......	33
6.	.018	18	19	10-15	....	38	......	52
7.	.022.	22	23	15-20	....	38	......	51
8.	.025	25	26	20-25	....	44	......	......
9.	.056	56	59	42-48	64	68	82	88
10.	.040	40	42	40-45	54	59	67	78
20.	.062	62	67	66	....	77	83	93
29.	.036	36	38	44	46	59	64	78
32.	.0003	0	0	0	0	13	21	44
39.	.042	42	44	39	....	59	70	80

It will be seen from the foregoing table that the free HCl as determined by Günzberg's reagent is very close to the actual concentration of the hydrogen ions as determined by the gas-chain method. Congo red gives always much more than the free hydrochloric acid and dimethyl-amino-azo-benzene (Töpfer) somewhat more. If the concentration of the hydrogen ions is low, however, Günzberg used in this way gives too low results. Pure, human appetite juice contains .12 NH ions.

In the case of pure gastric juice, when there is no admixture of stomach contents, congo red gives results more in harmony with the hydrogen ion titration. For then there is no admixture of food to bind some of the acid of the gastric juice. This fact is shown in the following determinations of various samples of gastric juice from a case of hypersecretion:

No. of sample	C_H by gas chain	C_{HCl} corrected for dissociation	Günzberg	Congo red	Phenol-phthalein
1.	.035	37	38	42	47
2.	.035	37	33	45	55
3.	.037	39	43	54	62
4.	.022	23	25	30	38
5.	10^{-8}	0	0	0	4
6.	.035	37	33	44	57
7.	.0073	77	77	83	88
8.	.056	59	58	64	71

With pure gastric juice there is not a very great difference between phenol-phthalein titration and that by Günzberg; in other words, there is very little difference between the free hydrochloric acid and the total acid, for in such juice when it is normal there is almost no lactic or organic acid and most of the acid is free and not bound to the organic matter in the juice. But as the protein admixture in the juice increases, or when by fermentation organic acids may be formed from the carbohydrates, then the difference between the two titrations may be very large. This fact is brought out if we compare the titrations of the stomach contents, first, of the pure juice, then of the contents after an Ewald test breakfast, which contains very little protein, and then after Bourget's breakfast which has more meat in it.

Titration Numbers.

	Günzberg's reagent	Congo paper	Phenolphthalein
Pure gastric juice	25	30	35
After Ewald's breakfast	25	45	65
After Bourget's breakfast ...	25	75	125

How the acidity of the juice may vary after various meals and at different times is shown by the following figures:

	Breakfast	Sample drawn hours after eating	C_H ions	C_{HCl}	Günzberg	Congo red	Phenol-phthalein
1.	250 grs. oatmeal soup .. 50 grs. meat 4 pieces white bread	3½	.032	34	36	64	91
2.	250 grs. oatmeal soup .. 100 grs. meat 4 pieces bread	2	.008	8	8	41	67
3.	250 grs. soup .... 100 grs. meat 2 pieces bread	3	.021	22	13	100	190
4.	250 grs. soup .... 100 grs. meat 2 pieces bread	2½	.062	67	71	91	117
5.	250 grs. bouillon ...... 100 grs. meat 2 pieces bread	2	.0007	1	—24	30	90

In the last experiment there was no free hydrochloric acid as shown by Günzberg; indeed, to get a positive Günzberg test it was necessary to add hydrochloric acid in considerable amounts, but nevertheless there was a normal total acidity and a normal acidity to congo paper.

The following table shows the number of c.c. of N/10 NaOH required to titrate 100 c.c. of unfiltered stomach contents after an Ewald test breakfast when various indicators were used:

Indicator	c. c. N/10 NaOH	
Tropaeolin	12 - 19	End point indefinite.
Methyl violet	15 - 25	End point indefinite.
Günzberg	25	
Boas	25	
Dimethyl-amino-azo-benzene	36 - 38	
Methyl orange	41 - 43	
Congo paper	43	
Alizarin	49 - 51	
Rosolic acid	51 - 53	
Litmus paper	56	
Phenol-phthalein	65	

Variation of hydrochloric acid in disease.—The determination of the secretion of hydrochloric acid is of considerable importance in the diagnosis of stomach disease. Thus in cancer of the stomach particularly, but also at times when the cancer is in some other part of the body, there is often a great diminution of the secretion of hydrochloric acid or its complete suppression. On the other hand, in ulcer of the stomach and in particular when the ulcer is in the pylorus, or duodenum, there is generally found hyperacidity. It may happen, however, that when an ulcer has a cancer grafted on it the secretion may be normal. In various neuroses the acidity and the pepsin content may be increased above the normal. In general, carnivorous animals have a more acid secretion than herbivorous, and a meat diet is supposed to increase the acidity; although no very convincing evidence of variations of acidity with diet has been found.

Theory of titration of stomach contents by indicators.—All the indicators employed for the titration of acids or alkalies are either acids or bases. Thus phenol-phthalein, dimethyl-amino-azo-benzene and congo red are acids. The color change is due probably to a rearrangement of the molecule to a colored form, usually a quinonoid, when in the salt form. This rearrangement is probably due to the dissociation of the molecule, the undissociated molecule not rearranging. Since the indicators are acids of different avidities, that is since they have different amounts of dissociation, some are weaker than others. Accordingly some are able to form salts in sufficient amounts to give a perceptible color in the presence of more acid than are others which are weaker. Thus congo red is a fairly strong acid and is able to take some of the base to itself and make a colored salt in the presence of some free acid, whereas phenol-phthalein is so weak an acid and its salts dissociate so much hydrolytically that it will only give an alkaline reaction, that is form enough salt to be seen, when there is quite a good deal of alkali present. The concentration of hydrogen ions at which the various indicators change their reactions is shown in the following table (see also page 544):

Indicator	C_H ions, at which the indicator changes. (Normal divided by 10 raised to the power indicated by the following figures. Congo red changes between $N/10^3$ and $N/10^5$ H ion)
Tropaeolin OO	1.4 - 2.6
Methyl violet	0.1 - 3.2
Dimethyl-amino-azo-benzene	2.9 - 4.0
Methyl orange	3.1 - 4.4
Congo red	3 - 5
Alizarin	5.5 - 6.8
Litmus paper	7
Neutral red	6.8 - 8.0
Rosolic acid	6.9 - 8.0
Phenol-phthalein	8.3 - 10.0

Günzberg's reagent.—The property of giving a red color under the conditions of the Günzberg reaction is not peculiar to hydrochloric acid, since sulphuric, nitric, phosphoric and boric acid give it also. Of these acids phosphoric and boric are weak acids, boric being very weak. But boric acid has the property of uniting by one of its hydroxyls with sugar and becoming thereby a much stronger acid, and it is possible that phosphoric acid has this same power less developed, since as the existence of phytin shows it has the property of combining with the hydroxyls of aromatic alcohols such as phloroglucin. Both these acids probably unite with the phloroglucin to make esters and their acidity is probably thereby much increased. Phosphoric acid is, for example, a much weaker acid than formic, which does not give the reaction. Oxalic, citric and tartaric acids give a positive reaction; succinic, propionic, lactic, acetic, butyric, benzoic, formic and phthalic are negative. No monocarboxylic fatty acid is known which is positive, but if there is more than one carboxyl present then it may be positive. Hydrochloric acid N/2500 still gives a noticeable reaction. It is clear that the reaction does not depend upon the number of hydrogen ions alone. For example, a mixture of glycocoll and hydrochloric acid having a concentration of hydrogen ions of $N \times 10^{-1.96}$, or roughly .01N, is just positive; while free hydrochloric acid N/2500 or $N \times 10^{-3.4}$ is positive. Christiansen concludes that the reaction depends on the nature of the acid and not on the H ion concentration. The probability is, however, that when glycocoll and hydrochloric acid are evaporated more and more of the acid is bound as the concentration increases. Consequently the ion concentration at the end may be far less than that indicated in the foregoing figures. Günzberg's reagent is certainly the most useful indicator for the determination of the free hydrochloric acid.

Origin of the hydrochloric acid.—What is the origin of this hydrochloric acid? How shall we picture the processes by which this acid in a concentration fatal to all animal cells, if once it enters them, is secreted by living matter from an alkaline fluid like the blood? In what part of the stomach is it formed and by what glands or cells? These are questions which are not yet solved. The problem is a difficult one.

The acid, unlike pepsinogen, is not stored in the cells of the stomach mucosa, for the aqueous extract of the mucosa is neutral, not acid, in reaction, and the mucosa contains only a small amount of chlorine, although two or three times as much as most of the other tissues of the body. Not only is the hydrochloric acid not stored, but neither are the chlorides stored or an organic chlorine compound, except in small amounts. The per cent. of chlorine in different tissues computed on the wet weight was found as follows by Nencki and Schumova-Simonowski:

Panniculus adiposus	0.076
Stomach mucosa	0.093
Liver	0.025
Bone marrow	0.034
Muscle	0.033
Kidney fat	0.032
Bones	0.033
Intestinal mucosa	0.040

The chlorine content of the mucosa is, therefore, somewhat greater than that of other tissues. But, owing to the small weight of the mucosa, the amount of chlorine stored is but a very small fraction of that excreted.

The hydrochloric acid is secreted chiefly by the fundus end of the stomach. The pyloric secretion is certainly less acid than the fundus secretion. It is found, too, that the fundus mucosa has a little more chlorine in it than the pyloric mucosa.

Cl Content of Fundus and Pyloric Mucosa. In Per Cent. of the Dry Weight.

Animal	Stomach Contents	Per cent. chlorine in dry weight of tissue		
		Fundus	Pylorus	Difference
Pig	....	0.87	0.67	0.20
Pig	Strongly acid	0.62	0.59	0.03
Dog	" "	0.72	0.68	0.04
Pig	Empty. Weak acid ..	0.75	0.44	0.31
Dog	Strongly acid	0.59	0.50	0.09
Pig	" "	0.60	0.50	0.10
Pig	Empty. Weak acid ..	0.83	0.62	0.21

The fundus has, then, more chlorine than the pylorus. The water content is about 85 per cent. in each.

The attempt was made by Heidenhain to determine whether the acid was formed in the pyloric or fundus part of the stomach by making a pouch of a portion of the stomach in each region. He found that the pouch at the fundus end was weakly acid and that at the pyloric end was alkaline. This experiment, however, is open to the criticism that the nerves were cut and the secretion consequently not normal. The pouch made by the Pawlow method takes in more of the fundus part of the stomach and it always secretes a strongly acid secretion. There is no doubt, therefore, that the secretion of this part of the stomach is certainly acid. Since in the mammalian stomach there is a marked difference between the character of the cells in the glands of the two regions, in that the so-called delomorphic, or parietal, or oxyntic cells are found in the glands of the fundus part of the stomach but not in the pyloric portion, the conclusion was drawn that the acid was secreted by these cells and they were named oxyntic cells in consequence. Many attempts have been made to discover some direct evidence that these cells secrete the acid, but the final result of such attempts has been to show beyond doubt that they do not secrete acid, but have an alkaline secretion.

Claude Bernard was one of the first who attempted to get some direct evidence of the place of formation of the hydrochloric acid. He injected into one vein of a dog potassium ferrocyanide and into another the lactate of iron. When these two reagents are brought together outside of the body, it is only in the presence of acid that they react to make a blue precipitate of Prussian blue. On killing animals after such an injection, he found the blue color only in the lumen of the stomach, and in the necks of the glands, but not in the mucous membrane. He concluded that the mucosa was alkaline in reaction and that the acid was formed only after excretion. Foster suggested that it was formed as an organic compound which after secretion decomposed, setting free the acid. All attempts to show the existence of such a compound in the mucosa have so far been fruitless. Fitzgerald recently repeated the work of Bernard, with the result that she found some of the parietal cells stained blue by the Prussian blue, but in general the color was in the lumen of the neck of the gland and in the stomach itself. Her results were interpreted to mean that the secretion of the parietal cells was acid. It has recently been shown by Harvey and Bensley that these conclusions are incorrect. A few of the parietal cells may take the stain, but the vast majority do not. Moreover, cells of the liver, and other tissues, may also stain and the blue deposit may be found in the lymph and blood where there is no possibility of the formation of acid. It is, therefore, clear that this method gives no reliable indications of the reaction of a cell or a tissue.

Macallum attempted to follow the matter further by means of a study of the distribution of the chlorine in the cells. By precipitating the chlorides with a solution of silver nitrate in dilute nitric acid and then reducing the silver chloride by exposing the section to the light, he was able to detect the presence of chlorine in cells. This method is open to the serious objection that perhaps other substances than chloride may fix the silver. Nevertheless, the method is better than none at all and enables a study of the silver-fixing elements of the cells. He found that both parietal and chief cells gave a strong reaction for chlorides, but that the parietal cells had the stronger reaction and he interpreted this finding as favorable to the view that the parietal cells secreted the acid. A repetition of this work by Lopez-Suarez gave the contrary result, that there was more chlorine in the chief cells and that the parietal cells were practically free from chlorine. That the acid is not formed by the parietal cells, but only in the neck of the gland and possibly in the lumen itself, is shown by the work of Harvey and Bensley. They found that cyanamine, which is blue when acid and red when alkaline, stains the secretion of the parietal cells an intense red in the living state. The secretion of these cells is not HCl, but it is full of organic matter. The secretion of the whole of a fundus gland is slightly alkaline or neutral until the foveola is reached. The gland contents in the foveola stain a blue color, indicating acid. From these observations it may be concluded that the acid is not secreted by the cells of the stomach at all, but is probably formed in the fovea by the reabsorption of some basic constituent, leaving the acid outside.

We may perhaps picture the formation of the acid of the gastric juice as follows: It is not formed in the cells, but in the cavity of the stomach and in the foveæ of the glands. The cells lining the mucous membrane are non-permeable to it. It is possible that the chlorine is secreted as ammonium chloride, or the chloride of some other weak base. It might also be secreted as an ester. When in the cavity of the stomach, perhaps as it passes along the lumen of the gland, or at any rate in the neck of the gland, a hydrolytic or other dissociation takes place, setting free hydrochloric acid.

$$NH_4Cl + H_2O \rightleftharpoons \underset{\text{Absorbed.}}{NH_4OH} + HCl$$

Either by some kind of selective absorption or adsorption the NH_4OH is removed by the cells of the neck of the gland, leaving the HCl behind. Such processes of an unknown nature, called selective adsorption, are believed to occur; but to fall back on this terminology here is in the nature of a subterfuge, since it simply puts the mystery under a new name. Possibly by a chemical reaction the ammonia is absorbed and converted into uramido compounds or carbamic compounds in the cell, keeping the pressure of the ammonia ion in the cell nearly zero. In every solution of NH_4Cl there are NH_4 and OH ions, and it is known that undissociated NH_4OH or NH_3 penetrates cells readily.

It may be urged in favor of this hypothesis that it is supported by all the direct evidence which we have. A similar formation of hydrochloric acid in just this manner occurs elsewhere in nature. Thus if the mould, penicillium, is grown in a medium containing ammonium chloride it absorbs the ammonia, leaving the hydrochloric acid outside. This is just the mechanism supposed in the foregoing theory for the stomach. Furthermore, the mucosa of the stomach contains more ammonia than any other tissue of the body. There are, for example, 40-52 mgs. in a hundred grams of stomach mucosa; whereas arterial blood contains about 1 mg. per 100 grams. The theory accounts, too, for the fact that the most acid juice is found in carnivora and in omnivora on a protein diet, a diet which by the deamidization of amino-acids sets free large amounts of ammonia. The theory has, therefore, much in its favor. The only great difficulty is the small degree of hydrolysis of ammonium chloride. The theory needs further confirmation before it can be wholly satisfactory. The fact that the mucous membrane or the cells of the mucosa, as long as they are alive and in good condition, have a resistance to the entrance of the acid cannot be doubted, since otherwise the acid would be neutralized by reabsorption. This impermeability is, however, of a very limited kind, and if the cell is weakened by disease or by partial occlusion of the arteries, or by great anemia, its resistance may be so lowered that the acid penetrates and digestion of the stomach wall begins. Particularly hyperacidity is dangerous, because the limit of resistance of the cells may be surpassed.

There have been two or three other suggestions of the nature of the process of the secretion of the acid which should be noted, although they are probably incorrect. The first is that of Maly, according to which by a double decomposition of the acid carbonates and alkaline phosphates and the chlorides some hydrochloric acid is formed in the cells of the stomach and the acid is then, in some unknown way, discharged in one direction and the alkali in the other. This seems highly improbable. In the first place, there is no evidence that acid is formed in the cells. In the second place, the amount of free acid thus formed would be extremely small, because the avidity of hydrochloric acid is so much greater than the very weak carbonic acid. Hydrochloric acid being stronger would take the sodium from the carbonate, not the other way around. In the third place, it explains one miracle by supposing a greater one in the separation of the free acid from an alkaline cell. Koppe has also made a suggestion that the hydrogen ions of the blood wander through the membrane and are exchanged for sodium ions which come in. The membrane is supposed to be non-permeable for negative ions like chlorine. The difficulty here is also to understand how such a membrane could be constructed, and the extremely small numbers of hydrogen ions in the blood. To get the acidity of the gastric juice as fast as it is secreted would be impossible. The explanation has the one merit of making the place of formation in the interior of the stomach rather than in the cells thereof. Otherwise it is no advance. As a matter of fact, the juice is secreted from the glands with full acidity and even in the absence of sodium chloride in the stomach.

One of the results of this acid secretion in the stomach is to render the whole blood and tissues of the body more alkaline, and this profoundly affects their metabolism. The sodium of the sodium chloride which has undergone decomposition finds its way into the blood as the alkaline phosphate or carbonate. The urine may even become alkaline during the eating of a meal and always its acidity is reduced. The change of alkalinity of the tissues increases the oxidation of the tissues and is responsible in part for the increased metabolism and heat production during digestion, observed by Lavoisier. It may very easily be a factor in the feeling of well-being following eating. Its importance for the body as a whole is generally overlooked.

Rennin and its action.—Gastric juice, or the aqueous infusion of the calf's stomach or of any other suckling mammal, clots milk. The essential fact in this clotting is that a soluble protein, casein (caseinogen, as the English call it), is transformed to an insoluble form called paracasein (English casein). The paracasein is so concentrated that it does not flock out of the solution as insoluble precipitates generally do, but it remains distributed all through the milk, and the water, with the fat, is held or entangled between the particles of paracasein so that a gel is formed. The substance in the gastric juice which causes this change in the casein is the rennin, or chymosin. If a little sodium oxalate or any other substance which precipitates calcium be added to milk, the milk will not clot on the subsequent addition of rennin. Calcium as well as rennin is, therefore, necessary to the clotting of milk. If after the addition of the oxalate and the rennin one waits for about half an hour or even a shorter period and then boils the milk so as to kill the rennin, the milk is still fluid, but if one now adds calcium chloride or some other calcium salt to the milk so that there is more than enough to combine with the oxalate, then clotting comes on at once. From this experiment it is clear that although in the absence of calcium no clotting has taken place, yet the rennin must nevertheless have acted on the casein and changed it so that, even in the absence of rennin, as soon as calcium is added the milk clots. Hammarsten found, also, that when milk clots and the whey or fluid portion is separated from the clot, a new protein, an albumose called *whey albumose*, appears in the whey. When milk clots, therefore, casein disappears and two new proteins appear, whey albumose and paracasein, the latter of which is found as a calcium compound in the clot; the former in the whey. Since a pure casein solution acts in the same manner, we may recapitulate all of these facts in the following explanation: Rennin splits casein into whey albumose (?) and paracasein, the paracasein thus formed is much less soluble than the casein as is shown by the fact that it is much more easily precipitated by acids, and its colloidal particles are of larger size. The paracasein forms with calcium an insoluble calcium salt, calcium paracaseinate, and this precipitate clots the milk. The molecular weight of casein, according to Van Slyke, is about 8,500, that of paracasein about 4,500. He believes that rennin splits casein into two molecules of paracasein.

The action of rennin appears, therefore, to be nothing more than the early stages of digestion or hydrolysis of the casein molecule. Casein clots when digested, for the reason that one of the early meta-protein hydrolytic products forms with calcium an insoluble compound. The action of rennin thus appears to be simply the action of a proteolytic ferment, and it is an interesting fact that all proteolytic enzymes appear to act similarly, although they differ considerably in their power. Thus not only does the gastric juice clot milk, but so will pancreatic juice, the juice of the intestine, the juice of the pineapple, of the cocoanut, the secretions of many bacteria and the extracts of some cruciferæ. In fact, wherever in nature one finds a proteolytic enzyme there one finds also a milk-clotting enzyme. It may be concluded from this either that the enzyme, rennin, is very widespread in nature, constantly accompanying proteolytic enzymes; or that the clotting of casein solutions is an indirect result of the proteolytic cleavage of the molecules of casein. The latter conclusion seems the more probable.

The foregoing conclusion does not mean, however, that rennin and pepsin are identical. This is a matter in dispute at the present time and a brief examination of the facts in the case will be worth while.

If it were possible to obtain a solution which had peptic but no clotting action, or vice versa, it might be inferred that the two substances were different. Hammarsten found that it is indeed possible to prepare solutions which have one but not the other action. By heating pepsin solutions for several hours at 40° C. they lose their rennin but not their pepsin action; if precipitated by lead acetate or magnesium carbonate, pepsin is completely precipitated, but a part at least of the rennin still remains in the solution. The most striking fact indicative of a difference between these two ferments is that peptic digestion takes place only in acid solution, whereas rennin coagulation will occur in neutral or amphoteric solution. These facts all look as if the ferments were two different substances. There is no doubt, however, that pepsin has a coagulating action on milk. Thus Hammarsten tried the following experiment: A pepsin solution, which had been heated 48 hours at 40° C. to make its rennin action very weak, after neutralization coagulated milk only after 6 hours and 10 minutes when added in the proportion of 1:5. By the addition of HCl the neutral solution was brought to an acidity of .1 per cent. HCl and it coagulated now in 12 minutes. A control showed that this coagulation was not due to the acid. Another solution of rennin, which according to Hammarsten was pepsin-free, was diluted with water so that it coagulated milk only after 45 minutes; it was then made acid to .1 per cent. HCl and it coagulated only after 20 minutes. The addition of acid, therefore, accelerated the action of the solution which contained much pepsin but little rennin far more than it did the solution containing rennin but no pepsin. The clotting in the first juice must, therefore, have been due to the pepsin.

Experiments similar to these have been carried out by Schmidt-Nielssen. He heated an acid extract of calves' mucosa so long at 40° C. that on neutralizing it with n/10 NaOH the neutral fluid added to milk in the proportion of 1:5 coagulated the milk only after 4-6 hours at 38°. This solution he called A. Another portion of the same extract was not heated, but was diluted with water and neutralized with n/10 NaOH until it had a rennin action approximately equal to solution A. These two solutions, A and B, when neutral had, then, the same coagulating action and presumably had the same content of rennin. If now there is only one ferment in the juice, then if acid is added to each of these solutions they ought to be affected to the same degree and they should have the same action on milk and fibrin. The results were as follows:

		Solution A. Warmed. Time for coagulation	Solution B. Not warmed. Time for coagulation.
1.	Neutral milk	370 minutes	355 minutes
	Acid milk	3 "	215 "
	Fibrin digestion	3 hours	80 hours
2.	Neutral milk	420 minutes	360 minutes
	Acid milk	18 "	250 "
	Fibrin digestion	3-4 hours	80 hours.

It is clear from these experiments that, although solution B had a little stronger action in the neutral solution and so presumably contained more rennin, yet the addition of acid affected the two solutions to very different degrees, A being enormously accelerated by the acid both in its coagulative and in its digestive power; whereas B was only slightly accelerated by the acid and had a very weak digestive power. The heating must have destroyed most of the rennin (or changed it to pepsin) and the coagulation in the acid solution must hence be due in chief measure to the pepsin, which is thus shown to have a clotting action. It has this clotting action only in an acid medium. There seems to be no escape from Schmidt-Nielssen's conclusion that the two ferments are not identical, or as he says, " that enzyme which coagulates neutral milk, chymosin, cannot be identical with pepsin."

A difference between the ferments is also shown by the time law of the coagulation. Thus for rennin in neutral solution the product of the time of coagulation by the ferment concentration is a constant; if one adds half the quantity of rennin, the time of coagulation is double, or C_f t=K. C_f is the concentration of the ferment. Schmidt-Nielssen found that the coagulation time in acid milk did not follow this simple proportionality law, as may be seen in the following experiment:

Milk containing 4 per cent. HCl.	Infusion warmed of calves' stomach	Cooked infusion for dilution	Concen. of enzyme	Coag. time—minutes	C_fT
10 c.c.	2	0	1	9.5	9.5
10 c.c.	1	1	.5	35	17.5
10 c.c.	.5	1.5	.25	300	75

An experiment showing the effect of acid in accelerating the clotting action of the pepsin-rich, rennin-poor solution is the following:

HCl in milk-enzyme mixture—per cent.	Coagulation time—minutes
0.42	3.5
0.25	10
0.17	42
0.08	280
0.00	Not determined

Other points of difference between the enzymes have been pointed out by Bang. Further evidence that there are at least two different enzymes is the fact that the infusion of calf's stomach has a powerful clotting action, but a weak digestive action on egg white; whereas the infusion of the pig's stomach has a powerful digestive but a weaker clotting action.

This view of the difference of the two enzymes does not stand unopposed. Pawlow and his pupils have established many facts which they have interpreted to mean that the two ferments are the same. Pawlow at first calls attention to the fact, so extremely significant, that one always finds a clotting action wherever one finds a proteolytic enzyme. This is true even in the case of plant enzymes which could never have been elaborated to act upon milk. This, he maintains, shows that all proteolytic enzymes coagulate milk and, vice versa, the coagulation of milk is caused by a proteolytic enzyme, a conclusion which has been supported by the discovery of the character of the change in the casein accompanying clotting. They were unable to prepare rennin solutions which were free from proteolytic power and pepsin solutions which were free from a rennin action. Pawlow and Parastchuk experimented with gastric juice of an adult dog obtained by sham feeding. They found that bread juice has a stronger milk-clotting action as well as a stronger proteolytic action. One of their experiments consisted in taking three samples of juice: 1, milk juice; 2, meat; and, 3, bread juice. These at the start had very different proteolytic actions. By dilution they were made of equal pepsin content and then they were tested to see whether they were now equal in clotting power. The experiment was as follows:

Milk juice		Meat juice		Bread juice	
Digestive power mm. egg alb. Mett's tube. 24 hours	Coag. time—minutes	Digestive power—mm.	Coag. time—minutes	Digestive power—mm.	Coag. time—minutes
1.8	50	3.6	5.25	5.8	2.5
1.85	30	4.05	3.0	5.8	0.75
2.45	12	3.8	3.2	6.4	0.75
2.9	35	4.0	9.0	6.4	5.5
2.5	26	3.6	6.5	6.6	2.5

It will be noticed that the bread juice coagulates in the shortest time and is the most powerful digestant, and the milk juice is the weakest. To show that when they have the same peptic content the coagulation time is the same, a sample of each kind of juice was taken and the amount of pepsin made the same in each by dilution, according to the law of Schütz and Borissow by which the amount of pepsin is proportional to the square of the egg albumin digested in a Mett's tube. At the start the bread juice digested 5.8 mm., the meat 2.8 and the milk 2.0 mm. of egg albumin in the same time. The pepsin in these samples was then in the proportion of the squares, or 33.64 :7.84 :4.0. By dilution with acid they were made equal in pepsin and acid. Equal amounts of the juices then coagulated milk respectively in 190 minutes, 190 minutes and 195 minutes. It will be observed that this is in an acid solution. It was found, also, that when the juice was kept in a thermostat it lost both actions at the same rate and both disappeared at the same time. Moreover, on heating five minutes at varying temperatures from 15° to 62°, the coagulation power and the pepsin action fell together, and both disappeared at the same temperature, namely with five minutes' heating at 62° C.

Although these experiments were interpreted by their authors as being in opposition to those of Hammarsten, they are in reality not so. Hammarsten's work was done on the neutral infusions of calf's stomach and the clotting of neutral milk was studied. What Pawlow and Parastchuk have really shown is that in dog's gastric juice pepsin has a rennin action; and, probably, that in the stomach of an adult dog gastric juice contains no specific rennin enzyme, the rennin effect being due to pepsin.

The experiments prove that in the stomachs of calves there is present in addition to pepsin an enzyme which will clot milk in neutral solution, and this enzyme was the one originally called rennin or chymosin.

The conclusion seems justified that the coagulation of milk is not a specific function of any single enzyme, but that it is a general property of all proteolytic ferments. The clotting is due to the fact that one of the first products of the decomposition of the casein forms an insoluble calcium compound which sets or gels. In the mucosa of the stomach there may be, and no doubt are, more than one proteolytic enzyme. There can hardly be a doubt that the mucosa of the stomach does produce different proteolytic enzymes in different portions of the stomach. The fundus glands produce pepsin, that is an enzyme active only in an acid medium and easily destroyed by slight alkalinity. The pyloric extract, on the other hand, is known to furnish an extract which is active not only in an acid medium but also in a neutral medium. It must, therefore, contain proteolytic enzymes more allied to trypsin and erepsin. It is not strange, therefore, that more than a single proteolytic enzyme should exist in the gastric juice and be present in varying proportions at different ages. It has been observed, indeed, that in the course of development ereptic enzymes which act on certain native proteins and albumoses appear in the tissues before pepsin, which is the specific enzyme of the stomach. A β-albumosecase (erepsin) has been found in gastric juice.

The facts clearly indicate that the coagulation of milk in a neutral solution, a power particularly developed in the stomachs of suckling animals such as the calf, is due to a particular proteolytic enzyme of the ereptic type active in neutral or very faintly acid solution. This particular ereptic ferment is rennin (chymosin). But in the adult stomach, particularly of carnivora, the amount of this enzyme is greatly reduced, or it may be absent, and the clotting of acid milk by the gastric juice of these animals is a result chiefly or entirely of the activity of pepsin. The clotting of milk is, then, no test for the presence of a specific ferment for casein, and, since other ferments than pepsin clot milk, the use of the time of clotting as a test for pepsin in gastric contents is entirely unwarranted. It is extremely significant in this connection that erepsin and ereptic ferments, although they do not normally digest native proteins, but only albumoses, do attack and digest casein, and that casein is one of the most unstable and easily hydrolyzed proteins known.

The question may be asked whether anything is gained by having in the stomachs of the new-born an enzyme which will clot and digest milk in a neutral or very faintly acid medium. The answer to this question is not difficult, for the advantages are obvious. The mucous membrane of a child's stomach is extremely delicate; it is very thin and so sensitive that it cannot bear even the firm clot of cow's milk. It cannot stand much acid. Milk is a liquid and liquids normally pass quickly through the stomach into the intestine. By having a proteolytic enzyme acting in a neutral or amphoteric medium on casein the milk is clotted in the stomach, even when the secretion is only faintly acid. The fluid portions are passed on, but the curds are held for digestion by the rennin and, as acidity develops, by the pepsin also.

Salivary and intestinal digestion in the stomach.—The food while in the stomach is acted upon not only by the juices secreted by this organ. Saliva swallowed with the food continues to act; and there is, at times, a regurgitation of digestive juices from the intestine so extensive that the stomach digestion may partake largely of the type of intestinal digestion.

The saliva continues to act in the fundus end of the stomach for a period longer or shorter, depending on the size and character of the meal and the amount of gastric juice. The center of the mass of food in the fundus end of the stomach does not become sufficiently acid to stop ptyalin action for about an hour, on the average, after the food is eaten (see page 335). During this time ptyalin is active and we now know that starch digestion can proceed well toward its end in the stomach. Moreover, we have always the action of the bacteria. These are killed for the most part when the juice is acid, but in hypochlorhydria, or when the motility of the stomach is depressed, the activity of the bacteria may be a source of much discomfort. By the fermentation by bacilli of a lactic acid type, or the butyric acid bacillus, or various sarcinas, lactic, butyric or acetic acids may be formed, and large quantities of hydrogen, or carbonic acid gas evolved. A sufficient amount of gastric juice to produce rapid acidity will prevent this bacterial growth. Hence, if much hydrochloric acid is formed, there is little lactic acid produced.

Still another factor enters into the gastric digestion: namely, the regurgitation of bile, pancreatic and duodenal juice into the stomach. This ordinarily occurs to a relatively slight extent, but under certain conditions it may result in the digestion becoming of a real intestinal nature. This fact, discovered by Boldyreff, has been taken advantage of by him for obtaining pancreatic juice for clinical examination. The presence of bile was observed in the stomach of Alexis St. Martin, but it does not seem to have occurred to physiologists that the pancreatic juice might come also. Boldyreff found that the presence of fat, and particularly fat with fatty acid in it, caused this regurgitation. Also, if the stomach is more than normally acid, there is a pouring-in of large quantities of intestinal juice to neutralize the hyperacidity.

Summary of gastric digestion.—The empty stomach is normally contracted on itself and its cavity obliterated so that its walls are in contact. Under these conditions the mucous membrane is white, or a pale pink, thrown into folds, or rugæ, and the surface is covered with a thin layer of mucus either alkaline, or neutral, or very faintly acid in reaction. When one is hungry the smell, or sight, or the taste of food is sufficient to start the secretion, so that before a meal is eaten there may be some gastric juice present in the stomach. This juice is acid in reaction, due to the presence in it of hydrochloric acid, and it has a strong solvent action on proteins, due to the presence of an enzyme called pepsin. When food is eaten the stomach slowly relaxes as the food is swallowed and thus adapts itself to the size of the meal which has been eaten. The food at first accumulates in the fundus end of the stomach in a large bolus, which retains a very faint acid or neutral reaction in its interior for a considerable period and in this bolus salivary digestion continues some time.

As soon as food enters the stomach, or even before it enters, the stomach begins to pour out the strongly acid secretion of the gastric glands. This juice attacks the exterior of this mass of food and slowly softens, digests and liquefies the peripheral portions of it. At the same time movements of the stomach begin and increase gradually in intensity. These movements are in the nature of peristaltic constrictions, which appear at about the junction of the pyloric and fundus regions of the stomach and then travel slowly toward the pylorus. By means of these movements some of the partially digested, softened portions of the periphery of the bolus of food are broken from the main portion and thoroughly mixed with the digestive juices.

At first when the peristaltic wave reaches the pylorus the latter does not open, but the wave is reflected back again toward the fundus end of the stomach, thus carrying the portions of the food back with it. By the combined action of the movements and the solvent action of the juice, the mass of food is slowly reduced to a state of fine subdivision and solution in an acid medium and this mixture of foods partly in suspension and partly in solution is called *chyme*. Gradually, as the free acidity of the juice increases, the vigor of the movements increases. As the peristaltic waves reach the pylorus the latter relaxes a little and some of the more fluid portions of the chyme are thus squirted into the intestine, where they cause the secretion of the juices of this part of the canal in the manner shortly to be described. In this manner the stomach slowly empties itself and the food is passed little by little toward and through the pylorus. It happens at times that larger masses are carried along in the peristaltic wave, but when these masses strike the pylorus the latter closes down upon them and does not let them pass.

The various foodstuffs are acted upon in such a way by the gastric juice and the saliva that when the contents leave the stomach the proteins have been reduced for the greater part to the state of acid albumin and proteoses. Some polypeptides, tri- and di-peptides are also in the chyme.

Some protein particles have not been digested, however, and meat fibers, as such, may pass the pylorus to be acted upon and digested in the intestine. The nucleins are not digested. The fats, if already saponified, have been in large part already converted by the lipase of the stomach into fatty acids and glycerol, but the fats which are not emulsified, such as the fat of meat, are passed through the pylorus still as neutral fat. The starches have been in large measure digested by the amylase of the saliva and changed to dextrins and maltose, so that in the chyme both of these are found; and some of the cane sugar, if that has been eaten, has been inverted by the action of the acid of the juice and converted into glucose and levulose, but most of the carbohydrates are passed on into the intestine only partially digested, to await their final digestion in that organ.

The bacteria and parasites swallowed by the individual have been for the most part killed by the acid of the juice, so that the strongly acid chyme is nearly or quite sterile.

The study of the origin of the pepsin and hydrochloric acid, which are the principal active constituents of the juice, has shown that the pepsin is formed in the cells of the glands of the stomach and particularly the cells of the fundus region. It exists in the cell presumably in an inactive form called pepsinogen, and is made active by the action of the acid of the juice. What cells form the pepsin is still uncertain. The origin of the hydrochloric acid is still very obscure. It is formed from the chlorides of the blood and by the fundus part of the stomach. The probability is that it is formed in the neck of the gastric glands, and perhaps all along the surface of the mucous membrane, by the reabsorption (selective adsorption?) of the basic part of the chloride (which is possibly ammonia) leaving the acid on the outside. The stomach wall is impervious to acid. It is certain that the acid is not formed by the parietal cells, as was at one time believed, but for which there never was any good, direct evidence. The formation of the acid is a problem of great importance which must be left to the future to solve.

The acidity of the juice formed is subject to wide variation in disease, and the determination of the amount and character of the acid is of diagnostic value in some stomach disorders. The methods for the investigation of this acidity have already been described. In carcinoma of the stomach the acidity is generally below normal or absent; in ulcer the acidity is generally above normal.

The resistance of the stomach to self-digestion is due to the fact that the acid is not able to penetrate the living protoplast of the cells of the membrane; but when they are dead or when their resistance is reduced, it does so penetrate and then the pepsin digests the dead organ. The cells contain, also, a substance which checks peptic action, an antiferment.

Not all of the products of digestion pass out into the pylorus. An unknown proportion, but probably a small proportion, of the food is absorbed directly by the stomach, according to the most recent investigation.

REFERENCES. STOMACH DIGESTION.

1. **General.** *Beaumont:* Physiology of Digestion.
Pawlow: The Work of the Digestive Glands. Translation by Thompson.
Bidder and Schmidt: Die Verdauungssäfte. Leipzig, 1852.

2. **Acidity of gastric juice.** *Christiansen:* Untersuchungen über freie und gebun. HCl in Mageninhalt. Biochem. Zeits. 46, 1912, pp. 24, 50, and 287.
Toepfer: Eine Methode zur titrimetrische Bestimmung der hauptsächlichsten Factoren der Magenacidität. Zeit. f. physiol. Chem., 19, 1894, p. 104. (Previous literature cited here.)

3. **Secretion in human stomachs.** *Richet:* Comptes rendus de l'Acad. de Sci. 84, Gaz. Hebd. 187, No. 10. *Carlson:* A. Jour. Physiol. (Various papers, 1912-14.)

4. **Formation of HCl.** *Harvey and Bensley:* Biological Bulletin, 23, p. 225, 1912.
Sjöqvist: Skandinävische Archiv f. Physiol. 6, 1895, p. 255.
Maly: Sitzungsber. d. Wiener Akad. 69. Abth. II and III.
Schultz: Die Zerlegung der Chloride durch CO_2. Pflüger's Archiv 27, p. 454.
Lopez-Suarez: Biochem. Ztschr. 46, 1912, p. 490.
Ritter: Reabsorption of NH_3 by moulds. Biochem. Zeits. 42, 1913. p. 2.

5. **Analysis of gastric contents.** *Panton and Tidy:* Quarterly Journal of Medicine, 4, 1910-11, p. 454.

6. **Pepsin. Method of determination.** *Gross:* Berl. klin. Woch., 1908, 45, p. 643.
Reicher: Wiener klin. Wochenschr. 48, 1907.

7. **Gastric juice. Composition.** *Rosemann:* Pflüger's Archiv f. d. ges. Physiol. 118, 1907, p. 467.

8. **Passage of duodenal juices into the stomach.** *Boldyreff:* Pflüger's Archiv 121, 1907, p. 13; 140, 1911, p. 436.

9. **Pepsin and rennin.** Not influenced by Roentgen rays. *Richter and Gerhartz:* Berl. klin. Wochen., 1908, 45.
Action of alkalies. *Ticomiroff:* Zeit. f. physiol. Chem., 55, 1908.
Law of action. *Schütz:* Wiener klin. Wochenschrift, 21, 1908, p. 729.
Various pepsins. *Wroblewski:* Zeits. f. physiol. Chem. 21, 1895, p. 1.
Identity of rennin and pepsin. *Schmidt-Nielsen:* Zeits. f. phys. Chem., 48, p. 92, 1906.
Hammarsten: Zeit. f. physiol. Chem., 92, p. 121, 1914.
Pawlow and Parastchuk: Zeit. f. physiol. Chem., 42, 1904.

10. **Pepsin. Colloidal nature.** Recherches sur le mode d'action de la pepsine dans la digestion de l'albumine. Arch. Intern. de Physiol. 13, 1913, p. 316.

11. **Pepsin. Composition.** *Nencki and Sieber:* Zeits. f. physiol. Chem., 32, p. 291, 1901.
Pekelharing: Zeits. f. physiol. Chem., 22, 1896.

12. **Energy transformations in enzyme reactions.** *Lengyl:* Pflüger's Archiv f. d. ges. Physiol. 115, p. 7, 190.

13. Acid control of the pylorus. *Cannon:* Amer. Jour. of Physiol., 20, 1907, p. 283.

14. Excretion of pepsin in urine. *Wilenko:* Berlin klin. Wochenschrift 22. 1908.

15. Pepsin in infants. *Rosenstern:* Berl. klin. Wochenschrift 1908, No. 11.

16. **Isolation of pepsin by elastin.** *Abderhalden and Strauch:* Zeits. f. physiol. Chem. 71, 1911, p. 315.

17. **Protection of stomach against digestion.** *Katzenstein:* Berl. klin. Woch. 39, 1908.

18. **Other proteolytic enzymes in the stomach.** *Hirayama:* Zeits. f. physiol. Chem. 65, 1910, p. 290.

19. **The chemical methods of gastric secretion.** *Edkins:* Journal of Physiology, 34, 1906, pp. 132-144.

20. **Favorable hydrogen ion concentration for gastric digestion.** *Michaelis:* Die Wasserstoffionen Concentration, p. 70, Berlin, 1914.

21. **Gastrin.** *Moronescu:* Inter. Beit. z. Path. u. Therap. Ernährungsstörungen, 1, p. 194, 1910.

Ehrmann: Ibid., 3, p. 382, 1911-12. *Emsmann:* Ibid., 3, p. 117, 1912.

Tomazewski: Cent. f. Physiol. 27, 1913. *Keeton and Koch:* A. Jour. Physiol., 37, p. 481, 1915.

CHAPTER X.

DIGESTION IN THE INTESTINE.

Duodenal secretion.—The discharge of acid chyme into the intestine causes the secretion of duodenal juice. We have now to inquire how it causes it and what are the properties and uses of the juice thus poured out. This brings us to one of the most interesting of the unsolved problems of physiology: namely, the true functions of the duodenum.

The duodenum (Latin, *duodeni*, twelve each), or the twelve-finger intestine, as the Germans call it (Zwölffinger Darm), because it is about eleven inches or twelve finger breadths in length, is that part of the intestine extending from the pylorus to the jejunum, or the empty intestine, and it receives the two very important secretions of the liver and the pancreas as well as its own peculiar one. It is lined throughout with a mucous membrane which contains a large number of small glands, the glands of Brunner.

The mechanism of secretion of these glands has not been sufficiently studied, but they pour into the duodenum a large amount of a strongly alkaline, albuminous juice. The alkalinity of the juice is due to carbonates and it effervesces strongly on the addition of acid. It has by itself very weak digestive powers. It contains some invertin, so that it will invert cane sugar; and some erepsin; but its main function is to increase the power cf, or to work in conjunction with, pancreatic juice and bile.

The quantity of duodenal juice secreted normally in man cannot be stated, but from a duodenal fistula of a loop cut off from the intestine and the stomach an extraordinarily large amount of a clear, colorless, alkaline (H ion content about 2×10^{-8} N) strongly albuminous fluid is discharged. In a dog weighing 5 kilos, 50 c.c. were secreted in 120 minutes. It does not seem possible, or probable, that the secretion can under normal circumstances be as large as this.

Enterokinase.—The duodenal juice is remarkable because of a substance in it called *enterokinase*, meaning literally the active substance of the intestine, which has the property, when mixed with pancreatic juice, of enormously increasing the action of the latter on proteins. If, for example, three samples are prepared in three test-tubes, one of pure pancreatic juice, one of pure duodenal juice and one of a mixture of equal parts of duodenal and pancreatic juice, and a piece of fibrin or other protein is introduced into each of the three, the mixture digests the protein at once, while the pure juice in each case leaves it almost unaffected.

Enterokinase is found in the mucous membrane of the duodenum and may be extracted with dilute bicarbonate solution. It is destroyed by heating. It is found in the intestines of all the vertebrates. In the dog-fish (Elasmobranch), Mustelus canis, one finds none of it in the stomach mucosa, but it is found all along the intestine mucosa nearly to the rectum, but the larger quantities are found in the upper part. It is supposed to act by converting an inactive trypsinogen in the pancreatic juice into active trypsin, but the evidence of this is still uncertain, although this interpretation is probable. It is claimed by some that enterokinase, or substances of a similar property, are found in leucocytes and other cells, but this is denied by Bayliss and Starling. It is certainly found nowhere else in the dogfish than in the mucous membrane of the intestine. A small amount of enterokinase is able to increase the digestive activity of the pancreatic juice on proteins very strongly, and for this reason it is thought to be a ferment itself.

Other enzymes in duodenal juice.—Besides containing enterokinase, the duodenal juice is important in digestion on account of its alkaline reaction, by which it aids in the neutralization of the acid chyme coming from the stomach, and because of the carbohydrate enzymes *invertin*, *maltase* and *lactase* it contains, which give it a powerful action on the disaccharides cane sugar, maltose and lactose, converting these to monosaccharides.

Other functions of the duodenum.—There are also certain facts about the juice which are well deserving of farther investigation. It seems to be necessary to life. Dogs live for long periods if the bile and pancreatic juice are diverted from the body by fistulas; but if the duodenal juice is completely diverted to the outside, they rarely live more than 48 hours. If, for example, a ligature is placed about the pylorus, so that duodenal juice cannot go back into the stomach, and the duodenum is cut away from the intestine about six inches from the pylorus, a gastro-enterostomy being made so that the food can pass from the stomach to the intestine, and if the duodenal sac thus made is drained to the exterior by a fistula, or if the duodenum be taken out entirely, no serious symptoms show themselves for 24 or 36 hours, but very shortly afterwards the dogs die. If, however, the duodenum is left in connection with the pylorus, being cut off at its lower end, so that the juice can get back into the stomach and so into the intestine; or if, after tying both ends of the duodenum, a rubber tube is connected with the duodenum and the jejunum so that the juice can pass along it, no serious results follow the operation. Indeed, extirpation of the first six inches of the dog's duodenum is invariably fatal in the experiments reported. Similar fatal results have been reported in human beings following resection of the duodenum. The cause of death is still obscure. Whether it is due to the rapid secretion of some necessary substance through the duodenum to the exterior; or whether it is due to the duodenum making something which is necessary to the functioning of the gut lower down, or whether it is due to some other cause, is quite unknown.

Excretory function of duodenum.—The duodenal juice may have in it not only its normal constituents, but many substances are excreted here. Thus sugars of various kinds put directly into the blood are excreted into the duodenum and reabsorbed farther down. This fact makes the interpretation of metabolism experiments in which sugars are injected directly into the blood very difficult, because one cannot easily tell whether the sugar is used directly, or only indirectly after its excretion and reabsorption. So, also, morphine is excreted here, and potassium ferrocyanide and many metals when they are injected subcutaneously, or intravenously; indeed, this part of the intestine is an important excretory organ, which functions especially in cases of renal insufficience.

Pancreatic juice.—Closely applied to the duodenum, or lying in the mesentery by its side, is the pancreatic gland, a digestive organ of the first importance which pours its secretion into the duodenum through two main ducts, the duct of Wirsung and the duct of Santorini. There may be, also, subsidiary ducts between these two. The latter of these ducts opens in human beings about four inches (9-10 cms.) below the pylorus; the former, in common with the bile duct, about five and three-fourths inches below the pylorus.

The pancreas, like the other digestive glands, does not secrete constantly, but only intermittently when its secretion is needed. Its strongly alkaline secretion gushes forth, together with the bile and the duodenal juice, when the acid chyme is squirted through the pylorus into the intestine, and it is continued until the acidity of this chyme is neutralized.

Composition of human pancreatic juice.—Human pancreatic juice has been obtained from artificial pancreatic fistulas. The juice is alkaline to litmus, due to the presence in it of sodium carbonate; it is as clear as water and coagulates on heating. The pancreatic juice of the dog obtained on stimulating the vagi or by the action of pilocarpine may contain so much protein that when heated it forms a solid coagulum. The freezing point of human pancreatic juice is above that of the blood: i.e., —0.42° to —0.49°. To neutralize it, using litmus as an indicator, by titration it requires for 1 c.c. of juice 0.1-0.15 c.c. of N/10 HCl. The neutralized juice becomes turbid at 47° C. and coagulates between 57° and 59° (Wohlgemuth). It is precipitated by an equal volume of saturated ammonium sulphate. The following is the composition of the juice examined by Glaesner and Wohlgemuth:

	Glaesner	Wohlgemuth
Water	98.72	98.70
Solids	1.27	1.30
Coagulable protein	.174	0.093
Nitrogen	0.0983	0.0813
Alcohol soluble substances	0.508	0.523
Specific gravity	1007.5	1007.13

Ash:

K	1.10%	Cl	50.75	P_2O_5	1.85
Na	36.65	SO_3	2.05	SO_2	0.34

Traces of Ca, Mg, Fe, SiO_2, CO_2 0.11%

This is the secretion from a permanent fistula. The secretion of the first juice secreted from the dog's pancreas on opening the duct is far more concentrated than this. It may contain nearly 10 per cent. of solids, 9 per cent. being organic matter. The concentration of hydroxyl ions in dog's pancreatic juice is equal to N/10,000 (Foa).

Control of the secretion of the pancreas.—How the secretion of the pancreas is controlled brings us to one of the most interesting of recently discovered facts concerning the chemical messengers of the body. To understand what follows we must have some knowledge of the structure, blood and nerve supply of this vital organ.

It is a relatively small organ weighing in the human adult about 87 grams. It is a tubular gland, the secreting cells being in acini and they are filled, or more than half filled, with granules which during secretion disappear from the cells and are replaced by a clear non-granular protoplasm containing mitochondria. Figure 42. The gland in the guinea pig and rabbit is very thin and spread out in the mesentery; and with a little care a loop of the intestine containing the organ in the mesentery may be so placed on a microscope stage that the gland may be watched secreting in the living state. The blood may be seen circulating about the base of the cells and the granules are clearly visible near the lumen. No structure could be seen by the author in the bases of these cells, but only a clear protoplasm, and the granules could not be seen to move in the cell itself. On stimulation of the vagus nerve the bases of the cells seemed to be a little more indented between the cells, but no other change was visible. Mitochondria may, however, be seen in the base of the living cell when appropriate methods are used.

The blood supply is from the pancreatic artery and the blood returns by the pancreatic vein into the mesenteric and ultimately into the portal vein, so that the blood after passing through the pancreas goes through the liver. There is a double nerve supply, fibers coming both from the vagi nerves, sometimes from one more than the other, and from the splanchnics. Stimulation of these nerves under certain conditions causes secretion from the pancreas. With this preliminary statement we may approach the problem of how the presence of chyme in the intestine causes the pancreas to secrete.

Secretin.—Since most organs are controlled by nerves, one naturally thinks of a nervous reflex. The acid chyme may be supposed either

FIG. 42.—Section of pancreas of mouse. Osmium bichromate (Mathews).

directly, or indirectly by changes it induces in the mucosa, to stimulate sensory nerve ends. Everyone knows how acids will stimulate such endings or such nerves in the skin and they may act in the same manner in the intestine. Such impulses may ascend the vagi nerves to the brain and be reflected back over the vagi, or splanchnics, or both, causing a dilation of blood vessels in the organ and a secretion of its stored juice. This explanation appeared quite satisfactory until it was discovered that after all nerves going to the pancreas were cut, putting acid into the duodenum still caused secretion from the gland. The first interpretation of this unexpected result was that there must be a local reflex mechanism; that the impulses did not need go to the brain, but were reflected through a local ganglion and so to the nerves of the organ. It is very difficult, if not impossible, to prove that this is not the case, but it was very soon found that certainly another mechanism different from this was called into play. Bayliss and Starling discovered that an acid infusion of the mucous membrane of the duodenum and the intestine when neutralized and injected into the blood caused secretion from the pancreas and liver, whereas acid infusions of other organs of the body had no such action. There was no doubt that a substance existed in the mucous membrane of the intestine, duodenum and jejunum, which could be extracted by boiling dilute acid, which was not protein or coagulable, which was soluble in alcohol, and which caused, when injected into the blood, secretion of the pancreas and liver. It caused, also, a marked fall of blood pressure and a diminution of coagulability of the blood, although these effects may be due to another substance than that causing secretion. This substance, stimulating secretion, they called "secretin" because of its action on the pancreas. This discovery suggested at once the possibility that the presence of acid chyme in the intestine sets free secretin in the mucosa; and that some of this secretin gets into the blood and makes the pancreas secrete.

The fact that such a substance exists in the mucosa does not necessarily mean that the substance is actively engaged in the normal secretion. To prove this it was necessary to prove that blood from animals actively secreting pancreatic juice really contained enough secretin to cause secretion. This proof was obtained by making a cross circulation in two dogs, by anastomosing the arteries with the veins. Cannulas were put in the pancreatic ducts of both dogs and then acid was placed in the duodenum of one dog. Under these circumstances both pancreases secreted. This experiment shows that secretin actually enters the blood when acid is put in the duodenum and that the amount is sufficient to cause the pancreas to secrete. Bayliss and Starling propose to call such chemical messengers as secretin, which arouse other organs of the body, "hormones," from the Greek, *hormōn,* "I rouse to activity." Adrenalin is another hormone which in some particulars resembles secretin.

It has been incorrectly inferred by some authors that this discovery means that the nervous system is not concerned in the secretion of the pancreas, but this is of course not true. It is unlikely that the nervous mechanism plays no part in the secretion. The existence of this mechanism can be proved by directly stimulating the nerves, causing thereby a copious secretion. Of course in this case, just as with the supra-renals and splanchnics (page 774), it is very difficult to prove that the secretion accompanying vagi stimulation is not due to the nerves acting on the duodenum or stomach so as to cause secretion of secretin, which then indirectly arouses the pancreas. A certain argument from analogy may be made for this conclusion. For example, it is known that when the sympathetic is cut and degenerates, organs supplied by this nerve become more than usually sensitive to the action of adrenalin, so that after division of the cervical sympathetic on one side the pupil of the eye on that side will dilate after amounts of adrenalin have been injected too small to affect the normal pupil. The pancreas shows a somewhat similar action. The vagus does not, in normal circumstances, cause secretion from the pancreas when it is stimulated, but if it is cut first and then stimulated after degenerating for two or three days, it often does cause a secretion on stimulation. It might be, therefore, that the gland after division of the vagus became hypersensitive to secretin. But while it is possible that the vagus causes secretion only indirectly in that it arouses the duodenum to secrete secretin, there is as yet no evidence that this is the case, and the analogy with adrenalin would indicate that secretin is probably acting to reinforce a nervous mechanism rather than to supplant it. There are several differences between the secretion due to secretin and that due to stimulation of the vagus. One is that, if the first dose of secretin injected is large, very little subsequent secretion is produced by subsequent doses. Nothing of this sort happens on stimulating the vagi. These cause secretion, under favorable circumstances, as often as they are stimulated and the secretion is copious. In this respect secretin resembles adrenalin, which in any but very small doses produces little effect on the second or third injection. Secretin causes secretion after atropin has paralyzed the gland so that nerve stimulation no longer causes secretin. The character of the secretion is different in the two cases also, the secretin secretion being more dilute with less organic matter than that obtained on nerve stimulation. In view of these facts, while no definite conclusions can be drawn without further investigations, it seems more probable that both chemical and nervous mechanisms are involved in the secretion of the pancreas and that part of the result of introducing acid chyme into the duodenum is due to a nervous reflex through the vagi, and part to the production of secretin by the direct action of the acid on the mucosa. There is, indeed, a remote possibility which has not been investigated so far as the author knows, namely, that secretin may produce its secretion from a different tissue of the gland than that of the nerve stimulation. There are two tissues in the gland, duct tissue and secreting acinary tissue. The cells of these tissues are different in their physical appearance and they react differently when the ducts are plugged. If a fat with high melting point is injected into the ducts, thus plugging them, the acinary tissue degenerates, but the duct tissue remains, so that the pancreas becomes, as Bernard says, like a tree without its leaves, all the small ducts and ductules standing out with all the acinary cells gone (Figure 63, page 782). It is not known whether the duct tissue contributes to the secretion and, if so, what constituents.

Secretin; chemical nature.—The chemical nature of secretin is still unknown. An acid extract of the mucosa contains, according to Dale, β-imidazolylethyl amine, a substance derived from histidine by splitting off carbon dioxide as follows:

```
HC—NH\            HC—NH\
||     >CH        ||     CH
C—N               C—N
|                 |
CH2               CH2
|                 |
CHNH2             CH2NH2
|                 β-imidazolylethyl
COOH                 amine.
Histidine.        Histamine.
```

According to some β-imidazolylethyl amine is identical with vasodilatin. It causes, certainly, a fall of blood pressure and a lowered coagulation of the blood, but this identity is denied by others. Whether secretin will act in the absence of any depressor substance in the extract is still uncertain, so that secretin and vasodilatin or imidazolylethyl amine are possibly not identical. Secretin is stable in acid solution, but readily oxidizes in neutral or alkaline solution and is destroyed. These are resemblances to adrenalin, which is apparently a body of the same kind: that is, a basic substance possibly derived from some amino-acid by the splitting off of carboxyl. It is not the only substance causing secretion of the pancreas. Thus pilocarpine has such an action and so also has Witte's peptone and curarin. It is asserted by Popielski that other tissues than the intestine will also yield extracts, vasodilatin, to acids which, when neutralized, will cause secretion of the pancreas on injection. It may be mentioned in this connection that, according to the author's observations, adrenalin causes secretion of the salivary glands, but not of the pancreas. Secretin may be extracted from the mucosa of the upper part of the intestine, after the mucosa has been hardened in $HgCl_2$, by boiling, rejecting the filtrate, extracting the residue with 2 per cent. acetic acid containing 1 per cent. of $HgCl_2$ and precipitating the filtrate by the addition of NaOH nearly to the neutral point. The white flocculent precipitate is treated with H_2S, filtered, the filtrate boiled to free from H_2S. The filtrate is very active (Dale and Laidlaw).

Digestive functions of the pancreas. Action on fats.—The greater part of our knowledge of the fundamental facts of the digestive action of the pancreas we owe to the great French physiologist, Claude Bernard, whose *Mémoire sur le pancreas* is one of the classics of physiology. Bernard undertook an investigation into the comparative physiology of digestion. His own words on the discovery of the action of the pancreas are as follows:

"During the winter of the year 1846 I studied the digestion of different alimentary substances in different carnivorous and herbivorous animals. Having fed fatty substances to dogs and to rabbits, I followed the physical or chemical changes fitting them for absorption that the different substances underwent in different parts of the alimentary canal. I perceived, in opening the intestines of animals, that in dogs the fatty substances were emulsified and absorbed by the lacteals from the commencement of the intestine and nearly from the pylorus; while in rabbits the phenomena became evident only very much farther down at a distance of 30 to 50 cms. from the pylorus, depending on the size of the animal. Struck by this difference, I sought with care to see if there was not some constant, particular, anatomical difference in this region between the two species. I determined in fact that in the dog the two pancreatic ducts opened very high up in the intestine, at the commencement of the duodenum, in the neighborhood of the choledochal canal; while in the rabbit, on the contrary, the principal pancreatic duct opened much lower than the bile duct and precisely at that point where I had seen that the emulsification of fatty matters commenced with great intensity. By this coincidence I was very naturally led to infer that the pancreatic secretion must have the property of so altering fats as to make them absorbable.

"It remained to test this hypothesis by experiments with the secretion of the pancreas. To this end I began experiments to obtain pancreatic juice and I was able to determine that the liquid possesses in truth the special property of emulsifying instantaneously the fats and of acting upon them chemically in a very remarkable manner."

Bernard found that pancreatic juice added to neutral olive oil formed almost instantaneously a permanent emulsion, the oil not separating as a layer; and that, if extract of litmus were added, the solution could be seen to have become distinctly acid. This acidity was due to the fatty acids which were set free from the neutral fats. The first property then to be noticed of pancreatic juice is *that it splits and emulsifies fats*. This property it loses if boiled first. It is due to an organic substance of an unknown nature contained in the juice which is called steapsin; and since it splits fats it is one of the group of fat-splitting enzymes called lipases, or lipolytic enzymes. The steapsin of the pancreas is the chief, or most important, enzyme in the digestion of fats. By it the neutral fats are split into glycerol and fatty acid by hydrolysis according to the following equation:

$$C_3H_5O_3(C_{18}H_{35}O)_3 + 3H_2O + \text{Steapsin} = C_3H_5O_3(C_{18}H_{35}O)_3.3H_2O.St =$$

Intermediate hypothetical stage.

$$C_3H_8O_3 + 3C_{18}H_{36}O_2 + \text{Steapsin}.$$

Glycerol. Stearic acid.

Steapsin. Origin.—The steapsin found in the juice is formed and stored in the tissue of the gland from which it is secreted. If a little piece of fresh pancreas is placed in a weakly alkaline $NaHCO_3$ emulsion of olive oil containing blue litmus, it will be seen under the microscope that very quickly the piece of tissue becomes surrounded by a red aureole or halo, due to the fatty acid set free from the fat. The piece of tissue may be put into a fat emulsion in gelatin colored as above. The lipase causes the gelatin to turn red. Pancreatic tissue differs from all other tissues in this property. The lipase may be extracted from the gland by glycerine or by water, but one of the best ways is by the use of 60 per cent. alcohol. The fresh gland shaken with alcohol of this concentration gives a maximum yield of lipase. It is probable that the steapsin is formed in the pancreas itself. It is supposed to be formed by the acinary cells, but decisive experiments in this regard do not seem to have been carried out.

Conditions of action of steapsin.—Pancreatic juice works usually in the presence of bile and it is not surprising, therefore, that it has been found that the addition of bile greatly increases the rate of splitting of fats by pancreatic steapsin. The following experiments show this fact: The figures give the c.c. of N/10 NaOH required to neutralize the fatty acid formed by the steapsin from the neutral fat with and without bile. The results also show the relative activity of glycerol extracts filtered or not filtered, or of the filter residue, and of suspensions of the gland. The extract was made by allowing glycerol to stand 14 days on pigs' pancreas. The filtration was through paper.

C.c. of N/10 NaOH required to neutralize equal mixtures of oil and pancreatic extracts.

	Without bile	With bile
Gland suspension	10.9 c.c.	21.7 c.c.
1st filtrate	7.4	18.0
Clear filtrate	6.9	14.6
Filter residue	14.2	32.6

In the presence of bile the amount of fatty acid formed was nearly doubled. The steapsin does not easily pass through a porcelain filter, at least the first filtrates are very poor in lipase. This is called an adsorption of the lipase by the filter. The filter soon becomes saturated, however, and thereafter the filtrate has a normal concentration of lipase. This is shown in the following experiment: A glycerol extract was made by extracting an alcohol-ether extracted pancreas with glycerol for 1-3 days. 100 c.c. glycerol was added to 1 gram of dried pancreas. This was filtered through paper until clear. The acidity was tested by taking 5 c.c. of the glycerol extract, 5 c.c. of a 1 per cent. solution of Platner's bile and 5 c.c. of olive oil, which were shaken and left for 20 hours. Then the acidity was determined by titrating with N/10 NaOH with phenolphthalein as indicator. The c.c. of NaOH required were the following:

1.	Filtered rapidly through paper	34.1 c.c.
2.	1st filtrate through Chamberlain filter	2.0
3.	2d " " " "	11.5
4.	3d " " " "	32.5

That the lipase of the pancreas differs from some other lipases is shown by the fact (discovered by Rosenheim and Shaw Mackenzie) that it attacks normal glycerides very much better than the lighter, normal ethyl butyrate. It is apparently made of two parts. If it be filtered through paper, it is said that the filtrate is inactive and the part on the filter paper is inactive. If they are reunited they become active. The co-enzyme is heat resistant, dialyzable, not soluble in alcohol or ether and its activity is lost on incineration. Sodium cholate activates the inactive enzyme. Glycerylphosphoric acid has also a remarkable action in activating, and this may account for the results of Hewlett, who ascribed such powers to lecithin.

Steapsin is not extracted by glycerine diluted with water from the pancreas which has previously been extracted by alcohol and ether and dried.

Inorganic salts do not increase, but instead they always decrease lipase activity, unless soaps are present in sufficient quantities to inhibit the steapsin. If this is the case, the addition of salt may appear to stimulate the action (Hanisik). The action is very dependent on the degree of alkalinity; the digestion goes best in a solution very faintly alkaline with a hydrogen ion content of $N \times 10^{-8}$, it is very poor in weak acid or stronger alkali. Cholesterol diminishes the action and the addition of hemolytic substances, such as the saponins, hastens the action of the lipase.

The favoring action of bile on the steapsin is not explained. It will be shown when discussing the bile that it is the bile salts which aid the lipase. One explanation of the manner of their action is that by the addition of these salts, which are the salts of weak acids, the change in acidity is reduced, the bile acting the part of a buffer so called, a substance like glycocoll which prevents any large change in the concentration of hydrogen ions. Were this the whole explanation of the action of bile acids, however, sodium acetate or sodium phosphate or glycocoll should act as well, since they have the same power of controlling acidity. But they do not accelerate lipase. Bile also acts by aiding the emulsification of the fat in the manner described under Bile, page 412. By this means the surface of contact between the fat and water is greatly increased, which has the effect of increasing the concentration of the fat. Lipase is certainly soluble in water containing glycerol and it either dissolves in the fat or unites with it, concentrating itself in the layer of separation of fat and water. It probably dissolves in the fat, since some may be extracted from the pancreas by ethyl butyrate.

Steapsin; its manner of action.—Steapsin, like most or all enzymes, becomes more resistant to heat, light and other reagents, in the presence of fat, its substrate. It may be inferred from this, and by the fact that fat will extract or shake it from water, that steapsin unites with neutral fat. The most probable point of union is with the double-bonded oxygen by means of the reserve or extra valences on the oxygen, as the oxygen of most esters is quadrivalent. This union protects the steapsin from

$$
\begin{array}{l}
\quad\;\; H \qquad\quad O = \text{Enzyme} \\
\quad\;\; | \qquad\quad\;\; \| \\
H-C-O-C-(CH_2)_{16}-CH_3 \\
\quad\;\; | \\
H-C-O-C-(CH_2)_{16}-CH_3 \\
\quad\;\; | \qquad\quad\;\; \| \\
\quad\;\; | \qquad\quad\;\; O \\
H-C-O-C-(CH_2)_{16}-CH_3 \\
\quad\;\; | \qquad\quad\;\; \| \\
\quad\;\; H \qquad\quad\; O
\end{array}
$$

Neutral fat-enzyme compound.

toxic influences. It leads to the breaking of the bond between glycerol and fat and the entrance of water. The action is hence an hydrolysis.

Action of pancreatic juice on carbohydrates.—Pancreatic juice, besides its emulsifying and splitting action on fats, has a very powerful starch-dissolving action. It has an amylolytic power. An aqueous or salt extract of the fresh pancreas of any vertebrate when added to a starch solution causes an almost instantaneous clearing of the solution. Instead of being opalescent, the starch becomes quite transparent; there is a marked fall in viscosity; and, if the solution is tested by Fehling's solution, it is found to have acquired a strong reducing action, due to the appearance of maltose. This action of the juice is lost if it is boiled; and a very small amount of the juice is able at ordinary temperatures to dissolve and digest a large amount of starch. We say, therefore, that the juice contains a catalytic agent, or enzyme, which converts starch into maltose. The enzyme is called "amylopsin."

Pancreatic juice is thus in most carnivora, in which the saliva has very little amylolytic power, the most important agent for the digestion of starch, or of glycogen in their food. The course of the digestion is much the same as in the case of ptyalin, but differs from it in certain particulars. The digestion goes best in both cases in very weak acid and the optimum temperature is about the same in each. Bile and duodenal juice appear to affect the amylolytic action of the juice very little, or not at all. The starch is converted into various dextrins, and these finally into maltose. Pancreatic amylopsin differs from ptyalin in the fact that with amylopsin the conversion of the starch into dextrin goes at a rate relatively greater than with ptyalin and the reducing sugar appears more slowly than with ptyalin. In fact, extracts of the pancreas are sometimes found which convert starch into soluble starch and dextrin with great speed, but which have little or no saccharifying power. For these reasons many observers are convinced that there are at least two different enzymes in amylopsin and perhaps in ptyalin, one an amylase which converts starch into dextrin, and one or more dextrinases which convert dextrin into maltose. These two enzymes are generally associated, but it may happen that they may exist separately or the dextrinase may be destroyed wholly or partly, leaving the other active. Similar facts are alleged for malt diastase. It is said that by heating malt diastase with $CaSO_4$ the dextrinases may be largely, or completely, destroyed, while the amylases remain, and this process has been patented for its application to brewing.

The pancreatic amylopsin is apparently the source of most of the amylolytic enzyme of the blood and urine, being either reabsorbed directly from the gland itself or from the intestine after secretion into the latter.

If dialyzed, pancreatic juice, like saliva, entirely loses its amylolytic action, but the addition of salts such as $CaCl_2$, NaCl, etc., to the juice restores its action completely. A possible explanation of this fact is that the active principle is of a globulin-like character insoluble in distilled water, but soluble in dilute salt solutions. The matter needs further investigation.

Preparation and properties of amylopsin.—Very active preparations of pancreatic amylase have been prepared recently by the following method (Sherman and Schlesinger): Mix thoroughly 20 grams of pancreatin powder with 200 c.c. 50 per cent. alcohol at 15-20°. Allow to stand 5-10 minutes, then filter, keeping the temperature below 20°. It takes 1-2 hours to filter. Pour the filtrate into 7 times its volume of a mixture of 1 part alcohol to 4 parts ether. Within 10-15 minutes the enzyme separates out as an oily solution. Decant supernatant liquid. Dissolve the precipitate in the smallest amount of pure water at a temperature of 10-15° C. and reprecipitate at once by pouring into 5 volumes of absolute alcohol. Allow to settle, keeping temperature low; filter, dissolve in 200-250 c.c. 50 per cent. alcohol containing 5 grams of maltose. Pour solution into 500 c.c. collodion sack and dialyze against 2,000 c.c. 50 per cent. alcohol at not above 20° C. and preferably not below 15°. Replace dialyzate twice after 15 hours and a second period of 8-9 hours with fresh 50 per cent. alcohol. Continue dialysis 40-42 hours. Filter. Pour clear filtrate into an equal volume of 1:1 alcohol-ether mixture. Filter in cold and place at once in a vacuum desiccator. The elementary analysis of the product on the basis that P was an integral part of the molecule was: C, 51.9; H, 6.6; N, 15.3; S, 1.0; P, 0.8; O, and undetermined, 24.4. The substance is evidently a protein. It coagulates on heating at 68-70°. .000,002,5 gram of the powder digested 5 c.c. of a 1 per cent. solution of Kahlbaum's soluble starch to a pure wine-red reaction with iodine in 30 minutes at 40°. This on Wohlgemuth's scale $D\frac{40°}{30'}$ =2,000,000. The weight of the starch digested was 20,000 times the weight of the enzyme. It later digested 800,000 times its own weight of starch in 30 hours to a red color, and in 72 hours to the achromatic point. It formed 455,000 times its own weight of maltose. In strong solutions it formed glucose as well as maltose. It acts only in the presence of electrolytes. It loses activity very fast on standing in pure water at room temperature, 45 per cent. activity being lost in 20 minutes. This preparation was not pure. It was more active proteolytically than the original pancreatin. It gave all the typical protein reactions, biuret, Millon, xanthro-proteic and tryptophane. After coagulation the coagulum took a violet color in the biuret test, the filtrate a rose-red color.

The pancreas appears not only to cause the digestion of starch by its own action, but it seems from the work of Roger and Simon to reactivate in the intestine the ptyalin which has been rendered inactive but not destroyed in its passage through the stomach.

Lactase and maltase.—The pancreatic juice acts also in children and young animals on lactose; it contains a lactase, an enzyme which splits lactose by the addition of water into galactose and glucose. This enzyme is absent in adult animals, except those like human beings and pigs which continue to ingest milk. The attempt to make the enzyme reappear in the glands of adult herbivorous animals which do not normally have milk by feeding milk has been thus far a failure. Invertin has sometimes been found in the juice, but at other times it has been missed. It appears then not to be a constant constituent.

Maltase, which converts maltose to glucose, is also present.

Action of pancreatic juice on proteins.—Not only does pancreatic juice act upon fats and carbohydrates in the chyme, but it also, when mixed with duodenal juice, has a powerful action on proteins, although, as secreted from the duct, it has very little action on proteins. Unmixed with succus entericus it will digest casein, but not coagulated proteins. But when it is mixed with duodenal or jejunal secretion, the mixture digests nearly all proteins at a very rapid rate and this mixture practically completes the digestion of the proteins which was begun by the stomach. Claude Bernard discovered that pancreatic juice would digest some proteins, but the action was very slow. He found that the digestion differed from the peptic digestion in two particulars: first, it went on best in a weakly alkaline medium and was stopped by acid;

and, second, in a tryptic digestion the digestive mixture in a day or so acquired the property of giving a violet color when bromine or chlorine water was added to it. He found no other secretion which gave this reaction, so that he considered it unique and characteristic of tryptic digestion. This reaction was called, therefore, the *tryptophane reaction* (Gr. *tripsis*, a rubbing; *phanos*, bright). It is now known to be due to the formation of free tryptophane, indol-amino propionic acid. These three discoveries of Bernard's were fundamental: first, that pure pancreatic juice was a weak digestant of proteins; second, that it worked in an alkaline medium; and, third, that by it other digestive products were formed than by peptic digestion.

After this work by Bernard further work appeared to show that the action on proteins was very intense. Thus, if the pancreas be dried, extracted with ether, and then treated first with salicylic acid solution for 24 hours at 38° C. and thereafter by Na_2CO_3 solution by Kühne's method, an extract very active on proteins was obtained. It was not until 1900 that Bernard's early observations were cleared up. It was found that the pure juice has indeed a weak action on proteins, as Bernard observed, but that if it is mixed with only a small amount of duodenal juice, or with an extract of the intestinal mucosa, the mixture becomes extremely powerful and digests proteins, even coagulated proteins, rapidly and so completely that monoamino-acids are formed and very little albumose or peptone remains. It is the enterokinase of the intestine which acts in conjunction with the pancreatic juice to digest proteins. The active principle of the pancreatic juice thus acting on proteins is called " trypsin " (Gr. *trypsis*, a rubbing; referring to method of preparation). It is probable that it is not a single enzyme, but that there are at least two, one acting on the complex proteins, for which the name trypsin may be retained, and the other of an ereptic nature, acting on the albumoses and some natural proteins like casein, and this might be called " pancreatic erepsin." It is probably this latter which has the power of clotting milk, or a rennin action, in the absence of intestinal juice.

Since pancreatic juice is inactive as it is secreted, but when mixed with intestinal juice digests rapidly, it is believed that trypsin does not exist as such in the gland, but in an inactive form called trypsinogen. It is ordinarily stated that the enterokinase acts by activating the trypsinogen or converting it into trypsin. It is well, however, to bear in mind the actual facts, rather than any interpretation of them, and the facts are that a mixture of pancreatic and intestinal juices digests proteins much more rapidly than either juice alone, and that the active principle found in the pancreas is called trypsin and that in the intestine is called enterokinase. But exactly how they support each the action of the other, cannot be stated at present and further work is required. The facts at present known, while not very convincing, indicate that enterokinase may be setting trypsin free from some union in which it is inactive, but other interpretations than this are possible. All observations on the digestive power of pancreatic juice made before the discovery of the rôle of enterokinase have to be interpreted with this fact in mind.

It is generally stated that digestion by the pancreas is most rapid in the presence of hydroxyl ions of a concentration of N/200 or in a slightly alkaline medium (.3 per cent. Na_2CO_3). It is a singular fact that, if this is true, the digestion must not take place in the alimentary canal under optimum conditions, for the reaction of the chyme is often weakly acid to litmus for a considerable distance along the intestine. It is possible that there are two or more enzymes of which the first, trypsin, working on coagulated protein, may be most active in a weakly alkaline medium, whereas the action on the albumoses by the erepsin may be best in a neutral or very faintly acid medium. This is a matter requiring further work. Rachford states that the bile, and particularly the chyme as it comes from the stomach, exerts a favorable action on tryptic digestion. He collected pancreatic juice from the pancreatic duct in rabbits. From his results, which antedated the discovery of enterokinase, it appears that the digestion is at an optimum under the conditions of the admixture of bile, duodenal juice and chyme with the pancreatic juice, such a mixture as exists in the intestine. The experiments made by Abderhalden and others on the digestion of artificial polypeptides have been made with juice containing both enzymes. Furthermore, natural, that is unactivated, pancreatic juice will hydrolyze peptone, casein and fibrin by means of the erepsin it contains. Besides enterokinase, it is maintained by Henri that pancreatic juice is activated also by calcium salts. The results on this are, however, diverse and from recent work it appears that calcium hastens the action of trypsin after it is formed and hastens the action of erepsin, but does not activate trypsinogen. The explanation of the action of the salt is, however, still largely guesswork.

Products of the action of trypsin-enterokinase on proteins.—The digestion of the natural proteins by the mixture of juices in the duodenum passes through much the same stages as peptic digestion, except that the decomposition of the peptones and albumoses into di-, tri- and mono-peptides takes place very much more rapidly than in the case of pepsin. The pancreas has an ereptic as well as a peptic power. As a result, if the digestive mixture is examined for the primary proteoses, very little will be found in the case of pancreatic digestion as compared with peptic. Furthermore, tyrosine, leucine and other amino-acids are produced in great amounts and crystallize out of the digesting mixture; these products are not formed by peptic digestion. It is uncertain whether these are produced by trypsin or by erepsin. No method of separating these enzymes is known.

The tryptic digestion differs, on the other hand, from the autolytic digestions of organs, that is from their self-digestion, in that little ammonia is formed in the course of tryptic digestion, whereas much is produced in autolysis. This, as has been pointed out, has the teleological explanation that acids set free in cells autolytic ferments and these form ammonia to neutralize the acid. There is no doubt that a good deal of ammonia is produced either by the action of digestive ferments or by the bacteria in the course of intestinal digestion, for the mucosa of the intestine contains much ammonia.

The digestion by trypsin-enterokinase does not, however, reduce all the polypeptides to amino-acids. There still remains in a trypsin digestion mixture a considerable number of carboxyl-amino linkages, since if a tryptic digestive mixture is titrated with formol in the Sorensen method before and after hydrolysis with acid, it is found that the number of free amino groups is considerably increased by the action of the acid, showing that some NH-CO linkages had not been split by the trypsin-enterokinase. This final step in the reduction of the proteins to the simplest building stones, the amino-acids, is brought about by the enzyme, erepsin, found in nearly all tissues, but in larger amounts in the intestinal mucosa and to some extent in the intestinal juice. Trypsin is, however, able to break many amino-carboxyl linkings, or, at least, activated pancreatic juice is able to do this, as is shown by the work of Abderhalden and his colleagues. They found the artificial poly peptides were digested or not digested by this mixture as shown in the table on page 404.

Whether the splitting was due to the trypsin-enterokinase or to the erepsin in the juice could not be positively stated. An examination of the appended list shows that the property of being digested by this mixture of enzymes depends on several properties of the molecule. In the first place (A) it depends on the structure of the molecule. Thus alanyl-glycine is hydrolyzed, but glycyl-alanine is not hydrolyzed; d-alanyl-l-leucine is digested, but alanyl-leucine (B), which was probably l alanyl-l-leucine plus d-alanyl-d-leucine, was not digested. The first peptide contains those active amino-acids found in nature. l-leucyl-l-leucine, the natural leucine, is digested, while l-leucyl-d-leucine and d-leucyl-l-leucine are not split. Dakin found, too, recently that casein and other proteins which have been racemized by dilute alkali are no longer digested by trypsin.

ARTIFICIAL POLYPEPTIDES HYDROLYZED AND NOT HYDROLYZED BY PURE ACTIVATED PANCREATIC JUICE (Fischer and Abderhalden).

Hydrolysed	Not hydrolysed
(r) Alanyl-glycine	Glycyl-alanine
(r) Alanyl-alanine	Glycyl-glycine
(r) Alanyl-leucine (A)	Alanyl-leucine (B)
(r) Leucyl-isoserine (A)	Leucyl-alanine
Glycyl-tyrosine	Leucyl-glycine
Leucyl-l-tyrosine	Leucyl-leucine
(r) Alanyl-glycyl-glycine	Aminobutyryl-glycine
(r) Glycyl-leucyl-alanine	Aminobutyryl-aminobutyric acid (B)
(r) Leucyl-glycyl-glycine	Aminobutyryl-aminobutyric acid (A)
Dialanyl-cystine	Glycyl-phenylalanine
(r) Alanyl-leucyl-glycine	Valyl-glycine
Dileucyl-cystine	Diglycyl-glycine
Tetraglycyl-glycine	Triglycyl-glycine
Triglycyl-glycine ester (Biuret base)	Dileucyl-glycyl-glycine
d-alanyl-d-alanine	d-alanyl-l-alanine
d-alanyl-l-leucine	l-alanyl-l-alanine
d-alanyl-l-tyrosine	l-leucyl-glycine
l-leucyl-l-leucine	l-leucyl-d-leucine
l-leucyl-d-glutamic acid	d-leucyl-l-leucine
l-leucyl-l-tyrosine	
l-prolyl-l-phenylalanine	
d-alanyl-diglycyl-glycine	

Peptides marked with an r are the racemic compounds.

In the second place (B) the power of being digested depends on the number of amino-acids present in the chain. Thus the tetraglycyl-glycine containing five glycine radicles was digested, while the diglycyl-glycine containing three was not digested.

Different proteolytic enzymes show quite different powers of hydrolyzing the different polypeptides, and the enzymes may be distinguished from each other in this way. It is besides in the highest degree significant that the enzymes found in the same organs in different kinds of animals show different digestive powers, and wide differences exist between the different organs of the same animal in its power of hydrolyzing these artificial polypeptides. Pepsin will not digest any artificial polypeptide. Fischer has suggested that this is because it has not yet been tried on a sufficiently long chain, but it may be due to other reasons. The various autolytic enzymes, or proteoclastic enzymes as they are also called, show very marked differences. But in general their powers of decomposition are greater than those of the pancreatic juice. Thus glycyl-glycine is hydrolyzed by the liver and muscle juice of the ox, rabbit, dog and by kidney press juice, but not by the serum or plasma of the horse. Ox plasma digests glycyl-dl-alanine, but not glycyl-l-tyrosine. In general, it may be said that when the racemic polypeptides are digested the naturally-occurring, optically-active, isomers are the ones digested and not their optical antipodes. Many of these peptides are split if injected directly into the blood of animals. Sometimes both the d- and l- forms are utilized, but one better than the other. In such cases it is possible that racemization is first produced, and the natural substance then consumed. Often the unnatural antipode is excreted more or less completely.

Sometimes when a tripeptide is exposed to pancreatic juice or other enzyme one of the amino-acids is split off very quickly, the other two being separated very much more slowly; or again, it may happen that one enzyme breaks the chain in one point while another breaks it at another, just as happens in the polysaccharides. In the tripeptide l-leucyl-glycyl-d-alanine, the proteases in yeast and saliva split this into leucine and the dipeptide glycyl-d-alanine; while pancreatic juice splits off d-alanine, leaving the dipeptide l-leucyl-glycine. Other examples have been studied by Koelker and Abderhalden. By this means differences between different proteases may be detected.

Law of the rate of tryptic digestion.—For the relation between the rate of proteolysis by trypsin and the amount of the enzyme present the same facts have been found as have been noted under pepsin. If the protein is in solution, if for example it be gelatin, there is an inverse simple proportionality between the amount of enzyme and the time required to digest to the point of liquefaction. That is, $dx/dt = KC_{ferment}$. Or $\log (A/(A-X)) = KC_{ferment}\, t$. In the gelatin experiment $\log (A/(A-X))$ is constant. If, however, the determination is made by Mett's tubes, values are obtained which are fairly in agreement with the Schütz-Borissow law, according to which the amount digested in a certain time is proportional to the square root of the amount of enzyme. Finally if fibrin colored blue with "spritblau" is the substrate, and the reaction is followed by the depth of color of the solution, the same law is found as Grützner found for peptic digestion under similar conditions: namely, that X, the amount digested, $= KC^{2/3}_{ferment}$. The amount digested is proportional to the two-thirds power of the concentration of the ferment (Palladin). Both of the latter relations are purely empirical.

Pancreatic juice dialyzed against water loses all its tryptic power, but it retains its ereptic power. It cannot then be activated by the addition of kinase. Kinase does not, then, hasten the ereptic action of dialyzed juice, nor will it activate trypsinogen in the absence of salts. Possibly trypsin is rendered insoluble by the dialysis, behaving as if it were a globulin. These experiments show that the juice contains at least two enzymes.

Pancreatic nuclease.—Common foodstuffs generally contain nucleic acid. This is not digested by proteolytic enzymes. There is in the pancreatic juice a nuclease which acts on the nucleic acids of the foods and reduces them probably to phosphoric acid, nuclein bases and carbohydrate. From the pig's pancreas (but the juice was not examined) *Jones* extracted a nuclease (tetranucleotidase) which had the property of splitting guanylic acid from yeast nucleic acid. The investigations of London show that nucleins are digested after reaching this portion of the intestine, and it is probable that this enzyme of the pancreas is the active principle. It is not impossible that the presence of guanylic acid in the pancreas, and presumably outside the nucleus, may be correlated with the presence of this enzyme. It is possible that the nuclease is in union with the guanylic acid of the gland which thus serves as a colloidal, non-digestible substrate.

Pancreatic juice. Summary.—The digestive action of the pancreatic juice may be summed up as follows: When mixed with bile and intestinal juice it very rapidly splits by hydrolysis and emulsifies fats; it reduces starches, dextrins and glycogen to maltose and glucose; it inverts cane sugar, at times, and in young animals it has the power of converting lactose into glucose and galactose; it attacks the partially digested peptones, albumoses, etc., of the chyme, very quickly hydrolyzing them into di- and tri-peptides and crystalline amino-acids which do not give the biuret reaction; it digests nucleic acid.

Pancreatic juice mixed with bile and intestinal juice often regurgitates into the stomach, as was mentioned when digestion in that organ was discussed. By stimulating this regurgitation it is possible to obtain samples for clinical examination. Boldyreff recommends the following method: The person should have eaten the day before only a small amount of easily digested food. The following morning 200 c.c. of olive oil containing 2 per cent. oleic acid is given on an empty stomach. After 1-1½ hours the stomach is emptied with a stomach tùbe. The contents settle in an oily upper layer and a watery lower layer. The latter is often alkaline in reaction and slightly brown in color. This can be tested for its tryptic activity by making slightly alkaline with sodium bicarbonate and adding fibrin. To see if it has pepsin it may be examined in acid solution. Digestion in the alkaline solution indicates the presence of pancreatic juice. The juice may also be investigated for its other pancreatic ferments.

THE BILE.—The third of the juices co-operating in duodenal digestion is the bile. The bile, or gall, the external secretion of the liver, made two (black bile and yellow bile) of the four cardinal humors of Hippocrates. It early attracted the attention of physicians and laity alike, partly because its formation was apparently the only function of one of the largest organs of the body, the liver; partly because of its green or golden color and viscidity; partly because of its extremely bitter taste and the yellow appearance of the skin and whites of the eyes when its outflow from the gall bladder was blocked; and partly because of the formation of gall stones and the extreme pain of their expulsion from the gall bladder. But while curiosity was very early excited as to the functions of this liquid, its real function has become known only within the past half century, and indeed it is still probably very imperfectly known. Its composition, also, is in many important particulars still obscure and there can be little doubt that many valuable discoveries remain to be made concerning its composition, function, formation and secretion.

The bile in human beings as it flows from a temporary biliary fistula is generally of a clear, golden yellow or brownish yellow color, but it is sometimes an olive green; it is very bitter and has a slimy consistence. The bile taken from the gall bladder after death is generally of a brownish green color, and is more viscid, containing more solids and more phosphoprotein (false mucin) than that flowing from a temporary fistula. The viscidity of the bile is a matter of importance, for the reason that the bile is secreted under so small a secretory pressure that the viscous bile may offer such a resistance as to block wholly or partially its outflow; and under these circumstances reabsorption takes place through the lymph vessels, leading to the appearance of bile pigment in the fluids and tissues of the body, and the excretions, and to the condition of icterus (Greek, *ikteros,* jaundice) or jaundice. The color of bile is generally yellow brown, but it may be green in its fresh state. If green, it readily changes by reduction, which may be produced by putrefactive changes or by the walls of the gall bladder, or possibly even by internal changes in the bile itself, into a reddish or golden brown. The freezing point of human bile is just about that of the blood, i.e., —0.55° to —0.61° C., so that its osmotic pressure is about the same or possibly a little greater than that of the blood. In this respect the bile differs from most other secretions which generally have a freezing point less than that of the blood.

Composition.—The composition of the bile is different in every species of animal examined, but all, or nearly all, biles contain three characteristic and very important substances: 1. The bile pigments; 2. the bile salts; and 3. cholesterol. The bile salts vary with the species. In the elasmobranch fishes they differ widely from those present in most animals. Most, if not all, biles contain also some phosphatide (phospholipin).

Human bile has been obtained for analysis either from biliary fistulas or from the gall bladder of recently executed criminals or persons dead by accident. The composition is indicated in the table. Bile is generally neutral or faintly alkaline in reaction to litmus.

COMPOSITION OF THE BILE.

Bladder Bile.

	Frerichs		v. Gorup - Besanez	
Water	86.00	85.92	82.27	89.81
Solids	14.00	14.08	17.73	10.19
Bile salts	7.22	9.14	10.79	5.65
Mucin and pigments	2.66	2.98	2.21	1.145
Cholesterol	0.16	0.26	4.73	3.09
Fat	0.32	0.92		
Inorganic	0.65	0.77	1.08	0.62

Fistula Bile (Hammarsten).

Water	97.48	96.47	97.46
Solids	2.52	3.53	2.54
Bile salts	0.93	1.82	0.90
Taurocholate	0.30	0.21	0.22
Glycocholate	0.63	1.61	0.68
Fatty acids	0.12	0.14	0.10
Mucin and pigments	0.53	0.43	0.52
Cholesterol	0.06	0.16	0.15
Lecithin + Fat	0.02	0.096	0.06
Soluble salts	0.81	0.68	0.73
Insoluble salts	0.025	0.05	0.02

The composition of bile flowing from a biliary fistula differs considerably from that of bile taken from the gall bladder. The fistula bile is less concentrated the more completely the bile has been cut off from the intestine. Thus bile from a complete fistula in a woman was found by Pfaff and Balch to contain only 3 per cent. of total solids, and Hammarsten found similar figures; whereas bile from the gall bladder contains from 10-20 per cent. of total solids. Moreover, of the total solids of fistula bile 30-50 per cent. are inorganic, whereas in human bladder bile only about 6 per cent. are inorganic. The freezing point of the bladder and of the fistula bile does not greatly differ, although that of the bladder bile is somewhat lower. The reason for this is that some of the organic constituents of the bladder bile are probably in colloidal solution, so that they do not much affect the freezing point. The greater concentration of the bladder bile is generally ascribed to a reabsorption of water in the gall bladder, but there is little evidence of this. It is more probably due in part to the fact that some of the organic matters, mucin, phosphoprotein and possibly some cholesterol, are added during the stay of the bile in the bladder, but it is chiefly due to the fact that when the bile is diverted from the intestine, as it is in most fistulas, there is no reabsorption of the bile and hence the bile salts are not re-excreted as they normally are.

Secretion of bile.—The bile is poured into the duodenum in man about 4-5 inches below the pylorus through the ductus choledochus. This duct opens very close to or in conjunction with one of the pancreatic ducts (Wirsung's), so that these two secretions are intimately mixed as they are poured out into the acid chyme coming from the stomach. They are also mixed with the duodenal juice, and it has been found that these three secretions mutually interact with each other and with the chyme and together form a digestive fluid very powerful and capable of digesting and dissolving all classes of foodstuffs.

The amount, character and time relations of the bile secretion can be studied by means of a fistula, either temporary or permanent. In a permanent fistula the gall bladder is opened, brought where possible to the surface, drained and fastened in the skin in such a way that an opening persists to the exterior. If it is desired to block completely all flow of bile into the intestine, it is necessary to resect a portion of the common bile duct (ductus choledochus), and even then the connection with the intestine may in time be re-established, so great are the reparative powers of animal tissues. As a rule, however, this connection is not re-established and the bile passes altogether to the exterior. If no resection of the duct is practised, some of the bile may continue to find its way into the intestine, and in human beings and dogs, if no means are taken to prevent it, the external opening will ultimately close and normal connections with the intestine will be re-established. The method usually employed in making such fistulas in dogs is that of Dastre. Such temporary fistulas are not uncommon in human beings following operations for gall stones, or for infection of the gall bladder.

The secretion of bile from such a fistula is found to be continuous, but by no means uniform. It is indeed subject to wide fluctuations in amount during the twenty-four hours, and the causes of these fluctuations are still quite obscure. While the formation of the bile and its flow from the fistula are thus continuous, the flow from the ductus choledochus into the intestine is intermittent, at least in those animals provided with a gall bladder, and occurs only when acid chyme is discharged from the pylorus into the intestine. The irregularity of the secretion of the bile from a fistula is illustrated in the following figures which give the secretion from a biliary fistula in a woman in the Massachusetts General Hospital, as recorded by Pfaff and Balch:

8 a.m. - 2 p.m.	149 c.c.
2 p.m. - 8 p.m.	84
8 p.m. - 2 a.m.	96
2 a.m. - 8 a.m.	136
8 a.m. - 2 p.m.	170
2 p.m. - 8 p.m.	129
8 p.m. - 2 a.m.	70
2 a.m. - 8 a.m.	105

The hourly secretion is subject to a wide diurnal variation, being least in the hours of the early morning from 2-5 A.M., rising rapidly after this time and culminating, as a rule, about 1-3 P.M., thereafter decreasing with some fluctuations. It thus runs a course not very different from the body temperature, which is at a minimum about 3-5 A.M. and at a maximum about 4-5 P.M. Other functions of the body also show such a diurnal or rhythmic variation. Similar variations in the flow of bile from a human fistula have been reported also by others. There are, besides the main variations in flow already mentioned, other variations apparently related in some way, not at present clear, to digestion. Thus in the case of Pfaff and Balch the great rise in the early morning coincided with breakfast and the next great rise with dinner, but sometimes the rise came before and sometimes after dinner, an hour or more, and the maximum outflow was not coincident with any ascertainable phase of digestion. In another case the flow often showed a remarkable fall immediately after a meal, particularly at the time of afternoon tea, and the authors suggested that possibly this sharp fall might be due to closure of the pylorus following the taking of food, thus diminishing the discharge of acid into the intestine and so reducing the stimulation by the secretin formed when acid is discharged. The flow from a dog's fistula shows somewhat similar variations.

Amount of bile secreted per day.—The amount of bile secreted per day by a human being is very difficult to estimate with any accuracy. Pfaff and Balch report in their case a secretion of 525 c.c. per day, and similar results have been reported by others. These figures, however, while they may give us some idea of the amount secreted, are not to be taken as a certain index of the amount normally produced, since they are all of them from permanent fistulas. It has been shown that, if bile is poured into the intestine, the secretion is increased both in concentration and amount. In these cases no bile was finding entrance to the intestine, and hence this would tend to diminish the secretion. On the other hand, the acid of the chyme acts as a stimulus to the formation and the discharge of bile from the liver. Now in the absence of the bile it is possible that the acid chyme remains acid for a longer time than normal and thus acts as a stimulus for a longer time. The absence of the bile in this way might act as a stimulus to the formation of more bile than usual. It is hence impossible to say whether 550 c.c. per day represents the secretion of bile in the human adult or not, but it is probable that in health and when the bile finds its way into the intestine the excretion is larger rather than smaller than this.

The amount of the bile formed is dependent also on various other factors, particularly on diet. There is a general consensus of opinion of all investigators that the secretion of bile is at a maximum on a meat diet and at a minimum on a carbohydrate diet. After hemorrhage the amount and solids of the bile are both reduced. Drinking water does not apparently influence the secretion. In starvation the bile continues to be formed, in dogs at least, up to the moment of death, and it does not materially change its composition, although it decreases in amount. The following experiment of Albertoni shows this fact:

BILE SECRETED ON SUCCESSIVE DAYS OF FASTING BY A DOG. WT. 21 KGS.

Days of fast	Fresh bile—grams	Dry bile—grams	N	Bile dry—per cent.	N—per cent.
1	74.75	3.45	0.115	4.61	0.153
2	40.65	3.34	0.118	8.21	0.290
3	31.25	2.48	0.08	7.90	0.256
5	40.25	2.10	0.123	5.21	0.305
10	33.25	2.01	0.080	6.04	0.240
17	23.00	2.11	0.079	9.18	0.343
21	25.00	3.05	0.120	12.2	0.480
23	24.05	1.75	0.073	7.20	0.303
27	16.00	1.37	0.260	8.50	1.625 Died.

Functions of the bile.—If the bile is cut off from the intestine and drained to the outside by a fistula, very serious disturbances of digestion follow. In dogs and human beings the feces lose their color, they become clay-colored; they increase greatly in amount and become very fatty if fat is taken in the food; constipation generally occurs; and the odor of the feces is very strong and offensive, putrefaction evidently being very marked. At the same time, although the person or dog may eat voraciously, there is a constant loss of weight, and dogs, at any rate, die in extreme emaciation in the course of four or five weeks, if they be prevented from licking the fistula and thus swallowing their own bile. Both human beings and dogs acquire a distaste for fats if the bile is cut off from the intestine and will not willingly eat fatty food.

The changes just described show that the absence of the bile is followed particularly by a failure to absorb fat, but the reabsorption of other foodstuffs, proteins as well, is also diminished. The direct examination of the bile shows that it has itself either no power of digestion at all or but slight action on either fats, starches or proteins. It is true that human bile is said to have some solvent action on fibrin but not on coagulated egg white; but on the whole the bile itself does not contain digestive enzymes. It has, however, very remarkable powers of aiding the digestive action of the pancreatic and intestinal juices and in helping the absorption of fats.

The power of the bile to aid the digestion of fats by the pancreas was observed by Claude Bernard, who clearly recognized the importance of the co-operation of the bile, pancreatic and duodenal juice in digestion. It was particularly studied by Rachford, who found that the power of the pancreatic juice of splitting fats was increased two to three times if bile were present in the digestive mixture, and it has been recently reinvestigated by von Fürth and Schütz, who found that the addition of 5 c.c. of bile to 20 c.c. neutralized olive-oil emulsion contain ing steapsin increased the amount of acid formed from 5-10 times over that formed by the steapsin alone.

Rachford did not find any accelerating action of the bile on amylase, but he thought that the digestion of proteins was accelerated. Other observers failed to find any accelerating action of bile on tryptic digestion, so that the matter has been uncertain, but recently this work has been repeated with pancreatic juice and bile from permanent fistulas of dogs. No accelerating action on the protein digestion of pancreatic juice and intestinal juice was observed, but there was a slight favoring action on amylopsin.

The substance in the bile which thus accelerates the fat-splitting action of the pancreas was found by Rachford to be the bile salts, sodium taurocholate and glycocholate, which were almost as favorable as the bile itself. Hewlett, who repeated his experiments, attributed the action of the bile to the lecithin it contained rather than to the bile salts. Hammarsten has recently shown that human bile has a very large amount of lecithin, or other phospholipin, in it, so that Hewlett's results seemed not unlikely, but von Fürth and Schütz in re-examining the whole matter and using synthetic bile salts, so as to exclude the possibility of the presence of lecithin as an impurity in the salts, have shown that the whole power of the bile is due to the bile salts and that sodium cholate acts as well as the glycocholate or taurocholate.

Circulation of the bile. Role in absorption.—But not only do the bile salts thus play a very important part in the digestion of fats; they play a not less important and certainly a very illuminating rôle in the absorption of fats. In the absence of bile the fatty matter in the feces is found to have been split into fatty acids, but these have not been absorbed. It might be, of course, that in the absence of the bile this splitting took place too far down the intestine for absorption to occur and the bile might by accelerating the splitting cause it to occur so far up in the gut that absorption could take place; but whether this is true or not, there can be no doubt that in quite another way bile influences absorption, if indeed it does not play the major rôle in the mechanism of absorption. When fats are being digested and absorbed, all the lymph vessels leading from the intestine across the mesentery to the receptaculum chylii and so to the thoracic duct are filled with a white milky lymph called chyle; and it is by this path mainly, and not by the portal circulation, that the fats are absorbed. Now, if the bile is cut off from the intestine of a dog and fat is fed the chyliferous vessels no longer are white, but filled with a lymph not more than opalescent. Evidently in the absence of bile fat does not pass into the chyle vessels. Furthermore, it is not absorbed even if it is given in the form of a soap, or as a fatty acid.

A further study of the manner in which bile affects the absorption has led to the discovery that bile is able to dissolve large amounts of fatty acids. Oleic acid is the more soluble in bile, but palmitic, stearic as well, acids which are quite insoluble in water, are soluble in bile. This solvent power of the bile for fatty acid is due at least in large measure to the bile salts. The probability is that the fatty acids form a loose union either with their amino groups or in some other way, possibly by cohesion. One or more molecules of fatty acid are perhaps thus united to the one molecule chemically bound to the bile salt. The fatty acids embark as it were on molecules of glycocholate and taurocholate and by this means they are ferried across the cell membranes and into the lymph channels. The bile salts thus reabsorbed are thrown into the blood, carried to the liver and are there re-excreted. There is thus a circulation of bile salts from the gall bladder to the intestine and from the intestine to the blood, by which they are carried to the liver and re-excreted. Each molecule of bile salt can thus function, theoretically at least, over and over again.

The circulation of the bile was first particularly emphasized by Schiff, although its possible occurrence had been suggested before that by Liebig. Schiff observed that the flow of bile from a biliary fistula was greatly increased, and in particular the bile solids were increased when bile was put into the intestine. He was struck by the fact that when a fistula is first made the bile flowing from it is far more concentrated than that coming after six hours or more. This decline in total solids and amount of bile and the immediate increase of both when bile was admitted to the intestine led him to conclude that the bile, or at least the bile salts, were reabsorbed and re-excreted and thus circulated. This fact has been established by subsequent observation, and particularly by the observations of Stadelmann, who found nearly the whole of the bile salts put directly into the intestine were re-excreted. How impossible it is for any process of diffusion to explain secretion is shown by this excretion of bile salts which are picked out from the blood, where they exist in the smallest proportions, and carried through the liver cells and into the bile, where their concentration is many thousand times as great as that in the blood.

Influence of bile on intestinal putrefaction.—Bile has a marked influence in reducing intestinal putrefaction and in stimulating peristalsis. It was at first thought that it must have antiseptic properties, but bile is itself easily putrescible, at least as long as it contains mucin. In fact, bile is a fair culture medium and it not infrequently happens that the gall bladder becomes infected, as in typhoid fever, and continues to serve as a permanent culture medium, constantly feeding bacteria into the intestine. The rôle of the bile in reducing putrefaction is, hence, probably an indirect one; by stimulating digestion and aiding absorption it prevents the accumulation of putrescible materials, and by its peristaltic stimulating powers it sweeps the bacteria out of the body and prevents constipation. Bile is, in fact, a natural laxative.

Summary of the functions of the bile.—The bile plays a very important part in the digestion and absorption of the foods. While of itself it has little digestive action, it stimulates greatly the digestion of the fats by steapsin and other lipases, and it holds in solution the fatty acids set free. By its alkalinity and because the bile salts, the taurocholate and glycocholate and other similar salts, are salts of weak acids, it helps neutralize the acidity of chyme and so makes a favorable medium for the action of trypsin and amylopsin. Even more important is its rôle in absorption, since in its absence absorption is much reduced. The fatty acids combine chemically or physically with the bile acids and the combination passes into the intestinal epithelial cells with ease. The bile salts are then freed from the fatty acids and carried to the liver, where they are re-excreted, thus forming a circulation of the bile. By stimulating peristalsis and absorption, putrefaction is much reduced. Bile is a natural laxative.

Bile acts also as a channel for excretion of various substances. Cholesterol is excreted in the bile and bile is the only fluid of the body which will dissolve this substance. In addition the bile pigments appear to be purely excretory substances formed by the decomposition of the blood pigment. So far as known they play no part in the physiology of the bowel. Other substances may also appear in the bile, such as toxins or metallic poisons of various kinds.

With this statement of function we may now consider the chemistry of the various constituents of the bile, the pigments, salts, lipins and mucin, and the manner and place of their origin.

Chemistry of the bile pigments.—The peculiar color of the bile is due to several pigments, of which the most important is bilirubin, which is the mother substance of most of the others. This is the pigment which gives the golden red color to bile; it is easily converted by oxidation into a green pigment, biliverdin; and by further oxidation it is converted into a series of colored substances, of which bilicyanin, a blue pigment, may be particularly mentioned. By reduction it forms urobilin, a pigment of the urine. Bilirubin is derived from the hemoglobin of the blood, and the amount of bilirubin, or the other pigments derived from it, in the bile may be increased by any means which causes laking of the red blood corpuscles and frees hemoglobin in the blood.

Bilirubin.—The chemical composition of bilirubin is still uncertain. It is a crystalline, organic, iron-free compound, of which the formula is either $C_{34}H_{38}N_4O_6$ or more probably $C_{33}H_{36}N_4O_6$, according to Willstaetter. The last formula makes bilirubin isomeric with hematoporphyrin. It yields on oxidation hematic acid, $C_8H_9NO_4$; on reduction with hydriodic acid, zinc or other reducing agents it yields hemopyrrol. This latter substance is a mixture of substituted pyrrols from which there have been isolated methyl, ethyl pyrrol and 2, 3-dimethyl, 4-ethyl pyrrol.

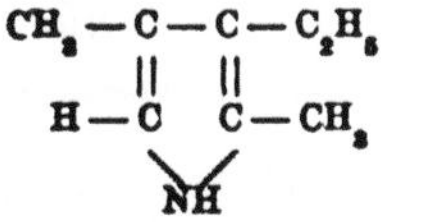

```
CH3 — C — C — C2H5            CH3 — C — C — C2H5
      ||   ||                        ||   ||
 H — C     C — CH3             CH3 — C     C — CH3
      \   /                          \   /
       NH                             NH
```

Dimethyl-ethyl-pyrrol. 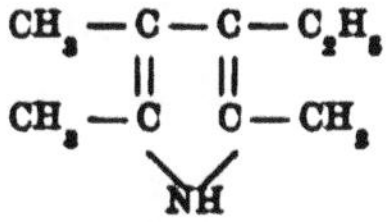 Phyllopyrrol.

Bilirubin is an acid and in the soluble form is present as the sodium salt. It probably contains four substituted pyrrol nuclei. The way they are put together is uncertain. The following formulas have been suggested for hematoporphyrin, and also of bilirubin, by Willstaetter and Fischer, although other formulas also have been proposed. This will serve to indicate the general nature of the compound.

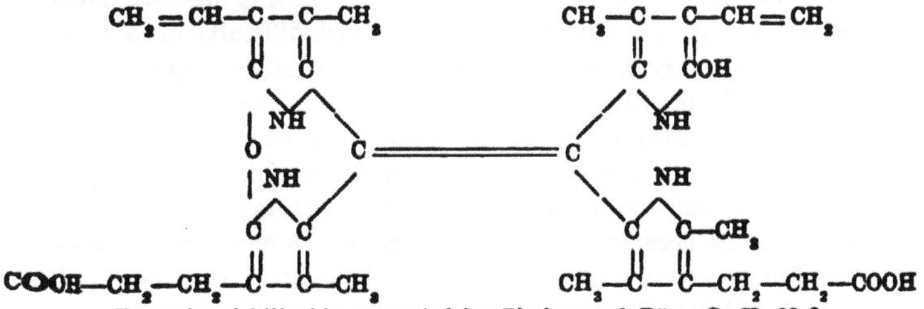

```
CH2=CH—C — C—CH3                      CH3—C — C—CH=CH2
       ||  ||                              ||  ||
       C   C                               C   COH
        \ / \                             / \ /
    |    NH   \                          /   NH
    O          C ======================= C
    |    NH   /                          \   NH
        / \ /                             \ / \
       C   C                               C   C—CH3
       ||  ||                              ||  ||
COOH—CH2—CH2—C — C—CH3              CH3—C — C—CH2—CH2—COOH
```

Formula of bilirubin suggested by Fischer and Röse, $C_{33}H_{36}N_4O_6$.

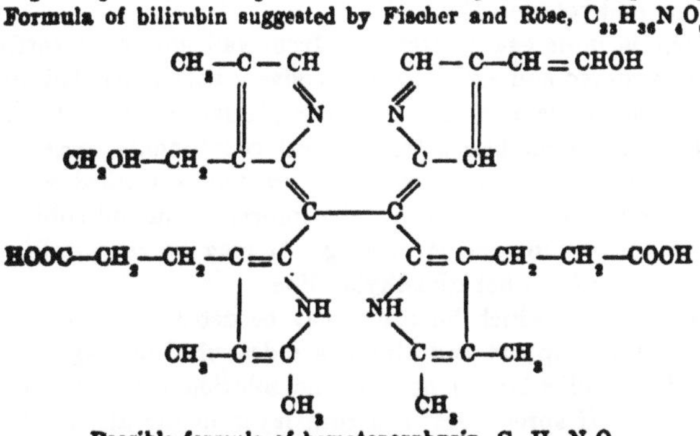

```
           CH3—C — CH        CH — C—CH = CHOH
               ||    \\      //    ||
               ||     N      N     ||
               ||    /        \    ||
 CH2OH—CH2—C — C              C — CH
                 \\          //
                   C ——————— C
                  /           \
HOOC—CH2—CH2—C = C             C = C—CH2—CH2—COOH
             |    \           /    |
             |     NH     NH       |
             |    /           \    |
        CH3—C = C              C = C—CH3
                |              |
               CH3            CH3
```

Possible formula of hematoporphyrin, $C_{33}H_{38}N_4O_6$.

The acid character and the substituted pyrrol nuclei are to be seen in this formula. In their chemical nature, then, the bile pigments are clearly oxidation products of hematin, the colored constituent of hemoglobin. Oxidation products similar to those of bilirubin and hematin are obtained from chlorophyll; all three yield hematic acid, and the two latter and mesobilirubin yield methyl-ethyl-maleic imide.

```
COOH—CH₂—CH₂—C = C—CH₃          CH₃—CH₂—C = C—CH₃
             |   |                       |   |
           O=C   C=O                   O=C   C=O
              \ /                         \ /
              NH                          NH
```

Hematic acid. Methyl-ethyl-maleic-imide.

In every spot of extravasated blood ("black and blue spot") this conversion of hematin into pigments analogous to, or identical with, the bile pigments may be seen. Pyrrol, which thus forms so important a part of the molecule of the pigments of bile, blood, urine and chlorophyll, is found in the protein molecule in tryptophane, and in the reduced form in proline.

Properties and preparation of bilirubin. Amorphous reddish-brown powder, or reddish-yellow or brown crystals. Crystallizes readily from chloroform solution by evaporation of the solvent. Long needles. Insoluble in water; soluble in warm chloroform, but with difficulty in cold; more soluble in alcohol; very slightly soluble in benzene, ether, amyl alcohol or glycerol. Soluble in dilute alkalies. The alkali salts are insoluble in chloroform. The solutions show no absorption bands. Solutions of bilirubin in dilute alkali are precipitated by milk of lime. An aqueous solution of alkali salt of bilirubin treated with ammonia and then with some zinc chloride becomes first deep orange, which changes to brownish green and finally green. The solution then shows absorption bands close to the C line and between C and D.

Bilirubin is most easily prepared from gall stones of cattle. The stones are powdered and extracted successively with ether, boiling water, 10 per cent. acetic acid, alcohol and hot glacial acetic acid. By these extractions cholesterol, bile salts, mineral constituents, green coloring matter, choleprasin are removed. The residue is washed with water, dried and extracted with boiling chloroform. The bilirubin crystallizes out of the chloroform on cooling. It may be recrystallized from hot chloroform or from hot dimethylaniline.

The reactions by which bilirubin may be detected, namely, those of Ehrlich, Gmelin, Huppert and others, are described on page 916.

Biliverdin.—Bilirubin in an alkaline solution has the property of auto-oxidation. If spread out in a thin layer in the air, it changes to a green pigment which is often present in the bile, known as biliverdin. The composition of this substance is probably $C_{32}H_{36}N_4O_8$. It does not crystallize. By further oxidation with nitric acid, or bromine water, it develops the series of colors noted for bilirubin. At times, in infants, especially in certain diarrheas, biliverdin may color the feces a bright green. Biliverdin may also be formed from bilirubin by gentle oxidation by sodium peroxide, or by Hübl's solution.

Biliverdin was found by MacMunn in the mesoderm of a sea-anemone (Actinia mesembryanthenum (?)). It is a significant fact that biliverdin, which contains no iron, combines spontaneously with oxygen and gives it up again almost as readily as oxyhemoglobin.

Properties. Amorphous. Soluble in alcohol, glacial acetic acid dilute alkalies; insoluble in water, ether, chloroform. The alkali salt is precipitated by heavy metals and alkaline earth salts. By reduction by putrefaction, or ammonium sulphide, it is converted back to bilirubin. By reduction with sodium amalgam it is converted into hydrobilirubin.

Urobilin, stercobilin, hemibilirubin and hydrobilirubin.—The bile pigments do not usually occur in the feces and urine, although they may at times be found there. The brown color of the feces is due to a reduced bilirubin called stercobilin or urobilin. These two are probably identical. Urobilin is one of the important coloring matters of the urine. It disappears from the urine when the bile is prevented from entering the intestine. It is formed from bilirubin by the reducing action of the bacteria of the gut. It is to be found also in blood serum and in bile itself. The composition of urobilin, or stercobilin, is still uncertain, for it is an amorphous body and it is impossible to know whether it has been prepared in a pure state. It is accompanied in the urine by a *urobilinogen*, a colorless substance which is converted into urobilin by the action of the oxygen of air. According to Fischer and Meyer-Betz, there are at least two of these urobilinogens: namely, hemibilirubin, to which they ascribe the formula $C_{33}H_{44}N_4O_6$, and an unknown substance. By the reduction of bilirubin by sodium amalgam a series of substances are obtained which resemble urobilin, but which are less stable. Among the reduction substances thus formed may be mentioned *hydrobilirubin*, which, according to Maly, has the formula $C_{32}H_{40}N_4O_7$. This formula calls for 9.8 per cent. of N. According to Garrod and Hopkins, however, urobilin as isolated by their method contained but 4.11 per cent of N. The relation of these bodies is, therefore, still uncertain. It is possible that what is known as urobilin is a mixture of various pyrrol derivatives, since all of the unstable pyrrols show color reactions such as the Ehrlich reaction and the fluorescence characteristic of urobilin.

Properties. Urobilin is soluble in water, but may be salted out by saturating its solution with ammonium sulphate. It is an amorphous, brownish substance which in dilute solution is reddish yellow or rose red, but is nearly colorless if slightly acid. It is distinguished by two properties: if a solution is made faintly ammoniacal and sufficient $ZnCl_2$ is added, the solution acquires a beautiful green fluorescence. In transmitted light it is a reddish color. Even very dilute solutions may be detected by this property. The other means of distinguishing urobilin is by the absorption spectrum. In a strong solution the light is absorbed from the violet end right up to Frauenhofer line b; but in a more dilute solution there is a well-defined band between b and F. If $ZnCl_2$ is added to the solution, another band appears between b and F, but nearer b than the former. Urobilin has the property of giving the Ehrlich reaction: that is, a brilliant red color on the addition of a little p-dimethylamino benzaldehyde to the acid solution. This reaction is given not only by urobilin, but by hemopyrrol and all pyrrols having an unsubstituted H atom on one of the carbon atoms of the ring (Fischer and Meyer-Betz). Urobilin is soluble in $CHCl_3$ and in acid alcohol. The method of separating it from the urine is given on page 761. Urobilinogen is said to give the biuret reaction.

Cholohematin.—Besides the pigments belonging to the bile, other pigments of an extraneous nature are sometimes found in it. Thus in herbivorous animals such as the rabbit, ox, hippopotamus, horse or goat, when fed on green fodder, the bile often contains a red pigment with very characteristic absorption bands called by its discoverer, MacMunn, cholohematin. The absorption bands are the following: Band 1, —λ638; band 2, λ604—582; 3, λ570—558; 4, λ530—515. There are two bands in the ultra violet. This substance is not a true bile pigment, but is derived from the chlorophyll the animal has eaten. This red pigment, *phylloerythrin,* as it is called, is found in the feces of cows. It is derived in some manner from chlorophyll, presumably by the action of the bacteria of the intestine; it is absorbed, carried in the blood to the liver and there excreted in the bile. If cattle are fed on dry fodder, the color disappears from the bile and from the feces. This fact illustrates again that the bile is not only a secretion, but an excretion, and toxins and poisons absorbed from the intestine may be excreted in it. Cholohematin is identical with bilipurpurin. It is extracted from the fresh feces of cattle fed on green grass. The solution in chloroform is cherry red. It is a weak base, insoluble in alkali. It is changed by HCl to a blue violet. It would be better to call it phylloerythrin.

Where are the bile pigments made?—Does the liver make the bile or does it simply secrete it from the blood as the kidney secretes urine? This question has been answered by seeing whether bile pigment and bile salts will accumulate in the blood after the liver has been cut out of the circulation. If the bile duct is tied, bile pigments and other salts accumulate in the blood of mammals and birds so rapidly that they may be easily detected in the blood serum after four to six hours. Now if the blood vessels, the portal vein and the hepatic artery are tied no such accumulation occurs, although animals may live 6 to 18 hours after the operation. Moreover, under such circumstances the bile pigments do not appear in the urine. In birds the liver may be extirpated, leaving only very small remnants about the vena cava, since in these animals a connection exists between the portal and kidney veins, by which the blood may return from the intestine into the general circulation. In birds, geese, ducks and hens, Minkowski and Naunyn found that there was no jaundice, nor did any bile salts appear in the urine, although the birds lived in some instances in a fairly good state for 12-19 hours after the operation. It is then clear that normally bile pigments and salts are formed in the liver itself, although every black and blue spot in a bruise shows that the bile pigments may be formed from extravasated blood in almost any location in the body.

The raw material from which the liver makes the bile pigments is the blood pigment, hemoglobin. Bilirubin and bemato-porphyrin are isomeric substances. The direct transformation of hematin of blood hemoglobin into bile pigments, bilirubin and biliverdin, is shown by the fact that any means which leads to the setting free of hemoglobin in the blood causes an increase in the bile pigments excreted. Thus inhalation of arseniureted hydrogen causes a great hemolysis, even leading to hemoglobinuria; and injections of toluylendiamine produce the same result. Bile secretion is greatly increased in animals poisoned by these substances. Injection of hemoglobin itself has a similar effect. Moreover, liver pulp rapidly destroys hemoglobin and, although it does not convert it into bilirubin, it is believed to make it into an antecedent of bilirubin. Microscopic examination of the liver shows the various steps in this transformation.

Transformation of hemoglobin into bile pigments.—We may now ask the question of the chemistry and method of the transformation of the blood into the bile pigments. The red blood corpuscles are constantly being formed in the cells of the red marrow of bones. It is clear, therefore, that they must somewhere be broken down and done away with or they would accumulate in the blood. All that is known indicates that it is in the liver that this destruction takes place, at least in the birds and amphibia, but that in mammals the spleen also plays an important, and perhaps the more important, rôle. The destruction of corpuscles in the liver cannot be demonstrated by the simple and direct expedient of counting the corpuscles in the portal and hepatic bloods. Counts of this sort reveal no appreciable differences between the blood entering and leaving the liver, nor is it to be expected that they would, since so rapid is the blood flow through this organ and so large is it, that in any single passage of the blood but a very few corpuscles can possibly succumb. Nevertheless, there is good reason to believe that they do thus die and degenerate in the liver tissue of both birds and amphibia, since both normally and when their decomposition is accelerated by blood poisons, such as those just mentioned, the process of decomposition is easily demonstrable by the microscope. Minkowski and Naunyn have particularly studied these processes in bird's liver and in rabbits after poisoning with arseniureted hydrogen, and Keyes has recently shown the same processes in normal livers of birds and amphibia. There are many questions which naturally arise in our minds as we approach this problem. Does the liver secrete into the blood a substance hemolytic in nature, which is not sufficient in amount to dissolve all the corpuscles but only the weakest ones? And does this hemolytic substance cause a discharge of the hemoglobin into the plasma of the blood from which the liver cells pick it up? or are the corpuscles engulfed as such by some of the phagocytic cells of the liver which by intracellular action destroy the corpuscles? If part of the hemoglobin is made into bilirubin, what becomes of the rest of the corpuscle? There are in the liver two quite distinct tissues: the endothelium of the blood vessels and the glandular tissue. Are both of these tissues or only one of them important factors in the production of bile from blood? While these questions cannot as yet be fully answered, certain facts have been discovered.

The endothelium of the blood vessels constitutes a great organ extending everywhere in the body, for which no function has as yet been found beyond that of serving as the lining of a tubular conduit of the blood. Is this its only function? or is it concerned also in the elaboration of the constituents of the blood; of its proteins, plasma and blood cells? Or does it play an important part in the regulation of the composition of the blood and the determination of the viscosity of the blood in the capillaries by the production of substances which act upon the colloids of the plasma, affecting frictional resistance? We cannot as yet answer these questions, for we know neither where the blood proteins arise nor the functions of the capillary endothelium. One thing, however, is clear and that is that these capillary cells must certainly play an important part in the control of lymph formation, to which they stand somewhat in the same relation as the glomerular cells of the kidney to urine formation. They are also endowed, in certain regions at least, with phagocytic powers and, indeed, it has recently been shown that the giant phagocyte cells found in tissues arise from the capillary endothelium. This phagocytic function is shown to a pre-eminent degree by the endothelium of the liver. Here and there in the liver these cells increase in size and when large are called Kupfer cells. These endothelial cells engulf bacteria. They also, according to Keyes, and as may be seen in the drawings of Minkowski and Naunyn, engulf the red blood corpuscles in birds and amphibia. One of these cells may take up many of these corpuscles which can be distinguished after engulfment by its outline and by the chromatin material, since the corpuscles of these forms are nucleated. These engulfed corpuscles go to pieces and may be seen with out their hemoglobin here and there in the Kupfer cells, and in case the destruction of the corpuscles is intense, a bright-green pigment, possibly allied to or identical with biliverdin, may be seen in them. At the same time the hemoglobin appears in the endothelial cells as masses of brown pigment. Similar masses appear shortly after in the liver cells proper and at the same time the cell body, particularly along the bile capillaries, becomes filled with fine brown granules, which stain a deep blue in potassium ferrocyanide, or brownish black in ammonium sulphide, showing that they contain iron. This iron has been set free from the hematin.

In the mammalian liver this phagocytic rôle of the endothelial cells has not been definitely established, but similar phagocytic cells were described in the finest capillaries in rabbit's liver after the injection of blood poisons by Minkowski and Naunyn. These cells were present, however, in mammals in much smaller numbers than in the bird's liver. There is also an accumulating mass of evidence that in mammals the spleen probably plays an important part in the destruction of the corpuscles. Cells like those phagocytic cells of the liver are found in numbers in the spleen containing engulfed corpuscles, and recent surgical work indicates that the spleen is active, in some pathological conditions at any rate, in blood destruction. Some kinds of anemia have been improved by extirpation of the spleen. If, however, the blood corpuscles are destroyed in the spleen, this cannot be the only place of their destruction, since the taking out of the spleen does not check the formation of bile in the normal animal, and the fact that bile pigments do not accumulate in the blood after an Eck fistula and ligation of the hepatic artery also shows that, if the spleen is active, only the first stage of the process can be taking place in that organ. The question of the destruction of the red corpuscles in mammals is, thus, in a very unsatisfactory state and more work must be done before definite conclusions concerning this very important question can be drawn.

All the evidence, however, points to the conclusion that the hemoglobin is not first discharged from the corpuscle in the blood stream and is then picked out by the liver, but that the corpuscle is engulfed as a whole and the separation of the hemoglobin occurs intracellularly in the phagocytic cells. It appears then that red cells in a peculiar condition, in the condition, possibly, which is produced by the first effect of a hemolyzing agent before hemolysis occurs, are engulfed by the phagocytic cells of the liver; these cells in birds and frogs, and possibly also in mammals, being in part at least the endothelial cells of the liver blood capillaries. Being engulfed, the hemoglobin separates from the stroma and is passed in some manner not known to the liver cells proper and there undergoes a transformation of such a kind that iron is split off and becomes readily demonstrable by microscopic methods; and ultimately the iron-free part of the hematin is further transformed in the liver cells into the bile pigments, bilirubin and biliverdin. It would be very interesting to know, in this connection, whether the endothelial cells of the blood change themselves so that the corpuscles more readily adhere to them leading to their engulfment, or whether they form a substance which, acting upon the corpuscles, makes them more easily engulfed, much as the bacteria are supposed to be acted upon by the opsonins. These and many other similar queries are among the most enticing questions of chemical physiology at the present time, since their solution may throw light on the important subject of the defense of the organism from bacteria.

It may be mentioned, in this connection, as possibly indicating one reason for the excretion of the bile pigments, that small amounts of hematoporphyrin occur normally in the blood. It has recently been found that hematoporphyrin has a very extraordinary effect in rendering animals sensitive to the action of light. If hematoporphyrin is injected into white mice, rats or guinea pigs, or if it is produced in the blood by the action of poisons, no toxic effects follow from the presence of the hematoporphyrin as long as the animal remains in the dark or in a dim light. But, if it is brought into direct sunlight, it presently begins to scratch vigorously, often rubbing the hair and skin off, it is very restless and it will die if not returned to the dark. Similar effects are produced in human beings by sunlight and hematoporphyrin, a rash and skin eruption appearing, followed by a tremendous œdema. The effects may persist for several weeks after taking hematoporphyrin before the sensitization of the body to light is lost. The mechanism of the action is not explained, but the observations are of very great interest. Animals with much pigment in the skin are protected from these effects. One of the functions of the liver is, then, to pick out the hematoporphyrin from the blood and to convert it to a harmless bile pigment.

The close relationship of the bile pigments to the blood pigments is shown by several facts. Thus if oxyhemoglobin is treated with dilute acid it splits into globin, a protein, and hematin, $C_{32}H_{32}N_4O_4Fe$; by strong acid hematin is converted into hematoporphyrin, and the iron of the hematin is set free as inorganic iron. Hematoporphyrin is supposed to have the formula, $C_{32}H_{36}N_4O_6$, or $C_{34}H_{38}N_4O_6$, and it is isomeric with bilirubin, $C_{33}H_{36}N_4O_6$.[1] On oxidation both give rise to hematic acids.

[1] The formulas of bilirubin and hematoporphyrin are still uncertain. The older formula called for 32 atoms of carbon. Küster gives 34 atoms as more probable,

$$\underset{\text{Hematin.}}{C_{32}H_{32}N_4O_4Fe} + 2H_2O = Fe + \underset{\text{Hematoporphyrin.}}{C_{32}H_{36}N_4O_6}$$

The bile pigments are also closely related to chlorophyll.

The relation of these three pigments may be illustrated by the following diagram:

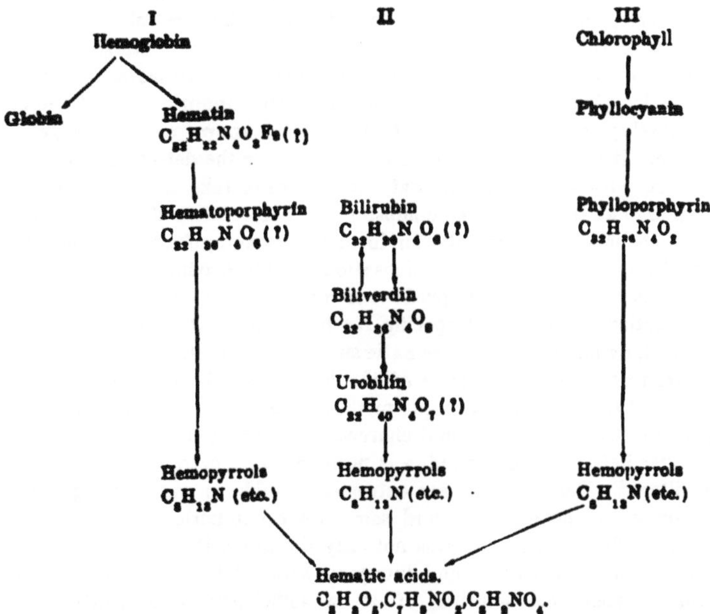

The bile salts.—These are the characteristic constituents of the bile which give to this fluid its properties of assisting in the digestion and absorption of fats. In the majority of mammals these salts consist for the most part of sodium salts of glycocholic and taurocholic acids, but in addition to these acids there may be present analogous acids such as taurocholeic and glycocholeic acids. The relative proportion of taurocholic and glycocholic acids differs in various animals; thus in dogs glycocholic acid may be almost or quite absent, although at other times it may be present. Its absence or presence may be shown readily by adding to the bile some neutral acetate of lead which precipitates the glycocholic acid but not taurocholic. The bile salts are salts of paired acids; that is, they split on hydrolysis into taurine, or glycocoll, and

whereas Fischer has recently expressed the opinion that there are 33 atoms of carbon and 38 of hydrogen.

cholic acid. The taurine and glycocoll are the same in all animals, but the cholic acid part of the molecule differs in different animals. The pig or the goose does not have the same cholic acid as the ox or man.

The bile salts appear very early in development, as soon in fact as the liver has been set apart; in the chick embryo they occur from the third day of incubation. They are not formed elsewhere in the body than in the liver, and in fact may be considered as the most characteristic products of the metabolism of this organ. They give a striking color reaction when they are mixed with a little sugar or formaldehyde and the solution placed in contact with concentrated sulphuric acid. A violet ring develops at the zone of contact. This is known as *Pettenkofer's* reaction. It is the same as Molisch's reaction for the detection of carbohydrates, with the exception that the bile salts take the place of the α-naphthol as the chromogenic substance. The reaction depends on the formation of an aldehyde-like furfural or oxymethylfurfural from the sugar by the acid and the condensation of this product with the bile acids, or some decomposition product of the latter, to a colored compound. The reaction is not at all specific for bile acids. It is given also by oleic acid, by many alcohols, aromatic substances and other compounds.

Preparation and properties of the bile salts. To make an impure solution of the bile salts, a simple method is to mix fresh bile with about 1 per cent. by weight of animal charcoal and to evaporate to dryness on the water bath. The dry residue is powdered and extracted with water and filtered. The bile salts are extremely soluble in water and go into solution while the pigments and some other impurities remain in the charcoal. The solution contains not only the bile salts, but some cholesterol, mucin, phosphatides and inorganic salts. The bile acids may be obtained from this solution, if it is sufficiently concentrated, by acidification.

Plattner's bile. Crystallized bile. This is prepared in the same way as the decolored bile just described, except that the dried residue containing the charcoal is extracted with boiling, absolute alcohol. The salts are very soluble in alcohol. By this means mucin, pigments and most of the inorganic salts are left behind. The alcohol after filtration through a dry filter into a dry flask, using care to prevent the entrance of water, is placed in a loosely stoppered flask and absolute ether run in until a precipitate begins to appear. It is then set apart in a cool place and the bile salts will crystallize out, provided the reagents have had little water in them. If water is present, they will come out as an oily liquid, which may later crystallize. The salts thus crystallized are very deliquescent and take up water rapidly from the air. They must be kept in a desiccator. They are known as Plattner's bile. From this partially purified product the acids may be separated. The cholesterol is separated by washing the salts with absolute ether. The salts generally have a bitter taste. They make a neutral solution. Some of them may be precipitated by saturating their solutions with ammonium sulphate.

Glycocholic acid.—$C_{26}H_{43}NO_6$. a. *Properties:* White, crystallizing in needles. Slightly soluble in cold water, more soluble in warm. Very soluble in strong alcohol, less in dilute. m.p. 193°. Shrinks 133-134° on rapid heating. Recrystallizes readily from dilute alcohol on cooling (Alcohol 10-30 per cent.) Slightly soluble in ether (1:1,000), and is precipitated from alcoholic solution by ether. Practically insoluble in benzol and chloroform. Rotatory power of alcoholic solution $(\alpha)_D^{18°} = +32.3°$ (Letsche). Na glycocholate, $(\alpha)_D^{13°} = +24.3°$, dissolved in water; $+27.8°$ in 90° alcohol (Letsche). The alkali and alkali earth salts are soluble in water; most others are insoluble. All of them are soluble in alcohol, except lead. Glycocholic acid is precipitated from solution by neutral acetate of lead. This distinguishes it from taurocholate, which is precipitated only by basic lead acetate. Its taste is bitter, with a sweet after taste. The affinity constant K is .0132. The acid is hence about as strong as lactic acid. Its salts do not make colloidal aqueous solutions.

b. *Decomposition by acids.* Cooked with acids it decomposes into glycocoll and cholic acid as follows:

$$\underset{\text{Glycocholic acid.}}{C_{26}H_{43}NO_6} + H_2O = \underset{\text{Glycocoll.}}{C_2H_5NO_2} + \underset{\text{Cholic acid.}}{C_{24}H_{40}O_5}.$$

By farther heating the cholic acid loses two molecules of water and is converted into *dyslysine*, $C_{24}H_{36}O_3$.

c. *Preparation.* From some biles containing principally glycocholic acid the glycocholic acid is very easily prepared by acidification after the addition of some ether. The bile of oxen often contains a good deal more of glycocholic acid than taurocholic. In some cases it is only necessary to filter the bile through charcoal, to partially or wholly decolorize it, slightly acidify to remove mucin, filter and then add a mixture of 5 parts HCl and 30 parts of ether to 100 parts of bile to have the glycocholic acid crystallize out. Generally, however, it will be found necessary to make first crystalline bile in the method described. This gets rid of the mucin. A strong solution of the crystalline bile is made after first freeing the crystals from cholesterol by anhydrous ether. Some ether is added to the solution, and then little by little a solution of sulphuric acid. This is added cautiously at first until crystallization begins and then more freely. The glycocholic acid crystallizes out, leaving the taurocholic acid in solution. This method does not always yield crystalline glycocholic acid, particularly if much taurocholate is present. It may be necessary to separate from taurocholic acid by precipitating with neutral lead acetate. This precipitates the lead glycocholate, leav-

amount of the sulphur as ethereal sulphate in human bile may vary from 6-17 per cent. of the total sulphur of the alcohol soluble solids.

It is possible by feeding animals cholic acid to increase somewhat the amount of taurocholic acid in the bile. The body seems to have a supply of taurine in reserve which it can unite with the cholic acid fed. This reserve of taurine is, however, soon exhausted and, if more cholic acid is given than enough to combine with the taurine, it appears in some other form than as taurocholic acid and chiefly in the form of glycocholic acid. The dog's body appears to form taurocholic acid by preference, but in the absence of sufficient taurine glycocoll is used instead. It thus happens that while taurocholic acid is generally present in large excess in the dog's bile, yet more or less glycocholic acid may be found there also. Indeed, after glycocholic acid is taken by the mouth, it reappears in the bile as such. But while the body has some reserve of taurine and a large reserve of glycocoll to pair with any cholic acid produced, there is no reserve of cholic acid to pair with the taurine. It has been found that, if cystine is given alone, there is no increase in the sulphur of the bile, the reason being that there is no reserve of cholic acid to pair with it; but, if cystine and cholic acid are given together, taurocholic acid is much increased.

Cholic acid.—$C_{24}H_{40}O_5$. This is also called cholalic acid. It is formed by hydrolysis from the conjugated acids just described. From 10 liters of ox bile Schryver obtained 225 grams of cholic acid, 75 grams of choleic and 40 grams of deoxycholeic acid in the crude crystalline form. Pregl and Buchtala obtained from ox bile 51.2 per cent. of the total fatty and bile acids as cholic acid.

a. *Preparation.* The following method of preparation of cholic and other acids of the bile was employed by Schryver (1912): 2.5 liters of fresh ox bile were mixed with 170 grams of NaOH dissolved in 300 c.c. of water and heated 30 hours in an iron digester with a reflux condenser. The mixture was then diluted with twice its volume of water and, while still warm, acidified with dilute HCl, vigorously stirring after each addition of acid. The crude acids separated as a viscid oil, pasty on cooling and sometimes granular. After standing overnight the mass was filtered, washed free from HCl by kneading in water, dried on the water bath and powdered. It was dissolved in an excess of dilute NH_4OH, so dilute that there was in solution finally not more than 5 per cent. of the ammonium salt, and some pigment was removed by boiling with animal charcoal. It was filtered, precipitated by dilute HCl, the precipitate washed with water, and dried in vacuo over $CaCl_2$ and soda lime. From time to time the surface lumps were removed and powdered, and finally it was a powder, which was recrystallized from hot acetone, filtered hot. On cooling, the crystals were filtered, using the pump, and washed in cold acetone until nearly free from mother liquor. The filtrates and washings united yielded a second crop of crystals when concentrated. The process was repeated until only a green mother liquor was left, from which no more crystals separated. All crystals were united. The separation of cholic, choleic and deoxycholeic acids was accomplished by the difference in behavior of the Mg salts. The magnesium salts of choleic and deoxycholeic acids are less soluble than cholic acid. The crystals are suspended in hot alcohol, a little phenolphthalein added and, while the alcohol is kept hot, NaOH is added until a faint alkaline reaction was obtained. The alcohol was then evaporated on the water bath, the sodium salts taken up in water so that 100 c.c. of solution corresponded to 1 gram of the crude crystals, filtered and made neutral to phenolphthalein with acetic acid. The solution was mixed with one-tenth its volume of 20 per cent. $MgCl_2$ and heated on the water bath. A bulky crystalline precipitate gradually formed. After heating 1 hour it was allowed to cool. The precipitate consisted of the Mg salts of choleic acid, deoxycholeic acids, with a little cholic acid. Cholic acid was obtained by acidifying the mother liquor. Choleic acid was separated from deoxycholeic by precipitating it as barium choleate, the barium salt being insoluble.

b. Prop*erties*. White, crystalline, very bitter substance. Almost insoluble in water, soluble in alcohol, but not very soluble in ether. It crystallizes from alcohol when water is added to the latter in the form of rhombic pyramids and tetrahedrons which contain a molecule of water. m.p. 198° (Bondi and Müller). Its solutions are dextro-rotatory $(\alpha)_D = +35°$. It gives Pettenkofer's reaction. It combines with water, with the halogen acids and potassium iodide. It is unsaturated. The alkaline salts and barium salt of cholic acid are very soluble in water, the other alkaline earths less soluble in water. On oxidation with nitric acid it forms first dehydrocholic acid, $C_{24}H_{34}O_5$, and finally oxalic acid, various volatile fatty acids and cholesterinic acid, $C_8H_{10}O_5$. Its structural formula is unknown, but it is related to cholesterol and the terpenes. A formula suggested is the following (Pregl, 1910):

$$COOH.CH \left\langle \begin{matrix} CH_2.CH_2 \\ CH_2.CH_2 \end{matrix} \right\rangle CH(CH_2)_5.CH \left\langle \begin{matrix} CH(CH_2OH).CH(CH_2OH) \\ CH_2 \text{———} CH \\ | \qquad\qquad | \\ CH_2.CH = CH.CHOH \end{matrix} \right\rangle CH_2$$

Another formula suggested by Panzer is given on page 430. It has a very characteristic iodine reaction, forming with the latter a blue-colored compound like starch. If 2 grams of cholic acid and 1 gram of iodine are dissolved in 40 c.c. of alcohol and to this is added 20 c.c. of a solution of KI containing 1 gram of I, a compound is formed which on the addition of water with constant stirring precipitates as a mass of bluish crystals. These are insoluble in water. They resemble starch iodide. The formula is supposed to be $(C_{24}H_{40}O_5)_4KI+nH_2O$. Other acids, such as deoxycholeic, hyocholic, bilianic, etc., do not give this reaction. When heated in water to 200° in an autoclave it forms the anhydride, *dyslisine*.

$$\underset{\text{Cholic acid.}}{C_{24}H_{40}O_5} = \underset{\text{Dyslisine.}}{C_{24}H_{36}O_3} + 2H_2O$$

Alkalies reconvert the dyslisine into cholic acid. The close relation of cholic acid to cholesterol and its probable derivation from that substance is shown by the fact that both cholesterol and cholic acid give the Lifschütz oxycholesterol reaction after oxidation with benzoylperoxide (see page 83) and that both yield on oxidation rhizocholic acid, $C_8H_8O_7$, which probably has the following composition:

```
   CH—C—COOH
   ‖   \
   ‖    C—COOH
   ‖   /
HOC—CH
     |
    COOH
```

Rhizocholic acid.

The same acid is obtained from camphor and oil of turpentine when oxidized, which shows that cholic acid and cholesterol are terpenes.

With HCl, cholic acid gives a color reaction (Hammarsten). Powdered cholic acid shaken with 25 per cent. HCl at room temperature colors the fluid at first yellow to green and then after several hours a bluish violet, which deepens for the next 24 hours. It shows an absorption band near D. The change takes place more rapidly on heating. Not all bile or cholic acids give this reaction.

```
     H H     H H      H OH    H H     H H
      \/      \/        \/     \/      \/
      C       C         C      C       C
     / \     / \  _____/ \    / \     / \
H2C      CH      CH        CH     CH     CH2
H2C      CH      CHCOOH    CH2    CH     CH2
     \/      \/                \/     \/
     CH2     CH                CH2    CH
             |                        |
             CH2OH                    CH2OH
```

Tentative formula of cholic acid (Panzer).

$C_{24}H_{40}O_5$

Glycocholeic acid.—$C_{26}H_{43}NO_5$. Besides the ordinary glycocholic acid found in ox and dog biles, a similar but somewhat different acid is also found, and it probably occurs elsewhere. It is present in smaller amounts than the ordinary glycocholic acid. It consists of glycocoll paired with choleic acid. Glycocholeic acid differs from glycocholic acid in the following points: It crystallizes in short, thick prisms; it has a very bitter taste, with only a weak sweet after-taste; it is soluble with difficulty in boiling water; the melting point is higher, namely, 175-176°; its aqueous alkali salt solutions are precipitated by the addition of barium, calcium or magnesium chlorides. In this way it may be separated from glycocholic acid. The sodium salt solution is far more easily precipitated by the addition of a saturated solution of NaCl than is the corresponding salt of glycocholic acid; and the pure solution of the alkali salt is precipitated by the addition of acetic acid, whereas the pure alkali salt solution of glycocholic acid is not precipitated by the addition of acetic acid; a solution of sodium glycocholate is precipitated by acetic acid, however, if free neutral salts such as NaCl are present in the solution (Wahlgren, 1902).

Glycocholeic acid may be separated from ox gall by Wahlgren's method as follows: The fresh bile is evaporated to a syrup and the mucin precipitated by the addition of alcohol. The alcohol is evaporated from the filtrate, the residue dissolved in water and precipitated by the successive addition of lead acetate, basic acetate and ammoniacal lead acetate. The first precipitate contains both glycocholate and glycocholeate. It is freed from lead by Na_2CO_3, filtered, the filtrate evaporated to dryness and the residue extracted with alcohol to free from carbonate. The glycocholate dissolves readily in the alcohol; the glycocholeate dissolves with difficulty. It may be separated by precipitating it with $BaCl_2$, which precipitates the choleate but not the cholate.

Taurocholeic acid.—This acid has been isolated from dog and ox bile. It is less abundant than the taurocholic acid. It consists of taurine and choleic acid. It contains more sulphur than taurocholic acid, namely, 6.25 per cent. instead of 5.94 per cent. in the sodium salt. It is separated from the taurocholate by being not so easily precipitated from its alcoholic solution by the addition of ether, as is the taurocholate, and when in aqueous solution it is precipitated by the addition of ferric chloride, whereas taurocholate is not precipitated (Gullbring, 1905). It is intensely bitter with no sweet after-taste. It has not been obtained in a crystalline form. Like taurocholic acid it is readily precipitated by NaCl, as a thick honey-like mass.

Choleic acid.—The probable formula for this acid is $C_{24}H_{40}O_5$ (Lassar-Cohn). It thus has the same formula as cholic acid, although its discoverer, Latschinoff, ascribed to it the formula $C_{25}H_{42}O_4$, for the water-free salt. Choleic acid is very much like cholic acid, but is less soluble in water, alcohol, and glacial acetic acid. The water-free acid melts between 185-190°; the water-containing acid melts at 135-140° and at 150° forms a homogeneous liquid. After repeated recrystallizations from glacial acetic acid it melts at 145°. This is probably the acetyl compound. This is the same melting point as deoxycholic acid, which is probably identical with choleic acid, according to Gullbring. (See, however, Schryver). It forms an insoluble barium salt by which it can be separated from cholic acid.

Other cholic acids.—In the bile of other animals than the ox and dog, other cholic acids and their conjugates are found. Thus in the pig hyoglycocholic acid occurs, and in the bile of geese chenotaurocholic acid. Hyoglycocholic acid has been given the formula $C_{27}H_{43}NO_5$. Chenocholic acid is $C_{27}H_{44}O_4$. There is little doubt that a number of these more or less closely isomeric cholic acids occur in different biles.

Soaps.—Bile contains small amounts of the sodium salts of various fatty acids (myristic, palmitic, stearic), among which sodium oleate may be especially mentioned. The presence of these soaps affects the ease of precipitation of the bile salts by neutral salts, the presence of sodium oleate particularly interfering with the salting out. From 10 liters of ox bile Schryver obtained the following amounts of fatty and bile acids:

Cholic acid	225 grams	350 grams of crude crystalline acids.
Choleic "	75 grams	
Deoxycholeic acid ...	40 grams	
Fatty acids	20 grams	65 grams from mother liquors of acetone crystallization.
Pigment (green) acid	12 grams	
Glassy acid	35 grams	

Pregl and Buchtala found in the bile of oxen of Gratz 10 per cent. of the total acids as fatty acids; 51.2 per cent as cholic; 11.9 per cent. choleic; 13.5 per cent. deoxycholeic; 12.6 per cent. non-crystallizable.

Cholesterol in bile.—Cholesterol has been found in the bile of all animals in which it has been looked for, with the exception of the hippopotamus. While the amount of cholesterol in human bile is not very great, its importance is increased by the fact that it is one of the chief constituents of gall stones. The amount in human bile is given as 1.6 parts per thousand by Frerichs; and as 1.00 per thousand in ox bile. In venous blood there is in the serum only 0.09 p.m. (Becquerel and Rodi), and in the whole blood only about .44-.75 p.m.

Cholesterol, as its name implies, i.e., solid bile, is one of the commonest constituents of gall stones of which it may form anywhere from 20-90 per cent. The general properties of this substance, so common in all cells, has already been given, page 81, and here only the amount in the bile, and its origin, function, and fate will be considered.

Cholesterol is entirely insoluble in water or in salt solutions whether neutral, acid or alkaline; its solution by the bile is, therefore, a singular fact. The bile is in fact able, as Moore and Roaf showed, to dissolve even more cholesterol than is ordinarily found in it. If solid pieces of cholesterol are put into bile they dissolve (Naunyn). This power of solution of the bile depends on the presence of the bile salts and specifically upon the cholic acid radicle of the salts. The bile salts themselves are extremely soluble. The cholic acid part of the salt is probably closely related chemically to cholesterol. It is probable that cholesterol unites either physically or chemically, in some way not yet known, with the cholic acid of the bile salts, and is thus held in solution. It will be recalled that the bile salts are hemolytic agents and in this respect act like the saponins and these saponins have been shown by Windaus and Yogi to form easily dissociable compounds with cholesterol. It is not improbable, therefore, that cholesterol unites similarly with the hemolytic group of the bile salts and the compound thus formed is soluble in the bile.

The origin of the cholesterol of bile is not yet certain. It may be derived in part from the food and in part from the cholesterol of the red blood cells, which are destroyed. A part of the cholesterol thus passing in might be changed to cholic acid, a part might be secreted unchanged. A part might be made in the liver from fat or sugar by the metabolism of that organ. Dorée and Gardner have found only traces of cholesterol in feces after feeding dogs various cooked foods, such as oatmeal and milk, beef and mutton, horseflesh or other foods. Stercorol was excreted on a diet of raw sheep's brains. If cholesterol is given in food to rabbits some is reabsorbed and finds its way to the blood, causing there an increase in both the free cholesterol and the cholesterol esters, as measured by the digitonin method. If phytosterol was fed to rabbits it, also, was absorbed in part, resulting in an increase of free cholesterol of the blood; but phytosterol did not appear as such in the blood. While the reduction of cholesterol to stercorin happens in the body, cholesterol is not reduced if added to the feces in vitro, possibly owing to lack of solubility. Dog's feces contain the unchanged cholesterol of the bile.

Since bladder bile is much richer in cholesterol than that from a fistula and the difference is much greater than the increase of concentration of the salts, it is generally concluded that cholesterol is added to the bile largely by the epithelium of the bladder. It is especially Naunyn who has defended this view and brought it into relation with the formation of gall stones. It is often the case that the kernel of the gall stone appears to be bladder epithelial cells and it is necessary to have some injury to the bladder, an inflammation or traumatism, before stones form. The experimental introduction of crystals of cholesterol into the bladder did not cause the formation of concretions unless infection of the gall ducts occurred also. The evidence does not appear to be at all conclusive that cholesterol is added to the bile in the gall bladder, and while it may well be that some of the cholesterol finds its way through the wall of the gall ducts or bladder, it would appear more probable that it is formed and secreted with the bile salts, to which it is so closely related chemically. It may be that the poor content of cholesterol in fistula bile, as compared with bladder bile, is due to a more complete conversion of the cholesterol into the bile salts in the former case when the bile is unusually poor in bile salts, than happens normally when there are bile salts being reabsorbed from the intestine. The presence of cholesterol in fistula bile in considerable quantities is evidence that certainly much of the cholesterol is secreted by the liver itself. It is, on the other hand, probable that cells of epithelium escaping into the bile may become impregnated with cholesterol and serve as the nucleus for the formation of a stone. Bile of the hippopotamus contains no cholesterol. It would be interesting to know whether the blood of the hippopotamus contains cholesterol. Perhaps some African physiologist may make some interesting discoveries by studying the bile and blood of this animal.

The functions of cholesterol have already been discussed. It has no function in digestion so far as we know. It is a singular fact that bile hastens the action of the lipase of the pancreas, whereas the cholesterol it contains is said to have the power of inhibiting lipolysis. There is a curious apparent contradiction in these statements. May the liver by excreting too much cholesterol reduce the power of the blood to withstand hemolysis, since cholesterol counteracts hemolysis? Does the liver make anything more out of cholesterol than the bile salts, if this is the origin of these substances? Are there any active oxidation products of the nature of bufonin made here in small quantities and are they active in the physiology of other parts of the body? These are some of the questions which must be left to the future for answers.

Whether the cholesterol secreted in bile is reabsorbed with the bile salts is not yet clear, since the metabolism of cholesterol has not yet been worked out. Certainly some of that taken in the food, and presumably some of that found in bile, is reabsorbed. But to what extent it is reabsorbed is uncertain; and whether the cholesterol of the body is formed in the body or absorbed from the food eaten is equally uncertain. Inasmuch, however, as the sterol of herbivorous animals is cholesterol and not phytosterol, the sterol of plants, and since cholesterol is found in all cells, it is probable that the animal body has the power of making

Stercorin. This is obtained from the dried and powdered feces by extraction with ether; the ethereal extract is decolorized with charcoal and evaporated. The residue is extracted with hot alcohol, the alcohol extract saponified with KOH, mixed with powdered salt and evaporated. The solid mass is extracted with ether, the ether washed with water until neutral in reaction, then the ethereal extract filtered and dried. The residue is extracted with boiling alcohol and on cooling stercorin crystallizes out in long fine needles radiating from centers.

Stercorin (Coprosterin) is soluble in $CHCl_3$, and the solution gives with concentrated H_2SO_4 at first a yellow color, which, by standing, changes to an orange and then a dark red. In the Liebermann test it gives at once a blue color, changing to a green. The most probable formula is $C_{27}H_{48}O$. Unlike cholesterol it does not take up bromine. The specific rotation $(\alpha)_D = +24°$.

In the feces of horses, cows, sheep and rabbits there is a *hippocoprosterol*, or hippostercorin, $C_{27}H_{54}O$, which is the phytosterol of grass, which has passed through the intestine unchanged. Gardner and Dorée have proposed to call it, therefore, *chortosterol* (Gr. *chortos*, grass). This substance melts at 78.5-79.5°. It is optically inactive and gives no cholesterol color reactions. It is stated by Bondzynski that dog's feces do not reduce cholesterol to stercorin.

The following experiment on the influence of diet on the excretion of stercorin was tried by Bondzynski and Humnicki. In five days on a normal diet the weight of stercorin in the feces was 4.30 grams. The next five days one gram of cholesterol was daily added to the food, making 5.0324 grams in all. In these five days there were excreted 5.835 grams of stercorin, and in the following five, when no cholesterol was added to the diet, 7.3694 grams, making a total increase of 4.629 grams. Of cholesterol they found only 0.5326 gram. In the following two days 1.4071 grams stercorin and in the second period of 1 gram cholesterol per day, for five days, 6.2812 grams stercorin were obtained. So that in one case only 10 per cent. and in the other only 3 per cent. of the cholesterol taken were refound in the feces, the rest having been transformed into stercorin. In human bile cholesterol has been found to be between .06-.6 per cent. It may be said then, roughly speaking, that about one gram a day of cholesterol may be discharged into the intestine from the bile. The feces contain, on the average, about 0.9 gram of stercorin per day.

Phospholipins in bile.—Bile contains also considerable quantities of lecithin and other phospholipins. In human and dog bile, the lecithin computed from the phosphorus is generally given as from 1-7 per cent. of the alcohol-soluble substances, but Hammarsten's investigations of human bile show that the amount of phospholipin is much greater than this and is indeed as high as that of the polar bear. The alcohol-soluble part of the bile contained 1.047 per cent. P, which calculated as lecithin would be 29.75 per cent. In the bile of the polar bear 23.5 per cent. of the alcohol-soluble solids were calculated as lecithin. Thudichum states that there is no lecithin in ox bile, but in place of it another phospholipin, in which the relation of P : N is as 1 : 4. Lecithin is certainly present in polar bear bile, according to Hammarsten, because a phospholipin was obtained in which P : N as 1 : 1 and from which stearic, oleic acids, glycerol and choline were isolated.

The amount of phosphorus in the alcohol-soluble solids of the bile of different animals and the amount of lecithin, computing the latter from the phosphorus on the supposition that all phospholipin is lecithin and contains oleic, palmitic or stearic acids and so 3.94 per cent. of P, is given in the following table:[1]

THE AMOUNT OF LECITHIN IN THE ALCOHOL-SOLUBLE SOLIDS OF VARIOUS BILES. (Hammarsten.)

	P— per cent.	Lecithin— per cent.
Polar bear	0.911 - 1.14	23.12 - 28.96
Man (bladder bile)	0.048 - 1.17	1.33 - 29.75
Man (fistula bile)	0.100 - 0.611	2.54 - 15.5
Dog	.768	19.50
Land-bear	.502	12.74
Orang-utang	.420	10.67
Pig	.334	8.47
Python	.332	8.43
Sheep	.289	7.35
Musk-ox	.272	7.04
Hippopotamus	.191	4.86
Ox	.181	4.60
Seal (Phoca Gr.)	.168	4.27
Goose	.162	4.10
Walrus	.043	1.08
Sea-wolf	.033	0.81
Haddock	Not determinable	...

There are probably other phosphatides (phospholipins) in the bile than lecithin, since one phospholipin had N : P as 3 : 5 and another N : P : : 2 : I. A similar phosphatide was obtained from brain by Thudichum and named sphingomyelin.

Nothing is certainly known concerning either the function, origin or fate of this bile phospholipin. Possibly it is derived, like the cholesterol, from the red blood corpuscles and it no doubt contributes toward giving bile its power of dissolving fats and cholesterol. Presumably it is split and digested in the intestine by the lipase of the intestinal secretions. The lecithin content is highest in the polar bear bile and this animal feeds on seals, which contain large quantities of fat. The great varia-

[1] Hammarsten, *Ergebnisse der Physiologie*, 4, 1905, p. 15.

tion in the biles of human beings would perhaps point to some variation in the diet as a cause. The phospholipins of the bile are combined, it appears, in part at least, with the bile salts, since they are precipitated in part with them by ether from the alcohol solution, and in part they remain in the ether and hold some of the bile salts in solution in a liquid in which, when pure, they are insoluble. It may be, however, we are dealing here with different phosphatides, one of which, like cephalin, may be insoluble in ether; the other, like lecithin, soluble. The fact that some of the phospholipin is precipitated with the bile salts by ether and that some of the bile salts remain in solution with the lecithin makes all the older analyses of the bile which did not take account of this fact unreliable.

Mucin.—Bladder bile is viscid, or slimy, due to its containing a mucin-like substance secreted by the walls of the gall bladder. Most of this mucin-like substance is not a true mucin, but a phosphoprotein (nucleo-albumin), which yields no sugar on hydrolysis and contains much more nitrogen (16.14 per cent.) than a true mucin. A small amount of true mucin may, however, be present in human bile, but according to Hammarsten this is secreted by the gall ducts and is certainly present in small amounts. This so-called slime or mucin can be precipitated by alcohol and it yields ordinarily a small amount of a reducing sugar on hydrolysis. Since many substances besides mucin, such as phosphatides, glycogen, dextrin, etc., have this same property, the appearance of a reducing substance under these circumstances is no proof that mucin is present. Wahlgren has obtained a secretion from the human gall bladder free from bile and finds it generally colorless and containing a nucleo-albumin, a little globulin and albumin. The presence of a phosphoprotein is, therefore, undoubted, but whether true mucin is present or not cannot be stated with certainty. No function has been found for the mucin-like bodies in the bile.

BACTERIAL DECOMPOSITION OF FOODS IN THE INTESTINE.

Both the unabsorbed food and the digestive juices are subjected in the intestine, and particularly in the large intestine, to the decomposing action of myriads of bacteria and the products thus formed are of the greatest importance to the organism. Many of them are toxic, producing headaches, drowsiness, or, at times, irritability, causing depression and a general feeling of malaise. By the destruction of red blood corpuscles they are one of the factors in anemia. That these products predispose, also, to various infections, particularly of the skin, but of other parts of the body as well, there can be no doubt. The study of these decomposition products and the discovery of methods for limiting them becomes, therefore, of great importance in hygiene. The toxic substances are derived in large measure from the proteins and are formed by putrefactive processes. The carbohydrates are decomposed by fermentation. But there is no essential difference in kind between fermentation and putrefaction, although, strictly speaking, a fermentation should involve the liberation of a gas.

The bacteria are found all the way from close below the pylorus to the rectum in constantly increasing numbers, but the numbers in the small intestine are small compared to the large. It is in the large intestine, the ascending, transverse and descending colon in which the intestinal contents, not reabsorbed, are stored for some time that the main putrefaction occurs. The chyme leaving the stomach is often sterile, but as acidity is neutralized a constantly increasing flora is found. The contents of the small intestine are sometimes weakly acid throughout and this reaction is unfavorable for the bacteria. The intestinal contents even at the junction of the colon with the ileum are not at all fecal-like. They are generally slightly acid in reaction, semifluid in consistence, and they are nothing more than the undigested remnants of the foods, cellulose, some starch, some meat fibers, seeds, particles of peas not fully digested and also, in considerable part, the unreabsorbed secretions of the alimentary canal. It sometimes happens that an artificial fistula must be made at the end of the small intestine because of tumor of the large intestine, and by this means knowledge has been obtained of the character of the material going into the large intestine and the time it takes to pass through the intestine. It has been observed that the discharge in such cases is not of a fecal nature. The time required for the food to traverse the stomach and the small intestine to such a fistula naturally varies somewhat with the character of the meal and the idiosyncrasy of the patient, but in round numbers it may be said that the first discharge from the end of the ileum takes place about four hours after eating and the discharge persists for about two hours. The amount discharged is small compared to the bulk of the food eaten; practically all water is absorbed and 90 per cent. of the solids of the foods.

Formation of the feces.—The transformation of the undigested remnants of the food and the secretions of the intestine not reabsorbed into the typical dark brown feces occurs in the large intestine, where during a period varying from ten hours to two days the remnants of the foods undergo putrefactive changes and fermentative decompositions produced by myriads of bacteria. Since the absorbing powers of the large intestine are very great, these products are reabsorbed and, coursing throughout the body in the blood, produce in it changes already mentioned, finally finding their exit in the urine, the perspiration or in the breath, which when this decomposition is unusually abundant may have a very marked fecal odor.

The number of bacteria in the feces is almost incredibly large. Something between one-half and a fourth of the dry matter of the feces consists of the bodies of bacteria. In a milligram of feces there are found by culture methods of living bacteria in normal men on a somewhat restricted diet 26,000,000 bacteria. There are in addition large numbers of dead bacterial cells, which can be computed by centrifuging an emulsion of feces so that the coarser fragments are thrown out and only the bacteria left in suspension, and then counting under the microscope the number of bacteria in a given volume. By this means the total number of living and dead bacteria excreted in the feces per day by healthy young men, on a restricted diet in a metabolism experiment, was found to be on the average 500×10^{10}.

The weight of dried feces passed per day is on the average in such conditions 15-25 grams, and of moist feces 80-120; but the amount may be much larger, particularly on a diet containing a good deal of indigestible substance such as cellulose. Of the 15-25 grams of dry matter 4-5 grams consist of bacterial bodies. Of the total nitrogen in the feces, which may be on the average about 1.5 grams, one-half is bacterial nitrogen.

The foregoing figures will give an idea of the enormous numbers of bacteria in the feces and will make one appreciate the probable very great importance of the products of their metabolism in human life.

The kind of bacteria which are present is, of course, of even greater importance than their numbers, since the character of the putrefactive and fermentative products formed is dependent on the species of bacteria. Ordinarily many different kinds of bacteria are found and the character of the flora varies from time to time. The greater number of the bacteria are bacilli of the type of the bacillus coli communis, but there are many races of this organism, some being much more toxic than others. To this same group of organisms belong the paratyphoid and the dysentery organisms and the connection between them and pathogenic forms, such as the typhoid bacillus, is still uncertain. The bacillus coli communis is a facultative anaërobe, that is it can grow both in the presence and in the absence of oxygen. It ferments glucose and lactose, forming lactic acid, alcohol and carbon dioxide; in anaërobic conditions, in the absence of sufficient carbohydrate, it forms from the proteins skatole and indole, the substances in the feces which produce the typical odor; and it will split off hydrogen sulphide from cysteine and cystine.

The study of the bacterial flora and in particular the careful study of its variation in health and disease and the different strains of the colon bacillus from different individuals is a matter of the greatest hygienic importance, in which as yet hardly more than a beginning has been made.

Bacterial decomposition of the carbohydrates. The bacteria form from the unreabsorbed carbohydrates and cellulose, carbon dioxide, butyric, lactic and other acids, alcohol, hydrogen, methane and other substances. Of these hydrogen, methane and carbon dioxide form a large part of the gases of the intestine. An analysis of human intestinal gases has shown the following composition in 100 volumes (Ruge):

Vegetable diet	CO_2, 21-34;	H_2, 1.5-4 ;	CH_4, 44-55;	N_2, 10-19
Meat diet	CO_2, 8-13;	H_2, 0.7-3 ;	CH_4, 26-37;	N_2, 45-64
Milk diet	CO_2, 9-16;	H_2, 43-54;	CH_4, 0.9 ;	N_2, 36-38

Lactic acid and alcohol are reabsorbed and burned by the body. That still other substances are produced from the carbohydrates is undoubted, but their nature and action are unknown. None of the products of carbohydrate bacterial decomposition is harmful so far as known, although the gas formation may be distressing.

Bacterial decomposition of the fats. The fermentation of the unreabsorbed fats and fatty acids seems to have been but little studied. The author has not found any data on the subject.

Bacterial decomposition of the proteins. It is, however, from the putrefactive decomposition of the proteins that the most toxic substances are produced. The amino-acids of the proteins set free by the intestinal enzymes are physiologically quite inert. They are absorbed and serve as foods. If they are introduced directly into the blood, they cause no physiological reaction whatever. The bacteria, however, like the cells of the body, have the power of tearing these amino-acids to pieces and some of the products are very toxic.

In the first place, it is certain that they have the power of deamidizing the amino-acids, setting free ammonia and leaving the fatty acid. It is not yet certain whether in the intestine this process involves the oxidation of the amino-acid to the ketonic acid first, as is the case in the oxidation of the amino-acids in the tissues of the body (see page 806), or whether the deamidization may be brought about by a hydrolysis. In any case, if the oxy-acids are thus formed, they are subsequently reduced so that from the amino-acids the fatty acids are produced. This may be illustrated by the decomposition of tryosine or phenyl alanine, which has been most carefully studied:

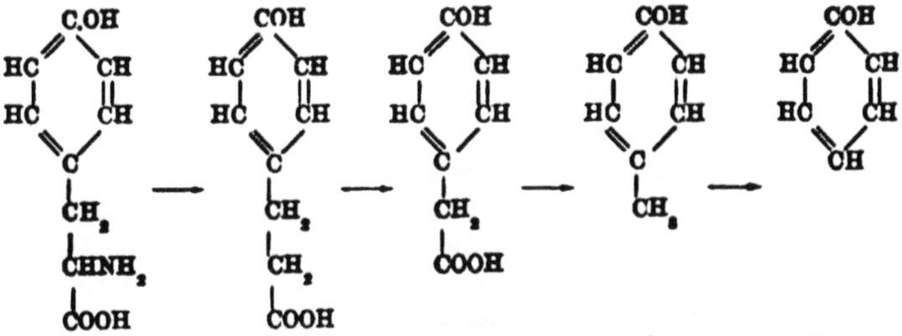

Tyrosine. → p-oxyphenyl-propionic acid. → p-oxyphenyl acetic acid. → Para-cresol. → Phenol.

The putrefaction of tryptophane probably is preceded by a deamidization:

Tryptophane. → Indole propionic acid. →

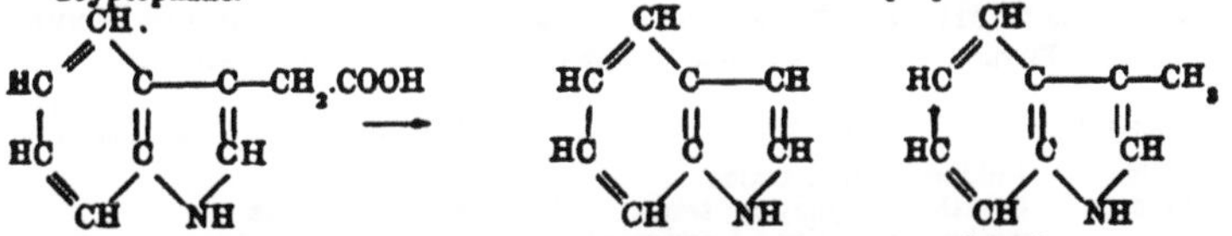

Indole-acetic acid. → Indole. Scatole.

A very important change is the splitting off of carbon dioxide by the action of so-called carboxylase bacteria. This may happen either before or after deamidization. If it happen before deamidization, amines of a highly toxic character are produced. Such amines produced by bacteria from amino-acids are called *ptomaines* by Gautier. As this is a process of great importance to the body we may examine it a little more at length. Thus from histidine there is formed imidazolylethyl amine; from alanine, ethyl amine; from tyrosine, phenyl-ethyl amine; from arginine, agmatine, or guanidine-butyl amine. Their formation is illustrated in the following reactions:

1. $CH_3.CH(NH_2).COOH = CO_2 + CH_3.CH_2(NH_2)$
   Alanine. Ethyl amine.
2. $C_6H_4(OH).CH_2.CH(NH_2).COOH = CO_2 + C_6H_4(OH).CH_2.CH_2.NH_2$
   Tyrosine. p-hydroxy-phenyl-ethyl amine.
3. $C_3N_2H_3.CH_2.CH(NH_2).COOH = CO_2 + C_3H_3N_2.\ CH_2.CH_2.NH_2$
   Histidine. Imidazolylethyl amine.
4. $NH_2—C(NH)—NH.CH_2.\ (CH_2)_2.CHNH_2.COOH =$
   Arginine.
   $CO_2 + NH_2.C(NH).NH.\ CH_2.(CH_2)_2.\ CH_2NH_2$
   Agmatine.

Such substances have been called by Kutscher, aporrhegmas (Gr. *apo*, from; *rhegma*, a fracture). Such aporrhegmas are the active principles of ergot. These substances have, in many instances, a powerful effect on the blood pressure and may be of great importance in the production of high blood pressure and the resulting thickening of the artery walls. They are very common in the metabolic products of plants and they are formed also in the metabolism of animals, since the power of splitting carbon dioxide from the amino-acids appears to be very general. Imidazolylethyl amine, or histamine, produces vasodilation, lowers the coagulability of the blood, and has been isolated from Witte's peptone and from the mucosa of the intestine. Adrenaline, the active principle of the supra-renal glands, is a substance of this nature, being a methylated ethyl amine derivative of tyrosine of the following composition:

$$C_6H_3(OH)_2.CH(OH).CH_2.NH(CH_3)$$

It can hardly be doubted that similar amines are produced from the other amino-acids, such as tryptophane, which gives rise to indole ethyl amine.

APORRHEGMAS.* ALL THE FRAGMENTS † OF THE AMINO-ACIDS WHICH ARE FORMED IN A PHYSIOLOGICAL WAY DURING THE LIFE OF PLANTS AND ANIMALS.

Amino-acid	Aporrhegma	Methylated aporrhegma	Methylated amino-acid
Histidine	Imidazolylethyl amine		
Arginine	Ornithine: agmatine; tetramethylenediamine; δ-aminovalerianic acid	Tetramethylputrescine	
Lysine	Pentamethylenediamine (Cadaverine)		
Glutamic acid	γ-aminobutyric acid	(γ-Butyro-betaine α-oxy-beta-butyro-betaine)	(γ-trimethyl amino butyric)
Aspartic acid	-alanine		
Glycocoll	Methyl amine		Sarcosine. Betaine
Alanine	Ethyl amine		
Leucine	Isoamylamine		
Proline	Pyrrolidine	N-methyl pyrrolidine	Stachydrine
Phenyl alanine	Phenyl-ethyl amine		
Tyrosine	p-oxyphenylethylamine	Hordenine	Surinamine
Tryptophane	Indole; scatole; indolethylamine		Hypaphorine (trimethyl tryptophane betaine.)

* Kutscher and Ackermann: *Zeit. f. physiol. Chem.*, 69, 1910, p. 265.

† It is better to limit the term to the nitrogen-containing products. A. P. M.

The decomposition of cysteine has not been fully worked out, but from the presence of mercaptans, such as methyl and ethyl mercaptans, $CH_3.CH_2SH$, in the feces, it is probable that there is a preliminary decarboxylation, with the formation of a thio-ethyl amine as an intermediate product:

$SH.CH_2.CHNH_2.COOH \longrightarrow CO_2 + SH.CH_2.CH_2NH_2 \longrightarrow SH.CH_2.CH_3 + NH_3$

Cysteine. Thioethyl amine, or amino-ethyl mercaptan. Ethyl mercaptan.

If the splitting off of the carbon dioxide occurs after the deamidization an amine cannot, of course, be formed, but the next lower carboxylic acid is produced by way of the aldehyde. Thus from tyrosine there may first be formed phenyl-pyruvic acid, which may be reduced to phenyl-lactic acid, reabsorbed and re-excreted in the urine; or the phenyl-pyruvic acid may be split into phenyl acetaldehyde and carbon dioxide, and the former be oxidized into phenyl-acetic acid, which is excreted in the urine. Phenyl-acetic acid is the mother substance of homogentisic acid and other aromatic compounds of the urine. Phenol or cresol may be set free.

These deamidized compounds have not as yet been shown to have any physiological action. But from tryptophane by a similar decomposition indole and scatole are set free and these bodies are toxic and give part of the bad odor to the feces. The formation of indole and scatole is shown in the reactions just given, although the intermediate products have not been thoroughly worked out.

By the decomposition of cysteine and cystine hydrogen sulphide is formed. This is reabsorbed readily and produces headaches and depression even when reabsorbed in small quantities and it is presumably one of the factors causing solution of the red blood corpuscles and so the anemia seen in those having chronic constipation. Mercaptans, that is ethyl or methyl sulphide, may also be formed and these are very ill-smelling compounds.

From the conjugated proteins the decomposition of the prosthetic group, such, for example, as that of the nuclein bases, may give rise to other products of importance to the body. But this matter has not yet been sufficiently investigated to permit of any definite statements being made.

It has been suggested that substances having the property of raising blood pressure may be produced from tyrosine in putrefaction and that these substances are active in causing arteriosclerosis and the ills which follow from this. Since a man is said to be as old as his arteries it would appear possible that intestinal putrefaction may be a factor in the production of the decrepitude of old age, as Metchnikoff has suggested. To what extent putrefaction produces premature decrepitude cannot be stated without more investigation.

There can, however, be no question of the importance of these putrefactive substances in the general well-being of the individual. We are all constantly exposed to food poisonings, which, when slight, are generally overlooked as the true cause of inefficiency, depression, sluggish mental processes, dissatisfaction or abnormal irritability. Particularly chicken, veal and pork are liable to contamination from some person handling these meats in the kitchen, who may carry a peculiar race of bacteria. Such food poisonings may come from infected meat-choppers, so that hashed meat is more apt to be a source of trouble than unhashed. If the bacteria are not killed in the subsequent cooking, and the hash is kept, as it is in restaurants, it may give rise to symptoms of food poisoning more or less marked. These symptoms generally come on in 1-5 hours after a meal, if they are due to ptomaines or toxic substances already formed in the meat before eating; but the bacteria themselves may develop in the intestine and form toxic substances. In such cases the onset of the symptoms is delayed coming 12-48 hours after the meal, varying with different individuals. If the symptoms are not well marked so as to lead to cramps, prostration and diarrhea, by which the bacteria are swept out of the system, only feelings of drowsiness, headache, migraine, lassitude or depression often preceded by mental or sexual excitement are produced, and the real cause of the trouble is overlooked. The symptoms are often extremely baffling and are referred to the nervous system, the ductless glands or any other rather than the true cause. Moreover, when once such a bacterium is lodged in the canal it may persist for a long time and the effects of a single food poisoning in an experiment on human metabolism were found to be clearly perceptible for a month or longer. These facts make it desirable, in case of headache, drowsiness or depression, even though there be no symptoms of deranged digestion or constipation, to try the effects of a good purge. The effect on the mentality is often remarkably prompt.

The putrefactive processes in the intestine have also a remarkable relation to the skin. In nearly all cases of excessive intestinal putrefaction the skin, and perhaps the whole organism, has its vital resistance lowered. Pimples, pustules, acnes and boils are constantly forming. The spotted skin is generally, although not always, a sign of intestinal putrefaction. This condition is generally referred by the laity to impure blood; and to cure it sulphur, burdock root and other local remedies are recommended as blood purifiers. Without exception all these blood purifiers, so called, are intestinal purgatives and their therapeutic action is due, in large measure at least, to their power of sweeping the intestine clean. Cold sores on the lips often have a similar origin. The increased sensitiveness of the skin to infection of the hair follicles is shown also in many cases of eczema, which may greatly improve if the diet is limited or if fasting is practised. Many hidden or chronic slight inflammations are often stimulated to become acute, or more active, by intestinal putrefaction. Thus colds may develop, catarrh become worse, chronic nephritis of a mild type become acute, erythemas develop in various parts of the body or even old healed wounds with bad circulation may inflame and break out afresh from this cause; and by attention to diet, or by the use of good purges, these results may be avoided.

The question is still unsolved whether besides their evil products the bacteria of the colon are also producers of some essential products. Bacteria are certainly not necessary for development, since the alimentary canal of the new-born is sterile. Chickens have been hatched and raised under sterile conditions; and guinea pigs removed by Cæsarean section have been found to grow and utilize their food when fed on sterile diets under aseptic conditions. As yet these bacteria have not been found to do good; but they certainly cause much evil. And while it will not do to condemn them completely without farther investigation, it would appear at the present time the part of wisdom to reduce putrefaction in the intestine as far as possible.

Methods of reducing putrefaction. The easiest method of reducing putrefaction is by a strict limitation of the diet, a reduction in the proportion of meat and a more thorough mastication of that which is eaten. By the careful use of these two precautions Mr. Horace Fletcher and others who have adopted his methods have been able to so reduce putrefaction in the intestine that the feces are entirely without odor. The general condition of these men has been greatly improved. Men and women of sedentary habits are very apt, particularly after the age of forty, to eat too much meat. The exact quantity and the particular quality of food most conducive to happiness, health, efficiency and retention of vigor has not yet been determined; but two facts have been found, namely, that the retention of vigor late in life, a green old age, generally goes with an abstemious diet; and in the second place, many have been greatly improved by a strong reduction in the amount of food eaten and in particular by a reduction in the amount of protein food. We cannot do better than to follow the wise advice of Don Quixote to his squire, Sancho Panza, as the latter was about to depart for the Island. "Eat little at dinner and less at supper; for the health of the whole body is tempered in the laboratory of the stomach."

Another method of reducing putrefaction is by eating carbohydrate food rather than protein, for the bacteria will not produce these putrefactive products from the proteins as long as there is carbohydrate food at their disposal. This peculiarity of the bacteria, which has been studied by Kendall, we shall have to return to under the chapter on metabolism, since it illustrates a general property of all living matter, namely, that the carbohydrates spare the proteins, and it is only in the absence of the carbohydrates that the proteins are attacked and torn to pieces to supply energy. By carbohydrate fermentation, also, acidity is increased and this also has a part in checking bacterial putrefaction.

since many bacteria cannot metabolize in the presence of acid. But the main action of the carbohydrate is that just stated, that as long as it is present it is utilized as a source of energy and the protein only to build the protoplasmic machine.

Summary of digestive changes.—The chief changes undergone by the foods in the intestinal tract by which they are made available to the body may be briefly summarized as follows:

The proteins by the action of pepsin hydrochloric acid of the gastric juice, by the intestinal and pancreatic juices containing enterokinase, trypsin and erepsin, are resolved into the amino-acids of which they are composed. The compound proteins, such as the nucleoproteins, are digested in part by the proteolytic enzymes and in part by the nucleases of the pancreatic juice, with the formation of phosphoric acid, nuclein bases, pyrimidin bases, and presumably carbohydrate, from the nucleic acid. Hematin is split off from hemoglobin. By the action of the bacteria some of these amino-acids are further decomposed, losing ammonia and being converted into fatty acids, or by decarboxylization being made into amines, some of which have a very powerful action on the blood pressure and other functions of the body. Other ill-smelling and harmful substances are produced, such as some of the mercaptans, sulphureted hydrogen, indole and scatole.

The carbohydrates taken as sugars and starches are digested by the ptyalin of the saliva, the amylopsin, maltase and lactase of the pancreas, and the invertin of the intestinal juice so that they are all reduced to the state of monosaccharides. Some of them are further broken up by the bacteria, with the formation of lactic acid, alcohol, marsh gas, hydrogen and butyric acid.

The fats acted upon by the lipase of the stomach and that of the pancreas and intestine are converted into fatty acids and glycerol, and by means of the bile salts are carried into the intestinal epithelial cells.

In short, the main object of the whole process of digestion appears to be to resolve the various food substances into those common building stones, amino-acids, monosaccharides and fatty acids, which are the common basis of all proteins, carbohydrates and fats. Thus each organism can use these building stones in the proportion and order it needs to construct its own proteins and organized matter, which in each organism has an architecture as distinct and characteristic as the form of the organism itself.

Finally by this decomposition into building stones but very little energy has been lost, and only a very small amount of heat set free. The advantage of this to the organism is obvious, for it means not only that energy has not been lost, but that very little energy will be required to reconstruct the tissues of the body.

We may now pass on to the consideration of the changes undergone by these building stones during absorption, and during their stay in the blood and the tissues.

REFERENCES. Secretion of Bile.

1. Human biliary fistula. *Pfaff and Balch:* Jour. of Expt. Medicine, 2, p. 49, 1897.
2. In fasting. *Albertoni:* Arch. Ital. de Biol. 20, p. 134, 1890.
3. In anemia. *Korentschewsky:* Arch. des scien. Biol. d. St. Petersburg, 16, p. 252, 1911. See also p. 490.

Composition of Bile

1. Reaction. *Quaglianello:* Atti R. Acad. dei Lincei, Roma, 20 (11), p. 312, 1911.
2. Osmotic pressure. *Dreser:* Archiv f. expt. Path. u. Pharm., 29, 1892. *Bonnani:* Bioch. Centrlblt., 1903.
3. Toxicity. *Bunting and Brown:* Jour. of expt. Med. 14, 1912, p. 444.
4. Iron in bile. *Beccari:* Archives Ital. de Biol., 28, 1897, p. 206.
5. Influence of cystine ingestion on taurocholic acid. *v. Bergmann:* Hofmeister's Beiträge 4, 1904, p. 192.
6. Cholesterol in bile. *Doyon and Dufourt:* Archives de Physiol. (5) 8, p. 587, 1896.
   Amount in human bile. *Pierce:* Archiv f. klin. Med. 106, 1912.
7. Conjugated sulphates in bile. *Hammarsten:* Zeits. f. physiol. Chem. 24, p. 322, 1898.
8. Stercorin or coprosterol. *Flint:* Amer. J. of Med. Sci. (N.S.) 44, 1862, p. 306 and following pages. Zeits. f. physiol. Chem. 23, p. 363, 1897. *Bondzynski and Humnicki:* Zeits. f. physiol. Chem. 22, 1896, p. 396.
9. Chortosterol. *Dorée and Gardner:* Proceed. Royal Soc. B, 80, pp. 212-26, 1908.
10. Phosphatides in bile. *Hammarsten:* Zeits. f. physiol. Chem. 36, 1902, p. 328.
11. Bile of hippopotamus. *Hammarsten:* Zeits. f. physiol. Chem. 74, 1911, p. 123.
12. General review. Article Bile. Dictionnaire de Physiologie, Richet, 2, p. 144, 1898. *Hammarsten:* Ergebnisse der Physiologie, 1905.
13. Preparation of unconjugated acids of ox bile. *Schryver:* Jour. of Physiol., 44, p. 265, 1912.
14. Taurocholeic acid of ox bile. *Gullbring:* Zeit. f. physiol. Chem., 45, p. 448, 1905.
15. Glycocholeic acid of ox bile. *Wahlgren:* Zeit. f. physiol. Chem., 36, p. 556, 1902.
16. Separation of taurocholate and glycocholate from ox bile. *Tengström:* Zeit. f. physiol. Chem., 41, p. 210, 1904. *Bleibtreu:* Pflüger's Arch., 99. *Letsche:* Zeit. f. physiol. Chem., 60, 1909, p. 462.
17. Choleic acid. Properties. *Lassar-Cohn:* Zeits. f. physiol. Chem., 17, 1893. *Latschinoff:* Ber. d. d. chem. Gesell. 18, p. 3039, 1885. *Fischer and Meyer:* Zeit. f. physiol. Chem., 76, pp. 95-8, 1912.
18. Acids of pig bile. *Jolin:* Zeits. f. physiol. Chem., 13, p. 205, 1889.
19. Cholic acid. Relation to cholesterol. *Schrötter and Weitzenböck:* Monatshefte f. Chem., 29, pp. 395-8, 1908. *Lifschütz:* Ber. d. d. chem. Gesell. 47, pp. 1459-60, 1914.
20. Cholic acid. Color reaction with HCl. *Hammarsten:* Zeits. f. physiol. Chem., 61, pp. 495-8, 1910.

21. Phocataurocholic acid and walrus bile. *Hammarsten:* Zeits. f. physiol. Chem. 61, p. 454, 1910.
22. Cholic acid. Iodine reaction. *Barger and Field:* Jour. Chem. Soc. 101, pp. 1394-1409, 1913.
23. Cholic acid. Composition. Structure. *Pregl:* Zeits. f. physiol. Chem. 65, p. 157, 1910.
24. Synthesis of taurocholic and glycocholic acid. *Bondi and Müller:* Zeits. f. physiol. Chem., 47, 1906, p. 499.
25. Lowering of surface tension by bile salts. *Billard et Dieulafé:* C. R. de la Soc. Biol. 54, p. 245, 1902.

Bile Pigments.

26. Composition of bilirubin. *Fischer and Röse:* Zeit. f. physiol. Chem., 89, 262, 1914. Some earlier literature cited. *Küster:* Zeits. f. physiol. Chem., 59, 1909, p. 63. Lit. before 1909 cited in this paper.
27. Bile pigments. Properties. *Jolles:* Arch. f. d. ges. Physiol., 75, p. 446, 1889.
28. Formation of biliverdin from bilirubin by Na_2O_2. *Hongouneng and Doyon:* Arch. de Physiol. (5) 8, p. 525, 1896.
29. Absorption spectra of bile pigments. *MacMunn:* Jour. of Physiol., 6, p. 22, 1885.
30. Cholohematin, bilipurpurin and phylloerythrin. *Marchlewski:* Zeit. f. physiol. Chem., 45, p. 466, 1905.
31. Urobilin. Spectrum. *Lewin and Stenger:* Arch. f. d. ges. Physiol., 144, 1912, p. 279.
32. Urobilin. Origin. *Fromholt:* Zeit. f. expt. Pathol., 20, 1911, p. 268.
33. Bile pigments in invertebrates. Molluscs. *Schulz:* Zeits. f. all. Physiol. 3, 1904, p. 91.
34. Origin in liver. *Naunyn and Minkowski:* Archiv f. expt. Path. u. Pharm., 21, 1886, p. 1. *Fiessinger and Lyon-Caen:* Jour. de Physiol. et Path. gen., 12, p. 958, 1910. *Weill:* Arch. Internat. de physiol., 1913, 13, p. 181.
35. Urobilin. Urobilinogen. *Fischer and Meyer-Betz:* Zeit. f. physiol. Chem., 75, 1911, p. 232. *Garrod and Hopkins:* Jour. of Physiol., 20, 1896, p. 112; 22, 1898, p. 451.
36. Chlorophyll composition and relation to bile and blood pigments. *Willstätter:* Jour. Amer. Chem. Soc., 37, p. 323, 1915.
37. Viscosity of bile. *Burton-Opitz:* Biochem. Bull., 3, p. 351, 1914.

Influence of Bile on Digestion and Absorption.

1. Influence of the bile on pancreatic digestion. *Rachford:* Jour. of Physiology, 12, p. 72, 1891. Ibid., Jour. of Physiol., 25, 1899, p. 165. *Martin and Williams:* Proceed. Roy. Soc., 48, p. 160 *Tschermak:* Cent. f. Physiol., 16, p. 329, 1902. *Bruno:* Arch. des Sci. Biol., St. Petersburg, 7. 1899. *v. Fürth and Schütz:* Hofmeister's Beiträge, 9, p. 28, 1907. (Earlier literature cited here.) *Hewlett:* Effect of bile on ester splitting action of pancreatic juice. Johns Hopkins U. Bull., 16, p. 166, 1906.
2. Effect of synthetic bile acids on pancreatic fat digestion. Magnus: Zeits. f. physiol. Chem., 48, 1906, p. 376.
3. Influence on absorption. *Pflüger:* Arch. f. d. ges. Physiol. 88, p. 431, 1901 *Moore and Rockwood:* Pro. Royal Society, 60, p. 439. Jour. Physiol., 21, p. 373, 1897.

4. **Influence on peptic digestion.** *Schiff:* Arch. f. d. ges. Physiol., 3, 1870, p. 200. Ibid., p. 613.
5. **Fat absorption.** , *Bidder and Schmidt:* Verdauungsäfte, p. 223. *Dastre:* Arch. de Physiol., 1890.
6. **Reabsorption of bile.** *Schiff:* Arch. f. d. ges. Physiol., 3, p. 598, 1870.

PANCREAS.

1. **General.** *Pawlow:* Work of the Digestive Glands.
2. **Discovery** of function in digestion. *Bernard:* Mémoire sur le pancreas. Paris, 1856.
3. **Composition of human** pancreatic juice. *Wohlgemuth:* Biochem. Zeit., 39, p. 321, 1912. (Other recent literature cited.)
4. **Clinical method of obtaining pancreatic juice from the stomach.** *Boldyreff:* Archiv f. d. ges. Physiol., 121, 1907, p. 13; 140, 1911, p. 436.
5. **Secretin. Discovery.** *Bayliss and Starling:* Journal of Physiology, 29 and 31, 1902. *Carlson et al.:* Jour. Amer. Med. Assn., 66, p. 178, 1916. Bibliography.

   Discharge of HCl into duodenum causes secretion of pancreas. *Popielski:* Cent. f. Physiol., 10, 1896, p. 405-9.

   Secretion of pancreas due to both humoral and nerve action. *Popielski:* Zeit. f. physiol. Chem., 71, 1911, p. 186. *Bylina:* Archiv f. d. ges. Physiol., 143, 1911, p. 531. *Popielski:* Ibid., 121, 1907-8, p. 239. *Mazurkiewicz:* Ibid., 121, 1907-8, p. 75. Good bibliography in Bylina's paper. p. 565.

   Secretin appears in blood when acid put in duodenum. *Enriquez et Hallion:* C. Ren. de la Soc. Biol., 55, 1903, p. 233.

   Secretin causes bile secretion. *v. Henry and Portier:* C. Ren. de la Soc. de Biol., 54, 1902, p. 620.

   Secretin inactivated by pancreatic extracts. *Lalou:* Jour. de Physiol. et de Path., 12, 1911, p. 343.

   Secretory action of curarin. *Czubalski:* Archiv f. d. ges. Physiol., 133, 1910, p. 225.

   Method of extraction of secretin from intestinal mucosa. *Dale and Laidlaw:* Pro. Physiol. Soc. Jour. of Physiol., 44, 1912, p. xi.

   Fate of secretin in pancreatic diabetes. *Evans:* Jour. of Physiol., 4, 1912, p. 461.

   Existence in mesenteric ganglia. *Delezenne and Frouin:* C. Rend de la Soc. Biol., 54, p. 896, 1902.
6. Enterokinase. Discovery. *Chepowalnikow:* Thesis, St. Petersburg, 1899. *Bayliss and Starling:* Jour. of Physiol., 30, 1903, p. 61.

   Supposed activation of trypsin by Ca salts. *Hekma:* Pro. Amsterdam Academy of Sciences, 13, (2), p. 1002. *Delezenne:* C. Rendu de la Soc. Biol. 1905, pp. 476, 523 and 612; 1906, p. 1070; 1907, p. 274. *Zunz:* Annal. de la Soc. Roy. des Sci. Med. et Nat. Bruxelles, 16, 1907.

   Conversion of trypsinogen to trypsin. *Mellanby and Wooley:* Jour. of Physiol., 45, p. 370-88, 1913.

   Time of appearance in fetal life. *Ibrahim:* Biochem. Zeits., 22, pp. 24-32, 1909.
7. Erepsin of pancreatic juice. *Schaeffer and Terroine:* Jour. de Path. et Physiol., 12, 1910, pp. 884 and 905.
8. Amylase of pancreas. Composition and method of obtaining. *Sherman and Schlesinger:* Jour. Amer. Chem. Soc., 34, 1912, p. 1104.

9. **Lipase of pancreas. Influence of dialysis on lipolytic activity.** *Bierry and Giaja:* C. Ren. de la Soc. Biol., 62, p. 432, 1907. *Slosse et Lambosch:* Biochem. Cen. 7, 1908, p. 510. *Hanisek:* Zeit. f. physiol. Chem., 71, 1911, p. 239. *Rosenheim and MacKenzie:* Jour. of Physiol., 40, 1910, pp. viii, xii, xiv.

10. **Lipase action on various esters.** *Morel et Terroine:* Jour. de Path. et Physiol., 14, 1912, p. 72.

11. **Lipase sensitive to heat.** *Visco:* Arch. Ital. de Biol., 54, 1911, p. 243.

12. **Alkalinity of pancreatic juice.** *Foa:* C. Ren. de la Soc. Biol., 59, 1905, p. 867.

13. **Trypsin and trypsinogen.** *Mellanby and Wooley:* J. of Physiol., 47, pp. 339-60, 1914. *Vernon:* J. of Physiol., 47, pp. 325-38, 1914.

14. **Erepsin. Conditions of action.** *Rona and Arnheim:* Biochem. Zeits., 57, 1894, p. 84.

15. **Intestinal Lipase.** *Falk:* J. Amer. Chem. Soc., 36, 1914, p. 1047.

16. **Action of pepsin and trypsin on each other.** *Edie:* Biochem. Jour., 8, 1914, p. 193.

17. **Optimum concentration of hydrogen ions for liquefaction of gelatine by trypsin.** *Palitzsch and Walbum:* Biochem. Zeits., 47, p. 1, 1914.

18. **Quantitative determination of trypsin.** *Wahlschmidt:* Arch. f. d. ges. Physiol., 143, pp. 189-229, 1911. Various methods are described.

19. **Ferments of Pancreas.** *Mellanby and Woolley:* Jour. Physiol. 49, p. 246, 1915.

20. **Pancreatic Insufficiency.** *Spriggs and Leigh:* Quart. J. of Med., 9, p. 11, 1915.

BACTERIA.

1. **Fecal bacteria of healthy men.** *MacNeal et al.:* Jour. Infectious Diseases, 6, pp. 123-169, 571-609, 1909.

2. **Method of estimation of fecal bacteria gravimetrically.** *Strasburger:* Zeit. klin. Med., 46, 413, 1902.

3. **Method by microscopic counting.** *Winterberg:* Arch. f. Hygiene, 59, p. 283, 1906. *Klein:* Centrlblt. f. Bakt. Abth. 1. Orig., 27, p. 834, 1900.

4. **General treatise on the feces.** *Schmidt and Strasburger:* Die Faeces des Menschen, Berlin, 1905.

5. **Bacteria and chemistry of feces of healthy men.** Studies in Nutrition by H. S. Grindley and W. J. MacNeal. Vols. 3, 4 and 5, University of Illinois, 1912.

CHAPTER XI.

ABSORPTION.

Absorption.—The foods in the alimentary canal are not yet, strictly speaking, in the body, since the cavity of the canal is in communication with the external world. How do the digested foods get into the body? Is it by a simple physical process of diffusion or osmosis occurring through the mucous membrane of the intestine, between the blood on the one side and the intestinal contents on the other? Or is it by the vital activity of the epithelial cells? The foods move in as though there were an attractive force exerted upon them by the hungry body cells. We do not believe any such force exists, but how is their entrance to be explained? Does anything happen to them during the process of absorption?

Absorption takes place throughout the length of the intestine. It is slight in the stomach; most of the absorption is in the small intestine. Only a small part of the food remains to be absorbed in the colon. The food in the small intestine is spread out as a thin layer over the surface of the gut. This surface is enlarged by the presence of finger-like processes, the villi, which dip down into the cavity of the intestine. Each villus is supplied with artery, capillaries and veins. It has in the center a lymph space called a lacteal, opening into the chyle vessels of the mesentery. Each villus has both longitudinal and circular muscular fibers so that it can expand and contract, taking up food like a sponge and being squeezed by the contraction. The mucous membrane, which lies between the food and the lacteal, is thin, only about 0.012 mm. in depth. It seems probable that a good deal of the absorption is due to purely physical processes of diffusion or osmosis of the amino-acids and mono-saccharides from the intestine, where they are present in quantities, into the blood in which they are present in very small amounts; but the problem is not so simple as this. This membrane of epithelial cells, although thin, is alive. The food matters must pass through the living protoplasm that is constantly altering its state. Such alterations in activity constantly change the rate of absorption; they change the permeability of the cells. Now substances do not pass through membranes by simply shooting through the holes. There is some kind of a union between the solvent and the solute and between the solvent and the membrane. Do the substances go through the cells because they disolve in the water which penetrates the cell? Or because they unite with he organic basis of the cell? One thing is probable, namely, that the activity of the cell in some way controls this process.

Absorption of fats.—That absorption is not due, primarily at any rate, to osmotic pressure is shown by the absorption of fat. Fat is entirely hydrolyzed in the intestine. The fatty acids thus set free are held in colloidal solution by the bile salts, probably by the union already discussed. The colloids have almost no motion of translation and very little osmotic pressure and yet the absorption of fats takes place with speed. The absorption is greatly aided by the bile. The bile salts are reabsorbed with the fatty acids. Glycerol also passes into the cells. The glycerol and fatty acid thus brought together in the living cell are resynthesized into neutral fat, which can be seen scattered about in the cell. Just how this resynthesis is produced it is at present impossible to say. Some have referred it to the reverse action of the lipase of the intestinal mucosa. But while it is true that lipase added to oleic acid and glycerol causes some resynthesis and this resynthesis is hastened by the bile, yet when water is present, as it is in the cells, the lipase produces only a very small amount of synthesis. Moreover lipase is not found precisely in those cells in which the synthesis is going on most rapidly, as in the cells of the mammary glands. All of these syntheses are dependent on a supply of oxygen and are inhibited by ether, which does not check lipase activity. The resynthesis is hence, probably, one of the synthetic powers of the cell correlated with cell respiration.

However the resynthesis of the neutral fats is brought about, it certainly occurs. The very finely emulsified fat is discharged in some way into the lacteal from which it is pressed along into the lymphatics of the mesentery in the form of a milk-white emulsion called chyle, into the receptaculum chyli and from here it passes by the thoracic duct, which lies just behind the pleura, not far from the median line posteriorly, up to the left shoulder, where it is poured into the blood at the junction of the jugular and subclavian vein. The chyle is forced up this duct partly by the negative pressure of the thorax, partly by the pressure due to inspiration on the abdominal viscera and partly by the pressure of the new chyle or lymph behind it, coming from the liver and other abdominal organs. Not all the reabsorbed fat is to be found in the chyle thus discharged into the jugular vein. A portion has disappeared, perhaps finding its way into the blood and so to the liver; or has been catabolized during its passage through the mucous epithelium; the fate of this remnant is not yet clear.

The chyle thus poured into the blood gives to the blood serum or plasma during the absorption of a fatty meal a milky appearance, and the amount of the fat in the blood may be so increased that it will rise on the plasma like so much cream. By the blood it is carried about the body, each tissue taking what it needs and any excess being picked out and stored in the fat depots of the body, the fat cells of the adipose tissue of the mesentery, the panniculus adiposus, or about the internal organs. How the blood fat finds its exit from the blood; how it gets through the vascular wall, whether by being split into soaps and glycerine or as neutral fat; and how it is accumulated in fat tissues, these are problems not as yet solved. There is a lipase in the red blood corpuscles and perhaps all the blood fat is rehydrolyzed before leaving the blood.

Absorption of carbohydrates.—What if anything happens to the carbohydrates during their passage through the mucous membrane is unknown. They are believed to pass directly into the blood capillaries as levulose, glucose or galactose. Sometimes maltose, lactose and cane sugar are absorbed as such and can be detected in the blood and urine, since the two last named sugars are not used if they enter the blood, but pass out in the urine. If sugars are injected into the blood they are in part excreted into the duodenum and presumably then reabsorbed. The sugars thus absorbed are found in the portal blood to the extent of 0.2-0.4 per cent. They pass to the liver, where a portion may be taken out and stored in that organ as glycogen; and in part they circulate to the other organs of the body, each of which removes from the blood what it needs. The sugar in the blood is generally glucose and it is in a form which may be dialyzed out of the blood by vivi-diffusion. Page 470.

If sugar in a concentrated solution is ingested it may cause vomiting, and in any case it is diluted in the bowel by the passage outward of water through the mucous epithelium. Ultimately, if the mucous membrane remains in a living, active form the sugar and water are reabsorbed.

Absorption of proteins.—The proteins are absorbed probably for the most part, if not altogether, in the form of amino-acids. It is certain that in the cephalopods, in which the conditions for studying this problem are very good, owing to the simple composition of the blood, they are absorbed in the form of amino-acids. There is now good evidence that they are absorbed in that form in the mammals also, since small amounts of the amino-acids are found in the blood plasma. (Page 472.) The quantity of non-protein nitrogen increases in the blood during the absorption of the proteins. The presence of erepsin in the wall of the intestine and the elaborate mechanism to secure the reduction of the proteins to the form of amino-acids is strong a priori evidence that they are absorbed in this form. This view is strengthened by the fact that the albumoses, when brought directly into the blood stream, act like foreign bodies, lowering the blood pressure and reducing the coagulability of the blood. There is no doubt that at times complex polypeptides are reabsorbed, since some individuals are found who have an idiosyncrasy toward some one protein, reacting by anaphylactic shock to the ingestion of this particular protein. Sometimes it is the protein of milk, sometimes of egg or some other protein of the food. This reaction is proof that at some time this protein in an unchanged form secured entrance to the blood. The very fact that this reaction is rare shows how unusual it is for the undigested protein to be reabsorbed.

There is no question that deamidization, that is the splitting off of amino groups, occurs to some extent during their passage through the wall or during their bacterial decomposition, for the intestinal mucosa contains more ammonia than any other tissue of the body with the exception of the stomach muscosa; and the blood of the mesenteric veins coming from the intestine carries from 6-10 times as much ammonia as any other blood in the body. A part of this ammonia is probably derived from the amide nitrogen split off by tryptic digestion, but a portion also comes, no doubt, from the amino groups. What proportion of amino-acids are thus deamidized it is impossible to say, but presumably not more than a third and it may be less than this.

The amino-acids which get past the intestinal mucosa are found in the blood rather than in the chyle. Their discovery here has been recently made. They do not accumulate, for they are removed from the blood by the tissues about as rapidly as they enter it. At any instant of time there are, then, only minimal amounts present.

For a long time it was believed that the amino-acids were resynthesized during their passage through the mucosa into some of the blood proteins and probably into serum albumin. This was supposed to be the food of the various tissues. This view, for which there never was any convincing or even strong evidence, has now been rendered very unlikely by the discovery of the constitution of the proteins and the presence of amino-acids in the blood.

Rôle of the white blood corpuscles in absorption. During the absorption of food and particularly of protein food, white blood corpuscles accumulate in the mucous membrane of the gut. They are to be found not only beneath the epithelial cells, but crawling between them and even out into the lumen of the gut. They may be seen also coming back into the chyle vessels. The rôle they are playing in absorption is still quite obscure. After the absorption of a meal they are found in the blood in unusually large numbers. There is, as it is said, a digestive leucocytosis. Schaefer described them as loading themselves with fat and carrying the fat to the chyle, where the fat was set free by the dissolution of the leucocyte. That fat may be found in the epithelial cells of the intestine indicates, however, that most of the fat goes through these cells rather than through the leucocytes. The leucocytosis is most pronounced during the digestion of proteins, and since the leucocytes show themselves to be positively chemotactic toward any decomposing tissue, or toward the products of digestion of proteins when the latter are in capillary tubes, it has been suggested by Hofmeister that most of the proteins are absorbed by uniting with the leucocyte body. By the decomposition of the whole or a part of the body of the cell they are then supposed to be set free. This view does not seem very probable in the light of the recent work, indicating that the proteins are absorbed as

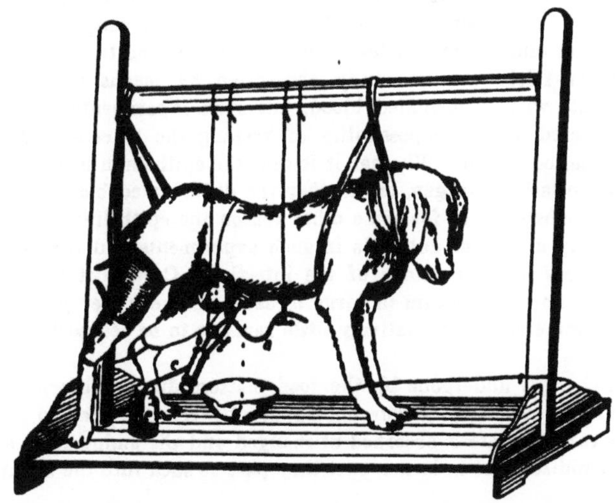

FIG. 43.—Polyfistula dog of London. The dog has three fistulas at different levels of the intestine. One is at *k* and the two others are discharging into the vessels *d* and *e*.

amino-acids and get into the blood in that form. It is objected also that the absorption is too rapid and too large to be accounted for by the relatively small weight of leucocytes found in the intestine during the absorption. Still there are many indications that the blood proteins may be formed by the leucocytes, and perhaps they are active in securing sufficient protein to supply the needs of the blood tissue alone. Further work is needed before any positive conclusion can be drawn concerning the rôle of the leucocytes during absorption.

The amount of absorption in different parts of the tract. By making fistulæ at various levels of the alimentary tract, London and his co-workers have studied the rapidity of absorption during the passage of food down the tract of dogs. They have found that the greater part of the absorption takes place in the upper parts of the intestine and that at the end of the small intestine only a small fraction of the total food remains to pass into the colon. At least three-quarters of the food is absorbed in the small intestine. Water passes very rapidly through the stomach and is absorbed in the intestine.

Conclusion. The absorption of the products of digestion is in part a physical process, but in a much larger degree it is a physiological process dependent upon the physiological activity of the cells of the intestinal epithelium. These cells have the power not only of reabsorbing food products, but of secreting into the bowel a good deal of water, salts and other substances. Moreover the solution of the problem of the nature of the processes involved in absorption is greatly complicated by the great sensitiveness of these cells to various conditions. It would be thought fairly easy to study absorption by means of perfusion experiments with defibrinated blood, but such experiments have hitherto shattered on the impossibility of keeping the mucous membrane in a normal condition. Whether it is that the epithelium is abnormally sensitive to lack of oxygen, or whether the defibrinated blood acts as a poisonous substance, or for some other reason the epithelium undergoes very readily a kind of autolysis in such experiments and the perfused blood escapes into the lumen of the intestine. Certain it is that the absorption normally occurs in large measure in the upper part of the small intestine and that relatively little remains to be reabsorbed in the colon.

The changes undergone by the food matters during absorption are still badly known, but the fats are apparently resynthesized to neutral fat, the carbohydrate is absorbed unchanged and the amino-acids are in part deamidized, but in larger part they pass as such into the blood.

REFERENCES. Protein Absorption.

1. *Abderhalden:* Zur Frage des Albumosengehaltes des Blutes und speziell des Plasmas. Biochem. Zeit., 8, 1908, p. 360.
2. *Abderhalden u. Rona:* Fütterungsversuche mit durch Pankreatin, durch Pepsinsalzsäure Pankreatin und durch Säure hydrolyzierten Casein. Zeit. physiol. Chem., 42, p. 528, 1904. Also 44, p. 198, 1905.
3. *Abderhalden and Oppler:* Weiterer Beitrag zur Frage nach der Verwertung von tief abgebauten Eiweiss im Organismus des Hundes. Zeit. physiol. Chem., 51, p. 226, 1907.
4. *Abderhalden u. Rona:* Same title. Zeit. physiol. Chem., 52, p. 507, 1907.
5. *Abderhalden and London:* Same title except investigation on an Eck fistula dog. Zeit. physiol. Chem., 54, p. 80, 1907.
6. *Abderhalden and Ollinger:* Same title as 4. Zeit. physiol. Chem., 57, p. 74, 1908.
7. *Abderhalden and London:* Weiterer Beitrag zur Frage nach dem Ab- und Aufbau der Proteine im tierischen Organismus. Zeit. physiol. Chem., 65, 1910, p. 251.
8. *Abderhalden and Suwa:* Same title as 4. Zeit. physiol. Chem., 68, p. 416, 1910.

9. *Abderhalden, Prym and London:* Ueber die Resorptionsverhältnisse von in den Magendarmkanal eingeführten Monaminosäuren. Zeit. physiol. Chem., 53, p. 326, 1907.

10. *Ascoli and Vigano:* Zur Kenntnis der Resorption der Eiweisskörper. Zeit. physiol. Chem., 39, p. 283, 1903.

11. *Bergmann:* Notiz über den Befund von Verbindungen im Blute, die mit Naphthalinsulfochlorid reagiren. Beitr. chem. Physiol. Pathol., 6, p. 40, 1905.

12. *Bostock:* On deamidization. Biochem. Jour., 6, p. 48, 1911.

13. *Brasch:* Weitere Untersuchungen über den Bakteriellen Abbau primarer Eiweissspaltprodukte. Biochem. Zeit., 22, p. 403, 1909.

14. *Cathcart and Leathes:* On the absorption of proteins from the intestine. Jour. Physiol., 33, p. 462, 1905.

15. *Cohnheim:* Die Umwandlung des Eiweisses durch die Darmwand. Zeit. physiol. Chem., 33, p. 451, 1901.

16. *Cohnheim:* Weitere Mittheilungen über Eiweissresorptionsversuche an Octopoden. Zeit. physiol. Chem., 35, p. 396, 1902.

17. *Cohnheim:* Versuche über Eiweissresorption. Zeit. physiol. Chem., 59, p. 239, 1909. See also 61, p. 189, 1909.

18. *Cramer and Pringle:* Assimilation of protein introduced enterally. Jour. Physiol., 37, p. 158, 1908.

19. *Embden and Knoop:* Ueber das Verhalten der Albumosen in der Darmwand, etc. Beitr. chem. Physiol. Pathol. 3, p. 120, 1903.

Halliburton: The absorption of proteins. Lancet, I, 1909.

20. *Hedin:* Trypsin and antitrypsin. Biochem. Jour., 1, p. 474, 1906.

22. *Henriques:* Die Eiweisssynthese im tierischen Organismus. Zeit. physiol. Chem., 54, p. 406, 1907.

23. *Hofmeister:* Das Verhalten des Peptons in der Magenschleimhaut. Zeit. physiol. Chem., 6, p. 69, 1882.

24. *Hopkins:* The utilization of proteins in the animal. Science Progress, 1, p. 156, 1906.

25. *Howell:* Amino-acids in blood and lymph. A. Jour. Physiol., 17, p. 273, 1907.

26. *Kühne:* Ueber die Verdauung der Eiweissstoffe durch den Pankreassaft. Arch. f. path. Anat. u. Physiol., 39, p. 130, 1867.

27. *Lang:* Deamidization in the animal body. Beitr. chem. Physiol. Pathol., 5, p. 321, 1904.

28. *Levene and Van Slyke:* Ueber Plastein. II. Bioch., Ztschr., 16, p. 203, 1909.

29. *London:* Various papers on digestion and absorption in the Zeit. physiol. Chem. from 1907 on. Especially 61, 69, 1909; and 60, 191 and 267, 1909.

30. *MacFadyen, Nencki and Sieber:* Die chemische Vorgänge im menschlichen Dünndarm. Arch. f. expt. Pathol. u. Pharm., 28, p. 310, 1902.

31. *Omi:* Resorptionsversuche an Hunden mit Dünndarmfisteln. Arch. f. d. ges. Physiol., 126, p. 428, 1909.

32. *Paton and Goodall:* Digestion leucocytosis. Jour. Physiol., 33, 20, 1905.

33. *Robertson:* Synthesis of paranuclein through the agency of pepsin. Jour. Biol. Chem., 5, p. 493, 1908.

34. *Salaskin:* Eiweissresorption im Magen des Hundes. Zeit. physiol. Chem., 51, p. 167, 1907.

35. *Taylor:* On the synthesis of protein through the action of trypsin. Jour. Biol. Chem., 3, p. 87, 1907. Ibid., 5, p. 381, 1908.

36. *Reid:* Intestinal absorption. Phil. Trans. Roy Soc., 192, p. 240, 1900.

CHAPTER XII.

THE BLOOD. THE CIRCULATING TISSUE.

Since each cell of the body needs food and oxygen and is injured by the accumulation of its own waste products, it is necessary, if a multicellular organism of a large size is to exist, that there be some means of getting food and oxygen to the cells in the interior of the mass, and some means of removing the waste matters. This problem was solved by one of the tissues, the blood, becoming liquid so that it could easily penetrate all the crevices of the body and come into intimate contact with every cell, and thus nourish it. There was finally developed, also, a system of tubes to contain the blood and a pumping organ, the heart, which drove it all over the body. Every cell of the body comes then into close contact with the circulating fluid and exchanges food and waste products with it. In some organisms this fluid bathes each cell directly, so that all cells of the body really live in the blood. This is the case, for example, in invertebrates and in the cortical parts of the suprarenal capsules of vertebrates where there are no capillaries. In these cases the blood passes directly from the ends of the arterioles into the tissue spaces and is collected from these spaces into the veins. But usually in vertebrate organs the blood is confined to the blood vessels. It has evidently been found on the whole to be better, for one reason or another, to confine the blood in this manner. The finer blood vessels, the capillaries, have walls so thin and so permeable, or are physiologically so constituted, that oxygen and other gases and water, salts and food matters pass easily through them, so that the tissue cells, while not directly in the blood stream, are in the lymph, which comes from the blood. The principal advantage of a closed vascular system is that the circulation is probably more easily controlled with such an arrangement than when the blood flows from the arteries into lacunar spaces in the tissues. By this device it is possible to drive the blood faster past the tissues and to provide more food and particularly more oxygen. The main difference in the metabolism of the higher and lower vertebrates is the much more intense oxidation in the higher. It is the nervous system which especially requires a very large supply of oxygen and it was probably primarily to supply this tissue, which has the keenest metabolism of all, that the many devices for improving the circulation have been evolved. It must be remembered that it is the nervous system which has been, on the whole, consistently selected in vertebrate and also in invertebrate evolution. The rest of the body really exists for it. Each cell of the body, then, even of the higher vertebrates, really lives in an aqueous medium just as do unicellular organisms, which live in water itself. And for this reason the blood has been called the internal medium of the body.

Lymph.—The spaces of the tissues are filled with a liquid called lymph, derived from the blood by secretion through the capillary walls. Lymph resembles, in many ways, the plasma of the blood, and contains some white corpuscles or cells in suspension in it. These are called lymphocytes. Some of this lymph, which is the immediate liquid environment of the tissue cells, passes back to the blood by absorption through the capillaries, but part of it is collected into thin-walled vessels called lymphatics, which resemble veins in structure, and after passing through lymph glands the liquid finally finds its way back to the blood by the thoracic duct and other lymphatics opening into the blood at the junction of the internal jugular and left subclavian vein. Food substances pass from the blood to the lymph and the cells take them from the lymph. The waste products pass in the opposite direction. One object of a separate lymphatic system may be to guard against infections, since when bacteria find access to the tissues they are carried by the lymph stream to the lymph glands, where they are generally digested and the infection is thus confined. There may be other advantages also in this separate lymphatic system.

Functions of the blood.—The blood has four great functions: 1. It carries food from the intestine to the tissues; and gaseous food, or oxygen, from the lungs to the tissues. 2. It removes waste products from the tissues and carries them to the kidneys, lungs, intestine, and skin, the excretory organs of the body. 3. It provides for the metabolic co-ordination of the body, in that it distributes the internal secretions from each organ to other organs which utilize them. It thus keeps tissues, which may be far separated, in metabolic co-ordination or exchange with each other; it is the internal medium of exchange. 4. It plays a very important part in the defense of the organism against the invasion of parasites.

Composition of the blood.—The blood is generally considered to be a liquid tissue lying within a system of tubes, the blood vessels, and the latter are not considered as part of the blood itself. More correctly, however, both blood and the endothelium of the blood vessels make part of a single system and it is impossible to alter one without at the same time altering the other. The liquid part of the blood, the plasma, is regarded as corresponding to the lifeless intercellular substance of connective tissue and cartilage; the cells or living part of this tissue being the red and white corpuscles. From this point of view the liquid part of the blood is a lifeless carrier of waste materials and food substances to the various other tissues of the body and only the cells of the blood are living. One can look at the blood, however, in a different way, as suggested by Wooldridge, and from this point of view it gains wonderfully both in interest and instructiveness. The blood as a whole, including the endothelial cells of the blood vessels, may be considered to be living matter, distinguished from most other living matter by its greater fluidity. There are, however, other kinds of living matter of a liquid kind. The protoplasm of the amœba and many plant cells is so liquid that it flows readily and is in reality a circulating liquid. The blood plasma may be regarded as a very liquid protoplasm formed essentially by the cells of the vascular endothelium, by the blood cells and the hematoblasts. The blood platelets and the red blood corpuscles may be regarded, from this point of view, as homologous with the granular inclusions of many cells. Blood separated from the endothelial cells dies and clots; in just the same way a peripheral nerve dies if severed from its nutritive center. We shall in this chapter adopt this point of view of Wooldridge and consider the whole blood, the more liquid portions together with the corpuscles both white and red, the platelets, and the cells lining the blood vessels, as consisting of a great mass of living protoplasm. The processes which occur in the blood are then in many important particulars probably identical with those occurring in living matter generally. Thus the clotting of the blood is not to be regarded as a separate and independent property peculiar to this tissue, but as typical of similar processes occurring in every form of living matter. For all living matter like the blood has the property of forming a gel or clotting. Blood, too, respires, like living matter. When the amino-acids, carbohydrates, etc., enter the blood by diffusion, or by secretion of the endothelial wall, they are not then entering an inert fluid, but they are entering real living matter and their subsequent fate in this fluid becomes extraordinarily interesting as showing the probable fate of similar substances in the body cells generally. The proteins of the blood plasma are not to be considered as inert simple proteins, as they are ordinarily considered, as so much globulin, albumin and fibrinogen in solution, but they are probably united in the blood plasma, at least to some extent, with phospholipins, just as they are in cells; and they probably make part of a complex substance, in reality an organized and very unstable substance, the blood plasma. The alkalinity of the blood is about the same as that of the tissues generally and the methods of maintaining its alkalinity are those employed by all forms of living matter. A study of the alkalinity of the blood, its variation both physiological and pathological and the means employed to hold it constant, throws light on the alkalinity of every cell.

We probably know more about the composition and the chemical and physical changes taking place in the blood than of any other tissue, but that our knowledge is still extremely fragmentary will appear in the pages which follow. The study of the blood is in many ways easier than that of any other tissue. Being liquid, it is readily obtained in large amounts for chemical and physical investigation, and by centrifugalization we are able to separate it, without inducing serious chemical changes, into portions of lighter and heavier specific gravity. Moreover, we can get such quantities of it that we are able to experiment with it more than with almost any other tissue of the body.

General composition of mammalian blood.—Mammalian blood as it circulates within the blood vessels consists of a more fluid part, the blood plasma, and this contains a large number of cells and organized structures in suspension. The cells are the leucocytes, or white blood cells; the special organized structures are the red blood corpuscles, or erythrocytes, and the blood platelets. The other cells belonging to this system, the endothelium of the blood vessels, are fixed, not circulating, cells and we shall for the present neglect them; but more than any other cells of the blood they probably control its composition and metabolism. The general composition of mammalian blood is as follows:

1. Blood (mammalian) consists of:
   - A. Plasma ..... 60-70% by volume; or about 55% by weight.
   - B. Corpuscles .. 40-30% " " " " 45% " "

A. Plasma.
(Specific gravity = 1.0237-1.0276 at 25°)

a. Water	90 - 92%
b. Solids	10 - 8
Organic	
Lipoproteins	5.5 -8.4
Carbohydrates	0.1 -0.2
Cholesterol	0.09-0.14
Fat and fat acids	0.3 -0.6
Extractives	0.05-0.2
Inorganic	1-2

B. Red corpuscles (Abderhalden).
(Specific gravity = 1.088)

a. Water	68.7-59.2
b. Solids	31.3-40.8
Organic	30.4-39.9
Hemoglobin	31.7
Stroma	
Phospholipin	0.37-0.39
Cholesterol	0.14-0.17
(Grigant et Huillier)	
Protein	5.7 -6.4
Inorganic (excluding Fe)	0.9

The following analysis of human blood was made by Carl Schmidt:[1]

BLOOD OF A MAN TWENTY-FIVE YEARS OF AGE

ONE THOUSAND GRAMS OF BLOOD CONTAIN

513.02 BLOOD CORPUSCLES

Water	349.69		
Substances not vaporizing at 120°	163.33		
Hematin (including 0.512 iron)	7.70		
"Blood casein," etc.	151.89		
Inorganic constituents (excluding iron)	3.74		
Chlorine	0.898	Potassium chloride	1.887
Sulphuric acid	0.031	Potassium sulphate	0.068
Phosphoric acid	0.695	Potassium phosphate	1.202
Potassium	1.586	Sodium phosphate	0.325
Sodium	0.241	Soda	0.175
Phosphate of lime	0.048	Calcium phosphate	0.048
Phosphate of magnesium	0.031	Magnesium phosphate	0.031
Oxygen	0.206		
		Total	3.736

486.98 PLASMA

Water	439.02		
Substances not evaporated at 120°	47.96		
Fibrin	3.93		
Albumin, etc.	39.89		
Inorganic constituents	4.14		
Chlorine	1.722	Potassium chloride	0.175
Sulphuric acid	0.063	Potassium sulphate	0.137
Phosphoric acid	0.071	Sodium chloride	2.701
Potassium	0.153	Sodium phosphate	0.132
Sodium	1.661	Soda	0.746
Calcium phosphate	0.145	Calcium phosphate	0.145
Magnesium phosphate	0.106	Magnesium phosphate	0.106
Oxygen	0.221		
		Total	4.142

SPECIFIC GRAVITY = 1.0599

1000 GRAMS OF BLOOD CORPUSCLES

Water	681.63
Substances not volatile at 120°	318.37
Hematin (including 0.998 iron)	15.02
"Blood casein," etc.	296.07
Inorganic constituents (excluding iron)	7.28

[1] C. Schmidt, "Charakteristik der epidemischen Cholera," pp. 29, 32: Leipzig and Mitau, 1850. Quoted from Bunge, *Physiologic and Pathologic Chemistry*, 2d edition, p. 212-213, Philadelphia, 1902.

Chlorine	1.750	=	Potassium sulphate	0.132
Sulphuric acid	0.061		Potassium chloride	3.679
Phosphoric acid	1.335		Potassium phosphate	2.343
Potassium	3.091		Sodium phosphate	0.633
Sodium	0.470		Soda	0.341
Calcium phosphate	0.094		Calcium phosphate	0.094
Magnesium phosphate	0.060		Magnesium phosphate	0.060
Oxygen	0.401			
			Total inorganic constituents (excluding iron)	7.282

SPECIFIC GRAVITY = 1.0886

1000 GRAMS OF PLASMA CONTAIN

Water	901.51
Solids non-volatile at 120°	98.49
Fibrin	8.06
Albumin, etc.	81.92
Inorganic constituents	8.51

Chlorine	3.536	=	Potassium sulphate	0.281
Sulphuric acid	0.129		Potassium chloride	0.359
Phosphoric acid	0.145		Sodium chloride	5.546
Potassium	0.314		Sodium phosphate	0.271
Sodium	3.410		Soda	1.532
Calcium phosphate	0.298		Calcium phosphate	0.298
Magnesium phosphate	0.218		Magnesium phosphate	0.218
Oxygen	0.455			
			Total of inorganic constituents	8.505

SPECIFIC GRAVITY = 1.0312

1000 GRAMS OF SERUM CONTAIN

Water	908.84
Solids not volatile at 120°	91.16
Albumin, etc.	82.59
Inorganic constituents	8.57

Chlorine	3.565	=	Potassium sulphate	0.283
Sulphuric acid	0.130		Potassium chloride	0.362
Phosphoric acid	0.146		Sodium chloride	5.591
Potassium	0.317		Sodium phosphate	0.273
Sodium	3.438		Soda	1.545
Calcium phosphate	0.300		Calcium phosphate	0.300
Magnesium phosphate	0.220		Magnesium phosphate	0.220
Oxygen	0.458			
			Total inorganic constituents	8.574

SPECIFIC GRAVITY = 1.0292

Corpuscles of the blood. White.—The white corpuscles are true cells. They have a nucleus; they are capable of spontaneous movement and reproduction. They are colorless, amœboid cells which have the power of phagocytosis; that is, of eating bacteria and other solid matters which enter the blood. Their specific gravity is less than that of the red blood corpuscles, so that they collect as a layer above the reds when the blood is centrifugalized. They have an active metabolism, and their composition is that of typical cells consisting of true nucleins, phospholipins, globulins and albumins, which are at least in part in union with phospholipin. They have the extractives usually found in cells. They are variable in size, but they are somewhat larger than the red blood corpuscles. In mammals' blood their diameter is about 4-13 μ. Their number is very variable. Normally there is about one leucocyte to 350-500 red cells in human blood, or about 7,000-15,000 per c.mm. of blood. The number is relatively larger in the young, and it is increased in many infectious diseases, by the injection of nuclein, by suppuration, and by a meat diet. In the pathological condition of leucæmia, or leucocythemia, the number may be greatly increased so that the blood has a creamy appearance, as if pus were mixed with it. The numbers may rise to 500,000 per c.mm. of blood. This disease may be analogous to a neoplasm involving the white cells or the tissue which produces them. The white corpuscles are of various kinds: a, polymorphonuclear; b, lymphocytes; c, eosinophile; d, basophile, etc. They are formed in the bone marrow, or in the lymph glands, and possibly in part in the spleen.

Motility of white cells. The white (and also the red) corpuscles when placed in a thin film of blood or Ringer's solution, and possibly under other circumstances as well, undergo very remarkable changes of form: not only are the white corpuscles amœboid, but they shoot out from themselves long, whip-like processes which extend many cell diameters and have an active movement. These processes move back and forth and they are very sticky, so that bacteria stick to them like flies on sticky fly-paper. The processes may be very quickly withdrawn within the cell and the latter resume its amœboid form (Kite). These processes may be seen with particular clearness with the dark field illumination. They seem to be due to some surface tension action. Very similar processes are found also on the red cells under certain conditions. By means of their motility the white cells can crawl out of the blood vessels into the tissues, and their main function appears to be to remove old, worn-out or diseased tissues and cells, and to digest or otherwise overcome parasites of all kinds attacking the body. Probably they help also to form the proteins of the blood plasma.

Red corpuscles. Erythrocytes.—The number of red corpuscles in mammalian blood is very great. In human blood it is normally between 5,000,000 and 6,000,000 per c.mm. of blood, but in leucæmia and in anemias of various kinds it may be reduced to half this number or even less. These corpuscles have normally in human blood the form of biconcave disks and they contain no nuclei. But in mammalian embryos, or in adults after hemorrhage when the blood is regenerating, larger, nucleated red cells may be present; and in all the vertebrates below mammals the corpuscles are red nucleated cells, oval in outline and biconvex. The mammalian red blood corpuscles, as they lack nuclei, are not complete cells. Their diameter in human blood is usually about 7-8 μ, but there may be some as small as 5 μ and larger ones up to 12 μ in diameter. Their numbers are increased by living at high altitudes, or by exposing an animal to a lowered pressure of oxygen. An animal responds to either of these conditions by the production of more corpuscles and more hemoglobin, so that the blood can carry more oxygen at a lower partial pressure of the oxygen.

The red corpuscles contain hemoglobin and their chief function is to carry oxygen to the tissues from the lungs. Besides the hemoglobin, they contain chiefly lipoprotein material. Their composition will be taken up presently. They are formed in the red marrow of the bones by cells called hematoblasts. How long they live is not known. Their form, while quite characteristic and apparently fixed in most preparations so that it is apt to give the notion of considerable rigidity, is in reality not so. They are soft, and their peripheral layer at least is apparently made of a sticky, fluid matter. Under certain conditions, when they are mounted on a glass slide at least, they undergo extraordinary changes of form. At times in a moment they all become covered with sharp projections like burs; they are then said to be crenated, and under high powers of the microscope, particularly in the dark field, it may be seen that fine protoplasmic processes may extend over a cell's diameter in length from the ends of the sharp processes. These processes may be in active movement. The processes may be withdrawn as rapidly as they are extruded and the cell round up to the typical shape. If stimulated by touching the corpuscle, the processes may be suddenly shot out, or as suddenly withdrawn, or by a slightly stronger stimulus the erythocyte may withdraw them, suddenly swell and undergo solution. In all these respects the erythocytes behave like many other cells, like the white blood cells, a great many kinds of egg cells and tissue cells of the lower animals, such as sponges. They resemble, too, some of the mitochondria (Kite; Oliver). These changes in form indicate very clearly that the corpuscles are irritable and at least the exterior substance of the cell is liquid, since only liquids have the power of molecular movement necessary for these rapid and extraordinary changes. The changes in shape have been variously interpreted. Some have regarded them as the beginning of death changes; but to the author, as to many who have watched them, they appear exactly analogous to the vital responses of all kinds of living matter. They show that the property of movement is not lost to the red cells, and that they are real living cells, albeit of a special kind, and not rigid, or semirigid, quiescent, dead structures. They lack nuclei and with this they have lost to a large extent, but not entirely, the power of respiration and all power of growth, or synthetic metabolism, but they are still reactive, although far less so than the white cells. They probably live for a certain time only and grow old like other cell structures. These are not the only cells without nuclei in the animal kingdom. In certain molluscs there are found in the seminal fluid gigantic anomalous spermatozoa in which the nucleus has totally degenerated and disappeared, and yet the cells continue to exist for a long time, and these cells without nuclei are capable of movement.

It is because of the presence of the corpuscles of the blood that even in thin layers the blood is not transparent, but opaque. A good deal of light is reflected from the surfaces of the corpuscles. If the corpuscles are dissolved, the blood takes a darker tint and becomes transparent. It is then said to be "laked." The darker tint is due to the fact that in laked blood there is a smaller admixture of white light reflected from the corpuscles.

Because of their small size, flat disk shape, and enormous numbers, the total surface of the red corpuscles in a human adult amounts to from 3,000-4,000 square meters. The blood makes about one-twelfth of the total weight of the body. In a man of 72 kilos weight there would be some 6 kilos of blood or something over 5.5 liters, since the specific gravity of the blood is about 1.055. The total number of red blood corpuscles in the body would be about 30,000,000,000,000. The surface area of a corpuscle is about 0.000,1 sq. mm. The large surface area of a corpuscle in comparison with its bulk facilitates the exit and entrance of oxygen.

Blood platelets.—The blood platelets are spherical, or disk-shaped, bodies found in the shed blood of mammals, but not in the blood of birds or other vertebrates lower than the mammals. The platelets are of somewhat irregular size, the diameter varying from 1.5-3 μ. Their numbers are also very variable, and it is a very suggestive fact that the more care that is used to keep the blood in a living state the fewer platelets there are. Injury to the wall of the blood vessels greatly increases their numbers. They are refractive bodies and apparently nearly or quite homogeneous. There is no evidence that they contain nuclei. The number of these bodies in shed blood, as has been said, is very variable, but they may be present in very large numbers: namely, from 300,000-800,000 per c.mm. of blood. Sometimes they contain hemoglobin, but they are generally colorless. The origin and nature of these bodies has been much disputed, but the following conclusions are those generally accepted. They are almost certainly derived from the white blood corpuscles, and in part at least from the red. In their chemical composition they appear to be identical with the stroma of the red corpuscles or the protoplasm of the leucocytes, and like undifferentiated protoplasm generally. By some they have been supposed to be composed of nuclein, but this is almost certainly incorrect and is due to the fact that they show some of the properties of nucleins, such, for example, as solubility and the property of yielding, when digested by pepsin hydrochloric acid, an insoluble residue rich in phosphoric acid. In reality the platelets consist of a phospholipin-protein compound. The phosphoric acid is in a phospholipin. The platelets consist of what Wooldridge calls A-fibrinogen. As we shall see when we come to the clotting of the blood, they play a very important rôle in this process and they yield fibrin, and also a substance called thrombin or fibrin ferment, and a proteolytic ferment.

It is very difficult to decide whether the platelets are preformed in the living, unchanged blood or whether they appear there with great ease when blood is disturbed. We now know that the chromosomes of the nucleus, which are certainly organized structural constituents, may appear with great rapidity when the nucleus is stimulated in any way (Chambers). It may be the same with the platelets. Although they have been seen in the living capillaries, yet it is significant that there are fewer the more care is taken to avoid injury to the capillaries and stasis of the blood. It is impossible to examine the blood without causing some injury to it. The platelets form the first beginnings of the thrombus in intravascular clotting and they appear in such numbers that it is impossible that they should have been preformed in the blood. If, for example, one ties off a portion of the carotid artery of the dog with its contained blood and then, the dog lying on its back, one injures the wall of the carotid on the upper (ventral) surface with silver nitrate or heat, a thrombus forms in the vessel at the injured point. It is found that the thrombus consists chiefly of platelets and there are vastly greater numbers than could have been present in the small amount of blood caught between the ligatures. Injury to the endothelial wall appears to call them forth. This observation is very strong confirmation of the view of Wooldridge that the platelets, in part at least, are in solution in the plasma in a supersaturated form and when disturbed they crystallize out. He regarded them as in the nature of imperfectly crystalline proteins. They appear to the author to resemble what are known as fluid crystals. Moreover, substances of this chemical nature are particularly apt to form fluid crystals. Schwalbe, who tried the experiment on thrombosis just cited, concluded that they must come from the red corpuscles, because some of the red corpuscles were found in various unusual forms as if they were decomposing. It is hard to see how without motion of the blood the platelets could accumulate at the upper part of the artery where they were found, since they are heavier than the blood plasma and would naturally sink. It is hence more probable that they are formed in the method described by Wooldridge by a process of crystallization from the plasma, the crystallization taking place first at the point of injury of the artery wall. It is probable that some are preformed in the blood as it circulates, but this would not at all militate against this view. Many authorities, however, are of the opinion that the platelets are almost wholly preformed, living elements (see Dietjen).

The platelets may be prepared by receiving blood in an equal volume of 0.2 per cent. sodium oxalate solution, or sodium metaphosphate or fluoride, so as to prevent coagulation and the dissolution of the platelets, and then by centrifugalization at a relatively slow speed to separate out the corpuscles. The plasma with the platelets is then poured off from the red and white corpuscles and centrifuged very fast for some time. The platelets are thus thrown out and collect as a grayish-white layer at the bottom of the tube. They are quite soluble in the plasma on warming when they are first separated (Wooldridge), but on standing lose their solubility. Their numbers are increased by cooling the plasma before centrifugalizing it. At first they dissolve readily also in salt solution containing a little alkali, but on standing rapidly change, becoming more insoluble and like fibrin. A suspension or solution of the washed, perfectly fresh, unchanged plates has the property of forming a typical clot, yielding fibrin. (Wooldridge; Schittenhelm and Bodong.)

Blood plasma.—Horse blood contains about 34.5 per cent. by weight of corpuscles; and 65.5 per cent. of plasma. In other mammals the proportions are not very different from this. In human blood the corpuscles make 35-40 per cent. of the blood by volume and nearly 50 per cent. by weight, since their specific gravity is greater than that of the plasma. The plasma contains about 10 per cent. by weight of solids, of which about 7-9 per cent. are proteins; 1 per cent. salts, and the rest lipin and various other substances, such as urea, creatine, amino-acids, dextrose, etc., present in small amounts. The composition of the serum is given in the following table:

Ox Blood (Abderhalden, Bunge's textbook).

Water	913.64
Solids	86.36
Proteid	72.5
Sugar	1.05
Cholesterol	1.238
Lecithin	1.675
Fat	0.926
Phosphoric acid as nuclein	0.0133
Soda	4.312
Potash	0.255
Iron oxide	.....
Lime	0.1194
Magnesia	0.0446
Chlorine	3.69
Phosphoric acid	0.244
Inorganic phosphoric acid	0.0847

The proteins which may be separated from it are 1, fibrinogen; 2, serum globulin; and 3, serum albumin. It is possible that the globulin is a mixture of two or more proteins. Another protein is present in the blood serum, but not in the plasma, called serum fibrinogen by its discoverer, Wooldridge, but more commonly fibrinoglobulin. The composition of the blood proteins will be taken up later on page 549. They are in part at least, as they exist in the plasma, in combination with phospholipins and possibly with cholesterol. The plasma is generally

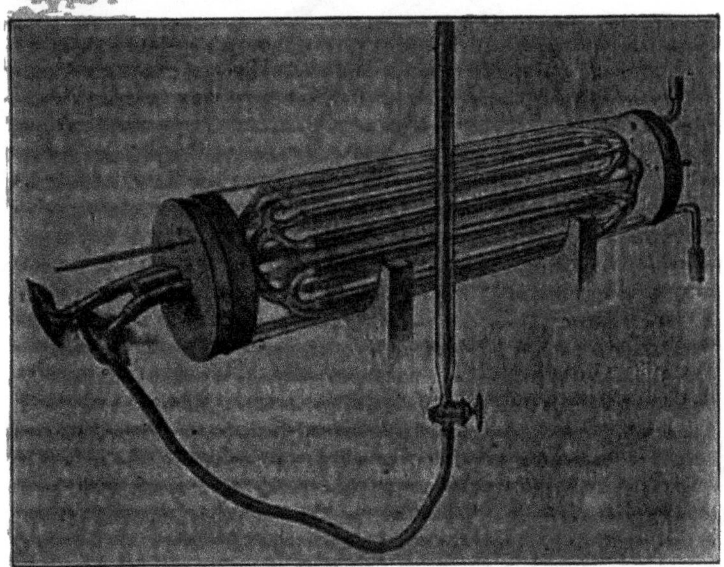

Fig. 44.—Abel's apparatus for the vivi-diffusion of the blood.

colored a light yellow, although in carnivorous animals it is at times almost colorless. The coloring matter in herbivorous animals is in part carrotin and is derived from green fodder, but in part it is due to urobilin. Plasma has a reaction alkaline to litmus, but acid to phenolphthalein, so that its hydrogen ion concentration is about 2×10^{-8} normal. It contains both carbonates and phosphates as well as chlorides. The greater part of the salt is sodium chloride.

With this brief statement of the structure and general composition of the blood we may proceed to discuss its functions.

1. **The blood as a carrier of food from the intestine to the tissues.**—The blood is to carry to the tissues the digested and absorbed foods. Water and salts; amino-acids and ammonia formed from the digestion

of the proteins; glucose and levulose from the digestion of carbohydrates; fats which have been split and then resynthesized during the process of absorption; all of these must find their way into the blood in order that they may be distributed to the tissues which need them. And all of them are to be found in the blood. Some hours after the ingestion of a fatty meal the blood contains so much of finely divided fat that the plasma may have a milky appearance and be no longer clear as it is

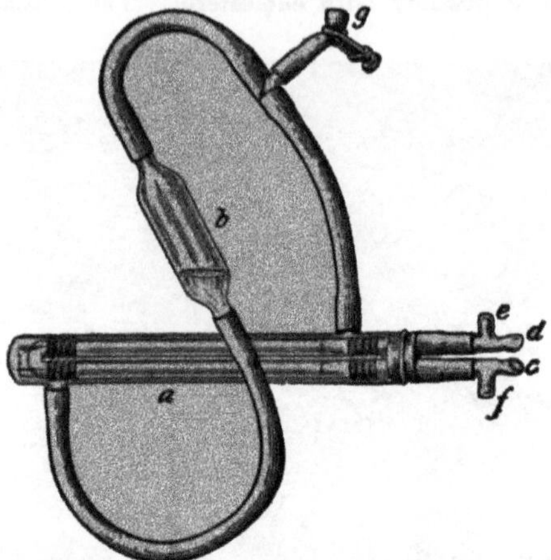

FIG. 45.—McGuigan and von Hess apparatus for vivi-diffusion. The blood enters and leaves at *d* and *c*, passing through collodion tubes inclosed in the diffusion jacket. By raising or lowering *b* diffusion is accelerated, and the pressure may be increased or diminished on the outside of the diffusion tubes through *g*.

in fasting animals. There may be so much fat present that it will rise on the serum separated from the defibrinated blood like so much cream. The amount of fat and fatty acid in the blood serum is normally 0.1-0.3 per cent., but after a fatty meal it may be from 0.6-1 per cent.

Dextrose is always present in blood in varying amounts. Normally the blood contains about 0.08 per cent. of dextrose, but it may in health rise to 0.15-0.2 per cent.; and in venous blood it may fall to 0.06 per cent. The dextrose is for the greater part in solution in the blood plasma in a dialyzable form. It may be dialyzed out of the blood by the method known as vivi-diffusion and, since this method is of the greatest importance in the investigation of the composition of the blood, it will be well to describe it at this point.

Vivi-diffusion method. The gist of the method consists in sending the blood from an artery through a collodion tube which is surrounded with physiological salt solution. Substances which are diffusible through collodion tubes will pass from the blood into the outer liquid and accumulate there until the concentration in the dialyzate is the same as that in the blood. After passing through the dialyzer, the blood returns to a vein. The dialyzer is thus a kind of artificial organ of excretion. The apparatus as described by Abel is shown in Figure 44. In this apparatus it is necessary to add hirudin or some other anticoagulant to the blood to prevent clotting. This introduced an element of complication in the interpretation of the results. The apparatus has been simplified and improved by McGuigan and von Hess, as shown in Figure 45. By avoiding all roughness in the glass and by keeping up a pulsation in the dialyzer they were able to prevent the blood from clotting without the use of any anticoagulant and greatly to hasten dialysis. The results obtained by the use of this method are very interesting, and the diffusion of dextrose is illustrated in the accompanying table (McGuigan and von Hess):

Dogs—anæsthetics used	Dialysis	Per cent. sugar in			
		Plasma	Dialysate	Plasma H_2O	Dialysate H_2O
Morphine	3.0. hrs.	0.135	0.139	0.147	0.141
Morphine	5.0 "	0.125	0.133	0.136	0.135
Morphine	6.0 "	0.088	0.099	0.096	0.100
Ether	4.5 "	0.083	0.087	0.090	0.088
Morphine and ether	5.0 "	0.073	0.082	0.080	0.083
Urethane	2.0 "	0.151	0.168	0.164	0.169
Urethane and ether	3.5 "	0.163	0.175	0.177	0.177
Urethane and ether	4.5 "	0.153	0.169	0.167	0.171
Urethane and ether	1.0 "	0.157	0.173	0.170	0.175

It is clear from the foregoing results that the sugar in the circulating blood is practically altogether in solution in the plasma and is not in union with a colloid, as had been suggested. In rabbits some glucose is found, also, in red cells (.05-.2 per cent.). Maltose also has been found in the blood plasma; and at times, during lactation or at the end of pregnancy, lactose in small amounts may be found there. It is surprising that the amount of dextrose in the blood does not increase more than it does during the absorption of a large amount of the products of starch digestion from the intestine. The reason is, probably, that the tissues take the dextrose out of the blood as rapidly as it enters, so that it does not accumulate in the blood. Liver and muscles have the power of storing a large amount of glucose as glycogen.

Water and salts are taken up by the blood from the intestine at a very rapid rate, but as the kidneys have the function of keeping the osmotic pressure of the blood approximately constant, when the tissues have taken out what they need any excess is rapidly eliminated in the urine. Nevertheless, foreign salts or the salts taken with the food can be shown to be present in blood. They are for the most part in solution in the plasma.

The end products of protein digestion are also found in the blood, but in very small amounts. After the injection of glycocoll into a loop of the intestine of a cat, there is an increase in the non-protein nitrogen, other than urea and ammonia, in the blood (Folin and Denis). A solution of glycocoll was injected into the ligated small intestine of a cat which had been fed 24 hours previously. The amount of non-protein nitrogen in the mesenteric and carotid blood before and after the injection was as follows:

Total non-protein N in 100 c.c.	portal blood before injection ..	30 mgs.
" " " " "	mesenteric vein 45 minutes after injection	85 "
" " " " "	carotid blood before injection ..	30 "
" " " "	" " 6 minutes after injection	34 "
" " " "	carotid blood 45 minutes after injection	57 "
" " " "	grams muscle before injection ..	250 "
" " " "	" " at end of experiment	346 "
" Urea N " "	grams muscle before injection ..	27 "
" " " "	" " at end of experiment	37 "

While the presence of amino-acids in the blood was made probable by these observations, by the discovery of their presence in invertebrate blood, and by the presence of these acids in the urine, their actual isolation from vertebrate blood has only recently been accomplished. They are present in very small amounts. A very large amount of blood was received from the slaughter-house (Abderhalden) and each liter poured into 15 liters of boiling water. After 15 minutes' boiling, 1 per cent. acetic acid was added, little by little, until the coagulation was complete and the solution became clear. It was filtered, the filtrate concentrated, and by the use of mercuric acetate and sodium carbonate the amino-acids precipitated. The precipitate contained proline, leucine, valine, alanine, glycocoll, aspartic acid, glutaminic acid, tryptophane, lysine, arginine and histidine. There is no doubt, therefore, that the amino-acids in extremely small amounts are to be found in solution in the blood plasma. Their presence may also be shown in the blood plasma by the vividiffusion method. They are diffused out of the circulating blood and may be detected in the dialyzate by the ninhydrin reaction and in other ways.

In the blood of a fasting dog Van Slyke and Meyer found in the femoral artery blood 4.4 mgs. of amino N per 100 c.c. blood; in the carotid artery 5.4-3.1 mgs., and in the mesenteric veins 3.9 mgs. per 100 c.c. blood.

While a very large amount of protein may be absorbed in the form of amino-acids by the blood, it is so rapidly removed therefrom by the tissues, or metabolized by the blood itself, that there is very little accumulation, but still there is some. From the figures just quoted from Folin and Denis it appears that amino-acids accumulate to some extent in the muscles after protein ingestion. The possibility exists that the amino-acids taken into the blood are in part synthesized into the proteins of this tissue, and by the digestion of these proteins again set free (Nolf); so that the amino-acids found in the blood need not be those immediately reabsorbed from the intestine. There is no good evidence, however, for any considerable transformation into blood proteins as an intermediary step.

2. **The blood as the carrier of oxygen from the lungs to the tissues.**—a. *The amount of different gases in the blood.* Venous blood as it leaves the tissues is purple in color. It returns to the right side of the heart and is driven through the pulmonary artery to the lungs. It there changes its color to a bright scarlet, and this scarlet, arterial blood is pumped by the heart to the tissues. The change in color of the blood in passing through the lungs is due to the fact that in these organs it takes on oxygen and loses carbon dioxide, so that arterial blood contains more oxygen and less carbon dioxide than venous; while in the tissues it loses oxygen and picks up carbon dioxide. The blood is then constantly carrying oxygen from the lungs to the tissues and carbon dioxide from the tissues to the lungs.

That arterial blood contains more oxygen and less carbon dioxide than venous blood may be shown by exposing the blood to a vacuum. The gases are given up to a vacuum and may be collected and examined. The blood yields all its oxygen to a vacuum and nearly all the carbon dioxide, but to get the last traces of carbon dioxide it is necessary to add to the blood some acid. This last portion is some of that combined as carbonate. In this and other ways it has been found that 100 c.c. of average human blood is able, if fully saturated, to take up from air or to give off to a vacuum between 18 and 19 c.c. of oxygen measured at 0° and 760 mm. pressure. Since arterial blood is usually only 96 per cent. saturated, the actual amount of oxygen recovered from blood under usual conditions is a little over 18 c.c. O_2 from 100 c.c. of blood. Venous blood yields generally about two-thirds as much oxygen as arterial, so that blood returns to the heart with about 12 per cent. by volume of oxygen still in it. In passing through the lungs 6 volumes per cent. of oxygen are restored to it.

The amount of carbon dioxide which can be pumped out of blood after the addition of acid, that is the total carbon dioxide, is about 40 c.c. from 100 c.c. of arterial human blood and about 48 c.c. from venous human blood. So that about 8 volumes per cent. are lost in passing the lungs.

The composition of cat's blood gases has been recently determined by Buckmaster and Gardner as follows, the oxygen capacity being somewhat less than that of humans:

ARTERIAL BLOOD. VOLUME IN C.C. AT 0° C. AND 760 MM. PRESSURE FROM 100 C.C. BLOOD.

	Total gas	CO_2	O_2	N_2
Carotid	39.34	25.81	12.70	0.83
"	28.59	15.79	11.97	0.83
"	33.73	18.54	14.18	1.00
"	38.41	22.11	15.13	1.17
Femoral	50.07	34.52	14.50	1.05
"	47.91	33.68	13.10	1.14
Mean	39.68	25.07	13.60	1.00

VENOUS BLOOD. ANESTHETICS AND HIRUDIN.

	Total gas	CO_2	O_2	N_2
Right auricle	56.66	44.24	11.31	1.12
" "	46.39	37.42	8.54	0.42
Mean	51.53	40.83	9.93	0.77

Besides carbon dioxide and oxygen, blood also contains small quantities of nitrogen. Nitrogen composes about four-fifths of the atmosphere. It is taken up by the blood in the lungs and exists in solution in all the tissues and fluids of the body. The amount of nitrogen gas given off by animals appears to be slightly greater than that inhaled, which would indicate the production of a small amount of gaseous nitrogen in the metabolism of the body. This result is not accepted by most observers, but it seems not unlikely, since by the action of bacteria in the alimentary tract nitrates are reduced to nitrites, some nitrites are constantly taken in the food and ammonium nitrite decomposes spontaneously, setting free nitrogen gas. The amount of nitrogen so formed is, however, very small. Nitrogen is very inert and exists simply in solution in the blood.

The amount of nitrogen taken up by the blood will depend on the pressure of nitrogen in the lungs. The amount absorbed is considerable in men working under compressed air in caissons, and as this nitrogen is released as a gas when the compression is suddenly removed it may collect as bubbles of gas in the blood vessels and by forming gas emboli be one of the causes of caisson disease. The amount of nitrogen in the blood of dogs, arterial and venous, for there is no difference usually between the amount in arterial and venous blood, was determined by

Bohr to be 1.2 c.c. in 100 c.c. of blood. Recent determinations in the blood of cats shows only 1.06 c.c. per 100 c.c. of blood. The air in the lungs, alveolar air, contains about 83.67 per cent. by volume of nitrogen. If this nitrogen pressure is reduced, the blood will lose N in passing the lungs and the venous blood may temporarily have more nitrogen than the arterial blood, due to the washing out of nitrogen from the tissues of the body. Nitrogen dissolves readily in fat, and at normal temperatures fat dissolves at least five times as much nitrogen as blood. The nitrogen in the blood and tissues is inert and probably plays no part in metabolism.

The amount of oxygen simply dissolved in the blood is small. It may be directly determined for the plasma and serum, but for the whole blood it is determined indirectly by finding the solubility of some inert gas like hydrogen, which does not combine with the blood corpuscles, and then multiplying the result thus found with the ratio of the solubility in water of hydrogen to that of oxygen.

The solubility of oxygen in ox serum at 29.7° is 2.47 per cent. by volume (0°, 760 mm.) or 94 per cent. that of its solubility in water. The solubility of hydrogen in ox serum is 1.56 per cent. by volume, or 95.5 per cent. of its solubility in water.

The difference in solubility of these gases in water and serum is due to the salts in the serum. A salt solution dissolves always less gas than an equal volume of water. Indeed, by the addition of salt, gases can be salted out of their solutions just as the proteins can be. Horse plasma dissolved 94 per cent. of the oxygen dissolved by an equal volume of water. The whole blood dissolved 91 per cent. as much hydrogen as an equal volume of water. The difference is due in part to the salts and in part to the volume of the dissolved protein and the solid matter of the corpuscles. It is evident, since the corpuscles make as a rule about 40 per cent. by volume of the blood, that hydrogen must dissolve in the water of the corpuscles as well as in the blood plasma.

b. *How are the gases carried in the blood? Oxygen.* Human arterial blood contains in 100 c.c. such an amount of gas that it will yield to a vacuum 18-19 c.c. of oxygen, 40 c.c. of carbon dioxide and about 1 c.c. of nitrogen, argon and other gases, all of these measured at 0° and 760 mm. pressure. Such an amount of gas is vastly more than can be dissolved in 100 c.c. of water, or in 100 c.c. of blood plasma. 100 c.c. of water at body temperature and the usual pressure of oxygen, that is a pressure of about one-fifth of an atmosphere, will absorb only 0.4 c.c. of oxygen, and serum will absorb about 94 per cent. of this amount. It is clear that oxygen must be combined chemically or physically with something in the blood so that its solubility is increased. It is, as a matter of fact, in greater part in union with the red coloring matter of the blood, hemoglobin, with which it forms oxyhemoglobin.

Carbon dioxide. Similarly 100 c.c. of water will absorb of carbonic anhydride at the temperature of the body and under the pressure of one-tenth to one-twelfth of an atmosphere, which is the pressure of CO_2 in the tissues, only about 10 c.c. Carbon dioxide is in part combined with the proteins of the blood plasma, and in part it is present in the plasma as the carbonate and bicarbonate of soda. There is in the blood plasma disodium hydrogen phosphate and sodium carbonate. When carbon dioxide comes into a solution of these salts, it combines with some of the sodium to make bicarbonate of sodium and is carried in the blood in large measure in this form. The proteins, such as the globulins, are also present in the blood as sodium salts. This sodium is removed from the globulin by the carbon dioxide. The corpuscles, too, act in the same manner as the globulins. They have sodium and potassium in them in organic union, and when carbon dioxide is given to the blood in the capillaries as it passes through the tissues, some alkali leaves the corpuscles to saturate the carbon dioxide so that the plasma has its total alkali increased by the action of carbonic acid, since the carbonates and bicarbonates have an alkaline reaction, or at any rate are titratable like alkalies. These various reactions by which the carbonates are formed may be written as follows:

$$Na_2HPO_4 + CO_2 + H_2O \longrightarrow NaH_2PO_4 + NaHCO_3$$
$$Na_2CO_3 + CO_2 + H_2O \longrightarrow 2NaHCO_3$$
$$\text{Na globulinate} + CO_2 + H_2O \longrightarrow NaHCO_3 + \text{Globulin.}$$
$$\text{Na lecithinate, etc., in the blood corpuscles} + CO_2 + H_2O \longrightarrow NaHCO_3 + \text{Organic compounds in a more acid state.}$$

All these changes take place when the blood is passing through the capillaries in the tissues. By this means it will be seen that the acidity of the corpuscles of the blood is increased, since they have lost alkali, Na and K, to carbonic acid. The result of this increase of acidity is that the power of the hemoglobin in the corpuscles to take up oxygen is greatly reduced, as we shall see presently, and the entrance of carbonic anhydride into the blood thus helps to turn the oxygen out of the blood and into the tissues. This factor is of great importance in cold-blooded animals where the affinity of hemoglobin for oxygen is so great, owing to the low temperature, that the pressure of the oxygen in the capillaries would be small. In the lungs, on the other hand, the opposite change takes place. With the passage of carbon dioxide outward alkali is set free again, is taken up by the corpuscles of the blood and their affinity for oxygen is so increased thereby that the blood saturates itself with oxygen very quickly in its passage through the lungs. This change is associated with a change of volume of the corpuscles. The volume in venous blood is larger than in arterial blood. Water passes into and out of the corpuscles.

Carbon dioxide is probably also carried in the blood in union with the proteins. When carbonic anhydride enters a solution of a protein which has free amino groups, the acid unites with these groups to make carbamino compounds (carbamino reaction of Siegfried). This reaction is the following:

$$\underset{\text{Protein.}}{R-\underset{COOH}{\overset{NH_2}{CH}}} + H_2CO_3 \longrightarrow \underset{\text{Carbamino compound of the protein.}}{R-\underset{COOH}{\overset{NH-COOH}{CH}}} + H_2O$$

These compounds are dissociable and the carbonic acid is easily recovered from this union. In all these ways, then, is the carbon dioxide carried back to the lungs. For the most part it is in solution in the plasma as carbonate, bicarbonate and protein compound, but some of it also is united with the corpuscles, presumably with the proteins of these structures (hemoglobin). A part of the carbon dioxide is dissolved as such in the blood. All these forms of carbon dioxide are in equilibrium with each other and with free carbonic anhydride, so that if the latter escapes more is set free to take its place, being dissociated from some of these unstable compounds. In the tissues the pressure of carbon dioxide is higher, 11 per cent. of an atmosphere, than in the blood, so that carbonic anhydride enters the blood until it is under equilibrium with this pressure of the gas. But when the blood reaches the lungs the pressure of carbon dioxide in the air in the lungs is low and carbon dioxide escapes from its solution and into the alveolar air. As soon as some of it escapes, more is set free from its union with alkali and protein in the blood. The pressure of carbon dioxide in the alveolar air in the lungs is about 5-6 per cent. of an atmosphere, whereas in the blood as it leaves the tissues it is about 8-10 per cent. Consequently in passing the lungs carbon dioxide is given up. The equilibrium may be represented as follows:

Tissues $CO_2 = 10\%$ of an atmosphere pressure. Alveolar air of lungs $CO_2 =$ 5-6% atmosphere.

1. $CO_2 + H_2O \rightleftharpoons H_2CO_3$
2. $H_2CO_3 + Na_2CO_3 \rightleftharpoons 2NaHCO_3$
3. $H_2CO_3 + Na_2HPO_4 \rightleftharpoons NaCHO_3 + NaH_2PO_4$
4. H_2CO_3 + NaProteinate $\rightleftharpoons NaHCO_3$ + H proteinate.
5. H_2CO_3 + Protein $\rightleftharpoons$ Protein COOH + H_2O
   Carbamino compound.

The reactions go in the right-hand direction in the tissues and in the left-hand direction in the lungs. Since the corpuscles and proteins have the power of combining with the sodium hydrate set free when the carbonic anhydride escapes in the lungs, they act as acids. It is for this reason that it is possible to pump nearly the whole of the carbon dioxide out of blood by means of a vacuum, whereas it is not possible to pump carbon dioxide to the same extent from a solution of sodium bicarbonate. From a solution of bicarbonate of sodium one can pump the carbon dioxide until sodium carbonate is formed. Thereafter the decomposition is almost immeasurably slow. A change occurs in the volume of the corpuscles when CO_2 enters them. They swell. This change is very significant, since it shows how small a change in acidity is required to cause swelling changes in vital structures. It is possible that it is the same process which is at the bottom of the contraction of muscle, as we shall later see. It is besides a true process of secretion of water into and out of the corpuscle, and is also a rhythmic process. When CO_2 increases in the corpuscle water enters it; when CO_2 is diminished water leaves, the sodium ion rè-enters and the corpuscle shrinks. There is an exchange of sodium ions and hydrogen ions back and forth between the corpuscle and the plasma.

c. *The mechanism of the entrance of oxygen into the blood and the passage of CO_2 outward. The exchange in the lungs.* We may now ask the question of the manner in which oxygen passes through the alveolar membrane into the blood. Is it by a simple physical process of diffusion, or is it by the active secretion of the lung tissue; or do both these processes occur?

The answer to this question, so important in medicine, cannot yet be given with certainty.

The walls of the alveoli of the lungs are extremely thin. They are composed of flattened plates. There is no doubt that in the lower forms, such as the amphibia, these plates are true cells and composed of living tissue; but there is a difference of opinion whether the plates in the mammalia are living or dead, and whether or not they have nuclei. Besides the very thin layer of alveolar plates, the gases must pass also the endothelium of the capillaries, which is also very thin, but certainly alive. The oxygen might enter either by diffusion, or one or both of these membranes might intervene actively in the process. If they do so intervene, they would probably be controlled by the nervous system.

The solution of this question of the method of entrance of the oxygen was approached in the following form: if the pressure of oxygen in the arterial blood as it comes from the lung is always lower than the pressure of oxygen in the alveolar air, then the process is probably one of diffusion and the membranes presumably do not actively intervene in it; if, however, the pressure of oxygen in the blood is ever higher than that of the alveolar air, then the exchange cannot be a simple physical process of diffusion. The first requisite for the solution of this problem was to find a method of estimating the oxygen tension of the blood in the arteries and of the air in the alveoli.

This tension is measured in two ways, a direct and an indirect. The direct method has been most frequently employed. The aërotonometer is an instrument designed for this purpose. The essential principle of the aërotonometer is the following: The blood is introduced directly from the blood vessels into an atmosphere of nitrogen, carbon dioxide and oxygen and allowed to remain in contact with the gas in a thin layer so that equilibrium is attained. The gas is then analyzed.

The principle of the method used by Krogh is to shake a small air bubble in a very small amount of the blood to be examined. As little as 1 c.c. blood may be used. The gases in the air bubble come quickly into equilibrum with those of the blood. The amount of oxygen is measured by the decrease in volume when the oxygen is absorbed by alkaline pyrogallate or some other oxygen-absorbing fluid, such as acid hyposulphite.

All measurements which were made with this instrument resulted uniformly in showing that the pressure of oxygen in the alveoli was slightly greater than the pressure in the arterial blood.

The matter thus seemed settled, but in 1890 Bohr [1] got the first definite evidence that the process was not one of simple diffusion. He observed in a few cases that the oxygen pressure in the arteries was higher than that in the alveoli. His results obtained by the aërotonometer method were so irregular as to suggest errors of manipulation and they have been seriously criticised by Krogh and by Haldane and Douglas. The aërotonometer is very sensitive to a change of temperature, and an accidental variation in this might have accounted for his results. Moreover, in one or two cases, as Haldane and Douglas point out, the results are so improbable as to indicate error very plainly. In spite of these defects, Bohr's paper served the end of reopening the question. The possibility of a definite secretion of oxygen by the lungs also gained in probability by the discovery of the high oxygen content of the air bladder of fishes, the organ from which the lungs were evolved. In 1907 Bohr afforded other evidence of the secretory activity of the lungs. He found that when pure air was breathed by one lung and air containing 8.8 per cent. by volume of CO_2 by the other, CO_2 was still given off from the lung breathing the CO_2 mixture, although the pressure of CO_2 in the venous blood from the right side of the heart was that of an atmosphere containing only 5 per cent. by volume.

[1] Bohr: *Skan. Archiv. f. Physiol.*, 2, 1890, p. 236.

It will be noticed that the pressure of CO_2 was determined in the heart blood, and the pressure in the lungs was supposed to be equal to this. However probable this assumption is, it weakens the proof.

Recently Haldane and Smith also got some evidence of the existence of a secretory activity of the lungs, but their work contained so many assumptions and possibilities of errors of fact and interpretation that not much weight can be given it. A more recent paper will be considered presently.

Krogh[1] has recently re-examined the whole question. He and Mrs. Krogh measured with great care by means of the micro-aërotonometer the pressures of oxygen and CO_2 in the lungs and the arterial blood. They found always that the pressures of CO_2 in the arteries and in the alveolar air were equal, the result which the diffusion theory demands. The oxygen pressure in arterial blood was always slightly less than the pressure in the alveoli, a result also in accord with the diffusion theory. These experiments give no evidence of a secretory activity on the part of the lung.

The matter is not yet settled, however, for Douglas and Haldane[2] have made a very complete study of the matter recently and obtained very interesting results.

Their method of measuring the arterial oxygen pressure was an indirect one. It consisted in partially saturating the blood with CO gas. When blood or hemoglobin is exposed to a mixture of O_2 and CO, the hemoglobin takes up some of each and the relative amount depends on the partial pressures of the two gases. But always far more CO than O_2 is held by the hemoglobin at the same pressures. If a person is made to breathe air containing a small per cent. of CO, the blood ultimately, after 30 minutes about, has taken up all the CO it will. If now the per cent. of saturation of the Hb by CO can be determined, then the amount of oxygen in the arterial blood can also be determined, since from the per cent. of saturation of hemoglobin by carbon monoxide the tension of oxygen necessary to prevent total saturation by carbon monoxide and to permit only the per cent. of saturation actually observed can be calculated. This calculated oxygen tension is assumed to be present in the blood. To determine the per cent. of saturation of the Hb by CO a sample of the blood is drawn and its tint, when diluted, is compared with a carmine solution which has previously been standardized against blood completely saturated with carbon monoxide. From the amount of dilution of the carmine solution the per cent. of saturation of the Hb can be calculated, Column 3, and from that the arterial pressure of oxygen can be computed, Column 5.

[1] Krogh: *Skan. Archiv f. Physiol.*, xxiii, 1910, p. 274.

[2] Douglas and Haldane: *Journal of Physiology*, 44, 1912, p. 305.

The results obtained by Douglas and Haldane are illustrated in the following protocol. Mice breathed air mixed with varying amounts of CO. They were then drowned and two drops of blood taken from the heart for analysis. The inspired air contained on the average 19.79 per cent. of O_2 and 0.29 per cent. CO_2; the alveolar air contained 14.06 per cent. O_2 and 5.64 per cent. CO_2.

1. Per cent. of CO in inspired air	2. Duration of expt. minutes	3. Per cent. saturation of Hb with CO	4.	5. Arterial O_2 tension per cent. of an atmosphere calculated from 3
		In vivo	In vitro	
0.016	60	26.2	17.2	12.2
0.018	45	26	18.5	13.6
0.046	40	29.1	22.7	15.0
0.053	40	37.7	30.2	16.2
0.100	32	45	43.0	19.3
0.129	31	50.4	56.3	20.8
0.213	13	59.1	75.5	44.7 (Animal died)
0.244	12	67.3	71.7	25.7 (Animal died)
0.262	20	66.4	73.7	28.2
0.275	25	66.5	76.9	35.9

The experiment shows that as long as hemoglobin was not more than 30 per cent. saturated with carbon monoxide, the pressure of oxygen in arterial blood (12-15 per cent.) was less than that in the alveoli; but when the per cent. of saturation of the hemoglobin with carbon monoxide was more than this, the calculated arterial tension was always higher (16-36 per cent.) and might be over 100 per cent. greater than that of alveolar air. The per cent. of saturation of the Hb by CO was less in vivo than in vitro.

A similar result was obtained for human beings when breathing air containing varying amounts of oxygen. The experiment was tried on a man breathing in a closed system. While resting and breathing air containing normal amounts of oxygen, the tension of oxygen in the alveolar air being from 12-15 per cent. and that of CO_2 5.6 per cent., the arterial blood had an oxygen tension varying in different experiments from 91.6-104.4 per cent. of the tension in the alveolar air. In other words, it was only once found to be higher than the tension in the alveoli, but was generally lower as the diffusion theory demanded. The same result was obtained when the subject was resting and breathing air containing more than the normal amount of oxygen. When, however, the amount of oxygen was reduced so that the per cent. of oxygen in the alveolar air was lower than normal, the pressure of oxygen in the blood was larger than the tension in the alveoli; the difference was particularly large when work was done.

Low oxygen pressure. Resting—arterial O_2 tension in per cent. of alveolar oxygen tension	O_2 tension alveolar	CO_2 tension alveolar
115.4	5.98	4.93
121.6	6.99	4.78
128.1	6.40	3.48
112.1	5.53	4.74
Moderate work. One arm		
124.8	12.82	5.39
Severe work. One arm		
131.7	15.11	5.33
135.0	15.49	5.69
128.0	11.64	5.22
128.0	19.38	4.43
Moderate work. One arm		
147.6	10.20	3.30

There is no doubt, therefore, that in normal circumstances during rest oxygen enters by a process of diffusion; or at least there is no evidence of any secretory activity by the alveolar endothelium. During work, however, or when there is a deficiency of O_2, the pressure of O_2 in the arteries rises far above, to 135 per cent., that in the alveolar air. It appears from these experiments, then, that the lungs may, when there is necessity, actively secrete oxygen into the blood. This discovery, if it be sustained, is evidently a very important one. The way in which the lungs are aroused to activity when the tissues need oxygen is still obscure. It may be either by way of the nervous system or by some metabolic products of the tissue activity. The authors state that it is certainly not by means of CO_2, or lactic acid, since these leave the process practically unaffected.

Conclusive though these experiments seem, they are not completely so, and the whole question must still be regarded as open. This is owing to the fact that in any indirect method of determining the pressure there are always many assumptions, some of which cannot easily be tested. In this method the following assumptions are made: First, that the colorimetric method of estimating the degree of saturation of the hemoglobin is reliable. A better method has been devised by Hartridge. But, even if the degree of saturation is correctly determined, the inference that the rest of the hemoglobin is combined with oxygen, or uncombined, is not proved. It is possible that hemoglobin unites with many other substances than gases. If so, these substances may be present in the blood in larger amounts than usual under the conditions of the experiment when there is partial asphyxia.

Another possible source of error in this indirect method of determining the oxygen tension of the blood is this. What is actually determined is the amount of carbon monoxide hemoglobin in the blood. From this figure one calculates how large the tension of oxygen must be in order to prevent the hemoglobin from taking up more carbon monoxide than it does. The assumption that is made in this is that the avidity of the oxygen and hemoglobin undergoes no change in the course of the experiment, but remains the same as in the experiments in vitro. This assumption may not be correct. In the experiment oxygen and carbon monoxide are quarreling for the hemoglobin. The power of the oxygen is measured by its success in the struggle under certain conditions. But let it be supposed that in times of stress, as in partial asphyxia, the body has the power of strengthening the hands of the oxygen; it might then wage a very much more successful struggle for the hemoglobin than before and displace more of the carbon monoxide. There are reasons for thinking that the body does possess just this power, because it forms oxidases which hasten oxidation when it needs oxygen. The formation of oxyhemoglobin is a process of oxidation. It is possible, therefore, that the smaller saturation of the hemoglobin by carbon monoxide in partial asphyxia is not due to the fact that the tension of the oxygen has been increased, but that the efficiency of that actually present has been increased in its oxidizing power by the oxidases. Perhaps the proportion of active oxygen molecules is increased. It would seem unlikely that oxidases should play no part in the oxidation of such an important substance as hemoglobin. This possibility should be investigated. That there is something in its favor is shown by the fact that a person exposed to a low oxygen pressure, if the pressure is not too low, shows a betterment of condition when slight work is done. The asphyxia seems somewhat relieved by the work. Perhaps by the activity of the tissues more of the oxidase is produced.

d. *Nature of the union of hemoglobin with oxygen.* There is little doubt that the union between oxygen and hemoglobin is chemical in nature. This opinion was almost universally held until W. Ostwald suggested that the union was one of adsorption. If the union is chemical, then if the per cent. of saturation of the hemoglobin is plotted along the ordinate and the tension of oxygen along the abscissa, as is done in Figure 46, the curve of saturation of the hemoglobin by oxygen should be a rectangular hyperbola. Bohr did not find this to be the case. The cause of the discrepancy was investigated by Barcroft, who found that if the hemoglobin solution was thoroughly dialyzed so as to rid it completely from salts, then the curve was a rectangular hyperbola, as the theory demanded.

That the union is chemical is shown also by Barcroft and Hill. The rate of reduction of HbO_2 by nitrogen was strongly influenced by temperature, going on at a much more rapid rate at a higher than at a lower temperature and the temperature coefficient between 38° and 18°

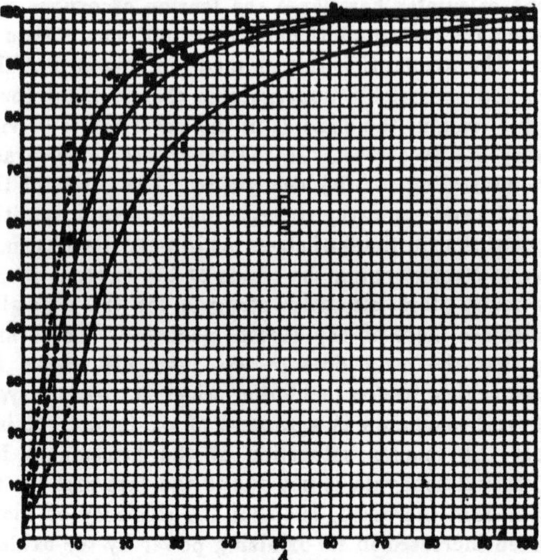

FIG. 46A.—Curve of dissociation of oxyhemoglobin showing the effects of various salts. I, 0.7% II, $NaHCO_3$; III, Na_2HPO_4. Bicarbonate and Na_2HPO_4 concentration equivalent to NaCl. Ordinate: % saturation of Hb. Abscissa: tension of O_2 in mms. Hg.

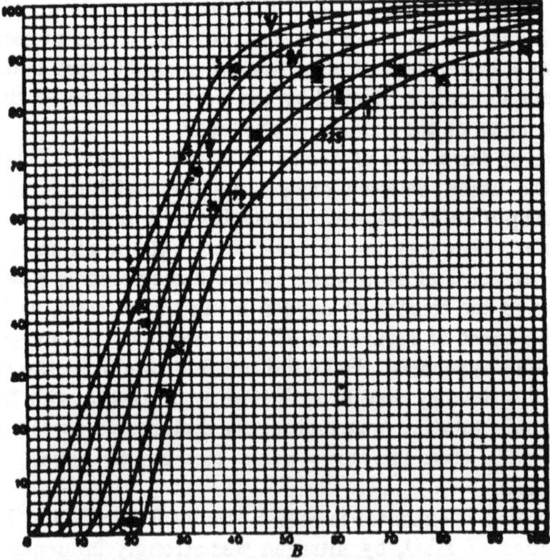

FIG. 46B.—Dissociation curves of sheep blood at various tensions of CO_2. I at 3; II at 10; III 20; IV at 40; and V at 80 mm. Hg. tension. Numbers below curves show actual CO_2 tensions observed. Temperature 37-38° C. Ordinate: % saturation of oxyhemoglobin. Abscissa: tension of O_2 mms. Hg (Barcroft and Camis).

is 3.7 for 10°. This indicates strongly a chemical union, for as a rule physical processes have much lower temperature coefficients than this, which is that of a chemical reaction. They also determined the heat set free when one gram of Hb was oxidized to HbO_2. They found for 1 gram Hb 1.85 calories. From this they calculated the molecular weight of the hemoglobin as 15,200, which is about that found by Hüfner, if one molecule of oxygen combines with one molecule of hemoglobin and if there is one atom of iron in the molecule. The process is evidently a limited, or partially consummated, oxidation process. One molecule of hemoglobin combines with one molecule of oxygen and a certain amount of heat is liberated in this process. We see, therefore, that to this extent at least Lavoisier was right and that some combustion really takes place in the lungs. There is some heat liberated there.

We have already considered on page 256 the physical chemistry of the process of oxidation, but we may at this point consider the application of those principles to the oxidation of hemoglobin. The velocity of the oxidation is then proportional to the concentration of the active oxidizing agent, to the concentration of the active reducing agent, and to the time required for the passing over of the positive charge from the oxidizing (O_2) to the reducing (Hb) body. This last factor, which varies so enormously in different oxidations, is in the case of hemoglobin in ordinary circumstances very long. It is much longer than oxyhemoglobin ordinarily exists before the reaction is reversed in the tissues. Hence HbO_2 is stable for a considerable period. This molecule has quite a long span of life. Nevertheless it is stable only under very narrow conditions and a change of alkalinity sufficiently great causes the consummation of the oxidation and the formation of methemoglobin, small amounts of which are present in normal blood and which under pathological conditions when nitrites, chlorates, aniline and many drugs are taken, is formed in large amounts. Hemoglobin is a substance combining easily with oxygen, but in which the oxidation does not go to a conclusion.

The velocity may be written in the form of an equation:

$$d(HbO_2)/dt = KC^{\cdot}_{O} \times C^{\cdot}_{Hb}$$

$C^{\cdot}_{O}$ is the concentration of the active oxygen, that is oxygen in a condition to unite; and $C^{\cdot}_{Hb}$ is the concentration of the active hemoglobin. The reaction is

$$1.\quad Hb \rightleftharpoons Hb''$$
$$2.\quad O_2 \rightleftharpoons O_2''$$
$$3.\quad Hb'' + O_2'' \rightleftharpoons HbO_2$$

The point of equilibrium and the velocity of the reaction will be determined not by the total concentration of the Hb and the O_2, but by the concentration of the active molecules present. Now most oxidations have these two peculiarities: They are accelerated by light, particularly by ultra-violet light, and they are all dependent upon water. In light, therefore, the per cent. of saturation of the hemoglobin at a given pressure of oxygen should be higher than in darkness and the velocity of the oxidation should be greater.

A second important fact in oxidation is the rôle of water. Substances do not oxidize in the dry state. The probable explanation of this fact, or at least one explanation, is that given in the case of bromide oxidations (p. 260). It is almost certain that when bromine oxidizes the active agent is not the bromine itself, but a positive bromine ion, which is formed by the interaction of the bromine and the water as follows:

$$1.\quad Br_2 + HOH \rightleftharpoons HBr + HOBr$$

$$2.\quad HOBr \rightleftharpoons \overset{+}{Br} + \overset{-}{OH}$$

The Br^+ set free is a powerful oxidizing agent and the speed of the reaction is probably proportional to its concentration. The oxidation by copper and oxygen is probably very similar as already discussed on page 260. We probably have the reactions:

$$3.\quad Hb'' + O^{++}(OH)_2 \rightleftharpoons HbO_2 + H_2O$$

Heat is liberated in the last reaction.

The condition of the hemoglobin must also be a great factor in the speed of the oxidation. All the evidence we have shows that reducing bodies are not always in a state to receive the oxidizing body and as a rule the condition of ionization of the

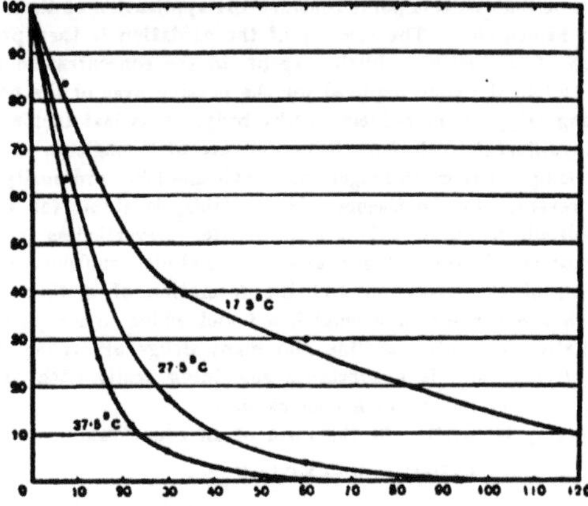

Fig. 47.—Effects of temperature on rate of reduction of sheep's blood by hydrogen. (Oinuma). Ordinate: per cent of saturation with oxygen; abscissa: time in minutes.

reducing body is of great importance. Now Hb is probably a salt of a metal, sodium or potassium, and the condition of the iron, with which the oxygen is combined, will probably be found to be a function of the particular metal in combination with the hemoglobin.

It will be seen then from the foregoing equations that the speed of oxidation and the per cent. of saturation of the hemoglobin with O_2 will depend in the first instance on the alkalinity, or number of hydroxyl ions. Hence an increase in the alkalinity of the blood will cause Hb to take up O_2 faster and hold a larger proportion of it since this increases the active mass of the oxygen; and acidity will have an opposite effect, causing the HbO_2 to give up O_2, hastening the reduction and lowering the point of equilibrium when saturation is reached. The effect of temperature, since the reaction is exothermic, will be to increase the dissociation as the temperature rises. Alkalies may also affect the active mass of the Hb. These theoretical conclusions are borne out by experiment.

e. *Factors influencing the dissociation of O_2 from HbO_2.* Bohr opened this subject by his discovery that carbon dioxide strongly influenced the dissociation curve of HbO_2. Hemoglobin takes up less O_2 as the tension of CO_2 increases; and HbO_2 gives off O_2 much quicker in the presence of CO_2. This is a matter of great importance in the body.

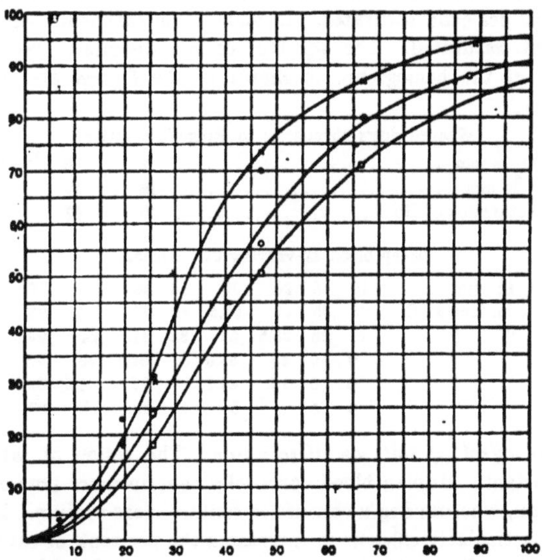

FIG. 48.—Curves illustrating the dissociation of oxyhemoglobin of sheep blood on the addition of various amounts of lactic acid (Barcroft and Orbell). The upper curve represents the per cent. of saturation of normal blood when the oxygen tension in mm. is that represented along the abscissa. The two lower curves show the effect of the addition of lactic acid.

The studies of Barcroft and his associates have shown the influence of alkalies, acids, salts, temperature and light on the dissociation of this oxide.

1. *Effect of temperature on dissociation.* The per cent. of saturation of hemoglobin in air and at different temperatures was determined by Barcroft and Hill as follows:

Dog's hemoglobin.

Saturation (%)	96	89	77	52
Temperature (°)	16	25	32	38

2. *Acids and alkalies.* Alkalies increase the speed and the per cent. of saturation under a given oxygen pressure and acids have the opposite effect. This is shown in the following figures from Barcroft and Camis and in figure 48 and 46A:

Tension O_2 in mm.	12.5	15.5	31	45	72
Per cent. saturation of hemoglobin in water	29	40	60	77.5	90.5
Per cent. saturation Hb alkaline with $(NH_4)_2CO_3$	78.5	79	92.5	97	99
Tension O_2 in mm.	20.5	21	30.5	37	59.5

It will be seen that in water at a tension of 50 mm. Hg hemoglobin is less than 80 per cent. saturated. In aminonium carbonate solution it is 98 per cent. saturated. The effect of carbon dioxide in reducing the saturation of the hemoglobin is shown in the figures and curves in Figure 46B:

Washed dog's corpuscles in Ringer's solution.

Tension of CO_2 in mm.	2	5	76	69
Tension of O_2 in mm.	18	19	17	18
Per cent. of saturation of Hb	55	57	6	9

Sheep blood at varying CO_2 tension.

Tension of O_2 in mm.		10	15	20	30	40	50	60	70	80	100
Per cent. saturation of Hb.	5 mm. tension CO_2	28	35	47	58.5	89	95	96	98	99	99.5
	10 " " "	11	26	38.5	63	83	91.5	94.5	96.5	97.5	98.5
	20 " " "	0	10	25	53.5		84.5	90	93	95	97.5
	40 " " "	0	0	11	42.5		77	83.5	88.5	93	95.5
	80 " " "	0	0	1	31		69.5	77	83	87.5	92.5

3. *The effect of salts on dissociation.* The action of salts is also very important. Human blood corpuscles contain more potassium salts than sodium, whereas dog's corpuscles contain more sodium. Potassium salts are particularly efficient in increasing the per cent. of saturation of the Hb. The salts in the corpuscles, or the nature of the base in union with the Hb, is, therefore, of considerable importance in this exchange.

At a tension of 50 mm. of oxygen the hemoglobin in solution in 0.7 per cent. NaCl is 85.5 per cent.; in 0.9 per cent. KCl it is 95 per cent. saturated; and in Na_2HPO_4 it is more than this. It is probable that this difference in salt content explains the difference in saturation of different bloods when exposed to the same tension of O_2.

4. *Other factors possibly influencing the dissociation of HbO_2.* There is, in addition to the factors already mentioned as controlling the union of Hb with O_2, one other which has been so far neglected, but which was mentioned on page 483. It may be that there are in the plasma or corpuscles substances which may facilitate the union of hemoglobin with oxygen. It would be interesting to see what influence small traces of iron might have on this process. It is said that small amounts of iron are constantly getting free from the Hb, particularly from reduced hemoglobin. Indeed, Bohr suggested that the iron was alternately set free and reunited with the Hb, but such is probably not the ease. The matter should be further studied.

f. Biological significance of factors influencing HbO_2 dissociation. The general biological significance of the facts thus described for blood is very great. We may indeed take hemoglobin as a type of a substance uniting with oxygen. Substances having the power of uniting with oxygen are found in all cells of the body, and it is probable, since the dissociation of the cell from oxygen is a matter of a good deal more difficulty than the dissociation of HbO_2, that these substances hold their oxygen a good deal more firmly than the oxygen is held by Hb. They do not easily give up their oxygen to a vacuum. The tension of oxygen in the tissues is very low. There are many reasons for thinking, however, that oxygen storage may occur there. The protoplasm is made up of reducing substances. We may be certain, in any case, that the oxidation of a cell, like that of hemoglobin, will be profoundly affected by sunlight, temperature, alkalinity and acidity and by salts. And indeed all our facts prove this to be the case. Oxidation is facilitated in the light; by fevers or high temperatures; by slight alkalinity; and by various salts.

The acidity produced by carbon dioxide and lactic acid is very important in turning oxygen out of its union with hemoglobin when the blood reaches the tissues. This must play a great part in cold-blooded animals, where at low temperatures the oxygen-hemoglobin compound dissociates very little. Carbon dioxide does not in the frog find its way out through the lungs, but through the skin. Perhaps it is kept in the body for this purpose.

g. *The exchange in the tissues.* While the exchange in the lungs has been supposed by some to involve secretory activity on the part of the plates of the lung endothelium, the exchange in the tissues is believed to be due only to processes of diffusion. The pressure of oxygen in the tissues is less than that in the capillaries and the pressure of carbon dioxide in the tissues is greater than that in the blood. So that there is no reason for supposing that any other factors than those of diffusion play a part in this exchange.

The blood as it leaves the tissues is still rich in oxygen. It is never, under ordinary circumstances, completely reduced. Indeed, venous blood still contains a large proportion of its oxygen. Analyses of blood coming from different organs show differences in this respect as might be anticipated, but the following table illustrates the composition of the gases of the venous blood from various organs; the figures are c.c. for 100 c.c. of blood:

Organ	Blood	CO_2	O_2	Observer
Submaxillary gland resting	Arterial	53.1	15.2	Chauveau and Kaufmann
	Venous	55.2	11.4	(Barcroft: Ergebnisse 7)
Leg of dog muscle resting	Arterial	21.92	14.4	Zunst
	Venous	36.32	1.2	

Organ	Blood	CO_2	O_2	Observer
Supra-renal gland	Arterial		21.79	Chassevaut and Langlois (Barcroft)
Brain	Arterial-carotid	40.86	16.81	Hill and Nabarre
	Venous (Torcula Heroph.)	44.74	13.39	(Mean values)

The following table from Barcroft shows the consumption of oxygen in c.cm. pro gram per minute by various tissues:

OXYGEN CONSUMPTION PRO GRAM PER MINUTE BY RESTING TISSUE

Name of tissue	Oxygen consumption	Animal	Observer
Skeletal muscle	.0037 c.c.	Horse	Chauveau and Kaufmann
Heart muscle	.010	Dog	Barcroft and Dixon
Salivary glands	.028	Dog	Barcroft and Dixon
	.023	Cat	Barcroft
Pancreas	.03-0.05	Dog	Barcroft and Starling
Intestinal canal	.023	Dog	Brodie, Halliburton and Vogt
Kidneys	.026	Dog	Barcroft and Brodie

h. *Is the respiratory pigment as it exists in the blood itself hemoglobin?* Does hemoglobin as it exists in the blood differ in its properties of oxygen absorption from isolated hemoglobin? Bohr thought that it did. He supposed that there were distinct differences between hemoglobin and the blood pigment, which he called hemochrome. The reason for this opinion was that solutions of hemoglobin in water were found by him to have a different curve of dissociation of the HbO_2 compound than the curve for the dissociation in blood itself. But Bohr overlooked the fact that the dissociation curve depends on the amount and character of the salts present. Barcroft and Camis showed that a solution of dog's hemoglobin, to which had been added the salts found in dog's corpuscles, gave a curve of dissociation like that of dog's blood; and that if to dog's hemoglobin salts, like those in human red corpuscles, were added a curve of dissociation was obtained like that of human blood. It appears, therefore, that there is no reason to suppose that the pigment as it exists in the blood is different from hemoglobin. There is no evidence, in other words, that the shape of the corpuscle, its wall or other properties play a part in the process of the union of hemoglobin and oxygen. Blood, in virtue of its alkali salts, has, however, a great advantage over a simple solution of hemoglobin in water. Thus at 30 mm. oxygen pressure hemoglobin in aqueous solution is only 62 per cent. saturated; whereas blood at the same pressure saturates itself to 69 per cent., owing to the potassium salts in the corpuscles. Hemoglobin in the corpuscles is, however, almost certainly in chemical union with the stroma, so that the oxygen-carrying substance in the blood is in reality stroma-hemoglobin compound and not free hemoglobin.

i. *Respiration of the blood itself.* Does the blood, then, not respire itself? Does it consume no oxygen and give off no CO_2? While the great bulk of the oxidation occurs in the tissues, there can be no doubt that a certain amount occurs in blood itself. Blood is a living tissue.

The white cells in it certainly respire, and the red corpuscles probably do also to a limited extent, since they contain in their stroma oxidizable substances. But the rate of their respiration is undoubtedly small. Not so, however, with the white or nucleated cells both white and red. Particularly after hemorrhage the blood shows a considerable power of oxidation of itself, and also during asphyxia, or whenever the tissue decomposition is greater than the tissue oxidation can burn, these substances get free in blood and in part burn there. Thus Bohr found that the ratio of O_2 to the iron of the blood underwent marked changes after hemorrhage, so that he suggested that there was more than one kind of hemoglobin in the blood. But it has since been shown (Douglas) that the oxygen capacity of the blood after hemorrhage is exactly proportional to its hemoglobin content, so that there is no change in the character of the hemoglobin. It has been found, too, that the blood of rabbits made anemic (Morawitz and Pratt) by repeated injections of phenyl hydrazine or repeated hemorrhage has a remarkable power of absorption of oxygen and production of CO_2. The following figures show the consumption of oxygen by rabbit blood after repeated hemorrhage:

	O_2 Per cent.	CO_2 Per cent.
After aëration	8.7	30.8
Incubated ¾ hour	6.0	34.8
" 1⅓ "	2.8	36.0
" 2¼ "	0.3	39.1
" 3¾ "		39.9

Respiratory quotient $CO_2/O_2 = .91$.

The main factors in this consumption are the white and the nucleated red blood corpuscles. That the oxygen-consuming power is found chiefly in the white corpuscles can easily be made evident by centrifuging defibrinated blood. It will generally be observed that the red color immediately beneath the layer of white corpuscles which rests upon the red corpuscles is that of reduced hemoglobin. In the blood of invertebrates the corpuscles are in many cases altogether white corpuscles. In Limulus, the king crab, there is a blood pigment which is blue when oxidized and colorless when reduced. This pigment corresponds to hemoglobin, but contains copper in place of iron. It is called hemocyanin. It will be observed in this blood if it is allowed to clot that the blood is white or reduced, except in the upper layers of the clot, where it comes in contact with the air.

j. *Evolution of hemoglobin.* The evolution of hemoglobin is of interest. Iron is found in all forms of living matter and in all it plays perhaps a predominant rôle in oxidation. In the course of evolution an iron compound was evolved which, while permitting the iron to take up oxygen, was not itself oxidized by it. It remained, therefore, an easily reduced, but otherwise fairly stable, oxide. This substance is hemoglobin. It is found very low in the animal kingdom in annelids, nemertines and mollusks. In the lowest forms and in its primitive condition it is a constituent of the muscles, just as it, or a closely allied substance, is found in many vertebrate muscles to which it gives a red coloration. Here it plays its primitive rôle of a storer of oxygen. The next step consisted in having it free in the circulating fluid as it occurs in the nemertines, so that it could obtain oxygen at the surface of the body and bring it back to the tissues. Finally we have it inclosed in corpuscles, where it may be surrounded by salts, which are particularly useful for its functions, but which, if at large in the blood stream, would be harmful to the organism. It is possible, also, that the concentration of hemoglobin in the blood can be increased by this means above that which is possible by simple solution; and finally it may be that the wall of the corpuscle has been particularly evolved to make a membrane which, like the gills of the fish, will let gases through readily, but will prevent the entrance of many substances which might combine with hemoglobin. It is possible that there are in the tissues other colorless protein or other compounds to which oxygen is more firmly attached than it is in the hemoglobin and which serve the purpose of storing oxygen in the tissues. The existence of such compounds can hardly be doubted; some have been described (Griffiths); and they are supposed by many physiologists to play a great rôle in anaërobic respiration.

k. *Other compounds of hemoglobin with gases.* Hemoglobin combines with many other substances than oxygen, and perhaps one advantage of placing it within corpuscles may be to protect it from such other substances. They may find difficulty in entering the blood cell. Indeed, it may be that the envelop of the red blood corpuscle has been devised to permit the easy passage of CO_2 and oxygen through it, but to resist most of the other food and metabolic products of the body. Among the substances readily penetrating the red blood corpuscles is carbon monoxide, CO, which is found in illuminating gas. This substance is either far more reactive than oxygen or else it forms a firmer compound with the hemoglobin. Probably the latter is the case. The HbCO compound, carbonyl-hemoglobin, dissociates less readily than oxyhemoglobin, or else the active mass of the CO, that is the proportion of molecules in a condition to unite with hemoglobin, is greater in it than in oxygen. When blood is exposed to a mixture of carbon monoxide and oxygen or air hemoglobin takes up by preference the carbon monoxide so that even though there is little carbon monoxide present, as compared with oxygen, blood saturates itself to a considerable extent with carbon monoxide. If hemoglobin becomes 50 per cent. saturated with CO, the life of an animal is endangered. In human blood 50 per cent. saturation of the hemoglobin with carbon monoxide occurs at room temperature in the presence of air containing about 0.05 per cent. by volume of CO gas. It is for this reason that the presence of even small amounts of carbon monoxide in

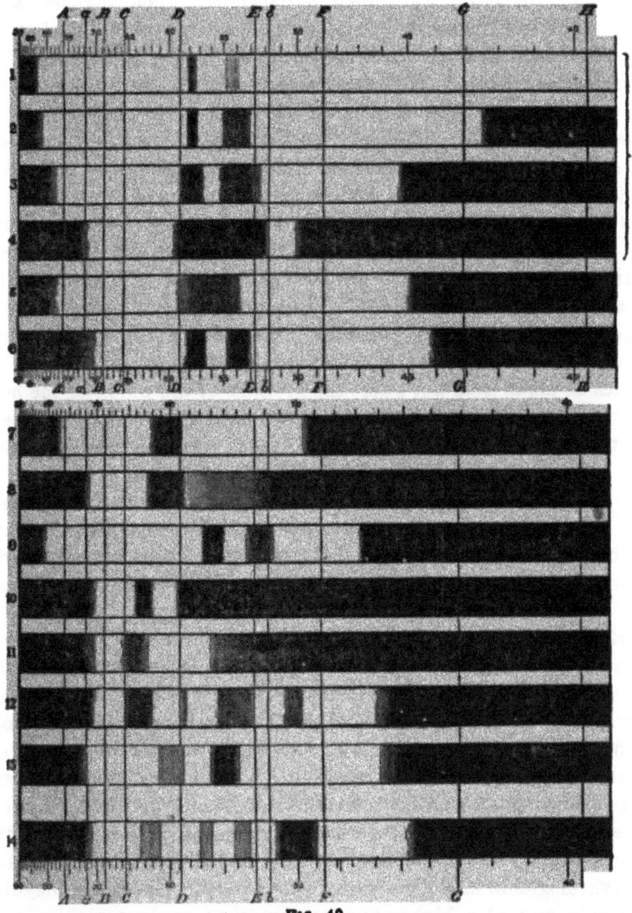

Fig. 49.

Explanation: Spectra 1, 2, 3 and 4, Oxyhemoglobin of various degrees of concentration; Spectrum 5, Hemoglobin; Spectrum 6, CO-Hemoglobin; Spectra 7 and 8, Hematin in alkaline solution of different degrees of concentration; Spectrum 9, Hemochromogen (Stokes' reduced hematin); Spectrum 10, Methemoglobin; Spectrum 11, Acid hematin (blood treated with acetic acid); Spectrum 12, Acid hematin in ethereal solution; Spectrum 13, Acid hematoporphyrin; Spectrum 14, Alkaline hematoporphyrin.

the air of houses is so detrimental to health. The bloods of different animals show different powers of saturation with carbon monoxide when they are exposed to the same mixture of air and carbon monoxide, and there is a variation also in different species of the same animal. This fact has not yet been explained. It may be due to the fact that the saturation of the blood by oxygen is dependent upon various factors which do not affect the carbon monoxide. Thus the percentage of saturation of the blood by oxygen is dependent upon carbon dioxide, alkalinity and lactic acid, whereas the dissociation of carbon monoxide hemoglobin does not appear to be much if at all affected by the presence of these substances. It may be, therefore, that the salts of the blood corpuscles being different in different animals determines that the per cent. of saturation of the hemoglobin with oxygen shall be different, and accordingly leave more or less hemoglobin free for carbon monoxide to unite with. Mouse blood, for example, is one-third less saturated with carbon monoxide than human blood when both are exposed to the same mixtures of O_2 and CO gases (Hartridge). The firmness of the union of carbon monoxide with hemoglobin makes it very difficult to replace it with oxygen and so to resuscitate a person poisoned by illuminating gas. Inasmuch as alkali does not affect the firmness of the union of the hemoglobin with carbon monoxide, but does increase the power of hemoglobin to unite with oxygen, it would appear wise, in cases of poisoning by coal gas, to give large amounts of sodium bicarbonate.

The spectrum of COHb differs slightly from that of oxyhemoglobin, Figure 49, the two absorption bands being shifted more to the right in the carbonyl hemoglobin. From this a very good method has been devised by Hartridge for estimating the per cent. of carbonyl hemoglobin in blood.

Carbon monoxide hemoglobin is more stable than oxyhemoglobin. It coagulates at a higher temperature and at a slower rate than HbO_2; carbonmonoxyhemoglobin is not attacked by ferricyanide of potassium in the dark, but is in the light. This shows that carbonmonoxyhemoglobin, or carbonyl hemoglobin, as it may be called, is dissociated by light, at least when in the presence of oxygen. If blood be exposed to the same mixture of oxygen and carbon monoxide in the dark and in the light the relative amounts of oxy- and carbonyl hemoglobin will be different in the two cases. It is probable that reduced hemoglobin is less stable than oxyhemoglobin and the reason that carbonyl hemoglobin coagulates more slowly and at a higher temperature than oxyhemoglobin is that there is less of the reduced hemoglobin formed by dissociation. Carbon monoxide probably unites like oxygen with the iron atom of the hemoglobin molecule. The union is of the nature of a carbonyl of iron, since this gas has the property of forming such carbonyl derivatives with the metals of the iron group.

Haldane and his co-workers have generally assumed that carbon monoxide unites only with hemoglobin and that it owes its toxicity solely to the fact that it thus interferes with the oxygen-carrying power of the hemoglobin. It is very doubtful whether this is the case. It is more probable that it unites with other oxygen receptors, as well as those of hemoglobin, and it may thus act directly on cells. It is found to be somewhat more toxic than an equivalent asphyxia for mammals, which would bear out this view. It is toxic toward some animals and plants which have no hemoglobin, but is not toxic for others.

Illuminating gas has in it another substance, ethylene, $CH_2:CH_2$, which toward many plants is vastly more toxic than carbon monoxide. Although ethylene is present in the gas in very small quantities it is the most toxic element of the gas for trees and various seedlings. The action of this gas on animals should be carefully investigated. It is possible that a part of the beneficial action of sleeping out of doors may be due to escaping the poisonous action of small amounts of illuminating gas, which penetrate from leaking pipes, joints and cocks into all dwellings.

Nitrous oxide hemoglobin. Nitrous oxide, or laughing gas, also forms a loose combination with hemoglobin. It is very suggestive that this mild and typical anesthetic is thus found to unite with an oxygen receptor in an easily dissociable union. It suggests that anesthesia may be produced by the saturation of the oxygen receptors of the protoplasm by anesthetics.

Nitric oxide (NO) hemoglobin is a firmer compound, and with this gas hemoglobin is easily oxidized to methemoglobin.

Hydrocyanic acid hemoglobin. Hydrocyanic acid, HNC, also unites with methemoglobin probably by means of its bivalent carbon atom H-N=C. There, again, is another typical respiratory poison and anesthetic occupying an oxygen receptor. It probably unites in the same manner in the cell and thus prevents union with oxygen. The union is in both cases dissociable.

Carbon dioxide unites with some part of the hemoglobin molecule, but it is more probable that it unites with the protein part of the molecule than with the iron.

Sulphureted hydrogen, H_2S, forms the compound HbH_2S. The union is probably with the extra valences on the sulphur, H_2S. Presumably, also sulphur, S_2, will unite with hemoglobin to give sulphur hemoglobin, but this compound does not seem to have been discovered.

To what extent hemoglobin unites with other substances has hardly been studied, but it will probably be found that many other substances will unite with it. For example, it is known that the anesthetics, such as ether and chloroform, when in blood, unite chiefly with the red blood corpuscles. It is believed by most observers that they form a loose union with the lecithin or other lipin of the corpuscles. Solutions of lecithin and cholesterin have the power of dissolving more anesthetic than water alone; but there may be in addition a union with the hemoglobin, which will retard its oxygen-carrying capacity, and thus play a part in anesthesia. Particularly chloroform from its greater chemical activity may be supposed to act in this way. The observations of Buckmaster and Gardener show that anesthetics in some way or other do lower the oxygen-carrying capacity of the blood.

l. *Summary of the oxygen-carrying capacity of the blood.* We may now briefly summarize the discussion in the previous pages. The blood contains in the red blood corpuscles a red pigment, hemoglobin, which is probably in union with the stroma. Hemoglobin has the property of uniting with molecular oxygen and giving it off again in a molecular form. In virtue of this property the blood is able to unite with considerable quantities of oxygen in the lungs to form oxyhemoglobin, which is of a scarlet color, and to carry oxygen to the tissues, which take the oxygen away in virtue of their reducing powers. In the tissues the pressure of the oxygen is extremely low, and in virtue of this fact oxygen dissociates from oxyhemoglobin and enters the tissues. The oxyhemoglobin is thus partially reduced and the blood changes to the purple color of venous blood, due to the presence in it of hemoglobin. The amount of oxygen in the arterial blood, as it leaves the lungs, is different in different individuals and in different animals, and it depends in the first instance on the amount of hemoglobin there is in one cubic centimeter of blood. But in general there can be extracted from 100 c.c. human arterial blood about 19-20 c.c. of oxygen gas, measured at 0° and 760 mm. of Hg pressure. From venous blood less oxygen can be extracted, the average amount being in human beings about 15 c.c. of oxygen, although it may under circumstances be less. The amount of oxygen taken up by the blood depends, in part, upon the partial pressure of the oxygen, but even when this partial pressure is reduced to only 13 per cent. of an atmosphere instead of the usual 20 per cent., the blood is still 93 per cent. saturated.

The per cent. of saturation of the blood by oxygen depends upon several factors; upon temperature, alkalinity or acidity of the blood, upon light and upon the presence of salts and of certain specific salts. Since these factors vary in different animals, the per cent. of saturation of their blood by oxygen when exposed to the same gas mixture varies also.

Blood also carries carbon dioxide from the tissues to the lungs, where it is given up. This carbon dioxide is in part dissolved as such in the water of the blood and the corpuscles, but in larger part it is present combined with other substances in solution in the plasma. In part it is there in sodium bicarbonate and in part in union with the proteins of the blood plasma. It is also carried in the red blood corpuscles, presumably united with the hemoglobin, but not united with the iron of the hemoglobin. The passage of carbon dioxide into blood from the tissues renders the blood in the capillaries more acid, or rather less alkaline, so that carbon dioxide in this way helps to turn the oxygen out of its union with hemoglobin and so make it available to the tissues. And when the lungs are reached the passage of carbon dioxide outward into the alveoli sets free the alkali to which the carbon dioxide had been attached. This facilitates the taking up of oxygen in the lungs. Other acids act in the same manner as carbon dioxide, so that the products of oxidation in the tissues, the organic acids, may thus assist in providing the tissues with oxygen by which these products may be oxidized.

The passage of oxygen into the blood and carbon dioxide out of the blood in the lungs is generally supposed to be due to processes of diffusion and to be thus a physical process. The pressure of oxygen in arterial blood is always, under ordinary conditions, lower than that in alveolar air; and the pressure of carbon dioxide is higher than that of the alveolar carbon dioxide. A few observations exist, however, which appear to indicate that in time of stress the lung epithelium may actively intervene in the process and actually secrete oxygen inward, so that the pressure in the arteries may be higher than that in alveolar air. The observations, however, upon which this conclusion of the activity of the lung or capillary endothelium depends are still open to other interpretations. They do not conclusively show that the oxygen is thus secreted. It is also unlikely that such very thin plates as the alveolar epithelium should have a secretory function, although the capillary endothelium might. It is better at present, therefore, to conclude that certainly diffusion is the principal factor concerned in the entrance of oxygen into the blood, but that possibly at times an active secretory process may also assist. Further work on this matter must be done before a definite conclusion can be drawn.

Hemoglobin unites, not only with oxygen, but with many other substances, such as coal gas, or CO. This union is firmer than the oxygen union with hemoglobin, and a part of the toxicity, probably the chief. and by some thought to be the total, action of the gas is to asphyxiate through its power of union with hemoglobin, so that the blood can no longer carry oxygen to the tissues.

Laking of the blood.—Blood may be laked, that is hemoglobin may be caused to pass out of the corpuscles into the plasma, by various agents. Many toxins lake the blood, particularly some of those of snakes and bacteria. Corpuscles are laked also by small amounts of alkali, by the addition of water to the blood, by the action of anesthetics, such as ether or chloroform, by bile salts and soaps. Blood may be laked also, by freezing and then thawing it. The explanation of this laking or hemolysis is still a matter of dispute. Some, perhaps the majority of observers, consider that hemoglobin is held in the corpuscles by the wall of the latter. This is not permeable to hemoglobin. The corpuscles are considered to be bags, or little cells, containing a concentrated solution of hemoglobin. When the membrane of the corpuscle changes its state it may happen that it becomes more permeable to the hemoglobin, which now diffuses out of the corpuscle. All these various laking agencies are said to act by affecting the permeability of the corpuscular membrane.

There are many reasons for doubting whether this explanation is correct. Hemoglobin may be held in the corpuscle by union with the stroma. It is true for all other cells, and probably it is true for the corpuscles, that they are not bags filled with fluid, but they are organized jellies. The corpuscles behave in many ways as if they were such jellies also. Hemoglobin does not escape, as one would expect it would if it were in solution, when the corpuscle is punctured or cut across, but it stays in the divided corpuscle. Moreover, when hemoglobin is set free in the corpuscle by some of these methods, particularly in the very large cells of Necturus, a tailed amphibian, the hemoglobin may crystallize in the corpuscle itself, which shows that it must be prevented in some way from crystallizing in the normal cell. Moreover, the concentration of hemoglobin in the mammalian corpuscle is greater than the solubility of oxyhemoglobin in an equal bulk of water. For these and other reasons some observers are of the opinion that hemoglobin is held in some kind of a loose chemical or physical union, presumably the former, with the stroma of the corpuscle and that the various hemolytic agents break this union. It is not at all impossible that the union is with certain reserve valences of the hemoglobin and the stroma and such unions are very unstable and easily broken. Stroma freed from its hemoglobin behaves as a poison, causing intravascular coagulation. The hemoglobin-stroma compound as it exists in the corpuscle is inert in this respect. Carbon dioxide protects the corpuscles from the hemolytic action of various hemolytic sera (Sawtschenko).

Whatever may be the explanation of hemolysis the process itself is of very great interest from a physiological point of view. It may be taken as a type of the processes which are occurring in all protoplasm. It is particularly instructive if the view be adopted that Hb and stroma are in union, for it indicates that similar loose unions may occur in protoplasm between other substances, for example between the fats and the proteins, or between lipins and proteins, and that the instability of protoplasm and its power of responding to stimuli of all kinds may depend in part upon this fact. It is well known that the presence of a certain amount of salt is necessary for the preservation of many cells and that if the salt is reduced in quantity the activity of the cell is profoundly affected. We see from the foregoing that the composition of the corpuscle, the hypothetical union of hemoglobin and stroma, depends on the presence of a certain amount of salt in the plasma, since diluting the plasma increases the tendency to hemolysis.

Composition of the red corpuscles.—The red corpuscles consist of hemoglobin and stroma. The latter contains a considerable proportion of phospholipin and cholesterol. 1,000 parts of erythrocytes contain, according to Abderhalden, the following amounts of lipins:

	Bull	Horse 1	Horse 2	Rabbit	Pig	Dog 1	Dog 2	Sheep 1	Sheep 2	Ox
Cholesterol.........	1.824	0.388	0.661	0.720	0.489	2.155	1.255	2.360	3.593	3.379
Phospholipin.	2.850	3.973	4.855	4.627	3.456	2.568	2.296	3.379	4.163	3.748
Total lipin.........	4.674	4.361	5.536	5.347	3.945	4.723	3.431	5.745	7.756	7.127
Phospholipin ÷ cholesterol. ...	1.563	10.240	7.345	6.426	7.067	1.193	1.829	1.428	1.159	1.109

The total lipins make, therefore, from 0.34-0.77 per cent. of the weight of the corpuscles.

1. *Hemoglobin. Chemistry.*—a. *Occurrence.* Hemoglobin is a pigment which is widespread in the animal kingdom and which is allied to chlorophyll of plants and to the pigments phycocyan and phycoerythryn found in algæ. It, or an allied pigment, occurs in many of the fixed tissues of animals, as well as in the blood, as, for example, in the striated muscle of most vertebrates, in heart muscle, in the pharyngeal muscles of many mollusks, such as Paludina, and in the pharyngeal muscle and ganglia of the polychete annelid, Aphrodite. Similar pigments, having the same power of combining with oxygen and giving the spectra of hematin and hemochromogen, have been found by MacMunn in the cells of sponges and echinoderms and he has called these pigments histohematins. While not identical with hemoglobin, they closely resemble it. The function of this pigment is apparently to serve as a storehouse of oxygen, and MacMunn has suggested that this was the original function of the pigment and that it was later developed into a means of transporting oxygen from the exterior to the tissues. Besides being in the tissues, hemoglobin is found in the blood or body fluids of a great variety of invertebrates and in all vertebrates with one or two exceptions. It occurs in these fluids either in solution, or confined to certain small bodies called erythrocytes, literally meaning "small red bodies." In all the vertebrates and in certain lamellibranch mollusks, i.e., Arca tetragona, Pecteniculus, etc.; in the polychete, Terebella; in various holothurians, i.e., Cucumaria Planci, etc.; in the worm, Thalassema erythrogrammon; in the polychete, Capitella, it is found in erythrocytes. It occurs in solution in the blood or body fluids of various Chætopoda,

crustacea, insects, leeches, and even in the echinoderm, Ophiactis virens. It is clear from this list that the power of forming hemoglobin must be very widespread in nature. Since the essential part of the molecule consists of various pyrrol nuclei, the power of making such pyrrol nuclei and uniting them to form hematin or chlorophyll must be a very general possession of protoplasm. Hemoglobin, or similar iron-containing pigments, are not the only oxygen-carrying proteins found in blood. In the blood of Limulus, and various crustacea and mollusks, there is found a copper-containing protein with a similar function, called *hemocyanin*. This is a blue pigment and these animals are the truly blue-blooded animals of the sea. Hemocyanin contains copper in place of iron. The composition of this pigment has not been investigated with the thoroughness of that of hemoglobin, but it resembles hemoglobin in its high histidine content and in some other properties. It is not so efficient an oxygen-carrier as hemoglobin and cannot carry nearly as much oxygen per gram of pigment as hemoglobin.

The development of hemoglobin in the blood has gone on *pari passu* with the development of the central nervous system. This system has a very great need of oxygen. It is more dependent on oxygen than any other tissue of the body, and its consumption of oxygen per gram of tissue appears to be larger. In the course of evolution the nervous system underwent a progressive development, presumably because animals have been selected chiefly for brain power. Hence as this system developed there developed the need of carrying large amounts of oxygen to it. The hemoglobin content of the blood increases more or less parallel with this growth of the nervous system. Thus man, with the largest nervous system, has the largest amount of hemoglobin in his blood. In human beings there is normally in the blood 20 per cent.; in dogs there is less; horses and sheep have still less, and in fishes and the lowest vertebrates the quantity is further reduced. The amount of hemoglobin in human blood is larger than could be held in solution. This large amount is made possible by placing the hemoglobin in the erythrocytes.

b. *Crystalline form*. There is not a single hemoglobin, but a whole series of hemoglobins, each animal probably having a kind differing from that of every other species. They all resemble each other in their main features, but they differ slightly in their composition, and above all, they differ in their crystalline form.

c. *Method of crystallization*. Oxyhemoglobin crystallizes with very great ease. In some animals in which hemoglobin is relatively little soluble, it is only necessary to lake the blood under the microscope to produce crystals. The crystals may even form in the corpuscles themselves, as in Necturus. Horse blood, guinea pig blood and squirrel blood crystallize most readily; ox blood with more difficulty. But all hemo-

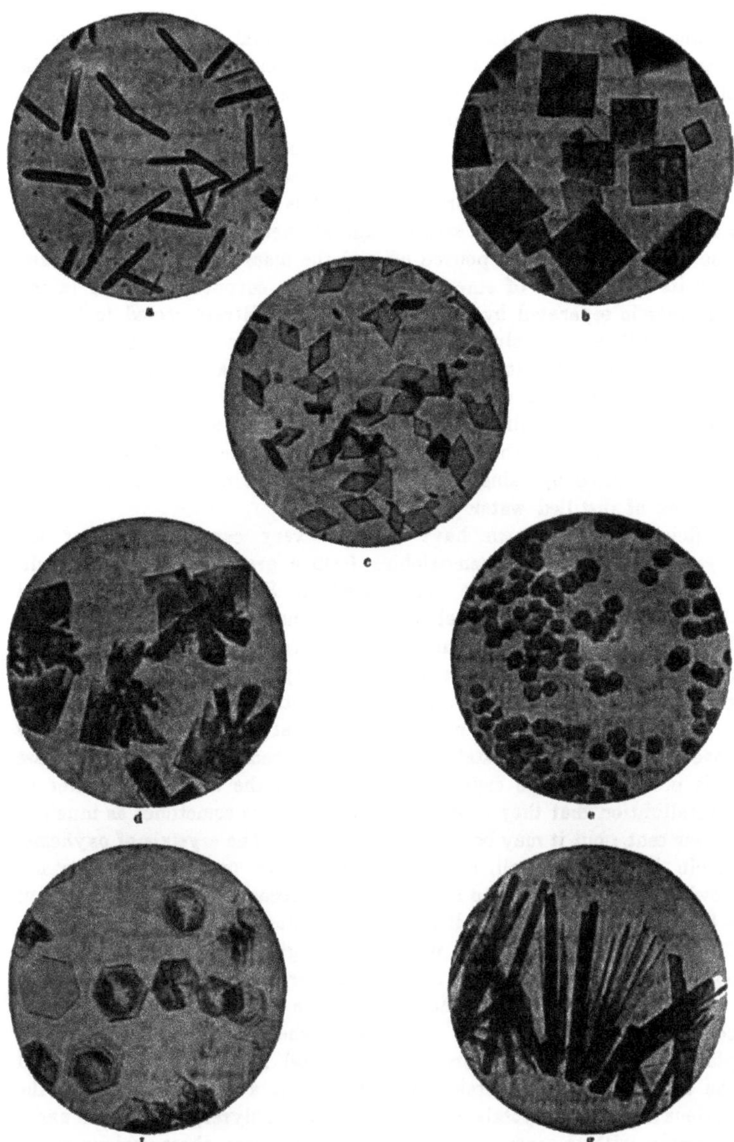

FIG. 50. Various forms of oxyhemoglobin crystals of different animals. a, Necturus maculatus; b, Trumpeter swan, Olor buccinator; c. Guinea fowl; d, Goose; e, Tasmanian wolf; f, Fox squirrel; g, Ground squirrel (Reichert and Brown).

globins may be crystallized by the use of special methods. To obtain large amounts of crystals in dog's blood it is only necessary to lake the blood corpuscles by shaking them with toluene and placing in the ice-box. The following methods are, however, better.

Hoppe-Seyler's method. The defibrinated dog or horse blood is diluted with 10 volumes of 3 per cent. salt solution, and the corpuscles allowed to settle. The supernatant liquid is poured off, the corpuscles washed twice with cold salt solution and allowed to settle in a cool place. The salt solution is then poured off and the mass of corpuscles is mixed with its own volume of ether. This lakes the corpuscles. After laking the ether is separated by rapid filtration, the filtrate cooled to 0° and diluted with a ¼th volume of absolute alcohol also cooled to 0°. It is kept at —5° or —10° until crystallized. The crystals separated by centrifuge or filtration are washed with cold, 25 per cent. alcohol, dried by pressure and recrystallized by dissolving them in water heated to 54°, cooling and adding ¼th volume of alcohol as before. Hüfner has shortened the method by using the centrifuge and laking pig's blood by the addition of distilled water.

Reichert and Brown have made a very careful study of the crystalline form of the hemoglobins from a great number of animals. Some of their figures are reproduced in Figure 50. They found that each species of animal had its peculiar kind of hemoglobin. The crystals of related animals were generally similar so that it was possible, the authors thought, to use the crystalline form of oxyhemoglobin as a means of aiding in the classification of animals and in discovering relationships. It often happens that one kind of animal may have more than one crystalline form of its hemoglobin. In such case it is possible that the crystals may differ in the amount of water of crystallization that they contain. This amount is sometimes as much as 11 per cent.; but it may be half this quantity. The crystals of oxyhemoglobin do not keep well, but even in a vacuum or when dry they are slowly converted in part to methemoglobin, become less soluble and have a brownish color. Nearly all the crystals belong to the rhombic system. The crystals when examined in polarized light are pleochroic. That is some of the crystals appear a brilliant scarlet; others have an orange color. This is due to the fact that in some of the crystals the light is coming through one axis of the crystal, whereas other crystals are so placed that light passes through the crystal in another direction and the refraction and dispersion of the light is a little different in the various axes. The crystals placed between an analyzer, Nicol prism and a polarizer, with the interposition of a gypsum plate, show various colors when the Nicol is rotated. The crystals show the same absorption spectrum as that of the solution, except that the distance between the two absorption bands is a little greater when the light comes through one axis of the crystal than when it passes through another axis. Reduced hemoglobin crystallizes with greater difficulty than oxyhemoglobin.

d. *Properties of oxyhemoglobin.* The solubility is increased by the addition of very small amounts of alkali. It is not precipitated by NaCl or $MgSO_4$ added to saturation; it is precipitated by $(NH_4)_2SO_4$ beginning to precipitate when about two-thirds saturated and continuing until saturation is reached. It resembles in this property the albumins. It is soluble in distilled water, but not soluble in alcohol or ether. Alcohol renders it insoluble and makes methemoglobin. It is a weak acid, goes to the anode on passing an electric current through the solution, and the isoelectric point, that is the point of minimum dissociation, is at a concentration of H ions of 1.8×10^{-7}. Acids, even when dilute, and alkalies decompose it into hematin and globin. It is not precipitated from solution when neutral by $CuSO_4$, $FeSO_4$, $AgNO_3$, $HgCl_2$, or by lead acetate. Methemoglobin is precipitated by lead acetate. Oxyhemoglobin is precipitated by $CHCl_3$ at 55° and the precipitate is insoluble in water. When heated oxyhemoglobin coagulates at 64° and decomposes, setting free hematin. An alkaline solution heated does not coagulate but begins to decompose at 54°. Oxyhemoglobin is dextro-rotatory, $(\alpha)_C = +10.4$ (Gamgee). The globin component is levo-rotatory $(\alpha)_C = -54.2°$

e. *Absorption spectrum.* The spectrum depends on the concentration. Strong solutions absorb from the violet end clear to the red; weaker solutions show two absorption bands between D and E. The centers of these bands are for α at 576 $\mu\mu$; β at 537 $\mu\mu$; there is also an absorption band in the ultra-violet, γ, at 414 $\mu\mu$. The relation of the width of the absorption band to the concentration is shown in Figure 51, from Rollet, in which the spectrum is plotted on the abscissa and the concentration on the ordinate. The depth of solution looked through is in each case 1 cm. Reduced Hb has a single absorption band between D and E.

It is the absorption of green and blue light which makes hemoglobin look red. It is possible that this absorption of blue light is of service to the organism in that it helps to protect the tissues from the more active blue light.

f. *Quantitative determination of the amount of hemoglobin and oxyhemoglobin.* Various clinical methods for the determination of the amount of hemoglobin in the blood are given in the practical part (page 922), and need not be repeated here. For the accurate determination of the amount of oxyhemoglobin present, even in the presence of other pigments which do not have the same absorption spectrum, the best method is that of the spectrophotometer. This, as its name implies, is a spectroscope so arranged as to measure the intensity of the absorption of any light ray desired, but a description may be omitted in a work of this character. The intensity of the absorption of the light is a simple function of the concentration of the hemoglobin. The product of the concentration by the depth of solution through which the light has to pass in order to produce the absorption of a given proportion of the incident light, say nine-tenths, is a constant. If c is the concentration of the hemoglobin, ϵ the reciprocal of the depth expressed in centimeters necessary for the absorption of nine-tenths of the light, ϵ is called the coefficient of extinction, then c/ϵ = a constant A. A being once determined if c or ϵ is known the other may be found.

g. *Composition.* Although the hemoglobins differ among themselves they are all alike in their general composition. They are all conjugated proteins, that is they break easily into a simple protein, called globin, and an iron containing radicle, which is hematin if oxyhemoglobin is broken up, or hemochromogen, if hemoglobin is broken up. Hemochromogen is reduced hematin. This decomposition is easily brought to pass either by the action of dilute acids or alkalies, or by boiling hemoglobin or by the action of digestive juices. The following table shows some of the results of the elementary analysis of hemoglobin by various observers:

Hemoglobin of	H_2O of cryst	C	H	N	S	Fe	O	P	Observer
Dog	3.4	53.85	7.32	16.17	0.43	0.39	21.84	0	Hoppe-Seyler
	11.39	54.57	7.22	16.38	0.568	0.336	20.93	0	Jaquet
		54.15	7.18	16.33	0.67	0.43	21.24	0	C. Schmidt
Horse	5	54.87	6.97	17.31	0.65	0.47	19.73	0	Kossel
		51.15	6.76	17.94	0.39	0.335	23.43	0	Zinoffsky
		54.81	7.01	17.06	0.6	0.468	19.86	0	Nencki
Beef	9.52	54.66	7.25	17.70	0.447	0.40	19.54	0	Hüfner
Pig	8.04	54.71	7.38	17.43	0.48	0.399	19.60	0	Hüfner
Fowl	9.33	52.47	7.19	16.45	0.858	0.335	22.5	0.197	Jaquet
Goose		54.11	6.83	16.58	0.65	0.51		0.0059	Abderhalden and Medigreceanu.

It will be noticed in the foregoing table that in dog's hemoglobin there are three sulphur atoms to one iron atom; while in horse hemoglobin there are two sulphurs to one iron. The sulphur of oxyhemoglobin, or of globin, is not split off by the action of alkali. Jaquet states that neither hen, horse nor dog hemoglobin contains sulphur in a form which can be split off by the action of alkali and lead acetate. If an alkaline oxyhemoglobin solution is cooked after the addition of a few drops of $MgCl_2$ solution the hematin split off is precipitated with the $Mg(OH)_2$, leaving an almost colorless solution. This alkaline solution of globin contains no sulphur precipitable by lead acetate as sulphide. The small amount of phosphorus in the hemoglobin of bird's blood is due to an impurity of nucleic acid, and is not part of the hemoglobin molecule.

h. *Composition of globin.* If the formula of hematin, namely, $C_{34}H_{34}N_4O_5Fe$, be subtracted from the formula of hemoglobin, $C_{758}H_{1203}N_{195}S_3FeO_{218}$, there is left $C_{724}H_{1169}N_{191}S_3O_{213}$, as the formula of globin. It might be that there were three molecules, each with one atom of sulphur,

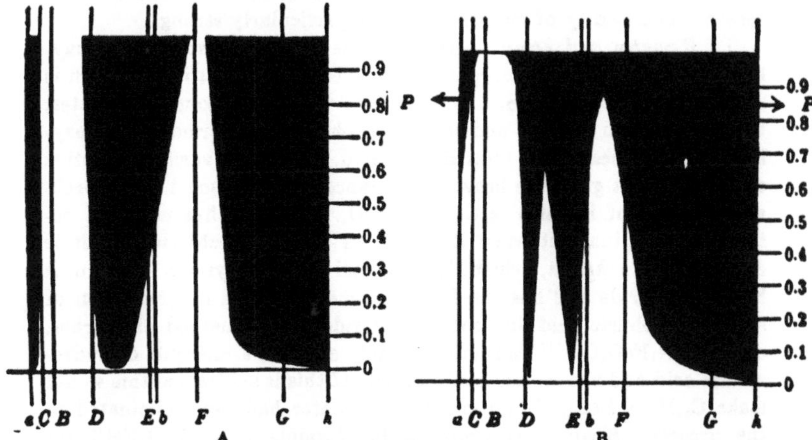

FIG. 51.—Absorption spectra of hemoglobin (*A*) and oxyhemoglobin (*B*). The abscissa shows the Frauenhofer lines of the spectrum; the ordinate represents the concentration when the depth is 1 cm. (Rollet).

attached to the hematin molecule in place of one very large molecule like the above. Globin is a histone-like body containing rather more nitrogen than most proteins, about 17 per cent., and somewhat more carbon. It is like histone in the fact that it is not dissolved by ammonia in the presence of ammonium chloride, and in the particular that it yields a very large amount of the basic amino-acids, namely, about 11 per cent. of histidine, 5.4 per cent. of arginine and 4.3 per cent. of lysine. For the preparation of histidine, blood corpuscles are most commonly used. Globin is insoluble in water, but readily soluble in alkalies and acids; it is coagulated by heat, but the coagulum is readily dissolved by acids (Schulz). It is interesting that the analogous blood pigment, hemocyanin, also yields a large amount of histidine on hydrolysis. The other amino-acids are those ordinarily found in proteins, as may be seen in the table on page 129, the percentage of leucine, namely 29 per cent., being rather larger than usual. It is rather interesting that, although globin is so strongly basic itself, yet it forms, when united with hematin, an electro-negative colloid. This is due to the fact that hematin is an acid. Concerning the nature of the union between hematin and globin nothing is known, but the great ease with which the bond is split both by acids, alkalies and by boiling water suggests that the union is simply that of a base and acid. It resembles, in this respect, such nucleoproteins as protamine nucleate, which is also a salt union, and which forms an electro-negative colloid. Hemoglobin, being an acid, is usually combined with some base and sodium or potassium is ordinarily found in it. It is more soluble in very dilute alkali than in water for this reason. The acidity of methemoglobin is particularly strong.

i. *Hematin and hemin.* Hematin, the colored constituent of oxyhemoglobin, is a brown, amorphous substance the constitution of which is not yet accurately known. It is an acid which is insoluble in water, but is soluble in alkalies and in acid alcohol. It is formed when oxyhemoglobin is heated, or treated with acid, alkalies or various digestive enzymes, and it gives the brown appearance to cooked meat. The probable formula of hematin is $C_{34}H_{34}N_4FeO_5$. Hematin has not been obtained crystalline, but when it is treated with hydrochloric acid it is converted into *hemin*, which forms small brown crystals. Hemin is hematin chloride and has the formula, when obtained by the action of alcohol, sulphuric acid and sodium chloride by the method of Mörner, of $C_{35}H_{35}N_4FeO_4Cl$. This hemin probably contains a molecule of hydrochloric acid and one molecule of alcohol. If this is subtracted this would make $C_{33}H_{32}N_4FeO_5$, if two molecules of water had been eliminated in the process. Küster has proposed the formula $C_{34}H_{33}N_4O_4FeCl$ for hemin, which by treatment with KOH would give for hematin the formula, $C_{34}H_{34}N_4O_5Fe$. This formula was that suggested by Hoppe-Seyler and is the most probable one. Hematin, when reduced, forms hemochromogen. Heated with strong acid it loses iron and is converted into hematoporphyrin, a possible graphic formula of which has been given on page 415.

$$\underset{\text{Hematin.}}{C_{34}H_{34}N_4O_5Fe} + 2HCl + H_2O = FeCl_2 + \underset{\text{Hematoporphyrin.}}{C_{34}H_{38}N_4O_6}$$

j. *Hematoporphyrin.* $C_{34}H_{38}N_4O_6$. This, the iron-free derivative of hemochromogen or hematin, is isomeric with bilirubin. It can be made by the method of Nencki and Zaleski, which consists in adding hemin, little by little, to glacial acetic acid, which has been saturated with HBr at 10° and stirring constantly. One adds 5 grams of hemin to 75 c.c. of acid and allows the mixture to stand at ordinary temperature for 3 to 4 days. The hemin having dissolved, the solution is diluted with water (1-2 liters) and filtered after some hours. The hematoporphyrin is then precipitated by the addition of just sufficient sodium hydrate to neutralize the hydrobromic acid. The precipitate is filtered, redissolved on the water bath by the addition of sufficient sodium carbonate and re-precipitated by addition of acetic acid. The precipitate is filtered, washed, sufficient water is added to make a thick suspension and HCl is added in small amounts until the hematoporphyrin has redissolved, the solution filtered and HCl added in excess, any resinous precipitate being filtered off. The filtered liquid is concentrated in vacuo until it crystallizes. The crystals are filtered, washed with 10 per cent. HCl and once recrystallized. These crystals are hematoporphyrin chloride. The free hematoporphyrin may be obtained by dissolving in water and neutralizing the acid with sodium acetate. It is a brown powder. The probable formula of this reaction is the following:

$$\underset{\text{Hemin.}}{C_{34}H_{32}N_4O_4FeCl} + 2HBr + 2H_2O = \underset{\text{Hematoporphyrin.}}{C_{34}H_{38}N_4O_6} + FeBr_2 + HCl.$$

Properties. Hematoporphyrin is a dark, violet-colored powder, having a green tint in transmitted light. It is insoluble in water and very slightly soluble in $CHCl_3$, amyl alcohol or ether; it is soluble in ethyl alcohol, and in water containing sodium hydrate or other alkali or dilute, strong acid. It is probably an amphoteric electrolyte. It is not very soluble in weak organic acids. The alcoholic solution, when neutral, is red, which becomes red-orange on the addition of a little acid, the color changing to a purple or violet by the addition of more acid. The solutions are very fluorescent and in the light have a strong physiological action. (See page 422.) Alkaline aqueous solutions are also red and have a different spectrum from acid solutions. (See the practical part, page 930.) Hematoporphyrin is an acid and forms definite salts, of which the sodium salt has been crystallized. The alcoholic and alkaline solutions have four absorption bands: in the red between C and D (621-610 $\mu\mu$), two between D and E (the first at D, 590-572 $\mu\mu$, the second near E, 555-528 $\mu\mu$), and the fourth from b to F (514-498 $\mu\mu$). For the spectrum, see page 493. The position of the bands depends somewhat on the concentration of the alkali, and their width on the concentration of hematoporphyrin. According to Schulz the third band can be resolved into three constituents. The spectrum of acid hematoporphyrin is different. There are two bands on each side of D: the first from 597-587 $\mu\mu$, the second at 541 $\mu\mu$. In both acid and alkaline solutions there is an absorption band in the ultra-violet between h and H (Gamgee). The acid salt dissolved in alcohol has the five-banded spectrum of alkaline hematoporphyrin, but by the addition of more HCl, the spectrum changes to the acid hematoporphyrin spectrum. Hematoporphyrin strongly resembles phylloporphyrin obtained from chlorophyll, but it contains two atoms of oxygen more than the latter. Heated with zinc and distilled both yield pyrrols.

Composition. When decomposed with hydriodic acid in glacial acetic acid, hematoporphyrin yields various hemopyrrols and hematic acids.

The composition of some of these bodies has already been indicated on page 416. It is the general opinion that the molecule of hematoporphyrin contains four substituted pyrrol nuclei, but the exact manner in which these are arranged is still uncertain. There is still some uncertainty about the molecular size, but the probability is that the molecule contains only four pyrrol nuclei. It is both an acid and a base, forming true salts with acids and bases. It is closely related to bilirubin in structure and it is the mother substance of the bile pigments. MacMunn says that hematoporphyrin is one of the pigments found in the shell of birds' eggs and that it is the pigment of Uraster rubens; of various mollusks, such as Limax and Arion; and of Lumbricus and Actinia. As a copper compound it is found as a pigment in the feathers (Laidlaw) of birds of the genus Mucophaga. It occurs in small amounts in the urine and in larger quantities after sulphonal ingestion and in some pathological conditions.

k. *Hemochromogen.* This is reduced hematin. It is formed by the action of acids or alkalis on hemoglobin, or it may be formed by the reduction of hematin. It oxidizes spontaneously in the air and forms hematin. It loses its iron more easily than hematin and forms hematoporphyrin. Like hemoglobin, one molecule unites with one molecule of CO and probably one of HCN. The union of hemochromogen with oxygen is apparently more stable than that with carbon monoxide, since oxygen will readily turn the latter out of its union with hemochromogen, but not with hemoglobin. This shows that the firmness of the union between the gas and the chromogenic radicle is affected by the presence of the globin radicle. There is still some question as to the amount of oxygen combined with hemochromogen to make hematin. It was observed by Ham and Balean that when hematin is formed from oxyhemoglobin by the action of strong acid some oxygen is set free, and the amount was about equal to one-half that which was set free from oxyhemoglobin by the action of ferricyanide of potassium. Oxygen, moreover, cannot be pumped out of the hematin solution to form hemochromogen. In this particular the state of the oxygen in hematin would appear to be more nearly that in methemoglobin. If carbon monoxide hemoglobin is acted upon by acid a carbonyl-hemochromogen is obtained. The union is one molecule of CO to one molecule of hemochromogen. The carbon monoxide can be displaced by a current of hydrogen. Nitric oxide also forms a molecular and firm, red-colored compound with hemochromogen. (Linossier).

Hemochromogen may be easily prepared from blood or hemoglobin by adding to solutions of the latter some solution of hydrazine hydrate, 5 per cent., in 10 per cent. sodium hydrate. The spectrum of hemochromogen appears. On shaking with air that of hematin temporarily appears, but is quickly reduced. The alkaline solution of hemochromogen is cherry red. It has absorption bands in the green between D and E nearer D (567-547 $\mu\mu$ mean at 559 $\mu\mu$) and one less well-defined from E to b (532-518 $\mu\mu$; mean at 525 $\mu\mu$). In the ultra-violet there is one between h and g at 420 $\mu\mu$ (Gamgee). In air solutions oxidize readily and take the brown color of hematin. It differs from hematin particularly in its sensitivity to acids. Even dilute acids transform it into hematoporphyrin, whereas hematin is decidedly resistant to acids.

3. **The blood as the carrier of waste substances.**—This is the third great function of the blood. All kinds of waste materials are found in the blood in solution in the plasma. The blood carries them from the tissues to the kidneys and other excretory organs. Some of these substances contribute to the composition of the medium in which the cells are accustomed to work and they are, therefore, important factors in their environments. Thus while urea and carbon dioxide are waste materials and generally regarded only as excretory substances some, and possibly all of the cells of the body, work better when they are present in the blood in the usual amounts. As they thus condition the activities of cells they may be regarded, also, as hormones. Thus carbon dioxide helps to rouse and regulate the respiratory center; and the heart beats with greater vigor, in some of the lower vertebrates at any rate, when there is a small amount of urea present. The amount and character of various waste products found in the blood is shown in the adjoining table. There are many which are not mentioned in this table, since nearly all the numerous substances excreted in the urine have been in the blood.

Accumulation of Nitrogenous Waste Products in Blood Under Various Conditions.*

Mgs. per 100 Grams Blood.

Condition of diet and health	Non-protein N	Urea N	NH_3 N	Uric Acid	Creatinine	Creatinine and creatine
Normal, purine-free, high N diet. Urinary N 24 grams	34	16	0.1	2.5	1.1	9.5
Normal(?) Urinary N 17 grams	37	18	0.1	3.0	1.2	8.5
Normal. Low N. Urinary N 6.2 grams	32	15	0.14	2.0	1.3	6.5
Normal. Urinary N 4.5 grams	24	11	0.11	[illegible]	1.[illegible]	6.0
Toxemia 3 weeks after delivery	26	12	0.1		0	
Typhoid fever. 104°. Second week	38	19	0.08		1	1
Uremia	288	222			31	
Uremia	284	228	0.06		26	46.0
Uremia, convulsions	200	160	1.0		0	20.0
Chronic nephritis	200	140			7	
Pneumonia (before crisis)	72	44		5.0	1	

* Folin and Denis: *Jour. Biol. Chem.*, 17, p. 488, 1914.

4. Blood as a distributor of internal secretions. Its co-ordinating function.—The blood plays a very important part in the metabolic co-ordination of the body. It carries from one organ substances elaborated there and which are necessary for the growth and normal life of tissues in other parts of the body. One of the first of these internal secretions (internal because secreted into the blood) to be recognized was that of the sexual glands. The metabolic products of these glands in some way or other determine the development, or aid in the control of, the so-called secondary sexual characters. Thus oxen are proverbially less fierce than bulls, their bodies have a different shape, their horns are larger. Many other organs besides the sexual glands produce internal secretions, which are necessary for the development of other organs. Thus the development of the long, silky hair and smooth skin of human beings, the development of the skull and intelligence, depend upon the thyroid gland. The metabolic products of this gland, whatever their nature, are supposed to find their way into the blood and to be distributed by that tissue to the tissues which need them. In this way the blood plays the part of a middleman, or a system of transportation in human society. The blood, after the nervous system, is the most important co-ordinating agency in the body. These internal secretions are probably present in the blood in very small quantities. They are probably in solution in the liquid part of the blood. Their nature will be considered in the chapter which deals with the glands which produce them. (Cryptorrhetic tissues. Chapter XVI.)

5. Physico-chemical factors of the blood important in its circulation. Its viscosity; coagulation.—Under this heading will be considered the physical chemical features of the blood, which are important in its functioning as a circulating fluid. The most important of these characters is its viscosity. The blood must be a liquid in order to penetrate freely all parts of the body. In order that the blood should be driven in a constant stream through the capillaries of the tissue it must be kept at a high pressure in the arteries. The pressure of the blood in the arteries is maintained by the contraction of the heart, which constantly pumps blood into the arterial system, and by the peripheral resistance in the arterioles and capillaries which prevents the blood running too rapidly through the capillaries. This peripheral resistance may be diminished or increased by widening or narrowing the path of the blood by dilating or contracting the arterioles; but it may also be changed by variations in its viscosity.

Viscosity of the blood. By the viscosity of a liquid is meant stickiness, or the resistance it opposes to flowing or changing its shape. A perfect gas has no viscosity. The molecules are so far apart that their size and mutual attractions in no way affect the freedom of their movements. In liquids the size of the molecules relative to the space between them is so large that the molecules get in each other's way and limit their freedom of movement. This resistance to freedom of movement due to cohesion and the space occupied by the molecules is known as the viscosity. In determining the viscosity of a pure liquid, or of a liquid containing molecules in solution, molecular size and cohesion are the determining factors. When the liquid contains substances in suspension in it if the amount of this material is small compared with the total space of the liquid there will be little change in the viscosity from that of a true solution, but when the amount of added material

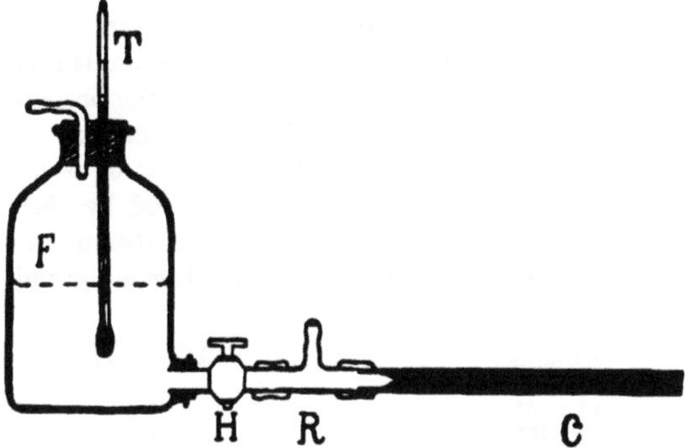

Fig. 52.—Viscosimeter (Burton-Opitz). C is the capillary through which the fluid in F flows.

becomes large enough to compare with the bulk of the liquid then the viscosity is increased. Water containing colloidal metals or clay in suspension has its viscosity very little changed until the amount of clay is large. If, however, the substances in suspension have a marked affinity for water so that they are united chemically or physically with several molecules of water, thus making what Naegeli called micellae, or what we may call hydrophile or emulsoid colloids, then the viscosity is more or less increased.

In blood the viscosity is affected by the corpuscles which are in suspension and by the proteins which are in solution. Blood consists to nearly half its volume of corpuscles, so that its viscosity is much greater than that of a salt solution, owing to this fact. And in the second place the proteins are markedly soluble. They are hydrophyl colloids. They may change their state so as to bind more or less water. There is about 8 per cent. of protein and this is sufficient to make the viscosity of the plasma greater than that of water.

Measurement of viscosity. The viscosity of a liquid is measured by an instrument called a viscosimeter. There are various forms of this instrument. One is figured in Figure 52. The method consists essentially in measuring the time required for a given bulk of a liquid to flow through a small opening or capillary tube under a constant temperature. One can use so simple a plan as the time required for a liquid to flow from a pipette and compare the time with that taken by water under similar conditions, the viscosity of water being taken as unity. If it takes twice as long for plasma to pass as for water, the viscosity of the plasma is said to be 2.

Viscosity of the blood.—The viscosity of mammalian blood is about 4.4-5.5. The viscosity of plasma or serum is much greater than that of water. At 38° it is 1.78-2.09 (Bence. Burton-Opitz). The viscosity diminishes with a rise of temperature until coagulation is reached. This high viscosity of the plasma and serum is due to the proteins in it. A dilute salt solution has about the same viscosity as that of water.

The viscosity of blood is much greater than that of serum, owing to the corpuscles in the blood and is much influenced by their tendency to clump. The viscosity is a function of the number of corpuscles, as may be seen in the following figures (du Pre' Demming and Watson):

No. of blood corpuscles	Viscosity of serum + corpuscles
0	1.9
3.2×10^6	3.3
6.3×10^6	4.9
12.6×10^6	15.6

The figures are for horse's blood at 32.2°. The corpuscle numbers are for cubic millimeters. The high viscosity of the last number is probably due to a partial blocking of the small capillary opening by the clumped corpuscles. The viscosity of the blood appears for this reason, unlike that of water or a true solution, to increase with the diminution of the size of the capillary. A capillary of 2 mm. diameter, a wide capillary, gives the more reliable figures. It is clear, however, that in the living body a narrowing of the capillary path will have the effect of apparently increasing the viscosity of the blood for the reason just stated.

The effect of temperature on the viscosity of the blood is marked. At 37° the viscosity is about 16 per cent. less than at 17° so that variations of temperature of 5°, which is within the limits of fever, may change the viscosity of the blood about 4 per cent. The viscosity of human blood in the mean is about 5.1 (Beech and Hirsch). It may vary between 4.73 and 5.89 at 38°. Since carbon dioxide causes an imbibition of water by the corpuscles and an increase in their size the viscosity is increased by carbon dioxide and diminished by oxygen. Hence the viscosity of venous blood is greater than that of arterial blood.

Viscosity of O_2 saturated blood	5.4
" after H_2 passed through it	5.88
" " CO_2 " "	6.57

In dyspnea the viscosity is increased. In dogs, hunger diminishes the viscosity (Burton-Opitz), a meat diet increases it most; a carbohydrate and fat diet produces a medium effect.

The blood contains a special mechanism for affecting its viscosity quite apart from those factors thus far mentioned. This is its power of clotting, or coagulation. By clotting the viscosity of the blood becomes so great that fluidity is lost and the blood is converted into a jelly-like solid. This power of clotting is of great service to the body in preventing hemorrhage, but it is not impossible that a partial coagulation takes place in certain regions of the circulation, causing a great increase in the viscosity and peripheral resistance in these regions. This would cause a very high intracapillary pressure and might contribute to a resulting edema. The tendency to clotting is always present. Even a sluggish movement of the blood in the veins may result, in certain special circumstances, in a premature permanent clotting, or thrombosis. This happens sometimes after anesthesia, which increases coagulability, or after parturition while the patient is lying flat in bed. Clotting is due to the conversion of one of the soluble proteins of the blood plasma, the fibrinogen, into an insoluble form of fibrin. This fibrin comes out in the form of a net which entangles the corpuscles and water and makes a jelly-like clot. Without considering at this point the nature of the change in the fibrinogen and the cause of the change, we may briefly examine the reasons for thinking that this change may perhaps normally occur to a slight extent and thus change the viscosity of the blood. The chief reasons are the following:

An extremely small amount of formation of fibrin from fibrinogen is sufficient to change enormously the viscosity of the blood. The conversion of .001 per cent. of fibrinogen into fibrin is able to change blood to a condition where it will hardly flow, but behaves as a gel. Even less than this will greatly affect its viscosity. If this happened in a certain part of the kidney, for example, the effect might be greatly to increase the pressure in the capillaries of the glomeruli and thus to increase the filtration. There is in the arterial blood entering the kidney generally about 0.1 per cent.-0.2 per cent. of fibrinogen. In the blood of the kidney vein there are always a few hundredths, or thousandths of a per cent., less than in the arteries. What becomes of this fibrinogen which thus disappears has never been entirely explained, although Nolf suggested that it deposited on the endothelial cells. It is not now believed that it is used as food, since the tissues are probably nourished by the amino-acids rather than by the blood proteins. It is possible that it has been converted (Nolf) into fibrin and possibly afterwards redissolved or partially digested, but not converted back to fibrinogen. A similar loss of fibrinogen occurs in nearly all the organs in the body, the kidneys perhaps being the most important of the organs destroying fibrinogen. The question may be raised whether the conversion of minute amounts of fibrinogen to fibrin may not take place in the peripheral capillaries, thus leading to an increased resistance and to a loss of fibrinogen. It may be that this was the original purpose of the fibrinogen. These are questions which cannot be answered positively without experiment. But the presence of this peculiar protein which has the function of enormously altering viscosity suggests that it may be a very important factor in the normal regulation of blood viscosity.

The great importance of the viscosity of the blood, and of the proteins of the plasma in determining it, is shown by the circumstance that if the blood is withdrawn from a dog's body and the corpuscles removed by centrifugalization and then the corpuscles suspended in a salt solution of the proper strength, such as Ringer's solution, and reinjected, the animal will not live. But it will live if gum arabic is added to the Ringer's solution in sufficient quantity to restore the viscosity to its normal amount.

Clotting of blood.—When mammalian blood escapes from the blood vessels into the tissues or when it is collected in a vessel, it changes in the course of two to ten minutes from a liquid to a jelly-like solid, which has the same volume as the blood in the liquid form. It clots. No heat is disengaged in this process of clotting. If the clotting takes place in a beaker or other vessel, the clot after a longer or shorter time, generally in the course of half an hour, begins to shrink and the clotted blood gradually separates into two portions, a clear, slightly straw-colored liquid, called the *serum;* and the solid, contracted clot consisting of the corpuscles, the greater part of the serum, and of an insoluble protein substance called fibrin. This process of contraction of the clot with the pressing out of the liquid serum is not peculiar to blood, but is a common property of many gels. It is called syneresis. The contraction of the blood clot is, however, more extensive than that of most other gels. The particular object secured by this process of conversion of the liquid blood to a solid clot is to stop bleeding and so to prevent fatal hemorrhage. While clotting is common to all forms of blood from that of the echinoderms to man, the process reaches its highest degree of perfection in the mammalia and birds, where the blood pressure is highest and the danger of fatal hemorrhage is correspondingly increased. We may now inquire more at length into the nature of this process of clotting. It may be stated at the outset that while no physiological phenomenon has been more extensively studied and for a longer time, yet we are still quite ignorant of the explanation of many of the steps of the process.

Clotting not due to contact with air. That clotting is not due to contact of the blood with air may be shown by the fact that the entrance of air into the veins does not cause clotting. In caisson disease, when men are too suddenly transferred from a compressed air to normal atmospheric pressure, there often occurs a disengagement of bubbles of nitrogen gas in the blood vessels of the body. These bubbles, while they may cause death by air embolism, do not cause any clotting about them in the blood. Moreover, if blood is received from an artery or vein into an evacuated vessel, or if it be collected over mercury without contact with air, clotting occurs as usual. Contact with the air is not, then, the cause of the clotting of blood as one might at first think.

Stopping of circulation not the cause of clotting. It might also be thought that the cessation of movement of the blood, its lack of agitation ensuing after its discharge from the blood vessels, is the cause of clotting, but this again can be shown not to be a sufficient explanation of the facts. If, for example, two ligatures be placed about a portion of the jugular vein of a horse so that some blood is imprisoned between the ligatures, it will be found on opening this vein after some time that the blood in it has remained quite fluid. It will, however, ultimately clot in the vein in these conditions, but its clotting takes very much longer. If the fluid blood is poured out of the vein into a glass vessel the blood clots at once. Moreover, if an attempt is made to keep blood in circulation through glass pipes at body temperature, it will be found that it clots in the pipes and fully as rapidly as if it remained at rest. Whipping the blood with glass rods increases the speed of clotting. The quick clotting of shed blood cannot then be due to the stoppage of the flowing motion of the blood, although this may, at times, in the blood vessels be a predisposing cause of clotting. Clotting is not due either to a lowering of temperature of the shed blood, since cooling usually retards the process of clotting, and a high temperature accelerates it.

Contact with the tissues greatly accelerates clotting. When blood is shed into the tissues, or when in the process of collecting it, it passes over the wounded surfaces of the tissues, its clotting is greatly accelerated. In fact, if blood of vertebrates lower in the animal scale than mammals, for example the blood of birds, reptiles, fishes and amphibia, be carefully collected by means of a cannula introduced into an artery, the greatest care being taken to avoid any contact with a wounded piece of tissue, the blood thus drawn will remain fluid for a long period of time, and if it is collected into vaselined glass vessels so that the blood does not touch the glass walls it will remain fluid almost indefinitely. If to such blood a small amount of the normal salt extract of any tissue of the body is added, clotting occurs very quickly. It is evident from this that there must be something in the wounded tissues which accelerates clotting of blood. In mammalian blood the conditions are essentially the same, except that clotting will occur without contact with the wounded tissue. Nevertheless such contact greatly accelerates clotting here also, and the addition of extracts of the tissues hastens the clotting of mammalian blood just as it does that of bird's blood. Among the substances of the tissues which have this power of accelerating clotting, phospholipins of the nature of cephalin, or unknown substances which accompany this fraction of phospholipin when the latter is separated from the tissues, have the power of acting in the same manner and are presumably the active substances in the tissues. It may be said here that these accelerating substances have a certain specificity of action in that the tissue of one kind of fish or other animal accelerates the clotting of its own kind of blood far more than it does that of blood of other species (Loeb, Nolf). The specificity is not, however, absolute, but is relative. The substances in the tissue extracts thus accelerating clotting are sometimes called thromboplastic substances, thrombokinase, tissue fibrinogens, or coagulins, as different writers have made different pictures of their method of acting.

Contact with foreign bodies greatly accelerates clotting. It has been found that not only will tissues hasten the clotting of blood. Any finely divided foreign substance, which can be wet by the blood, hastens clotting. Thus finely divided glass, glass wool, clay filters, feathers, or absorbent cotton, hasten clotting. It is not necessary for the blood to be shed in order to produce clotting by these means. If, for example, a needle be passed through the walls of an artery or vein it becomes covered with a coating of fibrin; and if the walls of a vein be injured, as by the action of an inflammatory process, by a hot needle or by caustics ($AgNO_3$), clotting more or less limited to the place of injury will be observed. Sometimes if the blood has been rendered abnormally easy to clot, as it is by anesthetics after an operation, or after hemorrhage, slight injury to a vein by an inflammatory process or mechanical process may lead to clotting in the vein and thrombosis. This is especially apt to happen, for example, after parturition. After any hemorrhage, including the more or less extensive hemorrhage of parturition, blood is more easily clotted, the circulation in the veins of the leg is slow and by tight bandages about the abdomen may be rendered slower than usual. Conditions somewhat similar, except for the hemorrhage, prevail after operations for appendicitis and clotting in the veins of the legs, particularly in the right leg after appendicitis operations, is not unusual. If these clots get loose in the circulation they may lodge in the pulmonary circulation and produce serious trouble or even death. If they remain fixed they disturb, in a painful way, the circulation of the leg. Milk leg is a result. Such clots are in part reabsorbed and in part are incorporated into the vein walls. While contact with many forms of matter produces clotting, yet blood received under paraffine oil and into smooth glass receptacles which have been oiled with paraffine oil has its clotting much delayed. Oftentimes the first signs of clotting in such blood occurs in the surface where dust particles may be present. Examination generally shows that the foci of such clots are some extraneous matters. Contact with foreign substances which may be wet with the blood is, hence, one of the determining factors of the process of clotting. Many inert solid substances thus have a thromboplastic action.

Physiological variations in tendency to clotting. It is observed that the tendency of blood to clot undergoes quite wide variations in different individuals. As has just been said, previous hemorrhage always greatly accelerates the tendency to clot. This is of course in the nature of an adaptive change to stop any recurrence of the hemorrhage. Certain individuals have blood which clots with much greater difficulty than normal. If this tendency toward delayed clotting is very pronounced, it may constitute a positive danger to life. Such people are known as bleeders; they have the disease known as *hemophilia*. This trouble is inherited and generally runs in families. In pneumonia and some fevers the clotting of the blood is somewhat slower than usual, and the settling of the corpuscles being rather more rapid, when the shed clotted blood of such individuals is examined it often happens that the upper parts of the clot may be found to be composed of plasma with few or no corpuscles, the corpuscles having settled before clotting occurred. The blood has a clear coat above. This is called the crusta inflammatoria. Horse's blood usually acts in this manner, the corpuscles settling very fast, particularly if the blood be cooled as soon as it is shed.

Ways in which the clotting of blood may be caused or prevented. Action of albumose. If a strong solution, 10-5 per cent., of albumose (Witte's peptone) in 0.9 per cent. NaCl be injected rapidly into the jugular vein of a dog in an amount of 0.2-0.3 gram albumose per kilo body weight, it is found that blood drawn from a few minutes to a couple of hours later has a greatly prolonged time of clotting. By the use of larger amounts of albumose the blood drawn 15 minutes after injection does not clot at all. Such blood is known as peptone or propeptone blood. This injection has no such pronounced action in rabbits, the injection of amounts of peptone which the animals will survive is followed by very little change in the coagulability of the blood. It has been found that, if the injection of peptone is made in a fasting dog, the results are different from those when the dog is in full digestion

(Wooldridge). The very first effect of the injection is a very brief period of quickened coagulation. Furthermore, if a small amount of peptone is injected first, so that the coagulability of the blood is only slightly depressed, then the subsequent injection of a large amount of albumose solution has no effect at all in changing the speed of clotting. The first injection appears to have made the dog immune to the albumose. What it is in the albumose which has this action is not known. It is perhaps significant that albumose prepared by the digestion of fibrin is most efficient in this action, but albumose from meat is also efficient, though to a less degree. Albumose added to blood outside the body does not have the effect of retarding coagulation. It somewhat accelerates the clotting. The effect of preventing clotting is, hence, not a direct action of the albumose on the blood, but on the tissues of the body. Some have thought that it is particularly acting on the leucocytes; others that it is acting on the liver. We shall come back to this presently.

Leech extract. If an aqueous extract be made of the head of the medicinal leech, it is found to have quite remarkable powers of delaying coagulation when injected into the body, or when added to the blood outside of the body. The active principle is called hirudin. It appears to be a deutero-albumose. The blood will not clot after leech extract has been added to it. Hirudin is an anticoagulant.

Tissue extracts. These have a most interesting action. If a normal salt extract be prepared of the thymus gland, spleen, testicle, brain, lymph glands, or other tissues, something goes into solution which is precipitated with dilute acetic acid. Redissolved in salt solution made slightly alkaline by sodium carbonate and injected into the vein of a dog, it will cause death if sufficient is injected. On post-mortem examination it will be found that intravascular clotting has occurred in the portal region. Sometimes the clot is confined to the portal vein and its branches entering the liver, but in digesting dogs it will be found that the clot may extend into the right heart and lungs. In rabbits by such tissue extract injections the whole circulation may be suddenly clotted. In dogs, however, it is found that a clot forms preferably in the portal circulation. If smaller amounts of the tissue extract are injected, it is observed that at first, immediately after the injection, there is an increased coagulability of the blood, and that there is a temporary cessation of the respiration and a sudden fall of blood pressure, but that in a few moments the dog recovers. Larger quantities of the tissue extract may now be injected without any obvious ill effects, but if the blood be drawn, it is found that it has lost wholly or partly its powers of spontaneous clotting. The addition, however, to this non-clotting blood of more of the same extract which had just been injected causes clotting at once outside the body, although it is ineffective within the body. If a dog which has received a dose of tissue extract not sufficient to kill him be killed in some other manner, it will generally be found on post-mortem examination that there are clots in the portal vein. These clots are already white, owing to the blood corpuscles having been washed out, and they ultimately undergo digestion and disappear. It is, hence, a very singular fact that these tissue extracts, which act in so many ways like peptone or albumose, produce within the body first a positive tendency to coagulation, followed by a negative reaction, whereas when added to blood outside the body they produce only the positive phase. This fact indicates that the negative phase of clotting, that is the tendency of the blood to remain fluid, is due to a reaction on the part of some tissues of the body to this dangerous substance, and not to a reaction on the part of the fluid blood itself.

Action of oxalates, fluorides and citrates. If blood is received into a solution of sodium or ammonium oxalate, fluoride or citrate so that it contains about 0.1 per cent. of the first two of these salts, it will remain indefinitely liquid. If, however, sufficient calcium chloride is afterwards added to such oxalate, or citrate blood so that there is an excess of calcium in the blood, clotting occurs normally. Fluoride blood does not clot spontaneously by the addition of an excess of calcium. Since these substances, except the citrate, have the power of precipitating calcium, and since the addition of calcium causes the blood to clot, it is concluded that the presence of calcium salts is necessary for the clotting of blood. While the citrate does not precipitate calcium, it is believed to unite with it and hold it in an un-ionized form so that its action also is supposed to be that of a decalcifier. The injection of calcium chloride solutions into the blood increases somewhat its tendency to clot, but does not in itself cause clotting, nor does it do so outside the body. Both strontium salts and barium salts will enable decalcified blood to clot, although the action is not so good as that of calcium. By collecting blood in oxalate solution and centrifuging the corpuscles out there is obtained a clear plasma, sometimes slightly colored by hemoglobin, which will not clot spontaneously, and this is called *oxalate plasma.*

Action of strong solutions of magnesium sulphate and sodium chloride. The clotting of the blood may be prevented *in vitro* by receiving the blood in an equal volume of 10 per cent. NaCl, so that the mixture contains 5 per cent. of NaCl. Such blood remains fluid for a long period. If the corpuscles are centrifugalized off, the clear plasma is called *sodium chloride plasma.* Such plasma will clot spontaneously if it be diluted with four volumes of water. Similarly, if blood be received into a saturated solution of $MgSO_4$ so that there will be three volumes of blood to one of the solution, the blood will not clot. If the corpuscles be centrifugalized off, then the addition of water to the magnesium sulphate plasma does not cause spontaneous clotting, but the diluted plasma will clot on the addition of some of the serum from clotted blood.

Explanation of the facts just cited. The facts which have just been stated about the clotting of blood, and some other facts which we shall now discuss, have been known most of them for a very long period, namely from 20-40 years, but the satisfactory explanation of these facts is still impossible. With these experimentally ascertained facts in mind we may now proceed to examine the nature of the processes involved in clotting. It is very important for the student to hold the facts themselves clearly in mind, rather than any explanation of them, for there are about as many explanations offered as there are investigators, and none of these explanations is as yet satisfactory. The great difficulty in the explanation is due to the complexity of the blood and our ignorance of the fundamental properties of solutions and the processes of crystallization, for the clotting of the blood, as we shall see in a moment, is at the bottom the crystallization of a supersaturated solution, the crystalline substance which separates being called fibrin.

Fibrin. Whatever may be the exact nature of the processes involved, all are agreed that the conversion of the blood from a liquid to a solid clot is due to the fact that a substance of an insoluble nature appears in it. This substance is called *fibrin*. It is a protein substance. It is deposited first as very fine acicular crystals lying between the blood corpuscles, and radiating from the blood plates or other finely divided foreign particles. (Figure 53.) These crystals shortly coalesce, or stick together, and form a network in the interstices of which corpuscles, both white and red, and the blood plasma are held so that the whole is in the nature of a solid clot. If blood as it is drawn be whipped with a glass rod, the fibrin as fast as it separates sticks to the rod in the form of tough strings in which some corpuscles are entangled. This is fibrin. The blood from which this fibrin has been thoroughly removed will no longer clot. It consists of a mixture of serum and corpuscles and is called *defibrinated blood*.

The question is, then, what is the origin of this fibrin? Does it pre-exist in the blood or does it appear only at clotting? Is it in the corpuscles or plasma and why does it appear when blood is shed? Some of these questions are not difficult; others of them are very hard to answer.

Fibrinogen. Origin of fibrin. The first question which may be asked is this: Does the fibrin come from the corpuscles or from the plasma? This question may fortunately be answered with certainty. The greater part of the fibrin comes from the plasma, not from the corpuscles, or at least not directly from the corpuscles. If blood is prevented from clotting by the action of any one of the anticoagulants just mentioned, such for example as strong NaCl solution, or leech extract, it is possible to remove all the corpuscles, both red and white, by centrifugalization. The clear plasma thus obtained, if it be peptone plasma or salt plasma, will clot if diluted with water. Fibrin appears in it. If oxalate plasma is obtained, it is necessary to add some serum or an extract from the alcohol-coagulated serum, or a calcium salt to make it clot, but then it clots and yields a typical fibrin. On the other hand, the blood corpuscles washed free from plasma by suspending them in physiological salt solution and recentrifugalizing will not clot and form fibrin under similar circumstances. This experiment shows clearly that the fibrin has come from the blood plasma. It is found that the amount of fibrin which a dog's blood will yield is about 0.1-0.3 per cent., but after suppuration or inflammation it may rise to double or treble this amount, i.e., to 1 per cent. The plasma yields naturally a larger proportion, since the corpuscles which make 30-40 per cent. by volume of the whole blood contain none. Plasma generally yields from 0.2-0.6 per cent. of fibrin. Since the fibrin once formed is, as such, quite insoluble or nearly insoluble in blood plasma, it is believed that the fibrin does not exist as fibrin in the plasma, but as a substance which gives rise to fibrin. This pre-existing substance is called *fibrinogen*. The essential act of clotting, therefore, appears to be the conversion of the soluble fibrinogen into fibrin. It may be said here, however, that besides that in solution in the plasma fibrinogen is also contained in the blood plates.

The optical phenomena of clotting. Before taking up the nature of the relation between fibrinogen and fibrin, we may for the moment turn aside to the microscopic examination of the clotting blood. The process of clotting of plasma or of blood itself can be best studied by means of the ultra microscope, that is the dark field microscope. If plasma or blood be watched as it clots, it will be found that fibrin comes out in the form of very fine, long, acicular crystals shown in Figure 53. These crystals are probably not solid crystals, but liquid crystals, since liquid crystals not infrequently take this form. Moreover, they have another property of liquid crystals, that of coalescing or sticking to things; it is because of this property that they will stick together to make thick, long strands of fibrin of which the crystalline nature could not be inferred in any way by microscopical examination except from their double refraction. That indeed shows that their molecules are oriented. The crystals appear first, as a rule, about little specks of dust, or, if blood platelets are present, they adhere to them and grow out from them. We shall later come back to the discussion of the relation between the platelets and coagulation. The fact is, however, that we have in fibrin formation a process of crystallization. The clotting of the blood is a crystallization of fibrin. This fact was clearly recognized by Wooldridge many years ago. The problem we have to answer is then this: Why does crystallization occur when blood is shed and why does it not occur in the blood vessels?

The relation of fibrin to fibrinogen. The first question which we wish to ask is then this: Is the fibrin preformed but held in solution in the plasma until clotting occurs; or is fibrin formed only at the moment of clotting from the fibrinogen which is itself soluble. If the latter hypothesis is true, namely, that fibrinogen is converted into an insoluble protein, fibrin, which crystallizes out, then the problem is essentially

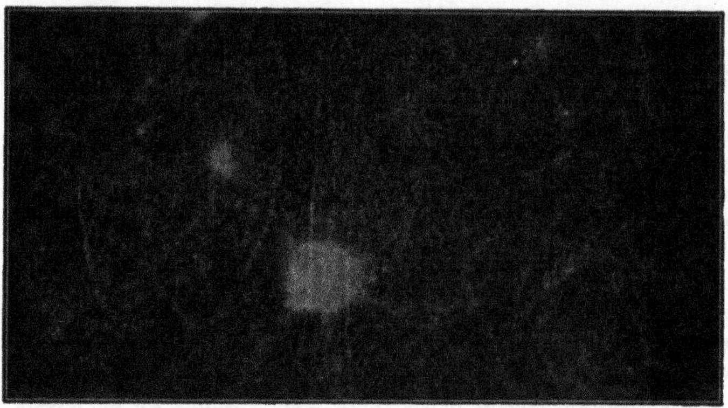

FIG. 53.—The clotting of fibrin showing the long, acicular crystals. A crystalline gel as seen in the ultra microscope (Stubel).

to learn the nature of the difference between fibrinogen and fibrin. The conversion of fibrinogen into fibrin is the essential thing. If, on the other hand, the fibrin is simply held in solution by union with some substance in the blood plasm and crystallizes out, when the influence of this substance is withdrawn, then the problem is quite a different one. The first question, then, to be discussed is the relation between fibrinogen and fibrin. What is the nature of fibrin? Now it unfortunately happens that we do not know what is the composition of fibrin. It is known that it is a protein. It is generally stated that it is a simple protein. The composition of it and of fibrinogen have been given by various authors. The figures of Hammarsten are as follows:

	C	H	N	O	S
Fibrinogen	52.93	6.90	16.66	22.26	1.25
Fibrin	52.68	6.83	16.91	22.48	1.10

It will be noticed that the analyses show no difference in composition between fibrinogen and fibrin. But, while these figures appear so clear-cut and decisive, their reliability is in fact an illusion. The substance which is analyzed as fibrin and called fibrin in the foregoing analysis is not the substance as it appears in the blood; and the substance which is analyzed as fibrinogen is not the substance which is obtained from blood by salting out, but each is a modified substance. Each of these substances before analysis has been extracted with alcohol and ether to remove the fat and lipoid. The substance remaining is a protein, of which the composition has just been given. Now it may be just in the lipin part of the molecule that the difference lies, and indeed Wooldridge maintains that this is precisely the case. It is true of all the proteins of the plasma that when they are precipitated from plasma by the addition of salt, the precipitate which is obtained always contains a considerable proportion of a phospholipin. Wooldridge showed that fibrin might contain as much as 14 per cent. of its dry weight as phospholipin. This phospholipin, or phosphatide, is often regarded as an impurity, but the probabilities are that it is a compound with the protein. The amount of phospholipin in the fibrin is less, Wooldridge thought, than that in fibrinogen, and this author, one of the ablest and keenest-sighted of all who have worked on this problem of coagulation, believed that the principal difference between these proteins consisted in the proportion of lecithin in the two cases. It may be mentioned in this connection that we have a very similar state of affairs in the case of hemoglobin and the red blood corpuscles. The amount of hemoglobin contained in the blood of most, or at least of many, mammals is very much more than can be dissolved when the oxyhemoglobin is free. It is held in solution by union with the phospholipin and protein of the stroma of the red blood corpuscles. When this union is broken hemoglobin crystallizes out, just as the fibrin crystallizes out. There are so many points of identity between laking blood and clotting that we may with some confidence believe that the processes are at bottom essentially of the same kind. There is at least one difference between fibrinogen and fibrin, which was emphasized by Wooldridge, but which seems to have been overlooked by all subsequent workers. If fibrinogen is obtained by Hammarsten's method of salting out by means of a saturated sodium-chloride solution, Wooldridge found that this fibrinogen when digested by pepsin hydrochloric acid left a large residue. This residue is by some authors (Pekelharing) called a nucleoprotein, or nuclein residue, owing to the fact that it is soluble in alkalies and contains a great deal of phosphoric acid; but it is incorrectly so regarded. The phosphoric acid containing part of the residue is soluble in alcohol. It is not, therefore, a nuclein, since nucleins are insoluble in alcohol. Practically the whole of the phosphoric acid is extractable with alcohol. This shows that fibrinogen as it exists in the plasma is a lecithoprotein, or at any rate that it contains a good deal of a phospholipin. Now Wooldridge found that fibrin did not yield nearly as abundant a residue of phospholipin when digested by pepsin-hydrochloric acid. It dissolved in the digestion mixture almost perfectly. The small residue it yielded Wooldridge believed to be due for the most part to some unchanged fibrinogen. These observations appear to indicate that fibrinogen and fibrin are not identical substances, but differ in the proportion of phospholipin in them, fibrin containing less.

Fibrinogen. The protein usually called fibrinogen, but which we may call Hammarsten's fibrinogen, may be obtained from centrifugalized oxalate plasma by adding to the plasma an equal volume of concentrated NaCl. None of the other proteins of the blood plasma precipitate with this treatment. The precipitate redissolves in dilute salt solution. It is insoluble in water and acts in this respect like a globulin. It may be redissolved in salt solution and reprecipitated several times so as to get it as free from the other proteins as possible. Thus obtained it is always contaminated with, or joined to, a phospholipin of some kind. It has a tendency to become insoluble and to be converted into a fibrinous, insoluble substance, apparently fibrin. If left long in the salt solution from which it is precipitated, it undergoes this change. Wooldridge called this a true conversion into fibrin. It is hard to know whether it is or not. Dog's fibrinogen particularly easily undergoes this change; horse's changes less readily. If we disregard the phospholipin and consider this only an impurity, the fibrinogen appears to be a simple protein, coagulating at 56° C. The kinds of amino-acids it has is shown on page 144. It has the property of clotting when in solution on the addition of thrombin or of serum. It is optically active.

We may sum up this discussion, then, by the statement that it is at present impossible to say whether fibrin pre-exists in the blood plasma being held in solution by its union with other proteins, or other substances, until the time of clotting; or whether an insoluble protein is formed from a soluble one, fibrinogen, the fibrin then crystallizing out. The latter view is the one at present held by nearly all physiologists, but it was strongly criticised and condemned by Wooldridge, and his objections, in the writer's opinion, have never been answered.

The rôle of the structural elements of the blood in clotting. When blood clots, then, we observe that the clotting is due to a protein substance which had been in solution in the blood plasma, crystallizing out of solution; while the exact nature of the changes involved in this protein which determines that it becomes insoluble or capable of crystallization is still unknown, we may proceed to examine the causes which are playing a part in the transformation. The first question which we shall ask is whether the red, the white corpuscles, or those more unknown elements, the blood plates, play any part in the process. This question is generally answered in the affirmative, namely, that both the white blood corpuscles and the platelets play a very important part in the initiation of the clotting process, but, as we shall see, the evidence that the white corpuscles are concerned in the process is at the best very dubious. On the other hand, there is no doubt that the elements called blood platelets are a very important factor in the clotting of the blood. Schmidt, who worked for many years on the clotting of the blood, and to whom most of the essential conceptions of the process are due, was of the opinion that the white corpuscles played a very important part. According to him, when blood is shed the white corpuscles break down in part and liberate a substance which he called *prothrombin*. This substance does not act, according to him, by itself, but in certain circumstances, i.e., in the presence of calcium salts, it is converted into a third substance, *thrombin*, or *fibrin ferment*, as he called it. This substance he believed to be a ferment, and by its action on the fibrinogen of the blood, with the co-operation of the paraglobulin, it formed fibrin. This view of Schmidt as regards the rôle of the leucocytes, the furnishing of prothrombin and thrombin and its action on fibrinogen, is accepted by the majority of workers on this subject. There is a doubt, however, whether it is essentially correct and well-founded. The evidence that Schmidt adduced to prove that the white corpuscles cause coagulation was this: If horse's blood is received directly from the artery or vein into a cooled glass vessel and the blood immediately cooled in a salt-water-ice mixture and kept cooled to about 0°, the clotting is so prolonged that time is given for the corpuscles to settle. The red corpuscles being more dense settle faster than the whites, and accordingly there is at the bottom of the tube a layer of red corpuscles, above this is a thin layer of white corpuscles and above this is the plasma, which is somewhat turbid, due to the admixture of some platelets and leucocytes. Now coagulation ultimately ensues in this blood and it is observed that the clotting begins always in the white layer, where the white corpuscles are most abundant, and that the plasma at the top and the red corpuscles underneath may not be coagulated, whereas the plasma just above and including the layer of the whites is coagulated. On examining this white layer under the microscope, Schmidt observed that there was among the white corpuscles a large amount of a granular precipitate looking like the débris of corpuscles which had gone to pieces. Schmidt interpreted the débris as being due to disintegrated leucocytes, although there was no evidence that it was so derived, and we now know that it has in part at least another origin. Schmidt concluded from these observations that when blood is shed many of the corpuscles disintegrate and set free a substance which causes clotting. This substance he called fibrin ferment. To support this conclusion Schmidt tried the experiment of injecting leucocytes derived from lymph glands, or débris of the latter, into the circulation and he believed that the intravascular clotting sometimes obtained was due to this fibrin ferment from the lymph glands. Now this conclusion of Schmidt, although it has persisted in the literature as a fact, is incorrect. Its incorrectness was shown by Wooldridge. So far as is known, the leucocytes of mammalian blood do not disintegrate to any considerable extent in shed blood. They persist and the débris which Schmidt observed did not come from them, but consisted of the substance now known as platelets, and the origin and nature of which we shall soon consider. All the evidence which Schmidt accumulated really applied to the blood plates, and it is these indeed which really inaugurate the process of clotting. That the leucocytes do not cause the clotting and are not necessary for it may be shown in several ways. Blood plasma quite freed from leucocytes and red blood cells by centrifugalization, such for example as salt plasma, or peptone plasma, will coagulate if simply diluted with water, or if a current of CO_2 is passed through the plasma. No corpuscles of any kind are necessary. Moreover, it can be shown that the leucocytes are quite inert. Wooldridge found that if lymph corpuscles are carefully washed with salt solution, they will cause clotting if added to peptone plasma, or if added to salt plasma, provided the substance of the platelets, which he called A-fibrinogen, is also present. But these same leucocytes do not clot the lymph in which they are. Exudates exist containing many leucocytes and fibrinogen, but these do not clot. Moreover, if the leucocytes which have been once in blood are collected by centrifugalization, washed and suspended in peptone plasma, they do not cause clotting of the plasma, although leucocytes of lymph glands, which have never yet been in the blood, will cause clotting. It is clear from this extraordinarily instructive and important observation that the leucocytes of the blood have not a power of inaugurating clotting. Wooldridge tried another very suggestive experiment. If very strong peptone plasma is used, a considerable amount of lymphocytes may be added without producing clotting. If now these added lymphocytes are centrifuged from the plasma and washed carefully in salt solution, so as to free them from adherent plasma, and are now introduced into a peptone plasma not so strong but one which will clot spontaneously in time, they do not clot this plasma, whereas a similar quantity of the same lot of lymphocytes which had not first been in the peptone plasma will clot it at once. This fact shows, according to Wooldridge, that leucocytes which have once been in the blood become inert toward it; whereas those which have not been in the plasma act like foreign bodies and hasten the clotting. Wooldridge was able to show, also, that it was the substance of the blood platelets which was the important substance in clotting. It is, however, possible to extract from the leucocytes a substance which hastens blood clotting. That the red corpuscles are not primarily concerned in the clotting is also clearly shown by the fact that the plasma has all the powers of clotting in the absence of the reds. But here again, just as with the whites, it is possible to isolate from the red corpuscles a substance which acts exactly like the substance from the whites and causes clotting. The fact, then, that it is possible to isolate from the white corpuscles a substance which hastens coagulation is no argument that the white corpuscles play a part in normal clotting, since exactly the same argument applies also to the reds, which are universally admitted to play no part in the process. It is a very interesting fact that while the intact red corpuscles are quite inert toward the clotting process, yet if the hemoglobin be separated from the stroma it is found that the stroma is very toxic, its injection produces intravascular clotting and when added to shed blood it greatly accelerates clotting. There is no good evidence, then, that either the reds or the whites are concerned in the normal clotting of blood, and we may now turn to the blood plates, the other morphological constituent of the plasma. Practically all authorities are of the opinion that these play a vital part in the clotting, since the addition of these to blood enormously accelerates the clotting process.

Rôle of blood plates in clotting. We now enter one of the most disputed fields of the morphology of the blood, for there is no unanimity of opinion as to the nature and origin of the blood plates, or indeed whether they pre-exist in the blood in the vessels. But whether they pre-exist or appear only after shedding, they play a part in clotting, and in fact their appearance is the first visible sign of the changes of the plasma resulting in clotting. The platelets are generally disk- or spindle-shaped bodies without color, about 1.5-3 μ in diameter. In other words, they are about one-third to one-seventh the diameter of a red blood corpuscle. They may be present in enormous numbers, as many as 800,000 per c.mm. of blood in dog, or human blood, or they may be quite absent, as in bird's blood and the blood of the lower vertebrates. They may be fixed by osmic acid. When first separated from oxalated blood by centrifugalization they redissolve readily in dilute alkaline salt solutions, but after a short separation they lose this power of dissolving and become converted into a fibrinous mass. It is very difficult to decide whether they are to be considered as morphological cells, as they are at present generally considered, or whether they are to be regarded as liquid crystals. It is well known that supersaturated liquids form crystalline deposits with the greatest ease, so that it is quite possible that we have in the blood plasma normally something in the nature of a supersaturated solution of the materials from which the plates are formed and that this deposits the plates very quickly on contact with foreign substances. Something of a very similar kind appears to happen in the cell nucleus and to result in the sudden appearance of the nuclear chromosomes. It has recently been shown by Chambers that the chromosomes may appear very suddenly in the clear homogeneous nuclei of the sperm mother cells of the testis of grasshoppers. The chromosomes appear whenever the cell is irritated, and their appearance inaugurates the phenomena of cell division just as the appearance of the plates inaugurates the process of clotting. The chromosomes are also organized substances of definite shape and, like the plates, are often disk-shaped. The phenomena of the appearance of the plates from the plasma has been particularly studied by Wooldridge. If horse's blood is cooled, the corpuscles separate and the plasma may be removed. The same result is had with dog's peptone blood. The corpuscles are separated by centrifugalization. The plasma which results is perfectly clear at the temperature of the body, but if it be cooled to about zero degrees there comes out of it a precipitate of platelets. This precipitate is not an amorphous precipitate, but is made up of clear, granular refractive disks, which look like imperfect protein crystals. On warming the blood plasma the platelets swell and redissolve and may be reseparated by cooling. The little plates thus separated from the plasma look exactly like very small colorless red blood cells and when slightly warmed they coalesce and may form larger disk-shaped, colorless structures of the shape and size of a red blood corpuscle. If after separating them in the cold, the precipitate with some plasma is warmed slightly, the disks lose their crystalline appearance and rounded or spindle shape and stick together in rows and clumps of granules which form fibrinous strings and become insoluble. They become changed to typical fibrin (Wooldridge, Schittenhelm and Bodong). The disks when they first appear look like the incompletely crystalline stage of proteins. Their behavior reminds one of the liquid crystals described by Lehman. Wooldridge concluded that they were in reality crystalline. By many authors they are supposed to be alive, because they show amœboid changes of form. But this again is a characteristic of some liquid crystals. Such crystals also have the power of changing their shapes and to make refractive myelin-like forms resembling protoplasm. It is possible, of course, that in some bloods they are present in such amount that they may be in part precipitated in the circulating blood. The plates have been collected by Wooldridge and their chemical composition seems to be identical with that of the stroma of the red blood cells and like the protoplasm of the whites. They resemble this stroma, also, in their physiological action, as we shall see. They contain a protein, fibrin and various lipins, such as a phospholipin, thrombin and a proteolytic ferment.

It was believed by Lillienfeld and Pekelharing that these plates were composed of nuclein. This belief was based on the fact that they were soluble in dilute alkalies, precipitated by dilute acids, were soluble in strong NaCl, contained phosphorus, stained with methyl green and basic dyes, and that they left a phosphoric acid containing residue when digested with pepsin HCl. Small amounts of nuclein bases may be obtained from them. But all these characteristics, except the last, are possessed by the lecithoproteins as well as the nucleins, and Wooldridge showed that the phosphoric acid in the residue was soluble in great part in alcohol. We know, besides, that particularly the phospholipins have this property of forming liquid crystals. While, then, the origin, and indeed the proper interpretation of the blood plates, is thus still in many points obscure, we may consider their relation to clotting. It may be said, in passing, that the interpretation of the clotting of the blood really rests on the determination of the origin and significance of the plates.

If any foreign substance wet by the blood is put into the blood stream, such as a needle or a thread put through a vein or artery, it is found that the foreign substance becomes covered with a deposit of plates and that this deposit becomes covered by, or is partially converted into, fibrin. Moreover, when platelets are added to plasma and the latter clots, it will be observed that the platelets are the centers from which the fibrin formation starts. The relation of the plates to clotting was also shown by Wooldridge as follows: Dog's blood, which is incoagulable from the injection of peptone, becomes coagulable if a stream of CO_2 passes through it. The same is true of the peptone plasma, provided the dose of peptone has not been too large. Peptone plasma will clot spontaneously if CO_2 is passed through it, or if acetic acid is added to the plasma, or if the plasma be diluted with two or three volumes of water, but it will not clot spontaneously if diluted with two or three volumes of 1 per cent. NaCl. Wooldridge found that if such coagulable plasma was cooled to a low temperature and the platelet stuff centrifugalized out of the blood and the plasma then filtered through several thicknesses of filter paper so that the plasma was perfectly clear and free from plates (A-fibrinogen, as he called it), the plasma had lost its power of clotting when diluted with water or when CO_2 was passed through it, but it reobtained its power if the platelets either suspended in a little plasma or redissolved in very dilute alkaline salt solution were added to it. Moreover, this plate material separated and dissolved in the manner indicated had the property when injected into the circulation of an animal, of causing intravascular clotting, in this respect resembling the stroma of the blood corpuscles or extracts of the various tissues. There can be no doubt, then, from these facts that the stuff known as blood platelets stands in some very important relation to clotting. They seem to act like the substances (fibrinogens) obtained from the tissues and from the wounded cells and tissues. A further very important fact is that the amount of fibrin recovered from plasma by the addition of plates to it increases with the amount of platelet stuff added, so that Wooldridge believed that the platelet stuff actually contributed to the substance of the clot, as well as acting as a starter of the process of clotting. This view has now been substantiated.

Prothrombin and thrombin. So far, we have seen that the plasma contains all the elements necessary for the clotting as long as the plasma contains platelets or the substance A-fibrinogen, which gives rise to the platelets. There remains, however, another factor to be considered. It has been found that, after plasma has clotted and the serum separated, the serum always, or nearly always, contains a substance which was not in the plasma and which has a remarkable action on the clotting of Hammarsten's fibrinogen. This substance is known as thrombin, or fibrin ferment. It is better to call it thrombin, since there is no evidence that it is a ferment, although it may be one. If blood *serum* is precipitated by adding to it four volumes of strong alcohol and the precipitate allowed to stand under alcohol for two to four weeks, all the proteins of the precipitate are coagulated by the alcohol. If now this coagulum be extracted with water, or dilute sodium chloride solution, an organic substance goes into solution and this solution added to a solution of Hammarsten's fibrinogen causes it to clot in a typical fashion. Furthermore, although its action is somewhat quicker in the presence of calcium salts and in the absence of oxalate, yet it will clot the fibrinogen solution in the presence of oxalate. The substance in the solution which has this property of clotting fibrinogen is destroyed by continued heating to 60°. It was called fibrin ferment by Schmidt, its discoverer, but is now generally called thrombin. Nothing is known about the nature of this substance. Lillienfeld and Pekelharing believed it to be a nucleoprotein, but it is probably not a nucleoprotein. There are reasons for thinking that it originates from the platelets, and these are almost certainly not nucleoprotein and contain none. Its nature is unknown. Now thrombin does not pre-exist in the blood. Like fibrin, it appears only if the clotting has occurred. It is an accompaniment of the clotting process, and, with as much justification as fibrin, is to be regarded as the result of the clotting process, but not its cause. It is, however, generally believed by most physiologists from Schmidt's time to be the cause of clotting. Both Wooldridge and Nolf have urged grave reasons for disbelieving that it is the appearance of this body which inaugurates the process of clotting normally. There is one reason which appears to the writer to be practically conclusive that the appearance of this substance is not the cause of normal clotting: that is the astonishing fact that enormous amounts of it may be injected into the circulation without causing any intravascular clotting. Now, if it be assumed that there is some anti-clotting substance normally present which is counteracting its action and keeping blood fluid, there cannot be the quantity required to hold back the action of so much of this substance. A very small amount of thrombin will cause clotting of a fibrinogen solution in the test-tube, but one can run in nearly as much thrombin as there is of fibrinogen in the blood without causing any intravascular clotting and without producing any change, or more than a slight change, in the clotting power of the blood. Now it is very astonishing, if thrombin is the cause of clotting and the formation of minute amounts of this substanec in the shed blood causes its clotting, that the injection of large amounts of thrombin into the blood, which contains already fibrinogen and calcium salts, has no effect at all. The result is the more astonishing since the injection of very little tissue extract, or blood platelet material, causes clotting both within and without the body with the greatest ease. Moreover, these same substances will clot peptone plasma outside the body, whereas thrombin does not clot it. These facts do not seem to have been properly appreciated by most writers who have accepted Schmidt's view. They constitute, in the writer's opinion, an insurmountable objection to the present view, that the first thing that happens in clotting is the formation of thrombin, and that this latter causes clotting by making fibrin out of fibrinogen. Wooldridge, because of these facts and others, held thrombin to be a result of clotting, not to be a ferment, and maintained that the fact that thrombin would not clot blood intravascularly showed that blood did not contain the substance known as Hammarsten's fibrinogen, but that the latter had been formed from the real fibrinogen of the blood, which he called B-fibrinogen, by the actions necessary to separate it.

Since thrombin did not exist in the plasma before clotting, but appeared after clotting, Schmidt and his pupils concluded that there must be an antecedent substance there which was called *prothrombin*, but which has also been named *thrombogen*, to indicate that it gave rise to thrombin. According to Schmidt, thrombogen was not present in the plasma, but was set free from the leucocytes when the latter disintegrated. This is the generally accepted belief at the present time, but it is not well grounded. There is no sufficient evidence that the leucocytes go to pieces in shed blood; Schmidt's evidence that the substance came from the leucocytes was vitiated by the fact that he neglected the platelets; it is the platelets, not the leucocytes, which added to peptone blood cause clotting, and it is from the platelets, not the leucocytes, that the prothrombin appears to be derived.

The rôle of calcium salts. There is no doubt that calcium salts play an important rôle in clotting (Green, Arthurs). They are not necessary for the clotting of fibrinogen solutions by thrombin, but such solutions will not clot on the addition of platelets unless calcium salts are present, and the calcium must be in an ionized form. The salts are supposed to act, hence, by converting thrombogen, or prothrombin, into thrombin.

Explanation of the clotting process by Morawitz and Fuld and Spiro. According to the work of Morawitz and Fuld and Spiro, the leucocytes contain proferment, or thrombogen. This thrombogen in the presence of calcium salts is converted by tissue extract, which they call *thrombokinase*, into an active thrombin. The thrombin, in some unknown way, converts the fibrinogen of the blood into fibrin. Blood remains fluid in the vessels of the body, because there is no thrombokinase present. This is supplied by the blood coming in contact with the tissues of the body. These authors, then, make the clotting of the blood analogous to the digestion of fibrin by pancreatic juice. The latter is supposed to be converted from trypsinogen to trypsin by the action of the enterokinase of the intestine in the presence of calcium salts. Thrombin, according to this view, is really an enzyme, and it is the thrombin of the blood which causes clotting. This view encounters two or three main objections. First, there is no evidence that the thrombin is a ferment, but on the contrary it can be shown that it unites with the fibrin; and the amount of fibrin formed is pretty strictly proportional, when the conditions are kept constant, to the amount of thrombin added. In the second place, the thrombokinase, which they supposed to be killed by heat, is in reality heat resistant, and Wooldridge showed that it was a phospholipin, but distinct from the usual lecithin of eggs. Howell has shown that it belongs to the cephalin group of phospholipins. There is no evidence that the conversion of the fibrinogen into fibrin is of the nature of a proteolytic digestion, although there is no evidence against this view, either. They cannot explain how it happens that the addition of very little of cephalin, or extract containing it, clots the blood far better than the addition of thrombin itself.

View of Nolf and Howell. To account for the fact that the blood remains fluid in the body, although it contains all the materials for clotting, the simplest explanation is that there is present some antagonistic substance which prevents clotting. This view has been suggested from various sources from Wooldridge down and has recently been advocated by Nolf and Howell. According to Nolf, the blood contains a substance called hepatothrombin to indicate that it is formed by the liver. This hepatothrombin unites with leucothrombin (the thrombogen of other authors) to form thrombin. Leucothrombin is supposed to come from the leucocytes, hence its name. According to Nolf, hepatothrombin by itself unites with fibrin and holds it in solution so that it cannot be precipitated. There is normally a slight excess of hepatothrombin in the blood so that it keeps the blood fluid, but when blood is shed there is a discharge of leucothrombin from the leucocytes. The leucothrombin then unites with the hepatothrombin, and the two together with calcium make thrombin; this then unites with the fibrinogen, and the three make fibrin. It happens that there is a slight excess of thrombin in the blood so that some remains in solution in the serum after the clotting has taken place. Nolf's hepatothrombin corresponds in a general way with Wooldridge's A-fibrinogen and with other tissue thrombins or fibrinogens. The hepatothrombin is what others have called antithrombin. In later work Nolf modified this view somewhat. According to Howell, the process is a little different. The blood remains fluid in the body because there is an excess of antithrombin there. This antithrombin is probably formed by the liver. The coagulative action of tissue extracts is due to the fact that they contain a cephalin-like body, as Wooldridge thought, but this cephalin-like body causes the clotting by uniting with the antithrombin and thus getting it out of the way. In the absence of antithrombin the thrombogen is converted into thrombin by the action of the calcium and unites with the fibrinogen to make fibrin. This view is in many ways like Nolf's, although the terminology is somewhat different. Both recall the work of Wooldridge.

Wooldridge's view. It will be obvious from this brief consideration how obscure the process is. Nearly all observers have neglected a possible change in the fibrinogen and have concentrated their attention on the thrombin and the changes it undergoes. Hammarsten's fibrinogen may not exist in the blood as such. The mere fact that there is a protein often present which will coagulate at 56°, the temperature of coagulation of fibrinogen, and that fibrinogen can be isolated from the blood, is not conclusive evidence that the fibrinogen exists there in an uncombined form as it does in the fibrinogen solution. It must be remembered that so-called tissue fibrinogens not clotting with fibrin ferment, but coagulable at 56°, can be isolated from a great variety of tissues of the body. In these circumstances of great confusion it is better, in my opinion, to go back to the work of Wooldridge, whose writings are the clearest and freest from preconceived ideas of any who have worked on this subject. A word may not be out of place here about this talented man, who was far ahead of his time in this work. His work has suffered a most unmerited neglect and his observations have been rediscovered since by those who were ignorant of his precedence. Wooldridge started with the Schmidt view of the importance of leucocytes in clotting, but he clearly showed in 1883-88 the untenability of this view. He showed the predominant place of the platelets in clotting and proved that these bodies were, in part at least, in solution in the blood and that they were to be considered as crystals rather than as morphological elements. It was Wooldridge who discovered the intravascular clotting by tissue extracts and that the acceleration of the clotting by these extracts and the plate material was due to the alcohol-soluble lecithin-like body in them. This body was not lecithin itself, since the lecithin from eggs had no such influence. It was he again who maintained at that time that the clotting of the blood was in the nature of the crystallization of some supersaturated liquid, and that view has now, within the past three years, been demonstrated to be correct. Wooldridge, a young man, was opposed to the views on coagulation of Schmidt and his school, views which were at that time, as at present, the dominant views, although he showed their weakness. This opposition secured for him the enmity or opposition of his older colleagues in England, and the Royal Society refused to publish the Croonian Lecture which he delivered before them, but buried it in their archives, and it was only published after his death. There is no doubt that that lecture was far ahead of any treatise at that time on the coagulation of the blood. This lecture, together with that entitled "Blood Plasma and Protoplasma," and his reports to the Grocers' Committee are among the most lucid and able treatises on coagulation and should be read by anyone who wishes to know the coagulation of the blood. The great difficulty in the way of the problem of coagulation at the present day arises from the attempt to harmonize the facts of coagulation with the facts of immunity and with those of the action of digestive enzymes. All of these subjects are in the greatest confusion from the complexity of the problems. It is probable that they all have the same basis. In all the same elements are found: namely, phospholipins, an enzyme-like substance, calcium salts and carbon dioxide.

Study of the coagulation of the blood will no doubt elucidate these other processes, also. Wooldridge very carefully avoided giving any names to his substances which might indicate a preconceived view of the rôle they were playing. According to his view, blood plasma contained two fibrinogens, A-fibrinogen and B-fibrinogen. C-fibrinogen was the ordinary fibrinogen of Hammarsten, but it did not pre-exist in the blood in more than very small amounts. A-fibrinogen gave rise to B-fibrinogen, and this in turn to C-fibrinogen, and this to fibrin. A-fibrinogen was the substance which constituted the blood plates. This was in solution in the plasma. Blood plasma was like protoplasma and was indeed nothing else than a dilute protoplasma. Now, in consequence of various actions when blood was shed, there was a change in this very unstable form of material. The first change was the precipitation of some of the A-fibrinogen. Whenever this was precipitated the equilibrium of the plasma was upset in some way and chemical changes were set up. The A-fibrinogen separated from the plasma became changed into fibrin, and one of the products of the change was thrombin. The action of calcium salts in hastening this change was unknown to Wooldridge. The main difference between the various forms of fibrinogen and fibrin was supposed to be a difference in the proportion of phospholipin they contained united with the protein. This view is, of course, inadequate, but it has the great advantage of being, in a sense, a physico-chemical explanation and is free from the preconceptions of complement, amboceptors and antibodies which enshroud most modern views in an impenetrable cloud of misunderstanding and confusion.

Summary of the clotting of the blood. The clotting of the blood is essentially the crystallization of a protein substance, fibrin, in the form of liquid crystals which coalesce to form fibrin strings. Whether the crystallization of this substance is due to the fact that the fibrin has been held in solution by its union with some substance which is dissociated from it at the time of clotting, or whether it may be that the fibrin does not pre-exist in the blood as such but in the form of a soluble precursor, fibrinogen, cannot be said with certainty. When blood is shed, and particularly when it comes in contact with wounded tissue, this crystallization occurs. Tissues contain substances which hasten the clotting. There is in blood plasma a substance which may appear in the form of the structural elements known as blood plates. This substance has the property of hastening clotting. Calcium salts are also necessary to hasten clotting. There appear in the blood as the result of clotting two new substances; one of these, called thrombin, has the property of clotting fibrinogen solutions but not circulating blood. The other is known as serum fibrinogen. Its relation to the clotting is not known. Clotting is very much hastened by and generally dependent upon contact with some foreign substance. Such substances appear to act as the points of deposition of the crystals of the blood plates, and from this separation the fibrin crystallizes out. The process of clotting is greatly accelerated by the intravascular injection of tissue extracts. How these various factors are interacting is still uncertain, but various theories have been given to explain them. None of these explanations are satisfactory. The process has many points of resemblance with the hemolysis of the red blood corpuscles. There, as in the blood, a crystalline substance comes out when its union with a lecithoprotein is broken. It is suggested that the lecithoprotein holding the fibrinogen in solution is the serum fibrinoglobulin which appears in the serum when the fibrin crystallizes out. The origin of the fibrinogens is still obscure, but the general resemblance in chemical composition of the plates and the stroma of red and white corpuscles would suggest the derivation of one from the other. The rôle of the liver in the formation of the fibrinogen would thus be explained, also. It is clear that the whole subject of the clotting of the blood is a very attractive field for study from the modern point of view of colloid chemistry and liquid crystals.

The following view is suggested very tentatively as harmonizing many of the facts of coagulation: Blood plasma is a very unstable, supersaturated solution corresponding to a very dilute protoplasm. Its instability may be upset in a variety of ways, these ways being generally those which we recognize as acting as stimulants of cells. Among the substances in solution in the plasma is a complex of enzymes, calcium, phospholipin and protein, a complex which probably corresponds pretty closely in nature to the clear, homogeneous substratum in cells. The blood plasma is supersaturated with this substance and under various conditions, particularly when removed from the blood vessels and brought in contact with foreign matter which it wets, it crystallizes out. These crystals are liquid crystals and, like many liquid crystals, they readily change their shape. They are known as blood platelets. When in solution they constitute the substance called by Wooldridge A-fibrinogen, and the part remaining in solution may be called B-fibrinogen. The composition of this material may possibly be of the following nature: *Protease-fibrin-phospholipin-thrombin-calcium-antithrombin*. This antithrombin is partially dissociable. The stability of the whole complex depends upon it. If a little cephalin, or tissue fibrinogen from injured tissues which contains cephalin (thrombokinase), is added to this mixture it combines with and removes some of the antithrombin. What remains is unstable and breaks apart between the phospholipin and the thrombin. The protease and fibrin crystallize out as *fibrin*, taking with them some of the phospholipin. The thrombin remains in solution. The *protease-fibrin-phospholipin* fragment may be what is called Hammarsten's fibrinogen, or C-fibrinogen of Wooldridge. The action of cephalin in offsetting the action of antithrombin would be explicable in the way suggested by Howell and others. The addition of thrombin to C-fibrinogen causes it to clot, but even in the absence of thrombin the dissociation of some of the phospholipin would lead to a slow conversion to fibrin. The addition of thrombin to blood would have no effect, since C-fibrinogen does not exist free in blood itself. The rapid digestion of the fibrin by the protease under appropriate conditions would be explained, also. It may be that the instability of the protease-fibrin-phospholipin-thrombin complex is much greater when this is present as the calcium salt, and perhaps greatest of all when carbon dioxide is added to the other bond of the calcium. That there are difficulties in the way of this view is self-evident and it is at the best but tentative. The advantage of it is that it agrees, on the whole, better with what we know of the general structure of protoplasm than the others suggested. But the further development of this subject cannot be taken up here.

Alkalinity of the blood.—Both the reaction of the whole blood, as well as that of the blood plasma, is alkaline toward litmus, but acid toward phenolphthalein. This means that there are very few hydroxyl ions present in the blood. But, while it is true that blood is thus almost neutral so far as its hydrogen ion content is concerned, it nevertheless has the power of neutralizing quite a large amount of strong acid before its reaction turns acid to litmus. The coincidence of these two properties means that blood has in it a good deal of some very weak alkali.

Total alkalinity. The power of the blood of neutralizing acid added to it without itself becoming acid to litmus is called the total alkalinity. It is the total titratable alkali. The amount of this titratable alkali varies with the amount of protein present and under other circumstances. In man it has been found to be equivalent to a sodium carbonate solution of 0.34 to 0.59 per cent. strength. In other mammals it generally lies between these extremes. In the dog it is 0.49, and in the horse 0.44 per cent. Na_2CO_3. It is somewhat reduced after violent exercise, in partial asphyxia and in diabetes. The alkalinity is said to diminish rapidly outside the body, perhaps owing to the production of acid in the blood itself, or perhaps for some other reason.

This total alkalinity of the blood is due to the presence in the blood serum or plasma and in the corpuscles of sodium carbonate, sodium bicarbonate and of Na_2HPO_4; and to the presence of the alkali salts of the proteins of both the corpuscles and the plasma. When a strong acid is added to blood it takes away the sodium and potassium from these weak acids, the carbonic, disodium hydrogen phosphate and the proteins, to make the neutral salts of the strong acid. All of these acids are very weak, and accordingly it takes a large amount of them before they furnish enough hydrogen ions to turn litmus red. The proteins of the blood will also, in another way, combine with the acid by means of their amino groups, so that in two ways proteins neutralize acids produced in the tissues and which find their way to the blood. This acid-combining power of the blood is of very great importance, for it is in virtue of it that the blood is able to pick up the carbon dioxide from the tissues and carry it to the lungs. It is also of very great value in caring for an excessive pathological acid production in the tissues, such as occurs in diabetes. In this disease large amounts of hydroxybutyric acid and aceto-acetic acid are set free in the tissues and the blood must carry them to the kidneys to get rid of them.

If, for example, HCl is added to blood, the strong acid takes first the sodium of Na_2CO_3, setting free the very weak acid H_2CO_3 and making NaCl. Similarly it dispossesses Na_2HPO_4 of its Na to form NaCl and NaH_2PO_4, and the latter is also a very weak acid. Finally the proteins exist both in the corpuscles and blood serum as salts, still weaker than these others. Na globulin and albumin at first give up their metals, leaving the free globulin, hemoglobin and albumin, and then, as more acid is added, they unite with the HCl to form globulin and albumin HCl, which dissociate few hydrogen ions. Moreover, the blood, like most tissues, possesses a power of ammonia manufacture by which small amounts of amino groups are split off from the proteins.

Hydrogen ion content of the blood. Blood, as so frequently emphasized, is a living tissue. The processes which occur in it, the processes of clotting and respiration, are processes probably like those taking part in all living tissues. Like the other tissues of the body, the blood to work at its best must be nearly neutral in reaction. It is normally very faintly alkaline, but when great quantities of carbon dioxide and other acid are poured into it by the tissues it may probably become nearly, if not quite, neutral. Normally, however, when tissues are thus pouring acid into the blood the circulation to that organ is so increased that sometimes the blood comes away from such a tissue less acid and less charged with carbon dioxide than when the tissue is at rest. The blood often issues from the vein of an organ doing work in an obviously arterial state. It has not been reduced.

The concentration of the hydrogen ions in the blood may be determined most accurately by means of the gas-chain method, but it may also be determined in blood plasma by means of indicators.

The gas-chain method. In this method a platinum electrode saturated with hydrogen gas is partially immersed in the blood, and another electrode also saturated with hydrogen gas is partially immersed in a weak solution of hydrochloric or some other acid of which the concentration of the hydrogen ions is known. The two electrodes are now connected through an electrometer, and the solution of acid and the blood are connected by putting in between them a saturated solution of KCl. The arrangement is shown in Figure 54. The hydrogen of the electrodes tends to go into solution as a positive ion, leaving the electrode in each case negative. There is thus set up at each electrode a difference of potential between the electrode and the blood or the solution in which the electrode is placed. The amount of the potential difference between the electrode and the solution is a function of the number of hydrogen ions in the solution. It is proportional to the logarithm of the ratio of the pressure P of the hydrogen in the electrode to the pressure of the hydrogen ions in the solution, π_H, or the electromotive force $E = K \log P/\pi_H$. If π_H is the pressure of the hydrogen ions in the blood and π_H'

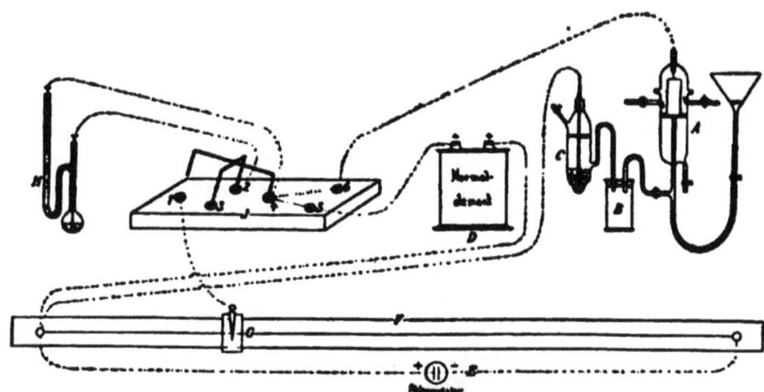

FIG. 54.—Arrangement of apparatus for the measurement of the hydrogen ion content of any fluid (Sörensen). *A* contains the fluid to be examined and the thin platinum electrode which barely touches the liquid and is surrounded with hydrogen gas; *B* contains a saturated solution of KCl; *C* is a normal calomel electrode, voltage 0.3377; *D* is the normal element which enables a determination of the potential of the accumulator *E*; *F* is a Wheatstone bridge; *H* is a capillary electrometer; *J* a key permitting short-circuiting of the capillary electrometer and the throwing into it either the current from the normal element or the gas chain. 4 is connected either with 5 or 6 at will by a stiff copper wire. The electrode *A* is shown somewhat enlarged in Fig. 55. This electrode is particularly designed for the investigation of the blood, as fresh blood is easily introduced without permitting air to mix with the hydrogen, and equilibrium between it and the CO_2 content of the hydrogen atmosphere above it is thus readily obtained.

that in the acid, then the potential difference between the blood and the electrode immersed in it would be

$$E = K \log(P/\pi_H)$$

and that between the acid and the other electrode would be

$$E' = K \log(P/\pi_H')$$

and the difference between the two electrodes, or the electromotive force of the circuit when the electrodes and the blood and acid are connected in the manner indicated, would be

$$e = E - E' = K(\log P/\pi_H - \log P/\pi_H') = K(\log \pi_H' - \log \pi_H) = K \log (\pi_H'/\pi_H)$$

In this equation the difference of potential between the boundaries of the solutions is disregarded, but this can be neglected when a strong connecting solution of KCl is used, since the two ions of this salt move with the same speed approximately and so no difference of potential is set up. Since the pressures of the hydrogen ions are proportional to the concentrations of the ions in the two solutions, we can put C_H and C_H' in place of π_H and π_H' and we have then:

$$e = K \log(C_H' / C_H)$$

The value of K is known to be .0577 volts when Briggs' logarithms are used. Hence, if we measure the electromotive force between the two electrodes and C_H' is known, that is the concentration of hydrogen ions in the acid solution of known strength, a simple calculation gives the value of C_H, the concentration of the hydrogen ions in the blood. Instead of using a second hydrogen electrode in an acid of known strength, a normal calomel mercury electrode of known electromotive force may be used instead. The current, or e, is measured by balancing a current of

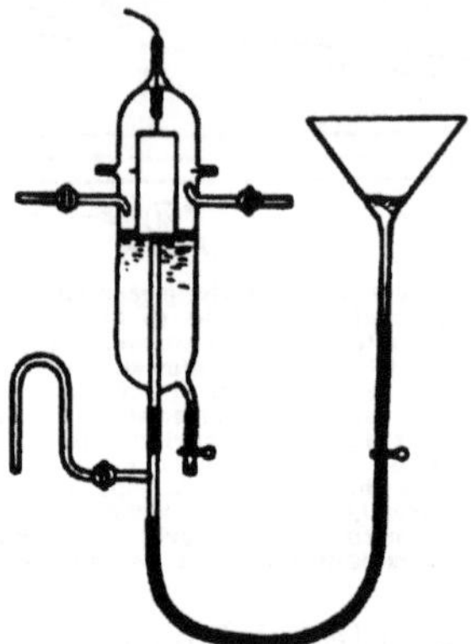

FIG. 55.—Enlarged electrode vessel for the examination of the hydrogen ion content of blood or other liquids containing CO_2. Capacity 15 c.c.; liquid 7 c.c. Metal spring clips hold the top on firmly.

known strength through the electrometer against the current between the electrodes. If the calomel electrode is used: $e = 0.3377 + .0577\, p_H$, p_H having the significance of page 541.

The indicator method of determining the hydrogen ions. The gas-chain method requires some experience and apparatus. A very useful substitute for it is the indicator method by which an approximation can be made to the actual concentration of the hydrogen ions. This method depends on the fact that various colored substances change their colors when they are made alkaline or acid, and the point at which the color changes in the different substances is at a different degree of alkalinity. If, then, it is known at what hydrogen ion concentration a color change of any substance occurs and we find that a solution just produces this color change, it must have at least this concentration of hydrogen ions. If another indicator can be found which does not change its color at this concentration of hydrogen ions, but requires a higher concentration of hydrogen ions, then, if the blood can change one indicator but not another, the hydrogen ion concentration of the blood must lie between these two concentrations. Two such indicators are litmus and phenolphthalein. Litmus changes to a blue violet when the concentration of the hydrogen ions are equivalent to a $N/10^7$ concentration. In concentrations of hydrogen ions greater than this litmus is red; in concentrations less than this it is blue violet or blue. Phenolphthalein is still colorless at this concentration of hydrogen ions. It only becomes pink when the concentration of hydrogen is less than $N/10^8$. Hence, as the blood is alkaline to litmus and acid to phenolphthalein, its hydrogen ion concentration must be between $N/10^7$ and $N/10^8$. By finding indicators which change between these two points and at other points, it is possible to find in this way the approximate hydrogen ion concentration of a liquid. The method may be still further refined by the comparison of the depth of color produced and the tint. For this purpose a set of tubes are made up of which the concentrations of hydrogen ions are known, having been determined by the gas-chain method or in other ways. These tubes may have different mixtures of Na_2HPO_4 and NaH_2PO_4. A drop of the indicator is added to these tubes, and then the indicator is added to the solution of unknown acidity and the unknown is matched against the colors of the known tubes until one is found to match. The concentration of the hydrogen ions in the unknown tube will then be equal to that in the known mixture, or it will lie between two of the standards. Of course, this method cannot be used with an already highly colored liquid like the blood, but it may be used with the urine or lymph, blood plasma or gastric juice or other body fluids little colored.

The following table shows the concentration of H ions at which the different indicators undergo their color changes. The figures under the heading p_H are the logarithms of the reciprocal of the concentration of the H ions. Thus when p_H is given as 0.11, it means that the concentration of the hydrogen ions is equal to $N/10^{.11}$ or $N \times 10^{-.11}$.

1. Hydrogen ion concentration of standard solutions of acids (Michaelis).

HCl at 18°			Acetic acid at 18°		
Concentration x normal	Concentration H ions x N	p_H	Concentration x N	Concentration H ions x N	p_H
1.0	0.78×10^{-0}	0.11	1.0	0.42×10^{-2}	2.38
0.1	0.91×10^{-1}	1.04	0.1	0.13×10^{-2}	2.89
0.01	0.96×10^{-2}	2.02	0.01	0.42×10^{-3}	3.38
0.001	0.98×10^{-3}	3.01	0.001	0.13×10^{-3}	3.89

2 Hydrogen ion concentrations of different salt mixtures for use in the indicator method.

Acetic acid and Na acetate 0°–50°				Primary and secondary Na phosphate 18°			
Acetic acid dilution Liters containing 1 gram mol)	Acetate dilution	Concentration H ions x N	P_H	NaH_2PO_4 m/3 dilution	Na_2HPO_4 m/3 dilution	Concentration H ions x N	P_H
1	32	5.76×10^{-4}	3.24	1	32	0.64×10^{-5}	5.19
1	16	2.88×10^{-4}	3 54	1	16	0.32×10^{-5}	5.49
1	8	1.44×10^{-4}	3.84	1	8	0.16×10^{-5}	5.80
1	4	0.72×10^{-4}	4.4	1	4	0.80×10^{-6}	6.10
1	2	0.36×10^{-4}	4.44	1	2	0.40×10^{-6}	6.40
1	1	1.80×10^{-5}	4.74	1	1	0.20×10^{-6}	6.70
2	1	0.90×10^{-5}	5.05	2	1	1.0×10^{-7}	7.00
4	1	0.45×10^{-5}	5.35	4	1	0.50×10^{-7}	7.30
8	1	0.22×10^{-5}	5.66	8	1	0.25×10^{-7}	7.60
16	1	0.11×10^{-5}	5.96	16	1	0.12×10^{-7}	7.92
32	1	0.56×10^{-6}	6.25	32	1	0.61×10^{-8}	8.21
64	1	0.28×10^{-6}	6.55				

Primary and secondary sodium phosphate 37°			
NaH_2PO_4 m/3 dilution	Na_2HPO_4 m/3 dilution	Concentration H ions x N	P_H
1	32	0.77×10^{-5}	5.11
1	16	0.38×10^{-5}	5.42
1	8	0.19×10^{-5}	5.72
1	4	0.96×10^{-6}	6.02
1	2	0.48×10^{-6}	6.32
1	1	0.24×10^{-6}	6.62
2	1	1.20×10^{-7}	6.92
4	1	0.60×10^{-7}	7.22
8	1	0.30×10^{-7}	7.52
16	1	0.15×10^{-7}	7.82
32	1	0.75×10^{-8}	8.12

NH_4OH 18°			NH_4OH and NH_4Cl at 18°			
Concentration Nx	Concentration H ions	P_H	NH_4OH dilution	NH_4Cl dilution	Concentration H ions	P_H
1.0	1.8×10^{-12}	11.74	32	1	1.02×10^{-8}	7.99
0.1	0.57×10^{-11}	11.24	10	1	0.51×10^{-8}	8.29
0.01	1.8×10^{-10}	10.74	8	1	0.26×10^{-8}	8.59
0.001	0.57×10^{-10}	10.24	4	1	0.11×10^{-8}	8.88
			2	1	0.64×10^{-9}	9.19
			1	1	0.32×10^{-9}	9.49
			1	2	0.16×10^{-9}	9.80
			1	4	0.80×10^{-10}	10.10

NaOH at 18°

Concentration x N	Concentration H ions	p_H
1.0	0.754×10^{-15}	14.12
0.1	0.65×10^{-13}	13.19
0.01	0.61×10^{-12}	12.21
0.001	0.59×10^{-11}	11.23

The mean value of the hydrogen ion content of defibrinated mammalian blood at room temperature, i.e., about 18-20° C., has been found to be between 6×10^{-8} and 2×10^{-8} (p_H=7.2—7.7). If the dissociation constant of water at this temperature is 0.72×10^{-14}, the OH ion concentration of the blood would be 1.2×10^{-7} to 3.6×10^{-7}. The blood is, therefore, a very weakly alkaline fluid. The concentration of the hydrogen ions depends, however, on the amount of carbon dioxide in the blood. Thus at 38.5° C. for defibrinated ox blood Hasselbalch and Lundsgaard got the following values: At this temperature the dissociation constant of water, that is the product of the hydrogen and hydroxyl ion concentration, is 2.7×10^{-14}, so that p_H of water is 6.78.

CO_2 tension	p_H (mean value)
30 mm.	7.45
40 "	7.36
50 "	7.31

An increase in the CO_2 increases the hydrogen ion content.

The suspension of blood corpuscles has a higher concentration of hydrogen ions, the serum a lower, than the whole blood when under the same pressure of CO_2. Sörensen gives the following figures:

CO_2 tension M.m Hg.	p_H serum	p_H corpuscle suspension	p_H whole blood
13.4	7.88		
19.7			7.55
29.5	7.68		
30.0			7.42
41.0		7.03	
41.7	7.63		7.31
53.5		6.96	
54.0	7.60		7.28

The meaning of this is that most of the acid neutralizing substances are in the serum. Venous blood contains at least twice as many hydrogen ions as arterial blood. Henderson showed that the concentration of the hydrogen ions is only slightly altered in passing from 18° to 38.5°. Since the value of the dissociation constant of water increases from $0.72—2.7 \times 10^{-14}$ within these same temperature limits, it is clear that the hydroxyl ion concentration of the blood must enormously increase in passing from 18° to 38°. The product of the concentration of the hydrogen and the hydroxyl ions is a constant. The concentration of the hydroxyl ions at 35° is at least twice or thrice that at 18°, and it increases 15-20 per cent. in rising from 38° to 42°. It is obvious, since

COLOR CHANGES OF INDICATORS (Salm).

Concen. H ions x N	Methyl orange	Mauve	Congo red	Sodium alizarin sulphonate	Rosolic acid	Phenol-phthalein	Naphthol benzene	Tropæolin O	Trinitro-benzene	Benzo-purpurin
2.0	Rose	Yellow	Blue	Yellow green	Yellow	Colorless	Brown yellow	Yellow	Colorless	Blue
1.0	....	Green	....		....	....		....	....	Blue violet
10^{-1}	....	Blue green	....	....	....	....	....	....	....	Violet
10^{-2}	....	Blue	....	....	....	....	....	....	....	....
10^{-3}	Orange red	Violet	Blue	....	Light brown	....	....	....	....	Red violet
10^{-4}	Orange	....	Violet	....	....	....	....	....	....	Rose
10^{-5}	Yellow	....	Scarlet	Brown	....	....	....	Greenish yellow	....	Yellow, red tint
10^{-6}	"	....	....	Red	Light brown	....	....	....	....	....
10^{-7}	"	....	....	....	Rose	....	....	....	....	....
10^{-8}	....	....	....	....	Red	Colorless	....	....	....	....
10^{-9}	....	....	....	....	....	Rose	....	....	....	....
10^{-10}	....	....	....	....	....	Red	Green	....	....	....
10^{-11}	....	....	....	Lilac	....	....	Greenish blue	Greenish yellow	....	....
10^{-12}	....	....	....	Violet	....	....	....	Orange	Colorless	....
10^{-13}	....	....	....	....	....	....	....	Orange red	Orange	Yellow red
10^{-14}	....	....	....	....	Red	Red quickly colorless	....	....	Orange red	Rose

oxidation is greatly dependent on the concentration of the hydroxyl ions, that combustion or respiration must be far more intense in fever than at normal temperatures. Blood is, moreover, but a type of a tissue. The elements controlling the alkalinity of the blood are the same as those which control the alkalinity of the cells. A rise of temperature in fever probably increases the hydroxyl ion concentration of the body cells in the same manner as that of the blood. The hydrogen ion coneen tration of the blood is not very different from that of sea-water, but as a rule sea-water is a little more alkaline, particularly in the south.

The number of H ions in the blood keeps remarkably constant. This is due to the co-operation of three factors: namely, the presence in the plasma of the salts of three weak acids—carbonic, phosphoric and protein. The first of these is present in the largest number of molecules probably and is the most important. When acid enters blood it reacts with the carbonates and phosphates to form carbonic acid and acid phosphates. The carbonic acid is removed through the lungs very quickly, and the kidneys pump out the acid phosphates, restoring the blood to its proper alkalinity. Any rise of H ions in the blood at once stimulates the respiratory center and leads to the elimination of carbon dioxide.

Osmotic pressure of the blood.—Since the relative osmotic pressure of the blood and the tissues helps determine whether liquid shall pass from the blood to the tissue, or vice versa, and since the activity of every body cell is dependent upon the amount of water in it, any change in the water content at once altering its activity, the osmotic pressure of the blood is of great importance in its functioning. It is obviously desirable that the osmotic pressure of the blood shall be kept as nearly uniform as possible, in spite of the considerable quantities of water leaving the body in the lungs, urine and through the skin, and the considerable income of water from foods and drink and from the oxidation of the hydrogen of the foods. It is one function of the kidneys to keep the osmotic pressure of the blood as constant as possible. The osmotic pressure of the plasma or the whole blood is determined by the freezing point method, which has already been described (page 201). The determination may be made with only a few c.c. of blood by the Wilson modification of this method. The freezing point of the blood of various mammals is as follows:

Mammal	Freezing point Δ	
Man	— 0.526°	(Varies .482-.605)
Ox	0.585	(" .543-.662)
Horse	0.564	
Rabbit	0.592	
Sheep	0.619	
Pig	0.615	
Dog	0.571	
Cat	0.638	

The freezing point of mammalian blood is, hence, about —0.6° C. This depression of the freezing point would mean an osmotic pressure of 0.6/1.85×22.4 atmospheres, or 7.3 atmospheres. This is about equal to a one-third molecular sugar solution. This osmotic pressure is subject to some variation even in the same individual. Thus Koeppe found in himself the freezing point to be as follows:

	Freezing point Δ		Freezing point Δ
9 A.M.	—.535°	Morning fasting 9 A.M.	—0.581°
12 M.	.558	11¼ A.M.	0.512
1½ P.M. (After dinner)	.585	1¼ P.M.	0.551
5¾ P.M.	.528	2 P.M. (After dinner)	0.617

The secretion of the gastric and intestinal juices thus increases the osmotic pressure of the blood. In a fistula dog this increase may be quite marked. (See page 346.)

The arterial blood has generally a slightly lower osmotic pressure than the venous blood. The difference is not marked (Nolf):

Carotid Δ	Jugular (Uncoagulated blood) Δ
—.574°	—.589°
.580	.587
.572	.576
.595	.597
.591	.593
.567	.565

The portal vein has a lower osmotic pressure than the hepatic vein (Fano and Botazzi):

Dog's blood

Portal vein Δ	Hepatic vein Δ
.692	.722
.617	.667
.602	.633

The osmotic pressure of the blood is due chiefly, but not exclusively, to the crystalloids it contains, to the salts, sugar, urea, etc., but the proteins also contribute somewhat to it.

In cholera, or in very hot, dry regions, the blood may have its osmotic pressure markedly increased. Its viscosity increases at the same time, hence the necessity of diluting it either with water or salt solutions.

The tissue fluids of invertebrates and some of the lower vertebrates have the freezing points shown in the accompanying table.

The great difference between the cartilaginous (Selachians) and bony fishes is seen in the table. The former have very little control over the osmotic pressure of their body fluids. Their blood is about the same freezing point as that of the sea-water. In teleosts partial control is attained, so that the osmotic pressure of their body fluids is lower than that of sea-water in the sea-fishes, and higher than that of fresh-water in the fresh-water fishes. This independence of the medium is in some fishes so complete that, like the salmon, they can pass from sea-water to fresh-water. It would seem that the covering of the gills of fishes must be of such a nature that it permits gases to pass but not water. In Selachians a considerable part of the osmotic pressure is due to the urea in the blood, which may be present to the extent of 1.5 per cent. The osmotic pressure due to the urea would be about 5.5 atmospheres. The total pressure is about 27.8 atmospheres. The osmotic pressure of human urine is about that of sea-water. $\Delta = -1.3\text{-}2.3°$.

FREEZING POINTS OF THE FLUIDS OF SOME INVERTEBRATES AND VERTEBRATES (Botazzi).

		Δ
Coelenterates:	Alcyonium palmatum	—2.196
Echinoderms:	Astropecten aurantiacus	2.312
Worms:	Sipunculus nudus	2.31
Crustacea:	Maja squinado	2.36
	Homarus vulgaris	2.29
Cephalopoda:	Octopus macropus	2.24
Selachians:	Torpedo marmorata	2.26
	Mustelus vulgaris	2.36
	Trygon violacea	2.44
Teleosts:	Charax puntazzo	1.04
	Cerna gigas	1.035
	Crenilabrus pavo	0.74-0.76
	Box salpa	0.82-0.88
Reptilia:	Thalassochelys carelta	0.61
	Fresh-water forms (Fredericq, etc.)	
Crustacea:	Astacus fluviatilis	0.80
Teleosts:	Anguilla vulgaris	0.58-0.69
	Barbus fluviatilis	0.475-0.558
	Lenciscus lobula	0.45
	Perca fluviatilis	0.512
Amphibia:	Rana esculenta	0.465
	Salamandra maculosa	0.479
Reptilia:	Emys europea	0.474

Conductivity of the blood.—The conductivity of the blood is of interest mainly because it enables a computation of the volume of the corpuscles in the blood. The conductivity is due to the salts in the plasma. The corpuscles occupy a certain amount of space in the plasma, but have almost no conductivity, so that the conductivity of the plasma is greater than that of the blood. If the conductivity of the plasma is determined on the one hand and that of the blood on the other, the volume of the corpuscles may be calculated by the formula (Stewart):

$$V_K = \frac{\lambda_b}{\lambda_s}(180 - \lambda_b - \sqrt{\lambda_b})$$

V_K is the volume of the corpuscles in the blood volume of 100; λ_b the conductivity of the blood, and λ_s the conductivity of the serum.

Enzymes of the blood.—The blood plasma contains many different enzymes. Indeed, the blood plasma may be regarded as a very dilute, liquid, not organized cell protoplasm, having in it many of the cell substances and showing many of the processes of cell metabolism. The blood plasma, like the plasma of cells, contains enzymes, and among these bodies are some proteolytic enzymes of which the importance in immunity and to the body is fundamental.

1. *Amylase.* There is always present a very small quantity of an enzyme which converts glycogen or starch to a reducing sugar. This enzyme is present in very small amounts. It is increased if dextrins, or starch or glycogen, are injected into the blood, or if these are fed in large amounts. It is increased very much if the pancreas ducts be ligatured. It is believed that this enzyme comes, at least in part, from the pancreas, but it is not impossible, since such enzymes are present in many other tissues such as the liver, the salivary glands, the white blood cells, that the enzyme is derived, in part, from other sources.

2. *Invertin.* This is also present in small quantities, but increases when large amounts of cane sugar are fed, or when cane sugar is injected directly into the blood. The amount present, however, is extremely small, the inverting power of the serum being very slight. Slight acidification greatly increases the activity of this enzyme.

3. *Glycolytic enzyme.* The blood plasma always has some power of destruction of glucose. This is ascribed to the presence of a glycolytic enzyme. What is made of the glucose, whether it is converted into isomaltose, or whether alcohol, lactic acid or other substances are formed from it, is uncertain. The amount of the glycolytic power is said by Lepine, who has principally studied this question, and by Slosse to be reduced in diabetes.

4. *Lipases* are also present. These have their origin perhaps in the pancreas.

5. *Proteolytic enzymes.* These are found normally in the plasma, together with their antibodies: that is, substances which prevent or inhibit their action. Thus there is always present in serum or plasma an anti-pepsin, anti-rennin and anti-trypsin. These digestive enzymes are thus rendered inactive. Where they come from, whether they are reabsorbed from the intestine in digestion or from the glands in which they are formed is still uncertain. Blood platelets contain or yield a good deal of proteolytic enzyme (Abderhalden and Deetjen). A very fundamental observation has recently been made by Abderhalden, an observation which may go far toward clearing up some obscure facts of immunity. He has found that the injection into the blood of strange proteins of any kind leads to the appearance in the blood within 24 hours of enzymes which will digest the albumoses, formed by acid hydrolysis from those proteins injected, and they will not split other albumoses. This fact he has applied to the detection of pregnancy. It was found that the blood serum of a pregnant woman, or other mammal,

has the property of digesting the albumose-peptone mixture made by hydrolyzing the placental tissue of that mammal with sulphuric acid. The albumose is prepared in the following way: The tissue or protein to be tested, in this case placental tissue, is ground fine in a meat-chopper and then allowed to stand 48 hours in 50 per cent. sulphuric acid at room temperature. The material is diluted, neutralized, filtered, boiled, filtered, and the albumoses precipitated by saturating with ammonium sulphate. It is freed from sulphate by dialysis. One c.c. is then mixed with 1 c.c. of blood serum, diluted to fill the tube of a polariscope and placed at 35° C. Any digestion occurring is shown by the change in rotatory power. This change is very slight and the observation must be carefully controlled, but there appears to be no doubt of its existence.

Another method consists in allowing the digestion to take place in a collodion tube. The products of the digestion are dialyzed out, the dialyzate is concentrated and the presence of amino-acids or proteins shown by the ninhydrin reaction.

By the optical method it has been possible to differentiate pregnancy from various tumors, and the method promises to be of value in diagnosis as well as of great theoretical interest. It has been found that any kind of protein, when injected, produces an enzyme in the blood which splits the albumose from that protein, but not from others.

What then is the explanation of this extraordinary power of the body to make a special enzyme which will digest the albumoses of the kind of protein which calls it into existence, but no others? How is it possible for the body to know at once how to make an enzyme which fits the particular protein injected, but no other, and to do this at the first attempt for a protein it and its ancestors have probably never met before? Possibly the enzymes are in each case formed from the proteins themselves, and hence resemble the proteins from which they came so closely that they fit them best.

6. *Cholesterases.* These split the cholesterin esters. They are present in the blood corpuscles.

Proteins of the blood plasma.—The plasma of mammalian blood is obtained by centrifugalizing blood which has been rendered non-coagulable by potassium oxalate, sodium fluoride, hirudin or other means. Such mammalian blood plasma contains normally 5-8 per cent. of coagulable proteins. These proteins are serum albumin, or seralbumin as it is called; serum globulin, or serglobulin; and fibrinogen. They are separated from each other by their varying ease of precipitation with acids or neutral salts. The relative amounts of these three substances vary under different conditions, but are approximately as follows:

Fibrinogen	0.15-0.6
Serum globulin	3.8
Serum albumin	2.5

The fibrinogen is subject to the widest variation, since whenever there is prolonged leucocytosis or suppuration anywhere in the body, the fibrinogen increases and may double or quintuple or even increase to eight times its normal amount, reaching as much as 0.9 per cent. of the whole blood or approximately 1.6 per cent. of the plasma. (Author's observations.) It is the impression of the author, although no definite study of this matter has been made, that young animals generally have more fibrinogen in their blood than old animals.

The relative proportions of serum globulin and serum albumin are reported to be different in different animals and in the same animal under different conditions, but there is no method which permits a sharp separation of the two bodies, hence all observations of their relative amounts are open to serious question.

Fibrinogen. Fibrinogen is the least soluble of the three proteins. It is almost completely precipitated by saturating the plasma with sodium chloride; or by the addition of a very small amount of acetic acid. It is also easily precipitated by water. It coagulates, also, at the lowest temperature, becoming insoluble at 56°-60° C. under the usual conditions of the plasma. It is rendered insoluble and converted into fibrin, undergoing some change as yet unknown, by various agents generally supposed to be catalytic agents, or enzymes, and found in all cells. These agents, whatever their nature, are called fibrin ferment or thrombin. It has not yet been shown that they are enzymes, and strong reasons have been given by Howell for doubting that they actually are.

Serum globulin. Although the globulins are ordinarily defined as being insoluble in water and precipitated from their salt solutions by dialysis, it is not possible to separate the globulin of the blood from the albumin in this way. Only a small fraction of the globulin is separated by dialysis of the serum. This small fraction is sometimes treated as a separate protein and called eu-globulin (*eu* meaning *well*); the name signifying that it is a typical globulin. A much larger amount of globulin is precipitated by diluting the serum several times with water and then mixing it with an equal volume of saturated solution of ammonium sulphate. This fraction, which is not precipitated by water, but is by half saturation by ammonium sulphate, is called pseudo-globulin. It is still uncertain whether they are distinct proteins, or whether the pseudo-globulin fraction continues to give small quantities of eu-globulin. The ease with which the proteins change their solubilities makes it very difficult to settle a point of this nature. The content of amino-acids in the two fractions is approximately the same. The globulin may be protected from precipitation by dialysis by the presence of some other colloid. The precipitate obtained by salting out the globulin is always strongly impregnated with a phospholipin (Hardy). To separate this it is necessary to extract it with alcohol.

Serum globulin is a simple, white, coagulable protein, coagulating in the plasma or in 3 per cent. salt solution at 75°. It is soluble in salt solutions, but is partially precipitated by a small quantity of carbonic or acetic acid. It is electro-negative for the most part, and probably exists in the plasma as the sodium salt. Its name comes from its supposititious origin from the white blood globules (A. Schmidt).

The difference in the composition of the globulin and albumin may be seen by comparing their basic amino-acids. The albumin contains far more of the basic amino-acids than the globulin.

In 100 Grams Ash-Free Protein (Lock and Thomas).

	Serum albumin (average)	Serum globulin 1	Serum globulin 2	Fibrin
Histidine	3.48	1.45	1.74	2.85
Arginine	4.67	4.51	4.07	5.52
Lysine	11.08	6.75	6.72	7.40
Total	19.23	12.71	12.53	15.77

Serum albumin. This simple protein resembles serum globulin closely, it coagulates at about the same temperature, but it is not precipitated from the serum by saturation of the latter with magnesium sulphate, or by half saturation with ammonium sulphate, but it precipitates if the filtrate from the globulin, which is half saturated with ammonium sulphate, is acidified with acetic acid. The precipitate consists of a sulphate of the albumin.

Origin and function of the plasma proteins.—The function of the blood proteins is still a matter of investigation. They were formerly supposed to be the protein food of the tissues, but this does not seem so probable since the presence of amino-acids in the blood has been shown and the radical differences between blood and tissue proteins have become clear. There never has been any evidence of the consumption of these proteins by the tissue cells. The whole matter of their nutritive value must be left to the future to determine. One function they undoubtedly have: they give to the plasma one of its most important properties, its viscosity. This is the function of the proteins in the other tissues of the body, for it is the proteins which give the structure, the jelly nature, to protoplasm; in other words, it is the proteins which determine the viscosity of the protoplasm. Similarly in the blood, which is a living tissue, the proteins contribute to the viscosity, a matter of much importance in determining the peripheral resistance. The proteins of the plasma resemble the proteins of the tissues of the body in other ways. As they occur in the blood, they are in union with phospholipins, and there is reason for thinking that the proteins of the body cells generally and of all forms of protoplasma are similarly joined to lipin. Just as the blood changes its viscosity with the greatest ease, at one time becoming more fluid and at another more solid, so do we see the same properties in the cell. The cell protoplasm clots like the blood. The blood, when shed, becomes more acid, just as the protoplasm when it dies becomes more acid. Moreover, there is a marked resemblance between the character of the proteins in the two cases. All tissues of the mammalian organism, or at least the majority of the tissues, contain a protein which resembles, but is probably not identical with, fibrinogen, a protein which coagulates at 56°. Many of them, also, contain globulins and albumins, which resemble, but are not identical with, the proteins of the plasma. Another important function of these proteins is in regulating the alkalinity of the blood. They neutralize in the ways described on page 545 the acids formed by the tissue cells, in part by yielding up to the acids the alkali metals they carry, and in part by uniting directly with the molecules of acid. Both these functions are in common with the proteins of the cell protoplasm. In fact we shall probably not go far wrong if we consider the blood plasma as a very dilute protoplasm. The processes which occur in it are probably the mirror of the processes which occur in all forms of living matter.

The origin of the plasma proteins is still uncertain. Schmidt, who worked for many years on this problem, believed that they came from the white blood corpuscles on the disintegration of the latter. Similar but not identical proteins are found in a wide variety of cells. Probably more work has been done upon the origin of fibrinogen than upon any other member of the group. This substance, because of its peculiar property of clotting and the ease with which it can be removed from the blood, is most easy to study, but the conclusions arrived at by various observers are nearly as diverse as the number of investigators. Some have concluded that the fibrinogen is formed in the liver, others from the white blood corpuscles, others from the bone marrow.

The problem has been attacked in the following way: If a dog or cat be anesthetized, a cannula put in the femoral artery and another in the femoral vein, between one-third and one-half of the blood may be withdrawn from the artery without killing the animal, if the blood is quickly reinjected. The whole blood is estimated at about one-thirteenth the body weight. This blood is kept warm and defibrinated by whipping, and after filtering through a cloth is reinjected slowly into the body. The injection must be slow or intravascular clotting may occur. If it does occur the clots generally appear in the heart, being attached to the valves and cordæ. After allowing 5 minutes for the reinjected blood to mix well with the blood in the body and the circulation to recover itself, the same amount of blood is again withdrawn and whipped and reinjected.

By repeating this process 5-6 times the whole time of defibrination taking from 1½-2½ hours, about nine-tenths of the fibrin of the blood is removed and so little remains, i.e., .002-.004 per cent., that the blood either will not clot at all, or at the most gives but a very weak jelly. The amount of fibrin thus recovered is generally a little less than, about nine-tenths of, the amount which was calculated to be present by a quantitative determination in a sample of blood taken at the outset of the defibrination. This fact shows that neither any pronounced destruction nor reformation has taken place during the process of defibrination, otherwise, unless both processes occurred equally, the amount recovered would be less or more than that found. The defibrinated animal reforms the fibrin and in the course of 24-48 hours the normal fibrinogen content of the blood is restored. This reformation of fibrinogen takes place equally well whether the animal is fed or is fasting, and it is clear that it does not come from the proteins of the food. Whence then does it come? Experiments have shown that the reformation takes place with normal, or even increased, speed after extirpation of the kidneys, spleen, pancreas, or after ligating off the brain and most of the nervous system. It also occurs normally after making an Eck fistula which cuts off all the portal blood flow through the liver, but leaves the hepatic artery open. Neither spleen, pancreas, kidneys, reproductive organs, lymph glands of the mesentery, nor nervous system is then necessary for the reformation. Perfusion experiments by which defibrinated blood is perfused through the legs, intestine and kidneys have shown no formation of fibrinogen in these organs. Perfusion through the liver is also negative in the experiments of some, and positive in others. The fibrinogen is reformed normally if the liver and intestine are intact, but the reformation is greatly reduced if the intestine is absent. An examination of the fibrinogen content of the blood in the arteries and veins in different parts of the body shows that in all regions of the body but the intestine the venous blood has less fibrin than the arterial. The portal or mesenteric veins alone show generally, but not always, a larger amount of fibrinogen than the arterial blood. From these observations it would appear that fibrinogen is taken out of the blood in its passage through most of the organs of the body and in particular in the kidneys, and that it is added to the blood during its passage through the vessels of the intestine or the liver, since some observations indicate that the amount of fibrinogen in the hepatic blood is greater than in the portal. While none of the methods of the quantitative determination of fibrinogen are quite satisfactory the results are so uniform as to indicate that the differences obtained between the arterial and venous blood really represent actual differences. The fibrinogen appears, then, to be added to the blood in the portal area and chiefly in the intestine. Whether the liver is active or not in the process is undecided. It was not supposed that the intestine formed the fibrinogen from its own peculiar tissues, except, perhaps, the lymph tissues, but that the fibrinogen came probably from the white blood cells and that the place of their destruction was probably the intestinal area.

The only method known to increase the amount of fibrinogen in the blood is by suppuration. Any prolonged suppuration anywhere in the body, and however produced, is accompanied by a great increase in the fibrinogen content of the blood. This increase may be enormous, more than eight times the normal amount being present. The reason for this great increase in fibrinogen has never been explained. It is possible that it is part of the defensive mechanism of the body against infections and fever, the viscosity of the blood being increased thereby. The blood in all such cases nearly always shows that the red corpuscles agglomerate and sink much more rapidly than normal so that the blood may have a coat of serum or plasma above the clot free from corpuscles, the so-called buffy coat, or crusta inflammatoria, which had been observed by the physicians when bleeding was common. This process of agglutination of the corpuscles must also increase the viscosity of the blood and would tend to reduce the speed of blood flow and perhaps make conditions favorable for the passage of the white blood cells out into the tissues where the infection is. Inasmuch as any great or prolonged leucocytosis, as in infections, is accompanied by an increase in fibrinogen and it is also accompanied by an increase in the decomposition of the leucocytes, the author drew the conclusion that probably the fibrinogen, and the other blood proteins as well, originated in the white blood cells.

Opposition to this view has, however, been fairly widespread. Both Nolf and Doyon believe that the liver forms the fibrinogen. The reason for this conclusion is that the liver, like many other organs, contains a protein which may be extracted and which coagulates at the same temperature as fibrinogen. It has never been shown, however, that this protein is fibrinogen and that it has the property of clotting on adding fibrin ferment. Another fact supporting the view that the liver forms fibrinogen is that when one poisons the liver with phosphorus or chloroform, so that there is extensive degeneration of this organ, then the amount of fibrinogen in the blood is reduced. These latter observations, however, do not show that the fibrinogen is formed in the liver. It is not to be supposed that the liver is the only organ affected in phosphorus or chloroform poisoning. The disappearance of the fibrinogen might be due to the fact that the consumption rose above the power of production, or that directly or indirectly other tissues which produced the fibrinogen were affected. It may be that the liver forms an enzyme, which may digest the fibrinogen and that on its destruction this enzyme gets loose in the blood.

There is still another possibility of the origin of the fibrinogen. It may be formed in the bone marrow. It is said, indeed, that the amount of activity of this tissue is greater after defibrination. And that is not improbable. It is said also that fibrinogen can be isolated from the bone marrow. Since this is the blood-forming tissue in the adult body it is most probable that the proteins are derived from the bone marrow or from the blood cells. The blood is a living tissue. Every living tissue has proteins peculiar to that tissue. Were the proteins the same, the chemical processes in different tissues would be the same. But we know that they are different. Every tissue forms its own peculiar proteins. It is probable, therefore, that the blood is formed by itself. It makes its own proteins. The place where it is made is either in the bone marrow, which is the blood-forming organ in the body, or by the disintegration, that is the further differentiation, of the blood corpuscles, and in particular the white cells. It is certain that the white cells, and possibly the reds as well, are constantly giving off substance to the blood. They disintegrate. Their surfaces, as Kite has shown, are sending out constantly great streamers of protoplasm into the plasma. It is probable, then, that they are constantly contributing to the protein constitution of this tissue and that they keep its constitution fairly constant. It is possible that the endothelial cells of the blood vessels also play a part in this process. This possibility should be carefully investigated, but the fact that the reformation of fibrinogen has never been described in the great number of perfusion experiments of living organs with defibrinated blood indicates that more than one tissue is concerned in this process. It may be that defibrinated blood does not reform its proteins on perfusion, because the raw materials, the leucocytes, are lacking. A circulation through the bone marrow and then through the intestinal area would possibly be more successful. The origin of fibrinogen and other proteins is then still obscure and must be left for future investigations, but the evidence favors its origin from the blood cells, the white blood corpuscles, and possibly the blood-forming organs, the bone marrow.

Experiment has shown that when the fibrinogen is reformed after defibrination, there is no increase in the per cent. of the other proteins of the blood. That is, the reformation of the fibrinogen does not involve a simultaneous increase in the globulin and albumin, as would be expected if all three were formed simultaneously by the decomposition of some cells such as the leucocytes. However, this fact is not conclusive evidence against the common origin of all three of the proteins, since some regulatory mechanism might remove these proteins as rapidly as they were formed. There is often a slight decrease in the other blood proteins coinciding with the reformation of the fibrinogen, from which it might be inferred that the fibrinogen was derived from the globulin, as Schmidt thought. But against this conclusion is the fact that the decrease takes place during the defibrination and not during the reformation. It looks rather as if some of the serglobulins contributed to the formation of the fibrin and this is not impossible. Schmidt found that the weight of the fibrin recovered from serum was increased by the addition of paraglobulin to the serum. The relation of the fibrinogen to the other proteins is still, therefore, a matter to be investigated.

The source of the other blood proteins is still more obscure than that of fibrinogen. The same method may be used in their study. The defibrinated blood may be centrifugalized, the corpuscles suspended in Ringer's solution and reinjected. It is necessary to add to the Ringer's solution some gum arabic to make the viscosity right. After several drawings and reinjections the blood is nearly freed from the proteins. The course of their reappearance can then be watched. The method is tedious and time-consuming. It is found that the paraglobulin and albumin will be reformed in about the same time as is required for the reformation of the fibrinogen. Nothing is known of the origin of these proteins.

Function of the endothelium of the blood vessels.—It is the purpose of this book to raise questions and if possible to raise more, by far, than it answers. The question must occur to every student of physiology what is the function of the endothelium of the blood vessels? These are generally thought of as passive tubes for the transportation of the blood, but we must now consider them as living things. They have a good nerve supply. What is the function of these nerve fibers distributed to the capillaries of the body? The vascular and lymphatic endothelium is a great organ, a living tissue, penetrating all the organs and parts of the body. It is probably not passive since it certainly plays a part in the coagulation of the blood. It must be constantly interacting with the blood, changing its composition, possibly affecting its viscosity, controlling the secretion of lymph, and possibly contributing hormones of importance to the body. What substances does it require for its nutrition? Does it, like the corpuscles of the blood, send out into the stream fine processes? Does it take fibrinogen out of the blood and put it in? If it controls the viscosity, as its relation to the clotting implies, how does it do it? Is it by shedding fibrin ferment? By secreting enzymes specifically suited to the blood proteins? Is it in these cells that the strange proteins injected into the blood are imprisoned and do they form the specific enzymes which appear in the blood when strange proteins are injected? Do they form the precipitins and immune bodies? Is their activity controlled by the nervous system? Are they actively phagocytic? Do they change their adhesiveness as the corpuscles do, so that bacteria and other cells will stick to them? Here is an organ coextensive with the body of which we know very little. Who knows but that some great gaps in our knowledge of the most fundamental questions may not be solved by its study. In the text-books of the future it may be that more than a chapter must be given to this tissue of whose chemistry, metabolism and function we are so profoundly ignorant.

REFERENCES. Coagulation of Blood.

1. *Wooldridge:* Chemistry of blood. (Collected Papers.) Report to the Scientific Committee of the Grocers' Assn. I, pp. 201-249; II, p. 266 et seq. Croonian Lecture, p. 141. Reply to Halliburton, Jour. of Physiol., 10, 1889, p. 329; Protoplasma and blood plasma; Arris and Cale Lecture, Chemistry of Blood, pp. 174 et seq. Various papers in the Proceedings of the Royal Society, 36, 1884, p. 417, and succeeding years to 1889.
2. *Nolf:* Archives internat. de Physiol., 9, 1910, pp. 407-459; 204-261. References here to earlier papers on coagulation by the same author.
3. *Archard and Ayrand:* Action des anticoagulants sur les globules. C. R. Soc. de Biol., 1908, 1, pp. 898-900; 2, pp. 724-725.
4. *Zak:* Relation of plasma lipoids to coagulation. Archiv f. expt. Path. u. Pharm., 70, p. 274, 1912.
5. *Gengou:* Adhesion moléculaire et phénomènes biologiques. Arch. internat. de Physiol., 7, pp. 1-87; 115-210, 1908-9.
6. *Fuld and Spiro:* Influence of coagulation inhibitors on bird plasma. Beiträge z. chem. Physiol. u. Path., 5, p. 171, 1903-4.
7. *Morawitz:* Beiträge zur Kenntnis der Blutgerinnung. Beiträge z. chem. Physiol. u. Path., 5, p. 133, 1903-4; 4, p. 381, 1903. See also Morawitz: in Oppenheimer's Handbuch der Biochemie.
8. *Schittenhelm u. Bodong:* Coagulation and especially the action of hirudin. Archiv f. allg. Path. u. Pharm., 54, pp. 217-244, 1906.
9. *Schmidt:* Ueber die Beziehung der Faserstoffgerinnung zu den farblosen Elementen des Blutes. Arch. ges. Physiol., 11, p. 515, 187.
10. *Popielski:* Importance of the loss of coagulability of the blood for the activity of the digestive glands. Archiv f. d. ges. Physiol., 144, pp. 135-151. Bull. internat. de l'acad. de Sci. de Cracovie, 1912, pp. 755-681; 1137-1196. Maly's Jahresber., 42, 178, 1912.
11. *Collingwood and MacMahon:* Anticoagulant substance in blood serum. Jour. of Physiol., 45, p. 119, 1911-12.
12. *Haycraft:* Action of a secretion from the medicinal leech on coagulation. Pro. Royal Soc., 36, p. 478, 1884.
13. *Wooldridge:* Relation of blood coagulation to the vascular wall. "On auto-infection in cardiac disease." Pro. Royal Soc. 1889, 45, p. 309. Du Bois' Archiv f. Physiol., 1888, p. 174.
14. *Bruce:* On the disappearance of leucocytes from blood after peptone injection. Pro. Roy. Soc., 55, p. 295, 1894.
15. *Wright:* Pro. Roy. Soc., 52, 1892-3, p. 364.
16. *Franz:* The nature of hirudin. Arch. f. expt. Pharm. u. Path., 49, pp. 342-366, 1903.
17. *Green:* On certain points connected with the coagulation of the blood. Jour. Physiol., 8, 354, 1887.

18. *Ringer and Sainsbury:* The influence of certain salts upon the act of clotting. Jour. Physiol., 11, p. 369, 1890.
19. *Arthus et Pages:* Nouvelle théorie chimique de la coagulation du sang. Archives de Physiol., 1890.
20. *Wright:* A study of intravascular clotting. Pro. Royal Irish Acad. 3d series, 2, p. 117, 1891-3.
21. *S. B. Jones:* Presence of prothrombin in blood platelets. Amer. Jour. Physiol., 30, p. 74, 1912.
22. *Davis:* Intravenous injection of thrombin. Amer. Jour. Physiol., 29, p. 160, 1911.
23. *Howell:* Rôle of antithrombin and thromboplastin in the coagulation of blood. Amer. Jour. Physiol., 29, p. 187, 1911.
24. *Howell:* Thrombin. Ibid., 1910, 25, p. 453. Nature and action of thromboplastic (zymoplastic) substances of the tissues. Amer. Jour. Physiol., 31, 1, 1912.
25. *Howell:* Prothrombin. Ibid., 35, 1914, p. 474.
26. *Retger:* Mechanical factors in clotting. Ibid., 24, p. 429, 1909.
27. *Howell:* Optical phenomena of coagulation. Ibid.
28. *Modrakowski:* Action of vasodilatin on coagulation. Archiv f. d. ges. Physiol., 133, p. 301, 1910.
29. *Burker:* Apparatus for determining time of coagulation. Arch. f. d. ges. Physiol., 118, p. 452, 1907.
30. *Nias:* Relation of salts given by the mouth to the coagulation of blood. Lancet, 174, p. 96, I, 1908.
31. *Fox:* Coagulability of blood in pregnancy and puerperal state. Lancet, 1908, p. 99, 1908.
32. *Lesourd et Pagniez:* Jour. de Physiol. et Pathol. gen., 1909.
33. *Wright:* Influence of CO_2 and O_2 on coagulation of blood in vivo. Pro. Royal Soc., 55, p. 279, 1894.
34. *Tait:* Types of Crustacean Blood Coagulation. Journal of Mar. Biol. Assn., Plymouth, 9, pp. 191-198, 1911.
35. *Hardy:* Journal of Physiol., 13, 1892, p. 165.
36. *Fredericq:* Arch. de Zool. Expt. 2d, 1, p. 413, 1883.
37. *Bordet et Gengou:* Coagulation of blood and the genesis of thrombin. Annales de l'Institut Pasteur, 26, p. 657, 1912.

Blood Platelets.

1. *Wooldridge:* Chemistry of Blood. Report to the Scientific Committee of the Grocers' Association.
2. *Schwalbe:* Thrombose u. Gerinnung. Ergebnisse der allg. Path. u. Path. Anat., 11, 2, 1907, pp. 901-927. (Literature.)
3. *Schwalbe:* Bau u. Entstehung der Blutplättchen. Ibid., 8, 1902, p. 192.
4. *Bürker:* Blood platelets and coagulation. Arch. f. d. ges. Physiol., 102, pp. 36-94, 1904.
5. *Pascucci:* Composition of platelets. Beiträge z. chem. Physiol. et Pathol., 6, pp. 540-552, 1905.
6. *Hayem:* Récherches sur l'évolution des hématies dans le sang de l'homme et des vertébrés. Archives de Physiologie, 1878-9.
7. *Meves:* Thrombocytes of Salamander blood and their relation to coagulation. Archiv. f. Mik. Anat., 68, pp. 311-358, 1906.

8. *Kreidl u. Neumann:* Ueber das Vorkommen von Ultramikroskopischen Teilchen im fötalen Blute. Zent. f. Physiol., 24, 54. 1910.

PROTEINS OF BLOOD PLASMA.

1 *Wooldridge:* Chemistry of Blood. (Collected papers.)
2. *Iscovesco:* Etude sur les constituents colloides du sang. Le fibrinogène. C. R. Soc. Biol., 60, 1906, pp. 923-5.
3. *Hardy and Mrs. Gardiner:* Jour. of Physiol., 40, 1910, p. 68. Pro. Physiol. Soc.
4. *Freund and Joachim:* Nucleoproteid of blood serum. Zeits. f. physiol. Chem., 36, 407.
5. *Lock and Thomas:* Content of blood proteins in basic amino-acids. Zeits. f. physiol. Chem., 86, 1913, p. 74.
6. *Huiskamp:* Fibrinoglobulin question. Zeits. f. Physiol. Chem., 44, pp. 182-197, 1905.
7. *Schaefer:* Proteids of chyle and the transference of food materials. Proc. Roy. Soc., 38, p. 89, 1885.

SUGARS IN BLOOD.

1. *Höber:* Distribution of sugar between corpuscles and plasma. Biochem. Zeits., 45, p. 207, 1912.
2. *Spallita:* Sur la nature du sucre du sang. Arch. Ital. de Biol., 53, 1910, p. 356.
3. *Scott:* The content of sugar in the blood under common laboratory conditions. Amer. Jour. Physiol., 34, p. 271, 1914. Literature cited here.
4. *Von Hess and McGuigan:* Sugar content determined by vividiffusion. Jour. Pharm. and Expt. Ther., 6, 1914.

FAT IN BLOOD.

1. *Hermann and Neumann:* Ueber den Lipoidgehalt des Blutes normaler u. schwangere Frauen sowie neugeborene Kinder. Biochem. Zeits., 43, p. 47, 1912.

VIVIDIFFUSION METHOD OF STUDYING BLOOD.

1. *Abel:* On the removal of diffusible substances from the circulating blood of living animals by dialysis. II. Jour. Pharm. and Expt. Ther., 4, 1914, p. 611.
2. *Von Hess and McGuigan:* See above, Sugars in blood, 4.

FERMENTS IN BLOOD.

1. *Moeckel and Post:* Diastase in different animals. Review of literature. Zeits. f. physiol. Chem.. 67, 1910, p. 433.
2. *Gould and Carlson:* Relation of pancreas to serum and lymph diastases. Amer. Jour. Physiol., 29, p. 174, 1911.
3. *Cytronberg:* Cholesterase of blood corpuscles. Biochem. Zeits., 45, p. 281, 1912.
4. *Haberlandt:* Existence of a leucocyte diastatic ferment. Archiv f. d. ges. Physiol., 132, p. 175, 1910.
5. *Bial:* Diastatic ferment of lymph and serum. Archiv f. d. ges. Physiol., 52, p. 137, 1892.
6. *Tschereskoff:* Récherches sur le ferment amylolytique du sang. Archives d. Physiol. Path. et Norm., 1895, pp. 629-635.
7. *Bainbridge and Beddard:* The diastatic ferment in the tissues of diabetes mellitus. Biochemical Journal, 2, p. 89.

8. *Carlson and Luckhardt:* Amer. Jour. Physiol., 23, p. 148, 1908.
9. *Arthus:* Sur la ferment glycolytique. C. R. Soc. Biol., 43. p. 60, 1891.
10. *Lépine:* Le diabète et les lésions du Pancréas. Revue de Médec., 12, 1892, I, p. 402; II, p. 481.
11. *Lépine et Bonal:* C. R. Soc. Biol., 44, p. 271, 1891.
12. *Abderhalden u. Kopfberger:* Invertin in blood serum. Zeit. f. physiol Chem., 69, p. 23, 1910.
13. *Abderhalden u. Rathsmann:* Serological studies with the aid of optical methods. Zeits. f. physiol. Chem., 71, p. 367, 1911.
14. *Abderhalden u. Schilling:* Ibid., 71, p. 385, 1911.
15. *Abderhalden u. Kampf:* Ibid., 71, p. 421, 1911.
16. *Abderhalden u. Fodor:* Studien über die Spezifizitat der Zellfermenten mittels der optischen Methode. Ibid., 87 and 88, 1913.
17. *Abderhalden u. Schiff:* The speed of appearance of protective ferments after repeated action of substrates foreign to blood plasma. Ibid., 87, p. 225, 1910.
18. *Abderhalden:* Abwehrfermente, 1914 *Van Slyke et al.:* J. Biol. Chem., 23, p. 377, 1915.

Leucocytes and Red Cells.

1. *Patella:* Origin in endothelium of blood vessels. Arch. Ital. de Biol., 54, p. 213, 1911.
2. *Morawitz and Pratt:* Münchener med. Wochen., 55, p. 1817, 1908.
3. *Morawitz:* Arch. f. expt. Path. u. Pharm., 60, p. 298, 1909.
4. *Lohner:* Erythrocyte membrane question. Arch. f. Mik. Anat., 71, p. 129, 1908.
5. *Deetjen:* Arch. f. Path. Anat. u. Physiol., 165, p. 280.
6. *Meltzer:* Effects of shaking on red blood corpuscles. J. H. U. Hospital Reports, 9, p. 135.
7. *Oliver:* Science, N. S., 40, p. 645, 1914.
8. *Kite:* Journal Infectious Diseases, 15, p. 319, 1914.
9. *Jaeger:* Destruction of red blood corpuscles. Arch. f. d. ges. Physiol., 94, p. 65, 1903.

Inorganic Constituents.

1. *Bertrand et Medigreceanu:* Sur le manganese du sang. C. R. Acad. Sci., Paris, 154, p. 941, 1911.
2. *Henze:* Vanadium compounds of blood cells. Zeits. f. physiol. Chem., 72, p. 494, 1911.

Alkalinity of Blood.

1. *Friedenthal:* Reaction of blood serum. Criticism of vital stains as tests for reaction. Zeits. f. allg. Physiol., 4, p. 60-61, 1904.
2. *Lundsgaard:* Reaction of blood. Biochem. Zeits., 41, p. 247, 1911.
3. *Peters:* Combined tonometer and electrode cell for measuring the H ion concentration of reduced blood at a given tension of CO_2. Jour. Physiol., 48, Pro. Physiol. Soc. VII, 1914.
4. *Hasselbalch:* Biochem. Zeits., 49, pp. 449, 457, 1913; 30, p. 317, 1910; 46, p. 403, 1912.
5. *Kormkoff:* Ibid., 51, p. 202, 1913.
6. *Sörensen:* Ergeb. d. Physiol., 12, 1912, p. 527. Literature. The measurement and significance of the H ion concentration for biol. processes.
7. *Höber:* Physical chemistry of blood. Oppenheimer's Handbuch der Biochemie, II, 2, 1914. Literature.

8. *Michaelis:* Die Wasserstoffionenkonzentration. Ihre Bedeutung für die Biologie und die Methoden ihrer Messung. Berlin, 1914.

CONDUCTIVITY OF BLOOD.

1. *Bugarsky and Tangl:* Archiv f. d. ges. Physiol., 72, 1898, p. 531.

OSMOTIC PRESSURE.

1. *Botazzi:* La Pression osmotique du sang des Animaux marins. Archiv. Ital. de Biol., 28, p. 61, 1897.
2. *Höber:* Arch. f. d. ges. Physiol., 70, 1898, p. 630.

RESPIRATORY FUNCTION OF BLOOD.

1. *Douglas and Haldane:* Jour. of Physiol., 44, p. 338, 1912.
2. *Peters:* Chemical nature of specific oxygen capacity of blood. Jour. Physiol., 44, p. 131, 1912.
3. *Buckmaster and Gardner:* Gases of arterial and venous blood of cat. Jour. Physiol., 41, p. 60, 1910.
4. *Barcroft and Hill:* Jour. of Physiology. 1912.
5. *Krogh:* The diffusion of gases through the lungs of man. Jour. Phys., 49, p. 271, 1915.
6. *Barcroft:* Effect of altitude on dissociation curve of blood. Jour. Physiol., 42, p. 44, 1911.
7. *Vernon:* Respiration of tortoise heart in relation to functional activity. Jour. Physiol., 40. p. 295. 1910.
8. *Vernon:* Conditions of tissue respiration. Action of poisons. Jour. Physiol., 39, p. 149, 1909.
9. *Hill:* Effects of aggregation of Hb on its dissociation curve. Jour. Physiol., 40, p. IV, 1911.
10. *Douglas:* Determination of total oxygen capacity of the blood at different altitudes by the CO method. Jour. Physiol., 40, p. 472, 1910.
11. *Hüfner:* Archiv f. Physiol., 1894, p. 130; 1904, Supp. 387.
12. *French, Pembrey and Riffle:* Compensatory increase in Hb and red corpuscles of blood in congenital heart disease. Jour. Physiol., 39, 1909. Pro. Physiol. Soc., p. ix.
13. *Firket:* Tension des Gaz du Sang arteriel. Arch. internat. de Physiol., 9, 1910, p. 291.
14. *Hüfner:* Composition of hemoglobin. Beiträge zur Physiol. Ludwig's Festgabe 70 Geburtstage, 1887.
15. *Willstätter:* Chlorophyll. This article contains much also on the composition of hemoglobin and references to literature. Jour. Amer. Chem. Soc., 37, p. 323, 1915.

VOLUME.

1. *Dreyer and Ray:* Blood volume of mammals as determined by experiments on rabbits, guinea pigs and mice and its relationship to the body weight and to the surface area expressed in a formula. Philosophical Transactions B, vol. 201, 1911, p. 133.

VISCOSITY OF BLOOD.

1. *Chick:* Viscosity of Blood protein solutions. Biochem. Journal, 8, p. 261, 1914.
2. *Burton-Opitz:* Viscosity of blood. Amer. Jour. Physiol., 35, p. 51, 1914. Arch. ges. Physiol., 82, p. 447, 1900. Jour. Physiol., 32, p. 8, 1904.

3. *Dupré-Demming and Watson:*
4. *DuBois-Reymond, Brodie and Müller:* Influence of viscosity on the circulation. Archiv f. Physiol. Supp., 1907, p. 38.
5. *Albutt:* Viscosity of the blood. A review. Quarterly Journal of Medicine, 4, p. 342, 1910-1911.
6. *White:* A new viscosimeter and its application to the blood and blood serum. Biochem. Zeits., 37, p. 489, 1911.

Hemolysis.

1. *Sawtschenko:* Action inhibitrice de l'acide carbonique sur l'hemolyse et la bacteriolyse. Annales de l'Instit. Pasteur, 26, pp. 1032-1035, 1912.
2. *Meyer:* Mechanism of saponin hemolysis. Hofmeister's Beiträge z. chem. Physiol., II, p. 357, 1907-8. Previous literature cited.

Amino-Acids.

1. *Abderhalden:* Der Nachweis von freien Aminosauren im Blute, etc. Zeits. f. physiol. Chem., 88, p. 480-1, 1913.
2. *Van Slyke and Meyer:* The amino-acid nitrogen of the blood. Preliminary experiments on protein assimilation, Jour. Biol. Chem., 12, p. 399, 1912.
3. *György and Zunz:* A contribution to the amino-acid content of blood. Jour. Biol Chem, 21, p. 511, 1915.
4. *Bock and Benedict:* Jour. Biol. Chem., 20, p. 47, 1915; *Greenwald:* Ibid., 21, p. 61, 1915.

CHAPTER XIII.

THE MASTER TISSUE OF THE BODY.

THE BRAIN AND NERVOUS SYSTEM.

CHEMISTRY AND METABOLISM OF THE MASTER TISSUE OF THE BODY.

The brain and nervous system control, either directly by nerve impulses or indirectly through the blood stream, the metabolism and activity of all the other tissues of the body. They are, hence, the master tissue of the body. While the greater part of the metabolism of the body is muscular metabolism, the muscles by their bulk dominating the character of the metabolism of the body as a whole, the superior reactivity or irritability of the nervous tissue enables it to control or to set the pace for all the other tissues.

The chemistry and metabolism of the nervous tissue is from almost every point of view the most absorbing and interesting of the problems of physiological chemistry. In the evolution of the vertebrates it seems to have been this tissue more than any other upon which the attacks of natural selection have been directed. Between the marsupials and the placental mammals the chief difference is not so much a change of form or structure of the body as it has been a change in capacity of the skull; a change in brain power. And the marsupials were, undoubtedly, in this respect, superior to the monotremes. It is brains which have won survival in the struggle for existence. Evolution, since the appearance of the vertebrates, is especially characterized by the steady development of the nervous system, its increase in amount and complexity. Concomitant with this perfecting has come the development of memory, self-consciousness and reason. It is, indeed, astonishing how extraordinarily small was the amount of nervous tissue in the gigantic Dinosaurs, reptiles which once must have dominated creation. But they were undoubtedly supplanted by their more clever though smaller relatives. The vertebrates differ from the invertebrates, also, as profoundly in the amount of nervous tissue they contain as in any other way. The circumesophageal ring of nervous matter of the invertebrate is replaced by the great ganglia of the vertebrate brain. In fact the whole of evolution is characterized by the steady development of the nervous system and by the steady development of no other tissue. The power of adapting the organism to a changing environment has been the problem nature has had to solve. It solved it by the development of a tissue of the body which should be most irritable, which should control the other tissues and which, having memory, could profit by experience. Adaptability of organisms was the end sought and this was obtained by the selection of the function of irritability. The nervous tissue and it alone shows, hence, a fairly steady progress from the lowest to the highest animals. It is by means of his nervous system and in that respect only that man stands at the summit of the animal world.

Not only has the nervous system been the point of attack of natural selection and so has played a predominant part in evolution, but it is also of the highest importance in embryological development. Since it is to control, directly or indirectly, the metabolism of all the other tissues of the body, it is almost or quite the first of the tissues to be set apart in embryogenic development. Its development appears, according to Child, to set the pace for the development of the other tissues which follow it. It is probable that, like the growing bud of the plant, the metabolism in the embryonic nervous system is higher than in any other tissue of the body. In the vertebrate it is the rudiment of the nervous system, the nervous folds of the cord and brain and the eye vesicles which appear earliest in development.

The function of the fore-brain and particularly of the cerebral hemispheres is memory and reason. This organ is, as it were, an epitome of the whole body, for it is brought into relation with every part by means of nerve fibers. No doubt this centralization of the body in the brain plays a part in the development and perfecting of self-consciousness and memory. Reasoning and all our psychic life are dependent, in some unknown way, upon this master tissue. For these reasons the study of the chemical composition and chemical transformations of the brain possesses a fascination above that of the metabolism of any other part of the body whatever.

Structure.—The nervous system is partly condensed into great ganglia constituting the brain and spinal cord, which are found in the cavities of the skull and vertebræ; and in part it is distributed through the body in the form of ganglia or nerve fibers or isolated nerve cells. Among these groups of ganglia, which are outside of the spinal canal, the ganglia of the sympathetic system found ventral to the vertebræ and in the abdomen are the most important. We shall treat, in this chapter, almost exclusively, the composition of the brain and cord, since these have been most studied.

The brain is composed chiefly of nerve cells with their processes, the dendrites and nerve fibers. There is also present, however, a small amount of connective tissue supporting the blood vessels; and a kind of supporting tissue composed of peculiar branching cells called neuroglia cells. These neuroglia cells appear to support the nerve cells. Nothing is more striking or significant in considering the metabolism of this tissue than the arrangements which insure a very large blood supply. The brain is supplied with blood from the external carotids, the internal carotids and the vertebral arteries; these pour their blood into a common series of great vessels about the base of the brain called the circle of Willis and from this the blood vessels are given off penetrating every portion of the brain substance. It thus happens that even if both external carotids are compressed, or otherwise rendered incapable of carrying blood, the other arteries are able to supply the needs of the brain. This remarkably copious blood supply, when taken in connection with the sudden change in activity of the brain when that supply is deficient, indicates, very clearly, that the brain must have a very intense metabolism.

The brain cells have usually two kinds of processes, the axon, so called, and the dendrites; the former generally constitutes the nerve fiber and in the vertebrates is generally surrounded by a special sheath, the medullary sheath. By means of these fibers the cells are brought into connection with other nerve cells at a distance, or with muscular and other tissues. One of these processes, the principal one, as has been said, is often surrounded by a peculiar cylindrical sheath of a glistening, white, fatty matter of a peculiar chemical nature. Nothing is definitely known of the function of this sheath, but it gives to nerves their glistening white appearance; and in the brain the nervous matter may be separated into white and gray matter, the difference between these depending on the relative amounts of medullated nerve fibers and cell bodies. The parts of the nervous tissue which consist chiefly of cell bodies and dendrites are gray; the fibers are white. The corpus callosum, the broad thick band of medullated fibers connecting the two cerebral hemispheres, is purely white matter; the great ganglia of the corpora quadrigemina and striata, and the cortex of the cerebrum are largely gray matter. These different parts of the brain, the white and gray matter, have different functions and different chemical compositions. The gray matter is, on the whole, more automatic, in that nerve impulses originate in it; the white matter is more purely a conducting tissue.

Chemistry.—The chemical composition of the brain can be best studied in the human brain, for not only is man's brain the largest, but the peculiar psychical processes correlated with, or dependent upon, the nervous system have in him reached their highest development. His brain is the most highly differentiated and appears to be the most perfect. We should expect to find in it, in the largest amounts, and in the purest unmixed state, the peculiar substances upon which the psychic processes depend. It is, moreover, relatively easy to obtain human brain material for analysis; the large number of accidents by which men in the prime of life are killed making available brain material, unaltered by disease.

The most striking point of difference between the chemistry of the brain and that of other tissues is the very large quantity of alcohol-ether soluble substances it contains; that is the very large proportion of lipins in it. No other tissue has anything like such a proportion, with the exception of fat tissue itself. The lipins of the brain, however, are almost entirely free from neutral fat. There is practically no neutral fat in the human brain; the lipins found there contain large amounts of phosphoric acid, or at least many of them do; they are phospholipins, glycolipins and cholesterol. That these lipins have a very important function in the physiology of the brain there can be no doubt, since they are present in very much smaller amounts in the embryonic nervous system and they develop *pari passu* with the development of the functions of the brain. Most of these lipins are in the white matter of the brain; but the proportion is also high in the gray. The proteins are far less prominent among the total solids than in muscle. In both white and gray substance, however, water makes by far the greater proportion of the weight. Thus, in the gray matter, it is 85.27 per cent.; in the white matter of the adult human brain it is 70.23 per cent. The proportion of water varies with the age, being largest in the youngest brains.

Our knowledge of the chemical composition of the brain is owing largely to Thudichum, a man of extraordinary care, accuracy, insight and industry, whose abilities were much underrated during his life. For, owing partly to an unusually combative nature, he alienated many of his colleagues and his work was long neglected. There is now, however, no question that he was far in advance of all others in this difficult field and his book, published in 1901, entitled, *Die chemische Konstitution des Gehirns des Menschen und der Thiere, nach eigenen Forschungen bearbeitet* is a monument to his ability and insight. A German by birth, he lived most of his life in England. He died in 1902.

Chemical Examination of the Brain. Separation of the Phosopholipins.—The method adopted by Thudichum for the chemical examination of the brain consisted in first freeing the brain from its membranes, the pia mater, etc., and the blood vessels, then drying and extracting it thoroughly with alcohol and ether. To get rid of the water in it the brain can either be ground, or cut into small pieces and dried in a current of air; or the pieces may be placed in 90 per cent. alcohol, at least three volumes of alcohol to each volume of brain. The latter method is the one described here as it is, on the whole, the better. The cold alcoholic extraction should be once repeated. This treatment coagulates all the proteins and takes in the alcohol most salts and so-called extractives. The hardened brain substance is then ground still finer and suspended in a small amount of fresh alcohol and put through a wire sieve of 144 meshes to the inch, by means of a stiff brush. This reduces the material to a fine purée, which can be thoroughly extracted. The material is then heated to 70°, with a large amount of 85 per cent alcohol, filtered hot through a cloth first and then a filter paper, and the residue reheated with fresh alcohol five times. To exhaust it completely it must be boiled at least 15 times more with 85 per cent. alcohol or absolute alcohol; even then some substances slightly soluble in alcohol, probably anhydrides of the lipins, remain behind in the proteins. By this repeated boiling with alcohol of 85 per cent., followed by absolute alcohol, practically all of the brain lipins are removed and the proteins left in a coagulated form. These united, hot alcohol extracts, if allowed to stand cool for 12 to 24 hours, separate out a large quantity of a white precipitate, which appears crystalline under the microscope, but is not homogeneous. This material may be called "*white substance.*" It is collected on a cloth, the mother liquor pressed out and serves for further fractioning into its constituents. It is sometimes called crude protagon. It contains nearly all the glycolipins (cerebrosides), cerebric acids, much cholesterol, cephalin (kephalin), various myelins and, if the alcohol solution was concentrated, some lecithin, amino-lipotides (amino-lipins) and small amounts of other substances. If the alcoholic filtrate from the white substance is concentrated by further evaporation until a test portion settles out a precipitate on cooling and is then allowed to cool, a precipitate comes out having, when pressed in a cloth, a soft buttery consistence and a yellow color. This precipitate, which may be called the "*buttery substance,*" is separated by filtration. It consists of much cholesterol, most of the lecithin and other phospholipins (phosphatides); the amino-lipins; but only traces of cerebrosides and cerebric acids.

The alcohol filtrate from the "buttery substance" is still further evaporated by distillation as long as alcohol, capable of burning, comes over. It is then evaporated further on the water bath in a porcelain dish. As soon as the last traces of alcohol are gone, there form, on the surface of the remaining water, oily drops uniting to masses, which hang to the sides of the dish. This forms the "*oily material* or *substance.*" It is separated hot, because if allowed to cool it mixes again to an emulsion with the water. The "oily substance" contains some cholesterol, but consists chiefly of amino-lipins (amino-lipotides) (bregenin) and phosphatides (kephalin).

The aqueous solution, which is left, contains the *extractives*, namely, inosite, lactic acid, salts, succinic acid, hypoxanthine, alkaloids, amino-acids and inorganic salts. This extract may be united with the extract obtained by evaporating the cold alcohol, which was used for dehydration. The latter has in it, also, some phospholipin. This may be removed most easily by shaking it up with water to make an emulsion, then adding some chloroform in small amount, shaking agsin and making it acid by the addition of hydrochloric acid. The lipins come out as an emulsion, which may be filtered from the water which contains the extractives.

The partition of substances sketched above may be summarized in the following table (Thudichum, Gehirn, p. 79):

1. *First extractive substances* in the cold alcohol for hardening and dehydrating.
2. *Insoluble protein and tissue residue* containing neuroplastin, protein, nuclein, phosphoproteins.
3. *White substance* containing:
   Kephalin with varieties and compounds.
   Lecithin with varieties and compounds.
   Paramyelin with varieties and compounds.
   Myelin with varieties and compounds.
   e. Amino-myelin with varieties and compounds.
   f. Cholesterol and Phrenosterol.
   g. Cerebrin mixture, mixture of cerebrosides (glycolipins), cerebric acids, cerebro-sulphatides and amino-lipotides with sphingomyelin and assurin.
4. *Buttery substance* containing:
   a. Kephaloidins with varieties and compounds.
   b. Lecithin.
   c. Paramyelin.
   d. Myelin.
   e. Amino-myelin.
   f. Sphingo-myelin, and assurin (small amount).
   g. Cholesterol and Phrenosterol.
   h. Phrenosin and other cerebrosides.
   i. Amino-lipins (Amino-lipotides).
5. *Oily substance* containing:
   a. Lecithin; b. Paramyelin; c. Oily liquid material of amino-lipotides.
6. *Last aqueous brain extract* containing:
   a. Alkaloids (hypoxanthine, etc.). b. Amino-acids. c. Inosite. d. Organic acids and salts. e. Inorganic acids and salts.

Fractioning the white substance. Both the white and buttery substances may be fractioned or separated into their constituents by the methods given in Thudichum. By extracting these substances with cold ether all the sterols and most of the phospholipins go into solution, leaving a white substance undissolved. This white substance consists chiefly of cerebrin and cerebrosides, cerebric acid and sulphatides. The kephalin is separated from the lecithin and some other phospholipins in the ether extract of the white and buttery substances by the addition of three volumes of absolute alcohol. This precipitates an impure kephalin. It may be purified by repeated solution and precipitation and finally by precipitation with cadmium chloride. Further details are given in Thudichum. We may now consider these various products, or educts, in detail.

Lecithin.—This is a phospholipin, or phosphatide, as it is called by Thudichum. The term phosphatide lays chief stress upon the phosphoric acid of the molecule; while phospholipin emphasizes its fatty character. Lecithin is found in the white substance in part, but in largest amount in the buttery substance. It is separated from the white or buttery substance by extracting these with cold ether. The kephalin is separated from the lecithin by the addition of three volumes of absolute alcohol to the ether. This precipitates nearly all of the kephalin, but leaves the lecithin in solution. The filtrate is precipitated by the addition of alcoholic ammoniacal lead acetate to get rid of the rest of the kephalin, kephalin being precipitated by this reagent, but lecithin not. The myelins, etc., are also precipitated; the filtrate is distilled to remove ammonia and ether until fairly pure cholesterol begins to come out. If some salve-like lecithin comes out, this is redissolved in 85 per cent. warm alcohol. To this warm solution a warm saturated solution of $CdCl_2$ in 85 per cent. alcohol is added, little by little, as long as a precipitate forms and then about as much more $CdCl_2$ as had already been added is poured in. The lecithin $CdCl_2$ compound crystallizes out as a white precipitate. This is washed by decantation with 85 per cent. alcohol. This precipitate is dried and freed from ether-soluble substances by prolonged extraction with boiling ether. Krinosin is the main impurity removed. The precipitate is next extracted with cold, water-free benzol to remove traces of the kephalin-$CdCl_2$, and finally is extracted with hot benzol, which dissolves the lecithin cadmium chloride, but leaves behind the para-and amido-myelin cadmium chlorides, which remain insoluble. The lecithin cadmium chloride compound is precipitated from the benzol by the addition of absolute alcohol and is recrystallized from hot alcohol.

There is thus obtained lecithin-$CdCl_2$, which crystallizes in spheres and stars of microscopic crystals. The purest lecithin cadmium chloride compound of ox brain had the composition: $C_{43}H_{84}NPO_8CdCl_2$. Lecithin may be freed from the cadmium chloride by suspending it in 85 per cent. alcohol and passing in H_2S. It is filtered through a hot funnel. On cooling a felt work of fine, needle-shaped crystals of lecithin chloride separate out. These crystals are microscopic plates, often hexagonal and very thin, so that they may be bent over to look like needles. They dry in vacuo to a white, easily-powdered mass of the composition:

$C_{48}H_{86}NPO_8Cl$; or $C_{42}H_{83}NPO_8Cl$.. HCl: N: P:: 1: 1.03: 1.09.

Pure lecithin. This is a white crystalline body, the crystals being thin plates, which, when pressed together, have a waxy consistence, but it is stickier than wax. It is very soluble in 85 per cent. alcohol and more soluble still in absolute alcohol. It is soluble in ether and chloroform, but does not crystallize from them on evaporation. It is not precipitated from its alcoholic solution by ammonia and lead acetate. When put in concentrated sulphuric acid it dissolves with a yellow color and if to this sugar solution is added, a deep purple red color develops (Pettenkofer's reaction). This is due to the oleyl group in the lecithin. The purple coloring matter is soluble in glacial acetic acid and shows absorption bands between D and E and in the blue. If lecithin cadmium chloride is suspended in water and dialyzed the lecithin remains in the tube; the cadmium chloride is dialyzed away.

The chief properties of lecithin are due, according to Thudichum, to the oleic-acid group it contains, but it is obvious that all of its constituents, and particularly the phosphoric acid, contribute to its properties. The oleic-acid radicle is somewhat more easily separated than the others. Thus it comes off readily when the $PtCl_4HCl$ lecithin compound is made. By decomposition with acids, or barium hydrate, lecithin yields neurine (choline according to most observers), oleic acid, palmitic acid or stearic and glyceryl-phosphoric acid. The formula for lecithin, given on page 90, is that of Diakonow, derived from the study of egg yolk lecithin, but it is uncertain what the formula really is, since Diakonow at no time probably had pure lecithin in his hands for analysis. How the various radicles are united to make lecithin is still uncertain except that glycerol is united with the phosphoric acid to make glyceryl-phosphoric acid. The chief lecithin of the brain is oleylpalmityl-glyceryl-neurylphosphatide. The other lecithins, containing stearyl in place of palmityl, are present only in small amounts. To indicate his belief that phosphoric acid is at the basis of the molecule, whence the name phosphatide, Thudichum represents the molecule as follows:

$$O=P\left\{\begin{array}{c} C_{18}H_{33}O_2 \\ C_{16}H_{31}O_2 \\ C_3H_7O_3 \\ C_5H_{11}N \end{array}\right\}=C_{42}H_{82}NPO_8$$

It is probable that the formula is really that given by Diakonow with the fatty acids substituted in the glycerol; and the choline, if present, united through the hydroxyl of the carbon, rather than that of the nitrogen, to the phosphoric acid. Thudichum states that he has very carefully examined the base of brain lecithin and that it is neurine and not choline. Unless the lecithin is suspended in water and shaken with hydrochloric acid it is impossible, he says, to free it from potassium.

which on hydrolysis precipitates with the platinum chloride-neurine compound, so that most of the cholines examined have been impure. If neurin pre-exists in the molecule it is very difficult to see how it can be attached, since the formation of the chloride of lecithin indicates, very clearly, that the hydroxyl of the nitrogen is free. It seems to the author more probably that the base in the lecithin is choline, from which neurine is formed on decomposition. There do not seem to be determinations of the molecular weight of lecithin. It is certainly colloidal in aqueous solution. It is possible that the formula is more complex than appears. Possibly the phosphoric acids may be joined much as they are in nucleic acid to give poly-phosphatides. The whole matter of the composition of the lecithin molecule is in need of investigation. Lecithin is often considered to be unstable, but Thudichum states that lecithin, in a dry state, or as the $CdCl_2$ compound, is so stable that it may be kept for years without change; and even as the hydrate when suspended in water it does not easily change. It is wet by water, does not float like fats, but sinks to the bottom, and swells to form an emulsion if it is present in less than 1 part to 100 of water. It does not dialyze. A very interesting fact is that, as ordinarily prepared, it contains some potassium. It probably exists in the cells in part as a potassium salt. It has the property of making myelin forms which are liquid crystals.

Kephalin.—Another mono-amino-mono-phosphatide is kephalin (Gr. *kephalos*, brain), or cephalin if the Latin spelling is used. This differs from lecithin, according to Thudichum, in that it contains another acid, kephalinic acid, which is an unsaturated acid of the linolinic acid series, in place of oleic acid. It differs, also, as we now know, in the character of the base it contains; it contains no choline, but in place of it amino-ethyl alcohol, or oxy-ethyl amine. This was isolated from it by Thudichum. Possibly other bases are also present, i.e., β-oxy-α-amino butyric acid (McArthur).

Preparation. It is isolated from the ether extract of the "white substance" or by extracting the dry brain with ether. The extracts are concentrated and freed from cholesterol by precipitation with acetone. The precipitate is redissolved in ether, if not clear, allowed to stand until any white matter has separated out, and the decanted clear solution precipitated by the addition of absolute alcohol as long as a precipitate forms. The kephalin is precipitated. After standing 24 hours in the cold the liquid is poured off from the kephalin precipitate. This precipitate is redissolved in ether and reprecipitated several times with alcohol to remove cholesterol and lecithin. It is then emulsified with 100 parts of water, allowed to stand, separated from any precipitate which may form by decantation and precipitated by the addition of hydrochloric acid just sufficient to precipitate it. The precipitate rises to the top and is lifted off and washed with water until by the separation of HCl it begins to get slimy. Then it is freed from water by alcohol, dissolved in ether, precipitated by alcohol, dried in vacuo and it is ready for analysis. The emulsification and precipitation of the kephalin, by acid, is necessary to free it from bases, Ca, K and Na, and phosphoric acid, which stick to it. There is always some ammonia separated in the water, and Thudichum states that the aqueous solution from the emulsification contains copper, giving, on evaporation, a deep blue solution with ammonia. This point should be reinvestigated to see whether this is in reality a normal constituent of all brains, or present only in human brains which happened to come for analysis. It is possible that the human brains he examined might have come from brass-workers or others exposed to copper poisoning. The calcium and potassium are attached, in part, directly to the kephalin molecule. These salts predominate over the others in the kephalin just as they do in the cell. It is not impossible that the greater predominance of potassium over sodium in the cell may be due to this firm union between kephalin and these bases. Similar salts are recovered from most phosphatides, particularly from myelin. Their possible rôle in the nerve impulse has been discussed by Pike.

Properties. The kephalin thus isolated is quite possibly still impure. It is at first a light yellow or white color, but in ether solution it changes rapidly to a red. It unites with water, forming an emulsion, the soluble kephalin becoming insoluble by heating the aqueous solution. A part of the kephalin, after precipitation with $CdCl_2$, when freed from $CdCl_2$ by dialysis in the manner described for lecithin, will diffuse through the paper. It makes a finer emulsion with water than does lecithin. It is soluble in water saturated with ether. 100 parts of boiling absolute alcohol dissolve 9 parts of kephalin; 2 parts come out on cooling, 7 parts remain in solution. It is much less soluble in alcohol containing water. Heated in water to 90-100° it melts to a thick, dark red oil. It forms a chloride with hydrochloric acid. This chloride is soluble in ether and is not precipitated by alcohol. It is precipitated from an ether-alcohol solution completely as kephalin $PtCl_4HCl$ by $PtCl_4$. It is precipitated also by $Ba(OH)_2$, and $Ca(OH)_2$ and $ZnCl_2$. The affinity of $CdCl_2$ for kephalin is less than that for lecithin. The Pettenkofer reaction is never so good as that of lecithin. In kephalin the nitrogen base splits off first on hydrolysis, like the base in sphingomyelin. Thudichum obtained by hydrolysis neurine, a second base probably amino-ethyl alcohol, which may be a decomposition product of neurine; a third base, of unknown nature; kephalinic acid, which is apparently an unsaturated partially oxidized palmitic or stearic acid; and glycerol. He isolated kephalylphosphoric acid, which is kephalin minus the neurine. More recent determinations of the composition of kephalin indicate that it contains no neurine. Koch found that when heated with hydriodic acid it gave rise to less isopropyl or methyl iodide than did lecithin, so that the base is certainly not neurine or a methylated base. Miss Foster has found that the iodide obtained by Koch was isopropyl iodide from the glycerol. No methyl iodide is formed. It is probable that the preparation of Thudichum was still not pure. There is no doubt that the chief base present is oxyethyl amine. Baumann obtained, like Thudichum, amino-ethyl alcohol and believes that this is the only base present; on the other hand, McArthur has isolated amino-oxy-butyric acid, serine, ethoxy amine and ammonia. How far these are decomposition products and how far they are preformed in the molecule is uncertain. No reliance can be placed on the results of the study of the decomposition products, unless the kephalin has been carefully separated from lecithin, myelin and other phosphatides, and emulsified and treated with acids to free it from extractives and salts. Neurine breaks up readily in alkaline solution, so hydrolysis in an alkaline solution yields results which are at the best difficult to interpret, but it is stable in acid. Either there are a number of kephalins with different nitrogen bases or products in various stages of decomposition have been analyzed. There is also unexplained the number of oxygen atoms which are not accounted for by the decomposition products isolated.

The composition of kephalinic acid is still uncertain, but it appears to be either $C_{18}H_{30}O_3$ or $C_{17}H_{30}O_3$. It is probably a mixture of very unsaturated acids of the linoleic or linolinic acid type and their partially oxidized derivatives. The constitution of kephalin was represented as follows by Thudichum, but it is probable that the neuryl radicle should be replaced by amino-ethyl alcohol.

$$O=P\left\{\begin{array}{ll}\text{Kephalyl} & C_{18}H_{24}O_3\\ \text{Stearyl} & C_{18}H_{35}O_2\\ \text{Glyceryl} & C_3H_7O_3\\ \text{Neuryl} & C_5H_{11}ON\end{array}\right\}=C_{43}H_{80}NPO_9$$

The manner in which these are united is unknown, as is also the molecular weight.

The extraordinary reducing powers of kephalin due to the unsaturated acids are extremely suggestive and interesting. In kephalin we have a body greatly more reactive than lecithin, capable of auto-oxidation and hence of respiration, and unstable. These phenomena, as pointed out on page 587, are possibly related to the phenomena of respiration and memory shown by the nervous system. This instability is one of the difficulties in the way of obtaining pure kephalin. Another possibility exists also: namely, the molecule may be formed of several phosphoric-acid groups to which various radicles are united. The further investigation of kephalin will probably yield results of great value.

The further study of the acids present indicates that they are stearic, as found by Thudichum, linolic, linolenic and carnaubic (McArthur).

Paramyelin.—This is another mono-amino-mono-phosphatide which is present in the white substance. It is precipitated by $CdCl_2$, but is separated from kephalin $CdCl_2$ by the solubility of the latter in cold, water-free benzol; and from lecithin $CdCl_2$ by its solubility in hot benzol. The paramyelin $CdCl_2$ is insoluble in hot benzol. The $CdCl_2$ salt is finally dissolved in boiling 85 per cent. alcohol, from which it crystallizes on standing. The $CdCl_2$ compound dissolved in hot alcohol, decomposed by H_2S and filtered hot, crystallizes out in white crystals as the chloride. When this is recrystallized from alcohol it loses HCl and crystallizes in rhombic and hexagonal plates like lecithin containing 4.31 per cent. of P and 2.06 per cent. of N. The computed molecular weight was 721. The formula: $C_{38}H_{75}NPO_9.CdCl_2$. On decomposition with Ba $(OH)_2$ it yields glyceryl-phosphoric acid, neurine, an acid giving Pettenkofer's reaction and another acid. It is a white, solid, crystalline substance, crystallizing from hot alcohol in plates and needles. It is a weaker base than lecithin or kephalin, easily losing the HCl. It is found both in the white and butter substance. This substance requires further examination. The character of the base is particularly doubtful.

Myelin.—This mono-amino-mono-phosphatide is found in the brain only in small amounts. It is sharply distinguished from paramyelin by its lead compound.

Preparation. If the white substance is extracted with cold alcohol and allowed to stand in the cold, myelin separates out mixed with sphingomyelin. The precipitate redissolved in hot alcohol is precipitated by alcoholic lead acetate; the lead compounds are extracted with boiling alcohol which dissolves the sphingomyelin. The lead compound is also insoluble in benzol. If it is suspended in hot alcohol, decomposed by H_2S, filtered hot, white myelin crystallizes out on standing. When recrystallized from alcohol it tends to form insoluble anhydrides. It is not precipitated in alcoholic solution either by $CdCl_2$ or by $PtCl_4$.

Properties. It crystallizes out of ether or alcohol on cooling the saturated solution either in spheres or very small microscopic crystals. In masses it is white like bleached ivory and when powdered it is entirely white. It dissolves in concentrated H_2SO_4 without color, but if sugar solution is added it gives a deep purple substance soluble in $CHCl_3$. The myelin lead compound is insoluble in ether; that of kephalin is soluble. Lecithin, paramyelin, amidomyelin and sphingomyelin give no lead compound. The analysis gave: C, 63.41 per cent.; H, 9.83 per cent.; N,

1.79 per cent.; P, 4.087 per cent.; O, 20.874 per cent., which corresponds to $C_{40}H_{75}NPO_{10}$. The decomposition products are unknown.

Diamino-mono-phosphatides.—There are also phosphatides with the nitrogen and phosphorus in the proportion of 2:1. These are di-amino-mono-phosphatides.

Amidomyelin. This is a diamino-mono-phosphatide of the formula, $C_{44}H_{92}N_2PO_{10}$, which is isolated as a $CdCl_2$ salt from the buttery substance. It crystallizes as the salt $C_{44}H_{92}N_2PO_{10}2CdCl_2$. This phosphatide has a very curious property: namely, it is completely soluble in cold water when freed from $CdCl_2$ by dialysis, but the solution gels on heating. The gel is not reversible, it does not redissolve on cooling. This property differentiates it from the other phosphatides. The free amidomyelin forms snow-white, microscopic needles and plates mostly arranged in stars. Over sulphuric acid it drys to a perfectly white powder. It gives Pettenkofer's reaction very quickly and intensely. Nothing is known about its decomposition products.

Sphingomyelin. This is the chief, but not the only, phosphatide left in the "white substance" when this is extracted with ether. The method of its separation is long and involved and will not be given here. It crystallizes out of alcohol in thick masses of needles, stars and six-sided plates. It is almost insoluble in ether even when HCl is added. It is thus separated from lecithin. It appears to combine with the cerebroside, kerasin, as a base in a weak union. It is precipitated by $CdCl_2$ from 85 per cent. alcohol solution. It swells in water and makes an emulsion. The $CdCl_2$ salt loses its $CdCl_2$ on dialysis. The analyses gave C, 65.37 per cent.; H, 11.29 per cent.; N, 2.96 per cent.; P, 3.24 per cent.; O, 17.14 per cent., which corresponds to the formula, $C_{52}H_{104}N_2PO_9H_2O$. On hydrolysis it yields no glycerol. After 5 hours' heating with $Ba(OH)_2$ only neurine is given off. The amount of neurine found as the platinum salt was 1.1 grams, where 1.3 grams was the theoretical. There remains sphingomyelinic acid, $C_{48}H_{95}NPO_{12}$. N:P::1:1. This yields sphingol, an alcohol, $C_9H_{18}O$, or $C_{18}H_{36}O_2$, soluble in alcohol and ether; sphingosin, a base, $C_{17}H_{35}NO_2$, and sphingo-stearic acid, $C_{18}H_{36}O_2$, probably an isomer of stearic acid melting at 57°. The $CdCl_2$ salts are beautifully crystalline and white. Sphingosin is apparently an unsaturated, mono-amino-dihydroxy alcohol. The structure is still unknown. The sulphate melts at 233-234° (uncor.) and the rotation is $(\alpha)_D^{20°} = -13.12$ (± 0.00) in a solution of .5304 gr. of sphingosin in 5 c.c. $CHCl_3$ and 1 c.c. glacial acetic acid. The recent analyses of sphingomyelin by Levene showed the composition, C, 64 per cent.; H, 11; N, 3.40; P, 3.60; inorganic bases, 3 per cent. N:P::2:1. It contains no free amino nitrogen. One nitrogen atom is in the form of choline; the nature of the other base is unknown; the acids split off on hydrolysis are lignoceric acid, $C_{24}H_{48}O_2$, and cerebronic acid, $C_{25}H_{50}O_3$ (α hydroxy pentacosanic acid), and sphingosin. The composition of sphingosin is still unknown, but its composition has been suggested by Levene to be $C_{12}H_{25}CH{=}CH.CHOH.CHOH.CH_2NH_2$.

Diamino-diphosphatides.—These contain two atoms of phosphorus and two of nitrogen to the molecule. They are present in small amounts only.

Assurin. This is found in the alcoholic extract of the cerebroside mixture after the separation of the myelin, sphingomyelin and kerasin. It is crystallized as the $PtCl_4$ salt, which is insoluble in boiling alcohol or ether and is either $2(C_{46}H_{94}N_2P_2O_9HCl)+PtCl_4$; or $2(C_{47}H_{101}N_3P_2O_9HCl)+PtCl_4$. There is apparently united with this body an amino lipotide phosphorus free, $C_{40}H_{81}NO_6$. It may be identical with bregenin.

The cerebrosides or galactosides.—In addition to the phospholipins, the brain is remarkable for the group of bodies called the cerebrosides, or glycolipins. These are not found in the embryonic brain, but develop as medullation comes on and are to be found chiefly in the medullary sheaths in the white matter of the brain. The most important of the cerebrosides are *Phrenosin* and *Kerasin*, or the cerebron of Thierfelder. The cerebrosides contain no phosphorus, but they all contain nitrogen, a sugar, first named cerebrose, but later shown to be for the most part galactose, and a complex fatty acid.

Phrenosin and *kerasin.* These two cerebrosides make the greater part of the white substance left behind on extraction of the phosphatides with ether. They are quite insoluble in ether, but are soluble in alcohol, particularly in warm alcohol. They crystallize very readily, although it is not so easy to get them quite clean. Phrenosin has also been named cerebron, by Thierfelder, but it is better to keep the original name. They are mixed in the white substance with another cerebroside, cerebric acids, from which they may be separated by precipitating the latter with lead oxide. They are prepared by exhausting the white substance by extracting with ether. If now they are redissolved and recrystallized from hot alcohol, the phosphorus content falls to about .8 per cent., but it is hard to get it lower. To purify further the mass is rubbed in a mortar with an alcoholic solution of lead acetate containing a little ammonia, and then, constantly stirring, it is poured into hot 85 per cent. alcohol. When all is transferred add an alcoholic solution of lead acetate and ammonia as long as a precipitate forms; filter hot; extract the precipitate repeatedly with hot 85 per cent. alcohol and collect the precipitates from all the hot alcohol extracts on cooling. Recrystallize from absolute alcohol and filter after 24 hours. The precipitate is a mixture of phrenosin and kerasin. On longer standing kerasin crystallizes out. The separation of these two is made by dissolving them in alcohol so that they are not too concentrated and allowing them to cool. Phrenosin comes out above 28°. Kerasin, which is more soluble, comes out below 28°. The sphingomyelin has to be separated by $CdCl_2$. Recrystallizing from glacial acetic acid also helps to eliminate the last traces of the phosphatide.

Phrenosin. Properties. The name is from the Greek, *phren*, brain. It is a white crystalline substance, which when warmed in water becomes neither doughy nor slimy, but floats in loose *flocculi* through the liquid. Rubbed with concentrated H_2SO_4, it apparently completely dissolves, then gradually develops a purple red color which is attached to particles in the acid. These are soluble in $CHCl_3$ or glacial acetic and show specific absorption spectra. The color reaction belongs to the sphingosin. No sugar need be added, since galactose is already there. Phrenosin has the composition $C_{41}H_{79}NO_8$. When boiled with acids it hydrolyzes and galactose is set free, the solution acquiring a strong reducing power in consequence. It yields *sphingosin*, $C_{17}H_{35}NO_2$, already described as a decomposition product of sphingomyelin; galactose; and a new acid, which Thudichum thought to be an isomeric stearic acid, but which is now known to be $C_{25}H_{50}O_3$, phrenosinic, or cerebronic, acid. On decomposition of phrenosin, the cerebronic acid breaks off first and there remains a compound of sphingosin and galactose of a basic nature named *psychosin*, $C_{23}H_{45}NO_7$. Out of hydrated phrenosin sulphuric acid splits off galactose, leaving *æsthesin*, $C_{35}H_{69}NO_3$(?), which is probably a compound of sphingosin and cerebronic acid. Sphingosin has a bitter taste and causes a burning sensation in the throat and a feeling of illness. Its pharmacology would perhaps be of interest.

Kerasin resembles phrenosin in most particulars, except it is optically different and it is more soluble in alcohol. If not more than one part is present in 321 parts of alcohol, it does not crystallize out above 28°. 100 c.c. of acetone dissolve 0.1576 gram of kerasin at 15°. It does not give the Pettenkofer reaction so easily as phrenosin. The differences between these two bodies have been shown to be due to the presence in phrenosin of cerebronic acid m.p. 84°; while in kerasin it is lignoceric acid, $C_{24}H_{48}O_2$. The cerebron of Thierfelder appears to be a mixture of both these bodies. Thudichum thought that the difference probably was that the cerebronic acid of the kerasin had about one carbon atom more than that of phrenosin. Kerasin is levo-rotatory. $[\alpha]_D = -2°$ in pyridine.

The phrenosin and kerasin spherocrystals are readily distinguished by microscopic examination with polarized light and a selenite plate (Rosenheim).

It is an interesting fact that the medullary sheath contains so large a proportion of galactose. It is possible that it is to supply this raw material to the brain at a time when medullation is proceeding at a rapid pace that the sugar of the mammary glands is lactose and contains, therefore, galactose.

Protagon. It was for a long time believed that the crystalline material falling out of the alcoholic extract of the brain constituted, when purified, a single substance composed of a compound of cerebrin and lecithin or kephalin. This was called "protagon." Since Thudichum succeeded in isolating 14 different substances of a complex nature from this mass, it is probable that protagon is a mixture and not a chemical individual. Nevertheless, it is not impossible that some of the compounds isolated may have been in weak chemical union. It is known, for example, that cholesterol forms a molecular, crystalline compound of definite composition with digitonin; but if the compound is extracted with ether it is decomposed, the cholesterol dissolving. The similar compound of cholesterol and saponin is stable in ether, but breaks up in hot alcohol.

The sulphur of the brain. Cerebro-sulphatides. Sulpholipins.—The cerebrosides while impure always contain some oxidized sulphur and, according to Thudichum, some unoxidized sulphur. The oxidized sulphur body was believed by Koch to be a union of sulphuric acid, a cerebroside and a phosphatide as follows:

$$\text{Cerebroside—O—}\overset{\overset{\displaystyle O}{\|}}{\underset{\underset{\displaystyle O}{\|}}{S}}\text{—O—Phosphatide}$$

Its nature is still entirely unknown. Taurine is found in the brain, but the unoxidized sulphur fraction of the lipins noted by Thudichum is still uninvestigated. It is probable that further study will show that sulphur is as important in the metabolism of the brain as phosphorus. The phospholipins from other cells not infrequently, also, contain sulphuric acid not extracted with water, but split off on hydrolysis with acid; and also a carbohydrate group. The sulphatide isolated by Thudichum from the brain contained 4 per cent. of sulphur.

Amido-lipotides. Amino-lipins.—These are found chiefly in the oily matter separating out at the end of the evaporation of the alcohol. They contain no phosphorus, but an amino group. When heated they collect as a thick oily mass on the surface of the water, or cling to the sides of the beaker, but on cooling they take up water again and swell to make a gelatinous mass. None of them are well characterized and they are present in relatively small amounts. Whether they are preformed or represent partially digested phospholipins cannot be said. Among these bodies two were isolated by Thudichum: *Krinosin*, $C_{38}H_{79}NO_{3}$, and *Bregenin*, $C_{40}H_{81}NO_{3}$. (From the Platt Deutsch word, *bregin*, brain.) If H_3PO_4 is added to this formula, we get the formula of the lowest lecithin.

Nothing is known of the composition of these bodies. Their physical properties are so interesting that they should be carefully studied. It is possible that they might throw light on the composition of the phosphatides. It is very hard to separate them completely from cholesterol. Cholesterol esters, also, have the property of taking up large amounts of water. See page 87.

Sterols.—The brain contains extraordinarily large amounts of cholesterol (1.6 per cent. of the wet weight of the brain), which appears to be for the greater part free cholesterol. This sterol is chiefly cholesterol melting at 145°; but there is also present another sterol melting at 137°, which is the melting point of iso-cholesterol. It was called by Thudichum phrenosterin, or as we should call it now, phrenosterol, to indicate its origin in the brain. The composition and characteristics of cholesterol have been discussed on page 82.

AMOUNT OF CHOLESTEROL IN DIFFERENT BRAINS (Rosenheim).

	Water—per cent.	Cholesterol—per cent. wet weight	Cholesterol—per cent. dry weight
Man	78.86	1.95	9.22
Child 3 months	85.80	0.69	4.89
Fœtus 36 weeks	90.29	0.39	4.07
Child 5 days	89.99	0.53	5.29
Dog	76.18	2.76	11.59
Cat	76.53	2.35	9.99
Ox	78.83	2.39	11.28
Sheep	79.50	2.13	10.37
Rabbit	77.86	2.12	9.57
Fowl	80.34	1.45	7.40
Codfish	84.03	1.92	12.02

The extractives.—When the lipins have been removed from the evaporated alcoholic extracts, cold and hot, there remain a group of water soluble substances, organic and inorganic, of which only a few have been isolated and identified. These are called the brain extractives. The solution is generally brown in color, with the smell of meat extract and a bitter, salty taste. Some of the organic extractives contain nitrogen and are precipitated by the alkaloidal precipitating agents like phosphotungstic acid. They are alkaloidsl in nature; others are of a carbohydrate nature; others are acids. It is very difficult to know whether these substances are normal constituents of the living brain, or whether they are formed by the hydrolytic decompositions which begin as soon as the circulation of the brain is interrupted. It is probable, however, that they pre-exist in the brain and that their accumulation there after death is due to the persistence of the vital processes in the absence of blood supply to remove the waste products. This view is supported by the relatively slight autolytic decomposition which brain material shows.

Alkaloid extractives. Of these *hypoxanthine* is the most important, but Thudichum isolated also another of unknown character having the formula for its gold salt $C_{15}H_{20}N_6O_5HClAuCl_3$, and another having a strong odor of human sperm. Hypoxanthine is discussed in connection with the occurrence of it in muscle. It is impossible to say whether it is formed from the nucleic acid of the brain by a partial hydrolysis, or whether it is found in the cell in a free state. Its function is unknown. It is interesting to recall that when caffeine is taken it accumulates in the brain tissue to a far greater degree than in any other tissue of the body.

Amino-acids. Among the amino-acids in the extractives Thudichum isolated tyrosine and a new leucine, a sweet leucine which he called glycoleucine, and correctly inferred it to be the normal caproic acid leucine having the formula:

```
    H  H  H  H  NH₂
    |  |  |  |  |
H—C—C—C—C—C—COOH
    |  |  |  |  |
    H  H  H  H  H
```

It has been recently isolated also by Abderhalden, who proposes to call it *caprine.* It is curious that the brain proteins should contain this normal leucine, whereas most other proteins have the iso-butyl-amino-acetic leucine. Caprine is less soluble in cold water and also the copper compound is less soluble than the ordinary leucine. Besides leucine and tyrosine, creatine and taurine have been found in the extractives. Creatine is methyl guanidine acetic acid, NH_2—C=NH.N(CH_3)—CH_2—COOH. The significance of its occurrence in the brain is unknown.

Organic acids. The principal organic acids found in the extractives are lactic, $C_3H_6O_3$, succinic, $C_4H_6O_4$, and also formic acid, CHOOH. These same acids are also the principal acid extractives of muscle. The lactic acid is the para, or meat, lactic acid.

Carbohydrates. In this group *inosite,* $C_6H_{12}O_6$, an optically inactive, sweet-tasting alcohol, not reducing Fehling's solution, is present in considerable quantities. This again is found in quantity in the muscles. Inosite is believed to be hexahydroxyhexamethylene. Its significance is unknown.

```
          H   OH
           \ /
     H      C      H
      \   /   \   /
       C         C
      /  |     |  \
    HO   |     |   OH
     H   |     |   H
      \  |     |  /
       C         C
      /   \   /   \
    HO      C      OH
           / \
          H   OH
```

Inorganic salts. The salts of the extractives are phosphates, chlorides, sulphates of calcium, potassium, sodium, magnesium and iron.

Distribution of substances between gray and white matter.—We may now take up the question of the distribution of these substances in the gray and white matter of the brain and their functions. While the methods of separation employed by Thudichum were far from quantitative, they give at least some idea of the distribution of many of the substances. Some analyses are given also by the improved methods of Koch.

Gray Matter of the Cortex. Dried First. (Thudichum.)

Water at 95°	85.27%
Solids[1]	14.73
Neuroplastin	7.608
Ether extract (kephalin, lecithin, cholesterol)	1.950
Cerebrosides, cerebric acid and myelin	0.424
Lecithin, kephalin, myelin from last oily matter	0.780
Inosite	0.193
Lactic acid	0.102
Alkaloids	.....
H_2SO_4 in neuroplastin	0.06
" " extract	.....
H_3PO_4	0.017
K	0.025
Na	0.092
Water extract	0.500

Loss by operations about 25% of the total solids.

Composition of the White Matter of the Corpus Callosum.

Water (dried at 95°)	70.23%
Neuroplastin	8.63
Ether extract (kephalin, lecithin, cholesterol)	11.497
Cerebrosides and myelin	6.91
Water extract	1.403
Lactic acid	0.0456
Inosite	0.2171
Alkalies or carbonates	0.1717

In contrasting these two tables it will be noticed that the white matter contains more solids and less water than the gray, and these solids are chiefly present in the ether extract and include the cerebrosides. Thus the cerebrosides make roughly 7 per cent. of the white matter and only 0.4 per cent. of the gray; and the kephalin, lecithin and cholesterol are 11.5 per cent. of the white and only 1.95 per cent. of the gray. The per cent. of protein is also higher in the white than in the gray. These figures indicate very clearly that most, if not all, the cerebrosides are in the medullary sheaths and that the greater proportion of phosphatides, myelins, etc., are there also. On the other hand, the proportion of these substances is very high even in the gray matter, the proteins making only about 50 per cent. of the total solids, the other 50 per cent. being extractives and ether-soluble material.

COMPOSITION OF THE SOLIDS OF THE HUMAN BRAIN.

Figures are per cent. of dry matter.*

	Whole brain (child)	Whole brain (adult)	Corpus callosum
Protein	46.6	37.1	27.1
Extractives	12.0	6.7	3.9
Ash	8.3	4.2	2.4
Phosphatide	24.2	27.3	31.0
Cerebrosides	6.9	13.6	18.0
Lipoid sulphur	0.1	0.3	0.6
Cholesterol	1.8	10.9	17.1

* Koch: Die Bedeutung der Phosphatide (Lecithane) für die lebende Zelle. II. Mittheilung. *Zeitschrift f. physiol. Chemie*, 63, p. 432, 1909.

Of the extractives, inosite forming 0.193 per cent. of the gray matter and 0.2171 per cent. of the white appears to be present in larger amounts in the latter. Lactic acid, on the other hand, is more abundant in the gray matter.

The specific gravity of the gray matter, 1.041, is somewhat greater that that of the white, 1.032; the specific gravity of the whole human brain is about 1.037, but there is some uncertainty about these figures. Computing from the specific gravity of the whole brain and the specific gravities of white and gray matter, the relative amounts of these two kinds of tissues were estimated as about 55 per cent. of white and 45 per cent. of gray matter. Tables giving a summary of the composition of the human brain have been published by Thudichum. More accurate analyses have been made by Koch. The adult human brain weighs about 1,200-2,000 grams; it contains approximately 78.9 per cent. of water and 100-120 grams of dry protein matter after all ether-soluble and alcohol-soluble matters have been extracted.

Proteins of the brain and nerve tissues.—There is nothing particularly striking or novel about the proteins of the brain which has been discovered so far, except the presence in the brain proteins, or some of them, of the normal amino-caproic acid. This is a matter probably of not very great significance and it is unlikely that this acid is only found in these proteins. The nervous tissues of frogs contain a globulin which coagulates at a very low temperature, 35°, and in the brains of mammals a globulin is present, Halliburton says, which is coagulated at about 42°. It is possible that the coagulation of this protein may be of importance in the pathology of heat stroke. A similar protein is found in the muscles of frogs, according to von Fürth.

Nucleic acid. The brain contains relatively little nucleic acid. This acid, isolated by Levene, had the usual structure of a typical animal nucleic acid. It yielded phosphoric acid, adenine, guanine or the oxidation products of these bases, hypoxanthine and xanthine. It yielded, also, levulinic acid and hence contained a hexose in the molecule like the thymus nucleic acid. Nothing is known of the character of the protein in the nuclei of the brain cells. The Nissl substance is supposed by some to be a nucleoprotein or a nucleoalbumin, but, since there is no certain way of distinguishing these bodies microchemically, it is impossible to be certain that the Nissl substance really contains phosphoric acid. The amount of purines isolated from the brain by Burian and Schur was small in amount, as was to be expected from the relatively small amount of nuclear matter.

Physical consistence of the protoplasm of the nerve cell.—It is very important to determine the physical consistence of the protoplasm of the ganglion cells. The experiments of Harrison, who cultivated embryonic nervous tissue under the microscope, showed that the fine ends of the growing axis cylinders were capable of spontaneous movement. This observation would indicate that the contents of the tips of the axis cylinder, at any rate, were liquid in nature and not jelly-like. It was found, also, by Heger and others in his Institute that the fine terminations of the dendrites in the brain were movable and capable of retraction. It was suggested as the result of this observation that the making and breaking of the connections by means of these processes accounted for the co-ordination in the brain and the psychical disorganization which sometimes occurs. On the other hand, Mott says that the substance of most ganglion cells is of a jelly-like consistence and that in the living cell there is no trace of the Nissl substance to be seen. It is possible that the cell bodies are jelly-like and the tips liquid. The observations of Kite certainly show that most tissue cells of the higher animals have a jelly consistence. Mitochondria, probably of a phospholipin nature, occur in nerve cells (Cowdrey).

Cerebro-spinal fluid.—This fluid was collected by Halliburton from a young woman in whom, owing to some malformation, there was a connection between one nostril and the ventricles of the brain so that the liquid dropped constantly from one nostril. The amount formed was 2.378 c.c. in 10 minutes when sitting quietly, but during straining or when the abdomen was compressed it might rise to 3 and 4 c.c. per 10 minutes. Analysis of the fluid showed that it was alkaline in reaction and of the composition:

Water	99.004%
Solids	0.966
Organic solids	0.118
Inorganic solids	0.878

This is a true secretion, as is shown by its low specific gravity, 1.004-1.007; it contains only a trace of proteid, of a globulin character. Fibrinogen and albumin are absent; it contains a reducing, non-fermentable substance which Halliburton thinks may be allied to pyrocatechin. It is hypertonic to lymph. It is probably formed by the secretory cells covering the choroid plexus. Its function is unknown. It is probable that the secretion of the pituitary passes into the cerebro-spinal fluid.

Trimethyl amine is normally present in cerebro-spinal fluid (Dorée and Golla).

Physiological interpretation of the chemical composition.—In the brain three fundamental properties of living matter are brought to their highest perfection: i.e., conduction, irritability and memory. One function, on the other hand, is practically in abeyance from an early age: that is, the function of growth. The brain early reaches its full development and thereafter division of the nerve cells does not occur, although there may be some increase in complexity and growth of the dendritic processes. We have before us, therefore, in the brain, in as pure a form as we can find it, a tissue specially built for the functions just mentioned, and we may infer from a study of its composition and a comparison with those tissues which are specialized for growth, or movement, what kind of protoplasm, or what substances in protoplasm, are particularly concerned in irritability.

The fact which at once strikes us when we undertake such a study is that the nerve fibers, which are pre-eminently conducting, are surrounded by medullary sheaths which consist chiefly of phospholipins, glycolipins (cerebrosides) and sulphatides. While it is undoubtedly true that conduction takes place in the axis cylinder, which is of the nature of gray matter, yet the gray matter also is unusually rich in these same constituents, or some of them, such as the phosphatides and cholesterol.

The composition and function of the medullary sheaths. The medullary sheaths consist for the most part of glycolipins (cerebrosides), phospholipins of various kinds and cholesterol. What the function of the medullary sheath may be is still uncertain. It has been suggested that it is in the nature of an insulating material which serves to prevent the spreading of the impulse from one fiber to the other; and, on the other hand, a nutritive function has been assigned to it. It is probable that the latter is the true explanation of its function. It may serve as a source of nutriment for the axis cylinder. The length of the latter is often so great that it would seem to be very difficult to control its nutrition from the cell body from which it is derived. Another fact which points in this same direction is that the medullated nerve fibers are almost indefatigable, that is they recuperate at once from fatigue, and the medullated nerves are less fatiguable than the non-medullated. The nerve fibers have been shown by Tashiro to have a very active metabolism, one of the most active of any tissue in the body, and the fact that they are not fatigued would indicate that they have a good source of supply of raw material. In the medullary sheath there is a large amount of inosite, galactose, fatty acids, various nitrogen bodies, phosphoric acid, sulphuric acid and of potassium, calcium and sodium. We find these same materials, or materials of a similar kind, in locations in which a very active metabolism is about to take place. They seem to constitute a peculiarly favorable material for the production of some of the active bodies of the cell. Thus they are found in the yolk of eggs, or in the eggs of lower forms of life, like the starfish. In the egg after fertilization there is a sudden outburst of metabolism at the expense of its stored materials. The composition, then, of the medullary sheath would lend some support to the view that the function of the sheath is nutritive. This view of its function has been greatly strengthened by Tashiro's recent discovery of the active metabolism of the nerve fiber, particularly of the non-medullated fiber. The axis cylinder appears to have a metabolism but little, if at all, less intense than that of the cell body from which it is derived. There are facts which he has found which indicate, also, that this keenness of metabolism is correlated with the rapidity of transmission of the impulse, being more intense in the fibers which transmit the impulse most rapidly, as in mammalian nerve. It is obvious that the power of immediate recovery from a stimulus, the short latent period and rapid transmission are very important in the functioning of the nervous system. If rapid transmission implies, as seems probable, a rapid metabolism also, a reserve supply of food readily assimilable by the fiber to replace at once that used up in the transmission of the impulse would seem to be necessary. Some mechanism similar to that of the medullary sheath would seem the most simple and direct way of solving the problem. The medullary cells on this view function as a kind of nurse cells, similar to the nurse cells of the eggs of many invertebrates. The laying down of this reserve substance within the living matter of the axis cylinder would obviously have many disadvantages and would probably reduce the rate of its metabolism and the rate of conduction, since that appears to be the result generally of an accumulation of foreign or lifeless matter in the living protoplast. The device adopted by the animal economy of having long nerve fibers, very thin, so that the surface is large in proportion to the bulk, and surrounded by a large number of special nourishing cells, would appear to be a most satisfactory solution of a very difficult problem which the organism had to face. If such rapid conduction were not needed, the problem might be solved by introducing relays of cells so that the length of any single axis cylinder was not beyond the powers of nourishment from the cell body; but this contrivance is not suitable where very rapid action is required, as in the brain and in those skeletal muscles by which an animal escapes death or secures its food by superior agility; for at every relay, or sinapse, there is an unavoidable delay in the rate of transmission of the impulse. In the internal organs where speed and agility, or indefatigability, are not the prime requisites, the nerves are still non-medullated for the most part and relays are used.

A second very striking and interesting fact in the chemistry of the brain, and one which may perhaps be correlated with the provision of reserve food in the medullary sheath, is the absence of reserve carbohydrate food in the brain, and the entire absence of neutral fat. Glycogen, which is found in so many tissues and which occurs in the young, undifferentiated nervous system, is absent in the adult brain. When medullation appears it disappears and no other carbohydrate is present in the adult brain, except inosite and galactose. The function of inosite is unknown, but it readily combines with phosphoric acid to make phytin and similar esters. Perhaps it plays the rôle of an intermediary substance. Inosite is readily destroyed when taken by the mouth. The absence of neutral fat while the phospholipins and other compound lipins are so abundant is indeed remarkable.

If the medullary sheath has this function of serving as a nutritive material, the question at once arises as to the constitution of the real living matter, the matter which has the properties of conduction, irritability and memory par excellence. How does it differ, for example, from living matter which is chiefly secretory, or metabolic, or contractile? The only facts to guide us in this dark region are analyses of the gray matter, for although it is impossible to get gray matter entirely free from medullary matter, the admixture in some of the regions of the brain is small. The main facts which may be gathered are these: The proportion of protein in the gray matter is high. The proteins make in the gray matter a larger proportion of total solids than in the white; the proportion of phospholipins is also high. Irritable living matter has, therefore, a larger proportion of water, 85 per cent., than contractile; it contains relatively less protein, though it is still protein which makes the main part of its solids; it contains a very large preportion of phospholipins and cholesterol, but no fat or carbohydrate. The irritable, conducting protoplasm is, therefore, decidedly aqueous and the colloids are protein-phospholipin colloids. These few facts tell us very little about this irritable matter, but so far as they go they indicate the importance in nervous activity of the chemical substances just mentioned. It may be significant that this substratum is of a kind which so readily forms liquid crystals: i.e., a kind whose molecules are readily oriented.

A third fundamental property of the human cerebrum is memory. It is possible to make some kind of an impression on the human brain of so persistent a nature that when made at from 4-6 years of age it may persist for from forty to eighty years. Of the physical-chemical basis of memory very little is known. When an impulse strikes into the brain something remains there, some trace, perhaps an enzyme. Something happens to the brain with every impulse fed into it. We may perhaps picture the process as follows in a tentative way, although the evidence is extremely meager that this picture at all represents the real process. There is a growing number of facts which indicate that when nerve impulses impinge on cells they do not at once arouse the peculiar activity of that kind of cell. There is always a latent period. It seems that under the influence of the nerve impulse a substance is produced which in its turn acts as the excitant of the cells' activity. This would explain the latent period of muscle and other cells. We may regard the gastric hormone or the secretin of the cells of the intestinal mucosa as such a substance. In muscle possibly lactic acid is the hormone. By a hormone is meant a substance which rouses to activity. Now in the nerve cells it may be that such memory hormones are produced and accumulate, or persist in the cells. There are only two facts of the chemistry of the brain, so far as the author sees, which may be brought into relation with this extraordinary power of memory. One of these facts is the great reduction of autolytic power of the brain tissue as compared with that of any other tissue of the body. If the stomach, the liver, the spleen, the muscles are finely hashed and kept sterile in water at 37° C., their proteins undergo rapid changes of autolytic digestion. The brain under similar conditions undergoes very little autolysis. In fact, the autolysis was at first overlooked, it is so slight, a very little alkali apparently checks it completely. What autolysis there is seems to occur in the first few hours after death. This great reduction of the brain's autolytic power points toward a stability of the brain's proteins, of an absence of wear and tear in the brain, which possibly, but only possibly, for in reality the matter is guesswork, but which involuntarily recalls the stability of impressions and memories in the brain. When autolysis does occur, or when nerve cells are destroyed in injury or by toxins, then memories disappear also, showing that there is some substantial basis of the memories. The stability of the nerve cells, the absence of autolysis, of cell division, all would appear to favor the stability of impressions of whatever kind which are made on the brain by the events about us. It is possible, also, that this absence of autolysis is correlated with the fact that the brain does not lose weight in starvation, but lives on those organs which autolyze the most readily.

The second fact of interest, although it may be simply a parallelism and not fundamentally connected with the process of memory, is the presence in the cephalin of the brain of very unsaturated acids of the type of linolinic acid. It has already been pointed out on page 67 that in the spontaneous oxidation of linolinic acid phenomena closely paralleling memory and learning occur. These are autocatalytic processes. In the case of linseed oil the memory, if it may be so called, consists in the persistence in the oil of an intermediary, catalytic oxidation product. If this substance could in any way be made stable, so that it would persist for a long period in the dark, linseed oil once illuminated would remember that illumination forever and would always respond differently, when exposed with oxygen to the light, from linseed oil which had not had the experience of a previous illumination. If it were possible that an impulse coming into certain cells caused in those cells the formation of a persistent autocatalytic, intermediary oxidation product, a physical basis of memory might be given. Our ignorance is so profound, however, that at present we can do hardly more than guess at this interesting problem and we must be on our guard against assuming that the basis of memory is in reality that sketched above. All experience shows how often we may be misled by seductive parallelisms of this nature into the conclusion that Nature is working in the manner we imagine her to be. To quote an old and homely, but true, saying: "Nature knows more than one way to kill a cat."

Having now considered the general composition of the nervous tissue and the connection between that composition and function, we may pass to the consideration of the metabolism of the brain. Is the respiration of the brain keen, or very low? What substances does it use as foods? What are its waste products?

The first question, that of the respiration or rate of metabolism of the brain, is very important for the light it will throw on the nature of the nervous mechanism. If those processes consist, as many have believed, in the simple dislocation of inorganic ions in the periphery of the nerve fiber, a dislocation which may be accompanied by a change of state in the nerve colloids,—if they are, in other words, physical processes,—one would expect the metabolism of the brain to be small, very small, for no material would be used up in this process, a temporary dislocation of the ions alone occurring. The only energy needed is the jar, or slight explosion, which starts it going, and if this took place in the sense-organs no energy would be required to be expended in the brain. If, on the other hand, the nerve impulse is in the nature of a chemical change taking place with great ease and swiftness, and if the velocity of conduction is proportional to the respiration, then the metabolism of the brain, as measured by its respiration, should be extremely high, higher than that of any other tissue of the body. The lack of cell division, the stability of the nerve cells, the large amount of neuroglia, the absence of autolysis, all point to a low formative metabolism, that is a low protein or growth metabolism after the adult condition is reached; but, on the other hand, the respiratory metabolism should be very high, if this is correlated at all with irritability. What, then, is known of the metabolism of the brain?

A few general facts may at once be noted. First, the blood supply to the brain is unusually large and every provision is made to supply the brain with a copious amount of blood. Not only are there the large direct carotids, external and internal, but there are also the indirect anastomoses through the vertebrals, which, in dogs at least, are able alone to maintain the life of the brain. This copious blood supply points unmistakably to a very vigorous metabolism of some kind which involves oxygen.

A second well-known fact is that a supply of oxygenated blood is absolutely necessary to the maintenance of consciousness in human beings. Consciousness cannot persist more than a few moments without it. Even though the tissues are saturated to their full extent with oxygen by inhalation of the latter and forced respiration, the supply cannot maintain consciousness for more than five minutes and, if the supply of oxygen is low, loss of consciousness comes on easily. Thus with a low blood pressure, as in fainting or syncope, the higher centers of the brain appear to pass out of action at once. Similarly swimmers under water holding their breath to the utmost sometimes become unconscious, and that it is the oxygen of the blood which is important is shown by the fact that if that oxygen be replaced in the blood by carbon monoxide, as in poisoning by illuminating gas, consciousness is lost. There is no question from this observation that it is only the oxygenated brain tissue which is irritable and conscious. In the absence of oxygen the brain loses its function more quickly than any other tissue of the body. These facts show that brain tissue either has a higher rate of consumption of oxygen, so that it exhausts its supplies more quickly; or that the nervous processes involve more oxygen than others; or that the brain has a smaller supply of stored oxygen than any other tissue of the body. The great dependence of the brain on atmospheric oxygen may be correlated, also, with the fact that there is no stored carbohydrate. The presence of stored carbohydrate like glycogen appears to increase the powers of tissues to live without atmospheric oxygen.

A third general fact which points unmistakably in the same direction is the great sensitiveness of the brain to cyanides and other drugs.

Cyanides certainly check respiration and undoubtedly combine with the cell protoplasm. The brain is most sensitive to cyanides, these drugs abolishing consciousness very quickly, and particularly the activity of the highest centers. Since it can be shown that many drugs accumulate in the nervous system, it must be that the latter has compounds with chemical bonds in a reactive or open form with which they can unite. It has been shown, too, that those tissues which are most active in respiration or have the highest rate of respiratory metabolism are the most easily affected by drugs like the cyanides. Hence, the sensitiveness of the brain to drugs also indicates clearly a higher rate of metabolism in the brain than in other tissues.

The persistence of the weight of the brain during starvation means either that the rate of formation of its substances must be very high or the rate of consumption extraordinarily low, since it is always the case that the more rapidly metabolizing tissue draws for its raw material upon the tissue having the less intense metabolism.

The blood coming from the brain also indicates metabolic changes in that organ. It is venous blood. Oxygen is consumed there and in large amount. That means that substances in the brain have been oxidized or burned, hence they must be constantly replenished or they would be used up. The blood coming from the brain has less sugar in it than in the arterial blood going to the brain. Evidently the brain consumes sugar. The blood of the jugular vein also contains less fibrinogen than the arterial blood of the carotid artery. But the interpretation of this fact is uncertain.

Direct determinations of the rate of respiration of the nerve fibers have now been made by Tashiro, who found indeed that the amount of carbon dioxide given off from the nerve fiber per gram of substance in 10 minutes was very high, higher than that of any other tissue examined. Thus the resting non-medullated claw nerve of the spider crab at 19° gave 6.7×10^{-7} grams CO_2 per cg. per 10 minutes; the frog's sciatic nerve, a medullated nerve, gave 5.5×10^{-7} grams CO_2 per cg. per 10 minutes, and the rate was about doubled when the nerves were stimulated by a weak electrical current. This rate of metabolism was probably somewhat larger in the cut than in the uninjured nerve, owing to the effect of the injury, but the amount is very high, higher than for the frog as a whole when equal weights are compared. The examination of the various parts of the nerve shows, also, that there is a gradient in its metabolism, the part of the nerve fiber nearer the source of the normal stimulus having a higher rate of metabolism than that farther from this source. Ganglia, too, seem to have a somewhat higher rate of metabolism than the nerve fibers, but the difference is not very great. Evidently the nucleus of the cell is not so important in respiration as had been thought. There is no doubt from these observations that nerve cells and nerve fibers produce carbon dioxide at a very rapid rate. If equal weights of nervous and other tissues are compared, the nervous tissues are found to give off more carbon dioxide per gram of tissue than any other tissues.

The examination of the oxygen consumption of the brain has been made by the method of Barcroft, which consists in estimating the amount of oxygen in the blood going to and coming away from the brain. The results of this study were as follows (Alexander and Cserna):

By comparing the oxygen content in the arterial blood with that of the venous blood of the sulcus longitudinalis and measuring the velocity of flow, it was found that in the waking animal the brain consumed 0.360 c.c. oxygen per gram per minute. This amount is enormously greater than that of skeletal muscle of 0.004 c.c. and of the salivary glands of 0.028 c.c. per gram per minute. In narcosis by ether the consumption sank in the mean 77 per cent.; the CO_2 output 59 per cent. In morphine narcosis the oxygen consumption sank 57 per cent.; the CO_2 output 61 per cent. Under ether the O_2 consumption could be reduced still farther. The first effect of the anesthetic is, however, to increase the consumption of oxygen. The initial action might increase the oxygen consumption 150 per cent.; the CO_2 production 359 per cent. This increased consumption of oxygen corresponds to the preliminary stimulation which is the first effect of the anesthetic. Tashiro and Adams found in the case of the nerve fiber and other cells that the first effect of other anesthetics was to increase the CO_2 production concomitant with the excitatory action, and that the second effect was to reduce the CO_2 output. By injection of $CaCl_2$, which aroused the brain from anesthesia by $MgSO_4$, the metabolism rose again to normal amounts. It is clear, then, that there is in the brain, as in the nerve fiber, a complete or wide-reaching parallelism between metabolism and the state of irritability. These results completely confirm the view of the chemical nature of anesthesia. They indicate clearly the conclusion that conduction and irritability are at bottom chemical processes, rather than physical, as they are sometimes pictured.

The rate of consumption of oxygen by nerve fibers has not yet been made so accurately and directly as the rate of carbon dioxide production, as the methods are badly in need of improvement. But there is no doubt from indirect evidence that the irritability of the nerve fiber is dependent upon a supply of oxygen and that it fatigues more rapidly in the absence of oxygen than when it is present. Thus, if a frog's sciatic nerve is stimulated in hydrogen or nitrogen, it becomes non-irritable more quickly than in oxygen and it recovers if returned to oxygen. The oxidation appears, however, to be in part independent of the oxygen of the atmosphere, or else there is some sort of storage of oxygen in the nerve, as the irritability of frogs' nerves will persist for some hours, although diminished in amount, in the absence of oxygen. The carbon-dioxide production does not appear to be due to a direct oxidation by the oxygen of the air. It may be that the oxygen in an available form is stored in some kind of a loose compound in the medullary sheath. Whether the nerve stores oxygen or not must be left for further investigation to decide. The carbon dioxide might be formed in various ways, either by carboxylase from the proteins or amino-acids, or by fermentation from the carbohydrates, or by indirect oxidation of the fatty acids, or by the decomposition of an unstable peroxide.

The attempt has also been made to get some direct measure of the metabolism of the brain in human beings by the use of the calorimeter, but so far without results. The brain in human adults weighs between 1,200 and 2,000 grams; it makes, therefore, between 2 per cent. and 3 per cent. of the weight of the body. On the other hand, the muscles make 42 per cent. of the body weight. If the brain could be brought into a condition of complete rest, the maximum change of metabolism of the body produced thereby would be only two per cent. But it is impossible to make the brain rest. The centers are always more or less active even during sleep, so that the difference in the metabolism to be expected would only be a per cent. or so of the total. Impulses are always coming into the brain over the sensory nerves from the skin or sense-organs or from the tendons and muscles. No difference in heat output, or in respiration, or metabolism could be detected by Benedict as a result of active intellectual work. There is, however, one clear and unmistakable evidence that intellectual work involves metabolism, presumably respiratory activity of the brain tissue: namely, keen intellectual work always causes a turgidity of the head and brain; blood increases in the head when one is working, as is evidenced by the flushed face, and if it does not increase keen intellectual work is impossible. In its absence one becomes sleepy, pale, and thoughts are sluggish, relations are not perceived. No change can be detected, so far, in the composition of the urine as a result of brain work. There is, in fact, no product of brain metabolism, that is, no substance whose presence indicates brain metabolism, which has been found as yet in the urine. Changes in the metabolism of the master tissue of the body cannot, therefore, as yet, be determined by any examination of the excretions of the body. It is, hence, impossible to say whether any drug, preservative or food substance has acted harmfully or beneficially on the brain or nervous system by an examination of the urine or excretions.

One of the most striking facts about the chemistry of the brain is the very large amount of phosphoric acid it contains. This fact was discovered very early, since in the ashing of the brain substance some of the acid is reduced and phosphorus formed. It has become so much a matter of common knowledge as almost to lose its significance. This phosphoric acid is in part free, as the salts of phosphoric acid; in part it is in combination in the nucleic acid and in the phosphoproteins; and in larger part it is present in the phospholipins. Liebreich crystallized the impression of the importance of phosphorus in the metabolism of the brain in the saying: " Ohne Phosphor keine Gedanken " (Without phosphorus no thoughts). It seems that phosphoric acid plays a highly important, though still obscure, rôle in the metabolism of all living matter. Not only do we find compounds of phosphorus most common in the protoplasm of the brain, and its importance was emphasized by Thudichum by the selection of the word phosphatide to indicate that the other radicles were grouped around it, but phosphorus occurs in large amounts in the nucleus and in phytin; it helps regulate the cell reaction and, combined with calcium and some organic radicles, it forms the basis of many structures like the bones; it has, besides, some peculiar relation to the alcoholic fermentation of sugars and other fermentations. Calcium metaphosphate may not be at the basis of life, but the importance of phosphoric acid is not to be denied.

Summary.—We may summarize the contents of this chapter very briefly, as follows:

In its chemical composition the human adult brain is characterized by the very large amount of alcohol-ether-soluble material in its solids; by its vigorous respiration; by its low rate of autolysis; by its lack of formative or growth metabolism. Of the solids the alcohol-ether-soluble material consists of a variety of phosphatides, or phospholipins, such as lecithin, kephalin, myelin, sphingomyelin, amino-myelin, paramyelin; of glycolipins or cerebrosides, such as phrenosin, kerasin, cerebron, homocerebrin, cerebric acids; of sulphatides; of cholesterol and phrenosterol, and of nitrogenous fats or amino-lipins (amino-lipotides). The brain contains no neutral fat. These substances are found chiefly, but not exclusively, in the medullary sheaths of the nerve fibers. Their proportion in the gray matter is also high. The protein material, or neuroplastin, contains only a small proportion of nucleic acid, but differs from other proteins in that it contains a normal amino-caproic acid or glycoleucine. The proteins of the brain have not in other respects been shown to be different from those of other tissues, although they are no doubt specific to the tissue. It has not yet been possible to examine the neurokeratin separately from the other proteins of the brain. There are also present extractives, among them inosite, lactic, succinic and formic acids, creatine, tyrosine, glyco-leucine, hypoxanthine and other nitrogenous bases of an unknown nature. On the whole, the brain extractives are very similar to those of muscle. Of the functions of these bodies nothing is known. Inorganic salts are also present.

The function of the medullary sheath is probably nutritive. The really active protoplasm of the nerve cell and body is characterized by its large content of water, at least 83 per cent. and probably more, by the absence of stored glycogen and neutral fat, and by the presence of large amounts of phospholipins and cholesterol. While the growth or formative metabolism of the brain is low, it has a very keen respiration and oxidative metabolism, the keenest of all tissues of the body, and the metabolism of the nerve fiber does not differ very much from that of the nerve ganglia in its intensity. The activity of the brain and nerve tissues in general is very dependent upon oxygen and they produce large amounts of carbon dioxide, the amount of which runs in general parallel to the state of activity of the tissue. It is probably its rapid and keen metabolism, which means that it has many reactive bonds in it, which makes the nervous tissue so very sensitive to drugs of all kinds.

We may infer tentatively from these facts that the lipins probably play a very important part in cell irritability without at this time venturing to define more in particular just the rôle they are playing. The phenomena of conduction appear to be closely correlated with the respiration of the tissue and are, hence, probably chemical in nature. More work along these lines is, however, necessary before any definite conclusion as to the nature of the nerve impulse can be drawn. At the present time it appears possible that it may be in the nature of an explosive decomposition of an organic oxide or peroxide which is propagated along the fiber. There are difficulties in the way of this view, one of them being the fact that the nerve seems to liberate no heat at the time it conducts. The chemical composition has so far thrown very little light on the chemical basis of memory, but it is suggested that possibly the presence in the brain of highly unsaturated acids of the linolinic-acid type may possibly be in some relation to this phenomenon, since in its oxidation this acid shows some of the phenomena we are accustomed to call memory, teaching and forgetting when they occur in living things.

We may close this chapter in no better way than in opening the question of the origin of the psychic qualities which are so related to the nervous system. Do these qualities arise *de novo* in the nervous system? Are they not found in their faintest form way down the slope of animal life? Do we not indeed see the beginnings of psychic life among the plants? And is it possible to start with the plants? Do not the foods every minute change into living matter in our bodies? Are not the atoms the same in the foods and living matter, and is it possible that they have different properties in the living and lifeless form? The atoms we now know are composed of electricity and the valences, or chemical bonds, are probably also electrical in nature. Are our thoughts also at the bottom electrical? Whenever a nerve impulse sweeps over a nerve it is accompanied by an electrical disturbance, and this disturbance is the surest sign of life. When the nerve impulses play back and forth over the commissures of the brain they are accompanied by this pale lightning of the negative variation. Is that pale lightning what we recognize as consciousness in ourselves? It would seem that there must be some psychic element in every electron if the atoms are made of electrons. There must be some psychic disturbance in every union of hydrogen and oxygen to make water and in every wave of the wireless telegraph. When an electron moves it generates a magnetic field; does it also generate a psychic field? How shall we escape the conclusion that there must be a psychic element in all matter both living and lifeless, since that matter is the same in the two forms? May it not be that just as magnetism, which is probably an attribute of all molecules, becomes most evident under certain conditions in certain substances, so the psychic life common to all matter shows its true character plainly only when organized as it is in the brain during its life? A magnet when heated loses its magnetism as surely as an organism when heated loses its vitality and its psychic life. In the case of magnetism all that has happened by the heating is that the orientation of the molecules has been changed so that the magnet is no longer an individual; in the case of the organism a similar loss of individuality results.

REFERENCES. BRAIN.

1. *Thudichum:* Die chemische Constitution des Gehirns des Menschen u. der Tiere. Tübingen, 1901.
2. *Rosenheim:* Purification of kerasin and phrenosin. Biochemical Journal, 8, 1914, p. 110.
3. *Dorée and Golla:* Trimethyl amine in cerebro-spinal fluid. Biochem. Jour., 5, 1911, p. 306.
4. *Koch:* Zur Kenntnis der Schwefelverbindung des Nervensystems. Zeits. f. phys. Chem., 53, p. 486, 1907.
5. *Simon:* Autolysis of brain. Zeits. f. physiol. Chem., 72, p. 463, 1911.
6. *Kutscher and Lohmann:* Autolysis of brain. Ibid., 39, p. 313, 1903.
7. *Pighini:* Indophenol oxidase in central nervous system. Biochem. Zeitschrift, 42, 1912, p. 124.
8. *Barbieri:* Catalase in cerebro-spinal fluid. Biochem. Zeits., 42, 1912, p. 138.
9. *Nizzi:* Absence of eterase and lecithase in nervous tissues. Biochem. Zeits., 42, 1912, p. 145.
10. *Masuda:* Cholesterol and fatty acid content of different parts of brain. Biochem. Zeits., 29, 1910, p. 161.
11. *Alexander:* Respiration of brain. Biochem. Zeits., 44, p. 127, 1912.

12. *Loeming and Thierfelder:* Cerebrosides. Zeits. f. physiol. Chem., 74, p. 282, 1911.
13. *Batelli and Stern:* Peroxidase of brain. Biochem. Zeits., 13, p. 44, 1908.
14. *Levene and Jacobs:* Sphingosine. Jour. Biol. Chem., 11, p. 545, 1912.
15. *Baumann:* Cephalin; nature of base in. Biochem. Zeits., 54, 1913, p. 30.
16. *MacArthur:* Cephalin. Constitution. Jour. Amer. Chem. Soc., 1914.
17. *Abderhalden:* Normal leucine in brain. Zeits. f. physiol. Chem., 84, 1913, p. 39.
18. *Frank:* Cephalin in liver. Biochem. Zeits., 50, 1913, p. 273.
19. *Schreiber and Lenard:* Oxycholesterol in brain. Biochem. Zeits., 49, p. 458, 1913.
20. *Kaufmann:* No free choline in ox brain. Zeits. f. physiol. Chem., 74, 1911, p. 175.
21. *Hannemann:* Inhibitory influence of the brain on body metabolism. Biochem. Ztschr., 53, p. 80, 1913.
22. *Abderhalden and Weil:* Amino-acids by hydrolysis of gray and white matter of brain. Zeits. f. physiol. Chem., 83, p. 425, 1913.
23. *Levene and Jacobs:* Cerebrosides of brain. Jour. Biol. Chem., 12, 1912, p. 389.
24. *Levene and Jacobs:* Cerebronic acid. Jour. Biol. Chem., 12, p. 381, 1912.
25. *Takaki:* Tetanus binding constituents of brain. Hofmeister's Beiträge, 11, p. 288, 1907-8.
26. *Halliburton and Dixon:* Choroid plexus and secretion of cerebro-spinal fluid. Jour. Physiol., 40, 1910, p. xxx, Pro. Phys. Soc.
27. *Koch and Koch:* Chemical differentiation of the brain. Jour. Biol. Chem., 15, p. 423, 1913.
28. *Koch, M.:* Brain of the albino rat at birth compared with that of fetal pig. Jour. Biol. Chem., 14, p. 267, 1913.

CHAPTER XIV.

THE CONTRACTILE TISSUES. MUSCLE.

Amount.—Muscle forms the greater proportion of the living tissues of the body. In an adult the weight of the muscle is about 42 per cent of the body weight. Of the whole metabolism during muscular rest about 50 per cent. is the metabolism of muscle; and during muscular activity probably nearly three-fourths of the chemical changes occurring are taking place in the muscle. These figures will make it clear that the muscles dominate, as it were, the metabolism of the body and their composition and the nature of the chemical changes taking place in them become of the first importance.

By far the greater proportion of the muscle is under the control of the will and is called voluntary muscle; but a smaller portion is not so controlled and is found in the heart, along the arteries, in the walls of the alimentary canal, in the uterus, bladder and other organs. This latter is involuntary muscle. These two kinds of muscle differ in their physiological properties. Thus the voluntary has a short latent period, it contracts and relaxes very quickly, whereas the involuntary muscle contracts much more slowly. They differ also in their optical appearance, in their structure and in their chemical composition. Heart muscle in all its properties takes an intermediate position between the other two. Thus voluntary muscle appears optically to be composed of alternate lighter and darker bands of matter. It is said to be cross-striated; its nuclear material is relatively small in proportion to the whole bulk; it contains a very high proportion of proteins and extractives and relatively little water. Involuntary muscle has relatively more nuclear material and it is not cross-striated, so that it is called smooth muscle. It is very faintly striated longitudinally in fixed preparations. All muscle begins development as smooth muscle. Cross-striated muscle may be regarded as smooth muscle which has been differentiated into a special structure, probably securing thereby greater speed of contraction. Having less nuclear material, it is possible that the striated material has a lower rate of nitrogen metabolism than the unstriated.

Structure.—A striated muscle consists of bundles of muscle fibers united or bound together by connective tissue. The amount of this connective tissue varies in different muscles and the more there is of it the tougher is the muscle. The connective tissue often contains deposits of fat. The blood vessels running in the connective tissues carry a large supply of blood to the muscles, the capillaries branching about the muscle fibers in such a way that they are in intimate contact with them. This arrangement brings directly to the muscle fibers a copious supply of oxygen and food and provides for the rapid elimination of wastes. During activity the arterioles dilate, owing in part to the action of chemical substances formed by the muscle, and a more than usually copious supply of blood is provided thereby. The nerves of muscle are both motor and sensory. The sensory are distributed at least in part to the tendon, but the motor fibers branch freely and a branch penetrates each muscle fiber, piercing the sarcolemma (the membrane surrounding the fiber) and ending in a swelling consisting of nuclei and granular protoplasm called the muscle plate and situated about the middle of the fiber. While the nerve fiber thus enters into a very intimate connection with the muscle substance so that nerve impulses pass freely to the latter and profoundly influence its metabolism, the nerve fibers are none the less nourished from the cells which originate them and they degenerate if they are cut off from these cells.

The muscle fiber itself is the real living, contractile part of the muscle. Such a fiber is an elongated, spindle-shaped cell. The structure differs in different kinds of voluntary muscle, appearing to have reached the highest perfection and differentiation in the wing muscles of insects in which the power of rapid contraction and relaxation is most developed, the cross-striation and fibrillar arrangement of the contents being most marked; in mammalian muscle, also, it is highly differentiated, but in the amphibia, and in particular in the very low vertebrate, Ammocetes, the fibers are of a more primitive structure and their cellular character is more plainly seen, the differentiation not being complete. In these latter the nuclei are arranged down the center of the fiber, surrounded by a larger or smaller part of undifferentiated protoplasm; but in mammalian and insect muscle the nuclei, of which there are several, are distributed through the contractile protoplasm and many lie close under the sarcolemma.

The exact structure of the differentiated contractile substance is still uncertain. It certainly consists of alternate disks or bands of different optical and staining properties. One of these disks is doubly refracting and more dense than the other and it stains more deeply in various stains like iron hematoxylin. In many muscles, and in particular in the insect muscle, these alternate disks appear to be arranged in fibrils which are called sarcostyles. There seems to be no doubt that in these muscles, at any rate, such a fibrillar arrangement really exists. Their exact structure in living muscle is still a matter of controversy. There are, besides, in the living muscle granules, some of which may be glycogen. It is generally supposed that there is also about the fibrils a more liquid portion called the sarcoplasm, but the existence of this is very uncertain. Some observers believe that there are networks of an elastic material extending in a regular pattern through the muscle and contributing to its relaxation. Such networks can be made out in the fixed muscle, but their existence in the living muscle has been disputed. Kite found that the resistance to the movement of a needle in the muscle of Necturus was about the same in all directions, and Kühne and Eberth observed in certain muscle fibers of frogs a minute parasitic nematode worm which moved through the content of the muscle fiber as if it were fluid.

The physical consistence of the contents of muscle fibers has been carefully examined by Kite by his very ingenious method of microscopical dissection. By means of very fine glass needles carried in a mechanical holder he cut the large muscle fibers of Necturus into pieces. He found the inside of the fiber, the protoplasm, the most "viscous, elastic and cohesive of the living gels examined." "The muscle substance sticks to a glass needle and can be drawn out into extraordinarily long threads, which when released almost regain their previous shape. The absorptive powers and turgidity of this substance are comparatively high." "When the whole or a piece of a muscle cell is stretched the striations become faint or disappear, only to reappear when the tension is removed." "If the point of a needle be pushed into a muscle cell, it can be moved in one direction about as easily as another." He goes on to say, "The optical appearance of a striped muscle is very misleading. Dissection has shown that the dark bands seen in living muscle are produced by concentrated areas of muscle substance, which absorb enough transmitted light of low intensity to appear as dark bands in the optical image. No fibrils could be dissected out. The substance lying between the concentrated regions and appearing as light bands is a highly viscous elastic gel, and has no physical properties that serve to distinguish it from the surrounding protoplasmic gel." "Hence absorption, diffraction, refraction and dispersion are involved in the formation of the optical image of striped muscle." The sarcolemma is extremely elastic. It is evident from this description of Kite that the muscle substance consists at least in part of alternate disks of more and less concentrated gel. It is, however, unlikely that this gel would split into the very regular and characteristic fibrils of the muscle of the water beetle as described by Schaefer if there was not in these latter, at any rate, a well-marked fibrillar arrangement into sarcostyles. The muscles of Necturus are possibly less differentiated than those of the insects. There is no doubt, however, that in all these cases the substance is highly elastic, and of a jelly consistence; it is probably not fluid as it was once supposed to be.

General composition.—Both striated and unstriated, mammalian muscle contains, in the adult, from 72-78 per cent. of water, and 22-28 per cent. of solids. In youth and in the fetus the per cent. of water is higher. The solids are very largely protein in nature, and in this respect the muscle shows a striking contrast to the nervous system in which the lipins hold so prominent a place. The total ether extract, lipin of all kinds, of the dog's heart is between 2.86-3.73 per cent. (Lederer and Stotte). The glycogen of the dog's heart is between 0.709 and 0.576 per cent. In the skeletal muscle it may be a good deal less than this, namely, 0.15 per cent. -0.3 per cent; 18-20 per cent. by weight of the muscle substance is, therefore, protein. The composition of ox meat, after extraction of the glycogen and fat, was found by Argutinsky to be as follows: C, 49.6 per cent.; H, 6.9 per cent.; N, 15.3 per cent.; Ash, 5.2 per cent.; O and S, 23.0 per cent. The extractives of muscle are also important. The proteins may be first considered. They are generally divided into two groups, the proteins of the muscle plasma and the proteins of the stroma. Part of the proteins are soluble in ammonium chloride solution. This protein is sometimes called myosin. In human muscle Danilewski found of 21.48 per cent. total solids, 3.68 per cent. extracted as myosin and 11.90 per cent. remaining insoluble as protein of the stroma. In most animals a somewhat larger per cent. of the protein may be extracted by ammonium chloride than this. In calves' muscle 15.6 per cent. was insoluble protein, 3.6 per cent. soluble. The composition of the muscle is strikingly different from that of the electrical organ which is formed from muscle. In the torpedo the electrical organ contains only about 8 per cent. of protein and over 90 per cent. of water. The development of a large amount of protein substance in muscle appears to be correlated, therefore, with the function of contraction.

Proteins of muscle.—The proteins of muscle are generally divided into two groups, those of the muscle plasma and of the muscle stroma. If a muscle is ground or hashed and then pressed in a hydraulic press, with very high pressures, 250 to 300 atmospheres, there is squeezed out a thick fluid, the plasma, which has the property of clotting spontaneously on standing. The amount of the plasma, which may thus be obtained from most muscles when they are living and subjected to the high pressures just mentioned, is about 60 per cent. of the weight of the muscle. The plasma has often a faint red color. After it clots the clot contracts somewhat just as does the blood clot, though not to the same degree, and the liquid which is thus expressed from the clot is called the serum. Muscle plasma may also be obtained by a method used first by the physiologist Kühne. The muscle is frozen, shaved fine in a knife machine or microtome, rubbed with sand while still cold and then pressed under high pressure. The part which remains behind in the press is called the stroma. Of course the proteins may be obtained directly from the muscle without this preliminary expression of the muscle juice or plasma, by the use of appropriate solvents. Thus either 5 per cent. sodium chloride or 5 per cent. ammonium chloride may be applied directly to the ground-up fresh muscle. By this means a portion of the proteins go into solution and these are supposed to come from the muscle plasma. The stroma, or the insoluble portion of the muscle, is insoluble in all neutral salt solutions and can only be dissolved by the action of strong acids and alkalies. It behaves like coagulated protein.

Proteins of the muscle plasma.—There is a great deal of uncertainty and contradiction in the literature in the account of the muscle proteins and much confusion in nomenclature. In these circumstances we shall follow the most simple account of von Fürth, who has given a great deal of study to this subject. It should be stated at the outset, however, that the conclusions of von Fürth are not accepted by all and have been contradicted in many important points recently. The difficulty arises from the uncertainty whether the change of a soluble to an insoluble protein involves a change in chemical composition, or is only a change in the state of aggregation. According to von Fürth there are three coagulable proteins, albumins or globulins in the narrow sense, in muscle plasma; these three are myogen, myosin and soluble myogen fibrin.

Myosin. This is the musculin of the older observers, or paramyosinogen of Halliburton. It is a globulin-like substance coagulating on quick heating between 44 and 50°, and it is characterized by its power of spontaneously coagulating on long standing, when it passes into an insoluble protein called myosin fibrin. Myosin is soluble in neutral salt solutions, is precipitated by dialysis and by water; is easily salted out of solution by one-half saturation with $(NH_4)_2SO_4$, which permits its separation from myogen; and it is easily precipitated by acids. Both in salt solution and when standing in water it becomes insoluble (myosin fibrin). Its per cent. of composition is as follows: (Kühne and Chittenden).

C, 52.79; H, 7.12; N, 16.86; S, 1.26; O, 22.97.

Myogen. This is another protein constituent of muscle plasma. It was formerly called, by Kühne, myosinogen, but this name von Fürth thinks should be dropped, since it implies that it gives rise to myosin and this is incorrect. There is about three to four times as much myogen in muscle plasma as myosin. Unlike myosin it is not precipitated by dialysis with water, nor by one-half saturation with ammonium sulphate. Thus it belongs to the class of albumins rather than to the globulins. It is precipitated by dilute acids like a globulin. The peculiar property which differentiates it from other proteins except myosin is its power of spontaneous coagulation. When in solution it passes first into a soluble metastable form called soluble myogen fibrin, and then into myogen fibrin. Some consider this simply a change in the state of aggregation of the same protein, the protein in solution passing slowly into an aggregated state. According to von Fürth, then, the spontaneous coagulation of muscle plasma takes place according to the following scheme:

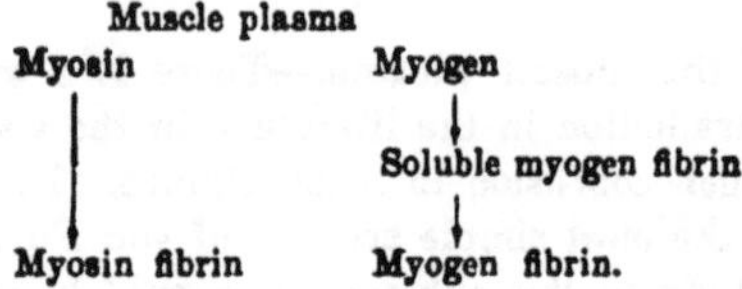

It is not impossible that this slow coagulation is due to the gradual hydrolytic decomposition of a salt of the protein, forming an insoluble free protein, but of the nature of the change involved nothing is certainly known. Halliburton believed that it was caused by an enzyme analogous to the fibrin ferment, but this view has not been accepted by von Fürth. The myosin fibrin and myogen fibrin are insoluble proteins behaving like coagulated proteins. They cannot be turned back into myosin and myogen. The change is not reversible.

Soluble myogen fibrin. This substance is easily detectable, either in solution or when present in muscle itself, by its extremely low temperature of coagulation. It coagulates at 30-40°, so that when soluble myogen fibrin is formed from myogen there is a fall in the temperature of coagulation of the solution from 55-60° to 30-40°. According to Botszzi, this apparent coagulation of the myogen is due to the heating hastening the spontaneous change of myogen to myogen fibrin, or in other words it is not a true coagulation, but the hastening of the process of aggregation already going on. The difficulty in this explanation is that it does not explain the change in the state of aggregation. According to von Fürth soluble myogen fibrin behaves like a globulin. It is precipitated by half saturation with ammonium sulphate or by dialysis, and is insoluble in water. It differs from myosin in its lower coagulation temperature. Stewart and Sollman believed that myosin and soluble myogen fibrin were identical, but von Fürth says that this is not so, for it is impossible to change a pure solution of myosin to soluble myogen fibrin coagulating 10° lower. Furthermore in frog's muscle both soluble myogen fibrin and myosin co-exist even in the living state, but in warm-blooded animals no soluble myogen fibrin exists in the muscle, for did it exist, it would be coagulated by the heat of the body. The ease of transformation of myogen into soluble myogen fibrin is thus described: "If one adds to a pure freshly prepared myogen solution some per cent. of a neutral salt and leaves it standing overnight, in the morning, although it shows no optical change to the eye, yet already a large portion of the myogen has been changed to soluble myogen fibrin."

Myoproteid. This is a non-coagulable protein found in the muscle plasma of the fish. It is not present in the muscle plasma of mammals. Its nature is still undetermined, but it is not an albumose.

Of these various proteins, myosin and myogen are found in all vertebrate voluntary muscles. Myogen has not been found in invertebrate muscle. Soluble myogen fibrin is found preformed only in the muscles of fishes and amphibia; while in reptiles, birds and mammals it occurs there only as a secondary decomposition product of the myogen. Myoproteid is found only in the muscles of fish in large amount, in the amphibia it is present only in traces and in reptiles, birds and mammals it is absent. In heart muscle there is twice as much myogen as myosin.

Proteins of the stroma.—The greater part of the proteins of the muscle are insoluble in ammonium chloride solution. These are the proteins of the sarcolemma and the stroma. It is stated that the structure of the muscle is not changed by the extraction of the myosin; the stroma is still doubly refracting. The amount of stroma insoluble in ammonium chloride is very hard to determine with any accuracy, owing to the tendency of the muscle proteins to become insoluble. From perfectly fresh living mammalian skeletal muscle it is said (Saxl) that 84-90 per cent. may be extracted by NH_4Cl, if rigor is prevented; whereas, when the muscle has stood for a short time only, 2-3 per cent. is extractable. According to Salkowski, of the nitrogen of flesh 77.4 per cent. was in the form of insoluble coagulable protein and 10.08 per cent. in the soluble coagulable form. 12.52 per cent. was non-protein nitrogen.

The stroma contains phosphorus. It contains a small amount of nucleoprotein. Two grams were obtained from 543 grams of dog muscle and this nucleoprotein contained 0.7 per cent. of phosphorus. Corresponding with this small amount of nuclear matter, it was found that the amount of purine bodies obtained by the hydrolysis of muscle was very much less than from the thymus and other organs. The result clearly indicates that the nucleus contains the nucleoproteins and that these are confined to the nucleus. It is possible that a nucleoprotein containing inosinic acid may be found in the cytoplasm, but the source of inosinic acid in the cell is uncertain.

The stroma contains also phospholipins (phosphatides) apparently in union with the proteins, although this cannot be said to be definitely proved. The phospholipins, which are present, are discussed on page 611. The protein matter of the stroma partakes of the character of coagulated or insoluble protein, myosin fibrin and myogen fibrin or albuminoid. It is either an albuminoid or metaprotein. It is partially digestible in pepsin hydrochloric acid and, if so treated and the extract neutralized with sodium carbonate, a precipitate called myostromin, containing phosphorus and probably a nucleoprotein or lecithoprotein, was obtained.

From the fact that the stroma retains its optical and structural properties after the extraction of some of the myosin Danielewski supposed that the stroma consisted of a doubly refracting network or spongy substance in the interstices of which the myosin was in solution. It seems probable from the general phenomena of the squeezing out from gels on contraction of the latter by the process called syneresis, of a solution containing the same matter as the gel, but in different concentration, that the plasma may, in reality, contain the same proteins as the stroma, but in more dilute form. The myosin is itself, when in strong solution or partially dried, doubly refracting like the stroma (liquid crystals). The fact of double refraction would indicate that the molecules of protein in the muscle are not arranged haphazard, but they must be oriented in some way similar to their orientation in a crystal whether liquid or solid. Otherwise there could be no definite double refraction. As a matter of fact there is no double refraction as long as the myosin is in solution; in other words, as long as the molecules are so far apart that they can take any position freely and have their axes directed in all directions. If the myosin is in the form of a gel, the gel as it exists in the muscle must be so concentrated that the molecules are oriented and have their freedom of movement somewhat restricted as they do in a solid. Furthermore the tenacity of the pieces of muscle cut out by Kite and the remarkable elasticity which the substance showed indicate that the muscle substance has rather the properties of an elastic solid, hydrated perhaps, but no less a solid, than the properties of a gel. If muscle is a gel it is certainly not a homogeneous gel, but one which in some way is definitely organized so as to give the remarkable structural pictures of the muscle. It is more comparable to a liquid crystal than a gel.

The protein of the stroma substance appears, then, to be made of an albuminous, insoluble material possibly united with phospholipins to make lipo- or lecitho-proteins. This union, however, is not very firm, it is dissociable and some of the protein remains free. The molecules are definitely oriented. It is also quite possible that in the living muscle some of the extractives, such as creatine, may be united with this complex and that carbohydrate, either glycogen or glucose, is also joined to it to make a very complex, unstable protein molecule.

The proteins of the different kinds of muscles do not differ widely in the composition of the amino-acids that they yield by hydrolysis, as may be seen in the following table:

Perhaps the interesting fact here is the relative large amount of the basic amino-acids, since these occur also among the extractives.

AMINO-ACID CONTENT OF VARIOUS MUSCLES (Osborne and Heyl).

	Fish—Halibut	Chicken	Ox	Scallop
Glycocoll	0.00	0.68	2.06	0.00
Alanine	?	2.28	3.72	....
Valine	0.79	?	0.81	....
Leucine	10.33	11.19	11.65	8.78
Proline	3.17	4.74	5.82	2.28
Phenyl alanine	3.04	3.53	3.15	4.90
Aspartic acid	2.73	3.21	4.51	3.47
Glutaminic acid	10.13	16.48	15.49	14.88
Serine	?	?	?	?
Tyrosine	2.39	2.16	2.20	1.95
Arginine	6.34	6.50	7.47	7.38
Histidine	2.55	2.47	2.66	2.02
Lysine	7.45	7.24	7.59	5.77
NH_3	1.33	1.67	1.07	1.08
Tryptophane	Present	Present	Present	Present

Proteins of smooth muscle.—Smooth muscle of mammals or birds extracted with 0.7 per cent. NaCl also yields a muscle plasma, which spontaneously coagulates at room temperature. By dialysis a globulin separates, which coagulates at 54-60°, and in the filtrate from the globulin an albumin was found which coagulated at 45-50°. This latter substance was myosin(?). Swale Vincent and Lewis got from the smooth muscle of the sheep's stomach by extracting with .7 per cent. NaCl a plasma coagulating at 47°, but no other protein was present. If the extraction was made with 5 per cent. $MgSO_4$, the coagulation temperature of the extract was 56°. The coagulation temperature of proteins varies within several degrees, depending on the salt content, nature of the salts, and reaction of the solution. 56° is the temperature of coagulation of the fibrinogen of the blood and also of similar globulins found in liver and other organs. Smooth muscle contains a larger proportion of nuclear material than striated and the nucleoproteid is five times the proportion of that of striated muscle. From these observations it appears that smooth muscle contains myosin and nucleoprotein, but no myogen.

The extractives.—The extractive substances in muscle are those organic substances which may be extracted with boiling water. The great interest attaching to these substances is due to the circumstance that they represent products of muscular metabolism and thus throw some light on the nature of the chemical changes going on in it. Furthermore, since the muscles form nearly half the weight of the body these substances probably represent the precursors of some of the substances excreted in the urine. The total amount of extractives, including both inorganic salts and organic matter, obtained from perfectly fresh living muscle by boiling it with water amounts to about 2 per cent. of the weight of the muscle. Of this amount .7 per cent. is organic and 1.3 per cent. inorganic.

The nitrogenous extractives. In this group are found creatine, methyl guanidine, carnosine, inosine, carnitine, sarcosine, taurine, glycocoll, urea and hypoxanthine.

Creatine. Creatine, or methyl guanidine acetic acid, $C_4H_9N_3O_2$ $NH_2-C(=NH)-N(CH_3)-CH_2-COOH$, is described on page 702. The amount of creatine in voluntary muscle varies from 0.3-0.45 per cent. of the weight of the fresh muscle. Most of the creatine may be extracted by soaking the muscle in cold water, but a somewhat better yield is obtained by making a boiled extract. Creatinine is found only in small amounts in perfectly fresh muscle and it is doubtful whether it exists in the living muscle.

The creatine in muscle is probably in union with the colloids of the muscle, since it does not readily dissolve out of the muscle. Thus muscle contains in 100 grams 400 mgs. of creatine, whereas the blood contains only 2-3 mgs. per 100 grams. Creatine is given off only in small amounts to the blood by the normal beating heart or by the living muscle, and while this might be due to the fact that the sarcolemma was almost impermeable to creatine it is more probably due to the creatine being in a loose chemical or molecular union with the muscle substance.

The origin of the creatine in muscle has been much debated and no satisfactory source has yet been established. It might be derived from arginine, which is guanidine-amino-valerianic acid, or from some of the phospholipins, or by synthesis from urea and sarcosine. The matter is still under investigation and no definite conclusions have yet been reached. It appears probable that the muscle is able to methylate guanidine acetic acid or glycocyamine. Whether the creatine of muscle is increased or diminished by stimulation has been much disputed. The older statements in this regard are all open to question because of the methods of determining creatine, which were not quantitative; by the method now used of determining it, by conversion into creatinine, a more accurate determination is made, but one also open to many errors. If the muscle is stimulated in situ with its normal blood supply several observers have found that the creatine is diminished instead of increased by stimulation. Liebig reported that the muscles of a hunted fox contained several times the normal amount of creatine. According to recent work, contraction of muscle does not change its creatine content, but any means which diminishes the tonicity of the muscle does diminish the creatine content.

Iso-creatinine. This has been isolated from fresh fish muscle. Its nature is uncertain, but it differs from creatinine in the greater solubility of the base and picrate in water and by the fact that oxidation by potassium permanganate does not produce methyl guanidine, but ammonia. It may be that the nitrogen is bound to different carbon atoms or that it is a tautomeric form of creatinine.

Carnosine (*Ignotine*) $C_9H_{14}N_4O_3$, is a basic extractive precipitated by phosphotungstic acid and by silver and barium hydrate, according to Kossel's histidine method. On hydrolysis it yields histidine, $C_6H_9N_3O_2$, and what is believed to be β-alanine. If this is so carnosine is the first natural dipeptide which has been found to have in it a β-amino-acid. It has been suggested that it may arise from histidyl-aspartic acid by the action of a carboxylase, splitting off the last carboxyl group of the aspartic acid.

Carnosine has the following properties:

100 grams water at 24.9-25° dissolve 31 grams carnosine. It is precipitated from water by alcohol. $(\alpha)_D^{20°} = +20.77$. The nitrate melts at 222° (Thiele's block). Colorless needle-shaped crystals, m.p. 219° with decomposition. Carnosine tastes insipid and reacts strongly alkaline. It is not precipitated by K_4FeCy_6, lead acetate, acid or basic, nor by $HgKI_3$. Saturated picric acid does not precipitate it but tannic and phosphotungstic acids do. Gold hydrochloride gives a precipitate. It has been found in the extract of horse, ox and calf muscle, and of fish, crabs and oysters, but only in the muscle. It is not found in kidneys or in liver. In the hind leg of a fresh-killed horse from 6.4 kilos of muscle, 0.58 gram of free purines, 2.0 grams of carnosine and about 10 grams of methyl guanidine were obtained. The accompanying table shows the amounts of various extractives contained in one kilo of various muscles. The amounts are expressed in grams.

AMOUNT OF EXTRACTIVES IN GRAMS IN ONE KILO MUSCLE.

	Horse	Ox	Salmon	Maguro	Calf	Wild rabbit	Pig
Creatine	0.58		3.2	3.0		2.0	
Purines	0.09-.07					0.04	
Carnosine	1.82	1.30	0.55	2.0	1.76	2.23	1.95
Methyl guanidine	.083-.1				0.22		
Carnitine	.2-.17				0.19		0.3

Carnitine (*Novain?*). This has been found in horse, calf and pig muscle. It is a base precipitated by phosphotungstic acid, giving with corrosive sublimate a double salt, m.p., 196°: $C_7H_{15}NO_3.2HgCl_2$. Carnitine is probably γ-trimethyl-β-oxybutyrobetain, the hydroxyl probably being in the β position (or α).

$$(CH_3)_3N \begin{cases} O\text{——————}CO \\ \qquad\qquad\qquad | \\ CH_2\text{—}CHOH\text{—}CH_2 \end{cases}$$

It would seem not impossible that this base might be the mother substance of choline. It is conceivably formed from glutamic acid by splitting off carbon dioxide and methylation. This base is said, by Krimberg, to be identical with the *Novain* isolated from meat extract by Kutscher, but according to the latter it is rather an isomer of Novain. The *oblitin* of Kutscher is the diethyl ester of carnitine.

Taurine. Taurine is found only in small quantities in mammalian muscle, but it is the chief nitrogenous extractive in the muscles of various invertebrates. The fresh muscle of the cephalopod, Octopus, contains at least 0.5 per cent. of taurine, but no creatine or creatinine. This muscle contains about .03 per cent. of hypoxanthine and no glycocoll. Taurine is

$$NH_2.CH_2-CH_2-\overset{\overset{\displaystyle O}{\|}}{\underset{\underset{\displaystyle O}{\|}}{S}}-OH$$

Purines of muscle. The fact that muscle contains far less purines than other tissues was discovered by Kossel. There is less nuclear substance in muscle than in most other tissues. In the cells of the thymus glands, for example, the nuclei nearly fill the cells and these cells contain seven to eight times as much purine nitrogen as the muscles, in which the greater part of the cell is cytoplasm. The pancreas is intermediate in the amount of nuclear material and purine bases it contains. This fact indicates that probably the purine bases are confined to the nucleus. The amount of purines in various tissues is shown in the following table in which the purine nitrogen is expressed in per cent. of the weight of the fresh organ:

Total Purine Nitrogen in Per Cent. of the Wet Weight of the Organ (Burian and Hall).

Organ	Per cent.	
Horse meat	.055	
Ox meat	.062	
Veal	.071	
Thymus	.482	(by quick work)
"	.429	(by slow work)
Pig's pancreas	.2	
Ox pancreas	.183	

The hypoxanthine of muscle increases during muscular work, whereas if the perfused muscle is stimulated the purine base nitrogen falls from .06 to .04 per cent. Muscles give off purine bases to the blood stream when they do work; they also give off some uric acid. The purine base and uric acid in the blood perfused through muscle behave as follows:

	Total purine N mgs.	Uric acid N mgs.	Uric acid mgs.	Purine base N mgs.
Before perfusion	4.0			4.0
After 1 h. at rest—1000 c.c. blood perfused	5.2	1.6	4.8	3.6
2d period muscle at work—1000 c.c. blood	8.5	2.1	6.3	6.4
3d period muscle at rest 1 h.—1000 c.c. blood	10.5	5.7	17.1	4.8

Glycocoll and Urea. Large amounts of glycocoll are found in the muscle of pecten irradians, 0.39-0.71 per cent. being present. Urea is present in only small amounts in the muscle of mammals, .04-.08 per cent., but about 1-2 per cent. is present in the muscles of elasmobranchs, such as the sharks and rays. If elasmobranch muscle dries, some of the urea decomposes yielding ammonium carbonate, so that the muscle appears to have urease in it.

Inosine (*Carnine*). $C_{10}H_{12}N_4O_5$. This substance is a decomposition product of inosinic acid and consists of a compound of hypoxanthine and d-ribose, a pentose. Similar substances (vernin) have been found in plant cells by Schulze. Of its function or significance nothing is known.

Inosinic acid. $C_{10}H_{13}N_4PO_8$. This was discovered in muscle by Liebig, who quite overlooked its phosphorus content, but found that it yielded sarkine (hypoxanthine) and a carbohydrate. It has since been shown to be a compound of hypoxanthine, d-ribose and phosphoric acid and to have the probable formula:

```
                                          O=C—NH
       O       H  H  H  H  H              |   |
       ‖       |  |  |  |  |          ╱NH—C   CH
HO—P—O—C——C——C——C——C——C<        ‖   ‖
       |       |  |  OH OH |          ╲N  —C—N
       OH      H  O—————————
```

About .11 per cent. of inosinic acid is present in the muscles of fowls, whereas none is present in the muscles of pigeons. None was found by Schlossberger in human muscle. Inosinic acid is obtained from the alcohol-extracted and powdered meat by extracting one kilo of the dry residue with 2-3 liters of warm water. The phosphates are carefully precipitated with $Ba(OH)_2$, the solution neutralized with HNO_3 and the inosinic acid precipitated with $AgNO_3$. The precipitate is decomposed with H_2S and the filtrate evaporated to a syrup, which becomes powdery with alcohol. The free acid is insoluble in alcohol and ether. 1,000 parts of water at 16° dissolve 2.5 parts of the salt. Inosinic acid is a mono-nucleotide. Whether it is found in the nuclei or cytoplasm cannot be said. Its relation to the nucleic acid of the nucleus is unknown.

i-Inosite. $C_6H_{12}O_6$; $C_6H_6(OH)_6$. Hexahydroxyhexamethylene. Inosite is found in nearly all organs of the animal body, in the heart, skeletal muscle, liver, pancreas, spleen, kidneys, supra-renals, testes, spinal cord, brain and lungs of mammals. It does not occur in the normal urine of man or rabbits, but it has been found in the urine in diabetes insipidus and mellitus (Kulz). It is not constantly present in the urine in these diseases, for in eight cases of diabetes insipidus it was found only once. It is sometimes present in quantities in the urine.

One case of inosite-uria has been described in which 18-20 grams of inosite were excreted daily (Vohl). Inosite is also widely spread in the plant world and it occurs in especially large quantities in green beans. It appears to be oxidized, or destroyed, in the body, since Kulz gave a man 30-50 grams of inosite and recovered in the urine only 0.2-0.5 gram. The wide occurrence of inosite in all forms of living cells indicates its connection with some fundamental process in the living protoplasm. It is found united with phosphoric acid as phytin in bran. Phytin is probably

```
            H OPO(OH)₂
               C
          H  /   \  H
(HO)₂OPO C       C OPO(OH)₂
          H |     | H
(HO)₂OPO C       C OPO(OH)₂
             \   /
               C
            H OPO(OH)₂
```

Phytin.

Phytin has a marked laxative action on the bowels and inosite also, when taken by the mouth, produces diarrhœa.

There are in nature three forms of inosite: i-inosite, found in muscle, which is the meso-inosite and inactive; d-inosite and l-inosite. It crystallizes in large, rhombohedric crystals; it is very soluble in hot water, insoluble in absolute alcohol and ether; its solutions have a sweet taste, but it does not reduce Fehling's solution.

It is generally thought to occur free in the muscle, but some think that it is in union with some other substance. The amount in muscle is small, only about .003 per cent. being present. From the brain 10 grams were obtained frbm 50 pounds. Of its function nothing is known. Presumably it is formed from carbohydrates.

Mytilite.—This is a body related to inosite which is found in the aqueous extract of the muscle closing the shell of Mytilus edulis. It appears to be a penta alcohol, having five groups capable of acetylation. It contains a hexa carbon ring and gives Scherer's reaction. It is isomeric, or stereoisomeric, with quercite and isoquercite. It resembles an anhydride of inosite. Its formula is $C_6H_{12}O_6.2H_2O$. Among the other extractives of this muscle are betaine (Brieger; Jensen), taurine and 1.5% of glycogen.

Xylose. Xylose was found by Henze to make from 0.4-0.9 per cent. of the dry weight of the muscles of Octopus.

Carnic and phosphocarnic acid. This is a substance found in the extractives of muscle and obtained as a copper salt. Carnic acid is probably a dipeptide of the formula $C_{10}H_{15}N_3O_5$. This substance binds hydrochloric acid so firmly that it forms no precipitate with silver nitrate except on boiling. It gives the biuret reaction and is identical with anti-peptone. It is said to yield lysine and some ammonia on hydrolysis. Carnic acid exists also in combination with phosphoric acid, according to Siegfried, to form phosphocarnic acid, to which important functions were originally assigned. From further work it seems doubtful whether phosphocarnic acid is a pure substance. There may be a series of similar compounds of an analogous nature. It need not be considered more at length until further work has been done upon it.

Succinic acid. $C_4H_6O_4$. COOH. CH_2—CH_2.COOH. This acid is one of the more abundant, non-nitrogenous extractives of muscle. It is found in the fresh muscle as well as in meat extract. 1.5 kilos of dog's muscle two hours after death yielded 0.24 gram of succinic acid; from 1.8 kilos of beef muscle, 48 hours after slaughter, 0.133 gram were obtained; and seven days later 1.75 kilos of the same beef yielded 0.122 gram. The amount is so large that it cannot possibly all be derived from phosphocarnic acid, even if it be admitted that the latter is an individual. Succinic acid possibly plays a rôle in oxidation, since it was found by Thunberg to influence oxidation of the muscle in quite a special manner. This acid is also one of the constituents of the oxidizing ferment laccase, which may be of interest in connection with the results of Thunberg. Whether muscle can oxidize or destroy this acid is unknown. It is doubtful whether it is oxidized to malic acid. The rôle it is playing in muscle, the conditions of its formation, and its fate, if normally formed there, require further investigation. It is found also in the brain extractives.

Lipins of muscle.—These consist of neutral fat, cholesterol and phospholipins. The latter are very little known at the present time. The neutral fat probably comes, for the greater part, from the connective tissue; the phospholipins, however, are derived from the muscle itself. Of the skeletal muscle the phospholipin makes about 30 per cent., usually, of the total lipins, but in the heart the proportion of phospholipin may be 60-70 per cent. of the total lipin. The cholesterol is about 0.2 per cent. of the dry substance. The phospholipins of the heart and skeletal muscle have been investigated recently by Erlandsen. In the heart there is found a peculiar lipin like cephalin, having highly unsaturated fatty acids, which has already been described on page 95. There are also present other lipins more of the nature of lecithin. Since these bodies have probably not been obtained as yet in a pure state and have been but little investigated, we need not discuss them further here. The total lipin content of the muscle is between 2 and 3 per cent. of the wet weight of the muscle. Figure 56.

Inorganic constituents of muscle.—The principal points of interest in the inorganic constituents of muscle are the relatively large proportion of potassium as compared with sodium and the considerable quantity of phosphoric acid present. The phosphorus is mainly in the inorganic form. The total amount in muscle is about 0.2 per cent. Of this amount in ox flesh, Constantino found 81 per cent. as inorganic and 19 per cent. as organic phosphorus. Grindley got 93 per cent. in the form of inorganic phosphorus in the beef he examined. Of the organic phosphorus, 16 per cent. is present as phosphatide and 3 per cent as other organic phosphorus compounds, nucleins, etc. In heart muscle, the inorganic is only 40 per cent. of the total and the organic 60 per cent. (Constantino). In smooth muscle 50-70 per cent. of the total phosphorus is organic, but the non-phosphatide, organic phosphorus is larger, probably

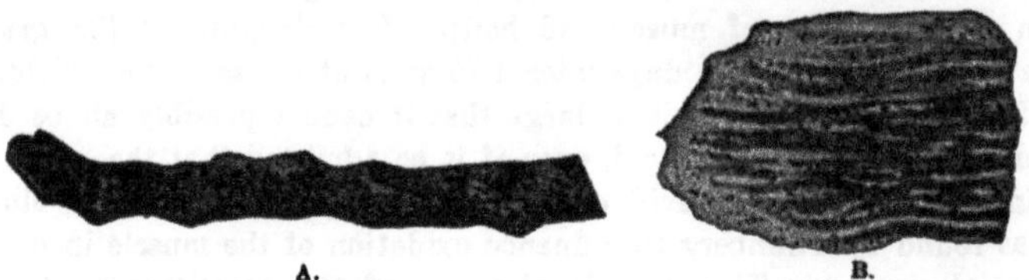

FIG. 56.—Distribution of lipin in muscle cells (Noll). *A*, Lipin stained with osmic acid in pectoralis of pigeon. *B*, Osmic acid preparation of rabbit smooth muscle.

owing to the larger amount of nuclear material present. Thus in the heart there were 42 per cent. of phosphatide phosphorus and 20 per cent. of non-phosphatide, organically bound phosphorus. In smooth muscle of the 50 per cent. organic phosphorus, 27-42 per cent. were in the form of non-phosphatide phosphorus organically combined.

In striated muscle there are from 2-3 times as many atoms of potassium as of sodium present. The atomic ratio K : Na is 2.60—3.34 : 1. In smooth muscle the ratio is variable. In the retractor penis it is very nearly 1 : 1. But in the stomach it is about the same as in striated muscle. In striated muscle of mammals there is about 0.32-0.42 per cent. of potassium and 0.04-0.07 per cent. of sodium computed on the wet weight; of smooth muscle that of the stomach has about the same proportion of sodium and potassium as striated, but the retractor muscle of the penis and smooth muscle of the uterus have about 0.26 per cent. of potassium and 0.15 per cent. of sodium (Constantino. See also Fahr and Meigs).

The internal secretion of muscle.—Like all other tissues of the body, the muscles are constantly giving off to the blood substances which are to be eliminated, or which affect more or less profoundly other organs. Knowledge of the substances thus given off is still uncertain, but the finding in the urine in small quantities of most of the extractives of muscle (Kutscher) shows that these substances are probably given to the blood by muscle. It has been found by perfusion experiments that carbon dioxide, creatine, methyl guanidine, hypoxanthine, lactic acid and acetone are, sometimes at any rate, given off from contracting muscle. Some of these substances are of importance in stimulating the respiratory center. This is true of the carbon dioxide and the lactic acid; others produce dilation of the arterioles and in this way serve the muscle; the acids, also, have the very important function, particularly in amphibia and cold-blooded animals, of helping to dissociate oxygen from oxyhemoglobin, thus turning the oxygen out of the blood so that it can be utilized by muscle. (See page 487.) Blood of tetanized dogs is said to be toxic and on transfusion to other dogs causes in them disturbances both of the circulation and respiration. (Compare parathyroid tetany, page 652.) The substances so acting are said to be soluble in alcohol. There is no doubt that much remains to be done in this branch of the subject. Methyl guanidine, which is found in some muscles, is decidedly toxic, as is also oblitine. These toxic bases have been found in the urine of dogs after parathyroid extirpation.

The metabolism of muscle.—It is extremely difficult, from the account which has just been given of its composition, to form any probable picture of the nature of the metabolism of muscle and the changes occurring in it, particularly during contraction or muscle rigor. By many authors an attempt has been made to distinguish between that part of the metabolism which is concerned with the formation of the muscle cell itself, its formative metabolism so called, and the metabolism which is involved in its activity during contraction, or its energy metabolism. It is doubtful to what extent this separation of the two phases is justified, but as it is not unlikely that the character of the chemical processes involved in forming the protein machinery of the muscle are different from those involved in furnishing the energy for contraction, the total metabolism may be discussed under these two headings.

The formative metabolism of muscle.—The formative metabolism of muscle, like that of all other cells, is that metabolism in virtue of which the substance of the muscle is produced. Its general features are those common to all cells. The proteins of the muscle are characteristic and, indeed, the muscle of each species of animal has its own specific protein. Probably were our methods fine enough, we should find the proteins of each individual of the same species to be different from the corresponding proteins in other individuals. These proteins are probably made from the amino-acids brought to the muscle from the digestive tract by the blood. The synthesis of the amino-acids into the specific proteins is, hence, a part of this formative metabolism. It is unknown how each tissue picks out or uses just those amino-acids in the right proportion to form its own specific proteins. Possibly it depends, in part, on a slightly different rate of destruction of the different amino-acids in different muscle cells. But not only has muscle the power of forming proteins from amino-acids; it is also able to destroy these proteins and reduce them to amino-acids. It has the property, in other words, of auto- or self-digestion. This happens in starvation; in wasting diseases, when fever exists; and it probably happens, also, in the wasting of a muscle from disuse. The fundamental protein metabolism of muscle is not especially keen. On the contrary, there are many indications that muscles are not remarkable for their power of growth and repair. The fundamental formative metabolism goes on at a faster or slower rate, depending on circumstances, such as temperature; and possibly upon the number and character of the nerve impulses reaching it; or upon age, being more rapid in youth than later in life; or upon health or sickness, a rapid catabolism taking place under conditions of starvation, etc.

But muscle is more than a simple protoplasmic cell. Its living material has either been transformed in large part into a machinery capable of producing changes in shape of the muscle at a rapid rate, or the living material has fashioned in itself a machinery, possibly not a truly living machinery, which is the machinery of contractility. It is impossible to say at present whether the contractile machinery is truly self-perpetuating or not; whether, in other words, it is living or not, or whether it is not to be regarded as a piece of lifeless protein machinery more like connective tissue, or the catgut fibers of Englemann's artificial muscle, which works under the influence of the chemical changes occurring in the living material which has formed and which surrounds it. Whatever may be the truth of this matter it seems possible that the muscle has quite a special metabolism, distinct from its formative metabolism, and that this second metabolism, the energy metabolism, is concerned with its functions as a motor, or contracting engine. This second metabolism, that of contraction, is thought to be, as it were, superimposed upon, or separate from, the fundamental metabolism of the muscle cell proper. It is a destructive metabolism, or catabolism, by which heat and energy are set free sufficient to move the machinery and produce contraction.

The nitrogen output of muscle is probably not so keen during rest as that of many other tissues of the body. There is a general law which prevails throughout the whole animal and plant kingdom that that organ which has the most powerful formative metabolism always draws upon and reduces the metabolism of other organs of a less intense metabolism. Thus the apical buds of plants inhibit the development of buds farther down the branch. So in times of fasting or starvation the muscle protein is torn to pieces and converted into amino-acids, which, passing from the muscle to the blood, are carried by that internai medium to those organs of which the metabolism is keener, to the brain and nervous tissues and the heart, and serve to nourish those organs at the expense of the muscles. Thus during fasting the voluntary muscles waste away, but the brain and the heart keep their weight almost unchanged, and those muscles which are of the most importance to the animal and are the most exercised draw upon those of less importance. This conversion of the muscle proteins, or self-digestion, is accomplished by means of *autolytic enzymes;* a plentiful supply of food and oxygen normally holds these enzymes constantly in check. Hence a rapidly contracting muscle like the heart is able to maintain its material intact, whereas a resting muscle gradually wastes away. Acids are particularly prone to set free the autolytic enzymes of cells.

The formative metabolism of muscle is also dependent directly, or indirectly through the blood supply, on the innervation. Muscle of which the nerve has been cut degenerates, the cross-striations break down, fat staining in osmic acid appears and there is a marked reduction in the bulk of the muscle. This degeneration may be the result either of the falling away of the nerve impulses of a heat-producing or tonic kind, which may normally be constantly impinging on the muscle, impulses which do not cause muscle contraction but keep it in a state of tone; or it may be due to the lack of control of the blood supply of the muscle. Ordinarily the metabolic demands of the muscle are very carefully met by the vascular system; but in the absence of this control, no regulation of supply and demand any longer exists. Which of these mechanisms determines metabolic control cannot be decided. It would seem possible, also, that the nerve impulses play an important part in the organization of the muscle substance. The re-establishment of the connection of the nerve with the muscle, after the nerve has been cut and degenerated, leads to the reorganization of the muscle substance and its restoration to a normal state. It would seem possible that this organization proceeds outward from the nerve terminals and is produced by the nerve impulses, just as the brain material is organized by the impulses coming into it. It is not possible to say whether the growth of a muscle accompanying its use is due to a stimulation of the formative metabolism coincident with a stimulation of the energy metabolism, or whether it is an indirect result of the stimulation of the latter. It is, on the whole, probable that it is a direct result of the energy stimulation and that a nerve impulse striking the muscle, not only causes the great catabolism which liberates energy and produces contraction, but also directly stimulates the other chemical changes of a formative nature. Something similar occurs in nerve tissue, in which it has been shown that the formative metabolism, which causes the development of the centers of the brain, is thus influenced by nerve impulses.

For example, if the eyelids of a puppy are sewed together at birth, so that they cannot be opened and no stimulation of the retina occurs, it has been found that the cells of the brain ganglia concerned with vision do not develop as they normally do, but remain in an infantile state. In the case of muscle it seems probable, as has already been stated, that a constant stimulation of the muscle through the nerve occurs. This stimulation does not cause muscle contraction, because it is not sufficiently intense, but it is sufficient to stimulate the formative metabolism and perhaps to maintain a certain tonus in the muscle.

In the course of the formative metabolism of muscle creatine, among other things, is produced. Nothing definite is as yet known of the real function of creatine, but according to the work of van Hoogenhuyze and Verploegh creatine is produced in the course of that metabolism which is concerned with tonus. These authors have shown that tonic contraction of muscle increases the creatine output in the urine and increases the creatine content of muscle. Some authors have reported an increase in the creatine content of muscle as a result of nerve stimulation. While this is not improbable, the evidence is as yet unsatisfactory. That muscle has in it the precursors of creatine is probable from the fact that on autolysis an increase of the total creatine and creatinine of the muscle occurs. On the other hand, Mellanby has failed to get any such increase when bacteria were absent. But blood certainly contains such a precursor. It is stated that the addition of gelatin to autolyzing muscle increases the creatine formed on autolysis. A study of the creatine content of muscle during disease also indicates the connection of creatine with the formative metabolism. Thus in diseases involving the wasting of muscle abnormally large amounts of creatine appear in the urine (Shaffer), if the methods used for detecting creatine are reliable, which seems somewhat doubtful.

Nucleins also play a part in the formative metabolism of muscle. The uric acid excretion undergoes a remarkable rise following muscular work. This rise lasts only for a couple of hours and is then followed by a fall so that the total excretion is not changed by work. Muscular work appears, in the first place, to increase the hypoxanthine content, and this is then oxidized to uric acid. It has been suggested that this oxidation happens only as the hypoxanthine is about to leave the muscle and is due to the fact that the hypoxanthine oxidase is found only in the periphery of the muscle. It is probable, since muscle is formed in young children when on a purine-free diet, that muscle has the power of making purines from non-purine precursors.

The formative metabolism includes the power of making glycogen and possibly of making carbohydrate from non-carbohydrate material, such as the proteins. Glucose brought to muscle is, in part, stored as glycogen. The origin of the d-ribose in inosinic acid is unknown, nor can anything be said concerning the origin of the phospholipins.

On the whole, then, the formative metabolism of muscle presents just those problems which all other tissues present; problems of which the solution remains for the future.

The energy metabolism of muscle.—The energy metabolism of muscle is, as it were, superimposed upon the formative metabolism. This metabolism is peculiar in that it does not directly involve any increase in protein catabolism. Muscular work does not increase the nitrogen output of the body or of muscle. This metabolism involves primarily the carbohydrates and we may now consider the changes in composition of muscle produced by muscular work.

1. *Change in glycogen content.* That muscles contain glycogen in considerable quantities was discovered shortly after Bernard found glycogen. The amount of glycogen in different muscles and in the corresponding muscles of different animals is variable. Thus horse muscle contains an unusually large quantity of glycogen, 1-2 per cent., and since glycogen can be detected in muscle microscopically by the brown reaction it gives with iodine, it is possible in this way to detect the presence of horse meat in sausage, or other food products in which it has been used. Another muscle having a large amount of glycogen is that of the scallop, or the shell-closing muscle of Mytilus edulis, which contains also about 1.5 per cent. Ox flesh and other forms of muscle contain less. Thus it has been found that perfectly fresh ox muscle contains, on the average, only 0.3 per cent. of glycogen. The glycogen content of muscle is subject to variation with the diet, but the variation is less than that of liver glycogen. Thus, in fasting, glycogen disappears more rapidly from the liver than from muscle. A large intake of glucose or other carbohydrate increases the glycogen of the liver more than that of the muscle, but still glycogen increases also in muscle in such cases. The heart muscle normally contains fully as large an amount of glycogen as the skeletal muscle, but glycogen disappears after death more rapidly from the heart than from skeletal muscle, so that many observers have found less glycogen in the heart than in other muscles.

The localization of glycogen in the muscle has been repeatedly investigated. It is found in the form of granules in the living matter itself (Arnold), in the sarcoplasm, but not in the fibrils. According to Arnold glycogen is best fixed by corrosive sublimate containing some glucose.

The glycogen of muscle undergoes a rapid decomposition after death, and in beef which has hung for some time the amount of glycogen is much reduced. In hen's muscle 30 to 60 minutes after death, 25-58 per cent. of the glycogen is lost and rabbit's muscle in 4 hours at 40° may lose 88.8 per cent. of its glycogen (Boruttau). This behavior is similar to that of the liver and, like the liver, it is due to the presence in muscle of an enzyme, an amylase or glycogenase, which rapidly converts glycogen into glucose when the reaction is slightly acid. There may also be found some iso-maltose in the muscle extract, which probably also comes from glycogen, but there can be no doubt that most of the sugar formed from the glycogen is glucose.

Quantity of Glycogen in Various Dog's Muscles, Showing the Quantity of Glycogen in Corresponding Muscles on Opposite Sides of the Dog's Body, the Effect of Fasting and of Age (Maignon).

Dog	Muscle		Glycogen per cent. of wet weight.
Very old. Fat.	Semi-tendinosus	Right	0.146
		Left	0.229
	Biceps femoralis	Left	0.235
4 years old. Fat.	" "	R	0.550
		L	0.516
	Semi-tendinosus	R	0.446
	" "	L	0.669
	Triceps brachialis	R	0.365
3 years old.	Biceps femoralis	R	0.527
		L	0.417
1 year old.	" "	R	0.077
		L	0.109
1 year old.		R	0.660
		L	0.559
	Fasting		
Old dog.	Temporal	R	0.134
		L	0.161
4 years old.	Semi-tendinosus	R	0.441
		L	0.544
	Semi-membranosus	R	0.679
		L	0.706
Old dog. 5 days fasting.	Semi-tendinosus	R	0.221
		L	0.200
Old, fat dog. 14 days fasting.	Biceps femoralis	R	0.220
		L	0.255
2 years old. 5 days fasting.	" "	R	0.280
		L	0.255

Besides having the power of converting glycogen into glucose, living muscle has a very remarkable power of destroying glucose, and even autolyzing or hashed muscle has this power to some degree, although it is soon lost after death. This glycolytic power, as it is called, it is needless to say is the most important problem in the metabolism of the carbohydrates and it has been keenly investigated of recent years. We will leave it for the moment to consider the change in the amount of glycogen, which accompanies muscular contraction and work.

Glycogen is much reduced when muscles do work. Thus the following results were obtained when dogs were anesthetized with ether, morphine and chloroform, and after the ligation of both the external iliac arteries and compression of the dorsal aorta the crural nerve on one side was stimulated as long as the muscle would contract. The stimulated muscle was then stimulated directly as long as it would respond. The two corresponding muscles of the two sides, stimulated and unstimulated, were then removed and their glycogen determined by the not very accurate Brücke process. The unstimulated muscle contained from .532-.684 per cent. of glycogen; while the active contained 0.116-0.194 per cent. Irritability of a muscle is lost long before its glycogen is exhausted. On a less-extended stimulation there was found in the unstimulated 0.716-0.560 per cent. and in the stimulated 0.440-0.112 per cent. On the other hand, division of a motor nerve increases the glycogen content of that muscle, as compared with the corresponding muscle of which the nerve has been left intact. In frogs the increase may be 20-30 per cent. of the original amount; in cats and rabbits the change is not so marked. Tetanus of frog's legs caused a diminution of 24-50 per cent. of the glycogen of the stimulated, as compared with the unstimulated side. In prolonged stimulation and tetanus by strychnine, glycogen undergoes a great diminution.

From the foregoing results there is no doubt that during muscular work there is a great diminution in the glycogen content of muscle. Glycogen is used up, presumably converted first to glucose and then oxidized. The energy used in muscular work may come, therefore, from this combustion of the glycogen. The question arises whether this is the sole source of the energy. Is there enough glycogen in the body to do the whole work which the muscle does? This question is very difficult to answer. A comparison of the amount of energy used in the work done with the amount which would be set free from the combustion of the glycogen which had disappeared during the working period has given a very bad agreement between the two values. Seegen found that the work done by an extirpated muscle represented 2-11 per cent. of the energy presumably available by the burning of the glycogen. From this factor he calculated that there was not enough glycogen in the body to yield the energy of muscular work, but Schenck has very justly criticised these conclusions because the muscle was working under unfavorable conditions, because in the body glycogen can be rapidly reformed, and because there was no proof that the disappeared glycogen had actually been completely burned. There is no doubt from the persistence of the power of muscular contraction in severe diabetes, where the power of combustion of carbohydrates has, presumably, been completely lost, that certainly muscle has the power of burning other foods, fats for example, as a source of its energy.

The fasting heart holds its glycogen longer than the body muscles, for it draws on the voluntary muscle for its nourishment during starvation. The heart has the power of nourishing itself even from a very greatly impoverished blood. In rabbits after six days of fasting the muscles still contain 0.04-0.06 per cent. of glycogen; cats retain 0.05-0.07 per cent. after 12-14 days of fasting; the muscles of doves after 2-8 days of fasting contain from 0.07-0.32 per cent. The most surprising result is found in horse muscle, which still contains from 0.98-2.43 per cent. of glycogen after 9 days of fasting (Aldehoff).

The glycogen in muscle, like that in the liver, is under the control of the internal secretions of the body, and particularly under that of the supra-renal gland. The observation was made nearly forty years ago that if cats were simply tied down on an operating board they showed after half an hour a very marked glycosuria, and that if they remained there the animals died after 36 hours, with a great fall of blood pressure and body temperature. The muscles of these animals proved to be completely free from carbohydrates and the liver lost its glycogen before the muscles. This disappearance of glycogen is probably due to the discharge of adrenaline from the supra-renal glands into the blood; a discharge which accompanies emotions such as anger and fear and the object of which may possibly be to promote the liberation of the carbohydrate so that it may be in a condition for rapid burning. Thus the chances of escape of an animal might be facilitated (Cannon). By causing the transformation of glycogen into glucose there are put at the disposal of muscle large stores of fuel, thus making possible a supreme effort. In this case the skeletal muscles are affected by nerve impulses to the supra-renals, these nerve impulses acting, as it were, at a distance, that is indirectly on muscle. The impulses impinging on the supra-renal glands cause these to set free substances which profoundly affect the metabolism of the muscle; may it not be possible that the effect of the nerve impulse to the muscle itself is to set free substances which cause the contraction of muscle? The nerve impulse may not be a direct, but an indirect, cause of the contraction. The latent period may have this explanation. The muscles have much less glycogen in pancreatic and phlorizin diabetes, but it does not completely disappear. Accompanying this loss of glycogen there is a marked weakening of muscular powers. Arsenic, also, causes a disappearance of muscle glycogen in cats. As adrenaline causes a discharge of glycogen from muscles so there may also be substances which cause a retardation of the transformation of glycogen to glucose, and it is not impossible that certain antipyretics, such as acetanilide or antipyrin, may have some such action. This possibility should be investigated.

From these observations it appears that glycogen is contained in muscle in considerable quantities. In a man's body, if his muscles contain 0.3 per cent. of glycogen and weigh 30 kilos, there would be roughly 90-100 grams of glycogen. This is partly consumed during muscular work, the transformation into glucose being caused by an enzyme similar to the glycogenase of the liver. The glycogen of muscle is less affected by diet than that of the liver, but still it is affected somewhat. In starvation it diminishes, and on carbohydrate food it increases. After death it disappears with greater or less speed, and under the influence of adrenaline or phlorizin it is greatly diminished. We have now to inquire concerning the further decomposition of the glucose set free from the glycogen.

The glycolytic power of muscle. Most of the sugar which is burned in the body is burned in muscle. It is a veritable conflagration which is there going on, a conflagration which warms, animates and invigorates us, but a conflagration giving a light too subtile to be seen. What is the nature of this conflagration, of this wonderful combustion in an aqueous medium, which causes heat and electricity but so seldom produces light? The true solution of this problem still eludes physiologists. Do the sugar molecules burn as such, or are they first torn to pieces by chemical union with some catalytic substance? It may be imagined that they unite with some of the substances in living matter and when the union is made the sugar molecule is no longer stable but breaks into fragments. When a piece of wood burns it is not the wood which burns, but the pieces of the molecules of wood which have been dissociated by the heat and which have such an affinity for oxygen of the air that they inflame spontaneously. And it is probably so in muscle, also, only it is not heat which fragments the sugar molecule in muscle, but a substance which unites with the sugar. Protoplasm is itself the torch.

The property of thus fragmenting the sugar molecule, is it a property of the living protoplasm only, or is it due to a substance which may be isolated from the protoplasm and which will have the same properties outside the cell as in it? Many men have sought for such a substance. The property of thus fragmenting the sugar molecule seems to be a property of living muscle only. Dead muscle no longer has this power. It is true that ground muscle has, also, some glycolytic power, but this power is soon lost; it inheres only in fragments of the living matter itself; it is never very great. Extracts of muscle do not have the power of glycolysis. It is impossible to say whether this power is due to some substance which is very unstable and thus hard to isolate, or which is destroyed by the action of other ferments such as the digestive ferments of muscle, or whether it is only the organized biogenic molecules which have the power. The course of development of the science in the past leads one to hope that the isolation of this substance will ultimately be made, but if it is made it will be one of the most fundamental discoveries in the history of science, for everything indicates that this fragmentation of the foods is a preliminary to their whole metabolism and not alone to their oxidation. It is a problem to be actively prosecuted by every means. Lepine and the French and Belgian physiological chemists with characteristic genius have most keenly attacked this fundamental problem. They have found that not only muscle but blood and other tissues have the power of making glucose disappear, but they have not shown that the disappeared substance has been oxidized or destroyed. It may have been resynthesized into maltose or some other substance having a lower reducing power than glucose. From yeast a substance has been isolated which does fragment the sugar molecule into carbon dioxide and alcohol. This substance is zymase. Possibly a similar enzyme fragmenting the sugar molecule in the same or slightly different way may exist in muscle. Definite statements cannot yet be made, but the presence of small amounts of alcohol in muscle tissue is held by some to be significant.

What, then, is the character of the fragmentation of the muscle sugar? This question cannot be answered. By some it is suggested that the glucose is fragmented by a process analogous to or identical with the fermentation by yeast into alcohol and carbon dioxide and that the alcohol is then oxidized to CO_2 and H_2O. But, while small amounts of ethyl alcohol have been recovered from muscle and alcohol is readily burned in the body, it is not probable that the whole metabolism goes in this direction. Lactic and succinic acid are produced during muscular work and acetone in large amounts when work is done during fasting. The fact that muscle has the power of methylating many substances, such as glycocyamine, would indicate that some formaldehyde was formed during the process of fragmentation. The ready synthesis of imidazole from pyruvic aldehyde and ammonia according to the reaction on page 185 and the presence of considerable amounts of histidine in muscle might suggest the formation of pyruvic aldehyde as an intermediate substance. Nothing is certainly known of the character of the first products of the fragmentation of the sugar molecule. The solution of this problem is of the highest importance for understanding muscular contraction and for the treatment of diabetes mellitus. While it is generally assumed that the fragmentation of the glucose molecule precedes its oxidation, it is not impossible that this view is incorrect and that primarily an oxidation to an osone or other ketone occurs and that this compound is then fragmented by some constituent of the muscle. These partially oxidized sugar molecules would probably be far less stable than the normal molecule, just as the partially oxidized fatty acid molecules are less stable than the saturated, unoxidized molecules. According to the view of Hermann, glucose is built into the nitrogen-containing complex molecules of the cell; this is then oxidized, the carbohydrate burned and the inogen molecule, so called, is reconstituted, the nitrogen part being retained in the cell, so that the total nitrogen excreted is not increased.

Lactic acid. Du Bois-Reymond found that muscle is normally alkaline in reaction to litmus, but becomes acid when it works. The acidity is very slight and not demonstrable if the muscle is well supplied with oxygen and blood, but it becomes noticeable if the circulation is interrupted and the muscle stimulated. The acid formed is in part the dextrorotary lactic acid, or sarco-lactic acid, CH_3.CHOH.COOH. It is still undecided whether lactic acid is formed exclusively from carbohydrate or not, but the best evidence is that most of it is derived from glucose. It is usually removed, if oxygen is present, as rapidly as it is formed. It was formerly thought to be oxidized, but Hill thinks it is resynthesized into muscle substance. Any accumulation is thus prevented, unless the entrance of oxygen is restricted. The development of acidity in muscle can be made very strikingly evident by an experiment devised by Dreser. If acid fuchsin is injected under the skin of a frog, it is carried by the blood all over the body, but it will not stain the tissues and muscles while these are alkaline. At best only a little superficial pinkness appears in parts of the connective tissue. After giving time for this distribution, the blood supply of both legs is interrupted by ligating the blood vessels and the sciatic nerve going to one leg is tetanically stimulated for some minutes. If, then, the skin is removed from the hind legs, it will be found that the muscle which has been stimulated is intensely red, while the resting leg is at most a faint pink. If the blood supply remains intact, stimulation of the muscle does not have this effect, for the acid is neutralized and removed as rapidly as it is formed.

The acidity of muscle may also be shown by immersing two gastrocnemius muscles in sodium chlorate, n/8 solution, and stimulating one. Sodium chlorate in neutral, or slightly alkaline, solutions is not toxic, but in the presence of only a little acid it produces rigor of the muscle. The tetanized muscle will be found to go into rigor while the other remains living.

Various mechanisms exist in muscle for the maintenance of its neutrality. These mechanisms have already been discussed on page 247. They are (1) the oxidation of the acid to very weak acid, such as CO_2, which will easily escape from muscle or combine with the amino groups of the proteins. (2) The neutralization of the organic or mineral acid produced either by sodium carbonate or sodium phosphate. By this the neutral salt of the strong acid is formed and carbonic and NaH_2PO_4 being very weak acids are but little ionized; (3) there is a deamidization of some amino-acid setting free ammonia which neutralizes the acid and then the carbon part of the amino-acid is oxidized to CO_2 and H_2O.

Every acidosis is accompanied by a larger or smaller increase in ammonia. It is reported that a working perfused muscle adds a little ammonia to the perfused blood. This is probably greater when the respiration is in any way reduced. (4) It is possible that other bases may be formed also. Muscle has the power of forming various diamines such as putrescine. These are formed by the splitting off of carboxyl from the amino-acid. Such bases occur in muscle and they may be formed in small quantities as part of the neutrality mechanism. They are certainly less important, however, than any of the other neutrality factors mentioned. (5) Another mechanism which might have some importance is the conversion of creatine to creatinine. The latter being the stronger base has much greater powers of neutralizing acid than has creatine. To what extent creatine thus plays a part in regulating the acidity of muscle cannot be said. If it does play any considerable part in this way, it would be an interesting example of how the balance of living matter is preserved or, as Hering puts it, how the need produces the satisfying of the need; of how, in other words, living matter shows its wonderful powers of adaptation to circumstances. The acid creates the base which saturates the acid.

Does the nucleus play a part in contraction? A very interesting but entirely unsolved problem is the possible involvement of the nucleus in the processes of contraction and energy metabolism. On stimulation of the muscles of amphibia it is said that there is a fall in the total purine nitrogen of the muscle of about 9-17 per cent. of that present at the start (Scaffadi). Burian states that on stimulation the hypoxanthine at first increases. These facts, scanty as they are, indicate the possibility that the nuclei are not the passive spectators of muscular contraction that they are often imagined to be. Perhaps they actively intervene. Histological studies of the great cells of Necturus might throw light on this problem.

The mechanism of the contraction.—During contraction of muscle carbohydrates disappear, heat is liberated, carbon dioxide is given off and lactic acid appears. No change in the bodies containing nitrogen is known to occur, except a slight increase in the ammonia and the changes in the purines which have been described. Work does not increase the nitrogen output in the urine. We may now ask the question: How is the contraction produced? Like all the other fundamental questions of physiology, this question cannot be completely answered at the present time. Indeed, we can do but little more than conjecture. The suggestions which have been made are the following:

The machinery or contractile engine in the muscle is certainly protein in nature and doubly refracting. The muscle sarcostyles consist of a very dense or concentrated gel and these gel particles are in very small units, so that their surface is very large compared to the bulk. Protein gels have the peculiarity of swelling greatly and very rapidly in the presence of acid. It has been suggested that the gel particles in protoplasm are arranged mechanically in very fine fibrils and that a more fluid gel or a liquid, the sarcoplasm, is about them. When the nerve impulse sweeps over the muscle it causes in some unknown way the combustion of the carbohydrate, which in part at least is probably built into the machinery, and this combustion produces acids such as lactic acid and succinic acid. Heat is liberated at the same time. Under the influence of the heat and the acidity these gel particles at once absorb water from their surroundings and swell. Being confined in tubes, or the molecules being oriented in a certain way, as in a crystal, the swelling does not take place uniformly in all directions, but they broaden and shorten in length. They swell in a direction transverse to the long axis of the muscle and thus cause a shortening in the other direction. This produces the contraction. But almost at once the lactic acid is burned or recombined and the carbon dioxide escapes from the muscle; the acidity is quickly reduced; the affinity of the contractile elements for water returns to the normal; and water passes out of the sarcostyles, which thus return to their normal size. The elastic properties of the muscle, which Kite has described, help it to regain its normal condition. On this view, then, the essential cause of the shortening is the production of acid. Everything follows as a result of this. Of course, the heat liberated will also assist in the process. There are various additional circumstances which may be urged in support of this view. If it is correct, then if the oxidation or removal of the lactic acid could be prevented the muscle should remain permanently shortened. Just this is what happens, according to many observers in "*rigor mortis.*" Muscles after death go into a condition of extreme contracture and hardness. They become rigid and this rigor persists for some time. A muscle which has been stimulated for some time just before death becomes rigid at death much quicker than one which has been at rest. Certain poisons of the picric acid group, which cause great fever in mammals, cause rigor to appear immediately after death. If a little acid is added to a fluid which is being perfused through muscle, its rigor comes on much earlier than normal; whereas, if alkali is added, rigor is delayed or may be prevented entirely. The beginning of rigor can be checked or entirely prevented by alkaline Ringer's solution. Furthermore, since oxygen in the presence of the muscle destroys lactic acid or causes it to recombine, oxygen should prevent rigor, and such is the case. If muscles are placed in atmospheres of compressed oxygen so that a plentiful supply of oxygen is present all the time, they take much longer to become rigid or they may not go into rigor at all. On the other hand, the muscles of frogs placed in hydrogen become rigid unusually soon. All of these facts agree with the hypothesis that acid is the direct cause of the muscle contraction and that in contraction we are dealing with a swelling or imbibition phenomenon of protein colloids. Another fact which points in the same direction is that muscles placed in salt solutions containing small amounts of acid take up more water and increase in weight much faster than if the muscle is in a neutral salt solution.

On the other hand, there are certain objections which may be urged to this view. The first objection is the very great speed of the contraction and relaxation of the wing muscles of insects. These may contract (bees) as many as 400 times per second. It has seemed to many physiologists that it is extremely unlikely that any such rapid imbibition and syneresis could occur. They have suggested as an alternative view that the change can only be one which affects the surface of the fibril, like a change in surface tension, so that the change, whatever its nature, need not penetrate any distance. The insurmountable objection to this view, or at least it seems insurmountable at the present time, is that the interior of the muscle where these changes are taking place is probably not a liquid, but is a gel and elastic and tough. (Perhaps more of the nature of a liquid crystal.) It is impossible to suppose that the phenomena of rapid change of form under the influence of surface tension can occur in such a solid material where the molecules are not free to move readily. It is only in mobile liquids that surface tension changes of form can occur. The objection that the rapidity of the process is too great in insect wing muscles to permit of this explanation may be met as Hofmeister met it, when he tried his first experiments in this direction, by the reply that, if the elements are small enough, the imbibitiou can take place very rapidly. The velocity is a function of the amount of surface. Now it is just in these insect muscles that the structure of the muscle has reached its highest perfection and the sarcostyle elements are extremely small. On the other hand, histological research has not yet been able to demonstrate clearly and in a convincing manner that the necessary machinery really exists in muscle which is required by this explanation.

Another suggestion was made by Englemann, who laid stress on the fact that double refraction and contractility go together. He suggested that the muscle fibrils absorbed water because of the heat of the combustion. It is possible to make an artificial muscle by taking a catgut string, which is also doubly refracting, passing a wire about it and suspending it in water or salt solution. If a current is sent through the wire so that the water is warmed about the string, it at once takes up water and contracts. On cooling it expands again. The phenomena of the contraction resembled in many particulars that of muscle. The difficulty in this theory in the minds of many is the improbability of the differences in temperature within the muscle in minute dimensions being sufficiently great to account for the whole process, but undoubtedly the heat of the contraction does facilitate the contraction. There is no contradiction between this view and that of imbibition due to acid. Very great differences in temperature within minute distances may momentarily occur, since temperature is only molecular kinetic energy.

It seems that the imbibition theory is the most probable theory of muscle contraction which we have. It is very suggestive that the myosin which is obtained from muscle is extremely sensitive to very minute amounts of acid. Thus one of the best solvents of it is ammonium chloride, which by hydrolytic dissociation produces small amounts of acid. One reason why this explanation of muscle contraction is satisfactory is that other living phenomena, such as those of secretion, may be explained in the same manner. In secretion we have also, probably, a rhythmical imbibition and syneresis on the part of the secreting cell, only in this case the cell acts as a whole and when the water leaves the protoplasmic gel it leaves it in a different direction from that in which it entered it. There is thus a regular pumping of water through the cell, such as occurs, for example, in the absorption of water from the intestine, or in the secretion of liquid from the salivary and other glands. In this case, too, acid may be at the bottom of the process. One could picture the process of secretion as follows: Let us suppose that a nerve impulse first impinges on the base of the cell and passes through it to the lumen end. As it passes down the cell it causes an explosive wave of oxidation by which acid is set free. This acid at once directly, or indirectly, increases the affinity for water on the part of the cell colloids. The protoplasm at the base of the cell first takes the water from the capillary and the lymph space about it. It swells. The oxidation of the acid in this part of the cell causes it to lose its affinity for water, but the next succeeding segment of the cell is now acid and has a great affinity for water so that it imbibes that which is lost from the base. The water is thus handed on from one part of the cell to the next until it reaches the lumen end of the cell, where it escapes more or less loaded with soluble substances into the gland lumen.

Not only are secretion and muscle contraction shown to be identical in principle upon this hypothesis, but other processes, such as ciliary movement and even amœboid movement, can be explained in the same way. By the production of acid at some point in the interior of an amœba water is absorbed here, the protoplasm becomes more mobile or fluid, the internal or osmotic pressure is increased and the firm ectosarc yielding at some point it bursts and the fluid contents stream in this direction. Thus the amœba actually appears to flow through itself. There are facts which indicate that the movements are not due to surface tension, as they were once supposed to be, but that the beginning of the process is in the interior. And if in a cilium either in the basal granule or along the cilium there was a rhythmic absorption by one side of the granule or cilium its bending might be understood. There is no doubt that in red blood corpuscles the slight change of acidity produced by the entrance of so weak an acid as carbonic acid is sufficient to cause the corpuscle to swell. On losing carbonic acid it shrinks again. Perhaps the sarcostyles act in the same manner.

But while so many facts find on this hypothesis a natural, simple explanation, so far without serious contradiction, yet experience teaches us to be cautious and the theory cannot be regarded as proved. It is not proved that the change in acidity is sufficient to account for the change in imbibition. It is rather at present to be looked upon as a guiding hypothesis helping us to see and experiment. It must not be forgotten that while acids do much, enzymes do more. Hydrochloric acid alone will cause the swelling and solution of fibrin, but hydrochloric acid plus pepsin causes a much more powerful effect. It is, hence, not improbable or unlikely that if the theory prove true in its essentials, namely, that the contraction is of the nature of an imbibition of liquid by an organized protein substance, it may be found that the imbibition is not induced directly by the acid, but that this only indirectly, by setting free enzymes which may cause a reversible change in the proteins, affects the imbibition. Or it might be that the organized living matter itself may in some other way have its affinity for water directly changed. It might be, for example, that living matter itself was composed of large colloidal aggregates, or biogens which were really the irritable substances. These biogens may be considered to consist of protein, oxygen, carbohydrate and possibly lipin material, and to have a certain affinity for water with which they are in union. On stimulation it might happen that the oxygen combines with and burns the carbohydrate, and by this process the biogen itself at once, by its change in composition, might acquire a greater affinity for water and absorb more of the latter. The biogen molecule at once reconstitutes itself, uniting with more oxygen and carbohydrate, thus changing to its former water-holding power. The irritable complex is re-established and relaxation occurs. According to this view, the essential explanation of the imbibition as the cause of the contraction remains unaltered, but acid no longer plays the part of the sole or chief cause. This view is in its essentials that of Hermann, who proposed it many years ago; the hypothetical substance here called a biogen being called by him *"inogen."* By the former hypothesis the contractile substance is in the nature of a lifeless protein similar to so many connective tissue fibrils which absorb and lose water under the influence of acid; by the latter view the contractile matter, while still colloidal and having the properties of colloids, is the living matter itself. The former view has a certain naïveté which is at the same time its charm and its weakness. In the former view the toiling spirit of life weaves the contractile web, remaining itself outside of and unaffected by the fate of the web; and, according to the latter view, the web is itself the organized living, self-perpetuating material which contracts because of its physical properties. Ignorance does not permit, as yet, a choice between these divergent views.

REFERENCES. Muscles.

1. *Lederer and Stolte:* Composition of human and dog hearts. Biochem. Ztschr., 35, pp. 108-112, 1910.
2. *Argutinsky:* Ueber die elementare Zusammensetzung des Ochsenfleisches. Archiv f. d. ges. Physiol., 55, pp. 345-365, 1894.
3. *v. Fürth:* General summary on the composition of muscle. Literature. Ergebnisse der Physiologie, 1, Abth. I, p. 110, 1902; Ibid., 2, I, pp. 575-606, 1903.
4. *v. Fürth:* Archiv f. expt. Path. u. Pharm., 37, 1896, p. 389; Hofmeister's Beiträge, 3.
5. *Stewart and Sollman:* Proteins of muscle. Journal of Physiol., 24, 1899.
6. *Winterstein:* Physiological nature of rigor mortis. Literature. Archiv f. d. ges. Physiol., 120, p. 225, 1907.
7. *Fletcher:* Relation of oxygen to the survival metabolism of muscle. Journal Physiol., 28, p. 474, 1902.
8. *Baglioni:* Comparative chemical study of the muscle, the electrical organ and blood serum of Torpedo ocellata. Beiträge z. chem. Physiol. u. Path., 8, 1906, p. 456-472.
9. *Weyl:* Physiology and chemical composition of electrical organ of Torpedo. Zeits. f. physiol. Chem., 7, p. 541, 1883.
10. *Buglia and Constantino:* Urease in muscle of Scyllum catulus. Zeits. f. physiol. Chem., 86, 1913, p. 137.
11. *Gautier and Landi:* Gas exchange of sterile muscle. Anaerobic respiration. Maly's Jahresbericht, 24, p. 410, 1895.
12. *Wilson:* Partition of water-soluble non-protein nitrogen. Jour. Biol. Chem., 17, 1914, p. 385.
13. *Meyers and Fine:* Creatine in muscle. Jour. Biol. Chem., 1913.
14. *Hill:* Oxidative removal of lactic acid. Jour. of Physiol., 48, 1914, p. x, Pro. Phys. Soc.
15. *Peters:* Lactic acid and heat. Jour. Physiol., 47, p. 270, 1913.
16. *Hill:* Total energy available in isolated muscle kept in oxygen. Jour. Physiol., 48, 1914. Pro. Physiol. Soc., xi.
17. *Botazzi and Quagriello:* Constitution physique et les propriétés chimicophysiques du suc des muscles lissés et des muscles striés. Archives internat. de Physiol., 12, p. 409, 1912.
18. *Galeotti:* Condensation of amino-acids with formaldehyde. (Methylation in muscle.) Biochem. Ztschr., 53, pp. 474-492, 1913.
19. *Gulewitsch:* Constitution of carnosin. Zeits. f. physiol. Chem., 73, 1911, p. 43.

20. *Thunberg:* Muscle respiration. Skan. Archiv f. Physiol., **25, 1911, p. 37.** Previous communications cited in this paper.

21. *Jona:* d-alanyl-d-alanine anhydride isolated from Liebig's extract. Zeits. physiol. Chem., 83, 1913, p. 458.

22. *Locke and Rosenheim:* Consumption of glucose by the mammalian heart. Journal of Physiol., 31, 1904.

23. *Mosso:* Die Ermüdung, Leipzig, 1891.

24. *Contardi:* Gazetta chimica Italiana, 42, pp. 408-412, 1912. Synthesis of phytin from inosite and phosphoric acid.

25. *Cabella:* Creatine content of different kinds of muscles. Zeits. f. physiol. Chem., 84, p. 29, 1913.

26. *Embech:* Succinic acid in meat extract and fresh muscle. Zeits. f. physiol. Chem., 87, 1913, p. 145.

27. *Jansen:* Mytilite in muscle of mytilus. Zeits. f. physiol. Chem., 85, p. 231.

28. *Siegfried:* Carnic acid. DuBois Archiv f. Physiol., 1904, pp. 401-418.

29. *Roaf:* Lactic acid theory of contraction. Pro. Roy. Soc. London (B), **88, pp. 139-150, 1914.**

30. *Scaffadi:* Relation of purine bases of muscle during work. Biochem. Ztschr., 30, pp. 473-480; 33, pp. 247-251, 1911.

31. *Pekelharing:* Nucleo-proteid of muscle. Zeits. f. physiol. Chem., **22, p. 245,** 1896-7.

32. *Burian:* Action of muscle work on purine excretion. Zeits. f. physiol. Chem., 43, p. 532, 1905.

33. *Buglia and Constantino:* Total amino nitrogen titrateable with formol in various kinds of muscle. Zeits. physiol. Chem., 81, pp. 109-112, 1912.

34. *Gulewitsch:* Extractive substances. Carnosin. Zeits. physiol. Chem., **87, 1913,** p. 1.

35. *Arnold:* Position of glycogen in muscle. Arch. f. Mik. Anat., 73, **1908-9, p. 265.**

36. *Maignon:* Influence of hunger on glycogen in muscle. Journal de physiol., **10.** pp. 203-221, 1908.

37. *Preti:* Formation of acetone in muscle. Biochem. Ztschr., 32, p. 231, 1911.

38. *Widmark:* Chemical conditions necessary for preservation of normal structure of cells. Skan. Archiv f. Physiol., 24, 1911, p. 339.

39. *v. Fürth and Leuch:* Rigor mortis an acid swelling and not a coagulation. Biochem. Ztschr., 33, 1911, p. 341.

40. *Jensen and Fischer:* State of water in living and dead muscle. Zeits. f. all. Physiol., 11, 1910, p. 23.

41. *Fletcher:* Lactic acid formation in autolysis. Journal of Physiol., **43, 1911,** p. 286.

42. *Feldman and Hill:* Influence of oxygen on lactic acid formation. Jour. of Physiol., 42, 1911, p. 439.

43. *Krimberg:* Carnitine. Zeits. f. physiol. Chem., 53, p. 514, 1907.

44. *Mendel and Leavenworth:* Composition of embryonic muscle and nerve. Amer. Jour. Physiol., 21, p. 9, 1908.

45. *Henderson and Brink:* Compressibility of muscle. Amer. Jour. Physiol., 21, 1908, p. 248.

46. *Fischer:* Analogy between absorption of water by fibrin and by muscle. Archiv f. d. ges. Physiol., 124, p. 69-99, 1908.

47. *Fahr:* Sodium content of skeletal muscle of frogs. Zeits. f. Biol., 52, p. 72, 1909.

48. *Constantino:* Potassium, sodium, chlorine, etc., in muscle. Biochem. Ztschr., 37, 1911, p. 52; 43, p. 165, 1912.

CHAPTER XV.

THE CONNECTIVE, OR SUPPORTING, TISSUES. THE BONES. CARTILAGE. TEETH. CONNECTIVE TISSUE.

The chemistry and metabolism of the supporting tissues of the body present several points of general interest. In plants these tissues are carbohydrate in nature. In the invertebrates the harder parts of the supporting tissues are chitin, which is a polymerized acetylated glucosamine and the composition of which has already been discussed on page 325. In the vertebrates these tissues are composed of an organic matrix which is always of a protein nature and in it is often deposited more or less inorganic material of the nature of phosphates and carbonates of calcium and magnesium. The connective tissues consist of both living and lifeless matter. These tissues are true tissues consisting of cells, but these cells have formed a large amount of intercellular, protein, lifeless matter which is generally of a fibrous nature and always tough and resistant to most reagents. There are several kinds of connective tissue proper and we may distinguish the yellow connective tissue, which occurs, for example, in the ligamentum nuchæ of the ox; white connective tissue, of which the tendon of Achilles is a good example; and reticular connective tissue found in lymphatic glands and elsewhere in the body. The yellow elastic tissue contains elastic, tough, yellowish fibers. It occurs not only in the tendons, but in the walls of the blood vessels, in the lungs and elsewhere in the body. It consists largely of a protein called *elastin*. The white fibrous tissue is also found generally throughout the body in the connective tissues as well as in the tendons of muscle. Bone also contains fibers of white tissue. Bone is indeed connective tissue in which mineral salts have been deposited. The character of the cells, and of course the character of the groundwork in which the fibers are embedded, differ in the various connective tissues. Cartilage is closely related to the white connective tissues. Our knowledge of the composition of these structures is largely owing to the work of Gies and his collaborators in this country and to Mörner.

1. **Composition of white connective tissue. Tendo Achillis.**—The tendo Achillis consists chiefly of white connective tissue and its composition is given by Buerger and Gies as follows:

COMPOSITION OF TENDO ACHILLIS OF THE OX.

Constituents	Fresh tissue		Dry tissue		Ash
	Calf	Ox	Calf	Ox	Ox
Water	67.51%	62.870%			
Solids	32.49	37.130			
Inorganic matter	0.61	0.470	1.88	1.266	
SO_3	....	0.031	.....	0.084	6.65
P_2O_5	....	0.039	.....	0.106	8.34
Cl	....	0.147	.....	0.397	31.27
Organic matter	31.88	36.660	98.12	98.734	
Fat (ether-sol. matter)	....	1.040	.....	2.801	
Albumin, globin	....	0.220	....	0.593	
Mucoid	....	1.283	....	3.455	
Elastin	....	2.633	....	4.398	
Collagen (gelatin)	....	31.588	....	85.074	
Extractives and undetermined	....	0.896	....	2.413	

2. Composition of yellow connective tissue. Ligamentum nuchae.—The ligamentum nuchæ of the ox consists for the most part of yellow connective tissue and its composition was found by Vandegrift and Gies to be as follows:

COMPOSITION OF LIGAMENTUM NUCHÆ OF THE OX. IN PER CENT.

Constituents	Fresh ligament		Dry ligament		Ash
	Calf	Ox	Calf	Ox	Ox
Water	65.10	57.570			
Solids	34.90	42.430			
Inorganic matter	0.66	0.470	1.90	1.100	
SO_3	......	0.026	....	0.062	5.64
P_2O_5	.....	0.035	....	0.081	7.39
Cl	......	0.136	....	0.318	28.95
Organic matter	34.24	41.960	98.10	98.900	
Fat (ether-sol. matter)	......	1.120	.....	2.640	
Albumin, globulin	......	0.616	.....	1.452	
Mucoid	......	0.525	.....	1.237	
Elastin	.....	31.670	.....	74.641	
Collagen (gelatin)	......	7.230	....	17.040	
Extractives and undetermined	......	0.799	....	1.883	

By an inspection of the foregoing tables it will be seen that the greater part of the dry matter of the tendons consists of protein material and that the per cent. of inorganic matter is low. The dry matter in each case consists of from 1-3 per cent. of mucoid. In white elastic tissue 85 per cent. of the dry matter consists of collagen which yields gelatin on heating with water; in the yellow elastic tissue, on the other hand, there is only some 17 per cent. of collagen in the dry matter, but the greater part of the dry matter, namely 74.64 per cent., consists of elastin.

The composition of the mucoid from the tendons has already been discussed on page 324. This mucoid is obtained by cutting the tendon or ligament into small parts and then boiling with half-saturated $Ca(OH)_2$ and precipitating the mucoid by acidification with acetic acid. This mucoid differs from true mucin apparently in several properties. It does not form the very stringy, sticky masses characteristic of mucin; it is less soluble in 0.1 per cent. HCl; and it contains about twice as much sulphur. The per cent. of sulphur in tendon mucoid is about 2.3 per cent., whereas in submaxillary mucin it is a little less than 1 per cent. Moreover, the mucoid of tendon contains chondroitic acid which relates it to the constitution of cartilage. Mucin and mucoid resemble each other in the fact that both yield a reducing sugar, either glucosamine or galactosamine, on hydrolysis. Both are glycoproteins. It would, perhaps, be well to call these mucoids *chondroproteins*, as Hammarsten does, to indicate that they contain a conjugated chon[illegible] acid. This acid will be discussed presently.

The composition of the gelatin, or collagen from which it is derived, has already been given in the table on page 129 and on page 110. It is remarkable, first, for its powers of gelatinization. It is a protein which contains a very large amount of glycocoll and which lacks tyrosine and tryptophane. Collagen yields with tannic acid an insoluble resistant substance, which does not putrefy readily, and this peculiarity is the basis of the tanning of leather.

Elastin is a very resistant, elastic protein, of which the composition is given on page 129. It consists of at least 25 per cent. of glycocoll and differs from collagen in the very much larger proportion of leucine in it, namely 21.38 per cent., as contrasted with 2.1 per cent. The three simple amino-acids, glycine, alanine and leucine, make over 50 per cent. of the molecule. Elastin is easily digested by pepsin. It contains 0.1-0.3 per cent. S, but none which is split off by alkali.

Elastin is readily prepared from the ligamentum nuchæ. It is first thoroughly extracted with 5 per cent. NaCl under thymol for 3-4 days to remove albumins and globulins, and then repeatedly boiled with water to convert all the collagen into gelatin and remove it, so that the filtrates give only a very slight turbidity with tannic acid. The undissolved residue is thoroughly washed with water, extracted with alcohol and ether and dried at 110°.

Both of these tissues yielded some extractive matter. Creatine and purines were recognized qualitatively in the extractives.

The general composition of several connective tissues is given in the following table compiled by Buerger and Gies:

Constituents	Tendon		Ligament		Vitreous humor	Costal cartilage	Bone with marrow	Adipose tissue kidney
	Calf	Ox	Calf	Ox				
Fresh tissue								
Water	67.51	62.87	65.10	57.57	98.64	67.67	50.00	4.30
Solids	31.88	37.13	34.90	42.43	1.36	32.33	50.00	95.70
Organic	0.61	36.66	34.24	41.96	0.48	30.13	28.15	95.51
Inorganic	32.49	0.47	0.66	0.47	0.88	2.20	21.85	0.19
Dry tissue								
Organic	98.12	98.71	98.10	98.90	35.29	93.20	56.30	99.80
Inorganic	1.88	1.29	1.90	1.10	64.71	6.80	43.70	0.20

II. Composition of cartilage.—Cartilage forms the internal skeleton of the lowest or cartilaginous fishes and in the higher forms it is generally laid down first and is later changed into bony tissue. In the cartilages of the trachea, and in particular of the larynx, it never undergoes transformation to bone, but remains cartilage throughout life. Cartilage is not found in any of the invertebrates, except in the cephalopods and the arthropods. The cartilage of these forms is, however, related to chitin rather than to true cartilage.

Histology. Cartilage consists of cells which are embedded in a homogeneous, or slightly fibrous, matrix. This matrix has in it some blood vessels for the nourishment of the cells and channels for the penetration of food materials to the cells. But the amount of blood supply must be extremely little and the penetration of the lymph would appear to be very slight. Most of the food materials find access to the cells possibly by soaking through the intercellular substance. This intercellular substance is secreted or formed by the cartilage cells. The cartilage cells arise from the mesodermal layer in embryonic development.

Chemistry. If cartilage is cut into small pieces and boiled in water for a long time, or, better, heated in water under pressure, the organic matrix of the cartilage slowly dissolves and there is left a mass of cell bodies and connective tissue fibers and blood vessels. The cell bodies consist, for the most part, of coagulable protein so that they are resistant to this treatment. By a careful study of the organic matrix Mörner found that it consisted, or rather that it yielded on decomposition, the following substances:

1. Chondromucoid.
2. Collagen.
3. Albuminoid.
4. Chondroitic acid
5. Inorganic salts.

The chondromucoid is in all essential respects like that isolated by

Gies and Buerger from tendons and it has already been described. Chondromucoid has the composition C, 47.30; H, 6.42; N, 12.58; S, 2.42; O, 31.28. It contains sulphur both in an oxidized and an unoxidized form. A part is split off as sulphide when boiled with alkali, a part as sulphate. This chondromucoid yields chondroitic acid. It is a white, acid-reacting substance insoluble in water, but soluble in very dilute alkali and precipitated by acetic acid. It is precipitated by iron alum, lead acetate and ferric chloride, but it is not precipitated by tannic acid. It gives the usual reactions of the proteins.

2. Chondroitic acid, or cartilage acid, is sometimes called chondroit-sulphuric acid, but as chondroitic acid is shorter and was the name first given, it should be adopted and the awkward name of chondroit-sulphuric acid dropped. This acid has already been described in part. It will be recalled that it splits on hydrolysis into sulphuric acid, a hexosamine (chondrosamine), acetic acid and glycuronic acid. It is a paired sulphate. Some of it is apparently free in the cartilage or attached through bases to the protein, since it may be extracted easily from the cartilage by Mörner's method by extracting with alkali and removing the proteins by neutralizing and precipitating the peptone by tannic acid. The salts of chondroitic acid are nearly all soluble. The composition of chondroitic acid is not yet definitely settled, the statements in the literature still being somewhat contradictory. (Fraenkel, *Annal. d. Chem. u. Pharm.*, 351.) It will be recalled that the composition of this acid is of considerable interest, since the presence in it of an acetylated hexosamine allies it at once with chitin.. It is either identical with, or very similar to, the glucothionic acid isolated from tendomucoid by Levene. A possible graphic formula is given on page 325.

3. Albuminoid.

Chondroalbuminoid is very closely similar to, but not identical with, the osseoalbuminoid of bones. It is the albuminous substance which remains after extraction of the osseoalbuminoid and the nucleoproteins, and the prolonged treatment of the cartilage with hot water. It is soluble in boiling 0.1 per cent. KOH and also in boiling 5 per cent. KOH. By the former it is hydrolyzed into various derived products. This substance is present in the cartilage in very small amounts. It is believed to form the lining of the tubules in the cartilage and bones. By some authors it was supposed to resemble keratin, but it is more closely related to elastin. It was prepared in the following way, according to Hawk and Gies:

The cartilaginous portions of the nasal septum of the ox were used. Several pounds of these were hashed in a machine after removal of the outer membranes; the hash washed in running water; mucoid, nucleoproteins, etc., were extracted by several treatments with dilute alkali after a preliminary treatment with 0.1-0.2 per cent. HCl.; the alkali washed out and the residue hydrated in boiling water for several days. The final product was extracted first with 0.1 per cent. sodium carbonate and then 0.5 per cent. HCl in which it was insoluble. The albuminoid remains undissolved. The composition of this substance and that derived from bone, or osseoalbuminoid, was as follows:

	C	H	N	S	O
Chondroalbuminoid	50.46	7.05	14.95	1.86	26.86
Osseoalbuminoid	50.16	7.03	16.17	1.18	25.46

The greater proportion of the organic matrix of the cartilage is collagen, but I have been unable to find any quantitative determination of the amount present.

III. The bones.—The bones also consist of cells and intercellular substance, but as in cartilage the intercellular substance is far in excess and determines the character of the tissue. The intercellular substance is formed by the cellular. It consists of two parts: of an organic basis of an albuminous nature in which a great amount of inorganic matter is deposited. The cells, it will be recalled, have to be nourished and the nourishment is arranged for by a series of channels, the cells being grouped in concentric rings about the blood vessels and having fine branching channels penetrating the bone in all directions.

The organic intercellular substance. This resembles strongly the organic matrix of cartilage. It consists of osseomucoid, the composition of which has just been given; and of osseoalbuminoid, which is present in only small amounts, and which is almost identical with the similar substance in cartilage and is supposed to be the lining of the Haversian canals. The greater part of the organic matter here as in cartilage and in white connective tissue is collagen, which yields gelatin on cooking with water.

The bones are many of them hollow and contain marrow in their centers. This marrow is of two kinds, being in part yellow in color and consisting of a large proportion of fat; and in part it is red marrow, the red color being due to the presence of a large number of red corpuscles in the erythroblasts.

The composition of the osseomucoid and some other mucoids is given in the following table taken from Cutter and Gies:

	C	H	N	S	O
Chondromucoid (Mörner)	47.30	6.42	12.58	2.42	31.28
Tendomucoid (Chittenden and Gies)	48.76	6.53	11.75	2.33	30.63
Tendomucoid (Cutter and Gies)	47.47	6.68	12.58	2.20	31.07
Osseomucoid (Hawk and Gies)	47.07	6.69	11.98	2.41	31.86

The composition of the gelatin from bone was found by Fremy to be C, 50.0 per cent.; H, 6.5 per cent.; N, 17.5 per cent.; O and S, 26.0 per cent. The following table is given by Richards and Gies:

	C	H	N	S	O
Ligament gelatin (Richards and Gies) ..	50.49	6.71	17.90	0.57	24.33
Tendon gelatin (Van Name)	50.11	6.56	17.81	0.26	25.24
Commercial gelatin (Chittenden and Solly)	49.38	6.81	17.97	0.71	25.13

The inorganic constituents of bone. These make about 40 per cent. of the dry residue of the bone, 60 per cent. being organic. In bone without the marrow the relations are just about reversed. The inorganic material is chiefly calcium phosphate and carbonate, but there is a little magnesium and there is also present a trace of fluoride and chloride. The composition of the inorganic materials is about as follows:

Calcium phosphate	85%
Magnesim phosphate	1.5
Calcium fluoride	0.3
Calcium chloride	0.2
Calcium carbonate	10.0
Alkali salts	2

Hoppe-Seyler pointed out that the relation of Ca to phosphoric acid is about the same as in apatite, namely, 10 Ca: $6PO_4$. There are about three molecules of tricalcium phosphate to one molecule calcium carbonate. The relationships remain the same, although the bone may be gaining or losing its inorganic compounds.

Practically nothing is known concerning the nature of the processes by which the salts are deposited in the organic basis of bone. It seems not impossible either that the conditions are such as to cause the precipitation of this double phosphate and carbonate here, or else that either the phosphoric acid or the calcium is united by one bond to the organic matrix, while with the other they unite with phosphoric acid or calcium brought from the blood so as to precipitate and so hold it. The attempt has been made to explain the deposition on the basis of an adsorption, but the relations are so regular, quantitatively, that it seems perhaps as probable that the union is a true chemical union.

The growth of the bones is in some way dependent upon the internal secretion of both the thyroids and the hypophysis. In case the hypophysis is extirpated in puppies, the bones do not grow normally, they remain in the state of youth and the epiphyses remain open. Normal growth of the bone is absent if the thyroids do not develop properly. The bones are also closely related to the reproductive organs. Castration of young animals nearly always causes the production of large-boned animals; and in osteomalacia of women it is not uncommon to remove the ovaries, the condition of the bones changing at that time. Changes occur in the bones and teeth in pregnancy. In fact, the bones are by no means a fixed tissue. They are alive with their own metabolism and the calcium may be laid down or again removed. We cannot for lack of time enter here into the work which has been done on the pathological changes which occur in bones and the attempts which have been made to produce rickets or other bone diseases artificially. That the proper formation of both bones and teeth is dependent on a proper supply of calcium in the diet is perhaps self-evident. How largely the composition of the bones may be affected by a change in the amount of calcium in the diet is shown in the following figures from Rohloff:

In growing dogs fed one on calcium rich, the other on calcium poor food, the composition of the shoulder blade was as follows, the amounts being in 10 c.c. of bone:

Diet		Dry substance	Ash	CaO	Phosphoric acid
Ca poor.	Shoulder blade	2.678	1.178	0.638	0.499
Ca rich.	" "	9.021	5.240	2.861	2.211

Similar and more extensive figures are given by Aron and Sebauer.

Teeth.—Teeth are composed of several parts: namely, enamel, dentine and cement. The cement is identical with bone in its structure and composition. The dentine makes the greater part of the tooth and in its chemical composition is practically identical with bone, although its structure is somewhat different. The enamel, however, is the product of an epithelial tissue. It is the hardest substance in the body and contains the smallest per cent. of water. It contains about 5 per cent. of water. The composition of the enamel of human and calves' teeth is as follows (Bertz):

	Human	Calf
Organic substance	6.82	16.56
CaO	50.22	44.24
MgO	0.73	0.96
P_2O_5	40.69	37.02

91 per cent. consists, therefore, of calcium phosphate. It is extraordinary to what widely different uses this interesting and common material, calcium phosphate, is put in the body. Here we find it called on to make the hardest and most resistant structure of the body, while it was but a short time ago that we found it assisting in the coagulation of the blood and in activating the enzymes, invertin and trypsinogen.

REFERENCES. Connective Tissues.

1. *Oswald:* Ueber Myxomucin. Zeit. physiol. Chem., 92, p. 146, 1914.
2. *Levene:* Preliminary communication on the chemistry of mucin. Jour. Amer. Chem. Soc., 22, p. 80, 1900.
3. *Hawk and Gies:* Chemical studies of osseomucoid, with determination of the heat of combustion of some connective tissue glucoproteids. Amer. Jour. Physiol., 5, p. 388, 1901.
4. *Vandergrift and Gies:* The composition of yellow connective tissue. Amer. Jour. Physiol., 5, p. 288, 1901.
5. *Richards and Gies:* Chemical studies of elastin, mucoid, and other proteids in elastic tissue, with some notes on ligament extractives. Amer. Jour. Physiol., 7, p. 93, 1903.

6. *Richards and Gies:* Methods of preparing elastin, with some facts regarding ligament mucin. Amer. Jour. Physiol., 5, 1901, p. xi.
7. *Gies and Cutter:* The composition of tendon mucoid. Amer. Jour. of Physiol., 6, p. 155, 1901.
8. *Hawk and Gies:* On the composition and chemical properties of osseoalbuminoid, with a comparative study of the albuminoid of cartilage. Amer. Jour. Physiol., 7, p. 340, 1902.
9. *Posner and Gies:* Do the mucoids combine with other proteins? Amer. Jour. of Physiol., 11, p. 340, 1904.
10. *Seifert and Gies:* On the distribution of osseomucoid. Amer. Jour. of Physiol., 10, p. 146, 1903.
11. *Buerger and Gies:* The chemical constituents of tendinous tissue. Amer. Jour. Physiol., 6, p. 219, 1901.

CHAPTER XVI.

THE CRYPTORRHETIC TISSUES. THE THYROID. PARATHYROID. SUPRA-RENAL. HYPOPHYSIS. REPRODUCTIVE GLANDS.

The cryptorrhetic tissues and organs.—There are quite a number of organs in the body which do not play a part in movement and, while they are more or less glandular in nature, they either have no ducts for the discharge of their secretions to the exterior or, if they have such ducts, there is abundant evidence that they form a hidden secretion which is passed back to the blood. Of course all organs form such substances which come back to the blood, but in the cryptorrhetic organs there are reasons for thinking that the formation of such substances is their main function. They are sometimes called glands of internal secretion. I have given them the name "Cryptorrhetic organs," from two Greek words—*kryptos*, concealed; and *rhoia*, flow. They are the tissues of hidden flowing. Among these organs the thyroid, parathyroids, ovaries and testes, pancreas, hypophysis and supra-renals are the most important.

THE HYPOPHYSIS.

Among the organs more particularly designated as those of internal secretion, the hypophysis is certainly among the most interesting and important.

Structure.—In human beings this organ is about as large as a pea, lying at the base of the brain with the optic chiasma just in front of it. The hypophysis cerebri (G. *hypo*, below) is also often designated as the pituitary body. It is connected with the third ventricle of the brain in the thalamus by a narrow stalk called the infundibulum, which is continuous with the epithelium lining the ventricles. The hypophysis proper is easily separated into three main parts: an anterior lobe, composed of glandular tissue and which is the larger part; a posterior lobe, composed of nervous tissue, and a pars intermedia, which resembles the infundibulum (?). These different parts have quite different structures, compositions and functions.

Anteriorly, the hypophysis is closely adherent to the *tuber cinereum* of the brain, and its separation from this in operations is a matter of great difficulty. This fact greatly complicates operations on the hypophysis, since any injury to this part of the brain is very serious.

By means of the infundibulum the organ is in direct connection with the brain ventricles, its secretions, if any, may be discharged into them, and when cut the ventricular fluid may escape from it. The hypophysis is thus essentially a downgrowth from the central nervous system toward the palate.

Situated as it is in the oldest part of the nervous system, it is not surprising that this organ, both phylogenetically and embryologically

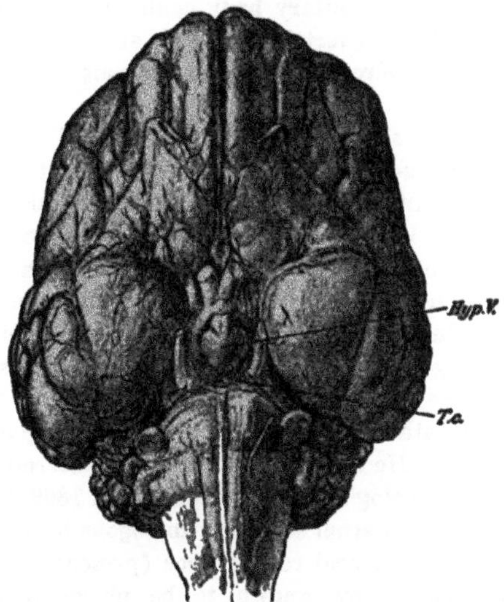

Fig. 57.—Under surface of the dog's brain showing the hypophysis, *Hyp. V.*, just behind the optic chiasma (Aschner).

is found to be very old. It is found in all the vertebrates without exception, except possibly in amphioxus, in which indeed the glandula subneuralis is supposed by some to be homologous with it. Even in the invertebrates a somewhat similar tissue has been described in worms, mollusks, and even in Echinoderms, although whether this tissue is homologous in nature and analogous in function is quite unknown.

Embryologically the organ is derived in part (posterior lobe) from the nervous system by a downgrowth of the floor of the third ventricle, and in part from the alimentary canal (anterior lobe) by a proliferation of cells from the dorsal side of the buccal cavity. It arises very early in embryological history. From all these facts it is evident that we have to do with a very old organ and one probably correlated with the most fundamental processes of the body.

The phylogenetic explanation of this organ generally accepted is that formerly the neural canal connected at this point with the alimentary canal. A probable and almost the only explanation of this, though an explanation almost universally rejected by zoölogists, is that of Gaskell, who has maintained that the vertebrate alimentary canal is a new structure, and that the old invertebrate canal is the present neural canal. The infundibulum, on this view, would correspond to the old invertebrate œsophagus, the ventricle of the thalamus to the invertebrate stomach, and the canal originally connected posteriorly with the anus. The anterior lobe of the pituitary body could then correspond to some glandular adjunct of the invertebrate canal, and the nervous part to a portion of the original circumœsophageal nervous ring of the invertebrates.

The hypophysis has always, even from the time of Aristotle, attracted attention and many guesses were early made as to its nature, some of them very near the truth. Thus its connection with the third ventricle led Diemerbroeck in 1686 to the conclusion that it made a secretion which was poured into the third ventricle, a view on the whole substantiated by modern work; another view was that it was the source of the cerebro-spinal fluid; another that it secreted the cerebro-spinal fluid into the blood. Engel called attention in 1839 to the pathological correlations between the hypophysis, pineal gland and the thyroid and to the commonness of pathological changes in the sexual glands when these glands were diseased. He called the hypophysis a secreting gland. The American-French physiologist, Brown-Séquard, in 1869 definitely placed it among the organs of internal secretion analogous to the thyroid. This he did because of its structural resemblance (presence of colloid) to the thyroid, a resemblance now known to be physiological as well as anatomical.

The modern view of the function of the pituitary dates mainly, however, from the work of Pierre Marie and Marie and Marinesco in 1886-1889. Marie, studying the disease known as acromegaly, a form of gigantism of the extremities (Gr. *akron*, extremity; *megas*, large), called attention to the invariably pathological state of the hypophysis in this disease and in true gigantism, and concluded that the abnormal hypophysis was the primary cause of the growth changes involved in acromegaly. A very large amount of work has since then been done by pathologists which has confirmed Marie's findings; though opinions are not unanimous whether the connection between the glands and the bones and other growing tissues is direct, or indirect through its influence on the thyroids, thymus and sexual organs. Pathologists have also shown that in several true dwarfs the gland is rudimentary or almost lacking. The relation to the sexual organs is very close. Thus during pregnancy the hypophysis undergoes a characteristic metamorphosis, in common with the thyroid and some other organs.

The pathological correlation between the hypophysis and disturbances of growth, and in the sexual organs (*dysplasia adiposo genitalis*) in the deposition of fat, many years ago led to the inference that the hypophysis was concerned in the regulation of the growth and metabolism of the body, but it was impossible to say whether it intervened through the nervous system, or directly by an internal secretion. The general conclusion was the latter and that it was the anterior lobe which particularly intervened in growth. The exact function of the organ could only be determined by experiment.

Such experiments were early undertaken, but, owing to the great difficulties of operating, the evidence obtained has been, until very recently, contradictory and confusing. The experimental difficulties arise from the inaccessibility of the organ; from the great difficulty of freeing the gland from the brain without injury to the tuber cinereum and surrounding parts, since a very slight injury here may be fatal; to the contraction of scar tissue putting tension on these parts which may result in killing the animal after several months; in part to the danger of too great loss of spinal fluid and fatal results from opening the third ventricle; and in part also to the danger of infection. The resistance to infection, particularly to skin infections, is said to be somewhat reduced and animals not infrequently die from these infections or from pneumonia.

Many of the symptoms which were obtained in the early experiments were undoubtedly due to injury of the brain, rather than to absence of the hypophysis. In nearly all of the early experiments extirpation of the organ was followed by the death of the animal within two to three days or weeks; and death was accompanied by pronounced nervous symptoms: cramps, coma, etc. A few animals lived without any apparent change for several months; but in these instances it was impossible to know whether the survival was due to some tissue being left in. Thus in Vasale and Sacchi's experiments the gland was destroyed either by thermocautery or acid. The animals, cats or dogs, always died in a few days, or hours, showing psychic depression, apathy, motor disturbances, fall in body temperature, polyuria, cramps, polydipsia, anorexia and loss of weight. Paulesco and Horsley got practically the same results. These symptoms are now known to be due to injury to the brain tissue, not to the hypophysis. On the other hand, Kreidel and Biedel in 1898 and Friedman and Mass conclude that animals can live a long time without an hypophysis. The American surgeon, Cushing, was one of the first to prove that definite metabolic changes occur after the partial extirpation of the anterior lobe of the hypophysis in dogs, particularly in young dogs. There was a great increase in fat in the body everywhere, fatty degenerations of the internal organs,

response. If the operation takes place after 10 weeks of age, the retardation of growth is still marked, but not so extraordinary as when performed on younger dogs.

It seems, therefore, from Cushing's and Aschner's experiments and the pathological evidence, that the hypophysis, like the thyroid, is nec-

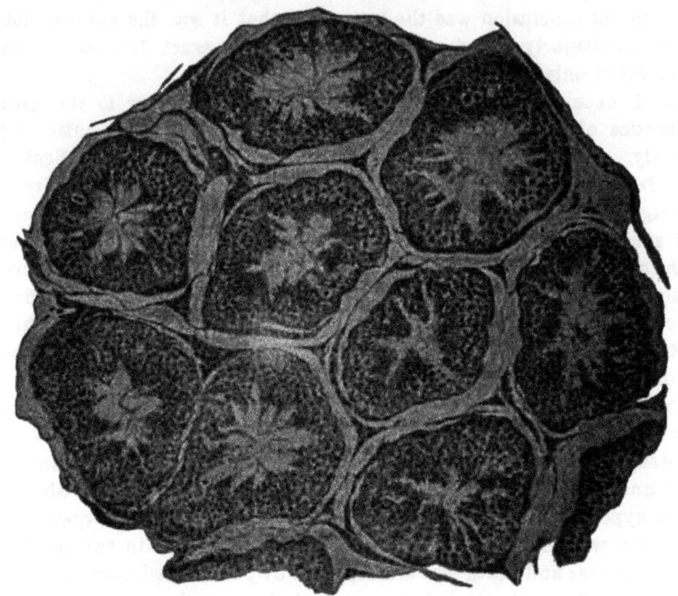

FIG. 59.—Section of the testis of normal dog at ten months of age. Notice the numerous sperm (Aschner).

essary for the growth and development of animals. The body weight, the development of the sexual organs, bones, etc., may be called secondary characters of the anterior lobe of the hypophysis.

The experiments of feeding or injecting extracts, or preparations, of the anterior lobe of the gland (Pituitrinum glandulæ) confirm these results. Schaefer has obtained a slight increase of growth by feeding such tablets to normal puppies; Magnus Levy and Solomon reported an increased gas exchange after taking hypophysis tablets; Falta found that an injection of the extract caused an increase of protein decomposition and a greatly increased glycosuria and increase of other reactions following adrenaline.

We may sum up the facts just presented as follows: In the absence of the anterior lobe in young animals growth is checked, an increase of fatty deposit occurs and a persistent infantilism, such as infantile sexual organs, persistent thymus, long retention of puppy hair and milk teeth; the epiphyses of the bones remain open; there is a marked diminution in susceptibility to adrenaline; lack of brain development and intelligence; a lowered body temperature; a lowered respiratory exchange;

FIG. 59A.—Section of the testis of the brother of the dog whose testis is shown in 59. This dog had the hypophysis extirpated early in life. Note the absence of spermatozoa. These two dogs were the same age.

a lowered nitrogen decomposition of the tissues. In adults there is no marked change, except an increase in fat; a lowering of temperature; lowered nitrogen output; and lowered sensitiveness to adrenaline. There is also an increase in the eosinophile cells of the blood and a diminution of vital resistance, particularly on the part of the skin. Feeding tablets of the hypophysis brings about the reverse change. These conclusions are supported by pathology. Hyperfunction of the anterior lobe of the hypophysis is accompanied by gigantism, true or acromegalous. Hypofunction is accompanied by persistent infantilism.

From these experiments we may conclude as probable that the anterior lobe furnishes an internal secretion necessary for normal growth and metabolism and which acts much as the thyroid acts. How this internal secretion acts, and where; whether by direct action on the tissues, or indirectly by trophic changes of the brain, thymus, sexual organs or in some other manner, is still an open question. Castration of the young generally leads to the production of an abnormally large animal. Castrated animals deprived of the hypophysis are less retarded in growth than those deprived of the hypophysis alone. The nature and point of attack of the secretion, whether it goes into the blood, lymph or cerebro-spinal fluid, are questions needing further examination.

The *posterior lobe* is quite distinct anatomically, phylogenetically, embryologically and physiologically. No disturbance of growth takes place when it alone is taken out. The sole change noted was one by Cushing: i.e., the increase of sexual desire. This, however, was not observed by Aschner in any of his experiments and may possibly have been due to slight injury to the brain parts immediately adjacent to the hypophysis. Indeed, it must be again emphasized that these parts are extremely important, injury to them being followed by glycosuria, polyuria, and changes in the sexual organs; and stimulation of them by contraction of the intestines, bladder, etc. The connection with the vagus is very close. Experiments by S. A. Matthews show that in very careful extirpation of the posterior lobe by Aschner's method there is often an enormous polyuria, or diabetes insipidus. A dog of 4 kilos may secrete as much as 3 liters per day. This diabetes insipidus is not accompanied by any lowering of the sugar tolerance, nor is the operation followed or accompanied by glycosuria.

The posterior lobe contains a substance, pituitrine, which raises the blood pressure like adrenaline, but its effect is not so transient as adrenaline, the blood pressure remains high for a considerable time; it may cause polyuria; and, most remarkable of all, it is one of the few substances which causes a flow of milk from the mammary glands. This last observation of the American physiologists, Ott and Scott, is extremely interesting, since only one or two other substances have such an action: namely, the extract of the reabsorbing uterine wall and the extract of the corpus luteum. If extract of the hypophysis is injected into a pregnant or parturient cat, a continuous stream of milk may often be expressed from the lacteal glands. The total quantity of milk is not increased. The substance raising the blood pressure has not yet been chemically isolated and identified. Its presence allies this lobe, perhaps, to the chromaffine tissues of the body. In other respects it acts like adrenaline. Its preparation should provide a valuable remedy to medicine. The presence of pituitrine in the body and the resemblance of its actions to those of adrenaline makes it always somewhat doubtful whether such a substance in the blood has come from the hypophysis or the supra-renals.

Among its other physiological actions may be mentioned its action on the uterus and bladder. It causes in each prolonged tonic contractions. It seems to act directly on smooth muscle regardless of the innervation, in this respect differing from adrenaline.

The composition of the active principle of the gland, or pituitrine, is not yet known. It is apparently a base and is precipitated by phosphotungstic acid (Engeland and Kutscher). The solutions of the active principle give a very strong histidine reaction with p-diazobenzene sulphonic acid (Fuhner; Aldrich). It would appear from this to be possibly a histidine derivative and in this connection it is of interest that β-imidazolylethylamine also causes contraction of the uterus like pituitrine. It is destroyed by trypsin, but slowly by pepsin, from which it would appear to be a polypeptide. It is very unstable in alkali and liberates a volatile amine. Aldrich obtained a crystalline picrate.

Pituitrine, whatever its nature, is apparently allied in its action to the digitalis group. A substance of somewhat similar properties has been found in the skin glands of toads. In the latter case the substance is an oxycholesterol. There is also a pressor substance in the anterior lobe, but it is in small amounts and its presence is masked by a depressor substance which may be separated by its solubility in ethyl alcohol.

The facts that are here presented show that in its actions on general metabolism the anterior lobe of the hypophysis strongly resembles the thyroid. Its action on metabolism is less pronounced. It is clear, from its relation to glycosuria, the chromaffine tissue and the sexual organs, that the hypophysis, thyroids, sexual glands, adrenals and thymus are a closely related series of organs which mutually influence each other's growth. The attempt has been made to bring these into a general scheme among themselves by Eppinger, Falta and Rudinger and later by Aschner. Such a scheme is that below.

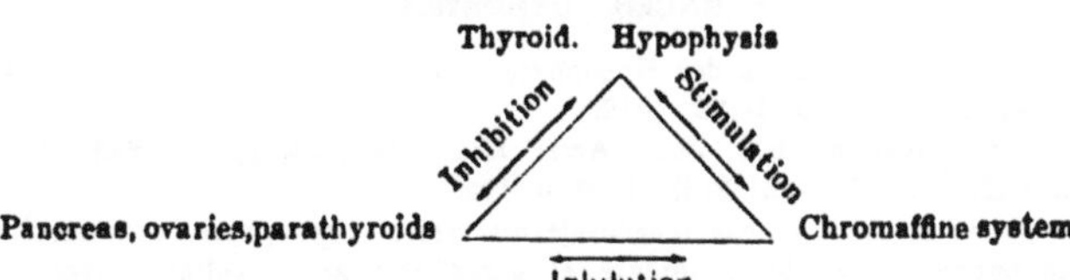

Thus the thyroid and hypophysis stimulate the chromaffine tissue, the latter inhibits the pancreas, ovaries and parathyroids; the latter inhibit, or are inhibited by, the thyroids and hypophysis. Such a scheme, while useful, is too restricted. It must be remembered that the metabolic co-ordination or correlation of the body embraces every organ in it. The body is an organic whole and these so-called cryptorrhetic organs, organs of internal secretion, are not unique, but the bones, muscles, skin, brain and every part of the body are furnishing internal secretions necessary to the development and proper functioning of all the other organs of the body. Such a scheme, to be complete, must embrace every organ; only the barest beginning has been made in this study so important, so necessary for the understanding of development and inheritance. Problems of development and inheritance cannot be solved until these physiological questions are answered.

The following table expresses the action of some of these organs:

On protein metabolism.

Stimulating	Inhibiting
Thyroid	Pancreas
Hypophysis	Parathyroids
Chromaffine system	
Ovaries, Testes.	

On calcium retention:

Favorable to	Inhibit
Hypophysis	Sexual glands
Thyroids	
Parathyroids.	

The internal secretions appear to the author to constitute strong evidence against the existence of such things as unit characters and inheritance by means of structural units in the germ, which represent different characters. We see in the internal secretions that every character in the body involves a large number of factors. The shape and size of the body; the coarseness of the hair; the persistence of the teeth; tendency toward fatness, may easily depend on the hypophysis, on the thyroid and the sexual organs; and these, in their turn, are but the expression of other influences played upon them by their surroundings and their own constitution. An accurate examination shows the untrustworthiness of any such simple or naïve view as that of unit characters.

REFERENCES. Hypophysis.

1. *Aschner:* Ueber die Funktion der Hypophyse. Full bibliography. Archiv f. d. ges. Physiol., 146, p. 1-126, 1912.
2. *Cushing:* Hypophysis cerebri. Jour. Amer. Med. Assn., 53, pp. 249-255, 1909. Johns Hopkins Hospital Bull., 1909, p. 105.
3. *Sotti et Soteschi:* Sur un cas d'agénésie du système hypophysaire accessoire avec hypophyse cérébrale intègre et gigantisme acromégalique, avec infantilisme sexuel. Archives Ital. de Biol., 57, p. 22, 1912.
4. *Miller and Lewis:* The effect on blood pressure of extracts made from the various anatomical components of the hypophysis. Transactions Chicago Pathological Soc., 8, pp. 133-134, 1911.
5. *Wells:* Presence of iodine in human pituitary glands. Jour. Biol. Chem., 7, 1910, pp. 259-261.
6. *Aldrich:* On feeding young pups the anterior lobe of the pituitary gland. Amer. Jour. Physiol., 30, p. 352, 1912.
7. *Schaefer:* Pro. Roy. Soc., 81, p. 442, 1909.

8. *Engeland and Kutscher:* Ueber einige physiologische wichtige Subetanzen A. Die physiologisch wirksamen Extraktivstoffe der Hypophyse. Zeit. f. Biol., 57, p. 527, 1911.
9. *Ott and Scott:* The action of animal extracts on the secretion of the mammary gland. Therapeutic Gaz., 35, pp. 689-691, 1911.
10. *Schaefer:* On the effect of pituitary and corpus luteum extracts on the mammary gland in the human subject. Quarterly Jour. Expt. Physiol., 6, pp. 17-19, 1913.
11. *Schaefer and Herring:* Action of pituitary extracts upon the kidney. Phil. Trans. Roy. Soc., 199, B, pp. 1-29, 1906.
12. *Hammond:* The effect of the pituitary extract on the secretion of milk. Quarterly Jour. Expt. Physiol., 6, pp. 311-338, 1913.
13. *Guggenheim:* Proteinogene amine. Peptamine: Glycyl-p-oxyphenylethylamine, etc., Biochem. Ztschr., 51, pp. 369-387, 1913.
14. *Fühner:* Pharmakologische Untersuchungen über die wirksame Bestandteile der Hypophyse. Zeits. f. d. ges. expt. Medizin, 1, pp. 397-443, 1913.
15. *Fühner:* Das Pituitrin und seine wirksame Bestandteile. Münch. med. Wochenschr., 59, pp. 825-853, 1912.
16. *Dale and Laidlaw:* A method of standardizing pituitary extracts. Jour. Pharm. and Expt. Ther., 4, pp. 75-95, 1912.
17. *Barger:* The simpler natural bases. Active principle of the pituitary body. p. 108. Monographs of Biochemistry. Longmans, Green and Co., 1914.
18. *Klotz:* Experimentelle Studien u. die blutdrucksteigende Wirkung des Pituitrins. Arch. f. expt. Pathol. u. Pharm., 65, p. 348, 1912.
19. *Schaefer:* Die Funktionen des Gehirnanhanges. Zent. Physiol., 15, p. 427, 1911.
20. *Hamburger:* The action of extracts of the anterior lobe of the pituitary gland upon the blood pressure. Amer. Jour. Physiol., 26, p. 178, 1911.
21. *Pankow:* Ueber Wirkungen des Pituitrin auf Kreislauf u. Atmung. Archiv f. d. ges. Physiol., 147, p. 89, 1912.
22. *Simpson and Hill:* Effects of repeated injections of pituitrine on milk secretion. Amer. Jour. Physiol., 36, p. 347, 1915.
23. *Weed and Cushing:* Effects of pituitary extract upon secretion of cerebrospinal fluid. Amer. Jour. Physiol., 36, p. 77, 1915.

PARATHYROIDS.

The parathyroids, or epithelial bodies, as they have been also rather badly named, are some small, pale, glandular masses either embedded in the thyroids, or lying close to them or attached to the thymus gland. There are in many cases four of these bodies, two within the thyroid called the internal parathyroids, and two without, or external parathyroids; but particularly in herbivora there are often accessory parathyroids.

For a long time these bodies were confused with the thyroids and results of their extirpation were attributed to the thyroids. Their relation to the thyroid is not yet clear, but it is certain that the tetany and death after 9-10 days originally ascribed to thyroidectomy are in reality due to the extirpation of the parathyroids.

Complete extirpation of the parathyroids is generally followed by a remarkable set of symptoms. Morel has described these very well: A dog 2-3 years of age which has undergone parathyroidectomy at first presents no appreciable symptoms. The symptoms begin on the second day. The animal then becomes sad and restiess; it moans and moves about. It eats no more, but drinks abundantly. From the second to the third day it seems stiff in its movements; a fleeting twitching of its muscles analogous to that of a horse bitten by a fly may be observed. There is fibrillary twitching of the tongue. On the 4th day it is worse. The animal makes a miserable appearance. It retires to a dark corner; it seems to suffer much and cries out at the least touch. Stiffness or rigor increases so as to interfere with its movements; the tremblings become general, more ample and less discontinuous. Convulsions throw the dog on the floor, with feet in extensor tetanus and head bent back; at times disordered movements come on, the feet beating the air and the face grimacing. At the same time respiration is quickened, the heart accelerated and the temperature increases. The attack is prolonged several minutes; then the dog comes to itself, gets on its feet and staggers about. The attack reappears after some hours, either spontaneously or under some stimulus such as a light, touch or noise. The attacks increase in number and intensity and tend to become continuous. Refusing all food, or vomiting the little it has taken, the animal rapidly becomes cachexic and gives out a putrid odor. Toward the 8th day it enters on the terminal phase of the malady, the convulsive seizures are repeated, but their violence diminished. Thin and without strength, the dog lies on one side. He asphyxiates little by little and the temperature falls. The respirations become few, deep and irregular. Then the animal dies. Death comes on the 9th or 10th day.

But while death generally follows, certain cases have been described in cats and other animals, in which recovery took place and no symptoms followed parathyroidectomy; and on autopsy no accessory parathyroids could be found. Such cases are most common in cats. Whether they are to be explained by undiscovered accessory parathyroids, by the getting into function of some other organ usually working with the thyroids, for example the hypophysis, or whether there is spontaneous cure by immunity, cannot be positively stated.

The tetany can be entirely done away with or greatly reduced by the giving of sodium bicarbonate or alkalies, or by bone injury, or by grafting parathyroid into a bone, or by calcium salts, but the length of life is little if at all prolonged thereby, provided all the parathyroids have been taken out. If, however, a small amount has remained in, but not enough to keep the animal alive and free from tetany, then the injection of calcium salts is said to permanently save the animals. It appears that the tetany is not the cause of death, but an indirect result of some profound change underlying it.

With the parathyroids, as with all other organs of internal secretion, we have to ask the question whether they are acting by an internal secretion, or by a process of detoxication. Of course both processes may coexist. In this case there is no doubt that the tetany is due to a poison of some kind circulating in the blood of the parathyroidectomized animals. It has been shown that bleeding and the injection of fresh blood or salt solution relieve the tetany. Furthermore the injection of the serum of animals in tetany may produce similar symptoms in other animals. The suggestion has been offered by some that the normal function of the gland is to remove this substance from the blood. Such a conclusion, however, is not justified. It may very easily be that this poison is not normally found in the body when the parathyroids are there. It may be formed as the result of their falling out of function. Methyl guanidine, a poison, has been isolated from the urine of dogs dead of tetany. Both guanidine and methyl guanidine produce tetany.

The symptoms of this tetany are similar to those which sometimes follow an Eck fistula when a dog is fed on meat. A bread and milk diet relieves them; meat predisposes to them. Substances appear to be formed in the digestive tract, either as the result of bacterial decomposition of meat or of digestion, or to pre-exist in the meat, which produce tetany and which the liver ordinarily removes. Similar acting substances of unknown nature and unknown origin appear also during parathyropriva and it has accordingly been suggested by Morel that the parathyroids are necessary for the liver to function normally; in their absence the liver is no longer able to detoxicate the body. Others have attempted to show that the liver has lost the power of holding back ammonia and that the tetany is an ammonia tetany, but their methods have generally been too inexact and their conclusions accordingly not dependable. It is not sure, also, whether the substances causing the tetany are the cause of death or an indirect result of the morbific process. It has also been suggested that an acidosis follows parathyropriva. Among the other suggestions made, again badly founded, was that in parathyropriva there was a hypersecretion of calcium and tetany as the result of this. This, however, is not the case, although there is no doubt that by calcium salts, as by several other ways, the tetany may be prevented. The way in which the parathyroids act is, therefore, still obscure. It is certain that there are serious disturbances of digestion.

One very interesting result of parathyropriva is the action on the dentine of the teeth and on the bones. Neither bones nor teeth calcify properly, so that in rats conditions analogous to rachitis and osteomalacia are produced.

THYROID.

The thyroid gland is a small, highly vascular, ductless gland, lying in the neck close to the larynx. It gets its name from the large cartilage of the larynx, the thyroid, or thyreoid cartilage, so called from its shield-like shape. (Gr. *thyreos*, shield.) In most of the higher vertebrates the gland is composed of three parts, two lobes lying close to the larynx one on each side and a little to the front of the trachea, and an isthmus of connecting glandular tissue which extends across the trachea in front. The organ is largest relative to the body weight in the mammalia and smallest in the fishes. The gland is found in all classes of vertebrates, but in the fishes and cyclostomes it is represented only by some small diffuse glandular patches little larger than the heads of pins, which extend along the aorta on the ventral side and out a little way along each gill arch. The gland is larger and fairly compact in amphibia and reptiles, but still small. The histological structure in all forms from the fishes up is the same and quite characteristic. In amphioxus, one of the lowest of the vertebrates, the gland is said to be represented by the neural groove, but there is some doubt about this homology. The only organ of the invertebrate which has been homologized with it is a gland of scorpions and arachnids. This gland has a structure very similar to that of the thyroid of the cyclostomes (Gaskell). It is interesting to notice that if this homology is correct, and few if any zoölogists would admit its correctness, the vertebrate thyroid represents an accessory sexual organ of the invertebrate. This is of interest on account of the close connection physiologically between the sexual organs and the thyroids of vertebrates of which we shall presently speak.

The thyroid is a very vascular gland. Relative to its weight a very large amount of blood passes through it, and in goiter this vascularity becomes still greater and the carotid arteries supplying the gland are generally enlarged, sometimes to double their normal size. So great is the blood supply that dilation of the arterioles of this gland alone may lower the general blood pressure many millimeters of mercury.

The nerve supply of the thyroid is from the superior and inferior laryngeal nerves. These nerves carry vasomotor and probably afferent and secretory fibers for the gland.

The thyroid is, therefore, pre-eminently a vertebrate gland. It has developed side by side with the vertebrate characteristics of a dry, hairy skin instead of a moist, mucus-bearing or chitinous skin; with a great development of the brain and intelligence; with an internal bony skeleton and a large skull. It may almost be called the vertebrate gland *par excellence*, for as we shall presently see it is intimately connected with these three fundamental vertebrate characteristics. Indeed, it appears to be related in a causal way with these organs, so that these typically vertebrate characters might almost be called secondary thyroid characters, just as the horns of cattle may be secondary sexual characters. Injury to the thyroid, particularly in growing animals, is associated with profound retrogressive changes or cessation of development in skin, internal skeleton and nervous system.

Embryology. The thyroid gland, although ductless in its adult state, arises as an outgrowth of the ventral side of the embryonic œsophagus.

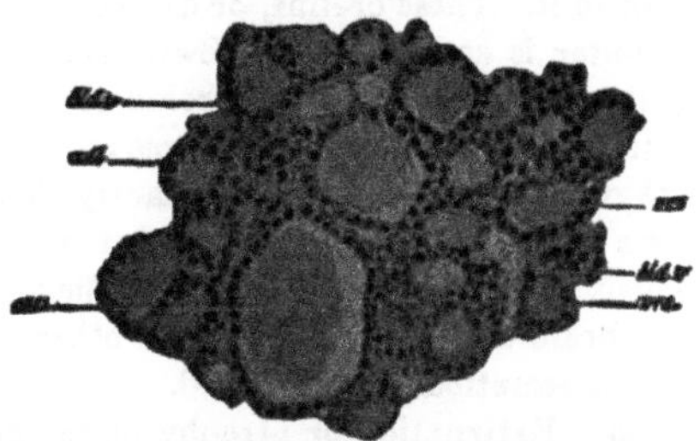

FIG. 60.—Section of the normal thyroid of a cat (Vincent and Joly). Note the low epithelium with the colloid material in the center of the alveolus.

It is hence hypoblastic in origin. Another such outgrowth is the lungs. It corresponds on the ventral side to the anterior lobe of the hypophysis, which grows out of the dorsal side, and there is some connection and similarity between the functions of these two organs. Very soon, however, connection of the thyroid with the œsophagus is lost; the gland is encapsulated with connective tissue and has no duct for the discharge of its secretion.

Histology. Histologically the gland consists of closed alveoli containing a single layer of epithelial cells and with the center of the alveolus (Figure 60), consisting of colloid material, which stains strongly in eosin. This colloid is believed to be the active principle of the gland. A similar colloid is found in the hypophysis formed by the pars intermedia and discharged into the third ventricle of the brain by means of the infundibulum. How the colloid escapes from the thyroid, whether it goes into the lymph or into the blood, or whether it is dissolved by some ferment which splits it into a soluble substance, osmosable and passing into the blood, is still unknown. It is, perhaps, most likely that such an enzyme exists in the gland and the colloid does not, as such, find its exit from the gland.

Function. The function of this gland was until recently obscure and even now not all the details of its action are known. The gland was classed as an organ of internal secretion by the great physiologist, Brown-Séquard. Knowledge of its function was derived first from pathology and then from surgery. It was found by surgeons that extirpation of the gland was followed by very grave symptoms in human beings; and the pathologists observed its connection with certain peculiar changes in the skin and skeleton.

(a.) *Cretinism.* If the gland fails to develop or atrophies or develops insufficiently in embryonic life, a cretin is the result. Growth is retarded, and a dwarf is produced, with a protuberant abdomen, low intelligence, or imbecility; the hair is coarse and scanty; the skin thick and dry with mucin in it. These cretins, or dwarfs, are found generally in regions where goiter is endemic, as in Switzerland and Styria. If, while very young, they are given sheep's thyroids, or extracts of thyroids, growth begins again and the abnormal symptoms more or less completely disappear. The thyroids must be taken constantly, however, or myxœdematous symptoms recur. It is then clear that arrested mental and physical development is a result of thyroid insufficiency. The proper development of the brain and skull and various other structures of the body depend upon the secretions of this gland.

(b.) *Myxœdema.* Extirpation or atrophy of the gland in adults is followed by the symptoms of myxœdema, so called from the deposit of mucin-yielding material in the thickened skin. (Gr. *myxa,* mucin; *oidema,* swelling.) The skin becomes thick, dry, the connective tissue is increased and yields mucin when extracted with alkali. The hair becomes scanty and coarse; there is a marked falling off of the sexual function; intellectual apathy and disinclination to exertion of any kind result. A general obesity may also develop. The body temperature falls about 1° below normal and the oxygen consumption and nitrogen metabollsm are reduced. (Compare with results of extirpating anterior lobe of hypophysis.)

(c.) *Basedow's disease.* On the other hand, if the thyroid is too active, exophthalmic goiter, so called from the protuberance of the eyes which commonly occurs, or Basedow's disease, results. In this case the symptoms are generally just the reverse of those in thyreopriva, or diminished function. The heart is very fast and often irregular (tachycardia); instead of a low temperature the temperature is high, one or two degrees above the normal; there is great nervous irritability in place of nervous sluggishness; nitrogen metabolism and oxygen consumption are increased, instead of diminished; and patients are thin instead of fat. Moreover, the symptoms are ameliorated by reducing the activity of the gland by tying some of its arteries or partial extirpation. The gland may or may not be larger than the normal, although it is generally somewhat enlarged, but it is generally found to contain but little colloid and more of the secretory epithelium.

From this pathological evidence it is clear that this small organ is enormously important in the proper development of bone, teeth, skull, brain and skin of the body; that its absence is attended by retrogressive changes in these tissues and organs; and that when it overfunctions it stimulates metabolism and nervous activity. The gland obviously has a very close relation to the nervous and the vascular systems throughout the vertebrate phylum.

(d.) *Extirpation of the gland in animals.* The results of extirpation of the gland differ in different animals. This is owing to the fact that the thyroid contains often in it the parathyroids, for example in dogs and cats; whereas in rabbits and the herbivora generally there are external parathyroids lying outside of the thyroids. The results of taking out the parathyroids are very different from those of taking out the thyroids alone. Since, in the beginning, the importance of the parathyroids was not recognized, many apparent contradictions resulted in the effects of the operation. Here only the effects of taking out the thyroids proper will be considered.

If the thyroids alone are taken out, or if the thyroids and internal parathyroids only are extirpated, leaving the external parathyroids intact, no very obvious symptoms follow at once. But in adults disturbances in metabolism shortly begin to appear and the symptoms of myxœdema develop. The animals live for months or years (dogs, cats, rabbits). The hair falls out. They are apt to become mangy; their nitrogen metabolism diminishes; they do not so easily develop adrenaline glycosuria when adrenaline is injected; the excitability of the vagus, depressor and other nerves is diminished; animals become apathetic and tend toward obesity, somewhat as they do after extirpation of the hypophysis. Their sexual life is also less intense. If the parathyroids are also removed tetany and death in 9-10 days nearly always result.

In young animals the consequences of extirpation of the thyroids are far more serious. Growth is checked; myxœdematous symptoms develop and intelligence remains very low.

(e.) *Effect of extirpation on the excitability of nerves.* It was mentioned that one of the results of extirpation of the gland is that the sympathetic nerves of all kinds, and in general what are known as the autonomic nerves, namely, those which regulate the visceral as distinct from the voluntary system of the body, are less active on stimulation than they were before. Now, in the case of the supra-renals there is no doubt that the action of the glands is due to an internal secretion, a substance of known composition, adrenaline, which comes from the gland to the blood. It is, hence, not unlikely that the thyroids also have an internal secretion. Physiologists have, accordingly, sought for evidence that such an internal secretion is produced by the gland. Such experiments were tried by von Cyon many years ago. He proved that stimulation of the thyroid branches coming from the superior and inferior laryngeal nerves caused a great dilation of the blood vessels of the gland, and that as a result of this stimulation, or of the injection of thyroid extracts, a remarkable change took place in the irritability of the vagus nerve and the depressor nerves. Stimulation of the vagus was far more effective in stopping the heart; and the depressor effect was far more sensitive after stimulation of the gland nerves or after the injection of the extract of the gland than before. On the other hand, when the thyroid was reduced in activity, as in some goiters, or when it was extirpated, there was a remarkable diminution in excitability of the vagus and depressor mechanisms. Both of these nerves are inhibitory nerves and are acting in an antagonistic manner to the sympathetics. These results have been confirmed by several observers. They appear to be well substantiated. It is also stated that the power of the splanchnics to raise blood pressure on stimulation by weak currents is also greater when the thyroid nerves are stimulated at the same time. This experiment is of importance, since the short time necessary to produce an effect, i.e., ninety seconds, would indicate that the thyroid secretion did not find its way into the lymph, but directly into the blood. A direct action of the thyroid is thus shown to be upon the vasomotor apparatus and sympathetic system, including the internal muscles of the eye. This effect is presumably partly peripheral and in part direct. This is of great importance in interpreting the symptoms of exophthalmic goiter, where such hypersensitiveness is marked. In thus increasing the excitability of the autonomic system the internal secretion of the thyroids appears to act somewhat like that of the supra-renals. But the thyroid extract does not itself cause a rise of blood pressure; it is much less powerful than that of the latter organs.

(f.) *Action of extracts of the thyroid. Blood pressure and vascular system.* While nearly all are agreed that extracts of the thyroid, including thyreoglobulin, which is obtained from the colloid matter, and iodothyrin, formed from the globulin by hydrolysis, influence the excitability of the vasomotor apparatus and the sympathetic system, very contradictory effects on blood pressure have been obtained as a direct result of the injection of extracts of the gland. These contradictory results appear to be due (a) to the use of anesthetics, curare and morphine in some cases but not others; and (b) to different methods of making the extract. Drugs often cause very different actions on the circulations of anesthetized and non-anesthetized animals. Thyroidectomized animals generally have, after a time, a low blood pressure. The serum of such animals is also hypotensic. Jeandelize and Perisot obtained the following results: Rabbits were operated on when 8 days old and tested after several months. The blood pressure in the controls and the thyroidectomized animals compared as follows:

Time from operation	Blood pressure in cms. Hg.	
	Controls	Thyroidectomized
2 months 18 days	11.5	9.6
6 "	11	6.5

The effect of injecting extracts of the gland is not uniform. Von Cyon and Asher got no direct action of extracts of the gland on blood pressure. Nearly all observers, however, have obtained a fall in blood pressure following the injection of aqueous, glycerol or acid extracts of the gland. It was suggested that this was due to choline. Popielski showed that if the depressor substance was first removed from the gland extract, a rise in blood pressure resulted on the subsequent injection of the extract. The amount of choline in the extracts is said to be too small to produce the result. The depressor substance in the acid extract of the gland is vasodilatin, according to Popielski. This extract causes all the typical symptoms produced by this substance, namely, secretion of the pancreas, fall in blood pressure, abolition of the coagulability of the blood. This substance would appear to be either imidazylethylamine or some similar amine. The recent work of Hunt on the great activity of some of the esters of choline suggests that possibly some of these may be present.

As regards the nature of the active principle which produces the change in excitability of the vagus and sympathetic nerves, experiments indicate that it is either the colloid, namely the iodine containing thyreoglobulin, or some derivative of this, such as iodothyrin, or some proteose from the globulin. The experiments of Popielski indicate the probability that there are also in the gland basic substances which may act directly on the blood pressure. The whole matter is, however, in such an uncertain state that more work will have to be done before the relation of the gland to the vascular system, a relation which no one doubts, is cleared up.

(g.) *Effect of feeding thyroids or extracts of thyroids on metabolism.* The relation of the thyroid to metabolism is shown very clearly by the cessation of growth and proper development on its removal. Moreover, feeding experiments give very positive results. If human beings or carnivora (the effects on herbivora are doubtful) eat thyroids or take extracts of the gland, there results a stimulation of their nitrogen metabolism. It is as if the protein oxidation had been stimulated. The temperature of the body is raised, just as it is by the ingestion of protein. The protein is rapidly oxidized. There is an increased consumption of oxygen and production of CO_2 and nitrogen excretion is increased. A diminution of the fat in the body results, so that thyroid tablets are recommended or used often for reducing fatness. The nervous and heart symptoms produced are, however, unpleasant. Very similar effects are produced on metabolism by the ingestion of very large amounts of protein.

The nature of the active principle of the gland, which is thus active is still uncertain, but it appears to be correlated with the colloid matter in the cells. It may be, however, that the nervous symptoms and the metabolic symptoms are due to different substances. The way in which the active principle acts is still quite obscure; it is uncertain whether it acts directly on the cells of the body, or indirectly by its action on the nerves. Further work in this direction will be needed. The indications are, however, from the work which follows, that it is the iodine-containing substance which is active.

(h.) *Nitrile reaction.* A very remarkable and curious reaction, making an almost incredibly delicate biological test for thyroid tissue, is the nitrile test of Reid Hunt. If so little as 0.1 mg. of dried thyroid tissue, or an equivalent amount of the thyroglobulin, per gram of body weight is fed to a white mouse on a cracker diet each day for 10 consecutive days, no visible change occurs except a loss of weight; but it is found that such a mouse will survive as much as 10 times the amount of acetonitrile given subcutaneously, which would have killed his brother or sister under conditions identical with the foregoing, except the thyroid feeding. White rats, on the other hand, are rendered far more sensitive to acetonitrile by this treatment. This test is the most delicate test for thyroid tissue which we have. All of this activity of the gland is found in the thyroglobulin of the gland and also in the metaprotein derived from it by acid hydrolysis; also in the iodine containing protalhumose formed from the metaprotein by acid hydrolysis. The amount of activity computed on the iodine present in these preparations remains what it was in gland substance containing the same amount of iodine. By further hydrolysis iodothyrin is formed, and while this is active, it is not so active per mg. of iodine as the original substance. The extreme sensitiveness of this reaction and the fact that the nitriles generally kill by their action on the respiratory center strongly indicate that the thyroid extract produces, in mice and rats, a change in this center, in the one case causing an increase, in the other a decrease in its sensitiveness to acetonitrile.

(i.) *Nature of the internal secretion of the thyroid.* This problem is not yet solved. The thyroid is peculiar among all the glands of the body because it normally contains iodine. No other tissue normally contains iodine in more than traces, except possibly the pituitary, and here the evidence is doubtful. The fact that the thyroids contain iodine naturally leads to the conclusion that the peculiarity of the gland's action is due in some way to this fact and that the active principle is also an iodine compound. There are strong reasons for believing this to be a fact, and some reasons for doubting it. The connection of iodine with the thyroid was long known empirically. It was found that iodide taken internally, or iodine painted on the outside, was beneficial in goiter, and reduced the goiter. It was early stated that the gland contained iodine and that regions in which goiter was endemic were marked by less than the normal amount of iodine in the water. These results, however, were not generally accepted until Baumann, in 1895, showed that iodine was present in the gland.

The amount of iodine in the gland is extremely variable and is much increased by various forms of medication by iodine, iodides or iodine-containing materials. The gland has the power of picking out iodine from the blood and storing it in organic union. Iodine is found in combination with a globulin which occurs in the colloid of the gland and is called *thyreoglobulin*. All of the iodine present is in this substance and certainly the effect of the gland on the nitrile susceptibility of mice is due exclusively to this substance. Glands which contain no colloid contain no iodine either; those which have much colloid usually contain also much iodine. As a rule the thyroids of persons living in goiterous regions contain, per gram of dry substance, less iodine than is common in other localities. The rule has, however, many exceptions. In a lot of human thyroids from the Stiermark (Styria), a goiterous region, von Rositzky found that the fresh weight of the glands of adults varied from 29 to 231 grams, and the dry substance from 5.2-67 grams. In 1 gram of dry substance the amount of iodine varied from 0.077 to 3.85 mgs. of iodine. The last figure was obtained in a case of tuberculosis. The total amount of iodine in the whole gland varied from 0.99 to 54.67 mgs. In children from birth to 1½ years of age, Baumann found from 0.09 to 0.342 mg. of iodine in the whole glands. In animal thyroids the amount of iodine usually runs from 1-1.5 mgs. of iodine in a gram of dry substance, but pigs may have less. In sheep there appears to be a seasonal variation in the amount of iodine, perhaps correlated with the character of the food. In Sweden in human adults the average amount found by Jolin was 11.20 mgs. of iodine in the whole glands. It is clear from these figures that the iodine content of human thyroids is subject to very wide fluctuations and these may be due in part to medication, disease, diet and other unknown causes. In exopthalmic goiter, when the colloid may be much reduced in amount, the iodine is reduced also. Some have interpreted these facts to mean that the gland simply stores iodine to get rid of it, as some organs store copper and arsenic. But this view is not borne out by the studies of Kocher, Oswald, Hunt and F. C. Koch.

Thyreoglobulin. This is the iodine-containing constituent of the gland and is the principal constituent of the colloid matter. It may be extracted by carefully drying the ground glands at room temperature, after previously separating them as far as possible from connective tissue, extracting the dried glands with gasoline to remove the lipins, removing the gasoline by evaporation at room temperature, and grinding the lipin-free material in a mill and passing the powder through a fine sieve. The globulin is then extracted from the fine powder with dilute salt solution. The globulin is purified by the usual methods of precipitating with ammonium sulphate, resolution, and reprecipitation by dialysis. It is a white, amorphous powder. The per cent. of iodine is variable, generally ranging from 0.34-1 per cent. Oswald found 1.6 per cent. The activity is proportional to the iodine content. It is not impossible that some of the symptoms of exophthalmic goiter may be due to some other constituent of the gland. The iodine-free globulin is inactive on the blood pressure, and does not affect the irritability of the vasomotor apparatus and sympathetic systems. The iodine-containing globulin is active in these particulars and Koch found that all the activity of the gland in the nitrile susceptibility was due to this constituent. There is then no doubt that thyreoglobulin is one of the active substances of the gland. Whether, however, it is the only active substance and whether it gets into the blood as such is doubtful. Koch found that full activity persisted in some of the first digestive products of the globulin. It is, therefore, possible that there is in the gland an enzyme, which, under nerve influence or other stimulation, digests some of the protein, setting free in the blood, not thyreoglobulin, but an iodine containing digestion product and this produces the action on the nervous system and possibly on the other cells of the body. It has not yet been shown, however, that any such enzyme exists in the gland and is active in this way and it is not known in what form the active principle is discharged from the gland. If such a process as that just mentioned occurs the gland would resemble the liver, which stores glycogen; the latter is not poured into the blood as such, but, by digestion, glucose is formed from it. Basedow's disease might, on this supposition, be due to the too great conversion of thyreoglobulin into its absorbable digested products and as a result the gland strives to manufacture more, just as the body makes glucose in diabetes. It is as if a brake had been removed. The small content of colloid, the activity of the gland, the thyroæmia all point in this direction. The condition would be analogous to diabetes, only it is a protein diabetes, not a carbohydrate.

The other constituents of the gland are those of cells in general, such as lipins, nucleic acid, probably guanylic acid, cholesterol, choline, albumins and extractives. The normal gland is said to contain small amounts of arsenic, but whether this plays any rôle in the gland, or whether it is useful or harmful, is unknown. We are constantly taking minute amounts of arsenic in the foods and it is possible that its presence in the gland is accidental. On the other hand, it is not certain that it may not be useful.

Iodothyrin. By acid hydrolysis of the gland Baumann and Ross isolated a brownish, amorphous substance containing from 2-14 per cent. of iodine. This iodothyrin gives no protein reaction. It is soluble in dilute alkali but not in weak acids, and, indeed, looks and behaves somewhat like melanin. It has been found that this substance has an action on goiters, cretins, myxœdematous patients and in modifying the irritability of nerves like the action of the gland itself but weaker than the latter, if the amount of action is compared with the amount of iodine in the iodothyrin and in the dried-gland substance. That is an amount of iodothyrin containing a milligram of iodine is less active than an amount of the dried gland containing an equal amount of iodine. Computing its activity in comparison with the iodine content it is only about one-tenth as strong as thyreoglobulin in rendering mice resistant to nitrile poisoning (Koch).

SUPRA-RENAL CAPSULES.

Anatomy.—The supra-renal capsules are two small glandular organs lying like caps on the upper poles of the kidneys, whence their name. They are richly supplied with blood vessels and nerves and together they weigh in an average adult about 6-7 grams. In youth and fetal life, in common with most of the glandular organs of the body, they are relatively larger. They are supplied with blood from two small arteries from the dorsal aorta and wide, short veins empty directly into the inferior vena cava to which the organs are closely attached. The amount of blood passing through them is very large compared to their size. The nerves which govern both the blood vessels and the tissue of the gland itself, being both secretory and vasoconstrictor, come from the two splanchnics, the left splanchnic supplying only the left gland, but the right sending a branch to the left as well as to the right.

Histology.—The gland is composed of two different kinds of tissue, making respectively the cortex and the medulla. The boundary between these layers is not sharply delimited, strands of cortex tissue penetrating here and there down into the medulla, particularly along the sympathetic nerves. These two tissues differ in their histological, chemical and physiological characters and in their embryonic origin.

Cortex. The cortex is composed of three regions, an external zone, a reticular and a glomerular zone, the differences between them being not sharp. The cells are arranged in a reticulum with blood spaces between. There are no endothelial cells on these cell strings and the circulation is of that extra capillary kind described by Minot in the liver and characteristic for most invertebrates. Hence the blood comes directly in contact with the cells of the tissue. The cortex does not stain in bichromate of potash and the cells are simply protoplasmic cells, containing some vacuoles and granules. During pregnancy and particularly just after parturition, this layer undergoes, in the guinea pig, a rapid cell proliferation and degeneration, the significance of which is not yet understood.

Medulla. The medulla is composed, in part, of nerve cells, but in large measure of granular protoplasmic cells, which are distinguished by their staining yellow when the gland is fixed in a solution of bichromate of potash and formol. This yellow stain is taken by some other cells also and tissue so staining is called *chromaffine* tissue. The staining power appears to be dependent upon, or correlated with, the presence of adrenaline, the active principle of the organ, since the content of adrenaline and the depth of stain rise and fall together. The stain is not, however, a peculiar characteristic of adrenaline, for other aromatic reducing bodies will probably be found to give the reaction also. However adrenaline content and chromaffine reaction often go together. Thus such tissue is found in the poisonous skin glands of toads and these contain an unusually large amount of adrenaline. The supra-renals have no ducts. If they be called glands their secretion is discharged (cryptorrhesis) into the blood stream, so that they are truly organs of hidden flowing.

Comparative anatomy.—The supra-renals are found in all classes of vertebrates and besides the organs themselves chromaffine tissue is found outside the glands. In all the mammals the glands are well-developed and in monkeys and human beings they are surrounded with strong capsules. In fishes, in which the pronephros or head kidney persists, chromaffine tissue resembling that of supra-renals is found in the head end of the kidney. In the frog, in which the mesonephros persists, the supra-renals are represented by small masses of tissue just above the kidneys. In mammals, besides the supra-renals, little accessory masses of chromaffine tissue are found along the trunk of the great sympathetic, sometimes in the kidneys also, and one large mass in particular is called Zuckerkandl's paraganglion. Nothing definite is known of the function of these accessory masses, but they are believed to act as accessory medullary substance. Whether chromaffine tissue exists in the so-called carotid gland is doubtful. The quite general occurrence of chromaffine tissue outside the supra-renals would not be surprising if it be true that many cells normally form small amounts of adrenaline when the nerve impulses strike them, as has been suggested by Sherrington.

Embryology.—The cortex and the medulla have different sources in the embryo. The cortex is derived from the mesoderm which gives rise to the kidney; the medullary portion contains many sympathetic cells and appears to be a modified mass of sympathetic nerve cells.

Functions.—The functions of these small organs have only been partially discovered and that within the past few years. As was the case in so many of the cryptorhetic tissues, attention to them was directed by Brown-Séquard, who classed them as organs of internal secretion many years ago and stated that they were essential to life. Pathology first indicated their function. It was found that in Addison's disease the autopsy constantly showed lesions of the supra-renal capsules. These lesions are generally tubercular in nature. The symptoms of this disease are a peculiar bronze coloration of the skin; a marked lassitude; lowered blood pressure; and emaciation. In no other disease is this pigmentation found, except in connection with the supra-renals. By this observation a connection was supposed to exist between skin pigmentation, the general tone of the body and the supra-renal capsules.

Extirpation. The consequences of extirpating the glands differ in different animals. In man the quick death following lesions, hemorrhages, pus formation, etc., in the glands leads us to believe that their absence could not be long survived. They have been extirpated in monkeys. The following protocol of one of Kahn's experiments will show the general course of events after adrenalectomy. The extirpation is easier in monkeys than in most animals because, in monkeys, the glands do not lie so near the vena cava and they are encapsulated. A female Macacus monkey, weight about 2,000 grams and fed on fruit. Right supra-renal removed under ether November 9, 1911: The wound heals quickly. On the 4th December, 25 days after the operation, when the weight was 1,850 grams, took out the left supra-renal. On the 5th the animal is very well and eats heartily. On the 6th she eats with normal appetite. Is active. There is a little œdema at the edges of the wound. On the 7th normal. On the 8th normal, but appetite a little less. On the 9th at 9 A.M., is fairly weak, lies stretched out on the bottom of the cage; no appetite; wound in best state. At 9:12 A.M. great increase in prostration. Electrical stimulation of the points of exit of the sciatics produces no effect. Direct stimulation of muscle effective. Apathetic. The eyes are open and look about. At 1:45 P.M., being nearly moribund, was killed with chloroform and the liver examined for glycogen, only a trace being found. The animal lived 5 days. For

4 days it could not be told from a normal animal. The sudden onset of the symptoms of extreme depression has the appearance of an intoxication.

Rabbits have an accessory paraganglion on the aorta and they survive total extirpation for at least a year without any symptoms. Rats also will survive for about three weeks, but dogs and cats rarely live more than 24 hours, the symptoms being again those of intense depression. It will be noticed that the carnivorous animals die most rapidly, and that the herbivora die with the same symptoms when their supplies of glycogen are exhausted and the animal must presumably call upon its protein for food.

The cause of this sudden death is by no means clear. The survival of monkeys for 4 days quite normally would seem to indicate that death was not due to the sudden withdrawal of something necessary for the body. The indications are, rather, that in the absence of the suprarenals, poisons, possibly protein degradation products, are formed or not destroyed. But whatever the explanation, there is no doubt of the fact that the organs are essential to life in most animals. The symptoms following adrenalectomy are little, if at all, alleviated by the injection of extracts of supra-renals. The extirpation is followed by the almost complete disappearance of glycogen in rats, dogs and monkeys from both liver and muscles. This is a point of some interest in connection with diabetes.

The glands have a particular relation to blood pressure. Oliver and Schaefer discovered that an extract of these glands made with 0.7 per cent. NaCl, when injected intravenously into cats or other mammals, causes a very quick, intense rise in blood pressure, the effect passing off rapidly. The substance producing this action was found to be in the medulla and not in the cortex. It was isolated by Abel as a benzoyl compound, which he called epinephrine, by extracting the glands with weak acid and benzoylating the extract. Unfortunately the crystalline substance Abel obtained was not the active principle itself, but the benzoylated active principle. A benzoyl group remained attached to it. Aldrich and Takamine, who followed Abel, obtained the free base which was named adrenaline. Adrenaline may be readily prepared by the following method (Abel):

The ground organs are extracted with an equal amount of 3.5% trichloracetic acid in absolute alcohol, shaken and filtered. The concentrated filtrate forms a microcrystalline precipitate of adrenaline on the addition of ammonia. The precipitate contains about 10% of ash. To purify it the crystals in a little water are treated with some oxalic acid and then alcohol-ether mixture added. The oxalate remains undissolved. It is converted into the acetate, taken up in alcohol-ether and precipitated with ammonia. v. Fürth has another method.

Chemical nature and properties of adrenaline. Adrenaline or suprarenin or epinephrine, as it is also called, is dihydroxyphenyl-hydroxyethyl-methyl amine. It is a derivative of pyrocatechin, with the following formula:

```
     HOC = CH       H  H
    /        \      |  |
HOC<          >C—C—C—NH(CH3)
   \\        //     |  |
     C  —  C       OH  H
     H     H
```

It is seen to be related on the one hand to choline, or oxyethyl amine, and on the other hand to tyrosine, of which it is possibly a derivative. The natural adrenaline is levo-rotatory and is far more toxic and active than the dextro-synthetic product. It has been synthesized. $(\alpha)_D^{20^\circ} =$ —50.72° (Abderhalden and Guggenheim); —51.30 (Abel and Macht). It gives an emerald green in an acid solution with ferric chloride, and carmine red in an alkaline solution. It is crystalline and precipitated from solution by ammonia.

Adrenaline is an unstable, weak base, decomposing rapidly when in solution in the free form, but being fairly stable as the dry free base and as the hydrochloride, in which form it is generally sold. Its decomposition is hastened by alkalies and light. It is a reducing substance and oxidizes spontaneously, if left in an alkaline solution exposed to oxygen. The oxidation products are physiologically inactive. The d-adrenaline is more stable. By oxidation with iodine, persulphate, mercuric chloride and other oxidizing agents, a red color is formed. It is distinguished readily by the green coloration it gives with ferric chloride. It is probably the substance which gives the yellow color to the medullary substance when the organs are fixed with bichromate. It gives a red color with I_2KI and $HgCl_2$. It is thermostable in an acid solution. It dissolves in strong alkali, NaOH, and probably forms a sodium salt.

Quantitative estimation. The amount of adrenaline in the glands may be estimated chemically by colorimetric and by physiological methods. The latter methods are the more delicate and extremely small amounts may be detected. For the colorimetric method, see page 1010, experiment 224. For the physiological methods, i.e., the frog eye, blood-pressure, uterus strip methods, reference must be made to other works. The free base is so active that as little as .000002 gram intravenously will produce a noticeable effect on blood pressure.

Amount of adrenaline in the glands. Bullock's supra-renals contain about 0.3 per cent. (Abel) or 0.25 per cent. (Hunt) of adrenaline. Folin, Cannon and Denis determined the amount in the adrenals of cats to be about 0.15 per cent.; in cattle and rabbits, 0.3 per cent.; in dogs and monkeys, 0.2-0.25 per cent. From human adults suddenly killed the supra-renals contained about 0.1 per cent.

Functions of adrenaline. The discovery of the active principle of the gland has given to the physician one of the most valuable of drugs. It is one of the most powerful local styptics. It causes contraction of the arterioles and it has proved invaluable in diminishing inflammation when a weak solution is dropped in the eye; in reducing hemorrhages, as in operations on the nose; in checking oozing hemorrhages in the mucosa of the stomach and bowels; in cases of shock with low blood pressure it stimulates the heart and raises blood pressure both by its central action on the central nervous system and by its peripheral action on the blood vessels. When subcutaneously injected in human beings it has been found of great value in asthma. Here it causes relaxation of the muscles of the bronchioles, upon which it has a remarkable action. It has besides many other actions. Injected intravenously, and sometimes when injected subcutaneously, it causes glycosuria; in very minute doses it checks the rhythmic peristalsis of the intestine. It causes contraction of the pregnant cat's uterus, but not of that of the virgin cat. In large doses in cats it causes the erection of the hair, relaxation of the bladder, dilation of the iris and all the symptoms which ordinarily accompany fright or anger. It will cause secretion of saliva and tears, but not of the pancreas or stomach. In fact we may at once say that it acts pre-eminently on all the functions innervated by the sympathetic system. Its action on any organ is always the same as is produced by stimulation of the sympathetic nerves to that organ and in different animals the effect may vary, since in some stimulation of a sympathetic nerve may cause activity, while in others it causes inhibition of activity.

But while adrenaline thus acts on all the sympathetic end organs, its action is not limited to these. It is, as a matter of fact, a general protoplasmic stimulant and in large doses it acts on all cells of all kinds of animals. Thus it checks the development of eggs of all kinds and in smaller doses produces monsters; it stops ciliary movement; and may poison the heart. It may make processes of repair slow. But, nevertheless, while thus acting, its main effect in small doses is, as stated, on the organs innervated by the sympathetic nerves. It has, however, an action also on skeletal muscle, helping it to recover from fatigue and in the perfused heart it strengthens the beat.

One of the most interesting of the effects of adrenaline is its power of producing hyperglycæmia, which is nearly always accompanied by glycosuria. This is the case in nearly all mammals. At first it was believed that the action was on the pancreas, a gland most intimately connected with the carbohydrate metabolism of the body. Herter reported that painting adrenaline on the pancreas produced glycosuria very readily and he and Richards described changes in the islands of

Langerhans as a result of this application. He and Wakeman found also that the glucosuria following pancreatectomy disappeared after extirpation of the adrenals. The general result of all the work has been, however, to indicate clearly that the substance acts by producing a stimulation of the liver and other cells analogous to that produced by stimulation of the sympathetic nerves (Figure 61). Claude Bernard

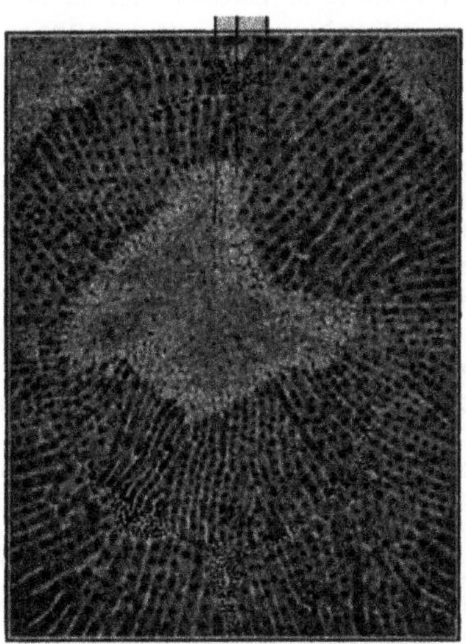

FIG. 61.—Section of a lobule of mammalian liver showing central necrosis of the lobule produced by repeated injections of adrenaline (Drummond).

discovered long ago that puncture of the floor of the fourth ventricle of the brain, the so-called sugar puncture, was followed by an immediate glycosuria, particularly apparent in rabbits; and shortly afterwards, Eckhard proved that this only happened if the splanchnic nerves were intact and if there was glycogen in the liver. It has been found by MacLeod that stimulating the splanchnic nerves increases the per cent. of sugar in the blood. It was not for a long time imagined that the sugar puncture could involve the supra-renals, but it was recently suggested that possibly the splanchnics caused glycosuria, not by direct action on the liver, but indirectly through the action of the nerves on the supra-renals, causing them to discharge more adrenaline into the blood and so to produce an adrenaline glucosuria. Such, however, is probably not the case. The amount of adrenaline in the blood is too small to produce such a glucosuria. Furthermore, MacLeod found that stimulation of the hepatic branches of the splanchnics still caused hyperglycæmia, whereas when the plexus about the veins was destroyed, stimulation of the splanchnics no longer had this effect. It appears from these experiments that two factors enter into this hyperglycæmia: stimulation of the splanchnics acting directly upon the liver and causing glycogenolysis; and, in the second place, setting free adrenaline. Adrenaline appears to be necessary for stimulation of the liver tissues by the nerves, either because it acts upon the sympathetic nerves themselves, or their terminations, increasing their excitability or their effectiveness; or because by direct action on the gland cells the latter are rendered so much more sensitive that they will respond to nerve stimulation; or because in some other way the nerve is modified. It is not impossible, as suggested elsewhere, that normally the nerve excites its terminal organ by causing the formation in that organ of either adrenaline or some similar hormone and that the amount formed by direct action of the nerve is not quite effective without the addition of that derived from the adrenals. MacLeod suggests that the adrenaline acts in the same way as the nerve stimulation.

One of the most interesting facts bearing upon this problem of the manner in which adrenaline stimulates is that tissues supplied by the sympathetic system and which have been denervated by cutting the nerves, or extirpating the ganglia, become hypersensitive to the action of the hormone. This is the case, for example, in the pupil of the eye and is the explanation of the so-called " paradoxical " dilation of the pupil. Stimulation of the cervical sympathetic nerve causes dilation of the pupil by stimulation of the dilator or radiating fibers of the iris. If the superior cervical ganglion is removed on one side, say the left, the pupil on that side at once contracts, owing to the tonic action of the sphincter muscle. After a few days, however, if a little adrenaline is injected subcutaneously, or intravenously, it will be found that the pupil of the operated side suddenly dilates, whereas that of the normal side does not. Furthermore, if the cat is excited, or frightened, or etherized, dilation of the pupil on the operated side occurs before that of the other side. This experiment was at first supposed to mean that the sympathetic carried inhibitory as well as stimulatory fibers to the iris, but was recognized by Langley as due to hyperexcitability and has been carefully studied by Anderson. The hyperexcitability of denervated tissue is not confined to the iris, but all denervated tissue ordinarily supplied by the sympathetic becomes hypersensitive to adrenaline after degeneration of its nerves. This fact has not yet been explained, but it may possibly be due to lack of function rendering the cell temporarily more unstable. There is perhaps an accumulation in the cell of substances with which the adrenaline usually unites to produce its action, and hence the cell is more than usually sensitive. Possibly the so-called paralytic secretions which occur after division of gland nerves may have a similar explanation.

This fact of the hypersensitiveness of denervated tissue may be made use of in detecting the presence of minute quantities of adrenaline in the blood. Thus, during anesthesia of a cat, the pupil of the side on which the cervical ganglion has been previously taken out dilates widely, whereas the normal one does not. This indicates the presence of adrenaline in small amounts in the blood at the beginning of anesthesia, a deduction confirmed by other experiments of which we shall shortly speak. Section of the splanchnics a few days before renders rabbits more easily glycosuric on adrenaline injection. Similarly the difference in behavior of the virgin and pregnant uterus may also be thus explained. Adrenaline added to blood intensifies the absorption bands of oxyhemoglobin. (Menten.)

The parathyroid glands have functions in many ways complementary to the supra-renals. The absence of the parathyroids renders the organism very resistant to adrenaline. No longer will glycosuria appear on its injection and the various symptoms of adrenaline poisoning require much heavier doses for their production.

That the supra-renal glands are constantly discharging adrenaline into the blood stream is now hardly doubtful. Thus Dreyer first showed that stimulation of the splanchnics made the blood from the supra-renal veins more active in raising blood pressure than it was before; this observation has now been confirmed both by direct and indirect experiment. During anesthesia the glands are very largely drained of their adrenaline, as is shown by making extracts of the gland at the end of anesthesia and comparing the amount of adrenaline with that found in similar normal non-anesthetized glands, and it is also shown by the loss of the chromaffine reaction in the anesthetized gland. This discharge of the adrenaline, which is physiologically so important, is due to a central stimulation, since after section of the splanchnic nerves it no longer takes place. During anger or fright, in cats at any rate, a substance having the properties of adrenaline is discharged into the blood in more than normal amounts and may be detected in the vena cava blood at the level of the supra-renal veins. The dilation of the pupil, erection of the hairs, etc., are all produced by adrenaline. The discharge may be so pronounced that a glucosuria may be produced, a glucosuria which does not occur if the glands are extirpated, or if one gland is extirpated and the veins of the other temporarily compressed, so that the secretion cannot get into the general blood stream. These observations are of the greatest interest as indicating the remarkable participation of the glands in the emotions. It must not be incorrectly inferred, however, that the emotion of fright depends on the supra-renal glands, or that the manifestation of this emotion necessarily so depends. The fact is probably quite otherwise. The central nervous system is responsible both for emotion and for the stimulation of the sympathetic system, only the sympathetic stimulates the supra-renals to secrete adrenaline, which in its turn makes the sympathetic innervation more efficacious. It is like a process of autocatalysis; the sympathetic system, as it is stimulated, automatically raising the efficacy of its own stimulation. Similar processes probably obtain in the nervous system itself, and are concerned in the matter of learning and habit formation, processes which also resemble autocatalysis.

The amount of adrenaline in the supra-renals varies at different ages and in different conditions of health. It is found in large amounts in those young animals born in an advanced age, such as lambs, colts and calves. These animals are able to run about when born and their adrenals contain normal quantities of adrenaline. In human beings this is not the case. Some observers have not found adrenaline in children's adrenals except in very small amounts up to one year of age. But it must be remembered that disease may, in these cases, have caused a discharge, or diminution, before birth or before death. A man and a horse are not at the same physiological age at birth. In comparison with their total span of life a horse is as old at birth as a human being at two years. And a child at birth is as old physiologically as a horse fetus at four months.

The amount of adrenaline in the glands is greatly reduced in poisoning by phosphorus, arsenic, and mercury.

Conclusion.—The supra-renal glands then have the very important function of manufacturing and secreting into the blood stream the active principle adrenaline. This substance is 1, 2, dibydroxyphenyl, 4, hydroxyethyl methyl amine.

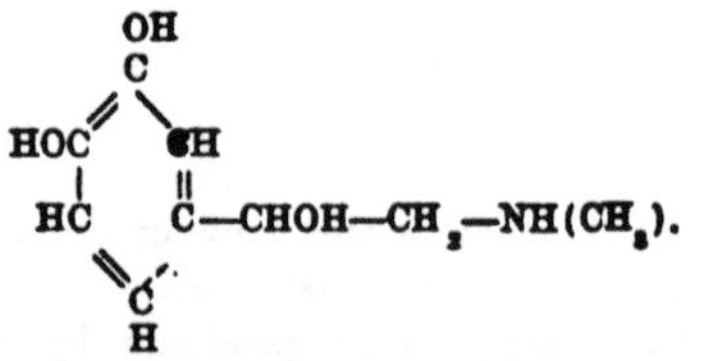

```
          OH
          C
        //  \
    HOC      CH
     |       ||
    HC       C—CHOH—CH₂—NH(CH₃).
       \\   /
          C
          H
```

This substance is presumably formed from tyrosine, or phenyl alanine, though there is as yet no direct proof of this. A 3.4 dioxyphenyl alanine has recently been isolated by Guggenheim from Vicia faba. Substances of a similar nature are found elsewhere in the animal kingdom than in the supra-renal glands. Chromaffine tissue is found in the skin glands (parotid gland) of poisonous toads and much adrenaline is found there also (Abel and Macht in Bufo agua). In ergot one of the active principles is hydroxyphenylethylamine.

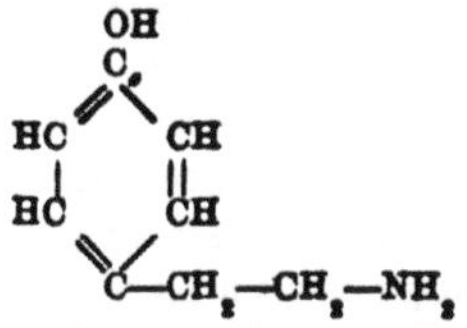

```
        OH
        C
      //  \
    HC     CH
    |      ||
    HC     CH
      \\  /
        C—CH₂—CH₂—NH₂
```

This would appear to be an intermediate product in the formation of adrenaline. This substance is formed from tyrosine by carboxylase splitting off carbonic acid. By oxidation and methylation this substance may be made into adrenaline. Similar substances, called by Kutscher by the group name *aporrhegmas*, are formed from other organs of the body and many of them are physiologically very active. It has been suggested that adrenaline, or related substances, may be formed in all cells on the advent of the nerve impulse and that it is by means of these active, or hormone, substances that the cell is roused to activity. Adrenaline is an unstable, easily oxidized substance, passing readily into a reddish and brown pigment matter. It is secreted into the blood stream from the glands and its function appears to be to reinforce the action of the sympathetic nervous system, thus increasing vasomotor tone, sympathetic tone, the tone of the digestive system, the bladder and uterus, and all other structures innervated by the sympathetic. It may also act on tissues not innervated by the sympathetic, but in small amounts its action is confined chiefly to that system. It also affects the central nervous system and it may play a part in the saturation of oxyhemoglobin.

The action of the supra-renals cannot, however, be said to be completely elucidated by these striking facts. We have as yet no satisfactory explanation of the sudden death of carnivorous animals following their extirpation and the later death of other animals. These have all the symptoms of an intoxication, and the injection of adrenaline has very little action in prolonging their life. It is not probable that death is the result of the lack of production of adrenaline. The complete loss of general irritability, skeletal as well as that of the sympathetic system, is too great to be thus explained. The result might be indirect. It is certainly possible that other factors come into play; the glands may have a detoxicating action. Further work on the cause of death following adrenalectomy is needed; nor is there any explanation of the pigmentation of the skin in Addison's disease.

The functions of the cortex of the gland are still entirely obscure. It has only a slight action on blood pressure, extracts causing a light fall. In fetal life the cortex is most developed and in the guinea pig it undergoes marked hypertrophy during pregnancy, followed by rapid retrogression after birth. It has been suggested that it elaborates the early stages of adrenaline, but this is not substantiated by any evidence. Nothing, in fact, is known of its function. It is, of course, not impossible that the adrenaline is made in the cortex and stored in the medullary cells. These medullary cells are modified sympathetic nerve cells and they may have been elaborated to act simply as storehouses for adrenaline, this substance having always a marked predilection for sympathetic tissue. Something analogous to this appears to be the case in the hypophysis. Pituitrine is found chiefly in the posterior lobe, which is nervous tissue, but there are some reasons for thinking that it may be made in the anterior lobe, which is glandular in nature. This matter, however, will need further investigation. Pituitrine resembles epinephrine closely in its action, but appears in many ways more nearly like d-adrenaline than the physiological l-adrenaline.

Besides producing adrenaline the supra-renals are remarkable chemically because of their very large lipin content. They contain more phospholipins, and other lipins of the general nature of those found in the central nervous system, than any other non-nervous tissue of the body. They have also a very intense metabolism.

THE SEXUAL GLANDS.

That the ovaries and the testes have a very marked effect on the growth and metabolism of the body has long been known. Indeed, these were the first tissues of which the cryptorrhetic powers were generally recognized. Castration has been practised for a very long time with the purpose of modifying the body form and temperament. That the effects were not due simply to cutting the nerves was easily shown. The reproductive organs give off either from the germ or interstitial cells substances which, circulating in the blood, change the growth of the horns and bones in cattle, the development of the hair and mammary glands and genitals in human beings, and profoundly modify the general psychological properties. There are a large number of so-called secondary sexual characters which depend upon the internal secretions of these organs.

The nature of the secretions thus formed is unknown, and this book is already so long that we can give no more here than a brief notice of some of the recent interesting work in connection with the corpus luteum. It will be remembered that when an ovum is discharged from the Graafian follicle of the ovary there is formed what is known as the corpus luteum, a yellow, glandular structure which after a time atrophies. It has been found that the extract of this corpus luteum determines the growth of the mammary glands which occurs during pregnancy. In some animals the ova are only discharged and corpora lutea formed after copulation. If copulation is performed by a sterilized male, so that pregnancy does not follow, the corpora lutea are formed as usual. The other symptoms of pregnancy follow, although there is no pregnancy. Thus dogs and cats will in such cases develop milk in the mammary glands, often they will proceed to prepare a nest for the litter. Marsupials, in similar conditions, show some of the same phenomena; they have been observed to clean out the pouch at the time when the young should normally appear. If, on the other hand, the corpora lutea be taken out, then these symptoms do not follow.

It seems probable that the corpora lutea play a very important part also in preparing the uterus for the fixation of the ovum. It is possible that extracts of the corpus luteum might be of value in checking abortion, or in aiding the implantation of the ovum. It has no relation to estrus.

That the ovary has also a very close relationship to the skeletal system is indicated by the healing effects of ovariotomy in osteomalacia, and by the large size of the bones of castrated cattle.

Owing to the uncertainty of the interpretation of many of the facts, however, a further discussion of this extraordinarily important subject will not be taken up here. It is not impossible that the further study of this subject may confirm or demolish some of the present theories of inheritance.

THYMUS GLANDS.

These glands also are cryptorrhetic organs, but their function is still unknown, or at least there is little that is clear-cut and definite to say about them. By many observers they have been brought into relation with the growth of the bones and the calcium excretion, but the evidence is contradictory. From a chemical point of view they are remarkable for their richness in nuclear material. The chemistry of the nucleic acid and histone found in them has already been discussed elsewhere. That they have an important function in the young animal can hardly be doubted.

PINEAL GLAND.

This is also a rudimentary organ in the brain corresponding in a general way on the upper surface to the hypophysis on the lower. The function of this organ is still unknown, but it appears to be closely related to the sexual organs. Tumors involving the pineal body are often accompanied by an extraordinarily early development of sexual maturity with all the secondary sexual characters. Somewhat similar observations have been made for tumors involving the cortical parts of the supra-renals. Feeding pineal glands to young chicks or guinea pigs accelerates their development (McCord).

REFERENCES. Sexual Organs.

1. *Steinach:* Willkürliche Umwandlung von Säugetier-Männchen in Tiere mit ausgeprägt weiblichen Geschlechtskarakteren und weiblichen Psyche. (Eine Untersuchung über die Funktion und Bedeutung der Pubertätsdrüsen. Archiv f. d. ges. Physiol., 144, p. 71, 1912.
2. *Bouin and Ancel:* Récherches sur la signification physiologique de la glande interstitielle du testicule des mammifères. Jour. de Physiol. et Pathol. gen., 6, p. 1012, 1904.
3. *Lederer and Pribram:* Experimenteller Beitrag zur Frage über die Beziehung zwischen Placenta und Brustdrüsenfunktion. Archiv f. d. ges. Physiol., 134, p. 531, 1910.
4. *Ribbert:* Ueber Transplantation von Ovarium, Hoden und Mamma. Archiv f. Entwickelungsmechanik, 7, p. 688, 1898.
5. *Clayton and Starling:* Pro. Roy. Soc., 77, p. 505, 1906.
6. *Marshall:* Phil. Trans. B 198, 1905.
7. *Meyns:* Ueber Froschhodentransplantation. Archiv f. d. ges. Physiol., 132, p. 433, 1910.
8. *Gudernatsch:* Fütterungsversuche an Amphibienlarven. Zent. f. Physiol., 26, p. 323, 1912.
9. *Oliver:* On the question of an internal secretion from the human ovary. Jour. Physiol., 44, p. 355, 1912.
10. *Paton:* The thymus and sexual organs. III. Their relationship to the growth of the animal. Jour. Physiol., 42, p. 267, 1911.
11. *Marshall:* The ovarian factor concerned in the recurrence of œstrus. Jour. Physiol., 43, p. xxi, 1911.
12. *Marshall:* Physiology of Sex. (Literature.)
13. *Marshall:* A note on the chemistry of the vesicular fluid of the hedgehog. Jour. Physiol., 43, p. 259, 1911.
14. *Marshall:* The male generative cycle in the hedgehog; with experiments on the functional correlation between the essential and accessory sexual organs. Jour. Physiol., 43, p. 247, 1911.
15. *Scherbok:* Versuche über innere Sekretion der Brustdrüse. Wiener klin. Woch., 25, p. 199, 1911.
16. *Steinach:* Geschlechtsbetrieb und echt sekundare Geschlechtsmerkmale als folge der innersekretorischen Funktion der Keimdrüsen. Zent. f. Physiol., 24, p. 552, 1910.
17. *Marshall and Runciman:* On the ovarian factor concerned in the recurrence of the œstrus cycle. Jour. Physiol., 49, p. 17, 1914.

Supra-renals.

A. General.

1. *Barger:* The Simpler Natural Bases, pp. 81-105, 1914. Monographs on Biochemistry.
2. *Bayer:* Die normale u. pathologische Physiologie des chromaffinen Gewebes der Nebennieren. Ergebnisse der pathol. Anat., 14, 1910.
3. *Biedel:* Innere Sekretion, ihre physiol. Bedeutung f. d. Pathologie. Berlin, 2d ed., pp. 313-521, 1913.
4. *Rolleston:* Suprarenal bodies. Goulstonian Lecture. British Med. Journal, I., pp. 629-634; 687-691; 745-748, 1895. Earlier literature cited.

5. *Vincent:* Innere Sekretion u. Drüsen ohne Ausführungsgang. Ergebnisse der Physiol., 9, pp. 509-586, 1910. Literature.

B. Pharmacy of adrenaline.

1. *Beckwith:* The pharmacy of adrenaline. Journal of American Pharmaceutical Assn., vol. III, pp. 1547-1554, 1914.

C. Physiology of glands.

1. *Elliott:* The control of the suprarenal glands by the splanchnic nerves. Jour. of Physiol., 44, p. 374, 1912.
2. *Cannon and Hoskins:* The effect of asphyxia and sensory stimulation on adrenal secretion. Amer. Jour. Physiol., 29, p. 274, 1911.
3. *Cannon and de la Paze:* Emotional stimulation of adrenal secretion. Amer. Jour. Physiol., 28, p. 69, 1911.
4. *Cannon, Sholl and Wright:* Emotional glycosuria. Amer. Jour. Physiol., 29, p. 280, 1911.
5. *Wertheimer and Battez:* Sur les nerfs glyco-sécréteurs. Archives internat. d. Physiol., 9, p. 363, 1910.
6. *Mayer:* Sur le mode d'action de la piqûre diabétique. Rôle des capsules surrénales. C. R. Soc. Biol., 60, p. 1123, 1906; 219, 1908.
7. *Dreyer:* On secretory nerves to the supra-renal capsules. Amer. Jour. Physiol., 2, p. 203, 1899.
8. *Joseph and Meltzer:* Effect of stimulation of the peripheral end of the splanchnic nerves upon the pupil. Amer. Jour. Physiol., 29, p. xxxiv, 1912.
9. *Paton and Watson:* The action of pituitrin, adrenaline and barium on the circulation of the bird. Jour. Physiol., 44, p. 413, 1912.
10. *Oliver and Schaefer:* The physiological effects of extracts of the suprarenal capsules. J. Physiol., 18, pp. 230-276, 1895.
11. *MacLeod:* The relation of the adrenal glands to sugar production by the liver. VIII. Amer. Jour. Physiol., 29, 1912, p. 419.
12. *Kahn:* Archiv f. d. ges. Physiol., 144, p. 396, 1912; 146, p. 578, 1912.

D. Adrenaline. Chemistry and physiology.

1. *Abel:* Further observations on the chemical structure of the active principle of the suprarenal capsules. Johns Hopkins Hospital Bulletin, 9, pp. 215-218, 1898.
2. *Abel:* Further observations on epinephrine. Johns Hopkins Hospital Bulletin, 12, pp. 80-84, 1901.
3. *Abel:* On epinephrine and its compounds with especial reference to epinephrine hydrate. Amer. Jour. Pharm., 75, pp. 301-325, 1903.
4. *Abel and Macht:* The poison of the tropical toad, Bufo agua. Jour. Amer. Med. Assn., 56, 1911, pp. 1531-1536.
5. *Abel and Macht:* Two crystalline pharmacological agents obtained from the tropical toad. Jour. Pharm. and Expt. Therapeut., 3, 1912, pp. 319-377.
6. *Aldrich:* A preliminary report on the active principle of the suprarenal gland. Amer. Jour. Physiol., 5, p. 457, 1901.
7. *Aldrich:* Adrenalin the active principle of the suprarenal gland. Jour. Amer. Chem. Soc., 27, pp. 1074-1091, 1905.
8. *Takamine:* Adrenalin the active principle of the suprarenal glands and its mode of separation. Amer. Jour. Pharm., 73, 1901, pp. 523-531.
9. *Takamine:* The isolation of the active principle of the suprarenal gland. Jour. Physiol., 27, p. xxix, 1901.

10. *Schultz:* Quantitative pharmacological studies: Adrenalin and adrenalin-like bodies. Bulletin Hygienic Laboratory, 55, Washington, 1909.

11. *Schultz:* Quantitative pharmacological studies: Relative physiological activity of some commercial solutions of adrenalin. Bulletin 61, 1910, Hygienic Laboratory, Washington.

12. *Stewart:* The alleged existence of adrenalin in pathological sera. Jour. Expt. Med., 15, pp. 547-569, 1912.

13. *Folin, Cannon and Denis:* A new colorimetric method for the determination of epinephrin. Jour. Biol. Chem., 13, pp. 477-483, 1913.

14. *Friedmann:* Die Konstitution des Adrenalins. Beiträge z. chem. Physiol., 8, pp. 95-120, 1906.

15. *Ewins:* Some color reactions of adrenine and allied bases. Jour. Physiol., 40, pp. 317-326, 1910.

16. *Ewins and Laidlaw:* Alleged formation of adrenine from tyrosine. Jour. Physiol., 40, 275-278, 1910.

17. *Fenger:* On the presence of active principles in the thyroid and surprarenal glands before and after birth. Jour. Biol. Chem., 11, 1912, pp. 489-492; 12, pp. 55-59.

18. *Barger and Jowett:* The synthesis of substances allied to epinephrine. Jour. Chem. Soc., 87, pp. 867-974, 1905.

19. *Blum:* Ueber Nebennierendiabetes. Deut. Archiv f. klin. Med., 71, 146-167, 1901.

20. *Abderhalden and Thiess:* Weitere Studien über das physiologische Verhalten von l, dl and d, Suprarenin. Zeit. physiol. Chem., 59, pp. 22-28, 1909. Earlier papers on subject cited here.

21. *Abderhalden and Guggenheim:* The action of tyrosinase from Russula delica on tyrosin containing polypeptides and suprarenin. Zeits. physiol. Chem., 57, 1908, pp. 329-331.

22. *Dakin:* On the physiological activity of substances indirectly related to adrenalin. Pro. Roy. Soc., B, 76, pp. 498-503, 1905.

23. *Gottlieb and O'Connor:* Ueber den Nachwies u. die Bestimmung des Adrenalins im Blute. Handbuch biochem. Arbeitsmethoden, VI, pp. 585-603, 1912.

24. *Guggenheim:* Dioxyphenylalanin, eine neue Aminosäure aus Vicia Faba. Zeits. physiol. Chem., 88, pp. 276-284, 1913.

25. *Kolmer:* Relation of adrenals to the sexual function. Archiv f. d. ges. Physiol., 144, p. 361, 1912.

26. *Elliott:* Some results of excision of the adrenal glands. Jour. Physiol., 49, p. 38, 1914.

Thyroid.

1. *Thompson:* On the thyroid and parathyroids throughout vertebrates, with observations on some other closely related structures. Phil. Trans. Roy. Soc. B, 201, p. 91, 1911.

2. *Koch, F. C.:* On the nature of the iodine-containing complex in thyreoglobulin. Jour. Biol. Chem., 14, p. 165, 1913.

3. *Oswald:* Neue Beiträge zur Kenntnis der Bindung des Jods im Jodthyreoglobulin nebst einigen Bemerkungen über das Jodothyrin. Arch. f. expt. Path. u. Pharm., 60, p. 115, 1908.

4. *Hunt:* The influence of thyroid feeding upon poisoning by acetonitrile. Jour. Biol. Chem., 1, p. 33, 1905.

5. *Hunt and Seidell:* Bulletins 48 (1908) and 69 (1910). Hygienic Laboratory, Washington.

6. *Baumann:* Ueber das normale Vorkommen von Jod in Thierkörper. Zeits. f. physiol. Chem., 21, p. 487, 1896; 22, p. 1, 1896.
7. *von Cyon and Oswald:* Ueber die physiologischen Wirkungen einiger aus der Schilddrüse gewonnener Producte. Archiv f. d. ges. Physiol., 83, p. 1901.
8. *von Cyon:* Metholodische Aufklärungen zur Physiologie der Schilddrüse. Archiv f. d. ges. Physiol., 138, p. 575, 1911.
9. *Claude and Blanchetiere:* Sur la teneur en iode de la glande thyroid dan ses rapports avec la constitution anatomique de l'organe. J. de Physiol. et Path. gen. 12, p. 563, 1910.
10. *Oswald:* Ueber den Jodgehalt der Schilddrüsen. Zeits. f. physiol. Chem., 23, 1897. Eiweisskörper der Schilddrüse. Ibid., 27, p. 14, 1899.
11. *Krause and Cramer:* On the effects of thyroid feeding on nitrogen and carbohydrate metabolism. Jour. Physiol., 44, p. xxiii, 1912.
12. *Gley:* Récherches sur la pathogenese du goitre exoptalmique. Jour. d. Physiol. et Pathol., 13, p. 955, 1911.
13. *Jeandelize et Parisot:* La pression artérielle après la thyroidectomie chez le lapin. J. de Physiol. et Pathol. gen., 12. p. 331, 1910.
14. *Rossi:* Sur les effets de la thyreo-parathyreoidectomie chez les animaux de la race ovine. Archives Ital. de Biol., 55, p. 91, 1911.
15. *Vincent:* Innere Sekretion u. Drüsen ohne Ausführungsgang. Ergebnisse der Physiol., 9, 1910. Literature.
16. *von Fürth u. Schwartz:* Ueber die Einwirkung des Jodothyrins auf den Zirkulationsapparat. Arch. f. d. ges. Physiol., 124, p. 113, 1908.
17. *Modrakowski:* Ueber die Identität des blutdrucksenkenden Körpers der Glandula thyroidea mit dem Vasodilatin. Arch. f. d. ges. Physiol., 133, p. 201, 1910.
18. *Bircher:* Weitere Beiträge zur experimentellen Erzeugung des Kropfes. Zeits. f. expt. Pathol. u. Ther., 9, 1911.
19. *Justchenko:* Die Schilddrüse u. die Fermentativen Prozesse. Zeit. f. physiol. Chem., 75, p. 141, 1911.
20. *Fenger:* On the presence of active principles in thyroid and suprarenal before and after birth. Jour. Biol. Chem., 12, p. 55, 1912.
21. *Koch:* On the occurrence of methyl-guanidine in the urine of parathyroidectomized animals. Jour. Biol. Chem., 12, p. 313, 1912.
22. *Greenwald:* Further experiments upon parathyroidectomized dogs. Jour. Biol. Chem., 14, p. 363, 1913; 14, p. 369, 1913.
23. *Fenger:* On the iodine and phosphorus contents, size and physiological activity of fetal thyroid gland. Jour. Biol. Chem., 14, p. 397, 1913.

PINEAL GLAND.

1. *Vincent:* Internal secretion and ductless glands. 1912.
2. *Baily and Jelliffe:* Tumors of the pineal body. Archives of Internal Medicine, 8, p. 851, 1911.
3. *McCord:* The pineal gland in relation to somatic, sexual and mental development. 65th Session Amer. Med Assn., June, 1914.
4 *Dana and Berkely:* Medical Record, 83, p. 835, 1913.
5. *Foa:* Hypertrophie des testicules et de la crête après l'extirpation de la glande pinéale chez le coq. Arch. Ital. de Biol., 57, p. 233, 1912.
6. *Berkely:* Med. Record, New York, 85, p. 513, 1914.

PARATHYROID.

1. *Carlson:* Condition of the digestive tract in parathyroid tetany in cats and dogs. Amer. Jour. Physiol., 30, p. 309, 1912.
2. *Cooke:* Changes in nitrogenous metabolism after parathyroidectomy. Jour. Expt. Med., 13, p. 439, 1911.
3. *Greenwald:* The effect of parathyroidectomy upon metabolism. Amer. Jour. Physiol., 28, p. 103, 1911.
4. *Ver Eecke:* Étude de l'influence de la sécrétion interne du corps thyroid sur les échanges organiques. Archives internat. de Pharmacodyn., 4, p. 81, 1898.
5. *MacCallum and Voegtlin:* On the relation of tetany to the parathyroid glands and to the calcium metabolism. Jour. Expt. Med., 11, 1909, p. 118. Jour. Pharm. and Expt. Ther., 2, 1911, p. 421.
6. *Berkely and Beebe:* A contribution to the physiology and chemistry of the parathyroid gland. Jour. Med. Research, 20, p. 149, 1909.
7. *Morel:* Les parathyroides dans l'ontogenèse. C. R. de la Soc. Biol., 67, 1910.
8. *Gley:* De l'exopthalmie consécutive à la Thyroidectomie. C. R. Soc., Biol., 68, p. 858, 1910.
9. *Falta and Kahn:* Klinische Studien über Tetanie mit besonderer Berücksichtigung des vegetativen Nervensystems. Zeits. f. klin. Med., 124, p. 1.
10. *Wiener:* Ueber die Art der Funktion der Epithelkörperchen. Archiv f. d. ges. Physiol., 136, p. 107, 1910.
11. *Vincent:* Some observations on the functions of the thyroid and parathyroid glands. Jour. Physiol., 32, p. 65, 1905.
12. *Arthus and Schapermann:* Parathyroidectomy et sels de chaux chez le lapin. Jour. de Physiol. et Pathol. gen., 12, p. 177, 1910.
13. *Morel:* L'Acidose Parathyroprive. Jour. de Physiol. et Path. gen., 13, 1911, p. 542.
14. *Meltzer and Joseph:* The inhibitory action of sodium chloride upon the phenomena following the removal of the parathyroids in dogs. Jour. Pharm. and Expt. Therapeut., 2, p. 361, 1911.
15. *Paton, Findlay and Burns:* On guanidin or methyl guanidin as a toxic agent in the tetany following parathyroidectomy. Jour. Physiol., 49, Proc. Physiol. Soc., p. xvii, 1915.

CHAPTER XVII.

THE EXCRETIONS OF THE BODY.

THE URINE.

It will appear from the account which has been given of the chemistry of the different tissues of the body how fragmentary is our knowledge of the chemical transformations undergone by the food materials after their absorption. For example, the simple proteins of the foods are reduced by digestion to the amino-acids. Some of these amino-acids are deamidized by the action of bacteria, or the enzymes of the alimentary mucous membrane. The amino-acids enter the blood of the portal system partly as amino-acids, partly as ammonia and ketonic acids, and to some extent possibly in other decomposition products not yet recognized. In the blood they circulate dissolved in the plasma and as they pass the various tissues each tissue removes, it is believed, those amino-acids in the quantity which it needs for its own metabolism; and it builds up its own protein from them. It remains for physiological chemistry to follow the subsequent fate of these amino-acids through the complex chemistry of each organ of the body, discovering what proportion is synthesized into the cell proteins; how rapidly they are broken up and what decomposition products arise from them; and following, also, the fate of the amino-acids not synthesized into protein, but broken up directly by the cell into products of great importance to the cell or its neighbors. This task, however, is beyond our powers at present and is indeed a vastly difficult task, since each organ has a different metabolism and must be studied by itself. We know scarcely anything, for example, about the chemical composition of many of these organs and particularly little about the master tissue of the body, the nervous system, so that we cannot proceed directly to each organ and describe just what chemical changes occur in it. We must of necessity take a roundabout course and follow the path which the science has actually followed in its development. Having followed the foods into the blood, the internal medium of the body, we turn to see in what form the various substances, nitrogen, carbon, hydrogen and oxygen, leave the body; and by working backward from the excretions gradually narrow down the problem to the metabolism of the various organs. We shall turn now to study the composition of he various excretions of the body.

These excretions are those of the kidneys, the skin, the lungs and the alimentary canal.

The total amount of the excretory material varies with the amount of food and drink consumed and the amount of work done. By reference to the metabolism experiment on page 292 it will be seen that a man doing pretty hard muscular work, so that he expends 5,500 calories of energy per day, excreted about 1,000 c.c. of water per day in his urine, about 240 c.c. per day in the feces. and about 4,000 grams per day in the respiration and perspiration. The total output of water amounted, therefore, to about 5,240 grams per day. For a person not working the amount is less than this and about 3,000 c.c. The relative loss through the lungs and skin can be determined by the very ingenious method of Lombard. He placed a man naked on one pan of a delicate balance, the opposite pan of which had a lever attached to it so that it traced a curve on a moving drum covered with smoked paper. Under these circumstances, of course, since the man is constantly losing weight by the evaporation of water from the skin and by the exhalation of water and carbon dioxide by the lungs, he grows constantly lighter and the pointer traces a fairly uniform falling line on the revolving drum. The man visibly evaporates before one's eyes. The loss in this case is both by the skin and lungs. The slope of the curve tells how rapidly the change of weight is occurring. Now while this curve is being taken, if the man simply holds his breath there is no longer any loss through the lungs, but there is a continuous loss through the skin. The slope of the curve for this period shows how rapid the loss is through the skin alone. By this means and by other experiments it has been found that by respiration there is lost about 30 per cent. of the total water, by the skin 17 per cent. and by the urine about 50 per cent. These figures are subject to wide variation. In the experiment just quoted the loss through the skin and lungs was at least double that through the urine, and it may be a still larger proportion during exercise on a hot day. In the lungs there is lost, also, about 1,700 grams of carbon dioxide per day. Water and carbon dioxide pass out, then, in large measure through the lungs. The skin carries out chiefly the water in the perspiration; but there are also present some salts and some nitrogen substances, such as urea. The amount of nitrogen lost through the skin is variously estimated. It is generally determined by collecting the perspiration in some form of cotton underclothing, washing this out and determining the amount of nitrogen in the wash water. The method is not exact. The nitrogen loss through the skin is undoubtedly larger in hot, dry climates than in moist, cold ones. It may be estimated as about one-half gram of nitrogen daily, but it may be nearly a gram a day. The feces have been considered already. They contain on the average about 1-2 grams of nitrogen per day. All the rest of the nitrogen excreted, except small portions lost in hair and the wear of the skin, passes through the kidneys into the urine. It is this fact which makes the urine of such great interest. The urine also contains various carbon compounds free from nitrogen and, in small quantities, a great number of substances which have escaped from the metabolism of the body. The composition of these substances is often of great interest because of the light they throw on the intermediate metabolism of the body.

The urine.—The urine is secreted by the kidneys. In man and mammalia these are two organs, more or less ellipsoidal in shape, weighing in male adults about 130 grams each, situated on each side of the spinal column, behind the peritoneum at about the level of the first three lumbar vertebræ. They are of mesodermic origin and develop in embryonic life from the genito-urinary ridge, a ridge of mesoblastic tissue extending from head to tail. They are very long organs in the fishes, extending nearly the length of the body cavity from far forward in the head; and are called pronephros or head kidneys. In amphibia the middle region of the ridge gives rise to the kidneys, which are called the mesonephros, and in mammalia only the posterior region or caudal portion thus develops into a more compact organ, the metanephros or true kidneys. In the invertebrates somewhat similar excretory organs are the segmental nephridia of the annelids and arthropods. The kidneys lie very close to the dorsal aorta, receiving an unusually large amount of blood by very large, short, straight arteries which deliver the blood under high pressure into a number of capsules of capillaries, each capsule inclosed in a very fine endothelial membrane. The resistance to the flow in the capillaries must be high. The glomerulus is supposed to function as a filtering apparatus (see, however, the recent work of Brodie) and from it a large amount of liquid containing salts and other soluble constituents, and at times protein, is believed to be filtered into the head of the kidney tubule. From these glomeruli the filtrate passes along contorted tubules, the kidney tubules, making up the greater part of the kidney substance and which are lined by epithelial cells, which are supposed to secrete the various nitrogenous and other constituents of the urine and, perhaps, to reabsorb some of the water. It cannot be said, however, that the mechanism of secretion of the urine is as yet in many particulars clear. The secretion is under the control of the blood flow; a more rapid blood flow, a higher pressure in the glomeruli, other conditions remaining normal, is accompanied by a higher secretion of urine.

Amount. The amount of urine secreted per day is extremely variable for different individuals, depending on diet, external temperature, amount of water drunk and on the constitution of the individual. Thus in warm weather, or when a restricted diet is being taken or little water drunk, it may be as little as 600-700 c.c. per day. In cold weather it is always larger and also on a heavy diet. An average abstemious American student in the Northern States secretes about 1,200-1,500 c.c. per day. The drinking of tea, coffee or beer increases the amount in part because of the fluid they contain, and in part because of the presence of diuretics in these beverages. Some individuals, generally of a nervous temperament, secrete larger amounts, from 2,000-3,000 c.c. The cause of this mild polyuria (diabetes insipidus) is still obscure, but it may be produced artificially by some injury to the part of the brain about the infundibular body. It may be due to slight pressure in this region.

Specific gravity. The specific gravity varies usually inversely with the amount secreted from 1.008-1.030. The color is generally a darker yellow, or brown the more concentrated the urine. In polyuria, hysteria and some nervous affections the urine is often almost colorless.

Reaction. Normal human urine is usually acid to litmus, though it may at times be alkaline. It is most acid on a meat diet; on a vegetable diet it may be neutral or alkaline. Its acidity is greatly reduced during the secretion of the gastric juice in the stomach during digestion, so that shortly after a meal an alkaline reaction may be had. There is at that time an alkaline tide in the body due to the separation of the hydrochloric acid (page 376). The urine of carnivorous animals is usually acid; that of herbivorous, such as cows and horses, is alkaline except during fasting, when it may be acid. The reason why the urine is acid when on a meat diet and alkaline on a herbivorous is that in protein there occur sulphur in an unoxidized condition, and esters of phosphoric acid in nucleic acid, phosphoproteins and in lecithin. During metabolism these are oxidized to sulphuric acid or set free as phosphoric acid. It is this acidity which accounts for the acidity of the urine. Nitrogen from the proteins is eliminated in the unoxidized form as urea, for the most part, but this is a very weak base, practically neutral in reaction, and hence not capable of neutralizing the acid produced. There are also some organic acids formed from the oxidation of parts of the protein, such as phenyl acetic acid, benzoic acid, uric acid, urocanic acid and so on, which also contribute to the urine acidity.

The alkaline reaction of the urine of herbivorous animals is owing to the fact that vegetables and fruits contain acid salts of dibasic acids, or polybasic acids, or other carboxylic acids, such as acid potassium malate, citrate, acetate, tartrate and so on. On oxidation in the body these are burned to carbonates. Some of the carbonic acid finds its exit through the lungs, leaving the associated base, generally sodium or potassium, with the very weak acid, carbonic acid, to find its way into the urine. The carbonates have an alkaline reaction owing to their forming some free alkali by hydrolysis. It seems paradoxical at first sight that by drinking an acid citrate the urine should be made alkaline, but it is explained in the way just stated.

The amount of acid which can be titrated in the urine by alkalies with phenolphthalein as an indicator amounts in a day to the equivalent of from 150-400 c.c. of N/10 acid. It is about equivalent to N/40th acid by titration. It varies greatly and is often less than this.

The real acidity of the urine, that is the number of the hydrogen ions it contains, can be determined either by the gas-chain method described on page 539 or by the indicator method. A recent extensive investigation of the acidity by the latter method by Henderson has shown that the acidity is about equal to $N \times 10^{-6}$ hydrogen ion. It may, for short periods, rise to about $N \times 10^{-5.1}$ or become as alkaline as $N \times 10^{-7.4}$. As a rule the total acidity as determined by titration and the concentration of the hydrogen ions go parallel. This may be seen, for example, in the following case:

Case No.	Amount c.c. urine	H' (P_H)	Acid c.c. N/10	NH_3 c.c. N/10	Total acidity c.c N/10 acid	Acidity ÷ NH_3
3	1000	6.0	222	360	582	0.62
	820	5.5	410	600	1010	0.68
	1100	6.3	195	390	585	0.50
	1500	5.7	315	535	850	0.59
	1140	6.0	150	290	440	0.52

It will be observed that when the hydrogen ion concentration is greatest and $N \times 10^{-5.5}$ the total acidity is highest, i.e., 1010.

There are no characteristic variations of the acidity in disease thus found, except that in cardio-renal disease the mean acidity is somewhat higher than normal, H ion$=N \times 10^{-5.38}$, but this is not higher than may occur in normal urine. It appears to be easier to develop high acidity in disease than greater alkalinity. This may be due to the fact that in serious disease the diet is generally restricted and the patient may become more purely carnivorous, the body calling on its own reserves. While all the acids of the urine contribute to this acidity, phosphoric acid, on account of its greater quantity, predominates in producing the effect. It is a weak acid and such strong acids as sulphuric will satisfy themselves at the expense of any alkali metal bound to the phosphoric acid.

In case the total acidity is increased, the ammonia generally rises, but not always proportional to the amount of acid produced. The amount of protein eaten, or other variations in the diet, such as the amount of fruit and so forth, determine whether the acid produced will be neutralized more by ammonia or by the fixed alkalies.

By the ingestion of acid and bicarbonates the acidity of the urine may be either increased or diminished. Four hours after the ingestion of 10 grams of monosodium phosphate the acidity of the urine had risen from $N \times 10^{-6.7}$ to $N \times 10^{-5.3}$ in an experiment by Henderson and Palmer. In one experiment 12 grams of sodium bicarbonate were ingested at 10.00 A.M. when the acidity of the urine was $N \times 10^{-5.80}$. At 12 o'clock the acidity had fallen to $N \times 10^{-8.70}$. The urine was then strongly alkaline. It remained at this alkalinity for several hours.

Osmotic pressure of the urine. The osmotic pressure of the urine is much higher than that of the blood. This means that the kidney must do work in the secretion of the urine and the amount of work done can be determined from the difference of osmotic pressure. It is evident that the secretion is not a simple filtration, but that the vital activity of the cells enables them to do this work. The work is done either in the reabsorption of water by the tubules, thus concentrating the filtrate from the glomeruli, or more probably in the secretion of the urea and other substances into the urine. The osmotic pressure is generally determined by the freezing-point method described on page 201. The freezing point of the urine varies from about —0.65° to —2.71°. The latter represents an osmotic pressure (see page 201) of 32.4 atmospheres. The freezing point of the blood is about —0.6°, which is an osmotic pressure of only 7.23 atmospheres. If the difference in osmotic pressure of the blood and the urine amounts to 25 atmospheres and there is secreted 1 liter of urine per day, the kidneys would have to do 25 liter atmospheres of work. 1 liter atmosphere is equivalent to 1.0132×10^{9} ergs, or 2.4211×10 small calories (15°). In a day, therefore, the kidneys would do 25 liter atmospheres or 605 small calories of work. This would be 0.605 large calories. As the total output of energy of the body per day is about 2,500 to 3,000 large calories, the work of the kidney represents a very small proportion of it. The burning of something less than 0.7 gram of glucose would yield this amount of energy. The osmotic pressure of the urine approximates more closely to that of the blood the more rapid the secretion; during exertion on very hot, dry days the freezing point may be still lower than —2.71° and may go below —3.0°. Bugarsky has found a formula of fairly general applicability expressing the relation between the osmotic pressure (freezing point or Δ) and the specific gravity: $\Delta = 75(s-1)$, where s is the specific gravity. In certain circumstances the urine may have a freezing point of only —0.2°, as after drinking large quantities of beer.

General composition of the urine.—The following table illustrates the average composition of human urine on an average diet containing 120 grams of protein per day. The amount of solids and water are of course widely variable.

TOTAL URINE PER DAY 1500 c.c. TOTAL SOLIDS 61 GRAMS.

Organic solids	38.2 grams.	Inorganic solids	22.7 grams.
Urea	32	Sodium chloride	14 grams.
Uric acid	0.7	H_3PO_4	2.0
Creatinine	1.8	H_2SO_4	2.6
Ammonia	0.7	K_2O	3.0
Hippuric acid	0.8	MgO and CaO	0.9
Residual organic	2.2	Residual inorganic	0.2
	38.2		22.7

Nitrogenous constituents of the urine.—The nitrogenous constituents found in the urine are urea, uric acid, ammonia, creatinine and creatine, amino-acids, allantoine, hippuric acid and a great number of basic and other nitrogenous substances found in very small quantities. But while present in small quantities many of these are of very great interest, since they undoubtedly represent intermediate products in the course of the transformations of the amino-acids in the body. In addition there may be found in pathological conditions proteins, or derived proteins, of various kinds. In human urine urea makes by far the larger proportion of the nitrogenous substances. Normally from 85-92 per cent. of the nitrogen in the urine is in the form of urea nitrogen; but on a diet containing a minimum quantity of protein food the proportion of urea nitrogen falls, so that it may be not more than 68-80 per cent. of the total urinary nitrogen.

In a group of about 25 normal young men in the course of an extended metabolism experiment, the average excretion of nitrogen per 24 hours was as follows:

Body-weight —kilos	Ingested nitrogen —grams	Total urinary N g.	Urea N g.	NH_3 N g.	Creatinine N g.	Uric acid N g.	Undetermined N g.	Total urinary N as per cent. of ingested.
71.0	13.43	11.30	9.30	0.49	0.659	0.198	0.65	84.12
Per cent. of total N in urine			82.29	4.35	5.87	1.76	5.73	

Urea. Chemistry.—Urea is the diamide of carbonic acid, $CO(NH_2)_2$:

$$O=C(NH_2)_2$$

Urea.

It may be regarded as formed from carbonic acid as follows

$$O=C(OH)_2 + NH_3 \longrightarrow O=C(NH_2)(OH) + H_2O$$

Carbamic acid.

$$O{=}C(NH_2)(OH) + NH_3 \longrightarrow O{=}C(NH_2)(NH_2) + H_2O$$

Carbamic acid. Urea.

According to this formula urea should be a very faintly basic substance, and acids should have some power of union with the amino groups, as they do have. If a urea solution is heated, however, it sets free some hydrogen ions and behaves as an acid. This is probably due to a rearrangement of the molecule in the direction of the imide form, and some molecules having the imide constitution probably exist in all solutions of urea. The imide form is the following:

$$O{=}C(NH_2)(NH_2) \rightleftharpoons HO{-}C(NH_2){=}NH \underset{+H_2O}{\overset{-H_2O}{\rightleftharpoons}} HN{=}C{=}NH \rightleftharpoons NH_2{-}C{\equiv}N$$

Urea. Imide form. Carbodiimide. Cyanamide.

Urea may be synthesized by heating ammonium cyanate which undergoes a rearrangement into urea:

$$NH_4{-}O{-}C{\equiv}N \rightleftharpoons NH_2{-}CO{-}NH_2$$

Ammonium cyanate. Urea.

Urea has no taste. It is colorless and odorless and crystallizes in long prisms, m.p. 132-133°. It may be purified by recrystallization from amyl alcohol. It is extremely soluble in water and in alcohol, but with difficulty in cold acetone. It is insoluble in ether. It forms salts with acids, and the oxalate and nitrate, $CH_4N_2O.HNO_3$, are much less soluble in water than urea itself. By heating with acids or alkalies it is split into ammonia and carbonic acid and the same change occurs on heating with water. Dry urea heated forms biuret, cyanuric acid and the amide of cyanuric acid, or ammelide:

$$2\,CO(NH_2)_2 - NH_3 \longrightarrow NH_2{-}CO{-}NH{-}CO{-}NH_2 \quad ; \quad 3\,CO(NH_2)_2 - 3NH_3 \longrightarrow C_3N_3(OH)_3$$

Urea. Biuret. Cyanuric acid.

Oxidized by hypobromite or nitrous acid, it is decomposed into carbon dioxide, nitrogen gas, some carbon monoxide and nitric oxide. The greater part of the decomposition, however, leads to the formation of nitrogen gas and carbon dioxide. The latter reaction is used for the clinical estimation of urea.

$$CO\,(NH_2)_2 + 2HNO_2 = CO_2 + 2N_2 + 3H_2O$$
$$CO\,(NH_2)_2 + 3NaOBr = CO_2 + 3NaBr + 2H_2O + N_2$$

Urea will also, like the amino-acids, form molecular compounds with salts. Thus, if a solution is evaporated with sodium chloride, prisms of $CO.(NH_2)_2.NaCl.H_2O$ crystallize out. Urea is decomposed into ammonium carbonate by an enzyme, urease, found in the Soy bean, in several bacteria and in other plant and animal tissues.

Amount.—The amount of urea secreted by the kidneys per day is very variable and depends upon the amount of protein in the food. For a person eating an average diet containing about 120 grams of protein per day, the urea excretion will be in the neighborhood of 30 grams; but on a low protein diet the amount of urea is greatly reduced. With a diet containing 50 grams of protein per day, or 8 grams of nitrogen, the urea will be only about 8 to 10 grams a day. No constituent of the urine is more variable than the urea. Ordinarily, when the protein intake is high, about 90 per cent. of the nitrogen of the urine is in the form of urea nitrogen; but on a low protein diet the proportion of urea nitrogen falls in the urine to about 60 per cent. of the total. This is shown in the table on page 750 taken from one of Folin's experiments.

The explanation of this variation of urea with the diet is that when more protein is eaten than is necessary to replace that decomposed in the vital processes in the body, the body does not store the excess, since there is no provision for the storage of an excess of protein, except in relatively small quantities. Instead of storing the excess, the nitrogen is split off from the amino-acids, converted into urea and excreted, while most of the carbonaceous part of the amino-acid molecule is converted into glucose, or fat, and stored in that form. Hence, when a very heavy protein diet is consumed, the amount of urea increases enormously and proportional to the protein consumption.

The fact that there is only a very limited storage of protein in the body is of very great significance. It indicates that the proteins play a different rôle from the fats and carbohydrates, probably because the proteins make part of the living matter of the cells. It is impossible to increase very much this vital matter without at the same time increasing the surface of the body, increasing its supply of oxygen and in other ways providing for its needs. Lifeless protein for storage evidently does not exist in the body, except in limited amounts: for example, in the connective tissue fibers, possibly the proteins of the blood, and some other supporting tissues.

Origin of the urea in mammals.—What is the origin of the urea found in the urine? In what organ is it formed? One turns first to the kidneys. Do the kidneys form the urea or do they only excrete it?

A definite answer to this question is obtained by extirpating the kidneys and examining the urea content of the blood before and after extirpation. If the kidneys are the chief or sole producers of the urea, then there will be no accumulation of urea in the blood after the extirpation of these organs; if, on the other hand, they simply excrete the urea which is formed elsewhere in the body and brought by the blood to them, then urea will accumulate in the blood, if there is no other organ which can take over the excretion. These experiments were tried by von Schroeder and some other experimenters. They showed that there was always an accumulation of urea in the blood in cats, dogs and other animals after kidney extirpation. Mammals survive loss of the kidneys for about three days. The cause of death which ensues is not yet certain. Von Schroeder and others got the following results:

Observer	Animal	Per cent of urea in the blood	
		Before kidney extirpation	After extirpation
v. Schroeder	Dog	0.045	0.208
Prevost and Dumas	Dog	....	0.830
" "	Cat	....	1.040
Hammond	Dog	0.026	0.083 (24 hours) 0.093 (48 ") 0.097 (61 ")
Voit	Rabbit	....	0.388 (60 ")

These experiments showed that in the absence of the kidneys urea accumulates rapidly in the blood, and, up to a certain point, the longer the animal lives the greater the accumulation becomes. A similar accumulation has been observed in human beings who have had anuria following or accompanying nephritis. When the kidneys fail to eliminate urea and it accumulates in the blood, the amounts in the other secretions greatly increase. Thus it may crystallize out on the skin on the evaporation of perspiration; urea goes also into the saliva, and particularly into the secretions of the duodenum and the intestine, since the intestine is one of the most important excretory organs of the body. In prolonged diarrhea or vomiting in nephrectomized animals the amount of urea in the blood may be considerably reduced.

From these observations we may conclude that the kidneys are not the chief organs for the production of urea and that it must be formed elsewhere in the body.

It would seem at first glance easy to solve a problem of this sort, since it would appear only necessary to examine the amount of urea in the blood before and after passing an organ in order to tell whether urea was produced there. If the blood coming away from the organ has more urea than that going to it, it must be formed in it. Actually, however, this method is seldom feasible, since the blood goes so rapidly through an organ as to take up at any one passage a very minute amount of the substance sought, an amount so small as to lie within the limits of experimental error. But while the amount taken away from any organ at any one circuit may be small, the total for the day may be large. Nor can we solve the problem by simply analyzing the organs and determining the amount of urea they contain, concluding that the organ with the highest urea content probably is the source of the substance. In the first place, the determination with accuracy of minute amounts of substances in such an albuminous fluid as the blood, or in any organ, is beset with many difficulties. Moreover, with such a soluble substance as urea which passes in and out of cells with ease, and which does not apparently form a union with the colloids of the cell, there is little accumulation, the stuff being excreted as rapidly as it is produced. But could we send the blood repeatedly through a single organ, the blood passing through the organ again and again, we might finally secure a considerable accumulation of the metabolic products of that organ in the blood. This we may do in the following way by perfusion: By taking the still living organ out of the body and establishing through it an artificial circulation of defibrinated, warm, arterial blood, the same blood passing through the organ again and again, we may finally secure so great an accumulation in the blood of any metabolic substances the organ may form that their nature may be determined. For such observations the organ to be examined is placed in a warm, moist chamber and as quickly as possible after removal from the body an artificial circulation of defibrinated blood, which has been arterialized by shaking with air, is sent through it. The blood is collected from the vein, shaken with air, warmed and injected over and over again, and this is repeated for many hours. We can, if desired, add certain substances to the blood and see how they are affected in passing through the organ and thus get an idea of the chemical powers of the organ.

This method, which looks so simple, is by no means without its drawbacks and difficulties. In the first place, many organs withstand very badly deprivation of their normal blood supply even for so short a time as twenty minutes. The intestines appear to be particularly sensitive in this regard, and the brain is also. A far more serious trouble comes from the fact that the organs are cut off from the nerve impulses which normally constantly impinge upon its cells and adjust the amount of blood coming to each part of the organ to the needs of that part. For no organ functions as a whole. Every gland has usually some alveoli in activity; some at rest. The circulation is adjusted to the needs of each alveolus. The same is true of smooth muscle which is constantly active, such as that of the intestine. The fibers take turns working and resting, so that at any instant of time some are at rest, others at work. Now in an organ taken out of the body, the blood vessels, as a rule, dilate and there is no control of the blood supply. The gland or other organ swells; it is apt to become œdematous; the circulation becomes smaller and smaller and finally stops entirely, or there may be extravasations of blood into the tissue. It has been found advantageous, also, to cause a rhythmic motion in the perfused blood to imitate as nearly as possible the natural conditions. Then, too, defibrinated blood is abnormal blood. It is not impossible that one of the main functions of the fibrin of the blood may be to regulate the viscosity of the blood in its passage through an organ, and this regulation will be lacking in the blood when it is defibrinated. For all these reasons the perfusion method, which at first glance promises so much for the study of the metabolism of individual organs, is subject to serious drawbacks and requires to be perfected.

With all its drawbacks, however, the method has thrown some light on the chemical possibilities of many organs of the body and will no doubt do far more when it is made a method of as great precision as it is capable of being made. Perfusion experiments have been carried out by von Schroeder, Salomon and others for the purpose of determining the origin of the urea. The following experiment illustrates the results obtained:

Blood Per Cent. of Urea.

	Before perfusion	After perfusion
Kidneys	0.0402	0.039
Muscles	0.014	0.0137
Liver	0.045	0.0812
	0.0538	0.1177 (3 hours perfusion)
		0.1253 (5 hours ")
Liver	0.0193	0.0236 (No $(NH_4)_2CO_3$ added)
		0.0599 ($(NH_4)_2CO_3$ added)

From these figures it is seen that neither the kidneys nor the muscles added urea to the blood during perfusion; but the liver did, and particularly when ammonium carbonate had been added to the blood before perfusion. The blood after passing the liver is uniformly richer in urea than when it entered it. This experiment shows very clearly that the liver has the power of forming urea and that it is capable of transforming ammonium carbonate into urea.

One cannot conclude from this experiment that the other organs have not the power of making urea. A positive result is convincing, but a negative result may mean many things. It might be, for example, that the liver happened to have a larger amount of precursors of urea in it than the other organs and that, if the proper forerunners of urea had been added to the blood, the other organs also would be found to have this power of urea formation. As a matter of fact, there is reason for believing, as we shall see in a moment, that other organs than the liver can make urea.

A second method of attacking a problem of this kind, and a very important method in physiological chemical research, consists in leaving the organ in the body, but in some way shunting the blood about it so that very little or no blood enters the organ and then examining the subsequent changes in the composition of the blood, or the excretions of the body. For the liver such experiments are difficult to perform. If, for example, the liver is taken out of the body of a mammal it dies in the course of a few hours. The cause of the death is still unknown and would probably well repay study. The liver has a double blood supply, getting arterial blood from the hepatic artery and venous blood from the portal vein, which has gathered the blood from all the intestine and its glandular annexes. About one-third of the blood from the liver is estimated to go through the hepatic artery, and the other two-thirds through the portal system. The first question which we will ask ourselves is: What will be the effect on the urea excretion of cutting off the blood supply coming through the portal vein? This supply cannot be cut off in mammals by simple ligature of the vessels and have the animals live for more than three or four hours. The blood accumulates in the intestinal area, there being no way for it to get back to the heart. It is necessary to so arrange the blood vessels as to send the blood around the liver and back into the circulation. Such a result can be accomplished by the so-called Eck fistula.

Eck fistula. The portal vein and the inferior vena cava run side by side before the former enters the liver. If one makes a slit in the adjoining sides of the veins and sews together the edges of the slits, the two veins are united and the blood can pass from one to the other through the hole, or fistula. This is called the Eck fistula, from its inventor. With modern technique it is not a very difficult operation to perform in dogs. After the fistula is made the portal vein beyond the fistula is ligatured and now the blood from the intestinal region no longer passes through the liver, but crosses through the opening into the inferior vena cava and back to the heart through that vessel. Animals so operated upon may live for months or years. The liver still has its circulation through the hepatic artery. The results of this operation on the liver cells is striking, since they shrink in size and look quite abnormal under the microscope.

As a result of this operation the blood from the intestine no longer is subjected to the action of the liver, but passes directly to the body cells and to the kidneys. The urine should show some change in composition. If the urea is formed in the liver and in the liver alone from raw materials brought from the intestine, marked changes in the excretions should result. Pawlow and Nencki and Hahn, who early studied this question, reported a marked change in the proportion of urea and ammonia nitrogen in the urine as the result of making an Eck fistula. They recorded the following ratios of ammonia and urea nitrogen in various experiments:

RATIO OF NH_3 TO UREA IN DOG'S URINE.

Before the Eck fistula	After the Eck fistula
1 : 73	1 : 33
	1 : 24.5
	1 : 16.1
	1 : 7.6
	1 : 9.9

There was, in other words, a great decrease in the urea nitrogen and an increase in the ammonia nitrogen of the urine. They found in the blood and urine of their Eck dogs ammonium carbamate, NH_4—O—CO—NH_2. There seemed, in other words, to be a failure of the body to convert ammonium compounds and carbamic acid into urea. In some of the experiments the changes were not so pronounced as those given in the foregoing table. The operated dogs when fed meat seemed to be poisoned. They became excitable, attempted to climb up the sides of the cages and showed symptoms like those of ammonia poisoning. The following experiment quoted from these observers will illustrate this:

An Eck fistula was made in a bitch on the 23d of January, 1892. She weighed 23.674 kilos. On March 11 her weight was 14.169 kilos, a loss of 9 kilos. She had been on a diet of bread and milk. On March 1 she was given 1,200 c.c. of milk and 200 grams meat powder. A stage of excitation ensued on the evening of this day; she turned constantly in the cage and tried to climb up the wall and bit objects near her. She seemed to be blind. The next day she received no meat. The weakness of the muscular system continued and the uncertain and irregular gait. The animal appeared to feel no pain. In three days on a bread and milk diet the pathological symptoms disappeared, but on repeating the experiment the animal died. The urine was very alkaline.

The authors go on to say: "We have observed, therefore, an undoubted and very characteristic fact: dogs in whom the blood has been shunted around the liver by means of the Eck fistula, so that it flows directly from the intestine without going through the liver, cannot stand meat without showing serious disturbance of the nervous system which may often result in death. It appeared from these experiments that one function of the liver was to detoxicate the blood; to take out of it ammonia which if it passed the liver would be extremely harmful." These experiments also indicated, like those of von Schroeder, that the liver was the chief source of the urea of the urine, since, when the blood supply was reduced, urea diminished and its place was taken by the carbamate of ammonium. Eck fistula dogs do not always show these symptoms. Sometimes they live hearty and well for months, particularly if care is taken to provide them with bones. A diet of bread and milk is not normal for a dog; they develop diarrhea on it. Dogs may live normally without loss of weight, or only slight loss after the Eck fistula if fed bread, meat and bones. Adhesions always form in these operations, and in the adhesions small blood vessels may grow into the liver from the pancreas or intestine, and it is possible that the nerves may at times be involved in the adhesions so that the interpretation of results is not always easy. As far as they go, however, the experiments cited bear out the conclusion that the liver forms a good part of the urea of the body.

Urea formation after corrosion of the liver and disease. If the liver cells could be injured, the metabolism might be changed in such a way as to throw some light on the question whether the liver is the sole source of the urea. The liver cells may be thus damaged in various ways either by disease or by the injection of corrosive or poisonous substances into the bile ducts. In acute yellow atrophy of the liver, interstitial hepatitis and cirrhosis of the liver, there is a very extensive degeneration of the liver cells. In all of these cases there is a reduction of the amount of urea in the urine, and an increase in the ammonia content. The change, however, is not so great as one would expect were the liver the sole source of the urea. In phosphorus poisoning one may have a great increase in the ammonia and a decrease in the urea. But in all these cases it is difficult to distinguish cause and effect. For whenever there is a production of acid in the body, there is always an increase in the ammonia of the urine, even though the liver cells be intact. In phosphorus poisoning there is such an acidosis. In acute yellow atrophy the nitrogen of the ammonia may amount to as much as 70 per cent. of that of the urea, whereas normally it is not more than 7 per cent. These experiments also indicate, then, that the liver is an important source of urea.

Experiments have also been tried of giving ammonium carbonate or citrate to patients suffering from liver disease, but the results have not been concordant. It seems, however, that except in the last hours of life in such cases the body is still able to convert the ammonia thus ingested into urea.

Experiments in the corrosion of the liver by the injection of sulphuric acid into the bile ducts of dogs have been tried by Pick. The liver cells are injured, but the circulation persists. Dogs thus treated remain apparently normal for 24 hours; they then become comatose and die in some six hours more. An analysis of the urine of dogs thus treated gave the following results:

Before corrosion		After corrosion	
Urinary N per cent. as HN_3 N	Per cent. as urea N	Per cent. as NH_3 N	Per cent. as urea N
1.51	98.23	3.93	82.6
		5.21	81.88
2.17	75.42	3.62	84.12
5.01	76.91	5.4	76.65
4.59	81	9.37	60.29
5.6		3.36	
2.95		1.51	

As a rule, the per cent. of urinary nitrogen as ammonia increased after the corrosion, but the difference was not marked.

Extirpation of birds' livers. Now, although in mammals the liver cannot be removed without causing speedy death, nor can the liver cells be killed with sufficient accuracy to yield sharp and certain results, it is possible in birds and reptiles to remove the liver without leading to the immediate death of the animal. It happens that in birds there is a connection between the portal and the renal circulation so that blood can go back to the circulation from the intestine after removal of the liver. Geese thus operated upon recover very quickly. They set at once to smooth their feathers and eat readily. After fifteen to twenty hours, however, or even earlier, symptoms of serious trouble set in. The animals stagger, become comatose and soon die. They live long enough, however, to answer the question as to what happens to the metabolism when the liver is gone. In birds and reptiles the urea of the urine is replaced by uric acid. There is a very small amount of urea excreted, but the urine consists of damp masses of crystallized urates. It has been shown by other observers that the uric acid of birds is formed from the same substances as the urea of the mammal. Thus, if urea is fed to birds, it is excreted in the form of uric acid. Ammonium compounds also are formed into uric acid. The uric acid of birds is, then, fairly commensurate in metabolism with the urea of the mammal, and presumably it is formed by the same organs of the body. Minkowski extirpated the livers of geese and studied the changes produced in the urine with the following result:

	Per cent. of total N as Uric Acid N	Ammonia N
Before	60-70	10-18
After	3-6	45-60
Before	50	
After	3.5	

From the results of these experiments it is seen that after removal of the liver from birds, uric acid disappears from the urine to a small remnant and is replaced by ammonia. The small amount of uric acid which appears after the extirpation may be the residue which had accumulated in the body and is slowly excreted. There is no doubt, from this experiment, that uric acid is formed in the bird's liver and not elsewhere in the body in any amount. Since the uric acid of birds clearly corresponds to the urea of mammals, this is an additional indication of urea formation by the liver.

Other sources of urea. The liver is not the only source of the urea, however, for dogs having an Eck fistula and with the hepatic artery ligated still have the power of increasing their urea output when amino-acids are injected under the skin (S. A. Matthews). Some of the other organs, perhaps all of them, certainly have the power in these animals of forming urea. Nevertheless, the evidence is that the liver is the main source of the urea of the urine.

The precursors of urea.—From what substances, then, is urea formed? In the first instance it is undoubtedly formed from ammonia. Not only does ammonia appear in the urine in larger amounts than normal in the total or partial absence of the liver, but direct determination of the ammonia of the blood before and after passing the liver shows that this substance is removed from the blood by this organ. It will be remembered that in the course of protein absorption and digestion some ammonia is split from the proteins. A part of this ammonia is split from the amide linkings of the proteins during digestion and by the action of the bacteria in the intestine; a part is formed from the partial deamidization of the amino-acids during the process of absorption; still another portion may arise in the liver itself by the action of the deamidizing enzymes which probably occur there. The blood of the portal vein is rich in ammonia. This ammonia comes in part from the intestine and in part from the pancreas, which also has a very large amount of ammonia in it. The relative amounts of ammonia in blood of the portal and hepatic veins are shown in the following table:

Mgs. NH_3 per 100 Grams Blood. (Pawlow, Nencki, Sieber.)

Portal vein	Hepatic vein
3.5	1.5
8.4	. .
5.6	1.1
12.0	1.9
4.0	1.8
3.5	1.9

The amount of ammonia in the portal blood is very much more than that in the hepatic blood, showing that the liver removes ammonia from the blood.

Various tissues had the following amounts of ammonia in mgs. per 100 grams of fresh tissue:

Lungs	Muscle	Liver	Intestine mucosa	Stomach mucosa	Intestinal contents	Stomach contents	Pancreas
1.1	0.2	22.8	23	37.1	42.6	10.4	8.8
	19.4	21.2	48.9	43.2	22.4	9.9	7.9
			41.7	52.8	40.2	24.3	16

The very high figures for the mucosa of the stomach will be noticed. It always contains more than the contents. The intestinal contents generally have more ammonia than the mucosa. Blood from various parts of the body had the following amounts of ammonia:

MGS. AMMONIA IN 100 GRAMS OF BLOOD OR ORGAN.

Hepatic vein	Portal vein	Pancreatic vein	Liver	Stomach mucosa
1.8	4.0	8.2	12.2	44.0
1.5	3.5	0.25	13.7	31.8

The carotid artery blood of a dog on a meat diet contains 1.5 mgs. of ammonia per 100 grams of blood. On fasting it may fall to 0.38 mgs.

Nencki, Sieber and Pawlow conclude their article with the following words: "The liver is, therefore, the true guardian of the organism, which changes the poisonous substances coming from the alimentary canal into harmless substances; since that which is true of ammonia may be assumed to be true also for the substituted ammonias, such as various plant alkaloids, bacterial poisons, etc."

From the foregoing it is apparent that all the organs contain more ammonia than the blood. A part of the nitrogen probably escapes from them in this form to the blood. Experiments show that when the ammonium salts of organic acids are fed to animals they increase the urea excretion. But ammonia is not the only substance which can serve as a precursor of urea. The same is true of any amino-acid and many other substances which contain amino nitrogen. The following results were obtained by von Knierem in the excretion of uric acid in hens when they were on a constant diet and fed an additional ration of various substances shown in the table:

Substance added to the diet	Uric acid excreted in grams	
	Before	After adding to diet
Asparagine	0.7601	2.4382
Aspartic acid	0.9673	1.5596
Glycocoll	1.3012	1.8010
Leucine	1.001	1.6282
NH_4Cl	1.8000	1.9728
$(NH_4)_2SO_4$	1.140	1.190

Very closely similar results were recorded by Schulzen and Nencki after feeding dogs glycocoll or leucine. The urea excreted rose from 3.8 grams per day to 8.3 grams per day after glycocoll was fed; a similar rise occurred with tyrosine and ammonium carbonate and formate.

Another precursor of urea in the mammal is arginine. This amino-acid may be split by an enzyme, arginase, found in the liver and intestine, but not in muscle, into urea and ornithine.

$$NH_2—\underset{\underset{NH}{\|}}{C}—NH—CH_2—CH_2—CH_2—CHNH_2—COOH + H_2O \longrightarrow$$

Arginine.

$$\underset{\text{Urea.}}{NH_2—\underset{\underset{NH_2}{|}}{CO}} + \underset{\text{Ornithine.}}{NH_2CH_2—CH_2—CH_2—CHNH_2—COOH}$$

In some animals, too, although it is doubtful whether it occurs in man, there is an oxidation and hydrolysis of uric acid to urea. This is probably, if it occurs, an unimportant source of human urea.

Summary. In the process of digestion and absorption of protein food large amounts of ammonia are set free. This ammonia is liberated in part as the result of the action of the enzymes of the digestive juices; in part by the action of the bacteria in the intestine, and in part, during the process of absorption, from the amino-acids of the proteins of the food. The ammonia thus liberated accumulates in the mucous membrane of the stomach and intestine, whence it is gradually removed by the blood and carried to the liver in the portal blood. The liver converts at least part of it into urea and the kidney eliminates it from the body. Another portion of ammonia is set free in the liver itself from the decomposition of the amino-acids brought to that organ in the portal blood, and the carbon moieties of those molecules are converted in part into glucose and glycogen. A relatively large fraction of the nitrogen of the food thus never becomes part of the living protein of the body at all, but is converted into urea and thrown out of the body. On a meat diet the amount of nitrogen thus converted into urea, so called superfluous nitrogen, is relatively large. It is not certain just how large it is, but it probably is at least a third of all the nitrogen in the urea, since on fasting the urea drops at once to about two-thirds its former value. Certainly with an output of 35 grams of urea per day this digestive nitrogen will represent at least ten grams of urea.

But this is not the only origin of the urea. The tissues are also capable of tearing their protein to pieces and deamidizing the amino-acids. For example, if a muscle has not a sufficient amount of carbohydrate to supply its energy, it will tear its protein to pieces to get the energy. This is always accompanied by a process of oxidation which forms a ketonic acid from the amino-acid and ammonia is set free in the way mentioned on page 123. Lactic acid is formed in muscles during their work; some ammonia probably escapes from the muscles in the form of ammonium lactate and this is in part made into urea. The endogenous urea has, therefore, a very varied origin, coming from the different organs in the form of a variety of nitrogenous substances which are somewhere, possibly in the liver, made in part into urea. Even on the minimum protein diet of only three or four grams of nitrogen intake per day the urea still makes 60 per cent. or more of the total nitrogen in the urine.

It is a very interesting observation that when large amounts of benzoic acid are fed to animals the excretion of urea is reduced and in place of the urea there appears the glycocoll of hippuric acid. Now the fact that under these circumstances glycocoll takes the place of urea has led some to think that possibly ammonia is synthesized into glycocoll normally before it goes into urea. This is not at all improbable. Glyoxal, COH.COH, and glycolaldehyde, $CH_2OH.COH$, are probably formed in the decomposition of the carbohydrates. With the former ammonia will condense as follows to form glycocoll:

$$\underset{\text{Glyoxal.}}{\begin{array}{l}H—C=O\\ \quad|\\ H—C=O\end{array}} + NH_3 \longrightarrow \begin{array}{l}H—C=NH\\ \quad|\\ H—C=O\end{array} ; + O \longrightarrow$$

$$\begin{array}{l}H—C=NH\\ \quad|\\ CO.OH\end{array} ; + H_2 \longrightarrow \underset{\text{Glycocoll.}}{\begin{array}{c}H\\ |\\ H—C—NH_2,\\ |\\ COOH\end{array}}$$

Other processes which may lead to the formation of the urea have, however, been suggested. Drechsel suggested that it is formed in part from ammonium carbamate by an alternate oxidation and reduction and he synthesized it in this manner by means of a very rapidly alternating electrical current.

$$1.\ NH_4—O—\overset{\overset{O}{||}}{C}—NH_2 + O \longrightarrow NH_2—O—CO.NH_2$$

$$2.\ NH_2O—CO.NH_2 + H_2 \longrightarrow \underset{\text{Urea.}}{NH_2—\underset{O}{C}—NH_2}$$

Salkowski has suggested that it is formed by the transformation of cyanamide:

$$\underset{\text{Cyanamide.}}{NH_2—C\equiv N} + H_2O \longrightarrow \underset{\text{Urea.}}{NH_2—CO.NH_2}$$

It is probable that it originates in a number of different ways, and in other tissues as well as the liver.

Physiological action of urea. Urea has several quite important physiological actions. In the first place, it is a natural diuretic and whenever there is an increase in the urea excretion there is always, other things being equal, an increase in the amount of urine. A person on a heavy protein diet will excrete per day between 1,000 and 2,000 c.c. of urine, except in very hot weather, when there is a great loss of water through the skin. On the other hand, a person having a small intake of protein, and so a small excretion of urea, will secrete not more than 300-500 c.c. per day. When one observes in a patient a small secretion of urine, it is always desirable to inquire into the nature of the diet before concluding that the kidneys need stimulating and giving a diuretic. Many people secrete this small amount of urine for years without any indication of uræmia or other symptoms indicative of abnormal secretion. Whether there is any advantage in a low or a high secretion of urine has not yet been determined. It is possible that by stimulating the secretion of the urine the drinking of more water results and a kind of catharsis of the cells of the body might be produced which might be either advantageous or disadvantageous to them.

Urea has also a very definite function in the cells of the elasmobranch fishes and possibly in the mammalia also. It is one of the normal constituents of the cell and of the blood and other fluids of the body, and since these cells have for long years of time been selected to work with the highest degree of efficiency in this urea-containing medium, it is found that the addition of a little urea to artificial perfusion solutions, when one is perfusing the heart or other organs, is as a rule advantageous. The effect of such an addition to the salt solutions used to sustain the heart-beat of fishes, both teleosts and elasmobranchs, is very marked. Tissues live much longer in the presence of some urea than in its absence. The effect on mammalia is much less marked, since the amount of urea in the blood of mammals is very small, .02-.04 per cent. The effect produced by the addition of urea to the artificial salt solutions in the elasmobranchs is that of a stimulation and the same effect may be produced by the addition of small amounts of ammonium carbonate. It is possible that in these animals there may be some conversion of urea into ammonium carbonate, or vice versa, and the effect may be due to the action of the ammonium carbonate in neutralizing acids. Urea is not, then, entirely inert.

Creatine and creatinine in the urine.—One of the most interesting of the nitrogenous substances found in urine is creatine and its anhydride, creatinine, creatine being methyl-guanidine acetic acid. In the urine of the average human adult there is present between 0.8 and 2 grams of creatinine a day. Creatine occurs in the urine of healthy men only in very small amounts, or it may be entirely absent if no creatine is taken in the food, but it is found in larger quantities in the urine of women and children even when they are taking no creatine in the food, and in men's urine the amount is increased by fasting and in disease. Both creatine and creatinine occur in the urine of other mammals. The amount of creatinine excreted daily by human beings varies with the weight, muscular development, state of health, sex and age, and slightly with the creatinine and creatine intake, *but is almost or quite independent of the protein intake of the body.* These facts show that creatine and creatinine have a very different significance from urea and that they stand in a very special relation to the fundamental metabolism of the body.

Chemistry. Creatine. Creatine (Greek *kreas*, flesh) is methyl-guanidine-acetic acid and has the formula $C_4H_9N_3O_2$ or:

```
NH2—C —N —CH2—C—OH
    ||  |      ||
    NH CH3     O
```

Creatinine is the anhydride of this: i.e., $C_4H_7N_3O$, or

```
        NH—CO
        |   |
  NH = C    |
        |   |
     CH3N —CH2
```

Creatine is thus related on the one hand to arginine, which is δ-guanidine-α-amino-valerianic acid; from which it might possibly be derived by oxidation and methylation; and on the other hand creatinine may be regarded as an imidazole derivative. It is thus allied to histidine and to the purines, both of which contain imidazole rings. Creatine may also be considered as a ureide of methyl glycocoll (amino-acetic acid), or sarcosine, which relates it to the betaines, methylated glycocoll derivatives found in many plants and represented in an extreme form of methylation by choline in animals. Creatine is also methyl-glycocyamine, glycocyamine being guanidine-acetic acid. It may be synthesized with great ease by the direct union of cyanamide, $NH_2.CN$, with methyl-amino-acetic acid, or sarcosine. If a strong aqueous solution of amino-acetic acid and cyanamide is made slightly ammoniacal at room temperature, glycocyamine crystallizes out. The reaction is as follows:

$$\underset{\text{Cyanamide.}}{NH_2.CN} + \underset{\text{Glycocoll.}}{H_2N.CH_2{-}COOH} \longrightarrow \underset{\text{Glycocyamine.}}{NH_2{-}C({=}NH){-}NH{-}CH_2{-}COOH}$$

The same reaction carried out with methyl-amino-acetic acid[9] gives creatine.

$$NH_2.CN + HN(CH_3)—CH_2—COOH \longrightarrow$$
$$NH_2—C(=NH)—N(CH_3)—CH_2—COOH$$

Creatine.

It will be noticed that cyanamide is an anhydride urea and can be prepared from urea by the action of sodium; and cyanamide yields urea when treated with 50 per cent. sulphuric acid. It would seem possible that urea might give a similar synthesis with amino-acetic acid:

$$1.\quad NH_2—CO—NH_2 \longrightarrow NH_2—C(OH)=NH$$

$$2.\quad NH_2—C(OH)=NH + NH(CH_3)—CH_2—COOH \longrightarrow$$

Urea. Sarcosine.

$$NH_2—C(=NH)—N(CH_3)—CH_2—COOH + H_2O$$

Creatine.

It is by no means impossible that small quantities of cyanamide may be present in urea solutions and this synthesis is one which might possibly occur in the body where both urea and sarcosine are to be found.

Creatine is a colorless, bitter, biting substance crystallizing in rhombic prisms with one molecule of water, which is lost by heating to 100°, the crystals then becoming white and opaque. It dissolves in 74.4 parts of water at 18°, but is much more soluble in hot water, almost insoluble in alcohol and insoluble in ether. Its aqueous solutions are neutral in reaction. It occurs not only in the urine, but in most organs and especially in the voluntary and involuntary muscle of vertebrates and some invertebrates. It has been isolated also from the blood, brain, liver, testis, transudates and the amniotic fluid. It has a stimulating action on the central nervous system, and methyl-guanidine, one of its decomposition products, is very toxic.

By hydrolysis with boiling barium hydrate creatine yields methyl-hydantoic acid, urea, methyl-amino-acetic acid and carbon dioxide. Probably it forms first by hydrolysis of an amino group the intermediary substance, $HO—C(=NH)—N(CH_3)—CH_2—COOH$, which by molecular rearrangement is transformed into methyl-hydantoic acid, $NH_2—CO—N(CH_3)—CH_2—COOH$. By oxidation by permanganate or hydrogen peroxide it forms urea and methyl-amino-acetic acid, or sarcosine. Arginase does not decompose it. By boiling with mercuric oxide it reduces the latter to metallic mercury and the creatine is oxidized to methyl-guanidine and oxalic acid. Creatine has, therefore, some reducing action, but it is far less marked than that of creatinine. It is precipitated in neutral solution by mercuric nitrate. It is not precipitated by cadmium chloride, phosphotungstic acid, or by lead acetate. Formaldehyde converts creatine to dioxymethylene creatinine.

Creatinine. Creatine goes over spontaneously into its anhydride form of creatinine, the reaction being similar to the transformation of glycocyamine to glycocyamidine.

$$NH_2—C(=NH)—N(CH_3)—CH_2—COOH \longrightarrow \begin{array}{c} NH—CO \\ NH=\overset{|}{C} \quad | \\ CH_3—\overset{|}{N}—CH_2 \end{array} + H_2O$$

This reaction is hastened by various tissue extracts, but it occurs if an aqueous solution of the creatine is boiled, or by the action of acid. A solution of creatine in half-normal hydrochloric acid heated in an autoclave for 30 minutes to 117° goes over almost quantitatively into creatinine (Benedict). The synthesis of creatinine from creatine is a synthesis of an amino and carboxyl group similar to the synthesis of a polypeptide from the amino-acids. The power of making such syntheses is an attribute of all living matter without exception. This synthesis is also instructive as illustrating the formation of a ring compound from an aliphatic. A similar synthesis, that of proline from glutamic acid, has already been discussed on page 124. In the latter case there is the formation of the pyrrollidine ring from glutamic acid by a union of a carboxyl and amino group, followed by the reduction of the ring which thus is rendered stable. The proline ring, like that of glycocyamidine, is also often methylated to form alkoloids of the type of stachydrine,

$$\begin{array}{ccc} CH_2 & — & CH_2 \\ | & & | \\ CH_2 & & CH—CO \\ & \diagdown\ \diagup & | \\ & N & —O \\ & \diagup\ \diagdown & \\ CH_3 & & CH_3 \end{array}$$

Stachydrine.

which occur in plants. Whether it is possible to reduce the creatinine ring in the body to form the imidazole ring is still unknown. The creatinine ring being thus an oxidized ring is unstable and creatinine in an alkaline solution is less stable than creatine from which it is derived.

Creatinine crystallizes in colorless, monoclinic prismatic crystals. It is readily soluble in water. It forms with zinc chloride a good crystallizing, not very soluble compound by which it may be identified, $(C_4H_7N_3O)_2ZnCl_2$. It is a strong base, the aqueous solution being alkaline; stable in acid solution, but not in alkaline. It is a reducing substance, reducing alkaline solutions of copper, silver and mercuric salts, but not reducing bismuth oxide. It reduces picric acid to the reddish picramic acid in alkaline solution, and it is determined quantitatively colorimetrically by means of this reaction. See page 962.

Origin and significance of the creatine and creatinine of the urine. The origin and significance of the one or two grams of creatinine and the few milligrams of creatine secreted daily have been keenly investigated of recent years, but have not yet been brought to a complete solution. Creatine is found in a great many organs of the body. It was discovered by Liebig as a constant constituent of voluntary muscle. It and creatinine are found in Liebig's beef extract, there being usually about twice as much creatine in it as of creatinine. Recent analyses by Beker have yielded the figures shown in the table.

AMOUNTS OF CREATINE AND CREATININE FOUND IN VARIOUS ANIMALS.

Animal	Organ	Mgs. creatine and creatinine computed as creatinine per 100 grams of the organ
Ox	Liver	25.3-37.24
Calf	"	26.54
Rabbit	"	18.9-21.2
Cat	"	20.6
Goat	"	11.2
Pig	"	15.72-17.46
Ox	Pancreas	12.51-19.62
Dog	"	12.9-16.13
Pig	"	10.69-14.2
Ox	Thyroid	11.4
Ox	Kidneys	12.26-17.6
Dog	"	10.28-16.34
Pig	"	15.2
Ox	Spleen	14.67
Dog	"	13.28-19.5
Calf	Thymus	9.76
Bull	Testis	76.4-97.2; 181, dog testis, Janney and Blatherwick.
Ox	Brain	51.4-63; 112, dog brain, Janney and Blatherwick.
	White matter	47.9-56.2
	Cerebellum	64.2-71.3
Dog	Brain	54.6-57.5
Ox	Voluntary muscle	403 (Average of 20 determinations)
Dog	"	314 " " 6 "
Pig	"	338 " " 3 "
Rabbit	White muscle	451 " " 4 "
"	Red muscle	326
Cat	"	354
Goat	"	316
Ox	Heart	220
Dog	"	228-257.4
Rabbit	Intestinal muscle	
"	Large intestine	32.5
"	Small intestine	23.4
Ox	Uterus	29.9-43
Pig	"	29.9-31.2
Ox	Blood	1.93-2.68
Dog	"	1.86-2.44
Pig	"	2.00-2.08

A fundamental and significant fact concerning the excretion of creatine and creatinine is that the amount of these excreted in the urine per day is independent of the amount of protein taken in the food and is wonderfully constant for each individual. This fundamental observation by Folin has been the starting point of the modern investigation of these substances. It shows clearly that creatinine has a different significance in metabolism from urea, of which the excretion varies directly with the protein intake. A second fundamental fact is that the excretion of creatine and creatinine in adult men is almost independent of the creatine intake, and that creatinine taken in the food appears in large measure in the urine as such. The third fact is that age, sex, state of health and starvation profoundly affect the excretion of these bodies, and the amount of creatinine excreted varies directly with the weight of the body—at least within certain limitations this is true. The significance of this latter fact will be apparent if it be remembered that nothing of the sort is true of urea, since a very small person, if he eats much meat, may excrete more urea than a large one on a more restricted diet. These facts show that creatine and creatinine are products of the endogenous metabolism of the body.

That the excretion of creatinine in male adults is independent of the amount of the protein ingested is shown by the following observations by Folin. The figures represent the excretion of an individual on two occasions, once when on an ordinary protein diet containing about 120 grams of protein per day; and the other time on a restricted protein diet when only about 20 grams of protein are ingested.

	Usual protein intake	Low protein intake
Vol. of urine	1170	385 c.c.
Total N	16.8	3.60 gr.
Urea N	14.7	2.20
Creatinine N	0.58 (3.60%)	0.60 (17.2%)

The creatinine excretion remained practically the same, although the protein was reduced so enormously. The contrast with the urea excretion is profound. The creatinine nitrogen made in the one case 3.6 per cent. of total nitrogen; and 17.2 per cent. in the other.

But, while the total amount of creatinine secreted is independent of the protein intake, it is closely dependent on the bodyweight and chiefly upon the muscular development of the body, although other factors affect it also. The average creatinine excretion of Dr. H., weighing 87 kilos, was 1.6 gram; Dr. A., weighing 56 ks., eliminated 1.15 gram; and F., who weighed 70 kg., but is rather short and corpulent, excreted 1.4 gram. The analytical data indicate that moderately corpulent persons (adults) eliminate per 24 hours about 20 mgs. creatinine per kilo bodyweight; while lean persons yield about 25 mgs. per kilo. The number of milligrams of creatinine excreted in 24 hours per kilo bodyweight is called the *creatinine coefficient*. Sometimes the term is applied to the milligrams of creatinine nitrogen per kilo bodyweight. Shaffer found that the creatinine coefficient in the latter sense had a value of between 8.1-11; for children it usually lies between 3.3 and 6.5.

But while the creatinine excretion is independent of the protein of the food it is affected by the intake of creatine or creatinine. The ingestion of creatinine is followed by the reappearance of most of it in the urine. The following figures from Shaffer show both how constant the creatinine excretion is from hour to hour and how it is increased by the ingestion of creatinine. The urine was passed each two hours. The average excretion in grams of creatinine per hour in successive two-hour periods was as follows: .066; .062; .057; .065; .068. From 6.30 to 12 at night the average was .053; from 12 to 9 A.M., while sleeping, it was .056; from 9 to 11.15 A.M. it was .062. The amount, therefore, is less at night than in the daytime. Creatinine does not then undergo the wide hourly variation which uric acid does. At 10.40 A.M., while his hourly excretion was .067 gr. per hour, he took 250 c.c. of water containing 0.70 gram of pure creatinine. From 10.15-11.15 the hourly excretion of creatinine rose to .084; 11.15-12.15 it was .174; 12.15-2.15, .138; 2.15-5.15, .101; 5.15-12.06 A.M., .084; 12.06-7.20 A.M. it was .073. In this experiment 76 per cent. of the ingested creatinine had been excreted in 21 hours.

The ingestion of creatine leads to a very small increase in the creatinine and is generally, but not always, followed by the excretion of a small amount of creatine. This point has been the subject of a good deal of controversy. Folin made the extraordinary observation that ingested creatine did not increase either the creatinine, creatine or total nitrogen of the urine. The creatine disappeared and he believed it to be metabolized in the body and retained there. It was afterwards suggested by Mellanby that it had been retained by the bacteria of the intestine and he isolated from the intestine a bacterium which would metabolize the creatine of a solution. It is, however, not yet certain what has become of the ingested creatine. Further experiment has shown that there is a marked difference between children and women, on the one hand, and men on the other, as regards their power of metabolizing ingested creatine. Men usually destroy nearly all, or all of that ingested, so that no increase of the creatine, or creatinine, occurs. In some cases, however, there is a very small increase in the creatinine. This is perhaps the rule, and occasionally some creatine appears in the urine unchanged.

Amount of creatine excreted under various conditions. Any discussion of the excretion of creatine in the urine must be prefaced by the statement that the creatine is not determined directly, but by difference.

The preformed creatinine is determined by Folin's colorimetrical method, and then the creatine present, if any, is converted into creatinine by heating the urine with acid, and the creatinine is then redetermined. This is called the total creatinine. It is sometimes larger than the figures first obtained and the difference is supposed to be due to the conversion of some creatine to creatinine. The difference is hence called creatine. It is obvious, however, that there is a considerable uncertainty about determinations made in this way, and it involves the assumption that the heating with acid has not in any way changed the urine, so that it will affect the creatinine determination, except by the conversion of creatine to creatinine. When it is remembered that the quantitative determination of the creatinine itself is made by measuring its reducing action, a property which is not peculiar to it, it is clear that inferences as to the presence or absence of creatine in the urine must be very cautiously drawn. As a matter of fact it has recently been alleged that the appearance of the presence of creatine is really due to the presence of small amounts of acetoacetic acid, which in the unheated urine make the creatinine determination somewhat lower than it should be. This acid is destroyed by heating the urine with acid, so that the second determination of creatinine is larger. It is just this difference which is usually called creatine. It is a very suspicious circumstance that most of the methods which are supposed to increase creatine excretion are just those which are known to increase acetoacetic acid excretion. As it is at present controverted whether the acetoacetic acid present will account for the whole of the so-called creatine or not, the excretion of creatine is treated here as if it certainly occurred. It may be that the so-called creatine is really creatine.

EXCRETION OF CREATINE IN HEALTHY CHILDREN. GRAMS PER DAY.

Sex	Age	Urine c.c.	Total N.	Creatinine N.	Creatine N.
Boy	5	620	6.10	.112	.025
Boy	8	710	7.20	.163	0
Boy	15	1150	11.97	.378	0
Boy	11	460	5.31	.157	0
Boy	12	680	4.19	.086	0
Boy	2	300	3.00	.025	.023
Boy	3	580	5.48	.057	.022
Boy	16	625	8.10	.268	0
		1120	11.70	.374	0
Girl	5	500	4.20	.069	.006
Girl	6	500	2.33	.032	.003
Girl	7	830	9.75	.157	.066
Girl	10	825	8.50	.147	.020
Girl	12	510	8.68	.201	.011
		1000	6.70	.224	.042

Creatine is found in the urine of most if not all mammals. In human beings it occurs in large amounts, relative to creatinine, in the urine of children, but in small amounts only in the urine of adults, except under special conditions of sickness, or diet, when the amount of creatine may be at least half of the total creatinine and creatine together. The results included in the table on page 708 show the excretion of creatine and creatinine in healthy children (Krause) when on a creatine-free diet.

From this table it is seen that on a creatine-free diet boys stop secreting creatine at about seven years of age, and that before this age the amount of creatine secreted, which in the first years of life has been about equal to the creatinine, becomes relatively and absolutely smaller until it disappears. Girls, on the other hand, continue to secrete creatine until puberty, and with women it reappears in the urine at each monthly sexual cycle. It will be noticed, furthermore, that the absolute amount of creatinine secreted increases with the age of the child and there is also an increase in the amount of creatinine nitrogen in mgs. per kilo bodyweight. This last factor, the " creatinine coefficient," in adults on a creatine-free diet is 8.1 to about 11. In the experiments just quoted the creatinine coefficient is higher in children of 12 than in those of 5. The coefficient at ages of 13 to 16 is only 3.6 to 4.1. In infants, Amberg and Morrill have found creatinine coefficients still lower. At 7 to 14 days old the coefficient was 1.8 to 3.3. The increase in the creatinine coefficient thus goes *pari passu* with the relative development of the musculature. In infants the brain and liver and other organs make relatively a larger proportion of the body weight than later in life. Not only does creatine appear in the urine of children, but it has been shown (Krause) that their power of metabolizing creatine is less than that of adults. By feeding creatine to children on a creatine-free diet it was found that the younger the child the less ingested creatine was retained.

Sex	Age	Per cent. ingested creatine excreted as creatine
Girl	6	56
Boy	8	43
Girl	11	31

In the case of women creatine is often absent from the urine, but reappears during menstruation, and during pregnancy and after delivery it occurs there for several days. Benedict and Meyers found considerable quantities of creatine in the urine of insane women when on a milk diet, or a creatinine and creatine-free diet. The ages of the patients ranged from 19 to 95 years and the amount of creatine excreted per day ranged from 70-200 mgs., while the creatinine was from 292 to 700 mgs. per day.

The ingestion of creatine causes in children and women always an increase in the creatine of the urine and a slight increase in the creatinine. In men, however, large quantities of creatine may be taken by the mouth without causing any increase in the creatine in the urine or indeed in the total nitrogen excreted. This is particularly noticeable when the amount of protein in the food is low. This fact was discovered by Folin, who on successive days took several grams of creatine with no change in the creatine or creatinine in the urine.

Krause in various experiments on two male adults of 26 and 36 on a creatine-free diet and with a total nitrogen in the urine of from 3.27-13.44 grams, gave .052-.126 grs. creatine nitrogen per day without any creatine appearing in the urine. The creatinine was only very slightly increased, and in some cases not at all increased. The total nitrogen in the urine not only did not increase, but seemed, on the contrary, to decrease. Thus it fell in three of the four experiments from 3.27 to 3.14, from 13.44-12.55 and from 13.30-12.07. Similar results were obtained in the case of children, except that in them a part of the creatine ingested reappeared as such in the urine.

Weber ingested meat extract and found a greater increase in the creatinine than was taken in the extract, so that some of the creatine had gone over into creatinine. Van Hoogenhuyze and Verploegh, Plummer, Dick and Lieb always got a small part of the creatine as creatinine. Pekelharing and van Hoogenhuyze found that, if they gave dogs and rabbits creatine subcutaneously in small amounts, it always increased both the creatine and creatinine, but if they gave it all at once they got an increase only in the creatine. Towles and Voegtlin also found a transformation of some of the creatine to creatinine. The importance of these observations arises from the fact that they show that creatine is turned into creatinine in the body and hence indicate that the creatine of the body is probably the source of the creatinine.

Fasting causes a decrease in the creatinine, but an increase in the creatine, so that the total is very little affected. Even in men who do not normally excrete creatine in the urine it appears there if they fast. In women the amount normally present is increased. A very interesting case of this sort was reported by Benedict and Diefendorf. The patient was an elderly woman having a religious mania so that she fasted periodically. One such period is that contained in the following

Date	Total N	Total creatinine —gr.	Preformed creatinine—gr.	Preformed creatine—gr.
Nov. 18	6.56	0.50	0.50	0.00
19	6.49	0.66	0.61	0.05
20	6.65	0.63	0.60	0.03
21	6.09	0.67	0.61	0.06
22	6.45	0.68	6.05	0.03
Begins to fast. First three days no water.				
23	4.19	0.65	0.61	0.04
24	6.05	0.66	0.57	0.09
25	6.38	0.61	0.54	0.07
26	6.93	0.50	0.44	0.06
27	6.16	0.65	0.49	0.16
28	4.41	0.49	0.34	0.15

End of fast.

Date	Total N	Total creatinine —gr.	Preformed creatinine—gr.	Preformed creatine—gr.	
Nov. 29	(3.04)	(0.34)	(0.23)	(.11)	Some loss
30	4.87	0.67	0.56	.11	No food
Dec. 1	3.17	0.55	0.50	0.05	
2	5.09	0.64	0.61	0.03	
3	6.05	0.61	0.60	0.01	

The great increase in the creatine on the 27th and 28th will be noticed as well as the wonderful constancy of the total of creatinine and creatine. The falling off in the creatinine is illustrated also in the following results on the professional fasting woman, Flora Tosca (van Hoogenhuyze and Verploegh) :

Total N	Creatinine per day in grams
Before the fast	
13.99	1.087
Fasting	
8.76	0.904
8.38	.577
10.73	.581
9.40	.634
7.87	.633
7.73	.590
6.11	.469
Fasting continued	
6.70	.689
7.35	.715
6.80	.602
6.14	.453
6.97	.566
5.62	.548
4.08	.426
4.38	.715
Broke fast	
	1.028

(In this case the creatine was not determined.)

The reappearance of the creatine in the urine of fasting men and the corresponding decrease in the creatinine seems to be an interesting casting back of their metabolism to the type of the juvenile metabolism. Regarded from this point of view it would appear that the metabolism is rejuvenated by fasting. It would be interesting to see if other signs of rejuvenescence are to be found; to see, for example, if allantoine would reappear in the urine of fasting men. Fasting rabbits also increase the creatine at the expense of the creatinine. The creatine quite disappears if the animals are given carbohydrate food, but not if fed fat or protein. It may be that the change in metabolism has affected chiefly the liver and muscles and particularly the former and that the exhaustion of the carbohydrate of the liver has affected its power to change creatine into creatinine. By some authors the increase in the creatine is supposed to indicate an unusual protein catabolism, but the small amount of the total nitrogen excreted and the slight decrease in the total creatine and creatinine is against such an interpretation.

Influence of carbohydrate food on creatine excretion. It has been found that a deficiency of carbohydrate food in the diet produces creatinuria. Various suggestions have been made to account for this fact. Thus it is supposed by some that creatine appears as the result of the abnormal catabolism of flesh in the body consequent on the withdrawal of carbohydrate; or owing to an impairment of the function of the liver. Thus creatinuria often accompanies diabetes both natural and phlorizin. The amount of creatine and creatinine together in these cases, however, is not larger than normal, so that creatine appears as the result of the loss of power of the body to change creatine to creatinine. As was stated at the outset, however, the reported creatine may not have been creatine, but acetoacetic acid.

Influence of the thyroid on creatine excretion. The thyroid has a marked but unexplained influence on the metabolism of the body. Injection of thyroid extract causes a stimulation of the protein catabolism of the body. It is interesting that it also causes an increase in the creatine excretion (acetoacetic acid?). Just how this result is produced it is impossible to say. It might be that the thyroid extract caused directly or indirectly the discharge of creatine from the muscles so that the excretion is increased. It may be that the creatine acts as a brake on the protein catabolism and its removal results in an abnormal catabolism of proteins.

Origin of creatinine. The foregoing facts show that the creatine and the creatinine of the urine have the same origin, and that creatinine is derived from the creatine of the tissues of the body. There is in the body of an average adult something more than 120 grams of creatine. That this creatine gives rise to the creatinine of the urine is indicated not only by the facts cited, but also by the following observations of Meyers and Fine:

	Creatine—muscle per cent	Creatinine—coefficient	Ratio—muscle creatine	Ratio—coefficients
Rabbit	0.52	14.3	1.4	1.7
Man	0.39	9.0	1.05	1.07
Dog	0.27	8.4	1.0	1.0

Origin of creatine. From what substances then does the body form the creatine found in the tissue? Where is the creatine changed into creatinine? How much of the creatine and creatinine is destroyed?

There are several possible sources of creatine. It might be formed from arginine, which is guanidine-amino-valerianic acid. By oxidation this might be turned to guanidine-amino-acetic acid and by methylation this changed to creatine. An ingestion of arginine does not increase the output of creatinine. But Inouye found in autolysis of the liver to which he added some arginine that there was an increase in the creatine and creatinine. The change was, however, small. Jaffé showed that the ingestion of guanidine acetic acid is followed by a slight increase in the excretion of creatine and this has been confirmed by others. He found, also, that the ingestion was followed by an increase in the creatine content of muscle. The greater part of the guanidine acetic acid was excreted unchanged, only 5-8 per cent. having been changed over. The ingestion of guanidine caproic acid and guanidine butyric gave no better results. It has not been possible to find glycocyamine in the metabolism of the body and it is accordingly very doubtful if it is an intermediate product in the formation of creatine. Another possibility is that creatine is formed from choline or some similar product of the phospholipins. Koch's experiments gave no increase in creatinine after the ingestion of large amounts of lecithin or choline. Another possibility is that the synthesis might be made with methyl guanidine as the intermediate body. Methyl guanidine has been found both in muscles and in the urine (Kutscher). Experiment has shown, however, that this is a very toxic substance and Jaffé and others have got no increase in the creatinine after injecting it. The synthesis might be made using urea and methyl amino-acetic acid. There is no evidence of this beyond the fact that the addition of urea to autolyzing liver completely prevents the destruction of creatine by the liver and the same is true of other organs. It would appear that in some way urea is concerned in the metabolism of creatine. The ultimate origin of creatine is thus uncertain. The presence of sarcosin and glycocoll so frequently in the muscles of invertebrates may be significant.

The transformation of creatine into creatinine is probably carried out in many different tissues. For example, the creatine of the blood slowly changes to creatinine, and the liver, kidney, muscle, spleen and lungs have the same power, due possibly to an enzyme they contain. It is even possible that this enzyme occurs in the urine since creatine turns to creatinine in the urine more rapidly than the number of hydrogen ions present would lead one to expect. Perfusion experiments also show clearly that the liver and muscles have the power of destroying and of making creatinine out of creatine. Creatine is destroyed also in many tissues, but it is not known what becomes of it. It has been suggested that it is first turned into creatinine and then destroyed, but this is not certain. If one extirpates the kidneys the amount of creatine in the blood increases from about 2 mgs. per hundred grams to 8 in the course of 48 hours. In six hours there is no noticeable change, a fact that indicates that the power of the body to destroy creatine is very great. That one place of transformation of creatine to creatinine is in the liver is indicated by the fact that if the liver is poisoned by phosphorus or hydrazine, creatine replaces creatinine in the urine. An Eck fistula, however, does not cause creatine to appear in the urine. Even heated blood serum is more powerful in transforming creatine to creatinine than is water. Blood contains the following amounts of creatine and creatinine in 100 grams:

Carotid defibrinated blood	Creatinine—mgs.	Creatine—mgs.
Cat (young female)	0.85	2.44
Dog (young female)	0.72	1.54
Adult (male dog)	0.38	2.92

Methylation. Creatine is a methylated amino-acid. This is a point of much interest, since none of the amino-acids of the proteins are known to be methylated. The question at once arises as to the mechanism of methylation, the place where it occurs, its significance and what substances may be methylated in the body. This general problem of methylation may very well be treated here in connection with creatine and creatinine which are the most important methylated substances in the urine.

The power of forming methyl-amino derivatives is very widespread in nature and many of such derivatives are of very great physiological and pharmacological importance. Choline and its more active relatives, neurine and muscarine, are methyl derivatives, the former being the trimethyl-oxyethyl ammonium hydroxide; neurine, trimethyl-vinyl-ammonium-hydroxide and muscarine is supposed to be the aldehyde of choline. Neurine and choline occur in the animal organism. Adrenaline is a methylated amine, being dioxyphenyl-oxyethyl-methylamine. In plants the betaïnes and similar compounds are common, and trigonellin, and stachydrin are examples of this class of bodies.

It has been found that human and other mammalian organisms have the power of introducing methyl groups into various substances ingested. The methyl derivatives are found in the urine. Thus ingested pyridine is excreted by dogs as methyl pyridine (His); tellurium salts are excreted as methyl tellurides (Hofmeister); and naphthalene also appears as the methyl derivative (Cohn). But rabbits do not methylate pyridine but excrete it unchanged, the carnivora and herbivora showing important differences in this respect.

The exact manner in which this methylation is produced is not clear, but it has been suggested that, in plants at any rate, it comes from the union of formaldehyde with amino groups. It will be recalled, from the discussion of the Sorensen titration method, that formaldehyde unites with amino groups to form methylene derivatives. By reduction these methylene amines can be changed to methyl amines. This seems the probable method of their formation. It has not been shown that formaldehyde is an intermediary product of metabolism in animals, but in the decomposition of the carbohydrates it is not impossible that small quantities may arise and that this formaldehyde is the source of the methylation. This view, however, can have no solid basis until the presence of formaldehyde in animal tissues is unequivocally established.

Concerning the place in which such formation of formaldehyde might occur one turns naturally to the two organs of the body most concerned in carbohydrate metabolism, the liver, where the carbohydrates may be formed from proteins and stored, and the muscles, the tissues in which carbohydrate is torn to pieces and the fragments burned for the production of muscle energy. It is, perhaps, more likely that it is in the muscles, where decomposition of carbohydrate occurs, that small amounts of formaldehyde may be produced and there the synthesis of methyl glycocoll take place. Certainly in some of the invertebrates glycocoll is found in large amounts, not methylated, in the muscles. Formaldehyde is itself very toxic and possibly glycocoll, which serves in other instances as a means of making toxic substances harmless, may also thus function in the muscle, thus forming sarcosine.

Relation of creatine and creatinine excretion to muscular metabolism. But while the muscles are the organs in which creatine is found in the largest amounts, and while the muscles are able to form creatine, it is probable that the creatine metabolism does not play a part in muscular contraction. While the evidence is somewhat contradictory on the behavior of muscle creatine in tetanus or during muscle contraction, it is certain that doing muscular work does not increase creatinine excretion. On the other hand, it is true that creatinine excretion is larger during the day than at night when the muscles are relaxed. It has been suggested that creatine is concerned in the tonic contraction of muscles. Thus standing in a rigid military position is said to increase creatinine excretion. It is doubtful, however, whether there is this fundamental distinction in kind between what is known as tonic contraction and the ordinary contraction of muscle. On the other hand, it has been suggested that creatine is evolved in the course of the formative metabolism of muscle which involves the making of muscle substance and its machinery. It represents rather the wear and tear of the machine than consumption for purposes of obtaining energy. This is, perhaps, the most reasonable view of creatine and it explains why all organs have and produce creatine. It is evidently not a substance having to do with a particular function, like that of contraction, but some function, like growth, common to all cells.

It is probable that creatine of muscle is not free, although it is so easily extracted with water, but that it is combined with the colloids of the muscle. Otherwise it would most probably escape from the cells. How much of it is actually given off from the muscles in the course of the day cannot be said. Safe conclusions cannot be drawn from the fact that it is given off from the muscles to perfusion liquids, for the condition of perfused tissues is generally far from normal. Most observers are of the opinion that the greater part of the creatine is destroyed in the body and that only a small portion escapes in the form of creatinine.

Summary. The principal results and conclusions of the chapter may be summarized as follows:

There is excreted in the urine of human male and female adults about 1 to 2 grams of creatinine a day. The amount is entirely independent of the protein intake, but is increased by the ingestion of creatinine, most of which reappears unchanged in the urine. The ingestion of creatine increases creatinine only very slightly and it may not increase it at all; it may increase somewhat the creatine excretion. What becomes of the greater part of the creatine thus ingested is unknown. Creatine is found only occasionally and in small quantities in the urine of adult men, a few mgs. per day, but it occurs, or is supposed to occur, in larger amounts, 60-100 mgs. per day, in the urine of women during menstruation, in the first few days after childbirth, and during pregnancy. It is usually present in children's urine, particularly in girls', and may make from one-tenth to one-third the amount of the creatinine.

The amount of creatinine excreted in adults both male and female is roughly proportional to the body weight; about 7-11 mgs. of creatinine nitrogen per kilo body weight being excreted in 24 hours. This figure is called the creatinine co-efficient. The combined creatine and creatinine undergoes little change during fasting, but the amount of creatinine diminishes and that of creatine increases, if the method of determination of creatine is correct. The same result occurs in carbohydrate starvation, in diabetes, both natural and phlorizin, and in the early stages of wasting diseases such as fevers, muscle-atrophy, etc. These facts all show that creatinine has an endogenous origin and that unlike urea it is unaffected by the intake of protein.

There is found in the voluntary muscles about 0.35-.48 per cent. of creatine and large amounts are found also in the liver, the involuntary and heart muscle, in the brain, testes and other organs. The total amount of creatine in the body is thus about 120 grams. This creatine of the body is probably the origin of the creatine and creatinine of the urine. The muscles have the power of forming creatine from glycocyamine and they, and other organs, can turn creatine to creatinine. Extracts of the organs have this same power which is probably to be ascribed to a ferment. Probably the kidneys are active in changing the creatine arriving by the blood to creatinine, but it is certain that they are not the only organs having this property. The tissues are also able to destroy creatine, but what is formed from it is unknown.

The increase of creatine in the urine in diabetes and carbohydrate starvation, or after the injection of thyroid extract, is ascribed by Folin, Shaffer and others to the increased catabolism of the tissues, but it is also possible that it is due to the setting free of the creatine from the combination in which it is in the muscles. Nothing definite is known of the function of creatine in the muscles and the other organs; whether it is primarily a waste product, or whether, by its presence, it affects the metabolism of the tissues. Further investigations are very necessary on this point. While all the really important questions about the origin and significance of creatine and creatinine still remain unanswered it appears that these substances hold a very important place in metabolism and the constancy of the excretion of creatinine indicates that it is, as Folin suggested, an index of the real catabolism of the vital machinery of the body proper, in distinction from that catabolism which increases the free energy.

Purine bodies and allantoine in the urine.—The most important purine found in human urine is uric acid, but there is also present about 30-50 mgs. of purine bases, xanthine, hypoxanthine, guanine and adenine. The first two are the more abundant. Allantoine is present in adult human urine in extremely small amounts, not more than 14 mgs. per day; but in other mammalia it may be present in very much larger amounts, 25-30 grams a day being secreted by cows. While allantoine is not, strictly speaking, a purine, it represents in many mammalia the end product of purine metabolism and must, therefore, be considered with the purines.

Uric acid.—Probably no nitrogenous substance in the urine has been more studied than uric acid. It is present in human urine only in small quantities, the excretion varying from 0.3-1.2 gram per day in the human adult, the average amount being about 0.6 gram. The amount varies with the diet, state of health and individual idiosyncrasy. Of the total nitrogen of the urine between 5 and 10 per cent. is excreted as uric acid nitrogen. But though present in small quantities, this acid has long attracted the attention of physicians and physiologists because of the problems involved in its origin, its significant variation in disease, its insolubility, which causes it to form part of the sediment of the urine, and above all because of its deposition in the joints in the form of crystalline insoluble salts in arthritis and gout.

While in the mammalia, amphibia and fishes the quantity of uric acid excreted in the urine is small and is but a small proportion of the total nitrogen outgo, in birds and reptiles this substance takes the place of urea and in these animals most of the nitrogen excretion is in this form. In the invertebrates, also, in the arthropods and particularly in the mollusks, the excretion of nitrogen, to a considerable extent at any rate, is in the form of the purine bases and uric acid.

There can be little doubt that the substitution of uric acid for urea as the form of nitrogen waste in reptiles and birds is an adaptation fitting them to a dry climate, the particular object of the change being the conservation of the water of the body. Uric acid has very little affinity for water and is almost insoluble; and the acid salts are also insoluble; hence uric acid is not excreted in bird's urine in solution, but in the form of masses of crystals, very little water being excreted at the same time. Urea, on the other hand, is a very soluble substance; it has a very great affinity for water and takes a good deal of it out of the body when it is excreted. It is a good diuretic. The birds were evolved from the reptiles and the reptiles from the amphibia which have urea as the nitrogen end product. The reptiles were probably formed or evolved in some arid region, since in all the particulars of their bodies they show this same adaptation. Thus they have replaced the moist skin of the amphibia by the scaly, hard, dry covering of the reptile, which prevents loss of water through the skin. Possibly if their lungs were examined they would be found to allow of less water passing through than is the case with the amphibian or mammalian lung.

Chemical nature. Uric acid is 2, 6, 8 tri-oxy purine.

```
H—N—C=O                          N=C—OH
    |   |                            |   |
O=C   C—NH                (2) HO—C   C—NH
  |    ||   \                    ||   ||   \
  |    ||    C=O                 ||   ||    C—OH (8)
  |    ||   /                    ||   ||   //
H—N—C—NH                         N—C—N
Uric acid.  Lactam form.               Lactim form.
```

The formula is usually written in the form first given, but another form is possible and accounts for the fact that it has an acid nature and that it forms two series of salts. This second form, the lactim form, is probably in equilibrium with the ordinary or ketone form. According to the latter formula the acid should be a tribasic acid, but only two series of salts are known. It is probable that the third hydrogen ion dissociation would be very weak. Intermediate forms between the lactam and the lactim probably exist in which only one hydroxyl is present. The lactam form is the less stable. The hydrogen in the 2 and 8 positions may be substituted, making possible two series of salts. Of these salts the acid salts are the less soluble and particularly the free acid and the mono-ammonium salts are very insoluble. The solubility of the free acid in water is one part in 39,480 parts at 18° and 1 in 15,505 parts of water at 37° (Gudzent).

It is a white, tasteless powder or crystalline substance, composed of rhombic prisms or plates (Figure 62). As it comes down in the urine it is combined with or associated with a red coloring matter, uroerythrin, and the crystals are colored red, or brown. The forms are very various, the so-called whetstone shape being common. Dumb-bell and other shapes occur. The acid urates also are, for the most part, very little soluble, but the dibasic salts of the alkali metals are more soluble. The acid may precipitate in the urine in the form of the acid sodium or ammonium salt in balls of needle-like crystals, of a brownish red color, or having irregular shapes. The solubility of the monobasic sodium salt is 0.8328 gram of the salt in a liter of water at 18°, and 0.4141 gram of the ammonium salt. At 37° it is respectively

FIG. 62.—Crystals of uric acid.

1.5043 and 0.7413 grams per liter. According to Gudzent these solubilities are only true of the fresh solution, sibce the solubility gradually diminishes due to the transposition of the lactam to the less soluble lactim form. One part of the normal sodium salt dissolves in 77 of water at 18°. The normal potassium salt dissolves in 44 parts of cold water; the normal calcium salt in 1500 parts and the acid lithium salt in 60 parts of water (Ralfe). The acid piperazine salt is much more soluble than the alkali salts. It dissolves in 50 parts of water at 17° C.

$$C_5H_4N_4O_3.NH\left\langle\begin{matrix}CH_2-CH_2\\ \\ CH_2-CH_2\end{matrix}\right\rangle NH$$

The methylglyoxalidine salt dissolves in 6 parts of water (Ladenburg. Ber. 27, 2952).

In acid solution uric acid is very stable. It may be dissolved in concentrated sulphuric acid without being destroyed, and from this solution it may be precipitated by addition of water. In alkaline solution, on the other hand, it is very unstable, breaking up rather rapidly and in the presence of oxygen oxidizing itself. Folin and Denis state that 0.5 per cent. Na_2CO_3 solution boiled three minutes with 10 mgs. of uric acid in 20 c.c. destroyed 12 per cent. of the acid present. In alkalies it probably breaks into dialuric acid and urea.

```
HN—CO                              HN—CO
|   |                              |   |
OC  C—NH                           OC  CHOH  +  H2N
|   ||   \                         |   |          \
|   ||    CO   + 2 H2O ——→         |   |           CO
|   ||   /                         |   |          /
HN—C—NH                            HN—CO         H2N
Uric acid.                         Dialuric acid.  Urea.
```

Uric acid, being auto-oxidizable in alkaline solution, is a reducing substance and reduces Fehling's solution, ammoniacal silver nitrate, phosphotungstic acid and other oxidizing substances. It is readily oxidized also by permanganate and it can be titrated and its amount quantitatively determined in this way. See page 964. When oxidized it forms various substances, such as alloxan, or oxaluric acid, urea, oxalic acid, carbonic acid, tartronic acid, allantoine and uroxanic acid, $C_5H_8N_4O_6$.

```
                                                   NH—C=O
  NH—C=O              NH—C=O                       |   |  OH
  |   |               |   |                        |   | /
O=C   C=O           O=C   CHOH                   O=C   C—NH
  |   |               |   |                        |   |    \
  NH—C=O              NH—C=O                       |   |     C=O
Alloxan (Mesoxalyl urea).  Dialuric acid.          |   |    /
                    (Tartronyl urea.)              NH—C—NH
                                                       \
                                                        OH
                                                   Intermediate form.
            O=C—OH
              |
            H—C—OH
              |
            O=C—OH
          Tartronic acid.
```

Reactions of uric acid. Murexide reaction. The crystals moistened with nitric acid and evaporated to dryness on the water bath on a porcelain plate at first dissolve and are partially oxidized, a red residue being finally obtained. On moistening this, after cooling, with ammonia a purple-red develops, due to the formation of ammonium purpurate, or murexide. It is called the murexide test because ammonium purpurate is the scarlet substance in the dye obtained from the sea-snail, *murex*. If caustic soda is used in place of ammonia a deeper blue is obtained, and the color disappears quickly on warming. Some other purines give this reaction. With xanthine and guanine the color does not disappear on heating. The formation of purpurate of ammonia is probably as follows:

By the hydrolysis and oxidation of uric acid, dialuric acid and alloxan are formed. These condense to form alloxantin which, in the presence of ammonia, forms ammonium purpurate or murexide. The reactions are as follows[1]:

[1] The formulas of alloxantin and purpuric acid are still uncertain. It is probable that those given here are not correct, but they are simpler than some others

```
HN—CO           HN—CO          HN—CO       OC—NH
|   |  H        |   |          |   |  OH   |   |
|   | /         |   |          |   | /     |   |
OC  C—OH   +    OC  CO  ——→    OC  C———————C   CO
|   |           |   |          |   |    /  |   |
|   |           |   |          |   |  HO   |   |
HN—CO           HN—CO          HN—CO       OC—NH
```

Dialuric acid. Alloxan. Alloxantin.

```
                             HN—CO        OC—NH
                             |   |   NH   |   |
                             |   | /    \ |   |
Alloxantin + NH3  ——→        OC  C————————C   CO
                             |   |        |   |
                             HN—CO        OC—NH
```

Purpuric acid.

Origin of uric acid. It was long believed that uric acid in mammalian urine was an intermediary product of protein metabolism, which was usually almost completely oxidized by the body to urea. The occurrence of more than the normal amount of uric acid in the urine was, hence, supposed to mean that the oxidative powers of the body were impaired in some way. While this view has certain elements of truth in it, it was in its essentials quite erroneous, as we now know. Uric acid does not come from the ordinary protein metabolism. The end product of that metabolism is urea; but it comes from the metabolism of the nucleins, both of those of the food and those of the tissues. The nucleins, it will be remembered, contain nucleic acid, and nucleic acid contains purines, which, when oxidized, form uric acid. Uric acid is, hence, of very particular interest because of its relation to nuclear metabolism.

That uric acid came from the nucleins either of the food or the tissues followed directly from Kossel's discovery that the nucleins were the mother substances of the nuclein, or xanthine, bases as they were called, adenine, guanine, xanthine and hypoxanthine. These bodies are now called purines following the suggestion of Emil Fischer, being regarded as all derived from purine. Uric acid was known to belong to the same group of substances and to be simply oxidized xanthine. Kossel suggested, therefore, that uric acid in the mammalian urine did not come from the proteins in general, but only from the nucleins. When nucleic acid was discovered by Altmann, this theory was made still more precise, the uric acid coming from this constituent of the nucleins.

The fact that uric acid comes from the metabolism of nucleins was shown in the first instance by feeding experiments. If one determines

which have been proposed. The union between alloxan and dialuric acid may be through an oxygen atom, being of the nature of an oxonium salt (Richter, Stieglitz and others).

the amount of uric acid in the urine it is found that the quantity increases when there is an increase in the nucleic acid ingested, but it does not increase nearly as much if there is an increased intake of proteins which do not contain nucleic acid. Glandular organs generally contain a good deal of nucleic acid, so that a diet of such organs in place of meat means an increase in the nuclein intake. In all such diets the excretion of uric acid is increased. This is shown in the following protocols (Jerome):

	Uric acid excreted per day—grms.
Usual diet	0.554
" "	0.480
" "	0.590
Nuclein diet. Testicles of herring	0.740
" " " " "	1.010
" " " " "	0.754
After period. Usual diet	0.452
" " " "	0.462
Nuclein diet. Pancreas period	0.606
" " " "	0.820
" " " "	0.612
After period. Usual diet	0.446
" " " "	0.474
Nuclein diet. Thymus period	1.546
" " " "	0.740
Diminished nuclein period	0.398

The usual diet was a fairly hearty diet. The breakfast consisted of two eggs, bread and butter, porridge, malt coffee, milk and saccharine. The dinner consisted of meat, potatoes, vegetables, milk or rice pudding, bread and butter, fruit, and two dessertspoonfuls of whisky, or a pint of champagne; lunch of fish, bread and butter, apples.

It will be observed that whenever glandular organs rich in nucleins were ingested then the excretion of uric acid was increased. On a nuclein-free diet, a starch and cream and egg diet, for example, the excretion fell to a minimum of about 0.4 gram per day. The same result has been obtained by many observers. There is no doubt that the ingestion of nucleins increases the excretion of uric acid; and the elimination of nucleins from the diet decreases the excretion to a certain minimum, but does not abolish it entirely. For a nuclein-free diet eggs, milk, sugar, starch and cream furnish an admirable diet almost purine free. On a diet of cream and starch Folin reduced his daily output of uric acid to about 0.3 gram and others under his direction did the same.

The increase in the uric acid excreted after the ingestion of nucleins might be either a direct or indirect result. That is the nucleins of the diet might directly in the course of their decomposition give rise to the uric acid of the urine, or indirectly they might stimulate uric acid production. It is believed that they act in the first manner and we accordingly say that some of the uric acid of the urine has an *exogenous* source, meaning that it comes from outside, from the nucleins of the food, which have not been incorporated in the nucleins of the body, but are decomposed and part of the molecule split off and excreted as uric acid. The details of this process, however, are very badly known. It is known that the nucleins of the food are digested by the juices of the intestine and its accompanying glands, such as the pancreas, and the purine bases are set free. In most tissues, taken as foods, the bases in the nucleins have already been partially oxidized, while still in the nuclein molecule, by the action of the auto- digestive and oxidizing enzymes of the tissue, so that they get free in the intestine in a partially oxidized form; in the form of hypoxanthine and xanthine, as well, probably, as guanine and adenine. These bodies are at least in part absorbed. Their absorption is followed by the appearance of some uric acid in the urine, but if it be asked where the oxidation of these substances occurs, whether in the intestinal mucosa, or whether in the liver or some other tissue, and whether they have, or have not, been part of the living matter when they were oxidized, or whether they were only dissolved in the cell sap and never incorporated in the living matter of the cell; and whether, indeed, they may not have acted by displacing some of the bases already in the cell,—to these questions very imperfect answers can be given, or none at all.

It is, however, worthy of note that of the total purine ingested, but a small part reappears in the urine as uric acid. Some of the remainder is probably destroyed in the intestine by the action of the bacteria, but some is probably destroyed in the tissues or retained there for the time being. The per cent. of purine nitrogen in the thymus gland, according to Burian, is 0.482. This would correspond to about 1.2 grams of purine in 100 grams of the fresh tissue. If one eats two hundred grams of sweetbreads it should, therefore, increase the uric-acid excretion by 2.4 grams, making in all nearly 3 grams per day of uric acid. The actual increase of uric acid in the urine is, however, not more than half this required amount, showing that the purines either had not been absorbed, or that they had been retained, or that they had been destroyed. This fact of the destruction of uric acid in the human body is well illustrated in the following experiment of Taylor and Rose. For three days the subject was on a purine-free diet, consisting of milk, eggs, starch and sugar; then for three days a part of the total nitrogen, 10 grams per day, was substituted in the form of sweetbreads, so that of the 10 grams, 7 grams were in egg and milk and 3 grams in the sweetbreads; for the next four days, six grams of nitrogen of the eggs and milk were replaced by sweetbread nitrogen; and for the next four days the purine-free diet, containing 10 grams of nitrogen, was restored.

	1st period Purine-free diet	2nd period	3rd period	4th period Purine-free diet
Total urinary N	8.9	8.7	9.1	8.8
Urea N and NH_3	7.3	7.1	7.1	7.05
Creatinine N	0.58	0.55	0.56	0.47
Purine N (Total)	0.11	0.17	0.26	0.10
Uric acid N	0.09	0.14	0.24	0.07
Remainder N	0.91	0.88	1.18	1.18

The intake of purine N in the 2d period was 0.17 and in the 3d, 0.34 grams per day. The increase in uric acid excreted accounted for less than half of that ingested. The purine base N in the urine remained constant. That much the larger portion of the purine nitrogen is either not absorbed, or else is retained or destroyed, is shown by Weintraud, who calculated that the amount of uric acid which was excreted after a nuclein diet was not sufficient to cover more than one-fifth of the amount computed that there should be from the increase in the phosphoric acid excretion.

Time of excretion. The study of the output of uric acid from hour to hour has led to the discovery of some very curious and unexpected facts. Hopkins and Hope found that taking food, even when it was free from nuclein, led in an hour or two to a great increase in uric acid excretion, which was at a maximum 3-4 hours after the meal, while the urea maximum was about 6-7 hours after eating. After fasting 6 hours a meal of bread and potatoes was eaten at 1:30 and the urea and uric acid measured in the urine each hour.

Time	Urea—grams	Uric acid—mgs.	Amount of urine—c.c.
10—11	1.07	26	175
11—12	1.13	27	118
12—1 P.M.	1.07	24	164
1—2 (meal)	0.64	21	60
2—3	1.12	22	43
3—4	1.16	38	41
4—5	0.84	40	53
5—6	1.16	56	59
6—7	1.20	39	56
7—8	1.37	30	95
8—9	1.47	33	183
9—10	1.33	24	155
10—11	1.33	23	190

These results have been confirmed by Smetanka, who showed that eating purine-free meat and even carbohydrates causes, 2-3 hours later, a marked increase in uric acid excretion. This increase is not due to the daily variation in the uric acid excretion which occurs even in fasting, the morning excretion being always greater than the afternoon, as in the experiment following from Smetanka:

HOURLY EXCRETION OF URIC ACID IN MGS.

Time	When fasting	When at 9 P.M. 330 grams of casein eaten	Increase
6—7 P.M.	9.6	10.2	
7—8	11.7	9.7	
8—9	12.2	9.6	
9—10	11.6	9.7	
10—11	10.7	17.9	7.2
11—12	10.5	19.7	9.2
12—1 A.M.	10.6	19.7	8.6
1—2	11.4	19.2	7.8
2—3	11.7	19.2	7.5
3—4	11.7	17.5	5.8
4—5	13.5	17.5	4.0
5—6	12.9	17.5	4.6
		Total	54.7 mgs.

What is the cause of this increase? It cannot come from the diet. It might come from synthesized uric acid, but Smetanka believes that it comes, as Mareç thought in 1887, from the work of the gastric and intestinal glands. It would seem not impossible that it might be due to a reabsorption of uric acid precursors from the intestine, due to the decomposition of the bacteria there and the increased blood supply, causing an increased reabsorption when digestion begins.

Endogenous uric acid. But by cutting the nucleins completely out of the food, or by starving, it is not possible to suppress the uric acid excretion entirely. There is still excreted about .3-.5 gram uric acid per day. The amount varies in different individuals. This residual uric acid evidently must have its origin either in the body tissues or else in the bodies of the bacteria of the alimentary canal. Since there is reason for thinking that the nuclei of the body cells are undergoing metabolism, it is generally believed that this uric acid takes its origin, largely at least, from the nucleins of the tissue nuclei, and the bacteria of the intestine are not supposed to play any important part in its formation. At the same time the increase in uric acid excretion, which accompanies the activity of the intestine, an increase just noted, would make a careful investigation of this possible source of uric acid desirable. The uric acid which is still produced in the body after the nucleins have been cut out of the foods is called *endogenous* uric acid, meaning formed, or generated, within.

Variation with disease. The endogenous uric acid will be, then, an indication of the nuclear metabolism of the body and may be expected to increase when that catabolism increases, and to decrease when it decreases. Thus it would be expected that the uric acid would increase during embryonic development, or during growth, when much nuclear material is being formed. That this is the case is shown by the excretion of uric acid per gram body weight from pregnant women, and also from children at different ages. The per cent. of uric acid nitrogen in the urine, computed on the total nitrogen, is also greater at this time, showing that the nuclear metabolism is greater relative to other metabolisms.

In diseases involving tissue decomposition uric acid excretion is also increased. Thus after the crisis in pneumonia, when the exudate of the lungs containing a large amount of leucocytes is being digested by autolysis and reabsorbed, there is a great increase in uric acid excretion. A similar increase is seen in leucæmia (leucocythemia), a disease in which the number of white blood cells is much increased above the normal and their decomposition is probably greater than the normal amount. After extensive burns of the skin causing marked destruction of tissues and their reabsorption, there is, similarly, an increase of uric acid.

Source of the endogenous acid. We have now to inquire what are the steps, what the processes and in what organs the endogenous acid is produced. The first steps in the solution of this problem were taken by Mareç and Horbaczewski. The latter first succeeded in deriving uric acid from a mammalian tissue by autolysis. By grinding dog's spleen in a meat chopper and then with sand, mixing the spleen pulp with blood well aërated and kept at body temperature, he was able to show that uric acid was formed from some elements of the tissue or from nucleic acid added to the pulp.

From these experiments Horbaczewski concluded that uric acid was produced by the oxidative decomposition of the body nucleins. Of the various cells of the body undergoing decomposition the leucocytes or white cells of the blood were those most obviously disintegrating. Horbaczewski suggested that most of the uric acid, namely, that following digestion, pneumonia, leucocythemia, came from the nucleins of these cells and the uric acid excretion was an index mainly, but not exclusively, of leucocytic decomposition. He suggested, also, that the rise in uric acid following nuclein ingestion was not due to the direct transformation of nucleins of the food into uric acid, but was an indirect result of the digestive leucocytosis and decomposition. Subsequent research has shown this view not to be strictly true. Leucocytosis may occur for a short period without an increase in uric acid, but it is none the less true that a long continued leucocytosis, involving as it does an increased decomposition of leucocytes, always increases uric acid. The parallelism is between the amount of leucocytic decomposition and uric acid excretion, not between the number of leucocytes in the blood and uric acid excretion.

Further investigation of the nature of the decomposition of nucleic acid in the body has led to a more precise knowledge of the various steps in its formation in mammals. There are in nearly all cells, and possibly in all cells, autolytic or endocellular enzymes, or nucleases, which decompose nucleic acid into its various constituents. The decomposition of the nucleic acid may occur in various ways. 1. There may be a cleavage into mono-nucleotides, such as guanylic acid. This acid consists of guanine, d-ribose and ortho-phosphoric acid. Such nucleases exist probably in yeast where guanylic acid and adenosine have been found. They occur also in the pancreas of the pig (Jones). The enzyme which thus splits nucleic acid into horizontal slices, as it were, is called polynucleotidase. 2. A second nuclease, phosphonuclease, splits off phosphoric acid either from the mono- or polynucleic acids, leaving the nuclein base and the carbohydrate united as they are in guanosine and adenosine. 3. Still another cleavage separates the nuclein bases from the molecule, leaving the phosphoric acid joined to the carbohydrate radicle.

These various cleavages appear in the autolysis of different cells and are believed to be due to different nucleases. At any rate the purine bases are set free in most autolyses. Before being set free, however, they may be oxidized or deamidized and then split free from the sugar. Thus in some organs during autolysis guanine is not set free as such, but as xanthine by an hydrolysis as follows:

```
    HN—C=O                                          HN—C=O
    |  |                                            |  |
NH=C   C—NH   +  H2O  +  (Guanase.) ——→   O=C   C—NH        + NH3
    |  ||  >CH                                  |  ||  >CH
    HN—C—N                                      HN—C—N
    Guanine.                                    Xanthine.
```

Adenine may be converted into hypoxanthine by hydrolysis by adenase and then by oxidation to xanthine. The xanthine may then, if oxygen is present, be converted by oxidation through the agency of the ferment xanthineoxidase to uric acid.

```
N=C—NH2      HN—C=NH                                    HN—C=O
|  |         |  |                                       |  |
HC  C—NH ——→ HC  C—NH  +  H2O  +  (Adenase.) ——→        HC  C—NH  +  NH3
||  ||  >CH  ||  ||  >CH                                ||  ||  >CH
N—C—N        N—C—N                                      N—C—N
Adenine.     Imide form.                                Hypoxanthine.

   HN—C=O                                               HN—C=O
   |  |                                                 |  |
   HC  C—NH  +  O  +  (Hypoxanthineoxidase.) ——→        OC  C—NH
   ||  ||  >CH                                          |  ||  >CH
   N—C—N                                                HN—C—N
   Hypoxanthine.                                        Xanthine.
```

```
HN — C = O                                                  HN—C=O
 |     |                                                     |   |
OC    C — NH  +  O  +  (Xanthinoxidase.)  =                 OC   C — NH
 |     ||    >CH                                             |   ||      C = O
HN — C — N                                                  HN — C — NH
Xanthine.                                                   Uric acid.
```

The following list of enzymes concerned in nucleic acid decomposition has been given by Jones and Amberg:

1. Phosphonuclease.
2. Purine nuclease.
3. Guanosine desamidase.
4. Adenosine desamidase.
5. Adenase.
6. Guanase.
7. Xanthosin hydrolase.
8. Inosine hydrolase.
9. Xanthinoxidase.
10. Uricase.

The phosphonuclease splits off phosphoric acid from nucleic acid; while the purine nuclease splits off the purines, leaving the phosphoric acid and sugar group united. This is quite similar to the splitting of raffinose by the two enzymes emulsin and invertin. Raffinose is fructose-glucose-galactose. Invertin splits off fructose, leaving melibiose, or glucose-galactose. Emulsin splits off galactose, leaving saccharose. By the action of these various enzymes uric acid will be formed from the catabolized nucleic acid.

The distribution of these various enzymes in different organs differs in different animals. They are found, however, for the most part, in the liver, the spleen, the pancreas and thymus. It is, on the whole, probable that some members of the group of enzymes are found in all tissues of the body, since the partial destruction of nucleic acid and the conversion of guanine and adenine in their nucleic acids to hypoxanthine and xanthine on autolysis appears to be a very common, if not a universal phenomenon. The determination of the presence or absence of these various enzymes in different organs in vitro is subject to various sources of error. Thus there may be inhibitory substances present, or the enzymes may be present at times but not at others, or diet may play a part in their appearance. The statements in the literature are, therefore, in part contradictory. Negative evidence is not worth a great deal. Wells has compiled the results and from his statement the following excerpt has been made:

1. Nuclease. Present in all cells investigated.

2. Adenase. Present in all cells and tissues examined, including bacteria, except human spleen, liver, pancreas, kidney and lung, the human fetus of three months, tissues of the fetal dog until birth.

3. Guanase. Present in tissues and cells investigated, except human spleen, spleen and liver of the pig and the pancreas of the dog.

4. Xanthine-oxidase. Present in the spleens of dogs, ox, horse, but absent from the spleen of man and the pig; present in the liver of men, cows, pig, rabbit, and possibly the dog; present in bovine muscle,

DIAGRAM REPRESENTING THE HYDROLYSIS AND OXIDATION OF NUCLEIC ACID. TYPICAL NUCLEIC ACID REPRESENTED AS DINUCLEOTIDE (Dakin).

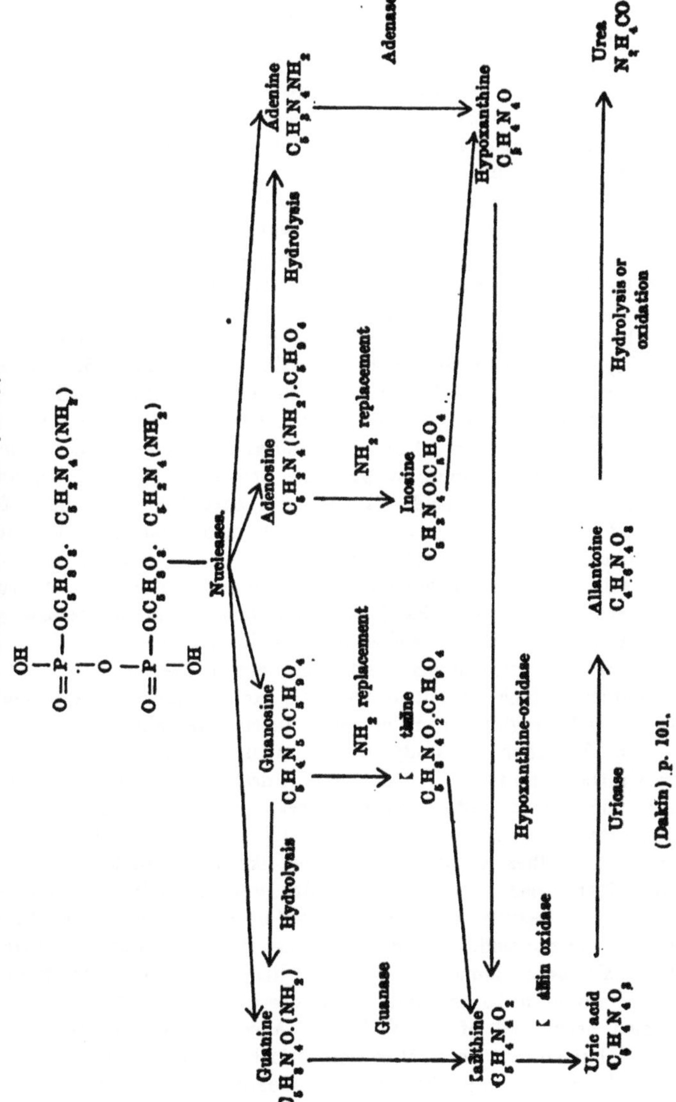

(Dakin) p. 101.

intestine and lung, but not in the thymus and blood of cattle, nor in the lungs, pig's pancreas, dog's pancreas and human placenta. It is lacking in the chief human tissues except the liver.

Destruction of uric acid. Uricolysis. Uric acid is an easily oxidized and hydrolyzed substance. It is not surprising, therefore, that what appears in the urine is only that portion which has escaped destruction. By oxidation allantoine is easily formed and this is certainly one of the substances formed in the bodies of most mammalia as a result of uricolysis. The human body alone and that of the chimpanzee appears to have lost the power of destroying uric acid. If uric acid is given by the mouth only a small portion of it reappears in the urine as such. For a long time it was thought that the remainder had been destroyed by the tissues, but it now seems more probable either that it has been destroyed by the bacteria in the intestine or not absorbed. Its fate is unknown. In most of the mammalia the purines are excreted chiefly in the form of allantoine, uric acid is the next most important substance and the bases are excreted in very small amount; but in other mammals the bases may, at times, surpass the uric acid excretion. In human beings, as has been said, all observers except Croftan have failed to find any uricolysis by the extracts of human organs and allantoine is present in such small amounts in human urine that the opinion is generally accepted that human tissues have lost the power of destroying uric acid. On the other hand allantoine, taken by human beings by the mouth, does not appear as such in the urine (Minkowski).

This destruction of uric acid is brought to pass chiefly in the liver or kidneys, which contain in mammals except man and the primates a uric acid-destroying enzyme or uricolytic enzyme, called *uricase.*

The first thorough study of the destruction of uric acid was made by Croftan, who found that uric acid is destroyed chiefly in the liver by carnivora, in the kidney by herbivora, and by both these organs in omnivorous animals, men and pigs. He succeeded in isolating a substance from these organs, an albumose-like body and a nuclein, which were inert when separate, but which were actively uricolytic when united. His results have been criticised by various workers to the effect that he did not sufficiently guard against the decomposition of uric acid by alkali and air alone, but in view of his controls the criticism appears to the author to be unfounded and subsequent investigations have confirmed nearly all of his findings. According to Schittenhelm and Wells, human liver does not contain uricase. It has been found in monkey's liver, but not in that of the chimpanzee. Croftan's positive finding of uricase in human liver remains, as yet, unexplained. It is possible that under different conditions of disease, or possibly of diet, the uricase may vary in amount.

The following are some of the results of Croftan in the destruction of uric acid by the ground-up dried organs after they had been extracted with alcohol and ether. The uric acid was dissolved in weak sodium carbonate and the organ powders were suspended in this and a stream of air passed through.

Animal	Organ	Flask 1. Uric acid determined at once	Flask 2. Boiled and then 48 hours at 38°	Flask 3. Not boiled at 38°	48 hours Loss per cent.
Dog	Liver	0.327	0.325	0.225	31.1
	Kidney	0.319	0.319	0.312	2.4
	Muscle	0.330	0.326	0.303	8.2
	Blood	0.321	0.317	0.313	2.5
	Spleen	0.327	0.320	0.319	2.4

Similar results have been obtained by Schittenhelm, Wiechowski and other observers.

Chemistry of the destruction of uric acid. Allantoine. In most mammals allantoine is formed by the oxidation of uric acid. Certainly in the dog all of the uric acid appears to go into allantoine, but whether this is always the case in other mammals or not is very doubtful. In most experiments in which uric acid has been ingested or injected only a portion of the uric acid thus ingested has been recovered as allantoine. What becomes of the rest is unknown. In human beings the allantoine is not increased by uric acid ingestion or injection, but the uric acid is in part excreted as such. In fowls it appears from the work of Ascoli that certainly a portion of uric acid is hydrolyzed in the liver to form dialuric acid, as shown on page 736. Perhaps this happens in other animals. This question must be left for further investigation.

According to Sundwik, the oxidation to allantoine by permanganate probably goes through uroxanic acid as follows:

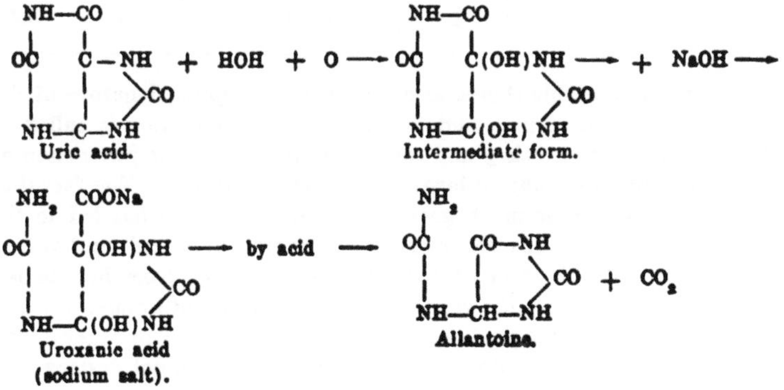

```
NH—CO                                     NH—CO
|   |                                     |   |
OC  C—NH  + HOH + O ——→ OC   C(OH)NH ——→ + NaOH ——→
|   |   \                                 |   |     \
|   |    CO                               |   |      CO
|   |   /                                 |   |     /
NH—C—NH                                   NH—C(OH) NH
Uric acid.                                Intermediate form.

NH2  COONa                          NH2
|    |                              |
OC   C(OH)NH ——→ by acid ——→        OC    CO—NH
|    |     \                        |     |    \
|    |      CO                      |     |     CO + CO2
|    |     /                        |     |    /
NH—C(OH)NH                          NH—CH—NH
Uroxanic acid                       Allantoine.
(sodium salt).
```

The oxidation in dogs and other mammals may follow a similar course.

Distribution of nitrogen between different purine bodies in different mammalia. In most of the mammalia 79-98 per cent. of the uric acid formed in the body is converted into allantoine. In man 80-100 per cent. appears to escape destruction. The proportion of uric acid and bases also varies widely. The following table, taken from Hunter and Givens' work, illustrates this variability in different mammalia. The figures under total purine nitrogen are simply average figures added from their other tables to give some idea of the total nitrogen appearing per day in the form of purines, including allantoine. This total is very variable and depends in part on diet, since many vegetables contain allantoine.

Orders and species	Total purine nitrogen—grms.	Per cent. of purine-allantoine nitrogen			Uricolytic index	Purine coefficient
		Allantoine	Uric acid	Bases		
Marsupialia						
Opossum.....	0.04	76.0	19.0	6.0	79	4.1
Rodentia						
Rabbit.......					95	26.0
Guinea pig...		91.0	6.0	3.0	94	27.0
Rat..........		93.7	3.7	2.7	96	37.0
Ungulata						
Sheep........	0.2-0.6	64.0	16.0	20.0	80	8.0
Goat.........	1.0	81.0	7.0	12.0	92	17.0
Cow..........	8.0	92.1	7.3	0.7	93	18.0
Horse........	1.6	88.0	12.0	0.5	88	3.7
Pig..........	0.3	92.3	1.8	5.8	98	12.0
Carnivora						
Raccoon......		92.6	5.4	2.0	95	16.0
Badger.......	0.25	96.9	1.9	1.2	98	28.0
Dog..........	0.1-0.3	97.1	1.9	1.3	98	29.0
Coyote.......	0.15	95.6	2.6	1.8	97	23.0
Primates						
Monkey......	0.045	66.0	8.0	26.0	89	4.5
Chimpanzee..					0	
Man.........	0.2	2.0	90.0	8.0	2	2.5

The foregoing table shows, at a glance, the exceptional nature of the purine metabolism of man and the chimpanzee. Attention is called to the fact that in the monkey and some other mammalia the proportion of purine base nitrogen may be larger than that of uric acid. The fact that in man uric acid is so much greater than the allantoine has led to the conclusion that man has no power of destroying uric acid, and this is in harmony with the fact of the absence of uricase from his tissues. But, on the other hand, the very small purine coefficient arouses the suspicion that some of his purine catabolism is represented in other forms of nitrogen, and the question whether man does or does not destroy uric acid or other purines cannot be said to be definitely settled. The purine coefficient in the above table represents the milligrams of purine-allantoine nitrogen secreted per day per kilo body weight; the uricolytic index is the ratio of allantoine nitrogen to the sum of allantoine and uric acid nitrogen only. "It is taken as the measure of the animal's capacity to oxidize uric acid arising intermediarily."

With the discovery of the origin of uric acid we have not, by any means, exhausted the subject. Uric acid is but one of the purines, purines are important constituents of the most important constituent of living matter, namely, the chromatin of the cell nuclei. The very pregnant question remains behind, namely, can the animal organism make its purines from other nitrogenous material of a non-protein kind, or must it depend entirely on purine materials in the food? If it does make purine from amino-acids, why may not some of the uric acid have this origin? Why must it be formed altogether from the nucleins of the food or from those of the body cells? The question thus raised is susceptible of but partial answer at this time. A recent observation of Taylor and Rose, in which a man on a purine-free, egg, starch and sugar diet increased the nitrogen intake from about 6 to 40 grams of nitrogen per day with an accompanying increase of uric acid excretion from about 0.3 to 0.82 gram per day would indicate that the uric acid of the urine may be synthesized in part from non-purine precursors. On the other hand, this rise may be due to a stimulated nuclein catabolism.

Synthesis of uric acid in birds and reptiles. There is no doubt that the sauropsida, the birds and reptiles, are able to form purines and uric acid from non-purine forerunners. Thus, if they be fed on proteins poor in nuclein, the greater portion of the nitrogen appears in the urine in the form of uric acid. Similarly, if their livers be perfused with blood containing various amino-acids, uric acid is formed. They convert ammonium lactate into uric acid. Purine synthesis is for them very easy of accomplishment. Moreover all the invertebrates, so far as they have been examined, are found to secrete their nitrogen largely in the form of purine nitrogen. They must manufacture their uric acid also from non-purines. All plants have this power. Assuredly so general a property of living matter is not lacking in the mammals. All mammals live during the first months of life chiefly on milk, which is almost free from purines, and at this very time they are manufacturing and catabolizing nucleins at a very rapid rate. In the developing bird's egg the purine-free proteins of the yolk and white are in part converted rapidly into nucleic acid.

The power of synthesizing purines from non-purine precursors appears, therefore, to be a universal attribute of living cells. It is not at all probable that this power is lost in the adult mammalia, for no evidence has thus far been obtained that purines are necessary in the food to make good purine waste. With the mammalian organism having this power of synthesis of purines which are converted so readily by the purine oxidases into uric acid, it would appear surprising if some of the uric acid was not formed directly from purine thus produced, before it has been incorporated into the nucleic acid. We have no evidence, however, that it is so produced, and it might be that the synthesis took place only in the nucleus of which the membrane might permit the passage inward of raw materials, but prevent the passage outward of the synthesized purines before they were incorporated in the chromatin or nuclein. The fact that the nuclear wall is, in many cases, derived from or composed, in part at least, of chromatin, through which the purine must pass before escaping to the cytoplasm, might be a device responsible for the failure of uric acid to be set free from these purines. The purine oxidases, perhaps, are in the cytoplasm, rather than in the nucleus, and so act only on those purines which escape from the nucleus.

Various attempts have been made to discover what the raw materials are from which pyrimidine and purine might be formed. This matter has been discussed already on page 183. The presence in the sperm head of such large amounts of the basic amino-acids, histidine, lysine, and arginine, draws attention to these substances as the possible precursors. Moreover, arginine has guanidine already in its molecule.

Not only does the dog and ox liver have the power of destroying uric acid, but it will also resynthesize it if the conditions are changed (Ascoli and Izar). The same fact is true of the livers of birds, although their behavior in this respect appears to vary with the diet. Thus hens fed in the laboratory resynthesized uric acid without difficulty (Izar), and behaved both in regard to uricolytic powers and resyntheses like dog's liver, whereas little resynthesis was obtained from the livers of hens bought in the market. Brunton and Bokenham showed long since that the power of resynthesis was lacking in livers of dogs which have fasted 72-192 hours; such dogs have also a much-reduced power of uricolysis, a fact of importance in understanding the contradictory results obtained with human livers. If blood was added which had been taken from a dog fed shortly before, the power of synthesis returned. The following protocol illustrates this.

Experiment Illustrating the Dependence on Diet of Uric Acid Destruction and Synthesis by Dog's Liver:

160 grams of sieved liver pulp of a dog 5 days fasting plus 1400 c.c. 0.85% NaCl plus 948.0 mgs. uric acid in 200 c.c. Li_2CO_3 (1:90) solution. 3 days autolyzed with air drawn through and then divided into 4 equal parts.

		Uric acid recovered
A. Coagulated immediately		149.5 mgs.
B. Added 100 c.c. NaCl solution	then 72 hours autolysis under CO_2	147.3
C. " " " defibrinated blood of dog fasting 72 hours		164.3
D. Added 100 c.c. defibrinated blood of dog fed 12 hours before		226.22

It will be seen that under CO_2 the autolysis for 72 hours caused no reformation of uric acid except when blood had been added from a fed dog. The uric acid increased in this case 77 mgs.

Uricolysis and Regeneration of Uric Acid in the Livers of Hens, Geese and Turkeys.

Liver of	10 per cent. sieved liver pulp c. c.	Hours fasting	Added uric acid—mgs.	Uric acid found after 72 hours autolysis under air	Uric acid found after a further 72 hours autolysis under CO_2
Hen 1	180	3	441	60.2	322.8
" 2	160	48	"	223.4	242.4
" 3	190	72	"	210.3	334.9
" 4	290	6	391.2	30.2	314.7
" 5	220	48	417.0	254.6	309.0
" 6	350	3	790	256.4	635.8
Turkey 1	1250	10	640.7	67.8	556.1
" 2	1010	120	817.4	752.6	740.2
" 3	1140	4	767.2	136.6	741.8
" 4	1380	10	970.1	186.0	814.2
" 5	1170	96	832.1	654.2	708.2
" 6	1240	48	847.4	550.2	622.8
Goose 1	1430	10	811.1	34.2	738.8
" 2	1000	10	817.4	168.6	784.2
" 3	1750	48	1014.0	966	982.2
" 4	1739	96	890	830.0	804.6
" 5	1540	48	1034.0	802.0	879.4
" 6	1150	102	913	876.0	868.0

These experiments show, first, that birds' livers have great powers of uricolysis and this power is enormously reduced by previous fasting; and, second, that the uric acid reappears if autolysis is continued for 72 hours under CO_2. If the uric acid is really destroyed, it would appear that when oxygen is present uric acid is destroyed or hydrolyzed and resynthesized by reduction or when CO_2 is abundant. The quantity reappearing is in all cases proportional to that which disappears. The resynthesis depends here, also, on the presence in the blood of a thermolabile enzyme, and an alcohol-soluble, heat-stable component in the liver, but this coferment is not present in the kidney. Hen's blood alone destroys uric acid very fast. The ferment is not specific, that is the ferment in dog's blood will act in the case of the hen's liver. A further investigation showed that in the presence of CO_2 liver forms uric acid out of dialuric acid and urea. On the other hand lactic, paralactic, tartronic, acrylic, oxalic, mesoxalic acids or their salts caused no uric acid formation. Allantoine had no effect. Izar was unable to isolate the intermediary substance. It would seem probable from these observations that the liver decomposed uric acid to dialuric acid and urea and resynthesized them under conditions of reduction and possibly of a change in reaction due to the CO_2. Recent work makes these results doubtful.

```
    NH—C=O                             NH—C=O
    |  |                               |  |
O=C    CHOH       NH₂    ——→       O=C    C—NH
    |  |      +      >CO               |  ||    >CO  +  H₂O
    NH—C=O        NH₂                  NH—C—NH
 Dialuric acid.   Urea.                Uric acid.
```

Allantoine.—$C_4H_6N_4O_3$. Allantoine is the diureide of glyoxylic acid. This acid will unite with urea through each of its hydroxyl groups to form a diureide according to the following equation:

```
NH₂       HO—CH—OH        H₂N             NH—CH—NH—CO—NH₂
|             |           |               |   |
CO   +        |      +    CO   ——→   O=C      |                + 2H₂O
|             |           |               |   |
NH₂       HO—C=O          H₂N             NH—C=O
Urea.   Glyoxylic acid.   Urea.           Allantoine.
```

Boiling with alkalies decomposes it by hydrolysis into urea and glyoxylic acid.

Allantoine is easily produced by the oxidation of uric acid with lead peroxide and the allantoine of the urine arises certainly in part from the oxidative decomposition of the purines as already discussed. Carbon dioxide is produced at the same time.

By boiling water allantoine is hydrolyzed into urea and allanturic acid as follows:

```
 NH—CH—NH—CO—NH₂                      NH—CH—OH
 |   |                                 |   |
OC   |                + H₂O  ——→      OC   |        + H₂N—CO—NH₂
 |   |                                 |   |
 NH—CO                                 NH—CO
Allantoine.                         Allanturic acid.     Urea.
```

If allanturic acid is reduced it yields hydantoine, $C_3H_4N_2O_2$, which when hydrolyzed goes over into hydantoic acid (glycol uric acid) and finally to glycocoll and ammonium carbonate.

```
NH—CH—OH                      NH—CH2
|    |                        |    |
OC   |      + 2H   ——→        OC   |      + H2O
|    |                        |    |
NH—CO                         NH—CO
Allanturic acid.              Hydantoine.
```

```
NH—CH2
|    |
OC   |  + H2O  ——→  NH2—CO—NH—CH2—COOH  ——→
|    |                          NH2—CH2—COOH + NH3 + CO2
NH—CO
Hydantoine.         Hydantoic acid.     Glycocoll.
```

It will be noticed that the ring in hydantoine, except for the double bonds, is the imidazole ring. Allantoine is thus related to the base histidine. It is also related to cyanuric acid, $C_3H_3N_3O_3$. Allantoine crystallizes from urine in sheafs of plate-like crystals, but when pure in clumps of prisms, m.p. 231-232.

Allantoine has been found in the urine of herbivorous and carnivorous animals; in calves', cows' and sheep urine and in that of the dog, cat and monkey. It was first found in the allantoic fluid of the calf, whence its name. It is not found in the urine of human adults, except that it has been, at times, found in the urine of pregnant and nursing women. It is said to occur, however, in the urine of children in the first week of life.. Human adults are said to be able to destroy allantoine taken by the mouth, but the lower animals, such as monkeys and the carnivora, are not able to do so, nor can a young child. It is doubtful whether human adults really have the power of destroying the substance.

In the lower animals and monkeys allantoine appears to be a terminal product of the purine metabolism and allantoine, given by the mouth, is excreted unchanged in the urine. In man, however, allantoine taken by the mouth does not appear as such in the urine (Minkowski). It is apparently oxidized but the products of its oxidation have not been found. It is possible that the destruction occurs in the intestine. Allantoine may be prepared from the urine of cows which secrete 20-30 grams a day. The method for its quantitative separation from the urine is complicated and described on page 970.

Since purines are found in plants as well as animals it is interesting to note, as showing how closely similar the chemical processes are in the two divisions of the living kingdom, that allantoine is also found there. It occurs in sprouting wheat seedlings and is a constituent of the bruisewort or slippery root, Symphytium officinale.

Hippuric acid.—This acid is found in the urine of herbivorous animals, such as horses or cows, in large amount, but only about .7 gram per day occurs in human urine. It is of especial interest because, unlike all the substances discussed hitherto, it represents one of the chemical methods of defense of the organism against toxic substances and is formed in the kidney itself, or at least part of it is. From both these points of view it will well repay a careful study. The name connects it with horse urine. (Gr. *hippos*, horse; and *ouron*, urine.)

Chemistry. Hippuric acid is benzoyl-glycin, $C_9H_9NO_3$,

```
          O
          ‖
     C  —C—NH—CH₂—COOH
   //  \
 HC     CH
  |     ‖
 HC     CH
   \\  /
    CH
```

Hippuric acid.

It is easily decomposed by alkalies, by bacterial action, by boiling acids, or by an enzyme of the kidney (histozyme) into glycocoll and benzoic acid. It crystallizes in long, fragile, rhombic, four-sided prisms, which dissolve in 600 parts of cold water, more easily in hot, readily in alcohol, slightly in ether, but readily in acetic ester; but which are not soluble in benzene, petroleum ether or carbon bisulphide. The melting point is 190.2° (187.5°?) and by farther heating the crystals form a red mass, which decomposes with the formation of an odor of hay and then of hydrocyanic acid and benzoic acid. Hippuric acid may be recognized by the odor of nitro-benzene obtained on evaporating it with nitric acid and heating the residue, a reaction given also by benzoic acid. It is differentiated from benzoic acid by the insolubility of hippuric acid in petroleum ether. The alkali and alkaline earth salts are soluble in water; the silver salt is less readily soluble.

Occurrence. Hippuric acid occurs, not only in the urine of cows, horses, pachyderms, carnivora and man; it has been found in the sweat after heavy doses of benzoic acid; in the urine of turtles and some insects; but not in bird's urine. In the urine of birds one finds in place of hippuric acid *ornithuric* acid, a compound of benzoic acid with diamino-valerianic acid, or ornithine, $C_5H_{12}N_2O_2$, in place of glycocoll. It is said by Baumann that hippuric acid completely disappears from dog's urine if there is no putrefaction in the intestine.

Amount. The amount contained in the urine varies with the diet. On a diet containing much fruit or vegetables the excretion by human beings may rise to two grams a day. Herbivorous animals excrete a great deal more than carnivorous. Its variation in disease has not yet been studied.

Origin. The wide variation of the excretion with the diet indicates at once that part at least of the hippuric acid must be derived from the food and this has been confirmed by experiment. An increase in the intake of benzoic acid, or of substances which can form benzoic acid in the body, causes a marked increase in the hippuric acid excreted. An increase in glycocoll intake, however, produces no change in the output of hippuric acid. Glycocoll is supplied by the body, benzoic acid mainly from the foods. Since benzoic acid is more toxic than hippuric acid, the conversion is evidently a process of detoxication, a means of defense of the organism against poisons. We find, indeed, that not only does the organism defend itself against benzoic acid by pairing it with glycocoll, but also against other aromatic substances, such as phenyl acetic acid, cresole and phenols, the same means of defense is used. The formation of glycocholic acid in the bile may be a similar process, since cholic acid is decidedly toxic. It may be mentioned also, in this connection, that other substances than glycocoll may be used for pairing purposes, namely, sulphuric and glycuronic acids.

Toxic substances owe their toxicity to two peculiarities: first, they possess a large amount of available potential energy, being generally unstable compounds which liberate energy on decomposing; and, second, they are abnormal substances not used in the normal metabolism of the cells in question. The organism protects itself against such substances in several different ways. It may oxidize them and thus make them more stable; or by making them unite with other compounds which are stable and inert they are rendered indifferent to the body. It may happen that oxidation renders substances unstable rather than more stable and an organism may in this way increase the toxicity of substances. Benzene, for example, which is not very toxic, is converted by oxidation into the toxic phenol; and some nitriles, like the propionitrile, may be oxidized to the lacto-nitrile, which is much more reactive and poisonous.

We may take advantage of this property of the body of pairing some of its metabolic substances with unstable toxic substances to study the intermediate metabolism of the body, as will be shown in discussing cysteine (page 815). By giving a toxic substance there may be combined with it, and thus brought into the urine, an intermediate substance not normally found there or found in very small amount. Glycuronic acid and cysteine are substances of this kind.

Place of origin of hippuric acid. The formation of hippuric acid from benzoic acid and glycocoll is of the general type of amino-acid condensations, like that of creatinine, and the power of making such condensations is an attribute of all living matter. The work of Bunge and Schmiedeberg proves that in the dog the perfused kidney can bring about this synthesis, but in the rabbit, and probably other animals as well, other tissues may do it also (Salomon). The synthesis depends like all such syntheses, upon a plentiful supply of oxygen and Drechsel considered it to be probably an oxidation-reduction synthesis.

Source of the benzoic acid. We not only take benzoic acid itself in small amounts in fruits and berries, particularly in cranberries, but grass and vegetables often contain other aromatic compounds, such as quinic acid, which by digestion, fermentation by bacteria or by oxidation yield benzoic acid. It is for this reason that a fruit or vegetable diet increases the hippuric acid secretion. Benzoic acid or benzoates are often used, also, as preservatives in canned fruit, catsup, or other food products, and even in milk, so that the consumption of such preserved foods leads to the consumption of benzoic acid. But the proteins themselves may also, by bacterial decomposition, give rise to benzoic acid. Thus phenyl alanine by oxidation goes over into phenyl pyruvic acid, $C_6H_5.CH_2.CO.COOH$, which by further oxidation is converted into phenyl acetic acid and this into phenyl carbonic or benzoic acid. It is probable from Baumann's observations that this latter process only occurs with the intermediation of the putrefactive bacteria in the intestine.

Source of the glycocoll. The question naturally arises how much glycocoll the body has at its disposal to neutralize toxic matters like benzoic acid and what is the origin of this glycocoll. Does it come from the protein or is it synthesized in the body? Concerning the first question of the amount of glycocoll which the body can supply for purposes of detoxication, experiment has shown that the herbivora have quite remarkable powers in this respect. Thus Ringer found that a goat might take 25 grams of benzoic acid a day and excrete it as hippuric acid. Magnus-Levy found in sheep and rabbits that after ingesting benzoic acid 27.8 per cent. of the urinary nitrogen might appear in the form of hippuric acid nitrogen; and Wiechowski that 50 per cent. might thus appear. There is only a slight diminution in the other nitrogen constituents of the urine, except possibly a diminution in the uric acid (Weiss and Levin). This would indicate that the glycocoll nitrogen was supplied in addition to that which would normally have been eliminated. In carnivora, and probably in man, the conditions appear to be different. No more glycocoll is in them available for pairing than can be accounted for by the decomposition of their body or food protein. Abderhalden, Gigon and Strauss found that in carnivora, herbivora and hens the entire amount of glycocoll in the whole body, exclusive of fat, feathers and intestinal contents, was only 2.33-3.34 per cent. of the total proteins. This amount is far too little to admit of the explanation that the glycocoll in herbivora is derived simply from the body proteins. It might come in small part from the purines which may yield glycocoll on certain decompositions and it will be remembered that glycocoll is present in considerable amounts in the muscle of pecten and other mollusks. It would seem probable either that glycocoll is synthesized in the body from ammonia and glyoxylic acid by reduction, or else the hippuric acid may be formed in part by the benzoic acid pairing with other amino-acids of a longer carbon chain, as happens in birds in ornithuric acid, and these longer chains are afterwards partially oxidized to amino-acetic acid. The former explanation is, perhaps, the more probable, since glyoxylic acid, or its aldehyde, is easily derived from the carbohydrates and the synthesis of ammonia and the aldehyde to glycocoll very probably occurs in the body, although demonstrative proof of this has not yet been found. It is not impossible that glycocoll may be formed in this way normally as one of the precursors of urea in the transformation of ammonia to urea. Whether the use of glycocoll to detoxicate benzoic acid reduces the amount of glycocholic acid in the bile should be investigated.

No definite answer can, however, be given as yet to the question of the origin of the large quantities of the glycocoll in herbivora until the matter has been more extensively studied.

Method of isolation. Hippuric acid is readily obtained from fresh horse or cow urine by first boiling it with milk of lime, filtering off the phosphates, evaporating to about half its original volume, cooling and adding strong hydrochloric acid to a plainly acid reaction. The hippuric acid crystallizes out. The crystals are separated by suction, pressed as dry as possible in filter paper, redissolved in milk of lime and recrystallized by the addition of acid. They may then be recrystallized from hot water, being decolorized if necessary by charcoal.

Quantitative determination. The method formerly used is that of Bunge and Schmiedeberg, which is very cumbersome and by no means exact. The urine, very slightly alkaline with sodium carbonate, is evaporated nearly to dryness. The residue is extracted thoroughly with strong alcohol, the alcohol evaporated on the water bath, the dry residue dissolved in water, transferred to a separatory funnel, the solution acidified with sulphuric acid and extracted by thorough agitation repeatedly (five or more times) with acetic ether. The acetic ether solution of hippuric acid is now shaken repeatedly with water in a separatory funnel, the acetic ether evaporated and the residue extracted with petroleum ether to remove benzoic acid, fats, oxy acids, phenols, etc. The hippuric acid remains undissolved. The residue is now dissolved in a little warm water and evaporated at 50-60° to crystallization. The crystals are weighed in a small weighed filter. The mother liquor is extracted with acetic ether, the ether evaporated and the weight of the residue added to that of the crystals. The better method of Folin and Flanders is given on page 968.

Ammonia.—Blood contains small amounts of ammonia and some of this is excreted in passing through the kidneys and appears in the urine. The amount of ammonia in the urine is greatly increased in any condition in which larger than normal amounts of acid are produced. Ammonia is one substance which is used to neutralize the acid formed in the course of cell metabolism, as has already been discussed on page 248. The amount of ammonia normally present in the urine of human adults is about 0.7 gram per day. The amount and the relative proportion it makes of the total nitrogen of the urine may be increased by the ingestion of mineral acids. In diabetes or in fasting, where there is an abnormal formation of acetoacetic acid, ammonia is also increased. On a high and a low protein diet, page 706, Folin found the total amount not much changed, but the relative proportion of ammonia was greatly increased on low protein. Directions for the determination of the ammonia are given on page 961.

Other nitrogenous substances present in small quantities.—*Amino-acids and peptides.* Normal urine contains small amounts of these substances and under pathological conditions the amount of amino-acids may increase. Thus after phosphorus poisoning, or in cirrhosis of the liver and in some other conditions, amino-acids such as tyrosine, leucine, glycocoll, etc., have been isolated from the urine. Normally, however, these substances are present in very small amounts indeed. A number of substances have been isolated, however, which are probably peptides or partially oxidized fragments of the protein molecules which have escaped the metabolism of the body. Such bodies are the *oxyproteic acid* of Bondyzynski and Dombrowski; *antoxyproteic acid* of the same authors and *alloxyproteic* acid; and uroferric acid of Thiele. The total amount of N in these substances found by Ginsberg and Gawinski in the urine of a man on a mixed diet amounted to 3-6.8 per cent. of the total nitrogen of the urine. This, it will be observed, is just about the amount of N of unknown nature in the urine of an average adult. The relative amount is greatly increased in phosphorus poisoning and in various conditions when body protein is being decomposed and the intake of protein is low.

	C	H	N	O	S	Reactions
Oxyproteic acid	39.62	5.64	18.08	35.54	1.12	No Ehrlich, biuret, or xanthoproteic.
Antoxyproteic acid ..	43.21	4.91	24.4	26.33	0.61	Ehrlich diazo positive. Others negative.
Alloxyproteic acid ..	41.33	5.70	13.55	37.23	2.19	Biuret and Ehrlich negative.
Uroferric acid					3.46	Biuret, Millon, Adamkiewicz negative.

It is doubtful whether these substances are unitary substances.

Basic substances. Small quantities of basic substances corresponding in a general way to those found in meat extract have been isolated from the urine by Kutscher and his colleagues and by other observers, particularly by French physiological chemists. Among these are *trimethyl amine; methyl guanidine; novain; reductonovain; dimethylguanidine; gynesin*, $C_{19}H_{22}N_2O_3$, from female urine; *mingin*, $C_{13}H_{18}N_2O_2$; *vitiatin*, a homologue of choline; histidine; imidazole-acetic acid; and methyl-pyridine chloride. This last is probably derived from tobacco or coffee and is not a natural product of the body metabolism. In addition *putrescine* and *cadaverine*, the former tetramethylendiamine and the latter pentamethylendiamine, were isolated from the urine by Baumann and von Udransky. Griffiths and Bouchard have particularly studied the ptomaines of urine. They generally isolate them by making the urine alkaline and shaking it out with ether. The bodies thus isolated have not been, for the most part, identified. They are said to be toxic to animals, and Bouchard and other French observers have standardized the toxicity of the urine by injecting it into rabbits. The urotoxic coefficient Bouchard defines as the weight of rabbit in kilos which is killed by the amount of urine secreted by 1 kilo of the body weight of the individual whose urine is being investigated. A part of the toxicity of human urine is generally ascribed to the potassium salts it contains, but probably not all of it can be thus accounted for.

A very interesting urinary constituent is that of *urocanic acid* found in the urine of dogs but not thus far isolated from human urine, although other imidazole substances have been found there. Urocanic acid has recently been found to be imidazolyl-acrylic acid and is, therefore, a decomposition product of histidine. It was found by Hunter in the digestive products of casein when digested for a long time by a pancreas mixture. The conditions of its appearance in such mixtures have not been determined. The formula is as follows:

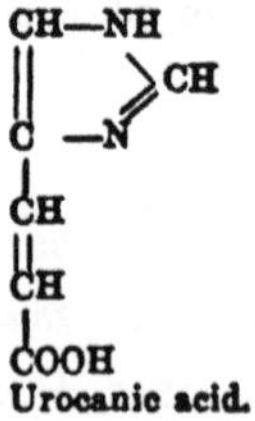

```
CH—NH
||    \
||     >CH
||    //
C  —N
|
CH
||
CH
|
COOH
```

Urocanic acid.

The acid forms very curious, sickle-shaped crystals.

Aromatic oxy acids of the urine.—These include *phenol, indoxyl, scatoxyl*, and *phenyl acetic, paraoxyphenyl propionic, oxymandelic* and *homogentisic* acids. They are all derived either from tyrosine, tryptophane, phenylalanine or other unknown phenyl derivatives of the proteins. The first group includes the ethereal sulphates or conjugated sulphates.

Ethereal sulphates. By the decomposition of the aromatic amino-acids of the proteins, phenols and indoles are produced. They are chiefly formed in the putrefactive decomposition of the proteins in the intestine. In their passage through the body they are oxidized to indoxyl, scatoxyl or hydroxyphenol, and then are paired with, or conjugated with, sulphuric acid to form what are known as the ethereal or conjugated sulphates. They are excreted in the urine for the most part as the potassium or sodium salts of these bodies. The amount excreted per day varies from 0.1-0.6 gram sulphuric acid. Urinary indican, that is the potassium salt of indoxyl-sulphuric acid, is such a substance. The place where pairing with sulphuric acid occurs is supposed to be the liver.

Phenol, C_6H_5OH, and *cresol,* methyl phenol, $C_6H_4(CH_3)OH$. The mother substance of phenol and cresol is tyrosine and phenylalanine, and possibly other aromatic amino-acids if any exist, other than tryptophane. The greater part of the phenol and cresol is excreted as conjugated sulphate, but a small portion is free in the urine, and a part is conjugated with glycuronic acid. If phenol is ingested, only a portion reappears in the urine. A part is evidently destroyed in the body. Probably like benzene some is destroyed by the rupture of the ring. The source of the phenol and cresol of the urine is believed to be the putrefactive decomposition of tyrosine and phenylalanine in the alimentary canal. The reason for this view is that the amount in the urine is much increased by excessive intestinal putrefaction and may be reduced to a minimum by a milk diet and by the ingestion of carbohydrates, a procedure which reduces putrefaction; or by the use of cathartics, and particularly such as have an antiseptic action such as calomel. These bodies are supposed, then, not to come from the metabolism of the tyrosine in the tissues, but entirely to be derived from the putrefactive decomposition in the intestine. For this reason the determination of the conjugated sulphates is supposed to give an indication, albeit a very uncertain one, of the amount of intestinal putrefaction. It is much easier, however, to determine putrefaction by means of the indican test to be spoken of in a moment.

The exact manner in which the tyrosine and phenylalanine are decomposed in putrefaction is not certainly known. But the transformation is believed to be as follows, all of these steps occurring in the intestine:

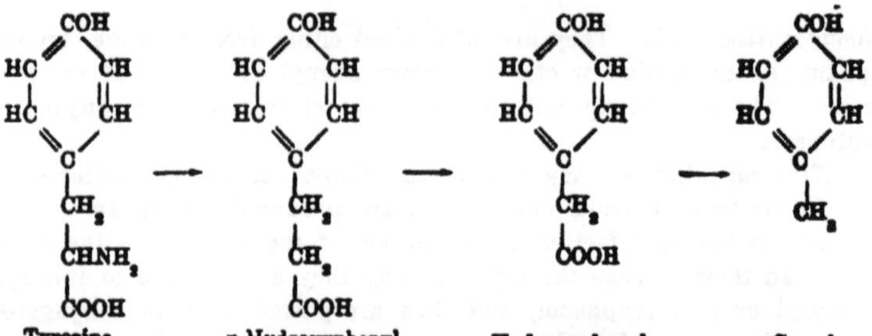

Tyrosine. → p-Hydroxyphenyl propionic acid. → Hydroxyphenyl acetic. → p-Cresol.

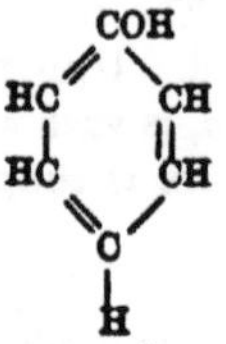

Phenol. 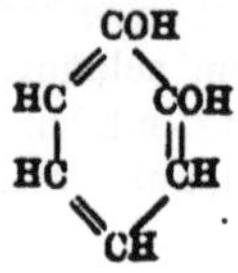 Pyrocatechol. 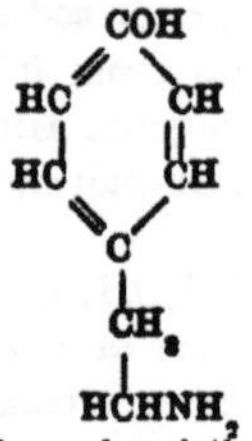 p-Hydroxyphenylethylamine.

It is possible, though perhaps not very probable, that p-hydroxyphenylethylamine is first formed which is then oxidized to hydroxyphenyl acetic acid. After the administration of phenol the urine may become dark colored, owing to the formation of hydrochinon, p-dioxybenzene and pyrocatechol. These in the air undergo spontaneous oxidation with the formation of dark coloring matters. They are reducing bodies, reducing Fehling's solution and other metallic oxides. If pyrocatechol is added to a very dilute ammoniacal solution of ferric chloride containing tartaric acid, the solution is colored a violet, or cherry-red color which is changed to a green on the addition of sufficient acetic acid. This same reaction is given by adrenaline. The oxyphenols are fairly stable in acid reaction, but very unstable in alkaline.

Indoxyl-sulphuric acid. Indican **of the urine.** This is found in the urine as the potassium salt. Its formula is as follows:

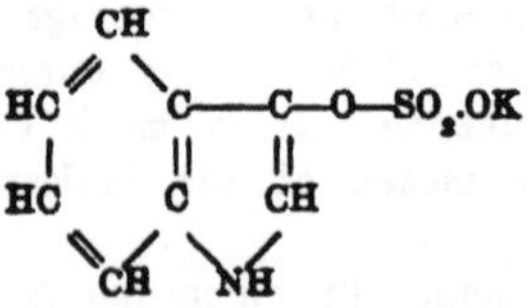

Indican.

Indole is formed by the putrefaction of tryptophane in the intestine. The indole thus formed is oxidized to indoxyl during its passage through the body and is paired for the most part with sulphuric acid, but in part also with glycuronic acid, presumably in the liver.

```
    CH                                          CH               CH
  //  \                                       //  \            //  \
HC      C——C—CH2—CHNH2—COOH              HC      C——CH    HC      C——COH
|       ||  ||                  ——>      |       ||  ||——>|       ||  ||
HC      C   CH                           HC      C   CH   HC      C   CH
  \\  /  \ /                               \\  /  \ /       \\  /  \ /
    CH    NH                                 CH    NH         CH    NH
   Tryptophane.                               Indole.          Indoxyl.

            CH                          CH
          //  \                       /  \\
        HC      C——CO        OC——C        CH
        |       ||   |        |   ||       |
        HC      C    C=========C   C       CH
          \\  /  \  /          \  /  \  //
            CH    NH            NH    CH
                      Indigo blue.
```

The steps in the transformation of tryptophane to indole are not entirely certain. It is probable that deamidization happens at first and then the carbon side chain is oxidized off, indole-propionic, indole-acetic acids and scatole being intermediate products (p. 441); or that by decarboxylization the indolethylamine is first formed, which is later converted into indole by oxidation. The important fact is, however, that the formation of indole does not occur in the course of the metabolism of tryptophane in the body, or if it does the indole so formed is destroyed. The presence of indican in the urine shows, therefore, that indole is being formed in the intestine. If more than the normal amount of it is present, it indicates the occurrence of an abnormal amount of intestinal putrefaction, or of putrefaction elsewhere in the body, as for example in decomposing abscesses.

The test for indican in the urine is very simple and is described in the practical exercises. It consists essentially in oxidizing the indoxyl in an acid solution by means of hypochlorite or ferric chloride to indigo blue and shaking out the indigo blue in chloroform. If the chloroform is more than a light blue, it means an abnormally large putrefaction. A satisfactory method for the determination of the amount of indican in the urine is that of Jolles. It was suggested by Folin that the color be compared with a standard of Fehling's solution in a colorimeter. The method is not very satisfactory. The total conjugated sulphuric acid may be estimated accurately by the gravimetric method, but this is too difficult a method for clinical use. A very good idea of the relative amount of putrefaction can usually be obtained by a little practice in making the test for indigo blue, so that an abnormally large putrefaction can easily be detected.

The amount of the putrefaction is greater on a heavy meat diet. It may be so great that the chloroform becomes almost black. On the other hand, it is often extremely faint. It increases after obstruction of the small intestine.

It must not be supposed that the indican test, if it is negative, means necessarily that intestinal processes are normal. It might be that the intestinal products contained little tryptophane, or that the bacteria present did not form indole, or that the indole absorbed was in part destroyed, or that the absorption was defective. But when the test is positive the conclusion is justified generally that there is abnormal putrefaction in the intestine, or rather an abnormally large absorption of putrefactive products, due possibly to constipation. According to Jaffé, the amount of indigo in the urine of a healthy man is between 5 and 20 milligrams in 24 hours.

Scatoxyl-sulphuric acid. Scatole (Gr. *skor*, excrement) is a crystalline, fecal-smelling substance, C_9H_9N, formed by the putrefaction of tryptophane by certain kinds of bacteria. It occurs in the feces. Some is absorbed from the large intestine and when passing through the body it is oxidized to scatoxyl and paired like indole, presumably in the liver, with sulphuric acid and excreted as the salt of this substance. In constipation so much of this substance may be absorbed as to give to the breath and exhalations of the body a very distinct fecal odor. Its presence in the urine, therefore, is an indication of intestinal putrefaction.

```
        CH                              CH
      //  \                           //  \
   HC      C———————C—CH₃           HC      C———————C—CH₃
   |       ||      ||              |       ||      ||
   HC      C       CH              HC      C       COH
     \\   / \     /                  \\   / \     /
       CH     NH                       CH     NH
   Scatole.                        Scatoxyl.

                    CH
                  //  \
               HC      C———————C—CH₃
               |       ||      ||
               HC      C       C—O—SO₂.O Na
                 \\   / \     /
                   CH     NH
               Sodium scatoxyl sulphate.
```

The constitution of the scatole of the urine is unknown, and also that of scatoxyl. The ingestion of scatole produces in the urine a chromogen which gives a bright red color when the urine is made strongly acid with hydrochloric acid. This color is sometimes referred to as *urorosein*. It is probable that urorosein is in reality indole acetic acid.

Scatole is probably formed from indolacetic acid, and may be one stage in the formation of indole from tryptophane.

Parahydroxyphenylacetic, parahydroxyphenylpropionic and parahydroxyphenylglycolic acid (oxymandelic acid). These occur in small quantities in the urine, the last, particularly in acute yellow atrophy of the liver. They all represent intermediary products of the oxidation of tyrosine and possibly of phenylalanine.

```
    COH              COH              COH              COH
   //  \            //  \            //  \            //  \
 HC     CH        HC     CH        HC     CH        HC     CH
  |     ||         |     ||         |     ||         |     ||
 HC     CH        HC     CH        HC     CH        HC     CH
   \\  /            \\  /            \\  /            \\  /
     C                C                C                C
     |                |                |                |
    CH2              CH2              CH2              CH(OH)
     |                |                |                |
    CHNH2            CH2              COOH             COOH
     |                |
    COOH             COOH
```

p-hydroxyphenyl-aminopropionic acid. p-hydroxyphenyl-propionic acid. p-hydroxyphenyl-acetic acid. p-hydroxyphenylglycolic acid.

These acids all give the Millon reaction. They are soluble in water and in ether in the free state. The melting point of p-oxyphenyl-propionic acid is 128° and that of paraoxyphenylacetic acid is 148°. The oxymandelic acid crystals melt at 162° C.

Homogentisic acid. Dioxyphenylacetic acid. $C_6H_3(OH)_2CH_2$ COOH. Hydrochinone acetic. This acid is the peculiar acid found in alcapton urine. It oxidizes spontaneously in the air and the urine turns black, at first in its upper layers where the oxygen has entrance. This acid is hydrochinone acetic acid.

```
        OH
        |
        C
      //  \
    HC     CH
     |     ||
    HC     C—CH2—COOH
      \\  /
        C
        |
        OH
```

Homogentisic acid.

This substance is formed from tyrosine and phenylalanine. The amount formed in some cases may be as large as 16 grams a day, usually, however, it is less than this, i.e., 3-5 grams. This substance is probably formed by a disturbed tyrosine metabolism, as discussed on page 809. The cause of alkaptonuria is still quite obscure. It appears to run in families. Baumann thought it was due to an abnormal intestinal flora, but that opinion is now abandoned. According to Garrod, it occurs most commonly where there is blood relationship of the parents. It is more common in males than in females. It is still doubtful whether the normal course of metabolism of tyrosine carries the latter through homogentisic acid, alkaptonuria being due to more of the intermediate substance escaping in the urine; or whether it is due to an abnormal manufacture of homogentisic acid in certain individuals. The work of Dakin indicates the latter.

Homogentisic acid is a strong reducing agent, reducing Fehling's solution, alkaline silver solutions and other metals. It does not reduce bismuth subnitrate. It crystallizes in large clear prisms. m.p. 146.5-147° C. It is soluble in alcohol, water and ether, but only slightly in benzene. It gives with ferric chloride a transient blue color, a reaction not specific but given by many other reducing substances such as cysteine, other phenols, etc. It is inactive. (For its preparation from the urine see Orton and Garrod, J. *of Physiology*, Vol. 27.)

Sulphur of the urine.—From 75 to 80 per cent. of the total sulphur of the urine is in the form of inorganic sulphate, when a person is on an ordinary diet of about 14-16 grams of nitrogen per day. The remainder of 20-25 per cent. is in organic union; a part as ethereal or conjugated sulphates, a small portion as taurine, a portion in the oxyproteic acids of the urine, some of this sulphur being unoxidized sulphur; and a part is present as cystine or cysteine.

Amount. The total amount of sulphur excreted per day depends upon the sulphur intake, rising and falling with this. Sulphur enters the body chiefly in the unoxidized form of protein sulphur and the total sulphur of the urine excreted per day is about 0.7-0.85 gr. The amount present as inorganic sulphate is about 0.5-0.6 gram per day. The remainder of 0.1-0.2 gram is organically bound. The methods of determining these different fractions in the urine are given in the practical part, page 981. The inorganic sulphates, the conjugated or ethereal sulphates and the neutral or unoxidized sulphur, are generally determined. The last group is a very heterogeneous group and has been very little studied. It is determined by difference.

Effect of diet on the distribution of sulphur. Sulphur is ingested chiefly as protein sulphur. There are, however, small amounts of inorganic sulphate, of sulphuric acid paired with phenols, or present as sulpholipins, and of taurine in all tissues, so that many different sulphur compounds are ingested. The ingested protein sulphur is for the most part excreted as inorganic sulphate, since protein ingested is not stored, but is burned in the body. Hence on a high protein diet not only does the total sulphur increase, but also the proportion of inorganic sulphur as well. On the other hand, on a low protein diet the total amount of sulphur is much reduced and the proportion present as neutral sulphur is increased. Since on a low protein diet the putrefaction in the intestine is reduced, there is also a reduction of the conjugated or ethereal sulphate. The following figures from a metabolism experiment by Folin[1] will show the change in the distribution of urinary sulphur when one changes from a normal to a reduced protein diet.

	Normal protein diet July 18	Reduced protein diet July 20
Volume of urine	1170 c.c	385 c.c.
Total nitrogen	16.8 gr.	3.60 gr.
Urea nitrogen	14.70 " = 87.5%	2.20 " = 61.7%
Ammonia nitrogen	0.49 " = 3.0	0.42 " = 11.3
Uric acid nitrogen	0.18 " = 1.1	0.09 " = 2.5
Creatinine nitrogen	0.58 " = 3.6	0.60 " = 17.2
Undetermined nitrogen	0.85 " = 4.9	0.27 " = 7.3
Total SO_3	3.64 "	0.76 "
Inorganic SO_3	3.27 " = 90.0	0.46 " = 60.5
Ethereal SO_3	0.19 " = 5.2	0.10 " = 13.2
Neutral SO_3	0.18 " = 4.8	0.20 " = 26.3

In these results it may be observed that the proportion of ethereal sulphate rose on the low protein from 5.2 per cent. to 13.2 per cent., although the absolute amount excreted diminished. But the greater relative rise was in the neutral sulphur, which rose from a proportion of 4.8 per cent on the normal protein diet to 26.3 per cent. on the reduced protein. It is very suggestive, also, that the absolute amount of the neutral sulphur excreted did not diminish at all on the low protein diet, but on the contrary rose slightly, the neutral SO_3 on the normal protein diet being 0.18 gram and on the low protein diet being 0.20 gram. It is probable from these facts that the neutral sulphur corresponds to the endogenous wear and tear, whereas the other two fractions are largely derived from the protein of the food torn to pieces by the oxidation of the body.

Variation of the sulphur under other circumstances. The per cent. of neutral sulphur in the urine is also dependent upon various drugs. Thus cyanides and nitriles when ingested leave the body in mammals largely in the form of sulphocyanate in the urine. Hence, when any substance is ingested which can form hydrocyanic acid in the body, it will increase the neutral sulphur. There is, also, an increase of neutral sulphur under the influence of chloroform and other anesthetics. In cystinuria also the proportion of this sulphur increases.

Phosphorus in the urine.—*Form.* Phosphorus is found in the urine

[1] Folin: *American Jour. Physiol.*, 13, p. 118, 1905.

wholly in the oxidized form of orthophosphoric acid. It is present as the disodium, and monosodium phosphate and free phosphoric acid, the relative amounts of these substances present depending on the character of the diet. On a heavy protein diet the urine is acid, due in a large measure to the sulphuric and other acids formed from the meat; under such circumstances there will be more free phosphoric acid and of the monosodium salt; on a vegetable diet, or when salts of acids which are burned to carbonic acid in the body are ingested, the proportion of disodium phosphate is larger. The bases of the urine distribute themselves among the various acids in proportion to the strength of the acids, the stronger acids taking the stronger bases. Phosphorus in an unoxidized form is toxic, and not present in any food. The phosphorus of the foods is present always as phosphoric acid, which is either inorganic or it may be in an ester form, as it is in nucleic acid, in phospholipins, phytin, etc.

The amount in the urine. The amount of phosphorus excreted per day depends on the amount ingested in part, but it is still more dependent upon the condition of the bowels. In other words, it depends upon the absorption of phosphates. The total amount of phosphorus ingested per day in the foods is between 1.2 grams and 2.0 grams in an adult on an ordinary diet. The proportion of this which goes out in the urine is very variable. In human beings from 50-65 per cent. of the income is found in the urine, and 30 per cent. to 50 per cent. in the feces. In constipation the proportion in the urine increases; in diarrhea the proportion in the feces increases and that in the urine diminishes. An examination of the phosphorus excretion in the urine alone in a metabolism study is quite worthless, since so large a proportion of the phosphorus is passing out through the feces. Both these exeretions must be examined in any metabolic experiments on phosphorus metabolism. The total amount of phosphorus in the urine of human adults is between 0.5 and 1.2 gram per day.

Variation in disease, and under various conditions. The excretion of phosphoric acid is increased during the catabolism of nucleins in the body, as for example in leucæmia, and during the reabsorption of pneumonic exudates. It is increased by the ingestion of nucleins. During starvation the bones are drawn upon for fuel and there is an increase over the amount of phosphorus excreted when the food contains sufficient fuel matter, but little phosphate. There appears to be a disturbance in the phosphate excretion after parathyroidectomy. There is an increase of phosphoric acid in the urine of dogs after feeding thyroid glands (Roos. Cauter). Ver Eecke found after complete thyroid and parathyroidectomy in rabbits that the secretion of urine was diminished by 30 per cent., the urea diminished 33.7 per cent., but the P_2O_5 of the urine was diminished by 61.7 per cent., while the NaCl augmented 60.8 per cent. In another experiment the P_2O_5 excretion diminished 72 per cent., while the NaCl augmented 164 per cent. Greenwald has confirmed these findings after parathyroidectomy in dogs on a carefully controlled diet. The urinary P as PO_4 fell on the day following the operation from 0.257 to 0.029 gram per day. It kept at this level two days and then increased as tetany came on to the normal of 0.254 gram. There was no corresponding increase in the phosphorus of the feces. It would appear, then, that in some way extirpation of the parathyroids caused phosphorus retention. There is as yet no explanation of this conservation. It may be correlated with the great changes in digestion which follow this operation.

The dependence of the phosphorus excretion on the diet is well illustrated in the experiments of Folin. On a diet containing 16 grams of nitrogen per day the excretion in the urine was 4.1 grams P_2O_5 in 24 hours. A week later on a diet of cream and starch containing 3.9 grams total nitrogen, the urinary P_2O_5 was 1.1 gram. The relation of the phosphoric acid retention to calcium intake and to bone deposition is not yet understood.

Chlorides.—The amount of chlorides, chiefly sodium chloride, excreted per day is dependent upon the food chlorides. The amount is very variable, but generally lies between 10-15 grams. Some people ingest very large amounts of salt with their food. This salt is absorbed and passes rapidly through the kidneys into the urine.

The chlorides may, under certain circumstances, almost completely disappear from the urine. This is the case, for example, in pneumonia during the formation of the exudate in the lungs. During its reabsorption, on the other hand, the chlorine reappears. It is not yet known how this failure to excrete chlorides is to be explained. A part of it is no doubt due to the great diminution in the intake, but this is not the whole explanation. The chlorine is in such cases held more firmly in the tissues. It is not to be explained by the formation of so much exudate. Further work on this problem is necessary.

It is a singular fact that no organic chlorine compounds other than chlorides are known to occur in the body. Whether there are any such in the urine is still disputed, but it is possible. In the tissues there is in lecithin or similar phospholipins a place for the attachment of chlorine to the choline radicle where it may replace hydroxyl. It is possible that under the conditions of disease this or some similar union becomes firmer, the chlorine is not dissociated as before and is so retained. It is not probable that all the chlorine of the tissues is in the form of inorganic chlorides.

The method for the determination of the chlorides of the urine is given on page 982.

Calcium and magnesium.—These occur in the urine in small amounts as phosphates, about 1 gram of these earthy phosphates being secreted per day. A great deal of work has been done on the calcium metabolism of the body, particularly from the point of view of explaining the defective formation of bone and other hard parts of the body. Calcium leaves the body largely in the feces and to a less extent in the urine. As these studies have not yet led to definite conclusions, further discussion of the subject will be omitted.

Pathological constituents of the urine.—*Protein.* Various forms of protein may appear in the urine under pathological conditions, but normally the urine is free from any protein matter. The forms which may appear are: 1. Coagulable protein, generally derived from the blood and being either serum globulin, albumin or fibrinogen. 2. Proteoses, formed from some hydrolysis in the tissues, as in pus. 3. Special proteins, such as Bence-Jones' protein, a kind of hetero-proteose found in disease of the bone marrow, myelomas and possibly osteomalacia. 4. Proteins derived from the urinary passages or glands such as mucin, from the bladder, or spermatozoa. The significance of these is of course very different. It is the coagulable or blood proteins which are the commonest forms and these occur in kidney inflammation, or nephritis, both acute and chronic. Small amounts of these proteins may be present without nephritis, but as a result of serious disturbances of the circulation, such as a dilated heart and a low blood pressure. Thus a certain number of individuals whose urine is free from protein normally may develop an albuminuria of slight degree following some very great exertion. During a Marathon race of about 23 miles on a cold day against a damp head wind nearly all the runners, whose hearts and urines had been perfectly normal before the race, developed albuminuria with heart murmurs and dilated hearts at the end of the race. Similar observations have been made on soldiers after exhausting marches. In all ordinary circumstances the appearance of appreciable amounts of coagulable protein in the urine is abnormal and should arouse suspicion of nephritis.

The presence of proteoses in the urine is not rare during the absorption of exudate in pneumonia or any other absorption of partially digested pus.

The Bence-Jones protein, described on page 939, is found during myelomas and possibly osteomalacia. It is apparently due to a decomposition of the bone marrow.

Carbohydrates.—*Dextrose.* Very minute amounts of dextrose are present in normal urine. The kidney is not a perfect filter and small amounts of glucose can penetrate it. The appearance of enough dextrose to give a precipitate of cuprous oxide in the Fehling test is abnormal. Dextrose may appear in the urine under several different circumstances. We accordingly distinguish the glucosurias as follows:

1. *Alimentary glucosuria.* Due to the ingestion of more dextrose than the body can store for the time being. Different individuals have different powers of utilizing carbohydrates and the power of the body may be increased by previous reduction in the amount of dextrose taken in the food.

2. *Diabetes mellitus.* This is a disease in which large amounts up to 200 grams of dextrose a day may be excreted in the urine. The word diabetes signifies an increased flow of urine; mellitus (from *mel*, honey) that sugar accompanies the polyuria. The urine is increased in quantity so that it may be four or even 10 liters per day. The urine is generally very light colored, but its specific gravity is about normal, due to the dextrose in it. There are all grades of the trouble, from a slight excretion of dextrose when carbohydrate food is ingested to a complete failure to burn any of the ingested carbohydrate, or that formed in the body. It is believed to be correlated often with disease of the pancreas.

3. *Emotional glucosuria.* A slight glucosuria appears in some individuals as a result of strong emotions such as anxiety. This glucosuria is believed to be due to the secretion of more than a normal amount of adrenaline by the supra-renal glands.

4. *Glucosuria due to drugs.* Dextrose may appear in the urine after the ingestion of certain drugs, of which phlorizin is an example. This glucosuria is accompanied by a hypoglycæmia. It is generally attributed to an impairment of the kidney which lets more than the normal amount of dextrose pass into the urine.

These various forms of glucosuria and the interpretation of them are discussed in Chapter XVIII on the carbohydrate metabolism of the body. Methods of identifying and estimating the glucose are given in the practical part, pages 940 and 984.

Lactose. Lactose has been obtained from the urine of women shortly before or immediately after childbirth. It is derived from lactose reabsorbed from the mammary glands. Its significance is of course totally different from that of dextrose. Lactose reduces Fehling's solution just as dextrose does, so that if such a reduction is obtained under circumstances in which lactose might be formed it is necessary to identify the sugar more carefully. This can be done most readily by the fermentation test with yeast. Dextrose ferments with ordinary yeast, but lactose does not. Lactose reduces bismuth in the Nylander test. It is also dexro-rotatory as is glucose.

Pentoses. Pentoses may occur in the urine in the condition which is called pentosuria. There is still some doubt as to the nature of the urinary pentose. Neuberg isolated i-arabinose; Luzzatto, l-arabinose. Pentoses may occur in the urine after the eating of fruits and fruit juices. They are also supposed to be diagnostic of pancreatic disease. The pancreas contains a pentose in its guanylic acid which is d-ribose. It is probable that it also contains in the amylolytic ferment, or the gum to which it is attached, an arabinose. Pentoses were first isolated from the urine of a person addicted to the morphine habit. They are identified by the orcine or phloroglucine tests, or by making their ozazones which melt at about 156-160° C. They reduce Fehling's solution, but do not ferment with yeast. The tests are described on page 946.

Glycuronic acid. This acid, see page 945, occurs in the urine in small amounts in the paired or conjugated form. It is levo-rotatory in the conjugated, but dextro-rotatory in the free state. It is evidently a normal intermediary product of metabolism, but unless it is united with something else it is burned in the body. The ingestion of various drugs increases the quantity in the urine. Thus camphor, chloral hydrate, many aromatic alcohols or phenols, and morphine appear in the urine conjugated with this acid. They are reducing substances and may be confused with dextrose or the pentoses. They may be distinguished from dextrose in that the latter is dextro-rotatory, whereas the paired glycuronic acids are levo-rotatory. The free glycuronic acid gives the orcine test and phloroglucine tests like the pentoses, but the conjugated acid does not give the orcine test. It may be distinguished from the pentoses best by the bromphenylhydrazine compound. This when dissolved in a mixture of alcohol and pyridine is very strongly levo-rotatory and may be distinguished in this way.

The origin of the glycuronic acid is still uncertain. It is clear from the fact that the aldehyde group is not oxidized that probably the oxidation of the glucose to the acid took place when the glucose was in a glucoside union, so that the aldehyde group was protected. After oxidation the acid was split off by hydrolysis.

Acetone and diacetic acid.—These substances appear in the urine together with hydroxybutyric acid in severe diabetes mellitus, or after the prolonged administration of phlorizin to dogs, or during various diseases when there is a deficient nourishment of the body.

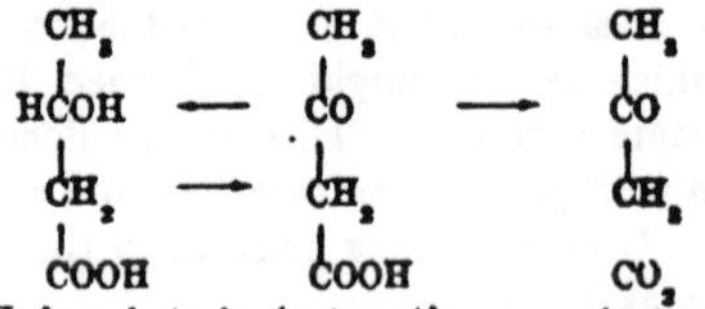

CH_3		CH_3		CH_3
HCOH	⟵	CO	⟶	CO
CH_2	⟶	CH_2		CH_3
COOH		COOH		CO_2
β Hydroxybutyric acid.		Acetoacetic acid.		Acetone + Carbon dioxide.

The amount of these bodies in the urine may be as large as 250 grams a day. They produce a veritable acidosis in severe forms of diabetes.

The acetone is formed from diacetic acid by splitting off of carbon dioxide. Usually very little of this transformation occurs, but in diabetes a considerable amount of dissociation of this sort takes place so that the urine may have the odor of acetone. Acetone, or some similar smelling substance, is in the urine and excretions in quantities in the disease of milk sickness, contracted from milk-sick cattle. There do not appear to have been any determinations of the amount of acetone given off in this disease.

The origin of the acetoacetic acid is in part from the fats and in part from the proteins. Several of the amino-acids, as for example leucine, tyrosine and histidine, give rise to acetoacetic acid when perfused through a dog's liver or when administered to dogs. This origin of acetoacetic acid is discussed on page 811. A more important source of acetoacetic acid is from the fats. By the oxidation of butyric acid, acetoacetic acid is formed in the body. The administration of fats containing butyric acid particularly increases the secretion of this substance in diabetics; but acetoacetic acid is also formed as the last term of the oxidation of the other fats, since it is probable that in all of them the fatty acids are oxidized in the β carbon, two carbon atoms at a time, leaving at last butyric acid, which then goes over into acetoacetic acid. It appears either that the power to oxidize this substance is reduced in diabetics or else that the amount formed is increased beyond the power of the body to destroy it. It will be recalled that in diabetics a large proportion of the energy requirement of the body must be covered either from fats or from proteins, both of which lead to this substance.

The source of the oxybutyric acid is not now doubtful. It is formed by the reduction of the acetoacetic acid. It was for a time thought that the oxybutyric acid was formed first and that the acetoacetic acid was derived from this. But it is impossible to increase the excretion of acetoacetic acid by giving oxybutyric acid. On the other hand, the excretion of oxybutyric acid is increased by giving acetoacetic acid, from which it is inferred that the ketonic acid is first produced and the alcoholic acid by reduction from this.

It is probable that in the normal body, and perhaps in the diabetic as well, the oxidation usually goes to the next step in the oxidation of the acetoacetic acid to two molecules of acetic acid, both of which are then burned to carbonic acid. This decomposition appears to fail largely, or completely, in diabetics.

The methods for the detection and the quantitative estimation of these acetone bodies are given on pages 943 and 1006.

The appearance of these substances in the urine means always that an acidosis is taking place. They accumulate in the blood. They combine with the alkali of the blood, thus setting free carbon dioxide and greatly diminishing the amount of this substance in the blood. The result of this is that the elimination of carbon dioxide is interfered with. In such cases the administration of sodium carbonate or bicarbonate is often very beneficial.

Metabolism of various substances not foods.—A great deal of information concerning the probable course of transformations of the foods, and concerning the chemical powers of the body, has been obtained by giving to animals substances of known chemical composition and then by an examination of the urine determining whether they have been affected at all in their passage through the body, and if they have been affected what the body has been able to do to them. A very large number of experiments of this kind have been tried and it is not possible in a book of this character to deal with all of them. Nor indeed would it be desirable, since it would burden the mind to no good end. But it is worth while to consider the general principles which have been worked out by experiments of this nature. Many different animals have been used for these purposes, often the substances have been ingested by the experimenters themselves when it is certain, or fairly certain, that they are not toxic, but as a rule the experiments have been tried on dogs, cats or rabbits. The chemical powers of different kinds of animals of course vary, else there would not be various species of animals.

A great many substances which are not foods, and many of them substances which the organism has had no experience in handling before, are burned wholly or partially, or otherwise affected by the cells of the body. This fact is of itself very interesting, for it indicates very strongly that at least one way of burning substances is while they are in solution in the water of the protoplasm and not when they are in the living protoplast itself, if there be such a thing. It is hardly likely that such materials as brombenzene are first built up into a living molecule before being oxidized. It is far more probable that they are, as it were, burned in the liquid of the cell while in solution there. This point, it will be recognized, is a matter of very great interest from a theoretical point of view, and we shall have occasion to come back to it in considering the combustion of the excess protein and amino-acids. It is in general harmony with the view that there are two kinds of combustion going on in a living cell: first, the oxidation of the living matter itself; and, second, the oxidation of substances not strictly a part of the living matter, but dissolved in the water which penetrates it. In most, if not all, oxidations where there is a spontaneous oxidation there is, as has been pointed out already in discussing the physical chemistry of oxidation, a formation of hydrogen peroxide accompanying the main oxidation. Now it is a peculiarity of many of the oxidations of living matter that they may be closely simulated outside the cell by oxidation with hydrogen peroxide at ordinary temperatures. This point has been particularly established by Dakin, who has made many beautiful discoveries concerning the nature of the oxidations in the body by following out this idea. Hydrogen peroxide is probably formed in the course of the vital reactions or oxidations. There is in practically all cells catalase for the purpose of destroying the peroxide. It would be strange if this ferment was so generally present and if it had nothing to do. It is probable, then, that there is produced alongside of the vital oxidation, or respiration, a secondary oxidation of a purely chemical kind of the materials in solution in the water of the protoplasm by the hydrogen peroxide formed. It is possibly this oxidation which gets hold of the substances as they enter the cells and partially or wholly oxidizes them. This oxidation is not at all specific and it may happen, then, that the cell has, in general, powers of oxidizing things it has never met with before. There is, however, another oxidation which is specific, as has been pointed out.

Substances which appear in the urine paired with glycuronic acid. One of the first general principles which was found was that the substance ingested after undergoing a partial oxidation, so that it obtained a hydroxyl group at some point, combined with glycuronic acid in an ester union and was excreted in the urine as the glycuronic acid ester. A very large number of different kinds of substances are thus excreted. Not always does all of the compound ingested reappear. Generally some, or most of it, is destroyed beyond recognition, but some of it by uniting with glycuronic acid becomes so stable that it is no longer oxidized or otherwise metabolized and it appears in the urine united with it. Just whère it finds the glycuronic acid, in what organ, whether the liver or the muscles or somewhere else, that is not definitely known; but it is probably one of these two organs, for it is probably an organ which has a supply of carbohydrate on hand from which the glycuronic acid can be formed. Among the substances which thus appear in the urine conjugated with glycuronic acid are a great number of aromatic substances which contain the phenol group, such as phenol, cresol, camphor, cyclic terpenes such as borneol, menthol, thymol, naphthalene, antipyrine, oil of turpentine, oxyquinolines, orthonitrotoluene and many other substances of this class. In addition many aliphatic alcohols also combine with glycuronic acid and appear in the urine in that form. Small quantities of the ester appear after the ingestion of isopropyl alcohol, amyl alcohol and many ketones and aldehydes which are probably first reduced to alcohols in the body. Chloral hydrate appears in the urine, after its ingestion, as trichlorethylglycuronic acid, $C_2Cl_3H_2.C_6H_9O_7$.

Substances which appear in the urine paired with glycine. Glycocoll, or glycine, is another substance which is very often united with ingested substances and when so united protects them from further decomposition. Hippuric acid is a glycine conjugate, but there are many others. Thus in the bile we have glycocholic acid, a conjugate of cholic acid and glycine. The body is apparently able to form very large amounts of glycine and thus to protect itself against many different foreign substances which by their decomposition might give rise to waste or active products different from those in whose presence the protoplasm is accustomed to work. The aromatic substances which by oxidation give rise to benzoic acid in the body appear in the urine, in part, as hippuric acid, although they are in part often oxidized further. In the table on page 810 it will be seen that phenyl-amino-acetic acid, phenylpropionic, phenylcinnamic acids, phenylserine, phenylglyceric and phenyl-b-hydroxypropionic acids undergo oxidation in the body to benzoic acid and leave the body in part as hippuric acid, being paired with glycine. Phenylethyl alcohol and phenylacetaldehyde are oxidized to phenylacetic acid and paired with glycine to form phenaceturic acid in dogs but not in man. In small part, too, they pair with glycuronic acid. Xylene is oxidized to toluic acid, $C_6H_4(CH_3).COOH$, and paired with glycine is excreted as toluric acid; mesitylene, $C_6H_3(CH_3)_3$, is oxidized to mesitylenic acid, $C_6H_3(CH_3)_2.COOH$, and after pairing with glycocoll is excreted as mesityluric acid, $C_6H_3(CH_3)_2.CO.NH.CH_2.COOH$; cymene, $(CH_3)_2CH.C_6H_4.CH_3$, is oxidized into cumic acid, p-isopropyl benzoic acid, and excreted as the glycocoll conjugate, cuminuric acid, $(CH_3)_2.CH.C_6H_4.CO.NH.CH_2.COOH$.

Substances which appear in the urine conjugated with sulphuric acid. We have already considered the principal substances of this group. They are for the most part aromatic alcohols, such as phenol, indoxyl, scatoxyl, pyrocatechol and the substituted members of the group. The formation of these conjugates is still essentially unexplained. It is not impossible that they may pair with taurine and the organic part of the molecule be split off. They may, however, be formed from sulphuric acid itself. The amount is in any case not very large.

Substances which are paired with ornithine. In the bird's body there does not appear to be on hand a stock of glycine for pairing purposes. Substances which in the mammals leave the body paired with glycocoll appear in birds' urine paired with ornithine to make ornithuric acid. This is not the only difference between the birds' metabolism and that of mammals. Nitriles which leave the mammal's body paired with sulphur, as sulphocyanate, do not appear in this form in birds' excretions. Ornithine is diamino-valerianic acid, $NH_2.CH_2.CH_2.CH_2.CHNH_2.COOH$.

Uramido acids. Attention has already been called to the fact that some substances leave the body conjugated with carbamic acid, $NH_2.CO.OH$. The formation of substances of this sort indicates that Siegfried's carbamic reaction occurs in living matter. Perhaps there first occurs a union of carbonic acid with the amino group, followed by its subsequent union with ammonia through a second hydroxyl of the CO_2. Thus in rabbits m-amino-benzoic acid appears in part in the urine as uramino-benzoic acid, $NH_2.CO.NH.C_6H_4.COOH$. Reactions similar to these may be at the bottom of the synthesis of uric acid, or of allantoine. These may undergo dehydration and appear in the urine as substituted hydantoines. Hydantoine is

```
NH—CO
|    |
OC   |
|    |
NH—CH2
```

Hydantoine.

Nitriles and cyanides. These appear in the urine of mammals as the sulphocyanates, having somewhere in the body picked up an unoxidized sulphur atom. In birds this union does not occur, or if it does the substances do not appear in the urine in this form. It has not been determined, so far as I can find, what becomes in birds' bodies of ingested cyanides and nitriles.

Processes of reduction in the body. Whenever they enter protoplasm substances are exposed both to reductions and oxidations. Probably the oxidations occur chiefly in the part of the cell where the oxygen is most plentiful: namely, that near the blood vessel. It probably happens, therefore, that the first exposure is to strong oxidation, but on penetrating into the interior of cells where processes of strong reduction occur it is not surprising that many substances are reduced as well as oxidized. This happens, for example, to the foods, since from the oxidized carbohydrates the reduced fats are formed. But many substances besides the carbohydrates are reduced in the body. Thus aldehydes and ketones are not infrequently reduced in part to alcohols, while a part is oxidized to acids. This happens, for example, to acetoacetic acid, which is reduced in part to β-hydroxybutyric acid. m-Nitrobenzaldehyde is in part carried over into m-acetylaminobenzoic acid. This is a reaction which involves an acetylation which is discussed later. (See page 820.)

Methylation and demethylation. Some substances such as pyridine and tellurium compounds are methylated in passing through the body, and some other substances lose their methyl groups perhaps by oxidation. The matter of methylation has been considered already under creatinine. The methylated xanthines, such as caffeine and theobromine, are in large measure demethylated somewhere in the body, since they appear in the urine in small part as the dimethyl or monomethyl purines.

Conclusion. The general conclusion from this work, besides the establishment of the general character of the decompositions of various compounds, both foods and non-foods, is that in their passage into the various cells of the body the ingested molecules of all kinds come into contact with a great variety of other molecules. Some of these molecules are of such a character that union takes place. The subsequent fate of the union depends upon its stability. If it is more easily oxidized than the substance which found entrance to the cell, decomposition and oxidation occur and we say that that particular cell contains an oxidase which is capable of hastening the oxidation of the ingested matter; if, on the other hand, the union is less easily oxidized, or otherwise decomposed, than the original substance, the compound thus formed takes its exit from the cell and eventually appears in the urine in the form of a paired substance. In such case we say that conjugation, or pairing, has preserved the substance and perhaps protected the body from the action of a toxic substance. For every substance which is thus rendered inert by the compound being more stable it is possible, however, that there are other substances which by union are rendered less stable and made more toxic. Thus some nitriles, such as benzonitrile, may be rendered far more toxic by partial oxidation, for the oxidized nitriles set free hydrocyanic acid more easily.

Urinary pigments.—Normal human urine has a yellow color. This yellow color, which may deepen on exposure to light and air, is due to certain pigments of which urochrome and urobilin are the best characterized. In addition urine when made strongly acid becomes a dark red color, due to the development of a pigment called uroerythrin. This red color is probably derived from the scatole of the urine. The composition of urobilin and urochrome is still uncertain. They are probably derived from the bile pigment, bilirubin, and urobilin appears to be identical with stercobilin, a reduced bilirubin found in the feces. It is quite possible that this reduction of bilirubin to urobilin can take place elsewhere than in the alimentary canal, since urobilin is found in the bile and blood. Both urochrome and urobilin yield pyrrol derivatives. They are probably derived in the long run from the hematin of the blood as discussed under the bile. As the chemistry and origin of these compounds is still so uncertain, further discussion may be postponed until methods of purifying them, particularly the urochrome, shall be perfected.

REFERENCES. UREA.

1. *von Schroeder:* Ueber die Bildungsstätte des Harnstoffes. Arch. f. expt. Path. u. Pharm., 15, p. 364, 1882.
2. *Gottlieb:* Ueber die quantitative Bestimmung des Harnstoffes in den Geweben und den Harnstoffgehalt der Leber. Arch. expt. Path. u. Pharm., 42, 1899, p. 239.
3. *Baglioni et Fredericq:* L'azione fisiologica dell urea sul cuore dei vertebrate. Zeit. f. allg. Physiol., 6, p. 482, 1907.
4. *Baglioni:* Der Einfluss der chemischen Lebensbedingungen auf die Tätigkeit des Selachierherzen. Zeit. f. allg. Physiol., 6, pp. 71-91, 1907; 6, pp. 213-216.
5. *Schultzen and Nencki:* Die Vorstufen des Harnstoffes im tierischen Organismus. Zeits. f. Biol., 8, p. 124, 1872.
6. *Macleod and Haskins:* Contributions to our knowledge of the chemistry of carbamates. Jour. Biol. Chem., 1, p. 319, 1905-6.
7. *v. Knierem:* Beiträge zur Kenntnis der Bildung des Harnstoffes im tierischen Organismus. Zeit. Biol., 10, p. 263, 1874.
8. *Salaskin:* Ueber die Bildung von Harnstoff in der Leber der Säugethiere aus Amidosäuren der Fettreibe. Zeits. physiol. Chem., 25, p. 128; 25, p. 449, 1898.
9. *Richet:* Compt. rend. Acad., 118, 1894; Compt. rend. Soc. Biol., 49.
10. *Stolte:* Ueber das Schicksal der Monoaminosäuren im Tierkörper nach Einführung in die Blutbahn. Hofmeister's Beiträge, 5, p. 15, 1903.
11. *Nencki, Pawlow and Zaleski:* Sur la richesse du sang et des organes en ammoniaque et sur la formation de l'urée chez les mammifères. Arch. d. sci. biol. de St. Petersburg, 4, p. 197, 1896.
12. *Kossel and Dakin:* Ueber die Arginase. Zeit. physiol. Chem., 41, p. 321; 42, p. 181, 1904.
13. *v. Jaksch:* Weitere Mittheilungen über die Vertheilung der stickstoffhaltigen Substanzen im Harne des kranken Menschen. Zeit. klin. Med., 50, p. 167, 1903.
14. *Folin:* Laws governing the chemical composition of the urine. Amer. Jour. Physiol., 13, p. 66, 1905.
15. *Osterberg and Wolff:* Day and night urines. Jour. Biol. Chem., 3, p. 165, 1907.
16. *v. Schroeder:* Ueber die Harnstoffbildung der Haifische. Zeit. physiol. Chem., 14, p. 576, 1890.
17. *Nencki and Hahn:* La fistule d'Eck de la veine cave inférieure et de la veine porte, etc. Arch. sci. Biol. de St. Petersburg, 1, p. 447, 1892. *Massen et Pawlow:* Ibid., p. 401.
18. *Abel and Muirhead:* Ueber das Vorkommen der Carbaminsäure im Menschen- und Hundeharn nach reichlichem Genuss von Kalkhydrat. Arch. expt. Path. u. Pharm., 32, p. 467, 1893; 31, p. 15.
19. *Salaskin u. Zaleski:* Ueber den Einfluss der Leberexstirpation auf den Stoffwechsel bei Hunden. Zeits. physiol. Chem., 29, p. 517, 1900.

URIC ACID.

1. *Dakin:* Oxidations and reductions in the animal body. Monographs on Biochemistry. Longmans, Green and Co., London.

2. *Jones:* Nucleic acids. Their chemical properties and physiological conduct. Monographs on Biochemistry. London.
3. *Izar:* Beiträge z. Kenntnis der Harnsäurezerstörung u. Bildung. Zeit. physiol. Chem., 73, p. 319, 1911.
4. *Horbaczewski:* Monatshefte f. Chemie, 10 p. 614, 1889; 12, p. 221, 1891.
5. *Hopkins and Hope:* Uric acid and diet. Jour. Physiol., 23, p. 271, 1898-9.
6. *Jerome:* The formation of uric acid in man and the influence of diet on its daily output. Jour. Physiol., 22, p. 146, 1897.
7. *Minkowski:* Untersuchungen zur Physiologie und Pathologie der Harnsäure bei Säugetieren. Archiv f. expt. Path. u. Pharm., 41, p. 375, 1898.
8. *Loewi:* Beiträge zur Kenntnis der Nucleinstoffwechsel. Archiv f. expt. Path. u. Pharm., 44, p. 5, 1900; 45, p. 157, 1901.
9. *Smetanka:* Zur Herkunft der Harnsäure beim Menschen. Archiv f. d. ges. Physiol., 149, p. 275, 1912-13.
10. *Wendt:* Skan. Archiv Physiol., 17, p. 231, 1905.
11. *Arnold and Larrabee:* The purin content of common articles of food. Jour. Amer. Med. Assn., 58, p. 18, 1912.
12. *Mares:* Sind die endogenen Purinkörper Produkte der Tätigkeit der Verdauungsdrüsen? Eine Antwort auf die Frage Sivens. Archiv f. d. ges. Physiol., 149, p. 275, 1912-13.
13. *Siven:* Ueber den Purinstoffwechsel des Menschen. II. Sind die endogenen Purinkörper Produkte der Verdauungsdrüsen? Archiv ges. Physiol., 146, pp. 499-516, 1912.
14. *Mares:* Zur Theorie der Harnsäure Bildung im Säugethierorganismus. Sitz. d. Kais. Akad. Wien, p. 101, 1892.
15. *Mares:* Der physiologische Protoplasma stoffwechsel u. die Purinbildung. Archiv ges. Physiol., 134, pp. 59-102, 1910.
16. *Brunton and Bokerham:* Zent. f. Physiol.
17. *Ascoli and Izar:* Quantitative Rückbildung zugesetzter Harnsäure in Leberextrakten nach vorausgegangener Zerstörung. Zeit. physiol. Chem., 58, p. 529, 1909.
18. *Crofton:* Synopsis of experiments on the transformation of circulating uric acid in the organism of man and animals. Medical Record, N. Y., p. 6, 1903.
19. *Crofton:* Zur Kenntnis der Harnsäure Umwandlung im Tier u. Menschenkörper. Archiv ges. Physiol., 121, p. 377, 1907-8.
20. *Milroy and Malcolm:* Metabolism of the nucleins. Jour. Physiol., 25, p. 105, 1899.
21. *Amberg and Jones:* Ueber die bei der Spaltung der Nucleine in Betracht kommenden Fermente mit besonderer Berücksichtigung der Bildung vom Hypoxanthin in der Abwesenheit von Adenase. Zeit. physiol. Chem., 73, p. 407, 1911.
22. *Jones:* Ueber die Beziehung der aus wässerigen Organextrakten gewonnenen Nucleinfermente zu den physiologischen Vorgängen im lebenden Organismus. Zeit. physiol. Chem., 65, p. 383, 1910. (References to earlier literature by same author. For general summary of work on this subject see Jones, Nucleic acids.)
23. *Voegtlin and Jones:* Ueber Adenase u. ihre Beziehung zu der Entstehung von Hypoxanthin im Organismus. Zeit. physiol. Chem., 66, 1910, p. 250.
24. *Taylor and Rose:* Uricolysis in the human subject. Jour. Biol. Chem., 14, p. 419, 1913.

25. *Schittenhelm:* Ueber den Nucleinstoffwechsel des Schweines. Zeit. physiol. Chem., 66, p. 53, 1910. (Earlier papers cited here.)

26. *Tollens:* Gicht und Schrumpfniere. Ausscheidung von Harnsäure u. Purinbasen im Urine u. im Kote des Gichtkranken bei Nierenstörungen. Zeits. physiol. Chem., 53, p. 164, 1907. (Historical account.)

27. *Van Noorden:* Handbuch d. Pathol. d. Stoffwechsels. II. Die Gicht.

28. *Block:* Die Herkunft der Harnsäure im Blute bei Gicht. Zeits. physiol. Chem., 51, p. 472, 1907.

29. *Brugsch:* Zur Stoffwechselpathologie der Gicht. Zeits. f. expt. Path. u. Therapie 6, 1909, p. 278.

30. *Miller and Jones:* Ueber die Fermente des Nucleinstoffwechsels bei der Gicht. Zeit. physiol. Chem., 61, p. 395, 1909.

31. *Biberfeld u. Schmidt:* Ueber den Resorptionsweg der Purinkörper. Zeit. physiol. Chem., 60, p. 292, 1910.

32. *Mendell and Mitchell:* The enzymes involved in purine metabolism in the embryo. Amer. Jour. Physiol., 20, p. 97, 1907.

33. *Preti:* Beiträge zur Kenntnis der Harnsäurebildung. IV. Zeit. physiol. Chem., 62, p. 354, 1909.

34. *Schittenhelm:* Ueber die Umsetzung verfütteter Nucleinsäure beim Hunde unter normalen u. pathol. Bedingungen. Zeit. physiol Chem., 62, p. 80, 1909.

35. *Bezzola, Izar and Preti:* Wiederbildung zerstörter Harnsäure in der künstlich durchbluteten Leber. Zeit. physiol. Chem., 62, p. 229, 1909.

36. *Ascoli and Izar:* Harnsäure Bildung in Leberextrakten nach Zusatz von Dialursäure u. Harnstoff. Zeits. physiol. Chem., 62, p. 347, 1909.

37. *Frank and Schittenhelm:* Ueber die Umsetzung verfütterten Nucleinsäure beim normalen Menschen. Zeit. physiol. Chem., 63, p. 269, 1909.

38. *Wiechowski:* Ueber die Zersetzlichkeit der Harnsäure im menschlichen Organismus. Archiv expt. Pathol. u. Pharm., 60, p. 185, 1909.

39. *Schittenhelm u. Weiner:* Ueber das Vorkommen u. die Bedeutung von Allantoin im menschlichen Urine. Zeit. physiol. Chem., 63, p. 283-88, 1910.

40. *Wiechowski:* Das Vorhandensein von Allantoin im normalen Menschenharn u. seine Bedeutung für die Beurtheilung des menschen Harnsäurestoffwechsels. Biochem. Zeits., 19, p. 368, 1909.

41. *Schmid:* Der Abbau methylierte Xanthine. Zeits. physiol. Chem., 67, p. 154, 1910.

42. *Schittenhelm:* Die Nucleinbasen der Feces unter dem Einfluss anhaltender Faulniss. Zeit. physiol. Chem., 39, p. 199, 1903.

43. *Levinthal:* Zum Abbau des Xanthins u. Caffeins im Organismus des Menschen. Zeit. physiol. Chem., 77, p. 259, 1912.

44. *Krüger u. Schittenhelm:* Die Menge u. Herkunft der Purinkörper in der menschlichen Feces. Zeit. physiol. Chem., 45, p. 14, 1905. Method.

45. *Krüger u. Schmidt:* Zur Bestimmung der Harnsäure u. Purinbasen im menschlichen Harn. Zeits. physiol. Chem., 45, p. 1, 1905.

46. *Bennett:* The purines of muscle. Jour. Biol. Chem., 11, p. 221, 1912.

47. *Burian:* Die Herkunft der endogenen Harnpurine bei Mensch u. Saugethier. Zeits. physiol Chem., 43, p. 532, 1904-5.

48. *Sachs:* Ueber die Nuclease. Zeits. physiol. Chem., 46, p. 337, 1905.

49. *Kojo:* Ueber Unterschiede im Harnbefunde beim Gesunden u. Carcinomatosen. Zeits. physiol. Chem., 63, p. 416, 1911.

50. *Ackroyd:* The presence of allantoin in certain foods. Bioch. Jour., 5. p. 400, 1911.

51. *Wiechowski:* Das Schicksal intermediarer Harnsäure beim Menschen und der Allantoingehalt des menschlichen Harns; nebst Bemerkungen über Nachweis und Zersetzlichkeit des Allantoins. Biochem. Zeits., 25, p. 431, 1910.
52. *Jones and Rohde:* The purin ferments of rats. Jour. Biol. Chem., 7, p. 237, 1910.
53. *Fine and Chace:* The influence of salicylates upon the uric acid concentration of the blood. J. Biol. Chem., 21, p. 371, 1915.

Creatinine and Creatine.

1. *Jaffé:* Untersuchungen über die Entstehung des Kreatins im Organismus. Zeit. physiol. Chem., 48, p. 430, 1906.
2. *Dorner:* Bildung von Kreatin u. Kreatinin im Organismus besonders Kaninchen. Zeits. physiol. Chem., 52, p. 225, 1907.
3. *Folin:* Laws governing the chemical composition of the urine. Amer. Jour. Physiol., 13, p. 66, 1905.
4. *von Hoogenhuyze u. Verploegh:* Beobachtungen über die Kreatininausscheidung beim Menschen. Zeit. physiol. Chem., 46, p. 415, 1906; 57, p. 161; 59, p. 101.
5. *Pekelharing and Harkink:* The excretion of creatinin in man under the influence of muscular tonus. Proc. Amsterdam Acad., 14, p. 310, 1910.
6. *His:* Ueber das Stoffwechselprodukt des Pyridins. Arch expt. Path. u. Pharm., 22, p. 253, 1887.
7. *Cohn:* Ueber das Verhalten einiger Pyridin u. Naphthalinderivaten im Tierischen Stoffwechsel. Zeit. physiol. Chem., 16, p. 112, 1894.
8. *Hofmeister:* Ueber Methylierung im Tierkörper. Arch. expt. Path. u. Pharm., 33, p. 198, 1894.
9. *Abderhalden, Brahm u. Schittenhelm:* Vergleichende Studien über den Stoffwechsel verschiedener Tierarten. Zeit. f. physiol. Chem., 59, 1909, p. 32.
10. *Inouye:* Ueber die Entstehung des Kreatins im Tierkörper. Zeit. physiol. Chem., 81, p. 79, 1912.
11. *Folin:* Hammarsten's Festschrift, 1906.
12. *Mellanby:* Creatine and creatinine. Jour. Physiol., 36, p. 447, 1907-8.
13. *v. Klercker:* Beitrag zur Kenntnis des Kreatins u. Kreatinins im Stoffwechsel des Menschen. Biochem. Zeits., 3, p. 45, 1907.
14. *Benedict and Meyers:* Determination of creatine and creatinine. Amer. Jour. Physiol. Chem., 18, p. 397, 1907; 18, p. 377; 18, p. 406.
15. *Cathcart:* Ueber die Zusammensetzung des Hungerharns. Biochem. Zeits., 6, p. 109, 1907.
16. *Jaffé:* Ueber den Niederschlag welchen Pikrinsäure in normalen Harn erzeugt und über eine neue Reaction des Kreatinins. Zeit. physiol. Chem., 20. p. 391, 1886.
17. *Achelis:* Ueber das Vorkommen von Methylguanidin im normalen Menschenharn. Zent. Physiol., 20, p. 455, 1906.
18. *Kutscher and Lohmann:* Der Nachweis toxischer Basen in Harn. III. Zeit. physiol. Chem., 49, p. 81, 1906.
19. *Engeland:* Ueber den Nachweis organischer Basen im Harn. Zeit. physiol. Chem., 57, p. 49, 1908.
20. *Meyers and Fine:* Jour. Biol. Chem., 21, 1915. Various papers.
21. *Traut and Mellanby:* Jour. Physiol., 44, p. 43, 1913.

Hippuric Acid.

1. *Baumann:* Zur Kenntnis der arom. Substanzen d. Tierkörper. Zeits. physiol. Chem., 7, p. 282, 1883.
2. *Baas:* Ueber das Verhalten des Tyrosins zur Hippursäurebildung. Zeits. physiol. Chem., 11, p. 485, 1887.
3. *Salkowski:* Ber. deutsch. Chem. Gesell. 11.
4. *Wiechowski:* Die Gesetze der Hippursäuresynthese. Ein Beitrag zur Frage der Stellung des Glykokolls im Stoffwechsel. Hofmeister's Beiträge, 7, p. 204, 1906.
5. *Lewinski:* Ueber die Grenzen der Hippursäurebildung beim Menschen. Zugleich ein Beitrag zur Glykokollfrage. Arch. expt. Path. u. Pharm., 58, p. 397, 1907-8.
6. *Abderhalden, Gigon and Strauss:* Studien über den Vorrat an einigen Aminosäuren im Organismus bei verschiedenen Tierarten. Zeits. physiol. Chem., 51, p. 311, 1907.
7 *Magnus-Levy:* Ueber die Neubildung von Glykokoll. Bioch. Zeits., 6, p. 523, 1907.
8. *Schmiedeberg and Bunge:* Ueber die Bildung der Hippursäure. Arch. expt. Path. u. Pharm., 6, p. 233, 1876.
9. *Schmiedeberg:* Ueber Oxydationen und Synthesen im Tierkörper. Arch. expt. Path. u. Pharm. 14, p. 288, 1881.
10. *Folin and Flanders:* A new method for the determination of hippuric acid in urine. Jour. Biol. Chem., 11, p. 257. 1912.
11. *Kingsbury and Bell:* The synthesis of hippuric acid in nephrectomized dogs. Jour. Biol. Chem., 21, p. 297, 1915.

Indican.

Jolles: Ueber eine neue Methode zur quantitativen Bestimmung des Indikans im Harne. Zeit. physiol. Chem., 94, p. 79, 1915.

CHAPTER XVIII.

THE METABOLISM OF THE BODY CONSIDERED AS A WHOLE. CARBOHYDRATES.

In the immediately preceding chapters there have been considered the chemistry and metabolism, so far as they are known, of the different organs and tissues of the body. Since each of these organs is normally working in conjunction with others, being co-ordinated either by the nervous system or the blood in the manner described, their normal metabolism can only be studied while they are in situ. The possibility of studying each organ separately is greatly limited by this fact. Much, however, can be learned by studying the metabolism of the body as a whole, without inquiring at first just where the various steps of that metabolism occur. In this method of study the organs are all working in their normal relations and at their maximum efficiency. Moreover, the fundamental processes of metabolism probably do not differ very widely in different organs, although, of course, the metabolism differs in particulars in each, so that the constitution of each is specific. The general course of the change which the carbohydrates, proteins and fats undergo in the course of their passage through the body, from the time of their absorption to their excretion in the form of various fragments, may now be examined.

The metabolism of sugar.—The carbohydrates are absorbed chiefly as glucose, levulose and galactose and about 500 grams of carbohydrate are normally consumed per day. We have now to follow, so far as we are able, the sugar thus absorbed in its course through the body. From the intestine it passes to the blood, where it is found in the blood plasma in a concentration varying from .08-.15 per cent. of the whole blood. The glucose thus present in the plasma is dialyzable and the greater part of it is free and not united to any colloidal matter, as was at one time suggested (McGuigan and von Hess).

In taking up this subject of the carbohydrate metabolism it is perhaps most interesting and instructive to follow the historical method. About the middle of the nineteenth century it was known from the work of Lavoisier that the carbohydrates, like other foods, were burned or oxidized in the body in large measure to carbon dioxide and water; that by means of this combustion heat was liberated to just the same amount as would have been produced by burning so much sugar outside the body; that of this heat part was consumed, or rendered latent, in the evaporation of water, or in doing work, and the greater part supplied the heat which the body was constantly giving off as if it were a stove. But where this combustion took place, how it could take place in an aqueous medium, and whether the sugar underwent intermediary transformations before oxidizing, as was indicated by the appearance of fat after carbohydrate ingestion, these were matters quite obscure.

A great step forward in this dark field was taken in 1843 by the French physiologist, Claude Bernard, a man whose name should be remembered for his striking discoveries, ingenious and skillful experiments, his clear thoughts, lofty imagination and the beautiful, simple and luminous style in which his books and papers were written. Since it was impossible to follow the course of the sugar directly after its entrance into the blood, Bernard turned his attention to those pathological cases in which sugar appears in the urine. Disease tries many experiments which we cannot as yet with our clumsy vivisectional methods hope to imitate, and for the wise man who can read the experiments aright pathology reveals many secrets of metabolism. It was known to Bernard that at times large amounts of glucose may appear in the urine of human beings, although it normally is present only in traces, .04 per cent. The urine is then greatly increased in amount, it is often sweet to the taste, very light colored, having a high specific gravity and often a sweetish or aromatic odor. Such pathological urine excretion is called glycosuria, or glucosuria, or diabetes mellitus (*dia*, through, and *mel*, honey; literally, honey diabetes). In severe forms of this disease with a fatal ending in coma, 60-200 grams of glucose might be eliminated per day. Nothing was known of the cause of the disease so grave in its prognosis; or of the origin of the glucose which appeared in the urine.

Bernard tried to produce the disease artificially. He found that injury to the floor of the fourth ventricle of the medulla oblongata in dogs and rabbits produced almost invariably a discharge of glucose in the urine. This puncture is known as the sugar puncture. This glucosuria was, however, temporary and never had a fatal ending. It was found on examining the matter farther that the puncture only produced glucosuria if the animals had been fed on carbohydrate, or were well fed. Starving animals produced little or no sugar. It did not occur either if the splanchnics, two sympathetic nerves supplying the abdominal viscera, had been divided before the puncture (Eckhard). Bernard did not succeed in producing artificially a severe glycosuria with the accompanying symptoms of the human disease, nor were physiologists more successful until about 1890, when von Mering and Minkowski discovered that extirpation of the pancreas would produce it.

Bernard attempted to find where the sugar came from which appeared in the urine after the sugar puncture. He examined the glucose content of the blood in different regions of the body. He found that the blood of the jugular and femoral veins contained less glucose than the corresponding arterial blood:

V. jugularis	.08%	Femoral artery	0.12%
Carotid artery	.12	Femoral vein	0.08

Evidently glucose was taken from the blood in passing through the tissues supplied by these blood vessels. Next, by means of a sound passed into the jugular vein, he drew a sample of blood from the right auricle. This contained as much glucose as the arterial blood. It was clear that somewhere in the body sugar must be added to the blood to make good the loss of that taken out by the tissues. A sound passed through the heart and down into the inferior vena cava to a level just above the kidneys showed that the blood of that vein contained 0.08 per cent. of sugar. The sound was now drawn carefully back to the level of the opening of the hepatic vein. The blood from here contained 0.14 per cent. or more of glucose, or more than the arterial blood. Evidently, then, somewhere in the portal circulation glucose was added to the blood. An examination of the blood of the portal vein showed generally that the portal vein blood contained less glucose than the hepatic blood. This proved that sugar was added to the blood during its passage through the liver and an examination of this organ resulted in the discovery of the following facts: If the liver was taken as rapidly as possible after decapitation out of the body of a rabbit and, excluding the gall bladder, was thrown into boiling water and cut up in it while boiling, only very small amounts of a reducing sugar could be extracted from it. Evidently the liver did not contain more sugar than might be attributed to the blood in it. But there went into solution in the water a substance which made an opalescent solution and which when boiled with acid or acted upon by saliva or malt diastase quickly set free a reducing sugar. Moreover, if the liver was left in the body after death, then it was subsequently found to contain considerable quantities of glucose and maltose. The living active liver, then, contained little or no glucose, but it contained a considerable quantity of a material from which glucose could be and was made, and this substance was called by Bernard for this reason *glycogen,* the glucose-maker.

Thus was made one of the first or fundamental discoveries in our knowledge of sugar metabolism and it placed the liver in an entirely new light. It had before that been known only as a digestive gland; it was now seen that it played another and possibly far more important rôle in the carbohydrate metabolism of the body and was one of its most important chemical factories.

Further study by Bernard showed that the glycogen which he had discovered was very generally found in the animal kingdom, where it played the rôle that starch plays in the plant world. He showed that it was present in muscle cells, in pus cells, in embryonic tissues, in fly larvæ, in earthworms and in other invertebrates as well as vertebrates, and it was present, though in varying amounts, in the livers of nearly all vertebrates. It became at once one of the fundamental constituents of animal cells.

Glycogen. $(C_6H_{10}O_5)_n$. The glycogen thus discovered is a colloidal polysaccharide which is digested by ptyalin into maltose and by the liver tissue and muscle tissue into glucose and maltose. When inverted by acid it yields dextrose. It is entirely resistant to 30 per cent. KOH and its quantitative determination depends upon this fact. It is prepared by boiling the tissue with 30 per cent. KOH until it dissolves; filtering through asbestos or glass wool and precipitating with alcohol. All protein and reducing sugars are destroyed by the strong alkali. It is snow-white in color, soluble in water to an opalescent solution. It is dextro-rotatory: $(\alpha)_D=+196.63°$. It does not reduce Fehling's solution; with iodine-potassium iodide it gives in the cold, if salt is present, a port-wine color and this is one of the means of its detection. By this reaction it may be detected in sections, and it can be seen in the liver tissue often deposited in radiating masses or granules about the nucleus. Some of the mitochondria described by histologists are possibly glycogen. It is, as will be seen from its resistance to alkali, an inert substance. It is colloidal and it is plainly a reserve material of the cell.

Origin of glycogen. The question of the origin of the glycogen of the liver was at once attacked. Bernard found that it not only appeared in large quantities when carbohydrate food was taken and hence was made presumably from the carbohydrate of the food, but it appeared also when a purely meat diet was consumed. Putrefying muscle contains almost no carbohydrate, but the larvæ of flies which have fed upon this muscle and on nothing else contain large amounts of glycogen in their bodies. Evidently they must be able to make glycogen from protein and since dogs fed exclusively on lean meat also have livers containing large amounts of glycogen they too must be able to make carbohydrate from protein. *A second fundamental discovery in the metabolism of the sugars was thus made, namely, that animals can make carbohydrate from protein.* This showed that animals had powers of synthesis which it had been believed were peculiar to plants.

To try these experiments properly it is necessary to get the liver practically glycogen free at the start of the feeding. This can, perhaps, be done best by Lusk's method of the injection of phlorizin of which we shall presently speak, but before the properties of this drug were known recourse was had to other methods. Animals were made to fast for considerable periods, forced to exercise by being put in a revolving cage, exposed to cold, which causes consumption of glucose, and finally given small doses of strychnine to produce mild tetanus. By these means the liver may be made almost glycogen or sugar-free and the amount in the muscles is reduced to a minimum.

Origin of glycogen from protein and carbohydrates. That the glycogen is formed from protein and carbohydrate food is illustrated by the following experiments by Külz and C. Voit. Külz fed hens casein, serum albumin and egg albumin, control hens being sacrificed at the start of the feeding to determine the amount of glycogen in the liver.

Substance fed	Glycogen content of the liver—grms.	
	Before	After feeding
Casein	1.013	1.85
Serum albumin	0.917	1.56
Egg albumin	1.016	1.78

C. Voit starved hens and rabbits for five days which reduced the glycogen to a minimum. He then fed various carbohydrates and killed the animals after six hours.

Sugar fed	Animal	Amount fed	Glycogen in liver
1. Dextrose	Cock	50 grams	15.34%
	Rabbit	80 "	16.85 (9.269 grams in liver; 8.972 grams in muscle)
2. Sucrose	Hen	100 c.c. 50% solution.	3.75% (1.215 grams)
3. Levulose	Cock	54.89 grams	10.50% (3.99 grs.; 3.562 grs. in rest of body)
4. Maltose	Rabbit	130 c.c. 54.8 grams	9.08% (5.26 grams)
	Cock	60 grams	10.43 (4 grams)
	Rabbit	60 "	9.822
5. Galactose	Cock	55 grams	1.29% (0.6716 gr.)
	Rabbit	68.2 grams	1.53 (0.871 gr.)
6. Lactose	Hen	16 grams	0.19%
	Rabbit	16 "	1.70
	"	50 grams	0.69 before; 3.61% after.

From these results it appears that feeding glucose, levulose, sucrose or maltose leads to a loading of the liver with glycogen. Lactose and galactose are utilized very little, if at all, by rabbits or hens. They are, however, utilized by dogs, human beings and other animals.

As regards the origin from carbohydrates Voit (Otto), after four days' fasting, gave a rabbit 80 grams of glucose in solid form and killed it after 8½ hours. The liver weighed 55 grams; he obtained 9.269 grams of glycogen or 16.85 per cent. In the rest of the body there were 8.972 grams of glycogen. The urine contained during this period 0.816 grams of nitrogen equivalent to 5.10 grams of albumin which at a maximum could only have yielded 2.42 grams of glycogen. There is, then, no doubt of the origin of glycogen from carbohydrate of the food.

Origin of glycogen from fat. While glycogen can thus be readily formed from protein and other carbohydrate very little is formed from fat. It is found by feeding glycerol to animals rendered artificially diabetic, or to diabetic human beings, that the glycerol of the fat molecule, if it get free, may be turned over into carbohydrate. But if fats are fed to diabetic animals or men they do not materially change the excretion of dextrose. It appears from this that the fatty acids do not go over into carbohydrate in the body, although the reverse formation, the change of carbohydrate to fat, goes with the greatest ease, and may very possibly be a part of every anaërobic respiration. It seems probable that the dextrose molecule in its decomposition may furnish oxygen to the living matter and be itself reduced to fat or other less oxidized substances. In this way in the absence of sufficient oxygen it aids in the respiration of the cell and it is itself stored as fat to be used when the supply of oxygen is more plentiful. The oxidation of the fatty acids has already been briefly discussed on page 75, and it will be recalled that they split into acetic acid and the fatty acid poorer in carbon by two carbon atoms. Acetic acid does not go over into glycogen but is oxidized to carbonic acid.

Origin of glycogen from other substances. Further study has shown that the liver can convert many other substances into glycogen, some of them very simple substances. Thus it can convert lactic acid, pyruvic acid, glycerol, propyl alcohol, glycerol aldehyde, aspartic acid, alanine, and some other amino-acids to carbohydrate. The power of the body to convert other substances into carbohydrate is most conveniently studied by Lusk's method of using phlorizinized dogs.

Fasting dogs, when injected subcutaneously with 1 gram of finely powdered phlorizin suspended in 7 c.c. olive oil develop a severe glycosuria lasting for several days. The ratio of dextrose to nitrogen, D : N, in the urine after a day or so reaches the value of 3.5-3.7. If now these dogs are fed or injected with substances which are converted into glucose in the body, the extra glucose thus formed is excreted in the urine and the amount may be determined. In this way it may be found whether animals have the power of converting any given substance into glucose. The following table has been compiled from Lusk's work; the table showing that the first four amino-acids are clearly capable of being turned either wholly or in part into glucose in the dog's body. The amount of these acids in meat will account for about 50 per cent of the glucose which is formed when meat is fed phlorizinized dogs. Probably serine, arginine, lysine, and valine also go into glucose in whole or in part. Leucine probably does not form glucose. Since these amino-acids thus go into glucose in the phlorizinized dogs it is probable that they have this power in the normal animal and thus contribute to the formation of glycogen.

Amino-acid fed	Amount—grams	Extra glucose excreted—grams	Theoretical if all amino-acid converted into glucose—grams	Per cent. of amino-acid made into glucose
Glycocoll	20	14.8-15.8	16	100
Alanine	20	18.76	20.2	100
Aspartic acid	20	13.4-14.9	13.52	100
			(If 3 C atoms into glucose.)	
Glutamic acid	20	13.16	12.24	100
			(If 3 C atoms to glucose.)	
Tyrosine		0		0

The place of the transformation of amino-acids to glucose is certainly in part in the liver. Glucose and glycogen can be made by the liver from protein. It will be recalled when discussing digestion that some deamidization of the amino-acids took place in the canal wall or before absorption. The amino-acids are carried also as such to the liver and this organ has the power of deamidizing some of them. This process has been studied by Knoop and Neuberg. Alanine, for example, when perfused through the liver loses its amino group and passes into pyruvic acid; glycocoll goes into glyoxylic acid:

$$CH_3.CHNH_2.COOH + O \longrightarrow CH_3.CO.COOH + NH_3$$

Alanine. Pyruvic acid.

$$CH_2.NH_2.COOH + O \longrightarrow CHO.COOH + NH_3$$

Glycocoll. Glyoxylic acid.

Not all of the amino-acids which reach the liver are thus converted into glucose. Some, and probably a large part of them, escape the gauntlet and circulating in the blood reach and nourish the tissues of the body. The power of oxidizing them in the way just indicated is probably an attribute of all living matter, although possibly especially developed in the liver. Folin and Denis have questioned whether the liver really deamidizes most of the amino-acids at all. Perfusion experiments appear, however, to be positive that some deamidization occurs there.

Conversion of glycogen into glucose. The circumstances and exact control of the conversion of glycogen into glucose by the liver are still a matter of investigation and they cannot be said to be clear in all their details. Its great importance arises in part from the fact that the process is typical of what goes on in cells in general, for stored foods are not uncommon. It may be said at the outset that the interruption of the blood flow through the liver, or depriving it of oxygen, always leads to the conversion, more or less complete, of glycogen into sugar. The liver can only form glycogen if it be supplied with sufficient oxygen. During digestion, when vasodilation occurs in all the intestinal organs, the liver receives blood by the portal vein, which is only slightly poorer in oxygen than arterial blood. At this time, therefore, glycogen is formed in abundance. But during digestive rest the liver receives blood by the portal vein which is already strongly venous; and the supply of arterial blood is relatively small by way of the hepatic artery, so that at this period the liver must be on the verge of asphyxiation. Under these circumstances glycogen will be converted into glucose.

Sugar puncture leads to the disappearance of glycogen from the liver; and the conversion of glycogen into glucose can be hastened by the stimulation of the splanchnics, or of nerves accompanying the blood vessels into the liver. Carbon monoxide poisoning leads to the discharge of the glycogen; curare injection has a similar result because, owing to the paralysis of the skeletal muscles, the respiration is reduced and partial asphyxia results. Chloroform and some other anesthetics easily produce glycosuria in rabbits and in dogs and human beings, also, if the body be chilled during the anesthesia. A very important discovery was that the injection of adrenaline, the internal secretion of the supra-renals, greatly increased the blood sugar, provided the liver contained glycogen, and hence produced hyperglycemia and glycosuria. So that the supra-renal glands were thus brought into relationship with the glycogen conversion. The question has now arisen whether the conversion of glycogen to glucose by stimulation of the splanchnics, or by sugar puncture, is due to the direct effect of the stimulus on the cells of the liver, or whether it is the indirect effect of stimulation of the supra-renal capsules, since these nerves supply these organs also.

Further examination of this problem has shown that if the supra-renals be extirpated, and rabbits will live for months without these organs, stimulation of the splanchnics no longer causes a conversion of glycogen into glucose. On the other hand, the amount of adrenaline discharged into the blood is not sufficient in itself to cause glycogenolysis. MacLeod is probably right in his opinion that stimulation of the splanchnics does two things: it directly stimulates the liver cells to convert glycogen into sugar; and it indirectly facilitates this result by setting free adrenaline, which carried to the liver greatly increases or re-enforces the effect of the stimulation, just as it does that of other sympathetic nerves. Strong emotion, either of fright or anxiety, causes hyperglycemia and in some human beings glucosuria. By some authors this result is supposed to be due to the indirect action of adrenaline, which is known to be discharged into the blood in large amounts in emotional excitement. It is, however, more probable that we are dealing here with a double action, the stimuli coming from the brain under emotional strain pass over the splanchnic nerves in part directly to the liver, arousing the cells to glycogenolysis, and in part indirectly they increase the efficiency of this stimulation by action on the supra-renals, causing a discharge of adrenaline.

The transformation of glycogen into glucose in the liver is then, in a condition of health, the product of at least two, and probably more, factors. In the first place there is the effect of nerve stimulation which in some way accelerates this transformation; and in the second place this stimulation is rendered far more powerful by the increase in the adrenaline content of the blood by the nervous impulses impinging on the adrenals. By the co-operation of these two actions glycogen is changed over into glucose and hyperglycemia is produced. Indirectly, also, either as the result of the hyperglycemia alone, or because at the same time the nerves going to the kidney influence the state of that organ, there may result an emotional glycosuria.

The hyperglycemia thus produced is undoubtedly in the nature of an adaptation. The object of the mechanism is to supply to the muscles, which under normal circumstances in nature may be called upon to undergo great exertions as the animal flees or attacks or defends itself, an abundant supply of glucose from which they can sustain their energy. Experiments have clearly shown that in the normal animal maximal contractions are made possible by an increase in the sugar of the blood. The heart, also, nourishes itself from this food and in its presence tissues are able to work with less oxygen than in its absence.

Just as the emotions may stimulate glycogenolysis, so also does exposure to the cold. If rabbits are simply cooled, or if dogs are cooled and anesthetized, or if human beings are similarly treated, hyperglycemia and discharge of glucose from the liver results. This again is in the nature of an adaptive change since the muscles, the great thermogenic tissues of the body, require the glucose for fuel.

The fact that emotions or cooling have this effect explains the negative results of several observers who have attempted to increase the glycogen content of the liver by the direct injection of glucose into the mesenteric or other veins. Thus Croftan got quite a negative result in such experiments unless glucose was given by the intestine, whereas positive results were obtained provided care was taken to prevent the access of stimuli from the nervous system to the liver.

It has been remarked that stimulation of the splanchnic nerve causes a change of glycogen to glucose. How is this produced? Does the nerve stimulation in some way set free a glycogenolytic enzyme in the tissue; or does it in some other way, for example by stimulating the active discharge of glucose from the cells, upset the equilibrium between glucose and glycogen and thus cause more glycogen to change to glucose? Is there in fact such an equilibrium? The same question arises in considering the physiology of muscle where also following nerve stimulation contraction takes place, glucose is destroyed and glycogen consumed. Just how does the nerve impulse act? This question cannot at present be answered definitely. MacLeod has especially studied it. There are many possibilities besides those just mentioned. Thus it might be that always a small amount of free glycogenase exists in the cell, but it is prevented by colloidal membranes from reaching the glycogen. It might be that the nerve impulse altered the state of these membranes so that the enzyme might get access to the glycogen. Another possibility is that the enzyme is anchored to some of the colloids of the cell, and the union is broken on stimulation. Or by the nerve impulse an oxidative change might result forming acid, this acid then might set free the enzyme from its substrate and enable it to attack the glycogen. Or perhaps the glycogen is being constantly attacked by the glycogenase and digested, the glycogenase being always present in the cell, but usually the sugar thus set free is at once resynthesized by the vital activity of the cell. The nerve stimulus might act by inhibiting in some way this synthetic activity, so that the action of the enzyme proceeds unchecked; or it might be, as already suggested, that the nerves simply stimulated the cells to discharge or secrete their glucose and automatically the cells produced more to make good the loss. MacLeod, in studying this question, found that no increase in glycogenase could be detected either in the tissue or in the vein as a result of nerve stimulation. On the other hand, an increased alkalinity of the tissue prevented the nerve stimulation from being effective.

The transformation of glycogen to glucose goes on rapidly in the liver after death, at least it goes rapidly in the first few hours. Thereafter it goes more slowly. It acts as if it might be a vital transformation. It is natural to suppose that this transformation is due to an endocellular enzyme, of the nature of amylase, which becomes active under the post-mortem acidity of the tissue. Extracts of liver tissue, however, have little more amylolytic power than those of the blood (Eves; Dastre); and Dastre thought the conversion not due to an enzyme. The conversion is greatly accelerated by chloroform. Paton gives the figures on the next page illustrating the speed of this conversion of glycogen to glucose (dog's liver).

The figures show that the conversion is very rapid in the first half hour and thereafter goes on more slowly. It is also clear that there is practically no post-mortem destruction of carbohydrate, or glycolysis. If the liver is thoroughly ground in a mortar, so as to destroy the cells as completely as possible, the conversion goes on much more slowly, but is not completely stopped. Chloroform has practically no influence on such a finely ground liver. This transformation is greatly retarded or stopped by an alkaline reaction and accelerated by small amounts of acid. Perhaps the slowing of the reaction is due to the destruction of the enzyme, or to the slowing of the reaction by the development of alkalinity, owing to the formation of ammonia by autolysis.

Time after death:	Two minutes	24 hours
Glycogen	5.88 grams	0.55 grams
Glucose	Trace	5.29
Total carbohydrates	5.88 grams	5.84

Time after death:	Two minutes	45 minutes		815 minutes	
		With $CHCl_3$	Without $CHCl_3$	With $CHCl_3$	Without $CHCl_3$
Glycogen	7.09 grams	5.68 grams	6.23 grams	4.60 grams	5.42 grams
Glucose	0.23	1.39	0.98	2.53	1.88
Total carbohydrates	7.32	7.07	7.21	7.18	7.20

The influence of the pancreas. Still another factor in the control of this transformation of glucose to glycogen and of glycogen to glucose has been discovered and it is one of the most important factors in the sugar control. It is the pancreatic gland. That this gland plays a very important part in this as well as other aspects of the control of the body carbohydrate metabolism was discovered by von Mering and Minkowski in 1889, a discovery not less pregnant and important than the discovery of glycogen itself. These authors found that if the pancreas be completely extirpated, hyperglycemia, glucosuria, and the complete loss of all power to burn glucose or other carbohydrate resulted. The animals showed all the symptoms of the severe form of human diabetes mellitus. They emaciated, were very thirsty and voracious, their muscles were weak, they passed urine containing not only glucose but also acetone, acetoacetic acid and β-hydroxybutyric acid, and they died in coma invariably in four to six weeks.

$$CH_3-\overset{}{C}(=O)-CH_3 \qquad CH_3.CO.CH_2.COOH \qquad CH_3.CHOH.CH_2.COOH$$

Acetone. Diacetic acid. β-hydroxybutyric acid.

Another very interesting and significant fact was that their vital resistance to infection was enormously reduced so that it was extremely difficult to avoid infection in the operation, or afterwards, and wounds healed slowly. A similar susceptibility to blood poisoning, boils, etc., in other words to the attacks of various streptococci, is shown by human beings with a diabetic trait. The possible significance of this fact in considering the cause of diabetes should not be overlooked.

Results similar to these have been obtained in all classes of vertebrates; in frogs, selachians (dogfish), birds, snakes and mammals.

The total extirpation of the pancreas is always followed by death and usually by a loss of power of burning carbohydrates.

By this discovery the problem set by Claude Bernard which he failed to solve, namely, the experimental production of the symptoms of diabetes mellitus, was solved.

The results which follow pancreas extirpation are not due, as was at first thought, to the extensive division of the nerves and traumatism of the sympathetic ganglia involved in the operation, but must be ascribed primarily to the actual taking out of the pancreas tissue. This was proved by Hedon, who made the following experiment: The abdomen of a dog was opened, a portion of the pancreas was extirpated and the branch of the gland, which extends into the mesentery, was brought to the surface and fastened into a pouch made underneath the skin of the abdomen. The blood vessels and nerves of this portion were left intact. After several weeks, during which no glycosuria appeared, the blood vessels of the skin enter into union with the vessels of the gland graft. Then another operation was performed and the grafted piece of pancreas was cut loose from the stalk connecting it with the abdomen. If the operation had been successful, if the blood connection in the skin had been established, no symptoms of diabetes resulted. The juice secreted by this grafted piece might discharge outwardly, or make a tumor, but there was no glucosuria as long as the graft remained. Some time later the graft was either removed, or at times it degenerated, and when this happened glycosuria began and continued until shortly before death.

These experiments of Hedon's were not very numerous but they were decisive. They showed clearly and conclusively that the nervous lesions, or circulatory disturbances, of this operation were not the cause of the glycosuria. The presence of the grafted tissue even under the skin was sufficient to prevent hyperglycemia and to enable the body to carry on its ordinary carbohydrate metabolism. There were, or seemed to be, but two possible explanations, namely, either the pancreas adds something to the blood or lymph, which is necessary for normal carbohydrate metabolism, or it takes something out, which if not removed causes glycosuria. Most investigators preferred the first of these possibilities and believed the pancreas must have an internal secretion even more important than its external and vitally necessary for the carbohydrate metabolism of the body. But the second or detoxication theory has also had its advocates.

This discovery of von Mering and Minkowski and Hedon led to an enormous amount of work to prove the existence of this internal secretion. This work extended over twenty years before a decisive result was obtained. It was found, however, that neither feeding the gland itself nor injecting the gland extracts caused any definite and clear-cut change in the course of the disease or of pancreatic diabetes. There were, it is true, many apparently successful experiments indicating that pancreatic extracts might have some temporary or trifling action in reducing the glycosuria; but other observers either failed to get any action at all, or the glycosuria might even be increased by feeding pancreatic tissue to depancreatized dogs or to human beings with the severe form of the disease. The ease with which a temporary relaxation of glycosuria might be produced was a special cause of error. In many instances, indeed, feeding pancreas after pancreatectomy so far from diminishing seemed really to increase the glycosuria. It appeared in 1910-1911 that if the pancreas had any internal secretion it did not accumulate, or was not stored in the gland as are all the other internal secretions known, and investigators began to turn to the study of other possibilities such as the detoxication theory, believing that the internal-secretion theory must be wrong. Within the past two years, however, more favorable results have been obtained. The most decisive experiments are those of Starling and Knowlton made with an acid extract of the pancreas. Their results in the following figures show that the addition of a neutralized acid decoction of the pancreas to diabetic blood which is being perfused through the heart taken from a depancreatized dog causes the heart to beat much stronger and to consume glucose added to the blood. The consumption of the glucose before the addition of the pancreas extract had been reduced to the vanishing point.

DIABETIC HEART FED WITH DIABETIC BLOOD.

(Sugar consumption in mg. per gram heart muscle per hour.)

First hour with blood alone	Second hour after addition of pancreatic extract
1.5	4.3
0.5	3.0
0.5	2.8
0.5	3.6

The normal heart of the dog when fed on normal blood under these circumstances consumes per hour per gram of heart muscle 3.5-5 mgs. of dextrose. It is clear from these experiments that the diabetic heart fed with diabetic blood has lost its power of burning glucose to a very considerable degree, and that the addition of the pancreas decoction has restored this power.

More recent experiments have indicated that the respiratory quotient of the diabetic heart is not always changed by the addition of pancreatic extract as one would expect if the heart really had acquired the power of decomposing glucose. Moreover both alkalies and adrenaline are said to have an equally beneficial effect. It was not shown in the experiments that the glucose had been actually burned. It might have been converted to glycogen. In view of these considerations it is still doubtful whether the results obtained really show the existence of an internal secretion.

These experiments and others indicate, however, that normal blood has in it a very small amount of a substance which enables glucose to be burned; this substance is derived from the pancreas. It is its internal secretion. If to an extract of muscle juice pancreatic extract is added, Cohnheim found the glycolytic power increased, but McGuigan and others have been unable to confirm these observations and Cohnheim himself in repeating them generally had negative results. Evidently some accidental circumstance in a few cases caused a greater disappearance of the sugar. The manner in which the extract acts is quite obscure. It may be that in the experiments cited the muscle, instead of destroying the sugar, may have converted it into maltose. The internal secretion might act by changing the permeability of the muscle membranes or by combining with glucose, acting as a between or intermediate body by which it is anchored to the protoplasm; or it might be the raw material from which the glycolytic substance of the muscle is made. Since a decomposition of the sugar almost certainly precedes its combustion, the muscle has probably lost not the power of burning the decomposition products, but rather the power of decomposition itself.

The internal secretion of the pancreas, like that of the supra-renals, is either rapidly eliminated or destroyed in the blood. It may possibly be a basic body very unstable in a free form, but more stable as a salt. At any rate blood contains no great amount of it and it quickly disappears when the pancreas is eliminated. It is not impossible that it is normally destroyed or eliminated by the kidney, as is adrenaline, and one of the causes of human diabetes may be that the destruction or elimination of this body may at times be greater than its rate of formation, so that a partial or total loss of combustive power ensues.

This internal secretion of the pancreas plays a part also in the control of the glycogenic function of the liver. In the absence of the pancreas, although the blood contains two to three times its normal supply of glucose, the liver forms very little glycogen. If, however, the liver be perfused with blood containing glucose, or if glucose is injected directly into the circulation, it stores more glycogen if at the same time there is added to the blood an extract of the pancreas. How this extract acts, however, is still quite uncertain. It may be acting on the ends of the sympathetic nerves reducing or counterbalancing the action of adrenaline, or it may be acting directly on the liver cells themselves. It will, no doubt, seem peculiar, at first glance, that the body should still be able to form small amounts of glycogen even when the further oxidation of glucose is impossible. The probable reason for this is suggested by the experiments of Lobry de Bruyn and also of Nef. For the complete oxidation of the glucose molecule a more extensive decomposition of the molecule is necessary than for the synthesis of the monosaccharides to the disaccharides. For the latter it is only necessary to produce a rearrangement of the last two carbon atoms of the chain; for the decomposition it is necessary to break the chain also. If, for example, sugars are put into very weak alkali, such as milk of lime, synthesis into disaccharides and mutual transformations of one sugar into another occurs, but the molecules do not break into fragments; it is only in the presence of stronger alkali that the decomposition of the molecule takes place (see page 31). The diabetic body, then, seems able to produce the easiest transformation of the carbohydrate, namely, that involved in the transformation of one kind of sugar into another, levulose into glucose, for example, or the synthesis of the monosaccharides in part to glycogen, although it has lost the power of breaking the molecule far enough to oxidize it.

Recently some results have been obtained by Dakin which may ultimately throw light on the relation of the pancreas to sugar metabolism. Dakin has found that nearly all tissues of the body, i.e., liver, thymus, thyroid, supra-renal, pituitary, kidney, spleen, heart muscle, skeletal muscle, lung, brain, etc., contain an enzyme known as "glyoxalase." This enzyme has the property of converting a simple or substituted glyoxaldehyde into the corresponding glycollic acid.

$$\underset{\text{Glyoxal.}}{\text{R. CO. CHO}} + H_2O \longrightarrow \underset{\text{Glycollic acid.}}{\text{R. CHOH.COOH}}$$

It is a very suggestive fact that the pancreas is the only organ, except the lymph glands, which lacks this enzyme, and in the pancreas alone is found a thermolabile substance, which antagonizes the glyoxalase, an antiglyoxalase. This antiglyoxalase is not found in the blood, it occurs in small amounts in the external secretion of the pancreas and probably, Dakin thinks, may be secreted internally. Whether this be the case or no, the formation of antiglyoxalase appears to be a specific function of the pancreas. (Another very interesting fact is that these glyoxals will be transformed into amino, as well as hydroxy-acids, when perfused through the liver (Dakin and Dudley)). While the relation of these results to the carbohydrate metabolism is still obscure, we have at any rate for the first time a specific chemical property of the pancreas obviously closely related to carbohydrate metabolism.

Internal secretion. Where produced in the pancreas. There are at least two distinct kinds of cells in nearly all glands, namely, the secreting cells of the acini and the cells of the ducts. These cells evidently differ in their chemical nature and their physiological function, but practically nothing is known in any gland of the functions of the duct tissue beyond its function of conducting the secretion. We do not know in the pancreas whether all the digestive enzymes are secreted by the acinary cells or whether some are secreted by the duct cells. It is known that the secretion obtained by the injection of acid in the duodenum often differs markedly in composition from that produced by secretin; but whether these two come from the same or different tissues

FIG. 63.—Islets of Langerhans in the guinea-pig pancreas. The islets are the dark cell masses attached to the ductules which grow out of the larger ducts. The acinary tissue has degenerated as a result of blocking the ducts (Bensley).

of the gland is unknown. There have also been described in the pancreas small clumps, or groups of cells which differ in their staining reactions and appearance from the acinary cells, but are in connection with the ducts. These are known as the *Islets of Langerhans*. Until very recently no methods were known for the certain identification of islet tissue, but not long ago Bensley found that in the guinea pig they stain more readily in neutral red, injected intra vitam, than do the acinary cells. Here and there, however, outside the islets proper, similar cells are to be seen. According to Bensley ligation of the duct of the pancreas in the guinea pig leads to the destruction of the acinary tissue, while the much branching duct tissue with the islets as masses of duct cells at the extremities of the outgrowths of the ducts remain. See Figure 63.

It was Bernard who first succeeded in showing the difference in reaction and function of the duct and the acinary cells. He found in dogs that injection of the ducts with a fat of high melting point, solid at the temperature of the body, produced a complete atrophy of the acinary cells, whereas the ducts remained intact, " like a tree which had lost its leaves." If such dogs developed diabetes Bernard did not observe it. If they did not, this experiment shows very clearly that duct tissue alone is sufficient to preserve the life of the animal in the absence of the acinary cells.

Some years ago Opie reported that in many cases of diabetes lesions were apparent only in the islands of Langerhans. He, accordingly, suggested that the internal secretion of the pancreas was due to the islets and not to the acinary cells. Unfortunately for this theory many cases of diabetes show lesions in neither tissue. The opinion has, however, become widespread that the internal secretion is due exclusively to the islets. The evidence for this is still very doubtful. Both in Bensley's guinea pigs and in Bernard's dogs there is no suggestion of diabetes. This does not show that the acinary cells form no internal secretion, but only that the duct tissue, including the islets, is alone sufficient to maintain life. Possibly both tissues form the secretion. The contrary experiment of the destruction of the duct and the survival of the acinary tissue has not been, and seemingly cannot be, tried. We cannot at present say, therefore, whether the internal secretion of the pancreas is due to the duct cells and the islet tissue alone, or whether the acinary tissue also contributes to it. The categorical statement often seen in text-books and papers that the internal secretion is supplied by the cells of Langerhans is entirely unjustified.

Cause of diabetes. Finally what is the cause of diabetes in human beings? Nothing is known of this whatsoever. The facts just stated lead pathologists generally to attribute the disease often to a lesion in the islets of Langerhans in the pancreas. Whether all diabetes have this origin is not certain. Nothing is known of the cause of the lesion if any such exists. Whether it is due to the absorption of a poison from the intestine, as some have thought, or whether it is due to a bacterium or other parasite, has been very little investigated and so far without result. Attention has been focused entirely on the symptoms of disturbed metabolism which accompany this disease, rather than on a search for the real cause. The fact that diabetics generally show such a lowered resistance to the attacks of the streptococci of blood poisoning arouses the suspicion that the trouble might be due to an infection of the islets by this organism which thus produces diabetes, just as it produces inflammation of the valves of the heart when it strikes them, or of the kidneys when it is located in those organs. Perhaps a strain of this protean organism will be found which will act on, or show a specific affinity for, the islets. At any rate the search for the cause of this disease, rather than a search for a method of alleviating its symptoms, would appear to be the part of wisdom.

Summary of the rôle of the liver in carbohydrate metabolism. We may now summarize the rôle of the liver in carbohydrate metabolism of the body. Carbohydrate food after digestion finds its way into the blood, where it is to be found in small amounts: 0.1-0.3 per cent. in the blood of the portal vein. The sugar thus circulating in the blood is free. It is diffusible and may be removed from the blood by the process of vividiffusion. The portal blood at the same time carries, also, any metabolic products or internal secretions of the spleen, pancreas, and intestine. Brought thus to the liver, this organ during digestion, when it is supplied with much oxygen, picks out from the blood some of the glucose, levulose and galactose passing through it and converts these into a colloidal polysaccharide, glycogen, which is laid down in such quantities in the liver tissue that on a rich carbohydrate diet the liver may contain 12-15 per cent. of glycogen.

The liver acts, thus, in the first instance, as a storehouse, or reserve depot of carbohydrates, in which surplus carbohydrate is stored until it is needed by the other tissues. It functions, then, in this particular like the fat tissue which stores superfluous fat.

Between meals, on the other hand, when no glucose is arriving from the intestine, this stored glycogen is called upon and more or less completely consumed; the liver of a fasting animal containing much less glycogen than that of a well-fed animal.

This conversion of glucose into glycogen and of glycogen into glucose depends on various factors. If the glucose in the blood surpasses 0.1 per cent., if the liver is well supplied with oxygenated blood, and with the internal secretion of the pancreas, the conversion of glucose to glycogen is at least as rapid as that of glycogen to glucose and no diminution of glycogen occurs. If, however, the content of glucose in the blood coming to the liver falls below 0.1 per cent., or if oxygen supply is reduced; or if impulses come into the liver over the splanchnic nerves; and particularly if there is at the same time an abundance of adrenaline in the blood due to the excitation of the supra-renal glands by the splanchnic nerves following emotional or other excitement, then glycogen is converted into glucose and this will take place even though the amount of glucose in the blood is already above the normal and as much as 0.2-0.3 per cent. It is uncertain how this glycogen is converted into glucose, but it is probably due to an endocellular glycogenase. Whether stimulation of the nerve causes an increase of glycogenolysis, or a decrease of conversion of glucose to glycogen, it is as yet impossible to say, but there seems to be no doubt that normally nerve stimulation is re-enforced by the concomitant stimulation of the supra-renals, causing them to throw into the blood an increased amount of adrenaline. There is no evidence of the setting free of glycogenase during the stimulation.

The liver is not the only storehouse of glycogen. The muscles contain an amount little, if any, less than that of the liver, and even in the absence of the glycogenic function of the liver, as, for example, when an Eck fistula is made, the eating of carbohydrate food seems no more liable than usual to produce glucosuria, nor does there seem to be any reduction in the sugar tolerance of the organism. This fact has not yet been explained, but it is probably to be ascribed to two or three circumstances. One is the storage power of the muscles; another is that possibly the liver, after the Eck fistula is made when its cells are greatly altered in their appearance and size and probably in function, is no longer able to form glucose from protein so that the production of glucose in the animal itself is reduced; and finally the nutrition of animals with the Eck fistula is often poor and possibly there is less complete absorption of carbohydrate and other foods.

But the liver is not only a magazine of carbohydrate; it is also a factory where proteins, amino-acids, glycerol, lactic and pyruvic acids, dioxyacetone and other substances are converted into dextrose. Many, if not all, of these syntheses involve oxidations and for their accomplishment the liver must have a supply of oxygenated blood. In some of these syntheses ammonia is split off and replaced by oxygen. This ammonia serves a double purpose: in part by its alkalinity it reduces acidity and hence checks glycogenolysis; in part it is changed to urea, a substance of importance in the maintenance of the activity of other organs; i.e., the heart. By this synthetic power of the liver not only is the body able to form dextrose from protein and other foods, being able indeed to live on protein alone, but many of the partially oxidized metabolic products of other tissues, such as lactic acid, are saved and resynthesized into glucose to be used over again. The liver in this respect works over waste products.

Finally the liver is not only a producer and storer of dextrose; it is also a consumer. Like all cells and tissues of the body it has the power of consuming carbohydrate. It respires and burns some of this carbohydrate, though how much it is impossible to say. Not all pyruvic aldehyde or acid is converted into dextrose, but a part at least is oxidized, burned to carbon dioxide and water; a part may, by reduction, be reunited with ammonia to form some of the simpler amino-acids, so that the reverse process of protein synthesis can also occur. In the bird's liver some of the products of carbohydrate destruction are used in synthesizing uric acid; and in the mammalian liver we have a formation of glycuronic acid so important as a means of neutralizing poisons.

No doubt, also, a ready transformation to fat occurs. But we have as yet a most imperfect picture of the chemical transformations occurring in this most important glandular organ.

But while we picture the liver as acting thus as a storehouse for glycogen with which it parts when glycogen is needed by the other tissues, there are reasons which make it probable that glycogen is stored in the liver primarily for the benefit of the liver itself. Glucose has a very particular relationship to anaërobic respiration. It is already partially oxidized and it can furnish energy for various decompositions and activities of the body, being itself reduced to alcohols and fatty acids at the same time. All tissues live very much longer without oxygen, if they are supplied with glucose. The liver is peculiar in that it must carry on its activities while its blood supply is very largely venous in character. It must be always nearer asphyxia than most of the tissues of the body. There can hardly be a doubt that the primary object of this stored supply of glycogen in the liver is to enable the liver to function properly in the presence of very little oxygen. It is found, indeed, that the liver in the absence of glycogen is far more liable to necrosis than when it is present. Particularly in chloroform anesthesia such necrosis is apt to occur and it shows itself first in the center of the lobules where, presumably, the oxygen need is greatest (Graham). This necrosis is very much less apt to occur if an animal has been well fed on carbohydrate before being anesthetized. It is possible that the giving to patients about to undergo operation considerable amounts of carbohydrate so as to load the carbohydrate reservoirs of the body would be a wise precaution, provided other unlooked-for results do not develop. If glycogen is discharged from the liver by adrenaline, central necrosis also occurs. (Figure 61, page 669.) The liver then stores glycogen not only that that glycogen may later be available for the other tissues of the body, but probably because glycogen is very necessary to its own safety in time of stress and oxygen want.

The further fate of glucose. The liver supplies glucose to the blood. What then is the ultimate fate of this glucose? The tissues which consume most of the carbohydrates are undoubtedly the muscles, because of their bulk; if for no other reason, this would be the case. They make about 50 per cent. of the weight of the body. They are always in activity and they produce much heat. The muscles contain a considerable quantity of glycogen. Indeed the glycogen may persist in the muscles after it has disappeared from the liver. That glycogen disappears from a muscle, or is consumed during muscle contraction, is shown by the determination of the glycogen content of muscle before and after prolonged tetanus, as shown by the figures on page 619.

Moreover that some non-nitrogenous substance furnishes the energy for muscular work is shown by the fact that physical work does not increase the nitrogen outgo of the body, but only that of carbon dioxide and water. We may, therefore, conclude that the sugar thus circulating in the blood is picked out by the muscles, which store it as glycogen, depositing it in their tissue, and that it is oxidized to carbon dioxide and water or to lactic acid during muscle work. The muscles are the great carbohydrate-consuming tissues of the body.

The question now arises concerning the consumption of dextrose by the muscle. How is the sugar consumed? Is it oxidized directly as glucose? Or is it decomposed first into simpler substances which then undergo oxidation? Do the muscles burn glucose by themselves or do they need the assistance of other tissues or glandular organs? Does this power of destruction depend on the vitality, structural integrity of the muscle, or does it occur also in hashed muscle? Is it due to an enzyme which is found in the muscle, or is it a vital process, that is one involving the structural integrity of the cell? Is the glucose burned in the muscle sap or must it first be broken and then built into the living tissue itself? How are we to explain the explosive decomposition which occurs when a nerve impulse sets up a muscular contraction? These and many other questions press at once for solution, but to only a few of them can we give a complete answer.

Conditions of sugar burning in muscle. The same factors which are active in the sugar metabolism of the liver play a rôle here in the metabolism of muscle. We may consider first the question whether the muscle is able by itself to burn sugar brought to it from the exterior, or only when it is assisted by the co-operation of other organs. To this question a definite answer may be given. The muscle cannot utilize extrinsic glucose as food and as a source of energy except in the presence of the pancreas. If the pancreas be completely extirpated in any vertebrate so far studied, the power of consuming glucose is completely lost by all the tissues of the body. If a portion of the gland be left in the body, the power of sugar consumption persists more or less completely. If glucose or glucose-producing food is given to a depancreatized animal it appears in the urine practically quantitatively.

But after pancreatectomy not only is ingested glucose completely eliminated, but also all that glucose normally produced in the body from protein is excreted also. It thus happens that even on a carbohydrate-free diet the excretion of glucose continues. It is as if the body were turning to sugar.

There are two aspects of this failure of the muscles and other tissues of the body to burn carbohydrate in the absence of the pancreas, which are truly remarkable. The power of burning carbohydrate and in particular the power of burning or fermenting glucose is practically universal. All kinds of plants, probably all plants, including the bacteria, have this power. Glucose is found in all the invertebrates in one form or another. Now it certainly is an almost incredible fact that this fundamental property of living matter should, in the vertebrate organism, be so completely lost, that no glucose at all should be burned in the absence of the pancreas. That the organs should have come to lean on the assistance of the pancreas would not be surprising, but that they should have totally lost one of the most fundamental properties of living matter so that they can only exist in the presence of the pancreas is so astounding as to arouse the greatest suspicion of the truth of the conclusion. It appears from the facts, however, that there is no escape from this conclusion, however unlikely it may appear.

The other aspect of the affair which is of great interest is this. Sugar is constantly being made in the diabetic organism. Indeed, the power of sugar formation is certainly not diminished and many have thought that it is stimulated in the diabetic organism. This glucose may be made out of all the substances from which it is made in the normal organism, from alanine and other amino-acids, from glycolaldehyde and other very simple substances. Now if the failure to burn dextrose is due to the fact that the molecule can no longer be fragmented, as is assumed by some investigators, then we should expect the fragments to burn as well as ever if these fragments were fed. No one has been able to discover any such fragments of the dextrose molecule, which can be burned by the diabetic organism when the power of burning glucose is lost. Of course there may be fragments which are normally burned and which will be found in the future. But it certainly is singular that all the search for such combustible fragments is still unsuccessful in spite of the considerable number which have been tried. Furthermore it is clear that if the hypothesis is correct that the failure of the tissues to burn the molecule is due to the fact that the preliminary fragmentation is wanting, these combustible fragments cannot be any of those from which the carbohydrates are formed. In other words, the molecule must break down in its fragmentation into substances of another kind than are used in the synthesis. For example, since the diabetic organism will carry lactic acid into sugar and not burn it, it is clear that the combustive decomposition of the carbohydrates is not through lactic acid, if the theory as to the reason of non-combustion is correct. It might perhaps be worth while to reexamine with great care the evidence upon which the conclusion is based that no sugar is burned in the diabetic organism to see whether this view is correct. It might be that the sugar formation was stimulated to such an extent that all that could be burned was being burned. Any excess accordingly appears in the urine. In other words, perhaps the sugar-burning powers of the body of the diabetic are already at a maximum without the introduction of glucogenetic foods from the exterior. The power of glucose storage is certainly greatly reduced, and perhaps this stimulates glucose production.

D:N ratio. If the carbohydrate is withdrawn as completely as possible from a depancreatized animal or from a diabetic individual, dextrose continues to be secreted in the urine. This dextrose, since it has not come from carbohydrate food, must have been made in the body itself, and since its amount rises with the protein and not with a rise in the fat intake, it must be derived from protein food. How extensive this manufacture of glucose from protein may be is illustrated by experiments by Minkowski and Lusk. If the protein was split quantitatively into glucose and urea there should be obtained, since 100 grams of protein contains 50 grams of carbon and 16 of nitrogen, about 175 grams of glucose. Not all of the protein, however, is converted into glucose. Some of it appears as acetone, diacetic acid, and other decomposition products. Some is burned to carbon dioxide and water so that actually the maximum amount which can be made into glucose in a dog's body in a condition of complete glucose intolerance is 60 grams. We have, then, since 100 grams of protein yield roughly 16 grams of nitrogen in the urine, a ratio of D:N, dextrose to nitrogen, of 60:16= 3.7. Higher ratios than this may easily be obtained on feeding food containing carbohydrates, but on a strictly protein and fat diet in a condition where glucose cannot be utilized at all, this ratio has been obtained by von Mering and Minkowski in depancreatized dogs, in diabetic human beings, and in dogs made completely sugar-intolerant by phlorhizin by Lusk. In an experiment by Reilly, Nolan and Lusk a fasting phlorhizinized dog was given 500 grams of meat without altering the D:N ratio. The figures are for the urine of 12-hour periods.

	Dextrose—grams	Nitrogen—grams	D:N
Fasting	23.87	7.0	3.41
500 grams meat	49.59	14.0	3.54
Fasting	25.36	7.1	3.56

This D:N ratio of 3.3-3.7 Lusk calls the fatal ratio, since its appearance in human diabetics when on a strictly carbohydrate-free diet means that none of the sugar can be consumed.

From the foregoing experiments it appears that at a maximum 60 grams of glucose may be made from 100 grams of protein. Several questions at once arise concerning this point. In the combustion of proteins in the body does it always happen that 60 per cent. are converted into glucose and are burned in that form? Or are we to assume that in the condition of carbohydrate starvation prevailing in diabetes, the process of manufacture of glucose from proteins is greatly stimulated in an endeavor to adjust the disturbed metabolism of the body and to refill the depleted carbohydrate reservoirs? There is not at present any way of deciding this question, but many facts indicate, as we shall see in studying protein metabolism, that certainly a very considerable part of the protein is normally made into dextrose.

Decomposition of sugar in the muscle. Fermentation. One of the most important questions in the combustion of sugar in the muscle is whether the dextrose is burned as such or is first decomposed into various splitting products which then are oxidized. Some light may be thrown on this question by a study of the spontaneous oxidation of glucose. In discussing the chemistry of the carbohydrates it was shown that oxidation did not occur directly but in large measure indirectly, the glucose molecule breaking up in alkaline solution and the particles thus produced then either oxidizing themselves spontaneously, or if sufficient oxygen was not present, the particles oxidized and reduced each other, or else condensed to form various other products. There are many indications that dextrose is not burned as such in the body, but by the action of the protoplasm decomposes, much as it does in alkali, to form various decomposition products which either oxidize if oxygen is present, or are reduced or condense to form fats or other metabolic products. The processes may very possibly be similar to the alcoholic fermentation of glucose. This decomposition takes place in the absence of oxygen, very little energy being set free; by the subsequent oxidation of alcohol energy is obtained. The exact nature of this alcoholic decomposition is not clear, but it may pass through glyceric aldehyde.

$$
\begin{array}{lclcll}
CH_2OH & & CH_2OH & & CH_2OH & \\
| & & | & & | & +\ 3\,O_2 \longrightarrow 2CO_2 + 3H_2O \\
CHOH & & CHOH & = & CH_3 & \\
| & & | & & & \\
CHOH & \longrightarrow & COH & & CO_2 & \\
| & & & & & \\
CHOH & & COH & & CO_2 & \\
| & & | & & & \\
CHOH & & CHOH & \longrightarrow & CH_3 & \\
| & & | & & | & +\ 3\,O_2 \longrightarrow 2CO_2 + 3H_2O \\
COH & & CH_2OH & & CH_2OH &
\end{array}
$$

Glyceric aldehyde.

As a matter of fact muscle and other tissues are able to burn alcohol readily and alcohol is found in small amounts in normal tissue; it is, however, very unlikely that in the combustion of glucose in animals alcohol is an intermediate product, since its toxic actions are too intense. The exact course of the destruction of the glucose in muscle is still entirely unknown.

The liver and the muscles are not the only tissues which need and consume dextrose, although they are by far the largest. The heart, the intestine, both support their movements by burning dextrose. Thus the addition of glucose to an artificial circulating fluid like that of Locke or Ringer or Tyrode restores or quickens their contractions.

Altogether aside from muscle tissue, however, there can be no doubt that the metabolism of other organs also requires dextrose and in its absence other sources of raw materials for energy and substance must be found. Particularly the relation of the kidneys to sugar metabolism needs careful investigation. There can be little doubt that the kidneys must have some sort of an affinity for glucose to enable them to secrete it from the blood to the urine, where at times it is in a much higher concentration. This power is greatly stimulated by phlorhizin; it seems also to be reduced in pancreatic and human diabetes, since the kidneys are no longer able to prevent an accumulation of sugar in the blood and a hyperglycemia. It is, of course, possible that the kidney is only able under the best of circumstances to secrete a certain amount of glucose, and that the hyperglycemia means that the kidneys are overwhelmed and unable to reduce the blood sugar to its normal level.

Phlorhizin diabetes.—Hitherto we have considered two experimental methods for producing glycosuria and more or less serious disturbances of carbohydrate metabolism. A third method was discovered by von Mering and has been particularly developed by Lusk. This method consists in the injection of the drug phlorhizin. Phlorhizin, a glucoside derived from the bark of the roots of the plum, apple, cherry and pear trees, as the name signifies (Gr. *phloios*, bark; *rhiza*, root), has the remarkable property of inducing glycosuria which is not accompanied by a hyperglycemia, but rather by a hypoglycemia. The method of employing the drug as developed by Lusk is by subcutaneous injection of 2 grams per day, one gram twice a day dissolved in a little Na_2CO_3. Another method of injecting 1 gram suspended in 7 c.c. olive oil is also used. The drug thus given, or when given in large doses by the mouth, causes a very great glycosuria, and if the injections are continued the usual symptoms of severe mellituria follow, namely, besides the polyuria, muscular weakness, acidosis, acetonuria and death in coma. If, however, the injections are stopped the animal recovers. The glycosuria is accompanied by a hypoglycemia. The dextrose content of the blood falls from 0.12-0.08 per cent. Nevertheless the kidneys continue to excrete glucose. The action of the drug appears to be, therefore, primarily on the kidney, causing it to secrete glucose more rapidly than usual and hence to keep the level of the dextrose in the body below the normal. As a result of this impoverishment of the blood, the liver and the muscles not in activity give up glucose to the blood in order to supply that organ (the kidney) whose consumption has thus reduced the whole supply. But as this happens to be the kidney, the result is simply like pouring water into a sieve. The dextrose is drained out of the body.

That phlorhizin thus acts on the kidney cells primarily, though not exclusively, is shown by an experiment of Levene. An anesthetized dog had cannulas in each ureter. Into the kidney artery of one side was then injected a small dose of phlorhizin. This kidney secreted glucose at once and the first appearance of glucose in the other kidney did not take place for two minutes later.

The drug causes also marked degenerative changes in the kidney epithelium leading ultimately to its complete destruction. It is sometimes stated that phlorhizin increases the permeability of the kidney epithelium to sugar, as if the kidney acted as a filter which normally held back the glucose and by the action of the drug was made more permeable so that glucose went through. While this may be the means of its action it seems more probable that the secretion of glucose is an active process and that this process is in some way stimulated by the phlorhizin. It is indeed probable that not only does phlorhizin increase the secretion of glucose by the kidney, but under its action glucose appears also in the bile. The secretion of urea is also increased directly or indirectly by the phlorhizin. The nitrogen output of fasting dogs is increased 3-5 times by a dose of phlorhizin. It has also been suggested that the glucose is in combination with some of the colloids of the blood and that the kidney under the action of the drug is able to make the glucose free, which now escapes and is excreted. The evidence for this is, however, extremely meager. It has been shown that the sugar in the blood is capable of diffusing by the method of vividiffusion and that its concentration in the dialysate is about that calculated to be in the blood. It is extremely hard to see why even if glucose is set free from such a hypothetical colloidal union, it should diffuse from a region where it is present in only 0.08 per cent. to one in which it is present to the extent of 2 per cent. Any such a concentration as this by a reabsorption of water by the contorted tubes of the kidney would be impossible. In any case the secretion of the glucose must be an active process quite analogous to that of the secretion of the bile salts from the blood by the liver cells. Phlorhizin produces microscopic changes in the pancreas.

REFERENCES. Carbohydrate Metabolism.

The literature of this subject is so enormous that no attempt will be made to give more than a few of the recent and some of the classical papers on this subject. A list of some 1,200 references to the literature will be found in Allen: Glycosuria and Diabetes, Boston, 1913.

Books.

1. *Allen:* Glycosuria and Diabetes, Boston, 1913. 1180 pages. This book has a fairly complete and exhaustive examination of our knowledge of diabetes mellitus and insipidus and various glycosurias, together with a very large number of original experimental observations.
2. *Lusk:* The Science of Nutrition, 2d edition, 1912. The writer considers this book to be the best general treatise on the subject of nutrition and a large amount of space is given in it to carbohydrate metabolism. References are given, the most important facts are cited in an interesting way and the book is not swamped by details.
3. *Pflüger:* Glykogen. Archiv ges. Physiol., 96, pp. 1-398, 1903. This has a very complete critical summary of the work done up to 1903.
4. *von Noorden:* Metabolism and Practical Medicine. Chicago, 1907.

References.

1. *Bernard:* Leçons sur la diabète et la glycogenèse animale, Paris, 1877.
2. *Bernard:* Leçons sur la physiologie et la pathologie du système nerveux. Vols. 1 and 2. Paris, 1858.
3. *Bernard:* Comptes-rendus de l'Acad., Paris, p. 884, 1859.
4. *v. Mering:* Ueber diabetes mellitus. Verhandl. d. Kong. f. inn. Med., 6, pp. 349-358, 1887.
5. *v. Mering:* Ueber diabetes mellitus. Zeits. klin. Med., 16, pp. 431-446, 1889.
6. *v. Mering and Minkowski:* Diabetes mellitus nach Pancreasextirpation. Arch. expt. Path. u. Pharm., 26, p. 371, 1889-90.
7. *Hedon:* Arch. de Physiol., 1894, p. 269.
8. *Külz:* Beiträge zu der Lehre von der Glykogenbildung in der Leber. Arch. ges. Physiol., 24, p. I, 1881.
9. *Külz:* Beiträge zur Kenntnis des Glykogens. (Beiträge zur Physiologie. Ludwig.) Marburg, 1891, pp. 69-121.
10. *Voit, C.:* Ueber die Glykogenbildung nach Aufnahme verschiedener Zuckerarten. Zeit. f. Biol., 28, p. 245, 1891.
11. *Minkowski:* Untersuchungen über den Diabetes mellitus nach Extirpation des Pankreas. Leipzig, 1893.
12. *Minkowski:* Ueber die Zuckerbildung im Organismus beim Pankreasdiabetes. Arch. ges. Physiol., 111, p. 13, 1906.
13. *v. Mering:* Zeit. f. klin. Med., 16, p. 437, 1889.
14. *Lusk:* The influence of cold and mechanical exercise on the sugar excretion in phlorizin glycosuria. Amer. Jour. Physiol., 22, p. 163, 1908.
15. *Lusk:* Ueber phloridzin diabetes. Zeit. Biol., 42, p. 31, 1904.
16. *Lusk:* Metabolism after the ingestion of dextrose and fat, including the behavior of water, urea and sodium chloride solutions. Jour. Biol. Chem., 13, p. 27, 1912.
17. *Lusk and Stiles:* On the formation of dextrose in metabolism from the end products of a pancreatic digest of meat. Amer. Jour. Physiol., 9, p. 380, 1903.
18. *Ringer and Lusk:* Ueber die Entstehung von Dextrose aus Aminosäuren bei Phlorhizinglykosurie. Zeit. physiol. Chem., 66, p. 106, 1910.
19. *Ringer:* Protein metabolism in experimental diabetes. Jour. Biol. Chem., 12, 1912, p. 431.
20. *Opie and Alford:* The influence of diet on hepatic necrosis and toxicity of chloroform. Jour. Amer. Med. Assn., 62, p. 895, 1914.

21. *Schöndorff:* Ueber den Maximalwerth des Gesammtglykogengehalts von Hunden. Arch. f. d. ges. Physiol., 99, p. 191, 1903.
22. *Whipple:* Insusceptibility of pups to chloroform poisoning during the first three weeks of life. Jour. Expt. Med., 15, p. 259, 1912.
23. *Graham:* The resistance of pups to late chloroform poisoning in its relation to liver glycogen. Jour. Expt. Med., 21, p. 185, 1915.
24. *Starling and Evans:* The respiratory exchanges of the heart in the diabetic animal. Jour. Physiol., 49, p. 67, 1914.
25. *Dakin and Dudley:* Glyoxalase. Part IV. Jour. Biol. Chem., 16, 1913-14, p. 505.

CHAPTER XIX.

PROTEIN METABOLISM OF THE BODY.

In the chapters on digestion and absorption, the course of the protein taken in the food was traced through the processes of digestion and into the blood. The simple proteins, it will be recalled, find entrance to the blood, in large measure at least, in the form of amino-acids, the primitive building stones of which the protein material of the body is to be constructed. Whether some of these amino-acids are synthesized to protein in passing through the wall of the intestine cannot be positively denied, but certainly the evidence that any such synthesis occurs, except for the building up of the proteins of the epithelial cells, is extremely unsatisfactory. Some of the amino-acids have been destroyed by the action of the bacteria of the tract and some have lost amino groups and been changed into ammonia and a carbon residue, possibly ketonic acids like pyruvic acid, during absorption. It is probable, however, that most of the amino-acids get into the blood as such. In the blood itself they are found in very minute amounts, but most of the important amino-acids have been found there in small quantities. So rapid is the circulation of the blood and so admirable are the mechanisms for maintaining its composition that the amino-acids are removed from the blood almost as rapidly as they find entrance to it; there is not, under normal circumstances, any accumulation of amino-acids in the blood. It has been shown, however, by Folin and Denis that there is always some increase, and not an insignificant increase, in the non-protein nitrogen of the blood after the ingestion of protein foods. The non-protein nitrogen includes amino-acids, urea and ammonia, among other constituents. We may now ask ourselves the question concerning the farther fate of these amino-acids.

We have already at various times touched on these questions and have considered at length certain aspects of protein metabolism when dealing with the origin of the nitrogenous substances in the urine. Thus we have already discussed the purine metabolism, the origin of urea, the formation of ammonia, the transformation of amino-acids to sugars, the significance of the creatine and creatinine excretion, and the origin of various other urinary constituents which arise from the proteins. Something has been said, too, about the influence of the thyroid gland on protein catabolism. In this chapter we shall only touch briefly on certain general questions of protein metabolism: What quantity of amino-acids are destroyed per day? What is the course of the transformations they undergo when they are decomposed and destroyed in the body? Whether they are also synthesized in animals as they are in plants. Has the body any power of storing protein when more protein is ingested than the organism needs at the time? How much protein per day must a person take and what are the consequences of taking more or less than enough?

Amount of protein needed per day by a human adult.—Few questions of recent times have been more debated than this: How much protein food must we eat a day in order to keep in the highest state of efficiency. This is a question of the highest importance in human nutrition. The proteins are the most expensive foods that we consume. It is nitrogen that is expensive. We may say at the outset that the quantity of protein needed will not be independent of the character of the protein, since the amino-acids are the substances which are really needed, rather than protein as such, and those proteins which have all the amino-acids in about the same proportions as they are found in the body as a whole will probably be more efficient than those which have an excess of one kind or another. We shall come back to this question presently. The minimum amount of protein required by the average human adult was stated a few years ago by Voit to be about 120 grams of protein per day. This amount was arrived at by measuring the amount which people in moderate circumstances consumed. The idea was that the human race had been for generations experimenting in order to arrive at this minimum. Proteins are expensive and difficult to get. In the struggle for existence which presses so hard on human beings as upon all animals, it is to be supposed that this amount of food which was so hard to get would be the minimum upon which a high state of efficiency, sufficient to conquer in the struggle for existence, could be maintained. If carbohydrates and fats, which are much easier to get and much cheaper, could give a more efficient individual, the persons who ate more of them than of proteins would, in the course of generations, have survived and supplanted their less sensible brothers. As a matter of fact, it was found that the races which used less protein, which were chiefly vegetarian races, were on the whole less active, vigorous and progressive. They were the Bengalis of India and races generally regarded as somewhat inferior and retrogressive. It seemed, then, that the point of view of Voit was well taken and that the minimum requirement for efficiency was about 120 grams of protein per day for an adult.

This idea was seriously attacked by an American, Mr. Horace Fletcher, some ten years or more ago and his work gave rise to a discussion of the whole matter which has done much to clarify our views of the rôle of protein in the animal economy. Mr. Fletcher being past middle life, and being refused life insurance on account of his poor condition, went seriously to work to regulate his diet so as to improve his condition. In this he was imitating the similar conduct of Louis Cornaro, an Italian of the fifteenth century, who in similar circumstances acted in the same way. After some experimentation both cut down their diets until they were eating far less food than before, and Mr. Fletcher particularly cut down his protein consumption. Cornaro took about 12 ounces of food per day. The physical condition of both Cornaro and Fletcher greatly improved. A bad catarrh and liability to catch cold which had troubled Mr. Fletcher quite disappeared. His general condition was so greatly improved that he became an extremely active man and was able to do exercises of a physical kind which only young men in good physical training can do without great fatigue and lameness. These results were so remarkable that he has devoted himself since then to teaching the great value of a restricted diet, particularly for men over forty years of age. The results in Cornaro's case were no less remarkable. He lived to be 102 years of age and at 82 and again at 94 he wrote treatises on the art of living long. The sum and substance of his prescription was temperance in all things. Neither Mr. Fletcher nor Cornaro restricted their diet to one kind of food. Cornaro is not very specific as to his exact diet, but apparently he partook of the ordinary foods, except fish and some things that did not agree with him; he drank wine temperately; and certainly Mr. Fletcher takes whatever he feels a desire for. In each case there has been a great diminution in the quantity of food taken.

The results were of such a nature that a careful investigation was undertaken in this country by Chittenden, a squad of soldiers volunteering to serve for the experiment to see what the effect would be of limiting protein consumption. Mendel, Folin and many others have contributed to this study.

The general result of this work has been to show that it is possible to live for a considerable period, at any rate, and apparently in a state of good health and without loss of weight, on far less protein than the Voit standard demanded. 120 grams of protein requires a nitrogen outgo of some 19 grams of nitrogen per day. Most of this of course will go in the urine, but some will be in the feces. The amount of nitrogen in Fletcher's urine was about 6 grams per day; in the soldiers' in Chittenden's experiments it ranged from 6-10 grams per day, and in himself and some of his colleagues and students it fell to a similar figure. Van Sommeren, a son-in-law of Fletcher, lived on an amount of protein food so small that his urinary nitrogen was only 4-6 grams per day and presumably it remained there for a long period, although he was actually under observation for a short time. Folin, by eating a diet containing chiefly starch, cream and sugar, reduced his nitrogen in the urine to about 6 grams a day for several days; and Thomas, in Rubner's laboratory, reduced his urinary nitrogen on a starch and cream diet, when a large amount was taken so as completely to cover the energy requirement, for various short periods during two years, to as little as 2.2 grams per day. This amount of nitrogen corresponds to a protein intake of only 15-20 grams per day.

It is clear from these experiments that it is possible to maintain the weight of the body and carry out ordinary exertions, to establish nitrogen equilibrium, at a far lower level than the Voit standard required. So perfect is the mechanism of the body that the utilization of the protein taken in the food is at a maximum under these conditions. It is almost completely absorbed and utilized, putrefaction in the intestine being reduced to a minimum. Furthermore, the general health in many of these individuals was better than it would have been under their former régime. It is, therefore, clear that the total nitrogen waste of the body may be reduced to a very low figure, and it must be concluded either that the proteins in the body are being torn to pieces very little, if at all, in metabolism, or else that the pieces into which they are torn are carefully saved and used over again. Which of these points of view is correct it is very hard to say, but perhaps modern work has emphasized the latter possibility rather than the former.

In order to reduce the amount of protein intake to a minimum while nitrogeneous equilibrium is maintained, that is while the outgo and income of nitrogen balance each other, it is necessary to cover the energy requirements of the body by eating carbohydrates and fats, for if sufficient energy-yielding food is not eaten, then the body tears its own tissues to pieces to secure the fuel necessary. Furthermore, the quantity of non-protein food eaten must be more than sufficient to cover the energy requirement, since there are reasons for believing that the carbohydrates in particular have the additional virtue of enabling a partial synthesis of at least some, and perhaps of many, of the amino-acids in the body from carbohydrate decomposition products and ammonia or other nitrogen derivatives of protein catabolism. For this reason they assist in keeping the nitrogen in the body. The total effect of the ingestion of carbohydrate is, therefore, to save the proteins of the body and they and fats are said to have a *protein-sparing function* in metabolism. The explanation of this action is not certainly known, but it may be in part that they are so much more easily oxidized that they protect the proteins from oxidation in this way; or they may, in the manner just cited, make possible the resynthesis of amino-acids from ammonia and other decomposition products of protein metabolism; or they may be important aids in the anaërobic respiration of cells which presumably occurs about the nucleus. Thomas found that in order to keep his nitrogen outgo down to 2.2-4.63 grams per day large amounts of carbohydrate had to be eaten. If fat were substituted for carbohydrate, the amount of nitrogen in the urine was somewhat increased. Perhaps this was due in part to the slight acidosis which generally occurs in the metabolism of large amounts of fats. The fats do not burn so easily and completely as the carbohydrates to carbon dioxide and water, but fragments of their molecules, such as acetoacetic acid, are apt to escape unburned in the urine. This acid is neutralized in part with ammonia and when it appears it carries out some ammonia in the urine, thus increasing somewhat the nitrogen outgo.

Since it is the amino-acids which are used to synthesize the proteins of the body it is necessary, if a real physiological minimum is desired, that just the right amount of each particular kind of amino-acid shall be eaten. Since the different proteins contain quite different proportions of the amino-acids, it makes a great difference to the body which protein is eaten. Dog flesh nourishes dogs with less waste than any other kind of protein. For example, some of the proteins lack completely certain amino-acids, and if the animal organism is incapable of manufacturing these acids in sufficient amounts to cover its needs, it will be impossible to maintain nitrogen equilibrium when that particular protein is used as a food. For example, gelatin lacks both tyrosine and tryptophane and it has been found impossible to nourish completely any mammal when gelatin is the sole protein food in the diet. However much gelatin may be taken, and however much carbohydrate be added to it, there is a slow loss of nitrogen to the body resulting eventually, if the diet is not changed, in death. Evidently it is impossible for the animal body to manufacture the lacking amino-acids from the food supplied in amounts sufficient to cover its requirements. Thomas found a considerable difference in the power of the different proteins to supply in the most efficient manner the nitrogen needs of the body. Meat and milk protein could replace the protein consumed with the greatest efficiency. It was necessary to eat least of these in order to supply the 2.2 grams of nitrogen which was the minimum outgo. Some of the vegetable proteins were far less efficient. If the protein minimum was covered by them, it was necessary to take far more of the protein. Indeed, of the total nitrogen taken in the form of vegetable protein, sometimes 60 per cent. was wasted: that is, that proportion of nitrogen was not in a form to cover the nitrogen minimum of the body. Potato protein was better for the physiological minimum than either peas or beans or wheat.

Recent work on the necessity of other constituents than protein in the foods (see page 833) makes the interpretation of these experiments somewhat obscure, since it is possible that the greater efficiency of milk and meat might be due to the presence in them of some non-protein constituent necessary to the body but not found in the vegetables consumed. But there is no doubt of the fact that the body can get along for a considerable time and often with advantage on less protein than is usually consumed. Rubner in 1883 expressed the opinion that not more than 5 per cent. of the energy requirement of the body had to be in the form of protein. The 2.2 grams of nitrogen in the urine of Thomas, Rubner suggests, came from the bacteria of the intestinal tract and from the blood decomposition. When doing very hard, muscular work while on this diet Thomas raised his nitrogen to 2.6 grams per day. This shows that the muscle substance does not wear out rapidly. The machinery does not wear out. Rubner thinks that there is a minimum decomposition of 2.0-3.0 grams of protein per day. Since the whole amount of protein in a man's body is about 2,000 grams, only 0.1 per cent. goes to pieces daily. If the loss were equally distributed, this would mean that the protein was renewed once in 5 years. The actual necessary wear and tear, he thinks, is less than this.

It appears, then, that the amount of nitrogen wear and tear of the body is not necessarily very great. This may mean one of two things. First, that the protein is metabolizing at a very slow rate indeed; that the protein makes actually a machinery of a very stable kind which moves and organizes the cell, but which does not itself burn, or metabolize at a rapid rate, but is moved by the energy set free from the combustion of the carbohydrates and fats; or, second, it may mean that the body is able to save and use over again the waste products of the protein metabolism so that it saves its ammonia. The first possibility is of considerable interest, since it may be that the stability of the brain proteins makes possible the stability of the memories of the body. This possibility is discussed under the chapter on the brain. The second possibility, however, has much in its favor. It is now certain that the body has the power of manufacturing some amino-acids from some of the products of carbohydrate metabolism and ammonia, and it is possible, hence, that the nitrogen is thus saved to the body and remade into amino-acids which are used to make good the protein wear and tear.

Is minimum protein desirable? It is possible to live for several years on less protein than is ordinarily consumed. Is it desirable that the bulk of the population should reduce their protein consumption so as to approach the minimum? Most physiologists are of the opinion that it is undesirable and that it is safer to provide for a certain excess above the minimum requirement. In the first place, it is certain that growth is very dependent upon protein food. For the proper development of the body it is necessary that protein should be eaten in considerable quantities. There is hardly a doubt that the increase in the average stature of the population of this country is due, in part at least, to better nourishment of the children. Growth is stunted by too little food. For growth protein is needed. Nature probably had to solve this problem by blind experimentation and the food provided for the rapidly-growing young is always protein food. Milk contains as much protein as it does carbohydrate or fat; young birds are fed on worms larvæ and insects, even though the adults may be graminivorous. But there are reasons even in adults for the excess of protein consumption above the minimum. While there is no protein storage in a narrow sense, there is certainly a reserve power which a well-fed person has and which an ill-fed one lacks. The muscles and cells of the body full of living matter have certainly a greater vitality and a greater resistance to disease than when they are depleted. Experiments have shown that the resistance of rats and other animals to snake venom is greater when they have been fed protein than when they have not been fed protein. It may be that the difference is due not to the protein, but to other constituents of the diet, but in our ignorance of what those constituents are it would appear wiser to eat the food which contains them. The whole matter is, hence, in an unsettled state. We are confronted, on the one hand, with the fact that long life is usually accompanied by a temperate disposition, and temperance in eating and drinking; and that many people, particularly those past middle life, are benefited by reducing their protein; on the other hand, peoples of great vigor are generally heavy protein consumers, and for the young, certainly, a plentiful protein diet seems to have been that elaborated by nature after many experiments.

Will the body store protein? The human body has the power of storing both carbohydrate and fat. If one eats more carbohydrate food than is necessary to cover the energy requirements of the body, it is not at once completely burned and got rid of, but up to a certain point it is stored either as glycogen in the liver, muscles and some other tissues, or it is converted into fat, and deposited as such in the great fat reservoirs of the body, which lie under the skin or about the internal organs. With proteins the matter is quite different. It is true that many plants have the power of storing proteins in their seeds and in other tissues. Nuts generally contain stored protein. The proteins in these cases are dead, reserve proteins. They are more stable than the usual proteins and they are often laid down in crystalline deposits in the cells. Some animals also store a small amount of protein in eggs to serve as food for the developing embryo. But it is always found that the cells which thus store protein have a metabolism slower than usual. Either they store protein because their metabolism is small, or else the accumulation

of this mass of inactive protein checks their metabolism. In the ad
human being there is certainly a very limited storage of protein. If
doubles the amount of protein necessary to replace the wear and t
of the protein of the body, in a well-fed person the sole result is
increase the output of nitrogen. If we eat 200 grams of protein
day, the body does not retain this, but it is at once oxidized and
rid of. The nitrogen is increased in the urine to the same extent
it has been increased in the food. It is not retained in the body. It
only after a long fast, or particularly after a prolonged low prot
diet, when the body has been covering its needs with carbohydrates
saving its proteins, that there is a retention of nitrogen in the bod
And this power of retention is not very great. It is, for example, af
wasting diseases when there has been a great loss of muscle substa
or after hemorrhage when there must be a rapid reformation of blo
that protein storage occurs.

In fact, so far is it from being the case that eating protein leads
protein storage, that the reverse is true. A large protein diet far
excess of the protein requirements leads to a consumption of fat,
that the body is thin and may actually lose weight. Proteins have
certain specific action. They stimulate heat production. If one ea
more protein, it is not as it is with the fats that the excess is stored
but it is burned at once, so that a large protein diet means an increa
output of heat. The heat production is at a minimum on a low prot
diet. In rest it may then sink to 2,000 calories per day; whereas at
high protein diet even at rest it rises to 3,000-3,500 calories per d
It is as if the fats burned in the heat of the proteins, for one way
getting thin is to eat large amounts of protein (Banting cure).

The explanation of this peculiarity of the proteins when contra
with the fats and carbohydrates has not yet been given in its entir
but it is not impossible that it has the following teleological explana
The proteins make part of the real living matter. It is impossible
increase the living matter of the cell, or the living matter of the b
as a whole, beyond the powers of the blood to supply oxygen to k
it alive. If more living matter is formed than can be supplied w
oxygen, hydrolytic or autodigestive processes are set at work w
digest the protein and thus tear down that which has been formed. T
amount of the living matter is evidently limited by the ratio of b
to surface and to the possibility of supplying oxygen. This may be t
way in which the amount of living matter is limited in the body. Pr
tein cannot be laid down in the protoplasm in the form of dea
reserve material without seriously checking the metabolism of the c
Stable, inert proteins are found only in those cells where the metabo
is not very intense. If it be asked how it happens that protein is

deposited in cells in spite of the possibly deleterious result of such a deposition, we have to confess that we know very little about it. The experiments of Ascoli and others on uricase and its variation in the liver of fasting and well-fed individuals appear to be full of significance. It will be recalled that in bird's and dog's liver the enzyme to destroy uric acid disappears when the diet is restricted and reappears when the diet is plentiful. This is evidently in the nature of an adaptive metabolic change. When the diet is restricted, or during fasting, it is possible that the uric acid is needed to replace the waste of the nuclein material. The activity of the uricase disappears under these circumstances. Feed the body well and destruction of uric acid results. The simplest explanation suggesting itself is that some of the food decomposition products give rise to the enzyme which destroys uric acid. But whether this explanation is correct cannot be said. Perhaps it is the same with the amino-acids. After fasting we know that the destruction of amino-acids is greatly restricted and the organism builds them over into protein to replace that which has been used up. It may be that the enzymes which destroy or hydrolyze the proteins, or oxidize the amino-acids, are reduced in quantity, so that the speed of the destruction of the amino-acids is reduced. On the other hand, when there is a luxus consumption of the amino-acids, perhaps some of their decomposition products are converted into catalytic agents which hasten the decomposition of the amino-acids. Consequently the oxidative decomposition is greatly increased and the heat of the body increased. Whatever may be the exact mechanism by which a storage of protein is prevented, there is no doubt of the fact that very little storage occurs, but that excess protein is torn to pieces, and the nitrogen eliminated as urea. Heat production is at the same time increased, and there appears to be a stimulated decomposition of the fats. Of the non-nitrogenous part of the protein molecule a portion at least is converted into glycogen, as has already been discussed on page 771, and may be stored as glycogen.

Catabolism of proteins.—The question we have now to ask is a very fundamental one and, like most fundamental questions, we cannot answer it. The question is this: What is the course of the metabolic decomposition of the proteins of the cells of the body? These proteins are complex, conjugated, colloidal proteins. Do they undergo oxidation or deamінization while they are in this form; or is the first step in their catabolism a digestive process which results in setting free the amino-acids and other constituents? And are these fragments then oxidized, or first fragmented by fermentation and the fragments oxidized? It will be recalled at the outset that the proteins with which we are dealing, that is the real, organized, protein basis of living matter, is not a simple protein, but it contains in its molecule certainly possibly various enzymes and carbohydrate and other mat it inorganic. Its composition is probably illustrated by th of the blood platelets or the stroma of the red blood corp might even say of the red blood cells as a whole, while tl in them hemoglobin. Presumably in the red blood cells tion of the protein is represented by a compound of hem pholipin-protein-lipase-cholesterol-potassium. This compou it be a chemical compound, is known to be very unstable variety of agents cause it to decompose. On the whole, is favorable to the view that something similar happens in that the first step in the catabolism of the proteins is a c of this complex, and the digestion of its constituents. Thu found that in all cells there ensues on death a digestion o material with the appearance of the splitting products of tl conjugated proteins. This digestion is known as *autolysis.* tissues, as soon as they die, there is a digestion of the adenine and guanine, ammonia is set free and the dean hypoxanthine and xanthine, are formed; the bases may free from their union with the carbohydrate or phosphori Lipases also become active and a digestion more or less the fats and phospholipins occurs; the simple proteins are the appearance of albumoses and amino-acids and other p have, then, in cells after death the appearance of diges which deamidize and decompose, or hydrolyze, a great mai constituents and among them the proteins. Nearly all ce teolytic enzymes of the erepsin type and of the deam Nearly all of these digestive actions occur best in a very medium and they are checked or prevented by the additi bicarbonate of soda. This is particularly true of the prot been suggested, and seems on the whole very probable, that t do not begin their work only at the moment of death when of the cell has become acid, but that they are more or less time, but that normally their activity is reduced either by of antibodies, or else by the reaction, or else that the syr of the cell is so great that in spite of their action th destroyed. It would seem probable that the first step in ca then, a hydrolysis of the proteins and that the oxidation or of the products set free succeeded this. This view, how been universally accepted. It is a very singular fact tha sible to isolate these digestive enzymes before autolysis h is the same difficulty which was noted in the case of the c of the glycogen in the liver; the enzyme appears only i amounts and after digestion has begun. A potato, although it undergoes hydrolysis of its starch very easily after lying for a time, seems to have no diastase in it when the potato is green. In other words, the diastase appears when the digestion begins. A quite similar fact is noted in the clotting of the blood: no thrombin can be isolated from the plasma, but only from the serum after clotting has occurred. The question in the clotting of blood is the same as in the digestion of the proteins, Is the enzyme which appears a result or a cause of the clotting or digesting process? No doubt it appears most probable that the enzyme is present and active even during life, but its activity is checked in some of the ways stated. Tissues waste away in starvation, even though they live, and this is probably due to a partial hydrolysis. There are not wanting those who maintain, however, that the enzyme is not the primary cause of the catabolism in the living tissue. The reason why the enzyme is not to be found in living active cells may be illustrated by the phenomena of the clotting of the blood. The blood platelets, according to Wooldridge, are of the nature of crystalline products. Now, if they are examined fresh, no fibrin ferment can be extracted from them, but if they are allowed to clot first then they yield fibrin and thrombin. They also yield an hydrolytic enzyme which has the power of digesting the fibrin and producing fibrinolysis. The probability seems to be that the enzymes in cells are in union with the substances upon which they act, they are in union with their substrates and so cannot be extracted; under certain conditions this union is stable and the substrate is not affected by the enzyme, but under certain other conditions, and a very slight reduction in alkalinity appears to be one of them, the reaction is consummated, the compound is hydrolyzed and both the hydrolytic product and the enzyme appear free at the same moment, just as fibrin and thrombin appear simultaneously. If this view is correct, the protoplasmic proteins already have combined with them the various enzymes which under different circumstances decompose them by autolysis. Perhaps under certain conditions these same enzymes have been responsible for the synthesis of the proteins which they digest under other conditions.

By this autolysis of the proteins amino-acids are produced. These amino-acids are either fermented, carbon dioxide and amines being formed from them; or they are oxidized with the formation of certain products, among them ketonic acids, ammonia and aldehydes. Ultimately the nitrogen residues escape from the body chiefly as urea; while the carbon residues are in part at least changed to glucose and glycogen in the manner indicated in the previous chapter.

Course of the oxidation of various amino-acids in the body. The course of oxidation of various amino-acids in the body has been studied by Neubauer, Knoop, Embden, Dakin and others by perfusing the liver with blood containing the amino-acids, by obtaining their decomposition products from the urine and by oxidation with hydrogen peroxide.

The general course of the oxidation of the simple amino-acids is first to form by oxidation the ketonic acid and ammonia. A subsequent oxidation converts them into the acid of the next lower series by the loss of carbonic acid. According to Dakin the aldehyde is formed as an intermediate product. The course of the oxidation of the simpler acids is shown by the following reactions. In not all cases has it been actually shown that the decomposition follows this rule, but it has in many of them and it is probable for the others.

$$\underset{\text{Glycocoll.}}{NH_2CH_2\text{—}COOH} + O \longrightarrow \underset{\text{Glyoxylic acid.}}{CHO\text{—}COOH} + NH_3$$

The reaction probably goes in two stages (Knoop and Neubauer), the oxyamino-acid forming first:

$$\underset{\text{Glycocoll.}}{NH_2CH_2\text{—}COOH} + O \longrightarrow \underset{\text{Hydrated imino, intermediate stage.}}{NH_2CHOH\text{—}COOH} \begin{cases} \longrightarrow \underset{\text{Imino-acid.}}{NH = CH\text{—}COOH} + H_2O \\ \longrightarrow \underset{\text{Glyoxylic acid.}}{O = CH\text{—}COOH} + NH_3 \end{cases}$$

The hydroxy amino compounds may be regarded as hydrated imino-acids. It will be remembered that oxygen and NH are very similar in many of their properties and mutually replace each other in compounds.

Alanine on oxidation forms pyruvic acid, thus:

$$\underset{\text{Alanine.}}{CH_3\text{—}CHNH_2\text{—}COOH} + O \longrightarrow \underset{\text{Pyruvic acid.}}{CH_3\text{—}CO\text{—}COOH} + NH_3$$

On further decomposition this acid yields by oxidation acetic acid and carbon dioxide:

$$\underset{\text{Pyruvic acid.}}{CH_3\text{—}CO\text{—}COOH} \longrightarrow \underset{\text{Acetic aldehyde.}}{CH_3\text{—}COH} + CO_2;\ CH_3\text{—}COH + O \longrightarrow \underset{\text{Acetic acid.}}{CH_3\text{—}COOH}$$

Relation to hydroxy acids. The amino-acids are converted often into hydroxy acids. Thus in perfusing the liver with blood containing alanine, lactic acid was obtained. Similar hydroxy acids have been obtained also from other acids. The hydroxy acid is not formed by a direct replacement of the amino group by hydroxyl, but by the reduction of the ketonic acid which is formed by the oxidation of the amino-acid and the subsequent elimination of ammonia. From alanine pyruvic acid is formed in the way described and this is by reduction converted into lactic acid.

$$CH_3{-}CO{-}COOH + 2H \longrightarrow CH_3{-}CHOH{-}COOH$$
Pyruvic acid. Lactic acid.

Lactic acid may, therefore, arise from the proteins as well as from the carbohydrates.

The evidence that lactic acid is not first formed and then oxidized to pyruvic acid is the fact that a substituted lactic acid, such for example as p-hydroxyphenyl-lactic acid, does not yield homogentisic acid when administered to an alcaptonuric, whereas p-hydroxyphenyl-pyruvic acid does yield it. This observation shows that the hydroxy acid cannot be in the normal course of oxidation of tyrosine; and that p-hydroxyphenyl pyruvic acid is probably in the chain of normal oxidation. The relation between the hydroxy acids, the ketone and amino acids may be represented as follows:

$$
\begin{array}{ccc}
 & & CH_3{-}CHOH{-}COOH \\
 & & O\downarrow\quad\uparrow H_2 \\
CH_3{-}CHNH_2{-}COOH & \underset{H_2}{\overset{O}{\rightleftharpoons}} & CH_3{-}CO{-}COOH + NH_3 \\
 & & \downarrow O \\
 & & CH_3{-}COOH + CO_2
\end{array}
$$

It is not impossible, however, that the exact course of the transformation may not be correctly represented by the foregoing scheme. It may be that an unsaturated acid is formed at the outset of the reaction by a deamidization and that this unsaturated acid is subsequently oxidized, to the hydroxy or oxy acid. An amino group behaves in general like a hydroxyl group. Hydroxy acids, like lactic acid, easily lose water and are transformed into the unsaturated acids like acrylic acid. However the β-hydroxy acids undergo this transformation more readily than the α-acids. The reaction would be as follows:

$$CH_3{-}CH.NH_2{-}COOH \longrightarrow CH_2 = CH{-}COOH + NH_3$$
Amino propionic acid. Acrylic acid.

By subsequent hydration or oxidation acrylic acid might be converted into the hydroxy or the ketonic acid:

1. $CH_2 = CH{-}COOH + HOH \longrightarrow CH_3{-}CHOH{-}COOH$
Acrylic acid. Lactic acid.

. $CH_2 = CH{-}COOH + O \longrightarrow CH_3{-}CO{-}COOH$
Acrylic acid. Pyruvic acid.

Such an unsaturated amino-acid has been obtained recently by Hunter from dog's urine, namely, urocanic acid. It has also been obtained from a pancreatic digest. It is derived from histidine by deaminization:

```
CH—NH\
||      CH
C  —N//
|
CH
||
CH
|
COOH
```

Urocanic acid.
(Imidazolylacrylic acid.)

```
C6H5
|
CH
||
CH
|
COOH
```

Cinnamic acid.

The corresponding derivative from phenyl alanine, called cinnamic acid, is found in oil of cinnamon, balsam of Tolu and elsewhere. An isomeric cinnamic acid, the alpha-phenyl acrylic acid, atropic acid, is obtained from the alkaloid, atropine. The evidence at present, however, favors the Knoop view of a preliminary oxidation before the desaturation.

Tyrosine and phenyl alanine. By oxidation tyrosine is converted probably in the first place to para-hydroxyphenyl pyruvic acid, and phenyl alanine to phenyl pyruvic acid. The isolation of these acids has not been accomplished from the urine after ingesting tyrosine, but indirect evidence shows their formation. Their further fate is very interesting. Under certain not well understood conditions a peculiar acid, homogentisic acid, appears in the urine. This acid is a dihydroxy-phenyl acetic acid and has the property of turning the urine dark on standing by the formation of melanin by the spontaneous oxidation of the acid. That this substance is derived from tyrosine is shown by the fact that the administration of tyrosine to alcaptone patients increases the amount of homogentisic acid. This acid differs from tyrosine in that the hydroxy group and the acetic acid radicle are not in the para positions to each other, so that to understand the formation of homogentisic acid from tyrosine it must be surmised that a rearrangement of the hydroxy groups takes place. This will be obvious from the following formulæ:

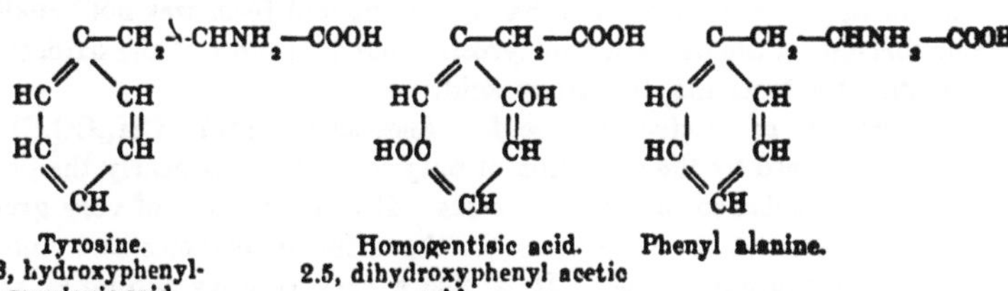

Tyrosine.
3, hydroxyphenyl-propionic acid.

Homogentisic acid.
2.5, dihydroxyphenyl acetic acid.

Phenyl alanine.

A similar transformation has, however, been observed by Bamberger and others, showing that such rearrangements are not uncommon and that they are due, probably, to the intermediate appearance of quinonoid derivatives formed by oxidation. Neubauer's explanation of the formation of homogentisic acid includes this explanation and is illustrated as follows:

```
      C—CH2—CHNH2—COOH         C—CH2—CO—COOH            OH
     //  \                    //  \                     |
   HC     CH                HC     CH                   C—CH2—CO—COOH
   |      ||      ——→        |      ||      ——→         /  \
   HC     CH                HC     CH                HC     CH      ——→
    \\  /                    \\  /                   ||     ||
     COH                      COH                    HC     CH
                       para-hydroxyphenyl pyruvic      \  /
                                                        C=O
  Tyrosine.                      acid.             Quinonoid form.

        OH                                 OH
        |                                  |
        C                                  C
      //  \                              //  \
    HC     C—CH2—CO—COOH               HC     C—CH2—COOH + CO2
    |      ||                ——→       |      ||
    HC     CH                          HC     CH
     \\  /                              \\  /
      C—OH                               C—OH
                                     Homogentisic acid.
 2.5 dihydroxyphenyl pyruvic      2.5 dihydroxyphenyl acetic
          acid.                            acid.
```

Homogentisic acid is possibly a normal product of tyrosine oxidation, which is usually further oxidized. On the other hand, it is possible that it is usually formed only in small amounts, and that normally most of the oxidation goes either directly from the quinonoid to the further decomposition of the tyrosine and relatively little is first carried over to homogentisic acid, or it does not go through the quinonoid form at all. Dakin found that when para-methyl-phenyl alanine and para-methoxyphenyl alanine were administered to alcaptonurics they did not go over into homogentisic-acid derivatives, but were completely oxidized. These substances cannot form the quinonoid intermediate products. From this it would appear more probable that the quinonoid form was not usually gone through in the oxidation of tyrosine, but that most of the oxidation went directly from the dihydroxy acid.

Formation of acetoacetic acid. Acetoacetic acid, $CH_3.CO.CH_2.COOH$, is formed by the oxidation of butyric acid and is nearly the final stage in the oxidation of the fatty acids. It is a substance of very great interest because of its appearance in the urine in diabetes and under various other circumstances. It is also a very reactive substance and acetoacetic esters form the starting point of some of the most fundamental syntheses of organic chemistry. It is extremely interesting that acetoacetic acid is produced, also, from the proteins and from several amino-acids. It has been shown by Dakin to be formed by the oxidation of tyrosine and phenyl alanine. In this case two of the carbons

	Substance	Fate in normal organism	when given to alcaptonuric	perfused thru surviving liver.
1.	Phenylethylalcohol, $C_6H_5.CH_2.CH_2OH$	$C_6H_5.CH_2.COOH$	—	
2.	Phenylacetaldehyde, $C_6H_5.CH_2.COH$	$C_6H_5.CH_2.COOH$		
3.	Phenyl acetic acid, $C_6H_5.CH_2.COOH$	Not oxidized	—	
4.	o-, m-, and p-hydroxyphenylacetic acids	" "		
5.	Homogentisic acid, $C_6H_3(OH)_2.CH_2.COOH$	Complete oxidation	+	+
6.	...acetic, $C_6H_5.CHNH_2.COOH$	$C_6H_5.COOH$	—	—
7.	...pionic acid, $C_6H_5.CH_2.CH_2.COOH$	$C_6H_5.COOH$	—	—
8.	...mic acid, $C_6H_5.CH = CHCOOH$	$C_6H_5.COOH$	—	
9.	Phenylalanine, $C_6H_5.CH_2.CHNH_2.COOH$	Complete oxidation	+	+
10.	Tyrosine, $C_6H_4(OH).CH_2.CHNH_2.COOH$	" "	+	+
11.	3.5, diiodotyrosine, $C_6H_2I_2(OH).CH_2.CHNH_2COOH$	" "		—
12.	p-methylphenylalanine	" "	Complete oxidation	+
13.	p-methoxyphenylalanine	" "	"	+
14.	Phenyl-β-alanine, $C_6H_5.CHNH_2.CH_2.COOH$	Not oxidized readily		
15.	Phenyl serine, $C_6H_5.CHOH.CHNH_2.COOH$	$C_6H_5.COOH$		
16.	Phenyl-a-lactic acid, $C_6H_5.CH_2.CHOH.COOH$	Complete oxidation	+	+
17.	p-hydroxyphenyl-a-lactic acid	Oxidation difficult but probably complete	—	Doubtful
18.	o-, and m-hydroxyphenyl-a-lactic acid	" "	—	
19.	Phenyl-glyceric, $C_6H_5.CHOH.CHOH.COOH$	$C_6H_5.COOH$	—	
20.	2.5 dihydroxyphenyl-a-lactic acid	Complete oxidation	+	
21.	Phenyl-b-hydroxypropionic acid	$C_6H_5.COOH$		
22.	o-, m-, p-hydroxyphenylpropionic acids	$C_6H_4(OH).COOH$		
23.	Phenyl pyruvic acid, $C_6H_5.CH_2.CO.COOH$	Complete oxidation	+	+
24.	p-, o-, m-hydroxyphenyl pyruvic acids	" "	o- and m- — p-, +	p-, +
25.	p-methylphenylpyruvic acid	" "		+
26.	p-methoxyphenylpyruvic acid, $C_6H_4(OCH_3).CH_2.CO.COOH$	" "		+
27.	2.5. Dihydroxyphenylpyruvic acid, $C_6H_3(OH)_2.CH_2CO.COOH$	" "	+	

* Dakin: Oxidations and reductions in the animal body. 1912, p. 71. Longmans, Green & Co. (Substances excreted in the organism in combination with glycine are recorded in the uncombined form.)

of the acetoacetic acid are derived from the benzene nucleus of the aromatic acids. The researches of Jaffé show that when benzene is given to dogs small amounts of muconic acid may be isolated from the urine, thus showing that the benzene ring is broken in the course of metabolism. The reaction is as follows:

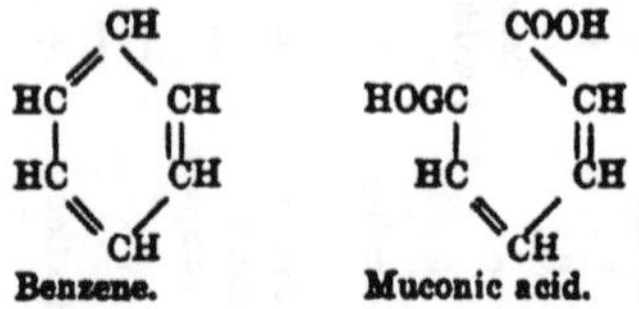

```
        CH                  COOH
      //  \                    \
    HC     CH         HOOC      CH
    |      ||           |       ||
    HC     CH          HC       CH
      \\  /              \\   /
        CH                 CH
    Benzene.          Muconic acid.
```

Dakin gives the following scheme for the possible formation of acetoacetic acid from phenyl alanine:

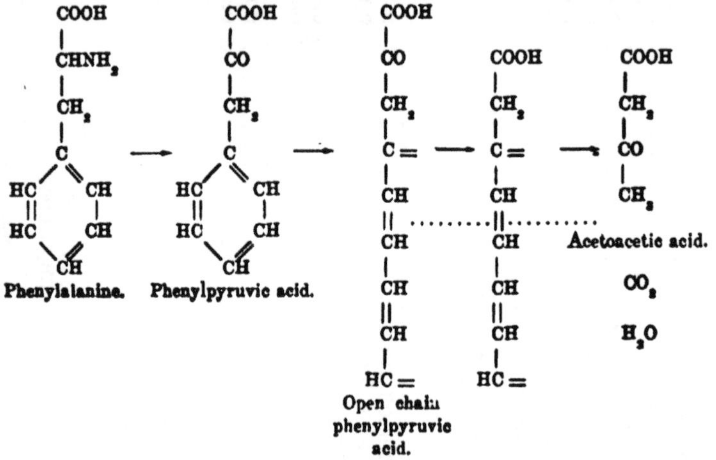

```
      COOH              COOH             COOH
      |                 |                |
      CHNH2             CO               CO          COOH          COOH
      |                 |                |           |             |
      CH2               CH2              CH2         CH2           CH2
      |                 |                |           |             |
      C        —→       C        —→      C=    —→    C=     —→     CO
    /  \\             /  \\              |           |             |
 HC      CH        HC      CH            CH          CH            CH3
 ||      |         ||      |             || ........||..........
 HC      CH        HC      CH            CH          CH        Acetoacetic acid.
   \   //            \   //              |           |
     CH                CH                CH          CH            CO2
Phenylalanine.  Phenylpyruvic acid.      ||          ||
                                         CH          CH            H2O
                                         |           |
                                        HC=         HC=
                                      Open chain
                                     phenylpyruvic
                                         acid.
```

The formation of acetoacetic acid from histidine has already been mentioned.

Decomposition of arginine. The easiest and most direct decomposition of arginine is the splitting off of urea from it by the action of the enzyme, *arginase*, found in the liver and various other tissues by Kossel and Dakin:

```
NH2—C—NH—CH2—CH2— CH2—CHNH2—COOH + H2O—→
    ||
    NH              Arginine.
CO(NH2)2 + NH2 CH2—CH2—CH2—CHNH2—COOH
  Urea.                Ornithine.
```

The further fate of the ornithine and the arginine is unknown. The possibility that creatine may result from the intermediate formation of guanidine butyric acid has already been discussed on page 712 in connection with the origin of creatine. There is no satisfactory evidence that arginine forms creatine in this way in the body, although it is not improbable. The further fate of this most interesting substance should be studied. Its close relation to the cell nucleus in protamine makes its fate of particular interest. In the urine of some people having an abnormal secretion of cystine, in cystinuria, Baumann found unusual quantities of the ptomaines, *cadaverine* and *putrescine*. These bodies almost certainly come from the amino-acids lysine, arginine and ornithine by a process of decarboxylization:

$$\underset{\text{Lysine.}}{NH_2.CH_2.CH_2.CH_2.CH_2.CHNH_2.COOH} \longrightarrow \underset{\substack{\text{Pentamethylenediamine} \\ \text{(Cadaverine).}}}{NH_2CH_2.CH_2.CH_2.CH_2.CH_2NH_2} + CO_2$$

$$\underset{\text{Ornithine.}}{NH_2CH_2.CH_2.CH_2.CH\ NH_2.COOH} \longrightarrow \underset{\substack{\text{Tetramethylenediamine} \\ \text{(Putrescine).}}}{NH_2.CH_2.CH_2.CH_2.CH_2NH_2} + CO_2$$

It is unlikely that more than a small part of the arginine or lysine decomposition normally takes this direction.

Sulphur metabolism of the body.—*Decomposition of cystine.* While the attention of physiologists and physiological chemists has been directed in the main to nitrogen in connection with protein metabolism, the possible rôle of sulphur which makes a part, albeit but a small part, of the protein molecule has of recent years come more into the foreground. Nearly all the proteins of the body, both the living proteins and those of the circulating liquids, contain a small amount of sulphur and this sulphur is in an unoxidized form. It is in the form of sulphide sulphur. Most proteins contain from 1-2 per cent. of this sulphur. It occurs in the protein molecule either in cystine or cysteine and possibly in some other similar acids, although no others have been positively identified.

The sulphur income of the body is almost altogether in the form of the sulphur of the proteins. Some of the foods, indeed the majority of them, contain small amounts of sulphates, but these sulphates do not, so far as we know, enter into the living matter of the body, although in small amounts they are to be found there. The per cent. of sulphate in the blood and tissues is small. It is not certain that sulphur taken in the form of sulphate is reduced in the body. It is more probable that it remains oxidized and while it may contribute something to the formation of the paired sulphates of the urine, even this is uncertain. There is taken on the average about 110 grams of protein per day. In this protein, sulphur makes about 1.2 per cent. This would mean 1.32 grams of sulphur income per day in the form of unoxidized sulphur.

Sulphur leaves the body for the most part in the form of oxidized sulphur. The amount of sulphuric acid excreted per day is about 2.5 grams. This is excreted as the sodium or potassium salt, principally the first. About two-thirds of the sulphur leaves in this form. The other one-third is excreted for the most part as so-called neutral sulphur in the form of conjugated sulphates, being united with indole, scatole, cresol, or phenol, as the case may be. There is also present a small amount of unoxidized sulphur, consisting of cystine, or polypeptides containing cystine, or some other sulphur compound.

In the intermediary metabolism of the body, that is the metabolism of the tissue, sulphur probably plays a very important rôle. This is shown not only by the fact that it is absolutely necessary for the continued existence of the body, as necessary as nitrogen or any of the other elements, but also by the fact that it is one of the most labile elements of the protein molecule. No other element is split off from the proteins with greater ease than this. It is, indeed, the labile element *par excellence*. Moreover cysteine, which is one of the amino-acids, readily oxidizes itself. It is a reducing body. It oxidizes spontaneously and there are many points in its oxidation which strongly resemble the processes of respiration. Thus the most favorable concentration of hydrogen ions for the oxidation of cysteine is the same as that in protoplasm; both cysteine and protoplasm are poisoned by many of the same substances, such as the nitriles, the cyanides, acids, and the heavy metals; their oxidations are catalyzed or hastened in the same manner by iron, arsenic and some other agents. For these reasons it has been suggested by Hefter and the author that there is more than a superficial connection between the oxidation of cysteine and the respiration of the cell.

If we try to follow the course of the absorbed protein sulphur through the body, we find that the fate of the cystine set free from the proteins by the digestive action of trypsin and erepsin is not known in all its details. Some is decomposed by the bacteria of the alimentary tract, forming hydrogen sulphide, a toxic and ill-smelling gas, which, when absorbed in quantities, dissolves red blood corpuscles and contributes, no doubt, to anemia; some cystine is probably absorbed as such. In the liver there is a quantity of taurine in taurocholic acid. Taurine is produced from cystine, or rather from cysteine, by a carboxylic decomposition. Perhaps thioethyl amine may be formed first and then the sulphur oxidized to sulphuric acid, or oxidation may occur first and decarboxilation second.

$$\underset{\text{Cysteine.}}{HS{-}CH_2{-}CHNH_2{-}COOH} \longrightarrow \underset{\text{Thioethyl amine.}}{HS{-}CH_2{-}CH_2NH_2} + CO_2$$

$$\underset{\text{Cysteine.}}{HS{-}CH_2{-}CHNH_2{-}COOH} + 3O \longrightarrow HO_3S{-}CH_2{-}CHNH_2{-}COOH$$

$$HO_3S{-}CH_2{-}CHNH_2{-}OOOH \longrightarrow \underset{\text{Taurine.}}{HO_3S{-}CH_2{-}CH_2NH_2} + CO_2$$

A portion of the cysteine probably passes the liver and is picked out by the tissues of the body. It must circulate in the blood as cystine, since it would at once be converted to cystine by the oxygen of the blood. When it enters the cells of the tissues it is probably, in part, built up into the proteins of the tissue: in part it is probably carried over into taurine since taurine is found in a great variety of tissues. It is an abundant constituent of the extractives of molluscan muscle, particularly of the muscles of cephalopods; it is found also in the mammalian brain and indeed there are few tissues without some. Whether all the decomposition of cystine takes place through taurine is unknown. Possibly a deamidization may occur, leading to a ketonic acid, but no such compound has as yet been found. When cyanides or nitriles are given to mammals they appear in the urine in the form of sulphocyanates, from which it may be inferred that they unite in the cell with the unoxidized sulphur, which in some way they cause to be detached. This union with sulphur does not seem to occur in birds. From the fact that the cyanides have such a remarkable inhibiting effect on respiration it has been inferred by some that sulphur must be very important in respiration.

It occasionally happens that some individuals excrete more cystine than normal. They have a cystinuria. This cystine comes from the tissues of the body, since cystine given in the food does not increase the amount excreted under such conditions. The cause of this metabolic anomaly is still completely unknown, but it may be due to a great hydrolysis of some of the proteins of the body, since there often are found in such cases more than the normal amounts of the other amino-acids, such as tyrosine and leucine, and since the bases, cadaverine and putrescine, may occur in the urine at that time.

Cystine was first discovered, as its name implies (*kystis*, bladder) in bladder stones, which often contain cystine or are composed of it. The meaning of the excretion of this substance was long obscure and it is worth while to examine how information was obtained that cystine was a normal product of the body metabolism, since the method employed for the solution of the problem was a general one and often used for the solution of similar problems. The question was whether the cystine which appeared at times in the urine was a normal constituent of the body metabolism, usually present in very small amounts, but now increased, or whether it represented a wholly new and strange metabolism. This problem was settled by Baumann, who at the same time introduced a very valuable method for the study of the intermediary metabolism of the body.

In the course of metabolism many substances of a very unstable nature are produced. They have a very temporary existence, since, being unstable, they are normally quickly oxidized and we can only guess at their existence, or infer from general principles that they may be formed. Now it is exactly these intermediary substances which are of the greatest importance in metabolism. From one such substance it is often possible by taking slightly different courses to go to several different end products; and we must know what these substances are in order to understand the nature of the metabolism of any single substance. One way of finding out what these substances are is by combining them with something so as to make them stable and thus cause them to escape, or to pass unscathed through the fire of metabolism, coming like Shadrach, Meshach and Abednego to testify to the truth or falsity of our faith. Baumann discovered that cysteine was such an intermediary metabolic product. He found that when brom- or chlor-benzene was given to dogs there appeared in the urine a very remarkable complex, namely, brom-phenyl-mercapturic-glycuronic acid. In this complex was cysteine. The composition of this mercapturic acid was as follows:

$$\begin{array}{l} BrH_4C_6\text{—}S\text{—}CH_2\text{—}CH\text{—}COOH \\ \qquad\qquad\qquad\qquad\quad | \\ \qquad\qquad\qquad\qquad\quad NH\text{—}CO\text{—}CH_3 \end{array}$$

Bromphenylmercapturic acid
(Bromphenylacetylcysteine).

The bromphenyl had united with the sulphur of a sulphur complex which was acetylated and appeared then conjugated with glycuronic acid. Here we have, in this complex, three substances of intermediary metabolism: cysteine, acetic acid and glycuronic acid. The acetylation has already been discussed. It may come from the union of ammonia with a molecule of pyruvic acid and possibly a molecule of the ketone acid, corresponding to cysteine, namely:

$$HS\text{—}CH_2\text{—}CO\text{—}COOH$$

Thiopyruvic acid.

The reaction might take this course:

$$\begin{array}{c} HS\text{—}CH_2\text{—}CO\text{—}COOH \\ + \\ NH_3 \\ + \\ CH_3CO\text{—}COOH \end{array} = \begin{array}{l} HS\text{—}CH_2\text{—}CH\text{—}COOH \\ \qquad\qquad\quad | \\ \qquad\qquad\quad NH \\ \qquad\qquad\quad | \\ \qquad\qquad\quad CO\text{—}CH_3 \end{array} + CO_2 + H_2O$$

This discovery proved that cysteine, which did not appear at all in dog's urine, or was not recognized at that time, but which did appear as cystine in certain cases, was a normal intermediary product of metabolism.

We do not know the method of the decomposition of cystine in the metabolism of the various tissues. Is the sulphur of cysteine oxidized while the latter is still a component of the protein molecule, giving a sulphuric-acid group attached to the protein? Can it be for this reason that so small a proportion of the total sulphur of most proteins is to be recovered in the form of cysteine? (See page 151.) Is the sulphur split off as sulphureted hydrogen, which is later oxidized to sulphuric acid, the cysteine persisting as serine? Or is the cysteine split out of the protein molecule by the action of autolytic enzymes, then losing carbon dioxide and with the sulphur oxidized to sulphuric acid, giving taurine? These questions are not yet answered. One thing at least is certain, namely, that in human beings most of the sulphur is oxidized to sulphuric acid before its excretion, four-fifths of it at least leaving the body in the oxidized form. It is, in fact, in large measure owing to the sulphuric acid formed in the course of protein oxidation that the urine of carnivorous animals is acid in reaction. Moreover, this oxidation takes place anywhere in the tissues.

The excretion of sulphuric acid is largest on a meat diet and the proportion of oxidized sulphur is also largest. The excretion of sulphur on a cream and starch diet is much reduced and its distribution in the urine is changed, a larger proportion appearing as unoxidized sulphur. This would indicate that cysteine or cystine of the diet is either largely decomposed in the alimentary canal, or that a great proportion of it is changed in the liver to taurine or sulphuric acid; whereas that set free in the tissues is relatively less oxidized and hence more of it escapes in the urine.

The sulphur compounds are often ill-smelling. The active principle of the odoriferous gland of the skunk is n-butyl mercaptane, C_4H_9SH. Ethyl sulphide is found in dog's urine. Its origin is unknown. Neubauer observed that on feeding ethyl sulphide to dogs it was methylated and excreted as diethylmethyl-sulphonium hydrate, $CH_3{-}SOH{=}(C_2H_5)_2$. Allyl sulphide and other unoxidized sulphur-containing compounds are found in mustard oil and in many other plants, i.e., oil of garlic. Ethyl mercaptane, C_2H_5SH, smells very bad, but the diethyl sulphide, when pure, has an ethereal odor. When impure its odor is disagreeable. The source of these compounds in animal metabolism is still uncertain. Presumably they are derivatives of cysteine. When sulphur is thus combined with alkyl radicles it has a strongly basic character, so that diethylmethyl-sulphonium hydrate, just mentioned, is a strong base.

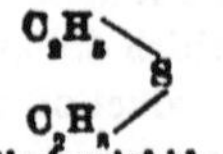

$$\begin{array}{l} C_2H_5 \\ \quad\quad\;\; \rangle S \\ C_2H_5 \end{array}$$

Ethyl sulphide.

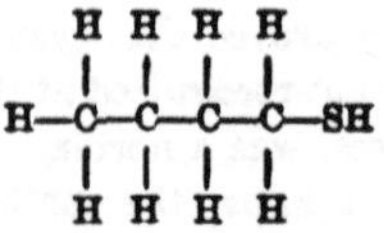

$$\begin{array}{c} \;\;H\;\;H\;\;H\;\;H \\ \;\;|\;\;\;|\;\;\;|\;\;\;| \\ H—C—C—C—C—SH \\ \;\;|\;\;\;|\;\;\;|\;\;\;| \\ \;\;H\;\;H\;\;H\;\;H \end{array}$$

N-butyl mercaptane.

In dogs a great part of the sulphur fed as cystine appears in the urine as sulphuric acid, but a part also, as thiosulphate, which indicates that thiosulphuric acid, $H—\overset{}{\underset{O}{S}}—OH$, is an intermediate product. This is confirmed by the fact that in rabbits taurine, taken by the mouth, appears, for the most part, as thiosulphate in the urine. In human beings taurine is excreted in part as taurocarbamic acid.

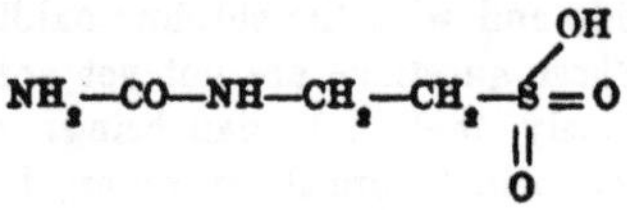

$$NH_2—CO—NH—CH_2—CH_2—\overset{OH}{\underset{O}{\overset{/}{\underset{||}{S}}}}=O$$

Taurocarbamic acid.

This is another example of the carbamino reaction in the body. The formation of thiosulphuric acid would require a reduction. It is not improbable that it is this thio acid which unites with the cyanides and nitriles in mammals to form sulphocyanates. Many bacteria, the so-called sulphur bacteria, have the power of reducing sulphates to sulphides. Nothing appears to be known about the sulphur metabolism of birds.

Ethereal sulphates. The quantity of ethereal sulphate in the urine of men in 24 hours is widely variable. v. d. Welden gives the limits of 0.094-0.620 gram. Sulphuric acid is used by the organism to pair with, and thus render less toxic, various aromatic decomposition products, most of them intestinal putrefactive products of the proteins. These compounds are called ethereal, or conjugated sulphates. They are in reality esters of phenol, cresol, paracresol, scatole and indole. All of these are the products of the putrefactive decomposition of tyrosine, tryptophane and phenyl alanine, and since this putrefaction takes place generally in the intestine the amount of the conjugated sulphates, or the amount of ethereal sulphuric acid, is an indication, although not a very good one, of the degree of intestinal putrefaction. The putrefaction may, however, take place elsewhere in the body, and putrefying and decomposing pus in old abscesses will also cause an increase in these bodies in the urine.

On the other hand, the elimination of ethereal sulphate can be greatly reduced by giving calomel or by reducing the protein diet, or by any other method which reduces intestinal putrefaction. Some proteins, for example casein, contain a great deal more tryptophane than others, so that the amount of ethereal sulphate will depend, too, on the character of the protein of the food, as well as upon its amount.

Benzene is itself a nearly inert substance, but by oxidation in the body it is converted into the reactive, unstable, toxic, convulsant phenol or carbolic acid. Fortunately phenol pairs very readily with sulphuric acid. In the human organism it appears to encounter most frequently sulphuric and glycuronic acids, with which it unites to form a non-toxic stable compound eliminated in the urine. The place in which this pairing occurs is probably the liver.

Synthesis of amino-acids in the animal body.—All plants have the power of synthesizing the amino-acids from ammonia and the products of carbohydrate fermentation. Have animal cells the same power or must they be fed on amino-acids already formed? This is obviously a very important question. A few years ago it was believed that animal cells differed from plant primarily in their powers of synthesis, animal cells being chiefly catabolic and plant cells anabolic. Further experience has served to correct that view. We now know that animal cells are able to synthesize carbohydrates from very simple substances, possibly even from formaldehyde, and almost certainly from glycolaldehyde; they can make fats from carbohydrates; nucleic acid from non-nuclein material; and in fact bring about a great many other syntheses so that animal cells certainly lack little of the powers of synthesis possessed by plants. Nevertheless the great fact remains that plants are able to nourish themselves from very simple sources of nitrogen, such as ammonia and nitrates or asparagine, whereas animals require, or at any rate eat, ready-made proteins. It may be observed, in passing, that the synthetic power of plants does not depend directly upon chlorophyll or light, since moulds and bacteria are able to construct their own particular kinds of proteins, which possess all the various sorts of amino-acids, from a single source of nitrogen, such as asparagine, if they at the same time have carbohydrate food. These plants contain no chlorophyll. The problem of how far animals have the power of making amino-acids has been attacked in various ways. Experiment has shown that at least glycocoll can be synthesized in large amounts in the animal body. If benzoic acid is fed mammals it leaves the body chiefly in the form of hippuric acid, having united with glycocoll somewhere in the body. It has been found by giving large amounts of benzoic acid that herbivora have the power of supplying glycocoll in far larger amount than is present in the proteins of the body. The solution of the question whether the other amino-acids can be formed has been sought by feeding proteins which lack some specific amino-acid. These experiments have generally been tried on young rats, since these animals grow very rapidly and are easy to keep. It was found by Hopkins, and Mendel and Osborne, that young rats would grow and develop normally when fed on a ration of butter fat, lard, some carbohydrate, such salts as are present in milk, and with a single pure protein added, provided that protein had in it the various amino-acids found in the body. Thus casein, edestin, serum albumin, were each sufficient to supply the protein needs. But if zein was the only protein in the diet, then the rats would not continue to grow, or indeed to live for long. Zein lacks both lysine and tryptophane. If these were added to the zein, then the diet became sufficient. We have already seen that gelatin is not in itself able to supply the whole of the protein requirement of the human body, and gelatin lacks both tyrosine and tryptophane. It was found also in the course of the experimentation that the amount of protein which it was necessary to add to the diet for the purposes of maintenance of the body weight or growth differed in different proteins. The minimum amount necessary appeared to be determined by the amount of some amino-acid which was required by the body and which was present in small amounts. In some proteins it was cystine which set the minimum; it was necessary to supply a certain amount of cystine per day and enough protein had to be fed to supply this amount of cystine. This emphasizes the point, made some time ago, that it is not protein as such but amino-acids which the body requires. It appears from these experiments that these animals, at any rate, cannot make sufficient tryptophane, tyrosine, lysine and cystine to supply their needs, but that these amino-acids must be present in the diet. This important field of investigation has just been opened and much more work must be done before we shall know how far different animals can synthesize different amino-acids. It is possible that the bacteria in the intestine may be playing a very important part in this process; they can synthesize many different amino-acids from a single one. By digestion these amino-acids might be set free and made available to the body, and thus the bacteria might be of considerable use to us. Perhaps the preliminary transformation of some of the amino-acids to amines by the intestinal bacteria may also be of value in the formation of certain hormones.

The formation of amino-acids from ketonic acids. A very fundamental fact was discovered by Knoop. Not only have animals the power of converting amino-acids into ketonic acids, and some of these latter into carbohydrates and indirectly into fats, but the reverse process is also possible. Out of carbohydrates ketonic acids are formed and the ammonia salts of these acids may, in some instances at any rate, be converted into amino-acids by reduction. Thus from ammonia and carbohydrate the body may have the power of making its proteins. It is not by any means possible, however, to cover all its protein needs by this reaction and it is doubtful how extensive this process may be in the animal body.' There is no doubt that the liver, and possibly other tissues as well, may carry out this synthesis which occurs probably in nearly all plant tissues.

The reaction is the reverse of that for the formation of ketonic acids and probably goes through the imino stage:

$$CH_3—CO—COOH + NH_3 \rightarrow \underset{\text{Iminopropionic.}}{CH_3—\underset{\underset{NH}{\|}}{C}—COOH} + H_2O \quad \overset{CH_3—CHNH_2—COOH}{+H_2}$$

Pyruvic acid. Iminopropionic.

This reaction being a reduction reaction consumes energy, it is endothermic. It may occur by the utilization of some of the energy set free by the oxidation of the system of which it is part. Another way in which the reaction might go is by means of acetylation. It will be recalled that when brombenzol is given dogs it is excreted in part as an acetylated derivative of cysteine and glycuronic acid. Acetylation of amino-acids is not unusual in the animal body. Knoop found that the phenyl-α-ketobutyric acid was excreted as the acetylated-amino-butyric acid. It has been suggested (Knoop) that the acetylation is brought about by a process analogous to the Canizzaro reaction. If pyruvic acid and ammonium carbonate interact there is a rise of temperature, the whole reaction is exothermic and acetylalanine and CO_2 are produced.

$$\begin{matrix} CH_3—CO—COOH \\ CH_3—CO—COOH \end{matrix} + NH_3 \longrightarrow \begin{matrix} CH_3—CH—COOH \\ | \\ NH \\ | \\ CH_3—CO \end{matrix} + CO_2 + H_2O$$

Pyruvic acid. Acetyl alanine.

It is not improbable that the acetylations in the body are thus produced from pyruvic acid and ammonia. In this case it will be noticed that one molecule of pyruvic acid is oxidized by the other. The total energy of the system is reduced, but the energy of one molecule of alanine is greater than of one molecule of pyruvic acid. It is clear from this that the amount of energy set free by the splitting off by oxidation of one carboxyl from the chain of carbon atoms is greater than the amount set free by the formation of the ketonic acid by oxidation from the amino-acid.

REFERENCES. Books.

Cathcart: The Physiology of Protein Metabolism. Monographs on Biochemistry, 1912.

Dakin: Oxidations and Reductions in the Animal Body. Monographs on Biochemistry. London, 1912.

Leathes: Problems in Animal Metabolism. Philadelphia, 1906.

Lusk: Science of Nutrition. 2d edition.

Fletcher: The A, B, Z of Digestion.

Cornaro: The Art of Living Long. Translation. Milwaukee.
Chittenden: Physiological Economy in Nutrition. 1905.
Voit: Hermann's Handbuch der Physiologie, VII, Leipzig, 1881.
Rubner: Die Gesetze des Energieverbrauchs bei der Ernährung. Berlin, 1902.
Mendel: Theorien des Eiweissstoffwechsels nebst einigen praktischen Konsequenzen derselben. Ergeb. d. Physiol., 11, pp. 418-525, 1911.

1. *Abderhalden and Steinbeck:* Weitere Untersuchungen über die Verwandbarkeit der Seidenpeptone zum Nachweis peptolytische Fermente. Zeit. physiol. Chem., 68, p. 312, 1910.
2. *Abderhalden, Einbeck and Schmidt:* Studien über den Abbau des Histidine im Organismus des Hundes. Zeit. physiol. Chem., 68, p. 395, 1910.
3. *Abderhalden and Suwa:* Weiterer Beitrag zur Frage nach der Verwertung von tief abgebauten Eiweiss im tierischen Organismus. Zeit. physiol. Chem., 68, p. 416, 1910.
4. *Abderhalden and Pincussohn:* Serologische Studien mit Hilfe der optischen Methode. Zeit. physiol. Chem., 64, p. 100, 1910; ibid., 66, p. 88, 1910.
5. *Abderhalden and Frank:* Weiterer Beitrag zur Frage n. d. Verwertung v. tief abgebauten Eiweiss im tierischen Organismus. Zeit. physiol. Chem., 64, 1910, p. 158.
6. *Abderhalden and Rona:* Ibid., Zeit. physiol. Chem., 67, p. 405, 1910.
7. *Abderhalden and Massuri:* Ueber das Verhalten von mono-palmytyl-l-tyrosin, distearyl-l-tyrosin und p-amino tyrosin im Organismus des Alkaptonurikers. Zeit. physiol. Chem., 66, p. 138, 1910.
8. *Abderhalden et al.:* Weitere Studien über die Verwertung verschiedener Aminosäuren im Organismus des Hundes unter verschiedener Bedingungen. Zeit. physiol. Chem., 74, p. 481, 1911.
9. *Abderhalden and Lampé:* Weiterer Beitrag zur Frage nach der Vertretbarkeit von Eiweiss resp. einem vollwertigen Aminosäuregemische durch Gelatine u. Ammonsalze. Zeit. physiol. Chem., 80, p. 160, 1912.
10. *Abderhalden:* Anaphylactic reaction with a synthetic polypeptide. Zeit. physiol. Chem., 81, p. 322, 1912.
11. *Abderhalden and Hirsch:* Fortgesetzte Versuche den Eiweiss bedarf des Hundes durch Ammonsalze u. ferner durch einzelne Aminosäuren ganz oder teilweise zu decken. Zeit. physiol. Chem., 80, 1912, p. 136.
12. *Abderhalden:* Weiterer Beitrag zur Frage nach der Verwertung von tief abgebauten Eiweiss im tierischen Organismus. Zeit. physiol. Chem., 57, p. 348, 1908; same title, ibid., 61, p. 194, 1909; 44, 1905, p. 198; 51, p. 226, 1907; 52, p. 507, 1907; with *London:* ibid., 54, p. 80, 1907; with *Messner and Windrath:* ibid., 59, p. 35, 1909; with *Frank and Schittenhelm:* ibid., 63, p. 215, 1909.
13. *Abderhalden:* Zur Frage des Albumosengehaltes des Blutes und speziell des Plasmas. Biochem. Ztschrft., 8, 1908, p. 360. See also Zeit. physiol. Chem., 42, 1904, p. 155.
14. *Abderhalden and Rona:* Die Zusammensetzung des "Eweiss" von Aspergillus niger bei verschiedener Stickstoffquelle. Zeit. physiol. Chem., 46, p. 179, 1905.
15. *Abderhalden and Schittenhelm:* Studien über den Abbau racemischer Aminosäuren im Organismus des Hundes unter verschiedener Bedingungen. Zeit. physiol. Chem., 51, p. 323, 1907.
16. *Abderhalden and Block:* Protein metabolism of an Alkaptonuriker. Zeit. physiol. Chem., 53, p. 464, 1907.

17. *Abderhalden, Bergell and Dorpinghaus:* Verhalten des Körpereiweisses im Hunger. Zeit. physiol. Chem., 41, p. 153, 1904.
18. *Abderhalden, Funk and London:* Weiterer Beitrag zur Frage nach der Assimilation des Nahrungseiweisses im tierischen Organismus. Zeit. physiol. Chem., 51, p. 629, 1907.
19. *Abderhalden, London and Reemlin:* Weitere Studien über die normale Verdauung der Eiweisskörper im Magendarmkanal des Hundes. Zeit. physiol. Chem., 58, p. 432, 1908.
20. *Ackermann and Kutscher:* Ueber die Aporrhegma. Zeit. physiol. Chem., 69, p. 265, 1910.
21. Ueber ein neues auf bakteriellem Wege gewinnbares Aporrhegma. Zeit. physiol. Chem., 69, p. 273, 1910.
22. *Argutinsky:* Muskelarbeit und Stickstoffumsatz. Arch. f. d. ges. Physiol., 46, p. 652, 1890.
23. *Ascoli and Vigano:* Zur Kenntnis der Resorption der Eiweisskörper. Zeit. physiol. Chem., 39, p. 283, 1903.
24. *Benedict:* The influence of inanition on metabolism. Carnegie Institution Report, Washington, 1907. See also Bulletin 203, 1914.
25. *Benedict:* The nutritive requirements of the body. Amer. Jour. Physiol., 16, p. 409, 1906.
26. *Benedict and Diefendorff:* The analysis of urine in a starving woman. Amer. Jour. Physiol., 18, p. 364, 1907.
27. *Bergell and Brugsch:* Ueber Verbindungen von Aminosäuren und Ammoniak. Zeit. physiol. Chem., 67, p. 97, 1910.
28. *Bergmann and Langstein:* Ueber die Bedeutung des Reststickstoffs des Blutes für den Eiweissstoffwechsel unter physiologischen und pathologischen Bedingungen. Zeit. chem. physiol. Pathol., 6, p. 27, 1905.
29. *Bickel u. Pawlow:* Untersuchungen über pharmakol. Wirkung des p-oxyphenylethylamins. Biochem. Zeits., 47, p. 345, 1912.
30. *Biedel and Winterberg:* Beiträge zur Lehre von der Ammoniak-entgiftenden Function der Leber. Arch. f. ges. Physiol., 88, p. 140, 1902.
31. *Billard:* Sur la valeur nutritive des albumines étrangères et spécifiques. C. R. Soc. Biol., 68, p. 1103, 1910.
32. *Blum:* Ueber das Verhalten des p-aminophenylalanins beim Alkaptonuriker. Zeit. physiol. Chem., 67, p. 192, 1910.
33. *Borchardt:* Ueber das Vorkommen von Nahrungsalbumosen im Blut und im Urin. Zeit. physiol. Chem., 57, p. 305, 1908.
34. *Boehm:* Ueber den feineren Bau der Leberzellen bei verschiedenen Ernährungszustände. Zeit. Biol., 51, p. 409, 1908.
35. *Brown and Cathcart:* The effect of work on the creatine content of muscle. Biochem. Jour., 4, p. 420, 1909.
36. *Brugsch:* Ueber die Rolle des Glykokolls im intermediaren Eiweissstoffwechsel beim Menschen. Zent. ges. Physiol. Pathol. d. Stoffw., 8, p. 529, 1907.
37. *Brugsch and Hirsch:* Gesammtstickstoff und Aminosäureausscheidung im Hunger. Zeit. f. expt. Path. u. Ther., 3, p. 638, 1906.
38. *Burckhardt:* Beiträge zur Chemie u. Physiologie des Blutserums. Arch. f. expt. Path. u. Phar., 16, p. 322, 1883.
39. *Busquet:* Contribution a l'étude de la valeur nutritive comparée d'une albumine spécifique et d'albumines étrangères chez la grenouille. C. R. Soc. Biol., 65, p. 652, 1908.

40. *Caspari:* Physiologische Studien über Vegetarismus. Arch. ges. Physiol., 109, p. 473, 1905.
41. *Cathcart:* The Physiology of Protein Metabolism. Monographs on Biochemistry. London, 1912. Extensive literature cited in this.
42. *Cathcart:* The influence of carbohydrates and fats on protein metabolism. Jour. Physiol., 39, p. 311, 1909.
43. *Cathcart and Leathes:* On the absorption of proteins from the intestine. Jour. Physiol., 33, p. 462, 1905.
44. *Chittenden:* Physiological Economy in Nutrition. 1905.
45. *Chittenden:* The Nutrition of Man, 1905.
46. *Cohn:* Zur Frage der Glykokollbildung im tierischen Organismus. Arch. f. expt. Path. u. Pharm., 53, p. 435, 1905.
47. *Cohnheim:* Die Umwandlung des Eiweisses durch die Darmwand. Zeit. physiol. Chem., 33, p. 451, 1901.
48. *Cohnheim:* Weitere Mittheilungen über Eiweissresorptionsversuche an Octopoden. Zeit. physiol. Chem., 35, p. 396, 1902.
49. *Cohnheim and Makita:* Zur Frage der Eiweissresorption. Zeit. physiol. Chem., 61, p. 189, 1909.
50. *Cramer:* On the assimilation of proteins introduced parenterally. Jour. Physiol., 37, p. 146, 1908.
51. *Cramer and Pringle:* The assimilation of protein introduced enterally. Jour. Physiol., 37, p. 158, 1908.
52. *Cremer and Henderson:* Ein experimenteller Beitrag zur Lehre vom physiologischen Eiweissminimum. Zeit. Biol., 42, p. 612, 1901.
53-54. *Dakin:* The products of the proteolytic action of an enzyme contained in the cells of the kidney. Jour. Physiol., 30, p. 84, 1904.
55. *Dakin and Dudley:* Conversion of α-amino acids to α-ketonic aldehydes and their relation to hydroxy acids. Proceed. Chem. Soc., 29, p. 192, 1913.
56. *Dakin and Wakeman:* The catabolism of histidine. Jour. Biol. Chem., 10, p. 499, 1912.
57. *Dakin:* Racemization of proteins and their derivatives. Jour. Biol. Chem., 13, p. 357, 1912.
58. *Ehrlich:* Ueber die Entstehung der Bernsteinsäure bei der alcoholischen Gärung. Bioc. Zeit., 18, p. 391, 1909.
59. *Ellinger and Kotake:* Synthese der p-oxymandelsäure und ihre angebliche Vorkommen im Harn bei akuter gelber Leberatrophie. Zeit. physiol. Chem., 65, p. 402, 1910.
60. *Engeland and Kutscher:* Ueber ein methyliertes Aporrhegma des Tierkörpers. Zeit. physiol. Chem, 69, p. 282, 1910.
61. *Embden and Almagia:* Ueber das Auftreten einer flüchtigen jodoformbildenden Substanz bei der Durchblutung der Leber. Beit. chem. physiol. Path., 6, p. 59, 1905.
62. *Embden and Knoop:* Ueber das Verhalten der Albumosen in der Darmwand und über das Vorkommen von Albumosen im Blute. Beit. chem. Physiol. Path., 3, p. 120, 1903.
63. *Embden and Schmitz:* Ueber synthetische Bildung von Aminosäure in der Leber. Biochem. Zeits., 29, p. 423, 1910.
64. *Folin:* A theory of protein metabolism. Amer. Jour. Physiol., 13, p. 117, 1905.
65. *Epstein and Bookman:* Formation of glycocoll in the animal body. Jour. Biol. Chem., 13, p. 117, 1912.

66. *Frank and Schittenhelm:* Beitrag zur Kenntnis des Eiweissstoffwechsels. Zeit. physiol. Chem., 70, p. 98, 1910.

67. *Frank and Trommsdorf:* Der Ablauf der Eiweisszersetzung nach Fütterung von abundanten Eiweissmengen. Zeit. Biol., 43, p. 117, 1902.

68. *Fraser:* Rendering animals immune against the venom of cobra and on the antidotal properties of the blood serum. Brit. Med. Jour., I, p. 1309, 1895. II, p. 416; II, 1896, p. 910.

69. *Funk:* The nitrogenous constituents of lime juice. Biochem. Jour., 7, p. 81, 1913.

70. *Frentzel:* Ein Beitrag zur Frage nach der Quelle der Muskelkraft. Arch. ges Physiol., 68, p. 212, 1897.

71. *Freund and Popper:* Ueber das Schicksal von intravenös einverleibten Eiweissabbauprodukten. Biochem. Zeit., 15, p. 272, 1909.

72. *Freund and Toepfer:* Ueber den Abbau des Nahrungseiweisses in der Leber. Zeit. f. expt. Path. Ther., 3, p. 633, 1906.

73. *Friedlander:* Ueber die Resorption gelöster Eiweissstoffe im Dünndarm. Zeit. Biol., 33, p. 64, 1896.

74. *Glaessner:* Ueber die Umwandlung der Albumosen durch die Magenschleimhaut. Beit. chem. physiol. Pathol., 1, p. 228, 1901.

75. *Grafe:* Beiträge zur Kenntnis des Stoffwechsels im protrahierten Hungerzustande. Zeit. physiol. Chem., 65, p. 21, 1910.

76. *Halliburton:* The absorption of proteins. Lancet, I, 1909.

77. *Halpern:* Beitrag zum Hungerstoffwechsel. Bioc. Zeit., 14, p. 134, 1908.

78. *Hopkins and Willcock:* The importance of individual amino-acids in metabolism. Observations on the effect of adding tryptophane to a dietary in which zein is the sole nitrogenous constituent. Jour. Physiol., 35, p. 88, 1907.

79. *Haskins:* The effect of transfusion of blood on the nitrogenous metabolism of dogs. Jour. Biol. Chem., 3, p. 321, 1907.

80. *Hamill and Schryver:* Nitrogenous metabolism in normal individuals. Jour. Physiol., 34, 1906, p. x.

81. *Hawk, Howe and Mattill:* Fasting Studies. III. Jour. Amer. Chem. Soc., 33, p. 568, 1911.

82. *Hedin:* Trypsin and antitrypsin. Biochem. Jour., 1, p. 474, 1906.

83. *Hedin:* Investigations on the proteolytic enzymes of the spleen of the ox. Jour. of Physiol., 30, p. 155, 1904.

84. *Heidenhain:* Neue Versuche über die Aufsaugung im Dünndarm. Arch. ges. Physiol., 56, p. 584, 1894.

85. *Heilner:* Ueber die Wirkung grosser Menge artfremden Blutserums im Tierkörper nach zufuhr per os und subcutan. Zeit. Biol., 50, p. 26, 1908.

86. *Henderson and Dean:* On the question of protein synthesis in the animal body. Amer. Jour. Physiol., 9, p. 386, 1907.

87. *Henriques:* Die Eiweisssynthese im tierischen Organismus. Zeit. physiol. Chem., 54, p. 406, 1907.

88. *Henriques:* Lässt sich durch Fütterung mit Zein oder Gliadin als einziger stickstoffhaltiger Substanz das Stickstoffgleichgewicht herstellen? Zeit. physiol. Chem., 60, p. 105, 1909.

89. *Henriques and Hansen:* Weitere Untersuchungen über Eiweisssynthese im Tierkörper. Zeit. physiol. Chem., 49, p. 113, 1906.

90. *Hirschfeld:* Ueber den Einfluss erhöhter Muskeltätigkeit auf dem Eiweiss-

stoffwechsel des Menschen. Arch. f. Path. Ant. u. Physiol., 114, p. 301, 1889.

91. *Hofmeister:* Das Verhalten des peptons in der Magenschleimhaut. Zeit. physiol. Chem., 6, p. 69, 1882.

92. *Hopkins:* The utilization of proteids in the animal. Science Progress, I, 1906, p. 159.

93. *Hoogenhuyze and Verploegh:* Creatinine excretion in man. Zeit. physiol. Chem., 46, p. 415, 1905.

94. *Hooven and Sollman:* A study of metabolism during fasting in hypnotic sleep. Jour. Expt. Med., 2, p. 405, 1897.

95. *Howell:* Presence of amino acids in blood and lymph. Amer. Jour. Physiol., 17, p. 273, 1906.

96. *Inaba:* Ueber die Zusammensetzung des Tierkörpers. Arch. f. Physiol. 1911, p. 1.

97. *Kauffman:* Ueber den Ersatz von Eiweiss durch Leim im Stoffwechsel. Arch. f. ges. Physiol., 109, p. 440, 1905.

98. Kayser: Ueber die eiweisssparende Kraft des Fettes verglichen mit derjenigen des Kohlenhydrats. Archiv f. Physiol., p. 371, 1893.

99. *Knoop:* Ueber den physiol. Abbau der Säuren und die Synthese einer Aminosäuren im Tierkörper. Zeit. physiol. Chem., 67, 1910, p. 489.

100. *Knoop and Kertess:* Das Verhalten von α-Aminosäuren und α-Kentonsäuren im Tierkörper. Zeit. physiol. Chem., 71, p. 252, 1911.

101. *Kotake:* Ueber l-oxyphenylmilchsäure und ihr Vorkommen im Harn bei Phosphorvergiftung. Zeit. physiol. Chem., 65, p. 397, 1910.

102. *Kovalevsky and Markewicz:* Ueber das Schicksal des Ammoniaks im Organismus des Hundes bei intravenoser Injektion von kohlensäuren Ammoniak. Biochem. Zeit., 4, p. 196, 1907.

103. *Kiesel:* Ueber den fermentativen Abbau des Arginins in Pflanzen. Zeit. physiol. Chem., 75, pp. 169-197, 1911.

104. *Krogh:* Ueber die Bildung freien Stickstoffes bei der Darmgährung. Zeit. f. physiol. Chem., 50, 289, 1907.

105. *Lang:* Ueber Desamidierung im Tierkörper. Beitr. chem. physiol. Path., 5, p. 321, 1904.

106. *Leathes and Cathcart:* On the relation between the output of uric acid and the rate of heat production in the body. Proc. Roy. Soc., 79, p. 543, 1907.

107. *Lehmann, Muller, Munk, Senator and Zuntz:* Untersuchungen an zwei hungernden Menschen. Arch. f. Path., Anat. u. Physiol., 131, Suppl. 1, 1893.

108. *Levene and Meyer:* The elimination of total nitrogen, urea and ammonia following the administration of some amino acids. Amer. Jour. Physiol, 25, p. 214, 1909.

109. *Loewi:* Eiweisssynthese im Tierkörper. Arch. f. expt. Path. u. Pharm., 48, p. 303, 1902.

110. *Lusk and Ringer:* Ueber die Entstehung von Dextrose aus Aminosäuren. Zeit. physiol. Chem., 66, p. 106, 1910.

111. *Luthje:* Zur Frage der Eiweisssynthese im tierischen Körper. Arch. f ges. Physiol., 113, p. 547, 1906.

112. *Lusk:* The fate of the amino acids in the organism. Jour. Amer. Chem. Soc., 32, p. 671, 1910.

113. *McCay:* Standards of the constituents of the urine and blood and the bearing of the metabolism of Bengalis on the problems of nutrition. Scientific Memoirs, Govt. India, 34, 1908.

114. *Lusk:* The Science of Nutrition, 2d edition.
115. *McCollum:* Nuclein synthesis in the animal body. Amer. Jour. Physiol., 25, p. 120, 1909.
116. *Magnus-Levy:* Ueber die Neubildung von Glykokoll. Bioc. Zeit., 6, p. 523, 1907.
117. *Mendell and Rockwood:* On the absorption and utilization of proteins without the intervention of the alimentary digestive processes. Amer. Jour. Physiol., 12, p. 336, 1905.
118. *Michaud:* Beitrag zur Kenntnis des physiologischen Eiweissminimum. Zeit. physiol. Chem., 59, p. 405, 1909.
119. *Morawitz:* Beobachtungen über den Widerersatz der Bluteiweisskörper. Beit. chem. physiol. Path., 7, p. 153, 1906.
120. *Medwedew:* Ueber Desamidierungsvorgänge im Blute normaler u. Schilddrüsenloser Tiere. Zeit. f. physiol Chem., 72, p. 410, 1911.
121. *Mörner:* Zur Chemie des Alkaptonharns bezw. der Homogentisinsäure (nebst einigen ihrer Verwandten). Zeit. physiol. Chem., 69, 1910, p. 330.
122. *Murlin:* The Nutritive Value of Gelatin. Amer. Jour. Physiol., 19, p. 285, 1907.
123. *Neubauer:* Ueber den Abbau der Aminosäuren im gesunden u. kranken Organismus. Deut. Arch. f. klin. Med., 95, p. 211, 1908.
124. *Neubauer and Gross:* Zur Kenntnis der Tyrosinsabbaus in der künstlich durchbluteten Leber. Zeit. physiol. Chem., 67, p. 219, 1910.
125. *Neubauer and Fischer:* Beiträge zur Kenntnis der Leberfunktionen. Zeit. physiol. Chem., 67, p. 230, 1910.
126. *Neuberg and Langstein:* Ein Fall von Desaminierung im Tierkörper. Arch. f. Physiol., 1903, Supp., 514.
127. *Oppenheimer:* Ueber die Anteilnahme des elementaren Stickstoffes am Stoffwechsel der Tiere. Bioc. Zeits., 4, p. 328, 1907.
128. *Romburgh:* Hypaphorine and the relation of this substance to tryptophane. Pro. Amsterdam Acad., 13 (2), p. 1177, 1911.
129. *Paton:* On Folin's theory of protein metabolism. Jour. Physiol., 33, p. 1, 1905.
130. *Paton and Goodall:* Digestion leucocytosis. 33, p. 20, 1905.
131. *Paton and others:* On the influence of muscular exercise, sweating and massage on the metabolism. Jour. Physiol., 22, p. 68, 1897.
132. *Pekelharing and v. Hoogenhuyze:* Die Bildung des Kreatins im Muskel beim Tonus und bei der Starre. Zeit. physiol. Chem., 64, p. 262, 1909.
133. *Pflüger:* Die Quelle der Muskelkraft. Arch. ges. Physiol., 50, p. 98, 1891.
134. *Pflüger:* Glykogen. Arch. ges. Physiol., 96, p. 381, 1903.
135. *Robertson:* Synthesis of paranuclein through the agency of pepsin. Jour. Biol., Chem., 5, p. 493. 1908.
136. *Röhmann:* Ueber künstliche Ernährung. Klin. ther. Woch., 1, 1902.
137. *Rubner:* Theorie der Ernährung nach Vollendung des Wachstums. Arch. f. Hyg., 66, p. 1, 1908.
138. *Rubner:* Verluste und Wiederneuerung im Lebensprozess. Archiv f. Physiol., p. 39, 1911.
139. *Rubner:* Die Beziehungen zwischen dem Eiweissbestand des Körpers und der Eiweissmenge der Nahrung. Archiv f. Physiol., p. 61, 1911.
140. *Rubner:* Ueber den Eiweissansatz. Archiv f. Physiol., p. 67, 1911.
141. *Salaskin:* Ueber das Ammoniak in physiologischer und pathologischer Hinsicht und die Rolle der Leber im Stoffwechsel stickstoffhaltiger Substanzen. Zeit. physiol. Chem., 25, p. 449, 1898.
142. *Salkowski:* Ueber Autodigestion der Organe. Zeit. f. klin. Med., 17, p. 77, 1890.

143. *Schryver:* Chemical dynamics of animal nutrition. Biochem. Jour. 1, p. 123, 1906.

144. *Schulze and Winterstein:* Studien über die Proteinbildung im reifenden Pflanzensamen. Zeit. physiol. Chem., 65, p. 431, 1910.

145. *Schwarz:* Ueber Verbindungen der Eiweisskörper mit Aldehyden. Zeit. physiol. Chem., 31, p. 460, 1901.

146. *Seitz:* Die Leber als Vorrathskammer für Eiweisstoffe. Arch. ges. Physiol., 111, p. 309, 1906.

147. *Shaffer:* Diminished muscular activity and protein metabolism. Amer. Jour. Physiol., 22, p. 445, 1908.

148. *Shaffer and Coleman:* Protein metabolism in typhoid fever. Arch. of Int. Med., 4, p. 538, 1909.

149. *Siven:* Zur Kenntnis des Stoffwechsels beim erwachsenen Menschen mit besonderer Berücksichtigung des Eiweissbedarfs. Skand. Arch. f. Physiol., 11, 1901, p. 308.

150. *Starling:* Die Resorption vom Verdauungskanal. Handbuch der Biochemie. (Oppenheimer) III, p. 240, 1909.

151. *Stepp:* Versuche über Fütterung mit lipoidfreier Nahrung. Biochem. Zeit. 22, p. 452, 1909.

152. *Stolte:* Ueber das Schicksal der Monaminosäuren im Tierkörper nach Einführung in die Blutbahn. Beit. chem. physiol. Pathol., 5, p. 15, 1904.

153. *Straub:* Ueber den Einfluss des Kochsalzes auf die Eiweisszersetzung. Zeit. f. Biol., 37, p. 527, 1899.

154. *Tallqvist:* Zur Frage des Einflusses von Fett und Kohlenhydrate auf den Eiweissumsatz des Menschen. Arch. f. Hygiene, 41, p. 177, 1902.

155. *Taylor:* On the synthesis of protamine through ferment action. Jour. Biol. Chem., 5, p. 381, 1908.

156. *Taylor and Cathcart:* The influence of carbohydrates and fats on protein metabolism. The effect of phlorizin glycosuria. Jour. Physiol., 41, 1910, p. 276.

157. *Stein:* Ueber die Bildung von Milchsäure bei der antiseptischen Autolyse der Leber. Biochem. Zeits., 40, p. 486, 1912.

158. *Thomas:* Ueber die Zusammensetzung von Hund u. Katze während der ersten Verdoppelungsperioden des Geburtsgewichts. Arch. f. Physiol., 1911, p. 9.

159. *Thomas:* Ueber das physiologische Stickstoffminimum. Arch. f. Physiol., 1911, Supp. 249.

160. *Totani:* Ueber das Verhalten der Phenylessigsäure im Organismus des Huhns, Zeit. f. physiol. Chem., 68, 1910, p. 75.

161. *Thompson:* The nutritive effects of beef extract. Proceedings British Assn., 1. 1910.

162. *Toepfer:* Ueber den Abbau der Eiweisskörper in der Leber. Zeit. f. expt. Path. u. Ther., 3, p. 45, 1906.

163. *Tsuchiga:* Ueber den Umfang der Hippursäure synthese beim Menschen. Zeit. f. expt. Path. u. Ther., 5, p. 737, 1909.

164. *Veraguth:* The effect of a meal on the excretion of nitrogen in the urine. Jour. Physiol., 21, p. 112, 1897.

165. *Vernon:* The rate of tissue disintegration and its relation to the chemical constitution of protoplasm. Zeit. f. allg. Physiol., 6, 1907.

166. *Voit:* Die Grösse des Eiweisszerfalls im Hunger. Zeit. f. Biol., 41, p. 167, 1901.

167. *Voit and Korkunoff:* Ueber die geringste zur Erhaltung des Stickstoffgleichgewichtes nöthige Menge von Eiweiss. Zeit. f. Biol., 32, p. 58, 1895.

168. *Voit and Zisterer:* Bedingt die verschiedene Zusammensetzung der Eiweisskörper auch einen Unterschied in ihrem Nahrwert. Zeit. f. Biol., 53, p. 457, 1910.
169. *Weinland:* Ueber die Ausscheidung von Ammoniak durch die Larven von Calliphora und über eine Beziehung dieser Tatsache zu dem Entwickelungsstadium dieser Tiere. Zeit. f. Biol., 47, p. 232, 1906.
170. *Weiss:* Untersuchung über die Bildung des Lachsprotamin. Zeit. f. physiol. Chem., 52, p. 107, 1907.
171. *White, Hale and Spriggs:* On metabolism in forced feeding. Jour. Physiol., 26, p. 151, 1901.
172. *Wolf and Osterberg:* Protein metabolism in phlorizin diabetes. Amer. Jour. Physiol., 28, 71, 1911.
173. *Zisterer:* Bedingt die verschiedene Zusammensetzung der Eiweisskörper auch einen Unterschied in ihrem Nahrwert? Zeit. Biol., 53, p, 157, 1910.
174. *Zuntz:* Ueber die Bedeutung der verschiedenen Nahrstoffe als Erzeuger der Muskelkraft. Arch. f. Physiol., p. 541, 1894.
175. *Winterstein and Kung:* Ueber das Auftreten von p-oxyphenylethylamine im Emmenthaler Käse. Zeit. f. physiol. Chem., 59, p. 138, 1909.

CHAPTER XX.

METABOLISM UNDER VARIOUS CONDITIONS. RESPIRATION. VITAMINES. CONCLUSION.

Metabolism during starvation and under various conditions.—When the body is starving or supplied with insufficient nourishment to cover its wear and tear and energy requirements, the tissues are themselves consumed, although to a very different degree in the various organs. A condition of temporary starvation must have been one of the commonest vicissitudes of life among our progenitors, who probably lived very much as the Patagonian savages described by Darwin. These savages are able to go for long periods without food, and then when they find a stranded whale or some other abundant source of food, they gorge themselves for days in succession. It is only after agriculture was cultivated that man began to be superior to these accidents and to have a regular and even food supply. These periods of fasting having probably been common, it is not surprising that we find that the body has a mechanism to provide for them. When abundant food is eaten it is stored in part, and during fasting these stores are drawn upon. Foods are stored pre-eminently as fat, and principally as the saturated fats in the great fat reservoirs of the body. These reservoirs are the fat tissues under the skin, the panniculus adiposus, the fat about the intestines and internal organs and in the connective tissue about the heart and muscles and even in the muscle cells themselves. The brain alone of all the organs appears to be free from fat. A second reserve of food, but one of far less importance and weight, is the glycogen which is stored in the liver and muscles. For the protein there is also a certain amount of reserve, since the muscles contain a small amount of amino-acid nitrogen and the proteins may be increased in the body up to a certain point, but nothing comparable to that of the fats or carbohydrates.

The first effect of starvation, or fasting, is to set free these reserve foods. The fats are first called upon to supply the needs of the body for energy or fuel. In fasting animals the adipose tissue almost entirely disappears, 93-97 per cent. of it being consumed. Glycogen is also greatly reduced, but a little remains in the liver and muscles even in a very advanced state of starvation. But the fats and carbohydrates cannot supply the protein decomposition of the body, and since there is very little protein reserve it is necessary for the body to conserve this in every way possible. The mechanisms by which the nitrogen is conserved in the body are still somewhat obscure, but we know now that the body has some power at least of resynthesizing the ammonia, which may have been set free, into some amino-acids which can be used for the resynthesis of the proteins. When fats or carbohydrates are eaten or drawn from the body's stores, not so much protein is decomposed as when they are absent. We say accordingly that these substances have a *protein-sparing action*. Possibly they act by furnishing the energy for the resynthesis and so aid the retention of the amino-acids; perhaps when they are oxidizing the activity of autolytic enzymes, which may attack the protein, is reduced; or it may be that in some other way they check the excretion of nitrogen and the destruction of protein. Certain it is that they do thus act. Whatever the explanation, it is found that in starvation the nitrogen output is reduced to a minimum. The lowest figures reported are those of Cetti, a professional faster, whose N output on the 25th day of fasting had fallen to a little more than 2 grams, and Thomas reduced his on a starch and cream diet to 2.2 grams. The normal excretion is approximately 12-16 grams. But while the nitrogen loss is thus reduced to a minimum, the total calories given off by the body per kilo body weight are not reduced in anything like the same proportion. The normal output of heat from a human adult, when very little active muscular work is done, is about 35 calories per kilo; it falls on fasting to about 30-32 calories per kilo.

The fact that the nitrogen excretion is so reduced in fasting has led several to the conclusion that this represents the necessary wear and tear of the tissues and the normal output, which is usually so far in excess of this, is due to a luxus consumption of protein. But this minimum may rather be regarded as the maximum reduction of nitrogen waste of which the body is capable. It is probable that this is the best which can be done under the most favorable conditions. It is quite possible that the amount of protein catabolized is far greater than this, but that the nitrogen is saved and resynthesized and used over again, the energy for the resynthesis coming from the carbohydrates or fats. It would be quite erroneous to conclude, as is sometimes done, that of the proteins of the body only an amount equivalent to this 2.5 grams are being catabolized per day. It is at least possible that this figure represents only the amount of net catabolism. How large the real catabolism and anabolism may actually be we have at present no means of knowing.

During fasting or starvation the various organs of the body lose weight. Human beings may fast so that their body weight is reduced to about two-fifths of the average weight and fully recover from the fast. Children, however, have a higher metabolism than adults, owing in part to their greater heat loss, as the surface of the body is larger relative to the body weight in them than in adults, and in part to the fact that their protoplasm is also younger and catabolizes faster. They can starve for much shorter periods without dying. The amount of loss of weight of the different organs of the body in fasting is given by Voit for a male cat as follows:

Adipose tissue ...	97%	Kidneys	26%	Bones	14%
Spleen	67	Skin	21	Heart	3
Liver	54	Intestine	18	Nervous system ..	3
Testicles	40	Lungs	18		
Muscles	31	Pancreas	17		

The heart and nervous system resist starvation better than any other organs. They lose but a small per cent. of their weight. The liver, which stores glycogen and fat; the spleen, which has to do with the blood destruction and formation; the testicles and the muscles, these besides the fat are most reduced. It is their material which is being used for the metabolism of the other tissues. It is not to be supposed that because the heart and the nervous system lose least that their metabolism is less than the other tissues. The direct contrary to this is probably the fact. It is those tissues of the most intense metabolism which preserve themselves best. We know that the heart has such an intense metabolism. It must continue at work whatever happens; and so must the nervous system. It is probable that the preservation of these tissues is brought about in the following way. By the autolysis of the other tissues a certain amount of amino-acids and other substances are set free. The metabolism in the heart and nervous system is so intense that in these organs the amino-acids and other products are in part catabolized and in part built up into the tissue substance. To maintain the equilibrium new food substances pass into these organs from the blood to make good the loss of the catabolized products; and from the other tissues amino-acids and other products pass to the blood to make good the loss from that tissue. Thus that organ with the highest rate of metabolism will call upon the other tissues of the body which have the lowest rate of metabolism; they will waste away at its expense. It is thus probably that a cancer impoverishes the other tissues.

The preservation of the brain and heart in this way at the expense of other less vital tissues of the body is evidently a measure of adaptation. These are the vital or master tissues and it is absolutely necessary that they be preserved. From its earliest origin the nervous system appears to have the most active metabolism and to dominate that of the rest of the body in the way so conclusively shown by Child.

Effects of fasting. The general effects of fasting are extremely inter-

eating. There is no doubt that in many cases fasting is very ben
to the general health. Chronic diseases, such as catarrhs and pi
boils, etc., are said often to have been permanently cured by this
expedient. The first few days of the fast may be, and generall
trying; there is not infrequently considerable nausea for a few
but thereafter the deprivation of food does not appear to caus
very painful sensations beyond great hunger. The total effect c
body is to sweep out of the protoplasm all the deposited waste or r
material. At least half of the muscle substance has to be regen
and, according to Child, fasting is, in its essence, of the nature
regeneration or rejuvenation. This is certainly the case in so
the lower animals, in particular in the flat worms, Planaria, and
animals upon which he worked. In these forms it is possibly to
that the fasting animals are in reality rejuvenated and emerge
the fast with all the characteristics of young animals, including a
lated metabolism, heightened respiration and so on. Whether the h
being is capable of a certain degree of rejuvenation by this same p
is not yet certain, but there are some indications that some
venescence is possible. It seems generally true that the depositi
colloidal matter in protoplasm is one of the conditions of senescenc
is possible that such depositions increase the difficulty of pass
certain substances into and out of cells, or they interpose barriers
way of a free exchange of material between the different parts o
and so ultimately break down the co-ordination of cell metabolis
this explanation of senescence, which has been proposed by Chi
true, then fasting would appear to be a means of combating the p
to some degree.

There are other changes in the excretions accompanying f
Thus the proportion of urea nitrogen in the urine instead of being
85 per cent. falls to 55 per cent. or 60 per cent. of the total nitroge
page 750). The quantity of ammonia increases slightly, due to the
acidosis produced by the burning of fat and protein; acetone and di
acid appear; and there is a relative increase in the neutral sulphur
corresponding decrease in the inorganic sulphates of the urine.
are the same changes observed by Folin and others in low protein di

The blood holds its composition fairly uniform in fasting, alt
there is some sinking in the per cent. of protein present in the pl
This falls from 6 per cent. to about 4-5 per cent. of the weight.
is a considerable reduction in the amount of blood in the body, a
per cent. of fat in the plasma may in the early days be somewhat l
than the average, as the fat reserves are mobilized.

The number of cells of the body does not decrease so much
size of the cells. The general course of a fasting experiment l

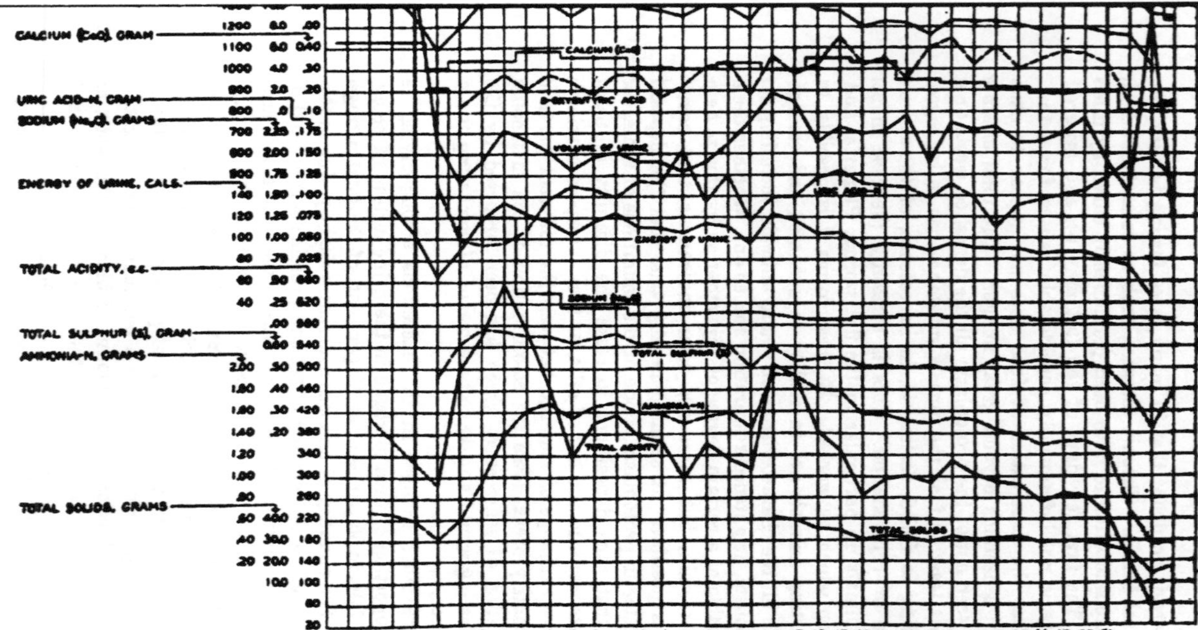

From Publication No. 203 of the Carnegie Institution of Washington.

FIG. 68A.—*Metabolism chart of the most important factors measured on subject L. throughout the fast at the Nutrition Laboratory, Boston, Massachusetts.*

Although the table gives the exact mathematical expression of the quantities of materials involved in the different measurements of the metabolism of this subject, a visualization of the relationship between the various factors of metabolism is best secured by a series of curves. A chart containing curves for all of the factors measured on this subject would be of such great size as to preclude convenient inspection. Hence we have selected here only the most important or more generally observed factors. Although no specific reference is made to these curves for any comparisons in the text, it is obvious that a constant

31 days recently carried out by Benedict is shown in the accompanying chart.

Lack of water and mineral substances. The body can go without food for 20-30 days, but not without water. Water is constantly escaping from the body in the urine, in the sensible and insensible perspiration and in the lungs. And the amount of water which is formed in the body by the combustion of the hydrogen is by no means sufficient to make good this loss. The loss of water makes the viscosity of the blood greater, the resistance to its flow increases; water passes from the tissues to the blood to make good the loss from the latter and this reduces the amount of water in the tissues, giving rise to excessive thirst which is almost intolerable. As the physical activities of the tissues are dependent upon the viscosity of the protoplasm, and the chemical activities on the water present, both the physical and chemical processes in the cells suffer. There is at first a reduction in the metabolism, which may be followed by an increased catabolism. An adult cannot live more than three or four days without water, the time depending naturally on the external temperature, amount of water lost by evaporation and so on. Death is the result probably of the increased viscosity of the blood.

Mineral substances are also necessary to life, and the result of keeping them out of the food is disastrous. In the perspiration and urine salts of various kinds are constantly being lost. Food as free as possible from mineral substances produced disturbances in the muscular system in Taylor's experiments on himself; disturbances of the nervous system have also been noted by Forster. A sufficient supply of phosphates and calcium are essential to the development of the bones and teeth. Herbivorous animals constantly have a diet poor in sodium and relatively rich in potassium. Such animals require from time to time some sodium chloride added to their ration. Carnivorous animals require no salt, since the salts in their prey are about those of their own bodies.

Vitamines. Substances of an unknown nature necessary for the nourishment of the body.—As has been so often remarked, nearly all of the really fundamental facts in nutrition remain still to be determined. This is illustrated in no more striking fashion than by the discovery in the past few years of the specific action of foods in nourishing the body quite apart from their protein, carbohydrate and fat content. A great field has thus been opened which promises to yield many valuable discoveries. It seemed a few years ago as if with the discovery of the fuel value of a food, of how many calories of energy it contained which were available to the body, and with the estimation of the grams of fat, carbohydrate and protein in it, all that was necessary to determine its food value was known. How far that was from being the truth will appear from what follows. We now know that the character of the fat, protein and carbohydrate is of as great importance as the amount. It is by no means the same whether one eats cane sugar or lactose, although they resemble each other so closely in calories and composition. It has been found, for example, that there is in the brain a large amount of the sugar, galactose. This occurs in the cerebrosides in the medullary sheaths of the nerves. We do not know whether the body, particularly in youth, has the power of making galactose from other carbohydrates. We see, in fact, that Nature, which has had charge of the rearing of children for millions of years, has provided in the mammary glands an organ for the manufacture of galactose, so that the child during the period of the medullation of the nerves of the cerebrum, when cerebrosides may be produced in great abundance, that is in the third to sixth month, may have a nourishment which contains quantities of galactose. The sugar of milk is not cane sugar; it is not maltose; it is not dextrose or levulose, or ribose; but lactose, a sugar containing half its weight of galactose, that important sugar of the brain. It may be almost stated as a truism that had it been more advantageous to have dextrose in milk than lactose, the sugar found there would have been dextrose. The suggestion, therefore, advocated by some physicians, to substitute for the lactose of the milk either cane sugar or dextrose, in artificial feeding of children can only be regarded with misgivings. It is wiser to accept the conclusions of nature which has tried, no doubt, many thousands of experiments of which we are ignorant, and which has provided lactose in the food of infants only after a prolonged research.

Neither is the fat consumed a matter of indifference, if only it be fat. It is not enough that the fat shall burn and liberate a certain amount of energy in the body. The nature of the fatty acid is important. We know indeed that the character of the fat laid down in the cells is somewhat dependent upon the character of the fat eaten. In the fat of dogs, an abnormal quantity of mutton fat appears when after fasting a dog is fed on large amounts of these fats. It was found, too, by Herter that the character of the fat laid down in the tissue of growing pigs was altered when the pigs were fed large amounts of acetic acid. The power of making the necessary fats peculiar to the body of the animal in which they are found, and indeed peculiar to the tissue, is so great that the fats may easily be formed from the carbohydrates. Nevertheless the melting point, and probably the ease of oxidation of the fats in the cells, is not entirely independent of the character of the fat fed.

Above all, the character of the protein is not a matter of indifference. Hitherto the protein of the food has been calculated by deter mining the nitrogen and multiplying by the factor 6.25 to give the amount of protein in the food. Sometimes only the coagulated extracted food has its nitrogen thus measured, sometimes it is raw food, and lipin nitrogen and nuclein nitrogen have been included. This method it is recognized is very crude. Not all nitrogen is protein nitrogen. There is the nitrogen in the phospholipins, in the amino lipins and in nucleic acid and other nitrogen-containing substances. Furthermore, it makes a difference what amino-acids are present in the proteins, and the various amino-acids we now know have specific functions which cannot be replaced by other amino-acids. This point has already been considered.

But beyond all these qualities of the food, evidence is accumulating that the foods must contain other bodies not protein or carbohydrate or fats or minerals, but of an organic nature and without which the organism will surely perish. These bodies are required apparently in very small quantities, but they are of vital importance. They have been named, therefore, *vitamines* by Funk, who has worked a good deal upon them, to indicate his opinion that they are necessary and that they contain basic nitrogen. While the whole matter is still under investigation and it is too soon as yet to draw any positive conclusions, the results obtained have been so extraordinary and so interesting that they should surely be considered here, even though opinion is not unanimous as to their nature.

Beri-beri. There is a curious metabolic disease, called beri-beri, found among Eastern peoples such as the Filipinos, the Japanese and East Indians, peoples who have a very restricted diet, of which rice is the main staple. This disease is characterized as follows:

It begins with a feeling of lassitude accompanied by numbness, stiffness or cramps in the legs. There is œdema of the ankles and face. In its further progress the patient loses the power of walking, there is partial paralysis of the leg muscles and other muscles, accompanied by anesthesia in the affected areas and often by pains and tingling sensations in the feet; the œdema becomes more general and breathlessness and palpitation may come on. There are neither fever nor brain symptoms. The symptoms are those of a peripheral neuritis which may involve the pneumogastric and phrenic nerves, but which generally begins in the regions farthest from the nerve centers. There is degeneration of the muscles. The mortality may be as high as 50 per cent. This disease has been variously explained in the past, some considering it due to spoiled rice, others to an infection of some kind. It has become possible to study it through the discovery of a similar disease in fowls which may be produced artificially, this is the polyneuritis of birds.

FIGS. 64 AND 65.—Early and late stages of polyneuritis in fowls after eating *exclusively* polished rice (Funk).

If fowls are fed exclusively on polished rice, that is the white rice the reddish exterior having been polished away, no changes are apparent for several weeks, but suddenly the symptoms appear and in the course of a couple of days or more they go rapidly to a fatal ending. The fowls become unable to walk about, then they become weaker, lie over on their sides and will surely die if the diet is not changed. Figures 64 and 65. If, however, the fowls are fed rice to which a little of the bran has been added, or if they are fed unpolished rice, they are not subject to the disease. They recover even when very ill by the injection of the extract of the bran. It seems, therefore, to be clear that there is something in the bran of rice, or in the outer layer of the kernels, which is absolutely necessary for the nourishment of the body of the fowl when

FIG. 66.—Degeneration in the peripheral nerves of fowls with polyneuritis (Funk).

it is on a rice diet. This substance is present in very small amounts. An amount of solid extract weighing a few mgs. is sufficient to cure a fowl when very ill. Concerning the nature of this substance, it may be said that it is not protein, it is soluble in alcohol, it probably contains nitrogen, it is organic in nature. Funk thought that it was allied to the pyrimidins, but the evidence is very unconvincing. Whatever its nature, it would seem that it must be present in the foods. Fat, protein, carbohydrate, mineral matters and energy are insufficient to nourish the body in its absence. Its lack appears to affect the peripheral nerves first, leading to their degeneration. Figure 66.

Inasmuch as human beings who take *unpolished* rice seem to be free from beri-beri, the conclusion has been drawn that beri-beri is also due to the lack of this vitamine which is found in the bran of rice, of wheat and in many other foods. It must be remembered, however, that peripheral neuritis may be produced in many different ways. It occurs in arsenical and lead poisoning, also in pellagra and in alcoholics. It is perhaps not entirely certain that beri-beri is caused directly by polished rice, although that seems at present most probable.

Further studies into the nature of the curative substance have shown that it is extractable from a great variety of tissues, both plants and animals, by alcohol. It follows the lipin fraction, but is probably not itself a lipin. Lecithin, cephalin, cerebrin, protagon, cholesterol, choline, nicotinic acid, guanidine and other substances isolated from the lipoid fraction are without any curative action. The substance contains no phosphorus. It probably contains nitrogen, since it is precipitated by phosphotungstic acid. It is precipitated from its alcoholic solution by ether. It is insoluble in acetone, benzene, chloroform and ether. It is soluble in water and is destroyed by boiling for some time, and by alkalies, even weak alkalies like ammonia. It is more stable in acids. It loses its activity on standing in a desiccator. Solutions which have a curative action generally, if not always, give a blue color with phosphotungstic acid, both the uric acid reagent and the polyphenol reagent of Folin and Denis. It is hence apparently a reducing substance. The substance is found not only in bran, but in milk, in muscle of all kinds, in the alcoholic extract of the brain, in autolyzing yeast and in other locations. It is apparently very unstable, particularly in light and in the presence of oxygen. It is probable that the results obtained by Stepp, who found that food substances thoroughly extracted with alcohol would no longer permit of growth and properly nourish animals, were due to the absence of these vitamines and not to the absence of the lipoid, as he supposed. The relation of vitamines to growth will be discussed in a moment.

It appears, then, from these experiments that birds, and human beings as well, require in their food certain unknown substances of an organic nature which are absolutely necessary to life. The evidence points, on the whole, to Funk's conclusion that they are pyrimidin derivatives. Possibly they are allied to alloxan or alloxantin, both of which are unstable. The discovery of the nature of these substances, or of this substance, is a very important matter. Nicotinic acid is found in the partially purified product, but nicotinic acid is itself inactive. Funk has suggested that possibly a mother substance of nicotinic acid is the active principle. All of the purified substances extracted from bran have been found to be inactive.

Pellagra. Another disease with some points of resemblance to beri-beri is pellagra. The name means *rough skin,* the skin, particularly on the backs of the hands and about the neck, being thickened and rough. Neuritis occurs here also and the symptoms of disturbance of the central nervous system are more pronounced. This disease was long ascribed by the Italian investigators to spoiled maize. It is common in the Southern States and in Italy. Other investigators have recently ascribed it to an infection carried by a fly or insect-carrier. Whatever may be the explanation of the cause of this disease, there can be no doubt about the fact of the impairment of the nutrition of the nervous system; but whether this is due to a poison elaborated by a parasite of some kind either in the body or in maize, or whether it is due to the lack of some substance in the diet, cannot be positively stated. It occurs generally among those having a very restricted diet, but other conditions of an unsanitary nature have usually been present, making the determination of the etiology difficult.

Scurvy. Scurvy is a disease of malnutrition, very common in the past when fresh or canned vegetables were rare and salt and preserved meats were eaten. The cause of the disease, what particular articles of diet are responsible for it, is not certain. But it is known that various foods have an anti-scorbutic action, so that when they are taken people do not get scurvy. Among these foods the juices of limes, oranges, lemons are particularly active, but it is not certain what the active substances in these foods are. In children an exclusive diet of boiled or sterilized milk sometimes produces scurvy. Orange juice appears to prevent this.

Importance of vitamines for growth. There appear to be then in milk, in orange juice and in various articles of diet, such as bran, substances which are not proteins, fats or carbohydrates, which are generally soluble in alcohol, and which have a very extraordinary influence in stimulating growth and in maintaining normal nutrition, particularly of the nervous system. What is the nature of these substances?

One would naturally look for such substances in milk and possibly in eggs, since these two foods have been provided especially to serve the needs of the rapidly-growing organism. If any substances stimulatory of growth are to be found anywhere, one would naturally look first in those foods which we know to be particularly good for growing animals and particularly good for the rapidly-growing nervous system. They ought to be found in human milk, since this food has to meet the requirements of a very rapidly-developing nervous system. Such substances have been found in milk. Hopkins, McCollum and Davis, Hopkins and Nevill, and Osborne and Mendel found, in testing the efficacy of various pure proteins and inorganic salts in promoting the growth of young white rats, that artificial diets containing some protein, such as edestin, albumin or casein, some inorganic salts like those of milk, some starch, and lard nourished the animals for a time, but that sooner or later they ceased to grow, so that they rarely attained more than two-thirds of the weight normal for rats of their age. If at this time some butter fat was substituted for a portion of the lard, the remainder of the diet being the same and the total energy not changing, the rats began to grow again and very rapidly reached their normal weight. Furthermore, milk itself has in it all the substances necessary for the growth and maintenance of rats. These experiments are illustrated in the

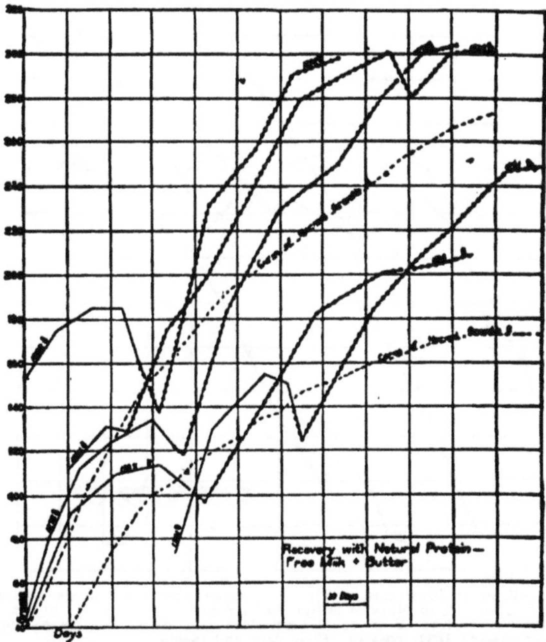

FIG. 67.—Curves of body weights of rats which have ceased to grow and have declined on foods containing the natural "protein-free milk" and have recovered when 18 per cent. unsalted butter replaced the same quantity of lard in the diet, as indicated by the interrupted lines (—o—o—o—). Rats 1204, 1281, 1202 had casein: rats 1268, 1276 had ovalbumin as the sole protein. Ordinates represent grams body weight; abscissas 20-day intervals. The diet was: Purified protein, 18 per cent.; starch, 26 per cent.; protein-free milk, 28 per cent.; lard, 10 per cent.; butter, 18 per cent.

curves in Figure 67. Sterilization of the milk did not in any way interfere with the value of the fat as a growth stimulant.

The experiments of Hopkins and Nevill were as follows: Twenty-four rats from various sources weighing from 50-60 grams each were placed on a diet containing protein and starch, which had been carefully extracted with alcohol; lactose, which had been repeatedly crystallized and extracted with alcohol; and a salt mixture similar to that used by Mendel and Osborne. Every rat, although eating well and taking sufficient food to cover its energy needs, rapidly ceased to grow, some on the sixth day, some on the ninth and all before the fifteenth day. A short period followed when they kept their weight, and to this succeeded a period of gradual decline in weight. The diet was kept the same for 18 of them up to death. Fourteen of them died about the fortieth day. In the case of six of the same set of rats after the decline had begun, 2 c.c. of milk per diem were added to the ration. An immediate betterment of the general condition was observed, growth was re-established and health was maintained. Another lot of six rats were given a little milk in addition to the ration of alcohol-extracted foods and these grew normally.

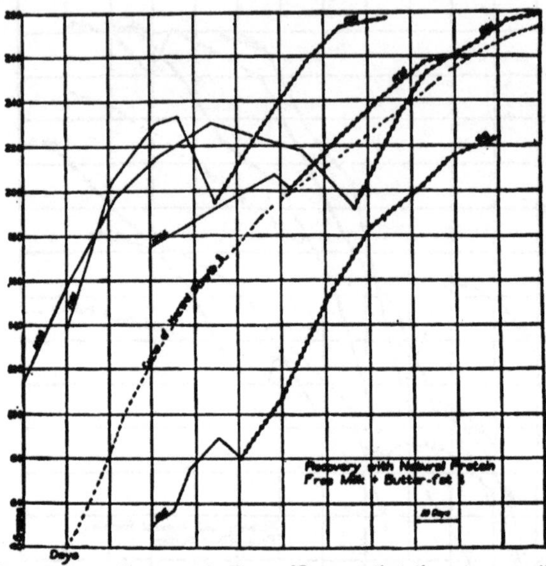

Fig. 67A.—Same as illustrated in Figure 67 except that the rats were all males and 18 per cent. of butter fat, in place of butter, as in Figure 67, was added to the diet, after decline had set in, in place of an equal amount of lard which was discontinued. Rats 1224, 1235 had casein; 1391, edestin; 1616, zein and casein.

The active substances in the butter fat are still undetermined. Cod-liver oil will act like the butter in promoting growth. Olive oil does not. Cholesterol is ineffective. The butter fat used by Osborne and Mendel contained neither ash, phosphoric acid nor nitrogen. The active substance for growth, they conclude, is probably not a glyceride of the ordinary fatty acids, nor a phospholipin, nor a vitamine in the sense of Funk. What it is must be determined by experiment.

Some experiments by Carrel may also be mentioned in this connection. Carrel has been growing tissues taken from the living organism in artificial culture media by Harrison's method of tissue culture. He found that if he placed these growing tissues in blood serum taken from adults, they continued to live for a long time, but they did not grow. The cells did not reproduce. If, however, the culture was made in the blood serum of young and growing animals, then the growth took place vigorously. It seems that there is some substance in the blood of young and growing animals which is not present in that of adults. What the nature of this substance is, is still unknown. If this experiment shall prove to be generally successful with many tissues and with the young and old sera, it would seem that we might at last be on the track of the substances of youth.

Tissue respiration.—Combustion is the source of all the energy of the human body in whatever form that energy may show itself. This combustion occurs in the living matter; it is the process of respiration. We cannot do better than to close this account of the principles of chemical biology by a brief examination of this most fundamental process. It is the breathing brain which is conscious. When oxygen unites with carbon and hydrogen in the brain under certain conditions, it produces or is accompanied by the psychic processes. The nature of respiration appears then to be the problem of all others of a chemical nature which it is desirable should be solved. Unfortunately we are still far from understanding the nature of this process, and there are in fact two main views as to its nature, and these two views are on the whole very antagonistic. They have been expressed by various observers under various guises, for the problem remains the same, although its outward form may change. The problem is this: Do the substances which enter the living protoplasm become part of the living protoplast before they burn, or not. What conception shall we have of the living matter with which the chapters of this book have dealt? There are two possibilities: one is that the cell is a kind of factory, the walls separating the rooms of which are made of colloidal matter which is relatively inert. The chemical changes which occur in the cell are due to the changes which are occurring in these separate rooms. This may be called the compartment theory. These compartments may be very small. Each may be imagined to have one or several enzymes or catalytic agents in it. These produce chemical decompositions, by purely chemical processes, in the amino-acids, sugars, or other substances which penetrate the compartments. They may undergo a process of fermentation by which CO_2 is produced. They may be oxidized, or hydrolyzed or otherwise changed. Or they may be combined with the walls of the compartment, either physically by adsorption, or by chemical union, and thus contribute to the chemical constitution of the colloids. According to this view, the oxidations in the cell are in all respects like those outside the cell. They are produced by active agents, oxidases of various kinds, which can be isolated from the cell and found outside to carry on the activities of protoplasm. By the oxidations of these substances acids or other compounds are produced, or heat is set free, and these compounds, or this heat, cause the change in the amount of water in the cell and so the physical changes of the lifeless colloidal material of the cell. These find their expression in the various forms of vital actions. This view was advocated by Hofmeister, among others, and it has been defended by Hopkins. According to this view, there is no vital matter in the cell; and the chemical transformations do not involve any large biogene molecule, but only such transformations as those with which we are all familiar, of relatively simple compounds in solution. The evidence in favor of this view is the fact that we are constantly succeeding in isolating from cells catalytic agents which cause in aqueous solution, in the beaker, oxidations, fermentations and hydrolyses identical in their main features with those in living matter. There is no denying the fact that there are many powerful evidences for the truth of this view. But it cannot be denied that there are also grave difficulties in its way. According to this view, the cell is not a unit; it is a large collection of separate enzymes, confined to separate parts, or compartments, of the cell. It gives no explanation of the apparent unity of the cell and the organism. The shapes of organisms are as characteristic as those of crystals. In crystals the matter is organized. In living matter, also, the matter is certainly organized. What is the organizing force in living matter? How shall we explain irritability and the phenomena of narcosis or anesthesia on the basis of the view just stated? The hypothesis that these drugs are acting by altering the permeability of the walls of the compartments has not yet been substantiated by any evidence which is really conclusive. It is certain that they influence the chemical processes profoundly since growth and respiration are annihilated by them; but anesthetics do not materially alter most of the oxidations outside the cell.

For these reasons another view has been proposed which has certain merits of its own. This view has been expressed by Pflüger, Verworn and other physiologists, often in slightly different forms, but in its essentials the same. The essential basis of this view is that the organizing property of the cell is made the point of departure. Living matter organizes the food that it receives and keeps what it needs; it makes always the right kind of living matter. Now there are two ways in which we may picture this organization force. The cell may be pictured as a kind of a kitchen in which there is a maid of all work, as Du Bois Raymond puts it. She receives the provisions, cooks the meals, throws out the wastes and keeps all in order. This maid of all work has received various names. Sometimes she is called entelechy, or vital force, but whatever her name her functions and activities are pictured by all who would employ her in their scheme of cell physiology, as being essentially the same. As long as the cell lives she is there; when she departs, things go at sixes and sevens and the machinery no longer works and the cell dies. The only escape from this conception, which recently has been again strongly advocated by Sir Oliver Lodge in his Presidential address before the British Association for the Advancement of Science, is to ascribe the organizing forces of the cell to the molecules of which it is composed. There are only two possibilities: either the molecules organize themselves as they do in a crystal or else something else organizes them, for that they are organized into a definite and characteristic form there is no doubt. Accordingly it is assumed that it is the molecular forces in the biogenes or large molecules of the cell. These large molecules are themselves the living, respiring units. It is the living matter, the organized matter itself, which is primarily burning in the protoplasm. But this is only part of the combustion. When combustion occurs in an auto-oxidizable matter, such as a biogene, there is formed, according to nearly all observers, a portion of hydrogen peroxide. That hydrogen peroxide is also formed in the course of living oxidations is indicated by the fact that there is present in all forms of living matter a special enzyme, catalase, which has the property of decomposing it and setting free oxygen. It is hardly probable that this catalase would be present in all forms of living matter without exception, unless it had some function there. We may assume that this hydrogen peroxide burns some of the molecules in solution in the cell. This is the other part of the combustion. Respiration consists on this view of two processes, the physiological or auto-oxidation of the real living protoplast or biogenes, and the oxidation, by means of the hydrogen peroxide, which is a secondary product of the primary respiration, of amino-acids and other fragments present in the solution. It is this oxidation which Hopkins has described. But this oxidation on this view is not the essential and primary oxidation. It is certainly a suggestive fact that the oxidations of this nature, of the amino-acids in cells and various fatty acids and other substances, are identical in character with those which are produced by hydrogen peroxide, as Dakin has shown.

We may, therefore, make the following picture of the fate of substances, such as amino-acids for example, on entering protoplasm. The amino-acid may enter the cell in solution, or it may ultimately find itself in solution in the water in the cell. Before it is combined with the protoplast or otherwise made over into the colloidal material of the cell, it has to run the gauntlet of the hydrogen peroxide and the oxidases.

A portion of the amino-acid is decomposed by this means and converted into ketonic acid and ammonia. The ketonic acid may be subsequently burned or it may be synthesized into the colloidal substratum of the cell. A portion of the amino-acid is caught up as such and synthesized into the protoplast. Oxygen when it enters the cell combines in the first instance with water and the living protoplast to make the irritable substratum of the cell. It is this substratum, a molecular union of oxygen, water and protoplast, which conducts, respires, contracts and is anesthetized by ether. The protoplast in its turn consists of unions of protein, phospholipin and various enzymes.

Whether this picture of the process is correct or not, there is no doubt that the respiration of the cell involves in its totality some oxidases, hydrogen peroxide, catalase, iron or manganese, and substances, presumably the living protoplast, which have the power of uniting with or decomposing the food matters and thus increasing their power of combustion.

This brief summary will serve to show how meager, as yet, is our knowledge, how impotent our attempts, to give any explanation of the great and fundamental problems of physiology and physiological chemistry: the nature of consciousness, of animal and plant forms and of the fundamental properties of respiration, irritability and growth.

REFERENCES. Vitamines and Growth Substances.

General.

1. *Funk:* Ueber die physiologische Bedeutung gewisser bisher unbekannter Nahrungsbestandteile, der Vitamine. Ergeb. d. Physiol., 13, pp. 125-205, 1913. Good literature lists for beri-beri, scurvy, pellagra, rickets and growth substances.

Special.

1. *Cooper and Casimir Funk:* Experiments on the causation of beri-beri. Lancet, p. 1266, 1911.
2. *Funk:* On the chemical nature of the substance which cures polyneuritis in birds induced by a diet of polished rice. Jour. Physiol., 43, p. 395, 1911.
3. *Funk:* The preparation from yeast and certain foodstuffs of the substance the deficiency of which in diet occasions polyneuritis in birds. Jour. Physiol., 45, p. 75, 1912.
4. *Funk:* An attempt to estimate the vitamine fraction in milk. Biochem. Jour., 7, p. 211, 1913.
5. *Drummond and Funk:* The chemical investigation of the phosphotungstate precipitate from rice polishings. Biochem. Jour., 8, p. 598, 1914.
6. *Funk and Macallum:* On the chemical nature of substances from alcoholic extracts of various foodstuffs which give a color reaction with phosphotungstic and phosphomolybdic acid. Biochem. Jour., 7, p. 356, 1913.
7. *Barger:* The simpler natural bases. Monographs of Biochemistry. London, 1914.
8. *Cooper:* The preparation from animal tissues of a substance which cures poly-

neuritis in birds induced by diets of polished rice. Bioch. Jour., 7, p. 268, 1913.
9. *Cooper:* Curative action of autolyzed yeast against avian polyneuritis. Bioch. Jour., 8; p. 250, 1914.
10. *Cooper:* The relation of vitamine to lipoids. Bioch. Jour., 8, p. 347, 1914.
11. *MacLean:* Purification of phosphatides. Bioch. Jour., 6, p. 355, 1912.
12. *Funk:* Is polished rice plus vitamine a complete food? Jour. Physiol., 48, p. 228, 1914.
13. *Edie, Evans, Moore, Simpson and Webster:* The antineuritic bases of vegetable origin in relationship to beri-beri. with a method of isolation of torulin, the antineuritic base of yeast. Bioch. Jour., 6, p. 234, 1912.
14. *Suzuki, Shimamura and Odake:* Ueber Oryzanin ein Bestandteil der Reiskleie und seine physiologische Bedeutung. Bioch. Ztschr., 43, p. 89, 1912.
15. *Funk:* The effect of a diet of polished rice on the nitrogen and phosphorus of the brain. Jour. Physiol., 44, p. 51, 1912.
16. *Chamberlain, Bloombergh and Kilbourne:* A study of the influence of rice diet and of inanition on the production of multiple neuritis of fowls and the bearing thereof on the etiology of beri-beri. Philipp Jour. Sci., 6, p. 177, 1911.
17. *Holst and Fröhlich:* Experimental studies relating to ship beri-beri and scurvy. Jour. Hyg., 7, p. 634, 1907.
18. *Funk:* Nitrogenous constituents of lime juice. Bioch. Jour., 7, p. 81, 1913.
19. *Lane-Claypon:* Value of boiled milk as a food for infants and young animals. Report to the Local Govt. Board, N. S., p. 63, 1912.
20. *Brachi and Carr:* Infantile scurvy in a child fed on sterilized milk. Lancet, 662, 1911.
21. *Lane-Claypon:* Observations on the influence of heating upon the nutritive value of milk. Jour. Hyg., 9, p. 233, 1909.
22. *Stepp:* Weitere Untersuchungen über die Unentbehrlichkeit der Lipoide für das Leben. Ueber die Hitzezerstörbarkeit lebenswichtiger Lipoide der Nahrung. Zeit. f. Biol., 59, p. 336, 1912.
23. *Stepp:* Experimentelle Untersuchungen über die Bedeutung der Lipoide für die Ernährung. Biochem. Ztscher., 22, p. 452, 1909.
24. *Osborne, Mendel and Ferry:* Feeding experiments with isolated food substances. Carnegie Instit., Washington. Nr. 156.
25. *Osborne, Mendel and Ferry:* Feeding experiments with fat-free food mixtures. Jour. Biol., Chem., 12, p. 81, 1912.
26. *Hopkins:* Feeding experiments illustrating the importance of accessory factors in normal dietaries. Jour. Physiol., 44, p. 425, 1912.
27. *Hopkins and Neville:* A note concerning the influence of diets upon growth. Bioch. Jour., 7, p. 97, 1913.
28. *Osborne, Mendel and Ferry:* Maintenance experiments with isolated proteins. Jour. Biol. Chem., 13, p. 233, 1912.
29. *Macallister:* A new cell proliferant: its clinical application in the treatment of ulcers. British Med. Jour., Jan., 1912.
30. *Macallister:* The action of Symphytum officinale and allantoin. British Med. Jour., Sept., 1912.
31. *Coppin:* The effect of purin derivatives and other organic compounds on growth and cell division in plants. Bioch. Jour., 6, p. 416, 1912.
32. *Carrel:* On the permanent life of tissues outside of the organism. Jour. Expt. Med., 15, p. 516, 1912.

33. *Carrel:* Artificial stimulation and inhibition of the growth of normal and sarcomatous tissues. Jour. Amer. Med. Assn., 55, p. 32, 1911.
34. *Umnus:* Die photobiologische Sensibilierungstheorie in der Pellagrafrage. Zeits. f. Immunitätsf., 13, p. 461, 1912.
35. *Hunter:* The sand-fly and pellagra. Jour. Amer. Med. Assn., 58, p. 547, 1912.
36. *Aldrich:* The feeding of white rats on the pituitary body. Amer. Jour. Phys., 31, p. 94, 1912.
37. *Orgler:* Die Kalkstoffwechsel des gesunden und des rachitischen Kindes. Erg. d. inn. Med. u. Kinderheil., 8, p. 142, 1912.
38. *Mohr:* v. Noorden's Handbuch der Pathol. d. Stoffwechsels, 2, p. 865.
39. *McCollum and Davis:* The necessity of certain lipins in the diet during growth. Jour. Biol., Chem., 15, p. 167, 1913.
40. *Osborne and Mendel:* The relation of growth to the chemical constituents of the diet. Jour. Biol. Chem., 15, p. 311, 1913; 16, p. 423, 1914; 17, p. 401.
41. *Benedict:* Experiment on a man fasting 31 days. Publication 203, Carnegie Institution, Washington, 1915.
42. *Funk and Macallum:* Die chemischen Determinanten des Wachstums. Zeit. physiol. Chem., 92, p. 13, 1914. *Funk:* Ibid., 88, p. 352, 1913.
43. *Folin:* On starvation and obesity with special reference to acidosis. Jour. Biol. Chem., 21, p. 183, 1915.
44. *McCollum and Davis:* The influence of the composition and amount of the mineral content of the ration on growth and reproduction. Jour. Biol. Chem., 21, p. 615, 1915; The influence of certain vegetable fats on growth. Ibid., 21, p. 179, 1915; Dietary deficiencies of rice. Ibid., 23, p. 181, 1915; The essential factors in growth. Ibid., 23, p. 231, 1915.
45. *Funk and Macallum:* On the probable nature of the substance promoting growth in young animals. Jour. Biol. Chem., 23, p. 413, 1915.
46. *Osborne and Mendel:* The resumption of growth after long continued failure to grow. Jour. Biol. Chem., 23, p. 439, 1915.
47. *Hunter, Givens and Lewis:* Preliminary observations on metabolism in Pellagra. Hygienic Lab. Bull. No. 102, U. S. P. H. Service, 1916.

PART III.

PRACTICAL WORK AND METHODS.

A. Equipment of the laboratory.

1. Desk *reagents.*

		Strength			Strength
1.	Hydrochloric acid. Conc. ..	37%	11.	Ammonium oxalate	3%
2.	Hydrochloric acid. Dilute ..	10	12.	Copper sulphate	2
3.	Sulphuric acid. Conc.	95	13.	Ferric chloride	2
4.	Sulphuric acid. Dilute.	10	14.	Potassium ferrocyanide	1
5.	Nitric acid. Conc.	70	15.	Millon's reagent.	
6.	Acetic acid. Dilute.	10	16.	Magnesium sulphate	Solid
7.	Lead acetate	2	17.	Ammonium sulphate	Solid
8.	Sodium hydrate	10	18.	Litmus paper. Red.	
9.	Sodium carbonate	10	19.	Litmus paper. Blue.	
10.	Ammonium hydrate	10			

A Word to the Student.—Laboratory work is of value only when it is done with a thorough understanding of what one is trying to do; otherwise it is of no more value to a student than any other form of mechanical exercise. An experiment should never be begun until the whole experiment has been carefully read through and a clear conception had of the object of the experiment. If you do not understand the object of the experiment and cannot discover it from reading the directions, ask the instructor. If he cannot tell you the object do not try the experiment, or if you do, try to find out from it what the object was. The experiments in this book all have a particular object in being there. Most of them are to enable you to see for yourselves what is described in the text; but some are also put in to show you methods of solving problems in chemical physiology or in clinical diagnosis. One generally remembers better what one sees than what one simply hears about. The experiments should be done while the text bearing on the matter is clearly in mind. Read over the work to be done and the text bearing on the general problems before coming to the laboratory. This will make the laboratory work far more valuable, and also remove from it much of the drudgery. In the quantitative work, if you cannot get a result you expect, or which you think is right, tell the instructor frankly about it and ask his advice. Perhaps your expectation may be wrong and your results are really right; or it may be that you have made some error in technique for which he can suggest a remedy. It is no disgrace to fail in getting an accurate result the first time a new quantitative method is tried. Even skilled chemists often fail in this same way. But by repeating the work one will become skillful. It is better to do one thing well than a dozen things badly. Remember that you are working for your own benefit, and that honesty is the first requisite for success in all walks of life, but above all it is the foundation stone of science.

2. *Side-shelf reagents.*

	Strength
1. Acetic Acid	Glacial
2. Acetic Acid	N
3. Alcoholic Alpha Naphthol	5%
4. Alizarin Red	5%
5. Ammonium Hydrate (0.90)	——
6. Ammonium Molybdate in HNO_3	——
7. Ammonium Sulphate	Sat.
8. Ammonium Sulphide	——
9. Aniline Acetate	50%
10. Barfoed's Reagent	——
11. Barium Chloride	5%
12. Benedict's Alkaline Tartrate	——
13. Benedict's $CuSO_4$	——
14. Benedict's K_4FeCn_6	——
15. Bismuth Subnitrate	Solid
16. Boas' Reagent	——
17. Bromine Water	Sat.
18. Calcium Chloride	5%
19. Calcium Hypochlorite	——
20. Casein	Solid
21. Charcoal (animal)	——
22. Chloroform	——
23. Congo Red	——
24. Cotton (absorbent)	——
25. Dextrin	Solid
26. Dimethylamidoazobenzol	0.5%
27. Egg White	Dried
28. Esbach's Reagent	——
29. Fehling's Solution No. 1	——
30. Fehling's Solution No. 2	——
31. Ferric Ammonium Sulphate	N/2
32. Folin-Shaffer Reagent	——
33. Formaldehyde	1:5000
34. Furfural	M/100
35. Gelatin	Solid
36. Glucose	Solid
37. Glycerol	95-100%
38. Glyoxylic Acid	——
39. Guaiacum Tincture	1%
40. Gum Arabic	Solid
41. Günzberg's Reagent	——
42. Haines' Solution	——
43. Hubl's Iodine Solution	——
44. Hydrochloric Acid	N
45. Iodine in KI	N/10
46. Lactic Acid	N
47. Lactose	Solid
48. Lead Acetate, basic	——
49. Levulose	Solid
50. Magnesia Mixture	——
51. Magnesium Chloride	1%
52. Maltose	Solid
53. Mercuric Chloride	1%
54. Methyl Orange	——
55. Nitric Acid	N
56. Olive Oil	——
57. Orcin	50% Sat.
58. Oxalic Acid	N
59. Para-dimethyl-amido-benzaldehyde	——
60. Paraffin	Solid
61. Peptone, Witte	Solid
62. Phenol	2%
63. Phenolphthalein	0.5%
64. Phenylhydrazine HCl	Solid
65. Phloroglucin in alcohol	2%
66. Phosphomolybdic Acid	2%
67. Phosphotungstic Acid	2%
68. Picric Acid in water	Sat.
69. Potassium Bichromate	Com'l.
70. Potassium Bichromate	N/2
71. Potassium Bisulphate	Solid
72. Potassium Chromate	Sat.
73. Potassium Iodide	1%
74. Potassium Mercuric Iodide	Sat.
75. Potassium Oxalate	Solid
76. Potassium Sulphocyanate	2%
77. Resorcin in Alcohol	0.5%
78. Saccharose	Solid
79. Silver Nitrate	1%
80. Soap Solution	1%
81. Sodium Acetate	Solid
82. Sodium Alcoholate	10%
83. Sodium Carbonate	N
84. Sodium Chloride	Solid
85. Sodium Chloride	N
86. Sodium Hydrate 40 gms. in 100 c.c.	——
. Sodium Nitrite	5%
Sodium Nitroprusside	Solid
. Sodium Thiosulphate	1%
Starch	Solid
. Sulphanilic Acid	5%
. Talcum Powder	Solid
87. Tannic Acid in alcohol	20%
[illegible]. Thymol in alcohol	10%

	Strength		Strength
95. Toluene	100%	98. Vaseline	——
96. Tropaeolin	0.5%	99. Zinc Sulphate	Solid
97. Turpentine	——	100. Zinc Sulphate	Sat.

3. *Special apparatus for general use.*

a. Blast of air and special apparatus carrying aëration tubes for ammonia determinations by aëration. Each apparatus has places for 12 simultaneous aërations. See Figure 72.
b. Kjeldahl digestion apparatus.
c. Block tin condenser still with 12-20 places for ammonia distillations.
d. 2 oil baths with tubes for use in Benedict's urea determination. Figure 71.
e. Long copper lined trough with running water for dialysis experiments, 8" x 8" x 15 ft.
f. Burettes connected with large bottles containing N/2 and N/10 H_2SO_4 and NaOH for quantitative titrations; also with saturated $Ba(OH)_2$ solution.
g. 40% solution of soda lye containing Na_2S for Kjeldahl determinations. To 10 lbs. NaOH (Commercial powder; soda lye) add *while stirring* 11, 320 c.c. water and then 250 grams powdered K_2S or 180 grams Na_2S. *Stir until all is dissolved.* Allow to settle out for 2-3 days and siphon off the clear supernatant liquid. For use in Kjeldahl-Gunning nitrogen determination. Take 100 c.c of this solution to neutralize the 20 c.c. concentrated H_2SO_4 used in the digestion. This soda lye solution is best kept in a large bottle sitting on a large, enameled-ware plate to protect from the drip, and provided with a bicycle pump or compressed-air connection by which air pressure can be made on the solution so as to drive some over from the delivery tube. The cork has two holes, one connecting with the compressed air and provided with a side tube to remove the air pressure when the required amount of liquid is discharged; the other carrying the delivery tube which opens near the bottom of the soda lye bottle.
h. The laboratory must of course have other common pieces of apparatus such as thermostats, polariscope, spectroscope, colorimeters, balances, etc.
i. Woolen blanket, pail of sand, pail of water, fire extinguishers.

4. *Desk outfit for each student.*

One of the blanks and the outfit is in the desk when the student enters the course. He checks off the apparatus, signs the blank and returns blank to the storeroom. The blank is as follows:

PHYSIOLOGICAL CHEMISTRY OUTFIT. (University of Chicago.)

This list must be checked, signed, and returned to the storeroom before the student can receive materials from the storeroom.

Not returnable goods in the drawer:—

Filter paper, 3 sheets.	Test-tube brush.
Matches (safety).	Towel.
Sapolio.	Wire gauze.

Returnable goods in the drawer:—

1 Buchner funnel with stopper.
1 Burette with pinch cock, 50 c.c.
2 Erlenmeyer flasks, 150 c.c.
2 " " 300 c.c.
3 Fermentation tubes.
2 Florence flasks, 250 and 500 c.c.
2 Glass rods.
1 Graduated cylinder, 5 c.c.
1 Graduated cylinder, 100 c.c.
1 Graduated pipette, 5 c.c.
1 Horn spoon.
1 Mortar and pestle.
1 Pipette, 25 c.c
1 Porcelain perforated filter plate.
1 Test-tube holder.
1 Thermometer.
2 Watch glasses, 2 inch.
1 Watch glass, 3 inch.

Returnable goods in the cupboard:—

3 Beakers, 75 c.c.
2 " 350 c.c.
1 Beaker, 1000 c.c.
3 Bottles, 300 c.c.
1 Bunsen burner.
1 Desiccator.
4 Evaporating dishes, 7, 10, 15, 21 cm.
1 Filter flask, 500 c.c.
2 Funnels, 5, 10 cm.
1 Iron clamp and holder.
3 Iron rings, 3 sizes.
1 Iron stand.
2 Reflux condensing tubes.
10 Test-tubes.
1 Test-tube rack.
1 Tripod.
1 Wash bottle.
1 Wire basket.

From the storeroom:—1 key.

Received these articles in good condition.

Signed ....

University Address ....

Home Address ....

Desk number

Date **Checked by** ....

Course

B. **General directions for work.**—The general aim should be to make as much of the work accurate and quantitative as possible. A quantitative experiment teaches all that a qualitative experiment teaches, and much besides; in addition it requires careful and accurate manipulation and cleanliness, qualities so essential to a physician or chemist. In the experiments which are described it is essential when quantities of certain materials are mentioned that they be accurately measured and all the conditions of the experiment carefully noted and fulfilled. The results of many of the experiments are only instructive when they have been accurately performed, under given conditions and with suitable controls.

The first requisites to proper laboratory manipulation of any kind, whether chemical, physical or surgical, are neatness and order. Neat workers only can be accurate manipulators. Orderly workers usually make fewer mistakes in laboratory manipulation than do those who go over the work hurriedly and without system or plan.

Apparatus found in the desks should first be cleaned, rinsed with distilled water and allowed to drain. Each time, immediately after use, the apparatus should be cleaned and kept ready for use. Most of tl apparatus soiled in this work is best cleaned by rinsing well with war water, then washing with soap and water and brush or cloth, then aga rinsing well with warm water and finally rinsing with distilled wat In every case friction must be used by rubbing with a brush. Nev dry the inside of a beaker or other chemical vessel with a towel. If must be dried quickly, rinse with alcohol and ether and dry with a cu rent of air; or dry by gently warming after rinsing in distilled water.

Test-tubes should never be rinsed only, but should be washed means of a brush and then should be well rinsed. In most cases clea ing mixture is not necessary, especially as usually employed by the ine perienced. Use judgment in cleaning your apparatus; for instance, not try to remove fats by rinsing with cleaning mixture and do n try to remove barium salts by means of cleaning mixture. Always u the appropriate solvent first, then remove the excess of the solvent a continue with the washing as above. Pipettes, after the prelimina cleaning with the appropriate solvents and rinsing as above indicat often require further treatment with hot cleaning mixture. Finall rinse well with warm water and distilled water.

In measuring reagents, do not insert a pipette into the reage bottle, but pour off about the amount desired into a dry vessel and th measure therefrom by the pipette. *Never return solid or liquid reage to the stock bottles.*

In taking samples or reagents with a pipette be sure to have yo pipette clean and dry, or rinsed with a part of the solution at lea twice before measuring it off. After rinsing with a part of the solutio hold the pipette between the thumb and second finger, draw up t liquid, place the index finger on the mouthpiece and by turning t pipette between the thumb and second finger allow the lower part the meniscus to come on a level with the mark. Then observe that drops adhere to the outside of the pipette and transfer the liquid to t proper vessel. Allow to drain by touching the side of the vessel wi the tip of the pipette, *but never blow into the pipette.*

Above are given two lists of reagents arranged in the order they a found on the shelves. Do not insert pipettes, glass rods or other utensi into these reagents. Do not lay the stoppers on the desk or other su faces; learn to hold the stopper in your hand while pouring from t bottle. Do not moisten litmus paper by means of the stoppers. PLEA RETURN EACH BOTTLE TO THE PROPER PLACE IMMEDIATELY AFTER USE.

Your note-book must contain for each experiment the following dat 1. The object of the experiment: the question to be answered by th experiment. Do not begin the experiment until you know this. 2. A bri but accurate statement of the methods employed for its solution. 3. A results of weighings, readings of burettes, measurements, etc., at the time they are made. All calculations should be given. 4. A concise critical statement of the conclusion.

Filtering. Nearly all slimy or flocculent precipitates are best filtered by folded or creased filters without the use of suction. Crystalline or granular precipitates may be filtered to advantage with suction. If it is desired to save a precipitate, select the size of the filter with regard to the size of the precipitate, not to the size of the filtrate. Albuminous precipitates which stick to the filter are best taken from the filter paper while they are still somewhat moist. Use filter plate and small filter paper where possible.

General principles. Avoid using excess of reagents; keep the number of different kinds of substances in a solution as small as possible.

CAUTION. Acetone, alcohol, ether, benzine, glacial acetic acid are inflammable. Never heat any of them over the free flame. Heat them on the steam or water or electric bath. Have no lights near when pouring ether, acetone or benzine from one vessel to another. In case of fire, if in a dish or flask, cover it with a wet towel, thus suffocating it. Water may be added to an alcohol or acetone fire. Blankets are kept in the laboratory for use if clothing catches fire.

I. CARBOHYDRATES.

In the work outlined below make your tests comparable in quantities and conditions. The stock solutions of the mono- and di-saccharides are M/5 in strength and are preserved by toluene. In making the dilutions called for, please do not use more than necessary for a day's work. Before coming to the laboratory you should have fully studied the part to be considered that day.

* 1. Physical characters of the carbohydrates.

* a. *Test the solubility of starch, saccharose and dextrose in ether, alcohol and water.*

Into clean, perfectly dry test-tubes introduce an amount of finely divided carbohydrate about equal in volume to the size of a wheat kernel. Now add 5 c.c. of the solvent. Allow to act with frequent agitation for 15 minutes, then filter through dry filter papers and funnels and apply the α-naphthol test (Molisch) on the *clear* filtrates, as outlined below. Expt. 25, page 864. Test the solubility of each of the three carbohydrates mentioned in both hot and cold alcohol, ether and water, warming the alcohol and ether on the water bath. In testing the filtrates, when alcohol and ether have been used as solvents, it is necessary first to remove the solvent by evaporation on the water bath. CAUTION! DO NOT WORK NEAR A FLAME. Then add 4 c.c. distilled water to the residue in the test-tube, cool and apply the Molisch test for carbohydrates. Make a blank control test on 4 c.c. distilled water. Note the differences in the intensities of the reaction in the various tubes.

* b. *State of aggregation of carbohydrates in solution. Crystalloids and colloids.*

Determine which of the carbohydrates given below diffuse through parchment paper. Use M/25 solutions of glucose and saccharose, 0.7 per cent. starch paste and a 0.6 per cent. solution of gum arabic. In making the starch paste, weigh off 0.35 gram of dry starch into a small porcelain dish, add 5 c.c. distilled cold water, stir until well suspended and then while continually stirring pour into 45 c.c. boiling distilled water. Boil the mixture for 2 minutes and finally add distilled water to make the total volume 50 c.c.

Prepare the parchment tubing by allowing it to remain in distilled water for 10 minutes. Fill with distilled water and see that the tubing does not leak. Empty out the distilled water and introduce into the tubing 10 c.c. of the carbohydrate solution to be tested and set the tubes in 75 or 100 c.c. beakers containing 20 c.c. distilled water, the ends of the parchment tubing being kept above the level of the water in the beaker so that no admixture of water and contents of the tube can occur except through diffusion. After standing one hour test 5 c.c. of the solutions in the beakers by the Molisch reaction. Why are some carbohydrates diffusible, others not?

c. *Precipitation of starch, a colloid, by salts.*

Saturate 10 c.c. of a 1 per cent. starch solution with powdered ammonium sulphate. Note the precipitate. Filter off and test the filtrate for starch by adding to it a drop of I_2KI solution. In the presence of starch a blue color.

* 11. Reactions of carbohydrates of biochemical interest.

A. ACTION OF STRONG ACIDS ON CARBOHYDRATES. MONOSACCHARIDES.

* 2. On Pentoses. (a) Formation of furfural from pentosans.

The distillations below may be conducted in groups of three: (a) being distilled by one, (b) by another and (c) by a third student. Compare the contents of each other's distilling flasks from time to time, also the odors of the distillates.

To 10 grams of wheat bran in a 250 c.c. Florence flask add 70 c.c. of a mixture of equal volumes of conc. HCl and water. Connect with a water-cooled, glass condenser; heat gradually with a flame, keeping the flame in motion until distillation has begun; then distill until about 30 c.c. has been collected. Shake the distillate with 15 c.c. portions of ether four times, removing and saving the ether layer each time. This shaking is best done in a separatory funnel, but may also be done in a large test-tube and the ether removed each time by means of a pipette.

Combine all the ether extractions in a beaker and allow to evaporate slowly on the steam bath. DO NOT EVAPORATE ON THE GAS-HEATED BATHS. When all the ether has evaporated, add 100 c.c. distilled water and stir until all is dissolved. Set aside for the comparative tests later.

Over the mouth of the flask, while still boiling very gently, place a piece of anilin-acetate paper. To prepare this place a drop of 50 per cent. anilin-acetate solution on a small piece of filter paper and allow to become partially dry. Furfural turns it pink.

* 3. On hexoses. (b) Formation of hydroxymethylfurfural, humic acids and levulinic acids from glucose.

Proceed in the same manner as given under (a) above, using 10 grams glucose instead of bran. Note and explain the color of the solution very soon after distillation has begun and at the completion of the distillation. Proceed with the ether extraction as above, but instead of dissolving the residue left on evaporating the ether extract in 100 c.c. water, use only 10 c.c. Test acidity; reduction of silver nitrate (formic acid); try iodoform test.

The residue in the flask contains humic acid, furfurals, levulinic acid and some formic acid. The formation of levulinic acid in this process is characteristic for hexoses and those polysaccharides which yield hexoses on hydrolysis. After removing the greater part of the furfurals by distillation, and the humic acids by filtration, one may extract the levulinic acid from the filtrate by ether. The ether solution on evaporation yields crude levulinic acid, which even in the cold readily forms iodoform when made alkaline with NaOH and a few drops of I in KI solution (No. 45) added. What other common substances readily yield iodoform by similar treatment? Are these substances likely to be present here? To further identify levulinic acid, it is usually separated as the zinc salt and then identified as the silver salt. (*Annalen*, 206, pp. 207 and 226.) Carry out the anilin-acetate test and compare with 2 (a).

* 3. (c) Formation of hydroxymethylfurfural, humic and levulinic acids from levulose.

Repeat 3 (b) above on 10 grams levulose, but before beginning the distillation add a few pieces of broken glass to the material in the distilling flask. Carry out the anilin-acetate test and compare with 2 (a) and 3 (b).

* 3. (d) Examine the three distillates from a, b and c above for the presence of furfural by the reactions on page 863.

Note any differences of color and the relative intensities of the reaction. Try the Molisch, orcine and phloroglucine reactions on the distillates. From which sugar do you get the most furfural? Write the reaction in your note-books, showing the production of furfural from a pentose. Write the structural formulas of levulinic acid and formic acid.

*B. ACTION OF STRONG ACIDS ON DISACCHARIDES.

*4. Inversion of cane sugar.

a. To 10 c.c. of a M/2 solution of saccharose add an equal quantity in a test-tube of $N/2H_2SO_4$ and immerse in boiling water for 45 minutes. While this is heating take in another test-tube 10 c.c. of M/2 saccharose, dilute it with an equal quantity of water and determine its specific rotatory power at 20° C. by means of the polariscope (see page 25). At the end of the 45 minutes of heating the original solution, cool it to the temperature of the solution you have just examined, about 20° C., and determine the specific rotatory power of this and compare it with the other. If the solution is still dextro-rotatory, return it to the water bath for longer heating. Explain the change in the rotatory power produced by the acid. Write the reaction.

b. After recording the polariscope reading of the solution which has been heated with acid, neutralize the acid exactly by adding carefully from a burette N/2NaOH or cold, saturated barium hydrate, using litmus as an indicator. Do not add an excess of alkali. Filter from the barium sulphate, if $Ba(OH)_2$ has been used, and test the reducing power of the solution with Fehling's and Barfoed's reagent. Has the reducing power been increased or diminished by the acid treatment? Explain. Test some of the filtrate for the presence of levulose.

5. Hydrolysis of lactose and maltose. (a) Repeat experiment B4.a. with lactose and maltose.

6. Comparative ease of hydrolysis of saccharose, maltose and lactose by acids.

Place in three separate 30 c.c. test-tubes 5 c.c. of M/5 solutions of lactose, maltose and saccharose. Add to each tube 15 c.c. $N/2H_2SO_4$. Immerse all three in boiling water for 45 minutes; transfer the solutions to 100 c.c. beakers and carefully neutralize to litmus by adding from a burette drop by drop, while stirring, cold saturated $Ba(OH)_2$ solution. About 17 c.c. of $Ba(OH)_2$ solution will be required. Avoid adding an excess of alkali, as the solution must not become alkaline and an excess may precipitate a part of the carbohydrates. Finally dilute each to 50 c.c. with distilled water, mix well, heat to boiling and set each aside to settle out the $BaSO_4$. Take 5 c.c. of the clear supernatant liquid from each into separate tubes, add to each tube 5 c.c. of Barfoed's solution and observe the relative speeds of reduction when immersed in boiling water. Which sugar hydrolyzes the most rapidly by acid? Test another 5 c.c. of each of the solutions for levulose by the Seliwanoff method. (Experiment 27.) Write the structural formula of lactose and show how the acid probably attacks the molecule. (Page 57.)

* C. ACTION OF STRONG ACIDS ON POLYSACCHARIDES.

* 7. Prepare 1 per cent. solutions of starch and gum arabic by mixing the powdered carbohydrate with a little cold water and pouring it while stirring into about 75 c.c. of boiling distilled water. Dilute the solution to 100 c.c. Test the reducing action of these solutions toward Fehling's solution in the manner described on page 860 and record your results. Also add to a portion of each solution a drop of I_2KI solution. Observe the blue color which develops in the starch solution if it is cold. Now place in two 150 c.c. Erlenmeyer flasks 0.1 gram of starch and gum arabic respectively; add 5 c.c. distilled water and 0.5 c.c. concentrated H_2SO_4 (measured by a 5 c.c. graduated cylinder). Fit the flask with a reflux condenser and heat on the water bath for four hours. Neutralize the solution as above with saturated $Ba(OH)_2$ solution. About 40-45 c.c. will be required. When neutral dilute to 50 c.c. Heat to boiling and allow to settle as above. Test 5 c.c. of the clear filtrate for starch with the I_2KI solution. Is any color formed? Record it. On other portions of each filtrate make the Barfoed, Seliwanoff and Tollens phloroglucin reaction. Record the results in a tabular form. What sugars are formed from starch by the action of the acid? What from the gum arabic? Do either of these carbohydrates contain pentoses?

D. DISSOCIATION AND DECOMPOSITION OF CARBOHYDRATES BY ALKALIES.

* 7(a). Caramel and humus formation. (Moore's test.)

Some of the carbohydrates are unstable in alkaline solution and decompose with the formation of caramel and humus, dark-colored substances.

To 1 c.c. of the M/25 solutions of saccharose, maltose, levulose, glucose, arabinose, 0.70 per cent. dextrin and 0.60 per cent. gum arabic solution, each in a separate test-tube, add 1 c.c. 10 per cent. NaOH solution. Immerse in boiling water and note the changes in the color and odor developed. Heat thus for 15 minutes. The dark color developed on heating with NaOH is known as Moore's test for carbohydrates. Most colloidal polysaccharides are not attacked even by strong alkali. Save the contents of the tubes and proceed with experiment 10. This experiment is the proof that cane sugar is not hydrolyzed by alkali.

* 8. An experiment showing how the reducing power of carbohydrates is increased by treatment with alkalies.

The object of this experiment is to show how the different components of Fehling's solution, namely, the alkali and the cupric salts, are acting, and also to show that the alkali greatly increases the reducing power of the sugar so that it becomes capable of reducing even an acid cupric sulphate, or methylene blue solution. Methylene blue becomes colorless when it is reduced.

a. Prepare the following solutions in test-tubes, measuring off the solutions in the order indicated and measuring carefully by means of a graduated pipette:

	A	B
M/5 glucose	5 c.c.	5 c.c.
N/2 NaOH	4 c.c.	0
N/2 H_2SO_4	0	5.5 c.c.
Immerse in boiling water for 10 minutes then add		
N/2 H_2SO_4	5.5 c.c.	0
N/2 NaOH	0	4 c.c.

Mix the contents of each tube well. Both tubes now contain the same concentrations of Na_2SO_4, H_2SO_4 and carbohydrate, but in tube A the carbohydrate has been acted upon by alkali, whereas B has not been exposed to alkali.

b. Measure off into tubes No. 1 and No. 2, 2 c.c. samples of A and B respectively. Now add to each, 2 c.c. of a dilute $CuSO_4$ solution (dilute Fehling's No. 1 $CuSO_4$ solution, No. 29, with 9 volumes distilled water). Mix the contents of each tube well and immerse in boiling water for 5-10 minutes. Note the difference in the colors of the solutions. The faint bluish color will disappear in one of the tubes. To each tube now add, after cooling, 2 c.c. strong NH_4OH (No. 5). Note the difference in color now. Cupric salt solutions become a deep blue color on the addition of ammonia. Are any cupric ions left in the solution which was heated while it was alkaline? Explain the changes in color and why solution A should act differently from solution B.

c. Repeat as in (b) above, but instead of adding the $CuSO_4$ solution add 1 c.c. of a 0.005 per cent. methylene blue solution to each and then immerse in boiling water. Which one decolorizes the methylene blue and why?

*** 9. The change of aldoses to ketoses and vice versa in alkaline solutions.**

Dilute alkali causes a rearrangement of the atoms of the first two or three carbons producing aldoses from ketoses and a variety of sugars from glucose.

Prepare tubes as indicated below:

A 1 c.c. saturated $Ba(OH)_2$ sol. and 1 c.c. of		B 1. c.c. H_2O and 1 c.c. of	
No. 1a	M/5 levulose	No. 1b	M/5 levulose
No. 2a	" glucose	No. 2b	" glucose
No. 3a	" arabinose	No. 3b	" arabinose

Cover the solution in each case with a layer of toluene about one-fourth inch thick and set aside overnight. The next day take 1 c.c. from each tube, neutralize, add 5 c.c. of Seliwanoff's reagent (in experiment 27) and immerse all the tubes in boiling water at the same time. Compare the speed of development and intensities of the colors in the different tubes. Seliwanoff's reaction is given most promptly by the ketoses. Explain the transformation of an aldose to a ketose. Why must the reaction be alkaline for the transformation to occur? See page 32. Write structural formulas showing the transformation of an aldose to a ketose.

To make this experiment of any value the solutions must be accurately measured and all tubes treated in precisely the same manner; heated for the same length of time, etc.

* 10. An experiment showing that cane sugar, starch and gum arabic are not hydrolyzed by alkalies.—Neutralize the contents of the tubes of experiment 7 exactly with N/2 H_2SO_4, add to each tube an equal volume of Fehling's *mixture,* and immerse in boiling water for 4 minutes. Do you obtain any reduction in the cane sugar, starch and gum arabic tubes? Are these carbohydrates attacked by alkalies? Why are they not attacked? Had these sugars been hydrolyzed by the alkali they would all have developed a brown color by the decomposition of the monosaccharides set free.

E. REDUCING REACTIONS OF CARBOHYDRATES.

11. Monosaccharides when treated with alkalies will spontaneously oxidize themselves from the oxygen of the air.

a. Place 2 grams of levulose in a 100 c.c. flask, add 50 c.c. N/2 NaOH, stopper the flask tightly with a rubber stopper which has a glass tube through it to which a short piece of rubber tubing is attached, the rubber tubing being closed with a pinch cock. Shake the flask vigorously for fifteen minutes to thirty minutes. Attach a short glass tube to the end of the rubber tubing, being careful not to open the pinch cock. Place the end of the glass tube under water and open the pinch cock. If there is a negative pressure in the flask, the water will rise in the tube. What has happened to develop a negative pressure in the flask?

* 12. Monosaccharides and some disaccharides reduce solutions of cupric salts.

* a. *Solvent action of carbohydrates, sodium potassium tartrate and glycerol on cupric hydrate. To prove that the cupric hydrate combines with the reducing body before it is reduced.*

The experiment which follows not only proves the union of oxidizing and reducing body, but also illustrates the function of the sodium potassium tartrate in Fehling's solution.

To 5 c.c. Fehling's $CuSO_4$ solution (No. 29) gradually add, while stirring well, 5 c.c. 10 per cent. NaOH. Measure off four 2 c.c. portions of this mixture into different test-tubes. Then proceed as follows:

Add to
Tube No. 1 about 1 gm. solid saccharose.
Tube No. 2 about 1 gm. sodium potassium tartrate.
Tube No. 3 5 c.c. glycerol.
Tube No. 4 nothing.

Boil and note in which tubes the precipitate disappears. Many other neutral substances possess this solvent action toward cupric hydrate. Polyhydric alcohols and salts of many polyhydroxy acids as well as amino-acids have this property. The same power appears in the solvent action of Schweitzer's reagent toward cellulose.

* 13. Reduction of Fehling's solution.

Prepare 20 c.c. of Fehling's solution by mixing thoroughly 10 c.c. of Fehling's solutions No. 1 and No. 2 respectively. Measure off eight 2 c.c. portions into different test-tubes and add thereto respectively 2 c.c. of M/500 solutions of arabinose, levulose, glucose, galactose, maltose, lactose, saccharose and water. Mix well and immerse in boiling water at the same time and heat for 15 minutes. Note the time when the first sign of reduction of the copper becomes manifest in each tube. A red precipitate of cuprous oxide will be formed and settle to the bottom of the tube.

* 14. Reduction of Barfoed's solution.

Barfoed's reagent is acid, not alkaline like Fehling's. It is prepared (No. 10) by dissolving 13.3 grams of crystallized neutral cupric acetate in 200 c.c. distilled water. This is filtered if necessary and to each 200 c.c. filtrate are added 5 c.c. of a 38 per cent. acetic acid solution.

Determine the speed in which the following solutions reduce 5 c.c. of the reagent furnished. Use 5 c.c. of M/100 solutions of arabinose, glucose, levulose and galactose, also M/100 and M/25 solutions of maltose, lactose and saccharose. Immerse in boiling water and continue heating therein for 25 minutes. Why is the reduction of an acid solution of a cupric salt slower than that of an alkaline solution? Barfoed's solution is sometimes recommended for distinguishing disaccharides from monosaccharides; is this a legitimate recommendation in the light of your experiment?

b. Other reduction tests for carbohydrates.

15. Haines' Solution.—Haines' solution (No. 42) is prepared by dissolving 8.314 grams copper sulphate in 400 c.c. water, adding 40 c.c. glycerin and then 500 c.c. 5 per cent. KOH solution. To 5 c.c. Haines' solution (No. 42) add one drop M/5 glucose solution and heat to boiling.

16. Pavy's, Purdy's and Long's solutions.—These are similar to Fehling's and Haines' solutions, but contain an excess of NH_4OH. To 1 c.c. of a mixture of equal parts of Fehling's solutions No. 1 and No. 2 add 1 c.c. strong NH_4OH (No. 5) and 1 c.c. M/5 glucose solution. Boil for a few seconds.

* 17. **Benedict's solutions for the quantitative volumetric estimation of "reducing" carbohydrates.**—The reagent here employed is made by mixing equal volumes of solutions A, B and C below.

Solution A contains 69.3 grams $CuSO_4$ in 1 liter.

Solution B contains 346 grams sodium potassium tartrate and 200 grams Na_2CO_3 (anhydrous) in 1 liter.

Solution C contains 30 grams K_4FeCN_6, 125 grams KCNS and 100 grams Na_2CO_3 (anhydrous) in 1 liter.

The principle of this method is that in presence of sufficient of the sulphocyanate the white cuprous sulphocyanate is precipitated as soon as reduction of the cupric ions has taken place. To 5 c.c. of the Benedict mixture (heated until free from precipitate) add 1 c.c. M/5 glucose solution and heat again.

* 18. **Reduction of bismuth salts to metallic bismuth (Böttger's, Nylander's and Almen's tests).**—The principle involved here is that an alkaline solution of "reducing" carbohydrate reduces bismuth ions to insoluble black metallic bismuth.

To a pinch of bismuth subnitrate add 1 c.c. N/2 NaOH and 1 c.c. M/5 glucose. Heat to boiling. How do sulphides interfere with this test? How would you remove sulphides?

* 19. **Reduction of mercury salts to metallic mercury. (Knapp's and Sachsse's reactions).**—To 1 c.c. of the potassium mercuric iodide solution (No. 74) add 1 c.c. N/2 NaOH and 1 c.c. M/5 glucose solution. Heat to boiling. Black precipitate.

* 20. **Reduction of silver salts to metallic silver.**—To 1 c.c. of $AgNO_3$ solution (No. 79) add 1 c.c. NH_4OH and 1 c.c. M/5 glucose solution. Warm gently and set aside. Note the mirror formed on standing.

* 21. **Reduction of molybdic acid to molybdous acid.**—To 1 c.c. phosphomolybdic acid solution (No. 66) add 1 c.c. N/2 NaOH and 1 c.c. M/5 glucose solution. Boil about a minute, then acidify with dilute H_2SO_4. The greenish color is due to molybdous acid.

* 22. **Reduction of methylene blue.**—See experiment 8.

For information on the behavior of carbohydrates in alkaline solution and on the use of Barfoed's solution, study the following:—H. McGuigan: *Am. Jour. Physiol.*, Vol. 19 (1907), p. 175; J. Nef: *Annalen der Chemie*, Vol. 357 (1907), p. 214; A. P. Mathews and H. McGuigan: *Am. Jour. Physiol.*, Vol. 19 (1907), p. 199; A. P. Mathews, *Jour. Biol. Chem.*, Vol. 6, p. 3, 1909; Nef: *Annalen der Chemie*, Vol. 403 (1913), p. 204; H. H. Bunzel: *Jour. Biol. Chem.*, 7, p. 157 (1910).

F. FERMENTATION OF CARBOHYDRATES BY YEAST.

* 23. **Action of living yeast on carbohydrates. Formation of carbon dioxide and alcohol.**—Yeast is able to ferment some sugars but not others. The process of fermentation is still obscure in its details,

but yeast owes its power to an enzyme called "zymase," which it contains. It is not improbable that animal tissues also break up carbohydrates in a manner similar to this fermentation; like yeast animal tissues are able to utilize some carbohydrates as foods, but not others, and in general they can utilize the same carbohydrates that yeast can ferment. Probably this fermentation plays an important part in respiration, since one of the products is gaseous.

The experiment may be done by a group of three students, each student testing the fermentability of three sugars. Those working together must select fermentation tubes of the same bore, so that the volumes of gas obtained may be easily compared.

Make a suspension of 3/4 inch cubed piece of compressed yeast in 10 c.c. of water by triturating very gently in a small mortar. To 15 c.c. portions of the carbohydrate solution (DO NOT USE THE TOLUENE PRESERVED SOLUTIONS HERE), add 1 c.c. of the well-mixed suspension, mix well and at once transfer to the fermentation tube. For carbohydrate solutions use M/25 solutions of arabinose, levulose, glucose, mannose, galactose, maltose, saccharose, lactose and 0.70 per cent. starch paste. Prepare the starch paste by the method outlined. Set aside the tubes and note the amount of gas liberated after 6-8, 10-12 and 18-24 hours. Note which of the sugars is fermented. What do the fermented sugars have in common which determines fermentability?

Products of fermentation. CO_2. The remainder of this experiment is to be continued by each student individually. Into a tube in which active fermentation has taken place introduce 1 c.c. strong NaOH solution (No. 86) into the closed end of the tube by means of a bent pipette. Next add enough water to completely fill the shorter end of the tube, place the thumb over the opening without introducing air and invert the tube two or three times without collecting any of the gas under the thumb in the shorter arm of the tube. Set the tube upright and note the suction on the thumb. When the thumb is removed the fluid rises in the closed arm, due to the adsorption of the CO_2 from the fermentation gas. Different kinds of bacteria and yeasts form gas with more or less H_2 intermixed with the CO_2.

Products of fermentation. Alcohol. Filter a portion of the solution in the fermentation tube and add 5-20 drops of I_2KI solution (No. 45) to 5 c.c. thereof. Sufficient I_2KI should be added to render the entire mixture light yellow for about one-half minute. Warm slightly, note the odor of iodoform, and set aside. Examine the separated yellow, iodoform crystals with a microscope. Acetone and some other substances will give the iodoform reaction with iodine.

Does yeast contain invertin? Does it contain lactase?

* 24. Condensation of carbohydrates with hydrazines. Formation

of osazones.—Weigh out 2.0 grams of phenyl hydrazine hydrochloride and 3.0 grams of sodium acetate on the torsion balance. Transfer to a clean mortar and grind together until the mixture is uniform in appearance. Now weigh a 0.5 gram portion into each of four dry test-tubes and to the tubes add respectively and accurately by a pipette 2 c.c. of M/5 solutions of glucose, mannose, levulose and lactose. Label carefully, and immerse all the tubes at the same time in boiling water. At intervals of one-half minute remove all from the bath at the same time, shake for an instant and return to the bath at once. Repeat this three times. Observe if any white crystalline precipitate, hydrazone, appears in any of the tubes. If it does remove a little with a pipette, observe under the microscope and draw some of the crystals. What is a hydrazone? Continue boiling the water for 25 minutes and note the time when precipitation occurs in the different tubes during that period. Now remove all the tubes from the water bath and, while cooling, note the time when a precipitate appears in those tubes which were clear. Examine a little of the precipitate from each tube on a slide under the microscope and draw the crystals of osazones formed. They are yellow in color. Mannose is at first precipitated as a colorless hydrazone, but later changes to the yellow osazone. The other hydrazones are soluble. Write the reactions involved in the experiment. Discuss the osazone formation; its value in identifying carbohydrates; explain the difference in the speed of reaction of the different sugars. Do mannose, glucose and levulose form the same osazone? What does this show as regards the constitution of the last four carbon atoms of these sugars? What is the object of adding the sodium acetate?

* G. METHODS FOR IDENTIFYING AND DETECTING CARBOHYDRATES.

*** a. Furfural reactions for detecting pentoses, hexoses and glycuronates.**

The following reactions all depend on the dehydrating action of strong acids on carbohydrates with the final formation of furfural or furfural derivatives and the interaction of these furfural derivatives with various amido- and hydroxy-aromatic substances to form characteristic colored substances. The pentoses and pentosans yield true furfural readily and almost quantitatively; the methyl pentoses, like rhamnose, yield methyl furfural and the hexoses (most readily the keto-hexoses) yield first 4-hydroxymethyl furfural which, when further acted upon by strong acids, breaks down into levulinic and formic acids. It is also believed by some that part of the 4-hydroxymethyl furfural obtained from hexoses is changed to true furfural in the distillation as usually carried out. It is not at all unlikely that products intermediate between the original carbohydrates and the furfurals are formed and that these reacting with the aromatic substances in presence of the strong acids usually employed give other colors than do the final products, the furfurals. This may account for the differences in the colors developed from time to time in the gradual heating of the carbohydrates with the acids in presence of the aromatic substances.

Numerous aromatic substances have been found to react with carbohydrates and furfural. Among those found most delicate and satisfactory are anilin acetate, alpha naphthol, resorcin, phloroglucin, orcin, diphenylamin and napththoresorcin.

* 25. Alpha naphthol reaction (Molisch reaction).—To the carbohydrate solution in a test-tube add 2 drops of a 5 per cent. solution of alpha naphthol in alcohol. Mix and then allow 5 c.c. concentrated H_2SO_4 to flow down the side of the test-tube so as to keep the solution and acid in separate layers. In some cases it may be necessary to agitate a little so as to have a slight mixing at the junction of the two liquids. Conduct this test on 1 c.c. each of M/50 solutions of levulose, glucose, arabinose, on 1 c.c. of the furfural distillates in experiments 2 and 3, on 0.35 per cent. solutions of starch and gum arabic as well as on distilled water. Note the colors of the rings, then mix the contents of each tube and again note the colors. Tabulate and discuss your results.

A negative result by this reaction is very good evidence of the absence of carbohydrates, but a positive one is simply an indication of the probable presence of carbohydrates. The test is probably of more value if used as modified by E. Pinoff (*Berichte* 38, p. 3308), who found that by heating in test-tubes definite quantities of carbohydrate, alcohol, sulphuric acid and 5 per cent. alcoholic alpha naphthol solution for definite lengths of time one obtains a colored solution, which, when examined for its absorption spectrum, gives characteristic bands with different classes of carbohydrates. Among the substances which give this test are various furfurals, glycuronic acid, glycoaldehyde, glycerose, erythrose, oxygluconic acid and glycerol aldehyde.

* 26. Blank or control tests showing action of acids on carbohydrates without chromogen addition (humic acid formation).—Dilute 15 c.c. concentrated HCl by distilled water to make a total volume of 45 c.c. Mix well and measure off 5 c.c. portions thereof into 7 different test-tubes; to the respective tubes now add 1 c.c. of M/5 solutions of arabinose, glucose, levulose and galactose, also 1 c.c. of the water solutions from the distillations (experiments 2 and 3 above). Mix the contents of each immediately after adding the carbohydrate solution. Immerse all the tubes in boiling water at the same time, continue the boiling

and note the changes. Heat thus for 30 minutes. Record your observations carefully as to time, color and intensity of color. These observations will serve as controls for the furfural reactions below, in which a chromogen, such as resorcin or α naphthol or orcin, is added to develop with furfural a special color.

* 27. **Resorcin HCl test for a ketose. Seliwanoff's reaction.**—Prepare 45 c.c. of the reagent by adding 4.5 c.c. of the 0.5 per cent. resorcin solution (No. 77) to 15 c.c. concentrated HCl and dilute to 45 c.c. with distilled water. Mix well and measure off 5 c.c. portions of this solution into 8 different test-tubes; to the respective tubes now add 1 c.c. of M/5 solutions (*mixing well immediately after the addition*) of arabinose, glucose, levulose and galactose, also 1 c.c. of the water solutions obtained from the distillation experiments 2 and 3 above. Also make one test using M/100 levulose solution. Immerse all the tubes in boiling water at the same time, continue the boiling and observe very carefully the changes taking place during 20 minutes' heating. In the presence of a ketose sugar the solution will quickly turn red and a red precipitate will form. Record the observations and compare with the blank tests made above as to humic acid formation by the action of the same strength of acid on the same concentration of carbohydrates.

This test is often described as a specific test for levulose. It is better to call it an indicative test for ketose, since other keto-hexoses, like sorbose, as well as keto-trioses, keto-tetroses, d-oxygluconic acid and formose give the test readily. All hexoses give the test very faintly and very slowly if proper precautions are observed as to concentrations of the acids and carbohydrates. The color developed here is due to the formation of 4-oxymethyl-furfural and the reaction of this with the resorcin in presence of a 12 per cent. solution of hydrochloric acid. The same reactions are involved in the Boas test for a certain hydrogen ion concentration in gastric juice. Boas' reagent is a solution of cane sugar and resorcin in 95 per cent. alcohol; the presence of a certain hydrogen ion concentration is indicated by the developing of a red color.

The Seliwanoff reaction has been modified in various ways, using lower concentrations of hydrochloric acid, alcoholic solutions of sulphuric acid, or a slightly ionized acid like acetic acid instead of the 12 per cent. hydrochloric acid solution employed above. In any case, due precautions as to length of time of heating and concentration of carbohydrate must be taken to make the reaction of diagnostic value. The test is carried out more satisfactorily when the precipitate, formed after heating, is filtered off, dissolved in 95 per cent. alcohol and this alcoholic solution then examined for its absorption spectrum.

For detailed information read: (1) Seliwanoff: *Berichte* 20 (1887), p. 181; (2) Ekenstein and Blanksma: *Ber.* 43, p. 2355, 1910; (3) Pinoff: *Ber.* 38, p. 3308,

1905; (4) Koenigsfeld: *Biochem. Ztschr.* 38, p. 310, 1912; (5) Ville and Derrien: *Bull. Soc. Chem.* 5, Ser. 4, p. 895, 1909.

***28. Phloroglucin-HCl reaction for pentoses.** (Tollen's phloroglucin reaction.)—Prepare 54 c.c. of the reagent by adding 6 c.c. of the 2 per cent. phloroglucin solution (No. 65) to 30 c.c. concentrated HCl and dilute to 54 c.c. with distilled water. Mix well and measure off five 9 c.c. portions of this solution into five different test-tubes. Then add 1 c.c. of the M/5 levulose, glucose, galactose and arabinose solutions and 1 c.c. of the water solution of the distillate from experiment 2 which contains furfural, to the respective tubes. Mix well immediately after adding the carbohydrate solution in each case. Immerse all the tubes in boiling water for the same time and record the changes in color which occur. Keep in boiling water for several minutes, 10-20. In the pentose tube the solution becomes a cherry red and finally a precipitate forms, which, dissolved in amyl alcohol, shows an absorption band between the D and E lines.

***29. Orcin HCl reaction for pentoses.**—This reaction is on the whole superior to the preceding for the recognition of pentoses. Prepare 54 c.c. of the reagent as in 28, using half-saturated orcin solution (No. 57) in place of the phloroglucin. Use the same sugars as in 28 and proceed in the same manner. Pentoses, when warmed with orcin and HCl, give a violet, followed by blue, red and finally a green color and a bluish green precipitate. This precipitate redissolved in amyl alcohol gives an absorption spectrum with band between C and D.

Note.—In both of the Tollen's reactions above it is very important to observe the concentrations of acid and orcin or phloroglucin, otherwise the velocity of the reaction and the colors obtained are very different from those described in the literature. Various modifications of these tests have been made, the most important being the shaking out of the colored solution with amyl alcohol and the examination of the absorption spectrum of the amyl alcohol solution after proper dilution. When thus employed one can more readily and accurately distinguish pentoses from hexoses; these reactions, however, do not distinguish pentoses from glycuronic acid.

By the use of naphthoresorcin and HCl, Tollens (*Berichte* 41, pp. 1783-87 and 1788-90, 1908) claims to be able to detect glycuronic acid with greater accuracy, in that the colored substances formed by the interaction of glycuronic acid with naphthoresorcin in presence of 18 per cent. HCl are readily soluble in ether, while the pentoses, hexoses and rhamnose yield colored substances difficultly soluble in ether. The ether solution of the former shows a characteristic absorption spectrum. (See experiment 185.)

The following articles deal with the above reactions: v. Udransky: *Zeit. Physiol. Chem.*, 12, pp. 355-95; Roos: *Ibid.*, 15, pp. 518-38; Treupel: *Ibid.*, 16, pp. 47-67;

1896; Fleig: *Ann. Chim. Analyt.* 13, p. 427, 1908.

***30. Detection of galactose. Mucic acid formation.**—Galactose, or polysaccharides containing it, yield, on oxidation with nitric acid, an insoluble, crystalline acid, mucic acid, COOH.CHOH.CHOH.CHOH.CHOH.COOH.

To 10 c.c. M/5 lactose solution in a 100 c.c. beaker, add 5 c.c. distilled water and 15 c.c. concentrated HNO_3. Heat on the water bath until concentrated to about 1/5th of the original volume. Now add 15 c.c. distilled water, mix well and set aside in the desk until the next day. Filter off the crystals of mucic acid, suspend in 10 c.c. water and add 1 c.c. strong NH_4OH (No. 5). Note if the solution has a slimy character. To the solution add 1 c.c. strong HNO_3 (1.42), stir again and set aside for crystallization. Examine the crystals under the microscope. Filter off the crystals, wash in cold water, dry over $CaCl_2$ in desiccator and hand in. Write the formula of the reaction. Glucose, and polysaccharides containing it, give, when similarly treated, an isomeric acid, saccharic acid, which is soluble.

***31 (a). Detection of starch by iodine.**—To 5 c.c. 0.7 per cent. starch solution, 5 c.c. of dextrin solution and 5 c.c. of glycogen solution, respectively, add a few drops of I_2KI solution. Starch gives a blue, dextrin a red and glycogen a brown-red color. Heat and observe the disappearance of the color on heating and its return on cooling. To different portions of the cooled solutions add, a few drops at a time, some alcohol, 10 per cent. NaOH and sodium thiosulphate solution (No. 89). Observe that each one of these reagents discharges the color. Explain. The blue color is due to an iodine starch compound which is dissociated on boiling. Iodine vapors are violet.

31. (b) Preparation of lactose from cow's milk.—To 100 c.c. skimmed milk add 300 c.c. distilled water. Mix well. Remove 40 c.c. of the mixture and add thereto by means of a pipette one drop of 10 per cent. H_2SO_4. Mix the 40 c.c. portion well and note whether a rather coarsely flocculent precipitate forms. In this way continue the addition of the 10 per cent. H_2SO_4 drop by drop until the precipitate separates in a coarse flocculent form. Now add this small portion to the larger volume of diluted milk and add 10 per cent. H_2SO_4 drop by drop in the same proportion as found necessary in the small portion tested. Allow to settle and decant the supernatant liquid into a 20 cm. porcelain dish. Save the precipitated casein and proceed therewith as directed under the preparation of casein (experiment 66). To the supernatant liquid now free from casein particles add saturated $Ba(OH)_2$ solution until it remains only slightly acid to litmus. (Why only slightly acid? Now heat to boiling for a moment to co lactalbumin. What are precipitated here?. Filter the hot s through a large creased filter until clear. Now concentrate the filt about half its original volume by boiling over a flame in a porcelai *but by using a small flame avoid baking the material on the side dish;* finally evaporate on the water bath to about 100 c.c., allow and filter. Again concentrate the filtrate to about 10-15 c.c., a cool and filter if necessary. To the filtrate add very gradually, whi ing, five volumes 95 per cent. alcohol; allow the precipitated lac separate out; decant the alcoholic solution and wash the lactose w per cent. alcohol four times. In this washing process be sure to kn pasty precipitate of lactose very thoroughly. After the final alcohol ing the lactose should no longer be tough and pasty, but hard and Finally wash once in the same way with ether, preferably by gr in a mortar and then adding the ether. (This preparation is impure lactose as proteins and inorganic salts are in part al cipitated by the alcohol.) Dry at room temperature by spreadi on clean filter paper. Then put in vial and label with your name a name of the substance. This and all other preparations are bottled and labeled in the same way and handed to the instructo completed. Test a small portion of your preparation by the M Fehling's and mucic-acid reactions.

QUESTIONS.

1. On boiling nucleic acid with 10 per cent. H_2SO_4 Kossel found levuli in the flask. What conclusion could he draw from this as to the presence or ab carbohydrate in the nucleic acid molecule?

2. On distilling nucleic acid from yeast with 10 per cent. H_2SO_4 h the distillate colored aniline acetate paper a cherry red and gave a blue col orcin. What conclusion could he draw from this as to the structure of the acid molecule?

II. LIPINS.

Neutral fats and oils. Phospholipins. Sterols.

Neutral fats mixed with other lipins may be obtained from and animal tissues both by physical and chemical means. The ph means are by warming and pressing. In this way oils and fats a tained from nuts, seeds and from animal tissues containing a amount of neutral fat, as in fatty tissue. The chemical mea by extraction with lipin solvents such as ether, carbon tetrachl alcohol, gasoline, benzene, etc. Whichever method is used resulting fatty matter is a mixture and generally contains, in ad to neutral fats and fatty acids, phospholipins and sterols. The n fats are separated from the phospholipins most readily by precipi of the ether solution by acetone. Soaps and phospholipins are insoluble in acetone; fats and sterols are soluble in acetone. The separation of cholesterol from neutral fat cannot be made without saponification of the latter and the isolation of cholesterol from the soap by ether.

1. Experiments illustrating the physical properties of fats and oils.

* 32. **Solubility.**—To one drop of olive oil in a test-tube add about 5 c.c. of the solvent. Allow to act in the cold. If solution is not evident from the usual physical changes, filter off the clear solvent and place a few drops on a piece of paper. In case some of the fat has dissolved, a translucent spot (grease spot) is left on evaporation of the solvent. Test in this way the solubility in hot and cold 95 per cent. and 70 per cent. ethyl alcohol; and in cold ether, chloroform, gasoline or naphtha, carbon tetrachloride, ethyl-acetate, water and acetone.

Test the solubility of castor oil in cold alcohol. How does castor oil differ in its composition from olive oil? What effect would this difference probably have on its solubility in alcohol, and why?

* 33. **Crystallization of neutral fats.**

Dissolve a little beef tallow in about 5 c.c. warm 80 per cent. alcohol, and allow to cool. Examine the precipitate which appears under the microscope to see if it is crystalline.

* 34. **Surface tension of neutral oils.**

Count the number of drops which will be formed from 1 c.c. of pure distilled water when issuing from a perfectly clean 1 c.c. pipette. Now dry the same pipette by alcohol and ether, fill it accurately to the mark with olive oil at room temperature and count the number of drops of olive oil delivered from the same pipette. For the relation between the number of drops and surface tension, see page 207. As performed in this way this method is only approximately accurate. The lower the surface tension the larger the number of drops.

* 35. **The surface tension of water is lowered by a small amount of oil.**

After the pipette used in experiment 34 has drained off its oil a very thin layer of oil still remains in it. Without cleaning it now fill it again to the mark with clean water and again count the number of drops issuing from the pipette. The number should be greater than before, for when the surface tension of the water is reduced, the film of water on the surface which supports the drop is no longer able to sustain as great a weight as before, and the drop falls when its weight is less and it is smaller.

* 36. **The surface tension of water is lowered by a very small amount of oil.**

Place in a perfectly clean beaker, that is a beaker which has been cleaned with soap and water, then by alcohol and ether, and then by cleaning fluid and finally thoroughly washed with distilled water, clean distilled water until half full. Sprinkle a few small pieces of camphor on the surface. If the water is perfectly clean the camphor will dart here and there on the surface. While it is moving, rub a glass rod on the side of the nostril and then touch it to the surface of the water. The fat of the skin is sufficient, usually, to add in this way enough oil to the surface of the water to make the camphor at once come to rest. If the camphor has already come to rest it will be observed usually that when the rod touches the surface of the water the camphor fragments move away from the rod in all directions. This is owing to the stretching of the surface due to the lowering of its tension at the point where the oil touches it. To get a clean surface of water in a laboratory is very difficult. There is generally some kind of oily matter in the air sufficient to contaminate the surface.

2. Experiments illustrating the chemical properties of oils and fats.

* 37. Free acid in oils.—Most commercial oils and fats undergo in the light and moisture a partial decomposition (rancidity) so that they have an acid reaction. To get a neutral oil precautions must be taken to separate the free acid thus formed. An experiment to illustrate this fact is the following:

To about 25 c.c. 95 per cent. alcohol in a 100-150 c.c. flask add about ½ c.c. phenolphthalein solution. Heat to boiling on a steam bath and gradually add from a burette N/10 NaOH until a faint pink coloration remains. Now add 5 c.c. olive oil by means of a pipette. Again heat to boiling, stopper the flask and shake gently; the alkaline alcohol becomes colorless. Continue the addition of N/10 NaOH and shaking until after a vigorous shaking the pink color in the alcohol layer just remains. Record the amount of NaOH required. Allow the two layers to separate in a large test-tube. Draw off, by means of a pipette, the oil layer for use as neutral olive oil in a subsequent experiment, and discard the alcoholic layer. Calculate the per cent. of acid in the fat, assuming the acid measured to be oleic acid and the specific gravity of the oil to be 0.90 at room temperature. All fats, even when freshly prepared, contain titratable amounts of acid. The amount of titratable acid increases slightly with development of rancidity. It is very doubtful, however, whether the acidities in the fresh and rancid fats are due to the same acids.

Experiment showing that neutral fats consist of glycerol and fatty acids; that they yield soaps and undergo a process of decomposition called saponification, when treated with alkali.

* 38. Saponification of fat.—Weigh off 10 grams mutton tallow, render it by heating gently and pouring off the melted fat, and transfer to a 500 c.c. flask, add 8 c.c. of the side shelf NaOH solution (40:100)

and 50 c.c. alcohol. Fit with a reflux air condenser, boil gently for an hour on the water bath and then evaporate to about 50 c.c. in a porcelain dish. Add 100 c.c. hot water and again evaporate to about 50 c.c. to get rid of alcohol. Next dissolve completely in about 500 c.c. water. Explain this process of fat decomposition, and write the formula of the reaction.

39. **Separation of the soap by salting out.**—Take 5 c.c. of the above solution, add 50 c.c. of water and saturate with NaCl. Filter off the precipitated soap. Dissolve in about 100 c.c. of water and test a part of the solution with a few drops of $CaCl_2$ and $MgSO_4$ respectively. What precipitates out? Explain the mechanism of the salting out process.

40. **Separation of the fatty acid.**—To the remaining 495 c.c. solution above, add concentrated HCl until the water solution is distinctly acid to litmus paper and a sharp separation of fatty acid occurs on heating on the steam bath. After heating set aside to cool, or draw off the lower layer by means of a separatory funnel, saving it and the upper layer of fatty acid. Wash the fatty acids twice with 100 c.c. hot water. Then remove the fatty acid layer. Dissolve a little in hot alcohol and show that it will neutralize NaOH and is acid to phenolphthalein. See experiment 37.

41. **Separation of glycerol.**—Filter the aqueous layer above and neutralize the filtrate with 10 per cent. NaOH solution. Evaporate to dryness, the final evaporation being carried on on the steam bath. Grind the residue to a fine powder, moisten with alcohol and again evaporate to dryness. Next add 25-50 c.c. alcohol, heat to boiling on the steam bath, filter and treat the undissolved residue once more with 25-50 c.c. alcohol in the same way. Combine the filtrates and evaporate to a syrup in a small dish on the water bath. The residue should be glycerol. It should have a sweet taste and give the acrolein test, as in experiment 42.

* 42. **Acrolein test for glycerol.**—To about 2 drops glycerol in a dry test-tube add an equal quantity of solid $KHSO_4$ (No. 71) or P_2O_5. Heat carefully and gradually, without charring, over a burner and note the penetrating odor of the vapor given off. This characteristic odor is due to the acrolein which is formed by the dehydrating action of P_2O_5 and heat on the glycerol. Perform the same test on the original fat and on the fatty acids obtained in 41. Write in the note-book the structural formulas of acrolein and glycerol.

* 42. (b) **Absorption of iodine by unsaturated fatty acids.**—Dissolve a small amount of the fatty acids in chloroform, then add thereto 2-3 drops of Hubl's iodine solution and shake. The iodine will be absorbed by the fatty acid and the solution decolorized if unsaturated

acids are present. Try this test also on the original fat as well as the fatty acid and make a control test, shaking the chloroform and iodine alone. Explain the reaction in detail. Define the iodine number of fats. Hubl's iodine solution contains 26 grams iodine and 30 grams $HgCl_2$ in 1,000 c.c. 95 per cent. alcohol.

3. Experiments showing how a neutral fat may be identified.

* **43. By the determination of the saponification number.**—Obtain a sample of oil, the saponification number of which is known to the instructor, from the storeroom. Read under general directions as to the use of pipettes. Accurately measure out 5 c.c. of the oil with a clean, dry 5 c.c. graduated pipette and transfer to a clean, dry 250 c.c. Erlenmeyer flask. (If facilities for weighing exist it is better to weigh about 5 grams of oil instead of measuring it. This can be done by placing a small amount of oil with a medicine dropper in a small flask, weighing all accurately, then transferring about 5 c.c. of oil to the Erlenmeyer flask and reweighing the medicine dropper and flask. The difference in the weighings gives the weight of oil taken.) Now add 50 c.c. of the special alcoholic KOH solution, this also to be measured accurately with a dry 25 c.c. pipette. Into another flask of the same kind also measure accurately 50 c.c. of the alcoholic KOH solution with the same 25 c.c. pipette. Fit each flask with a reflux air condenser and boil gently over a wire gauze for 30 minutes; next cool, add about 1 c.c. phenolphthalein solution and titrate with the N/2 H_2SO_4 solution. Titrate the blank flask in the same way.

By saponification number is meant the number of mgs. of KOH necessary to neutralize the fatty acids in 1 gram of fat, phenolphthalein being the indicator. Record your measurements and calculations in some such way as the following: (Specific gr. of oil may be obtained from instructor.)

Oil taken 5 c.c. sp. gr. = ... = ... Grams oil.
Burette readings in titrating the blank ....
Burette readings in titrating the saponified oil ....
c.c N/2 H_2SO_4 equivalent to the KOH used in the saponification ....
c.c. N/2 KOH required to saponify .... grams of oil ....
1 c.c. N/2 KOH contains .... grams of KOH
Saponification number of the oil = ...

44. By the determination of the iodine number.—The iodine number is the per cent. of iodine absorbed by the fat. That is it is the number of centigrams of iodine absorbed by 1 gram of fat. Neutral fats of saturated fatty acids absorb no iodine; but neutral fats, which contain some unsaturated acids, will absorb iodine at the unsaturated bond, two atoms of iodine being absorbed to each double bond.

The solutions needed in this determination are made up and titrated by the instructors and the titration values of the thiosulphate in mgs. of iodine and of the Wijs solution (iodine dissolved in iodine trichloride, ICl_3) are marked on the bottles. In accurate work the oil or fat must be weighed, but in the class work it may be measured by a pipette and the weight calculated from the specific gravity. (See page 66.)

Weigh a dry 100 c.c. flask. Add to it about 0.28 c.c. of olive oil and weigh again carefully so as to obtain the amount of oil used. Dissolve the oil in 10 c.c. of carbon tetrachloride and add after solution 25 c.c. of Wijs' iodine solution, stopper the flask and allow it to stand in the dark for from one to two hours. Then pour the contents of the flask without loss into a 750 c.c. Erlenmeyer flask, adding to the small flask 10 c.c. of KI solution to wash out the iodine left in it and wash the potassium iodide quantitatively into the 750 c.c. flask. The total volume of fluid should be about 300 c.c. Now titrate the iodine solution with the standard thiosulphate solution, using starch as the indicator. Do not add the starch until the amount of iodine remaining is sufficient to give only a light-brown color. The calculation should be recorded as follows:

Thiosulphate used in the titration .... c.c.
1 c.c. thiosulphate equals .... mgs. iodine
Amount of iodine not absorbed by the fat = ... x .... = ... mgs.
Amount of iodine in 25 c.c. Wijs solution .... mgs.
Mgs. of iodine absorbed by the fat = ...
Weight of fat taken = ...
Iodine number of fat = ... = ...

44 (b). *Standard thiosulphate solution for the iodine value of fats.* Dissolve 48 grams of sodium thiosulphate in 2 liters of water. Allow to stand about 24 hours and then standardize it according to Vollhard as follows: Dissolve 3.8657 grams of potassium bichromate in water and make up to 1 liter. 10 c.c. of 10 per cent. KI solution, 5 c.c. strong HCl and exactly 20 c.c. of the bichromate solution are mixed in a 750 c.c. Erlenmeyer flask and then diluted with about 300 c.c. of water. This amount of bichromate solution when acidified with HCl liberates exactly 0.2 gram of iodine from the KI. Now determine by titration how many c.c. of the thiosulphate solution it takes to reduce exactly this amount of iodine. To do this run in from a burette carefully some of the thiosulphate until the iodide solution is only faintly yellow from the iodine, and then add a little starch solution. Now carefully add the thiosulphate until the blue color of the starch is just lost. The number of c.c. of thiosulphate equivalent to 0.2 gram iodine is thus determined. From this, by division, the amount of iodine equivalent to 1 c.c. thiosulphate may be calculated.

44 (c). *Iodine solution of Wijs.* This is an iodine monochloride solution made as follows: Weigh into a 300 c.c. flask 9.4 grams of iodine

trichloride and then pour in about 200 c.c. of glacial
flask with a cork carrying a $CaCl_2$ tube and hea
till all is dissolved. Rub 7.2 grams iodine to a fine
and wash it with glacial acetic acid into a similar f
the same manner to solution. Pour the contents
stoppered graduated liter flask, add glacial acetic a
iodine and reheat until all the iodine has been diss
the liter flask. Stopper, allow to cool, make up ex
acetic acid and titrate on the following day by t
phate solution. To do this measure exactly with a p
the solution into a large Erlenmeyer flask, add 10
solution and dilute with 200 or 400 c.c. water.
burette the thiosulphate until pale yellow, then a
add thiosulphate carefully to the discharge of t
strength of the iodine chloride solution may alter
week, but thereafter, Leathes says, it remains very
some weeks. Restandardize from time to time. Tl
should be twice recrystallized and the mother liq
the crystals before use. Care must be taken to pr
of water.

Lewkowitsch modifies Wijs solution by adding t
trichloride in molecular proportions: 7.9 grams IC

44 (d). *von Hubl's iodine solution.* Dissolve
in 500 c.c. 95 per cent. alcohol mixed the day bef
an equal quantity of a solution of 30 grams mercuri
of 96 per cent. alcohol. Waller recommends addin
to 1 liter of the mixed iodine and mercuric chloride
its titer better as it prevents the reaction of the IC
H_2O=IOH+HCl). Von Hubl's solution gives l
iodine number in linseed oil and cholesterol than W

45. The emulsification of fats by soaps.
soaps.—In this experiment, use the oil layer, desc
ment 37, as the neutral oil; before using for the te
the water bath until it has become clear. For
use the olive oil on the side shelf. Prepare

Tube	Oil added	Other addition	State
a	3 drops rancid	5 c.c. H_2O	
b	3 " neutral	" " "	
c	3 " rancid	5 c.c. H_2O + 0.3 c.c. N/10 NaOH	
d	3 " neutral	" " " " " " "	
e	3 " rancid	5 c.c. 0.05% Soap Sol.	

Tube	Oil added	Other addition	State of emulsion after 30 minutes
f	3 drops neutral	5 c.c. 0.05% Soap Sol.	
g	3 " rancid	5 c.c. 10% Egg White	
h	3 " neutral	" " " " "	
i	3 " rancid	5 c.c. 10% Gum Arabic	
j	3 " neutral	" " " " "	

Shake all tubes equally thoroughly. Note the appearance immediately after shaking and 30 minutes later.

Write a brief explanation of the manner in which soaps emulsify fats. See page 222. How does soap reduce the surface tension at the boundary of the oil and the fat?

5. Experiments illustrating the method of obtaining and the physical and chemical properties of phospholipins and glycolipins.

* 46. Preparation of a glycolipin, cerebrin. (Phrenosin and kerasin mixture.)—Obtain 200 grams finely chopped calves', pigs' or sheep brains at the storeroom. Transfer to a 10-12-inch porcelain dish and gradually add, while stirring well, 200 c.c. 95 per cent. alcohol. Next transfer the mixture to a 750-1,000 c.c. flask; add 200 c.c. 95 per cent. alcohol, mix, fit flask with a reflux condenser, heat on the steam bath for ¾ hour, agitating the flask from time to time. Filter hot, through a dry, creased filter paper, into a dry flask, removing as much of the liquid as possible from the undissolved tissue. Again add 250 c.c. 95 per cent. alcohol to the residue which has been returned to the flask, boil on the steam bath as before and filter into the same filtrate as before. Return the residue to the flask and extract again with 250 c.c. fresh alcohol. Combine the three filtrates and set aside in a covered beaker in a cool place until the next day to permit of crystallization. Save the residue, return it to the flask and add 250 c.c. ether, allow to extract at room temperature for several hours, filter off the ether and extract once more with the same amount of ether. Save the residue for experiment 68, p. 893. Unite and save ether filtrates. The hot alcohol takes out cholesterol, lecithin, and the cerebrosides, besides other substances. The ether will take out an additional amount of cephalin.

Separation of cerebrin. The next day filter off the cold alcohol from the white crystalline precipitate which has separated out, using the Buchner suction funnel. The filter paper in the funnel must be first moistened with water, suction applied and then the water removed from the paper with alcohol. The filtrate contains phospholipin and may be united with the ether extract. The white precipitate consists of some phospholipin, cerebrin and some cholesterol and other substances. It is sometimes called impure protagon. Transfer the white material to a small beaker and dissolve in 100 c.c. hot methyl or ethyl alcohol. Filter hot. Allow the hot, clear solution to cool slowly and thoroughly. Impure cerebrin and cholesterol crystallize out. Filter off, adding the filtrate to the ether extract above. Wash the crystals in ether. Cholesterol dissolves in ether and will go into solution, while cerebrin is insoluble. Add the ether washings to the ether extract. Recrystallize the cerebrin crystals once more from hot methyl alcohol. The result is an impure phrenosin and kerasin (cerebroside). To remove the last traces of phospholipin it would be necessary to recrystallize in hot glacial acetic acid. The following experiments can be tried with the cerebrin thus obtained.

* 47. Hydrolysis of cerebrosides (glycolipin).—The object of this experiment is to show that the glycolipin from brain is hydrolyzed by acid and yields galactose.

To about 0.5 gram cerebrin in a 100-150 c.c. flask add about 50 c.c. distilled water and 0.5 c.c. con. H_2SO_4 (measured by a 5 c.c. graduated cylinder). Then proceed with the hydrolysis as in the hydrolysis of a polysaccharide, experiment 7. Do fatty acids form on the surface? Remove the sulphuric acid in the same way. Test a portion of the clear water solution by Fehling's solution in the usual way. A reduction should be obtained. Test another small portion for phosphoric acid. Evaporate the remainder to about 5 c.c. and proceed with the mucic acid test as given in experiment 30.

* 48. Preparation of phospholipins (phosphatides).—Concentrate the united ether extracts and cold alcoholic filtrates from experiment 46 in a porcelain dish on the water bath. Evaporate to the consistence of a salve or paste. Avoid overheating or baking. Add 25 c.c. 95 per cent. alcohol, moistening all the solid substance in the dish therewith. Try to redissolve the material as much as possible so as to cover less space in the dish. Again evaporate as above to a thick syrup. Again take up in 95 per cent. alcohol and repeat the evaporation. This is to remove the water. Allow to cool and immediately on cooling extract the residue in the dish three times, using 20 c.c. ether each time. Decant the ether off each time into a large test-tube. Be very careful to avoid getting water into the material in the dish and be sure to break up the lumps as much as possible while treating with ether. The milky ether extracts, combined in a large test-tube, stoppered, are set aside in a cool place to settle. After a day or two the clear supernatant liquid is poured into a 300 c.c. beaker and then, while stirring well, three volumes of anhydrous acetone added. Be sure to use acetone which has not been allowed to absorb moisture by standing exposed to the air or by being measured with vessels which contain water or alcohol. The precipitated phospholipin (lecithin, cephalin and various myelins) is

allowed to settle out and the supernatant acetone-ether solution poured into a 10-12-inch porcelain dish and allowed to evaporate spontaneously for the cholesterol preparation. (Experiment 50.) The phospholipin precipitate is now kneaded with a horn spoon or a glass rod until all the particles are collected into one mass. Knead this mass well with two 15 c.c. portions of acetone, adding the acetone washings to the acetone-ether solution, evaporating for cholesterol. The phosphatide preparation may be kept on a watch glass in a desiccator. Examine a portion of it to determine its composition as follows:

*** 49. Tests for glycerol, fatty acids and organic phosphoric acid in phospholipin.**—Take a portion of phosphatide about equal in bulk to a sphere with a diameter of one-eighth inch, place in dry test-tube and cover with a little P_2O_5 and heat gently in the flame, smelling from time to time as soon as white fumes come off from the fused phosphatide. Note the irritating, penetrating odor of acrolein, showing the presence of glycerol. Take another portion of phospholipin of the same amount, dissolve this in a test-tube in about 6 c.c. 95 per cent. alcohol. Heat to boiling in a steam bath and add an equal quantity of concentrated HCl and heat on the steam bath for 20 minutes. Transfer to a 100 c.c. beaker and evaporate until the alcohol has been boiled off. Then shake the residue in 20 c.c. ether in a separatory funnel, separate the ether layer from the other and evaporate both separately on the steam bath until the ether and HCl have been removed.

Note the character of the material, fatty acid, left from the ether. Dissolve this in the least quantity of dilute NaOH. Acidify again and heat to boiling. What separates?

The residue from the evaporation of the aqueous solution above should be tested for phosphoric acid. It is dissolved in a little water, transferred to a test-tube, most of the water evaporated and 10 drops of concentrated H_2SO_4 added to destroy organic matter. Heat in the flame until charred, then add one drop of concentrated HNO_3 and again heat until charred. *Be very careful to hold the mouth of the test-tube directed away from you when you add the nitric acid; in applying this test to fatty substances or to substances containing glycerol, alcohol or ether one must be very cautious, as a violent reaction is likely to take place at the beginning of digestion.* Repeat the addition of one drop of nitric acid and the heating until the material no longer chars on heating sufficiently to give off white sulphuric-acid fumes. Allow to cool, then carefully add about 5 c.c. distilled water and strong NH_4OH (No. 5) until faintly alkaline. Now again add just enough concentrated HNO_3 (one drop at a time) to render acid. To this solution warmed to about 70° C. add a few c.c. ammonium molybdate solution (No. 6). Agitate with a stirring rod. If phosphorus is present in the original substance it will be precipitated here as the yellow, ammonium phosphomolybdate. Write a graphic formula for lecithin.

* 50. **Preparation of cholesterol.**—To the residue left on evaporating the acetone-ether solution decanted above, add 25 c.c. 95 per cent. alcohol and 5 c.c. strong NaOH solution (No. 86). Mix well, transfer to a 150 c.c. Erlenmeyer flask, using 10 c.c. more 95 per cent. alcohol to rinse the dish with. Fit the flask with a reflux air condenser and boil on the water bath for 30 minutes. Then evaporate in a 10 cm. dish to remove most of the alcohol; neutralize to litmus by gradually adding, while stirring, concentrated HCl. Cool and extract with three 20 c.c. portions of ether. Allow the combined *clear* ether extracts to evaporate spontaneously to a small volume in a 100 c.c. beaker. The crystals of cholesterol may be recrystallized from 95 per cent. alcohol. They should be colorless plates.

* 51. **Cholesterol reactions.**—(a) IODINE-SULPHURIC-ACID TEST. Prepare a mixture of 5 c.c. concentrated H_2SO_4 + 1 c.c. water. Cool and add a few small white crystals of cholesterol. Allow to act at room temperature until the edges of the crystals are colored pink. Then add 3 drops of a *dilute* iodine solution (1 vol. No. 45 + 10 vol. water), agitate and note the colors developing. Gradually add 5 to 10 drops more of the same iodine solution, agitating after each addition. Observe under microscope.

(b) SALKOWSKI'S REACTION. Dissolve a small crystal of cholesterol in about 1 c.c. chloroform in a *dry* test-tube. To this solution at room temperature add an equal volume concentrated H_2SO_4. Do not agitate at once, but note the red color of the upper chloroform layer and the fluorescence of the sulphuric-acid layer after standing a minute or so. Agitate to mix the two layers and set aside. Note the intense cherry-red coloration of the upper layer on separating into two layers.

(c) LIEBERMANN-BURCHARD REACTION. In a *dry* test-tube dissolve a few small crystals of cholesterol in about 2 c.c. chloroform. Next add 15 drops acetic anhydride, mix and then add concentrated H_2SO_4, drop by drop, 2-5 drops. Shake the contents of the tube. Allow to act further without heating and note the changes in color. A deep blue color develops.

III. PROTEINS.

* 52. **Elementary composition of the proteins.**—Carefully heat a small amount of dried casein or other dried protein in a dry test-tube until the characteristic odor of burning hair is developed. Note the drops of water on the sides of the tube. The presence of this water, if the protein has been thoroughly dried at 100°, means that proteins contain hydrogen. The charred black mass shows the presence of carbon.

To show the presence of nitrogen now add about 1 c.c. of a mixture of equal parts of powdered magnesium metal and dry sodium carbonate. Again heat carefully until the material has burst into a flame. Now at once dip the end of the tube into 10 c.c. of distilled water in a mortar. The end of the tube should be broken by the water. Break up the material with a pestle and filter. To the filtrate now add a small crystal of ferrous sulphate and boil about one-half minute. By the heating with the magnesium, and by dry distillation of proteins, hydrocyanic acid is formed. The ferrous sulphate unites with the sodium cyanide to make ferrocyanide. Now add to the solution concentrated HCl until just acid. If the dark-blue ferric-ferrocyanide is not formed at once, add a drop of ferric chloride. The presence of nitrogen in proteins is shown by the formation of the cyanide. Many proteins contain also sulphur in an unoxidized form, which may be detected by the formation of lead sulphide as in experiment 53 K.

*53. **Methods of detecting proteins in solutions or solids. Color reactions of the proteins.**—For these experiments take portions of blood serum which has been diluted 10 times with water; or egg white, diluted with five volumes of water; or a 2 per cent. solution of commercial peptone.

**A. Biuret (Piotrowski's) reaction.* To 2 c.c. portions of the solutions of blood albumin, egg white and commercial peptone add 2 c.c. 10 per cent. NaOH and then one drop of a 0.5 per cent. $CuSO_4$ solution. Mix well and add more $CuSO_4$, drop by drop, until a distinct pink or violet color is developed or until a precipitate of $Cu(OH)_2$ is formed. If the rose or violet color which constitutes the biuret reaction does not come at once, allow to stand 15-20 minutes.

This reaction is interfered with if much ammonium salt is present. In its presence it is necessary to add a large amount of strong NaOH to get rid of the influence of the NH_4OH. If NaOH is added to a solution containing an ammonium salt, such as $(NH_4)_2SO_4$, sodium sulphate will be formed and ammonium hydrate. The ammonium hydrate is a very weak base, particularly in the presence of ammonium salts; it combines with the cupric ions, removing them from the solution, so that it is necessary to add NaOH until nearly the whole of the ammonium sulphate or other ammonium salt has been decomposed. To show this add to the peptone test above, drop by drop, a saturated solution of ammonium sulphate until the blue color of the ammonium cupric salt replaces the pink color, due to the biuret reaction. Now add the strong 40 per cent. NaOH (No. 86) until the pink color returns.

Read on page 144.

Repeat this test with biuret. Make the biuret from urea as follows:

**A. 1. Formation of biuret and cyanuric acid from urea.* Carefully heat over a low flame a layer of urea about one-eighth inch deep in a dry test-tube. The urea melts, then the liquid boils and finally a [illegible] solid remains on the sides and bottom of the tube. During the [illegible] note the odor of the vapor (ammonia) escaping from the tube. [illegible] cooling add about 5 c.c. of water, warm slightly and decant from [illegible] undissolved residue into another tube and then make the biuret test in the aqueous solution. A pink color should be obtained. Biuret is formed according to the following reaction:

$$\underset{\text{Urea.}}{NH_2{-}CO{-}NH_2} + \underset{\text{Urea.}}{NH_2{-}CO{-}NH_2} \longrightarrow \underset{\text{Biuret.}}{NH_2{-}CO{-}NH{-}CO{-}NH_2} + NH_3$$

The undissolved residue left in the tube is mainly cyanuric [illegible] Its barium salt is insoluble. To the white residue add about 1 [illegible] NH_4OH and warm. To the clear solution then add a few drops of $BaCl_2$ solution (No. 11) and acidify with HCl. The precipitate is barium cyanurate.

$$\text{Cyanuric acid is } C_3H_3N_3O_3\text{, or } \begin{array}{l} HO{-}C = N - C{-}OH \\ \quad\;\; | \qquad\quad \| \\ \quad\;\; N = C - N \\ \qquad\quad\; | \\ \qquad\quad OH \end{array}$$

Write all the reactions showing the various stages in the formation [illegible] biuret and cyanuric acid. Write the lactam formula for cyanuric acid (see page 718).

* *B. Xanthoproteic reaction* (see page 147). This reaction, the yellow protein reaction, is given both by solid and dissolved protein, but is most delicate when applied to the solid protein. To a small amount of the solid substance or the solution add about 1 c.c. concentrated nitric acid and boil until dissolved. Cool. Note the lemon-yellow color which develops in case protein is present containing a benzene nucleus. Cool the solution and add strong NaOH solution until the solution is slightly alkaline. The color should change to a deeper orange if the test is positive. See if solid gelatin, casein, solutions of the proteins, one drop of benzene, one drop of toluene, a few drops of 2 per cent. phenol solution and a little salicylic acid give the same reaction. This reaction is not given by proteins which lack the aromatic nucleus. Explain the reaction (see page 148). What amino-acids found in the protein molecule give this reaction?

* *C. Millon's reaction.* For the composition of Millon's reagent see page 147. When this test is applied to solutions one adds a few drops of the reagent to four or five c.c. of the solution and then heats to boiling. Generally there is first formed a white precipitate which turns pink or red slowly on heating. In case a pink or red color does not

develop it is best to add a few more drops of the reagent and again heat to boiling and allow to stand. The amount of reagent to be added varies considerably, depending on the concentration and nature of the solution. Try this reaction on 2 c.c. of the 2 per cent. gelatin and casein solutions, and on solid gelatin and casein and on the albumin solutions. Test also 2 drops of 2 per cent. phenol solution added to 3 c.c. of water and test also a few crystals of resorcin, phloroglucine, thymol, vanilline and tyrosine. For an explanation of the reaction see page 147.

a. The reaction is interfered with by the presence of sodium chloride. Repeat the phenol or any of the other reactions which have been positive, but add to the solution some solid NaCl first. It will be found necessary to add much more Millon's solution before a red color is developed if NaCl is present.

b. What groups in the protein molecule will give this reaction?

* *D. Tryptophane reaction. (Adamkiewicz reaction.)* To 5 c.c. glacial acetic acid add 3-10 drops of the 2 per cent. protein solution, mix well and pour concentrated H_2SO_4 down the side of the tube so that it forms a layer underneath the acetic acid. A violet ring should develop at the plane of contact if the test is positive. This generally develops after standing a few minutes. After it appears, or if the test is negative, agitate the tube so as to mix the sulphuric acid and acetic. The whole solution may become violet.

Try this test on gelatin and casein in particular. Is gelatin positive or negative? If the test is negative, it is well to repeat it by dissolving some of the dry protein in glacial acetic acid by warming, then cooling and adding the H_2SO_4. A violet ring or a violet solution should be obtained on mixing if the test is positive. If tryptophane and indole can be procured, repeat this test on small quantities of these substances or their solutions. For explanation of the reaction see page 148. What proteins will not give this reaction?

* *E. Glyoxylic acid reaction (Hopkins-Cole reaction).* To about 2 c.c. of glyoxylic acid reagent (No. 38) add 2 c.c. of the 2 per cent. protein solutions, or of the protein solutions to be tested, mix and pour concentrated H_2SO_4 down the side of the tube. A purple-violet ring at the plane of contact is a positive reaction. Repeat, mixing the contents of the tube. This reaction does not always come immediately, so if negative allow it to stand 15 minutes. If still negative, repeat, using a little of the solid protein matter dissolved in glyoxylic acid and sulphuric.

The glyoxylic-acid reagent is made in the following way:

In a 500 c.c. flask place 10 grams powdered magnesium, cover with distilled water and slowly add 250 c.c. of saturated oxalic-acid solution, cooling from time to time under the tap. Filter off the magnesium oxalate, acidify slightly with acetic acid, dilute to one liter and keep in a stoppered bottle, with a little chloroform added. The solution contains oxalic acid and glyoxylic acid, CHO.COOH.

This reaction, like the preceding, is for tryptophane. See page 148.

Cole states that this reaction is interfered with by the presence of chlorides in excess, and in the presence of chlorates, nitrates and nitrites. It is also important to use pure sulphuric acid.

• *F. Acree-Rosenheim formaldehyde reaction.* To 2 c.c. of the 2 per cent. solution add 3 drops of formaldehyde solution 1 : 5,000 (No. 33). Mix and pour concentrated sulphuric acid down the side of the tube, as in the preceding. Note the purple ring which forms after standing about 5 minutes if the test is positive. This is also a tryptophane reaction. For explanation see page 149. A trace of iron is necessary.

• *G. Liebermann's reaction.* This test can be made in two forms. Boil the solid protein with 5 c.c. concentrated HCl for about a minute with a few drops of dilute saccharose solution. A violet color develops if the test is positive. Test in this way dry egg albumin, casein and gelatin.

In the other way, treat the same proteins by boiling with alcohol and ether in the water bath first, pour off the alcohol and ether before boiling with concentrated HCl and do not add the saccharose. The solid protein often takes a beautiful blue coloration if it does not dissolve, and the solution becomes violet if it does dissolve. For an explanation of these reactions read page 149. In the first case the aldehyde is supplied by the decomposition of the saccharose; in the latter case it is present already in the alcohol or ether or both, so that the addition of glyoxylic acid or formaldehyde is unnecessary. Sometimes carbohydrate is already present in the protein molecule and it is only necessary to boil with hydrochloric acid to develop the violet color.

• *H. Ehrlich's p-dimethyl-amino-benzaldehyde reaction.* In place of formaldehyde, or glyoxylic acid or the aldehydes developed from the carbohydrates by the action of strong acids, we may use also aromatic aldehydes for the detection of tryptophane. Among these the p-dimethyl-amino-benzaldehyde is often used. This reagent has the formula:

```
      C—COH
     //  \
   HC     CH
    |     ||
   HC     CH
     \\  /
       C
       |
       N(CH3)2
```

To 1 c.c. of the protein solution add an equal volume of concentrated HCl and boil for one-half minute; note the color developed, if any (Liebermann reaction). Then add two drops of the 5 per cent. solution of p-dimethyl-amino-benzaldehyde in 10 per cent. H_2SO_4 (No. 59), mix and again note the color. A red to violet color develops in case tryptophane or other indole derivatives are present. A few drops of the 0.5 per cent. $NaNO_2$ (No. 87) solution added now changes the color to a blue. Another method of making this test is described on page 919. This reaction is also used for the detection of indole substances in the urine.

I. Ehrlich's diazo reaction. Diazo-benzene-sulphonic acid. This is a histidine and tyrosine reaction. The formula of diazo-benzene-sulphonic acid is $C_6H_5N_2SO_3H$. Take about 1 c.c. of the 0.5 per cent. sulphanilic-acid solution in 2 per cent. HCl (No. 91), add an equal volume of the 0.5 per cent. $NaNO_2$ (No. 87) solution. Mix well and after about one-half minute add 1 c.c. of the 1 per cent. protein solution. Again mix and then add enough NH_4OH or Na_2CO_3 to make the mixture distinctly alkaline. Histidine gives a red to orange color and tyrosine gives an orange color, but less intense. Tyrosine when converted into the benzoyl derivative no longer gives the test, but benzoyl histidine does (K. Inouye, *Zeits. f. physiol. Chem.*, 83, 1913, p. 79). Carry out this test with gelatin and casein solutions, with distilled water and with tyrosine and histidine, if they can be provided by the instructor.

• *J. Molisch reaction.* See the directions under carbohydrates. Try this test on solutions of casein, egg albumin, blood proteins, gelatin and Witte's peptone. Note which are positive. A positive reaction is an indication of the presence of carbohydrate in the protein molecule or solution.

• *K. Reduced sulphur reaction.* To the protein solution, and for this test it is better to take a fairly concentrated solution of the protein, add four volumes of 10 per cent. NaOH and boil for a few moments, 1-2 minutes. Then add a few drops, 3-10, of lead-acetate solution. A brown color or a black precipitate formed on the addition of Pb acetate shows the presence of unoxidized sulphur in the molecule. It is split off as the sulphide and precipitated as sulphide of lead. Test also gelatin, casein and Witte's peptone solution by this test. Which are positive? Has casein sulphur in it? If the instructor can supply some cystine, repeat this test with a little cystine dissolved in sodium hydrate containing some lead acetate.

• *Ninhydrin (triketohydrindenehydrate) reaction* (Ruhemann, *Trans. Chem. Soc.*, 97, p. 2025, 1910; Abderhalden and Schmidt, *Zeit. f. phys. Chem.*, 72, p. 37, 1911). Ninhydrin is triketohydrindenehydrate. This reaction is one of the most sensitive of the protein reactions. It is positive with proteins, peptones and amino-acids, with the exception of those which have an imino instead of an amino group. One carboxyl and one α-amino group must be free for protein or peptide to give a positive test. The reaction is made as follows: 0.1 gr. of the reagent is dissolved in 30-40 c.c. of water. One or two drops of this solution is added to 1 c.c. of the solution to be tested and heated a short time boiling. On cooling a more or less intense blue color develops, if test is positive. It is necessary that the fluid should be neutral in tion. If acid, the color is more red-violet and the reaction is retarded alkali. See page 150. The solution of ninhydrin does not keep well, so it is well to make it up in small amounts. It comes in the trade in 0.1 gram vials. In using this test for the presence of amino-acids dialyzate Abderhalden recommends that it be made as follows: To c.c. of the dialyzate 0.2 c.c. of the 1 per cent. solution of ninhydrin added. A boiling stick is then added and the solution boiled for exactly one minute from the appearance of the first air bubble on the side the test-tube. A control should be run at the same time, the control having no substance in it giving the test. A blue color develops on ing if the test is positive (Abderhalden, *Abwehrfermente des tierischen Organismus*, Berlin, 1912). Ninhydrin is a trade name for triketo-hydrindenehydrate.

* 54. **Determination of the amount of nitrogen in protein bodies by the Kjeldahl method.**—Follow the directions for the determination of nitrogen by the Arnold-Gunning modification of the Kjeldahl method. Determine the nitrogen in a sample of protein material given you the instructor. When you have made the determination, compare your results with the correct values to be obtained from the instructor. The sample given you is probably not a pure simple protein, but a mixture; possibly a ground-up, alcohol-extracted, dried tissue.

Procure a weighing tube of the unknown substance from the store-room. Transfer the contents to a clean, dry 75-100 c.c. beaker and dry in an oven at 100-105° C. for 1 hour; then remove from the oven, break up any caked masses which may have formed, transfer the finely divided material to the weighing tube and again dry for 1 hour at 100-105° C. In the meantime, if your desiccator is not already in good condition clean and dry same, then transfer thereto fresh granular $CaCl_2$ from the storeroom; also apply a very small amount of vaseline to the desiccator cover to make it tight. After drying in an oven for 1 hour, cool the loosely stoppered tube in the desiccator. Next weigh the tube and contents carefully on an analytical balance, then transfer, without loss, about 0.5 gram of powder to a clean, dry, 500 to 800 c.c. Kjeldahl flask, stopper the tube and weigh again carefully. Keep all these records in your note-book in ink. With a fine jet of pure water wash down the powder adhering to the neck of the flask (use as little water as possible). Add 5 grams potassium sulphate, or 10 grams crystalline Na_2SO_4 20 c.c. pure concentrated H_2SO_4 (Sp. G. 1.84) and 10 c.c. of the $HgSO_4+CuSO_4$ solution (containing 1 gram $HgSO_4$ and 1 gram $CuSO_4$ per 10 c.c.).

Heat on the digestion shelf until the water has been removed, then add 15 grams more of potassium sulphate or 30 of $Na_2SO_4.10H_2O$ and continue the digestion for 2½-3 hours. To have the digestion complete, the mixture must, however, have been boiling for at least 2½ hours and the hot solution left must not be colored yellow, but simply green, due to the $CuSO_4$. After the digestion is complete, remove from the digestion shelf and just as the contents of the flask begin to crystallize gradually add, while agitating 250 c.c. distilled water; stopper loosely and set aside until it is to be treated as indicated below.

While your digestion is under way, clean out the distilling apparatus as follows: To a clean 800 c.c. Kjeldahl flask transfer 300 c.c. distilled water, add a knife point of granular zinc, connect with the proper distilling bulb and distill about 150 c.c. into a clean receiver. Throw this away and then proceed with the distillation of the ammonia as follows:

Having cleaned and prepared the distilling apparatus, measure off exactly 25 c.c. $N/2H_2SO_4$ into a 250-300 c.c. conical (Erlenmeyer) flask by means of a *burette*. Place the flask so that the glass adapter end of the distilling apparatus is just sealed by the acid. Dry the Kjeldahl flask on the outside and carefully wipe the first inch of the inside of the neck of the flask with a piece of filter paper (this is done to insure a good tight joint with the rubber stopper later on). Add a small knife point of granular zinc to the contents of the flask to insure quiet ebullition and just before adding the alkali and steadily (not so rapidly as to mix the liquids) pour down the side of the neck of the flask (avoid moistening the first inch of the neck) 100 c.c. of the special $NaOH+K_2S$ solution in such a way that the two solutions form distinct layers. Now rapidly connect the flask with the distilling apparatus. In making the connection do not try to twist the stopper into the flask, but hold the stopper firmly and twist the flask on to the stopper. Now start the burner with a medium-sized flame and at once rotate the flask to insure *thorough mixing* of the contents. After the first 5-10 minutes' heating the liquid can be boiled rather vigorously until about 150 c.c. have distilled over. Then lift the end of the adapter out of the distillate by lowering the receiving conical flask and continue the distilling for about 5 minutes longer. Wash off the tip of the adapter with a jet of distilled water, the washings going into the distillate. Then remove the conical flask and *turn off the gas last of all.* Titrate the distillate with n/2 NaOH, using congo red as indicator. Make a correction of about 0.5 c.c. N/2 solution for the ammonia in a blank test. Calculate the per cent. of nitrogen in the dry substance. This has been determined to be about the correction necessary for an ordinary distillation carried out in this way. In accurate work it is necessary to make a blank distillation with the reagents used (H_2SO_4, etc.) to determine the amount of ammonia in them. 1 c.c. $N/2H_2SO_4$ = 1 c.c. N/2 NH_4OH. 1 c.c. N/2 NH_4OH contains .007 gram N.

The preparation and properties of various kinds of proteins.

I. Simple, coagulable proteins: albumins and globulins.

Albumins and globulins. Preparation. Methods. For this work we shall use blood serum which contains both these proteins. To obtain blood serum the ox blood is collected in pans. It clots and then the clot shrinks, squeezing out a clear, yellowish-colored liquid called the serum. This serum has in it two proteins, or groups of proteins, called serum globulin and serum albumin.

Obtain from the storeroom 100 c.c. of the serum. Test its reaction to litmus paper and to phenolphthalein. It is usually alkaline to litmus and acid to phenolphthalein. Test its specific gravity by means of a specific gravity bulb. For this, place the serum in a measuring cylinder and immerse a clean urinometer in the serum. Read the specific gravity from the neck of the urinometer. Record the results in your note-book for future reference.

* 55. Precipitation of the globulins by dialysis.—The serum contains salts (test a little for chlorides and phosphates) and these salts hold the globulin in solution. A portion of the globulin is precipitated if these salts are removed. Place 100 c.c. of the serum in a parchment dialyzing tube, examining the tube first to see that it does not leak, and dialyze the serum for 24 hours against running tap water. At the end of that time remove the tube, notice if there is in it a deposit of protein and pour the contents into a clean beaker. A portion of the protein should now be in suspension in the liquid. This portion is called *euglobulin*. Remove this by filtration, through a small folded filter. Save both the filtrate and the precipitate. After filtration take some of the precipitate, suspend it in distilled water. It will not dissolve. Now add to the suspension a crystal of sodium chloride. Both salt and euglobulin dissolve. This illustrates the fact that globulins are insoluble in distilled water, but soluble in neutral salts. Test the solution by the biuret test to make sure that the precipitate was protein in nature, and boil some after adding 1 drop of acetic acid to see that it is coagulated.

* 56. The precipitation of globulins by salts.—Globulins are not only insoluble in water; they are more or less completely precipitated by half saturating their solution with ammonium sulphate, and some are precipitated by saturation with sodium chloride. Fibrinogen is one of these, but serum globulin is not thus precipitated by NaCl. The filtrate from (a), that is the dialyzed serum, still contains some globulin, indeed the greater part, because there are substances in the serum (phospho-

lipins) which hold the globulin in solution in the absence of salt. The fact that this globulin is there can be shown by salting it out. To show this, dilute the filtrate with an equal amount of distilled water, and then add to the diluted serum an equal volume of saturated ammonium sulphate solution (No. 7). A precipitate of globulin appears and it is complete after standing for some hours. The solution may, if necessary, be left covered with a watch glass until the next morning, but not longer, since concentration of the sulphate by evaporation will precipitate some of the albumin. Filter off the precipitate of serum globulin, and wash the precipitate twice with half-saturated $(NH_4)_2SO_4$, saving both precipitate and filtrate. The filtrate contains serum albumin and is used in 58. Test the precipitate as follows:

57. Dissolve the precipitate on the filter with some distilled water. There is generally enough salt in the precipitate to bring it into solution. Make a biuret test in a little of the solution. Boil about 3 c.c. of the remainder. It should coagulate. Keep the remainder of the solution for comparison with the albumin solution.

* 58. **Preparation of serum albumin.**—The filtrate from the half-saturated, diluted blood serum in 56 contains still some protein, serum albumin. To separate this, saturate the solution by the addition of powdered ammonium sulphate. Saturating with ammonium sulphate in a faintly acid solution removes all proteins from solution, except the peptones. To the saturated solution add enough 10 per cent. acetic acid to have the resulting solution contain 1 per cent. acetic acid. In saturating a solution with a solid substance this substance must be added in a finely powdered form in small quantities at a time with very frequent (best continual) stirring or shaking until considerable of the solid remains undissolved. Each 100 c.c. of half-saturated $(NH_4)_2SO_4$ solution will dissolve about 35 grams $(NH_4)_2SO_4$. Allow to stand for 2 to 24 hours, then filter through a creased filter and allow to drain as well as possible. Transfer the precipitate to 50 to 100 c.c. distilled water, neutralize toward litmus by the addition of the 10 per cent. Na_2CO_3 solution and then dialyze until free from sulphates. Observe the same precautions in making the test here as above in the preparation of serum globulin. Remove the solution from the tube to a clean beaker. Test 3 c.c. of the solution by the biuret test and another 3 c.c. heat to boiling. If it does not coagulate, take another 3 c.c., add to it a little ammonium-sulphate solution, or a crystal of the solid, and repeat.

59. Having made the above tests, secure solid serum globulin and albumin in a coagulated form by precipitating the solutions with alcohol. To do this pour each solution into an equal volume of 95 per cent. ethyl alcohol. If the solutions are very poor in salts, the alcohol does not readily precipitate the proteins, but a milky solution may result. If

this happens, add to the alcohol 1 c.c. of 10 per cent. acetic acid to 100 c.c. of alcohol used. Allow to settle, filter the white precipitate through a creased filter, or through a small suction filter, drain well and either dry with suction, using alcohol and ether, or spread the moist precipitate in a thin layer on a watch glass and dry in a desiccator. Test the solubility of the dry powders in water and dilute salt. They will not dissolve. They have been coagulated, changed to *metaproteins* by the action of the alcohol. Test the two preparations for carbohydrate, tryptophane, oxyphenyl and unoxidized sulphur groups. Which substance gives the better tryptophane and Millon reactions?

* 60. **Heat coagulation of the proteins.**—For the study of the conditions governing heat coagulation, either egg white diluted with a double volume of distilled water or blood serum similarly diluted may be used. To show that not all proteins are coagulated by heat, a 2 per cent. albumose or gelatin or casein solution should be heated under similar conditions. Many of the proteins are denatured by heat. That is, they are rendered insoluble metaproteins. There are two distinct changes involved in this process: a chemical and a physical. The chemical change will occur in the absence of salt or electrolytes, but the physical change, the agglomeration into a coagulum, will only happen if electrolytes are present. The chemical change occurs best if the solution is very faintly acid, or neutral. The nature of the chemical change involved is unknown. If the solution of a protein is dialyzed or diluted with distilled water so that it contains no electrolyte, or very little, then if it is heated the solution generally becomes slightly opalescent, but there is no precipitation. If, however, salt is added to the previously heated but cooled solution, the protein now precipitates, although it would not precipitate before heating. The coagulation and precipitation are most complete if both salt and very little acid are present. If the reaction is very alkaline, no precipitation occurs, even though the salt be present, but the protein is converted by heating into a soluble metaprotein. If it is more than very faintly acid, also, the same result will be obtained. Both of these metaproteins, acid and alkali, will be precipitated if the solution is made exactly neutral. The dependence of the coagulation on salts, acids and alkalies is shown in the following experiments:

* 1. Heat to boiling 5 c.c. of the undiluted egg albumin and blood serum, after assuring yourself that they are faintly alkaline to litmus. They will be found to coagulate, although faintly alkaline.

* 2. Repeat, using the diluted serum and egg albumin. It will be found that the solution coagulates very imperfectly. Now add to the opalescent tubes some powdered, solid, sodium chloride, about the amount on a knife point. On standing the proteins are precipitated.

This shows that heating has changed the proteins and made them more easily precipitated by salt, although it has not coagulated them.

• 3. Now take four tubes containing each 5 c.c. of the diluted serum, or egg white, add to (1) a couple of crystals, about a centigram, of solid sodium chloride; to (2) add two drops of 10 per cent. acetic acid; to (3) add two drops of 10 per cent. NaOH; to (4) add a c.g. of NaCl and two drops of 10 per cent. acetic acid; immerse all four in boiling water and note which coagulate best, and how the coagulation compares with that in (b). Number 4 will generally coagulate most firmly and completely.

After heating and cooling neutralize (3) exactly with acetic acid. A precipitate of metaprotein, alkali albumin, is now obtained at the neutral point.

4. Is water involved in heat coagulation? Place a small amount of finely divided dry egg albumin into each of two *dry* test-tubes. To the one add 5 c.c. of a 1 per cent. NaCl solution. Immerse both tubes in a bath of boiling water and heat thus for 10 minutes, agitating frequently. Allow to cool and then add 5 c.c. of a 1 per cent. NaCl to the tube containing the dry egg albumin and allow action to take place for 10 minutes at room temperature, with frequent agitation. Next decant the clear liquid from each tube or filter if necessary and perform the biuret test in exactly the same way on equal volumes of each solution. To another portion of each solution add 1 drop 10 per cent. acetic acid and heat to boiling. Do not conclude that prolonged heating of the dry protein will give the same results, as the solubilities of heat-coagulable proteins are distinctly altered by heating the dry proteins at 100° C. for periods of one hour or longer. Similar changes are brought about more slowly at room temperature.

61. Precipitation by salts of heavy metals.—As most of the heavy metal ions are precipitated by alkaline solutions without the presence of other substances, it is necessary to make careful control tests when adding such solutions to alkaline protein solutions and to compare the amount and character of the precipitates obtained. Furthermore, as acids alone precipitate some of the proteins, a second set of control tests is necessary in such cases.

Make the following comparative tests. Add the solutions gradually and shake well after each addition.

	A	B	C	D
Distilled water	5 c.c.	4 c.c.	4 c.c.	4 c.c.
2% $FeCl_3$ solution	5 drops	0	5 drops	5 drops
Phenolphthalein solution	1 drop	1 drop	1 drop	1 drop
Protein solution	0	1 c.c. egg white	1 c.c. egg w.	1 c.c. peptone

Make each tube now very faintly alkaline by adding 1 per cent. NaOH solution and observe and record what happens; then add a little more alkali to make it distinctly alkaline and record what happens; then make each tube very faintly acid by adding 10 per cent. HCl, observe and then make distinctly acid with HCl. Does the $FeCl_3$ precipitate in acid or alkaline solution best? Observe the amount and character of precipitate obtained in each case.

Repeat the same tests as above, using 5 drops 2 per cent. lead acetate, 10 drops 1 per cent. mercuric chloride and 5 drops $CuSO_4$ (Fehling's No. 1) as precipitants, and as acids use 10 per cent. acetic, 10 per cent. HCl and 10 per cent. H_2SO_4 respectively. Discuss and explain your results.

62. Precipitation by various anions (alkaloidal reagents).—Here only one control test is necessary (A, below). Proceed as follows:

	A	B	C	D
Water	2 c.c.	1 c.c.	1 c.c.	1 c.c.
1% K_4FeCN_6	0	1 c.c.	1 c.c.	1 c.c.
Phenolphthalein .	1 drop	1 drop	1 drop	1 drop
Protein solution .	1 c.c. egg white	1 c.c. egg white	1 c.c. peptone	1 c.c. gelatin

Add in succession to each tube as in 61:

10% NaOH solution to render distinctly alkaline.
10% H_2SO_4 " " " very faintly acid.
10% " " " " distinctly acid.

Repeat the above using one protein only and instead of the 1 per cent. K_4FeCN_6 solution the following reagents: 2 per cent. phosphotungstic acid, 2 per cent. phosphomolybdic acid, saturated picric acid and iodine in potassium iodide.

II. Preparation of crystalline globulins, edestin and excelsin, and various types of proteins.

63. Preparation of edestin.—Grind 25 grams hemp seed to a fairly fine powder, transfer to a 250-500 c.c. flask and add 200 c.c. of a 5 per cent. NaCl solution, previously heated to 60° C. Do not heat above 65° or the edestin will be coagulated. Keep immersed in a bath at 60-65° C. for ½ to 1 hour with very frequent stirring. Prepare a creased filter, and just before beginning the filtration moisten the paper with 5 per cent. NaCl solution heated to 70° C. Filter the extraction until about 100-150 c.c. have been collected. Warm the filtrate to 60-65° C. and a milky solution should remain. Now cool rapidly by immersing and agitating in *cold* tap water or in ice water. Allow the crystals to settle out and decant the supernatant liquid. Filter off the

white or grayish solid, redissolve in the smallest volume of 5 per cent. NaCl solution kept at 60-65° C. and allow to cool very gradually by keeping the container immersed in a large bath kept at 60-65° C. and then allow bath and all to cool gradually. The next day, or when cooled to room temperature, examine the crystals with a microscope, filter off the main bulk of the crystals and wash with 95 per cent. alcohol. Spread out on a watch glass and dry in a desiccator. Test portions of the filtrate by boiling, by adding an equal volume of saturated $(NH_4)_2SO_4$ solution and by adding a few drops 1.0 per cent. acetic acid. Test a part of your preparation for "reduced" sulphur, for histidine and for tryptophane.

64. **Preparation of excelsin.**—Excelsin, a globulin from Brazil nuts, is very easily obtained crystalline (see Osborne, *The Plant Proteins*, 1909). The ground-up nuts are freed from fat by extraction with gasoline, or petroleum ether, or benzene. The dried powdered residues are then extracted with 10 per cent. NaCl. The excelsin is dissolved. Filter.

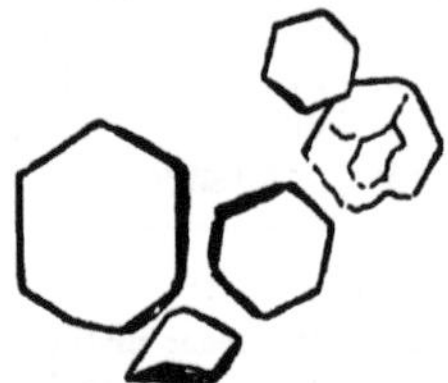

FIG. 68.—Crystals of excelsin.

Dialyze the filtrate against water. The excelsin will often crystallize out in the dialyzer. Crystals are small, hexagonal plates. Redissolve the precipitate from the dialysis tube in dilute ammonium sulphate and precipitate with an equal volume of saturated ammonium sulphate. Filter, redissolve in dilute ammonium sulphate and dialyze. The excelsin crystallizes out in the dialyzer. Figure 68.

65. **Preparation and properties of a prolamine. Gliadin.**

Preparation of gliadin. To 100 grams wheat flour gradually add enough water to make a thick dough. Then knead in the hand in a stream of cold water until all the starch is washed out. Now cut the protein into small pieces with a knife or scissors and extract twice (½ hour each time) with 200 c.c. portions of boiling 70 per cent. alcohol, by boiling in a flask on the steam bath. Filter hot each time and evaporate the alcoholic filtrates to about ¼ of the original volume, then allow to cool and add while stirring 10 c.c. 10 per cent. NaCl solution. Allow to settle out, filter off and dehydrate with cold 95 per cent. alcohol in the usual way.

Carry out the xanthoproteic, biuret, Millon and glyoxylic-acid tests on your preparation.

*** 66. Preparation and properties of a phosphoprotein, casein.**—To the crude casein precipitated in experiment 31 (b) above add 30 c.c. water, stir well and gradually add of a 5 per cent. NaOH solution until dissolved. Avoid an excess here. What is in solution now? Now add enough water to make the total volume about 400 c.c. Precipitate again by adding 10 per cent. acetic acid, allow to settle out, decant, redissolve the casein in the same way and reprecipitate twice in the same way. Wash twice by decantation with water, transfer to a 7 cm. Buchner funnel and suck dry by filter pump. Now transfer the casein to a mortar and macerate with 95 per cent. alcohol, filter through a 7 cm. Buchner funnel and repeat this alcohol treatment twice, removing the casein from the funnel each time and macerating thoroughly so as to remove the water as thoroughly as possible. Finally treat twice in the same way with ether. Remove the casein from the paper and spread out in a thin layer in a 3-inch watch glass. When perfectly dry and free from ether, bottle as directed above under lactose.

Organic phosphate test. Take a small amount of your casein in a test-tube, add ten drops concentrated H_2SO_4 and heat until charred, then add *one* drop of concentrated HNO_3 and again heat until charred. *Be very careful to hold the mouth of the test-tube away from you; in applying this test to fatty substances or to substances containing glycerol, alcohol or ether one must be very cautious, as a violent reaction is likely to take place at the beginning of the digestion.* Repeat the addition of *one* drop nitric acid and the heating until the material no longer chars on heating sufficiently to give off white sulphuric-acid fumes. Allow to cool, *then carefully* add about 5 c.c. distilled water and strong NH_4OH (No. 5) until faintly alkaline. Now again add just enough concentrated HNO_3 (one drop at a time) to render acid. To this solution warmed to about 70° C. add a few c.c. ammonium molybdate solution (No. 6). Agitate with a stirring rod. If phosphorus is present in the original substance, it will be precipitated here as the yellow ammonium phosphomolybdate.

*** 67. Preparation and properties of the nucleoproteins. Nucleic acid.**—Nucleic acid is, as its name implies, the acid of the nucleus. It constitutes the part of the chromatin of the nuclei which has an affinity for basic stains. It is, hence, in that part of the nucleus which appears as chromosomes in cell division, and it is this substance which the histologist by his basic dyes follows in his study of inheritance and mitosis. Nucleic acid has been described on page 163. It yields, it will be remembered, phosphoric acid, purine and pyrimidine bases and a carbohydrate group. Its structural formula is that indicated on page

170. The nucleic acid of yeast has a pentose in it, d-ribose. That of the thymus gland has a hexose sugar. Both are polynucleotides.

Preparation of nucleic acid from compressed yeast. To 150 grams finely divided compressed yeast in a liter beaker add 450 c.c. of a 1.5 per cent. NaOH solution. Stir well while heating on a boiling water bath and digest thus for 30 minutes. *Stir frequently.* While still hot filter through a creased filter in a 6-inch funnel. To the *cooled* filtrate obtained after 1 hour's filtration add glacial acetic acid, drop by drop, until faintly acid to litmus. Stir well and again at once filter through a creased filter. Now make this filtrate slightly alkaline by the addition of a 10 per cent. NaOH solution. Concentrate on the water bath in a porcelain dish to about one-half of the original volume. *Cool this solution* and again add glacial acetic acid until just acid to litmus. Filter rapidly again if necessary and then pour the solution, *while stirring,* into two volumes *cold* 95 per cent. alcohol containing 1 c.c. concentrated HCl per 100 c.c. alcohol. Transfer the milky solution at once to 1-inch test-tubes and allow to settle out therein. Decant the supernatant alcoholic layer from the separated nucleic acid as soon as possible and transfer all of this solid material to one of the tubes, using in all 50 c.c. 95 per cent. alcohol in so doing. Mix this well and again allow to settle out; decant the supernatant alcohol. Again add 50 c.c. 95 per cent. alcohol to the precipitate, mix well and allow to settle out. Repeat this treatment with 95 per cent. alcohol twice. Finally gradually add 50 c.c. ether to the precipitate in the tube, mix well, allow to settle out and decant as before. Again add 25 c.c. ether, mix well and transfer rapidly on to a suction filter of hardened filter paper moistened with alcohol. Wash twice on the filter with 25 c.c. portions of ether. Suck dry by suction. When dry remove from the paper and expose on a watch glass until all ether is removed. Note the solubility of your product in water, dilute acids and bases. Note that it does not coagulate on heating. Carry out the biuret test on your preparation. Heat a small amount of your product with 10 per cent. H_2SO_4 for a minute or two, then cool and apply the α-naphthol test. Test a small amount also for organic phosphorus, as given under casein, experiment 66. *Observe the precautions in the digestion with nitric acid.*

*** 68. Determination of phosphorus quantitatively in nucleoproteins.**—All cells contain nucleoproteins and also phosphoproteins, so that, if the residue after exhausting the tissue with alcohol and ether is examined, it will be found to contain organically bound phosphoric acid. Determine what per cent. of phosphorus there is in the brain residue after complete extraction by alcohol and ether. Obtain the sample of the tissue from the storeroom. This is the residue from completely extracted sheep's brains or other tissues and the phosphorus content have been determined by the instructor. The Pemberton-Neumann method is sufficiently accurate if all conditions are carefully observed. It is more rapid than the magnesium-phosphate method, but not so reliable. In practice it must be frequently controlled by gravimetric determinations. The point of difficulty seems to be in getting the molybdate-phosphoric-acid precipitate to have always the same composition.

Weigh off into a 300 c.c. Kjeldahl flask 0.4 to 0.5 gram of the substance previously dried as directed in the Kjeldahl method. To this then add 5 c.c. concentrated H_2SO_4 and heat under a hood over a free flame until well charred. From a separatory funnel, furnished for the purpose, now gradually add, *drop* by *drop*, concentrated HNO_3 until the char has disappeared. CAUTION! IN HEATING AFTER ADDING THE NITRIC ACID, OR WHILE ADDING THE ACID, DO NOT HOLD THE MOUTH OF THE FLASK TOWARD YOUR FACE! Heat again carefully until the nitric oxide fumes are boiled off. The mass will char again. Repeat in the same way the *gradual* addition of the nitric acid and the heating until, after the last addition and heating, white fumes of sulphuric acid appear without charring of the liquid contents of the flask. Heat on the digestion shelf for about 10 minutes longer. Then allow to cool. Explain the above process. What other method is used for decomposing the organic matter and retaining the phosphorus as phosphate?

Next transfer, without loss, the contents of the flask to a 250 c.c. beaker and also rinse the flask 5 times with 15 c.c. portions distilled water, adding each washing to the beaker. Gradually add, while stirring, concentrated NH_4OH until the solution is very slightly alkaline to litmus, then add sufficient HNO_3 to make the solution *slightly* acid. To this solution now add 15 grams NH_4NO_3 and heat to 65° C.

To 30 c.c. ammonium molybdate solution (see below) add 1.5 c.c. concentrated HNO_3 and filter, then add this to the above solution while stirring and keep in the bath at 65° for 15 minutes.

Filter this with suction through a prepared asbestos filter (p. 996) and wash beaker and precipitate three times with 10 per cent. HNO_3. Follow this by 5 washings with 2 per cent. NH_4NO_3 solution. All of this manipulation must be done neatly and without loss. See that your filtrates are clear.

Transfer the precipitate and asbestos by means of a glass rod from your funnel to a beaker in which the precipitation was conducted. Add exactly 20 c.c. N/2 NaOH to the mass in the beaker, stir with a glass rod and remove the adhering precipitate from the funnel by means of a drop of the mixture in the beaker. Finally rinse the funnel with a jet of distilled water, allowing the washings to flow into the beaker. Stir until all is dissolved and add enough water to make about 100 c.c. volume. Add phenolphthalein solution (1 c.c.) and titrate with N/2 H_2SO_4. Calculate the per cent. of phosphorus in the dry substance. The precipitate obtained above, if the instruction as to time, temperature and other conditions of precipitation are followed, is of the formula $(NH_4)_3PO_4.12MoO_3$. Each cubic centimeter of the N/2NaOH is equivalent to 0.674 milligram phosphorus. Explain the titration above.

a. The ammonium molybdate solution used in the above method is made as follows: Dissolve 100 grams of molybdic acid in 144 c.c. of ammonium hydrate, sp. g. 0.90, and 271 c.c. water; slowly and with constant stirring, pour the solution thus obtained into the mixture of 489 c.c. of nitric acid (sp. gr. 1.42) and 1,149 c.c. of water contained in a large porcelain dish. Keep the mixture in a warm place for several days, or until a portion heated to 40° deposits no yellow sediment, and preserve in glass-stoppered bottles. Before using a portion of this solution for the precipitation of PO_4, add 5 c.c. con. nitric acid (sp. gr. 1.42) to each 100 c.c., mix well and filter.

The solution may also be made by mixing 489 c.c. HNO_3 (sp. gr. 1.42) and 1,149 c.c. H_2O in a large porcelain dish (18-inch) and then pouring into this while stirring a solution made as follows: powder 121 grams ammonium molybdate, dissolve in 355 c.c. water, stirring frequently to hasten solution and then add 60 c.c. strong ammonium hydrate (sp. gr. 0.90). This solution is then poured *into* the nitric acid as indicated above.

*** 69. Preparation and properties of the secondary-derived proteins.**—In this group are the products of hydrolytic cleavage known as albumoses, or proteoses and peptones. These substances, it will be recalled, are soluble in water, not coagulated by heat, and all but the peptones are separated by saturating their solutions with ammonium sulphate. They may be prepared either by the hydrolytic action of sulphuric or other acids, or by digesting proteins with pepsin-hydrochloric acid. For the experiments which follow obtain the materials from the storeroom. Witte's peptone is a commercial product which consists chiefly of proto- and deutero-proteose. There is relatively little peptone in it. Armour's peptone consists chiefly of peptone, with little of the albumose.

* A. *Albumoses.* Obtain 5 grams of Witte's peptone and dissolve it in 100 c.c. distilled water by heating, slightly acidify with acetic acid and filter.

1. Take about 1 c.c. for each test and add about twice the volume of water and test the solutions for tryptophane, tyrosine, unoxidized sulphur and by the biuret and xanthoproteic and Molisch test. Record your observations.

2. To separate the various albumoses mix the filtrate in a beaker with an equal volume of saturated $(NH_4)_2SO_4$ (No. 7). A precipitate forms which sticks to the rod on stirring. Remove this from the solution. It is *primary* albumose and consists of a mixture of proto- and hetero-albumose. Save the filtrate which contains the secondary albumoses.

3. *Proceed with the primary albumoses.* Dissolve them in a little water.

a. Heat to boiling a portion of the solution. It does not coagulate.

b. Acidify slightly with acetic acid and add a drop of ferrocyanide of potassium. A precipitate is formed. Distinction from secondary albumoses.

c. Add a drop of copper-sulphate solution. A precipitate forms.

d. Add a few drops of concentrated nitric acid. A precipitate forms which dissolves on heating.

e. Add some tannic acid. A precipitate forms.

f. Add some lead acetate. A precipitate forms.

g. Add a few drops of sodium or potassium bichromate. A precipitate forms if the solution is acid, but not if it is neutral.

h. Make a test for unoxidized sulphur.

*4. *Secondary albumoses. Deutero-albumoses. Thio-albumose, synalbumose, etc,* To the filtrate from the separation of the primary albumoses add an equal volume of saturated ammonium sulphate in a beaker and stir. Thio-albumose (and possibly other albumoses) are precipitated. Remove the precipitate; save the filtrate.

Precipitate. Thio-albumose. Dissolve the precipitate in a little water and make the test for unoxidized sulphur. It should be particularly strong. Hence this is called thio-albumose. Repeat the tests a, b, c, d of the preceding experiment with this albumose.

Snyalbumose, etc. The filtrate or the solution after the separation of the thio-albumose is saturated while warm and stirring constantly by the addition of powdered ammonium sulphate. The remainder of the deutero-albumoses separate out and generally stick to the rod or the sides of the beaker. Pour off and keep the solution. Dissolve the precipitate of deutero-albumose in distilled water, and test small portions as follows:

a. Heat to boiling. No coagulation.

b. Biuret test. Pink and positive.

c. Acidify slightly with acetic acid and add a drop or two of potassium ferrocyanide. No precipitate. Difference from protoproteose.

d. Add a few drops of copper sulphate. No precipitate.

e. Add a few drops of nitric acid. No precipitate forms.

f. Add a few drops of tannic acid. A precipitate forms.

g. Add a few drops of lead acetate or mercuric chloride. A precipitate forms.

h. Acidify and add a few drops of potassium-bichromate solution. A precipitate forms.

i. Make a test for unoxidized sulphur, tryptophane and tyrosine. How do they compare with the similar tests of the other albumoses?

* B. *Peptone.* There is still left in the solution which has been saturated with ammonium sulphate some substances which give a biuret test. Test some of the solution for these substances. Use 40 per cent. NaOH after a drop or two of $CuSO_4$, since in the presence of so much ammonium sulphate a large amount of NaOH must be used. A pink biuret test shows the probable presence of peptone. To isolate the peptone is a matter of difficulty.

Peptone. Its properties. To study the properties of peptone take two grams of Armour's peptone from the storeroom and dissolve in 50 c.c. distilled water.

a. Dilute a small portion and make in it the usual protein reactions.

b. Saturate 5 c.c. with powdered $(NH_4)_2SO_4$. There will be little or no precipitate showing the absence of all albumoses.

c. Add a few drops of picric acid. There will be little precipitate showing the absence of albumoses.

d. Add a few drops of ferrocyanide to the solution slightly acidified with acetic acid. It is negative.

e. Add to the solution three volumes of alcohol. A precipitate shows that the peptones are not soluble in strong alcohol.

f. Add a few drops of lead acetate. A white precipitate. From these reactions it appears that the peptones differ from the albumoses in their solubility in saturated ammonium sulphate. They are still precipitated by lead acetate and tannic acid, and these reagents are among the best methods of separating the proteins completely from a solution.

70. Albuminoid reactions.—*A. Gelatin.* Gelatin is in reality a derived protein, being formed by the partial hydrolysis of the substance, collagen, found in white connective tissue and in the ground substance of cartilage and bone. A very pure gelatin may be obtained, also, from the scales of fishes after the removal of the other albuminoid, icthylepidin. Take some of the 2 per cent. solution of gelatin and make with it the following tests:

a. The various protein reactions. Observe whether the sulphur, tyrosine and tryptophane reactions are positive or negative.

b. Obtain some pieces of gelatin from the storeroom. Test the solubility of some small pieces in cold water. They swell but do not dissolve. Heat. They go into solution. Cool the tube under the tap. If the solution is concentrated enough, it will set or gel. This is the best test for gelatin. The connective tissues of the invertebrates do not yield any substance which will set or gel.

b. If the solution is very dilute, say 1 per cent., the solution will not gel. Make such a solution or obtain some of the stock 2 per cent. solution and make with it the following tests:

c. The various protein reactions. Observe whether the sulphur, tyrosine and tryptophane reactions are positive or negative.

d. Is gelatin precipitated by half saturating the solution with ammonium sulphate? By lead acetate? By tannic acid? By ferrocyanic acid? By picric acid?

e. If boiled with acid gelatin loses its power of gelatinizing. Make some of the solution distinctly acid by adding to 5 c.c. of a 10 per cent. solution 1 c.c. of 10 per cent. HCl, heat to boiling for 2 minutes and then cool under the tap. If it does not set, neutralize with 10 per cent. NaOH and test its setting powers. If it sets, reacidify and heat again for 2 minutes.

SALIVA.

Collect some saliva by chewing paraffine and expectorating into a beaker. Collect about 50 c.c. of filtered saliva.

*71. With this saliva carry out the following experiments:

a. Reaction. Test its reaction to litmus, phenolphthalein and congo red. About what is the concentration of hydrogen ions in saliva? See page 322.

b. Mucin. Precipitate 20 c.c. filtered saliva by adding 4 volumes of 95 per cent. alcohol. Allow to settle, then filter through a 7 cm. filter and identify the mucin in the precipitate by the following tests:

1. Dissolve the precipitate as fully as possible in a little water. To a portion of the solution add, drop by drop, dilute acetic acid. A stringy precipitate insoluble in acetic acid but soluble in 0.1 per cent. HCl indicates mucin. In a little dilute sodium-carbonate solution the precipitate from the acetic acid redissolves to a slimy solution.

2. Make the biuret, Molisch and Millon tests with the solution. The positive Molisch test shows the presence of carbohydrate in the molecule. Mucin is a glycoprotein.

c. Digestive action. Saliva has the property of dissolving starch and converting it to maltose, a reducing sugar. The active principle, or enzyme, which causes the transformation of the starch is called "*ptyalin*." For a fuller account of this substance and the conditions of its activity see page 331. To illustrate this action and to study the conditions of activity of ptyalin, prepare about 100 c.c. of starch paste by the method given on page 854.

72. Raw starch is not digested by ptyalin, except very slowly.—Place a little raw starch, as much as can be held on the point of a knife, in 5 c.c. of water in a test-tube and add 1 c.c. of the filtered saliva. Mix and allow the two to act together for 30 minutes or longer. At the end of that time test the reducing action of a portion of the solution by Fehling's solution as indicated in experiment 13. Compare the result with the action on boiled starch. The failure of the enzyme to digest the uncooked starch is due to the fact that the outside of the grain of starch is either cellulose or some other carbohydrate not easily penetrated by the ptyalin. If this is broken by chewing, etc., then the starch digests. To show this take some of the same kind of starch used in the preceding experiment, chew it thoroughly for a few moments, collect it then in a test-tube, dilute with a little water and test for a reducing sugar. The test should be positive.

*** 73. Starch paste is digested with great speed by ptyalin.**—Take 5 c.c. of the 1 per cent. starch paste in a test-tube, add about 1 c.c. of the filtered saliva, mix thoroughly by inverting the tube with the thumb held over the end. Then at once heat the solution to boiling in the Bunsen burner in order to stop the action of the ptyalin. Notice the very rapid clearing of the starch solution when the saliva is added, if the saliva is active. Now take about 3 c.c. of the solution, add 3 c.c. of the Fehling mixture and boil. A copious reduction to the red cuprous oxide is obtained if the saliva is normally active. This shows that saliva forms a reducing sugar from the starch at the very start of its action. Try to see how short the time of contact of the saliva and starch can be and still result in the formation of a reducing sugar, maltose. The action of the saliva can be stopped most easily and instantaneously by making the mixture fairly alkaline with NaOH. The great speed of the action on cooked starch contrasts strongly with the slowness of digestion of uncooked.

*** 74. Formation of dextrins from starch.**—Place on a white porcelain plate several drops of KI_3 solution. Now place in a test-tube 10 c.c. of starch paste and 2 c.c. of saliva, mix well and place in the water bath at 38-40° C. From time to time remove a drop on a glass rod and touch it to a drop of KI_3 on the plate. At first the color will be pure blue, that of starch itself. Gradually the color will change to a violet, then to a reddish or reddish brown and finally the red color will be given more and more faintly and ultimately it will disappear. The time when this happens should be noted. It is the *achromic point*. The time taken to reach this point under similar conditions of experimentation may be taken as a measure of the digestive strength of the saliva. The red color is due to erythrodextrin, or red dextrin, the colorless solution still contains a colloidal dextrin, achroodextrin.

* 75. **The action of the saliva is lost by boiling.—Heat some sali** to boiling and then mix it with starch paste under the same conditi as in experiment 74. From time to time test the iodine reaction s see if any reducing substance (maltose) appears in the mixture. T fact that the property of the saliva is lost by heating shows that t substance which is active is unstable, heat labile and, since a [illegible] it converts a great deal of starch, it is called an enzyme.

76. Conditions of activity of the ptyalin.—The digestion of stard by saliva goes fastest in very weak acid media. It is stopped by [illegible] acid and alkali. It is shortened by the addition of a little NaCl. [illegible] this experiment use diluted saliva. To 5 c.c. of fresh saliva add 45 [illegible] of distilled water and mix well. If this dilution takes too long to [illegible] to the achromic point, use with it a more dilute starch solution. [illegible] should not take much over 20 minutes at 40° to reach the achrom point.

Action of bases, acids and salts on the activity. Measure out the [illegible] lowing into test-tubes and determine the length of time for each to [illegible] to the achromic point. Add the solutions in the order given below:

Tube	Starch paste	Water	Other addition	[illegible]
1	2 c.c.	2 c.c.	0	2 c.c.
2	"	I "	1 c.c. N/10 HCl	"
3	"	1.5 "	0.5 " " "	"
4	"	1.9 "	0.1 " " "	
5	"	1.0 "	1.0 " " Na_2CO_3	"
6	"	1.0 "	1.0 " " NaCl	"
7	3 c.c. of A below + 3 c.c. of B below			
8	3 " " C " + 3 " " D "			

A. ...... 4 c.c. starch paste + 2 c.c. N/10 Na_2CO_3 kept at 40° C for 15 minut
B. ...... " saliva (diluted) + 2 " " HCl " " " " " "
C. .. . " starch paste + 2 " " " " " " " " "
D. ...... " saliva (diluted) + 2 " " Na_2CO_3 " " " " " "

Explain the results and their significance in connection with natur digestive processes.

77. Excretion of salts in saliva.—The saliva contains many sul stances which are adventitious, excretory substances. Iodides pa readily into the saliva. By swallowing iodides and noting the time which the iodine reappears in the saliva, an idea may be had of th motor activity of the stomach, since the absorption is largely from th intestine. The iodides must be passed through the stomach before the are reabsorbed.

GASTRIC DIGESTION.

On account of the difficulty of obtaining sufficient quantities of gas tric juice, the experiments which follow will be made with artifici gastric juice. In making this juice it is best, if the mucous membranes of the hog's stomach can be obtained, to make the juice by extracting the dried or undried mucous membrane with 0.4 per cent. HCl. A large amount of the mucous membranes can be dried on glass plates, then extracted with gasoline and ground and kept on hand. If mucous membranes cannot be obtained, a similarly acting juice may be made from commercial pepsin preparations, which are generally made from hog's stomach mucosa by allowing it to digest with acid, dialyzing out much of the protein digestive products and drying the pepsin solution on glass plates at low temperatures. This makes scale pepsin.

*** 78. Reaction of the mucous membrane.**—Test the reaction of the fresh mucous membrane of the hog's stomach to litmus. It will be found to be either very faintly acid or slightly alkaline. There is no storage of hydrochloric acid in the membrane, although it secretes a juice strongly acid with HCl.

*** 79. The active digestive principle of gastric juice is stored in the mucosa and digests proteins.**—That the mucous-membrane contains pepsin which digests proteins in an acid solution may be shown as follows:

Take 50 grams of hashed pig's stomach mucosa, add to it 250 c.c. of 0.4 per cent. HCl, place it in a flask and allow it to digest itself for 24 hours at 37-40° C. At the end of that time filter the solution from the undigested residue and test the solution for the metaproteins: acid albumin (precipitation on exact neutralization to litmus), protalbumose, deutero-albumose and peptone. How could you remove the peptone and peptides from the solution without injuring the pepsin?

*** 80. Existence of pepsinogen.**—*Principle and object.* This experiment constitutes the chief evidence of the existence of the pepsin in the mucous membrane of the stomach in an inactive, more resistant form, which has been called pepsinogen. The evidence consists in the fact, shown by this experiment, that a neutral infusion of the stomach mucosa can be made alkaline by sodium carbonate without permanently losing its activity. That is, digestive action returns when it is reacidified; whereas an acid infusion, or a neutral infusion which has been made acid by HCl and consequently contains active pepsin, cannot be made alkaline without losing the greater part of its activity. The pepsin is evidently very sensitive to alkalies; whereas the substance in the neutral infusion is not nearly so sensitive. This inactive precursor, or form, of pepsin is called pepsinogen, the theory being that it goes into pepsin by the action of acid.

To make the neutral infusion. To 0.5 gram of dried, fat-free, powdered hog stomach mucosa add 50 c.c. distilled water. Mix well and allow to stand for 20 minutes. Then strain through cheese cloth.

The experiment. Measure four 5 c.c. portions of this infusion into four test-tubes and label them (a), (b), (c) and (d).

(a) To (a) add 4 c.c. water and 1 c.c. N Na_2CO_3. Allow to stand for 15 minutes at room temperature, then make it acid by the addition of 3 c.c. NHCl and dilute to 20 c.c. This tube now contains N/10 HCl.

(b) To (b) add 2 c.c. N HCl and dilute to 20 c.c. This tube also contains N/10 HCl, approximately 0.36 per cent., but the infusion has not been alkaline at any time.

(c) To (c) add 0.6 c.c. N HCl and 0.4 c.c. H_2O, so that the total volume is 6 c.c. After standing 15 minutes at room temperature, add 1.6 c.c. N Na_2CO_3 and dilute to 10 c.c. Allow to stand for 15 minutes. This tube now contains the same concentration of NNa_2CO_3 that (a) did, but the solution has been exposed to acid for a time, so as to form pepsin. Now add to this tube, after standing 15 minutes, 3 c.c. N HCl and dilute to 20 c.c. with water. This tube also now contains N/10 HCl.

(d) To (d) add 1 c.c. N NaCl and dilute to 20 c.c.

Transfer 10 c.c. samples of these solutions to four dry test-tubes, label each accurately, add a piece of fibrin about the size of a pea to each and digest at 35-40° C. at the same time and for the same length of time. Observe the digestion in each tube. (b) should be the fastest, but compare especially the rate of digestion in (a) and (b).

Draw your inference.

*** 81. An experiment to show the optimum concentration of acid for peptic activity and the action of different kinds of acids of the same concentration.**—*Principle.* This experiment illustrates a very interesting fact which is apparent everywhere in nature: namely, that all kinds of living phenomena take place at their optimum under the conditions which prevail in nature. This is called "adaptation." It may be observed in this experiment that the digestion goes best at the concentration of HCl normally present in the stomach; and that the most favorable acid is HCl, the acid normally present.

Experiment. Preparation of artificial gastric juice. Digest 50 grams finely divided, fresh, or about 10 grams of the dried, hog stomach mucosa in a 350 c.c. beaker with 250 c.c. 0.4 per cent. HCl previously warmed to 40° C. Set aside in a bath at 35-40°, with frequent stirring until almost dissolved. Then transfer to a 500 c.c. flask, stopper loosely and again allow to digest for two or three days. Then filter and test a small portion of the filtrate for metaprotein, proteoses and peptones. (See page 896.) Save the remainder in a stoppered flask. Explain what happens in the above process.

Prepare the following mixtures in large test-tubes. For the pepsin solution take some of the artificial juice and add thereto Na_2CO_3 solution until the solution is no longer acid to congo red, but is still acid to litmus. After the contents of each tube have been well mixed, add to each an amount of fibrin equal in volume to a small pea. Shake each tube from time to time and digest at 35-40° C. for 10-20 minutes.

Tube	Neutralized juice c.c.	First addition c.c.	Second addition c.c.	Calculated HCl concentration	Relative activity
a	5	5 H_2O	5 H_2O		
b	5	9.8 "	0.2 N HCl		
c	5	9.5 "	0.5 " "		
d	5	9.3 "	0.7 " "		
e	5	9.0 "	1.0 " "		
f	5	8.5 "	1.5 " "		
g	5	8.0 "	2.0 " "		
h	5	6.0 "	4.0 " "		

The action of other acids in peptic digestion. Prepare a set of large tubes as above, one with the optimum amount of N HCl added and the others with the same volumes of N HNO_3, N H_2SO_4, N oxalic, N lactic and N acetic respectively. Then add fibrin and digest as above. Do all the acids act equally well? How about the H ion concentrations in the different solutions?

82. The effect of addition of bile, peptone and saccharose on peptic activity.—In the same way determine qualitatively the effect of 1, 2 and 5 c.c. of bile, 15 per cent. peptone and 5 per cent. saccharose solutions when acting in presence of the optimum concentration of HCl. Be sure to have the concentration of enzyme and HCl the same in each test. Tabulate your experiments and explain results.

83. Pepsin quantitative determination.—The quantitative determination of pepsin is made by one of the following methods. Most of these are adapted for clinical examination of the gastric juice. The activity of the juice is compared with the activity of a standard solution of pepsin.

a. *Jacoby's method.* 0.5 gram of ricin are dissolved in 50 c.c. of a 5 per cent. NaCl solution and filtered. To the opalescent liquid enough N/10 HCl is added to make it slightly cloudy, and 5 c.c. is placed in each of 10 test-tubes in the water bath at 38° C. The stomach juice, or the pepsin solution, is added in decreasing amounts to each of the tubes and the tubes are made up to the same volume by the addition of distilled water or cooked juice. The greatest dilution which after three hours in the thermostat clarifies the ricin shows the lowest concentration of juice which can fully digest this amount of protein. If 1 c.c. of juice diluted 100× is able to do this, it contains 100 pepsin units. If the total acidity of the juice is normal and equal to 40-60 c.c. of N/10 NaOH, the juice, if normal, should contain 100-200 peptic units. In hypo acidity the pepsin is generally reduced; in hyper acidity it remains about the same. In the absence of HCl there may still remain a very weak digestive action.

* b. *Mett's method.* Procure two glass tubes (2-3 mm. internal diameter and 7 cm. long) from the storeroom, draw up fresh egg white in the same until filled; then place horizontally in water which is boiling hot, remove the flame and allow the tubes to remain in the water for 10 minutes. Remove and allow to cool. When the solutions, etc., below are all ready for use, cut the capillary tubes into lengths of approximately 1 cm. each. The coagulated egg albumen must be free from air bubbles and the broken surfaces must be even with the glass ends.

Prepare two ¾-inch flat-bottom vials as follows:

(a) 0.5 c.c. 0.3 per cent. solution of 1:3,000 U.S.P. pepsin in 0.2 per cent. HCl plus 4.5 c.c. 0.2 per cent. HCl and 2 Mett's tubes.

(b) 1.0 c.c. unknown solution in 0.2 per cent. HCl plus 4.0 c.c. 0.2 per cent. HCl and 2 Mett's tubes.

Rotate gently, stopper lightly with cotton and set aside at 35-40° C. for 24 hours. Gas bubbles may cling to the ends of the tubes, be sure to brush these off with a fine wire before setting aside to digest. The Mett's tubes must lie flat on the bottom of the tube. Remove all the tubes at the same time and place on a millimeter scale, then by means of a lens measure the lengths of albumen dissolved at both ends of each tube. This gives four readings for each solution tested. Take the average of the four readings.

The above conditions are such that the Schütz law holds. What is this law and what are the conditions under which it holds? Calculate the per cent. concentration of the unknown solution in 1:3,000 U.S.P. pepsin. The U.S.P. standard of 1:3,000 means that 1 part of pepsin dissolves 3,000 parts of coagulated egg white in 2½ hours when tested by the U.S.P. assay method.

* 84. **The coagulation of milk by rennin. Rennin is activated by HCl, like pepsin.**—*Principle.* This experiment is designed to show that mucous membranes of calves' stomachs contain an enzyme or active principle which coagulates milk in a neutral solution, and that it exists in the mucosa in an inactive form, being activated by HCl.

(a) To 0.5 grams dried, fat-free, powdered calves' rennets add x c.c. of water. The value of x will be given you by the instructor. It has to be determined by a preliminary experiment, as the rennet activity is variable, and the condition of the milk varies. It should be so chosen that the time of coagulation in the following most favorable tube is between 5 and 10 minutes. After standing 15 minutes strain through cheese cloth. Take three 5 c.c. samples and treat as follows, making all measurements very carefully:

(b) To 5 c.c. add exactly 45 c.c. distilled water and mix well.

(c) To 5 c.c. add 1 c.c. N/10 HCl, and after 5 minutes *gradually add while stirring* 1 c.c. N/10 Na_2CO_3. Next add 43 c.c. distilled water and mix well.

(d) To 5 c.c. add 1 c.c. N/10 NaCl and after 5 minutes add 44 c.c. distilled water and mix well.

Now determine how long it takes for each (b), (c) and (d) to curdle milk when the following conditions are observed. Measure into three clean, dry, large (1-inch diameter) test-tubes (large tubes are used so as to permit of instantaneous mixing) 10 c.c. samples of sweet, well-mixed milk. Heat each to 40° C. When all the tubes are ready add 1.0 c.c. of (b), (c) and (d) to the respective tubes; mix *immediately* after the addition and accurately note the time to within 5 seconds.

Place all in the bath at 40° C. Now do not shake the mixture, but remove the tubes one by one from the bath from time to time (every 30 seconds), incline the tubes gently and note the time when the curd breaks away smoothly from the side of the tube. Explain the results.

85. Relation of the time of coagulation to the amount of rennin. Time law.—Take 5 c.c. samples of (c) above and prepare dilutions as follows:

(e) 5 c.c. plus 2½ c.c. water and mix well.

(f) 5 c.c. plus 5 c.c. water and mix well.

(g) Now determine how long it takes for 1.0 c.c. of (c), (e) and (f) to curdle 10 c.c. of milk at 40° C. What law does calves' rennet follow here?

86. Calcium salts are necessary for the coagulation of milk, but not for the action of the rennin on casein.—Prepare large test-tubes as follows:

Tube	Milk	Other addition
a	10 c.c.	2 c.c. 2% $(NH_4)_2C_2O_4$ Sol.
b	" "	4 c.c. H_2O
c	" "	2 c.c. 5% $CaCl_2$ Sol. + 2 c.c. 2% $(NH_4)_2C_2O_4$ Sol.

Warm to 40° C. and keep in a bath at 40° C. To each tube now add sufficient of solution (c), experiment 84 above, to cause the contents of the tube (b) to curdle in 7-12 minutes. As soon as (a) has stood with the enzyme solution the length of time it required (b) to curdle, at once add 2 c.c. 2 per cent. $CaCl_2$ solution to (a), mix well and note what happens after ¼ to 2 minutes. Explain.

87. The calf's stomach extract has a weak peptic action, but a strong rennin action; the conditions are reversed in the pig's stomach.—Prepare 50 c.c. 1 per cent. water extract of dried calves' rennets and

50 c.c. 1 per cent. extract of dried hogs' stomach mucosa. Allow each to digest for 15 minutes, then filter and measure off accurately (without contaminating one solution with the other) into *dry* vessels two 10 c.c. portions of each. To each 10 c.c. portion now add 2 c.c. N HCl. After standing 1 to 5 minutes, add carefully 2 c.c. N Na_2CO_3 to one tube from each set and, after mixing each well, compare these as to curdling action on milk.

After 5 to 30 minutes add to the other 10 c.c. portions of acid solutions 10 c.c. distilled water, mix each well, warm to 40° C. and add the same amount of well-washed fibrin to each. Digest at 40° C. and note from time to time the relative rates at which the fibrin is dissolved.

The acidity of gastric juice.—Natural gastric juice is obtained either from fistulas or clinically by eating a test breakfast of a roll and a glass of water, or weak tea, and 45 minutes later emptying the stomach by means of a stomach tube. As this is not a very agreeable procedure, skill in testing the contents had best be obtained by examining the artificial gastric mixture provided by the instructor and which contains the ingredients which one usually finds in such stomach contents. The most important part of the examination is the determination of the amount of hydrochloric acid present, since the pepsin usually goes parallel with the acid, and the acid is of considerable significance, being generally much reduced in cancers and some other digestive troubles and increased at other times, particularly in gastric or duodenal ulcer.

The total acidity of the juice may be due:

1. To free hydrochloric acid.
2. To hydrochloric acid combined with proteins.
3. To acid salts such as acid phosphate.
4. To organic acids such as acetic, lactic, butyric, etc., which have been formed by fermentative decomposition of the foods. These last are generally not found if the HCl is normal or hypernormal; for the presence of the acid keeps down the development of the bacteria and moulds. In the absence of HCl the acidity due to these other acids may be marked. The sum of 1, 2, 3 and 4 gives the total acidity of the juice. The sum of 1 and 2 makes the part of the acid which is normal and physiologically active.

It is necessary to determine the total acidity, and the presence and amount of the free hydrochloric acid, and the combined acid. Read pages 366-374.

* 88. Total acidity.—Titrate 10 c.c. of the unfiltered gastric contents with N/10 NaOH, using phenolphthalein as an indicator. The total acidity is often expressed as grams of HCl in 100 c.c. of juice. To get this number multiply the number of c.c. of NaOH required to neu-

tralize the juice with the factor 0.0365, since each c.c. of N/10 HCl would contain this number of grams of the acid. The acidity is sometimes expressed as c.c. of N/10 NaOH required for neutralizing 100 c.c. of juice. Why is phenolphthalein chosen as the indicator for total acidity? It is advisable not to filter the juice, although the particles in suspension in the juice may make a considerable part of its volume, since on filtering a slight loss occurs in the acid attached to the substances in suspension.

* 89. Free HCl.—The estimation and detection of free HCl is made by means of the *Günzberg test* as follows (see page 367 for the composition of the reagent): One drop of the reagent is placed on a white porcelain plate or evaporating dish and dried very cautiously over a free flame or better on a water bath. It must not be heated too high. Then add one drop of the filtered juice to the clear yellow spot left by the evaporation of the indicator and warm again carefully. If hydrochloric acid is present in the free state, a purplish red color develops on heating. If the spot is heated too much, it will be brown. On diluting the gastric juice and repeating the test, a limit will be found when the reaction is just perceptibly positive. At this dilution the juice contains about N/2,500 HCl.

a. Confirm the accuracy of this test for free HCl by repeating it, using N/300 and N/2,000 HCl; with a mixture of N/30 acetic acid in which a little NaCl has been dissolved; and with N/10 lactic acid. Which are positive?

b. To show that the presence of proteins reduces the amount of free hydrochloric acid, add to 10 c.c. N/30 HCl 10 c.c. of a 1 per cent. solution of Witte's peptone. Mix. Determine by the Günzberg reagent what the amount of free HCl is now in the solution.

c. Titrate 5 c.c. of the mixture from b with N/10 NaOH, using phenolphthalein as an indicator. Has the total acidity been reduced by the addition of the peptone?

d. Titrate another 10 c.c. with N/10 NaOH, using di-methyl-amino azobenzene as an indicator (Töpfer's test). How do these figures compare with the total acidity and with the acidity as determined by Günzberg?

e. In place of determining the free HCl by the dilution of the gastric juice, it can also be determined by titrating by the addition of N/10 NaOH and determining the end point at which the free HCl is neutralized by testing from time to time 1 drop of the liquid by the Günzberg method. This method is perhaps more tedious than the dilution method.

For clinical purposes the Günzberg, the total acidity, and the Töpfer titration are all that are required. By consulting the tables on page 368 it will be seen that the Töpfer titration, i.e., the dimethyl-amino azo benzene, shows an amount of acid between the free and the total acidity. It gives an imperfect notion of the amount of free HCl. It is not affected by organic acids, except when they are present in high concentration. The concentration of hydrogen ions at which it changes color may be seen in table, page 371, to be $N/10^{2.9}$—$N/10^4$. It requires a little higher concentration than congo red.

* 90. **Free HCl by Boas reagent. Tropæolin OO.**—The determination of the free HCl by Boas reagent gives the same results as Günzberg, but the latter is sharper. The Boas reagent has no points of superiority, except that it holds better. Some drops of saturated solution of tropæolin OO in 94 per cent. alcohol are placed on a porcelain plate and dried at 40° C. To the dried spot a drop of the liquid to be tested is added and again dried at 40° C. In the presence of as much as 0.006 per cent. HCl a violet spot is left when the liquid evaporates.

a. Control this test with dilute HCl, lactic acid, etc.

The student should read the action of indicators in one of the following: Stieglitz's *Qualitative Analysis*, Part I; Jones' *Elements of Physical Chemistry*; Smith's *Introduction to Inorganic Chemistry*. Other references on the use of indicators are: Abderhalden's *Handbuch der Biochem. Arbeitsmethoden*, Vol. I, p. 534; Vol. III, p. 1337; and Vol. V, pp. 500 and 1905.

91. **Hayem and Winter method for the determination of hydrochloric acid.**—This method requires the facilities of a chemical laboratory. It is based on the determination of chlorine. Three portions of 5 c.c. each of stomach contents are placed in three crucibles, a, b and c. To (a) is added an excess of sodium carbonate and all three are dried on the water bath and then in an oven at 100° until completely dry. The crucible (a) is then ashed over a free flame, very lightly, and the ash extracted with water. The chlorine is then determined in the water by silver nitrate titration. This gives the total chlorine, both acid, free and combined, and sodium chloride. The second crucible (b) as soon as completely dry has Na_2CO_3 added and ashed and the chlorine is determined as in (a). By heating on the water bath and at 100° the free hydrochloric acid is driven off. The difference between (a) and (b) gives, then, the free hydrochloric acid. The third crucible (c) is ashed directly without the addition of the carbonate. By this only the chlorine of the sodium chloride is left. The difference between (b) and (c) gives, then, the combined HCl. This method, however, is only approximate, since in the ashing some acid phosphate is formed which will expel some of the chlorine from the sodium chloride. Moreover, in the drying some of the combined hydrochloric acid is set free and goes off.

92. **Lactic-acid detection.**—By fermentation of the carbohydrates in

the stomach, when the hydrochloric acid is low, a considerable degree of acidity may be formed, due to lactic or other organic acids. It is sometimes desirable to examine the contents for lactic acid. The lactic acid may be detected either by the decolorization of phenol-ferric-chloride solution (Uffelmann's reaction) or by Hopkins' method. The latter is the better.

a. By Uffelmann's reaction. The reagent is made by taking 10 c.c. of 4 per cent. solution of carbolic acid, 20 c.c. of distilled water and 1 drop of a 2 per cent. ferric-chloride solution. This solution has a deep violet color, due probably to the partial reduction of the iron. The addition to this solution of a solution of lactic acid, even very dilute (1 part in 10,000), causes the color to change to a yellow. To 3 c.c. portions in test tubes add respectively 1 drop of N/10 and N/100 lactic acid and N/10 and N/100 HCl. Note yellow color with lactic. The reaction is not specific for lactic acid, but it is not given by dilute HCl. Now see if the filtered gastric contents discharge the color from Uffelmann. It is better to shake out the contents with ether, in which the lactic acid is soluble, evaporate the ether on the steam bath in a beaker, dissolve the syrup, if any, which is left in a little water and repeat the test with this. Keep a portion of the ether extract for testing by Hopkins' method, which is as follows:

* *b. Hopkins' method of detecting lactic acid.* This depends on the cherry-red color developed by thiophene.

Into a clean, dry test-tube place 3 drops of a 1 per cent. solution of lactic acid in alcohol, add 5 c.c. of concentrated sulphuric acid and 3 drops of a saturated solution of $CuSO_4$. Mix and place in a beaker of boiling water and heat for 5 minutes. Cool under the tap and add 2 drops of a 2 per cent. alcoholic solution of thiophene and shake. Replace the tube in boiling water. A cherry-red color develops.

c. Repeat by using in place of the lactic acid that recovered from the gastric contents by ether. It is not probable that this method either is specific for lactic acid, but it is the only substance in gastric contents which gives the reaction.

INTESTINAL DIGESTION.

A. Digestive actions of the pancreatic juice.—The three juices, bile, pancreatic and duodenal, are mixed with the chyme and digestion is the result of the mixture. The different juices will here be studied separately. Since it is impossible to provide large classes with sufficient pancreatic juice for experimental work, it is necessary to use extracts of the pancreas and the duodenal mucosa. These act, on the whole, like the secretions, since the active principles or enzymes are stored in the glands and may be extracted. In some particulars, however, the actions may not be strictly comparable, since by extraction substances may be obtained which are not normally secreted. Pancreatic juice digests starches and dextrins, proteins and fats.

***93. Amylolytic activity of pancreatic extracts.**—Qualitative detection. Prepare some 1 per cent. starch paste as in experiment (1). Get from the storeroom about 10 grams of fresh pig's pancreas which has been freed as far as possible from fat and been finely hashed in a meat chopper. Add 40 c.c. of 35 per cent. ethyl alcohol and grind and mix well in a mortar. Both the lipolytic and the amylolytic enzymes are readily soluble in dilute alcohol. Filter off the extract. Test its reducing action on Fehling's solution. It should be negative.

To 5 c.c. starch paste in a test-tube add 1 c.c. of the extract and mix. Observe the very rapid clearing of the starch due to the formation of soluble starch. Place in beaker of water heated to 40° C. At the end of 1 minute of action remove a drop to a drop of dilute I_2KI solution and test for starch. Remove also 2 c.c. and see if reduction of Fehling's solution occurs. After 10 minutes repeat both tests. The pancreatic extract contains an active principle, *amylopsin,* probably a mixture of enzymes, which converts starch to dextrins and maltose. It differs from the action of ptyalin in that the preliminary transformation of the starch to dextrins goes relatively more rapidly than, and the formation of maltose slower than, the salivary.

***94. Proteolytic activity of the pancreas.**—The proteolysis caused by extract of the pancreas differs from that produced by pepsin in that there are set free amino-acids and tri- and di-peptides. The amount of albumoses formed is relatively very little. One of the amino-acids thus set free is tryptophane, so that the pancreatic digest forms free tryptophane, and by this it is easily distinguished from the digest of the stomach. To show this add to 20 grams of hashed, fresh pig's pancreas 100 c.c. of 1 per cent. $NaHCO_3$ solution and 10 c.c. of a fresh 10 per cent. aqueous extract of duodenal mucosa, add toluene, shake well and allow to digest 24 hours at 40° C. At the end of that time filter off a little of the extract; make in a portion of it a biuret test; in another portion test for free tryptophane in the following way: Acidify about 5 c.c. of the filtered extract with acetic acid and add to it, drop by drop, a solution of bromine in water (No. 17). A violet and pink color develops in the presence of free tryptophane. When the maximum color is developed add 2-3 c.c. amyl alcohol and shake. The color dissolves in the amyl alcohol.

Permit the rest of the digestion mixture to digest and test it from time to time by the biuret test. After a few days this will be almost gone. When this is faint remove from the thermostat. Observe any white deposits (tyrosine or leucine) and examine under the microscope to see if they are crystalline. Test some of the white matter by the tyrosine test. Heat the remainder of the digestion mixture, filter while hot and evaporate on the water bath to a small bulk, 20 c.c. A crystalline crust and deposit will form if the digestion has been sufficiently prolonged. Identify tyrosine and leucine in the crust. Tyrosine crystallizes in balls of needle-shaped crystals; leucine in solid conglomerates having a radiate structure.

* 95. **Lipolytic activity.**—Take 2 c.c. of olive oil, 3 c.c. of water in a test-tube. Shake thoroughly. Then add 1 drop of litmus solution. The reaction will probably be slightly acid. If it is, add now from a pipette or burette enough N/10 NaOH just to make the reaction of the litmus blue. Now add to this tube 1 c.c. of the dilute alcoholic extract of the pancreas just prepared, shake well again and place in the beaker of water at 40° C. Shake from time to time. Observe if the litmus changes to an acid reaction, due to the splitting of the fats by the lipase, and also observe if on shaking the emulsion becomes more permanent, the fat not separating as at first. If the litmus becomes red, make it blue again by the addition of a little NaOH and place in the bath again. Write the reaction for the splitting of the neutral fat by lipase.

96. **The activity of both amylase and lipase depends on the presence of salts.**—To show this take the remnant of the alcoholic extract of the pancreas, place it in a small dialyzing, parchment tube and dialyze against running water for 24 hours. Then remove the contents of the tube to a beaker and test the activity on starch paste and on neutral fat suspended in distilled water. Make the tests just as in experiments 93 and 95, but neutralize the olive oil separately by shaking it with a little sodium hydrate, dilute and removing 2 c.c. of the neutral oil to a tube of distilled water. Repeat the experiment, having added a drop of $CaCl_2$ solution to each tube. The activity should have disappeared on dialysis and return on the addition of a little salt, such as $CaCl_2$ (No. 18).

97. **The Roberts method for amylase determination.**—Prepare 100 c.c. of a 1 per cent. starch paste as per directions previously given. Take 10 c.c. of this paste and add 90 c.c. boiling hot water. Cool to 40° C. and set aside in the water bath kept at 40° C. Now at a noted time add 1 c.c. of the enzyme solution and mix well by stirring. After acting for 3 minutes remove 1 c.c. of the mixture and add thereto 1 c.c. of the following iodine solution (dilute 0.5 c.c. of the N/10 iodine solution on the side shelf with 100 c.c. distilled water). Note the time when, on mixing the solutions as indicated, a colorless or yellowish solution remains. Repeat the digestion tests on 100 c.c. portions of the

diluted starch paste with varying quantities of the amylase solution until the achromic point is reached in 4 to 6 minutes after addition of the enzyme solution. The activity is then calculated by the formula $D=10/V\times5/n$ where V=volume of enzyme solution and n=time in minutes required to reach the achromic point. D then is the number of c.c. 1 per cent. starch paste hydrolyzed to the end point here taken, in 5 minutes, by 1 c.c. of the enzyme solution.

Determine the relative activities of two substances or solutions by the above method.

98. **Effect of hydrogen ion on pancreatic amylase action.**—Prepare an extract by grinding 2.5 to 5 grams fresh hog pancreas with a small amount of sand and gradually add thereto 100 c.c. distilled water. Strain through cheese cloth.

Note the reaction of this solution toward litmus, methyl orange and dimethylamidoazobenzene. To a 15 c.c. portion thereof add one drop of the dimethylamidoazobenzene, mix well and then add just enough of a N/20 HCl solution to render the solution faintly pink. *Measure the amount of N/20 HCl added by means of a burette.* Now prepare another 15 c.c. portion in the same way with the added indicator, but instead of adding the N/20 HCl add a volume of distilled water equal to the volume of acid added above. Mix each solution well and determine their relative activities by the Roberts method above. Tabulate your results.

99. **Effect of hydroxyl ion on pancreatic amylase action.**—To 10 c.c. of the 5 per cent. pancreas solution add just enough N/10 Na_2CO_3 to render it alkaline to phenolphthalein. To two other 10 c.c. portions add an additional amount of a Na_2CO_3 solution to make the concentration of additional $NaCO_3$ 0.01 per cent. and 0.05 per cent. respectively (for 20 c.c. volume). Dilute each solution to 20 c.c. and compare the three solutions with the original 0.50 per cent. pancreatin solution (diluted with an equal volume of water) for amylase by the Roberts method.

100. **Protection of pancreatic amylase by carbohydrates.**—Prepare an extract by grinding 5 grams fresh hog pancreas with a small amount of sand and gradually adding 100 c.c. water. Strain through cheese cloth.

From the undiluted pancreatic extract above prepare the following solutions:

(a) 5 c.c. extract plus 10 c.c. H_2O + toluene.
(b) 5 c.c. " " 10 c.c. 50% glucose solution + toluene.
(c) 5 c.c. " " 10 c.c. 50% saccharose solution + toluene.

Shake the contents of each tube well, then allow the toluene layer to separate and at once compare the activities of the aqueous solutions by the Roberts method, using 5 c.c. thereof for 10 c.c. starch paste. Set the remainder of the solutions aside at 35-40° C. for 24 hours and test again in the same way. Which is now the stronger? Explain.

* 101. **Enterokinase and trypsinogen.**—*Principle.* The object of this experiment is to show that a mixture of the extract of the intestinal mucosa and the extract of the pancreas digests proteins very much faster than either alone. This fact is interpreted to mean that there is in the intestinal secretion a heat-sensitive substance, which has been named enterokinase, which has the function of activating the inactive proferment of the pancreas so that it will digest proteins. The inactive proferment is called trypsinogen.

Experiment. Prepare 50 c.c. of a 1 per cent. water extract of dried fat-free hog pancreas and also 25 c.c. of a 1 per cent. water extract of dried, fat-free hog duodenal mucosa.

Instead of the above, one may prepare 10 per cent. water extracts of the fresh tissues as directed under amylase above.

Now prepare the following mixtures in large test-tubes.

Tube	Pancreas extract	Intestinal extract	Water	Treatment
a	10 c.c.	0	5 c.c.	Keep at 40° C for 20 min.
b	10 c.c.	5 c.c.	0	" " " " " "
c	0	5 c.c.	10 c.c.	" " " " " "
d	10 c.c.	5 c.c.	0	(Heat in boiling water for 5 min., cool to 40° C and keep at 40° for 20 min.
e	10 c.c.	5 c.c. (boiled)	0	Keep at 40° C for 20 min.

To each tube now add 5 c.c. 10 per cent. Na_2CO_3 solution and mix well *immediately* after the addition. Now add to each tube the same amount(about the size of a hazel-nut) of well-washed fibrin. Agitate well and digest at 40° C. Note the changes in 2 to 40 minutes. Tube b should digest very much faster than the others.

* 102. **Determination of the optimum alkalinity for tryptic digestion.**—*Principle.* The object of this experiment is to determine whether trypsin digests better in an acid or alkaline medium, and what the optimum concentration of alkali or acid is.

Experiment.

a. *The optimum Na_2CO_3 concentration for trypsin action.* Prepare 100 c.c. of an activated trypsin solution. Take 5 c.c. samples and add thereto 14.5; 14.0; 13.5; 13.0; 12.5; 12.0 and 11.5 c.c. water, and 0.5; 1.0; 1.5; 2.0; 2.5; 3.0 and 3.5 c.c. N Na_2CO_3 solution respectively. *Add the water first.* Shake well immediately after each addition and add to each a piece of well-washed fibrin about the volume of a hazel-nut.

Digest at 35-40° C. Calculate the concentration in Na_2CO_3 in each case above and note the optimum concentration for trypsin action.

b. *Effect of hydrogen ions on trypsin action.* In the same way as above determine the effect of adding to 15 c.c. of the activated trypsin solution 14.9; 14.8; 14.7; 14.6 and 14.5 c.c. water, and 0.1; 0.2; 0.3 0.4 and 0.5 N/10 HCl respectively.

103. **Protective action of hydrolyzed protein on trypsin.**—*Question:* Does trypsin unite with proteins when it is acting on them? To solve this we look to see if the presence of hydrolytic cleavage products makes the enzyme more stable. The experiment is obviously far from being conclusive. At the best it is but an indication. OH ions should be the same concentration in each tube.

Experiment. Prepare 50 c.c. of activated trypsin solution, take two 10 c.c. portions and add 0.4 c.c. N Na_2CO_3 solution to each. Next dilute one by adding 10 c.c. water and to the other add 10 c.c. Witte's peptone solution, also add a few drops toluene to each and mix well. Now test each qualitatively as in previous experiment, but set aside about one-half of the enzyme mixture at 35-40° C. for 12-24 hours. Then again test in the same way for tryptic activity. Discuss your results.

* 104. **Steapsin is destroyed by heat and its activity is greater in the presence of bile.**—Prepare 100 c.c. of a 10 per cent. water extract of fresh hog pancreas by grinding in a mortar with sand and gradually adding the water. Saturate with toluene and set aside for 24 hours. After standing for 24 hours, strain through cheese cloth and make the following tests, adding one drop toluene to each tube and digesting at 37-40° for 24 hours. Use large test-tubes all of the same diameter. After the digestion shake in ice water, add a drop of phenolphthalein and titrate with N/20 NaOH until a permanent pink remains. *Shake well while titrating.* Why shake well?

Tube	Olive oil	Water	Other addition	Enzyme solution
a	0.5	4.5	0	5 c.c.
b	0.5	4.5	0	" boiled
c	0	4.5	0	5 c.c.
d	0	4.5	0	" boiled
e	0.5	9.5	0	0
f	0.5	8.5	1 c.c. bile	0
g	0.5	3.5	1 " "	5 c.c.
h	0	8.5	1 " "	0
i	0	3.5	1 " "	5 c.c.

Record for each tube the amount of NaOH necessary to neutralize the fatty acid set free by the action of lipase.

105. **Following the course of a protein digestion by trypsin by formol titration (Sorensen method).**—*Object:* This experiment illus-

trates a very important method, that of formol titration of amino-acids (see page 363), and at the same time affords a proof that in the course of a proteolytic hydrolysis there is a steady increase in the number of free amino groups.

Experiment. The Sörensen method depends on the amino-acids having their basic character destroyed by the formaldehyde and thus bringing about a greater ionization into H^+ and RCHN (:CH_2) COO^-. Consequently the greater the concentration of the amino-acids produced by the hydrolysis of the protein the greater the acidity, and thus from a measure of the increased acidity we have a measure of the degree of hydrolysis.

Prepare 40 c.c. of a 10 per cent. extract of fresh hog pancreas. Also prepare 10 c.c. of a 10 per cent. extract of duodenal mucosa and mix the two.

Take 10 c.c. of the above and add to 100 c.c. of a 4 per cent. casein solution in 0.5 per cent. Na_2CO_3 solution. Immediately after mixing remove 25 c.c. by means of a pipette, then add 10 c.c. neutralized formalin (a 40 per cent. formaldehyde solution rendered neutral toward phenolphthalein by adding N NaOH), and set aside the remaining solution at 35-40° C. Titrate the 25 c.c. sample by N/10 NaOH or HCl, as the case may be, using phenolphthalein as indicator. At the end of ½, 1½ hours and 24 hours remove 25 c.c. samples and titrate these in the same way. Explain your results and write the reactions involved in the titration. Why not use congo red as indicator here?

106. **Invertin in intestinal mucosa.**—Prepare 100 c.c. of a 10 per cent. water extract of fresh intestinal mucosa in the same way as directed for pancreatic amylase above. Detect invertin in this solution by seeing if an addition of 2 per cent. of cane sugar to 10 c.c. of this solution and with the addition of one drop of 10 per cent. acetic acid will give rise to glucose.

107. **Erepsin in duodenal mucosa.**—To 10 c.c. of the above extract add in a flask 50 c.c. of a 2 per cent. peptone solution, add 1 c.c. toluene and shake. Prepare another mixture in the same way, but boil the intestinal extract first. This will be the control. At once remove exactly 10 c.c. from each flask, add phenolphthalein and titrate with N/10 H_2SO_4 or N/10 NaOH to the neutral point. Now add formalin as in experiment 105 and titrate the acidity which develops. Keep a record of your results. Place both flasks in the incubator at 37° and remove 10 c.c. portions and titrate this same way after a period of digestion of 3 hours, 24 hours and 48 hours. If erepsin is present, there should be a digestion of the peptone with the setting free of amino-acids and dipeptides, and these will be detected by the acidity which develops when formalin is added.

THE BILE.

The bile has very little digestive action, its main function being act in co-operation with the pancreatic and duodenal juices and to as in absorption, particularly of the fats. It is also an excretory su stance. The constituents which are important in digestion and absor tion are the bile salts, sodium glycocholate and taurocholate, and alkali, sodium carbonate. Its pigments make it very striking. far as known these are excretions and have no function. The bile co tains also some phosphatide and cholesterol.

* 108. **The physical properties of bile.**—Obtain 25 c.c. ox bi (Note—This is best kept and transported in the ligated gall bladder Note the specific gravity, the viscidity, slimy character, the taste, sm and color. Test its reaction to litmus paper and to phenolphthalein.

109. **Powers of solution.**—To 5 c.c. of bile in a test-tube add 0 c.c. of oleic acid and shake. The oleic acid partially dissolves and held in solution in the bile.

110. To 5 c.c. of fresh bile in a test-tube add a little dry, powder cholesterol, about the amount on the point of a knife. Warm for 2 minutes at 38°. The cholesterol dissolves. Bile is the only body flu which has the power of dissolving cholesterol.

* 111. **Reactions by which its pigments can be detected.**—Th bile pigments have the property when oxidized of passing through series of brightly colored compounds, finally becoming yellow. Th final yellow color is called choletelin. Bilicyanin, a blue color, an bilifuscin are intermediate oxidation products. The two principal pig ments of the bile are biliverdin, $C_{33}H_{36}N_4O_8$ (?), and bilirubin, $C_{33}H_{36}N_4O_6$ (?). Bilirubin by oxidation is converted into biliverdin.

* 112. **Gmelin's test for bile pigments.**—This test may be tried in various ways. It depends on the oxidation of the pigments by concen trated fuming nitric acid: i.e., nitric acid containing some nitrous acid.

a. Place a thin layer of bile on a porcelain dish and place on it a drop of fuming nitric acid. The drop becomes surrounded by a series of various colors, blue, pink, orange, etc.

b. Place 3 c.c. of fuming nitric acid in a test-tube and carefully place above it a layer of bile. On agitating very gently a series of col ors develops at the zone of contact. Repeat this test, diluting the bile many times with water.

* 113. **Huppert-Cole test for bilirubin.**—To about 50 c.c. of diluted bile add an excess of baryta water or milk of lime. The bilirubin is pre cipitated as an insoluble barium or calcium compound. It coheres. Allow to settle. Remove most of the supernatant liquid with a pipette transfer the remainder to a small filter Remove the precipitate from

he filter; place it in a test-tube; add about 5 c.c. of 95 per cent. alcohol, 3 drops of strong sulphuric acid and 2 drops of a 5 per cent. solution of potassium chlorate and boil for a minute. After the settling of the precipitate of calcium or barium sulphate, the supernatant liquid will be an emerald or blue-green.

114. **Hammarsten's reaction for bilirubin.**—Take one volume of the acid mixture (see below) and add to it 4 volumes alcohol. If a drop of a solution containing bilirubin is added to a few c.c. of this mixture, a beautiful, permanent green color at once develops. By the addition of more of the acid mixture to the green solution the other colors of Gmelin's test to choletelin can be obtained. The acid mixture should be prepared beforehand. It is made by mixing 1 volume of nitric acid and 19 volumes of HCl, each acid being about 25 per cent. in strength. It is used when it has become yellow by standing. It will keep at least a year.

THE BILE SALTS. SODIUM SALTS OF GLYCOCHOLIC AND TAUROCHOLIC ACIDS.

115. **Preparation of glycocholic acid.**—From some biles rich in glychocolic acid the acid will crystallize. Place 100 c.c. of ox bile in a stoppered cylinder, add 5 c.c. of con. HCl and 12 c.c. of ether and shake vigorously. Put aside in a cool place. In some cases the glycocholic acid will crystallize out in a few minutes or hours. If this does not happen, make the solution faintly alkaline with sodium-hydrate solution and precipitate the glycocholic acid by lead acetate or ferric chloride. Filter. Decompose the precipitate with a small amount of soda; filter; and treat the solution with HCl so that it is plainly acid. Glycocholic acid should crystallize out. It may be recrystallized from hot water.

*116. **Preparation of bile salts. Platner's bile.**—Mix 40 c.c. bile with enough powdered charcoal to make a thin paste, about 10 grams, and evaporate to dryness on the water bath. Grind the residue in a mortar, transfer it to a flask and extract it with about 75 c.c. of absolute alcohol by boiling it on the water bath for a few moments. Cool and filter through a dry filter into a dry beaker. Add ether to the alcohol until a permanent cloudiness results and put in a cool place, covering the beaker with a plate. The bile salts should crystallize out as long needles, or a thick hard mass. At times their crystallization is delayed. Filter off and dry in the desiccator.

Dissolve in water a little of these salts for the tests for bile salts which follow:

*117. **Pettenkofer's test for bile salts.**—To 5 c.c. of the solution of bile salts, or diluted bile, in a test-tube add a small crystal of cane sugar and shake until dissolved. Then holding the tube inclin concentrated sulphuric acid, about 5 c.c., to run down beneath salt solution. In the presence of the salts there develops at of junction a purple color. This reaction is the same as the Moli tion, only in this case we use the bile salts, the cholic acid pa molecule, as the chromogenic substance in place of α-naphth reaction is due to the production of methoxy furfural or furfu the sugar by the sulphuric acid and the union of this with the to give the color.

Note—This reaction is also given as Raspail's reaction by some fat namely, by those of sphingosine and lecithin.

* 118. Hay's test for bile salts.—Bile salts like soaps have t erty of greatly lowering the surface tension of water. This may as a very delicate test for their presence, although it is of co specific. Prepare two perfectly clean beakers and fill them ab full of perfectly clean, not greasy or soapy, water. To one add a little of the bile salt solution. Now sprinkle on the top some flowers of sulphur. If the water is clean, the sulphur w on the top of one, but will sink in the one to which the bile h added, owing to the lowering of its surface tension.

* 119. Bile salts precipitate proteins.—The bile salts have th erty, common to many acids, of forming insoluble precipitat the proteins. Slightly acidify a filtered solution of Witte's about 1 per cent. solution, and add to it a drop or two of bile salt or of diluted bile. A white precipitate should form. The bile p thus precipitates and unites with the proteins in the chyme wh discharged from the pylorus.

120. Cholesterol in bile.—Evaporate 10 c.c. bile to dryness water bath. Extract twice with small quantities of ether and ev the ether extracts to dryness in another dish. DO NOT WORK A FLAME WHEN USING ETHER. CARRY ON ALL SUCH ON THE STEAM BATH. Dissolve the residue left from the eth tion in about 2 c.c. chloroform and apply the Salkowski test.

INTESTINAL PUTREFACTION PRODUCTS.

The large intestine contains myriads of bacteria which act up food products and intestinal secretions which have escaped abso and produce, by fermentation and putrefaction, various products of which are offensive and harmful. Among these products whic to feces their characteristic odor are indole and scatole. The of separating these substances from the feces is by distillation. method and some of the properties of these bodies are illustrated following experiments, but instead of using feces a putrefying protein solution, blood, or fibrin, or albumose, is used instead.

* 121. Obtain 100 c.c. putrefied protein from the storeroom. Distill off about 50 c.c., using a water-cooled condenser. It may be necessary to coagulate and filter before conducting the distillation.

Acidify the distillate with hydrochloric acid and extract three times in a separatory funnel with 20 c.c. ether each time. Allow the ether extract to evaporate spontaneously. Save also the acid aqueous layer for the work below.

Dissolve the residue from the ether extract in about 10 c.c. water, filter if necessary and test for phenol and cresol and for indole and scatole. For the two former apply the Millon test. To another portion add saturated bromine water, a crystalline precipitate indicates tribrom-phenol and tribrom-cresol formation.

Indole gives a red coloration by the glyoxylic acid $+H_2SO_4$ test, and a violet blue with formaldehyde $+H_2SO_4$. Scatole requires the addition of a drop of very weak $FeCl_3$ solution (or other oxidizing agent) to give the test with formaldehyde $+H_2SO_4$. Apply these tests on 1 c.c. portions of your solution above (Dakin, *Jour. Biol. Chem.*, Vol. II., 1907, p. 289).

With para-dimethyl-amido-benzaldehyde indole gives a red color which becomes darker red on the addition of sodium nitrite. Scatole yields a blue-violet color before adding the nitrite and a deep blue after addition thereof. This blue color may be extracted by chloroform (Steensma, *Zeits. f. physiol. Chem.*, Vol. XLVII, 1906, p. 25).

The original acid-water solution above may contain sulphides and ammonia. Test directly for sulphides therein. Concentrate the remainder of the solution to about 5 or 10 c.c. and test the same for ammonia.

The graphic formulas of indole and scatole are given on page 441.

THE BLOOD.

Vertebrate blood consists of a liquid called the plasma, which holds in suspension a vast number of small bodies: the red corpuscles, the white corpuscles and the blood platelets. There are in 1 c.mm. of human blood about 5,500,000 red corpuscles and 8,000 whites. The plasma consists of water holding in solution about 8 per cent. of coagulable proteins, serum globulin, fibrinogen and serum albumin,—salts, a little glucose (about 0.1 per cent.) and a great number of other substances, amino-acids, urea, etc., which are present in very small quantities. The red corpuscles owe their color to hemoglobin, a protein having the power of forming a loose union with oxygen and which carries oxygen to the tissues. The number of white and red corpuscles varies under different conditions of health and their number often furnishes a valuable means of diagnosing disease and of showing the state of health or disease of the blood. For this reason simple and expeditious methods have been devised for the determination of the number of white and red corpuscles in the blood and of the amount of hemoglobin, and also of the relative amount of corpuscles and plasma. The study of the blood may be begun with the determination of these factors.

122. **Determination of the number of red corpuscles in the blood. Method of using the hemacytometer of Thoma-Zeiss.**—The apparatus consists of two pipettes for dilution and a slide ruled in squares. The pipette for the determination of the red corpuscles dilutes the blood 100 times; that for the white, 10 times. Prick the finger and when a good drop of blood has collected draw it up into the pipette for the reds to the mark 1. Wipe the blood off the exterior of the pipette and then draw up quickly some Hayem's solution to the mark 101. Mix the contents thoroughly at once in the bulb by shaking the pipette back and forth so that a uniform suspension of corpuscles is obtained. Allow some of the mixture to flow out of the pipette and then transfer a small drop to the ruled platform of the cell. But before doing this see that the cell is perfectly clean and that there is no dust either on the slide or the cover slip. The drop should be so large that when the cover slip is put on it does not go into the moat, but completely covers the ruled platform. Allow the slide to stand for a few moments for the corpuscles to settle and then count 16-20 of the small squares, beginning at one corner and going straight across the cell and then the next tier and so on. The average number of corpuscles per square is determined. The squares are ruled so that each square is one-four-hundredth of a square mm. and the depth of the space above the ruled platform is one-tenth of a mm., so that each square represents the number of corpuscles in one-four-thousandth of a c.mm. This number multiplied by 4,000 and by 100, since the sample of blood was diluted 100 times, gives the number of red corpuscles in 1 c.mm. of blood. Hayem's solution in the following: Na_2SO_4, 5 grams $HgCl_2$, 0.5 gram; NaCl, 1 gram; distilled water, 200 c.c.

The determination of the white corpuscles is made in a similar way, except that the other pipette is used which dilutes only 10 times. It is filled with blood to the mark 1 and then the point is wiped off and at once the solution of 0.5 per cent. acetic acid is drawn up to the mark 11, and the blood mixed with the solution by shaking and rotating the pipette. It is necessary to work rapidly to prevent the clotting and sticking together of the corpuscles. It is difficult to get a uniform distribution of the corpuscles. After thorough mixing place on the slide as before and count the corpuscles. The red corpuscles are dissolved by the acid and the whites are fixed. The number per square multiplied by 40,000 gives the number of corpuscles per c.mm. It is well to count the whole of a large square. This determination is a difficult one to make accurately.

123. Oliver's hemacytometer.—The Thoma-Zeiss method of directly counting the corpuscles is trying on the eyes and consumes much time. A simple, accurate clinical method of determining the number of red corpuscles is that of Oliver. The principle of this method consists in diluting a given amount of blood with a solution until the flame of a candle makes just the image of a line through it. The apparatus is simple. It consists of a measuring, capillary pipette, a pipette like a medicine dropper and a test-tube graduated and having a rectangular cross-section. The method of using this apparatus is described by Cabot as follows:

Clean and dry the capillary pipette by drawing through it a needle carrying thread or daining cotton saturated with water and then with alcohol and ether. The medicine dropper is then filled with Hayem's solution. The capillary pipette is then filled with blood in the usual way, any superfluous blood on the outside being quickly removed, the pipette connected at the blunt end with the rubber on the end of the medicine dropper and the blood washed out by means of the Hayem's solution into the graduated test-tube. If the previous hemoglobin estimation has shown nearly the normal amount of hemoglobin, the blood may now be diluted to about 80 with Hayem's solution. Hold it in the hand in the manner indicated, with the thumb and first finger extending up the sides of the tube, and holding the tube close to the eye in a dark room and about 9-10 feet from the small wax (Christmas) candle, see if the image of the candle suddenly appears as a bright line across the tube. Dilute by adding a few drops of Hayem's solution, hold the finger over the mouth of the tube and invert once or twice, wiping each time the thumb on the mouth of the tube so that the liquid sticking to it goes back in the tube, and continue diluting and testing until the bright line suddenly appears. This dilution is such that the solution contains approximately 52 corpuscles to 36 squares of the Thoma-Zeiss counter. If enough fluid has been added to make the volume in the tube 100, the blood contains 5,000,000 corpuscles per c.mm. If it has been necessary to dilute it to 80, it will contain 80 per cent. of 5,000,000 or 4,000,000; if only diluted to 50, the number is 2,500,000. The values obtained by this instrument with human blood after a little practice are usually correct within 1 per cent., unless a very great number, 100,000 per c.mm., or more, leucocytes are present when somewhat higher figures are obtained. The tube held in the hand in the manner indicated shuts off with the hand the light of the candle, except that coming through the tube.

124. Hematokrit method.—In this method the relative volume of plasma and corpuscles is determined by the separation of the corpuscles by rapid centrifugalization in a Daland hematokrit. The finger or ear is pricked and a sample of blood is sucked up into the capillary, graduated tube of the hematokrit. The tube must be exactly filled. To do this grease the first finger with a little vaseline and, as soon as the tube is removed from the blood-drop, slip the greased finger tightly over the end. Holding it there, detach the rubber tube from the other end and

slip the tube into the hematokrit, with the beveled end toward the axis of rotation. An empty tube has previously been put into the other arm of the centrifuge for balancing purposes. Now rotate the centrifuge two minutes at the rate of 70 revolutions of the handle per minute. If the work has been rapidly done, the blood is centrifuged before it has clotted, the corpuscles are packed as a solid plug at the outer end of the tube. They make, as a rule, in normal blood about 40 per cent. of the length of the tube. This corresponds to about 5,000,000 corpuscles per c.mm. Since many factors influence the volume of the corpuscles, the method is not accurate for determining the number of corpuscles. The centrifuge used for the purpose is noisy, crude and generally unsatisfactory.

Determination of hemoglobin.—The most convenient clinical method for this purpose is that of Tallqvist. The most elaborate is that of von Fleischl, but as the latter method is inaccurate, cumbersome and expensive, it possesses no superiority over and should be replaced by the Tallqvist and Dare methods.

125. **Tallqvist method.**—The advantage of this method is its cheapness, the Tallqvist scale costing but $1.25, and its extreme simplicity. A determination of the hemoglobin within at least 10 per cent. can be made at the bedside and in a few moments. A drop of undiluted blood is soaked into a piece of filter paper and after a moment or so, and before it has dried, it is matched with the color of the scale against a white background and in ordinary daylight. The scale has been carefully prepared by matching the tint of blood of various degrees of anemia and known hemoglobin content when soaked into filter paper. The colors were lithographed and the lithographed scale bound up with 50 pieces of filter paper. It can easily be slipped in the pocket. The errors are not greater than with the von Fleischl apparatus in the usual hands.

126. **Oliver's hemoglobinometer.**—This has, in general, the same principle as the foregoing, but uses diluted blood. The capillary pipette is filled with blood from a pricked ear or finger and washed as in the case of the hematocytometer into the cell by some water. The cell is then filled with water, the cover put on so that only a very small bubble remains under the cover slip, to show that it is not overfull, and compared with the scale. The color of the cell is matched with the colors of the scale. Two riders are supplied which, when placed above the scale, make the tint somewhat deeper and so increase the number of subdivisions. For example, if the sample has a tint betwen 80 and 90, a rider placed on 80 makes it 85. If the test comes between 85 and 90, it could be called 87.5. 100 per cent. is taken as the normal. This corresponds to 14 per cent. of hemoglobin in the blood. Reflected candle

light is used. By this a reading may be had to 2 per cent. A set of disks can be had adjusted to daylight readings.

127. Dare's hemoglobinometer.—This instrument is extremely convenient. It uses undiluted blood so that it avoids the errors due to dilution. It is compact, but it costs $20. It is more accurate than Tallqvist's method, but not so convenient. It is probably the best of the methods here given.

The film of blood is drawn between two plates of glass by capillarity, and the tint compared with a revolving scale of a circular disk of glass. The two colors are matched and the per cent. of Hb in terms of the normal read off. The colors are compared by candlelight and the observation made through a telescopic attachment so that light is cut off other than that transmitted. A dark room is not necessary. The instrument is simply pointed at some dark corner. The candle is attached to the instrument.

HEMOGLOBIN.

This, the red coloring matter of the blood, is confined in the vertebrate blood to the red blood corpuscles. If small quantities are present in the plasma, the amount is so small that at present we cannot detect it. The function of this red conjugated protein is to convey oxygen from the lungs to the tissues. It has the power of forming a loose, dissociable union with oxygen, the compound being known as oxyhemoglobin. It has also the power of combining with other substances. The absorption spectra due to the absorption of light of these various compounds differ somewhat and the compounds may be detected and distinguished by these spectra. We will consider, first, influences which cause the hemoglobin to leave the corpuscles and become free in the plasma. This process is called laking the blood; second, we will consider methods of crystallizing hemoglobin; third, the spectra of various compounds of hemoglobin with oxygen, carbon monoxide, of hemoglobin itself and methemoglobin and some other derivatives; and fourth, some of the decomposition products of hemoglobin.

Influences which cause the hemoglobin to leave the corpuscles. Laking the blood.—The separation of hemoglobin from corpuscles is a matter of very great interest because, owing to the color of the hemoglobin, the process can be very easily followed and the causes and nature of the processes involved can be studied. It is not probable that the passage out of hemoglobin is at all peculiar, but that all cells in a similar manner lose some of their constituents when subjected to these same processes. We cannot, however, so easily detect them, owing to the lack of color. In studying the laking of the blood, therefore, it is probable that the matter is of greater interest as a type of cellular reaction than it is for its immediate significance for the physiology of the blood. Laking may be produced by the following agents: by warming, or freezing and thawing blood; by the action of anesthetics of all kinds; by the pumping of the gases out of the blood, particularly by taking out the oxygen; by the action of very dilute alkalies; by the action of bile salts; by the action of specific hemolytic agents such as the hemolysins; by diluting the blood with water so that the blood is hypotonic; by condenser discharges through the blood; by various toxic substances, such as certain snake venoms, and saponins.

It will be observed that most of these agents cause either stimulation or depression of protoplasmic processes. Since the discharge of hemoglobin from the corpuscle means that a change in the distribution of the constituents of the protoplasm has occurred under the influence of the reagent, it has been suggested that the nature of the process of stimulation consists in the physical change in distribution of the constituents of the protoplasm caused by the laking agent or the stimulus. The hemoglobin is generally supposed to be held in the corpuscle by the limiting membrane, which is of such a nature that it cannot go through. Many authors accordingly speak of the hemolytic agents as affecting the permeability of the sheaths of the corpuscles. As pointed out on page 498, the phenomena may also, and perhaps more correctly, be interpreted on the hypothesis that the hemoglobin is in loose union with the constituents of the stroma of the protoplasm and these various reagents alter the stability of that union.

128. Influence of hypotonicity. Laking by dilution with water.—To illustrate this take 7 test-tubes and place in each 10 c.c. respectively of the following solutions: a, distilled water; b, 0.2 per cent. NaCl solution; c, 0.4 per cent. NaCl; d, 0.5 per cent. NaCl; e, 0.7 per cent. NaCl; f, 1.0 per cent. NaCl; g, 1.2 per cent. NaCl. Most easily prepared by running the proper amounts of 2 per cent. NaCl solution and water from two burettes. To each tube add from a pipette two drops of defibrinated blood. Mix well and allow to stand for a few minutes. Record the results. The laked blood is a darker color and more transparent. Which tubes lake the blood?

129. Laking by anesthetics.—Anesthetics unfortunately have the property of laking blood. Take two tubes containing 10 c.c. 0.9 per cent. NaCl solution. To each add three drops of defibrinated blood and mix. Now to one add 1 c.c. of the anesthetic and mix again. Allow both to stand. Observe the laking in the anesthetic tube. Test in this way ether, chloroform, naphtha, toluene, acetone and alcohol.

130. Laking by warming.—To 5 c.c. of 0.9 per cent. NaCl add 3 drops of defibrinated blood and warm carefully to about 50°, not higher, holding the blood for a few minutes at that temperature. Laking will

occur. Laking will also occur by freezing the blood and thawing. The laking in this case is possibly accelerated by the fact that on melting the ice crystals local differences of concentration occur. But this is not the only explanation.

131. Laking by bile salts.—To 10 c.c. of a 0.3 per cent. solution of bile salts in 0.9 per cent. NaCl add 3 drops of defibrinated blood. Observe the laking.

132. Laking by saponins.—To 10 c.c. of a 0.9 per cent. solution of NaCl containing 0.1 per cent. of saponin add 3 drops of defibrinated blood.

133. Laking by dilute alkalies.—Make a precipitate of sodium magnesium phosphate by adding to a solution of $MgCl_2$ some solution of Na_2HPO_4, and if necessary a little NaOH. Filter off and wash with water. Now suspend some of the precipitate in 10 c.c. of 0.8 per cent. NaCl, add 3 drops of defibrinated blood, mix well by inverting the tube twice and allow to stand. Have a control tube with the blood and sodium chloride alone. The hemolysis may be slow, so observe for some time.

134. Crystallizing hemoglobin.—Various bloods crystallize with different ease. Guinea-pig blood crystallizes with the greatest ease and crystals can be obtained by mixing a drop of the defibrinated blood with a drop of water on the microscopic slide and putting on a cover glass. The crystals form very quickly. Rat and mouse blood crystallize readily, and also the blood of the amphibian, Necturus. Dog's blood oxyhemoglobin crystals may be obtained by adding a few drops of ether, or better toluene, to 10 c.c. of defibrinated dog's blood in a test-tube and mixing until laked. Then add a little powdered ammonium oxalate, shake once or twice and place in the ice-box or a cool place for a few hours. Crystals of hemoglobin will generally be found. Examine them microscopically and draw some of them.

The forms of the crystals are, as shown by Reichert, characteristic for different animals. See Figure 50.

135. Spectra of various compounds of hemoglobin.—Hemoglobin, both solid and when in a solution, has a purplish-red color. The colors of the various compounds of hemoglobin differ somewhat from each other and from hemoglobin. Oxyhemoglobin is a bright scarlet; carbonylhemoglobin is more of a cherry red; methemoglobin has a brownish-red color; and so on. The difference in color is due to the fact that the different compounds absorb light of different wave lengths so that the light which comes through the solution, or the crystals, has lost vibrations of certain wave lengths and so no longer appears white but colored. Most of the red is obviously transmitted, but some of the green or blue must be absorbed. By passing the light which has traversed a hemo-

globin solution through a prism which spreads it into a spectrum it is easy to see which of the rays have been absorbed. An instrument for thus determining the absorption or emission spectra is called a spectroscope. These are of two forms, but the one of a very convenient form for such work is a direct vision spectroscope provided with a wave-length scale. The direct vision spectroscope of Zeiss is a convenient form.

The direct vision spectroscope. The instrument consists of a train of crown and flint glass prisms so arranged that the course of the rays, though scattered to form a spectrum, is not bent, but passes directly to the eye in line with the object looked at. At one end is the eyepiece. At the other end is a small slit or opening which can be narrowed by the use of a screw. At the side of the main tube is a small secondary tube which contains the wave-length scale which by means of a mirror is sent into the eyepiece so that the wave-length scale and the spectrum appear to lie side by side. In using the instrument, point it at the source of light, with the small tube at the left. Have a very narrow slit so that the spectrum is barely visible. Now draw out the tube gently until the spectrum appears to be traversed in a vertical direction with very fine, dark lines (Frauenhofer lines) ; adjust the slit and the focus of the tubes until these are very sharp. One very plain one will be noticed in the orange just to the left of the yellow. This is the D line of the spectrum and corresponds to the main line of sodium. The wave length of the light at this point of the spectrum is 0.59 μ or λ 589. By means of the screw adjust the wave-length scale so that the line of the scale 0.59, or λ 589, comes exactly above the D line of the spectrum. The wave length of any of the other dark lines can now be determined very readily by reading off the scale.

The sample of blood, hemoglobin or other substance of which it is desired to determine the absorption spectrum is placed in a test-tube or beaker, or better in a small flat-walled glass, and this is placed just in front of the slit and between it and the source of light. A bright daylight is good, but any other bright white light may be used. If the solution is too concentrated, too little light will come through, most of the field appearing dark. The more dilute the solution, the narrower will the absorption bands become. With a proper dilution of the oxyhemoglobin a broad absorption band will be seen in the violet and two narrow absorption bands in the green between the D and E lines of the spectrum. These bands are blackest in their centers and shade off somewhat toward the edges. In recording your observations, record the wave length read from the scale corresponding to the center of the band, and the wave lengths at the sides of each band.

The width of the absorption bands will vary with the concentration of the solution looked through. The absorption spectra are very commonly plotted, therefore, in the manner indicated in Figure 51, p. 505, the absorption being indicated by the shaded portions, the wave lengths being marked along the abscissa and the concentration of the solution along the ordinate. The solution is examined in a layer 1 centimeter thick. The point of maximum absorption is usually given by the tip of the plot.

In making the experiments which follow, it is not necessary to prepare the hemoglobin in a pure form first; for experiment has shown that the absorption spectra of the isolated hemoglobin and the hemoglobin of laked blood are the same. Make some laked blood by adding a little blood to distilled water in a test-tube, filtering the solution and then examining the absorption spectra of various dilutions. It will be observed that if the solution is strong enough the various absorption bands fuse together, giving for example one broad absorption band for oxyhemoglobin.

136. Spectrum of oxyhemoglobin.—Shake the laked blood with some air in a test-tube. Examine the spectrum. Make a scale in your note-books and plot on it the absorption bands. The bands of oxyhemoglobin can be distinguished in a solution which contains one part of oxyhemoglobin to 10,000 of water. The spectroscope could be used for the estimation of the amount of oxyhemoglobin in a solution. A 0.3 per cent. solution of oxyhemoglobin gives a good spectrum. The Hb in blood is 14 per cent. From this complete the dilution necessary to give 0.3 per cent. Map the absorption bands. The α band (the one toward the red end) has its center at λ 578, that of the β band is about λ 540.

137. Reduced hemoglobin.—Take 5 c.c. of a solution of oxyhemoglobin so dilute that the two absorption bands are fairly far apart and not fused in one. Add to this solution two drops of ammonium-sulphide solution and warm very gently, not over 55°. Observe if there is any darkening of the tint of the solution. Now without shaking the tube examine its spectrum. It should now show the single broad absorption band of reduced hemoglobin. The center is about λ 565. Instead of ammonium sulphide as a reducing agent, Stokes' reagent may be used. It is a freshly prepared solution of ferrous sulphate to which some tartaric acid has been added and which is made slightly alkaline with ammonium hydrate just before use. Map the spectrum of hemoglobin.

Stokes' solution is the following: 3 grams ferrous sulphate dissolved in cold water; add a cold aqueous solution of tartaric acid, 2 grams, and make up to 100 c.c. with water. Just before using add enough strong ammonium hydrate just to redissolve the precipitate which is

first formed. It rapidly oxidizes itself, so must be freshly pr
Its advantage is that it reduces the oxyhemoglobin without warmi

138. **Carbonylhemoglobin.**—This is the union of hemoglobi carbon monoxide in place of the oxygen. Its spectrum is like that hemoglobin, but the two bands are shifted slightly toward the end of the spectrum. Prepare a solution of carbonylhemoglobin b ing a little illuminating gas through a solution of laked blood the hood. Notice the change in tint of the blood. It has a bluis Dilute if necessary and examine the spectrum. The middle of 572 and of β at 535.

139. Carbonylhemoglobin is not reduced by ammonium su To show this repeat experiment 137, using the carbonylhemogl place of the solution of oxyhemoglobin. See that the spectrum r unchanged after warming with $(NH_4)_2S$.

140. Another very convenient way of distinguishing carbon oxyhemoglobin is to take two solutions of oxyhemoglobin and ca hemoglobin of about the same tint and dilute equally with wat will be seen that the solution of oxyhemoglobin becomes yellowish that of carbonylhemoglobin acquires a carmine tint. Try this.

141. Another test for carbonylhemoglobin in the blood is *Kat* Add 5 drops of blood to 10 c.c. of water. Then add 5 drops of colored ammonium sulphide. Mix and make faintly acid with acetic acid. Blood containing CO develops a rose-red color; blood, a dirty greenish-gray. The difference is perceptible with e of CO blood to 5 of normal blood.

142. **Methemoglobin.**—This is a union of oxygen and hemo but the union is firmer than oxyhemoglobin. The gas cannot be p out. Oxyhemoglobin passes with great ease into methemoglobin i slightly acid, or alkaline, or in the presence of various oxidizing such as ferricyanide, chlorates, etc. It is formed in the blood in and chlorate poisoning and by a great variety of other poisons s acetanilide.

The exact difference in composition between oxy- and methem is unknown, but probably oxyhemoglobin is a union of a mol nature, the union involving the residual valences of the oxygen. difference may be schematically represented as follows:

$Hb = O—O$	$Hb = O = O$	;	$Hb \langle {}^{O}_{O} \rangle$ (O—O bonded)	$Hb \langle {}^{O}_{O}$
Oxyhemoglobin if oxygen is univalent.	Oxyhemoglobin if oxygen is bivalent.		Oxyhemoglobin.	Methemogl

Probably none of these representations is correct, for the ex involves water, and probably the first union is a union between oxygen and hemoglobin.

143. To a few c.c. of water add three or four drops of blood. Mix. This lakes the blood. Now add two drops of a saturated solution of potassium ferricyanide. The blood color changes to a chocolate brown. Examine the solution spectroscopically. Methemoglobin has an absorption band in the red, λ 630. There is marked absorption of the blue end.

144. It is possible by reducing agents to restore the oxyhemoglobin and then to produce reduced hemoglobin. To show this take 5 c.c. of the methemoglobin solution just prepared and add to it a few drops of ammonium sulphide. It will be seen that the color changes to a red. Examine by the spectroscope and see that the red absorption band of methemoglobin disappears and the bands of oxyhemoglobin appear. On warming gently the bands of oxyhemoglobin disappear and reduced hemoglobin appears. Now shake strongly with some air. The bands of oxyhemoglobin reappear. It is probable that small amounts of methemoglobin appear at times in normal blood. This experiment is particularly instructive, as it shows that in the act of oxidation of substances several intermediate stages occur. Always a molecular union is made first.

Another very singular fact is that the addition of potassium ferricyanide to blood causes the liberation of an amount of oxygen equal to the oxygen combined as oxyhemoglobin in the blood. This fact is the basis of the method generally used at present for the estimation of the oxygen in blood. No air-pump is needed.

Haldane has given the following provisional equation to represent what happens, but it is not satisfactory:

$$HbO_2 + 4Na_3FeCy_6 + 4NaHCO_3 = HbO_2 + 4Na_4Fe\dot{C}y_6 + 4CO_2 + 2H_2O + O_2$$

Perhaps a molecular union of HbO_2 and ferricyanide occurs first and then this is rearranged, oxyhemoglobin being changed to methemoglobin, the ferricyanide reduced and oxygen set free.

145. Take 3 c.c. of defibrinated blood, lake by diluting with 3 c.c. water and warming gently. Then add 6 c.c. of saturated potassium ferricyanide solution, mix by inverting and observe the liberation of bubbles of gas. O_2.

Decomposition Products of Hemoglobin.

146. Hemin crystals (Teichman's crystals).—Prick the finger. Collect a drop of blood on a microscope slide. Dry. Put on a drop of glacial acetic acid, cover with a cover glass and boil. Examine for crystals of hemin. Dark-brown plates and prisms. In examining old blood stains by this method it is necessary to add a minute crystal of salt. There is salt enough in fresh blood. These crystals are acetylated hematin chloride. The hydroxyl of the hematin is replaced by chlorine

and one acetyl group enters from the acetic acid. Hematin is $C_{34}H_{32}N_4O_4Fe$.

147. Hematoporphyrin.—This is iron-free hematin. It is prepared by the action of strong acids or alkalies on hemoglobin.

Acid hematoporphyrin.—The strong acid at first splits off hematin from the globin and then this hematin loses its iron and is converted to hematoporphyrin. As reduced hemoglobin is far less stable than oxyhemoglobin, the iron is split off easiest from the reduced hemoglobin. Add a couple of drops of putrefying blood (the putrefaction reduces the hemoglobin) to a few c.c. of concentrated sulphuric acid and mix by gently shaking. The solution has a purple color. Examine spectroscopically. Center of absorption bands: α at λ 600 relatively fainter, and β at λ 554.

148. Alkaline hematoporphyrin.—Make a somewhat stronger solution of acid hematoporphyrin than the preceding, but in the same way, and pour it into 50 c.c. of water in a beaker. Stir well. Collect the precipitate which rises to the surface on a rod, transfer to a test-tube, add a few c.c. of alcohol and boil. Make alkaline with a few c.c. of NaOH. Examine the spectrum of alkaline hematoporphyrin thus obtained. 4 absorption bands: λ 622, λ 576, λ 539, and λ 504.

149. Acid hematin.—Hematin split off from oxyhemoglobin by the action of acids is soluble in ether. Heat a few c.c. of defibrinated blood with a drop of strong HCl and a few c.c. of acetic acid. Cool and extract with 5 c.c. of ether. Examine the spectrum of acid hematin in ether. Red, λ 638. On dilution other bands appear.

150. Hemochromogen.—A solution of oxyhemoglobin made alkaline with NaOH forms alkaline hematin (band from D to λ 630). Reduce this by adding a few drops of $(NH_4)_2S$. This forms the red hemochromogen. Bands in the green. α at λ 558; β at 520.

COAGULATION OF THE BLOOD.

Clotting of the blood depends on the interaction in the blood of four elements: Fibrinogen, calcium salts, prothrombin and cephalin.

For these experiments one can use ox or horse's blood rendered non-coagulable by receiving it into a saturated solution of sodium or potassium fluoride.

151. Fibrinogen.—Centrifugalize 20-30 c.c. of oxalate or fluoride plasma. Note the relative volume, approximately, of plasma and corpuscles. The plasma should be a light-yellow color and clean. Pipette it off from the corpuscles very carefully without getting an admixture of corpuscles. This is best done by connecting a test-tube with the suction water pump by a tube (a) which reaches just through the cork

Another tube (b) goes through the cork, but ends nearer the bottom. The other end of the tube (b) is attached to a flexible piece of rubber tubing. The pump is started going and one end of this rubber tubing is introduced into the plasma just below the surface of the plasma. The plasma is sucked over into the second tube. As it is drawn over follow it down with the end of the rubber until no more can be taken off without mixing with corpuscles.

152. Take 5 c.c. of the plasma in a test-tube and add an equal volume of *saturated* NaCl solution. Fibrinogen will be precipitated.

153. Take another 3 c.c. in a test-tube and immerse in a beaker partly filled with water. Heat slowly with a thermometer in the fibrinogen solution. The tube must not touch the bottom of the beaker. Note the temperature of coagulation. About 58°.

154. To another 5 c.c. of the plasma add sufficient $CaCl_2$ to precipitate all the oxalate or fluoride and leave a very small excess of calcium salt in the solution. Does the solution clot on standing? If not, add to it some serum or a few c.c. of 0.9 per cent. NaCl solution which has been rubbed in a mortar with the fresh fibrin of defibrinated blood. Allow to stand and observe clotting. By extracting the fibrin a solution of thrombin is obtained.

The power of fibrinogen solutions to clot is the real test for this substance.

155. Clotting is also prevented by receiving the blood at once in strong solutions of magnesium sulphate. A saturated solution of magnesium sulphate is placed in a bottle, or jar, and blood run in directly from the veins, mixing well as it flows, until the combined volume of blood and sulphate solution is about five times that of the sulphate solution alone. The blood is then centrifugalized to remove the corpuscles and the plasma drawn off and kept in a cool place. It is *salt plasma*. It may be used in place of the oxalate or fluoride plasma.

156. Dilute some of the plasma by adding about 4 times its volume of distilled water. Divide into two portions in two test-tubes. To A add nothing; to B add some of the fibrin ferment solution. Place both in the bath at 38° C. Observe that both clot, but that B clots the sooner.

THE URINE.

The urine is a most important excretion. In it the nitrogenous wastes and most of the mineral and some of the carbon wastes are excreted. The determination of the amount and character of these is of great importance in throwing light on the physiology and pathology of the body. Among the nitrogenous wastes urea, uric acid, creatinine, ammonia and hippuric acid are the more abundant, although a very large number of nitrogen-containing substances occur in the urine small amounts. Of the inorganic wastes, chlorides, sulphates and ph phates may be mentioned, and of the substances containing carb various aromatic compounds, such as phenyl-acetic acid, and aliph compounds such as acetone, hydroxybutyric acid, lactic acid, acetoac acid. There are, however, a large number of other substances pres in very small amounts.

QUALITATIVE EXAMINATION OF THE URINE.

* 157. Collect a sample of urine and note its color, odor, tra parency, reaction to litmus and to phenolphthalein and its spec gravity.

* 158. Make qualitative tests for the presence of chlorides, s phates and phosphates. For the chlorides acidify with nitric acid a add a few drops of silver nitrate. A white precipitate indicates presence of chlorides. For the sulphates, acidify with hydrochloric a and add a few drops of barium chloride. A white precipitate sho the presence of inorganic sulphuric acid. For the phosphates add the clear urine made alkaline by ammonia some ammonia-magnesia m ture used for the determination of phosphates. A white precipitate in cates phosphates. Or add to the urine some drops of nitric acid a then some ammonium-molybdate solution. A yellow precipitate phosphomolybdate occurs in the presence of phosphates.

* 159. Preparation of urea and uric acid from urine.—Evapora 500 c.c. of fresh urine in evaporating dish to dryness on steam ba Extract residue with three successive portions of 95 per cent. alcohol. acetone, using 25 c.c. each time and heating to boiling on the water ba to facilitate solution of the urea. Decant the hot alcohol or acetone fr the insoluble residue each time, pouring the hot solution containing t urea into a small evaporating dish. The uric acid is left in the resid in the large evaporating dish, while the urea, separated from most the salts, is in the alcohol or acetone. Save the residue with the uric ac and proceed with the preparation of the urea.

Preparation of urea. Evaporate the alcoholic or acetone urea sol tion to a syrup on the water bath, add 10 c.c. of water, and after thoro mixing add, little by little, while stirring and keeping the evaporati dish cold by floating it in cold water, half concentrated, white, not fu ing, nitric acid as long as crystals of urea nitrate continue to separate ou When no more crystals form, filter through a small suction filter, usi suction flask and the little filter plate, and suck the nitrate of urea dr Most of the color remains in the filtrate. To the filtrate add a litt strong nitric acid to make sure that all urea has been precipitated. The is danger of oxidizing the urea, so do not use fuming nitric acid; b sure to cool the urea solution between each addition of nitric acid. Ne transfer crystals of urea nitrate from the filter paper to small evapora ing dish. Add to the crystals in the evaporating dish 15 c.c. distill water and then, little by little, powdered barium carbonate, stirri between each addition, as long as effervescence lasts, and then add mo to make sure that all the urea nitrate is decomposed and that there is an excess of barium carbonate present. The solution will then be very faintly acid, due to the carbonic acid. By this one forms barium nitrate, carbon dioxide and free urea. The carbonate must be really in excess, or the urea will be oxidized on warming. Now add about half a level teaspoonful of good powdered bone black and heat on water bath for about fifteen minutes. Meanwhile prepare the small filter plate for suction filtering, placing a test tube in the filter flask to catch the filtrate, and at the end of fifteen minutes filter the hot solution and transfer the filtrate, which should be clear as water, to a small evaporating dish, and evaporate nearly to dryness on the water bath. Crystals of barium nitrate separate out. Extract the residue without separating the crystals, with two 10 c.c. portions of 95 per cent. ethyl alcohol, heating on the water bath each time and transferring the hot alcohol containing the urea by decantation to another small evaporating dish. Barium nitrate remains undissolved. Evaporate the alcoholic solution of urea on the water bath until it begins to crystallize and then remove it from the bath and allow it to crystallize at room temperature. Urea crystallizes in long prismatic crystals. The preparation should be white. If it is yellow it will not crystallize so well. If this should be the case dissolve in a little alcohol and decolorize again with powdered animal charcoal. The yield should be 6-10 grams.

Uric acid. To the residue in the evaporating dish from which the urea was extracted, add 50 c.c. distilled water distinctly acid with hydrochloric acid. All the inorganic salts dissolve and leave the uric acid as a small, reddish, granular, crystalline precipitate. Decant and discard the liquid but keep the crystals. Wash them once or twice with fresh water by decantation. Finally add about 30 c.c. of water and then, drop by drop, 10 per cent. Na_2CO_3 solution, stirring after each addition. Seven or eight drops are usually enough, and warm for a minute or so on the water bath to hasten solution. Add a pinch of animal charcoal. Filter through a small suction filter, transfer without loss to a small beaker and acidify with hydrochloric acid. The uric acid will crystallize out promptly as a sandy precipitate, perhaps not entirely white. After standing for half an hour or over night, filter through a small suction filter. Allow the crystals to dry on the filter, then remove them to a small vial. Make the various tests for uric acid with them. Yield about 0.2 gram.

Reactions for the identification of uric acid.

* 161. **Murexide test.**—See page 720 for the reaction involved. Place a few crystals of uric acid in a porcelain dish, moisten them with a drop or two of concentrated nitric acid and dry on the water bath until the nitric acid is completely gone. A reddish residue remains. Now cool this and moisten it with dilute ammonia solution. The residue becomes a violet or purple red, due to formation of ammonium purpurate. Add a little 10 per cent. NaOH. the residue becomes a bluish violet. On heating the color disappears. Of the other purine bases adenine and hypoxanthine do not give the murexide test. Guanine and xanthine give the

reaction forming the yellow nitroxanthine first, which turns violet or purple when moistened with sodium hydrate. The color does not disappear on heating.

* 162. Reducing reactions.—Uric acid is easily oxidized and it is accordingly a reducing substance. Many tests have been devised for its detection based upon this property. None of these are specific, since they are given by other reducing reagents. Uric acid reduces Fehling's solution. Take a few crystals, dissolve in a little sodium hydrate and assure yourself that they reduce Fehling's solution. To another portion of the sodium-hydrate solution add some powdered bismuth subnitrate and heat to see if this is reduced also (Böttger's glucose reaction).

163. Schiff's reaction.—Dissolve a few crystals of uric acid in a few c.c. sodium-carbonate solution (sodium hydrate cannot be used because it precipitates the silver as brown silver hydrate), the solution being distinctly alkaline. Pour a drop of the solution on a filter paper moistened with silver nitrate solution. A black spot will be formed in the presence of uric acid. This reaction depends on the power of the uric acid to reduce alkaline silver solutions.

* 164. Folin reaction.—This reaction depends on the power of uric acid to reduce sodium phosphotungstate solution. To a very small amount of uric acid in a beaker is added 20 c.c. saturated sodium-carbonate solution. It may be warmed to hasten solution. As soon as the uric acid is dissolved add 1 c.c. of sodium phosphotungstate reagent (Folin's). A blue color is obtained, due to the formation of a lower oxide of tungsten.

Folin's sodium phosphotungstate reagent is the following: 100 grams pure sodium tungstate, 80 c.c. 85 per cent. orthophosphoric acid and 750 c.c. distilled water are boiled gently or in a flask with a reflux condenser for 1½-2 hours. Cool and dilute to 1 liter. This solution is reduced by other compounds, for example by polyphenols. It is used in the microchemical estimation of uric acid.

165. The reduction by uric acid is most rapid in an alkaline solution. Benedict's solution, which is reduced by carbohydrates, is not so alkaline as Fehling's and is hence reduced by uric acid at a very much slower rate. Test the reducing action of a sodium-carbonate solution of uric acid on *Benedict's solution*. Experiment 17.

* 166. Precipitation reactions of uric acid.—The insolubility of the free acid has already been noted. The ammonium salt is also very insoluble, particularly in the presence of other soluble ammonium salts. This method is the basis of the quantitative method for the estimation of uric acid of Hopkins.

Make a saturated solution of uric acid by heating some crystals with 10 c.c. of 2 per cent. Na_2CO_3. After adding two drops of ammonia, saturate the solution with ammonium chloride. Note the white, amorphous precipitate of ammonium urate which forms. If desired, the precipitate may be identified as uric acid.

*** 167. Precipitation by ammoniacal silver solution.**—Add an excess of ammonia to the sodium-carbonate solution of uric acid, and then a few drops of silver nitrate. A white precipitate is formed. This is the silver compound of the uric acid. Other purines are precipitated by this method (see Salkowski's method).

168. Detection of uric acid in urine and other fluids.—To show the presence of uric acid in small quantities of urine, Folin's method may be used. Place 1-2 c.c. of urine in an evaporating dish, add 1 drop of saturated oxalic-acid solution and evaporate to complete dryness on the water bath. Cool, add 10 c.c. 95 per cent. alcohol and allow to stand for 5 minutes to extract phenols which will also, if present, give the reduction reaction with phosphotungstate. Pour off the alcohol. Add to the residue 10 c.c. of water and a drop of saturated sodium-carbonate solution. Stir to complete the solution of the uric acid; transfer to a small beaker; add 1 c.c. of Folin's reagent (sodium phosphotungstate) and 20 c.c. of saturated sodium carbonate. A blue color indicates the presence of uric acid.

169. Another method is the following (Cole): Take 50 c.c. of urine, add 2 drops of ammonia and then saturate with powdered ammonium chloride. Allow the excess of ammonium chloride to settle for 15 seconds and pour off into another beaker. Allow to stand. Note the gelatinous precipitate of ammonium urate. Filter: scrape the precipitate from the filter and transfer it to an evaporating dish. Add 3 or 4 drops of strong nitric acid and evaporate to dryness, then with ammonia make the murexide test. If urates are present in the precipitate, they will give a positive reaction.

Creatinine.

170. Preparation of creatinine. Zinc-chloride method.—Make 500 c.c. of urine alkaline with milk of lime and add $CaCl_2$ solution to completely precipitate the phosphates. Filter, acidify the filtrate with acetic acid and evaporate to a syrup. Extract the creatinine from the syrup by treating it with warm 95-99 per cent. alcohol, 100 c.c., in two 50 c.c. portions. Allow alcohol extract to stand 8-24 hours in a cool place. Filter. A little sodium acetate is added to the alcoholic filtrate to reduce the acidity and about 1 c.c. of strong, acid-free zinc-chloride solution, sp. gr. 1.2. Stir and allow to stand 2 to 3 days in a cool place. Creatinine zinc chloride crystallizes out as a sandy, yellowish powder

composed of fine needles in rosettes or balls. Collect by filtering on a small suction filter, wash with alcohol, suspend the crystals in a little (100 c.c.) warm water and add some freshly precipitated lead hydrate, or lead oxide, and filter. The precipitate is lead chloride. Decolorize the filtrate with animal charcoal, filter, evaporate the solution to dryness, extract with strong alcohol (creatinine dissolving, creatine remaining insoluble) and evaporate the alcoholic extract to beginning crystallization and then allow to stand and crystallize. Make some of the following tests with the crystals. The crystals when pure are colorless, monoclinic prisms.

* 171. Preparation of creatinine from the urine by the picrate method.—Add to 500 c.c. of urine 100 c.c. of a 5 per cent. picric-acid solution in alcohol. A crystalline precipitate of the double creatinine and potassium picrate is formed. Filter this off after standing overnight. While still moist, heat the crystals in 50 c.c. of water, to which is added a little potassium bicarbonate to decompose the crystals. Neutralize with sulphuric acid, add four volumes of alcohol to precipitate the sulphates. Filter. Precipitate the creatinine by zinc chloride and sodium acetate as in the preceding determination and proceed as in the foregoing experiment.

* 172. Reactions for the detection of creatinine. Weyl's reaction. Nitroprusside reaction.—Like some other reducing substances, creatinine gives a red color with sodium nitroprusside. See the test for acetone (Legal's). Cysteine gives a similar reaction. To 5 c.c. of urine in a test-tube add a few drops of a dilute, fresh solution of sodium nitroprusside and make alkaline with sodium hydrate. A ruby-red color which fades to yellow is the result. If this solution, cold, is treated with an excess of acetic acid, a precipitate of the nitroso compound ($C_4H_6N_4O_2$) results. If acidified with acetic acid and heated, the solution becomes green, then blue and Prussian blue separates. Dissolve one crystal of the creatinine in water and repeat the test with it.

* 173. Picramic-acid reaction. Jaffé.—Creatinine combines with picric acid. If made alkaline, it reduces the picric acid forming the red-colored picramic acid. See page 41. To 5 c.c. of urine in a test-tube add a few drops of picric-acid solution and make alkaline with sodium hydrate. A red color is produced. This reaction is the basis of Folin's creatinine quantitative method. It is not specific for creatinine, but depends on the reducing action of creatinine in alkaline solution.

Hippuric acid.

174. Preparation of hippuric acid from urine. Roaf.—To 500 c.c. of the urine of a horse, or cow, add 125 grams of ammonium sulphate

and 7.5 c.c. concentrated sulphuric acid. Hippuric acid crystallizes out. Filter. Wash the crystals with a little cold water, redissolve in a small amount of hot water, boil with animal charcoal to decolorize, filter and allow to cool. The hippuric acid crystallizes on standing. If some of the crystals are evaporated to dryness on the water bath with 1-2 c.c. concentrated nitric acid and then the residue heated by a flame, an odor of nitrobenzene may be perceived. What is the origin of hippuric acid?

* 175. **Detection of indican.**—Indican is the potassium salt of indoxyl-sulphuric acid.

```
    CH                          CH               CH
HC//  \C—C—O—SO3K          HC//  \C—CO   OC—C//  \CH
 |     ||  ||               ||    ||  |    |  ||    |
HC     C   C—H             HC     C   C == C  C     CH
  \\  /  \/                  \\  / \ /      \ /  \ //
    CH   NH                   CH   NH        NH   CH
     Indican.                         Indigo blue.
```

Indican is formed from the indoxyl which has been formed from indole by oxidation. Indole is derived from the tryptophane of proteins by processes of putrefaction. The presence of indican in larger than usual quantities in the urine generally means excessive putrefaction in the alimentary canal. In constipation the indican is usually increased. Indican is detected by converting it to indigo blue by oxidation. Various oxidizing agents may be used, but ferric chloride or hypochlorite is usually employed. The oxidation takes place in strong acid solution, the acid splits off sulphuric acid, leaving indoxyl which is oxidized to indigo blue.

* 176. **Jaffé's indican test.**—Take 5 c.c. of urine in a test-tube and add 5 c.c. of concentrated HCl and then 1 drop of a 3 per cent. $KClO_3$ solution. By the action of the acid hypochlorite is formed. Add a little chloroform, 2-3 c.c., and shake. If the chloroform settles out colorless, add another drop of hypochlorite and shake again. If indican is present, the chloroform should be colored blue. If thymol has been added to the urine for preserving it, the chloroform will be a reddish or violet color if indican is present. Keep on adding drop by drop of the chlorate and shaking until a maximum blue is obtained. The oxidizing agent must be added cautiously, since the addition of too much carries the oxidation to indigo white which is colorless. Calcium hypochlorite or bleaching powder may be used in this test in place of the chlorate. The result is the same.

* 177. **Detection of indican by ferric chloride. Obermayer's test.**—This may be performed either directly on the urine or after partial purification by precipitation with lead acetate. The test in the latter case is performed as follows: Add to 50 c.c. acid urine 5 c.c. basic lead

acetate. Filter. Take 10 c.c. of the filtrate in a test-tube, add an equal volume of concentrated HCl containing 2-4 grams $FeCl_3$ per liter, and 2-3 c.c. chloroform and shake thoroughly. The chloroform which settles out should be more or less blue if indican is present. There is not so much danger in this test of overoxidation as there is in the hypochlorite test.

DETECTION OF PATHOLOGICAL CONSTITUENTS OF URINE.

Detection of proteins.—Normal urine does not contain protein. In abnormal conditions various proteins may occur. Coagulable proteins derived from the blood plasma may be present in inflammation of the kidney, or nephritis; the nephritis being either chronic, due to a chronic infection of the kidney, or acute. Acute nephritis may accompany scarlet fever and other diseases. A deutero-albumose, generally called peptone, may be present particularly during the absorption of partially digested pus from abscesses or in the absorption of pneumonia exudate. An especial protein called the Bence-Jones protein appears in the urine in osteomalacia or multiple myeloma. This is more nearly related to a hetero-albumose. The nitrogen in the protein in the latter case may amount to as much as one-third of the total nitrogen of the urine.

* 178. **Coagulable protein.**—Very little of this may appear in the urine, even though pretty extensive chronic nephritis prevails. The presence of any in the urine should be regarded with suspicion. If the urine is clear and acid, place about 7 c.c. in a test-tube and, while holding the tube inclined, heat the upper parts of the tube to boiling. If coagulable protein is present, this part of the urine should appear cloudy as compared with the unboiled part. If it becomes turbid, add a drop of dilute acetic acid. The turbidity might be due to phosphates, but these dissolve in the presence of dilute acid. The protein coagulum does not dissolve. If the urine is alkaline, make it before boiling very faintly acid with dilute acetic acid.

* 179. **Heller's test.**—Place in the bottom of a test-tube about 4 c.c. nitric acid and holding the tube inclined pour carefully down the side of the tube some urine. The urine should float on top of the nitric acid. In the presence of albumin a white ring appears at the junction of the liquids. This is a very delicate test. It may happen that there is a ring of urea-nitrate or uric acid in concentrated urines. If the ring is due to these substances, it will disappear on diluting the urine and repeating the test. Urines preserved with thymol may give a ring of nitrothymol. The thymol may be separated from the urine by extracting it with a little naphtha. Resinous substances which appear in the

urine after a person has been treated with balsams also may produce a white ring. It is well, therefore, to further identify the protein, if the test is positive. The colored ring which develops due to the oxidation of various chromogens may be disregarded.

180. Various modifications of these tests have been proposed. Various salts such as sodium chloride are added to the urine, or to the nitric acid to increase the delicacy of the test. The Heller test is, however, already sufficiently delicate for all ordinary purposes. The identification of the protein can be confirmed by trying the precipitation test with potassium ferrocyanide and acetic acid. Make the urine faintly acid with acetic acid and add a drop of potassium ferrocyanide solution. In the presence of albumins or protalbumose a white precipitate appears.

A case came under the author's observation in which the urine contained so little protein that it was with great difficulty that its presence could be certainly established. At times the urine was free from protein. The person died not long afterwards of an acute infection, and on autopsy the kidneys were found to have been the seat of a long-standing and extensive nephritis.

* 181. Quantitative estimation of albumin present. Esbach's method.—*Principle.* The method consists in precipitating the protein with picric acid and estimating the volume of the precipitate. A graduated tube known as *Esbach's albuminometer* is used.

Process. Fill the tube with urine to the mark U and then add the picric-acid reagent to the mark R. Cork, mix by inverting a couple of times and then allow to stand upright for 24 hours. At the end of that time read off on the scale etched on the tube the height of the column of precipitate. The figures give the number of grams of dried protein in a liter of urine. If the amount is less than 0.05 per cent., it cannot be accurately determined in this way. If the precipitate comes above 5, it is better to dilute the urine once and repeat.

Picric-acid reagent. Dissolve 10 grams picric acid and 20 grams citric acid in 800 c.c. boiling water, transfer to a 1,000 c.c. volumetric flask, cool and make up to the mark.

182. The coagulable protein can also be directly determined by boiling 50 c.c. of urine, slightly acidified with acetic acid, till the separation is complete, filtering through a dried weighed filter paper, washing thoroughly with hot water and alcohol, drying and reweighing.

183. Bence-Jones protein.—This protein is detected by its peculiarity of becoming insoluble on heating to 60° and the precipitate redissolving on heating to boiling.

If the urine is acid, it may be heated directly; if alkaline, make it faintly acid with acetic acid. Heat slowly. The urine becomes turbid at about 45° and a precipitate appears at 60° which redissolves on further heating to boiling. On cooling the reverse phenomena appear.

* 184. Albumose.—This can be detected if in sufficient quantity by the rose-colored biuret test in urine which will not coagulate on heating. It may also be detected as follows: Remove the coagulable protein by heating to boiling the faintly acid urine and filtering. In the filtrate a biuret reaction or a ring test with Spiegler's reagent indicates the presence of albumose or peptone.

Spiegler's reagent. $HgCl_2$, 40 grams; tartaric acid, 20 grams; NaCl, 50 grams; glycerine, 100 grams; H_2O, 1,000 c.c. This reagent is a very delicate test for proteins, as both albumoses and peptones are precipitated by it. It is used in the same way as the nitric acid in Heller's test. 1 part protein in 250,000 can be detected by it. Control it by normal urine.

* 185. Detection of glucose in urine.—Glucose appears in the urine under the circumstances already described on page 754. If the urine is light colored, of large volume, three liters or more per day, and with a normal specific gravity, the presence of glucose is indicated. To detect glucose three or four methods may be used: i.e., reduction test with Fehling's or Benedict's solution, reduction of bismuth subnitrate, polarimeter examination, the fermentation by yeast, formation of phenylglucosazone. The best method is Benedict's or 189(a).

* 186(a). To 4 c.c. of urine add an equal quantity of Fehling's mixture, p. 860, and heat to boiling. Boil for a few moments. In the presence of glucose, lactose and some other substances a red precipitate of cuprous oxide is formed. Normal urine has some powers of reduction, but not sufficient to form a precipitate of cuprous oxide under these conditions. Caution: If chloroform has been added to the urine as a preservative, it will itself reduce Fehling's solution.

186(b). Benedict's qualitative reaction.—To 5 c.c. of the special Benedict reagent (see experiment 220) heated to boiling, add 8 drops of the urine and boil vigorously for two minutes. The fluid will contain a dense but very finely divided red, yellow or green precipitate if glucose to the extent of 0.05 per cent. or more is present. The green precipitate appears with the lower concentrations and the red with the higher. It is best ordinarily not to use more than 8 drops of the urine, although if the urine is very dilute as to uric acid, creatinine and phosphates, one may use as much as 16 drops of the urine to 5 c.c. reagent.

* 187. Many substances besides glucose reduce Fehling's solution. The reduction might be due to lactose, pentoses, aromatic bodies, urates, creatinine or glycuronic acid, for example. If glucose is present, the urine will be dextro-rotatory and will reduce bismuth subnitrate and ferment with yeast. Nylander's reaction. Bismuth subnitrate reaction.

To 5 c.c. of urine in a test-tube add a little powdered bismuth subnitrate, make the urine alkaline with sodium hydrate and heat to boiling for a few minutes. If glucose is present, the subnitrate will be reduced to the black bismuth.

* 188. Fermentation test.—Fill a fermentation tube with urine which has not been preserved by the addition of formol or other preservative, and in which some yeast has been suspended. Place in a warm place (40°). If the reducing body is glucose, the yeast will ferment and a gas will collect in the closed arm of the tube. Lactose, pentoses and glycuronic acid or aromatic substances will not ferment.

* 189. Form the osazone in the method described in experiment 24. If the osazone, the reduction and the fermentation tests are positive, it may be concluded that the substance is glucose or levulose. The rotation will distinguish between these two. The melting point of the osazone can also be determined.

189(a). The identification of small amounts of glucose and lactose in urine (Cole).—This is probably the best method of identifying small amounts of glucose and lactose in urine in the shortest time.

Principle. The use of Fehling's solution for the detection of glucose is open to the following disadvantages: 1. Other substances than glucose reduce this solution: namely, pentoses, glycuronic acid, lactose, urates, creatinine and some aromatic bodies. Moreover, creatinine holds the cuprous oxide in solution so that it does not precipitate. 2. The strong alkali used will destroy small amounts of glucose. To many samples of urine as much as 0.5 per cent. of glucose can be added without producing more than a greenish cloud with Fehling. Normal urine contains between 0.03 and 0.08 per cent. of glucose. Benedict's solution removes many of these difficulties because by the use of carbonate in place of sodium hydrate the reducing action of uric acid and creatinine is climinated. Cole proposes the following improvement on Benedict (Cole, *Lancet*, September 20, 1913). The principal improvement introduced is that glucose is not adsorbed by good blood charcoal in the presence of acetic acid, whereas other reducing substances, and particularly lactose, are. In the absence of acetic acid small amounts of glucose are adsorbed. In the actual reduction also sodium carbonate is used in place of sodium hydrate.

Procedure. " In a dry boiling tube or large test-tube place about 1 gram of Merck's pure blood charcoal. (A spatula about three-eighths of an inch broad, well piled up with the charcoal for just over an inch, carries about ½ gram. Merck's ' Blutkohle mit säure gereinigt ' is recommended.) Add 10 c.c. of the urine and shake from side to side to mix thoroughly. Heat to boiling point, shaking the whole time. Cool thoroughly under the tap and shake at intervals for about five minutes.

Filter through a small paper (9 to 11 cm. in diameter) into a rather wide test-tube containing about half a gram of anhydrous sodium carbonate. When the fluid has filtered through, add 6 drops of pure glycerine (the glycerine used must not itself reduce the copper), shake and heat to boiling. Note the time when boiling commences. Maintain active boiling for 50 seconds, shaking from side to side to prevent spurting. Immediately add 4 drops of a 5 per cent. solution of crystallised copper sulphate. Shake for a moment to mix the solutions, and allow the tube to stand without further heating for one minute. With normal urine the fluid remains blue, with a variable amount of a grayish precipitate of the earthy phosphates. If glucose is present to the extent of 0.02 per cent. or more, above the average normal amount, the blue color is discharged, and a yellowish precipitate of cuprous hydroxide forms. The rapidity with which the precipitate forms is a rough measure of the amount of glucose present. With 0.05 per cent. it appears in a few seconds. With 0.02 per cent. it may not appear till 50 seconds. A yellowish precipitate or coloration appearing after 60 seconds must not be taken as evidence of abnormal glycosuria. It may be due to the normal amount of sugar in the urine.

Remarks. The method is more sensitive than Benedict's and Nylander's. $CHCl_3$ does not give a positive result even when present in considerable excess. Urines of patients treated with chloral hydrate and containing glycuronates are also negative. There is no necessity to remove albumin, but it is advisable to do so by boiling and filtering before treatment with charcoal. If specific gr. of urine is more than 1,025, dilute with equal volume of water and take 10 c.c. of the diluted urine. With a modification of this method Cole detects one part of glucose in a million parts of water, when glucose is present alone.

* 189 (b). Method of distinguishing lactose and glucose in urine (Cole)—In the case of urine from a pregnant or nursing woman the following procedure should be adopted: "Treat 20 c.c. of the urine with 1 gram of charcoal in the method described in 189(a). Treat the whole of the filtrate with another gram of charcoal and repeat the process" (by this means the lactose is adsorbed by the charcoal). "To 5 c.c. of the filtrate from this add ½ gram of anhydrous Na_2CO_3 and 6 drops of glycerine and boil for 50 seconds. Now add to the hot solution 4 drops of 5 per cent. $CuSO_4$ solution and set the tube aside for 1 minute. A reduction appearing within the specified time indicates the presence of at least 0.04 per cent. of glucose in the urine.

If less than 0.3 per cent. of lactose is present, it is entirely adsorbed by 10 per cent. of charcoal, as in the routine method. By using 5 per cent. of charcoal, as in the special modification, a considerable amount of lactose is removed, and that left is entirely adsorbed by the second reatment. The addition of 1 per cent. lactose to normal urine fails give a positive test in the filtrate by this method, whereas 0.04 per ent. glucose added gives a distinct reaction. If this test gives a negaive result, and the original urine responds to Benedict's test, the urine lmost certainly contains lactose.

189(c). Identification of lactose in urine (Cole).—*Principle.* dsorption of lactose by charcoal and the formation of lactosazone in he presence of acetic acid.

Procedure. To 1 gm. of charcoal add 25 c.c. of suspected urine, mix y shaking, boil for a few seconds, cool thoroughly and shake at interals for 10 minutes. Filter through a small paper or use a filter pump. Vhen the charcoal has completely drained, transfer it to a porcelain dish ontaining 10 c.c. of water and 1 c.c. glacial acetic acid. This is best one by opening the paper, holding it by the clean half and moving it bout in the liquid. The greater part of the charcoal is thus removed rom the paper. Stir the charcoal with a glass rod and transfer the iixture to a boiling tube. Heat to boiling for about 10 seconds and lter the hot solution through a small paper into a test-tube containing s much solid phenyl-hydrazine-hydrochloride as will lie on a quarter, nd twice this amount of solid sodium acetate. Mix thoroughly and lter from any insoluble oily residue. Place the tube in a boiling water ath and leave it there for 45 minutes. Remove the tube and allow it o stand at room temperature for at least one hour. It is advisable to llow it to stand longer if possible. Pipette off a little of the deposit, f any, and examine it on a slide under the high power of the microscope. actosazone crystallizes in characteristic clumps with projecting spines "hedge-hog" crystals). If desired filter off crystals, redissolve in hot vater and allow to recrystallize. Melting point of crystals 200° C.

* 190. Detection of acetone in urine. Detection in the undisilled urine. Legal's nitroprusside reaction.—To 5 c.c. of urine in a est-tube add a few drops of a fresh dilute solution of sodium nitroprusside and then make the urine alkaline with sodium hydrate. A ruby-red olor results both in normal and abnormal urine, due to the creatinine f the urine. If the urine is now made acid with acetic acid, if creatinine lone is responsible for the color, the solution will become yellow, whereas f acetone is present the red color is intensified by the addition of cetic acid. Control this test with normal urine.

* 191. Rothera's nitroprusside reaction.—This test is better than Legal's. To 5 c.c. of urine add a little solid ammonium sulphate and 2-3 drops of a fresh 5 per cent. solution of sodium nitroprusside and 1-2 c.c. of concentrated ammonium hydrate. If acetone is present, a permanganate color develops.

Acetone may be more accurately detected in the distillate from urine.

Distill about 20 c.c. of liquid from 200 c.c. of urine acidifie addition of several drops of concentrated HCl. Use a good wi denser, as acetone is volatile. The distillate may be subjecte following examination.

* 192. Iodoform test.—*Gunning's.* To 5 c.c. of the distill distilled urine may be used) add a few drops of KI_3 solution = 4 grams iodine; 6 grams KI; 100 c.c. water) and enough (5-10 drops) to make a black precipitate of nitrogen iodide. ing this is changed to iodoform (CHI_3). Detect this by the by examining the crystals. They are yellow in color and form h plates and rosettes. Neither alcohol nor aldehyde form iodofor these conditions. Acetoacetic acid is decomposed by heating so that the distillate contains acetone from this as well as p acetone.

* 193. Detection of acetoacetic acid.—$CH_3.CO.CH_2.COO$ cetic acid is readily decomposed, forming acetone if the acid boiled. It gives a much more sensitive reaction with sodiu prusside than does acetone in the Le Nobel test. The test may in the following way (Harding and Ruttan, *Biochem. Journal* 445, 1912). Acidify the urine with acetic acid, add 0.5 c.c. N/1 nitroprusside (1 c.c. = 0.0298 grs. $Na_2Fe(CN)_5NO.2H_2O$) a overlie the solution with concentrated aqueous NH_4OH. A vi is produced. This is a modification of Taylor's method of carr the reaction. Acetoacetic acid gives this reaction far better than A solution of acetone 0.057 per cent. gives a very faint reddis ring after about 20 minutes. Acetoacetic acid gives it at once. acetic acid 1 pt. in 30,000 will give a positive reaction. It is fa delicate than the Gerhardt ferric-chloride test.

* 194(a). Gerhardt's reaction for acetoacetic acid.—To 5 urine in a test-tube add ferric-chloride solution, drop by drop, as a precipitate forms. An addition of more $FeCl_3$ to the filtra duces a Bordeaux red color if acetoacetic acid is present. 1 7,000 is the limit of the reaction. Many other substances give th tion with $FeCl_3$, such as salicylic acid, and substances excreted af ingestion of antipyrin, phenacetin and thallin. It is an enol re Acetoacetic acid may be extracted from the urine by acidifyin shaking with ether and the test made in the ether extract. Aceto not give this reaction.

* 194(b). Salicylaldehyde reaction for acetone.—To 5 c.c. (or better distillate) add 1 c.c. of a 10 per cent. alcoholic solut salicylaldehyde, shake well and introduce a piece of solid NaOH the size of a large pea. Set aside. Prepare a control tube with di water or normal urine in the same way. In case acetone is pres

the extent of 0.01 per cent. or more a deep orange-red ring is formed at the junction of the two layers. Normal urines and urines containing aldehydes give a yellowish to brown ring. The red-colored substance formed with acetone is 0.0-dioxydibenzolacetone.

* 194(c). **Black's reaction for beta-oxybutyric acid.**—Concentrate 15 c.c. urine to 5 c.c. on the steam bath; this is done to remove any acetoacetic acid which may be present. Now add two drops concentrated HCl and the least quantity of plaster of Paris necessary to form a solid mass. Pulverize this mass and extract twice with ether. Evaporate the ether extract without loss, add 5 c.c. water and a small amount of $BaCO_3$. To the almost neutral water layer now add 2 or 3 drops hydrogen-peroxide solution, shake and add a few drops of 5 per cent. $FeCl_3$. Set aside for a few minutes. If beta-oxybutyric acid was present, it has been oxidized to acetoacetic acid and gives the usual red color with ferric salts. Black claims this method will detect the acid in concentrations of 1:10,000.

195. **Detection of glycuronic acid.**—CHO.(CHOH)$_4$.COOH. See page 755. This acid gives many of the reactions of the pentoses. It may be differentiated as follows (*Tollen's reaction*): Add to equal quantities of urine and concentrated hydrochloric acid 0.5-1 c.c. of a 1 per cent. naphthoresorcin solution in alcohol. The solution is heated slowly to boiling and boiled one minute, shaking the tube constantly. Allow to stand and cool for four minutes, then cool under the tap and shake out with ether. The red or violet color goes into the ether if a glycuronate was present, but is insoluble in ether if the color is due to a pentose or hexose. The ether solution examined spectroscopically shows two bands, one on the D line and one between D and E.

Glycuronates reduce Fehling's solution, but they do not ferment with yeast. On heating with strong acid the glycuronic acid is set free. In the free state it is dextro-rotatory. In the paired condition it is levo-rotatory. It occurs in the urine normally in very small amounts; in larger quantities after certain drugs are taken (see page 755).

196. **Detection of glycuronic acid in urine** (Tollens-Neuberg and Schwket, *Biochem. Zeitschrift*, 44, 1912, 502).—In a small separatory funnel place 10 c.c. fresh urine with 2 c.c. dilute H_2SO_4. Add at once 10 c.c. ordinary alcohol and 20 c.c. ether. After several strong shakings, hasten the separation by the addition of a few c.c. of water or NaCl solution. As soon as the ether layer is separated, draw off the alcohol-water layer and shake the ether with 2-3 c.c. H_2O or NaCl solution. Separate the ether carefully and filter through a small, dry filter into a porcelain dish. Add 5 c.c. water and evaporate the ether on the water bath. Divide the fluid which remains, and which may have some oil drops or be slightly cloudy, into two parts and test one with orcin

and the other with naphthoresorcin. By using 10 c.c. of normal urine both tests are positive. By shaking out with ether the glycuronic acid esters are separated from the pentoses which may be present and which give similar reactions. If the amount of glycuronic acid is small, the amyl-alcohol extract of the orcin test only gives a plain absorption band after standing.

197. Detection of pentoses.—These occur in the urine in disease of the pancreas and in some other conditions, the amount depending on the amount eaten. There is still some doubt about the character of the pentose found in the urine in endogenous pentosuria. The presence of pentoses may be inferred if the urine reduces Fehling's solution, does not ferment with yeast, gives a positive reaction by Tollen's naphthoresorcin reaction, the color being insoluble in ether; and forms furfural when the urine is made strongly acid with phosphoric or hydrochloric acid and distilled. Furfural comes off both from pentoses and glycuronates. The other pentose reactions may also be tried: i.e., with phloroglucin and orcinol. To make the phloroglucin test, take 1 c.c. of 2 per cent. *phloroglucin* solution, 5 c.c. of concentrated HCl and 6 c.c. of urine, mix well and immerse in boiling water. A red color indicates pentose or glycuronic acid. In using *orcinol* proceed in the same way, but use half-saturated orcinol solution, 1 c.c., in place of the phloroglucin. A blue color and precipitate indicates pentoses (see page 866). Extract the color with amyl alcohol.

* 198. Detection of bile salts. Reduction of surface tension of urine. Hay's test.—Bile lowers the surface tension of water. This lowering may be detected by sprinkling some flowers of sulphur on the surface of the urine. In the presence of bile the sulphur falls at once to the bottom of the beaker. In normal urine it floats on the surface. Other greases act in the same way as bile. The reaction is not specific.

* 199. Pettenkofer's test for bile salts.—This has already been described in experiment 117, page 917. A red ring is formed at the junction of the sulphuric acid and the urine after the addition of a crystal of saccharose. Keep the temperature below 70° in making this test. A little furfural may be added in place of the saccharose. The test remains essentially the same.

* 200. Detection of bile pigments in urine.—Gmelin's test already described can be applied to the urine. Pour 5 c.c. of urine over some concentrated nitric acid in a test-tube, holding the tube inclined while pouring so that the liquids do not mix. At the zone of contact a series of violet, green-blue and red colored rings develop in the presence of bile. The blue-green ring is characteristic. Indoxyl and scatoxyl also produce colors in this test. Urobilin does not.

200(b). The bile pigments can also be detected by the Huppert-Cole reaction described in experiment 113.

200(c). Examination for other pigments.—(a) Urobilin. To 5 c.c. urine add an equal volume of a 10 per cent. solution of zinc acetate in absolute alcohol and filter. The filtrate should possess a reddish-green fluorescence. In case this gives a negative result shake 10 c.c. urine with 5 c.c. warm amyl alcohol after acidifying with hydrochloric acid. Remove the amyl-alcohol layer, filter if necessary, render it alkaline with NH_4OH and add ½ volume of the 10 per cent. alcoholic zinc-acetate solution. Shake. Green fluorescence if urobilin present.

(b) Bile pigments (Gmelin-Rosenbach reaction).—Saturate a piece of filter paper with urine by filtering much urine through it. Allow it to dry partially and drop thereon a few drops of concentrated nitric acid. Around each drop of nitric acid soon appears a series of concentric rings—yellow-red, red, violet, blue and green.

(c) Huppert-Nakayama reaction for bile pigments.—To 5 c.c. urine in a centrifuge tube add 5 c.c. 5 per cent. $BaCl_2$ solution, mix and centrifuge. Pour off the supernatant clear solution, add to the precipitate 2 c.c. of the special ferric-chloride reagent (4.0 grams $FeCl_3$ dissolved in 990 c.c. 95 per cent. alcohol + 10 c.c. concentrated HCl) and heat to boiling. The solution becomes green to bluish green and after adding a drop of concentrated nitric acid (yellow) the solution becomes violet and red. This reaction is stated to detect 1 part bilirubin in 1,000,000 parts urine.

Blood in Urine.

* 201. Hemoglobin derivatives.—(a) *Guaiac reaction.* To 5 c.c. acid (slightly) urine add about 0.5 c.c. of a freshly prepared tincture of guaiac, mix and then add about 1 c.c. old turpentine. Shake thoroughly and set aside for a few minutes. If a green to blue color has not developed, add a few c.c. of alcohol and again shake gently. In case the result is still negative one can conclude that hemoglobin is absent from the urine. A blank test, using water in place of the urine, should be made also. The turpentine can be prepared by exposing it in a flat dish for some hours to direct sunlight. Or in making the test add a drop of hydrogen peroxide. It is necessary that the turpentine be partially oxidized.

(b) *Teichmann hemin reaction.* In case (a) above is positive it is best to confirm it by this test. To 10 to 15 c.c. urine add 10 per cent. NaOH to make it strongly alkaline, heat to boiling and set aside to cool. Filter off the precipitate of phosphates, etc. Wash with distilled water and finally transfer the precipitate to a microscope slide, add a very small crystal of NaCl and very carefully evaporate to dryness over a low flame. Cover the *dry* material with a cover glass [illegible] allow to flow under it a drop or two of *glacial* acetic acid. Now again warm gently over a free flame until the acetic acid begins to boil gently. Allow to cool and examine under the microscope for the reddish [illegible] hemin crystals.

(c) *Spectroscopic examination.* Examine by a direct-vision spectroscope for oxyhemoglobin and methemoglobin.

QUANTITATIVE EXAMINATION OF THE URINE.

For the quantitative experiments all the urine for 24 hours must [illegible] carefully collected. Two kinds of urine are to be examined: (a) while the student is on a low protein diet and (b) while he is on a high [illegible]tein diet.

In the diet part of the work students may work in pairs, provided they alternate in making the various determinations. The urine must, however, be collected in both periods from the same individual and during 24 consecutive hours each time. The same amount and kind of [illegible]cise must be taken during both diet periods. Bottles provided for the urine are to be properly cleaned and rinsed with a saturated alcoholic solution of thymol before taking from the laboratory. During the first period the subject is to take a low protein diet for two days. On [illegible] second day of this period the urine voided before breakfast is to be discarded, but all passed during the next 24 hours is to be collected. Other hours may be more convenient, but *consecutive 24 hours should be taken each time.* Also have the urine as fresh as possible for analysis. In the same way collect a second sample of urine two weeks later on the second day of a two-days' diet on high protein. If possible weigh the food taken and from the record obtained calculate the caloric value in both diets. Submit your proposed diet to an instructor before beginning thereon.

* 202. Volume.—Determine the volume of the well-mixed 24-hour sample by means of a 1,000 or 2,000 c.c. cylinder.

* 203. Specific gravity.—Determine the specific gravity by means of a urinometer. Note the temperature and for every 3° C. over the temperature recorded on the urinometer add 0.001 to the observed specific gravity. For every 3° below subtract 0.001 from the observed specific gravity. It is better, however, to cool the urine to the temperature at which the urinometer is to be employed and then determine the specific gravity.

* 204. Odor, color and transparency.—Note these carefully and interpret your observations.

* 205. Total nitrogen by the Gunning-Arnold modification of the

Kjeldahl process.—Carefully measure into a long-necked 800-1,000 c.c. Kjeldahl flask 5 c.c. of urine by means of a 5 c.c. pipette. With a fine jet of water wash any urine in the neck of the flask down into it. Use as little water as possible. Add 5 grams potassium sulphate, or 10 grams sodium sulphate, 20 c.c. pure concentrated H_2SO_4 (Sp. G. 1.84) and 10

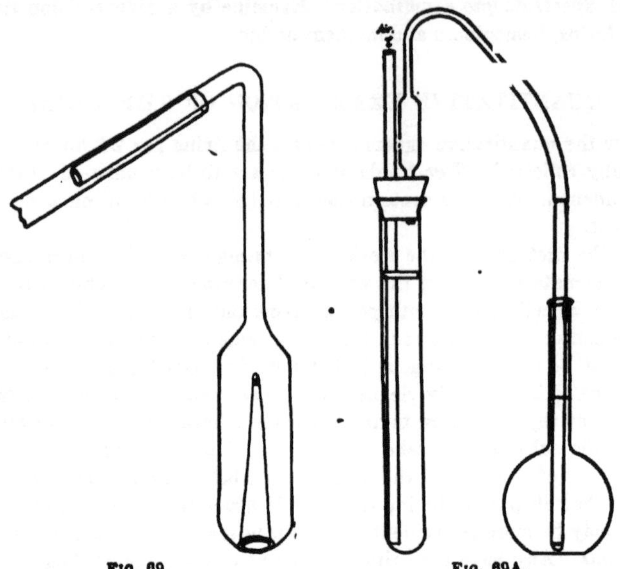

FIG. 69. FIG. 69A.

FIG. 69.—Adapter to be placed in the top of the Kjeldahl digestion flask or large test-tube to be connected with the suction for the removal of SO_2 or other harmful fumes (Folin).

FIG. 69A.—Apparatus set up for the aëration of the ammonia into the sulphuric acid. The test-tube contains the Kjeldahl digestion made alkaline. The air is blown through this and into the 100 c.c. volumetric flask containing the sulphuric-acid solution. A perforated cork disk is placed part way down the tube to prevent foam going over (Folin).

c.c. of the $HgSO_4+CuSO_4$ solution (containing 1 gram $HgSO_4$ and 1 gram $CuSO_4$ per 10 c.c.).

Heat on the digestion shelf until the water has been removed, then add 15 grams more of potassium sulphate or 30 grams $Na_2SO_4 10H_2O$, and continue the digestion for 2½ to 3 hours. To have the digestion complete, the mixture must, however, have been *boiling* for at least 2½ hours and the hot solution left must not be colored yellow, but simply green, due to the $CuSO_4$. After the digestion is complete, remove from the digestion shelf and, just as the contents of the flask begin to crystallize, gradually add, while agitating, 250 c.c. distilled water; stopper loosely and set aside until it is to be treated as indicated in experiment 54.

206. Determination of total nitrogen in urine. Mi
(Folin and Farmer, *Jour. Biol. Chem.*, XI, 1912, p. 493
urine are measured into a 50 c.c. measuring flask, if the sp
of the urine is over 1.018, or into a 25 c.c. flask if the speci
less than 1.018 (the urine should be so diluted that 1 c.c. c

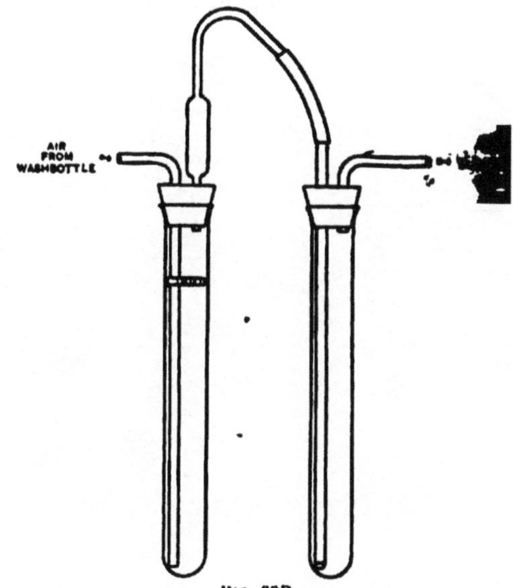

FIG. 69B.

FIG. 69B.—The same as Fig. 69, except arranged for suction in g
(Folin).

0.75-1.5 mgs. of N). Fill the flask to the mark with wate
a few times to secure a thorough mixing. 1 c.c. of this c
is measured into a large test-tube of Jena glass (size 20 t
200 mm.). To the urine in the test-tube add 1 c.c. of
sulphuric acid, 1 gram of potassium sulphate, 1 drop of
$CuSO_4$ solution and a small, clean quartz pebble to preve
Boil over a micro-burner for about 6 minutes: i.e., about 2 r
the mixture has become colorless. (Figure 69 and Figure
to cool about 3 minutes until the digestion mixture is l
become viscous, but it must not be allowed to solidify.
6 c.c. of water at first, a few drops at a time, then more r
to prevent the mixture from solidifying. To the acid solu
added an excess of sodium hydrate (3 c.c. of saturated s
the NH_3 is aspirated by means of a rapid air current (see
into a 100 c.c. measuring flask containing about 20 c.c. of

c.c. of N/10 H_2SO_4. The air current used for driving off the ammonia should be rather moderate for the first 2 minutes, but thereafter for 8 minutes it should be as rapid as the apparatus can stand. Compressed air is best used, but in the absence of this an air current may be drawn through by a pump. In that case the NH_3 is not received directly into the measuring flask, since the mouth of the latter is too narrow to take

FIG. 70.—Apparatus as set up in Folin's laboratory, showing the small Kjeldahl digestion flasks or test-tube at the right with the adapter above and micro-burner beneath. Other solutions giving off odors or irritating fumes are shown over the other burners. The large bottle contains an alkaline solution, and the whole is connected with the water pump at the left (Folin).

a double-bored stopper, but is received in a second test-tube and the contents of this afterwards washed into a 100 c.c. measuring flask.

Nesslerizing. After 10 minutes disconnect; dilute the contents of the flask to about 60 c.c. and dilute similarly the standard ammonium-sulphate solution, of which enough has been taken to contain 1 mg. of N, to about the same volume in a second 100 c.c. volumetric flask. Nesslerize both solutions as nearly as possible at the same time with 5 c.c. of Nessler's reagent diluted immediately beforehand with about 25 c.c. of water (5 c.c. of Nessler's reagent gives the maximum color with 1 to 2 mgs. of ammonia and when diluted, as indicated, turbidity is avoided). The color produced does not reach its maximum for about half an hour, but the increase is small and is immaterial when both are Nesslerized at the same time. Fill the two flasks at once to the mark with distilled water and, after mixing each, determine the relative intensities of color by a Dubosc colorimeter. Set the standard at 20 and match the unknown with it by moving the prism up or down. The calculation is easy. If the unknown is at 14 when the two have the same tint in the colorimeter, the amount of nitrogen is 20/14× 1 mg. in the amount of urine taken. It is important in reading the colorimeter to bring the two sides to equality of color, first from below and then from above, or vice versa, and to take the mean of several readings. In the case cited 1 c.c. contains 1.43 mgs. N. If the urine was diluted 10×, then each c.c. of the original urine contained 14.3 mgs. 100 c.c. of urine would contains 1.43 grams of N.

Preparation of the standard solution of ammonium sulphate. Ammonium salts generally contain pyridine bases. To prepare pure ammonium sulphate decompose a high-grade ammonium sulphate with caustic soda and pass the ammonia by means of an air current into pure sulphuric acid. Precipitate the salt so obtained by the addition of alcohol, redissolve in water and again precipitate with alcohol and finally dry in a desiccator over sulphuric acid.

9.4285 grams of ammonium sulphate are dissolved in water and the volume made up to 1 liter (stock solution).

100 c.c. of the stock solution are diluted to form 1 liter. This is the standard solution. 5 c.c. contain 1 mg. of nitrogen.

Nessler reagent. Dissolve 62.5 gm. KI in about 250 c.c. distilled water. Set aside a few c.c. and add gradually to the larger part a cold saturated solution of mercuric chloride until a faint permanent precipitate is produced (about 500 c.c. required). Add the reserve portion of KI and then more $HgCl_2$ solution gradually until a slight permanent precipitate is again formed. Dissolve 150 grams of solid KOH in 150 c.c. distilled water, allow to cool and add gradually to the above solution and make the volume finally to 1 liter with water. After settling, decant the clear liquid into another bottle and keep in the dark. Reagent improves on keeping.

Quantitative determination of urea.—The best method for the quantitative determination of urea is probably the urease method. The other methods which are given here are, however, good, particularly when applied to human urine which has little or no allantoine. It is the experience of some that the Benedict method is more reliable in the hands of students than that of Folin. The Benedict method determines, like the Folin, not only the urea, but about 50-70 per cent. of the allantoine. Allantoine is present in the urine of the lower animals in larger amounts than in human urine, so that the urease method is to be preferred in examining such urines.

207. Estimation of urea by the Folin method.—*Principle.* The urea is hydrolyzed to ammonium carbonate by the action of hydrochloric acid, the boiling temperature being raised by the addition of $MgCl_2$. The ammonia is then distilled as in the Kjeldahl method.

Method. Measure off 5 c.c. urine into a 1,000 c.c. Kjeldahl flask, add 20 grams magnesium chloride, a piece of paraffin about the size of a pea and 5 c.c. concentrated HCl (use from the special bottle provided for this purpose). Now fit the flask with a test-tube or Folin's condenser and boil on a wire gauze. Heat until the liquid bumps, then reduce the heat and continue the boiling for 1½ hours. Next add 300 c.c. distilled water and granular zinc, as in the Kjeldahl method. In case the $MgCl_2$ has crystallized out, it must be dissolved before adding the alkali as directed below. Next add 5.0 c.c. of the special NaOH solution (40:100), immediately connect with the distilling apparatus as in the Kjeldahl method and distill into 50 c.c. n/10 H_2SO_4 in the same way. Observe the same precautions here in drying the flask and the neck. Continue the boiling until about 200 c.c. distillate have been collected. Titrate as usual. Calculate the total grams of urea nitrogen in the 24-hour sample after allowing for a correction for NH_4OH in the reagents used. After determining the ammonia nitrogen in the urine, subtract this from the above figure and the difference is urea nitrogen. Explain this process throughout. Calculate the per cent. of the total nitrogen found as urea nitrogen, also the grams of urea nitrogen in the 24-hour sample and the grams of urea in the 24-hour sample.

* **208. Estimation of urea by the Benedict method.**—*Principle.* The urea is hydrolyzed to ammonium carbonate by the acid developed from $KHSO_4$ and $ZnSO_4$ by heating for a certain time at definite temperatures. The solution is then made alkaline and distilled as usual into a measured amount of sulphuric acid as in the Kjeldahl process.

Method. Weigh off 3 grams $KHSO_4$ and 2 grams $ZnSO_4$ and transfer to a large (1 inch) test-tube. Add 5 c.c. urine, a pinch of paraffin and of pumice. Boil almost to dryness in the oil bath kept at 130-150° C. Now transfer to another bath kept at 165-170° C. and heat thus for 1 hour. Allow to cool partly, then transfer the contents of the tube to an 800 c.c. Kjeldahl flask, using hot distilled water and diluting to about 400 c.c. with cold distilled water. Add a small amount of granular zinc and 5 c.c. of the 40:100 NaOH solution. Distill into 50 c.c. n/10 H_2SO_4 as usual. Use congo red as indicator. Correct for the ammonia in the urine and also make a blank test on your reagents and process. Calculate as directed under Folin's method above.

* **209. Quantitative determination of urea by the urease method (Marshall; Van Slyke and Cullen).**—*Principle.* The urea is decom-

posed to ammonium carbonate by the action of the specific enzyme urease, found in the castor bean. The dried urease is prepared and put on the market. This is in many ways the ideal method, for it requires no heating, no addition of salt and the action of the enzyme is specific, not affecting or hydrolyzing the other constituents of the urine. The ammonia nitrogen must be separately determined, as in the other methods, and the amount subtracted from the amount given by this

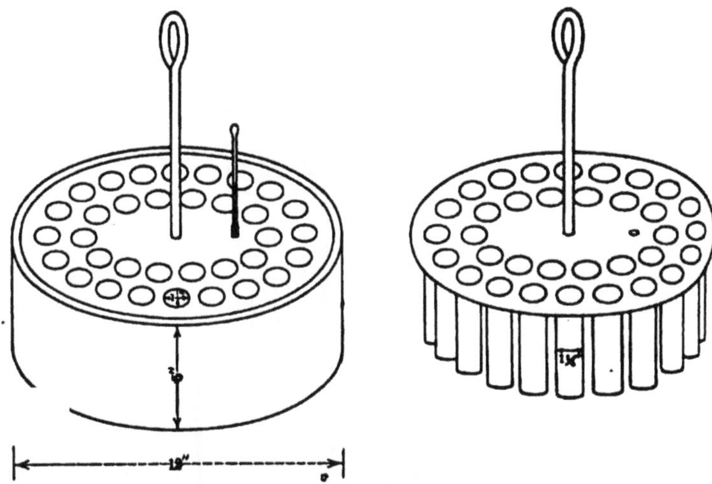

FIG. 71.—Copper baths designed by Koch for use in the Benedict method. They are of copper, and the outer is partially filled with paraffin. The test-tubes, with the urine, etc., go into the copper tubes, which they closely fill, but they do not come in contact with the oil. Two such baths are in the laboratory; one kept at 140-150° and the other at 165-170° C.

method. The ammonia set free by the urease is carried over by an air stream into standard acid and determined either by titration, if large amounts have been taken, or better by Nesslerizing if small amounts of urine are used.

Experiment: Determination of urea in urine. Into each of two clean absorption cylinders (Figure 72) measure off carefully by a burette 20 c.c. N/10 H_2SO_4, then add enough distilled water just to cover the bell of the absorption tube, B, when the same is attached to the Koch ammonia aëration tube, A (Figure 72). Having these prepared, now very carefully measure off into the aëration tube, A, 1 c.c. urine, 10 c.c. water and 5 c.c. of a 5 per cent. urease solution, mix well, add 5 c.c. paraffin oil and at once connect with the absorption tube. Prepare another aëration tube in the same way, using 1 c.c. water instead of 1 c.c. urine. After both tubes are connected, at once aërate at a fair rate for 30 minutes. Next rapidly add to each aëration tube 5 grams dry Na_2CO_3, again connect and again aërate vigorously (avoid foaming) for 1 hour. Titrate the acid solution in the absorption cylinder with N/10 NaOH, using congo red as indicator and leaving the absorption tube in the cylinder. Allow for the blank titration and calculate to urea

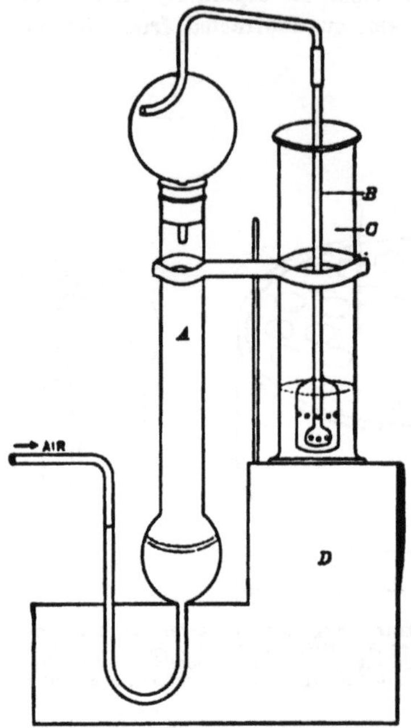

FIG. 72.—Koch's aëration tube for use in Folin's microchemical or macrochemical method of aërating ammonia from a solution to acid. *D* is a section of a wooden base containing, in the Chicago laboratory, 12 aëration apparatus, as indicated. Only half of the wooden base is shown in the cut. The other half is like this and faces it, and the air inlet is a metal pipe provided with cocks running between the two halves. *A* is the aëration tube containing the alkaline solution to be aërated and filled to the mark shown and covered with a little paraffin oil. *C* is a cylinder containing the acid, the amount of which is known; *B* is a Folin aëration tube. An upright metal rod carries metal clips shown, but not lettered, which support the tubes.

nitrogen after correcting for the ammonia nitrogen as estimated below. Calculate the per cent. of the total nitrogen found as urea nitrogen and also the grams of urea nitrogen and urea as such in the 24-hour sample.

209(b). Determination of urea by urease.—The following description of the method has been taken from the circular sent out with the commercial preparation " arlco-urease " and may be followed in the absence of the aëration device just described:

"*Standardization of the enzyme.* One-tenth gram of arlco-urease dissolved in 1 c.c. of water and added to 5 c.c. of 1 per cent. solution of pure urea will hydrolyze .0168 gram of urea, yielding .00953 gram of ammonia, which is equivalent to 28 c.c. of fiftieth normal acid, in 15 minutes at 25° C., or in correspondingly less time as the temperature approaches 50° C. as a maximum. Therefore, from these standardization data one can calculate the amount of arlco-urease required to decompose a given quantity of urea in a solution where the analysis is desired.

"*Determination of urea* (in urine). Dilute the urine ten times. Take 5 c.c. of the diluted urine for analysis. Add to this, in a 100 c.c. test-tube, a few drops of caprylic alcohol, or 1 c.c. of amyl alcohol or kerosene to prevent subsequent foaming; then add 1 c.c. of a 15 per cent. solution of arlco-urease. Close with stopper, and allow to stand until the reaction is complete. Allow 15 minutes at 20° C. (68° F.) or 10 minutes at 25° C. (79° F.); 3 minutes suffice at 50° C. (122° F.). The tube is connected through its stopper with a second tube containing 25 c.c. of fiftieth normal hydrochloric acid. Aërate the digestion mixture for half a minute after the reaction is complete and before the tube is opened, in order to prevent loss of any possible ammonia fumes in upper part of tube. Open tube, add dry potassium carbonate (4 to 5 grams), close quickly and aërate for about 15 minutes, as in Folin's micro-method for ammonia determination, or until all ammonia gas has been carried over into the standard acid. Then to complete the analysis, titrate the acid solution to neutrality with fiftieth normal sodium hydroxide, using as indicator 1 drop of a 1 per cent. sodium alizarin sulphonate solution. The number of c.c. of fiftieth normal acid neutralized by the ammonia from the urea is multiplied by .12 to give the per cent. urea in the urine, or by .056 to give the per cent. urea nitrogen. Under ordinary conditions, the total time for a complete analysis need not exceed 30 minutes, and it can be reduced considerably below this limit.

"(In blood). Urease is particularly valuable in permitting a simple and accurate determination of urea in the blood. Five c.c. of freshly-drawn blood are mixed in a 100 c.c. test-tube with 1 c.c. of 5 per cent. potassium-citrate solution to prevent clotting, a few drops of caprylic alcohol to prevent foaming during subsequent aëration and 1 c.c. of a 10 per cent. solution of arlco-urease. The remainder of the procedure is the same as described for urine, except that only 10 c.c. of fiftieth normal acid need ordinarily be used. In this case the number of c.c. of fiftieth normal acid neutralized is multiplied by .012 to give the per cent. of urea, or by .0056 to give the per cent. of urea nitrogen."

210. Quantitative determination of urea by Folin's microchemical method (*Jour. Biol. Chem.*, XI, 1912, p. 513).—*Principle.* The urea

is decomposed, forming ammonia, by heating with solid potassium acetate and a little acetic acid at 153-160° C. The ammonia is then liberated from the ammonium acetate by the addition of sodium hydrate and carried over by a strong air current into a measured amount of standard acid in a volumetric flask. The contents of the flask are filled up to the mark and the amount of ammonia determined by Nesslerizing as in the total nitrogen method.

Procedure. The urine is diluted so that 1 c.c. contains 0.75 to 1.5 mgs. of urea nitrogen. Dilutions of 1 in 20, 1 in 10 or rarely 1 in 5 are usually adequate for this purpose. One cubic centimeter of the diluted urine is then transferred by means of an accurate pipette (Ostwald) to a large, dry Jena-glass test-tube (200 mm. by 20 mm.) previously charged with 7 grams of dry potassium acetate (free from lumps), 1 c.c. of 50 per cent. acetic acid, a small sand pebble or better a little powdered zinc (not zinc dust) to prevent bumping during the boiling, and a temperature indicator. Close test-tube by means of a rubber stopper carrying an empty, narrow calcium-chloride tube without a bulb (size 25 cm. by 1.5 cm.) to act as a condenser. Suspend the tube by means of a burette clamp so that it can be raised or lowered and heat over a micro-burner. As soon as the acetate dissolves and the mixture begins to boil, usually in about 2 minutes, the indicator begins to melt, showing that the desired temperature (153-160°) has been reached. Continue gentle boiling for 10 minutes, when decomposition of the urea is complete. Remove apparatus from the flame and dilute the contents of the tube with 5 c.c. of water added by a pipette partly through the $CaCl_2$ tube so as to wash off the bottom of the rubber stopper and the sides of the tube. Use not more than 5 c.c. for this purpose. Add 2 c.c. of saturated NaOH or K_2CO_3 solution and drive the liberated NH_3 over into a 100 c.c. volumetric flask containing about 35 c.c. water and 2 c.c. N/10 sulphuric acid, by means of an air current as in the total nitrogen determination. (See Figure 69A.) 10 minutes is sufficient if a rapid air current is used. Determine the N in the flask after making up to 100 c.c. by the Nessler and colorimetric method as described under experiment 206.

This method determines both urea and ammonia nitrogen. Neither creatinine nor hippuric acid gives a trace of ammonia. Allantoine may give off about ½ its N in the presence of zinc, but in its absence it behaves like urea, provided its quantity does not exceed 0.5 mgs. of allantoine N. Uric acid may give a little N, but it is quantitatively imperceptible, in the presence of the urea.

It is important to have potassium acetate which is dry and free from ammonia. This is made by J. T. Baker Chemical Company, Phillipsburg, N. J. The best German product is generally dry enough, but it should be tested for ammonia. The American acetate, except that mentioned, is apt to have too much water in it. It may be dried by placing about a pound of it in a large porcelain dish and letting it stand on a warm plate (at about 115°) for about 24 hours. The plate must not be too hot or acetic acid will be driven off.

Temperature indicator. This consists of a small amount of HgICl sealed into a small glass bulb not over 1 mm. in diameter. They may be obtained from Eimer and Amend in New York. This salt is bright red at ordinary temperatures. At 118° it turns lemon yellow and at 155° melts to a clear dark red liquid. It solidifies on cooling at 148° C. and slowly takes again its red color in about 24 hours. The acetate begins to cake and solidify at 160° C. So this temperature must be avoided. The indicator is made by heating in a dry state intimately mixed molecular proportions of mercuric chloride and iodide at 150-160° for 6 to 8 hours. It is then cooled, powdered and kept dry until sealed up in bulbs for use. The advantage of this method is that a thermometer need not be introduced.

It is important in making the determination that no bumping or spattering occur. The test-tube must be dry. If too much heat is applied the acetate cakes at the bottom, if too little it cakes at the top. The bottom of the test-tube should be some distance above the micro-flame, which should be about 0.5 cm. long. It should boil gently without caking. Bottomless beakers make excellent wind shields for these micro flames.

Computation. If the colorimeter tube A, the standard, has been set at 20 and the other tube B reads 15 when both colors match, then the amount of nitrogen in the 1 c.c. of diluted urine taken would be 20/15 mg. If the urine was diluted 10 times, the amount of urea N in 1 c.c. of undiluted urine would be 200/15 mgs. And in 100 c.c. of urine it would be 100 times this amount.

*211. Estimation of urea by the hypobromite method (Krogh, *Zeit. f. physiol. Chem.*, LXXXIV, 1913).—*Principle.* The urea is oxidized by bromine, or more properly by sodium hypobromite, with the formation of carbon dioxide and nitrogen gas. The carbon dioxide is absorbed by strong alkali and the nitrogen collected and measured volumetrically. The reaction is as follows:

$$CO(NH_2)_2 + 3NaOBr = 3NaBr + CO_2 + 2H_2O + N_2$$

Even with pure urea and with temperature and barometric pressure correction this method gives inaccurate results, but in the urine as ordinarily tried it gives very inaccurate results. Not only urea but also ammonia salts and other nitrogen constituents of the urine give off their N so that the amount of N found is generally nearer the total N than it is the urea N. It gives, however, a rapid, not too difficult, approximate determination of the total N of the urine which can be used clinically. Some CO and NO gases are formed also.

Process. Use the apparatus figured in cut 73. Various forms of ureomoters have also been proposed, of which one is figured in Figure 74. They are very inaccurate, but convenient. A 50 c.c. burette (a) is

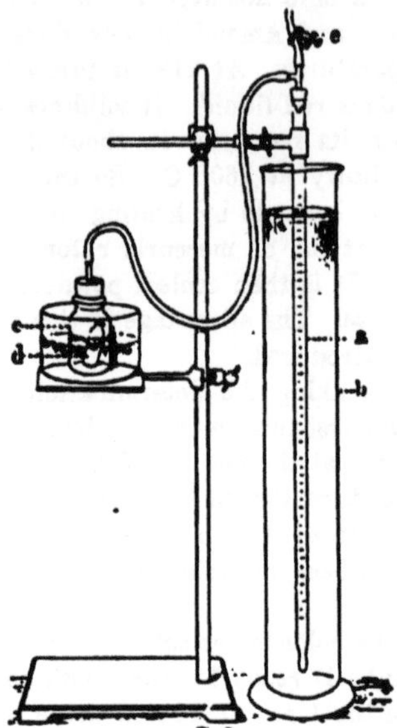

Fig. 73.—Arrangement for the determination of urea by the hypobromite method.

supported in a clamp attached to a lampstand and immersed in a tall measuring cylinder filled with water. The upper end of the burette carries a one-hole rubber stopper through which passes a T tube, the free end of the tube being closed by a rubber tubing clamped by a screw clamp (e). The other arm of the T is connected by a piece of rubber tubing to a glass tube which goes through the one-hole rubber stopper in the neck of the wide-mouth bottle C. The bottle C is supported in a crystallizing dish resting on a ring of the lampstand. Water is put in the crystallizing dish so as nearly to cover the bottle. A short glass vial (d) of about 10 c.c. capacity is placed in the bottle.

Place 25 c.c. of hypobromite solution in the bottle C. Into the vial place 5 c.c. of the urine. With forceps carefully place the vial upright in the hypobromite bottle so that the urine does not come in contact with the solution. While the screw clamp (e) is open place the stopper in the bottle and the bottle in the jar of water and leave for a few moments to acquire the temperature of the water. Adjust the burette so that the level of the water in the burette is a little below the beginning of the scale on the burette, screw the screw clamp tight shut and read the burette. The level of the water should be the same within and without the burette when the reading is made. Now raise the bottle

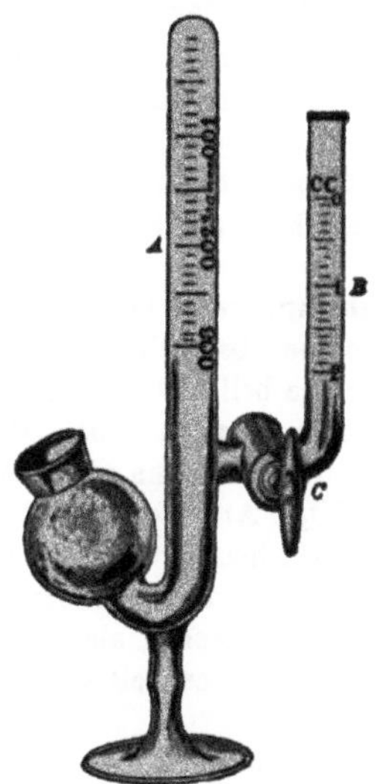

Fig. 74.—Doremus ureometer.

and capsize the vial and mix the hypobromite and urine thoroughly. Gas is given off. Place the bottle back in the water and after three minutes again adjust the burette by raising it so that the level of the water within and without the burette is again the same. Read the burette carefully and subtract from this the former reading. The result is the c.c. of N_2 gas which have been evolved.

Calculation. The reading is corrected for temperature and pressure and the pressure of aqueous vapor as follows: If v is the actual reading, if t is the temperature at which the reading was made, p' the pressure due to aqueous vapor and p the barometric pressure, then at 0° and 760 mm. barometric pressure the real volume of N gas evolved measured at 0° and 760 mm. is v'.

$$v' = \frac{v \times 273 \times (p - p')}{(273 + t) \times 760}$$

1 gram of urea evolves 357 c.c. of N, measured under standard conditions. v'/357 represents, then, the amount of urea in 5 c.c. of urine. And 20 v'/357, the grams of urea in 100 c.c. of urine.

Hypobromite solution. Add slowly 25 c.c. of bromine to the cooled solution of 100 grams NaOH in 250 c.c. of water. Cool under the tap while adding the bromine. The solution must be made shortly before use, since on standing a part is converted to bromate. The reaction is as follows:

$$2NaOH + Br_2 = 2NaOBr$$
$$3NaOBr = 2NaBr + NaBrO_3$$

* **212. Estimation of ammonia nitrogen by the Folin macrochemical method.**—*Principle.* The ammonia of the urine is freed by making the urine alkaline by the addition of Na_2CO_3. A strong current of air is then blown or drawn through the apparatus and carries the ammonia over into a second tube which contains a measured amount of N/10 H_2SO_4.

Procedure. Set up the apparatus as shown in Figure 72. Introduce 20 c.c. N/10 H_2SO_4 into the absorption cylinder and add enough distilled water to just cover the bell of the absorption tube when the same is attached to the ammonia aëration tube. Carefully measure off by means of a pipette 25 c.c. urine into the aëration tube, also 5 c.c. paraffin oil and 5 grams dry Na_2CO_3. At once connect with the absorption tube and aërate vigorously for 1½ hours. Titrate in the absorption cylinder as indicated above, using congo red as indicator. Calculate the total grams of nitrogen found as ammonia, also calculate what per cent. of the total nitrogen this represents, as well as the total grams of ammonia found in the 24-hour sample.

213. Microchemical method for the estimation of ammonia in urine (Folin and Macallum, *Jour. Biol. Chem,* XI, 1912, p. 523).—Into a large test-tube measure by means of an accurate pipette 1 to 5 c.c. of urine. The volume taken should give 0.75 to 1.5 mgs. ammonia nitrogen. With normal urines 2 c.c. will most often give the amount. With very dilute urines 5 c.c. will be necessary, and with diabetic urine with much ammonia 1 c.c. may be too much and the urine must be diluted. Add to the urine in the tube enough water to make the total about 5 c.c., then a few drops of a solution containing 10 per cent. of K_2CO_3 and 15 per cent. potassium oxalate, and a few drops of kerosene or heavy, crude machine oil to prevent foaming. Introduce a tube through a cork as in the determination of the total nitrogen and pass a strong current of air through the mixture for 10 minutes, or as long as it is necessary to drive off all the NH_3, and collect the NH_3 in a 100 c.c. measuring flask containing about 20 c.c. of water and 2 c.c. N/10 H_2SO_4. Nesslerize as in the total N method and compare the color with 1 mg. of nitrogen obtained from a standard ammonium-sulphate solution and similarly Nesslerized. This method is very convenient and rapid. The computation is made as in the other colorimetric determinations.

*** 214. Estimation of creatinine by the Folin colorimetric method**—The significance and chemistry of creatinine is discussed on page 702.

Principle. The principle of the method consists in the reduction of picric acid, $C_6H_2(NO_2)_3OH$, to picramic acid, $C_6H_2(NO_2)_2(NH_2)OH$. Picramic acid in alkaline solution has a reddish-brown color and its amount is determined colorimetrically by comparing it in a Duboscq colorimeter with a standard of N/2 potassium bichromate solution, or with a standard formed by reducing picric acid by a known amount of creatinine.

Process. Before making the estimation, learn to use the Duboscq colorimeter and compare two samples of N/2 potassium-bichromate solution with each other, setting the one at a depth of 8 mm. and adjusting the other until the two fields appear the same in intensity. Then read the scale; the variation should not be more than 0.2 mm. Continue this drill until you have accustomed your eyes to the instrument.

Into a 500 c.c. volumetric flask introduce 10 c.c. urine, measured by a pipette. Now add by means of a pipette 15 c.c. saturated (aqueous) picric-acid solution, mix well and then add 5 c.c. of a *clear* 10 per cent. NaOH solution. At once mix well, note the time and allow to stand at room temperature for exactly 5 minutes. In the meantime prepare the Duboscq colorimeter with a standard tube of N/2 potassium-bichromate solution and set the scale at 8 mm. After standing the 5 minutes, add enough water to make up to 500 c.c., mix well and compare the solution with the bichromate solution, taking at least three readings, but keeping the standard set at 8 mm. Match the colors both when approaching the correct point from above as well as from below. Slightly different values will be obtained. The three readings should not vary more than 0.2 to 0.3 mm. from each other and the color of the creatinine solution should be of such intensity that the reading on the colorimeter scale will be somewhere between 6 and 12 mm. If the reading is not between these extremes, it is necessary to repeat the test with more or less urine, depending on whether the color is too weak or too intense. Be sure to avoid having air bubbles under the surface of the movable glass prisms.

It has been determined by experiment that 10 mg. of creatinine treated as above and diluted to 500 c.c. gives a colored solution of such intensity that it requires a layer 8.1 mm. deep to give the same intensity as the 8 mm. layer of N/2 potassium bichromate. A layer of 8.1 mm. is then equivalent to 10 mg. creatinine in the original solution under the conditions given here, or the constant $8.1 \times 10 = 81$ holds for all dilutions of creatinine within the limits stated. If the reading X is known, then $81/X$ = mgs. of creatinine in the original urine taken. This method is very good for normal urines, but in certain abnormal urines where acetoacetic acid is present this may give low results. Dextrose up to 5 per cent. is said not to interfere (Greenwald, *Jour. Biol. Chem.*, XIV, 1913, p. 87). Calculate the grams of creatinine found in the total 24-hours' sample, also the grams of nitrogen corresponding thereto and the per cent. of this in terms of total nitrogen.

Note. The color involves the dissociation of the salt of picramic acid and of the bichromate. These do not change their color in the same way with the temperature. The standard bichromate solution has been made to compare with the picric acid-creatinine mixture at a temperature of about 20° C. The reading should be made at this temperature. At higher or lower temperatures the bichromate solution no longer corresponds to the 10 mg. of creatinine when set at 8.1. Each colorimeter, also, should be standardized to make sure that when set at 8.1 it really matches the bichromate solution and a standard creatinine-picric acid solution of the strength indicated. To avoid these difficulties Folin has recently suggested that for comparison a standard solution of creatinine should be treated as is the urine and compared in the colorimeter with it. Variations in temperature will not then be important. Probably picramic acid could itself be used to advantage. Bichromate has the great advantage of convenience and stability.

Bichromate solution. 24.55 grams of pure potassium bichromate dissolved in water and made up to 1,000 c.c.

215. Estimation of creatine by the Folin-Benedict method.—*Principle.* The preformed creatinine is separately determined in the manner just mentioned. Then the urine is heated with acid in an autoclave which converts all the creatine to creatinine. The creatinine is now redetermined. This figure gives the combined creatine and creatinine. The difference between this and the preformed creatinine is considered to represent the creatine present.

Process. Introduce 20 c.c. of urine, or less if the urine is concentrated and contains much creatinine, into a 50 c.c. volumetric flask, add an equal volume of N HCl and heat in an autoclave at 117-120° C. for ½ hour. Dilute the cool mixture to 50 c.c., mix well and introduce by means of a pipette 25 c.c. of the mixture into a 500 c.c. volumetric flask and proceed with the estimation of the creatinine as described in experiment 214. Calculate the amount of creatine present.

Caution. By this method the creatine is determined indirectly. The difference between the two readings is supposed to represent creatine. It has been found that when acetoacetic acid is present in the urine the first, or creatinine, determination is low. By heating with acid the acetoacetic acid is destroyed so that the second reading of creatinine is now higher. The difference would be counted as creatine, whereas it is due to another substance. Other oxidizing substances which are destroyed by acid may behave like acetoacetic acid. Reducing substances will increase the reduction of picric acid and make the creatinine appear high. The interpretation of results must be made with these [illegible] stances in mind.

* 216. Quantitative estimation of uric acid by the Folin-Shaffer method.—*Principle.* The uric acid is precipitated from the urine [illegible] the ammonium urate and the amount in the precipitate determined [illegible] titration with a standard potassium-permanganate solution in the [illegible] ence of acid. Phosphates and some organic matter are first [illegible] tated by uranium acetate and $(NH_4)_2SO_4$.

Process. Into a 350 c.c. beaker introduce 100 c.c. urine (more [illegible] the specific gravity is less than 1.020) and while stirring add by [illegible] of a pipette 25 c.c. Folin-Shaffer reagent (500 grams ammonium [illegible] phate and 5 grams uranium acetate in 710 c.c. of an 0.84 per [illegible] acetic-acid solution). Allow to stand until the precipitate has [illegible] rated fairly well. Then filter through a dry filter paper into a [illegible] beaker or flask. Now measure off 100 c.c. of the filtrate (or more if [illegible] urine is dilute) by means of a pipette into a 300 c.c. beaker. To [illegible] solution then add while stirring 5 c.c. of concentrated ammonium hydroxide and set aside for 24 hours. Filter through a well-washed, properly prepared asbestos filter in a Hirsch funnel or Gooch crucible and [illegible] rapidly with a 10 per cent. ammonium-sulphate solution until the [illegible] is free from chlorides. Now by means of 100 c.c. hot water transfer the asbestos and precipitate back into the 350 c.c. beaker. *After cooling* add gradually while stirring 15 c.c. concentrated H_2SO_4 and titrate at once with n/20 $KMnO_4$ solution from a glass stopcock burette.

Titrate to the point where, on vigorously stirring, the entire solution remains faintly pink for an instant after adding two drops of the standard permanganate solution. The end point is somewhat difficult to determine.

Each c.c. of the n/20 $KMnO_4$ is equivalent to 0.00375 gram uric acid. Calculate the grams of uric acid in the original urine sample taken, adding 0.003 gram to correct for the solubility of ammonium urate in 100 c.c. Calculate the total grams of uric acid, total grams of uric acid nitrogen and the per cent. of the total nitrogen found as uric acid nitrogen in the 24-hours' sample.

N/20 Permanganate solution. 1.581 grams of pure potassium permanganate are dissolved in pure water and made up to 1,000 c.c. Keep protected from light.

217. Quantitative microchemical determination of uric acid (Folin-Macallum, *Jour. Biol. Chem.,* 13, 1913, p. 363).—*Principle.* This method depends on the development of a blue color with phosphotungstic acid by uric acid and its quantitative colorimetric estimation by comparison with the color produced by a standard uric-acid solution. Other

substances which may give the same reaction are removed by extracting the evaporated urine with ether-methyl alcohol. These substances are believed to be chiefly polyphenols. (See 2d method, exp. 218.)

Process. From 2 to 5 c.c. of urine (depending on specific gravity) are measured into a 100 c.c. beaker and after adding a drop of saturated oxalic-acid solution evaporated to dryness on a water bath. To the dry, cool residue add 10-15 c.c. of a mixture consisting of 2 parts of pure dry ether (distilled over sodium) and 1 part of pure methyl alcohol. After standing for about 5 minutes carefully decant the solution and add another 10 c.c. of the ether-alcohol mixture, the residue being allowed to settle and again decanted. To the washed residue add 5-10 c.c. water and a drop of saturated sodium carbonate and stir to secure solution of uric acid. To this solution add finally 2 c.c. of the phosphotungstic acid reagent and then 20 c.c. saturated sodium carbonate. Transfer the blue solution to a 100 c.c. graduated flask, dilute to the mark with water and at once compare the color in a Duboscq colorimeter with a standard solution of uric acid and the reagent.

The standard solution is obtained by weighing out 250 mgs. of Kahlbaum's uric acid, transferring it to a 250 c.c. volumetric flask by means of 25-50 c.c. of water, then adding 25 c.c. of the 0.4 per cent. lithium carbonate solution and shake at intervals for an hour before diluting with water to 250 c.c.. 1 c.c. of this solution is carefully measured with a pipette to a 100 c.c. volumetric flask, 5-10 c.c. of water added and then 2 c.c. of the phosphotungstic-acid reagent and 20 c.c. of saturated sodium carbonate solution and the whole diluted to the mark with water. This standard uric-acid solution will not keep more than 5-6 days.

Since the color is not permanent but fades rapidly it is essential that the addition of the reagent to the standard uric-acid solution and to the urine solution which is being tested for uric acid should be made as nearly simultaneously as possible. Not more than 5 minutes at the outside should separate the two additions.

The two solutions are now placed in the two sides of the Duboscq colorimeter. The standard is set let us say at 20, the unknown is now matched with it. If the standard reads 15 and the unknown 20 then the concentration in the unknown is 15/20 of that of the known. The known contains 1 mg. uric acid in 100 c.c. The unknown then contains 0.75 mg. in the amount of urine taken. If this was 2 c.c. then 100 c.c. of urine would contain 37.5 mgs. of uric acid (0.75×50).

Phosphotungstic-acid reagent for uric acid.—See experiment 164, p. 934.

218. Quantitative microchemical determination of uric acid in urine. Second and improved method (Folin and Denis, *Jour. Biol.*

Chem., 14, 1913, 97).—*Principle.* Development of color in alkaline phospho-tungstic-acid solution after precipitation by ammoniacal silver nitrate and estimation colorimetrically in a Duboscq colorimeter by comparison with a standard. The color is not specific to uric acid.

Process: Measure 1-2 c.c. urine by an accurate pipette (Ostwald type) into an ordinary centrifuge tube. Add distilled water to make the volume in the tube 5 c.c., 6 drops of 3 per cent silver lactate solution two drops magnesia mixture and enough (10-20 drops) of concentrated NH_4OH to dissolve the AgCl. Centrifuge 1-2 minutes, pour off the supernatant liquid from the residue at the bottom of the tube which contains the. uric acid as a silver compound, and add 5-6 drops of freshly prepared saturated hydrogen sulphide solution and one drop of concentrated HCl. Place tube in a beaker of boiling water until all H_2S is driven off. Since H_2S also gives a blue color with the reagent it must be completely removed. To prove that it is removed add one drop of 0.5 per cent. lead acetate solution to the contents of the tube after the latter has been in the water bath about 5 minutes; if any H_2S remains a deep brown precipitate forms. If this is the case heat longer. When the tube is cool add 2 c.c. of the uric-acid reagent, 10 c.c. of saturated Na_2CO_3 solution, transfer to a 50 c.c. volumetric flask and make up to the mark with water. Compare the color as usual against the color obtained from 5 c.c. of the standardized uric acid-formaldehyde solution or a freshly prepared uric-acid solution as directed in experiment 217.

This method is better than the former method, since in many urines (rat and cat, and human urines containing albumin and sugar) the evaporation as formerly practised makes it impossible completely to extract the phenols. Urines containing albumin on addition of H_2S solution take a brownish tint. This difficulty is overcome by addition of 2-10 drops 10 per cent. sodium-acetate solution. The addition makes the results slightly low and should not be added if not necessary.

Standard uric acid-formaldehyde reagent. Dissolve 1 gram of accurately weighed uric acid in 200 c.c. of a 0.4 per cent. solution of Li_2CO_3 in a liter volumetric flask. Add to the solution 40 c.c. 40 per cent. formaldehyde solution, shake and allow to stand a few minutes. Acidify the clear solution by adding 20 c.c. N acetic acid and dilute to the mark with water. The solution should remain perfectly clear and the next day, but not before, standardize it against a freshly prepared lithium carbonate solution of uric acid (see experiment 217). The color produced by 5 c.c. of the solution corresponds very nearly to the color obtained from 1 mg. uric acid. The colorimeter reading obtained when compared with the uric-acid standard is to be used as that which corresponds to 1 mg. uric acid. The solution of formaldehyde-uric acid appears to keep its color-producing power unaltered for many months both in the light and in the dark.

This method gives very good results even in the presence of protein or sugar. Modification: Benedict, *Jour. Biol. Chem.*, 20, 1915, p. 629.

219. Quantitative microchemical determination of uric acid in blood (Folin and Denis, *Jour. Biol. Chem.*, 13, 1913, p. 469).—*Principle.* Phosphotungstic acid colorimetric method. With phosphotungstic acid an amount of uric acid as little as 1 part in 1,000,000 of water gives a positive reaction.

Method. Normal human blood contains about 0.25 mg. of uric acid in 15-25 c.c. of blood and this amount suffices for a quantitative determination. 15-25 c.c. of blood is drawn into a previously weighed, small, wide-mouth bottle containing about 0.1 gram of finely powdered potassium oxalate and shaken with the oxalate. The bottle is then reweighed and the difference gives the quantity of blood taken. An amount of N/100 acetic acid equal to five times the weight of blood taken is placed in an ordinary liter flask and heated to boiling. The oxalate blood is poured quantitatively into the boiling acetic-acid mixture and the heating is continued until the solution has begun to boil. The mixture is filtered while still hot. The coagulated material on the filter paper is transferred back into the flask by means of a spatula, about 200 c.c. of boiling water are poured over it and it is allowed to stand 5 minutes. This mixture is filtered through the same filter as was used for the first filtration. The filtrate in the receiving flask should be nearly as clear as water. To the filtrate add 5 c.c. of 50 per cent. acetic acid and evaporate over a free flame in a deep half globular 10 cm. evaporating dish until fairly concentrated and then on the water bath until a very small volume (3 c.c.). Pour the liquid into an ordinary small centrifuge tube and wash the dish with two successive portions of 0.1 per cent lithium carbonate solution, using about 2 c.c. for each rinsing, and adding the washings to the tube; any solid material adherent to the sides of the dish is removed by a rubber-tipped stirring rod. The liquid in the tube should not be more than 10 c.c. Add to it 5 drops of 3 per cent. silver lactate solution, 2 drops of magnesia mixture, and enough strong ammonium hydrate (10-15 drops) to dissolve the silver chloride. Centrifuge for 2-3 minutes, pour off the supernatant liquid and to the residue add 4-5 drops fresh, saturated H_2S water and 1 drop concentrated HCl. Place now the tube for 5-10 minutes in a beaker of boiling water to remove the excess of H_2S. To make sure that it is all gone add a drop of 0.5 per cent. lead acetate solution to the tube. If any blackening occurs heat the tube another five minutes in the water bath. It is necessary to remove the H_2S completely, since this gives a blue color with the phosphotungstic acid reagent. When all the H_2S is thus shown to be gone centrifuge the tube for one or two minutes. Transfer the supernatant liquid by decantation as completely as possible to a small beaker and wash the inside of the tube with a stream of water from a wash bottle, care being taken not to break up the precipitate in the bottom of the tube. The wash water (not over 5 c.c.) is then added to the liquid in the beaker, and to this acid solution containing the uric acid is then added 2 c.c. of the uric-acid phosphotungstic-acid reagent and 10, 15, or 20 c.c. of saturated sodium carbonate, depending on whether the color obtained requires a final dilution of 25, 50 or 100 c.c. Three volumetric flasks representing 25, 50 and 100 c.c. must be at hand and ready and the blue unknown solution is transferred to one of them and diluted to the mark with water. The standard uric-acid solution and the blue solution is made as directed in experiment 218. It contains 1 mg. in 1 c.c. and is diluted to 100 c.c. This must be made just before the addition of the sodium carbonate to the unknown. It is necessary sometimes to filter the latter before being transferred to the colorimeter for comparison. In all other ways the determination is the same as that for uric acid in the urine.

The calculation is as follows, giving the amount of uric acid in 100 grams of blood. V represents the volume (25, 50 or 100 c.c.) to which the unknown is diluted and W represents the weight of blood taken for the determination. 20 is the reading of the colorimeter of the standard solution and R the reading in millimeters of the unknown solution when its color matches the color of the known solution. Then

20 V/RW=milligrams of uric acid in 100 grams of blood.

220. Quantitative determination of hippuric acid in urine (Folin and Flanders, *Jour. Biol. Chem.*, 11, 1912, p. 257).—*Principle.* The hippuric acid is hydrolyzed to benzoic acid first by treatment of the urine with NaOH and then by boiling with strong nitric acid and a little $Cu(NO_3)_2$. The benzoic acid is shaken out of the mixture with chloroform, the chloroform washed and then the benzoic acid it contains determined by titrating with N/10 Na alcoholate, using phenolphthalein as indicator.

Process. Measure 100 c.c. of urine into a porcelain evaporating dish by means of a pipette. Add 10 c.c. of 5 per cent. NaOH and evaporate to dryness on the steam bath. Transfer the residue to a 500 c.c. Kjeldahl flask by means of 25 c.c. of water and 25 c.c. of concentrated nitric acid. Add 0.2 gram $Cu(NO_3)_2$, a couple of pebbles to prevent bumping and boil very gently four and one-half hours over a micro-burner. The necks of the flasks are fitted with Hopkins condensers, that is large test-tubes of Jena glass carrying a double-holed rubber stopper, through which one tube goes nearly to the bottom, the other to the top of the tube and a stream of cold water is kept circulating through the tubes.

After cooling rinse the condensers with 25 c.c. water and transfer contents of flask to a 500 c.c. separatory funnel by the use of 25 c.c. more water. The total volume of the solution is now 100 c.c. Just saturate the solution with powdered $(NH_4)_2SO_4$ (about 55 grams). Extract 4 times by shaking with freely washed chloroform, using 50, 35, 25 and 25 c.c. portions. The first two portions may be used to rinse out further the Kjeldahl flask before they are added to the separatory funnel. Collect the successive portions of $CHCl_3$ in another separatory funnel and wash the chloroform by adding to the combined extracts 100 c.c. of saturated solution of pure NaCl, to each liter of which has been added 0.5 c.c. concentrated HCl. Shake well and draw the chloroform, which contains the benzoic acid, into a dry 500 c.c. Erlenmeyer flask and titrate with N/10 sodium alcoholate, using 4-5 drops of phenolphthalein as indicator. The first distinct end point is taken, although the color may fade on standing a short time.

The sodium ethylate is made by dissolving 2.3 grams of cleaned metallic sodium in 1 liter of absolute alcohol. It is advisable that it be slightly weaker rather than stronger than N/10. It may be standardized against purified benzoic acid in washed $CHCl_3$. It may also be standardized against N/20 HCl provided the ethylate is free from carbonic acid. As a rule, some carbonate will be present and then the titrations in aqueous media will show the ethylate somewhat stronger than in the $CHCl_3$. The carbonate does not titrate in the latter case.

221. Determination of allantoine in the urine. Urease method.—Allantoine is most easily determined by determining the urea, allantoine and preformed ammonia by Benedict's urea method and then the urea and preformed ammonia by the urease method. Benedict's method detects not only the urea nitrogen, but about 70 per cent. of the allantoine nitrogen as well. The urease detects only the urea nitrogen.

Method. In one sample of urine the preformed ammonia is determined by Folin's method, experiment 212. In another sample the urea and allantoine are determined by Benedict's method, experiment 208. In a third sample the urea and preformed ammonia are determined by hydrolyzing the urea by the soy bean and drawing air through to remove the ammonia formed, as in Folin's method. To determine the urea and preformed ammonia by the soy bean urease introduce 5 c.c. of urine into a Folin ammonia cylinder, add 50-60 c.c. of water, and 1 gram of finely ground soy bean. Keep in the water bath at 35-40° with an air current going through for one hour. The air current must be freed from ammonia as usual. Then disconnect, add 1 gram anhydrous Na_2CO_3 to the cylinder, some liquid petroleum and determine the ammonia as described in experiment 200. Receive the ammonia in 25-50 c.c. n/10 H_2SO_4, of which the volume has been accurately measured.

Titrate with n/10 NaOH, using alizarin or congo red as an indicator. In dog's urine in 5 c.c. about 1.3-1.9 c.c. of N/10 acid is needed to neutralize the ammonia from the allantoine. The difference between the combined ammonia, urea and allantoine obtained in Benedict's method and the urease urea and ammonia gives the allantoine ammonia. Since this detects only about 70 per cent. of the allantoine the result must be multiplied by 100/70 to give the amount of ammonia from the allantoine in 5 c.c. of urine.

222. **Allantoine determination by the Wiechowski method.**—This is an accurate but rather long method. It gives the allantoine directly and is hence superior in accuracy to the method described in 221. (Hofmeister's *Beiträge,* 11, 1907-08, 109).

Principle. A dilute solution of mercuric acetate in the presence of a large amount of sodium acetate precipitates the allantoine as a white precipitate. The urine is first purified from basic substances, chlorine and ammonia, by phosphotungstic acid, lead and silver acetate; then sodium acetate is added and the allantoine precipitated by 0.5 per cent mercuric acetate.

Method for dog urine. (Hofmeister's *Beiträge,* 11, 129, 1907-08.) Allantoine is most easily prepared from dog urine in which it occurs in considerable amount. In human urine only very small quantities are present.

" Dog and rabbit urine are to begin with diluted. In general I have diluted the 24-hour urine of a rabbit to 150 c.c., that of a dog to 300 c.c., utilizing also the cage washings.—For the allantoine determination 100 c.c. are taken, 10 c.c. of 8 per cent. H_2SO_4 added, transferred to a volumetric flask of appropriate size, precipitated with the exactly necessary amount of phosphotungstic acid 10 per cent. (a preliminary test showing how much it is necessary to add) and the flask filled to the mark with water. After standing at least an hour it is filtered through a thick, folded filter into an evaporating dish, and the clear, generally dark-colored filtrate treated with lead carbonate and rubbed as long as CO_2 is evolved and until the liquid reacts weakly acid or neutral. Filter under sharp suction. As large a rounded amount as possible of the filtrate, which is often weakly blue but always neutral, is measured off into a volumetric flask of sufficient size, and precipitated, while avoiding an excess, by basic lead acetate. The amount necessary to add is determined by a preliminary test. The volume is made up to the mark with water. The filtrate from the lead precipitate is treated with H_2S and the filtrate from the lead sulphide freed from H_2S by a current of air. If chlorine is present an aliquot rounded portion of the acid filtrate is taken and precipitated in a volumetric flask with silver acetate solution and made up to the mark with water. The filtrate from the silver chloride is handled, in the same manner as that from the lead precipitation, with H_2S and the filtrate freed from H_2S with air. This filtrate is now tested in small portions to see that the precipitation with phosphotungstic acid, basic lead acetate and silver nitrate has been complete, and now the reaction is absolutely negative. If this is the case two rounded aliquot portions of the filtrate are taken after preliminary neutralization with chlorine-free NaOH (prepared from Na) and the allantoine precipitated by mercuric acetate and sodium acetate.

"The reagent is best made as follows: commercial mercuric acetate (Merck) is dissolved in water to 1 per cent. and pure sodium acetate added to saturation and diluted to such a degree with water that the content of mercuric acetate is 0.5 per cent. The completeness of the allantoine precipitation is determined by the addition of some of the reagent or allantoine to a small filtered test portion. After at least one hour standing the precipitate is brought on a filter and washed with water until the disappearance of a precipitate or of a yellow color on the addition of Na_2S. One of the precipitates is now taken and its N determined by Kjeldahl. The precipitate from the second portion is washed by a stream of water into a beaker, and while heated to boiling completely decomposed with H_2S. The mixture is then evaporated to dryness on the water bath, the residue digested with water, quantitatively transferred to a small measuring cylinder or volumetric flask and the volume increased with water to 25 c.c. generally, or 50 c.c. if much allantoine is present. The final filtration is made through very thick filter paper and repeated until the filtrate is quite clear. A rounded volume of the same is evaporated in a weighed glass dish, dried at 100° and reweighed. The remainder of the filtrate serves for testing the identity and purity of the allantoine. If the allantoine crystals are colored they may be decolored with H_2O_2 in the method indicated. Either at once or after one recrystallization the melting point of the crystals is determined (230-232° with decomposition) and a small portion burned on platinum foil (no residue). Finally the value of allantoine and allantoine nitrogen obtained is multiplied with the total volume numbers (volume of urine, volume after addition of phosphotungstic acid, volume after basic lead acetate, after addition of silver acetate and final volume after decomposition of the allantoine precipitate) and divided by the value of the aliquot parts taken, so as to get the value for the total urine. The determination takes not more than 6-12 hours." The large number of dilutions makes it necessary to be extremely careful with the measuring. Carefully tested and calibrated pipettes and flasks must be used. The precipitation of the allantoine is quantitative. The result is exact only to about 0.01 gram allantoine per day. For human urine the method must be changed as described in *Biochem. Ztschr.*, 19, 1909, 378.

223. Quantitative estimation of total purines and of purines other than uric acid. Salkowski-Arnstein.—*Principle.* The total purines are precipitated by ammoniacal silver nitrate, the precipitate washed and treated with magnesia to free from ammonia and the nitrogen determined in it by the Kjeldahl method. The uric-acid nitrogen is separately determined by Folin's method and subtracted from the total. The difference is computed as xanthine.

Process. Measure 200 c.c. of protein-free urine into a 300 c.c. volumetric flask, add 50 c.c. of magnesia mixture, make up to the mark with water, mix well and allow to stand for a little. Filter through a dry filter paper into a dry 500 c.c. flask. To 200 c.c. of the filtrate add 10-15 c.c. of a 3 per cent. silver nitrate solution. This precipitates all the purines, including uric acid. Filter and wash the precipitate thoroughly with water to free it from ammonia, finally place the funnel in the neck of a Kjeldahl flask, puncture the end of the filter and with a wash bottle wash the whole of the precipitate into the flask. Wash down the neck of the flask. Add to the flask a little magnesia and heat to boiling. Boil for 20 minutes to free from ammonia. Proceed to determine the nitrogen remaining by the Kjeldahl method, described in experiment 54. From the nitrogen thus determined calculate the total purine nitrogen in 100 c.c. of urine. Subtract from this the uric-acid nitrogen in 100 c.c. of urine, as determined by the Folin method. Compute the remaining nitrogen as xanthine, or leave it as purine base nitrogen.

Magnesia mixture. 100 grams of crystallized magnesium chloride are dissolved in water, ammonia added until liquid smells strongly of it, and then ammonium chloride sufficient to dissolve the precipitate. Make up to a liter.

* **224.** Total acidity of urine by Folin's method.—By means of a 25 c.c. pipette measure off 25 c.c. urine into a 250 c.c. conical flask, add 25 c.c. distilled water, 5 drops phenolphthalein solution and 15 grams *finely* powdered potassium oxalate. Rotate the contents of the flask for 2 minutes and titrate at once with n/10 NaOH to a pink tint. Note the number of c.c. required. Immediately after the reading continue with the titration as given under "Amino-acids" below. Calculate the total c.c. of N/10 NaOH required to neutralize the entire 24-hour sample of urine. Explain the method.

The end point is not sharp and it will vary with different observers. It is a good plan to have a standard color and go to this point for the different urines examined. The oxalate is added to precipitate the lime salts which otherwise precipitate the phosphates when the hydrate is added.

*225. **Determination of amino-acids by the formol titration of Sörensen.**—Immediately after the above titration, add to the sample you have just titrated 50 c.c. of a fresh formalin solution, made by taking 15 c.c. of formalin, 30 c.c. of water and enough N/10 NaOH to make it just very faintly alkaline to phenolphthalein. Mix and now titrate again, noting the reading of the burette when you start, by adding N/10 NaOH until a faint permanent pink remains. Each c.c. of the N/10 NaOH is equivalent to 1/10,000 of a gram equivalent of amino-acid. How much nitrogen does this represent calculated for a mono-amind-monocarboxylic acid? This titration (the last one) gives us a measure of the NH_4 and amino-acids together, according to the following equations:

$$4NH_4Cl + 6HCHO + 4NaOH \longrightarrow \underset{\text{Hexamethylenetetramine.}}{N_4(CH_2)_6} + 10H_2O + 4NaCl$$

$$\begin{matrix} R \\ | \\ CHNH_2 \\ | \\ COOH \end{matrix} + \begin{matrix} CH_2 \\ \| \\ O \end{matrix} + NaOH \longrightarrow \begin{matrix} R \\ | \\ CH{-}N = CH_2 \\ | \\ COONa \end{matrix} + 2H_2O$$

How do you account for the increased acidity developed after adding the formaldehyde? Calculate the grams of nitrogen in the form of amino-acid nitrogen in the total 24-hour sample, also calculate the per cent. of the total nitrogen found as amino-acid nitrogen. (See p. 363.)

226. **Amino-acid nitrogen in urine. Quantitative determination. Formol titration after removal of ammonia** (Benedict and Murlin, *Jour. Biol. Chem.*, 16, p. 386, 1914). *Principle.* Precipitation of ammonia and basic substances in the diluted urine by phosphotungstic acid; removal of excess of phosphotungstic acid with KCl and titration of the amino-acid by formol and N/10 NaOH.

Process. Measure into a 500 c.c. Erlenmeyer flask 200 c.c. of a 24-hour human urine diluted to 2,000 c.c. Add an equal quantity of 10 per cent. phosphotungstic acid (Merck) (note, Kahlbaum's preparation is a very different substance) in 2 per cent. HCl. Let stand at least 3 hours; better overnight.

Pour off 250 c.c. of the clear fluid; add 1 c.c. of a 0.5 per cent. solution of phenolphthalein and barium hydrate in substance until the whole fluid turns decidedly pink. The $Ba(OH)_2$ should be added a very little at a time. Let stand one hour. Filter off two 100 c.c. samples (=50 c.c. urine). Neutralize to litmus (Squibb's papers answer for all practical purposes) with N/5 HCl. Add 10—20 c.c. neutral formalin and titrate cautiously with N/10 NaOH, to a deep red color, i.e., until a drop produces no additional color.

Correct by deducting the amount of N/10 NaOH necessary to produce the same depth of color in an equal quantity of CO_2-free with the same quantity of neutral formalin added.

227. Determination of amino nitrogen by the Van Slyke me —(Van Slyke, *Jour. Biol. Chem.*, 12, 1912, p. 275; 16, **1913**, p.

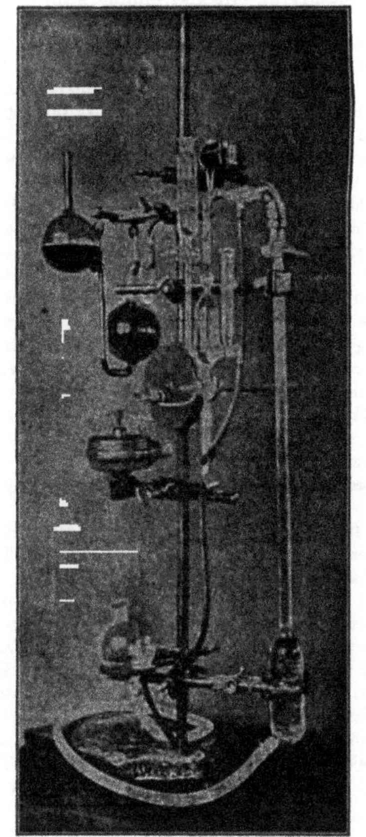

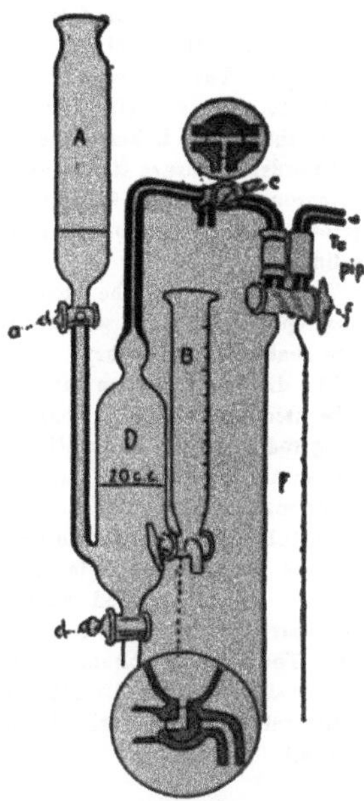

FIG. 75. FIG. 76.

FIG. 75.—Van Slyke animo-nitrogen apparatus.
FIG. 76.—Detail of Van Slyke apparatus (Van Slyke).

23, 1915, p. 408.) *Principle.* Nitrous acid acting on amino-nitro sets free nitrogen gas which is collected and its volume determined.

$$RNH_2 + HNO_2 = ROH + N_2 + H_2O.$$

There are two forms of apparatus, a larger and smaller, differing only in capacity.

Process. The following description of the method is taken from Van Slyke, *Jour. Biol. Chemistry,* 12, 1912, pp. 277-284:

"The structure of the apparatus and the manner in which it is set up are apparent from the accompanying cut and photograph.

D is of 40-45 c.c. capacity, *A* of about 35 c.c. and the burette *B* of 10 c.c. The wire from which the deaminizing bulb *D* is suspended should be fairly stiff, and rigidly fastened in position from above so that the loop about the capillary acts as a fixed center. A is then so placed that its center of gravity comes near this center and the shaking of *D* is accomplished with a minimum motion in *A*, and, consequently, without putting a dangerous strain on the tube which connects *A* with D. This tube is strong-walled and of 3 mm. inner diameter. It is essential that the bore of cock *a* should also be 3 mm. The reason for this is that during the analysis gas containing some nitrogen collects in the tube. Unless *a* is of as wide bore as the tube the liquid from *A* may flow around the bubble instead of forcing it into *D* at the end of the reaction. The cock *d* is also of large bore in order to facilitate emptying *D*. The neck connecting *D* and *B* must be of at least 8 mm. inner diameter in order to allow free circulation of the solution in *D* up to the cock B. The small bulb at the top of D keeps the reacting solution from splashing into the capillary.

In order to insure tightness of the cocks and to prevent their becoming loosened by vigorous shaking it is well to lubricate them with a paste made by dissolving together over a flame one part of rubber, one part of paraffin and two parts of vaseline.

The structure of the modified Hempel pipette is entirely apparent from the photograph. This form would undoubtedly facilitate absorption in all gas analyses where shaking is necessary.

The driving wheel, as can be seen from the photograph, is so arranged that it can be used alternately to shake the deaminizing bulb or the Hempel pipette. The driving rod is shown in position for shaking the deaminizing bulb. By lifting the rod from the shoulder of *D* and placing the other hook, at the end of the rod, over the horizontal lower tube of the pipette, the power is transferred to the latter. Rubber tubes drawn over the hooks at the end of the driving rod and those from which the Hempel pipette is suspended make the apparatus almost noiseless. For power one can use a good water motor. Still more convenient is a small electric motor, particularly when connected with a rheostat enabling one to regulate the speed. The gearing should be so arranged that the driving wheel, to which the driving rod is eccentrically attached, makes 300 to 500 revolutions per minute. The driving rod is attached about 1.5 cm., in no case over 2 cm., from the center of the wheel.

The manipulation is in principle the same as described in this *Journal,* ix, pp. 189-91. As there are slight variations due to the different form of the present apparatus, however, we describe the present technique, dividing the determination into three stages as in the original description.

1. *Displacement of air by nitric oxide.* Water from *F* fills the capillary leading to the Hempel pipette and also the other capillary

as far as *c*. Into *A* one pours a volume of glacial acetic acid suffici to fill one-fifth of *D*. For convenience, *A* is etched with a mark measure this amount. The acid is run into *D*, cock c being turned so to let the air escape from *D*. Through *A* one now pours sodium nitı solution (30 gms. $NaNO_2$ to 100 c.c. H_2O) until *D* is full of solut and enough excess is present to rise a little above the cock into It is convenient to mark *A* for measuring off this amount also. The exit from *D* is now closed at c, and, *a* being open, *D* is shaken for a seconds. The nitric oxide, which instantly collects, is let out at *c*, a the shaking repeated. The second crop of nitric oxide, which was out the last portions of air, is also let out at c. *D* is now connec with the motor and shaken till all but 20 c.c. of the solution have b displaced by nitric oxide and driven back into *A*. A mark on *D* in cates the 20 c.c. point. One then closes *a* and turns *c* and *f* so that and *F* are connected. The above manipulations require between one two minutes.

2. *Decomposition of the amino substance.* Of the amino solut to be analyzed 10 c.c. or less, as the case may be, are measured off in Any excess added above the mark can be run off through the outfl tube. The desired amount is then run into *D*, which is already c nected with the motor, as shown in the photograph. It is shaken, wl α-amino-acids are being analyzed, for a period of three to five minu With α-amino-acids, proteins or partially or completely hydrolyzed teins, we find that at the most five minutes' vigorous shaking compl the reaction.[1] Only in the cases of some native proteins which, wl deaminized, form unwieldy coagula that mechanically interfere with thorough agitation of the mixture, a longer time may be required. case a viscous solution is being analyzed and the liquid threatens to fo over into *F*, *B* is rinsed out and a little caprylic alcohol is added thro it. For amino substances, such as amino-purines, requiring a lon time than five minutes to react (cf. p. 191, former article), one mer mixes the reacting solutions and lets them stand the required length time, then shakes about two minutes to drive the nitrogen complet out of solution.

When it is known that the solution to be analyzed is likely foam violently, it is advisable to add caprylic alcohol through before the amino solution. *B* is then rinsed with alcohol and dri with ether or a roll of filter paper before it receives the amino so tion.

3. *Absorption of nitric oxide and measurement of nitrogen.* T

[1] Only 95 per cent. of the lysine nitrogen reacts in five minutes, but remaining one-twentieth of the lysine nitrogen is a practically negligible prop tion of the total nitrogen of a complete protein.

reaction being completed, all the gas in *D* is displaced into *F* by liquid from *A* and the mixture of nitrogen and nitric oxide is driven from *F* into the absorption pipette. Solution in absorbing pipette is $KMnO_4$ 50 grams + 25 KOH (or 18 grams NaOH) in one liter. The driving rod is then connected with the pipette by lifting the hook from the shoulder of *d* and placing the other hook, on the opposite side of the driving rod, over the horizontal lower tube of the pipette. The latter is then shaken by the motor for a minute, which, with any but almost completely exhausted permanganate solutions, completes the absorption of nitric oxide. The pure nitrogen is then measured in *F*. During the above operations *a* is left open, to permit displacement of liquid from *D* as nitric oxide forms in *D*.

Testing completeness of reaction. Particularly when the mechanical shaker is used, there is little danger of failing to obtain a complete evolution of nitrogen. The point may be tested, however, as follows. The nitrogen from *F* is driven out at *c*; *a* is closed and *D* connected with *F*. The gas which has formed in the nitrous-acid solution in *D* during the absorption of the nitric oxide and measurement of nitrogen is shaken out and driven over into *F* and then into the Hempel pipette as before. After absorption of the nitric oxide, the gas left should not measure more than that obtained in blank tests, usually less than 0.1 c.c. After the gas has all been forced from *D* over into *F* at the end of the reaction, the nitrous solution is run out from *D*, by opening *d*, through a tube leading to a drain. *B* is rinsed and dried with a roll of filter paper or with alcohol and ether and the apparatus is immediately ready for use again.

Blank determinations, performed as above except that 10 c.c. of distilled water replaces the solution of amino substance, must be performed on every fresh lot of nitrite used. The amount of gas obtained on a five-minute blank is usually 0.3 to 0.4 c.c., with very little increase for longer tests. Nitrite giving a much larger correction should be rejected.

The following determinations, performed with an $\frac{N}{10}$ solution of leucine, indicate the speed of the reaction. The correction applied for reagents was 0.40 c.c. Ten cubic centimeters of $\frac{N}{10}$ leucine solution, containing 14.01 mgs. of nitrogen, were used for each determination.

Time of reaction	N	Temperature	Pressure	N obtained	N obtained on second shaking of solution	Total N obtained
Minutes	*C.c.*	*Degrees*	*Mm.*	*Mgs.*	*Cc.*	*Mgs.*
2	24.38	23	762	13.71	0.45	13.97
3	24.65	22	762	13.93	0.20	14.03
4	24.80	22	762	14.01	0.00	14.01
10	25.07	24	762	14.03	0.00	14.03

The driving wheel was making 300 revolutions per minute. At speeds of 400 or 500 revolutions the reaction can be driven to completion in three, or, with higher room temperature, in two minutes.

The rate of reaction of *ammonia* is shown in the following table. Ten c.c. portions of $\frac{N}{10}$ ammonium sulphate solution, containing 28.0 mgs. of nitrogen each, were used.

Time of reaction	N	Temperature	Pressure	Weight of N	Per cent. of total ammonia nitrogen
Minutes	*C.c.*	*Degrees C.*	*Mm.*	*Mgs.*	
3	12.1	24	752	6.86	21.6
5	18.4	24	752	10.16	36.3
10	31.5	24	752	17.38	62.1

As pointed out before, ammonia reacts slowly compared with the amino-acids. *For accurate determination of NH_2 nitrogen in digestion solutions, etc., it is advisable to first remove the ammonia;* although good comparative results can be obtained, in the presence of the relatively small proportion of ammonia usually present, if reaction conditions of time, temperature and concentration of solutions are kept constant, so that the proportion of the ammonia decomposed is the same in each determination. The ammonia can be conveniently removed and determined by distillation with $Ca(OH)_2$ under diminished pressure, as described on page 21, Vol. X, of this *Journal*. After the distillation the excess Ca $(OH)_2$ is dissolved with acetic acid. It is essential that all the ethyl alcohol should be distilled off, as it decomposes nitrous acid with formation of large volumes of gases which can be removed with permanganate only with difficulty and by the use of perfectly fresh permanganate solution. The point at which the alcohol has all been boiled off is usually indicated when the solution begins to foam in the distilling flask.

The following results were obtained with lysine picrate. Lysine, as previously stated, reacts more slowly than the other amino-acids because it contains not only an α-amino group, but also an ω-amino group. In the fifteen- and thirty-minute determinations the solution was shaken only during the last five minutes.

Weight of lysine picrate	Time of reaction	N	Temperature	Pressure	NH_2-N found	NH_2-N calculated
Gram	*Minutes*	*C.c.*	*Degrees C.*	*Mm.*	*Per cent.*	*Per cent.*
0.200	5	25.4	24	764	7.13	7.47
0.200	15	26.7	24	764	7.49	7.47
0.200	30	26.7	24	764	7.49	7.47

Solutions to be analyzed should be free of ethyl alcohol and acetone. These substances when mixed with nitrous acid give off gases or vapors which are with difficulty absorbed by the permanganate.

Amyl alcohol, which in the original description of the amino method was recommended to prevent the foaming of viscous solution, must be replaced for this purpose by caprylic alcohol [Kahlbaum's "octyl-alkohol (sekundär) I"]. Amyl alcohol, boiling at 131°, has the disadvantage of a very noticeable vapor tension. Permanganate solution apparently possesses the power to absorb slight amounts of amyl alcohol vapor. Particularly on hot days, however, and when relatively much of the alcohol is used, it is necessary to change the permanganate with every analysis or else reduce the volume of gas observed by multiplication with an empirically determined factor.

For convenience in calculating results the following table is appended. The figures are calculated by dividing by 2 those for moist nitrogen given by Gattermann in the *Praxis des organischen Chemikers*, ninth edition. They represent the weights of amino-nitrogen in milligrams which correspond to 1 c.c. of nitrogen gas, obtained by the action of nitrous acid and measured over water, at the temperatures and atmospheric pressures indicated."

228. Van Slyke micro-apparatus for amino-nitrogen.—The smaller apparatus has the following dimensions: gas burette, 10 c.c., the upper part measuring 0.02 c.c. The deaminizing bulb is 11-12 c.c. and the 10 c.c. burette is replaced by one of 2 c.c., only 10 c.c. of nitrite solution and 2.5 c.c. acetic acid are required and the correction for the reagents is 0.06-0.12 c.c., according to the quality of the nitrite used. "With the micro-apparatus the error need not be more than 0.005 mg. of N when 2 c.c. or less of N gas are measured. One can analyze 1/5th of the amount of substance needed for the larger apparatus. 0.5 mg. of amino N can be measured to an accuracy of 1 per cent. The only change of procedure from that already described is in the speed at which the deaminizing bulb is shaken and the Hempel pipette. The deaminizing bulb should be shaken at a very high rate of speed; about as fast as the eye can follow, or an unnecessary amount of time is lost in freeing the apparatus from air. This stage is accelerated by warming the nitrite solution to 30° before it is used. During the absorption of the nitric oxide the Hempel pipette should be shaken not faster than twice per second. Because of the small amount of nitrogen to be measured it is especially necessary that the removal of air in the first stage be complete. This is assured by shaking the solution in the deaminizing bulb back each time in this stage until the bulb is two-thirds filled with nitric oxide. The hook or wire loop from which the deaminizing bulb is suspended should be perfectly rigid and hold the capillary outlet tube tightly. Binding the tube to the holder with a strip of rubber band is a satisfactory method of insuring a firmly held apparatus." "Clean whole apparatus with bichromate-sulphuric-acid mixture. For especially accurate results measure the amino-acid solution with an Ostwald 2 c.c. pipette graduated to deliver between 0.001 and 0.002 c.c. int 2 c.c. burette and wash the burette twice with 6 or 7 drops of wate tributed about the entire inner walls of the burette for each washin

MILLIGRAMS OF AMINO NITROGEN CORRESPONDING TO 1 C.C. OF NITROGEN G 11°-30° C; 728-772 MM. PRESSURE.*

t	728	730	732	734	736	738	740	742	744	746	748
11°	0.5660	0.5695	0.5710	0.5725	0.5745	0.5760	0.5775	0.5790	0.5805	0.5820	0.5840
12°	0.5655	0.5670	0.5685	0.5700	0.5720	0.5735	0.5750	0.5765	0.5780	0.5795	0.5815
13°	0.5630	0.5645	0.5660	0.5675	0.5695	0.5710	0.5725	0.5740	0.5755	0.5770	0.5785
14°	0.5605	0.5620	0.5635	0.5650	0.5665	0.5680	0.5700	0.5715	0.5730	0.5745	0.5760
15°	0.5580	0.5595	0.5610	0.5625	0.5640	0.5655	0.5670	0.5685	0.5705	0.5720	0.5735
16°	0.5555	0.5570	0.5585	0.5600	0.5615	0.5630	0.5645	0.5660		0.5690	0.5710
17°	0.5525	0.5540	0.5555	0.5575	0.5590	0.5605	0.5620	0.5635	0.5650	0.5665	0.5680
18°	0.5500	0.5515	0.5530	0.5545	0.5560	0.5580	0.5595	0.5610	0.5625	0.5640	0.5655
19°	0.5475	0.5490	0.5505	0.5520	0.5535	0.5550	0.5565	0.5580			0.5630
20°	0.5445	0.5460	0.5475	0.5495	0.5510	0.5525	0.5540	0.5555			0.5600
21°	0.5420	0.5435	0.5450	0.5465	0.5480	0.5495	0.5510	0.5525			0.5575
22°	0.5395	0.5410	0.5425	0.5440	0.5455	0.5470	0.5485	0.5500	0.5515	0.5530	0.5545
23°	0.5365	0.5380	0.5395	0.5410	0.5425	0.5440	0.5455	0.5470	0.5485	0.5500	0.5515
24°	0.5335	0.5350	0.5365	0.5380	0.5400	0.5415	0.5430	0.5445	0.5460	0.5475	0.5490
25°	0.5310	0.5325	0.5340	0.5355	0.5370	0.5385	0.5400	0.5415	0.5430	0.5445	0.5460
26°	0.5280	0.5295	0.5310	0.5325	0.5340	0.5355	0.5370	0.5385	0.5400	0.5415	0.5430
27°	0.5250	0.5265	0.5280	0.5295	0.5310	0.5325	0.5340	0.5355	0.5370	0.5385	0.5400
28°	0.5220	0.5235	0.5250	0.5265	0.5280	0.5295	0.5310	0.5325	0.5340	0.5355	0.5370
29°	0.5195	0.5210	0.5220	0.5235	0.5250	0.5265	0.5280	0.5295	0.5310	0.5325	0.5340
30°	0.5160	0.5175	0.5190	0.5205	0.5220	0.5235	0.5250	0.5265	0.5280	0.5295	0.5310
t	728	730	732	734	736	738	740	742	744	746	748

t	752	754	756	758	760	762	764	766		770	772
11°	0.5870	0.5885	0.5900	0.5915	0.5935	0.5950	0.5965	0.5980		0.6010	0.6030
12°	0.5845	0.5860	0.5875	0.5890	0.5905	0.5925	0.5940	0.5955		0.5985	0.6000
13°	0.5820	0.5835	0.5850	0.5865	0.5880	0.5895	0.5910	0.5930		0.5960	0.5975
14°	0.5790	0.5805	0.5825	0.5840	0.5855	0.5870	0.5885	0.5900	0.5915	0.5935	0.5950
15°	0.5765	0.5765	0.5795	0.5810	0.5810	0.5845	0.5860	0.5875	0.5890	0.5905	0.5920
16°	0.5740	0.5755	0.5770	0.5785	0.5800	0.5815	0.5830	0.5850		0.5880	0.5895
17°	0.5710	0.5730	0.5745	0.5760	0.5775	0.5790	0.5805	0.5820	0.5825	0.5850	0.5865
18°	0.5685	0.5700	0.5715	0.5730	0.5745	0.5765	0.5780	0.5795	0.5810	0.5825	0.5840
19°	0.5660	0.5675	0.5690	0.5705	0.5720	0.5735	0.5750	0.5765	0.5780	0.5795	0.5810
20°	0.5630	0.5645	0.5660	0.5675	0.5690	0.5705	0.5725	0.5740	0.5755	0.5770	0.5785
21°	0.5605	0.5620	0.5635	0.5650	0.5665	0.5680	0.5695	0.5710	0.5725	0.5740	0.5755
22°	0.5575	0.5590	0.5605	0.5620	0.5635	0.5650	0.5665	0.5680	0.5695	0.5715	0.5730
23°	0.5545	0.5560	0.5575	0.5595	0.5610	0.5625	0.5640	0.5655	0.5670	0.5685	0.5700
24°	0.5520	0.5535	0.5550	0.5565	0.5580	0.5595	0.5610	0.5625	0.5640	0.5655	0.5670
25°	0.5490	0.5505	0.5520	0.5535	0.5550	0.5565	0.5580	0.5595	0.5610	0.5625	0.5640
26°	0.5460	0.5475	0.5490	0.5505	0.5520	0.5535	0.5550	0.5565	0.5580	0.5595	0.5610
27°	0.5430	0.5445	0.5460	0.5475	0.5490	0.5505	0.5520	0.5535	0.5550	0.5565	0.5580
28°	0.5400	0.5415	0.5430	0.5445	0.5460	0.5475	0.5490	0.5505	0.5520	0.5535	0.5550
29°	0.5370	0.5385	0.5400	0.5415	0.5430	0.5445	0.5460	0.5475	0.5490	0.5505	0.5520
30°	0.5340	0.5355	0.5370	0.5385	0.5400	0.5415	0.5430	0.5445	0.5460	0.5475	0.5490
t	752	754	756	758	760	762	764	766	768	770	772

* *Journal of Biological Chemistry*, xii, No. 2. Van Slyke: The Quanti Determination of Amino Groups. II.

*** 229.** Quantitative estimation of total sulphates in urine (F *Jour. Biol. Chem.* 1, 1905, 6, 153). Gently boil in a 200-250 c.c. Er meyer flask for 20-30 minutes 25 c.c. of urine and 20 c.c. of dilute (1 part HCl, sp. gr. 1.20 to 4 pts. water); or 50 c.c. urine and 4 con. HCl. Keep covered with a small watch glass during boiling. Cool for 2-3 minutes under the tap and dilute with cold water to 150 c.c. To this cold solution add 10 c.c. of 5 per cent. $BaCl_2$ slowly drop by drop, without shaking or stirring the solution during the addition, and allow to stand without shaking for one hour. After one hour, or later as convenient, shake the mixture and filter through a Gooch weighed crucible with an asbestos mat, wash with about 250 c.c. cold water, dry and ignite, cool and weigh. Ten minutes' ignition is sufficient unless organic matter is present.

Preparation of the asbestos mat. Cut a good grade of fibrous asbestos with scissors into lengths of 50-70 mm. Place a few grams at a time in a cylinder with 300 c.c. of 5 per cent. HCl and pass a strong air current through. This separates the fibers. Keep ready for use in dilute HCl. From 50-100 mgs. of asbestos is used for each mat. By a good vacuum pump the asbestos mat is packed firmly in a thin but compact layer. It is washed finally with water with only enough vacuum to make the water go through slowly. Dry, ignite and weigh. The same mat may be used repeatedly until about 1 gram $BaSO_4$ has collected in it. It is somewhat difficult to prepare such mats, which show no loss on filtering and ignition. On igniting do not apply the flame directly to the bottom of the crucible, or losses will occur. To ignite, place the lid of a platinum crucible on a triangle and let the crucible stand upright on the lid. Another platinum lid should cover the crucible. The flame is applied to the lower platinum lid. Ten minutes' ignition is sufficient.

230. Estimation of inorganic sulphates of urine (Folin, *Jour. Biol. Chem.* 1, 1905, 151).—Measure 25 c.c. of urine, 100 c.c. of water and 10 c.c. of HCl (1 vol. concentrated HCl to 4 vols. of water) into a 250 c.c. Erlenmeyer flask. If the urine is dilute 50 c.c. may be taken instead of 25 and correspondingly less water. Add drop by drop, preferably by means of an automatic dropper, 10 c.c. of 5 per cent. $BaCl_2$ solution. The urine solution is not to be shaken or stirred during the addition of the $BaCl_2$ nor for an hour afterward. At the end of the hour proceed with the determination of the precipitated sulphates in the manner already described for the estimation of the total sulphates.

231. Estimation of the conjugated sulphates.—The conjugated sulphates are estimated by taking the difference between the total sulphates as determined in experiment 229 and the inorganic sulphates in experiment 230.

232. Estimation of total sulphur by the Benedict-Denis method.—

Reagent: 25 grams $Cu(NO_3)_2$
25 " NaCl
10 " NH_4NO_3
100 c.c. H_2O

Process. In a 4-inch porcelain dish evaporate 10 to 25 c.c. urine wit 5 c.c. of the above reagent. When dry, heat carefully over a free flam gradually raising the temperature until heated to a dull red heat. Kee at this temperature for 10-15 minutes. Cool and dissolve in 10-20 c. 10 per cent. HCl solution. Warm and dilute to about 150 c.c. Filte if necessary and determine the SO_4 as described in 230.

233. Ethereal and inorganic sulphates. Volumetric method (Ro senheim and Drummond, *Bioch. Journal,* viii, 143, 1914).—*Principl* Precipitation of the sulphates with benzidene hydrochloride and titratio of the acid in the benzidene sulphate with KOH.

Process. Make the benzidene (p-diamino-diphenyl, $NH_2.C_6H_4.C_6H_4.NH_2$) solution by rubbing 4 grams of Kahlbaum's benzidene t a fine paste with about 10 c.c. of water and transfer to a liter flask wit about 500 c.c. of H_2O; 5 c.c. con. HCl (sp. gr. 1.19) are added and th solution made to a liter with water. 150 c.c. of this solution, whic keeps indefinitely, are sufficient to precipitate 0.1 gr. H_2SO_4.

Inorganic sulphates.—Measure 25 c.c. of urine with a pipette into 250 c.c. Erlenmeyer flask, add dilute (1:4) HCl until the reaction i distinctly acid to congo paper. 1-2 c.c. of acid is usually enough. 10 c.c. of benzidene solution is then run in and the precipitate which form in a few seconds is allowed to settle 10 minutes. Filter the precipitat under suction, wash with 10-20 c.c. H_2O saturated with benzidene sul phate; transfer the precipitate and filter paper to the original flas with about 50 c.c. of water and titrate hot with 0.1 N KOH after addin a few drops of a saturated alcoholic solution of phenolphthalein. 1 c. N/10 KOH corresponds to 4.9 mgs. H_2SO_4.

Total sulphates.—Measure 25 c.c. urine into a 250 c.c. flask, add 2 c.c. dilute (1:4) HCl and boil for 15-20 minutes to hydrolyze the etherea sulphates. The authors state that a smaller amount of acid, namel 2-5 c.c., will do as well. If the larger amount of acid is used carefull neutralize the acid after boiling with KOH and then again add HC until the reaction is acid to congo red. Cool solution and precipitate a once with benzidene solution. Proceed from this on as in the inorgani sulphate determination. Ethereal sulphates are given by difference be tween the total and inorganic. Method is as accurate as the gravimetri and much easier.

* 234. Quantitative determination of the chlorides of urine. Voll hard's method.—*Principle.* All the chlorides in the urine are precipi tated as silver chloride by the addition of a known amount of silve nitrate to the urine after the addition of nitric acid. The excess of silve remaining in solution is then titrated with a standard solution of potassium sulphocyanate, using a ferric salt as an indicator.

Process. Introduce into a 100 c.c. volumetric flask 10 c.c. of urin with a pipette, add 4 c.c. concentrated nitric acid and 20 c.c. of a standard solution of silver nitrate. Fill the flask to the mark with distilled water, mix well, and filter through a dry filter paper into a dry beaker or flask. Take accurately 50 c.c. of the filtrate, add to it 5 c.c. of iron alum solution and titratê with the standard sulphocyanate solution until a permanent red color is obtained. Record the reading of the burette containing the sulphocyanate at the beginning and end of the titration.

Calculation. On standardization of the KCNS it was found that 1 c.c.$=x$ c.c. $AgNO_3$ solution. In the above titration y c.c. KCNS solution have been required to titrate the excess $AgNO_3$ in 50 c.c. of the filtrate. 100 c.c. or the whole filtrate would require $2y$ c.c. KCNS. This is equivalent to $2xy$ c.c. $AgNO_3$ solution. The whole amount of $AgNO_3$ solution added was 20 c.c., hence $20—2xy$ c.c. is the amount which has been used to precipitate the chlorides. 1 c.c. of the standard $AgNO_3$ is equivalent to .01 gram NaCl or .00606 gram Cl. The total Cl in 10 c.c. of urine is hence $(20\text{-}2xy) \times .00606$ grams. In 100 c.c. it is 10 times this amount.

Standard solutions. Silver nitrate. Dissolve 29.063 grams fused $AgNO_3$ in distilled water and make up to 1 liter. 1 c.c.=.00606 gr. Cl.

Potassium sulphocyanate. 8 grams of the salt dissolved in water and made to 1 liter. Standardize by titrating against $AgNO_3$ solution.

Iron alum. Concentrated solution. *Pure HNO_3* free from chlorine.

Standardization of sulphocyanate solution. Measure accurately with a burette or pipette 10 c.c. of the standard $AgNO_3$ solution into a clean 300 c.c. beaker or flask, add 5 c.c. of pure nitric acid, 5 c.c. of iron alum solution and 80 c.c. of water. From a burette run in the solution of sulphocyanate carefully until a permanent faint red tinge is obtained. Note the reading of the burette before and after the addition of the sulphocyanate to the silver nitrate. Two determinations should be made and the mean taken.

$$x \text{ c.c. KCNS} = 10 \text{ c.c. } AgNO_3 :: 1 \text{ c.c. KCNS} = 10/x \text{ c.c. } AgNO_3.$$

235. **Estimation of calcium.**—(a) *Solutions required.* 2.5 per cent. oxalic acid, 20 per cent. sodium acetate, 0.5 per cent. $(NH_4)_2C_2O_4$ sol.

(b) *Process.* Before measuring off a sample of urine, determine the reaction toward litmus paper. If neutral or alkaline, make slightly acid by the addition of conc. HCl. Mix well and filter a portion. Measure off 200 c.c. of the filtered urine; if only faintly acid to litmus paper add 10 drops conc. HCl; if the filtered urine is strongly acid make slightly alkaline with NH_4OH and then acidify as above indicated, finally adding 10 drops conc. HCl (sp. gr. 1.19) or 10 c.c. N/2 HCl.

Now add 10 c.c. of the 2.5 per cent. oxalic acid drop by drop to the cold solution; then in the same way add 8 c.c. of the 20 per cent. $NaC_2H_3O_2$ solution. Now allow to stand overnight or shake vigorously for 10 minutes.

Filter off on ashless paper, wash free from chlorides with 0.5 per cent. $(NH_4)_2C_2O_4$ solution. Ignite in a porcelain crucible and weigh as CaO. Ignite strongly until constant in weight. Or filter through asbestos, transfer washed precipitate and asbestos to a beaker with a little water, acidify with sulphuric acid and titrate with N/20 permanganate.

***236. Estimation of total phosphates by the uranium acetate titration.**—Measure off by a pipette 50 c.c. urine into a 150 to 300 c.c. beaker, add 5 c.c. sodium-acetate solution (1 liter contains 100 grams sodium acetate and 30 grams acetic acid) and heat to boiling. Now while boiling gently and agitating, add drop by drop of the standard uranium-acetate solution (35.461 grams made up to 1 liter) until one no longer is able to see the formation of a distinct precipitate as the solution is added. Now remove a drop on to a titration plate containing drops of 10 per cent. K_4FeCy_6 solution. The end point is reached when the mixture of the drops at once yields a brownish-red precipitate.

It may be necessary to repeat the titration to arrive at a more accurate end point. Calculate the total grams of phosphorus in the 24-hours' sample. Each c.c. of the uranium-acetate solution is equivalent to 0.002183 gram phosphorus.

This method is not so accurate as the gravimetric, but is sufficiently accurate for all ordinary purposes and much easier than the gravimetric.

237. Quantitative determination of glucose by Bang's method (*Bioch. Ztschr.* 2 and 11; and 49, 1913).—*Principle.* A standard solution of $Cu(CNS)_2$, cupric thiocyanate, made alkaline by K_2CO_3 and $KHCO_3$ is reduced by glucose to colorless CuCNS, cuprous thiocyanate. An excess of the reagent is used and the excess of the cupric salt is titrated to the colorless cuprous salt by a standard solution of hydroxylamine sulphate. From the amount of hydroxylamine required to reduce the excess of cupric salt the amount of cupric salt reduced by the sugar can be found and from this by reference to a table which has been prepared, by determining the reducing power of known amounts of dextrose, the amount of dextrose in the solution is determined.

Process. Measure from a pipette 10 c.c. of urine, which must not contain more than 0.6 per cent. of glucose, into a flat-bottomed 200 c.c. flask. If the urine contains more than this amount of sugar take a proportionately smaller amount of it, 2-5 c.c., and dilute with water so that the bulk of fluid added to the flask is 10 c.c. Add 50 c.c. of the copper solution and heat to boiling on a wire gauze and boil for exactly three minutes. Remove and cool at once under the tap to stop the reaction. Now run in slowly from a burette, the reading of which has been taken, the standard hydroxylamine solution, shaking the flask with each addition, until the solution loses its blue color and becomes colorless or has a slight yellow tint. The end point is not very sharp. An error of 0.5 c.c. of hydroxylamine is possible for beginners. From the amount of hydroxylamine solution required the amount of dextrose can be found by consulting the accompanying table:

Hydroxylamine solution	Sugar	Hydroxylamine solution	Sugar	Hydroxylamine solution	Sugar	Hydroxylamine solution	Sugar
C.c.	Mgs.	C.c.	Mgs.	C.c.	Mgs.	C.c.	Mgs.
43.85	5	29.60	19	17.75	33	7.65	47
42.75	6	28.65	20	16.95	34	7.05	48
41.65	7	27.75	21	16.15	35	6.50	49
40.60	8	26.85	22	15.35	36	5.90	50
39.50	9	26.00	23	14.60	37	5.35	51
38.40	10	25.10	24	13.80	38	4.75	52
37.40	11	24.20	25	13.05	39	4.20	53
36.40	12	23.40	26	12.30	40	3.60	54
35.40	13	22.60	27	11.60	41	3.05	55
34.40	14	21.75	28	10.90	42	2.60	56
33.40	15	21.00	29	10.20	43	2.15	57
32.45	16	20.15	30	9.50	44	1.65	58
31.50	17	19.35	31	8.80	45	1.20	59
30.55	18	18.55	32	8.20	46	0.75	60

(Bang.)

Standard solutions: Copper *solution.* Dissolve 12.5 grams of copper sulphate ($CuSO_4+5H_2O$) in 60 c.c. water, warming if uecessary, and after cooling to room temperature dilute to 75 c.c. Place in a large evaporating dish or beaker 200 grams of powdered potassium thiocyanate, 250 grams of potassium carbonate and 50 grams of potassium bicarbonate, and about 600 c.c. of water. Stir and if necessary aid the solution by gently warming on the water bath, but not over 50°. Cool the solution under the tap to the usual temperature, add the copper-sulphate solution very slowly, a little at a time while stirring. Transfer to a 1 liter volumetric flask, washing the evaporating dish or beaker out carefully into the volumetric flask and make up to 1 liter. The solution slowly changes on standing and in three months is useless. With a little care the solution may be made directly in the flask. The process outlined above must be followed exactly or a false titer will be found.

Hydroxylamine solution. Dissolve 200 grams potassium thiocyanate in 1,500 c.c. of water in a 2-liter volumetric flask. Add to it a solution of 6.55 grams hydroxylamine sulphate in a little water, carefully washing all the solution into the flask, and make up to 2 liters.

Standardization. These two solutions should exactly correspond. Titrate 50 c.c. of the copper solution, to which 10 c.c. water is added, with the hydroxylamine-sulphate solution, to make sure that they do correspond.

The advantages of this method are that the solution is less alkaline so that the sugar is not broken up readily by the alkali and no precipitate forms so that the end point is sharper. The reducing action of the uric acid and creatinine is small in this faintly alkaline solution, so that they introduce less error.

238. Quantitative determination of glucose by Bang's micromethod (Bang, *Biochemische Zeitschrift*, XLIX, p. 1, 1913; LVII, 301, 1913).—*Principle.* The macrochemical method is expensive on account of the cost of hydroxylamine and thiocyanate and the copper solution is not stable. By this method very minute amounts of glucose, even 0.1 mg., may be accurately determined. The copper solution is modified from the former method in that less carbonate is taken and potassium chloride is substituted for potassium thiocyanate. The copper is reduced in the absence of air and the reduced cuprous chloride, which is kept in solution by the large amount of KCl present, is titrated directly with N/100 or N/25 or N/10 I solution, soluble starch being used as the indicator. The iodine solution under these conditions oxidizes the cuprous chloride back to the cupric condition.

Process. Place in a 100 c.c. Jena-glass, flat-bottomed flask having a neck without a turned-over edge 0.1-2 c.c. of the sugar solution to be examined. The solution must not contain more than 10 mgs. of glucose, and it is better to have not more than 5 mgs. If the sugar solution is more dilute, more than 2 c.c. of the solution may be added. Then add 55 c.c. of the copper solution, measured with a 55 c.c. pipette, and draw over the neck of the flask a closely-fitting, good rubber tube, leaving about 2 cms. projecting above the neck of the flask. A good Mohr pinch cock is laid ready to put on the rubber tubing so as to close the flask air-tight at the end of boiling, or the special apparatus figured in Figure 77, which enables one to lift the flask from the flame and close the mouth at the same instant, is adjusted on the flask. Heat over a flame so high that it takes 3½ to 3¼ minutes to come to the boil and then boil vigorously for exactly 3 minutes. At the end of 3 minutes exactly, clamp air-tight the rubber tube on the neck of the flask, lift the flask from the flame and cool at once under the tap. The object of the clamp is to prevent any oxidation by air while cooling. After cooling to room temperature, open the rubber tube and add ½ to 1 c.c. of a 1 per cent. solution of soluble starch in a saturated KCl solution to the contents of the flask and titrate with N/100 or N/25 iodine solution run in from a glass-stoppered, small burette. At first the decolorization of the starch takes place at once and the iodine may be run in freely. When the blue color of the starch iodide appears, shake the flask gently once and allow it to stand. Near the end point the color persists 2-3 seconds and then disappears. Add iodine solution until the color persists

5-10 seconds. Particularly in titrating urine a slow decolorization of the starch due to the absorption of the iodine goes on. This is to be disregarded and the end point taken as that at which the color persists for 5-10 seconds.

Computation. There is a direct proportionality between the amount of sugar oxidized under these circumstances and the amount of iodine

FIG. 77.—Flask for use in Bang's micro-method.

solution used, at least up to 9.5 mgs. of sugar. It is only necessary, therefore, in order to find the mgs. of glucose in the amount of solution taken, to divide the number of c.c. of N/100 I used in the titration by the factor 2.7. If the titration is by N/25 I divide by the factor 0.7, and if N/10 I by the factor 0.285. Since in making the copper solution 16.5 c.c. of stock solution are diluted by the addition of 38.5 c.c. of KCl solution, it is possible to replace the 38.5 c.c. of KCl by 38.5 c.c. of the sugar solution to which has been added 11-11.5 grams of solid KCl. In this way 0.1 mg. of glucose may be accurately determined when present in a concentration of 1:300,000.

Copper solution. Dissolve first 160 grams $KHCO_3$, 100 grams K_2CO_3 and 66 grams KCl with about 700 c.c. water in a 1-liter flask. Since the bicarbonate is not readily soluble, pulverize and dissolve it first at a temperature of about 30° C. Then dissolve the KCl and finally, under cooling, the carbonate. Now add 100 c.c. of a 4.4 per cent. solution of $CuSO_4+5H_2O$ and fill to the mark when the slight evolution of CO_2 is at an end. Shake the solution only gently so as not to absorb much air. Use the solution on this account only after 24 hours' standing. This is the *stock* copper solution. To use, dilute 300 c.c. to 1,000 c.c. by the addition of saturated KCl solution. Shake this mixture also very little, and use only after standing some hours.

Iodine solution. This is a N/100 I solution. This may be made by diluting a N/10 I solution with boiled water. Such a solution in a dark-colored bottle keeps its strength unchanged for 3 months. One can make it also from iodide and iodate. Pour 1 c.c. of a 2 per cent. KIO_3 solution in a 100 c.c. volumetric flask, add 2-2.5 gr. KI and exactly 10 c.c. N/10 HCl. The acid sets free an equivalent amount of I. The reaction is instantaneous and the flask is filled to the mark with water.

Starch solution. 1 per cent. soluble starch in saturated KCl solution. This keeps perfectly.

Note. 55 c.c. of the copper solution is reduced completely by 10 mgs. glucose.

239. Microdetermination of glucose in blood (Bang, *Bioch. Zeits.*, XLIX, 1913, p. 23; LVII, 1913, p. 301.—*Principle.* Two or three drops of blood are received on a previously weighed, small piece of blotting paper and rapidly weighed. The blotting paper is then heated with slightly acid salt solution to coagulate the proteins, which remain in the blotting paper while the glucose goes into the salt solution. The glucose is then determined by the method given in 238, except that smaller amounts of copper solution and a weaker iodine solution, N/200, are employed. Microchemical methods are also given for the determination of water and chlorides in blood.

Process. Small pieces of a good, thick blotting paper are prepared about 16 x 28 mm. in area and weighing about 100 mgs. The paper must be extracted thoroughly first with hot water acidified with acetic acid to remove any iodine-binding substances, chlorides and reducing matters. (Papers all prepared, weighed and having attached to them a small wire for attaching them to the balance may be had from Warmbrunn and Quilitz, Berlin, who also have the other apparatus recommended.) The blotting paper is weighed beforehand. Two or three drops of blood weighing about 100-130 mgs. are drawn into the paper and immediately weighed. The weighing is very conveniently done on a small torsion balance of a special type recommended in the original article. Weighing to 1 mg. is sufficient. The blotting paper with the blood is now placed in a test-tube and 6.5 c.c. of boiling hot salt solution containing HCl is added and, after standing ½ hour, the clear liquid is poured carefully without loss into a 50 c.c. Jena glass flask with a neck without a rim and provided with a piece of rubber tubing like that described in the previous method—238. The salt solution with HCl consists of 1,360 c.c. of saturated KCl + 640 c.c. H_2O + 1.5 c.c. 25 per cent. HCl in a 2-liter flask. The salt solution may be measured into a test-tube with a mark at 6.5 c.c. in which it is boiled. The blotting paper is now washed out again with another 6.5 c.c. of hot salt solution and this is added to the 50 c.c. flask. The flask is cooled and after cooling add 1 c.c. of the copper *stock solution* described in 238. This is sufficient for at least 0.8 mg. of glucose. The flask provided with the rubber tubing and with the holder ready to clamp the tubing (Figure 77) is now heated to boiling so that it takes 1 min. 30 sec. to come to the boil (a variation of 5 seconds more or less permitted) and then boiled gently for exactly 2 minutes. At the end of 2 minutes exactly the rubber tube on the neck of the flask is clamped air-tight, the flask lifted from the flame and cooled under the tap. The reduced cuprous chloride is now titrated with N/200 iodine solution after the addition of 1-2 drops of starch solution and while the air in the flask is being driven out by a gentle stream of CO_2 which is carried to the bottom of the flask by a small tube. It is very necessary to prevent the entrance of air during the titration, as this oxidizes some of the cuprous chloride. The N/200 I is run in from a small, sharp-pointed, glass-cock burette of about 2 c.c. capacity, graduated to one-fiftieth of a c.c. One adds the iodine solution until the slight blue color persists for from 30-60 seconds. At first the color will disappear in 2-3 seconds as one draws near the end point.

Computation. Since, for some reason, the salt solutions and copper solutions have the power of binding about 0.12 c.c. of N/200 I, this amount is first subtracted from the amount of N/200 I consumed in the titration and the result divided by 4 gives the number of mgs. of glucose in the amount of blood or liquid taken. 0.01 mg. of glucose is equivalent to 0.04 c.c. of N/200 I solution. For example, if 0.72 c.c. of N/200 I has been used, the amount of glucose will be 0.72—0.12÷4, or 0.15 mg. of glucose. The results thus obtained with blood are a little higher than results given by Bertrand or the macro-method. This is due to the presence of some I-binding material in blood. Experience shows that this is about equivalent to 0.01-0.015 per cent. of glucose, so that this amount should be subtracted from the results obtained to give the true value of the glucose content.

Solutions. The N/200 I is made fresh since it is not stable. Into a 100 c.c. volumetric flask bring 1-2 c.c. 2 per cent. KJO_3, about 2 grs. KI, exactly 5 c.c. of N/10 HCl and fill to the mark with water. The copper solution has already been described in 238..

* 240. Glucose by Benedict's method (*Journal Amer. Med. Ass.*, LVII, 1193, 1911).—*Principle.* Complete reduction of a standard cupric sulphate to a colorless cuprous state by urine run in from a burette.

Process. 25 c.c. of the reagent are measured by a pipette accurately into a 25-30 cm. evaporating dish. Add 10-20 grams of crystalline Na_2CO_3, or half the quantity of the anhydrous salt, some powdered pumice or talcum, and heat to boiling over a free flame until the carbonate has completely dissolved. Dilute 10 c.c. of urine to 100 c.c., unless the amount of sugar in it is small, when it can be used without dilution, fill a burette with the urine, or diluted urine, to the zero mark and run into the boiling copper solution rather rapidly at first, then more slowly as the color grows less, then add but a few drops at a time until the solution is colorless. A white precipitate forms. If the mixture becomes too concentrated during the process, add water to replace that evaporated.

Calculation. The reduction of the 25 c.c. of the copper reagent is accomplished under these conditions by exactly 50 mgs. of glucose. The amount of urine run in from the burette contained this amount of glucose, provided of course there was no other reducing substance in it. If the urine was diluted 10 times, then the per cent. of glucose in the original urine was 0.050×1,000/x, where x was the number of c.c. used from the burette.

Standard solution. $CuSO_4$ (crystalline), 18.0 grams; Na_2CO_3 (crystalline), 200 grams; sodium or potassium citrate, 200 grams; KCNS, 125 grams; potassium ferrocyanide, 5 per cent. solution, 5.0 c.c.; distilled water to make a total volume of 1 liter. Dissolve the carbonate, citrate and thiocyanate in about 700 c.c. water and filter if necessary. Dissolve the copper sulphate in 100 c.c. water and add slowly little by little, stirring constantly, to the 700 c.c. Add the ferrocyanide and make up exactly to 1 liter. The copper salt must be weighed exactly; the other constituents need not be so exactly measured.

The advantage of this method over the original titration by Fehling's solution in a similar way is that the solution is less alkaline so that the sugar is less profoundly decomposed by the alkali and there is less danger of oxidation by the oxygen of the air. The end point in this method is sharp, whereas in the titration by Fehling's solution it is often very indefinite. The method is probably as accurate as any volumetric method. The drawback to the method lies in the changing concentration of the copper as the titration proceeds, so that relatively less copper is used to oxidize each molecule of glucose and the danger of oxidation by the air increases. It is, however, the most convenient and exact of the volumetric methods, except the method of Bang, which has some points of superiority, but is not so simple.

Other methods have been suggested, but they are inferior to those given here and are accordingly omitted. This second method is superior to Benedict's earlier method.

241. Quantitative determination of reducing sugars in urine and other fluids. Bertrand method (Bertrand, *Bull. de la Soc. Chimique de Paris*, XXXV, p. 1285, 1906).—In the absence of other reducing substances in the urine the amount of glucose, or other reducing sugar, can be best determined by the Bertrand method or by a combination of the Munson and Walker and Bertrand methods which follow (242).

Principle. The sugar containing liquid is boiled with a definite amount of Fehling's solution for a specific time. There must always be a large excess of Fehling's solution used, so that the blue color is little changed. The cuprous oxide formed is filtered off on an asbestos filter, and the amount determined by redissolving it in a solution of ferric sulphate containing sulphuric acid and titrating the amount of ferrous sulphate formed by standard permanganate of potassium solution. From the amount of copper thus determined one finds by means of tables the amount of glucose or other sugar present in the quantity of original solution used.

Process. Place in a 150 c.c. Erlenmeyer flask 20 c.c. of the neutral sugar solution (urine) to be determined. These 20 c.c. must not contain more than 100 mgs. of glucose or other sugar: in other words, it must not be more than 0.5 per cent. It is preferable to have between 10 and 90 mgs. Dilute the solution if more concentrated than this, but always take 20 c.c. Add 20 c.c. of the cupric-sulphate solution (1) and 20 c.c. of the alkaline liquid (2) and heat to boiling. Boil gently, so as not to concentrate the solution too greatly, for exactly 3 minutes. Withdraw the flask from the fire, and after waiting a moment for the cuprous oxide to settle, filter while still hot under suction through a previously prepared Gooch crucible with an asbestos mat. The way to make this filter is described in the following modification of this method, p. 996. Bring as little as possible of the cuprous oxide on the filter, but wash the cuprous oxide by decantation with distilled water, decanting through the asbestos filter each time. After several washings by decantation and suction, when the asbestos filter is free from blue color and any tartrate, remove and rinse carefully the filter flask so that it is quite clean and then replace the filter over it. Now dissolve the Cu_2O in the Erlenmeyer flask by adding successive portions little by little, about 5 c.c. at a time, of the ferric-sulphate solution (3). About 10-20 c.c. will be necessary. The precipitate changes to a blue black and dissolves to a green solution. When this is dissolved pour this solution carefully, using mild suction, through the asbestos filter so as to dissolve any cuprous oxide which may have been caught in the filter. It may be necessary to add a little fresh ferric-sulphate solution to bring this all in solution. Rinse the Erlenmeyer carefully with distilled water and pour the washings through the filter. Now titrate the filtrate in the suction flask by adding from the burette the potassium permanganate solution (4) until a single drop gives a permanent pink color. The end point is very

Solutions needed:

1.	Pure crystalline $CuSO_4$	40 g.
	Distilled water to	1 L
2.	Pure Rochelle salts	200 g.
	NaOH in sticks	150 g.
	Water to	1 L
3.	Ferric sulphate	50 g.
	Con. H_2SO_4	200 c.c.
	Water to	1 L
4.	Potassium permanganate	5 g.
	Water to	1 L

Determine the permanganate as follows: 0.250 g. ammonium oxalat is warmed with 100 c.c. water and 2 c.c. con. H_2SO_4 in a beaker to 60-80° Run in permanganate from a burette until a rose color remains (abou 22 c.c. will be needed). Multiply the amount of ammonium oxalate b $\frac{63.6 \times 2}{142.1}$ (=0.8951). This gives the amount of Cu which the amount o permanganate used in the titration represents. In round figures 1 lite of permanganate solution equals 10 gr. of copper. Reactions:

1. $Cu_2O + Fe_2(SO_4)_3 + H_2SO_4 = 2CuSO_4 + H_2O + 2FeSO_4$
2. $FeSO_4 + 2KMnO_4 + 8H_2SO_4 = 5Fe_2(SO_4)_3 + K_2SO_4 + 2MnSO_4 + 8H_2O$
3. $5C_2H_2O_4 + 2KMnO_4 + 3H_2SO_4 = 10CO_2 + 2MnSO_4 + K_2SO_4 + 8H_2O$

1 molecule $(NH_4)_2C_2O_4$ is equivalent to 2Fe.

Table for the Determination of Reducing Sugars by the Bertrand Metho

(Note.—This table can only be used when the directions of Bertrand as to the co centration of the Fehling's solution and the time of heating have been exactl followed. Different values are obtained if the heating is for another time an if the relative concentrations of sugar, copper sulphate and alkali are differe from those specified.)

Cu. in Mgs. for the Following Sugars:

Sugar in mgs.	Glucose Cu.	Invert sugar Cu.	Galactose Cu.	Maltose Cu.	Lactose Cu.	Mannose Cu.	Xylose Cu.	Arabinose Cu.
10	20.4	20.6	19.3	11.2	14.4	20.7	20.1	[illegible]
11	22.4	22.6	21.2	12.3	15.8			
12	24.3	24.6	23.0	13.4	17.2			
13	26.3	26.5	24.9	14.5	18.6			
14	28.3	28.5	26.7	15.6	20.0			
15	30.2	30.5	28.6	16.7	21.4			
16	32.2	32.5	30.5	17.8	22.8			
17	34.2	34.5	32.3	18.9	24.2			
18	36.2	36.4	34.2	20.0	25.6			
19	38.1	38.4	36.0	21.1	27.0			
20	40.1	40.4	37.9	22.2	28.4	40.5	39.6	[illegible]
21	42.0	42.3	39.7	23.3	29.8			
22	43.9	44.2	41.6	24.4	31.1			
23	45.8	46.1	43.4	25.5	32.5			
24	47.7	48.0	45.2	26.6	33.9			
25	49.6	49.8	47.0	27.7	35.2			
26	51.5	51.7	48.9	28.9	36.6			
27	53.4	53.6	50.7	30.0	38.0			

Cu. in Mgs. for the Following Sugars:

n	Glucose Cu.	Invert sugar Cu.	Galactose Cu.	Maltose Cu.	Lactose Cu.	Mannose Cu.	Xylose Cu.	Arabinose Cu.
	55.3	55.5	52.5	31.1	39.4			
	57.2	57.4	54.4	32.2	40.7			
	59.1	59.3	56.2	33.3	42.1	59.5	58.7	62.0
	60.9	61.1	58.0	34.4	43.4			
	62.8	63.0	59.7	35.5	44.8			
	64.6	64.8	61.5	36.5	46.1			
	66.5	66.7	63.3	37.6	47.4			
	68.3	68.5	65.0	38.7	48.7			
	70.1	70.3	66.8	39.8	50.1			
	72.0	72.2	68.6	40.9	51.4			
	73.8	74.0	70.4	41.9	52.7			
	75.5	75.9	72.1	43.0	54.1			
	77.5	77.7	73.9	44.1	55.4	78.0	77.3	81.5
	79.3	79.5	75.6	45.2	56.7			
	81.1	81.2	77.4	46.3	58.0			
	82.9	83.0	79.1	47.4	59.3			
	84.7	84.8	80.8	48.5	60.6			
	86.4	86.5	82.5	49.5	61.9			
	88.2	88.3	84.3	50.6	63.3			
	90.0	90.1	86.0	51.7	63.3			
	91.8	91.9	87.7	52.8	65.9			
	93.6	93.6	89.5	53.9	67.2			
	95.4	95.4	91.2	55.0	68.5	95.9	95.4	100.6
	97.1	97.1	92.9	56.1	69.8			
	98.9	98.8	94.6	57.1	71.1			
	100.6	100.6	96.3	58.2	72.4			
	102.3	102.3	98.0	59.3	73.7			
	104.1	104.0	99.7	60.3	74.9			
	105.8	105.7	101.5	61.4	76.2			
	107.6	107.4	103.2	62.5	77.5			
	109.3	109.2	104.9	63.5	78.8			
	111.1	110.9	106.6	64.6	80.1			
	112.8	112.6	108.3	65.7	81.4	113.3	113.2	119.3
	114.5	114.3	110.0	66.8	82.7			
	116.2	115.9	111.6	67.9	83.9			
	117.9	117.6	113.3	68.9	85.2			
	119.6	119.2	115.0	70.0	86.5			
	121.3	120.9	116.6	71.1	87.7			
	123.0	122.6	118.3	72.2	89.0			
	124.7	124.2	120.0	73.3	90.3			
	126.4	125.9	121.7	74.3	91.6			
	128.1	127.5	123.3	75.4	92.8			
	129.8	129.2	125.0	76.5	94.1	130.2	130.6	137.5
	131.4	130.8	126.6	77.6	95.4			
	133.1	132.4	128.3	78.6	96.6			
	134.7	134.0	130.0	79.7	97.9			
	136.3	135.6	131.5	80.8	99.1			
	137.9	137.2	133.1	81.8	100.4			
	139.6	138.9	134.8	82.9	101.7			
	141.2	140.5	136.4	84.0	102.9			
	142.8	142.1	138.0	85.1	104.2			
	144.5	143.7	139.7	86.1	105.4			
	146.1	145.3	141.3	87.2	106.7			
	147.7	146.9	142.9	88.3	107.9			
	149.3	148.5	144.6	89.4	109.2			
	150.9	150.0	146.2	90.4	110.4			
	152.5	151.6	147.8	91.5	111.7			
	154.0	153.2	149.4	92.6	112.9			
	155.6	154.8	151.1	93.7	114.1			
	157.2	156.4	152.7	94.8	115.4			

Cu. in Mgs. for the Following Sugars:

Sugar in mgs.	Glucose Cu.	Invert sugar Cu.	Galactose Cu.	Maltose Cu.	Lactose Cu.	Mannose Cu.	Xylose Cu.	[illegible]
88	158.8	157.9	154.3	95.3	116.6			
89	160.4	159.5	156.0	96.9	117.9			
90	162.0	161.1	157.6	98.0	119.1	163.3	164.2	172.[illegible]
91	163.6	162.6	159.2	99.0	120.3			
92	165.2	164.2	160.8	100.1	121.6			
93	166.7	165.7	162.4	101.1	122.8			
94	168.3	167.3	164.0	102.2	124.0			
95	169.9	168.8	165.6	103.2	125.2			
96	171.5	170.3	167.2	104.2	126.5			
97	173.1	171.9	168.8	105.3	127.7			
98	174.6	173.4	170.4	106.3	128.9			
99	176.2	175.0	172.0	107.4	130.2			
100	177.8	176.5	173.6	108.4	131.4	179.4	180.6	[illegible]

Table for Glucose by Bertrand for Smaller Amounts of Cu. (Moeckel [illegible] Frank).

Cu Mg.	Glucose Mg.	Cu Mg.	Gl[illegible]
1.1	0.5	11.5	
2.2	1.0	12.5	
3.3	1.5	13.5	
4.4	2.0	14.5	
5.5	2.5	15.5	
6.5	3.0	16.5	
7.5	3.5	17.5	
8.5	4.0	18.5	
9.5	4.5	19.5	
10.5	5.0	20.5	

242. Quantitative determination of reducing sugars in urine other fluids by a combination of the Munson and Walker and Bertra methods (Bertrand, *loc. cit.* *Munson and Walker: U. S. Department Agriculture,* Bulletin 107, 1912, p. 240; see also Circular 82).—In experience of this laboratory this method is the most accurate and co venient of all the methods of sugar analysis. It differs from the Muns and Walker method in that the cuprous oxide is not weighed but titrat by permanganate as in the Bertrand method. It differs from the Be trand method in that the composition of the Fehling solution is ferent and the time of heating is more precisely defined.

Principle. This is the same as the Bertrand method.

Process. Transfer 25 c.c. each of the copper (1) and alkaline ta trate solutions (2) to a 400 c.c. Jena beaker and add 50 c.c. of *neutral* sugar solution, or 10 c.c. if urine be used and add water make the final volume 100 c.c. Cover the beaker with a watch crys and heat on an asbestos gauze over a Bunsen burner so regulated th boiling begins in exactly 4 minutes, and continue the boiling for exac 2 minutes. Filter the cuprous oxide at once through an asbestos felt a porcelain Gooch crucible, using suction. (A device which has be found useful in this laboratory to control the time of heating is to a $CaCl_2$ solution, boiling at 112° C. This is brought to boiling while boiling the beaker with the sugar and Fehling's solution is immersed in it for exactly 6 minutes. It takes 4 minutes to come to the boil and boiling lasts 2 minutes. Another device has been suggested by Cole: namely, to connect a water manometer with the gas supply and, having once found what pressure of gas with a given burner will boil the beaker in 4 minutes, this gas pressure is obtained each time by regulating a thumbscrew.) Wash the cuprous oxide and the filter thoroughly with water at a temperature of about 60°, using suction. (At this point in the Munson and Walker method the precipitate is washed with alcohol and ether, dried at 100° and weighed as Cu_2O.) In the Bertrand modification proceed as follows: Transfer the asbestos and precipitate as soon as clean to the beaker in which the precipitation was conducted, suspend in water and then add 10 c.c. of the Bertrand ferric-sulphate-sulphuric-acid solution, or 10 c.c. of M/4 ferric-ammonium-sulphate-sulphuric-acid solution (4). Stir until all of the Cu_2O is dissolved, taking care to dissolve all the Cu_2O on the stirring rod or adherent to the inside of the crucible or the sides and lip of the beaker. A green solution is obtained. Titrate at once with continual stirring until the pink due to the permanganate persists for about 10-15 seconds. 1 c.c. N/20 $KMnO_4$ is equivalent to 0.00315 gram of Cu. Calculate the amount of Cu reduced and from this weight find the amount of reducing sugar to which it is equivalent in the appended tables (Bulletin 107, Dept. of Agriculture, pp. 243-251. For lactose see the correction in *Jour. Am. Chem. Soc.*, XXXIV, p. 202, or Circular 82). In the tables appended here the lactose correction has not been made.

Solutions and preparation of the asbestos filter. 1. Copper-sulphate solution (Soxhlet's modification of Fehling's):

$CuSO_4.5H_2O$ (Kahlbaum or Baker's best quality) ..	34.639 grams.
Water to	500 c.c.

2. Alkaline tartrate solution:

Rochelle salts	173 grams.
NaOH (best quality)	50 "
Water to	500 c.c.

3. Potassium permanganate solution: Instead of the solution recommended by Bertrand, a N/20 solution is generally used in this laboratory, as it is weaker:

$KMnO_4$	3.16 grams.
Water	2 liters

Dissolve the $KMnO_4$ in 2,000 c.c. distilled water, allow to stand in the dark for 2-3 days, then filter through asbestos or glass wool or decant the supernatant clear fluid into a clean, dry bottle. Keep in a bottle covered with black paint or black paper to protect from light. Standard ize against sodium oxalate. Dry overnight about 0.66 gram of sodium oxalate (pure) in a weighing tube in a steam oven and carefully weigh off three samples of 0.10-0.15 gram each of this dried sodium oxalate Dissolve each sample in 100 c.c. of water, add 5 c.c. of H_2SO_4 (1:1 by volume), warm to 70° C. and titrate the $KMnO_4$ against this. 1 c.c N/20 $KMnO_4$ is equivalent to 0.0035 gram sodium oxalate, or 0.0031 gram copper.

4. Ferric-ammonium alum and sulphuric acid: In place of the ferric sulphate recommended by Bertrand the ferric-ammonium alum may be used. The solution used is M/4.

$(NH_4Fe(SO_4)_2)_2.24H_2O$	240.5 grams.
Sulphuric acid (Con.)	200 c.c.
Water to	1 liter.

5. Asbestos filter: Digest the amphibole variety of asbestos, cut in pieces about 1 inch long, for 3 days with 1:3 HCl. Wash free from acid and digest for a similar period with soda solution, after which treat for a few hours with hot alkaline copper tartrate solution of the strength employed in sugar determinations. Wash the asbestos free from alkali, digest with nitric acid for several hours and after washing free from acid break into shreds by shaking for some time in water Folin suggests suspending it in water and forcing a brisk air current through it. Load the Gooch crucible with a layer of the washed asbestos about ¼ inch thick, wash it repeatedly with water to remove fine asbestos and then with alcohol. Dry at 100° if the gravimetric Munson and Walker method is to be used. With the Bertrand method it is not necessary to weigh or dry. The filters improve with use. In the gravimetric method the cuprous oxide may be washed out with nitric acid after each determination and the felts used over and over. In the Bertrand method the asbestos is filtered off after each titration, washed thoroughly and used for the next filtration.

Table for calculating dextrose, invert sugar alone, invert sugar in the presence of sucrose (0.4 gram and 2 grams total sugar), lactose (three forms), and maltose (anhydrous and crystallized). [For correction of lactose figures see Cir. 82.]

[Expressed in milligrams.]

Cuprous oxid (Cu_2O).	Copper (Cu).	Dextrose (*d*-glucose).	Invert sugar.	Invert sugar and sucrose.		Lactose.			Maltose.		Cuprous oxid (Cu_2O).
				0.4 gram total sugar.	2 grams total sugar.	$C_{12}H_{22}O_{11}$.	$C_{12}H_{22}O_{11}+\frac{1}{2}H_2O$.	$C_{12}H_{22}O_{11}+H_2O$.	$C_{12}H_{22}O_{11}$.	$C_{12}H_{22}O_{11}+H_2O$.	
10	8.9	4.0	4.5	1.6	....	3.8	3.9	4.0	5.9	6.2	10
11	9.8	4.5	5.0	2.1	....	4.5	4.6	4.7	6.7	7.0	11
12	10.7	4.9	5.4	2.5	....	5.1	5.3	5.4	7.5	7.9	12
13	11.5	5.3	5.8	3.0	....	5.8	5.9	6.1	8.3	8.7	13
14	12.4	5.7	6.3	3.4	....	6.4	6.6	6.8	9.1	9.5	14
15	13.3	6.2	6.7	3.9	....	7.1	7.3	7.5	9.9	10.4	15
16	14.2	6.6	7.2	4.3	....	7.7	8.0	8.2	10.6	11.2	16
17	15.1	7.0	7.6	4.8	....	8.4	8.6	8.8	11.4	12.0	17
18	16.0	7.5	8.1	5.2	....	9.1	9.3	9.5	12.2	12.9	18
19	16.9	7.9	8.5	5.7	....	9.7	10.0	10.2	13.0	13.7	19
20	17.8	8.3	8.9	6.1	....	10.4	10.6	10.9	13.8	14.6	20
21	18.7	8.7	9.4	6.6	....	11.0	11.3	11.6	14.6	15.4	21
22	19.6	9.2	9.8	7.0	....	11.7	12.0	12.3	15.4	16.2	22
23	20.4	9.6	10.3	7.5	....	12.3	12.7	13.0	16.2	17.1	23
24	21.3	10.0	10.7	7.9	....	13.0	13.3	13.7	17.0	17.9	24
25	22.2	10.5	11.2	8.4	....	13.6	14.0	14.4	17.8	18.7	25
26	23.1	10.9	11.6	8.8	....	14.3	14.7	15.1	18.6	19.6	26
27	24.0	11.3	12.0	9.3	....	15.0	15.3	15.7	19.4	20.4	27
28	24.9	11.8	12.5	9.7	....	15.6	16.0	16.4	20.2	21.2	28
29	25.8	12.2	12.9	10.2	....	16.3	16.7	17.1	21.0	22.1	29
30	26.6	12.6	13.4	10.7	4.3	16.9	17.4	17.8	21.8	22.9	30
31	27.5	13.1	13.8	11.1	4.7	17.6	18.0	18.5	22.6	23.7	31
32	28.4	13.5	14.3	11.6	5.2	18.2	18.7	19.2	23.3	24.6	32
33	29.3	13.9	14.7	12.0	5.6	18.9	19.4	19.9	24.1	25.4	33
34	30.2	14.3	15.2	12.5	6.1	19.5	20.1	20.6	24.9	26.2	34
35	31.1	14.8	15.6	12.9	6.5	20.2	20.7	21.3	25.7	27.1	35
36	32.0	15.2	16.1	13.4	7.0	20.9	21.4	22.0	26.5	27.9	36
37	32.9	15.6	16.5	13.8	7.4	21.5	22.1	22.7	27.3	28.7	37
38	33.8	16.1	16.9	14.3	7.9	22.2	22.8	23.3	28.1	29.6	38
39	34.6	16.5	17.4	14.7	8.4	22.8	23.4	24.0	28.9	30.4	39
40	35.5	16.9	17.8	15.2	8.8	23.5	24.1	24.7	29.7	31.3	40
41	36.4	17.4	18.3	15.6	9.3	24.1	24.8	25.4	30.5	32.1	41
42	37.3	17.8	18.7	16.1	9.7	24.8	25.4	26.1	31.3	32.9	42
43	38.2	18.2	19.2	16.6	10.2	25.4	26.1	26.8	32.1	33.8	43
44	39.1	18.7	19.6	17.0	10.7	26.1	26.8	27.5	32.9	34.6	44
45	40.0	19.1	20.1	17.5	11.1	26.8	27.5	28.2	33.7	35.4	45
46	40.9	19.6	20.5	17.9	11.6	27.4	28.1	28.8	34.4	36.3	46
47	41.7	20.0	21.0	18.4	12.0	28.1	28.8	29.5	35.2	37.1	47
48	42.6	20.4	21.4	18.8	12.5	28.7	29.5	30.2	36.0	37.9	48
49	43.5	20.9	21.9	19.3	12.9	29.4	30.1	30.9	36.8	38.8	49
50	44.4	21.3	22.3	19.7	13.4	30.0	30.8	31.6	37.6	39.6	50
51	45.3	21.7	22.8	20.2	13.9	30.7	31.5	32.3	38.4	40.4	51
52	46.2	22.2	23.2	20.7	14.3	31.3	32.1	33.0	39.2	41.3	52
53	47.1	22.6	23.7	21.1	14.8	32.0	32.8	33.6	40.0	42.1	53
54	48.0	23.0	24.1	21.6	15.2	32.6	33.5	34.3	40.8	42.9	54
55	48.9	23.5	24.6	22.0	15.7	33.3	34.1	35.0	41.6	43.8	55
56	49.7	23.9	25.0	22.5	16.2	33.9	34.8	35.7	42.4	44.6	56
57	50.6	24.3	25.5	22.9	16.6	34.6	35.5	36.4	43.2	45.4	57
58	51.5	24.8	25.9	23.4	17.1	35.2	36.1	37.1	44.0	46.3	58
59	52.4	25.2	26.4	23.9	17.5	35.9	36.8	37.7	44.8	47.1	59
60	53.3	25.6	26.8	24.3	18.0	36.5	37.5	38.4	45.6	48.0	60
61	54.2	26.1	27.3	24.8	18.5	37.2	38.2	39.1	46.3	48.8	61
62	55.1	26.5	27.7	25.2	18.9	37.8	38.8	39.8	47.1	49.6	62
63	56.0	27.0	28.2	25.7	19.4	38.5	39.5	40.5	47.9	50.5	63
64	56.8	27.4	28.6	26.2	19.8	39.2	40.2	41.2	48.7	51.3	64

Table for calculating dextrose, invert sugar alone, invert sugar in the presence sucrose (0.4 gram and 2 grams total sugar), lactose (three forms), and malto (anhydrous and crystallized)—Continued. [For correction of lactose figures s Cir. 82.]

[Expressed in milligrams.]

Cuprous oxid (Cu_2O).	Copper (Cu).	Dextrose (d-glucose).	Invert sugar.	Invert sugar and sucrose. 0.4 gram total sugar.	Invert sugar and sucrose. 2 grams total sugar.	Lactose. $C_{12}H_{22}O_{11}$.	Lactose. $C_{12}H_{22}O_{11}+\frac{1}{2}H_2O$.	Lactose. $C_{12}H_{22}O_{11}+H_2O$.	Maltose. $C_{12}H_{22}O_{11}$.	Maltose. $C_{12}H_{22}O_{11}+H_2O$.	Cuprous oxid (Cu_2O).
65	57.7	27.8	29.1	26.6	20.3	39.8	40.9	41.9	49.5	52.1	65
66	58.6	28.3	29.5	27.1	20.8	40.5	41.6	42.6	50.3	53.0	66
67	59.5	28.7	30.0	27.5	21.2	41.1	42.3	43.3	51.1	53.8	67
68	60.4	29.2	30.4	28.0	21.7	41.8	42.9	44.0	51.9	54.6	68
69	61.3	29.6	30.9	28.5	22.2	42.5	43.6	44.7	52.7	55.5	69
70	62.2	30.0	31.3	28.9	22.6	43.1	44.3	45.4	53.6	56.3	70
71	63.1	30.5	31.8	29.4	23.1	43.8	44.9	46.1	54.3	57.1	71
72	64.0	30.9	32.3	29.8	23.5	44.4	45.6	46.8	55.1	58.0	72
73	64.8	31.4	32.7	30.3	24.0	45.1	46.3	47.5	55.9	58.8	73
74	65.7	31.8	33.2	30.8	24.5	45.7	47.0	48.2	56.7	59.6	74
75	66.6	32.2	33.6	31.2	24.9	46.4	47.6	48.8	57.5	60.5	75
76	67.5	32.7	34.1	31.7	25.4	47.0	48.3	49.5	58.2	61.3	76
77	68.4	33.1	34.5	32.1	25.9	47.7	49.0	50.2	59.0	62.1	77
78	69.3	33.6	35.0	32.6	26.3	48.4	49.6	50.9	59.8	63.0	78
79	70.2	34.0	35.4	33.1	26.8	49.0	50.3	51.6	60.6	63.8	79
80	71.1	34.4	35.9	33.5	27.3	49.7	51.0	52.3	61.4	64.6	80
81	71.9	34.9	36.3	34.0	27.7	50.3	51.6	53.0	62.2	65.4	81
82	72.8	35.3	36.8	34.5	28.2	51.0	52.3	53.7	63.0	66.3	82
83	73.7	35.8	37.3	34.9	28.6	51.6	53.0	54.4	63.8	67.1	83
84	74.6	36.2	37.7	35.4	29.1	52.3	53.7	55.0	64.6	68.0	84
85	75.5	36.7	38.2	35.8	29.6	52.9	54.3	55.7	65.4	68.8	85
86	76.4	37.1	38.6	36.3	30.0	53.6	55.0	56.4	66.2	69.7	86
87	77.3	37.5	39.1	36.8	30.5	54.3	55.7	57.1	67.0	70.5	87
88	78.2	38.0	39.5	37.2	31.0	54.9	56.4	57.8	67.8	71.3	88
89	79.1	38.4	40.0	37.7	31.4	55.6	57.0	58.5	68.5	72.3	89
90	79.9	38.9	40.4	38.2	31.9	56.2	57.7	59.2	69.3	73.0	90
91	80.8	39.3	40.9	38.6	32.4	56.9	58.4	59.9	70.1	73.8	91
92	81.7	39.8	41.4	39.1	32.8	57.5	59.0	60.6	70.9	74.7	92
93	82.6	40.2	41.8	39.6	33.3	58.2	59.7	61.3	71.7	75.5	93
94	83.5	40.6	42.3	40.0	33.8	58.8	60.4	61.9	72.5	76.3	94
95	84.4	41.1	42.7	40.5	34.2	59.5	61.1	62.6	73.3	77.2	95
96	85.3	41.5	43.2	41.0	34.7	60.2	61.7	63.3	74.1	78.0	96
97	86.2	42.0	43.7	41.4	35.2	60.8	62.4	64.0	74.9	78.8	97
98	87.1	42.4	44.1	41.9	35.6	61.5	63.1	64.7	75.7	79.7	98
99	87.9	42.9	44.6	42.4	36.1	62.1	63.8	65.4	76.5	80.5	99
100	88.8	43.3	45.0	42.8	36.6	62.8	64.4	66.1	77.3	81.3	100
101	89.7	43.8	45.5	43.3	37.0	63.4	65.1	66.8	78.1	82.2	101
102	90.6	44.2	46.0	43.8	37.5	64.1	65.8	67.5	78.8	82.9	102
103	91.5	44.7	46.4	44.2	38.0	64.7	66.4	68.1	79.6	83.8	103
104	92.4	45.1	46.9	44.7	38.5	65.4	67.1	68.8	80.4	84.7	104
105	93.3	45.6	47.3	45.2	38.9	66.1	67.8	69.5	81.2	85.5	105
106	94.2	46.0	47.8	45.6	39.4	66.7	68.5	70.2	82.0	86.3	106
107	95.0	46.4	48.3	46.1	39.9	67.4	69.1	70.9	82.8	87.2	107
108	95.9	46.9	48.7	46.6	40.3	68.0	69.8	71.6	83.6	88.0	108
109	96.8	47.3	49.2	47.0	40.8	68.7	70.5	72.3	84.4	88.8	109
110	97.7	47.8	49.6	47.5	41.3	69.3	71.1	73.0	85.2	89.7	110
111	98.6	48.2	50.1	48.0	41.7	70.0	71.8	73.6	86.0	90.5	111
112	99.5	48.7	50.6	48.4	42.2	70.6	72.5	74.3	86.8	91.3	112
113	100.4	49.1	51.0	48.9	42.7	71.3	73.1	75.0	87.6	92.2	113
114	101.3	49.6	51.5	49.4	43.2	71.9	73.8	75.7	88.4	93.0	114
115	102.2	50.0	51.9	49.8	43.6	72.6	74.5	76.4	89.2	93.9	115
116	103.0	50.5	52.4	50.3	44.1	73.2	75.2	77.1	90.0	94.7	116
117	103.9	50.9	52.9	50.8	44.6	73.9	75.8	77.8	90.7	95.5	117
118	104.8	51.4	53.3	51.2	45.0	74.5	76.5	78.6	91.5	96.3	118
119	105.7	51.8	53.8	51.7	45.5	75.2	77.2	79.1	92.3	97.2	119

Table for calculating dextrose, invert sugar alone, invert sugar in the presence of sucrose (0.4 gram and 2 grams total sugar), lactose (three forms), and maltose (anhydrous and crystallized)—Continued. [For correction of lactose figures see Cir. 82.]

[Expressed in milligrams.]

Cuprous oxid (Cu_2O).	Copper (Cu).	Dextrose (d-glucose).	Invert sugar.	Invert sugar and sucrose.		Lactose.			Maltose.		Cuprous oxid (Cu_2O).
				0.4 gram total sugar.	2 grams total sugar.	$C_{12}H_{22}O_{11}$.	$C_{12}H_{22}O_{11}+\frac{1}{2}H_2O$.	$C_{12}H_{22}O_{11}+H_2O$.	$C_{12}H_{22}O_{11}$.	$C_{12}H_{22}O_{11}+H_2O$.	
120	106.6	52.3	54.3	52.2	46.0	75.8	77.8	79.8	93.1	98.0	120
121	107.5	52.7	54.7	52.7	46.5	76.5	78.5	80.5	93.9	98.9	121
122	108.4	53.2	55.2	53.1	46.9	77.1	79.2	81.2	94.7	99.7	122
123	109.3	53.6	55.7	53.6	47.4	77.8	79.9	81.9	95.5	100.5	123
124	110.1	54.1	56.1	54.1	47.9	78.5	80.5	82.6	96.3	101.4	124
125	111.0	54.5	56.6	54.5	48.3	79.1	81.2	83.3	97.1	102.2	125
126	111.9	55.0	57.0	55.0	48.8	79.8	81.9	84.0	97.9	103.0	126
127	112.8	55.4	57.5	55.5	49.3	80.4	82.5	84.7	98.7	103.9	127
128	113.7	55.9	58.0	55.9	49.8	81.1	83.2	85.4	99.4	104.7	128
129	114.6	56.3	58.4	56.4	50.2	81.7	83.9	86.0	100.2	105.5	129
130	115.5	56.8	58.9	56.9	50.7	82.4	84.6	86.7	101.0	106.4	130
131	116.4	57.2	59.4	57.4	51.2	83.1	85.2	87.4	101.8	107.2	131
132	117.3	57.7	59.8	57.8	51.7	83.7	85.9	88.1	102.6	108.0	132
133	118.1	58.1	60.3	58.3	52.1	84.4	86.6	88.8	103.4	108.9	133
134	119.0	58.6	60.8	58.8	52.6	85.0	87.3	89.5	104.2	109.7	134
135	119.9	59.0	61.2	59.3	53.1	85.7	87.9	90.2	105.0	110.5	135
136	120.8	59.5	61.7	59.7	53.6	86.3	88.6	90.9	105.8	111.4	136
137	121.7	60.0	62.2	60.2	54.0	87.0	89.3	91.6	106.6	112.2	137
138	122.6	60.4	62.6	60.7	54.5	87.7	90.0	92.3	107.4	113.0	138
139	123.5	60.9	63.1	61.2	55.0	88.3	90.6	93.0	108.2	113.9	139
140	124.4	61.3	63.6	61.6	55.5	89.0	91.3	93.6	109.0	114.7	140
141	125.2	61.8	64.0	62.1	55.9	89.6	92.0	94.3	109.8	115.5	141
142	126.1	62.2	64.5	62.6	56.4	90.3	92.6	95.0	110.5	116.4	142
143	127.0	62.7	65.0	63.1	56.9	90.9	93.3	95.7	111.3	117.2	143
144	127.9	63.1	65.4	63.5	57.4	91.6	94.0	96.4	112.1	118.0	144
145	128.8	63.6	65.9	64.0	57.8	92.2	94.7	97.1	112.9	118.9	145
146	129.7	64.0	66.4	64.5	58.3	92.9	95.3	97.8	113.7	119.7	146
147	130.6	64.5	66.9	65.0	58.8	93.5	96.0	98.4	114.5	120.5	147
148	131.5	65.0	67.3	65.4	59.3	94.2	96.7	99.1	115.3	121.4	148
149	132.4	65.4	67.8	65.9	59.7	94.8	97.3	99.8	116.1	122.2	149
150	133.2	65.9	68.3	66.4	60.2	95.5	98.0	100.5	116.9	123.0	150
151	134.1	66.3	68.7	66.9	60.7	96.2	98.7	101.2	117.7	123.9	151
152	135.0	66.8	69.2	67.3	61.2	96.8	99.3	101.9	118.5	124.7	152
153	135.9	67.2	69.7	67.8	61.7	97.5	100.0	102.6	119.3	125.5	153
154	136.8	67.7	70.1	68.3	62.1	98.1	100.7	103.3	120.0	126.4	154
155	137.7	68.2	70.6	68.8	62.6	98.8	101.4	104.0	120.8	127.2	155
156	138.6	68.6	71.1	69.2	63.1	99.4	102.0	104.7	121.6	128.0	156
157	139.5	69.1	71.6	69.7	63.6	100.1	102.7	105.3	122.4	128.9	157
158	140.3	69.5	72.0	70.2	64.1	100.7	103.4	106.0	123.2	129.7	158
159	141.2	70.0	72.5	70.7	64.5	101.4	104.1	106.7	124.0	130.5	159
160	142.1	70.4	73.0	71.2	65.0	102.0	104.7	107.4	124.8	131.4	160
161	143.0	70.9	73.4	71.6	65.5	102.7	105.4	108.1	125.6	132.2	161
162	143.9	71.4	73.9	72.1	66.0	103.4	106.1	108.8	126.4	133.0	162
163	144.8	71.8	74.4	72.6	66.5	104.0	106.7	109.5	127.2	133.9	163
164	145.7	72.3	74.9	73.1	66.9	104.7	107.4	110.2	128.0	134.7	164
165	146.6	72.8	75.3	73.6	67.4	105.3	108.1	110.9	128.8	135.5	165
166	147.5	73.2	75.8	74.0	67.9	106.0	108.8	111.5	129.6	136.4	166
167	148.3	73.7	76.3	74.5	68.4	106.6	109.4	112.2	130.3	137.2	167
168	149.2	74.1	76.8	75.0	68.9	107.3	110.1	112.9	131.1	138.0	168
169	150.1	74.6	77.2	75.5	69.3	107.9	110.8	113.6	131.9	138.9	169
170	151.0	75.1	77.7	76.0	69.8	108.6	111.4	114.3	132.7	139.7	170
171	151.9	75.5	78.2	76.4	70.3	109.2	112.1	115.0	133.5	140.5	171
172	152.8	76.0	78.7	76.9	70.8	109.9	112.8	115.7	134.3	141.4	172
173	153.7	76.4	79.1	77.4	71.3	110.5	113.5	116.4	135.1	142.2	173
174	154.6	76.9	79.6	77.9	71.7	111.2	114.1	117.1	135.9	143.0	174

Table for calculating dextrose, invert sugar alone, invert sugar in the presence sucrose (0.4 gram and 2 grams total sugar), lactose (three forms), and malt (anhydrous and crystallized)—Continued. [For correction of lactose figures Cir. 82.]

[Expressed in milligrams.]

Cuprous oxid (Cu_2O).	Copper (Cu).	Dextrose (d-glucose).	Invert sugar.	Invert sugar and sucrose.		Lactose.			Maltose.		Cuprous oxid (Cu_2O).
				0.4 gram total sugar.	2 grams total sugar.	$C_{12}H_{22}O_{11}$.	$C_{12}H_{22}O_{11}+\frac{1}{2}H_2O$.	$C_{12}H_{22}O_{11}+H_2O$.	$C_{12}H_{22}O_{11}$.	$C_{12}H_{22}O_{11}+H_2O$.	
175	155.5	77.4	80.1	78.4	72.2	111.9	114.8	117.7	136.7	143.9	175
176	156.3	77.8	80.6	78.8	72.7	112.5	115.5	118.4	137.5	144.7	176
177	157.2	78.3	81.0	79.3	73.2	113.2	116.1	119.1	138.3	145.5	177
178	158.1	78.8	81.5	79.8	73.7	113.8	116.8	119.8	139.1	146.4	178
179	159.0	79.2	82.0	80.3	74.2	114.5	117.5	120.5	139.8	147.3	179
180	159.9	79.7	82.5	80.8	74.6	115.1	118.2	121.2	140.6	148.0	180
181	160.8	80.1	82.9	81.3	75.1	115.8	118.8	121.9	141.4	148.9	181
182	161.7	80.6	83.4	81.7	75.6	116.5	119.5	122.6	142.2	149.7	182
183	162.6	81.1	83.9	82.2	76.1	117.1	120.2	123.3	143.0	150.6	183
184	163.4	81.5	84.4	82.7	76.6	117.8	120.9	123.9	143.8	151.4	184
185	164.3	82.0	84.9	83.2	77.1	118.4	121.5	124.6	144.6	152.2	185
186	165.2	82.5	85.3	83.7	77.6	119.1	122.2	125.3	145.4	153.0	186
187	166.1	82.9	85.8	84.2	78.0	119.7	122.9	126.0	146.2	153.9	187
188	167.0	83.4	86.3	84.6	78.5	120.4	123.5	126.7	147.0	154.7	188
189	167.9	83.9	86.8	85.1	79.0	121.0	124.2	127.4	147.8	155.5	189
190	168.8	84.3	87.2	85.6	79.5	121.7	124.9	128.1	148.6	156.4	190
191	169.7	84.8	87.7	86.1	80.0	122.3	125.5	128.8	149.3	157.2	191
192	170.5	85.3	88.2	86.6	80.5	123.0	126.2	129.5	150.1	158.0	192
193	171.4	85.7	88.7	87.1	81.0	123.6	126.9	130.1	150.9	158.9	193
194	172.3	86.2	89.2	87.6	81.4	124.3	127.6	130.8	151.7	159.7	194
195	173.2	86.7	89.6	88.0	81.9	125.0	128.2	131.5	152.5	160.5	195
196	174.1	87.1	90.1	88.5	82.4	125.6	128.9	132.2	153.3	161.4	196
197	175.0	87.6	90.6	89.0	82.9	126.3	129.6	132.9	154.1	162.2	197
198	175.9	88.1	91.1	89.5	83.4	126.9	130.3	133.6	154.9	163.0	198
199	176.8	88.5	91.6	90.0	83.9	127.6	130.9	134.3	155.7	163.9	199
200	177.7	89.0	92.0	90.5	84.4	128.2	131.6	135.0	156.5	164.7	200
201	178.5	89.5	92.5	91.0	84.8	128.9	132.3	135.7	157.3	165.6	201
202	179.4	89.9	93.0	91.4	85.3	129.5	132.9	136.3	158.1	166.4	202
203	180.3	90.4	93.5	91.9	85.8	130.2	133.6	137.0	158.8	167.2	203
204	181.2	90.9	94.0	92.4	86.3	130.8	134.3	137.7	159.6	168.0	204
205	182.1	91.4	94.5	92.9	86.8	131.5	135.0	138.4	160.4	168.8	205
206	183.0	91.8	94.9	93.4	87.3	122.1	135.6	139.1	161.2	169.7	206
207	183.9	92.3	95.4	93.9	87.8	132.8	136.3	139.8	162.0	170.5	207
208	184.8	92.8	95.9	94.4	88.3	133.4	137.0	140.5	162.8	171.4	208
209	185.6	93.2	96.4	94.9	88.8	134.1	137.6	141.2	163.6	172.2	209
210	186.5	93.7	96.9	95.4	89.2	134.8	138.3	141.9	164.4	173.0	210
211	187.4	94.2	97.4	95.8	89.7	135.4	139.0	142.5	165.2	173.8	211
212	188.3	94.6	97.8	96.3	90.2	136.1	139.6	143.2	166.0	174.7	212
213	189.2	95.1	98.3	96.8	90.7	136.7	140.3	143.9	166.8	175.5	213
214	190.1	95.6	98.8	97.3	91.2	137.4	141.0	144.6	167.5	176.4	214
215	191.0	96.1	99.3	97.8	91.7	138.0	141.7	145.3	168.3	177.3	215
216	191.9	96.5	99.8	98.3	92.2	138.7	142.3	146.0	169.1	178.0	216
217	192.8	97.0	100.3	98.8	92.7	139.3	143.0	146.7	169.9	178.9	217
218	193.6	97.5	100.8	99.3	93.2	140.0	143.7	147.3	170.7	179.7	218
219	194.5	98.0	101.2	99.8	93.7	140.6	144.3	148.0	171.5	180.5	219
220	195.4	98.4	101.7	100.3	94.2	141.3	145.0	148.7	172.3	181.4	220
221	196.3	98.9	102.2	100.8	94.7	141.9	145.7	149.4	173.1	182.2	221
222	197.2	99.4	102.7	101.2	95.1	142.6	146.3	150.1	173.9	183.0	222
223	198.1	99.9	103.2	101.7	95.6	143.2	147.0	150.8	174.7	183.9	223
224	199.0	100.3	103.7	102.2	96.1	143.9	147.7	151.5	175.5	184.7	224
225	199.9	100.8	104.2	102.7	96.6	144.6	148.4	152.2	176.3	185.5	225
226	200.7	101.3	104.6	103.2	97.1	145.2	149.0	152.9	177.0	186.4	226
227	201.6	101.8	105.1	103.7	97.6	145.9	149.7	153.6	177.8	187.2	227
228	202.5	102.2	105.6	104.2	98.1	146.5	150.4	154.3	178.6	188.0	228
229	203.4	102.7	106.1	104.7	98.6	147.2	151.1	154.9	179.4	188.8	229

Table for calculating dextrose, invert sugar alone, invert sugar in the presence of sucrose (0.4 gram and 2 grams total sugar), lactose (three forms), and maltose (anhydrous and crystallized)—Continued. [For correction of lactose figures see Cir. 82.]

[Expressed in milligrams.]

Cuprous oxid (Cu_2O).	Copper (Cu).	Dextrose (d-glucose).	Invert sugar.	Invert sugar and sucrose.		Lactose.			Maltose.		Cuprous oxid (Cu_2O).
				0.4 gram total sugar.	2 grams total sugar.	$C_{12}H_{22}O_{11}$.	$C_{12}H_{22}O_{11}+\frac{1}{2}H_2O$.	$C_{12}H_{22}O_{11}+H_2O$.	$C_{12}H_{22}O_{11}$.	$C_{12}H_{22}O_{11}+H_2O$.	
230	204.3	103.2	106.6	105.2	99.1	147.8	151.7	155.6	180.2	189.7	230
231	205.2	103.7	107.1	105.7	99.6	148.5	152.4	156.3	181.0	190.5	231
232	206.1	104.1	107.6	106.2	100.1	149.1	153.1	157.0	181.8	191.3	232
233	207.0	104.6	108.1	106.7	100.6	149.8	153.7	157.7	182.6	192.2	233
234	207.9	105.1	108.6	107.2	101.1	150.5	154.4	158.4	183.4	193.0	234
235	208.7	105.6	109.1	107.7	101.6	151.1	155.1	159.1	184.2	193.8	235
236	209.6	106.0	109.5	108.2	102.1	151.8	155.8	159.7	184.9	194.7	236
237	210.5	106.5	110.0	108.7	102.6	152.4	156.4	160.4	185.7	195.5	237
238	211.4	107.0	110.5	109.2	103.1	153.1	157.1	161.1	186.5	196.3	238
239	212.3	107.5	111.0	109.6	103.5	153.7	157.8	161.8	187.3	197.2	239
240	213.2	108.0	111.5	110.1	104.0	154.4	158.4	162.5	188.1	198.0	240
241	214.1	108.4	112.0	110.6	104.5	155.0	159.1	163.2	188.9	198.8	241
242	215.0	108.9	112.5	111.1	105.0	155.7	159.8	163.9	189.7	199.7	242
243	215.8	109.4	113.0	111.6	105.5	156.3	160.5	164.6	190.5	200.5	243
244	216.7	109.9	113.5	112.1	106.0	157.0	161.1	165.3	191.3	201.3	244
245	217.6	110.4	114.0	112.6	106.5	157.7	161.8	166.0	192.1	202.2	245
246	218.5	110.8	114.5	113.1	107.0	158.3	162.5	166.6	192.9	203.0	246
247	219.4	111.3	115.0	113.6	107.5	159.0	163.1	167.3	193.6	203.8	247
248	220.3	111.8	115.4	114.1	108.0	159.6	163.8	168.0	194.4	204.7	248
249	221.2	112.3	115.9	114.6	108.5	160.3	164.5	168.7	195.2	205.5	249
250	222.1	112.8	116.4	115.1	109.0	160.9	165.2	169.4	196.0	206.3	250
251	223.0	113.2	116.9	115.6	109.5	161.6	165.8	170.1	196.8	207.2	251
252	223.8	113.7	117.4	116.1	110.0	162.2	166.5	170.8	197.6	208.0	252
253	224.7	114.2	117.9	116.6	110.5	162.9	167.2	171.5	198.4	208.8	253
254	225.6	114.7	118.4	117.1	111.0	163.5	167.9	172.2	199.2	209.7	254
255	226.5	115.2	118.9	117.6	111.5	164.2	168.5	172.8	200.0	210.5	255
256	227.4	115.7	119.4	118.1	112.0	164.8	169.2	173.5	200.8	211.3	256
257	228.3	116.1	119.9	118.6	112.5	165.5	169.9	174.2	201.6	212.2	257
258	229.2	116.6	120.4	119.1	113.0	166.2	170.5	174.9	202.3	213.0	258
259	230.1	117.1	120.9	119.6	113.5	166.8	171.2	175.6	203.1	213.8	259
260	231.0	117.6	121.4	120.1	114.0	167.5	171.9	176.3	203.9	214.7	260
261	231.8	118.1	121.9	120.6	114.5	168.1	172.5	177.0	204.7	215.5	261
262	232.7	118.6	122.4	121.1	115.0	168.8	173.2	177.7	205.5	216.3	262
263	233.6	119.0	122.9	121.6	115.5	169.4	173.9	178.3	206.3	217.2	263
264	234.5	119.5	123.4	122.1	116.0	170.1	174.6	179.0	207.1	218.0	264
265	235.4	120.0	123.9	122.6	116.5	170.7	175.2	179.7	207.9	218.8	265
266	236.3	120.5	124.4	123.1	117.0	171.4	175.9	180.4	208.7	219.7	266
267	237.2	121.0	124.9	123.6	117.5	172.0	176.6	181.1	209.5	220.5	267
268	238.1	121.5	125.4	124.1	118.0	172.7	177.2	181.8	210.3	221.3	268
269	238.9	122.0	125.9	124.6	118.5	173.3	177.9	182.5	211.0	222.1	269
270	239.8	122.5	126.4	125.1	119.0	174.0	178.6	183.2	211.8	223.0	270
271	240.7	122.9	126.9	125.6	119.5	174.6	179.2	183.8	212.6	223.8	271
272	241.6	123.4	127.4	126.2	120.0	175.3	179.9	184.5	213.4	224.6	272
273	242.5	123.9	127.9	126.7	120.6	176.0	180.6	185.2	214.2	225.5	273
274	243.4	124.4	128.4	127.2	121.1	176.6	181.3	185.9	215.0	226.3	274
275	244.3	124.9	128.9	127.7	121.6	177.3	181.9	186.6	215.8	227.1	275
276	245.2	125.4	129.4	128.2	122.1	177.9	182.6	187.3	216.6	228.0	276
277	246.1	125.9	129.9	128.7	122.6	178.6	183.3	188.0	217.4	228.8	277
278	246.9	126.4	130.4	129.2	123.1	179.2	184.0	188.7	218.2	229.6	278
279	247.8	126.9	130.9	129.7	123.6	179.9	184.6	189.4	218.9	230.5	279
280	248.7	127.3	131.4	130.2	124.1	180.6	185.3	190.1	219.7	231.3	280
281	249.6	127.8	131.9	130.7	124.6	181.2	186.0	190.7	220.5	232.1	281
282	250.5	128.3	132.4	131.2	125.1	181.9	186.6	191.4	221.3	233.0	282
283	251.4	128.8	132.9	131.7	125.6	182.5	187.3	192.1	222.1	233.8	283
284	252.3	129.3	133.4	132.2	126.1	183.2	188.0	192.8	222.9	234.6	284

Table for calculating dextrose, invert sugar alone, invert sugar in the presence of sucrose (0.4 gram and 2 grams total sugar), lactose (three forms), and maltose (anhydrous and crystallized)—Continued. [For correction of lactose figures see Cir. 82.]

[Expressed in milligrams.]

Cuprous oxid (Cu_2O).	Copper (Cu).	Dextrose (*d*-glucose).	Invert sugar.	Invert sugar and sucrose.		Lactose.			Maltose.		Cuprous oxid (Cu_2O).
				0.4 gram total sugar.	2 grams total sugar.	$C_{12}H_{22}O_{11}$.	$C_{12}H_{22}O_{11}+\frac{1}{2}H_2O$.	$C_{12}H_{22}O_{11}+H_2O$.	$C_{12}H_{22}O_{11}$.	$C_{12}H_{22}O_{11}+H_2O$.	
285	253.2	129.8	133.9	132.7	126.6	183.8	188.7	193.5	223.7	235.5	285
286	254.0	130.3	134.4	133.2	127.1	184.5	189.3	194.2	224.5	236.3	286
287	254.9	130.8	134.9	133.7	127.6	185.1	190.0	194.9	225.3	237.1	287
288	255.8	131.3	135.4	134.3	128.1	185.8	190.7	195.5	226.1	238.0	288
289	256.7	131.8	135.9	134.8	128.6	186.4	191.3	196.2	226.9	238.8	289
290	257.6	132.2	136.4	135.3	129.2	187.1	192.0	196.9	227.6	239.6	290
291	258.5	132.7	136.9	135.8	129.7	187.7	192.7	197.6	228.4	240.5	291
292	259.4	133.2	137.4	136.3	130.2	188.4	193.3	198.3	229.2	241.3	292
293	260.3	133.7	137.9	136.8	130.7	189.0	194.0	199.0	230.0	242.1	293
294	261.2	134.2	138.4	137.3	131.2	189.7	194.7	199.7	230.8	242.9	294
295	262.0	134.7	138.9	137.8	131.7	190.3	195.4	200.4	231.6	243.8	295
296	262.9	135.2	139.4	138.3	132.2	191.0	196.0	201.0	232.4	244.6	296
297	263.8	135.7	140.0	138.8	132.7	191.7	196.7	201.7	233.2	245.4	297
298	264.7	136.2	140.5	139.4	133.2	192.3	197.4	202.4	234.0	246.3	298
299	265.6	136.7	141.0	139.9	133.7	193.0	198.0	203.1	234.8	247.1	299
300	266.5	137.2	141.5	140.4	134.2	193.6	198.7	203.8	235.5	247.9	300
301	267.4	137.7	142.0	140.9	134.8	194.3	199.4	204.5	236.3	248.8	301
302	268.3	138.2	142.5	141.4	135.3	194.9	200.0	205.2	237.1	249.6	302
303	269.1	138.7	143.0	141.9	135.8	195.6	200.7	205.9	237.9	250.4	303
304	270.0	139.2	143.5	142.4	136.3	196.2	201.4	206.5	238.7	251.3	304
305	270.9	139.7	144.0	142.9	136.8	196.9	202.1	207.2	239.5	252.1	305
306	271.8	140.2	144.5	143.4	137.3	197.5	202.7	207.9	240.3	252.9	306
307	272.7	140.7	145.0	144.0	137.8	198.2	203.4	208.6	241.1	253.8	307
308	273.6	141.2	145.5	144.5	138.3	198.8	204.1	209.3	241.9	254.6	308
309	274.5	141.7	146.1	145.0	138.8	199.5	204.7	210.0	242.7	255.4	309
310	275.4	142.2	146.6	145.5	139.4	200.1	205.4	210.7	243.5	256.3	310
311	276.3	142.7	147.1	146.0	139.9	200.8	206.1	211.4	244.2	257.1	311
312	277.1	143.2	147.6	146.5	140.4	201.4	206.7	212.1	245.0	257.9	312
313	278.0	143.7	148.1	147.0	140.9	202.1	207.4	212.7	245.8	258.8	313
314	278.9	144.2	148.6	147.6	141.4	202.8	208.1	213.4	246.6	259.6	314
315	279.8	144.7	149.1	148.1	141.9	203.4	208.8	214.1	247.4	260.4	315
316	280.7	145.2	149.6	148.6	142.4	204.1	209.4	214.8	248.2	261.2	316
317	281.6	145.7	150.1	149.1	143.0	204.7	210.1	215.5	249.0	262.1	317
318	282.5	146.2	150.7	149.6	143.5	205.4	210.8	216.2	249.8	262.9	318
319	283.4	146.7	151.2	150.1	144.0	206.0	211.5	216.9	250.6	263.7	319
320	284.2	147.2	151.7	150.7	144.5	206.7	212.1	217.6	251.3	264.6	320
321	285.1	147.7	152.2	151.2	145.0	207.3	212.8	218.3	252.1	265.4	321
322	286.0	148.2	152.7	151.7	145.5	208.0	213.5	218.9	252.9	266.2	322
323	286.9	148.7	153.2	152.2	146.0	208.6	214.1	219.6	253.7	267.1	323
324	287.8	149.2	153.7	152.7	146.6	209.3	214.8	220.3	254.5	267.9	324
325	288.7	149.7	154.3	153.2	147.1	210.0	215.5	221.0	255.3	268.7	325
326	289.6	150.2	154.8	153.8	147.6	210.6	216.2	221.7	256.1	269.6	326
327	290.5	150.7	155.3	154.3	148.1	211.3	216.8	222.4	256.9	270.4	327
328	291.4	151.2	155.8	154.8	148.6	211.9	217.5	223.1	257.7	271.2	328
329	292.2	151.7	156.3	155.3	149.1	212.6	218.2	223.8	258.5	272.1	329
330	293.1	152.2	156.8	155.8	149.7	213.2	218.8	224.4	259.3	272.9	330
331	294.0	152.7	157.3	156.4	150.2	213.9	219.5	225.1	260.0	273.7	331
332	294.9	153.2	157.9	156.9	150.7	214.5	220.2	225.8	260.8	274.6	332
333	295.8	153.7	158.4	157.4	151.2	215.2	220.8	226.5	261.6	275.4	333
334	296.7	154.2	158.9	157.9	151.7	215.8	221.5	227.2	262.4	276.2	334
335	297.6	154.7	159.4	158.4	152.3	216.5	222.2	227.9	263.2	277.0	335
336	298.5	155.2	159.9	159.0	152.8	217.1	222.9	228.6	264.0	277.9	336
337	299.3	155.8	160.5	159.5	153.3	217.8	223.5	229.2	264.8	278.7	337
338	300.2	156.3	161.0	160.0	153.8	218.4	224.2	229.9	265.6	279.5	338
339	301.1	156.8	161.5	160.5	154.3	219.1	224.9	230.6	266.4	280.4	339

Table for calculating dextrose, invert sugar alone, invert sugar in the presence of sucrose (0.4 gram and 2 grams total sugar), lactose (three forms), and maltose (anhydrous and crystallized)—Continued. [For correction of lactose figures see Cir. 82.]

[Expressed in milligrams.]

Cuprous oxid (Cu_2O).	Copper (Cu).	Dextrose (d-glucose).	Invert sugar.	Invert sugar and sucrose.		Lactose.			Maltose.		Cuprous oxid (Cu_2O).
				0.4 gram total sugar.	2 grams total sugar.	$C_{12}H_{22}O_{11}$.	$C_{12}H_{22}O_{11}+\frac{1}{2}H_2O$.	$C_{12}H_{22}O_{11}+H_2O$.	$C_{12}H_{22}O_{11}$.	$C_{12}H_{22}O_{11}+H_2O$.	
340	302.0	157.3	162.0	161.0	154.8	219.8	225.5	231.3	267.1	281.2	340
341	302.9	157.8	162.5	161.6	155.4	220.4	226.2	232.0	267.9	282.0	341
342	303.8	158.3	163.1	162.1	155.9	221.1	226.9	232.7	268.7	282.9	342
343	304.7	158.8	163.6	162.6	156.4	221.7	227.5	233.4	269.5	283.7	343
344	305.6	159.3	164.1	163.1	156.9	222.4	228.2	234.1	270.3	284.5	344
345	306.5	159.8	164.6	163.7	157.5	223.0	228.9	234.7	271.1	285.4	345
346	307.3	160.3	165.1	164.2	158.0	223.7	229.6	235.4	271.9	286.2	346
347	308.2	160.8	165.7	164.7	158.5	224.3	230.2	236.1	272.7	287.0	347
348	309.1	161.4	166.2	165.2	159.0	225.0	230.9	236.8	273.5	287.9	348
349	310.0	161.9	166.7	165.7	159.5	225.6	231.6	237.5	274.3	288.7	349
350	310.9	162.4	167.2	166.3	160.1	226.3	232.2	238.2	275.0	289.5	350
351	311.8	162.9	167.7	166.8	160.6	226.9	232.9	238.9	275.8	290.4	351
352	312.7	163.4	168.3	167.3	161.1	227.6	233.6	239.6	276.6	291.2	352
353	313.6	163.9	168.8	167.8	161.6	228.2	234.2	240.2	277.4	292.0	353
354	314.4	164.4	169.3	168.4	162.2	228.9	234.9	240.9	278.2	292.8	354
355	315.3	164.9	169.8	168.9	162.7	229.5	235.6	241.6	279.0	293.7	355
356	316.2	165.4	170.4	169.4	163.2	230.2	236.3	242.3	279.8	294.5	356
357	317.1	166.0	170.9	170.0	163.7	230.8	236.9	243.0	280.6	295.3	357
358	318.0	166.5	171.4	170.5	164.3	231.5	237.6	243.7	281.4	296.2	358
359	318.9	167.0	171.9	171.0	164.8	232.1	238.3	244.4	282.2	297.0	359
360	319.8	167.5	172.5	171.5	165.3	232.8	238.9	245.1	282.9	297.8	360
361	320.7	168.0	173.0	172.1	165.8	233.5	239.6	245.8	283.7	298.7	361
362	321.6	168.5	173.5	172.6	166.4	234.1	240.3	246.4	284.5	299.5	362
363	322.4	169.0	174.0	173.1	166.9	234.8	241.0	247.1	285.3	300.3	363
364	323.3	169.6	174.6	173.7	167.4	235.4	441.6	247.8	286.1	301.2	364
365	324.2	170.1	175.1	174.2	167.9	236.1	242.3	248.5	286.9	302.0	365
366	325.1	170.6	175.6	174.7	168.5	236.7	243.0	249.2	287.7	302.8	366
367	326.0	171.1	176.1	175.2	169.0	237.4	243.6	249.9	288.5	303.6	367
368	326.9	171.6	176.7	175.8	169.5	238.1	244.3	250.6	289.3	304.5	368
369	327.8	172.1	177.2	176.3	170.0	238.7	245.0	251.3	290.0	305.3	369
370	328.7	172.7	177.7	176.8	170.6	239.4	245.7	252.0	290.8	306.1	370
371	329.5	173.2	178.2	177.4	171.1	240.0	246.3	252.7	291.6	307.0	371
372	330.4	173.7	178.8	177.9	171.6	240.7	247.0	253.3	292.4	307.8	372
373	331.3	174.2	179.3	178.4	172.2	241.3	247.7	254.0	293.2	308.6	373
374	332.2	174.7	179.8	179.0	172.7	242.0	248.4	254.7	294.0	309.5	374
375	333.1	175.3	180.4	179.5	173.2	242.6	249.0	255.4	294.8	310.3	375
376	334.0	175.8	180.9	180.0	173.7	243.3	249.7	256.1	295.6	311.1	376
377	334.9	176.3	181.4	180.6	174.3	243.9	250.4	256.8	296.4	312.0	377
378	335.8	176.8	182.0	181.1	174.8	244.6	251.0	257.5	297.2	312.8	378
379	336.7	177.3	182.5	181.6	175.3	245.2	251.7	258.2	297.9	313.6	379
380	337.5	177.9	183.0	182.1	175.9	245.9	252.4	258.8	298.7	314.5	380
381	338.4	178.4	183.6	182.7	176.4	246.6	253.0	259.5	299.5	315.3	381
382	339.3	178.9	184.1	183.2	176.9	247.2	253.7	260.2	300.3	316.1	382
383	340.2	179.4	184.6	183.8	177.5	247.9	254.4	260.9	301.1	316.9	383
384	341.1	180.0	185.2	184.3	178.0	248.5	255.1	261.6	301.9	317.8	384
385	342.0	180.5	185.7	184.8	178.5	249.2	255.7	262.3	302.7	318.6	385
386	342.9	181.0	186.2	185.4	179.1	249.8	256.4	263.0	303.5	319.4	386
387	343.8	181.5	186.8	185.9	179.6	250.5	257.1	263.6	304.2	320.3	387
388	344.6	182.0	187.3	186.4	180.1	251.1	257.7	264.3	305.0	321.1	388
389	345.5	182.6	187.8	187.0	180.6	251.8	258.4	265.0	305.8	321.9	389
390	346.4	183.1	188.4	187.5	181.2	252.4	259.1	265.7	306.6	322.8	390
391	347.3	183.6	188.9	188.0	181.7	253.1	259.7	266.4	307.4	323.6	391
392	348.2	184.1	189.4	188.6	182.3	253.7	260.4	267.1	308.2	324.4	392
393	349.1	184.7	190.0	189.1	182.8	254.4	261.1	267.8	309.0	325.3	393
394	350.0	185.2	190.5	189.7	183.3	255.0	261.8	268.5	309.8	326.1	394

Table for calculating dextrose, invert sugar alone, invert sugar in the presence of sucrose (0.4 gram and 2 grams total sugar), lactose (three forms), and maltose (anhydrous and crystallized)—Continued. [For correction of lactose figures see Cir. 82.]

[Expressed in milligrams.]

Cuprous oxid (Cu_2O).	Copper (Cu).	Dextrose (d-glucose).	Invert sugar.	Invert sugar and sucrose.		Lactose.			Maltose.		Cuprous oxid (Cu_2O).
				0.4 gram total sugar.	2 grams total sugar.	$C_{12}H_{22}O_{11}$.	$C_{12}H_{22}O_{11}+\frac{1}{2}H_2O$.	$C_{12}H_{22}O_{11}+H_2O$.	$C_{12}H_{22}O_{11}$.	$C_{12}H_{22}O_{11}+H_2O$.	
395	350.9	185.7	191.0	190.2	183.9	255.7	262.4	269.1	310.6	326.9	395
396	351.8	186.2	191.6	190.7	184.4	256.3	263.1	269.8	311.4	327.7	396
397	352.6	186.8	192.1	191.3	184.9	257.0	263.8	270.5	312.1	328.6	397
398	353.5	187.3	192.7	191.8	185.5	257.7	264.4	271.2	312.9	329.4	398
399	354.4	187.8	193.2	192.3	186.0	258.3	265.1	271.9	313.7	330.2	399
400	355.3	188.4	193.7	192.9	186.5	259.0	265.8	272.6	314.5	331.1	400
401	356.2	188.9	194.3	193.4	187.1	259.6	266.4	273.3	315.3	331.9	401
402	357.1	189.4	194.8	194.0	187.6	260.3	267.1	274.0	316.1	332.7	402
403	358.0	189.9	195.4	194.5	188.1	260.9	267.8	274.6	316.9	333.6	403
404	358.9	190.5	195.9	195.0	188.7	261.6	268.5	275.3	317.7	334.4	404
405	359.7	191.0	196.4	195.6	189.2	262.2	269.1	276.0	318.5	335.3	405
406	360.6	191.5	197.0	196.1	189.8	262.9	269.8	276.7	319.2	336.0	406
407	361.5	192.1	197.5	196.7	190.3	263.5	270.5	277.4	320.0	336.9	407
408	362.4	192.6	198.1	197.2	190.8	264.2	271.1	278.1	320.8	337.7	408
409	363.3	193.1	198.6	197.7	191.4	264.8	271.8	278.8	321.6	338.5	409
410	364.2	193.7	199.1	198.3	191.9	265.5	272.5	279.5	322.4	339.4	410
411	365.1	194.2	199.7	198.8	192.5	266.1	273.1	280.1	323.2	340.2	411
412	366.0	194.7	200.2	199.4	193.0	266.8	273.8	280.8	324.0	341.0	412
413	366.9	195.2	200.8	199.9	193.5	267.4	274.5	281.5	324.8	341.9	413
414	367.7	195.8	201.3	200.5	194.1	268.1	275.2	282.2	325.6	342.7	414
415	368.6	196.3	201.8	201.0	194.6	268.7	275.8	282.9	326.3	343.5	415
416	369.5	196.8	202.4	201.6	195.2	269.4	276.5	283.6	327.1	344.4	416
417	370.4	197.4	202.9	202.1	195.7	270.1	277.2	284.3	327.9	345.2	417
418	371.3	197.9	203.5	202.6	196.2	270.7	277.8	285.0	328.7	346.0	418
419	372.2	198.4	204.0	203.2	196.8	271.4	278.5	285.6	329.5	346.8	419
420	373.1	199.0	204.6	203.7	197.3	272.0	279.2	286.3	330.3	347.7	420
421	374.0	199.5	205.1	204.3	197.9	272.7	279.8	287.0	331.1	348.5	421
422	374.8	200.1	205.7	204.8	198.4	273.3	280.5	287.7	331.9	349.3	422
423	375.7	200.6	206.2	205.4	198.9	274.0	281.2	288.4	332.7	350.2	423
424	376.6	201.1	206.7	205.9	199.5	274.6	281.9	289.1	333.4	351.0	424
425	377.5	201.7	207.3	206.5	200.0	275.3	282.5	289.8	334.2	351.8	425
426	378.4	202.2	207.8	207.0	200.6	275.9	283.2	290.5	335.0	352.7	426
427	379.3	202.8	208.4	207.6	201.1	276.6	283.9	291.1	335.8	353.5	427
428	380.2	203.3	208.9	208.1	201.7	277.2	284.5	291.8	336.6	354.3	428
429	381.1	203.8	209.5	208.7	202.2	277.9	285.2	292.5	337.4	355.1	429
430	382.0	204.4	210.0	209.2	202.7	278.5	285.9	293.2	338.2	356.0	430
431	382.8	204.9	210.6	209.8	203.3	279.2	286.5	293.9	339.0	356.8	431
432	383.7	205.5	211.1	210.3	203.8	279.8	287.2	294.6	339.7	357.6	432
433	384.6	206.0	211.7	210.9	204.4	280.5	287.9	295.3	340.5	358.5	433
434	385.5	206.5	212.2	211.4	204.9	281.2	288.6	295.9	341.3	359.3	434
435	386.4	207.1	212.8	212.0	205.5	281.8	289.2	296.6	342.1	360.1	435
436	387.3	207.6	213.3	212.5	206.0	282.5	289.9	297.3	342.9	360.9	436
437	388.2	208.2	213.9	213.1	206.6	283.1	290.6	298.0	343.7	361.8	437
438	389.1	208.7	214.4	213.6	207.1	283.8	291.2	298.7	344.5	362.6	438
439	390.0	209.2	215.0	214.2	207.7	284.4	291.9	299.4	345.3	363.4	439
440	390.8	209.8	215.5	214.7	208.2	285.1	292.6	300.1	346.1	364.3	440
441	391.7	210.3	216.1	215.3	208.8	285.7	293.2	300.8	346.8	365.1	441
442	392.6	210.9	216.6	215.8	209.3	286.4	293.9	301.4	347.6	365.9	442
443	393.5	211.4	217.2	216.4	209.9	287.0	294.6	302.1	348.4	366.8	443
444	394.4	212.0	217.8	216.9	210.4	287.7	295.3	302.8	349.2	367.6	444
445	395.3	212.5	218.3	217.5	211.0	288.3	295.9	303.5	350.0	368.4	445
446	396.2	213.1	218.9	218.0	211.5	289.0	296.6	304.2	350.8	369.3	446
447	397.1	213.6	219.4	218.6	212.1	289.6	297.3	304.9	351.6	370.1	447
448	397.9	214.1	220.0	219.1	212.6	290.3	297.9	305.6	352.4	370.9	448
449	398.8	214.7	220.5	219.7	213.2	290.9	298.6	306.3	353.2	371.7	449

Table for calculating dextrose, invert sugar alone, invert sugar in the presence of sucrose (0.4 gram and 2 grams total sugar), lactose (three forms), and maltose (anhydrous and crystallized)—Continued. [For correction of lactose figures see Cir. 82.] [Expressed in milligrams.]

Cuprous oxid (Cu_2O).	Copper (Cu).	Dextrose (d-glucose).	Invert sugar.	Invert sugar and sucrose.		Lactose.			Maltose.		Cuprous oxid (Cu_2O).
				0.4 gram total sugar.	2 grams total sugar.	$C_{12}H_{22}O_{11}$.	$C_{12}H_{22}O_{11}+\frac{1}{2}H_2O$.	$C_{12}H_{22}O_{11}+H_2O$.	$C_{12}H_{22}O_{11}$.	$C_{12}H_{22}O_{11}+H_2O$.	
450	399.7	215.2	221.1	220.2	213.7	291.6	299.3	306.9	353.9	372.6	450
451	400.6	215.8	221.6	220.8	214.3	292.3	299.9	307.6	354.7	373.4	451
452	401.5	216.3	222.2	221.4	214.8	292.9	300.6	308.3	355.5	374.2	452
453	402.4	216.9	222.8	221.9	215.4	293.6	301.3	309.0	356.3	375.1	453
454	403.3	217.4	223.3	222.5	215.9	294.2	302.0	309.7	357.1	375.9	454
455	404.2	218.0	223.9	223.0	216.5	294.9	302.6	310.4	357.9	376.7	455
456	405.1	218.5	224.4	223.6	217.0	295.5	303.3	311.1	358.7	377.6	456
457	405.9	219.1	225.0	224.1	217.6	296.2	304.0	311.8	359.5	378.4	457
458	406.8	219.6	225.5	224.7	218.1	296.8	304.6	312.4	360.3	379.2	458
459	407.7	220.2	226.1	225.3	218.7	297.5	305.3	313.1	361.0	380.0	459
460	408.6	220.7	226.7	225.8	219.2	298.1	306.0	313.8	361.8	380.9	460
461	409.5	221.3	227.2	226.4	219.8	298.8	306.6	314.5	362.6	381.7	461
462	410.4	221.8	227.8	226.9	220.3	299.4	307.3	315.2	363.4	382.5	462
463	411.3	222.4	228.3	227.5	220.9	300.1	308.0	315.9	364.2	383.4	463
464	412.2	222.9	228.9	228.1	221.4	300.7	308.7	316.6	365.0	384.2	464
465	413.0	223.5	229.5	228.6	222.0	301.4	309.3	317.3	365.8	385.0	465
466	413.9	224.0	230.0	229.2	222.5	302.0	310.0	317.9	366.6	385.9	466
467	414.8	224.6	230.6	229.7	223.1	302.7	310.7	318.6	367.3	386.7	467
468	415.7	225.1	231.2	230.3	223.7	303.3	311.3	319.3	368.1	387.5	468
469	416.6	225.7	231.7	230.9	224.2	304.0	312.0	320.0	368.9	388.3	469
470	417.5	226.2	232.3	231.4	224.8	304.7	312.7	320.7	369.7	389.2	470
471	418.4	226.8	232.8	232.0	225.3	305.3	313.3	321.4	370.5	390.0	471
472	419.3	227.4	233.4	232.5	225.9	306.0	314.0	322.1	371.3	390.8	472
473	420.2	227.9	234.0	233.1	226.4	306.6	314.7	322.8	372.1	391.7	473
474	421.0	228.5	234.5	233.7	227.0	307.3	315.4	323.4	372.9	392.5	474
475	421.9	229.0	235.1	234.2	227.6	307.9	316.0	324.1	373.7	393.3	475
476	422.8	229.6	235.7	234.8	228.1	308.6	316.7	324.8	374.4	394.2	476
477	423.7	230.1	236.2	235.4	228.7	309.2	317.4	325.5	375.2	395.0	477
478	424.6	230.7	236.8	235.9	229.2	309.9	318.0	326.2	376.0	395.8	478
479	425.5	231.3	237.4	236.5	229.8	310.5	318.7	326.9	376.8	396.6	479
480	426.4	231.8	237.9	237.1	230.3	311.2	319.4	327.6	377.6	397.5	480
481	427.3	232.4	238.5	237.6	230.9	311.8	320.0	328.2	378.4	398.3	481
482	428.1	232.9	239.1	238.2	231.5	312.5	320.7	328.9	379.2	399.1	482
483	429.0	233.5	239.6	238.8	232.0	313.1	321.4	329.6	380.0	400.0	483
484	429.9	234.1	240.2	239.3	232.6	313.8	322.1	330.3	380.7	400.8	484
485	430.8	234.6	240.8	239.9	233.2	314.4	322.7	331.0	381.5	401.6	485
486	431.7	235.2	241.4	240.5	233.7	315.1	323.4	331.7	382.3	402.4	486
487	432.6	235.7	241.9	241.0	234.3	315.8	324.1	332.4	383.1	403.3	487
488	433.5	236.3	242.5	241.6	234.8	316.4	324.7	333.1	383.9	404.1	488
489	434.4	236.9	243.1	242.2	235.4	317.1	325.4	333.7	384.7	404.9	489
490	435.3	237.4	243.6	242.7	236.0	317.7	326.1	334.4	385.5	405.8	490

243. Acetone. Quantitative determination. Folin method (*Jour. Biol. Chem*, III, 1907, 177).—*Principle*. The actone is carried over by a strong air current from the urine and is absorbed and changed to iodoform by an alkaline hypoiodite solution of known strength. The amount of iodine remaining is then titrated with a standard thiosulphate solution with starch as an indicator. The method is accurate, rapid and simple, and requires only 20-25 c.c. urine.

Process. Into an aërometer cylinder like that described in experiment 209 measure accurately 20-25 c.c. of urine with a pipette, add 0.2-0.3 gram of oxalic acid or a few drops of 10 per cent. phosphoric acid, 8-10 grams of solid NaCl and a little petroleum. The salt renders the acetone less soluble. Connect with the absorbing bottle which is like that described for the ammonia apparatus, experiment 209. In the absorbing bottle place about 150 c.c. water, 10 c.c. 40 per cent. KOH solution and an excess of standardized solution of N/10 I, the amount being carefully measured from a burette or pipette. An excess of 10-15 c.c. of the standardized iodine solution should be added. Hawk recommends that to get an idea of the amount of iodine solution necessary to add, take in a test-tube 10 c.c. of urine and add 1 c.c. of ferric-chloride solution (100 grams ferric chloride in 100 c.c. water). Compare the color with 10 c.c. of the original ferric-chloride solution in a similar test-tube. If the colors about match, then take 20 c.c. of the iodine solution. If the urine is darker so that it has to be diluted once with water, then take 35 c.c.; if still darker take more. It is important to connect the apparatus and begin the air current at once after adding the strong alkali to the iodine, since the hypoiodite goes over very rapidly into iodate. Run the air through briskly, but not so fast as for ammonia, for 20-25 minutes. Then disconnect, acidify the contents of the hypoiodite tube by the addition of 10 c.c. concentrated HCl for each 10 c.c. of strong alkali added at the start, and titrate the excess of iodine by means of the thiosulphate solution with starch as indicator. The sodium thiosulphate solution must have been titrated against the iodine beforehand. It must be kept in the dark. For standardization of the thiosulphate see experiment 44(b).

244. Determination of acetone and diacetic acid (Folin-Hart).—*Principle.* The acetone is first determined as in experiment 243, then the urine is acidified and heated while air carries over the acetone which is set free from the diacetic acid in these circumstances. The two determinations may be made as one, the total acetone being received in a single solution of hypoiodite, or the preformed acetone may be separately determined first.

Method. The arrangement of the apparatus is as in experiment 243, except that in place of the cylinder for holding the urine, the latter is placed in a large Jena test-tube which is supported in such a way that it can be heated. The air inlet tube dips to the bottom of the Jena glass test-tube. The air is passed through a hypoiodite solution first to remove any substance which may reduce iodine. Place in the large test-tube, which carries the stopper and tubes for aërating the urine, 20 c.c. of urine, and the phosphoric acid, oxalic acid, salt and petroleum as described in 243 and without heating the tube air is passed through for

25 minutes. The air with the acetone passes into a measured amount of fresh hypoiodite as described. Then disconnect the hypoiodite, place a fresh cylinder of a known amount of hypoiodite solution to receive the acetone and heat the tube just to boiling for 25 minutes while the air is going through. By this heating the acetoacetic acid is decomposed, acetone is liberated and absorbed by the second hypoiodite tube. The first hypoiodite tube being titrated in the manner indicated in 243 gives the preformed acetone; the second hypoiodite tube gives the acetone from the diacetic acid. The sum of the two gives the total acetone in 20 c.c. urine.

245. Quantitative determination of acetone and acetoacetic acid and hydroxybutyric acid in the same sample of urine. Method of Shaffer and Marriott (*Jour. Biol. Chem.*, XVI, p. 276, 1914).—*Principle.* The urine is distilled with acid after precipitation of sugar and other substances by basic lead acetate and ammonia. This separates the acetone, both that preformed and that arising from the acetoacetic acid. The acetone is titrated with standard iodine and thiosulphate. The urine filtrate, from which the acetone has been distilled, is oxidized by the cautious addition of bichromate to the boiling urine containing the oxybutyric acid. This oxidizes the oxybutyric acid to acetone, the acetone is distilled off and titrated in the same way as the preformed acetone. The yield is about 90-95 per cent. of the theoretical.

Process. From 25-100 c.c. of urine (usually 50 c.c.) are measured with a pipette into a 500 c.c. volumetric flask containing 200-300 c.c. of water. Add basic lead acetate solution (U.S.P.) in amounts equal to the urine taken and mix the liquid well. If the urine contains but little sugar, only half the amount, or less, of the lead acetate should be used. Next pour into the flask strong ammonia water equal to about half the volume of lead-acetate solution taken, dilute to the mark with water, shake and after a few moments' standing filter through a folded filter. Measure 200 c.c. of the filtrate into a round-bottom flask (800 c.c. or liter Kjeldahl), dilute with water to about 600 c.c., add 15 c.c. concentrated sulphuric acid and talc or boiling stone and distill the mixture until about 200 c.c. of the distillate has been collected. (Distillate A. This contains the preformed acetone and that derived from the acetoacetic acid.) The distilling flask must be fitted with a dropping tube and water run in from time to time to prevent the volume in the flask from becoming less than 400-500 c.c. The tube at the end of the condenser should dip below the surface water in the receiving flask, a second Kjeldahl, to prevent loss of acetone.

Distillate A is redistilled for about 20 minutes after the addition of 10 c.c. of 10 per cent. sodium hydroxide. If a high degree of accuracy is not required, distillate A may be titrated directly with standard iodine solution, N/10, and thiosulphate. The results are a little higher than after redistillation from an alkaline solution. The distillate from A, which may be called A_2, is titrated with iodine and standard thiosulphate.

After A has been distilled off, a new receiving flask is adjusted, the end of the tube dipping below the water in the receiving flask and bichromate solution as indicated below is slowly added to the distilling flask while the distillation is continued to give B.

The residue of the urine plus sulphuric acid from which distillate A was obtained is again distilled, adding water or bichromate solution when necessary to keep the volume between 400 and 600 c.c. From 0.5-1 gram of bichromate will usually be sufficient and not more than 1 gram should be added, unless the liquid turns green, indicating a great reduction to chromium sulphate: very rarely 2 or 3 grams of bichromate may be necessary, especially if the sugar has not been completely removed. To make the distillation proceed as follows:

A 10 per cent. solution of potassium bichromate is kept on hand and 10 c.c. of this, diluted to 100 c.c., are measured out for each determination. 20 c.c. of the dilute solution (0.2 gram $K_2Cr_2O_7$) are first added slowly through the dropping tube and then 10 c.c. portions every 15 or 20 minutes until the whole has been added. Should the liquid become markedly green the bichromate must be added at correspondingly shorter intervals and in amount sufficient to maintain a slight red-yellow color of the chromic acid, which may be detected even in the presence of the green. The distillation is continued with moderate boiling for from 2-3 hours. The distillate B, which should be collected in a liter flask to avoid transference, is again distilled for about 20 minutes after adding 10 c.c. of 10 per cent. sodium hydroxide and 25 c.c. of 3 per cent. H_2O_2. The flask must be heated cautiously until the peroxide has been decomposed. This distillate, B_2, is titrated with the standard iodine and thiosulphate.

Computation. 1 c.c. of N/10 iodine=0.968 mg. acetone=1.736 mg. oxybutyric acid, or

1 c.c. of 1.035/10 iodine (=13.13 mg. I_2)=1 mg. acetone=1.793 mg. oxybutyric acid.

Note.—This method gives results usually a little lower than the extraction and polarization method of Black, but is more accurate for small amounts of oxybutyric acid and somewhat more convenient.

In making this titration an excess of N/10 iodine solution is added and then the solution made alkaline by the addition of 10 c.c. 60 per cent. NaOH, the flask stoppered, shaken and allowed to stand 5-10 minutes, then acidified by the addition of 15 c.c. concentrated HCl and the liberated iodine titrated with standard thiosulphate in the usual way.

For the determination of acetone in blood, or when small quantities are present, see Marriott, *Jour. Biol. Chem.*, 16, p. 284 and p. 289, 1913. For determination of oxybutyric acid in blood and tissues, see Marriott, *ibid*, page 293.

246. Quantitative determination of saccharose in urine (Jolles *Biochem. Ztschr.* 43, p. 56, 1912).

Determine polarimetrically after treating the urine with 0.1 per cent NaOH 24 hours in the thermostat at 37°. All other mono- and disaccharides become inactive under these circumstances. Saccharose is not affected.

247. Quantitative determination of the hydrogen ion cencentration of urine. Sorensen indicator method. (Henderson and Palmer, *Jour. Biol. Chem.*, 13, 398, 1913).—*Principle.* Matching the color of the urine, diluted if necessary and containing a known amount of some indicator, with the color developed by a similar concentration of the indicator in a series of solutions containing known concentrations of hydrogen ions.

STANDARD SOLUTIONS: MIXTURES

No.	NaH_2PO_4	Na_2HPO_4	P_H+	Indicator	
1.		0.1000 N	9.27	Phenolphthalein (1–3)	
2.	0.0001 N	0.0480 "	8.7		
3.	0.0001 "	0.0120 "	8.0	Neutral red (3–6)	Sodium alizarine sulphonate (3–13)
4.	0.0166 "	0.0833 "	7.48		
5.	0.0010 "	0.0060 "	7.38		
6.	0.0010 "	0.0023 "	6.90		
	CH_3COOH	CH_3COONa			
7.	0.0009 N	0.0920 N		p-Nitrophenol (7–13)	
8.	0.0023 "	0.0020 "			
9.	0.0046 "	0.0920 "			
10.	0.0092 "	0.0920 "			
11.	0.0230 "	0.0920 "		Methyl red (11–13)	
12.	0.0460 "	0.0920 "	6.70		
13.	0.0920 "	0.0920 "	6.90		

Process. In a series of exactly similar 250 c.c. flasks place 10 c.c. of each of the standard solutions indicated above, make up the solutions to 250 c.c. with distilled water and add enough of an aqueous solution of alizarine sulphonate of sodium so that the concentration of the latter is about 0.0003 per cent. and is exactly equal in all cases. 10 c.c. of urine are now introduced into another 250 c.c. flask and the same amount of distilled water and indicator added. The color of the diluted urine with its indicator is now matched with one of the standard series. If the reaction as thus measured is more acid than $P_H=5.3$ (H ion$=N\times 50\times 10^{-7}$) further tests are made with methyl red; for the range 5.3—6.7 with p-nitrophenol and with neutral red; for more alkaline urines phenolphthalein is used.

(a) H ion concentrations greater than 50×10^{-7}. "10 c.c. por-

water and 20 c.c. ethyl ether and shake for 5 minutes. Add then 20 c 95 per cent. alcohol and shake again for 5 minutes. Allow to stand ur contents separate into two distinct layers. Remove the upper layer the pipette, as shown in Figure 81, into a weighed evaporating di Add 5 c.c. ether to the cylinder so as to wash down the sides of cylinder; remove by the pipette to the evaporating dish; repeat t process of washing 5 times, using 5 c.c. each time and removing e time to the evaporating dish. Evaporate the washings and upper la to dryness on the water bath, or cool over sulphuric acid in a desicca and dry at room temperature to constant weight. The weight of residue may be taken as very nearly that of the total lipin. It conta a slight amount of lactose, and the extraction of lipin material is quite so complete as the Soxhlet method.

INDEX.

INDEX. PRACTICAL WORK AND METHODS.

Lightning Source UK Ltd.
Milton Keynes UK
UKHW021936161118
332480UK00007B/75/P